Catalysis and Reaction Engineering Division

Core Programming Topic at the 2011 AIChE Annual Meeting

Minneapolis, Minnesota, USA
16-21 October 2011

Volume 1 of 2

ISBN: 978-1-61839-739-3

Printed from e-media with permission by:

Curran Associates, Inc.
57 Morehouse Lane
Red Hook, NY 12571

Some format issues inherent in the e-media version may also appear in this print version.

Printed by Curran Associates, Inc. (2012)

AIChE
3 Park Avenue
New York, NY 10016-5991

Phone: (203) 702-7660
Fax: (203) 775-5177

www.aiche.org

Additional copies of this publication are available from:

Curran Associates, Inc.
57 Morehouse Lane
Red Hook, NY 12571 USA
Phone: 845-758-0400
Fax: 845-758-2634
Email: curran@proceedings.com
Web: www.proceedings.com

TABLE OF CONTENTS

Volume 1

Volume 2

(13a) Mechanistic Modeling of Fast Pyrolysis of Cellulose to Predict Bio-Oil Composition

Vinu Ravikrishnan and Linda J. Broadbelt, Chemical and Biological Engineering, Northwestern University, Evanston, IL

Fast pyrolysis, a potential strategy for the production of transportation fuels from biomass, involves a complex network of competing reactions, which result in the formation of bio-oil, non-condensable gaseous species, and solid char. Bio-oil is a mixture of anhydro sugars, furan derivatives, and oxygenated aromatic and low molecular weight (LMW) compounds. Currently, the successful modeling of fast pyrolysis reactors for biomass conversion is hampered by lumped kinetic models, which fail to predict the bio-oil composition. Hence, a fundamental understanding of the chemistry and kinetics of biomass pyrolysis is important to evaluate the effects of process parameters like temperature, residence time and pressure on the composition of bio-oil.

We have recently developed a comprehensive mechanistic model to characterize the primary products of fast pyrolysis of pure cellulose. The model incorporates the following condensed phase reactions with ionic and electrocyclic fragmentation mechanisms for the formation of various LMW products in bio-oil: glucosidic bond scission, retro-aldol, retro-Diels-Alder, 1,2-dehydration, 1,3-dehydration, hydrolysis, cyclic Grob fragmentation and enol-keto transformation reactions. The kinetic rate coefficients for most of the above reaction steps were obtained from the literature, based on either experimentally determined Arrhenius parameters for cellulose and model compounds, or theoretically computed values of activation energies using quantum chemical calculations. A computational framework based on continuous distribution kinetics was constructed to solve the kinetic model. The model tracks 96 distinct polymer species and 25 LMW products, and incorporates over 350 reactions. The model predictions compare well with the experimental data (Patwardhan *et al.*, *J. Anal. Appl. Pyrolysis* **2009**, 86, 323-330) for the formation of important products like levoglucosan (60 wt.%), glycolaldehyde (6 wt.%), 5-hydroxymethyl furfural (3 wt.%), furfural (1 wt.%) and formic acid (6 wt. %). Levoglucosan was found to be formed by the initiation of ionic chain ends by mid-chain glucosidic bond scission, followed by end-chain ionic unzipping. Some of our efforts in extending the kinetic model of pyrolysis of pure cellulose to cellulose with mineral matter and hemicellulose will also be discussed in this presentation.

See more of this Session: Catalytic Conversion of Renewable Resources to Synthesis Gases and Pyrolysis Oils
See more of this Group/Topical: Sustainable Engineering Forum

(13b) Mixed Rare Earth Oxides (REOs) for Desulfurization and Tar Removal From Gasifier Effluents

Rui Li, Joseph Bridges, Weishi Kong and Kerry.M. Dooley, Chemical Engineering, Louisiana State University, Baton Rouge, LA

The removal of H_2S, NH_3 and tars from syngas generated in biomass or coal/biomass gasification plays an important role in further processing of syngas. We have developed materials capable of both H_2S adsorption and tar reforming at high temperatures (~600-650ºC). These are mixed rare earth oxides (REOs) such as CeO_2/La_2O_3, prepared by templated sol-gel methods, and then impregnated with transition metals (Mn, Fe, etc.). Desulfurization tests have been carried out using model syngas mixtures. Successive desulfurization-regeneration cycles some of these materials (e.g., $Mn/Ce/LaO_x$) demonstrated reasonable sulfur capacities in the presence of water and CO_2, along with regenerability. Naphthalene and propane have been used as models for tar reforming/cracking, and the effects of H_2S, CO and CO_2 on the model reactions were examined. H_2S and CO_2 are strong inhibitors, while CO is not. Fresh, used and regenerated sorbents and catalysts have been characterized by BET, TPR, XRD, XANES and EXAFS.

See more of this Session: Catalytic Conversion of Renewable Resources to Synthesis Gases and Pyrolysis Oils
See more of this Group/Topical: Sustainable Engineering Forum

(13c) Multi-Scale Study On the Pyrolysis of Sustainable Biomass Feedstock

Jessica D. Murillo, Chemistry, Environmental Sciences, Tennessee Technological University, Cookeville, TN, Joseph J. Biernacki, Chemical Engineering, Tennessee Technological University, Cookeville, TN and C. Pat Bagley, College of Agricultural and Human Sciences, Tennessee Technological University, Cookeville, TN

Multi-Scale Study on the Pyrolysis of Sustainable Biomass Feedstock

Jessica D. Murillo

Department of Chemistry, Environmental Sciences, Tennessee Technological University, Cookeville, TN 38505

Joseph J. Biernacki

Department of Chemical Engineering, Tennessee Technological University, Cookeville, TN 38505

C. Pat Bagley

College of Agricultural and Human Sciences, Tennessee Technological University, Cookeville, TN 38505

Pyrolysis of lignocellulosic material for the production of biofuels and value-added products offers a short- to long-term renewable energy option to help curb fossil fuel use, gain energy independence, and mitigate global climate change. The main components of lignocellulsic material are cellulose, hemicellulose and lignin; however, there are many issues with using such a complex chemical feedstock for the production of renewable fuels. During pyrolysis the main components degrade via primary and secondary reactions, producing a spectrum of products, including volatiles, char, and bio-oil. To gain a more complete understanding of the pyrolytic behavior of lignocellulosic biomass, a deeper knowledge of the morphology and chemical structure is required.

Preliminary meso-scale experimental results from the pyrolysis of tall fescue hay, sorghum-sudangrass, Bermuda grass, switchgrass, and alfalfa hay, have paved the way for study at microscopic lengths. The physical and chemical characteristics of each grass hay was studied using X-ray diffraction, mid-infrared spectroscopy (MIR) and electron scanning microscopy to determine the microstructure and to quantify the content of lignin, cellulose, and hemicelluloses. Comparison studies on the reaction kinetics of grass samples and isolated pseudo-components were performed using thermogravimetric analysis (TGA).

(13d) Catalytic Cracking of Oak Pyrolytic Vapors Via Bench-Scale In-Situ Fixed-Bed Catalysis

David J. Mihalcik, Akwasi A. Boateng, Charles A. Mullen and Neil Goldberg, Eastern Regional Research Center, Agricultural Research Service, USDA, Wyndmoor, PA

Pyrolysis oils derived from lignocellulosic feedstocks by way of fast pyrolysis offer the possibility of a distinct alternative to the use of petroleum-based fuels. However, an inherent drawback associated with pyrolysis oil its high oxygen content which is due to the nature of the lignocellulosic biomass from which it derives. Extensive efforts are being put forth to treat pyrolytic vapors in situ using zeolite catalysts. For our experiment, the catalytic conversion of oak pyrolysis vapors was performed at 500°C using a 2.5 kg/hr fluidized-bed fast pyrolysis reactor that was fitted with a re-heated (425°C) split-stream fixed-bed catalytic cracker containing zeolite-type catalyst. The placement of the split-stream fixed-bed catalytic reactor was varied within the condenser train to determine the effect of the removal of increasing amounts of water and oxygenated components from within the pyrolytic vapor stream. The cracked pyrolytic vapors resulting from the fixed-bed catalyst were collected by way of immediate contact with a dry-ice acetone bath followed by secondary collection using a methanol spray condenser. Karl Fisher titration and elemental and quantitative GC/MS analyses were performed to determine relevant trends within the experiment. The recovered liquid products that passed through the fixed-bed catalytic reactor displayed a remarkable decrease in the amount of oxygen containing species and a significant increase in the presence of aromatic hydrocarbons.

See more of this Session: Catalytic Conversion of Renewable Resources to Synthesis Gases and Pyrolysis Oils
See more of this Group/Topical: Sustainable Engineering Forum

(13e) A Dual-Stage Laminar Entrained Flow Reactor (LEFR) and Plug Flow Reactor (PFR) System for Studying Biomass Pyrolysis and Gasification

Mark W. Jarvis, Calvin Mukarakate, David Robichaud and Mark Nimlos, National Renewable Energy Laboratory, Golden, CO

An understanding of pyrolysis vapor cracking kinetics and mechanisms are important for developing approaches for reducing tars during gasification and ultimately reducing the cost of producing transportation fuels from gasification and fuel synthesis. In order to reduce tars using predictive design approaches such as Computation Fluid Dynamics (CFD), it is necessary to have robust chemical reaction models containing mechanism that lead to the formation of these species. Existing reaction models only contain limited, highly generalized mechanisms for the formation of tars and are not robust enough to be predictive. In particular, the initial steps of gasification are inadequately characterized. These steps are primarily cracking reactions that break down the pyrolysis vapor species into smaller molecules ultimately leading to syngas and tars. In this study, we have undertaken an experimental investigation of these types of cracking reactions. The solid phase vaporization reactions were separated from the gas phase cracking reactions by conducting experiments in a two-stage reactor. In the first stage, particles of biomass were pyrolyzed in a Laminar Entrained Flow Reactor (LEFR) at a low temperature (450°C) and in the second phase the vapors were cracked at variable residence times (0.2 to 1.5 s) and temperatures (430° to 680°C) in a Plug Flow Reactor (PFR). The tar species and other product gases were measured using a Molecular Beam Mass Spectrometer (MBMS). The data from the MBMS showed trends in the tar species concentrations as a function of residence time and temperature, which will allow the development of more accurate and robust chemical kinetics models of gasification. Eventually, these kinetic models will be used in CFD models of gasification in order to optimize it for reduced tar formation and increased syngas yield.

See more of this Session: Catalytic Conversion of Renewable Resources to Synthesis Gases and Pyrolysis Oils
See more of this Group/Topical: Sustainable Engineering Forum

(13f) Reforming of Residual Tars and Oils From Biomass Gasification

Lyman Frost[1], Elango Elangovan[1] and J. Hartvigsen[2], (1)Ceramatec, Inc, Salt Lake City, UT, (2) Electrochemical, Ceramatec, Inc, Salt Lake City, UT

One of the barriers to wider use of many kinds of biomass is the fact that they generate refractory tars and oils when gasified. These compounds cause difficulty when they are passed to downstream components in the system by causing plugging and drop in performance. The compounds can be condensed and removed but this involves extensive heat exchange, increased cost, and removal to a hazardous waste landfill.

Working as a subcontractor to Emery Energy, a Salt Lake City company with expertise in biomass gasification, Ceramatec will be applying its non-thermal plasma reformer technology to eliminate any refractory tars and oils that exit the Emery gasifier. Preliminary laboratory work indicates that it is possible for the plasma reformer to operate on the gas exiting the gasifier and convert these tars and oils to additional hydrogen and carbon monoxide. This work is being done at Western Research Institute in Laramie, Wyoming under a US Department of Energy award to Emery Energy.

This paper will present an overall process structure of the gasification process and the non-thermal plasma reformer in particular. Operational characteristics and laboratory results related to the reforming of the refractory tars and oils will be presented. The larger pilot sized facility at Laramie will also be described. Data is expected to be available from the larger pilot sized facility in Laramie by the time for the conference. The resultant gas stream is of a quality that could be used by a high temperature fuel cell.

See more of this Session: Catalytic Conversion of Renewable Resources to Synthesis Gases and Pyrolysis Oils
See more of this Group/Topical: Sustainable Engineering Forum

(13g) Low Temperature Gasification of Impregnated Biomass Feedstock

Foster Agblevor, Biological Engineering, Utah State University, Logan, UT, Ofei Mante, Biological Systsems Engineering, Virginia Polytechnic Institute and State University, Blacksburg, VA, Allen Aradi, Afton Chemical Inc, Richmond, VA and Tze-Chi Jao, Afton Chemical, Richmond, VA

Hybrid poplar biomass feedstocks were impregnated with catalysts and gasified at three different temperatures in a bubbling fluidized bed reactor. The impregnated biomass were gasified in a nitrogen atmosphere at 400 °C, 600° C, and 800 °C. The gas yield increased with increasing temperature as expected and the char and liquid yields decreased with increasing temperature. Te catalytic effect was observed at 400 C when the impregnated gasification was compared with the control that did not contain any catalyst. At 400 C, the impregnated sample had a gas yield of 51wt% compared to 41 wt% for the non-catalytic gasification at 450 C. The char yield was 11 wt% compared to 14 wt% for the non-catalytic run at 450 C. The char yield decreased from 11 wt% to 6.5 wt% when the gasification temperature was increased from 400 to 800 C. Although there was 88 wt% gas yield at 800 C, the thermal effect was very strong and the catalyst was not effective in cracking the lignin degradation products. Thus, this catalyst needs considerable improvement to effectively crack the lignin fraction of the biomass pyrolysis products.

See more of this Session: Catalytic Conversion of Renewable Resources to Synthesis Gases and Pyrolysis Oils
See more of this Group/Topical: Sustainable Engineering Forum

(18a) Database Driven Discovery of Materials and Catalysts

Thomas Bligaard, Center for Atomic-scale Materials Design, Technical University of Denmark, Kgs. Lyngby, Denmark

We will present our approach to data-rich discovery of materials.

See more of this Session: Computational Catalysis I - Databases and Informatics
See more of this Group/Topical: Catalysis and Reaction Engineering Division

(18b) Fuels and Chemicals From Biomass: Challenges and Opportunities From a Systems Perspective

Prodromos Daoutidis, Chemical Engineering and Materials Science, University of Minnesota, Minneapolis, MN

Biorefineries are envisaged to produce green transportation fuels and chemicals by upgrading biomass in a sequence of processing steps. Although considerable emphasis has been given so far to the "upstream" conversion of biomass to intermediate platforms (sugars or syngas), progress in "downstream" conversion to chemicals and intermediates is still lagging. Due to the oxygen present in biomass and the diversity of raw materials derived from biomass, the necessary downstream reaction and separation processes are generally different from existing ones based on fossil fuels. Furthermore, there is limited data available on physical properties of such molecules, and on their full array of chemical transformations, and their kinetics and thermodynamics. These challenges lead to several emerging opportunities for systems and computational research, spanning several scales: elucidation of the underlying chemistry at the mechanistic level, modeling and design at the reactor and process level, and an enterprise-level analysis of process economics and environmental considerations.

This talk will discuss recent results on: i) the elucidation of the chemical transformations involved in biomass conversion using RING, a recently developed reaction network generation and analysis tool, and ii) the design and optimization of novel reaction-separation processes for biomass-based chemical synthesis. It will also highlight several avenues for research in computational chemistry and chemical informatics, microkinetic and multiscale modeling, and process design, optimization and control, which can collectively have a major impact on the realization of the ambitious concept of an integrated biorefinery.

See more of this Session: Computational Catalysis I - Databases and Informatics
See more of this Group/Topical: Catalysis and Reaction Engineering Division

(18c) Materials Design Workbench: An Open-Access Database for Materials and Molecular Design

Glen Allen Ferguson[1], Jeff Greeley[2], Peter Zapol[1], Larry A. Curtiss[3], Svetlozar Evtimov Nestorov[4], Ian T. Foster[5], David Landis[6], Jens Strabo Hummelshj[7], Karsten W. Jacobsen[8], Thomas Bligaard[9] and Jens Nrskov[10], (1)Material Science Department, Argonne National Laboratory, Argonne, IL, (2)Center for Nanoscale Materials, Argonne, IL, (3)Materials Science Division, Argonne National Laboratory, Argonne, IL, (4)Computer Science, University of Chicago, Chicago, IL, (5)Mathematics and Computer Science and Department of Computer Science, Argonne National Laboratory and University of Chicago, Argonne, IL, (6)Department of Physics CAMD, Technical University of Denmark, Kongens Lyngby, Denmark, (7)SLAC National Accelerator Laboratory, SUNCAT, Menlo Park, CA, (8)Center for Atomic-scale Materials Physics, Dept. of Physics, Technical University of Denmark, Lyngby, Denmark, (9) Center for Atomic-scale Materials Design, Technical University of Denmark, Kgs. Lyngby, Denmark, (10)Chemical Engineering, Stanford University and SUNCAT, Stanford, CA

The use of modern computing resources and efficient computational chemistry software packages has resulted in the production an extraordinary amount of computational data related to materials and catalytic problems. Very few such data, however, are available for public browsing; they tend to be stored locally and are rarely publically disseminated. A significant reason for this is absence of an easy to use, open-access public database for computational materials and molecular data. To address this problem, we are implementing an open-access database, the Materials Design Workbench (MDW), that will store data from computational chemistry software. The data are in a form that can be conveniently searched, visualized, analyzed, and retrieved by users.

The implementation of the Materials Design Workbench is a collaboration between scientists at Argonne National Laboratory, the University of Chicago, the Center for Sustainable Energy through Catalysis at the Stanford Linear Accelerator Center, and the Technical University of Denmark. The implementation consists of several modules. The central piece is a MySQL database that stores information from computational chemistry software and associated experimental data. To facilitate entering computational data into the database, a stand-alone python interface named the Computational Materials Repository (CMR) has also been developed. One of the most important roles of the CMR is to convert the output from calculations into XML format files. These XML files can then be modified to add additional data and uploaded to the database. Two additional applications are under development for visualization and analysis of data that have been entered into the database. The first is a client-side application that runs in any HTML 5 enabled web browser. The second is a server-side web interface.

In this talk, we will discuss two important parts of the MDW. The first is how the MySQL database stores and organizes the computational data. The second, which will include a preview, is the server-side application. This application, which is in development at the University of Chicago and Argonne National Laboratory, currently consists of a web interface for viewing and retrieving structures from the database. A user can query the database using any of the stored criteria and can display the results as a single molecule or as a collection of molecules. These collections can be analyzed, saved for later analysis, or downloaded. Users can also attach citations or other any other desirable data to their calculations. It is also possible to restrict access to data sets to particular users and groups or to make the data globally readable. These features, combined with the powerful visualization and analysis software that is also being developed, make MDW an attractive solution for the storage, retrieval and analysis of materials and molecular data.

See more of this Session: Computational Catalysis I - Databases and Informatics
See more of this Group/Topical: Catalysis and Reaction Engineering Division

(18d) Model-Based Design of Catalysts

W. Nicholas Delgass[1], Mahdi Abu-Omar[2], James M. Caruthers[1], Fabio Ribeiro[3], William F. Schneider[4], Kendall T. Thomson[1], Grigori Medvedev[1], Jun Wang[3], John Clay[4], Mary Jones[2], Wen-Sheng Lee[1], Jorge Pazmino[3], Kaiwalya Sabnis[1], Mayank Shakhar[1], Stephen D. Stamatis[1], Jeffery M. Switzer[1], D. Keith Steelman[2], Nicholas Travia[2], W. Damion Williams[1] and Silei Xiong[1], (1)School of Chemical Engineering, Purdue University, West Lafayette, IN, (2)Chemistry Department, Purdue University, West Lafayette, IN, (3)Chemical Engineering, Purdue University, West Lafayette, IN, (4) Chemical and Biomolecular Engineering, University of Notre Dame, Notre Dame, IN

This team of researchers is working on a model-based approach to catalyst design called *Discovery Informatics*. In this concept, the first objective is to identify descriptors that that can be linked to the kinetic performance of a family of catalysts through a quantitative *forward model*. This demanding task requires a number of computational tools. Since a microkinetic description is essential for the forward model, we have focused on tools for model discrimination and accurate extraction of rate constants from kinetic data. For single site olefin polymerization, which is our homogeneous catalysis test bed, we have developed a parallelized population balance modeling tool which allows molecular weight distributions from gel permeation chromatography to be included in a multi-response data set that is used to verify mechanism and provide rate constants for each of the necessary elementary steps. For the water gas shift (WGS) reaction on supported metals, our heterogeneous catalysis test bed, we have included Bayesian statistical analysis that provides specific accounting for error and Markov Chain Monte Carlo analysis to give probability distributions for parameters and their correlations with each other. While descriptors to relate to rate constants can sometimes be identified through linear correlations from statistically designed experiments, we have turned in this developmental stage to density functional theory to help identify descriptors, to better define reaction pathways, and, particularly in the case of heterogeneous systems, to compute energies that are difficult to measure.

Progress in building forward models will be discussed in terms of hexene polymerization over non-Cp coordination complexes of Zr and Hf and WGS over supported Au and Pt. In the homogeneous polymerization system we have discovered that the pendant ligand, NMe_2 versus THF, has a large effect on rate constant for chain transfer, while the change from Zr to Hf, lowers the propagation rate constant by a factor of 20. We will interpret this data in light of DFT results and our previous finding that the ion pair separation energy is an important descriptor[i]. For the WGS catalysts, we have identified the corner and perimeter atoms in contact with the support as the active sites for supported Au, all surface atoms as active for supported Pt, and, in spite of these differences, that the support plays an important role for activation of water for both Au and Pt catalysts. Models integrating these findings will be presented.

The value of a successful forward model is that it can be used in a guided stochastic search of the descriptor space to identify optimum catalyst candidates. Tools and information needed and obstacles which must be overcome to close this *inverse* cycle will also be discussed.

[i] T. A. Manz, K. Phomphrai, G. Medvedev, B. B. Krishnamurthy, S. Sharma, J. Haq, K. A. Novstrup, K. T. Thomson, W.N. Delgass, J. M. Caruthers, and M. M. Abu-Omar, *J. Am. Chem. Soc.*, *129(13)*, 3776-3777 (2007).

See more of this Session: Computational Catalysis I - Databases and Informatics
See more of this Group/Topical: Catalysis and Reaction Engineering Division

(24a) Photocatalytic Degradation of Organic Molecules On Visible-Light-Active Sn(II)-Containing Photocatalysts

Bharat Boppana and **Raul F. Lobo**, Chemical Engineering, University of Delaware, Newark, DE

Development of visible light active photocatalysts is a challenge we need to overcome to harvest solar energy effectively considering our need for clean fuels. Doping of d0 (TiO_2) or d10 ($ZnGa_2O_4$) oxides to make them absorb visible light is inefficient because of cation mid-gap dopant states or due to inhomogeneous anion doping. Thus, these doped oxide catalysts generally show lower UV light activity compared with their un-doped counterparts because of this new presence of dopant recombination centers and also due to poor hole mobilities. To offset this, in this presentation, we will explore different ways to surface-modify native oxide and thereby red-shift their absorbance edge while not losing efficiency.Firstly, we focus on the visible light photocatalytic activity of Sn(II)-TiO_2. The catalysts are prepared from a simple reaction of titanium butoxide and several tin precursors at 80oC in aqueous solutions. Samples synthesized with $SnCl_2$ have lower band gaps (red-shifted to the visible region) with respect to anatase TiO_2 attributed to the presence of surface Sn^{2+} species comprised of 5s orbitals. The catalysts are isostructural to anatase TiO_2 even at the highest loadings of Sn^{2+}. When the precursor is changed to $SnCl_4$, rutile is the predominant phase obtained but no reduction in the band gap is observed further corroborating that the earlier observed red-shift is due to Sn^{2+}-5s orbitals. The majority of surface cation sites are tin, with a fraction of tin also incorporated in the bulk particles. The experiments also indicate the presence of chlorine in the samples, also influencing the optical and catalytic properties as confirmed by comparison to materials prepared using bromide precursors. These catalysts can photocatalytically degrade of organic molecules with rates higher than the standards (P25 TiO_2) and also evidenced from the generation of hydroxyl radicals using visible light. The analysis of the degradation of rhodamine B dye by a de-ethylation mechanism using multivariate chemometric methods will also be presented to explain the need to better understand a common test reaction for photocatalytic activity measurements but often incorrectly interpreted.Based on the successful results observed with Sn (II)-titania, this protocol is then extended to incorporate Sn^{2+} with other oxide semiconductors to prepare photocatalysts with interesting electronic properties. We chose a d10 photocatalyst, $ZnGa_2O_4$ previously shown to be active for CO_2 reduction to methane and water splitting, albeit in UV light as its band gap is 4.5 eV. In this case, the presence of a surface SnO_x species containing Sn^{2+} enhances both visible light and UV light activity unlike traditional cation of anion doped titania, wherein the new dopant states act as mid-gap electron traps. This improved activity is due to improved hole mobilities observed by running reactions conducted in the presence of scavengers. The results here expand the materials available for efficient visible light photocatalysis and suggest that an array of novel photocatalysts may be obtained from other d0 or d10 oxide precursors upon the presence of surface modified species.

(24b) Detailed Microkinetic Modeling for Emissions Oxidation From Engine Exhaust

Hom Sharma, Department of Chemical, Materials, and Biomolecular Engineering, University of Connecticut, Storrs, CT, Angela M. Moreno, Center for Clean Energy Engineering, University of Connecticut, Storrs, CT and Ashish B. Mhadeshwar, Department of Chemical, Materials, and Biomolecular Engineering and Center for Clean Energy Engineering, University of Connecticut, Storrs, CT

The precious metal-based Diesel oxidation catalyst(DOC) plays a fundamental role in reducing diesel fuel particulate matter andother harmful emissions such as hydrocarbons (HC), carbon monoxide (CO) and nitricoxides (NO_x), which have adverse impact on human health andenvironment. Despite the extensive research on these catalysts, comprehensiveand predictive kinetic models for simultaneous prediction of multiple emissionsoxidation are lacking. In this work, a detailed elementary step microkineticmodel for oxidation of CO, NO, other nitrogen containing emissions ($NH_3$and HCN) as well as toxic aldehydes (CH_2O) is developed. Thedetailed mechanism development is carried out using several parameterestimation techniques: semi-empiricalUnity Bond Index-Quadratic Exponential Potential (UBI-QEP), Transition State Theory(TST), quantum mechanical Density Functional Theory (DFT), and temperatureprogrammed experiments. Model predictions for catalytic oxidation of variousemission components will be discussed as a function of operating conditions(see Figure for HCN oxidation).

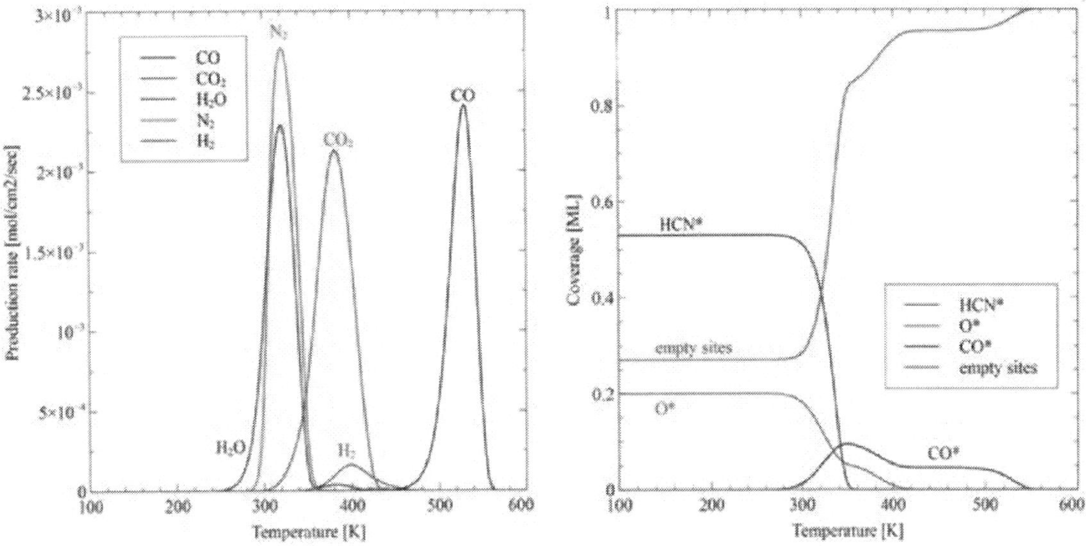

See more of this Session: Fundamentals of Environmental Catalysis
See more of this Group/Topical: Catalysis and Reaction Engineering Division

(24c) Influence of Co-Doped Metals On Mn/TiO2 Catalyst and Its Effect On Selective Reduction of NO with NH3 At Low-Temperatures

Thirupathi Boningari, Krishna Reddy Gunugunuri and Panagiotis Smirniotis, Chemical Engineering Program, School of Energy, Environmental, Biological and Medicinal Engineering, University of Cincinnati, Cincinnati, OH

In recent years, there is an increasing interest in developing a novel low-temperature (353–523 K) SCR catalyst for the removal of NO_x. In our earlier studies, manganese-based titania-supported catalysts have been established successfully for the low-temperature selective catalytic reduction (SCR) of NO. In continuation of our research in low-temperature SCR of NO, we prepared a series of Mn-Me/TiO_2 (Me = Cr, Fe, Co, Ni, Cu, Zn, Ce, and Zr) catalysts by adopting wet-impregnation method and investigated for the low-temperature SCR of NO with NH_3 by varying Me/Mn atomic ratio. The catalytic performance of these materials has been carried out with a high gas hourly space velocity (GHSV) 50,000 h^{-1}, at the temperature range 433-523K. Physico-chemical techniques such as XRD, N_2 physisorption, TPR, XPS, and *in-situ* FTIR were used to investigate the influence of co-doped metals. Our XPS results illustrated that the MnO_2 phase is extremely dominant over the Mn_2O_3 phase (Mn^{4+}/ Mn^{3+}=22.31, 96%) in our best catalyst Mn-Ni(0.4)/TiO_2 anatase, whereas Mn_2O_3 phase is in competition with MnO_2 in other catalysts (Mn^{4+}/Mn^{3+}=1.34-12.67). Increase in reducibility of manganese and absence of the high-temperature (736 K) peak in H_2-TPR studies, suggested that the dominant phase is MnO_2 in Mn-Ni/TiO_2 catalyst. The catalytic performance and characterizations of optimized materials will be discussed in the presentation.

See more of this Session: Fundamentals of Environmental Catalysis
See more of this Group/Topical: Catalysis and Reaction Engineering Division

(24d) Unusual Low-Temperature Ammonia Oxidation Behavior In Zeolite-Based SCR Catalysts

Krishna Kamasamudram, Aleksey Yezerets and Neal Currier, Corporate Research and Technology, Cummins Inc, Columbus, IN

Selective catalytic reduction of NOx with NH3 (SCR) has emerged in recent years as a primary technology for meeting the increasingly stringent environmental regulations for diesel exhaust emissions. This was enabled by a new class of SCR catalysts, introduced in the US in the recent years, based on metal-exchanged zeolites. These catalysts combine a number of beneficial characteristics, including excellent activity across a broad range of conditions and outstanding hydrothermal stability.

Our detailed characterization of the individual redox and acid-base functions of these catalysts showed that they possess poor activity for NH3 oxidation by O2, with no appreciable conversion up to ~ 400C, as expected from catalysts designed to selectively react NH3 with NOx in the presence of vast excess of O2. On the other hand, we discovered that, under the SCR reaction conditions unexpectedly large NH3 oxidation by O2 (parasitic NH3 oxidation) was observed, even at temperatures as low as ~250C. Further study of this unexpected phenomenon by systematically varying inlet gas composition has confirmed that such parasitic NH3 oxidation has only occurred in the presence of NOx in the gas feed. The data collected to date indicate that NOx modifies oxidation state of the exchange metal sites in zeolite, leading to their increased activity in the ammonia oxidation process.

See more of this Session: Fundamentals of Environmental Catalysis
See more of this Group/Topical: Catalysis and Reaction Engineering Division

(24e) Experimental and Kinetic Modeling Study of Selective Catalytic Reduction of NOx On Fe- and Cu-Based Zeolitic Monolithic Catalysts

Pranit S. Metkar, Michael P Harold and Vemuri Balakotaiah, Department of Chemical and Biomolecular Engineering, University of Houston, Houston, TX

The selective catalytic reduction (SCR) of NOx with NH_3 is considered to be a highly promising technique for the efficient reduction of highly detrimental NOx (to N_2) emitted from diesel engine vehicles. Amongst the various catalysts available for SCR, Fe and Cu-based zeolite catalysts are considered to be highly stable and efficient towards NOx conversion over a broader temperature range. Cu-zeolite has been found to be a very good low temperatures (< 350 °C) NOx conversion catalyst. However, Fe-zeolite has been found to be a better catalyst at higher temperatures (> 350 °C).

This is a comprehensive experimental study coupled with the kinetic modeling of key SCR reactions. A comparison is made between Fe- and Cu-zeolite catalysts and the observed differences between the two catalysts are explained with the kinetic model. These experiments included NH_3 and NOx uptake and temperature-programmed desorption (TPD), NO and NH_3 oxidation, standard SCR (NO+NH_3), fast SCR (NO+NO_2+NH_3), and NO_2 SCR (NO_2+NH_3).

The Cu-zeolite catalyst was seen to have higher ammonia and NO_2 storage capacities. It was insensitive to NO_2/NOx ratios at lower temperatures and gave very high NOx conversions. However, Fe-zeolite catalyst was found be a better NO oxidation catalyst and NO_2 was found to have a strong inhibition on NO oxidation reaction. NH_3 was seen to inhibit the standard SCR reaction on both Fe and Cu-zeolites at lower temperatures. But this inhibition was less pronounced for Cu and it disappeared at slightly higher temperatures. Ammonium nitrate was found to inhibit NOx reduction at lower temperatures in the presence of excess feed NO_2.

Monolith catalysts with different washcoat loadings, washcoat thicknesses and lengths were synthesized to study the mass transfer limitations in zeolite catalysts. A steady state experimental study of various SCR reactions showed the presence of washcoat diffusion limitations at intermediate temperature range and hence these diffusion limitations should be considered in any kinetic model and catalyst designs. A detailed analysis of the effect of temperature on the transitions between the controlling regimes (for various SCR reactions) is presented.

A kinetic model which includes the effect of washcoat diffusion limitations is developed on the basis of all the above experimental observations. A good agreement is obtained between the experimental observations and the model predictions.

Keywords: Selective Catalytic Reduction, Cu-zeolite, Fe-zeolite, Standard SCR, Fast SCR, Ammonia oxidation, NO Oxidation, Washcoat diffusion, Mass transfer, Kinetic model.

See more of this Session: Fundamentals of Environmental Catalysis
See more of this Group/Topical: Catalysis and Reaction Engineering Division

(24f) NOx Reduction Over Perovskite Based Lean NOx Trap

Qi, G., General Motors Global R&D

Conventional Lean NO_x Trap (LNT) catalysts contain high loadings of Pt in order to oxidize NO to NO_2, a critical reaction in LNT application, and NO_2 stored on NO_x trap material (such as BaO) to form nitrates/nitrites during a lean period and reduced to nitrogen during a shorter rich period. GM R&D has reported that high NO oxidation to NO_2 was achieved over a Mn based perovskite and a $LaMnO_3$ based, Pt free LNT catalyst showed comparable activity to a commercially available Pt based catalyst.

This report presents our investigation on the mechanistic aspects of the storage and reduction of NO_x over $LaMnO_3$, BaO/Al_2O_3, $LaMnO_3/BaO/Al_2O_3$ and $LaMnO_3/Pd/BaO/Al_2O_3$ catalysts. Our results showed that in the presence of oxygen, NO was effectively oxidized to NO_2 on $LaMnO_3$, while no NO_x ($NO+NO_2$) were stored; under the same conditions, a significant amount of NO_x were stored over $LaMnO_3/BaO/Al_2O_3$ indicating NO_x should be adsorbed on BaO site not $LaMnO_3$. Although NO_x can be stored over $LaMnO_3/BaO/Al_2O_3$ during the lean periods, it could not be reduced by CO. After combining Pd/Al_2O_3 and $LaMnO_3/BaO/Al_2O_3$ catalyst, it was found that NO_x could be stored during the lean conditions but also reduced by CO. Based on our experimental results it could be concluded that the Pd catalyst was mainly responsible for NO_x reduction during the rich conditions and plays a negligible role for NO oxidation under lean conditions; the functionality of $LaMnO_3$ and BaO/Al_2O_3 is for NO oxidation to NO_2 and NO_x storage under lean conditions, respectively.

(24g) Nitrate Formation/Decomposition On Ba-, and K-Based Model LNT Catalysts: Similarities and Differences

Do Heui Kim[1], Kumudu Mudiyanselage[2], Janos Szanyi[2], Ja Hun Kwak[1] and Charles H. F. Peden[1], (1) Institute for Interfacial Catalysis, Pacific Northwest National Laboratory, Richland, WA, (2)PNNL, Richland, WA

Diesel engines are a promising near-term solution for obtaining high energy efficiency because of their unparalleled power density and low CO_2 emissions. However, there are challenges for the catalytic aftertreatment systems, especially due to the necessity to selectively reduce nitrogen oxides (NO_x) in the net oxidizing environment. NO_x storage-reduction (NSR) catalysts are generally considered as one of the potential options for removing NO_x in the diesel engine exhaust. In the NSR technology, alkali or alkaline earth elements (K or Ba) take up NO_x under lean operation and store them as nitrates, which are subsequently reduced during rich operation. In this presentation, we compare the characteristics of Ba and K storage elements during NOx adsorption and desorption processes using multiple techniques, including TR-XRD, FTIR and NO_2 TPD. We find that, although both elements store NOx as nitrates, there are significant differences between the morphology changes of the Ba and K nitrates.

See more of this Session: Fundamentals of Environmental Catalysis
See more of this Group/Topical: Catalysis and Reaction Engineering Division

(37a) Mechanistic Study of Methane Hydrate Nucleation Under Realistic Conditions

Brandon C. Knott, Michael F. Doherty and Baron Peters, Chemical Engineering, UC Santa Barbara, Santa Barbara, CA

The formation of methane hydrate clathrates is of great interest in many scientific and industrial contexts, including as a hindrance in oil and natural gas pipelines and as a potential energy resource. We employ semigrand canonical Monte Carlo (SGCMC) simulations to study the nucleation of methane hydrate clathrates utilizing a coarse-grained model that represents each molecule as a single particle.[1] SGCMC simulations facilitate the study of nucleation at low (and constant) driving force. This is desirable because these are the conditions under which most scientific and industrial systems of interest operate. We seek to shed light on the hydrate formation mechanism under the conditions at which they form in nature and important industrial contexts. We study nucleation pathways along two distinct coordinates: 1) the clustering of methane molecules regardless of water structure and 2) the number of clathrates (water) cages formed.

[1]L.C. Jacobson and V. Molinero, J. Phys. Chem. B **114**, 7302 (2010).

See more of this Session: Multiscale Modeling I
See more of this Group/Topical: Catalysis and Reaction Engineering Division

(37b) Development of a Coarse -Grained Model, and Its Application to Free Energy Studies In Proteins

Spyridon Vicatos, Anna Rychkova, Anatoly Dryga and Arieh Warshel, Chemistry, University of Southern California, Los Angeles, CA

Simplified Coarse Grained protein models have been in use since 1975 and are being used extensively in folding studies. In this work, we provide a brief description of our Coarse Grained (CG) model and present its application on the free energy calculations of large biological systems. This model has been recently used for the study of the free energy associated with transmembrane (TM) helices insertions (Rychkova et al, PNAS 107,17598-17603). Our CG model, after recent modifications for the calculation of electrostatic interactions, was used for the study of TM/translocon system even though the structural characteristics are still not clear. The CG model has been used for the study of voltage activated ion channels. Meaningful converging microscopic simulations for ion channels are not yet available and macroscopic treatments involve major uncertainties in key features. The CG model has been extended, in order to perform simulations of membrane/protein systems under external potential (Dryga et al, BBA, under preparation). Finally, the CG model shows promising results, when used for the calculation of protein folding energies ΔG_{fold} and changes of folding energies upon mutations $\Delta\Delta G$.

See more of this Session: Multiscale Modeling I
See more of this Group/Topical: Catalysis and Reaction Engineering Division

(37c) Coarse-Graining Proteins Using Relative Entropy Theory

M. Scott Shell, Scott Carmichael and Aviel Chaimovich, Department of Chemical Engineering, University of California Santa Barbara, Santa Barbara, CA

The development of accurate coarse-grained molecular models is central to many efforts in soft-condensed matter, materials, and biophysics that seek to understand large length and time scale properties. We recently proposed a broad new statistical-mechanical framework for such problems based the relative entropy, a phase-space functional measuring the "information" lost upon coarse graining and hence the (inverse) fitness of a given coarse-grained model [1]. We suggest that minimization of the relative entropy provides a universal variational principle for coarse-graining.

Here, we show how this approach provides a quantitative and systematic approach to the development of coarse-grained models of proteins. We first describe the development of stable, efficient, and robust numerical coarse-graining algorithms based upon relative entropy minimization [2]. In particular, these approaches are able to locate optimal models in very high dimensional parameter spaces, here for one-, two-, and three-bead (per amino acid) models of peptides. We show that the relative entropy framework offers exquisite control over coarse-graining errors, and is able to generate coarse peptide models whose structural correlations and folding behavior are in excellent agreement with more expensive all-atom simulations. We also demonstrate how these models are well-suited to large-scale simulations of peptide self-assembly. Second, we show that the relative entropy approach provides an a priori prediction of coarse-graining errors (i.e., errors in the properties predicted by coarse-grained models), and gives a systematic strategy for designing coarse-grained potentials [3]. In particular, we use this theory to develop simple analytical energy-landscape models of protein folding. Here, an ensemble of predicted structures for a large protein is projected onto a coarse-grained funnel landscape. The relative entropy quantifies which structure, when considered at the bottom of the funnel, best describes the energy-structure relationships, in turn predicting which structure is nearest to the true native structure. We present results on several model protein systems and from a recent community-wide blind structure prediction competition.

[1] M. S. Shell, "The relative entropy is fundamental to multiscale and inverse thermodynamic problems," J. Chem. Phys. 129, 144108 (2008).

[2] A. Chaimovich and M.S. Shell, "Anomalous waterlike behavior in spherically-symmetric water models optimized with the relative entropy," Phys. Chem. Chem. Phys 11, 1901 (2009).

[3] A. Chaimovich and M. S. Shell, "Coarse-graining errors and optimization using a relative entropy framework," J. Chem. Phys. 134, 094112 (2011).

A Multi-scale Modeling Study of MDH-Catalyzed Methanol Oxidation: The Effect of the Ion in the Enzyme Active Site

Purnima Kharidehal and Daniela S. Mainardi*

*Institute for Micromanufacturing, Chemical Engineering Program, Louisiana Tech University, Ruston, LA 71272

ABSTRACT

A multi-scale modeling approach is presented here to investigate methanol oxidation by Ca^{2+}- and Mg^{2+}-containing Methanol Dehydrogenase (MDH) enzymes. A combination of tools involving Density Functional Theory (DFT), Transition State Theory (TST), and Kinetic Monte Carlo (kMC) are combined in order to model enzyme kinetics. Information regarding energy barriers and pre-exponential factors needed to determine reaction rates involved in each step of the Modified-Hydride Transfer methanol oxidation mechanism by MDH is obtained from DFT and TST respectively. This microscopic reaction rates are then provided as inputs in the kMC program, and the methanol oxidation process is modeled on a 2-D reactive surface representing the enzyme active site. A grid representing possible methanol adsorption, diffusion and reactions sites were mapped onto the projected active site image in a 2-D surface for a random substrate approach onto the reactive enzyme surface. Comparisons regarding enzyme kinetics were performed for methanol oxidation by Ca^{2+}- and Mg^{2+}-containing MDH.

INTRODUCTION

Methanol Dehydrogenase (MDH) enzyme is a bacterial quinoprotein that oxidizes methanol and other primary alcohols to their corresponding aldehydes [1]. The active site of MDH contains a Pyrroloquinoline Quinone (PQQ) cofactor, a divalent calcium cation, a catalytic base ASP-303 and twelve other amino acids, and water molecules[1]. It has been found that the Ca^{2+} ion in the active site of MDH is bonded to the nitrogen atom molecule (N6), the C5 quinone oxygen (O5), the oxygen of the carboxylate (O10) of the PQQ, the oxygen atoms (O12 and O13) from the carboxylate of Glu-177, and the oxygen atoms O11 and O14 of Asn-261 and Asp-303, respectively (Figure 1) [2-8]. Results from X-ray studies, distances from the Ca^{2+} ion to the N6, O5, and O10 atoms of PQQ are in the 2.32–2.47, 2.25–2.77, and 2.30–2.44 Å range, respectively[2-8] .Ca^{2+} ion in the active site holds the PQQ in place and also acts as a Lewis acid during the oxidation mechanism[5]. From previous studies, it was observed that the energy barrier for the oxidation of methanol was influenced by the atomic size of the metal ion in the active site of MDH[9].

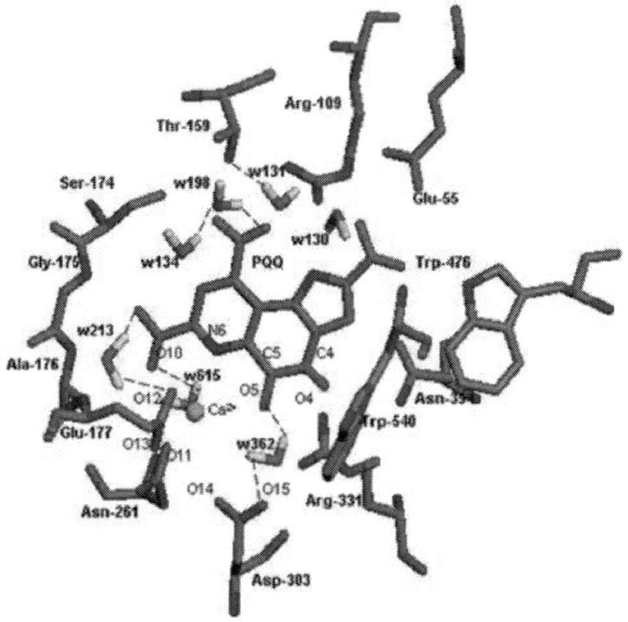

Figure 1: Active site of Methanol Dehydrogenase enzyme (MDH) [7].

Oxidation Mechanism for MDH

In the literature, two mechanisms for the oxidation of methanol by the naturally occurring Ca^{2+}-containing MDH have been commonly proposed; the three step addition-elimination (A-E) mechanism and the four step Hydride transfer (H-T) mechanism[6, 10]. The rate determining step for methanol oxidation by naturally occurring Ca^{2+}-MDH is the formation of formaldehyde[6, 10]. No agreement has been reached in the literature regarding which mechanism (A-E or H-T) actually operates in nature; however the Hydride Transfer (H-T) mechanism seems to be the preferred one. Hence, in this study the Hydride transfer mechanism is considered[11].

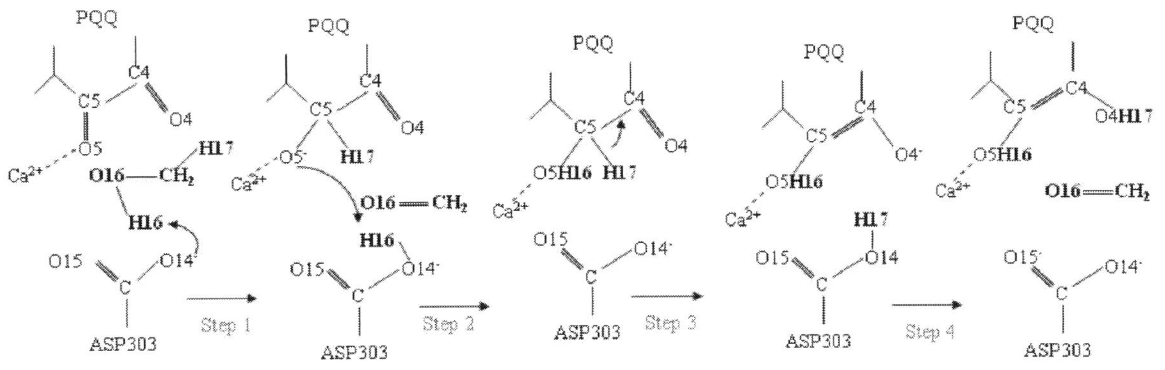

Figure 2: Four step Hydride-Transfer (H-T) methanol electro-oxidation mechanisms by Ca^{2+}-containing Methanol Dehydrogenase Enzyme[10].

Several experimental and theoretical studies have been reported that other bivalent metal ions can replace the natural ones in the active site of MDH enzymes[12]. Spectroscopic and semi-empirical calculations have shown that the binding constant of Sr^{2+} and Ba^{2+} to PQQ are weaker than that of Ca^{2+}; which can be attributed to the size of the calcium ion as that best fits in the binding pocket of PQQ in the active site. It has been found that the binding of the larger metal ions to the PQQ cause a distortion, making the coordination smaller than that of the calcium ion[12]. To observe the structural variations of the active site and kinetics of MDH enzyme on the methanol oxidation, the MDH active site Ca^{2+} ion was replaced with Ba^{2+} and Sr^{2+} and reported that there were no major differences between these enzymes in the interactions between PQQ and the metal ions; however, whe tested upon the oxidation of methanol, the activation energy decreased in both modified MDH enzymes[12]. Sr^{2+}-MDH showed a decrease in activation energy (7.6 kcal/mol) with respect to that of Ca^{2+}-MDH (8.5 kcal/mol); but the lowest activation energy was found in the case of Ba^{2+}-MDH (3.5 kcal/mol), which was unexpected since Barium ion (weaker Lewis acid) should decrease the activity of the enzyme and hence the activation should be larger[12]. Thus, it shows that as the ionic radii increased (Ba^{2+}(1.35Å)>Sr^{2+}(1.12Å)>Ca^{2+}(0.99Å)) the activation energy decreased ($E_{act}Ba^{2+}$-MDH < $E_{act}Sr^{2+}$-MDH < $E_{act}Ca^{2+}$-MDH).

Kinetic Monte Carlo

Methanol oxidation by MDH enzymes is investigated using a multi-scale approach, as implemented in a Dynamic (kinetic) Monte Carlo (kMC) program which permits the investigations of length scales up to meters and time-scales up to the range of seconds. The kMC program, Carlos

4.1, developed by John Lukkien and A. P. J Jansen is described here[13, 14]. The approach followed here starts by first describing the system of the catalytic reactions by a stochastic model where an MDH catalytic surface (active site of the enzyme) is created for methanol oxidation to take place, and the lattice-gas models are considered to define the catalytic surface. The surface is defined by a discrete, regular two dimensional array of lattice points called "sites" (Figure 3) [13-15]. These sites with characteristic values and the lattice together are termed as a "*configuration*" [13-15]. Thus, a change from one configuration to another can be seen as a reaction. The species can diffuse, react, associate and dissociate once they are added onto the lattice following a set of probabilities and rules. As the simulation proceeds, the system undergoes dynamic evolution until a quasi –equilibrium steady state is reached.

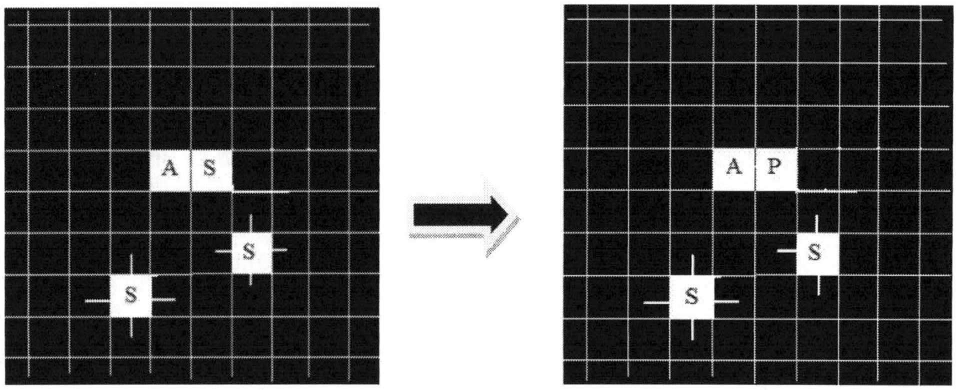

Figure 3: The sites occupied by the substrate molecules (methanol, S) on the Lattice (black grid) after following a random walk. When the substrate molecules reach the most reactive region of the enzyme active site (A), they get converted into product molecules (formaldehyde, P) [16].

Monte Carlo simulations can be implemented for steady-state as well as dynamic systems[13-15]. In a steady state system, where the system does not vary with time, a simple pattern is followed (Figure 3), where a lattice site is chosen randomly first and then the probabilities are calculated to determine whether the reaction at that site is possible. Once the reaction probabilities are calculated, the next site is chosen and the same procedure is repeated. After each grid site has been visited at least once on average, the simulation completes one Monte Carlo time step. In the case of dynamic systems (vary with time), there is an uncertainty in calculating the reaction probabilities with respect to the reaction steps. In order to simulate the dynamical interpretation, Monte Carlo methods by Fichthorn and Weinberg are used[17]. Microscopic reaction rates are needed as input, and these can be obtained either from experiments or *Ab Initio* calculations, and as a result, kMC methods yield time evolution of individual reaction rates (difficult to determine experimentally); which compete with each other as a function of external variables such as pressure and temperature. The rates are specified as probabilities and the time evolution of surface configuration is given by a Master Equation[14] (1), which is derived from first principles as Monte Carlo is a probabilistic approach.

$$\frac{dP(c,t)}{dt} = \sum_{c' \neq c} [P(c',t) r_{c',c} - P(c,t) r_{cc'}] \qquad \text{Eq. (1)}$$

where $P(c,t)$ is the probability of finding the system in configuration c, at time t, and $r_{cc'}$ is the microscopic rate of the reaction that transfer c into c'. There are methods which are used internally by the Carlos program to solve this equation: First-Reaction Method (FRM), Variable Step Size Method (VSSM), Random Selection Method (RSM) [14]. The First-Reaction Method (FRM) [14] is used here, which is appropriate for cases where reaction constants vary with time. According to this method, when the system is in a given configuration c, the set of all possible reactions is determined, and a time of occurrence $t_{c'c}$ is generated for each reaction i compatible with configuration c, according to equation (2).

$$\exp\left[-\int_{t}^{t_{c'c}} r^{i}_{c'c}(t') \cdot dt \right] = \delta \qquad \text{Eq. (2)}$$

In equation (2), $r^{i}_{c'c}$ is the time dependent rate of reaction i, and δ is a random number selected uniformly in the interval (0,1). Then, the reaction with the smallest time is selected, the configuration is changed accordingly, and the time is incremented to t. Finally, the set of possible reactions is generated according to the new configuration c', and the procedure is repeated. The microscopic rate r_i are related to macroscopic parameters as stated by equation (3),

$$r_i = v_i \exp\left[-\frac{E_{ai}}{RT} \right] \qquad \text{Eq. (3)}$$

where v_i is the prefactor, and E_{ai} is the activation energy of a reaction i.

Transition State Theory

Activation energies are required in order to calculate the rate constants for the processes to be modeled. For the MDH-assisted methanol oxidation study, the activation energies are obtained using Density Functional Theory (DFT) calculations on MDH models. The calculation of rate constants from the activation energies is conducted using Transition State Theory (TST) for the prediction of pre-exponential factors and Equation (3) [18]. According to the TST, the reactants need to overcome an energy barrier in order to form products and it postulates equilibrium (Boltzmann) energy distribution, which relates the reaction rate of a reactions with its Gibbs free energy by the following expression:

$$k = \frac{k_B T}{h} \exp\left(-\frac{\Delta G}{RT}\right) \qquad \text{Eq.(4)}$$

where k is the rate constant (s^{-1}), k_B is the Boltzmann's constant (3.29×10^{-24} cal/K), h is the Planck's constant (6.626×10^{-34} Js), T is the absolute temperature (298.15 K at room temperature), R is the universal gas constant (8.314 J/Kmol) and ΔG is the Gibbs free energy of activation[18].

METHODOLOGY

Density Functional Theory

The DMOL^3 module[19] of the Materials Studio® software by Accelrys, Inc. is used in this work[20]. It is a quantum mechanical package that models the electronic structure and energetics of molecules, solids and surfaces using Generalized Gradient Approximation (GGA) functionals within the Density Functional Theory formalism[21, 22]. The Becke-Lee-Yang-Parr (BLYP) exchange correlation functional[23, 24] and the double numerical with polarization (DNP) basis set, the best set available in DMOL^3, were used for performing the optimization calculations. The DNP basis set considers a polarization *d* function on heavy atoms and a polarization *p* function on hydrogen atoms. It can be compared to split-valence double zeta 6-31G** basis set in size, but is more accurate than the Gaussian basis sets of the same size. The errors are expected to be in the second decimal for calculated bond lengths (Å) and in the order of 2–5 (kcal/mol) in energies when BLYP/DNP theory level is used; which has been proven to be acceptable for the study of Ca^{2+}- and Ba^{2+}-MDH [25].

Small models of enzyme active sites are used to study the enzyme H-T reaction mechanism (Figure 2)[26-33]. Accordingly for treating the neglected enzyme environment, a homogeneous polarizable medium with some assumed dielectric constant is used, and density functional theory calculations are then performed to obtain reaction energy barriers[26-28, 30-33]. A homogeneous polarizable medium was considered with a continuum solvation model known as COSMO (conductor-like screening model) as implemented in the $DMOL^3$ module, to estimate solvation effects of the rest of the enzyme that are not included in the model[34-36]. This model works in such a way that the solute molecule forms a cavity within the dielectric continuum of permittivity that represents the solvent. To model the protein surroundings, the standard value for the dielectric constant ε was chosen to be 4 and so, that the final energy includes the $DMOL^3$/COSMO electrostatic energy, the geometry optimizations were performed in the presence of this solvation model[26-28, 30-33]. To confirm that the stationary points on the potential energy surface of the systems are in fact local minima (all real frequencies) or transition states (only one imaginary frequency), the harmonic vibrational frequency calculations were performed.

Kinetic Monte Carlo Modeling

To explore the kinetics and construction of the lattice representing the enzyme active site, it is important to discuss the assumptions and boundary conditions of the systems, since the reaction mechanics are relative to these aspects. The following assumptions are made to describe the model and to reduce complexity based on the assumptions reported [16]:

1. On the surface of the lattice, the substrate molecules follow a random walk.
2. The individual substrate molecules do not interact with each other.
3. There are no interactions between the substrate molecules and the obstacles.
4. The substrate molecules move towards the active site by the shortest available distance following the random walk motion.
5. The excluded volume condition is maintained i.e. at any instant of time one lattice site cannot be occupied by more than one molecule of the same or different species.

The well-known Michaelis-Menten equilibrium enzyme kinetics [37] equation is considered:

$$E + S \rightarrow ES \rightarrow E + P \qquad \text{Eq.(5)}$$

where E denotes the enzyme, S denotes the substrate (methanol), ES is the intermediate complex (enzyme-substrate), and P is the product (formaldehyde). Since the equation (5) is assumed to be diffusion limited, the formation of an intermediate ES is not taken into account and the substrate is almost instantaneously converted to product at the active site (A).Therefore, equation (5) reduces to a simpler reaction between the active site (A) and the substrate (S) molecule.

$$A + S \xrightarrow{k_{(t)}} A + P \qquad \text{Eq. (6)}$$

To carry out the simulations, a 64 × 64 lattice with cyclic boundary conditions was used. An important aspect to consider while modeling enzymatic reactions is the distribution of the substrate and obstacles. Previous works reported on enzyme kinetics modeling have used random sampling to distribute the species on the lattice and their reports showed that the enzyme kinetics were influenced by aspects like obstacle density. Thus, according to the approach described by Dandala *et al.*[16], a biased approach is used in sampling the lattice and to investigate its effect on the kinetics of methanol oxidation by MDH. The center of masses for all the amino acids in MDH are calculated and plotted with the active site amino acids as the center; which is in a way similar to the coarse graining method where the molecules are represented as beads. The representation of the MDH active site as beads is shown in Figure 4.

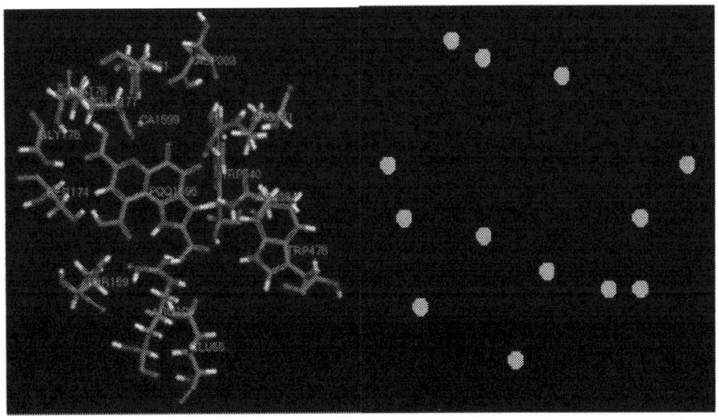

Figure 4. MDH enzyme Active Site (left) showing the various amino acids and the active site model showing the amino acids plotted as dots on the lattice according to their center of masses (right) [16].

The rate constants of H-T process are given as input at the start of each simulation and these are calculated from the activation energies obtained through the DFT calculations. The obstacle and substrate concentrations are specified at the start of each simulation and the simulation time is counted in Monte Carlo steps. For each simulation, the obstacle concentration is fixed while the substrate concentration decreases as the simulation proceeds due to the conversion of substrate molecules to product.

The simulation starts off with the substrate molecules (red) diffusing (making a random walk) on the lattice with the obstacles distributed (black) on it (Figure 5). Active site amino acids (green) are distributed at the center of the lattice and when the substrate molecules come in contact with any one of the active site amino acids, a product (blue) is formed which also diffuses on the lattice. The substrate molecule (red) entering the active site regime to react can be seen in Figure 5 [16]. Through this model, it is not only possible to visualize the enzymatic reactions that take place inside the enzyme but also to get useful information, such as the concentration profiles of substrate and product with respect to Monte Carlo time, obstacle density, reaction rate constants and the positions of obstacles on the lattice [16].

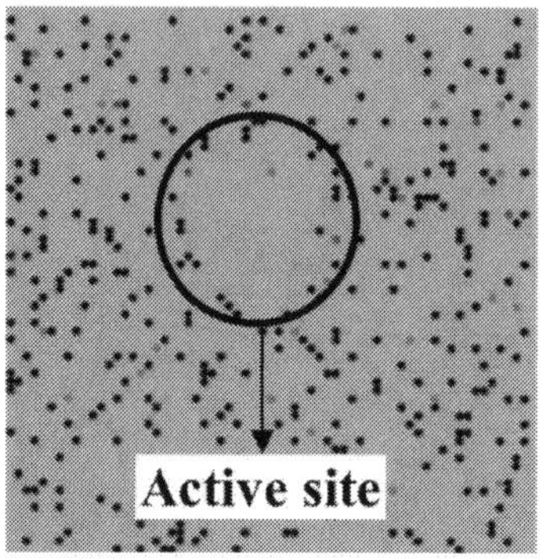

Figure 5. The green dots represent the active site cluster plotted on the two dimensional lattice with the Substrate (Methanol molecules) as red dots and Product (Formaldehyde Molecules) as blue dots making a random walk. The black dots represent the Obstacles (other amino acids in MDH) [16].

RESULTS AND DISCUSSION

DFT Calculations

In order to investigate the proposed H-T methanol oxidation mechanism, from previous studies Mg^{2+}-MDH active site models were selected[38]. The active site model consists of the PQQ molecule, Mg^{2+} ion, aspartic acid, glutamic acid, asparagine and three water molecules (Figure 6). According to the Himo *et al.* approach[39], unprotonated (charged) amino acids can be modeled by their corresponding protonated (neutral) amino acids. This was implemented and tested by Himo *et al.* [39] and well justified by the fact that proteins and enzymes have low dielectric constants ($\varepsilon \sim 4$), indicating that charge separation, if present, should be very small. These authors have shown that, by using these charge-neutral models, accurate enzymatic reaction energies and barriers were yielded. Thus, in our model we used the "Charge-neutral models" for amino acids[39]. The active site model corresponding to Mg^{2+} ion instead of the naturally occurring Ca^{2+} in the MDH active site was tested upon the H-T methanol oxidation mechanism. The reactant, intermediates and product for the four steps of the mechanism were optimized at the BLYP/DNP theory level and included solvation effects with a dielectric constant of 4 to represent the protein environment. The transition states were obtained for each step of the mechanism and the energies, representative bond distances, and energy barriers were compared to the ones obtained for the naturally occurring MDH.

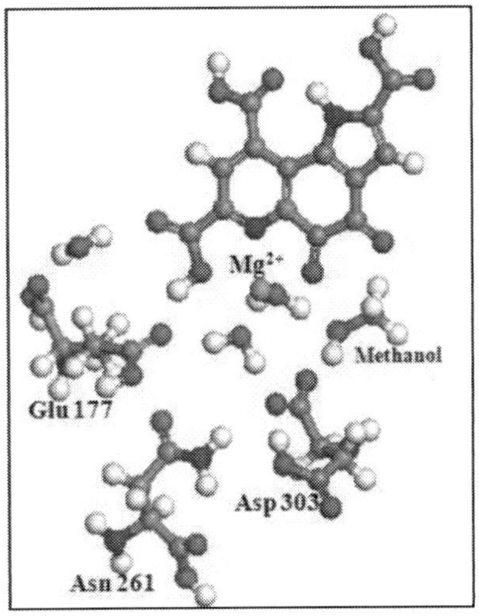

Figure 6: Active site model for Mg^{2+}-containing MDH.

To obtain the minimum energy configuration the reactant, intermediates and product were energy minimized and the ground state conformations were found. The energies obtained for the different orientations were compared and the minimum energy structure was considered for obtaining the transition states (Figure 7). The geometries of the reactant, intermediates and product of Mg^{2+}- containing MDH are in good agreement with the experimentally observed configurations of Ca^{2+}-MDH[38].

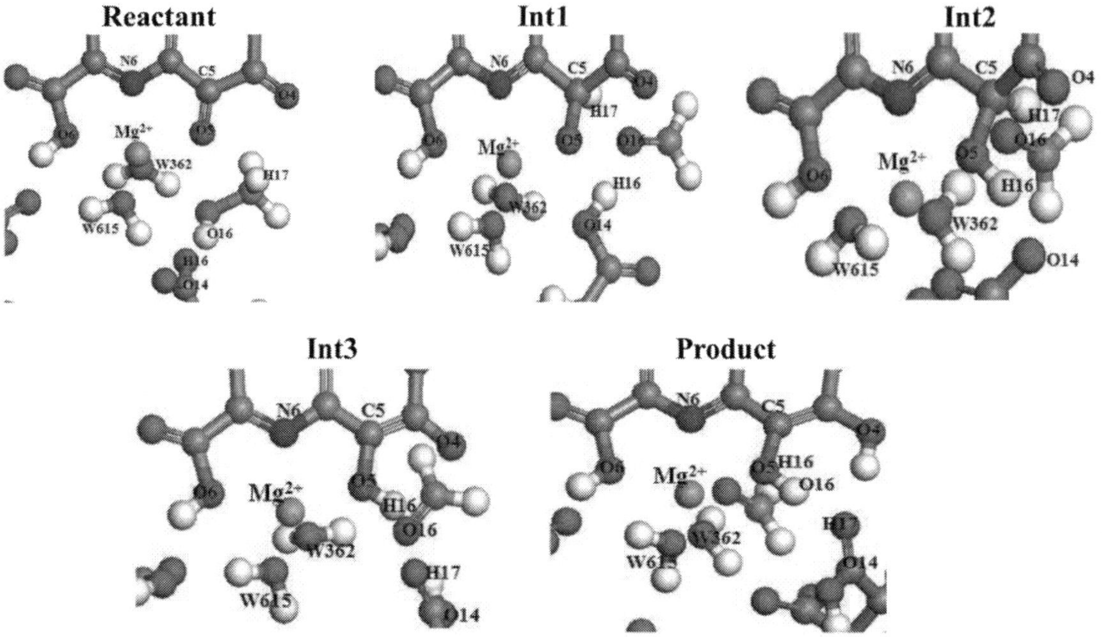

Figure 7: Minimum energy structures corresponding to Mg^{2+}-MDH models corresponding to the H-T methanol oxidation mechanism of Figure 2 obtained at BLYP/DNP theory level.

Rate Determining Step of the Hydride Transfer Mechanism by Mg^{2+}-MDH:

In this step, the hydride (H17) transfers from methanol to the C5 of PQQ and the H16 proton abstraction from methanol to Asp-303 occurs simultaneously forming formaldehyde (Figure 7). Thus, the intermediate (INT1) is evolved through the first transition state (TS1). At BLYP/DNP theory level on Mg^{2+}-MDH, the C5-H17 bond length was reduced from 3.63Å (Reactant) to 2.769Å (TS1) and finally to 1.121Å (INT1) and the bond length O14-H16 evolved from 1.768Å to 1.041Å (TS1) to 1.013Å (INT1) showing the abstraction of the proton H16 by Asp-303. For Mg^{2+}-MDH, the free energy barrier for TS1 was calculated to be 21.2 Kcal/mol (19.2 Kcal/mol for Ca^{2+}-MDH) [38] with respect to the reactant (Figure 8). Mg^{2+} did not polarize the C5-O5 bond as Ba^{2+}[38], thus resulting in less negative charge on the oxyanion O5 of PQQ. The hydrogen bonding (W362-O5) increased from 3.085Å (Reactant) to 3.347Å (TS1) to 3.169Å (INT1); hence the barrier is higher for this step compared to that of Ca^{2+}-MDH.

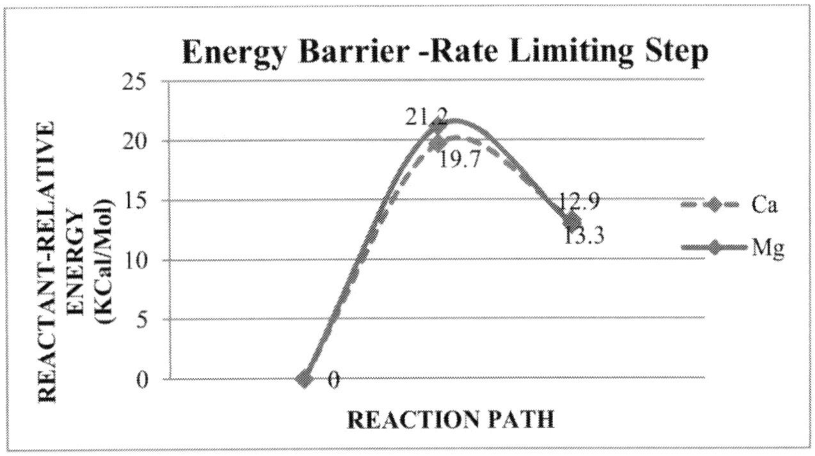

Figure 8: Energy Barriers for the Rate Determining Step in the H-T mechanism by MDH active site models with the two ions.

Kinetic Monte Carlo Simulations

Using Equation (4) and the free energy barrier values from the DFT calculations, the rates for the mechanism steps are obtained. The rate constant obtained for the case of Mg^{2+}-MDH is lower than the rate constant in the case of Ca^{2+}-MDH for the rate determining step as the energy barrier in the case of Mg-MDH is higher than that of Ca^{2+}-MDH. These rate constants are given as input and the substrate (methanol) and obstacle (amino acids that are not part of the active site) concentrations are also specified at the start of each simulation. The influence of the number of amino acids (obstacle density) on the diffusion of methanol towards the active site is investigated (Figure 9 for Ca^{2+}-MDH and Figure 10 for Mg^{2+}-MDH). This will provide useful information like the variations in the product formation rate or substrate consumption as the number of obstacle is changed.

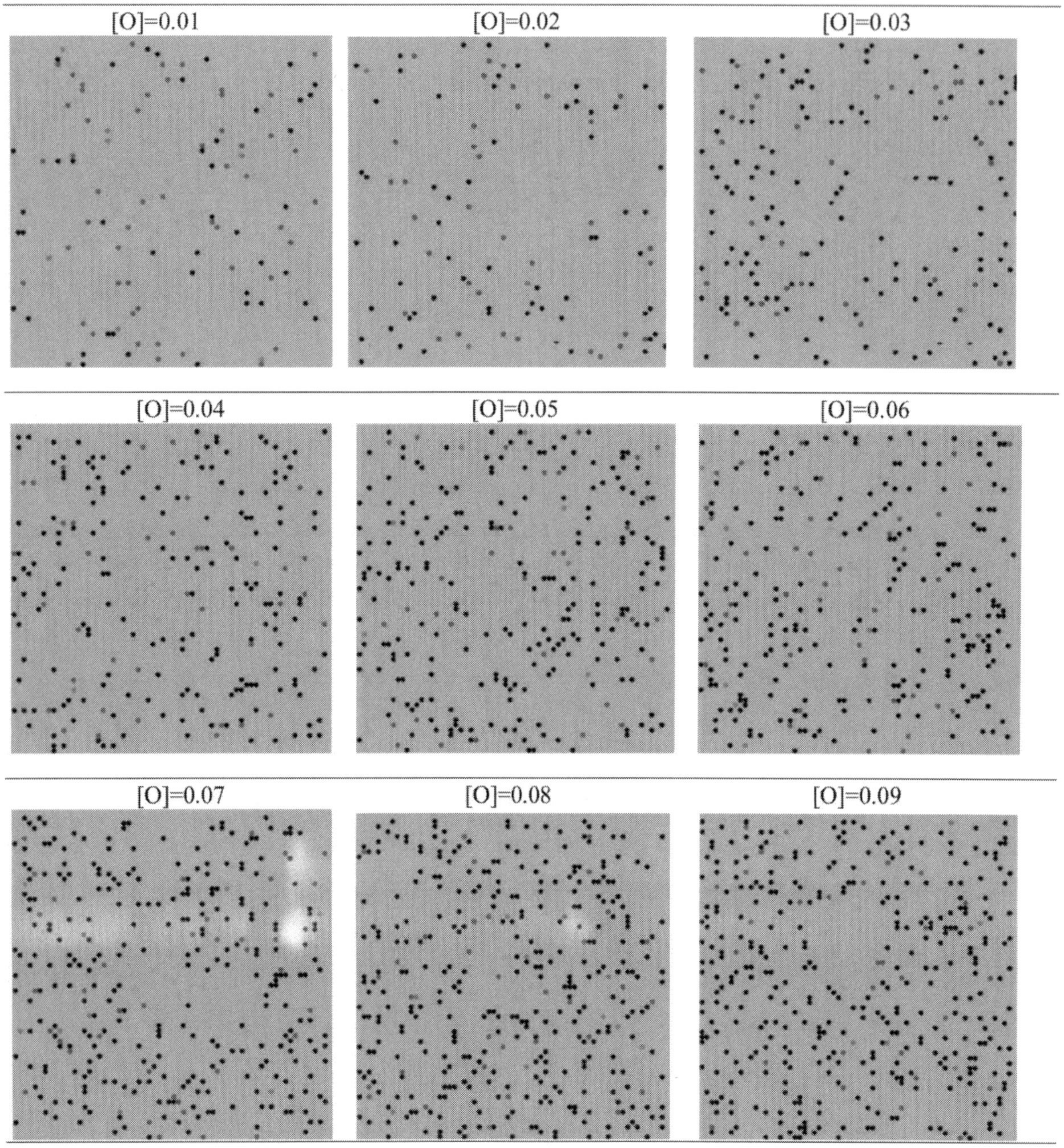

Figure 9. Lattice configurations showing the Substrate (Methanol molecules) as red dots and Product (Formaldehyde Molecules) as blue dots with increasing Obstacle concentration (other amino acids in MDH), [O], at t=1500MC steps for Ca^{2+}-containing MDH and the green dots represent the active site cluster.

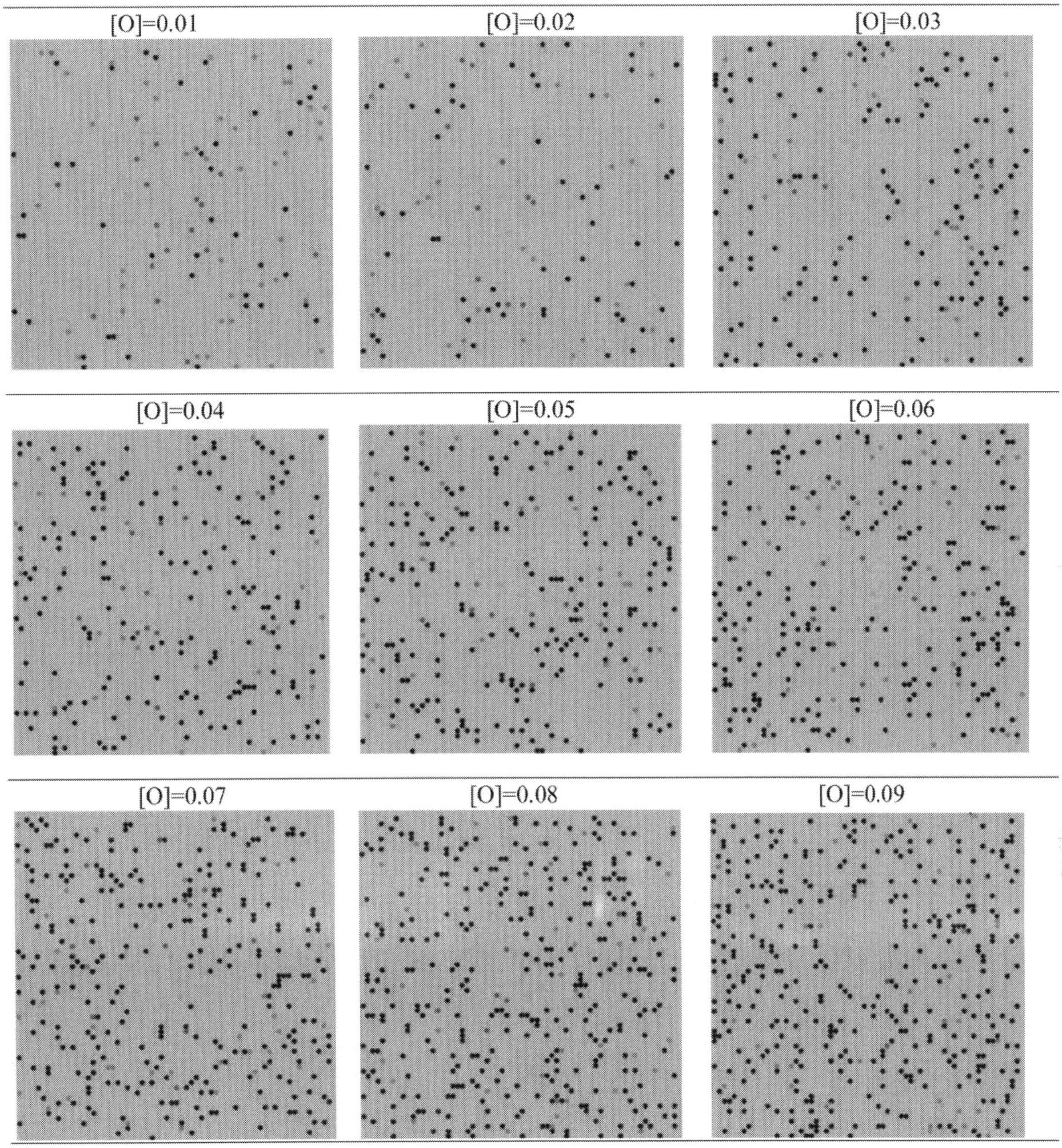

Figure 10. Lattice configurations showing the Substrate (Methanol molecules) as red dots and Product (Formaldehyde Molecules) as blue dots with increasing Obstacle concentration (other amino acids in MDH) at *t*=1500 MC steps for Mg^{2+}-containing MDH and the green dots represent the active site cluster.

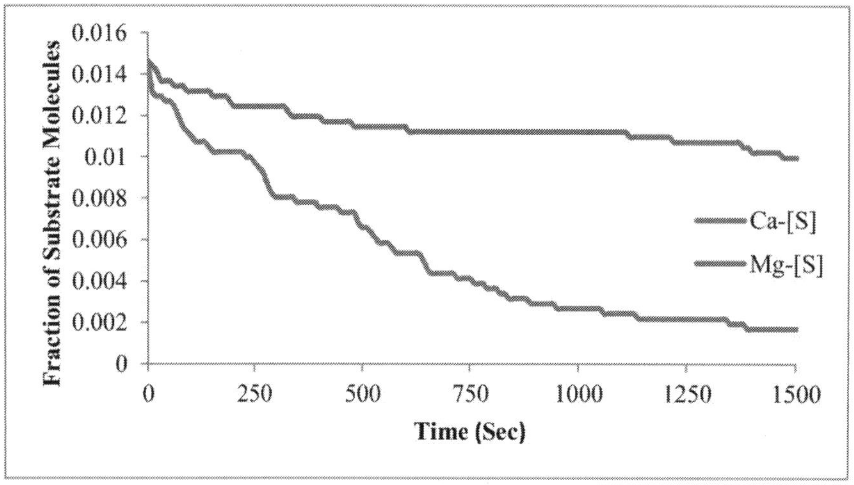

Figure 11. Fraction of Substrate (methanol) molecules formed for MDH active site models with the two ions with respect to time.

From Figure 11, the concentration of substrate molecules decreased with respect to time, implying $\frac{-d[S]}{dt}$=*rate*. For both the MDH active site models, the substrate consumption decreased. As observed from the simulations (Figure 9 and Figure 10) for increasing obstacle concentrations ([O]) i.e. the number of amino acids in MDH, the formation of formaldehyde is reduced in the case of Mg^{2+}-MDH and increased in the case of Ca^{2+}-MDH proving that Ca^{2+}-containing MDH is more reactive than Mg^{2+}-containing MDH. Thus, this implies that the amino acids around MDH also play an important role for the product formation (Figure 12). As shown in Figure12, after t=1500 MC steps for [O]=0.01 and [O]=0.09, the average of the fraction of product molecules is higher for the case of Ca-MDH.

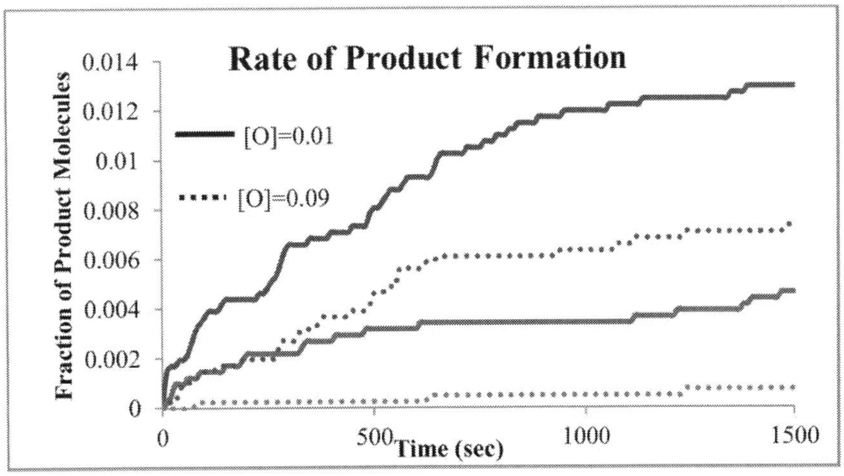

Figure 12. Rate of Product (Formaldehyde Molecules) formation for the two ions MDH active site models with respect to time for different Obstacle (amino acids) concentrations. Blue represents Ca-MDH and red represents Mg-MDH.

When the rate of formation of product molecules was plotted for increasing obstacle concentrations for both the ions MDH active site models, the Ca^{2+}-containing MDH has more product formation for [O]=0.01 and [O]=0.09 than the Mg^{2+}-containing MDH.

CONCLUSION

Using the generalized gradient approximation within the density functional theory formalism at the BLYP/DNP theory level methanol oxidation by a PQQ-containing MDH enzyme was investigated using Mg^{2+} in place of Ca^{2+} in the active site. The proposed hydride transfer mechanism was tested for Mg^{2+}-MDH active site model and it was found that the geometries of the intermediates, reactant and product agreed well with the experimental results. The free energy barrier for the rate limiting step (step1) for the Mg^{2+}-MDH was found to be 1.5kcal/mol more than the Ca^{2+}-MDH. The obtained energy barriers for the methanol oxidation reaction mechanism in Mg^{2+}-MDH were used as input for calculating the rate constants for the rate limiting step and these were used as input for the kMC program. The diffusion of methanol towards the active site is investigated for an increasing number of amino acids (obstacle density) and thus observed that Ca-containing MDH is more reactive than the modified MDH.

ACKNOWLEDGEMENTS

Financial support provided by NSF through the "Louisiana EPSCoR Research Infrastructure Improvement" grant (EPS 1003897) is acknowledged.

REFERENCES

[1] C. Anthony, *Methanol Dehydrogenase, a PQQ-Containing Quinoprotein Dehydrogenase*. New York.: Kluwer Academic/ Plenum Publishers, 2000.

[2] Z. X. Xia, Y. N. He, W. W. Dai, S. White, G. Boyd, and F. S. Mathews, "Detailed Active Site Configuration of a New Crystal Form of Methanol Dehydrogenase from Methylophilus W3a1 at 1.9 Å Resolution," *Biochem J.*, vol. 38, pp. 1214-1220, 1999.

[3] Z. X. Xia, Y. N. He, W. W. Dai, S. White, G. Boyd, and F. S. Mathews, "Determination of Gene sequence and Three-dimensional Structure at 2.4 Å resolution of Methanol Dehydrogenase from *Methylophilus W3A1*," *J. Mol. Biol.*, vol. 259, pp. 480-501, 1996.

[4] Z. X. Xia, W. W. Dai, J. P. Xiong, Z. P. Hao, V. L. Davidson, S. White, and F. S. Mathews, "The Three-Dimensional Structures of Methanol Dehydrogenase from Two Methylotrophic Bacteria at 2.6 Å Resolution," *J. Biol. chem.*, vol. 267, pp. 22289-22297, 1992.

[5] S. White, G. Boyd, F. S. Mathews, Z. X. Xia, W. W. Dai, Y. S. Zhang, and V. L. Davidson, "The Active Site Structure of Calcium Containing Methanol Dehydrogenase," *J. Biochem*, vol. 32, pp. 12955-12958, 1993.

[6] P.R. Afolabi, K. Amaratunga, O. Majekodunmi, S. L. Dales, R. Gill, D. Thompson, J. B. Cooper, S. P. Wood, P. M. Goodwin, and C. Anthony, "Site-Directed Mutagenesis and XRay-Crystallography of the PQQ-Containing Quinoprotein Methanol Dehydrogenase and Its Electron Acceptor, Cytochrome CL," *Biochem J.*, vol. 40, pp. 9799-9809, 2001.

[7] P. A. Williams, L. Coates, F. Mohammed, R. Gill, P. T. Erskine, A. Coker, S. P. Wood, C. Anthony, and J. B. Cooper*, "The atomic resolution structure of methanol dehydrogenase from Methylobacterium extorquens," *Acta. Crystall. Section D*, vol. D61, pp. 75-79, 2005.

[8] M. Ghosh, C. Anthony, K. Harlas, M. G. Goodwin, and C. C. F. Blake, "The Refined Structure of the Quinoprotein Methanol Dehydrogenase from Methylobacterium Extorquens at 1.94 Å," *Struc. (London)*, vol. 3, pp. 1771-1787, 1995.

[9] N. B. Idupulapati and D. S. Mainardi, "Coordination and binding of ions in Ca2+- and Ba2+-containing methanol dehydrogenase and interactions with methanol," *Journal of Molecular Structure: THEOCHEM*, vol. 901, pp. 72, 2009.

[10] C. Anthony and P. Williams, "The Structure and Mechanism of Methanol Dehydrogenase," *Biochimica. Biophysica. Acta*, vol. 1647, pp. 18-23, 2003.

[11] Y.-J. Zheng, Z.-x. Xia, Z.-w. Chen, F. S. Mathews, and T. C. Bruice, "Catalytic mechanism of quinoprotein methanol dehydrogenase: A theoretical and x-ray crystallographic investigation," *Proceedings of the National Academy of Sciences of the United States of America*, vol. 98, pp. 432, 2001.

[12] M. G. Goodwin and C. Anthony, "Characterization of a Novel Methanol Dehydrogenase Containing a Ba2+ Ion at the Active Site," *Biochem J.*, vol. 318, pp. 673-679, 1996.

[13] J. J. Lukkien, J. P. L. Segers, P. A. J. Hilbers, R. J. Gelten, and A. P. J. Jansen, "Efficient Monte Carlo methods for the simulation of catalytic surface reactions," *Physical Review E - Statistical Physics, Plasmas, Fluids, and Related Interdisciplinary Topics*, vol. 58, pp. 2598-2610, 1998.

[14] R. M. Nieminen and A. P. J. Jansen, "Monte Carlo simulations of surface reactions," *Applied Catalysis A: General*, vol. 160, pp. 99-123, 1997.

[15] A. P. J. Jansen, "Monte Carlo simulations of chemical reactions on a surface with time-dependent reaction-rate constants," *Computer Physics Communications*, vol. 86, pp. 1-12, 1995.
[16] A. P. J. J. a. D. S. M. N.K.R.Dandala, "A Multi-scale Modeling Approach for Studying MDH-Catalyzed Methanol Oxidation," in *Multiscale Modeling: From Atoms to Devices*, T. C. Pedro Derosa, Ed. Florida: CRC Press Inc, 2010, pp. 91-112.
[17] K. A. Fichthorn and W. H. Weinberg, "Theoretical foundations of dynamical Monte Carlo simulations," *The Journal of Chemical Physics*, vol. 95, pp. 1090-1096, 1991.
[18] J. B. Foresman, "*Exploring chemistry with electronic structure methods*," *Gaussian, Inc, Pittsburgh, PA,1996*, 1996.
[19] Accelrys Inc., "DMOL^3 User Guide," (San Diego, 2003).
[20] Accelyrs, "Materials Studio," 2006.
[21] R. M. Dreialer and E. K. U. Gross, *Density Functional Theory: An Approach to Quantum Many Body Problem*. Berlin: Springer, 1990.
[22] W. Koch and M. C. Holthausen, *A Chemist's Guide to Density Functional Theory*, Second ed: Wiley-CVH, 2001.
[23] C. Lee, W. Yang, and R. Parr, "Accurate and simple analytic representation of the electron-gas correlation energy," *phys. Rev. B*, vol. 37, pp. 786, 1998.
[24] A. D. Becke, "Density-functional thermochemistry III. The role of exact exchange," *J. Chem. Phys*, vol. 88, pp. 2547, 1988.
[25] N. B. Idupulapati and D. S. Mainardi, "A DMOL3 Study of the Methanol Addition-Elimination Oxidation Mechanism by Methanol Dehydrogenase Enzyme," *Mol. Sim.*, vol. 34, pp. 1057-1064, 2008.
[26] R. Zhen Liao, J. Yu, F. M. Raushel, and F. Himo, "Theoretical Investigation of the Reaction Mechanism of the Dinuclear Zinc Enzyme Dihydroorotase," *Chem. Eur. J.*, vol. 14, pp. 4287-4292, 2008.
[27] R. Sevastik and F. Himo, "Quantum chemical modeling of enzymatic reactions: The case of 4-oxalocrotonate tautomerase," *Bioorg. Chem.*, vol. 35, pp. 444-457, 2007.
[28] P. Velichkova and F. Himo, "Theoretical Study of the Methyl Transfer in Guanidinoacetate Methyltransferase," *J. Phys.Chem. B*, vol. 110, pp. 16-19, 2006.
[29] M. Leopoldini, N. Russo, and M. Toscano, "The Preferred Reaction Path for the Oxidation of Methanol by PQQContaining Methanol Dehydrogenase: Addition–Elimination versus Hydride-Transfer Mechanism," *Chem. Euro. j*, vol. 13, pp. 2109-2117, 2007.
[30] L. Noodleman, T. Lovell, W. Han, J. Li, and F. Himo, "Quantum Chemical Studies of Intermediates and Reaction Pathways in Selected Enzymes and Catalytic Synthetic Systems," *Chem. Rev.*, vol. 104, pp. 459-508, 2004.
[31] K. H. Hopmann and F. Himo, "Theoretical Study of the Full Reaction Mechanism of Human Soluble Epoxide Hydrolase," *Chem. Eur. J.*, vol. 12, pp. 6898-6909, 2006.
[32] K. H. Hopmann and F. Himo, "Quantum Chemical Modeling of the Dehalogenation Reaction of Haloalcohol Dehalogenase," *J. Chem. Theory Comput.*, vol. 4, pp. 1129-1137, 2008.
[33] P. Georgieva and F. Himo, "Density functional theory study of the reaction mechanism of the DNA repairing enzyme alkylguanine alkyltransferase," *Chemical Physics Letters*, vol. 463, pp. 214, 2008.
[34] B. M. R. Cammi., J. Tomasi, *J. Phys.Chem*, vol. 103, pp. 9100, 1999.
[35] B. M. R. Cammi., J. Tomasi, *J. Phys.Chem A*, vol. 104, pp. 5631, 2000.

[36] G. S. M. Cossi, N. Rega,V. Barone, *J. Chem. Phys*, vol. 43, pp. 117, 2002.
[37] A. Cornish-Bowden, *Fundamentals of enzyme kinetics*. London: Portland Press, Ltd., 2004.
[38] N. B. Idupulapati and D. S. Mainardi, "Quantum Chemical Modeling of Methanol Oxidation Mechanisms by Methanol Dehydrogenase Enzyme: Effect of Substitution of Calcium by Barium in the Active Site," *The Journal of Physical Chemistry A*, vol. 114, pp. 1887, 2010.
[39] F. Himo, "Catalytic Mechanism of Benzylsuccinate Synthase, a Theoretical Study," *The Journal of Physical Chemistry B*, vol. 106, pp. 7688, 2002.

(37e) Dynamics and Control of Aggregate Thin Film Surface Morphology for Improved Light Trapping: Implementation On a Large-Scale Kinetic Monte-Carlo Simulation

Jianqiao Huang, Xinyu Zhang, G. Orkoulas and Panagiotis D. Christofides, Department of Chemical and Biomolecular Engineering, University of California, Los Angeles, Los Angeles, CA

Photovoltaic (solar) cells are an important source of sustainable energy. Thin-film silicon solar cells are currently among the most important and widely used solar cells and their share of the overall solar cell market is steadily increasing (e.g.,[1]). Research on optical modeling of thin-film silicon solar cells indicates that the scattering properties of the thin film interfaces directly influence the light trapping process and the efficiencies of thin-film silicon solar cells (e.g., [2],[3]). However, no efforts have been seen in improving the conversion efficiency of thin-film solar cells via the regulation of the thin film surface morphology during the manufacturing process by appropriately tailoring the surface slope and roughness to desired specifications. Thus, it is desirable to develop systematic approaches to manufacture thin film solar cells with optimal conversion efficiencies via computational modeling and real-time model-based control of the manufacturing process. Recently, we have initiated an effort towards modeling and control of thin film surface morphology to optimize the light reflectance and transmittance properties of thin films. In this direction, we have studied the dynamics and lattice size dependence of surface mean slope [4] and have developed predictive control algorithms to regulate both surface roughness and slope of KMC simulators with small domain size in one [5] and two [6] spatial dimensions. However, control of thin film surface morphology within a domain comparable to the visible light wavelength has remained unsolved.

This work focuses on the application of a model predictive controller to a large-scale kinetic Monte-Carlo model. The dynamics of the evolution of the thin film surface height profile are modeled by an Edwards-Wilkinson-type equation (a second-order stochastic partial differential equation) in one spatial dimension. The thin film surface morphology is characterized in terms of surface roughness and surface slope. Analytical solutions of the expected surface roughness and surface slope are obtained by solving the Edwards-Wilkinson equation and are used in the controller design. The model parameters of the Edwards-Wilkinson equation are estimated from large scale kinetic Monte-Carlo (KMC) simulations. This parameter dependence on the deposition rate is used in the formulation of the predictive controller to predict the influence of the control action on the surface roughness and slope at the end of the growth process. The cost function of the controller involves penalties on both surface roughness and mean slope from set-point values as well as constraints on the magnitude and rate of change of the control action. The controller is applied to a large-scale one-dimensional KMC model with a domain size comparable to the wavelength of visible light. Simulation results show that the proposed controller successfully regulates surface roughness and slope to set-point values at the end of the deposition that yield desired levels of thin film reflectance and transmittance.

[1] M. A. Green. Thin-film solar cells: Review of materials, technologies and commercial status. Journal of Materials Science: Materials in Electronics, 18:15--19, 2007. [2] M. Zeman. Optical and electrical modeling of thin film silicon solar cells. Journal of Material Research, 4:889-898, 2008.
[3] A. Poruba and A. Fejfar. Optical absorption and light scattering in micro-crystalline silicon thin ˉ lms and solar cells. Journal of Applied Physics, 88:148-160, 2000.
[4] J. Huang, G. Hu, G. Orkoulas, and P. D. Christofides. Dynamics and lattice-size dependence of surface mean slope in thin film deposition process. Industrial & Engineering Chemistry Research, 50: 1219--1230, 2011.
[5] X. Zhang, G. Hu, G. Orkoulas, and P. D. Christofides. Predictive control of surface mean slope and roughness in a thin film deposition process. Chemical Engineering Science,65:4720-4731, 2010.
[6] X. Zhang, G. Hu, G. Orkoulas, and P. D. Christofides. Multivariable Model Predictive Control of Thin Film Surface Roughness and Slope for Light Trapping Optimization, Industrial & Engineering Chemistry Research, 49: 10510-10516, 2010.

See more of this Session: Multiscale Modeling I
See more of this Group/Topical: Catalysis and Reaction Engineering Division

(37f) A Predictive Control Approach for Thin Film Growth of Photovoltaic Systems

Ali Rahnamoun and **Antonios Armaou**, Chemical Engineering, The Pennsylvania State University, University Park, PA

A Predictive Control approach for thin film growth of photovoltaic systems

Ali Rahnamoun and Antonios Armaou

Department of Chemical Engineering, The Pennsylvania State University,University Park, PA 16802

When one considers the limited fuel resources of energy, the importance of photovoltaic systems in the future energy provision becomes obvious. The limited conversion efficiensy of these systems however severely limits their use currently. During the last two decades, there have been lots of studies to enhance the efficiency of photovoltaic systems. The efficiency of such systems is highly dependent on controling their microstructure within strict limits. Two important factors affecting this microstructure are the surface roughness and porosity of the produced thin film solar cells. Due to the lack of accurate sensors that can timely provide microstructure information, model based approaches to control the quality of these photovoltaic systems have been pursued. The models that predict film microstructure however, pose significant challenges both from an analysis and control point of view [1]. Such difficulties are attributed in part to the unavailability of closed form models to describe the process evolution at molecular-level detail and their computationally intensive nature that prevents their real time implementation.

To circumvent the mentioned limitations, one of the previous approaches includes identified stochastic partial differential equation models to design the controller [2]. This approach however assumes specific structure to the nonlinear stochastic terms. Another approach relied on the off-line and subsequently the on-line identification of bilinear models for the process, which are then used for the controller synthesis [3].

The objective of the present work is to develop a model predictive control algorithm to simultaneously regulate the surface roughness and porosity of thin film solar cell to obtain higher efficiency. To simulate the evolution of thin film growth, Kinetic Monte Carlo simulation is utilized. Two processes in this simulation include deposition with considering shadowing effect and migration of the deposited particles. For achieving the closed loop model of the simulated system, fuzzy system identification is utilized [4]. The proposed method explicitly accounts for the presence of stochastic terms in the process and the issue of noise in the measurements. The kalman filter is applied to the model to reach to modeling results with more accuracy. The developed model is then employed with a model predictive framework to drive the process at the desired process objective. The simulation results demonstrate the effectiveness of the nonlinear model predictive controller designed to regulate the surface roughness and porosity of the thin film solar cell to achieve higher efficiency.

REFERENCES:

[1] P. D.Christofides, A.Armaou , "Control of Multiscale and Distributed process systems-preface," *Computers &Chemical Engineering*., **29**,687-688, 2005.

[2] Y Lou, P. D.Christofides, "Nonlinear Feedback Control of Surface Roughness Using a Stochastic PDE: Design and Application to a Sputtering Process," *Industrial & Engineering Chemistry Research*., **45**, 7177-7189, 2006.

[3] A.Varshney, A.Armaou, "Multiscale Optimization using hybrid PDE /KMC process systems with applications to thin film growth", *Chemical Engineering Science*., **60**, 6780-6794, 2005.

[4] A.Rahnamoun, A.Armaou, "A Nonlinear Predictive Control of Processes with Multiscale Objectives using a Fuzzy-System Identification Approach",American Control Conference 2011.

See more of this Session: Multiscale Modeling I
See more of this Group/Topical: Catalysis and Reaction Engineering Division

(37g) Moment Closure for Chemical Reactions

Yiannis Kaznessis and **Fatemeh Ghasemi**, Department of Chemical Engineering and Materials Science, University of Minnesota, Minneapolis, MN

We present a systematic procedure for constructing moment closure functions of arbitrary order for the stochastic models of chemical reaction systems. In the stochastic formulation of chemical reactions, the dynamics of the first M-order moments of the species populations generally do not form a closed system of differential equations, in the sense that the time-derivatives of first M-order moments generally depend on moments of order higher than M. However, for analysis purposes, these dynamics are often made to be closed by approximating the needed derivatives of the first M-order moments by nonlinear functions of the same moments. These functions are called the moment closure functions. We will discuss a new approach to construct moment closure functions. Our method is based on calculating the correlation function of individual moments. We use several examples, with linear and nonlinear sets of reactions, to demonstrate the applicability and the limitations of the scheme.

See more of this Session: Multiscale Modeling I
See more of this Group/Topical: Catalysis and Reaction Engineering Division

(37h) Multiscale Analysis and Simulation of Bubbly Flow

Ning Yang[1], Zongying Wu[2], Jianhua Chen[1] and Jinghai Li[1], (1)State Key Laboratory of Multiphase Complex Systems, Institute of Process Engineering, Chinese Academy of Sciences, Beijing, China, (2) Department of Chemical Engineering, Xi'an Jiaotong University, Xi'an, China

Bubble columns are considered as favorable reactors in process engineering. Scale-up and optimization of these reactors require an extensive understanding of gas-liquid two-phase flow in order to gain further insight into mixing and transfer properties. However, the gas-liquid flow and/or gas-liquid-solid flow in bubble columns is quite complicated not only because of its versatile macro-scale flow regimes, but due to those intricate structures occurring at meso and micro scales, posing great challenges for CFD simulation of multi-phase flow in these reactors. Simulation based on Eulerian-Eulerian method, though usually employed in commercial CFD packages, is still not powerful enough to capture real physics due to our limited understanding on the multi-scale flow structure and hence inadequate development of closure laws including drag, lift and turbulence models. The aim of this study is to use the dual-bubble-size (DBS) model to develop a closure law for the ratio of effective drag coefficient to bubble diameter, and then apply this model in simulation of gas-liquid two-phase flow, mass transfer and chemical reaction for FT synthesis. In addition to mass and momentum conservation equations, the stability condition reflecting the compromise between two different dominant mechanisms serves as another constraint for the model, and therefore it is expected to offer a more accurate drag closure for CFD simulation. Prior to this integration, we investigate the respective effects of bubble diameter and correction factor and find that the effect of bubble diameter is limited, whereas the correction factor due to the bubble swarm effect is eminent and appropriate correction factor has to be selected for different correlations of standard drag efficient to be in accord with experiments. By contrast, the new model can well predict the radial gas holdup distribution, the total gas holdup as well as the two-phase flow field without the need of adjusting model parameters, and then it is used for a scale-up process of F-T systhesis, showing its great potential and advantage in understanding the complex nature of multiscale structure of gas-liquid flow in bubble columns.

See more of this Session: Multiscale Modeling I
See more of this Group/Topical: Catalysis and Reaction Engineering Division

(40d) Core-Shell Type Magnetically Active Fe3O4@Au Nanoparticles As Supported Nanocatalysts for CO Oxidation

Sarthak Gaur[1], Faruq Mohammad[2], Challa S.S.R. Kumar[2] and James J. Spivey[3], (1)Cain Department of Chemical Engineering, Louisiana State University, Baton Rouge, LA, (2)Center for Advanced Microstructures and Devices(CAMD), Louisiana State University, Baton Rouge, LA, (3)Louisiana State University, Baton Rouge, LA

Size-definedand thiol-ligated Au clusters vcan be used to prepare nanoparticles (NPs) withless than 2 nm in diameter. These materials exhibit distinct quantumconfinement effects that differ significantly from bulk Au, for exampleexhibitingdiscrete electronic electronic structure and molecular propertiessuch as HOMO-LUMO transitions and intrinsic magnetism. These properties mayhave an interesting effect on catalysis by these clusters.

Recently,dumbbell shaped Au-Fe_3O_4 nanoparticles prepared by colloidaldeposition and supported on TiO_2 have been successfully tested ascatalysts for CO oxidation[1]. The particle size of Au in that study was 2.5-3.5nm and that of Fe_3O_4 was about 15-16 nm, and the dumbbellshape ensured a strong interaction between the metals. In such NPs, a stronginteraction exists between Au and Fe_3O_4 due to theepitaxial growth of Fe_3O_4 on Au[1].

Here, wereport the catalytic activity of core-shell type SPIONs@Au supported on $TiO_2$for CO oxidation. The NPs were synthesized as described earlier[2, 3] by modifying a reported methodfor the synthesis of these particles[4]. The Au/Fe_3O_4 particles wereimpregnated on TiO_2 (P25, Degussa) to give Au/Fe_3O_4/TiO_2catalysts. These catalysts were treated in an oxidative environment at 300C,400C, and 500C, respectively before CO oxidation to remove the sulfur ligands.Fresh and treated catalysts were characterized by HRTEM, XPS, XRD, and by COoxidation. Results show that Au/Fe_3O_4/TiO_2 isan active catalyst for CO oxidation, only if calcined at 300C. Other treatmentmethods sintered the catalysts, which results in no catalytic activity.

References:

[1] H. Yin,C. Wang, H. Zhu, S.H. Overbury, S. Sun, S. Dai, Chemical Communications (2008)

4357-4359.

[2] F.Mohammad, G. Balaji, A. Weber, R.M. Uppu, C.S.S.R. Kumar, The Journal ofPhysical ChemistryC. 114 (2010) 19194-19201.

[3] C.S.S.R.Kumar, F. Mohammad, The Journal of Physical Chemistry Letters. 1 (2010)3141-3146.

[4] WangWang,J. Luo, Q. Fan, M. Suzuki, I.S. Suzuki, M.H. Engelhard, Y. Lin, N. Kim, J.Q.Wang, C.-J.Zhong, The Journal of Physical Chemistry B. 109 (2005) 21593-21601.

See more of this Session: Nanoscale Materials as Catalysts I
See more of this Group/Topical: Catalysis and Reaction Engineering Division

(40b) High Oxygen Reduction Activity From Electroless Co-Deposition of Ag and MnOx Nanodomains On Carbon

Daniel A. Slanac[1], Anthony Lie[1], Joel A. Paulson[1], Keith J. Stevenson[2] and Keith P. Johnston[1], (1) Chemical Engineering, The University of Texas at Austin, Austin, TX, (2)Chemistry and Biochemistry, The University of Texas at Austin, Austin, TX

Currently there is renewed interest in alkaline fuel cells (AFCs) as an alternative to acid-based proton exchange membrane fuel cells (PEMFCs) due to potentially faster oxygen reduction reaction (ORR) kinetics and improved catalyst stability. For AFCs less costly non-precious group metal (NPGM) catalysts supported on carbon can be highly effective. Manganese oxides (MnOx) on carbon have been widely explored as cathode catalysts in base, as the ORR reaction intermediate, the HO_2^- anion, disproportionates back into oxygen to yield a pseudo 4 electron reduction similar to that observed for Pd/C and Pt/C. However, the kinetics of the reaction remain sluggish, but may be improved by intimately combining the MnOx with metal nanoparticles (e.g. Ag) to gain a synergistic effect between the two for both ORR and HO_2^- disproportionation. The key challenge here is to synthesize monodisperse, nano-sized metal and MnO_2 particles with uniform dispersion over the carbon support for achieving high activity.

In this work a new electroless co-deposition scheme is developed whereby Ag and MnO_2 are co-reduced on vulcan carbon support for oxygen reduction in alkaline media. By employing the surface specific, limited reaction between the redox pairs of potassium permanganate/carbon and Ag/MnOx, thin domains of manganese oxide and Ag nanoparticles (<5nm) were deposited uniformly over the carbon surface. The Ag-MnO_2/C composite displays a mass activity of 0.98 A/mg), close to that of Pd/C (0.11A/mg) and 36% higher that MnOx/C (0.072 A/mg). Furthermore, the onset potential for oxygen reduction is 15 mV positive that of Pd/VC, suggesting facile kinetics. Mechanistically, the origin of the activity enhancement for Ag-MnOx/C is shown to result from fast kinetics for the chemical disproportionation of the superoxide anion (HO_2^-), an ORR intermediate formed by the partial 2 electron reduction of oxygen. The chemical compositions of the catalysts were determined with X-ray photoelectron spectroscopy and energy dispersive spectroscopy. Catalyst morphology was examined with high resolution electron microscopy to better understand the relationship between the catalyst structure and composition with respect to activity.

See more of this Session: Nanoscale Materials as Catalysts I
See more of this Group/Topical: Catalysis and Reaction Engineering Division

(40c) Monometallic Gold, Palladium and Bimetallic Gold-Palladium Catalysts for the Methanol Steam Reforming Reaction

Matthew B. Boucher, James Kammert, Chongyang Wang, Nan Yi and Maria Flytzani-Stephanopoulos, Chemical and Biological Engineering, Tufts University, Medford, MA

Abstract:

In this study we investigatethe active nature of nano-scaled gold, palladium andgold-palladium catalysts for the steam reforming of methanol (SRM). Mono- and bimetallic catalysts wereprepared by the DP-urea method both in the presence of metal oxide supports (ZnO, CeO_2) and in the absence of a support. Preparations in the absence of a supportwere carried out in the presence of PVA to avoid aggregation of mono- andbimetallic nanoparticles. Mono- andbimetallic colloids prepared in the presence of PVA were later immobilized oninert metal oxide supports (SiO_2, Al_2O_3). While the gold catalysts prepared in thestudy already exemplify remarkable SRM activity and selectivity to CO_2,low concentrations of palladium (Au:Pd= 10:1) were determined to have a promotional effect on the catalyticperformance.

See more of this Session: Nanoscale Materials as Catalysts I
See more of this Group/Topical: Catalysis and Reaction Engineering Division

(40a) "Finding the Right Fit": Catalysis and Confinement At the Nanoscale

Rajamani Gounder and Enrique Iglesia, Department of Chemical and Biomolecular Engineering, University of California at Berkeley, Berkeley, CA

Dispersion forces within voids of molecular dimensions solvate transition states that mediate chemical reactions. These interactions give rise to ubiquitous enthalpy-entropy compromises for confined species, which influence, in turn, the Gibbs free energy differences between transition states and reactants. Kinetically-relevant transition states for dimethyl ether carbonylation (438 K) are more effectively stabilized by van der Waals interactions within eight-membered ring (8-MR) pockets than in larger 12-MR channels of MOR zeolites. At low temperatures, enthalpic contributions dominate free energies and the tighter fit in 8-MR pockets decreases activation barriers and leads to higher turnover rates. Turnover rates for monomolecular alkane (C_3H_8. n-C_4H_{10}, i-C_4H_{10}) activation (748 K) are also higher on 8-MR protons in spite of higher barriers, which reflect less effective van der Waals contacts that result from partial confinement of larger transition states within shallow 8-MR pockets. Partially-confined transition states have higher entropies than their fully confined analogs, leading to higher rates because entropy effects contribute to free energies more strongly at higher temperatures. The preeminent effects of entropy and partial confinement at high temperatures are selective for reactions involving looser transition states; they are also consequential for C_3H_6 hydrogenation with H_2 and C_2H_4 alkylation with CH_4, which involve the same ion-pair transition states as their C_3H_8 monomolecular dehydrogenation and cracking counterparts. These routes occur at rates constrained by gas-phase thermodynamics and with the same site requirements as alkane activation, because catalytic surfaces remain vacant with varying reactant pressures and distances from chemical equilibrium. These findings highlight the role of temperature in mediating contributions of transition state enthalpy and entropy, which depend strongly on confinement, to free energies and turnover rates in catalysis.

See more of this Session: Nanoscale Materials as Catalysts I
See more of this Group/Topical: Catalysis and Reaction Engineering Division

(40e) Nanostructure Architecture of Pt Based Electrocatalysts for Energy Conversion Applications

Chao Wang, Nenad Markovic and Vojislav Stamenkovic, Materials Science Division, Argonne National Laboratory, Argonne, IL

Structure architecture at nanoscale is a challenging task to develop advanced nanomaterials. Particularly in the electrocatalyst design for chemical-electrical energy conversion, it becomes critical in order to achieve both the targets of high catalytic activity and long durability simultaneously. Platinum (Pt) in the form of nanoparticles supported on high surface area carbon matrix is widely used for catalyzing the hydrogen/methanol oxidation reaction at the anode and oxygen reduction reaction (ORR) at the cathode of fuel cells today. However, the precious metal Pt is not only expensive causing the high cost, but also scarce which limited the large scale application of fuel-cell technology. Moreover, Pt that is generally chemically inert becomes unstable when exposed to the hostile electrochemical environment where Pt surface atoms dissolve and migrate, resulting in aggregation of nanoparticles and losses of surface area, activity and power density. Particularly, the ordinary activity and degrading of Pt at the cathode represents one of the major limitations for commercialization of this technology.

Here I introduce our recent work on structure architecture of Pt-based nanoparticles for electrocatalytic applications. The particle size, shape, alloy composition and composite heterostructure were tuned with properties tailored for superior behavior in catalysis. The mechanisms of controlled nanoscrystal growth were depicted by molecular dynamic simulation and are thus able to be generalized to other system. Improvements in catalytic activity and stability were also explained by comparative studies on extended surfaces and first-principle calculations, which provided comprehensive understanding of the correlation between nanoparticle structure and catalytic performance. The developed strategy of combining state-of-the-art nanomaterial synthesis with fundamental study of model catalyst system and theoretical analysis represents a convictive approach toward advanced materials for catalytic and other applications.

See more of this Session: Nanoscale Materials as Catalysts I
See more of this Group/Topical: Catalysis and Reaction Engineering Division

(40f) Confinement-Induced Reactivity

Julibeth M. Martinez-De la Hoz, Chemical Engineering and Materials Science and Engineering Program, Texas A&M university, College station, TX and Perla B. Balbuena, Chemical Engineering, Texas A&M University, College Station, TX

Confinement-induced reactivity

Julibeth M. Martinez de la Hoz and Perla B. Balbuena

Department of Chemical Engineering and Materials Science and Engineering Program

Texas A&M University, College Station, TX, 77843

Important to the field of heterogeneous catalysis is the analysis of the ability of a given metal to adsorb different reactants, as well as the estimation of activation energies for the formation of intermediates and final structures on the surface. Also, it has been found that solid metal surfaces separated by small distances (3 – 10 Å) can induce new interesting phenomena, related to changes in the chemical, electronic and magnetic properties of the metallic material. Therefore, understanding how the spatial confinement of adsorbates between metallic surfaces influences their adsorption energies and energy barriers for dissociation, offers the possibility of using the degree and/or nature of confinement to facilitate specific reactions.

In this work, we use DFT methods with the aim to elucidate whether a specific reaction is facilitated by spatial confinement of the adsorbate among metal surfaces separated by a few Angstroms. We analyze two systems: the well known O_2 dissociation on Pt (111) and the the dissociation of NO_2 on Fe (111). We found that molecular oxygen chemisorbed among Pt (111) surfaces needs a significantly smaller activation energy to be dissociated at surface – surface separations of 4.7 Å, compared to that needed on a single Pt(111) surface. The difference between these activation energies seems to be related to a larger charge transfer from the substrate to the confined O_2 molecule, which induces larger changes in the electronic structure of the molecule and weakens the O – O bond strength. Similar qualitative results are found for the dissociation of NO_2 on Fe (111); however, there are quantitative differences between the charges transferred to the adsorbate as a function of the surface-surface separation, and no significant variation of the activation energies in confinement with respect to the single surface. Consequently, it is shown that the degree of facilitation provided by the confinement, for a given reaction also depends strongly on the nature of the surface.

Viral-templated palladium nanocatalysts for Suzuki coupling reaction

Cuixian Yang[1], Amy K. Manocchi[1], Byeongdu Lee[2] and Hyunmin Yi[1]

[1]Department of Chemical and Biological Engineering, Tufts University, Medford, MA 02155

[2]X-ray Science Division, Argonne National Laboratory, 9700 S. Cass Avenue, Argonne, IL 60439

Abstract:

We demonstrate and thoroughly examine tobacco mosaic virus (TMV)-templated palladium (Pd) nanocatalysts for the ligand-free Suzuki coupling reaction under mild conditions. The surface-assembled TMV templates allow for facile catalyst synthesis under mild aqueous conditions that leads to high Pd surface loading and stability. Further, the chip-based format enables simple catalyst separation and reuse as well as facile product recovery. We found that the solvent ratio played an important role in the Suzuki reaction selectivity; a higher water/acetonitrile ratio significantly facilitated the desired cross-coupling pathway. AFM, GISXAS and XPS results showed minimal particle size change and stability along with the overall stability of Pd-TMV complexes in the catalytic reaction conditions employed. The majority of both products (4-MB and BP) were formed within the first 5 h, with "apparent" reaction rate of 0.0165 mol/(min·g) based on this amount of Pd employed for each batch reaction. Preliminary reaction mechanism studies suggest that the Suzuki reactions may be catalyzed largely by trace amounts of Pd species (1–90 ppb) in the reaction solution. The TMV-templated Pd nanoparticles are thus considered as "pre-catalysts" of the catalytically active Pd species in the reaction solution. We envision that our viral template-based bottom-up assembly approach can be readily extended to other biotemplates, metal catalysts and organic reaction systems.

(41a) Keynote Speaker Abstract: Design of Hydrophobic Zeolites Containing Lewis Acid Active Sites for the Isomerization of Glucose In Aqueous Media

Mark E. Davis[1], Eranda Nikolla[1], Yuriy Roman[1], Manuel Moliner[2] and Sonjong Hwang[1], (1)Chemical Engineering, California Institute of Technology, Pasadena, CA, (2)Chemistry, Instituto de Tecnologia Quimica (UPV-CSIC), Valencia, Spain

We have reported that tin or titanium-containing zeolite beta catalysts (Sn-Beta or Ti-Beta) catalyze the isomerization of glucose to fructose in aqueous media at low pH in the presence of HCl and an inorganic salt (Proc. Natl. Acad. Sci., 107 (2010) 6164, ACS Catal, 1 (4), (2011) 408). Our current efforts are focused on obtaining a fundamental understanding of the nature of the active site in the Sn or Ti-Beta zeolite and the mechanism of the isomerization reaction of glucose to fructose. For example, we have utilized solid-state NMR experiments to probe the environment of tin in the zeolite. Sn-Beta zeolite is synthesized with labeled Sn-119 in order to provide a solid that can be effectively analyzed by solid-state NMR. We have also performed NMR experiments with deuterated and C-13 labeled glucose to study the interactions of the reactant with the tin active site in the zeolite. Detailed kinetic studies including isotope-labeling experiments have been conducted in order to obtain kinetic parameters such as activation energies and rate limiting steps for the isomerization of glucose to fructose. An overall picture of the state of the tin active center and the reaction pathway is emerging. In this presentation, we will discuss the important features of the catalytic materials that are necessary for obtaining reactivity. Control of the zeolite hydrophobicity and framework localization of either the tin or titanium are critical to achieving activity when the reaction medium is water.

See more of this Session: Novel Catalytic Materials I
See more of this Group/Topical: Catalysis and Reaction Engineering Division

(41b) Perovskites As Alternative Catalysts for Solid Oxide Fuel Cell Anodes: Effect of Dopants

Hyunkyu Choi, Chemical & Biomolecular Engineering, Ohio state university, Columbus, OH and Umit S. Ozkan, Chemical & biomolecular Engineering, Ohio state university, Columbus, OH

Perovskites as alternativecatalysts for solid oxide fuel cell anodes: Effect ofdopants

HyunkyuChoi, Umit S. Ozkan*

The OhioState University

HeterogeneousCatalysis Research Group

Departmentof Chemical and Biomolecular Engineering

Solid oxide fuel cells (SOFCs) show greatpotential for effectively generating clean power from a variety of fuels.Current state of the art anode catalysts based on Ni catalysts supported on yittria-stabilized zirconia (Ni/YSZ) deactivate in thepresence of coal-derived gas because they are highly susceptible to coking andsulfur poisoning. These processes decrease the anodic reaction rates, lower thepower densities, and lead to instability during long time operation. Therefore,the development of carbon and sulfur tolerant as well as highly activematerials suitable for anodes is essential to bring coal-gasfed SOFC systems closer to wide-spread application.

Ongoing research in our group examines developingperovskite-type catalysts by incorporating cerium tothe doped-lanthanum ferrites. The materials are prepared by sol gel techniques.The present work studies the kinetics of the hydrocarbon fuel oxidationreaction, the resistance to carbon coking and sulfur poisoning, the formationof oxygen vacancies, and the bulk and surface chemistry of cerium-doped perovskites. The phase transformation ofthese materials are characterized using in-situ X-ray diffraction (XRD).The activity as well as stability of these catalysts in the presence of $H_2S$is studied with time-on-stream reaction experiments for various anodereactions. Oxygen nonstoichiometry, an indicator ofoxygen vacancies, is vitally important for the diffusion of the oxide ion toreact in the three-phase boundary (TPB) and is determined by thermogravimetric analysis. Surface sites are probedusing methanol adsorption and diffuse reflectance infrared Fourier Transform(DRIFT) spectroscopy.

Session : Novel catalytic materials I

*Presenting author

331 KoffoltLaboratory•
140 West Nineteenth Avenue

Columbus, OH 43210

Phone: 614-292-4993

Fax: 614-292-3769

Email: choi.508@osu.edu

Umit S. Ozkan

333A KoffoltLaboratory•
140 West Nineteenth Avenue

Columbus, OH 43210

Phone: 614-292-6623

Fax: 614-292-3769

Email: ozkan@chbmeng.ohio-state.edu

See more of this Session: Novel Catalytic Materials I
See more of this Group/Topical: Catalysis and Reaction Engineering Division

(41c) Hybrid Organic-Inorganic Materials for Heterogeneous Catalysis: Development of Highly Structured Multifunctional Silicon Surfaces

Heidrun Gruber-Woelfler[1], Peter Feenstra[1], Georg J. Lichtenegger[1], Eleonora Polo[2] and Johannes G. Khinast[1], (1)Institute for Process and Particle Engineering, Graz University of Technology, Graz, Austria, (2)Dip. di Chimica, C.N.R.-ISOF Sez. Ferrara, Ferrara, Italy

Functionalization of single-crystal silicon surfaces is an area of intense research due to the potential use of hybrid organo-silicon materials as the basis for integrating various functionalities, such as molecular recognition devices and biomolecules, including proteins, DNA and carbohydrates. For instance, a wide variety of methods was developed to prepare organic monolayers of 1-alkenes and 1-alkynes on hydrogen-terminated silicon (H-Si).[1,2] Taking advantage of the strong Si-C bond, this kind of functionalization can also be used for the covalent immobilization of organometallic complexes onto silicon, thus leading to a powerful approach for the preparation of catalytically active materials, which can be repeatedly recycled.

We recently reported the covalent immobilization of a tethered ethylenebis(indenyl) (=EBI) ligand,[3] tethered EBI-zirconocenes[4] and of biomolecules[5] on H-terminated Si(111)-surfaces.

In this contribution we present the design of hybrid organic-inorganic silicon surfaces that include different kinds of organometallic catalysts in a highly structured arrangement.

The preparation of these multifunctional surfaces included in the first step the immobilization of an organic bis(oxazoline)-ligand on a H-terminated Si(111)-surface. The immobilization reaction was carried out via a UV-mediated hydrosilylation of the terminal double bond of the ligand and the Si-H group of the Si-wafer. In addition to being a clean and straightforward approach, this method lends itself to the sub-micron structuring of catalytic surfaces via photopatterning and photolithography. The immobilized ligand was then metalated with $Pd(OAc)_2$ to give a highly potential heterogeneous Pd-catalyst. On the same Si-wafer tethered titanocenes were attached to the surface applying again a hydrosilylation reaction. The resulting multifunctional surfaces were tested for the heterogeneous synthesis of active pharmaceutical ingredients, such as amines and substituted biphenyls.

(1) Buriak, J. M. *Chem.Rev.* **2002**, *102*, 1271-1308 and references therein.

(2) Scheres, L.; Giesbers, M.; Zuilhof, H. *Langmuir* **2010**, *26*, 4790-4795 and references therein

(3) Langner, A.; Panarello, A.; Rivillon, S.; Vassylyev, O.; Khinast, J.; Chabal, Y. J. *J.Am.Chem.Soc.*

2005, *127*, 12798-12799.

(4) Gruber-Woelfler, Rivillon, S.; Chabal, Y. J.; Schitter, G.; Polo, E.; Ringwald, M.; Khinast, J. G.. *Chem.Comm. (Hot Paper)* **2008**, 1329-1331

(5) Jeanquartier, C.; Schider, G.; Feichtenhofer, S.; Schwab, H.; Schennach, R.; Stettner, J.; Winkler, A.; Gruber-Woelfler, H.; Schitter, G.; Eder, R. J. P.; Khinast, J. G.. *Langmuir* **2008**, *24*, 13957-13961

See more of this Session: Novel Catalytic Materials I
See more of this Group/Topical: Catalysis and Reaction Engineering Division

(41d) Supports As Co-Catalysts for Activating Molecular Oxidation Catalysts

Nicholas Schoenfeldt[1], Andrew Korinda[1], Randall J. Meyer[2] and **Justin M. Notestein**[1], (1)Chemical and Biological Engineering, Northwestern University, Evanston, IL, (2)Department of Chemical Engineering, University of Illinois - Chicago, Chicago, IL

There is a wide class of non-heme oxidase enzymes thatconsist of histidine-coordinated first-row metal oxides, further activated bycarboxylates, and these structures have been extensively mimicked by solubleinorganic complexes. Many of these amine and pyridine-ligated complexes areactivated towards productive oxidation using H_2O_2 or $O_2$by coordination of carboxylates. As one example, dimeric1,4,7-triazacyclononane manganese oxides are known to be efficient and highlyactive homogeneous oxidation catalysts whose selectivity toward a number ofoxidation reactions (epoxidation / cis-dihydroxylation, alkane oxidation, alcoholoxidation) can be controlled by the addition of various acid co-catalysts.

Here we present a synthetic route to supported triazacyclononanemanganese oxides through an *in-situ* assembly method whereby the surfaceactive-site is synthesized under reaction conditions by simply combining apreformed Mn dimer and the carboxylate modified metal oxide co-catalytic supports.Self-assembly from MeCN/H_2O_2 solutions ontoorgano-functionalized oxide supports provides an inherent advantage bysupplying a co-catalyst which also acts as the support tether, thus eliminatingsynthetically cumbersome ligand alterations which can lead to differences inactive-site structure and reactivity or selectivity from the homogeneousanalogues. For example, we have shown sterically hindered surfaces to increaseselectivity toward *cis*-diol. Additionally, the high local concentrationsof surface carboxylates lead to increased reaction rates and shorter reactiontimes required for maximal productivity than is observed with analogoushomogeneous carboxylate co-catalysts. Recent work using other metal complexes (e.g.V), but the same general principles, are also described. Surface structures areunderstood through X-ray absorption fine structure, diffuse reflectanceUV-visible spectroscopy, and DFT modeling.

See more of this Session: Novel Catalytic Materials I
See more of this Group/Topical: Catalysis and Reaction Engineering Division

(41e) Unsupported NiMoW Sulfide Catalysts Prepared by Thermal Decomposition of Thiosalts for Hydrodesulfurization of Dibenzothiophene

Yanjiao Yi[1], Christopher T. Williams[2], Guang Xiong[1] and Changhai Liang[1], (1)School of Chemical Engineering, Dalian University of Technology, Dalian, China, (2)Chemical Engineering, University of South Carolina, Columbia, SC

More and more stringent environmental regulations and fuel specifications are demanding a significant reduction of sulfur and nitrogen containing compounds, as well as aromatics in fuels. Conventional hydrotreating processes are performed over alumina supported sulfide catalysts based on Mo or W and promoted with Ni or Co. These materials are produced by sulfidation of the corresponding oxides. It well known that incomplete sulfidation due to strong interaction between metal and support results in decreased hydrotreating activity [1]. Unsupported sulfide catalysts have attracted recent attention due to their high activity in the hydroteating reactions. The structures and catalytic properties of these sulfides obtained by comaceration [2], precipitation [3], and hydro/solvothermal route [4] depended strongly on their preparation atmosphere and thermal treatment conditions. Thiosalt decomposition provides a simple and reproducible method for obtaining unsupported sulfide materials with controlled stoichiometry and specific surface area [5]. Herein, a family of unsupported NiMoW sulfides with different ratios were prepared by thermal decomposition of tetramethylammonium thiomolybdotungstate and $Ni(NO_3)_2$. The catalytic materials were characterized by Fourier transformation infrared spectroscopy, ultraviolet-visible spectroscopy, thermogravimetric analysis, X-ray diffraction, N_2 physorption, scanning electron microscopy and transmission electron microscopy. The results show that structures and properties of unsupported NiMoW sulfides depend on nickel content and decomposition condition. The layering and stacking degree of the sulfide particles shows a decreasing trend as the nickel content increases in the NiMoW sulfides. This family of catalysts was evaluated for the hydrodesulfurization (HDS) of dibenzothiophene (DBT). Low nickel content NiMoW catalysts gave a positive effect on the HDS performance, while catalysts with high nickel content showed a negative effect. The $Ni_{0.5}MoWS_x$ catalyst showed the highest activity in the HDS reaction compared to other samples. The results suggest that unsupported NiMoW sulfide catalysts may be promising candidates for wide application in hydrotreating of fuel oil.

References

[1] R. Prins, *Adv. Catal.* 46 (2001) 399-464.

[2] G. Hagenbach, P. Courty, B. Delmon, *J. Catal.* 31 (1973) 264-273

[3] R. Candia, B. S. Clausen, H. Topsøe, *J. Catal.* 77 (1982) 564-566

[4] E. Devers, P. Afanasiev, B. Jouguet, M. Vrinat, *Catal. Lett.* 82 (2002) 13-17.

[5] X. Jin, C.L. Ma, Y.J. Yi, Q.M. Zhang, J.S. Qiu, C.H. Liang, J. Phys. Chem. Solids 71 (2010) 642-646.

See more of this Session: Novel Catalytic Materials I
See more of this Group/Topical: Catalysis and Reaction Engineering Division

(41f) Reactive Nanoparticles Immobilized In Hydrogel for Toxic Organics Degradation

Li Xiao and D.B. Bhattacharyya, Chemical and Materials Engineering, University of Kentucky, Lexington, KY

Nanoscale metallic particles have important applications in the environmentally important catalytic reactions due to the novel physical and chemical properties over bulk particles. Functionalized membrane supported metallic nanoparticles can be used to conduct reductive detoxification of chlorinated organics. In this research, a novel strategy was employed for stabilizing metalic nanoparticles with temperature responsive hydrogel (poly - N-isopropylacrylamide) as stabilizer. Hydrogel/Membrane immobilized nanoparticles possess many advantages: prevention of aggregation, tunable nanoparticle size and morphology, and potentially adjustable reactivity by the variance of hydrophobicity and hydrophilicity with the environmental change.

In this work, poly (N-isopropylacrylamide-co-acrylic acid) [p(NIPAAm-AA) immobilized metallic nanoparticles were synthesized with uniform size (around 30nm~50nm) and distribution immobilized in polymer domain. These iron nanoparticles containing hydrogel-composites were utilized as a reaction media for the reduction reactions of chlorinated organics. Also, this polymer network has tunable properties and it was found that reactivity toward TCE dechlorination with Fe/Pd (2 wt% Pd) increased three fold with a temperature change of only 4 °C. On the other hand, experiments without any hydrogel showed an increase of only two fold, showing the importance of hydrogel tunable properties. Hydrogel supported nanoparticles can be regenerated and reused only with 30% reduction in the initial reduction rate. The dechlorination of dichlorobiphenyl was also conducted with the system of bimetallic nanoparticles Fe/Pd (4.6 wt% Pd) nanoparticles immobilized in hydrogel. The batch reaction results showed the formation of biphenyl and complete dechlorination (as measured by chloride analysis) at 25 °C. Support of this research has been provided by NSF-IGERT and NIEHS-SRP.

See more of this Session: Novel Catalytic Materials I
See more of this Group/Topical: Catalysis and Reaction Engineering Division

(49a) Steam Gasification of a Cellulose Surrogate Over a Fluidizable Ni/-Alumina Catalyst: A Kinetics Model

Enrique Salaices, Chemical & Biochemical Engineering, The University of Western Ontario, London, ON, Canada

This study considers the catalytic steam gasification of a cellulose surrogate using a fluidizable Ni/α-alumina catalyst. Experiments were carried out in the CREC fluidized Riser Simulator On this basis a reaction network and a kinetic model for biomass catalytic steam gasification were proposed. This kinetic model was developed using a sound reaction engineering approach where reaction rates for various species are the result of the algebraic addition of dominant reactions. The modeling procedure also included the decoupled determination of intrinsic kinetic parameters and adsorption constants as allowed in the CREC Riser Simulator. The implemented approach eliminates overparametrization with successfully parameter correlation. Numerical regression of the experimental data led to intrinsic kinetic parameters with narrow spans showing that the proposed kinetic model satisfactorily describe the catalytic conversion of glucose under the selected gasification conditions.

See more of this Session: Reaction Engineering for Biomass Conversion
See more of this Group/Topical: Catalysis and Reaction Engineering Division

(49b) Review of Fast Pyrolysis of Biomass

Aditya Kashyap and Vijay Kumar Agarwal, Chemical Engineering, Indian Institute of Technology Roorkee, Roorkee, India

Renewable energy is of growing importance in satisfying environmental concerns over fossil fuel usage. Wood and other forms of biomass including energy crops and agricultural and forestry wastes are some of the main renewable energy resources available. The potential for biomass to supply much larger amounts of useful energy with reduced environmental impacts compared to fossil fuels has stimulated substantial research and development of systems for handling, processing, and converting biomass to heat, electricity, solid, liquid and gaseous fuels, and other chemicals and products. Greater use of biomass has also been motivated by improvements experienced in local and global environmental quality. The increasing dependence on imported oil as well as the urgency to reduce Greenhouse emissions abounds in justifying an energy policy that carefully considers the role of renewable sources as energy carrier.

Pyrolysis and other thermochemical conversion processes offer an important opportunity for the utilization of biomass and waste. Biomass fast pyrolysis is of rapidly growing interest all around the world as it is perceived to offer significant logistical and hence economic advantages over other thermal conversion processes. The principle of fast pyrolysis is rapid heating of relatively small biomass particles (> 5 mm) to approx. 500^0C in absence of oxygen. Fast pyrolysis is meant for the conversion of biomass to an intermediate, liquid bio-fuel that has a significantly increased volumetric energy density and much lower ash content than the original biomass.

The objective of this review is to analyze the design considerations faced by the developers of fast pyrolysis, upgrading and utilization processes in order to successfully implement the technologies. Aspects of design of a fast pyrolysis system include feed drying; particle size; pretreatment; reactor configuration; heat supply; heat transfer; heating rates; reaction temperature; vapour residence time; secondary cracking; char separation; ash separation; liquids collection. Each of these aspects is reviewed and discussed. The primary liquid product obtained is characterized by many properties that impact its usage in many applications. Applications of bio-oil in various fields have also been critically discussed.

See more of this Session: Reaction Engineering for Biomass Conversion
See more of this Group/Topical: Catalysis and Reaction Engineering Division

(49c) Cellulose Pyrolysis Residence Time Distribution and the Mechanism of Reactive Boiling Ejection of Aerosols From Biomass

Paul Dauenhauer[1], Andrew R. Teixeira[1], David P. Schmidt[2] and Kyle G. Mooney[2], (1)Chemical Engineering, University of Massachusetts Amherst, Amherst, MA, (2)Department of Mechanical and Industrial Engineering, University of Massachusetts Amherst, Amherst, MA

The pyrolysis of cellulose and lignocellulosic materials occurs through a molten intermediate liquid which thermally degrades to volatile components and produces primary aerosols. The generation of primary aerosols from biomass hinders the production of biofuels by pyrolysis, and intensifies the environmental impact of forest fires and the health implications associated with cigarette smoking. High speed photography is utilized to elucidate the ejection mechanism of aerosol particles from thermally decomposing cellulose at the timescale of milliseconds. Fluid modeling, based on first principles, and experimental measurement of the ejection phenomenon supports the proposed mechanism of reactive boiling ejection. Identification of the ejection mechanism of intermediate cellulose confirms the transportation of nonvolatile material to the gas-phase and provides fundamental understanding for predicting the rate of aerosol generation and the residence time distribution of reacting, molten biopolymers.

See more of this Session: Reaction Engineering for Biomass Conversion
See more of this Group/Topical: Catalysis and Reaction Engineering Division

INTENSIFICATION OF BIOFUEL SYNTHESIS BY MICROWAVE IRRADIATION

Armando T. Quitain,[1] *Shunsaku Katoh,*[2] *Mitsuru Sasaki*[1] *and Motonobu Goto*[1]
[1]*Kumamoto University, Kumamoto, JAPAN*
[2]*Kagawa Industry Support Foundation, Takamatsu, Kagawa, JAPAN*

Abstract

Intensification of heterogeneously catalyzed syntheses of two promising biofuels (*i. e.* biodiesel and *bio*ETBE) by applying microwave irradiation was investigated. Results showed a significant increase in the yield of biodiesel even at short time less than 1 min, in which the attained bulk temperature did not reach 60°C. For ETBE production, at a microwave power of 350W, and irradiation time of 1 min, the mixture reached solvothermal conditions obtaining EtOH conversion closed to 90%. The conversion and yield were found to be dependent on operating parameters such as temperature, microwave irradiation power, time and amount of catalysts. The proposed method offer benefits including energy efficiency, development of a compact process, rapid heating and instant on-off process (instant heating-cooling process).

Keywords: microwave, biodiesel, ETBE, heterogeneous catalysis, solvothermal

Introduction

The threats posed by the recent climate changes along with increasing global demand for energy call for environmentally benign and economically viable alternatives to fossil-derived fuels. Biodiesel and biomass-derivable ethyl *tert*-butyl ether (ETBE) are seriously being pursued as potential alternatives. In this regard, development of efficient and energy-saving methods for the production of these two most promising biofuels had been a subject of intensive global research in recent years. Literature searches on published articles having "biodiesel" on its title using JST and SCOPUS database resulted to heavy hits, clearly indicating a dramatically increasing global interest on research, development and analysis related to biodiesel production, especially during the past decade, due primarily to the motivation of reducing fossil-derived carbon dioxide emissions to the atmosphere.

Biodiesel is a mixture of fatty acid methyl esters produced from the transesterification of plant oils or animal fats with methanol over alkali or acid catalysts. As reported in the book edited by Saka (1), the conventional method for its production utilizes homogeneous alkali catalysts, such as NaOH and KOH, in a batch mode. Post treatment procedures after reaction require neutralization of catalysts and their removal from the products utilizing enormous amount of water. Alternative methods to avoid the problems and high costs of treating wastewater associated with the process are being explored. Moreover, the demand for biodiesel is highly expected to increase, and a more efficient continuous process is being sought with the purpose of reducing capital or production costs.

Aside from biodiesel, ETBE, which had been previously considered as a replacement for lead as gasoline octane booster, has also gained popularity over ethanol as biofuel due to its superior properties which blend well with gasoline. ETBE also outranks MTBE as an octane enhancer due to its low blending Reid vapor pressure, beside, ETBE is a better option because it is derived from ethanol

(EtOH) which can be obtained from renewable resources like biomass. ETBE is produced from the reaction of isobutene (IB) and EtOH, however, the supply of IB, which is mostly derived from non-renewable crude oil, may become limited, and for this reason, alternative routes for its synthesis are also currently being explored. *tert*-Butyl alcohol (TBA), a major byproduct of propylene oxide production from isobutane and propylene, can be employed instead of IB as a reactant (2). With suitable enzyme catalysts, TBA can also be possibly produced from biomass sources similar to that used for bioethanol production. The prospects are high for the use of these two biomass-derivable lower alcohols for the production of suitable fuel for conventional engines running on gasoline.

The synthesis route for ETBE production employing *tert*-butyl alcohol (TBA) instead of IB has long been investigated by Norris and Rigby using concentrated sulfuric acid as catalyst (3). Habenicht et al. also investigated the reaction at elevated temperatures and pressures (4). Yin et al. (5) introduced the use of heterogeneous catalysts such as ion-exchange resin and heteropoly acid. Matouq et al. (6) applied an advanced method of reactive distillation, utilizing low-grade alcohol catalyzed by potassium hydrogen sulfate ($KHSO_4$). Results showed that ETBE could be produced from the reaction, and that the reactive distillation column was a good choice to separate ETBE from the reacting mixture. In subsequent works, pervaporation technique was incorporated at the bottom of the column to remove the water byproducts, shifting the equilibrium forward, resulting to better yields (2). The works were further extended on the utilization of ethanol at a concentration as low as that obtained from the fermentation of biomass (about 2.67 mol% in aqueous solution). Almost complete conversion of TBA was obtained, with ETBE selectivity of about 36% (7).

In this work, the application of microwave irradiation to intensify syntheses of these two most promising biofuels by microwave irradiation will be discussed and summarized, focusing on our recent studies on heterogeneously catalyzed processes.

Experimental Methodology

Microwave-Assisted Synthesis of Biodiesel

In most of the experiments, rapeseed oil commercially available from Nacalai Tesque (Japan) was used. The average molecular weight of the oil was assumed to be 806 (8). Methanol (HPLC grade), $Ca(OH)_2$ and CaO (99.9%) were purchased from Wako (Japan), while other catalysts were purchased from Sigma-Aldrich (Japan). In some experiments, commercial slaked lime in pellet form supplied by Inoue Lime Industrial Company (Kochi, Japan) was also used.

All microwave-assisted batch experiments were performed using an in-house microwave apparatus working at 2.45 GHz frequency, with a power programmable from 0 to 700W. Temperature could be controlled, and the reactants could be mixed using a magnetic stirrer. Continuous experiments were performed in a similar apparatus (Shikoku Instrumentation Co. Ltd., Kagawa, Japan) design for microwave-assisted drying, but the power can be programmed to a maximum of 1500W.

In a typical batch experiments (as outlined in Fig. 1), about 11.5 g methanol and 48.5 g rapeseed oil (MeOH-oil molar ratio = 6) were placed in a three-necked round bottom flask, and heated either in an oil bath or in a microwave apparatus described above. In all runs, the MeOH-oil ratio was fixed at a commonly used molar ratio of 6, which is also the ratio being employed in industrial scale production of biodiesel. The amount of catalyst was varied from 1 to 20 g. The reaction temperature was set at 60 ^oC, unless otherwise specified. In experiments involving constant microwave heating power, the temperature was not controlled, but the maximum attained temperature was noted.

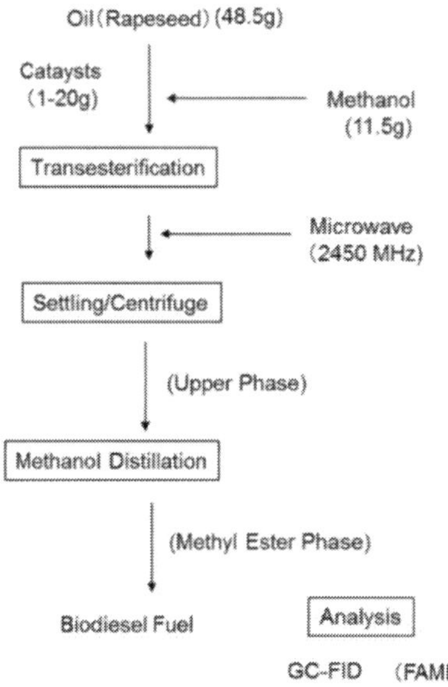

Fig. 1. Outline of experimental procedures for batch experiments on biodiesel synthesis

The products were collected, then centrifuged to separate the catalysts and the glycerin phase. The unreacted MeOH in the products was then removed using a rotary evaporator at 70 °C. The products were analyzed of its composition by a gas chromatography – flame ionization detector (GC-FID) apparatus (Shimadzu GC-14B) connected to a computer for data collection and analysis. Component separation was made in a 50m x 0.25mm CP Sil 88 capillary column (GL Science, Japan), tailor-made for FAME analysis using helium as a carrier gas. The column, detection and injection temperatures were set to 190, 300 and 270 °C, respectively. The sample injection volume was 5 μl and peak identification was made by comparing the retention time between the sample and the standard compound. FAME quantitative mixtures (GL Science, Japan) were used for peak identification and for quantitative analysis.

Preliminary experiments were conducted to evaluate catalytic activities of various solid catalysts such as Amberlyst 15, Amberlite-OH, Amberlite-Acid, zeolite, sulfated zirconia (in powder and pellet forms), $Ca(OH)_2$ and CaO. Among the catalysts investigated, $Ca(OH)_2$ showed to be the most active, while CaO also gave fairly good results. The use of these two relatively cheap catalysts showed potential for biodiesel production, thus $Ca(OH)_2$ was used in the succeeding experiments for biodiesel synthesis unless otherwise specified.

Microwave-Assisted Synthesis of ETBE

The method was also applied to the synthesis of ethyl *tert*-butyl ether (ETBE) from two biomass-derivable alcohols (ethanol and *tert*-butyl alcohol). Microwave-assisted experiments were performed using the same microwave apparatus used for biodiesel synthesis, working at 2.45 GHz frequency, with a power programmable from 0 to 1000W. In a typical run, about 0.25mol of EtOH and TBA, and 20 g of catalyst were placed in a reactor vessel, and heated using a microwave apparatus described above. GC-FID was used for the analysis of the products using an internal standard.

Preliminary experiments on the evaluation of catalytic activities of various solid catalysts such as Amberlyst 15JWET, sulfated zirconia, sulfated charcoal and zeolite showed Amberlyst 15JWET to be the most effective. The yield of ETBE using sulfated charcoal and and zeolite is almost negligible compared to that of Amberlyst 15JWET and sulfated zirconia. Thus, Amberlyst 15JWET was used in the experiments unless otherwise specified.

Results and Discussion

Microwave versus conventional heating

Microwave heating for the production of biodiesel in a batch mode was compared with that of the conventional. In case of the conventional method, the oil bath temperature was set at 60°C, and the mixtures of reactants and catalysts were heated for 1 min. Using microwave irradiation, the power was set at 700 W. Heating for 1 min, the bulk temperature of the mixtures did not reach 60°C in all runs.

Fig. 2 shows a remarkable increase in the yield of methyl esters using microwave heating compared to the conventional. The yield, corresponding to the amount of methyl esters in the oil phase, reached above 95% using 20g $Ca(OH)_2$. Even if the bulk temperature did not reach 60°C, it is likely that localized heating above 60°C occurred at the surface of the catalysts, which brought about a significant increase in reaction rate, resulting into high yield. This is advantageous especially on the viewpoint of equipment design as this entails less provision for heat and pressure-resisting reactor materials.

In this proposed method, the reaction time was reduced to less than 60 sec compared to 1 to 8 h using the conventional method. In addition, the use of solid catalysts avoids the rigors and complexities of dealing with post-reaction treatments (*i. e.* neutralization of homogeneous catalysts and washing of the products with water). Furthermore, with short reaction time, development of a continuous process is highly feasible thus reducing equipment costs.

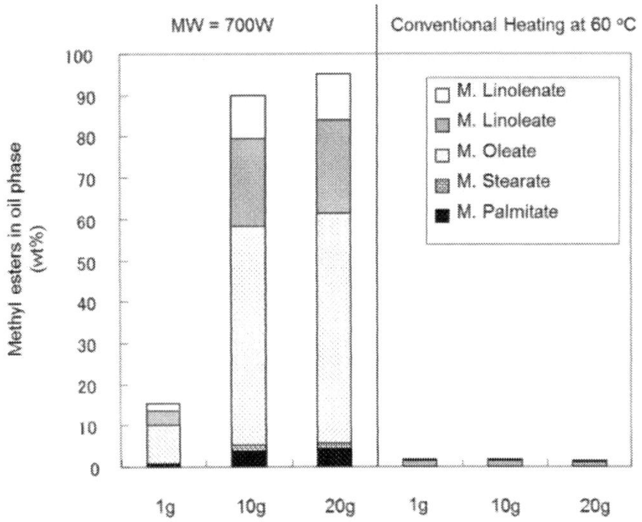

Fig. 2. Comparison of microwave and conventional heating (t = 1min, catalysts: $Ca(OH)_2$, MW = 700W)

Elucidation of the effects of operating parameters

Fig. 3 shows the effect of microwave irradiation power on the yield at various amounts of catalysts. No significant differences were observed at 140 and 350W, but the yields were comparatively high at 700W especially at 10 and 20g $Ca(OH)_2$. Using 10g $Ca(OH)_2$, the reaction time was increased to 5 min at the same microwave power of 700W. Results showed that while the temperature increased sharply above 110 °C in just 5 min, the yield decreased to 20%. It is likely that reverse reaction took place brought about by an increase in reaction temperature and the subsequent evaporation of MeOH from the reaction zone. The same results were observed in the works of Hernando et al. (9) on the batch tests performed with microwaves.

Application of the proposed method to other oil feedstocks

Rapeseed oil contains mostly oleic acid, while soybean and coconut oils are rich in linoleic and lauric acids, respectively. These differences might have an effect on the transesterification of the oils, thus the results obtained using rapeseed oil were compared with that of soybean and coconut oils. Almost similar results were obtained with different types of vegetable oils investigated, with the yield for coconut oil higher than the two other types of oil. This implies that the proposed method can be applied to any types of oil (or fat) feedstocks for biodiesel production.

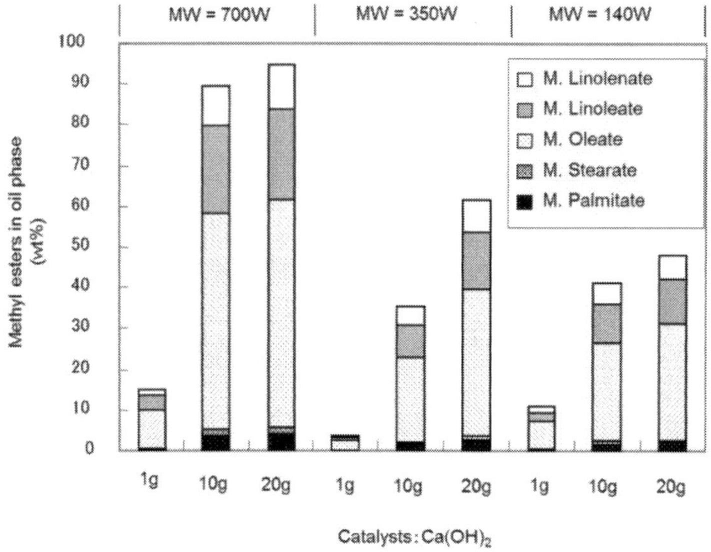

Fig. 3. Dependency of yield on microwave irradiation power at various amounts of $Ca(OH)_2$ catalysts (t=1min)

Commercial slaked lime in comparison to pure $Ca(OH)_2$ catalysts

For low-cost production of biodiesel, cheap and readily available catalysts for its production are being sought. The use of cheap commercial grade $Ca(OH)_2$ catalysts could be considered. For this purpose, the activity of the commercial slake lime pellets (supplied by Inoue Lime Co. Ltd., Kochi, Japan) containing 60% $Ca(OH)_2$ was compared with that of pure $Ca(OH)_2$. Results in Fig. 4 show that the yield was low at around 30% using the un-pretreated pellets. If pre-dried, the yield increased by more than 10%. After further pulverization, a three-fold increase in the yield was obtained using dried catalysts as a result of the increase in surface contact area. In the succeeding continuous-flow experiments, non-pulverized dried slaked lime pellets were used.

Continuous-flow methods and results

For continuous-flow experiments, about 120g of slaked lime pellets was placed in a 100-ml glass flask which served as a reactor. Slaked lime was selected based on the results of our previous work on the investigation of activities of various catalysts. Methanol and rapeseed oil (fixed at a molar ratio of 6) were vigorously mixed and allowed to pass through the reactor at various flow rates. The residence time was calculated based on the void space of the reactor after placing the catalysts.

Fig. 5 shows the typical experimental conditions and results of the continuous process for biodiesel production using microwave. The temperature was controlled at 60°C by supplying microwaves at a maximum peak of 30% corresponding to a power of about 300 W. The flow rates were varied from 12 to 50 ml/min corresponding to residence times of 7 to 1.5 min, respectively.

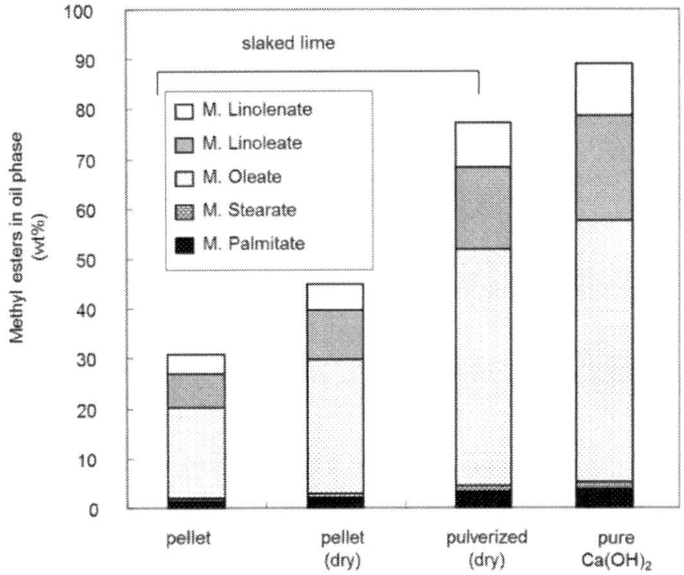

Fig. 4. Experimental results using commercial slaked lime (60% pure) compared to pure $Ca(OH)_2$ catalysts (t=1min, catalysts=10g, MW=700W)

In runs 1 and 2, high yields were observed initially when most of the microwave irradiation were supplied and absorbed by the reactants and catalysts. However, above the set reaction temperature of 60°C, the microwave irradiation automatically ceased, possibly causing an intermittent lowering of the yield. Below 60°C, the system automatically activated, supplying microwave back to the reactor. The reaction reached steady state after about 8 sampling runs corresponding to a run time of about 8 and 16 mins for runs 1 and 2, respectively. Other than the thermal effects, microwave effects were evident from the experimental results, and thorough investigation would be necessary to further validate this interesting phenomenon.

	Run 1	Run 2	Run 3
Flowrate (ml/min) :	50	25	12
Corresponding residence time(min):	1.5	3.5	7
Sampling period (min):	1	2	4

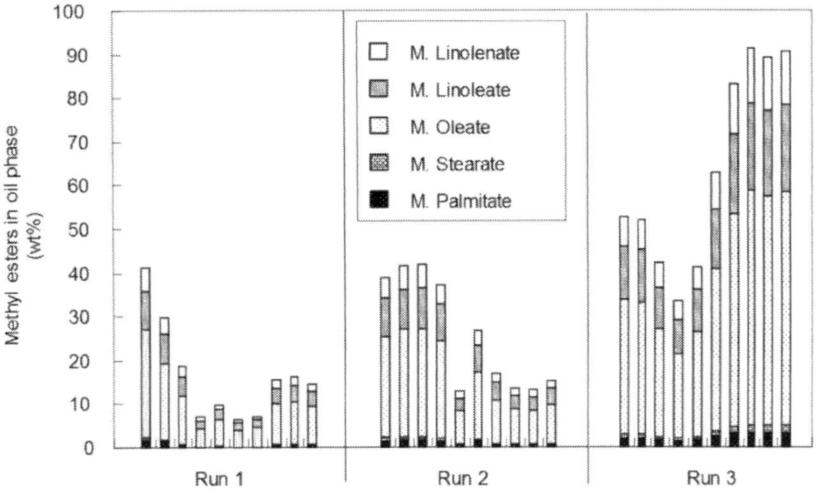

Fig. 5. Continuous-flow experimental results for biodiesel production under microwave irradiation

In run 3, steady state was also observed after about 8 sampling runs, and a yield above 90% was obtained in short residence time of 7 min. Mixing of the reactants and catalysts by stirring could improve the homogeneity of microwave absorption and could increase the yield. However, in the case of solid catalyst pellets, recirculation of the reactants or products back to the reactor could be a better alternative to stirring.

Esterification of free fatty acids present in oil feedstock

Most of the oil feedstocks for biodiesel syntheses contain relatively high amount of fatty acids especially the waste cooking oil. This has become a big hurdle for industrialization of the proposed process, because the presence of fatty acids significantly affects the solubility of Ca-based catalysts in the products. Government quality standards for biodiesel require the level of Ca to be below 5 ppm, while the fatty acid content should not exceed 1wt%. Thus, pretreatment of free fatty acids in oil is necessary prior to transesterification of the triglyceride contents. In this regard, microwave irradiation was also applied to convert free fatty acids into biodiesel. Results in Fig. 6 show an 88% conversion of fatty acids in waste cooking oil in 1 min of microwave irradiation at a power of 700W using ion exchange resin as catalysts. With these results, a two-step process is proposed for the conversion of waste oil, or any type of oil feedstocks containing high amount of fatty acids, to biodiesel fuel. The process consists of a first step of esterification of fatty acids followed by a second step of transesterification of the triglyceride. While the two-step process seems ideal for the treatment of free fatty acids in oil, this also minimizes the solubility of Ca-based catalysts as a result of the reduction of fatty acid contents.

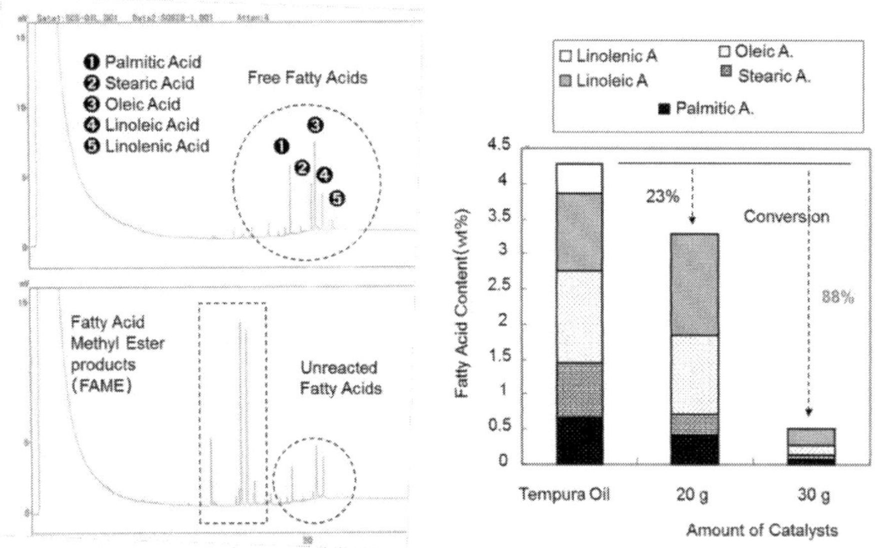

Fig. 6. Results of experiments on esterification of free fatty acids in waste oil using microwave (MWpower=700W, t = 1 min)

ETBE Synthesis by Microwave Irradiation

Experiments at atmospheric pressure using a batch reactor showed that the yield hardly increased beyond the 20% level. The experiments were extended to continuous flow at various conditions, but the yield did not exceed 35% as shown in Fig. 7. Almost similar results were obtained by other researchers (Japanese Patent JP2007-126450) (10), and the lower yield was likely due to the selective dehydration of TBA to IB, a highly volatile compound that could easily escape from the reaction zone. If IB could further react with EtOH to produce ETBE, better yield could be obtained. Thus, experiments using a sealed reactor vessel were conducted using another microwave apparatus

(Milestone Ethos). Fig. 8 shows the conversion of EtOH to ETBE using MW at various power and irradiation time. A maximum yield of about 87% was obtained at MW power of 350W at irradiation time of 1 min. At this condition, the attained temperature was around 87°C, higher than the boiling points of the two alcohols. The conversion was also found to be dependent on the amount of catalysts, reaction time and microwave power.

	Run 1	Run 2	Run 3	Run 4	Run 5	Run 6	Run 7	Run 8
Flowrate (ml/min) :	100	100	50	50	20	20	10	10
Residence time(min):	1	1	2	2	5	5	10	10
Sampling time (min):	1	1	2	2	5	5	10	10
MW duty (%):	13	20	13	20	13	20	13	20
Temperature (°C):	28	67	69	71	70	71	70	71

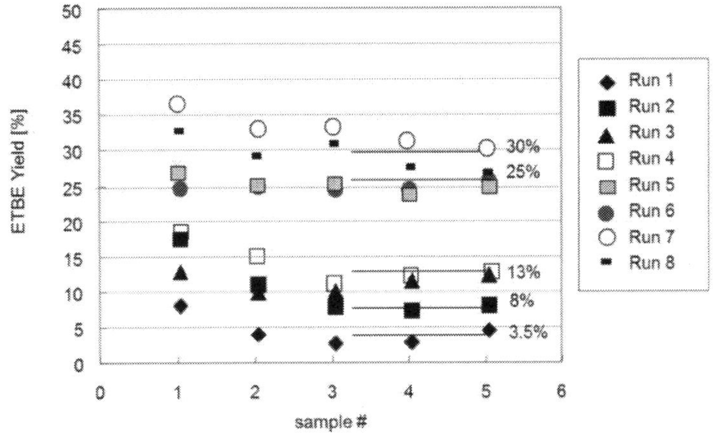

Fig. 7. Typical continuous-flow experimental results for ETBE production using A15JWET ion exchange resin catalysts under microwave irradiation

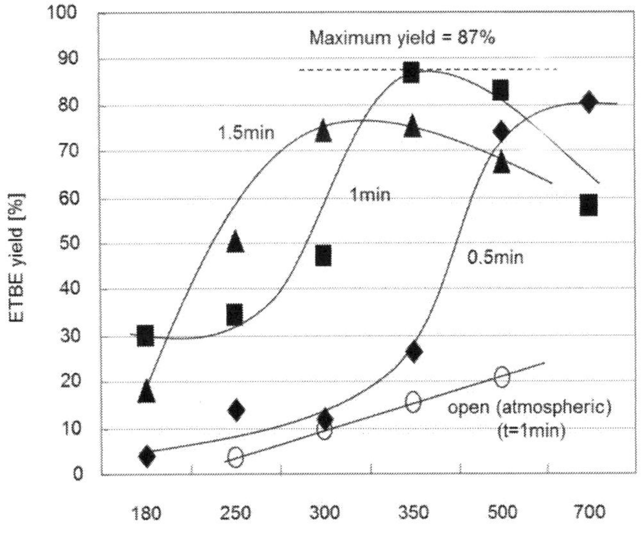

Fig. 8. Yield of ETBE at various microwave irradiation power and reaction time (closed system, TBA=EtOH=0.25mol, A15JWET = 20g)

While several methods reported a conversion in the range of 9 to 38 % using the conventional heating methods, we obtained a maximum conversion of about 87% under microwave irradiation at a power of 350 W for 1 min. At this condition, the bulk temperature reached about 86 ^{o}C, which was above the boiling points of the two alcohols. The results imply that solvothermal condition (closed system) is ideal for ETBE synthesis because the IB generated from the dehydration of TBA can further react with EtOH to produce ETBE, resulting to higher conversion.

Drawbacks of the Proposed Process

Microwave-assisted reactions offer several great advantages to the synthesis of biofuels, however, there are also some drawbacks associated with its use. Microwave could not work well with large quantities of materials, and thus could not be easily converted from laboratory to a multikilogram industrial-scale production. The penetration depth of MW irradiation into the absorbing materials is only a few centimeters, and this significantly limits scale up of the technology. Microwave irradiation is non-homogeneous and formation of "hotspots" is likely, thus control of reaction is too difficult. Mixing may improve homogeneity, however, with the use of solid catalysts, appropriate methods of mixing remains a challenge. Safety consideration is another factor for industrial utilization of microwave. The use of batch microwave reactors, for the processing of comparatively large volumes under pressure may not be safe because any malfunction or rupture of a large pressurized reaction vessel, which are usually made of Teflon or glass materials, may result into massive spillage causing significant operational damages to the working place and the environment.

Future Directions

As the demand for biofuels continue to increase in the near future, and while the search for an efficient and low-cost production process continues, the global outlook is positive for the use of microwave irradiation to the synthesis of two most promising biofuels - biodiesel and bioETBE. To overcome the limitations for scaling up microwave-assisted technology for biodiesel production, development of a continuous process is suggested, but still poses several challenges that require detailed investigation. The future also calls for the development of cheap, effective and stable solid catalysts for the synthesis of the abovementioned fuels. While the use of microwave irradiation offers great benefits with regards to rapid reaction or synthesis, safety is a big factor to consider in designing a large scale production plant. However, this can be avoided if multilayered compact reactors operating under microwave irradiation can be developed instead.

Conclusion

Microwave irradiation was applied to intensify syntheses of two most promising biofuels, *i. e.* biodiesel and *bio*ETBE. The application of microwave irradiation to the synthesis of the two above mentioned biofuels offer benefits including energy efficiency, development of a compact process, rapid heating and instant on-off process (instant heating-cooling process), among others. Unlike the conventional heating, the heat is generated within the material, thus rapid heating occurs. Besides, microwave effects on reaction also likely occur, thus obtaining dramatic increase in the yield even at low bulk temperatures. Results showed a significant increase in the yield of biodiesel even at short time less than 1 min, in which the attained bulk temperature did not reach 60°C. For ETBE production, at a microwave power of 350W, and irradiation time of 1 min, the mixture reached solvothermal conditions obtaining EtOH conversion closed to 90%. The conversion and yield were found to be dependent on operating parameters such as temperature, microwave irradiation power, time and amount of catalysts.

The benefits have been indicated by the above mentioned results. However, there are some drawbacks that would require thorough investigation prior to its commercialization. Although the field is in its infancy, the outlook is bright for the proposed methods due to foreseen high global demands for biofuels. The next few years should see development of continuous compact process, along with cheap, effective and stable solid catalysts.

Acknowledgements

The research works were mostly supported financially by the Japan Science and Technology. Funds for collaborative research between Kagawa Industry Support Foundation and Shodoshima Clean Service, Co. Ltd. for the investigation of the treatment of free fatty acids are also gratefully acknowledged. This work was also partly supported by Kumamoto University Global COE Program "Global Initiative Center for Pulsed Power Engineering".

References

(1) Saka, S. (1996). *All About Biodiesel*, IPC Publishing, ISBN- 9784901493567, Tokyo, Japan.

(2) Yang, B. & Goto, S. (1997). Pervaporation with Reactive Distillation for the Production of Ethyl *tert*-Butyl Ether. *Sep. Sci. Tech.*, Vol. 32, pp. 971-981.

(3) Norris, J. F. & Rigby, G. W. (1932). The Reactivity of Atoms and Groups in Organic Compounds. *J. Am. Chem. Soc.*, Vol. 54, pp. 2088-2100.

(4) Habenicht, C.; Kam, L. C.; Wilschut, M. J. & Antal, M. J. (1995). Homogeneous Catalysis of Ethyl *tert*-Butyl Ether Formation from *tert*-Butyl Alcohol in Hot, Compressed Liquid Ethanol. *Ind. Eng. Chem. Res.*, Vol. 34, pp. 3784-3792.

(5) Yin, X.; Yang, B. & Goto, S. (1995). Kinetics of Liquid-Phase Synthesis of Ethyl *tert*-Butyl Ether from *tert*-Butyl Alcohol and Ethanol Catalyzed by Ion-Exchange Resin and Heteropoly Acid. *Int. J. Chem. Kinetics*, Vol. 27, pp. 1065-1074.

(6) Matouq, M.; Quitain, A. T.; Takahashi, K. & Goto, S. (1996). Reactive Distillation for Synthesizing Ethyl *tert*-Butyl Ether from Low-Grade Alcohol Catalyzed by Potassium Hydrogen Sulfate. *Ind. Eng. Chem. Res.*, Vol. 35, pp. 982-984.

(7) Quitain, A. T.; Itoh, H. & S. Goto, S. (1999). Reactive Distillation for Synthesizing Ethyl *tert*-Butyl Ether from Bioethanol. *J. Chem. Eng. Japan*, Vol. 32, pp. 280-287.

(8) Kusdiana, D. & Saka, S. (2001). Kinetics of Transesterification in Rapeseed Oil to Biodiesel Fuel as Treated in Supercritical Methanol. *Fuel*, Vol. 80, pp. 693-698.

(9) Hernando, J., Leton, P.; Matia, M. P., Novella, J. L. & Alvarez-Builla, J. (2007). Biodiesel and FAME Synthesis Assisted by Microwaves: Homogeneous Batch and Flow Processes. *Fuel*, Vol. 86, pp. 1641-1644.

(10) Japanese Patent Application: JP2007-126450. ETBE Synthesis Using Microwave.

Hydrogenation of cellulose to sorbitol over sulfated zirconia supported Ru catalysts

*Yufei Niu, Hua Wang, Mingrui Liu and Jinyu Han**

Key Laboratory for Green Chemical Technology of State Education Ministry, School of Chemical Engineering & Technology, Tianjin University, Tianjin, P. R. China.

Introduction

Lignocellulosic biomass, composed of cellulose, hemicellulose and lignin, abundantly exists in nature and could be a promising alternative to fossil derived chemicals. Cellulose is a polymer of D-glucose units connected through β-1, 4-glycosidic bonds. Dueing to the highly crystalline structure with extensive intra- and inter-molecular hydrogen bonds, it is more resistant to depolymerization. Up to now, extensive efforts have been devoted to the conversion of cellulose in various reaction media including hot water, mineral acids and ionic liquids. It is of great significant and challenge to achieve direct conversion of cellulose to sugar alchohols, which combining hydrolysis and hydrogenation. Furthermore, previous studies have demonstrated the effectiveness of heterogeneous solid acid catalysts in hydrogenolysis of cellulose to hexitols.

Sulfated zirconia, as a typical solid acid catalyst, was widely used in reactions such as isomerization of isobutane, cyclopropane/n-hexane conversion, and recently in cellobiose hydrolysis. In this work, we prepared a series of sulfated zirconia supported Ru (SO_4^{2-}/ZrO_2-Ru) catalysts by precipitaion methods followed by impregnation in diluted H_2SO_4, and loading a proper amount of Ru. BET method, powder X-ray diffraction (XRD), and temperature programmed desorption of NH_3 (NH_3-TPD) were used for the catalysts characterization.

Characterization of catalyst

The results of temperature programmed desorption of NH_3 (NH_3-TPD) showed that the strength of acidity was effectively improved by sulfation: the weak acid site was enhanced with desorption peak shift to higher temperature; in addition, there was a broad desorption peak distributed on the region of 800~1000K, and the area of desorption peak increased remarkablely with the raising concentration of H_2SO_4 solution. XRD patterns revealed that the sulfated zirconia calcined at 773K was completely tetragonal phase which was responsible for the acidity; with the calcination temperature raised, component of monoclinic phase increased, thus the acidity was reduced. The concentration of H_2SO_4 solution and the kind of doping metal affected the BET surface area more than the amount of Ru loading on the sulfated zirconia.

** Supported by the National Natural Science Foundation of China (No.21076152) and Specialized Research Fund for the Doctoral Program of Higher Education of China (No. 20100032110018)

* *Corresponding author: hanjinyu@tju.edu.cn*

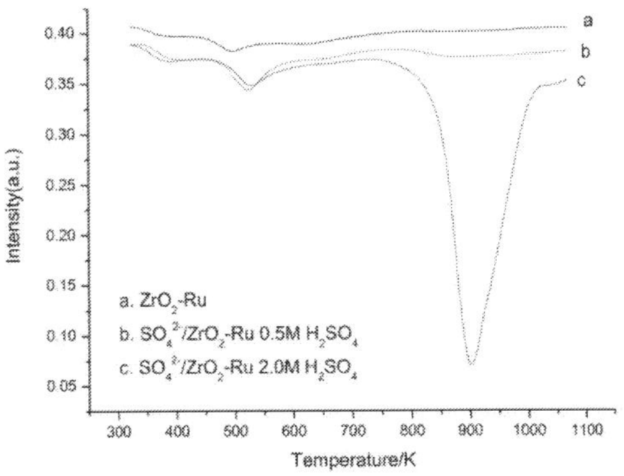

Figure1. NH_3-TPD profiles of catalysts

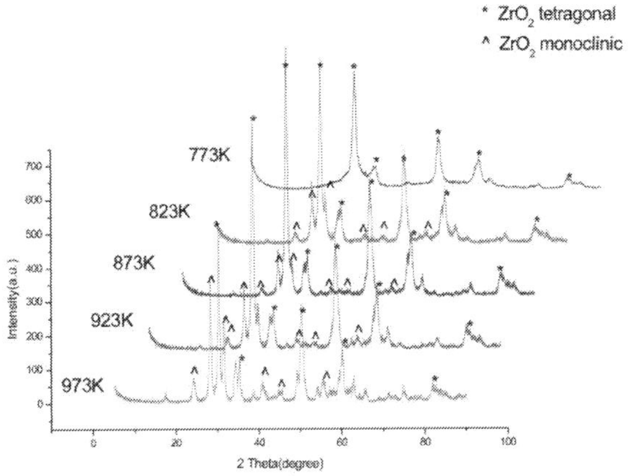

Figure2. XRD patterns of SO_4^{2-}/ZrO_2-Ru calcinated at different temperatures

Table1. Physical characterisitics of catalysts

Entry	Catalyst[a]	BET Surface Area m^2/g	V_{BJH} cm^3/g	Pore size nm
1	ZrO_2-Ru	38.05	0.1168	12.28
2	SO_4^{2-}/ZrO_2-Ru	84.89	0.0972	4.58
3	SO_4^{2-}/ZrO_2-Ru[b]	4.19	0.0071	6.77
4	SO_4^{2-}/ZrO_2-MgO-Ru	36.89	0.0651	7.06
5	SO_4^{2-}/ZrO_2-Al_2O_3-Ru	90.93	0.0738	3.25
6	SO_4^{2-}/ZrO_2-SiO_2-Ru	113.48	0.1419	5.00
7	SO_4^{2-}/ZrO_2-SiO_2-Ru[c]	151.16	0.2778	7.35
8	SO_4^{2-}/ZrO_2-SiO_2-Ru[d]	257.18	0.5565	8.65
9	SO_4^{2-}/ZrO_2-Ru 1%Ru	89.45	0.1045	4.67
10	SO_4^{2-}/ZrO_2-Ru 2%Ru	100.54	0.1036	4.12
11	SO_4^{2-}/ZrO_2-Ru 3%Ru	101.90	0.1020	4.05
12	SO_4^{2-}/ZrO_2-Ru 4%Ru	100.34	0.0988	3.94
13	SO_4^{2-}/ZrO_2-Ru 5%Ru	84.89	0.0972	4.58

[a] All catalysts were impregnated with 0.5M H_2SO_4 in the sulfation step and calcined at 773K; for entry 4,5,6, amount of metal introduced was in the molar ratio Zr: M= 100: 15;

[b] Catalyst was impregnated with 2.0M H_2SO_4 in the sulfation step;

[c] Molar ratio Zr: Si=50:50;

[d] Molar ratio Zr: Si=15:100.

Catalytic performance of sulfated zirconia supported Ru catalyst

The plausible reaction pathway was summerized in Scheme1. The reaction process is roughly divided into two steps: catalytic hydrolysis of cellulose by solid acid into monomeric sugars such as glucose and fructose, and hydrogenation of sugars to get hexitols. Selectivity loss of hexitols may be caused by reactions as follow: (a) degration of sugars to form furan compounds like 5-HMF, and then it was converted to levulinic acid and formic acid by dehydration, or hydrogened into DHMF and DHM-THF furtherly; (b) degration of sugar alcohols, including dehydration of hexitols into sorbitan which was further dehydrated to isosorbide, and hydrogenolysis of C-C and C-O bonds producing smaller polyols such as glycerol, ethene glycol (EG), and propanediols.

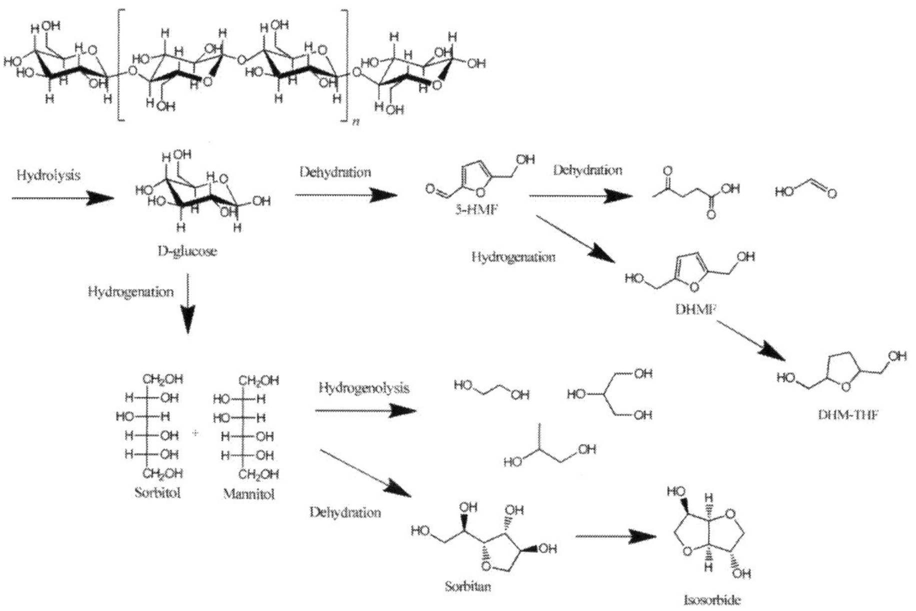

Scheme1. Reaction pathways for acid-catalyzed hydrogenation of cellulose

The activity of as-prepared catalysts for hydrogenation of cellulose was investigated in a lab-scaled batch reactor. The main products detected were sorbitol, slight amount of xylitol and C2~C3 polyols. Trace of glucose was detected in the liqiud product, verifying that cellulose was first hydrolyzed into sugars by the acid sites on the solid catalyst. No product generated by dehydration of C6 sugars like furfual, 5-HMF, levulinic acid and formic acid were obtained, indicating that sugars were hydrogened by Ru supported on the sufated zirconia without being dehydated. Mannitol, a common byproduct reported in most literatures generated from hydrogenation of fructose, was also not detected in the liquid mixture, revealing that isomerization of glucose into fructose in the hydrolysis process could be effectively prevented using sulfated zirconia catalyst.

The acidity strength of catalyst had a remarkable influence on cellulose conversion and the yield of sorbitol. The increase of acidity on catalyst surface promoted hydrolyzing of cellulose, thereby enhanced the conversion of cellulose.While dueing to the side reactions such as dehydration of glucose/hexitols and C-C cleavage of sorbitol, too dominant acid catalysts caused lower selectivity of sorbitol. The strength of acidity can be adjusted by changing of calcination temperature, concentration of H_2SO_4 and the amount of doping metal.

Calcination temperature was elevated from 773K to 973K. As the XRD patterns shown, elevating calcination temperature induced the phase transformation of tetragonal ZrO_2 into monoclinic phase which leads to the acidity loss, as a result the conversion of cellulose and yield of products decreased. In order to investigate the effect of concemtration of sulfuric acid on catalytic performance, an concentration of 0.5M, which optimized by Gonzalez et al. using the isomerization of n- butane as a reaction probe, was chosen. By comparison, a higher concerntration of 2.0M was also used. Although sulfating with 2.0M H_2SO_4 solution provided more acidic sites, thus promoted the hydrolysis of cellulose into liquid product, the yields of xylitol and C2~C3 polyols in liquid product increased duing to the C-C bond cleavage side reaction, and the yield of sorbitol declined significantly. Mg, Al and Si were introduced to sulfated zirconia supported Ru catalyst before precipitaion step respectively, in the attemp to improve the catalytic activity. The result showed that only doping with Si obtained higher cellulose conversion and sorbitol yield, which might due to the enhanced acidity as previous studies reported. Besides, the amount of doping Si affected the performance of catalyst. Yield of sorbitol was getting higher with the increase of Zr/Si molar ratio.

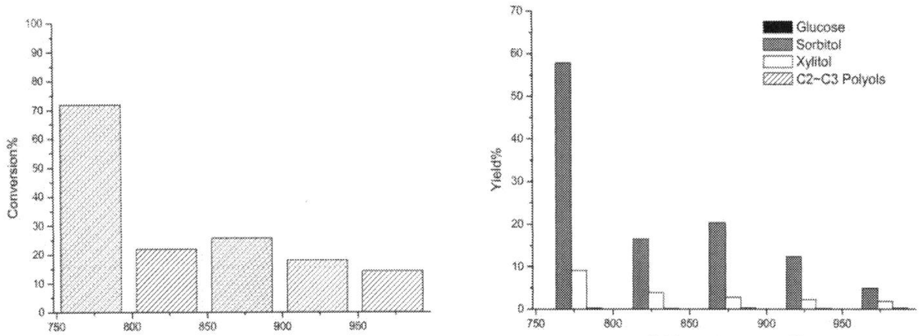

Figure3. Effect of calcination temperature on calalytic activity

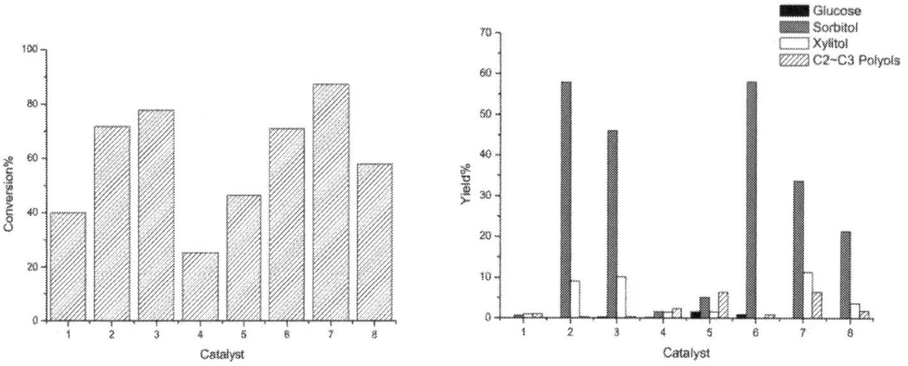

Figure4. Catalytic activity of catalysts

(1)SO_4^{2-}/ZrO_2-Ru; (2) SO_4^{2-}/ZrO_2-Ru, 0.5M H_2SO_4; (3)SO_4^{2-}/ZrO_2-Ru, 2.0M H_2SO_4; (4) SO_4^{2-}/ZrO_2-MgO-Ru; (5) SO_4^{2-}/ZrO_2-Al_2O_3-Ru; (6) SO_4^{2-}/ZrO_2-SiO_2-Ru, Zr: Si=100:15; (7) SO_4^{2-}/ZrO_2-SiO_2-Ru, Zr: Si=50:50; (8) SO_4^{2-}/ZrO_2-SiO_2-Ru, Zr: Si=15:100

In conclusion, for microcrystalline cellulose without pretreatment, 71.1% conversion and 58.1% yield of sorbitol were obtained under the mild conditions (at 473K after 10h, 5Mpa H_2 pressure; cellulose,0.25g; catalyst, 0.05g; distilled water 5g) catalyzed by SO_4^{2-}/ZrO_2-SiO_2-Ru catalyst (Zr/Si = 100:15, 5%Ru).

References

A. Fukuoka, and P. L. Dhepe, Catalytic conversion of cellulose into sugar alcohols, Angew. Chem. Int. Ed., (2006), 45, 5161 -5163

B. Tyagi, K.B. Sidhpuria, B. Shaik, and R.V. Jasra, Effect of Zr/Si molar ratio and sulfation on structural and catalytic properties of ZrO_2–SiO_2 mixed oxides, J. Porous. Mater, (2010), 17, 699-709

C. Luo, S. Wang, and H.C. Liu, Cellulose conversion into polyols catalyzed by reversibly formed acids and supported ruthenium clusters in hot water, Angew. Chem. Int. Ed., (2007), 46, 7636-7639

E. Zhao, Y. Isaev, A. Sklyarov, and J.J Fripiat, Acid centers in sulfated, phosphated and/or aluminated zirconias, Catal. Lett., (1999), 60, 173-181

H. Kobayashi, Y. Ito, T. Komaanoya, Y. Hosaka, P. L. Dhepe, K. Kasai, K. Hara, and A. Fukuoka, Synthesis of sugar alcohols by hydrolytic hydrogenation of cellulose over supported metal catalysts, Green Chem., (2011), 13, 326-333

I.A. Ignatyev, C.V. Doorslaer, P.G.N. Mertens, K. Binnemans, and D.E. de Vos, Reductive splitting of cellulose in the ionic liquid 1-butyl-3-methylimidazolium chloride, ChemSusChem, (2010), 3, 91-96

J. Geboers, S.V. de Vyver, K. Carpentier, P. Jacobs, and B. Sels, Efficient hydrolytic hydrogenation of cellulose in the presence of Ru-loaded zeolites and trace amounts of mineral acid, Chem. Commun., (2011), 47, 5590-5592

J.H. Wang, and C.Y. Mou, Alumina-promoted mesoporous sulfated zirconia: A catalyst for n-butane isomerization, Appl. Catal. A, (2005), 286, 128-136

J. N. Chheda, G. W. Huber, and J.A. Dumesic, Liquid-Phase catalytic processing of biomass-derived oxygenated hydrocarbons to fuels and chemicals, Angew. Chem. Int. Ed., (2007), 46, 7164 -7183

M.G. Cutrufello, U. Diebold, and R.C. Gonzalez, Optimization of synthesis variables in the preparation of active sulfated zirconia catalysts, Catal. Lett., (2005), 101, 1-2

N. Ji, T. Zhang, M.Y. Zheng, A.Q. Wang, H. Wang, X.D. Wang, and J.G. Chen, Direct catalytic conversion of cellulose into ethylene glycol using nickel-promoted tungsten carbide catalysts, Angew. Chem. Int. Ed., (2008), 47, 8510-8513

P.L. Dhepe, and A. Fukuoka, Cracking of cellulose over supported metal catalysts, Catal. Surv. Asia, (2007), 11,186-191

P. L. Dhepe, and A. Fukuoka, Cellulose conversion under heterogeneous catalysis, ChemSusChem (2008), 1, 969 -975

R. Barthos, F. Lonyi, J. Engelhardt, and J. Valyon, A study of the acidic and catalytic properties of pure and sulfated zirconia-titania and zirconia-silica mixed oxides, Top. Catal., (2000), 10, 79-87

R. Palkovits, K. Tajvidi, J. Procelewska, R. Rinaldi, and A. Ruppert, Hydrogenolysis of cellulose combining mineral acids and hydrogenation catalysts, Green Chem., (2010), 12, 972-978

R. Rinaldi, R. Palkovits, and F. Schuth, Depolymerization of cellulose rsing solid catalysts in ionic liquids, Angew. Chem. Int. Ed., (2008), 47, 8047-8050

S.V. de Vyver, J. Geboers, P. A. Jacobs, and B.F. Sels, Recent advances in the catalytic conversion of cellulose, ChemCatChem, (2011), 3, 82–94

V. Degirmenci, D. Uner, B. Cinlar, B.H. Shanks, A. Yilmaz, R.A. van Santen, and E. J.M. Hensen, Sulfated zirconia modified SBA-15 catalysts for cellobiose hydrolysis, Catal Lett, (2011), 141, 33-42

W.P. Deng, X.S. Tan, W.H. Fang, Q.H. Zhang, and Y. Wang, Conversion of cellulose into sorbitol over carbon nanotube-supported ruthenium catalyst, Catal Lett, (2009), 133,167-174

X. J. Shi, Y.L. Wu, P.P. Li, H. F. Yi, M.D. Yang, and G.H. Wang, Catalytic conversion of xylose to furfural over the solid acid SO_4^{2-}/ZrO_2-Al_2O_3/SBA-15 catalysts, Carbohyd. Res, (2011), 346, 480-48

X.R. Chen, Y.H. Ju, and C.Y. Mou, Direct synthesis of mesoporous sulfated silica-zirconia catalysts with high catalytic activity for biodiesel via esterification, J. Phys. Chem. C, (2007), 111, 18731-18737

Y. Wang, J.H. Ma, D. Liang, M.M. Zhou, F.X. Li, and R.F. Li, Lewis and Brønsted acids in super-acid catalyst SO_4^{2-}/ZrO_2–SiO_2, J. Mater. Sci, (2009), 44, 6736-6740.

(49f) Kinetic Modeling of Tandem Hydrogenolysis/ Reforming of Glycerol to 1,2-Propanediol with In Situ Formed Hydrogen In a Batch Slurry Reactor

Arely A. Torres, Debdut S. Roy, Bala Subramaniam and Raghunath V. Chaudhari, Department of Chemical and Petroleum Engineering, Center for Environmentally Beneficial Catalysis, University of Kansas, Lawrence, KS

Glycerol is viewed as an important feedstock due to its availability in large quantities, as a byproduct during biodiesel production and possible production from sugar derived polyols. The efficient conversion of glycerol into higher value added chemicals, such as 1,2-propanediol, ethylene glycol and lactic acid is necessary to achieve the economic viability of such processes. When using a supported bimetallic Pt-Ru catalyst, this reaction system offers the possibility of generating hydrogen *in situ* instead of providing it from an external costly source, greatly improving the economics of the process. Furthermore, since the reaction can run at milder conditions compared to traditional hydrogenolysis with externally added hydrogen at high pressures, it improves the inherent safety of the process. This tandem approach has been recently demonstrated in our group (Roy et al., Catalysis Today 156 (2010) 31–37) for glycerol to 1,2-propanediol using admixture of supported Ru and Pt catalysts. In this presentation, a detailed kinetic model of this reforming/hydrogenolysis reaction using supported bimetallic Pt-Ru catalyst will be addressed. We build on our previous efforts (Torres et al., I&EC Research, 49 (2010) 10826–10835) by including several additional reaction pathways discerned by performing experiments with selected intermediate products such as 1,2-propanediol, ethylene glycol, methanol, syngas, among others. The experiments were carried out in a batch slurry reactor, using a 2%Pt-2%Ru/C catalyst, in which both gas and liquid phase product distribution was measured. The effects of different glycerol concentrations, pressure, and catalyst loading over a temperature range of 453-493 K were studied. A detailed analysis of mechanistic Langmuir Hinshelwood type kinetic models following a rigorous model discrimination procedure will be presented along with statistical analysis of the model parameters. This study is useful in understanding the mechanism of the reactions involved in hydrogenolysis of polyols in general, as well as the modeling and scale up of slurry and fixed bed reactors.

See more of this Session: Reaction Engineering for Biomass Conversion
See more of this Group/Topical: Catalysis and Reaction Engineering Division

Green synthesis of platinum/gold bimetallic nanoparticles supported onto hydrotalcite surface as heterogeneous catalyst for selective oxidation of glycerol

Duangta Tongsakul, Chuchaat Thammacharoen, Sanong Ekgasit, Sensor Research Unit, Chulalongkorn University, Bangkok, Thailand*

Shun Nishimura and Kohki Ebitani, School of Materials Science, Japan Advanced Institute of Science and Technology, Nomi, Japan*

Abstract

Platinum/Gold bimetallic (Pt/Au) nanoparticles were successfully synthesized by green synthesis method in aqueous solution under mild condition using polysaccharide as a protecting agent as well as a reducing agent. The metal nanoparticles were supported on hydrotalcite (HT) surface by sol immobilization method and further utilized as heterogeneous catalyst. The surface morphology and compositions of supported platinum/gold (PtAu/HT) were analyzed by XRD and ICP-MS. The lattice parameter calculated from XRD diffraction pattern (111) of Pt_mAu_{100-m} exhibited linearly with their compositions only at low Pt concentrations (i.e., 20 and 40% Pt). However the result XRD pattern support that Pt/Au metal might be formed the partially alloy or partially phase segregation at high Pt concentrations, (i.e., 60, 80, and 90% Pt). The catalysis of Pt_mAu_{100-m}/HT (1 wt%) were evaluated for glycerol oxidation in base-free aqueous solution using molecular oxygen as an oxidant (glycerol/metal = 700 mol/mol). The moderate activities of Pt_mAu_{100-m}/HT also remain with increasing of Au amount to 20%, in addition, the increasing of Au amount more than this value resulting decreased of an activity. It is due to the Pt_mAu_{100-m}/HT particles size become larger with Au amount. However, the increasing of glycerol/metal ratio to 500 mol/mol was clearly seen that the presence of Au in Pt metal improved the selectivity toward glyceric acid, especially, $Pt_{80}Au_{20}$/HT catalyst. This phenomenon is explained by the synergistic effect of Pt/Au bimetallic resulting in the decreasing of the undesired product.

Introduction

Glycerol can be obtained from biomass as the by-product, the production of biodiesel by tranesterification of triglyceride causes the output of glycerol approximately 10 wt% of the total product. Therefore, it would be of significantly advantage if the glycerol could be transformed to value-added chemical. Glycerol is an intermediate in the synthesis of a large number of compounds used in industry, therefore, there are many catalytic processes which transformed glycerol to valuable products, e.g., hydrogenolysis, polymerization, esterification, dehydration, and oxidation.(1-3) Especially, all glycerol oxygenates derivatives are value-added products. However, the similarly reactive with molecular oxygen of both hydroxyl groups cause the poor selectivity. Therefore, supported metal such as Pt/AC, Pt/MWNTs, Au/C, Pd/C, Au/TiO_2, Au/Fe_2O_3, Rh/AC, Au/graphite, Pd/graphite, Au/TiO_2, Nb_2O_5, Au/$MgAl_2O_4$, and Pt/HT catalyst have been used for catalysis of the glycerol oxidation reaction toward desired product.(4-13) However, the drawback for utilizing of monometallic as a heterogeneous catalyst in the liquid phase with oxygen as an oxidant is the sensitivity in deactivation due to over oxidation and poisoning from by-product.(13)

Recently, the publications have been reported that the bimetallic structure is of interest in heterogeneous catalysis, since they have different activity and selectivity comparable to that of monometallic because the presence of foreign metal atoms at the surface change its electronic and structures.(14, 15) Several bimetallic catalyst have been exhibited more active than either of their individual metals. For instance, the bimetallic catalyst of Au and Pd exhibit very high alcohol oxidation rate and selectivity in both solvent-free and aqueous solution reaction.(16-17) In particularly Pt/Au bimetallic catalyst, it was reported that their bimetallic has proved to be beneficial, performing better than platinum alone in oxygen reduction at the fuel cell cathode and in selective oxidation of polyol.(18, 19)Hence, Pt/Au bimetallic could provide a synergistic catalytic effect that involves the suppression of adsorbed poisonous species and a change in electronic band structure for modifying the strength of the surface adsorption.(20)

In this work we focused on the development of a simple method for synthesis of Pt/Au bimetallic nanoparticles in aqueous solution by using polysaccharide as a reducing agent as well as protecting agent and further supported HT surface (PtAu/HT) by sol immobilization method. Hydrotalcite, $Mg_6Al_2(OH)_{16}CO_3 \cdot nH_2O$ (Mg-Al-CO_3), consists of a positively charged two dimensional Brucite-like layer, with anionic species in the interlayer to form neutral material. One of an interest in characteristic of HT is the surface basicity.(21) Furthermore, we evaluated the catalytic activity of an obtained PtAu/HT bimetallic heterogeneous catalysts by glycerol oxidation in base-free aqueous solution using molecular oxygen as an oxidant under atmospheric pressure. It is well-known that the oxidation reaction of primary alcohol requires the alkaline condition therefore the selection of hydrotalcite as a supporter has an advantage for carrying out the reaction under the absence of homogeneous base.

Experiment

Material: $H_2PtCl_6 \cdot 6H_2O$ (99.9%), $HAuCl_4 \cdot 4H_2O$ (99.9%) and polysaccharide were purchased from Wako Pure Chemicals. NaOH (90.0%) was obtained from Kanto Chemicals. Hydrotalcite (HT), Mg/Al = 5, was purchased from Tomita Pharmaceutical.

Preparation of platinum and gold monometallic supported on hydrotalcite catalyst (Pt/HT and Au/HT): Pt and Au nanoparticles were synthesized with chemical reduction method by using the polysaccharide simultaneously as protecting agent as well as reducing agent and were supported HT by sol immobilization method. The desired concentration (2.5 mM, 5 mL) of $H_2PtCl_6 \cdot 6H_2O$ and $HAuCl_4 \cdot 4H_2O$ were prepared in aqueous solution. $H_2PtCl_6 \cdot 6H_2O$ and $HAuCl_4 \cdot 4H_2O$ were pH adjusted with NaOH until the solution became neutral (pH 7), and then mixed with NaOH solution (0.05 M, 5 mL). The polysaccharide (4 wt%) were dissolved in hot water, then, that solution (5 mL) were mixed with NaOH solution (0.05 M, 5 mL). The above two solutions were heated at 100°C for 20 min. Then, their solution were mixed together and continuously stirred at 100°C for 10 min. Within 10 min after sol generation, HT (Mg/Al = 5, 0.49 g) was added into the solution under the continuously stirred and refluxed at 100°C for 1 h. The amount of HT required was calculated to have a metal loading of 1 wt%. After that, the solid solution were cooled down at room temperature, filtered and washed with water. The filtrated was dried overnight at 100°C.

Preparation of platinum and gold bimetallic supported on hydrotalcite catalyst (PtAu/HT): PtAu bimetallic nanoparticles were synthesized and supported HT with the same method of monometallic catalyst. The desired concentration (2.5 mM of total metal loading, 5 mL) of $H_2PtCl_6 \cdot 6H_2O$ and $HAuCl_4 \cdot 4H_2O$ were prepared in aqueous solution, the several metal contents in bimetallic nanoparticles were listed in **Table 1**. H_2PtCl_6 and $HAuCl_4$ were pH adjusted with NaOH until the solution became neutral (pH 7). The NaOH solution (0.05 M, 5 mL) was added into the mixed solutions of H_2PtCl_6 and $HAuCl_4$. In parallel, 5 mL of polysaccharide solution (4 g/100 mL H_2O) were also mixed with NaOH solution (0.05 M, 5 mL). The above two solutions were heated at 100°C for 20 min. Then, their solution were mixed together and continuously stirred at 100°C for 10 min. Within 10 min after sol generation, HT (Mg/Al = 5, 0.50 g) was added into the solution under the continuously stirred and refluxed at 100°C for 1 h. The amount of HT required was calculated to have a metal loading of 1 wt%. After that, the solid solution were cooled down at room temperature, filtered and washed with water. The filtrated was dried overnight at 100°C.

Catalyst characterization: Samples were characterized by X-ray diffraction (RINT-2000), and the metal contents in catalyst were analyzed by inductively coupled plasma atomic emission spectroscopy (ICPS-7000).

Catalytic testing: Catalytic test of Pt/HT and Pt_mAu_{100-m}/HT were evaluated for glycerol oxidation in base-free aqueous solution using molecular oxygen as an oxidant. The reaction of glycerol solution (0.5 mmol), 2 mL of water and 1 wt% of Pt/HT Pt_mAu_{100-m}/HT heterogeneous catalyst were carried out at 323 K for 5 h under oxygen flow (10 mL/min). All performed in a Schlenk tube attached with reflux condenser. After the reaction, the vessel was cooled down to room temperature and catalyst was separated by filtration. The filtrated were analyzed by using high performance liquid chromatography (HPLC) with a Aminex HPX-87H column from Bio-Rad Laboratories and refractive index (RI) detector. The analysis conditions were set as follows: the eluent was an aqueous solution of H_2SO_4 (10 mM), flow rate 0.5 mL/min and column temperature 323 K.

Results and Discussion

The results are divided into two sections, first section, we are discussing the morphology and compositions of Pt, Au mono- and bimetallic nanoparticles along with the support of their metal. Second section, we discuss the catalytic activity of supported-Pt mono- and bimetallic for glycerol oxidation.

1. Morphology and compositions of Pt, Au mono- and bimetallic nanoparticles

The bimetallic of Pt/Au was synthesized by green synthesis method using polysaccharide as a reducing agent as well as protecting agent. The various metals loading (Pt_mAu_{100-m}) were calculated in 1 wt% total metal loading with HT support. The calculations of metal loading for PtmAu$_{100-m}$/HT are shown in **Table 1.** along with the obtainable of their metal loading that were analyzed by ICP-MS.

Figure 1(A) shows the XRD pattern of HT supported-Pt_mAu_{100-m} nanoparticles. The diffraction patterns of Pt and Au metal could not be observed, since they overlapped with the HT peaks. Therefore, we are considering the diffraction patterns of Pt_mAu_{100-m} nanoparticles without supportive material that was shown in **Figure 1(B)**. There revealed that the diffraction pattern of Pt_mAu_{100-m} located between the patterns of Pt (2θ = 39.7) and Au (2θ = 38.3).(22) The peak position of the (111) reflection of Pt_mAu_{100-m} nanoparticles shifted to lower angle with increasing of Au amount, i.e., 2θ = 39.7, 39.6, 39.4, 39.1, 38.8, 38.5, and 38.3 for Pt_{100}, $Pt_{90}Au_{10}$, $Pt_{80}Au_{20}$, $Pt_{60}Au_{40}$, $Pt_{40}Au_{20}$, $Pt_{20}Au_{80}$, and Au_{100}, respectively. The broadening of (111) diffraction peak of

Pt_{100} becomes narrower and sharper with increasing of Au amount. It is well-known that the increasing in the particles size results in a narrowing of diffraction peak.

Table 1. Composition of Pt mono- and PtAu bimetallic nanoparticles.

Entry	Sample[(1)]	Metal loading		Nanoparticles[(2)]	
		Pt	Au (%)	Pt	Au (%)
1	Pt	100	-	90	-
2	$Pt_{90}Au_{10}$	90	10	84	10
3	$Pt_{80}Au_{20}$	80	20	72	20
4	$Pt_{60}Au_{40}$	60	40	57	34
5	$Pt_{40}Au_{60}$	40	60	41	47
6	$Pt_{20}Au_{80}$	20	80	24	65
7	Au	0	100	-	85

(1) Total amount of metal/HT = 1 wt%. (2) Analysis by ICP-MS.

Nevertheless, the lattice parameter calculated from XRD diffraction pattern of (111) reveals that the presence of Au increased the lattice distance of Pt. **Figure 1(C)** show the lattice parameter of Pt_mAu_{100-m} nanoparticles that exhibited linearly with their compositions at low Pt concentrations (i.e., 20 and 40% Pt). It indicated that the obtained Pt_mAu_{100-m} could be formed the completely bimetallic alloy only at low Pt concentrations. However, the result XRD pattern support that Pt/Au metal might be formed the partially alloy or partially phase segregation at high Pt concentrations, (i.e., 60, 80, and 90% Pt). Mott et al., proposed that bimetallic between Pt and Au could be formed not only completely alloy, but also partially alloy or partially phase segregation and completely phase segregation.(23)

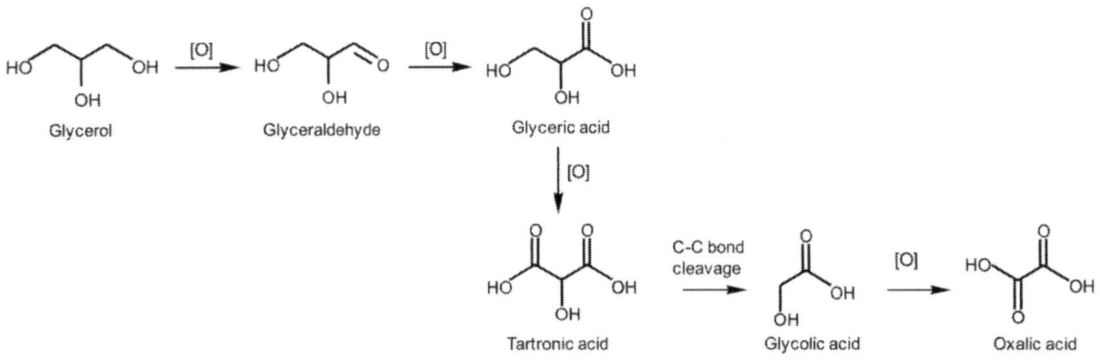

Scheme 1. A possible reaction pathway of glycerol oxidation catalyst by Pt_mAu_{100-m} heterogeneous catalyst.

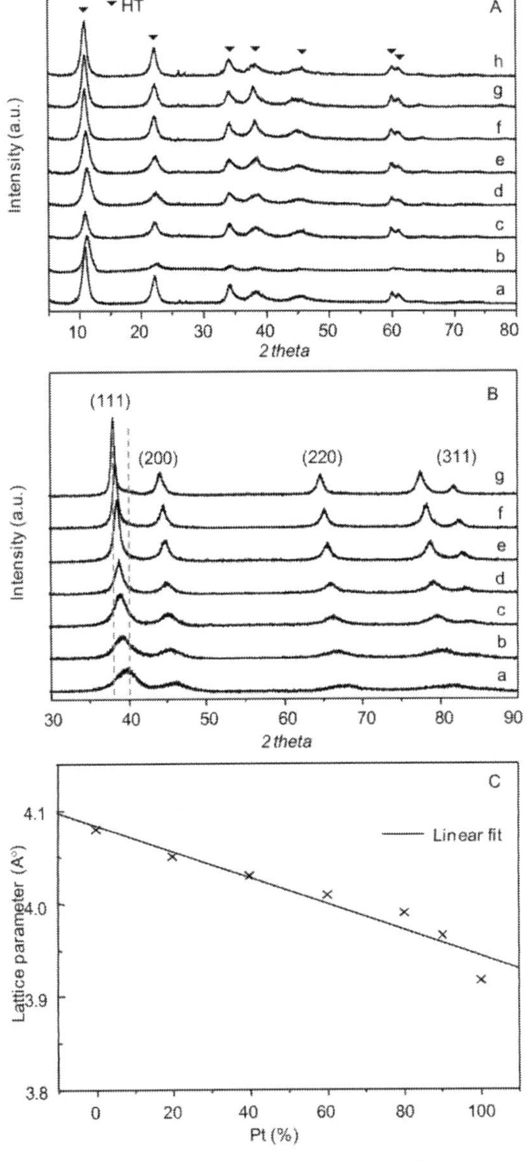

Figure 1. (A) XRD patterns of HT supported Pt_mAu_{100-m} nanoparticles: (a) Pt_{100}/HT, (b) $Pt_{90}Au_{10}$/HT, (c) $Pt_{80}Au_{20}$/HT, (d) $Pt_{60}Au_{40}$/HT, (e) $Pt_{40}Au_{60}$/HT, (f) $Pt_{20}Au_{80}$/HT, (g) Au_{100}/HT, and (h) HT (Mg/Al = 5). (B) XRD patterns of Pt_mAu_{100-m} nanoparticles (without support): (a) Pt_{100}, (b) $Pt_{90}Au_{10}$, (c) $Pt_{80}Au_{20}$, (d) $Pt_{60}Au_{40}$, (e) $Pt_{40}Au_{60}$, (f) $Pt_{20}Au_{80}$, and (g) Au_{100}. (C) Dependence of the lattice parameters for Pt_mAu_{100-m} nanoparticles on the relative composition of Pt% (cross mark on solid line).

2. Catalytic activity testing (glycerol oxidation)

The catalytic activity testing of Pt_mAu_{100-m} was carried out by glycerol oxidation in base-free aqueous solution under atmospheric oxygen pressure and the results were shown in **Table 2**. The possible reaction pathway of glycerol oxidation catalyst by Pt_mAu_{100-m}/HT catalyst was shown in **Scheme 1**. The reaction was accomplished with the oxidation at hydroxyl group (primary alcohol) of glycerol to glyceraldehyde further oxidized to glyceric acid (GA), tartronic acid (TA), glycolic acid (HA), and oxalic acid (OA). The catalytic activity of Pt_mAu_{100-m} for glycerol/metal = 700 mol/mol revealed that the glycerol conversion maintained with increasing of Au amount to 20% (entry 1, 3, and 5) and continuously decrease with increasing the amount of Au to 100% (entry 7, 8, 9, and 10). This results compatible with previously reported values that the catalytic activity showed a limitation when metal particle size increased.(24) However, the selectivity toward GA was boosted to 76 and 74% for 10 and 20% Au, respectively. In addition, it is obviously observed that not only the selectivity toward GA was increased, but also GA yields. The glycerol conversion did not occur for the reaction that was carried out without catalyst (entry 11). It indicated that the bimetallic of Pt and Au improved the selectivity toward GA, although, Au itself scarcely had selectivity toward GA. These improvements are clearly seen with increasing the amount of glycerol/metal = 500 mol/mol. In case of Pt/HT (entry 2), the glycerol conversion sharply increased to 80% along with the decreasing of selectivity toward GA to 60%. For $Pt_{90}Au_{10}$/HT(entry 4) and $Pt_{80}Au_{20}$/HT (entry 6), the increasing rate of glycerol conversion increasing smaller than that of Pt/HT. Especially $Pt_{80}Au_{20}$/HT catalyst, it was found that slightly increasing of the conversion (64 to 69%) was mostly transformed to GA (47 to 52% yield) while the selectivity toward GA was constant.

For further comparisons of the catalytic activity of Pt/HT and $Pt_{80}Au_{20}$/HT catalyst, the reaction time courses were performed by using glycerol/metal = 500 mol/mol, glycerol (0.5 mmol), H_2O (2 mL), under oxygen flow (10 mL/min), 323 K. The reaction time courses of glycerol oxidation catalyzed by Pt/HT and $Pt_{80}Au_{20}$/HT catalyst are shown in **Figure 2. and 3**. Their Figures presented of the two different important issues, the first one is the initial reaction rate and the second one is the occurrence of the undesired product. Firstly, we discussed the initial reaction rate, it was found that the reaction catalyzed by $Pt_{80}Au_{20}$/HT showed the higher initial reaction rate than that was catalyzed by Pt/HT.

Table 2. Results of glycerol oxidation catalyst by Pt/HT and PtAu/HT in base free aqueous solution using molecular oxygen as an oxidant.[(1)]

Entry	Catalysts	Conversion (%)	Selectivity (%)				GA Yield (%)
			GA	TA	HA	OA	
1	Pt/HT	63	68	10	16	6	43
2	Pt/HT[(2)]	80	60	14	17	9	48
3	$Pt_{90}Au_{10}$/HT	61	76	7	12	5	47
4	$Pt_{90}Au_{10}$/HT[(2)]	71	70	12	13	6	49
5	$Pt_{80}Au_{20}$/HT	64	74	8	14	4	47
6	$Pt_{80}Au_{20}$/HT[(2)]	69	75	10	8	7	52
7	$Pt_{60}Au_{40}$/HT	39	86	3	10	1	33
8	$Pt_{40}Au_{60}$/HT	12	87	-	11	2	11
9	$Pt_{20}Au_{80}$/HT	4	67	-	28	6	2
10	Au/HT	2	39	-	46	15	1
11	Blank	1	30	-	9	3	-

(1) Reaction conditions: glycerol (0.05 mmol), H_2O (2 mL), glycerol/metal = 700 (mmol/mmol), total amount of metal/HT = 1 wt%, Mg/Al = 5, under O_2 flow (10 mL/min), 323 K, 7 hr. (2) Glycerol/metal = 500 (mmol/mmol).

This phenomenon can explain in term of the electronic interaction between Pt and Au. Generally, the feature of the accepted mechanism for metal catalyzed dehydrogenation process in aqueous solution involves hydride subtraction and subsequent oxidation by oxygen.(18) Firstly, the hydride subtraction at alcohol position has occurred further toward alkoxide. After that, there are formed bonding with metal site toward to glyceraldehyde and undergo oxidized to GA. This step, the electronic interaction between Pt and Au atoms located in close proximity to each other cause a partially of electron transfer from Au to Pt, with the cause of, increasing of the back-donation of electrons from the filled *d* orbitals of Pt to the $-CO^{-}$.(25) Consequently, the desorption of alkoxide toward glyceraldehyde is faster, therefore, the initial reaction of glycerol oxidation catalyzed by PtAu/HT is faster than Pt/HT. Secondly, we discuss the synergistic effect of Pt and Au bimetallic to the lower production of undesired products (i.e., TA, HA, and OA) The possible effect of the lower production of undesired products revealed the changing of *d* spacing in crystal structure of metal particles. It has been reported for the Au-Pd system, Au acted as promoter to isolate Pd monomer sites thus also limiting oxygen coverage, in other words, Au improved the geometric of Pt surface. This means that the presented of Au could enhance the resistance to oxygen poisoning. Gold also reported to reduce Pt self poisoning.(18,26) As a result, we can say that the bimetallic nanoparticles in the presence of Au in Pt can improve the activity of PtAu/HT catalyst for glycerol oxidation by the synergistic effect electronic and geometric.

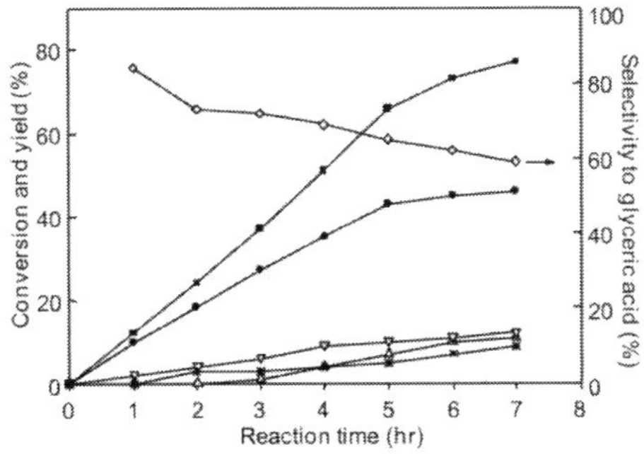

Figure 2. Reaction time profile of glycerol oxidation catalyst by Pt/HT in water used O_2 as an oxidant. Reaction condition: glycerol (0.5 mmol), H_2O (1.5 mL), 1wt% Pt/HT (Mg/Al = 5), glycerol/metal = 500 (mol/mol), under O_2 flow (10 mL/min), 323 K. Glycerol conversion (■ : close square), glyceric acid yield (● : close circle), tartronic acid yield (△ : open up triangle), glycolic (▽ : open down triangle), oxalic acid yield (* : star), and glyceric acid selectivity (◇ : diamond).

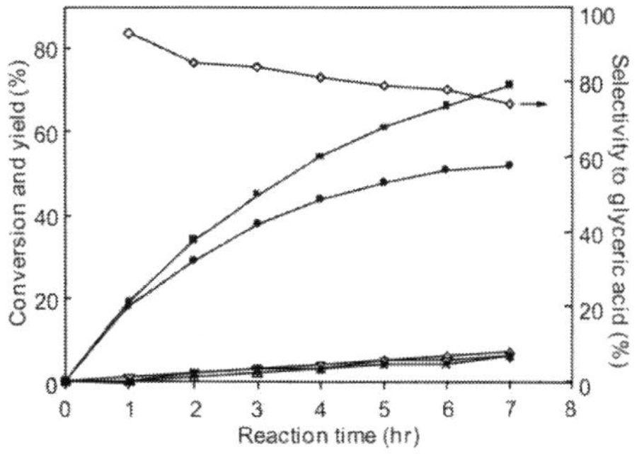

Figure 3. Reaction time profile of glycerol oxidation catalyst by PtAu/HT in water used O_2 as an oxidant. Reaction condition: glycerol (0.5 mmol), H_2O (1.5 mL), 1wt% PtAu/HT (Mg/Al = 5), glycerol/metal = 500 (mol/mol), under O_2 flow (10 mL/min), 323 K. Glycerol conversion (■ : close square), glyceric acid yield (● : close circle), tartronic acid yield (△ : open up triangle), glycolic (▽ : open down triangle), oxalic acid yield (* : star), and glyceric acid selectivity (◇ : diamond).

Conclusion

Green synthesis of HT supported-Pt/Au bimetallic nanoparticles was successfully synthesized by reduction of Pt and salt utilizing polysaccharide as a reducing agent as well as protecting agent. The analysis of metal morphology reveals that the Pt and Au formed a partially alloy onto the HT surface. The activity of Pt_mAu_{100-m}/HT as a heterogeneous catalyst for glycerol oxidation in base-free aqueous solution with atmospheric pressure O_2 flow as an oxidant showed that the glycerol conversion decreased with increasing of Au amount, since, the increasing of particles. However, the $Pt_{80}Au_{20}$/HT exhibited the best catalytic activity (considered both of conversion and selectivity) under these conditions. Even though, the presentation of Au in Pt is not formed the completely alloy, but the experimental result revealed that the PtAu/HT showed the synergistic effect to the improvement of selectivity toward GA.

References

1. C.-H. (Clayton) Zhou, J. N. Beltramini, Y.-X. Fan, G. Q. (Max) Lu (2008), *Chem. Soc. Rev.*, 37, pp. 527-549.
2. M. Pagliaro, R. Ciriminna, H. Kimura, M. Rossi, C. D. Pina (2007), *Angew. Chem. Int. Ed.*, 46, pp. 4434-4440.
3. A. Corma, S. Iborra, A. Velty (2007), *Chem. Rev.*, 107, pp. 2411-2502.
4. N. Dimitratos, J. A. Lopez-Sanchez, D. Lennon, F. Porta, L. Prati, A. Villa, *Catal. Lett.* 2006, 108, pp. 147-153.
5. J. Gao, D. Liang, P. Chen, Z. Hou, X. Zheng (2009), *Catal Lett.*, 130, pp. 185-191.
6. N. Dimitratos, A. Villa, C. L. Bianchi, L. Prati, M. Makkee (2006), *Appl. Cata. A.*, 311, pp.185-192
7. E. Taarning, A. T. Madsen, J. M. Marchetti, K. Egeblad, C. H. Christensen (2008), *Green chem.*, 10, pp. 408-414.
8. E. G. Rodrigues, S. A. C. Carabineiro, X. Chen, J. J. Delgado, J. L. Figueiredo, M. F. R. Pereira, J. J. M. Orfa ̃o (2011), *Catal Lett.*, 141, pp. 420–431.
9. N. Dimitratos, F. Porta, L. Prati (2005), *Appl. Cata. A.*, 291, pp. 210–214.

10. K. Musialskaa, E. Finocchiob, I. Sobczaka, G. Buscab, R. Wojcieszakc, E. Gaigneauxc, M. Zioleka (2010), *Appl. Catal., A.*, 384, pp. 70–77.
11. A. Villa, A. Gaiassi, I. Rossetti, C. L. Bianchi, K. v. Benthem, G.l M. Veith, L. Prati (2010), *J. Catal.* , 275, pp. 108–116.
12. A. Tsuji, K.T. V. Rao, S. Nishimura, A. Takagaki, K. Ebitani (2011), *ChemSusChem.*
13. N. Dimitratos, A. Villa, D.Wang, F. Porta, D. Su, L. Prati (2006), *J. Catal.*, 244, pp. 113–121.
14. Y. Luo, H. O. Seo, K.-D. Kim, M. J. Kim, W. S. Tai, M. Burkhart, Y. D. Kim (2010), *Catal Lett.*, 134, pp. 45–50.
15. K. Tekaia-Elhsissen, F. Bonet, P. -Y. Silvert, R. Herrera-Urbina (1999), *J. Alloys Compd.*, 292, pp. 96-99.
16. N. Dimitratos, J. A. Lopez-Sanchez, J. M. Anthonykutty, G. Brett, A. F. Carley, R. C. Tiruvalam, A. A. Herzing, C. J. Kiely, D. W. Knight, G. J. Hutchings (2009), *Phy. Chem. Chem. Phys.*, 11, pp. 4952-4961.
17. W. C. Ketchie, M. Murayama, R J. Davis (2007), *J. Catal.*, 250, pp. 264-273.
18. L. Prati, , A. Villa, C. Campione, P. Spontoni (2007), *Top. Catal.*, 44, pp. 319-324.
19. A. N. Grace, K. Pandian (2006), *Electrochem. Commun.*, 8, pp. 1340–134.
20. D. Mott, J. Luo, P. N. Njoki, Y. Lin, L. Wang, C.-J. Zhong (2007), *Catal. Today*, 122, pp. 378–385.
21. K. Kaneda, K. Ebitani, T. Mizugaki, K. Mori (2006), Bull. *Chem. Soc. Jpn.*, 79, pp. 981-1016.
22. M.-L. Wu, D.-H. Chen, T.-C. Huang (2001), *Chem. Mater.*, 13, pp. 599-606.
23. B. N.Wanjala, J. Luo, B. Fang, D. Mott, C.-J. Zhong (2011), *J. Mater. Chem.*
24. M.-L. Wu, D.-H. Chen, T.-C. Huang (2001), *Chem. Mater.*, 13, pp. 599-606.
25. L. B. Ortiz-Soto, O. S. Alexeev, M. D. Amiridis (2006), *Langmuir*, 22, pp. 3112–3117.
26. M. Chen, D. Kumar, C.-W. Yi, D. W. Goodman (2005), *Science*, 310, pp. 291-293.

(54a) Incorporation of AquaporinZ Via Binary-Lipid Langmuir Monolayers

Guo Fei Sun[1], Hu Zhou[2], Yi Li[1], Kandiah Jeyaseelan[3], Arunmozhiarasi Armugam[3] and Tai-Shung Chung[1], (1)Chemical and Biomolecular Engineering, National University of Singapore, Singapore, Singapore, (2)Biological Sciences, National University of Singapore, Singapore, Singapore, (3) Biochemistry, National University of Singapore, Singapore, Singapore

In this work, a new approach of incorporating transmembrane protein AquaporinZ (AqpZ) into the lipid bilayer has been developed with the aid of Langmuir-Blodgett technique. Aquaporin is a big family of transmembrane proteins that selectively conduct water or small solutes in or out of the biological membrane driven by osmotic pressures. The hydrophobic property of AqpZ makes the incorporation of membrane proteins with the LB method a challenging and intriguing work. The incorporation in this study is achieved by combining a pure binary-lipid monolayer with an AqpZ-associated binary-lipid monolayer and subsequent refolding of AqpZ in the bilayer. The binary-lipid monolayer is composed of (1) gel-phase lipids that prevent detergent dissolution and (2) nickel-chelating lipids that attach the AqpZ from the subphase. To gain better film integrity and enhance protein adsorption on the monolayer, detergent in the subphase is removed by BioBeads. It is shown that the removal rate can be determined by BioBeads amount and subphase circulation. Furthermore, the bilayer formed with reconstituted AqpZ is imaged by AFM and the incorporation mechanism is explained in a three-step process. The new approach of AqpZ reconstitution revealed by this work could potentially be applied in biomimetic membranes for water purification and biosensor applications.

Acknowledgement This work was financially supported by the National Research Foundation of Singapore, Environment and Water Industry Programme Office (EWI) project 0804-IRIS-01 (NUS grant number: R-279-000-293-272).

See more of this Session: Research Frontier of Water Sustainability
See more of this Group/Topical: International Congress on Energy 2011

(54b) Planar Biomimetic Aquaporin-Incorporated Triblock Copolymer Membranes On Porous Solid Supports

Phuoc H.H. Duong[1], Tai-Shung Chung[1], Kandiah Jeyaseelan[2], Arunmozhiarasi Armugam[2] and Minghui Hong[3], (1)Chemical & Biomolecular Engineering, National University of Singapore, Singapore, Singapore, (2)Department of Biochemistry, National University of Singapore, Singapore, Singapore, (3)Data Storage Institute, Singapore, Singapore

The aim of the present study was to demonstrate the preparation and characterization of planar biomimetic bacterial water-channel protein Aquaporin Z (AQPZ)-incorporated polymer membranes on porous solid substrates by vesicle spreading for water treatment. AQPZ -incorporated polymer vesicles were prepared from disulfide-functionalized poly(2-methyloxazoline)-block-poly(dimethylsiloxane)-block-poly(2-methyloxazoline) (disulfide-functionalized $PDMS_{20}$-$PDMS_{75}$-$PMOXA_{20}$) tri-block copolymer which was previously proven to be able to maintain protein activity after the incorporation as occurred in lipid vesicles. AQPZ -incorporated vesicle permeability was evaluated using stopped-flow spectroscopy. The results demonstrate that AQPZ could maintain its performance after inserting into polymer vesicles. The vesicles were designed with disulfide functional end groups to enhance vesicle spreading on gold-coated solid substrates by covalent interactions between polymer membranes and substrates. The morphology of polymer membranes on difference types of porous solid substrates were elucidated by using atomic force microscopy (AFM) and field-emission scanning electron microscope (FESEM). For the first time, AQPZ -incorporated polymer vesicles were demonstrated to be able to form planar AQPZ -incorporated polymer membranes on the porous substrates.

See more of this Session: Research Frontier of Water Sustainability
See more of this Group/Topical: International Congress on Energy 2011

(54c) Ultrafast Water Permeation of Biomimetic Membranes Embedded with Aquaporin for Water Reuse and Desalination

Honglei Wang[1], Tai-Shung Chung[1], Yen Wah Tong[1], Zaichun Chen[2], Minghui Hong[2], Kandiah Jeyaseelan[3] and Arunmozhiarasi Armugam[3], (1)Department of Chemical and Biomolecular Engineering, National University of Singapore, Singapore, Singapore, (2)Department of Electrical and Computer Engineering, National University of Singapore, Singapore, Singapore, (3)Department of Biochemistry, National University of Singapore, Singapore, Singapore

Biomimetic membranes embedded with Aquaporin have received worldwide attention for water reuse and desalination. Compared to commercially available polymeric membranes, biological membranes exhibit far superior water transport and regulation ability via transmembrane water channel proteins - aquaporin (Aqp). The water permeability of a single tetramer ranges from 6 to 24 $\times 10^{-14} cm^3/s$ depending on the source of the Aqp. To our best knowledge, for the first time, a proof-of-concept will be demonstrated in this presentation. The Aqp Z molecules are inserted into amphiphilic polymers in a functional folding and then coated on a porous substrate. The selective layer exhibits the world's highest water permeability with a reasonably low salt reverse flux. Both water and seawater are used as the feed in forward osmosis (FO) processes.

Acknowledgement
This work was financially supported by the Environment and Water Industry Programme Office (EWI) (NUS grant number: R-279-000-293-272) under the Singapore National Research Foundation (NRF).

See more of this Session: Research Frontier of Water Sustainability
See more of this Group/Topical: International Congress on Energy 2011

Forward Osmosis: Potential use in Desalination and Water Treatment

Dr. Ali Altaee, CEC's, Abu Dhabi University, UAE

Abstract:

There has been a recurring interest in using Forward Osmosis (FO) process in water treatment and desalination. The recent advance in membrane technology has encouraged further investment in the process technology. Despite the promising results from pilot and bench scale experiments the technology is still not commercialized yet. Obviously, this is due to the complicated nature of the process which usually involves multiple stages of treatment in addition to the FO membrane process. Unfortunately, most of the recent studies were focused on studying the FO process alone and didn't provide enough data about the actual cost of the process as whole which includes the osmotic agent regeneration stage/s. This issue resulted in some uncertainties about the total cost of the water treatment by the process. Furthermore, more data are required to evaluate the impact of the osmotic agent losses on the overall cost and efficiency. More data are required to evaluate the effect of salt diffusion from seawater to the draw solution on the product water quality. This is particularly important due to the salt accumulation in the system when the draw solution is recycled and reused. In case if the draw solution is regenerated by membrane treatment, a suitable membrane should be selected to ensure an optimal salt rejection. For power generation by Pressure Retarded Osmosis (PRO) process, there was an evident progress. However, the process is site specific; i.e. it is dependent of the availability of the draw and donor solution. This suggested that the process is applicable to certain areas but can't be generalized. A quick review on the FO processes applications in water treatment and power generation is covered here to assess its potential use based on the recent development that have been made so far.

Introduction:

Thermal and membrane technologies are the forefront processes for seawater desalination. Multi Stage Flashing (MSF) and Multi Effect Distillation (MED) are the leading processes in thermal technologies while Reverse Osmosis (RO) dominates the membrane desalination technologies [1-3]. The thermal processes are widely used in the Gulf region of the Middle East due the difficult nature of gulf water such as high salinity and concentrations of impurities. This trend has been changed in the last decade in favor of the RO process due to the high membrane performance and low power consumption. Worldwide, RO process is, almost, the only technology for seawater desalination [3, 4]. Mainly, this is because of its reliability and lower power consumption compared to the thermal processes [4, 5]. The cost of water desalination by the RO process is less than USD 1 $\$/m^3$ [6, 7]. Unfortunately, the RO cost for desalination is still relatively expensive and probably unaffordable to many countries. This encouraged the scientists and researchers to investigating cheaper processes for seawater desalination such as FO [6-8]. The attractiveness of the FO process consists in its low power consumption, and potential high recovery rate [6, 8, 9]. Freshwater transports across a semi permeable membrane from the low to the high concentration solution due to the osmotic pressure gradient. Different chemicals have been proposed as a potential osmotic agent such as table salt, magnesium sulfate, glucose, ammonia carbon dioxide, and magnesium chloride [9, 10]. In practice, the osmotic agent is recycled and reused to reduce the cost of desalination. This can be achieved by thermal processes or membrane filtration.

In the FO desalination process, there is a number physical and chemical processes which affect the efficiency and cost of the process such as salt diffusion from the seawater to the draw solution, osmotic agent diffusion across the membrane, and concentration polarization. In general these phenomena reduce the process efficiency and the cost of desalination [11]. Yet, there isn't enough information about the overall effect of salt diffusion in both directions on the product water quality and cost. Although osmotic agent can be regenerated by either thermal or membrane processes, neither of these processes are capable of completely recycling the osmotic agent. Of course, this has impact on the product water quality which depends on the process used for regeneration such as thermal or membrane processes. In case of the membrane processes, the regeneration process is also affected by the types of the osmotic agent and membrane used such as RO, Nanofiltration (NF), and membrane distillation (MD).

Another attractive and promising application of FO process is in power generation [9, 10, 12]; also know as PRO. The process was suggested by Loeb [13] and there is a wealth of literature describing its potential and application in power generation. The mechanism of operation is similar to that explained in FO water desalination but the product water leaving the FO membrane, preferably pressurized, is directed to a turbine to convert hydraulic energy to electrical power. There are a number of options in which the process can be designed but they are using the same concept. The breakthrough in PRO was made by StatKraft which built the first pilot plant for power generation in Norway [14]. The pilot plant used freshwater as donor solution and seawater as a draw solution. The optimal operating condition for PRO can be achieved when fresh water is used as a donor solution to avoid the internal concentrative concentration polarization. StatKraft plant for power generation by PRO is the largest in the world to demonstrate the feasible scalability of PRO from bench to pilot scale. To date, there is no data about the economical feasibility of the PRO in power generation. It should be stressed here that the type and concentration of the osmotic agent could have significant impact on the FO process as will be discussed later in this paper.

FO for Seawater Desalination:

Seawater desalination is an essential process for fresh water supply. The cost of desalination is still high and need to be reduced. FO, therefore, has received a lot of attention in the past decade of being a competitive process for seawater desalination. There are a number of technical and operating factors which affect the competency of the process. Technical factors such as internal and external concentration polarizations were found to have detrimental impact on the membrane flux and the recovery rate. The phenomenon of concentration polarization is associated with the membrane characteristics and it is mainly due to the relative salt concentration at the membrane surface [9, 11]. Depending on the flow mode of the draw and feed solutions, concentration polarization is classified into internal and external effects [11]. The latter effects, in turn, are classified into concentrative and dilutive concentration polarizations. Concentrative concentration polarization usually refers to the change in the concentration of feed solution and it is due to the increase in the concentration of feed solution leading to a decrease in the osmotic pressure gradients across the membrane [11, 15]. As a result, the fresh water flow across the membrane is reduced. On the other hand, the dilutive concentration polarization is associated with the draw solution which results in a reduction in the concentration of draw solution. In response to this effect the osmotic pressure gradient is decreased as well as the water flux. Both dilutive and concentrative concentration polarizations are divided into internal and external effects. The former effect occurs inside the lumen while the latter on the shell side or at the membrane selective layer side [11, 15]. The recent advance in the membrane manufacturing technology has dealt with the internal concentration polarization problem. This was achieved by reducing the thickness of the membrane support layer [16]. However, the effect of external concentration polarization is an inherent nature of the FO membrane process and can't be avoided.

The effect of internal concentration polarization was demonstrated to be more serious than the external concentration polarization (figure 1) [7]. This is especially important when the FO membrane is operated in the Reverse Osmosis mode or the draw solution in the lumen side of the membrane. Pilot and bench scale experiments showed a steep drop in the membrane flux under the effect of internal concentration polarization. Although the new FO membranes have successfully reduced the effect of internal dilutive and concentrative concentration polarization, it can't be completely eliminated. Therefore, the efficiency of FO process will still be affected by the internal concentration polarization phenomenon. The new experimental work showed that FO process is operated better when the draw solution is in the shell side of the membrane while the feed solution/donor solution is in the lumen side [12, 15]. Accordingly, FO process operates better in inside to outside flow mode or when the draw solution is in the shell side and the feed solution in the lumen side. But, it should be noted here this conclusion holds true when the FO process is applied for seawater desalination in the conventional way. In some cases, as will be discussed later, it is preferred to have the draw solution in the lumen not to improve the membrane flux but to reduce the membrane fouling propensity [15]. This is especially true if the donor solution has a high concentration of fouling agents such as wastewater effluent.

It is highly desirable to have a high recovery rate in the desalination process. Currently, the recovery rate in seawater RO desalination is less than 50% and in the thermal desalination processes (MSF and MED) is around 30% [4, 8, 17, 18]. The main mechanism by which fresh water transports across the membrane in the FO process is the osmotic pressure gradient. For a given feed water/seawater TDS, the recovery rate can be increased by increasing the concentration of draw solution. However, experimental work demonstrated that the relationship between the concentration of draw solution and the recovery rate is not linear (figure 2). Water flow across the membrane dilutes the draw solution at the membrane surface. This micro layer adjacent to the membrane surface will have a concentration lower than the bulk solution and hence reducing the driving force of water transport from the feed solution to the draw solution. As shown in figure 2, the increase in the osmotic pressure gradient resulted in an increase in the recovery rate. The relationship was linear at low osmotic pressure gradient but this trend changed at high osmotic pressure gradients. Primarily, this is due to the concentration polarization effect at the membrane-solution interface. It should be noted here that the energy requirement for fresh water extraction will be increased with increasing the concentration of draw solution. This is because of the increase in the concentration of diluted draw solution has to be regenerated (figure 2). Accordingly, a subsequent increase in the concentration of draw solution wouldn't necessarily lead to the desirable improvement in the performance of FO process.

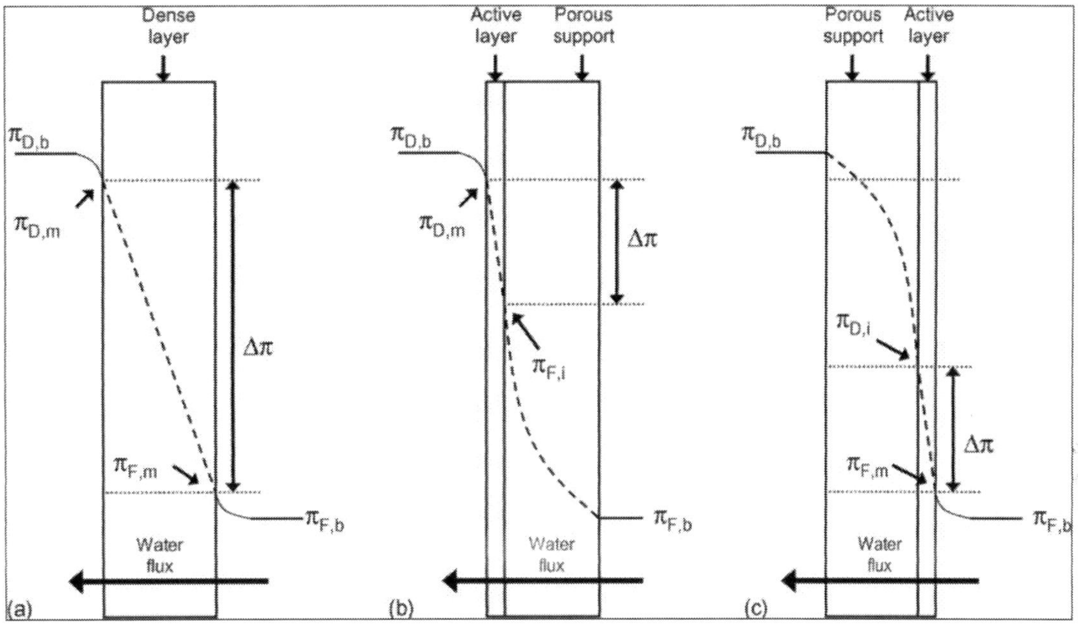

Figure 1: Concentration polarization in FO membrane (a) A symmetric membrane (b) An asymmetric membrane draw solution facing the active layer (c) An asymmetric membrane feed solution facing the active layer (Jeffery et. al., 2006)

The regeneration process of draw solution is the most expensive stage in FO desalination process. In contrast to the FO process which is driven by the natural osmosis phenomenon, the regeneration process is relatively more power intensive. Typically, the osmotic pressure of the diluted draw solution will be equal or higher than the osmotic pressure of feed solution. Both membrane and thermal processes were proposed for freshwater extraction and draw solution regeneration [9, 10]. NF, RO, and MD membranes were proposed for the draw solution regeneration and fresh water extraction [9, 12]. Each membrane has its own advantages and disadvantages which should be considered upon choosing the suitable membrane for FO process. In addition to the membrane processes, thermal processes such as MSF and MED were suggested for draw solution regeneration [20].

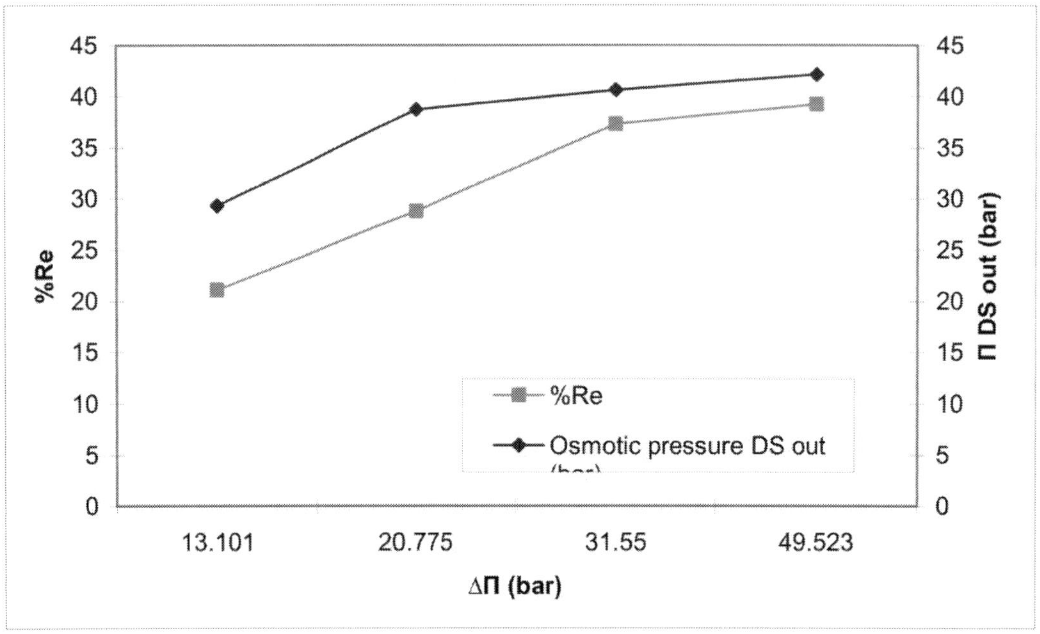

Figure 2: Effect of osmotic pressure on recovery rate and the concentration of diluted draw solution

Membrane Processes for Draw Solution Regeneration:

As mentioned previously, membrane processes such as NF, RO and MD were suggested for the draw solution regeneration and concentration. The application of membrane process is affected by:

1. Type of the osmotic agent used in the draw solution
2. The concentration of draw solution
3. Type of the feed water
4. Type of the membrane

For example, the permeability of NF membranes is higher than RO membranes but they exhibit a lower rejection rate compared to the RO membranes. NF membranes, therefore, are more suitable in the regeneration of low concentration than high concentration draw solution. This is because the maximum feed pressure recommended for the NF membranes shouldn't exceed 40 bar. Practically, this pressure is out of the range required in treating a diluted draw solution for seawater desalination; minimum osmotic pressure 27 bar. Otherwise the recovery rate of NF membrane, if used in seawater desalination, will not be economical. Therefore, NF membranes are more suitable in the regeneration of a draw solution having a lower osmotic pressure than seawater such as brackish water. For simplicity, it is assumed here that the osmotic pressure of the brackish water to be treated by FO process is 1.6 bar (feed TDS 2000 ppm NaCl) and the recovery rate is

75%. In ideal FO process the osmotic pressure of the diluted draw solution will be at least equal to that of the feed solution concentrate; i.e. 6.5 bar. ROSA software by Filmtec will be used in the simulation of the membrane performance in the regeneration process. Two different draw solutions were tested here (NaCl and $MgSO_4$) to investigate the effect of draw solution on the membrane performance and product water quality. The cost of the regeneration process and product water quality is also affected by the type of NF membrane in the regeneration process. For comparison purpose, NF and BWRO membranes were applied for the draw solution regeneration and concentration. The simulation results showed that the permeate TDS was lower when BWRO membrane was used for draw solution regeneration (figure 3). The permeate TDS was also affected by the type of draw solution (figure 3). For a given BWRO membrane, a lower permeate concentration was achieved when NaCl was used as a draw solution. This is because of the higher osmotic pressure generated from the same concentration of NaCl compared to MgSO4; hence a lower concentration of NaCl is used for the desalination of brackish water. It is well understood that the permeate concentration increases with the recovery rate [8, 19]. The required concentration of NaCl and $MgSO_4$ in the draw solution to generate an osmotic pressure equivalent to 6.5 is 8250 ppm and 29972 ppm respectively. As such, NaCl is likely to be more efficient osmotic agent than $MgSO_4$ if BWRO membrane is used for the regeneration. In case of NF membranes, it is preferable to use MgSO4 as an osmotic agent due to the high rejection rate compared to NaCl (figure 3). It should be noted that the higher rejection rate of BWRO membrane comes at the cost of higher energy consumption (figure 4). As expected, the specific energy consumption was higher in case of BWRO membrane especially when $MgSO_4$ was used instead of NaCl as a draw solution. This is because:

1. $MgSO_4$ has more affinity to the membrane surface because of the higher ionic charge compared to NaCl
2. The higher concentration of $MgSO_4$ required for the generation of osmotic pressure equivalent to that from NaCl Salt. Higher concentration causes higher concentration polarization.
3. Higher membrane rejection to $MgSO_4$

$$C_w = C_b * CP \quad [1]$$

Where C_w is the concentration at the membrane wall, C_b is the bulk concentration and CP is the concentration polarization factor.

In contrast, the specific energy consumption decreased when NF was used instead of BWRO membrane regardless the osmotic agent. The specific energy consumption increased slightly with the recovery rate increase which is typically observed in membrane filtration processes.

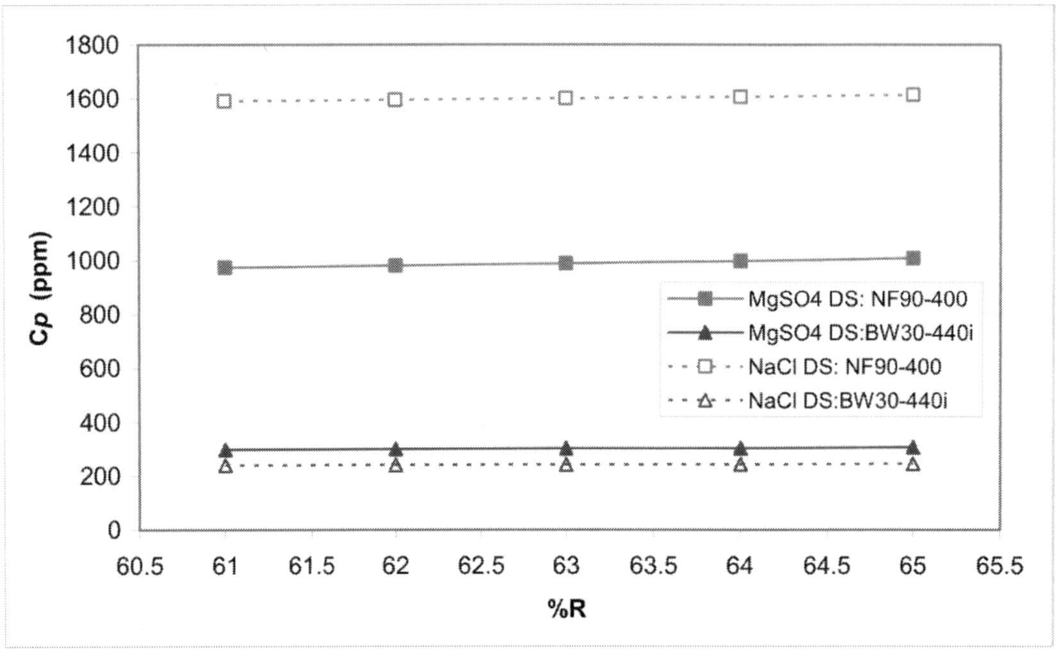

Figure 3: Effect of recovery rate on the permeate concentration

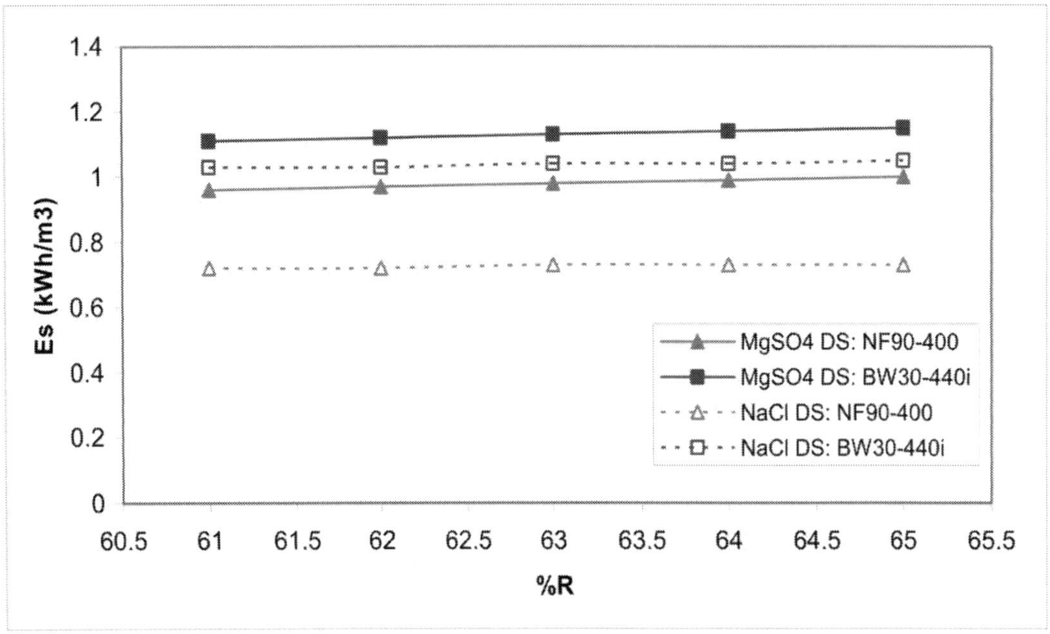

Figure 4: Power consumption as a function of recovery rate

BWRO membranes have a tighter structure than NF membrane which is why it required a higher feed pressure for draw solution treatment (figure 5). In fact the feed pressure was higher in BWRO than NF membranes for both osmotic agents $MgSO_4$ and NaCl. In case of NF membrane the power consumption decreased when NaCl was used as a draw solution because of the lower NF rejection rate to monovalent ions. As such NaCl is not recommended as an osmotic agent if NF membranes are used for the regeneration of draw solution. When BWRO membrane was used for draw solution regeneration, the feed pressure required for MgSO4 filtration was higher than that for NaCl (figure 5). The reason for that was due to the high rejection rate of $MgSO_4$ by BWRO membrane and hence the concentrate pressure was higher than that for NaCl (figure 6). Indeed, the concentrate osmotic pressure increased with increasing the recovery rate in both draw solutions; i.e. NaCl and $MgSO_4$. Therefore, it is preferable to use NaCl or any monovalent ions of high osmotic pressure in FO process if tight membranes such as BWRO/RO are used in the regeneration process.

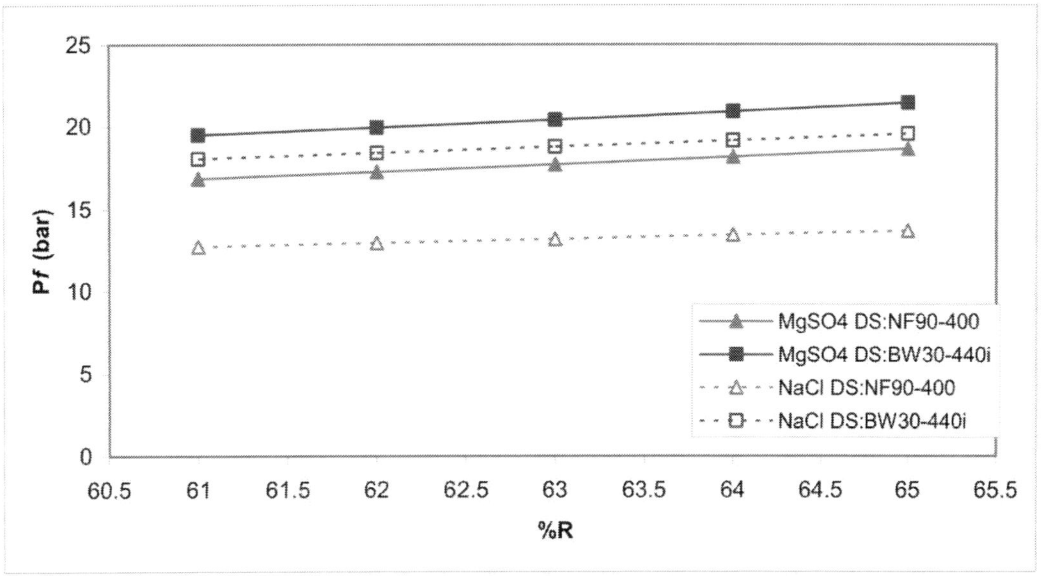

Figure 5: Effect of the recovery rate on the feed pressure for NF and RO membranes

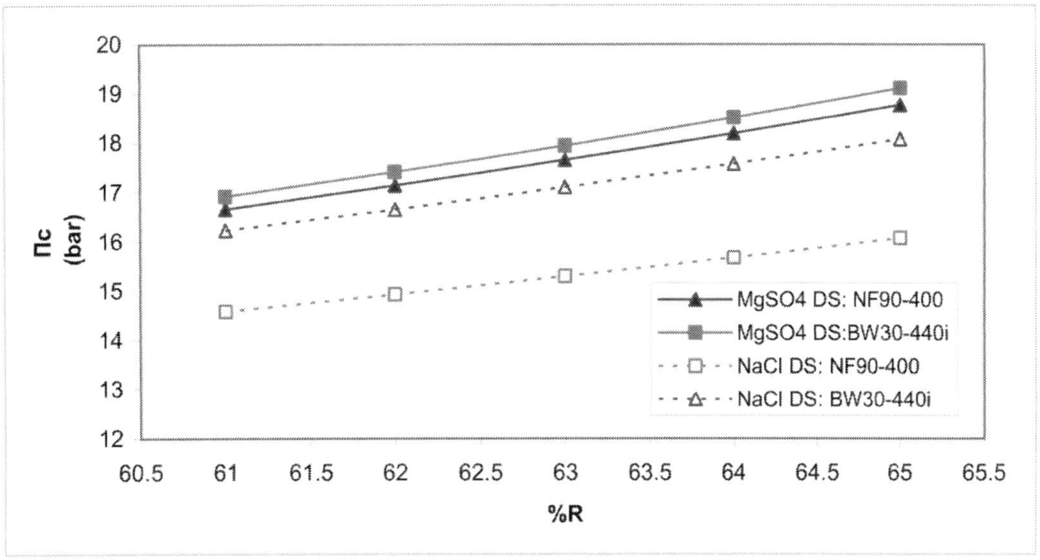

Figure 6: Concentrate osmotic pressure at different recovery rates

For seawater desalination, RO membranes should be used in conjunction with FO membranes. High pressure seawater RO membranes are preferably applied for water extraction and draw solution concentration. Once again ROSA was used in this study to simulate the RO membrane performance for draw solution concentration and regeneration. Three different types of osmotic agents were investigated; $MgSO_4$, $MgCl_2$ and NaCl and a number of recovery rates ranged between 42 and 54 at 3% intervals were examined. The results showed that the feed pressure requirement for membrane filtration was the highest incase of $MgSO_4$ followed by $MgCl_2$ and NaCl respectively (figure 7). Though, the difference between $MgCl_2$ and NaCl was insignificant. The reason why feed pressure was the highest in case of $MgSO_4$ was due to the high rejection and concentration of chemicals used to generate a desirable osmotic pressure for FO seawater desalination as mentioned above. Figure 8 shows that the concentrate osmotic pressure was the highest in case of $MgSO_4$ which was caused probably by the high membrane rejection to divalent ions and highest concentration polarization. This substantiated the fact why $MgSO_4$ draw solution required a higher feed pressure than $MgCl_2$ and NaCl. As a result of high feed pressure, the specific power consumption was the highest in case of the MgSO4 (figure 9). Furthermore, the diffusion of SO_4 ions is lower than Cl which may aggravate the intensity of concentration polarization of $MgSO_4$ draw solution [29, 30]. For MgCl2 and NaCl the difference in power consumption was insignificant.

It is evident from figure 9 that the specific power consumption was affected by the recovery rate which depends on the type of the osmotic agent in use. In case of MgSO4 draw solution, the simulation results show the optimal specific power consumption could be achieved at around 42% recovery rate. Then the power consumption increased with the recovery rates increase above 42% (figure 9). Therefore, the optimal recovery rate for $MgSO_4$ in the regeneration stage should be around 42%. When MgSO4 was replaced by $MgCl_2$, the optimal power consumption was achieved at a recovery rate about 48% (figure 9). Accordingly, a higher recovery rate can be achieved if $MgCl_2$ was used instead of $MgSO_4$ as a draw solution. Similarly, when NaCl was used as a draw solution the optimal power consumption was achieved at a recovery rate around 51% (figure 9). Based on these results, a higher recovery rate can be achieved when NaCl was used in the draw solution. This is probably one of the advantages of using low molecular weight osmotic agent in the draw solution.

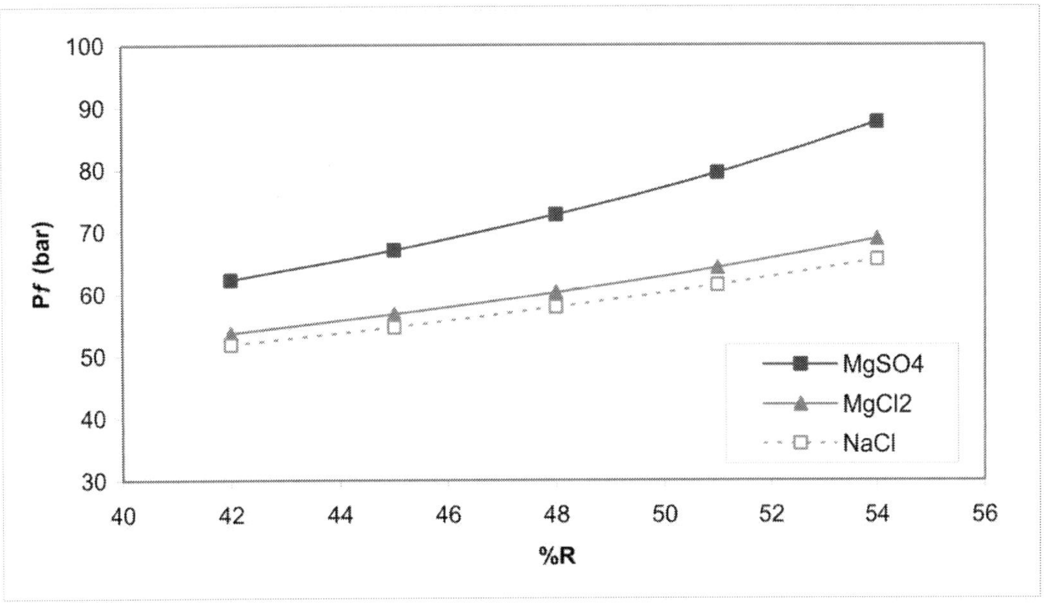

Figure 7: Effect of recovery rate on the feed pressure

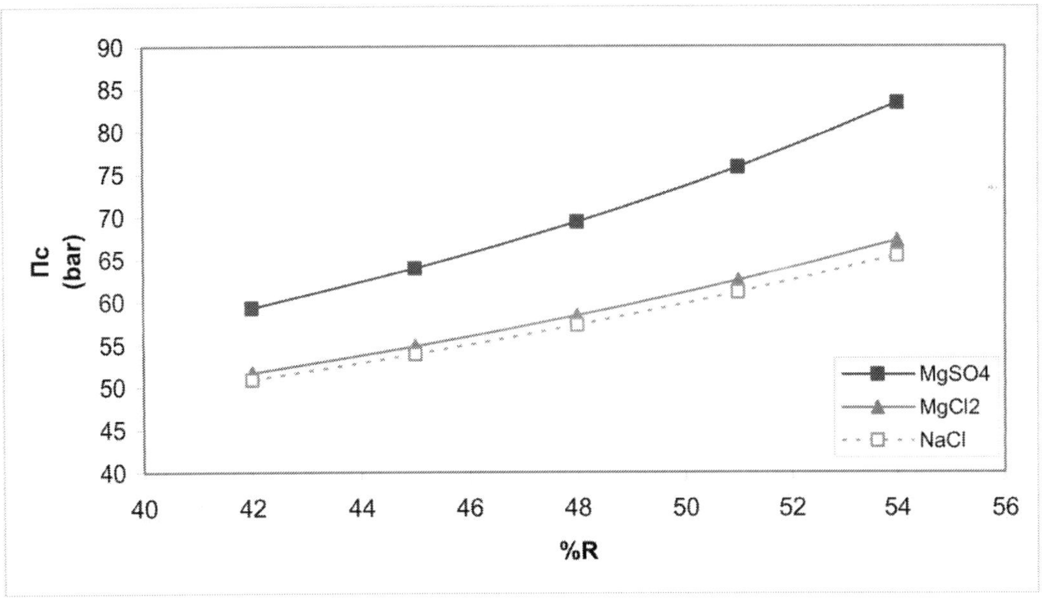

Figure 8: Effect of recovery rate on the concentrate concentration

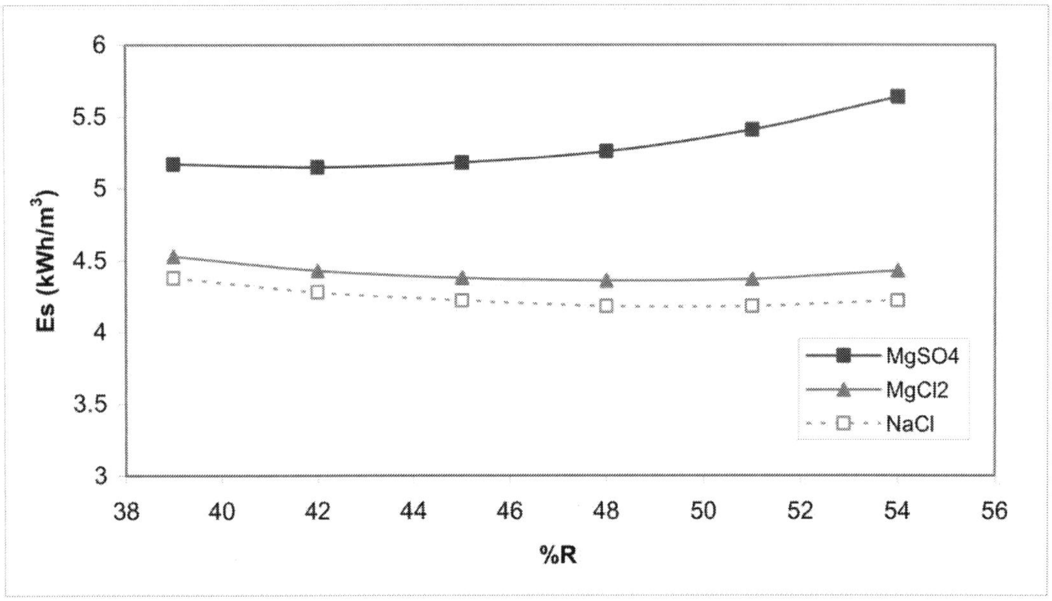

Figure 9: Effect of recovery rate on the Specific power consumption

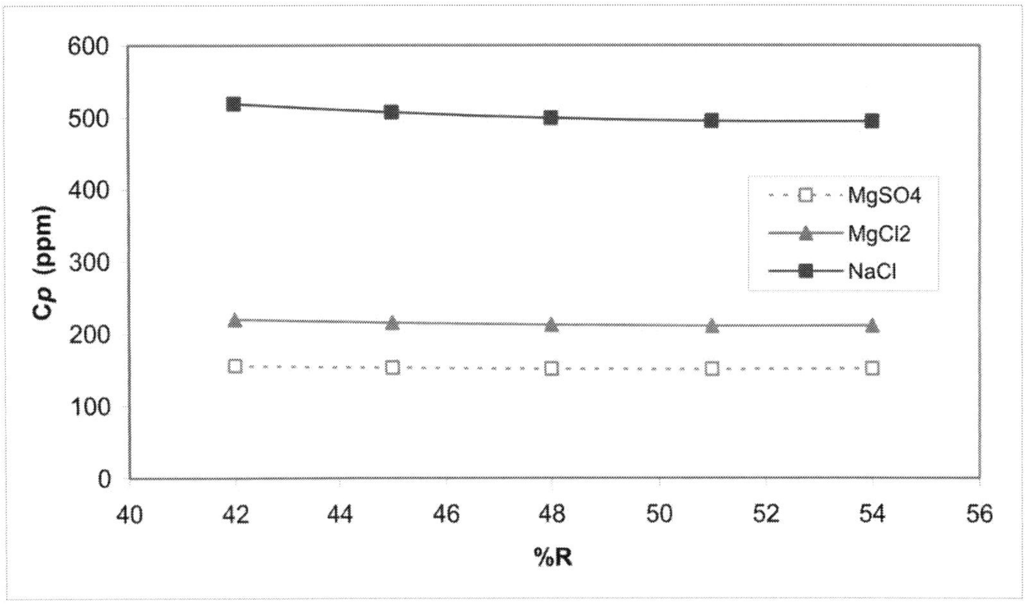

Figure 10: permeate concentration at different recovery rates

The TDS of permeate from $MgSO_4$ draw solution was lower than that from $MgCl_2$ and NaCl draw solutions (figure 10). Primarily, this was due to the large molecular size of $MgSO_4$ which was easily rejected by the RO membrane. The high permeate concentration will require an additional membrane filtration to reduce the permeate TDS leading to an increase in the treatment cost. The lowest permeate concentration was observed when $MgSO_4$ was used as a draw solution. Noting that the difference between the permeate TDS in case of $MgSO_4$ and $MgCl_2$ was insignificant. The permeate TDS, however, decreased with increasing the recovery rate (figure 10). This was due to increasing the permeate dilution factor with the recovery rate increase.

Although membrane hyperfiltration processes were suggested, so far, for the draw solution separation and regeneration, MD process can also be used for the regeneration of draw solution [20]. MD process was demonstrated to be economical in the regeneration of some draw solution such as ammonia carbon dioxide [7, 9]. In the latter process, the diluted draw solution is heated up to 60 ^{o}C before it is fed into the MD unit. Inside the membrane, ammonia carbon dioxide is evaporated and condensed in the permeate side of the membrane leaving behind fresh water in the concentrate side of the membrane. The concentration of ammonia carbon dioxide in the draw solution varies depending on the feed water solution but in general it is lower than the volume of distilled water. For instance, the FO membrane flux for seawater desalination using a draw solution contains 12% ammonia carbon dioxide is 10 L/h [7]. After leaving the MD unit, the concentrated ammonia carbon dioxide is mixed with distilled water to prepare a draw solution of desirable concentration. Despite the low cost of draw solution regeneration by MD, the process suffers from several draw backs [21]:

1. Membrane wetting which causes a reduction in the permeate flux and increasing the permeate TDS.
2. Low recovery rate
3. Small membrane area of the MD unit

In addition, the osmotic agent residues in the concentrate side of the MD membrane will affect the product water quality. According to the WHO, the concentration of ammonia in the drinking water should be less than 1 ppm. Similarly, upon chlorine reaction with ammonia to produce chloramine, the concentration of chloramine shouldn't exceed 1 ppm as recommended by the environmental and health agencies. A high concentration of disinfectant in the drinking water triggers the formation of disinfection by-products such as Trihalomethane (THM) which is a carcinogenic compound. The level of disinfectant should be adjusted to the desirable level; this in turn will increase the cost of treatment. It is also expected that the product water quality is affected by the salt diffusion (mostly NaCl) from the seawater to the draw solution side of the FO membrane. Salt concentration in the draw solution side is not affected by the MD process and it will remain in the product water stream. The concentration of salt in the product water will increase due to the draw solution recycling till it reaches the actual concentration level of salt diffusion through the FO membrane. The lower the membrane salt rejection rate the higher the salt diffusion. Practically, the salt diffusion in FO process using high salt rejection membrane (Rejection>99%) was 2051 ppm; feed water concentration 34590 ppm. The results in figure 11 show the increase of salt concentration in the product water as a function of the draw solution recycle. It is evident from figure 11 that the concentration of NaCl in the product water increased with the number of draw solution recycling. After recycled 15 times the concentration of NaCl in the draw solution reached 2000 ppm. The concentration of NaCl in the product water reached the actual salt diffusion concentration, 2051.093 ppm, after recycle number 60 and remained unchanged.

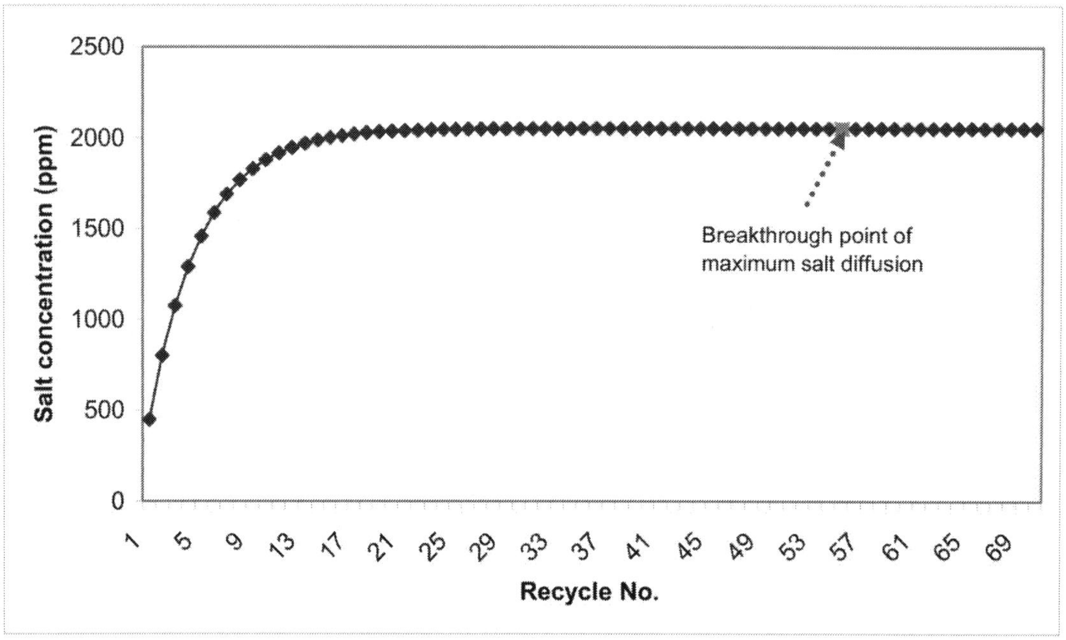

Figure 11: Concentration of NaCl salt in product water as function of draw solution recycle

The accumulation of NaCl in the product water renders it brackish. As a result it needs another membrane treatment to reduce the salt concentration to the desirable level. The additional membrane treatment may employ a low pressure BWRO membrane to adjust the concentration of permeate TDS and its cost should be added to the total cost of the seawater treatment. There are a number of parameters affecting salt diffusion across the FO membrane such as feed concentration, type of the membrane, feed and draw solutions flow rate and temperature which need a further investigation.

Thermal Processes for Draw Solution Regeneration:

Thermal processes such as MSF and MED were proposed for draw solution regeneration and recycle [20]. MSF was very popular in the Gulf region of the Middle East due to the high salinity and impurities in the Gulf water. At the beginning of the 21st century the new generation of MED was introduced and proven to be very competitive to the traditional MSF design [4, 22, 23]. MSF can be operated in either once through or brine recycle mode. The recovery rate in once through mode is 10% while in brine recycle is 33% [24]. Most of the current MSF plants are operated in brine recycle mode to achieve a high recovery rate [4, 24]. In contrast, MED plant doesn't have a flexible operating mode. A recovery rate of 33% can be reached in the MED plant. The high performance of MED attracted a lot of attention especially when the fuel price is high. The top brine temperature in MSF is 110 ^{o}C while in MED is 65 ^{o}C [25].

The cost of draw solution regeneration by thermal processes is expected to be higher than membrane technologies. The feed solution in the thermal processes (MSF and MED) is the diluted draw solution from the FO process. MSF and MED are more suitable for ammonia carbon dioxide regeneration than organic and inorganic salts such as sucrose, glucose, MgSO4, MgCl2, etc. because of the low temperature required for the evaporation of ammonia carbon dioxide. The following point should be observed upon using thermal processes for the regeneration of draw solution:

1. The potential of scale caused by the draw solution such as MgSO4 treatment by MSF
2. The diluted draw solution needs to be recycled more than often to achieve the desirable water recovery and draw solution concentration. As such the cost of regeneration will increase

As mentioned before, the expected concentration of ammonia dioxide in the draw solution is about 10% for seawater desalination, TDS 35000 ppm. In case of MSF process, once thorough operating mode could be adequate for ammonia carbon dioxide regeneration (concentration 10% in draw solution). The Gain Output Ratio (GOR) is 8.9 otherwise GOR of 8 is achieved in brine recycle. Usually, the GOR for MED is 8. If thermal processes are used in conjunction with FO for ammonia carbon dioxide regeneration, there will be salt residues in the fresh water as explained above in the FO-MD process. Therefore, an additional membrane treatment may be required for the remove of salt from the product water.

Forward Osmosis for power generation:

Power generation from renewable energy such as solar, wind, geothermal has received a lot of attention due to the continuous increase in the fossil fuel prices and environmental awareness about green house gas release. The idea of using FO in power generation was dated back to the 70's of last century [26]. Sidney Loeb was first who suggested using osmotic energy in what so called Pressure Retarded Osmosis (PRO) in power generation [26, 27]. There area analogy between the application of FO in seawater desalination and power generation. In the latter process, two solutions of different concentrations are fed into the FO membrane. The high concentration solution is known as the draw solution while the low concentration solution is the donor solution. Fresh water transports across the membrane barrier from the low to the high concentration solution due to the osmotic pressure gradient. After leaving the FO membrane, the diluted draw solution is fed into turbine for power generation (figure 12).

PRO can be used alone or combined with RO process for power generation and seawater desalination [27]. As such, seawater is applied into the RO membranes for desalination. Permeate from the RO system is the product water while the concentrate is the donor solution in the FO process. This process has the advantage of reducing the brine concentrate discharge from the RO to sea. Although PRO was suggested long time ago, its commercial application wasn't achieved till 2009 when StatKraft company built the world first power plant operates by osmotic energy (figure 12). The process was slightly different from that suggested by Sidney Loeb by incorporating an Energy Recovery Instrument (ERI) to enhance the overall performance of the process according to the following equation:

$$Es = \frac{PV}{\mu} \quad [2]$$

Where Es is the power generation from the PRO, P is the pressure of feed solution to the turbine, V is the volume of feed solution to the turbine, and μ is the pump efficiency. The power generation increases with increasing the volume and pressure of the feed solution to the turbine system. Earlier FO membrane exhibited a low flow due to the adverse impact of concentration polarization (figure 1). External or internal salt accumulation at the feed side of the membrane surface reduces the osmotic pressure gradient and hence the driving force across the membrane. New FO membranes have dealt with this problem by reducing the thickness of the support layer which resulted in reducing salt accumulation at the membrane surface (figure 14) [16]. This development has encouraged more research and pushed the PRO process a step ahead towards commercialization. These efforts have been culminated by building the world first power generation plant using the PRO process [14]. Fresh water was used as feed solution while seawater was the draw solution. Using fresh water as a feed solution will eliminate the effect of concentrative concentration polarization at the membrane surface. However, the process is site specific; i.e. it is dependent of the availability of draw and feed solutions. In many countries affected by water shortage it is rather impossible using fresh water as a feed solution for power generation by PRO process. Since water shortage problem has affected many areas around the world, seawater was proposed to be the feed solution provided that the draw solution must be of a higher concentration than seawater. In the latter design, the concentrative concentration polarization is more effective than the previous Norwegian design in which freshwater is used as a feed solution [15]. In matter of fact PRO process for power generation using fresh water as a feed solution and seawater as a draw solution is a site specific process and can't be generalized worldwide. It depends on the abundancy of the feed and draw solutions in a particular area.

Alternatively, wastewater effluent was proposed to replace the fresh water as a feed solution in the PRO process to overcome the fresh water shortage problem [10, 12]. Any impaired solution with low salinity could be used as a feed solution. The PRO process diagram using wastewater effluent and seawater as feed and draw solution respectively is shown in figure 14. Wastewater effluent contains some impurities comparing to fresh water but equal concentration of dissolved ions. Example of the impurities in wastewater is the organic matters, total nitrogen (T-N), total phosphorus (T-P) and suspended solids (TSS). Organic matter, in particular, presence in feed wastewater effluent increases the propensity of membrane fouling and affecting the overall water transport across the membrane [15, 28]. The coupled effect of organic matter and the concentration polarization effect were well investigated in the literature [15]. The experimental work showed that PRO operates better when the wastewater effluent facing the membrane surface while the draw solution facing the support layer [15]. Although such design reduces the osmotic driving force across the membrane but it is more effective in reducing the coupled effect of organic matter membrane fouling and concentration polarization.

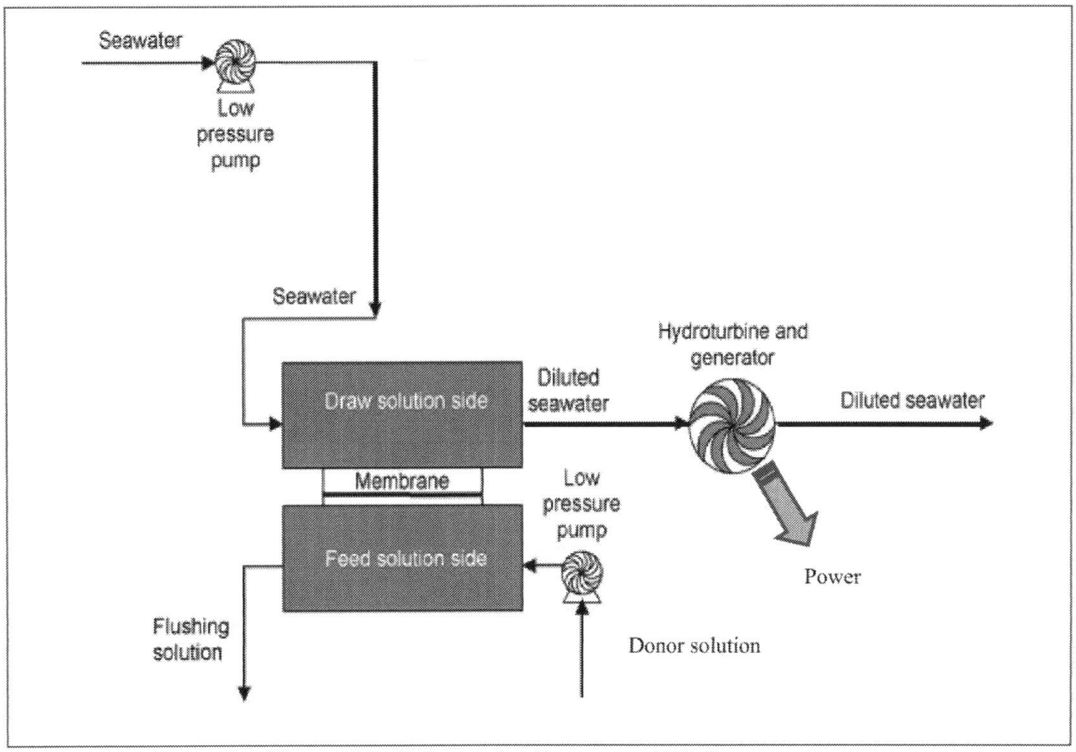

Figure 12: Schematic diagram of PRO process

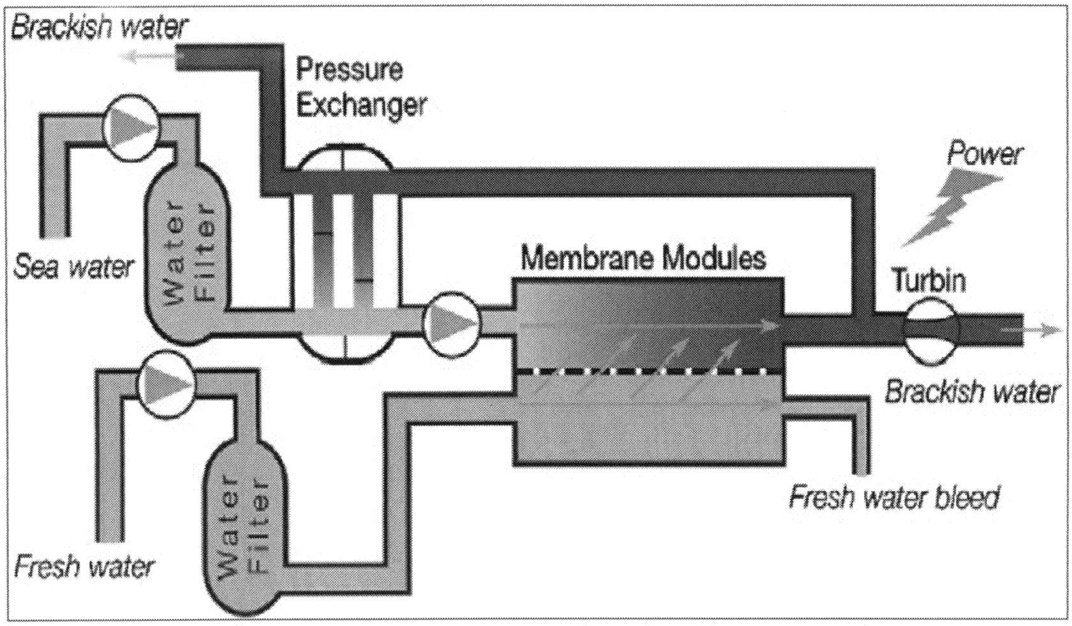

Figure 13: StatKraft PRO power generation plant (*from StatKraft website*)

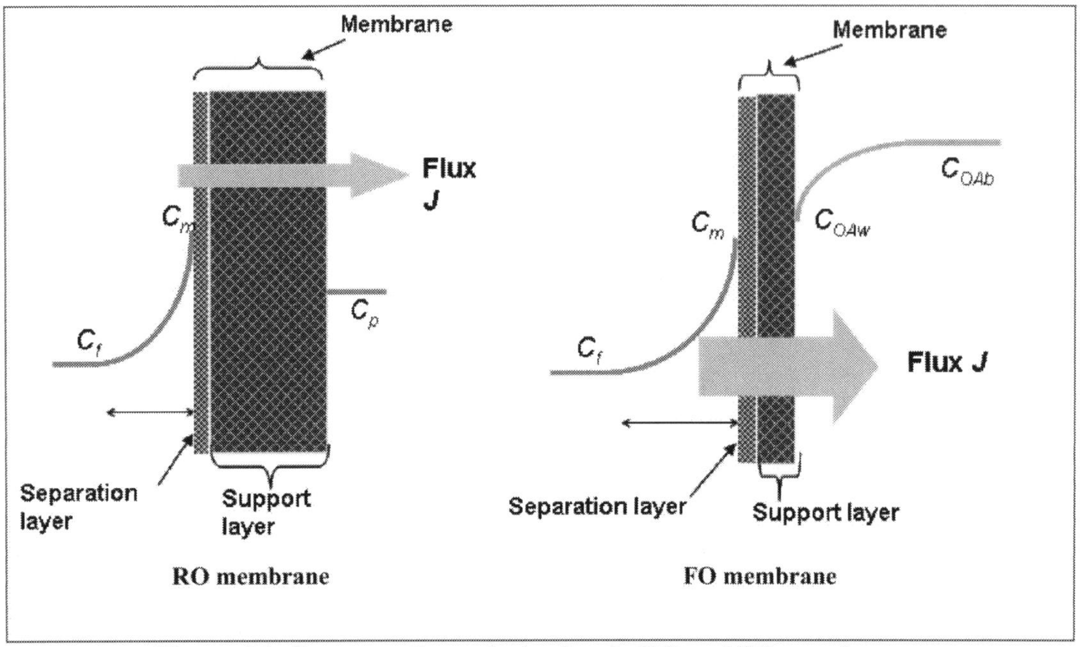

Figure 14: Concentration polarization in RO and FO membranes

A suggested PRO design by Shung et. al., wastewater effluent passed in the FO cell as a feed solution while seawater was the draw solution (figure 14). Part of the diluted draw solution is passed through a pressure exchanger for energy reuse and mixed then with the rest of the diluted draw solution before it is sent to a second FO membrane. In the latter membrane a costume design draw solution was used for water extraction from the diluted seawater. Although using wastewater effluent as a feed solution will overcome the problem of fresh water shortage, the process performance will be lower than the fresh water feed. Organic matter fouling is the main drawbacks of using wastewater effluent as a feed in PRO. To alleviate the effect of organic fouling, wastewater effluent should face the selective layer of FO membrane while the draw solution faces the support layer. This operating mode is renown of yielding a lower membrane flux but more effective in reducing the FO membrane fouling [15]. Practically, the concentration of organic impurities in wastewater effluent varies depending on the type and level of wastewater treatment.

The PRO design shown in figure 14 has two unit of FO membrane process which makes it rather complicated. The diluted seawater from the first FO process can be treated directly by thermal or membrane processes for fresh water extraction and draw solution recycle (figure 15). This will reduce the FO membrane and cost. Additionally, the plant foot plant will be more compacted. The wastewater concentrate leaving the FO membrane can be used for irrigation or discharge to a proper water system. Either design in figure 14 or 15 are an alternative to the use of freshwater and they need to be confirmed experimentally.

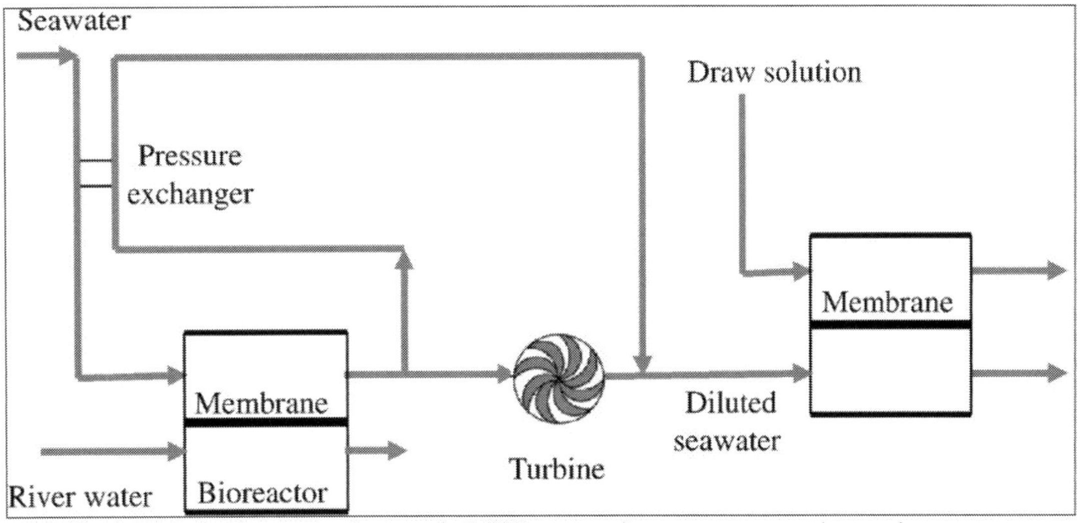

Figure 14: An integrated osmotic MBR, osmotic power generation and seawater desalination system (Tai-Shung Chung et. al., 2010)

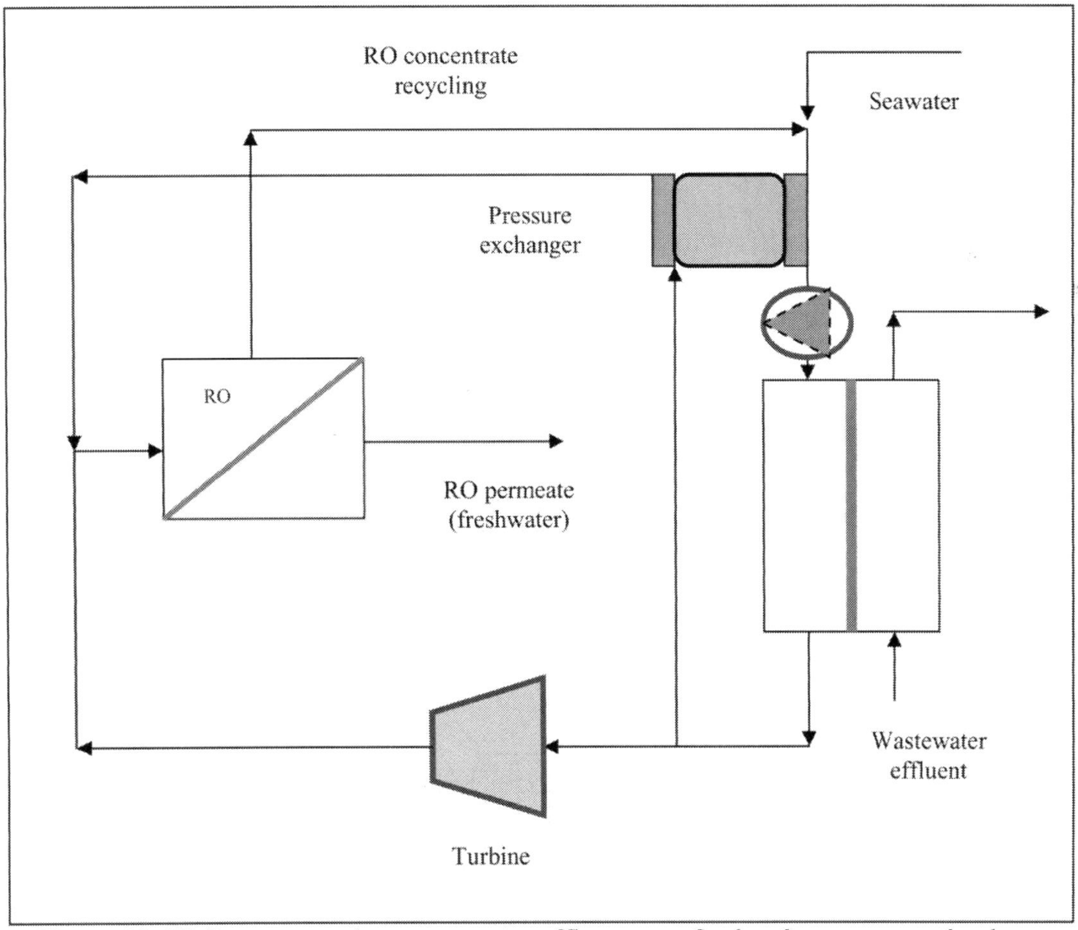

Figure 15: PRO process using wastewater effluent as a feed and seawater as the draw solution

Conclusion:

Despite the wealth of literature and experimental work conducted in FO membrane process, its application is still limited to bench and some pilot plant studies. In seawater desalination, the process is still under investigation. Its wide application in seawater desalination was hampered, at the beginning, due to the lack of appropriate membrane. Understanding the phenomenon of concentration polarization in the FO process has resulted in the development of a suitable FO membrane. The real challenge encountered the commercial application of FO process was the economic feasibility of the FO and if it can be competitive to the existing membrane desalination technologies such as RO. Any successful application of FO requires a cost-effective regeneration process. This is because most of the energy required in FO desalination is spent in the regeneration process. Results from previous research studies suggested using NF membrane in the regeneration of tailored design draw solution of large divalent ions such as MgSO4. Attention should be paid here that this process is more suitable for brackish water desalination as most of the available NF membrane cant tolerate feed pressure more than 40 bar. Different organic and inorganic salts were suggested to be used as draw solutions. The simulation results in this study showed that NaCl is more efficient than $MgSO_4$ and $MgCl_2$ due to the higher recovery rate can be achieved at lower power consumption. Osmotic agent of small molecular weight is more efficient draw solution than large molecular weight osmotic agent due to the higher osmotic pressure possessed by the former osmotic agent.

One of the inherent problems in FO is the salt diffusion from seawater to the draw solution side of the membrane. In particular, this is important when MD/thermal processes are used for draw solution evaporation and concentration such as in ammonia carbon dioxide. Low pressure BWRO membrane process can be used for salt removal from permeate to the desirable level. But the cost of this process should be added to the total desalination cost.

FO process has the potential of use in power generation by what so called PRO process. The only commercial application of such process in power generation is the pilot plant built by StatKraft in Norway. Such process is a site specific as it uses fresh water and seawater as feed and draw solution respectively. Wastewater effluent was proposed as a feed solution and hence the geographic application of PRO can be extended to include water shortage areas. However, membrane fouling due to the organic matters in the wastewater effluent should be investigated to reduce membrane fouling.

References:

1. Mohammad Al-Sahali, Hisham Ettouney, Developments in thermal desalination processes: Design, energy, and costing aspects, Desalination, 214 (2007), 227-240
2. Ioannis C. Karagiannis, Petros G. Soldatos, Water desalination cost literature: review and assessment, Desalination, 223 (2008), 448-456

3. Kah Peng Lee, Tom C. Arnot, Davide Mattia, A review of reverse osmosis membrane materials for desalination—Development to date and future potential, Journal of Membrane Science, 370 (2011), 1-22
4. Toufic Mezher, Hassan Fath, Zeina Abbas, Arslan Khaled, Techno-economic assessment and environmental impacts of desalination technologies, Desalination, 266 (2011), 263-273
5. Lauren F. Greenlee, Desmond F. Lawler, Benny D. Freeman, Benoit Marrot, Philippe Moulin, Reverse osmosis desalination: Water sources, technology, and today's challenges, Water Research, 43 (2009), 2317-2348
6. Robert L. McGinnis, Menachem Elimelech, Energy requirements of ammonia–carbon dioxide forward osmosis desalination, Desalination, 207 (2007), 370-382
7. Jeffrey R. McCutcheon, Robert L. McGinnis, Menachem Elimelech, Desalination by ammonia–carbon dioxide forward osmosis: Influence of draw and feed solution concentrations on process performance, Journal of Membrane Science, 278 (2006), 114-123
8. Kah Peng Lee, Tom C. Arnot, Davide Mattia, A review of reverse osmosis membrane materials for desalination—Development to date and future potential, Journal of Membrane Science, 370 (2011), 1-22
9. Tzahi Y. Cath, Amy E. Childress, Menachem Elimelech, Forward osmosis: Principles, applications, and recent developments, Journal of Membrane Science, 281 (2006), 70-87
10. Tai-Shung Chung, Sui Zhang, Kai Yu Wang, Jincai Su, Ming Ming Ling, Forward osmosis processes: Yesterday, today and tomorrow, Desalination, In Press, Corrected Proof, Available online 11 January 2011
11. Jeffrey R. McCutcheon, Menachem Elimelech, Influence of concentrative and dilutive internal concentration polarization on flux behavior in forward osmosis, Journal of Membrane Science, 284 (2006), 237-247
12. Andrea Achilli, Tzahi Y. Cath, Amy E. Childress, Power generation with pressure retarded osmosis: An experimental and theoretical investigation , Journal of Membrane Science, 343 (2009), 42-52
13. Sidney Loeb, Energy production at the Dead Sea by pressure-retarded osmosis: challenge or chimera?, Desalination, 120 (1998), 247-262
14. World Desalination Report, Vol 45, Issue 39, 26 October 2009
15. Chuyang Y. Tang, Qianhong She, Winson C.L. Lay, Rong Wang, Anthony G. Fane, Coupled effects of internal concentration polarization and fouling on flux behavior of forward osmosis membranes during humic acid filtration, Journal of Membrane Science, 354 (2010), 123-133
16. Sidney Loeb, Leonid Titelman, Emmanuel Korngold, Joseph Freiman, Effect of porous support fabric on osmosis through a Loeb-Sourirajan type asymmetric membrane, Journal of Membrane Science, 129 (1997) 243-249
17. M. A. Darwish, Hisham El-Dessouky, The heat recovery thermal vapour-compression desalting system: A comparison with other thermal desalination processes , Applied Thermal Engineering, 16 (1996), 523-537
18. Ian Lomax, Experiences of Dow in the field of seawater reverse osmosis, Desalination, 224 (2008), 111-118

19. Mark Wilf, Kenneth Klinko, Optimization of seawater RO systems design, Desalination, 138 (2001), 299-306
20. Abdulsalam Al-Mayahi and Adel Sharif, Solvent Removal Process, USPC Class: 210644
21. Mohamed Khayet, Membranes and theoretical modeling of membrane distillation: A review, Advances in Colloid and Interface Science, 164 (2011), 56-88
22. Narmine H. Aly, Adel K. El-Figi, Thermal performance of seawater desalination systems, Desalination, 158 (2003), 127-142
23. Jacques Andrianne, Félix Alardin, Thermal and membrane processe economics: Optimized selection for seawater desalination' Desalination, Volume 153, Issues 1-3, 10 February 2003, Pages 305-311
24. Ahmed M. Helal, Once-through and brine recirculation MSF designs — a comparative study, Desalination, 171 (2005), 33-60
25. Suresh Patel, Michael A. Finan, New antifoulants for deposit control in MSF and MED plants, Desalination, 124 (1999), 63-74
26. Loeb S., Norman R. S. (1975). "Osmotic Power Plants". Science 189 (4203): 654–655.
27. Loeb S, "Energy Production at the Dead Sea by Pressure-Retarded Osmosis: Challenge or Chimera?", Desalination 120 (1998), 247–262.
28. Sangyoup Lee, Menachem Elimelech, Salt cleaning of organic-fouled reverse osmosis membranes, Water Research, 415 (2007), 1134-1142
29. Bernard Vigneault, Peter G. C. Campbell, André Tessier and Richard De Vitre, Geochemical changes in sulfidic mine tailings stored under a shallow water cover, Water Research, 35 (2001), 1066-1076
30. Mustafa, Syed, Shah, Khizar, Naeem, Abdul, Waseem, Muhammad, Ahmad, Tauqeer, Khan, Sadullah, Water, Air, and Soil Pollution, 217 (2011), 57-69
31.

(54e) Influence of Sublayer Structure On Concentration Polarization and Membrane Performance In FO Processes

Jincai Su[1], Rui Chin Ong[1], Bradley J. Helmer[2] and Tai-Shung Chung[1], (1)National University of Singapore, Singapore, Singapore, (2)Eastman Chemical Company, Kingsport

Influence of sublayer structure on concentrationpolarization and membrane performance in FO processes

Jincai Su[1],Rui Chin Ong[1], Bradley J. Helmer[2], Tai-Shung Chung[1,*]

[1]*Departmentof Chemical & Biomolecular Engineering, National University of Singapore, 4Engineering Drive 4, Singapore 117576*

[2]*Eastman Chemical Company, P.O. Box 1972, Kingsport, Tennessee 37662*

**Corresponding author, Email: chencts@nus.edu.sg, Fax: (65)-67791936*

Key words: Forward osmosis;Hollow fiber membrane; Internal concentration polarization; Sublayer structure

Abstracts

Forwardosmosis (FO) has been considered as a promising technology for water-relatedapplications. Intensive studies have been conducted in the last several yearson the development of high performance FO membranes and novel draw solutions. Aserious problem of the FO process is concentration polarization (CP). Sincemost FO membranes are asymmetric, CP may occur at the surface of the membraneactive layer (external concentration polarization, ECP) or within the membranesublayer (internal concentration polarization, ICP). The influence of ECP canbe reduced to some extent through enhancing the turbulence of draw solutions andusing draw solutions with lower viscosity. However, these approaches are noteffective for suppressing ICP. Running the feed or the draw solution againstthe porous sublayer would result in the concentration of the feed or thedilution of the draw solution. Both cases would inevitably reduce the effectivedriving force, thus influencing the water flux.

Inorder to find out the relationship between ICP and the membrane sublayer, wehave prepared hollow fiber membranes with different sublayer structures andevaluated their FO performance with DI water feed. It is observed that varyingthe sublayer structure does not apparently affect the FO performance withrunning the draw solution against the selective layer but has significantinfluence on the performance if running the draw solution against the sublayer.Theoretical studies are carried to examine how different structural parameters,i.e. sublayer porosity, tortuosity and thickness, affect the FO performance. DesiredFO membrane should have no sublayer. Since phase inversed membranes always havea sublayer, the preferable sublayer should have high porosity, low tortuosityand small thickness. Based on experimental observation and theoreticalmodeling, desired sublayer structure are proposed for FO membranes.

See more of this Session: Research Frontier of Water Sustainability
See more of this Group/Topical: International Congress on Energy 2011

(62a) Optimization of a Louisiana Native Algal Co-Culture for Biofuel Production

Rong Bai[1], Maria Teresa Gutierrez-Wing[2], **Michael G. Benton**[1] and Kelly A. Rusch[2], (1)Cain Department of Chemical Engineering, Louisiana State University, Baton Rouge, LA, (2)Civil and Environmental Engineering, Louisiana State University, Baton Rouge, LA

Microalgae have been reported to possess significantly higher lipid productivity per unit area than traditional oil crops, a benefit for biofuel production. Microalgal strains such as *Nannochloropsis sp., Botryococcus braunii,* and *Chlorella vulgaris* have been reported to contain high lipid/oil content. However, these results are mostly at the lab scale. In large-scale, commercial culture systems, tolerance to local environmental conditions and invasive algal species significantly affect lipid/oil production. Therefore, native microalgal strains with high lipid productivity are likely better options for large-scale biofuel production.

A native fresh water microalgae strain from Louisiana was investigated. This work measures the impacts of irradiance and nitrate levels on total lipid yield and the fatty acid profile of this strain. The culture was harvested in the late exponential-stationary phase to maximize the lipid content, and Soxhlet extraction with chloroform/ methanol (2:1, v/v) was used for total lipid content evaluation. Microwaves and nanoparticle assisted Folch's extraction methods have been used to optimize lipid extraction. The fatty acid profiles of neutral lipids were analyzed by gas chromatography combined with mass spectrometry.

The results indicated the lipid yield and content at 400 and 800 mmol $m^{-2}s^{-1}$ was the highest and no significant differences (32.13 and 32.76 mg $L^{-1}d^{-1}$, 31% and 30%) between the two treatment levels were observed. Results also indicate that although the lower nitrate level increased lipid content, the increase did not compensate for a decrease in growth rate. The fatty acid profiles indicate the major components are 16- and 18-carbon fatty acids, and ~35% are saturated fatty acids. The fatty acid profiles do not vary significantly with the irradiance and nitrate levels.

See more of this Session: Advances In Algal Biorefineries I
See more of this Group/Topical: Sustainable Engineering Forum

(62b) Controls of Microalgal Biomass and Lipid Production In Municipal Wastewater-Fed Bioreactors

Belinda S.M. Sturm[1], Val H. Smith[2], Marie-Odile Fortier[1], Edward Peltier[1] and Frank Jerry deNoyelles[3], (1)Civil, Environmental & Architectural Engineering, University of Kansas, Lawrence, KS, (2)Ecology and Evolutionary Biology, University of Kansas, Lawrence, (3)Kansas Biological Survey, Lawrence

The use of wastewater as a nutrient source for algal growth as a biodiesel feedstock could increase the viability of this process and reduce wastewater discharge of N and P compounds. In this study, four algal bioreactors were fed continuously with municipal wastewater effluent over a six month period; these were monitored weekly to determine algal biomass and lipid production, nutrient removal, and carbon and nutrient availability. Algal biomass production ranged from 0.78 to 15.9 g dry weight m^{-2} d^{-1} over the experimental period. Algal reactors removed 19% of dissolved nitrogen and 43% of dissolved phosphorus from wastewater effluent. A stoichiometric analysis of particulates revealed that algal biomass had far lower C:P ratios (67:1) and N:P ratios (6:1), but higher C:N ratios (17:1), than Redfield values. These data suggest that the algae produced were rich in phosphorus, and that N was likely to be growth-limiting in these systems. Numerous studies also suggest that algal biomass can be enhanced by providing CO_2 in excess of normal atmospheric concentrations, and that algal biofuel production can contribute to CO_2 sequestration efforts. In order to optimize nutrient removal in wastewater-fed algal reactors, and to determine the potential for CO_2 sequestration, future reactor engineering research should continue to incorporate the expertise of ecological stoichiometry.

See more of this Session: Advances In Algal Biorefineries I
See more of this Group/Topical: Sustainable Engineering Forum

(62c) Effects of Total Inorganic Carbon and Nitrogen Concentrations On Lipid Formation In Chlorella Vulgaris

Jinsoo Kim, Department of Chemical & Materials Engineering, University of Cincinnati, Cincinnati, OH, Joo-Youp Lee, Department of Chemical and Materials Engineering, University of Cincinnati, Cincinnati, OH and Kaniz F. Siddiqui, The Metropolitan Sewer District of Greater Cincinnati, Cincinnati, OH

Effects of totalinorganic carbon and nitrogen concentrations on lipid formation in *Chlorella vulgaris*

Jinsoo Kim[1], Joo-YoupLee[1], Kaniz F. Siddiqui[2]

[1] Chemical EngineeringProgram, University of Cincinnati, OH 45221-0012

[2] The Metropolitan SewerDistrict of Greater Cincinnati, Cincinnati, OH 45204-2002

Microalgae have recently received greatattention as a promising alternative bioenergy source because they can produce asubstantial amount of triacylglycerols (e.g. 20-50 % dry cell weight) and canbe readily converted into biodiesel. *Chlorella vulgaris* has been selected forthis growth kinetic study since it is one of the fastest growing microalgae andcontains a reasonable amount of triacylglycerols (e.g. 14-22 % dry cell weight). Dissolved carbon dioxide (CO_2(aq)) and bicarbonate ion (HCO_3^-) areessential inorganic carbon forms required for autotrophic algal growth, but fewstudies have been systematically conducted on the impact of inorganic carbon onalgal growth. In this respect, thegrowth kinetics of *Chlorella vulgaris*was previously investigated dependent on the total inorganic carbonconcentration (i. e., CO_2(aq) and HCO_3^-) As a result, it was observed that the growthkinetics was increased when the total inorganic carbon concentration wasincreased in the culture medium.

Under many nutrient limiting conditions,lipids are accumulated as a means of storage when energy (i.e. light) andcarbon (i.e. CO_2 for autotrophic growth) sources are available. However, many of these nutrients areessential for photosynthesis responsible for metabolic flux generation intolipids formation. Thesenutrient-deficient conditions decrease biomass growth rate, and offset thebenefit of lipid accumulation. Thegrowth of algal biomass such as *Chlorella vulgaris* is dependent on bicarbonate ion concentrationsavailable in the culture medium. Bicarbonateion also plays a critical role as a catalyst in the first step of fatty acidsbiosynthesis for the synthesis of malonyl-CoA.In this study, a change in total lipids content has been investigatedusing the Nile red stain method when the total inorganic carbon concentrationin culture medium is increased. This isexpected to provide fundamental understanding of the utilization of bicarbonateions for the design and operation of bioreactors.

Keywords: *Chlorella vulgaris*, total inorganic carbon, nitrogen deficiency,lipid content.

See more of this Session: Advances In Algal Biorefineries I
See more of this Group/Topical: Sustainable Engineering Forum

(62d) Medium Optimization for Enhanced Lipid Production by Chlorella Protothecoides UTEX25

Kuan-Chen Cheng[1], Ming Ren[1] and Kimberly Ogden[2], (1)Chemical and Environmental Engineering, The University of Arizona, Tucson, AZ, (2)University of Arizona, Tucson, AZ

Algal biodiesel as the third generation biofuels has received considerable attention in recent years because of its renewable, biodegradable, and nontoxic fuel properties. Algal also exhibit several advantages over other energy crops, such as higher photosynthetic efficiency, higher biomass production and lipid content, and faster growth rate. The composition of algal-produced hydrocarbons vary among the microalgae species resulting in many different lipids, hydrocarbons, and other complex oils.

In the present study, a fresh water microalgae, *Chlorella protothecoides* (UTEX25), was evaluated for its lipid productivity through medium optimization. A one-factor-at-a-time approach was applied to determine the influence of each ingredient of proteose medium on both biomass and lipid production. $MgSO_4$°E7H_2O, $NaNO_3$, $CaCl_2$°E2H_2O, K_2HPO_4 and KH_2PO_4, and NaCl were removed individually from the culture medium, and the results were compared with the one obtained from the original medium. After 11-day cultivation, the results demonstrated that the removal of MgSO4°E7H_2O has positive effect on both biomass growth and lipid production. The biomass reached 0.95 g/L (the biomass production obtained from original medium was 0.54 g/L). The lipid content also increased to from 4.6 to 9.5 % when MgSO4 was removed. On the other hand, the removal of $NaNO_3$ and proteose adversely reduced biomass growth. The biomass production decreased from 0.54 to 0.37 and 0.29 g/L, respectively, when $NaNO_3$ and proteose was removed. Therefore, MgSO4°E7H_2O, $NaNO_3$, and proteose were chosen for further evaluation.

A statistical experimental design, response surface methodology (RSM), was applied to determine the optimal concentration of the three nutrients for both biomass and lipid production. The predicted results of these three nutrients are validated in comparison with experimental data. The change of the lipid profiles obtained from original and optimized medium are evaluated using GC-FID. Results are applied to minimize nutrient requirements in large-scale pond cultivation.

See more of this Session: Advances In Algal Biorefineries I
See more of this Group/Topical: Sustainable Engineering Forum

(62e) Single-Cell Growth Kinetics of Algae, Chlorela Vulgaris

Alim Dewan[1], Jihye Kim[2], Swastika S. Bithi[2], Marci Kerls[2], Siva A. Vanapalli[2] and M. Nazmul Karim[2], (1)Chemical Engineering, Texas Tech University, Lubbock, TX, (2)Chenical Engineering, Texas Tech University, Lubbock, TX

Single-cell growth kinetics of algae, *Chlorela vulgaris*

AlimDewan, JihyeKim, Swastika S. Bithi, Marci Kerls, Siva A.Vanapalli, M. Nazmul Karim*

Departmentof Chemical Engineering, Texas Tech University, Lubbock, Texas 79409, USA.

*E-mail: naz.karim@ttu.edu

Abstract

Traditionally,the growth kinetics of algae is investigated using cell culture equipment thatcontains large number of cells. The kinetic parameters are measured fromrepresentative samples of the cell culture and the measured parametersrepresent the statistical average of all the cells in the samples. In thisstudy, we have investigated the growth kinetics of single-cell of *Chlorellavulgaris* using droplet based PDMS microfluidic chip. The kinetic parameterswere determined by counting the total number of cells under investigationduring the entire period of the experiment. We found that there is significantheterogeneity in the growth and size of single cells and kinetic parameters aredifferent even if the initial cell numbers are the same. We found that thereis significant heterogeneity in the growth and size of single cells and kineticparameters are different even if the initial cell numbers are the same. Figure1exmplifies the heterogeneity of single-cell growth kinetics. We hypothesizethat the heterogeneity occurs due to the difference in initial cell conditionssuch as the age of the cell, and the size.

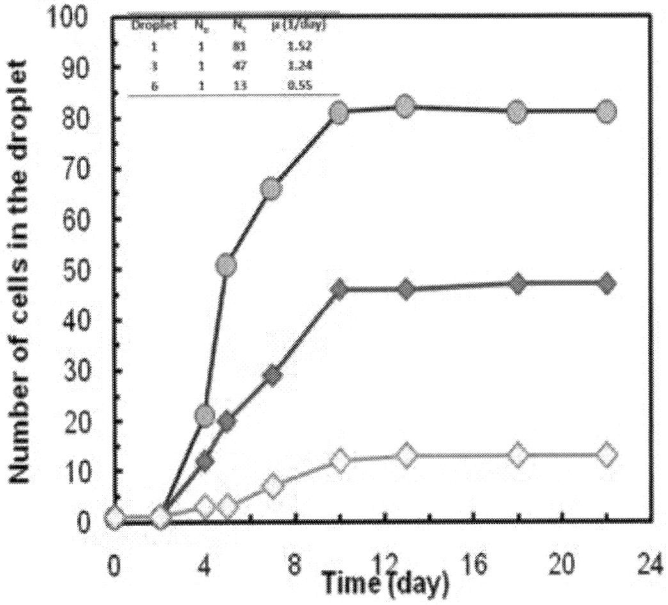

Figure1.Example of single-cell growth curves. Inset shows the initial cell number (N_o),final cell number (N_t) and the specific growth rate ().

See more of this Session: Advances In Algal Biorefineries I
See more of this Group/Topical: Sustainable Engineering Forum

(62f) Modeling and Optimization of Algae Based Biodiesel Production

Soumya Yadala, Chemical Engineering, The University of Tulsa, Tulsa, OK and Selen Cremaschi, Department of Chemical Engineering, The University of Tulsa, Tulsa, OK

Processingof biodiesel from algae biomass has attracted much interest in recent times dueto many reasons. Significant amongst them are global warming and scarcity infossil fuels. It has been suggested that this third generation biofuel, acarbon neutral alternative, is capable of meeting the existing demand fortransport fuels [1]. However, substantial challenges are encountered inderiving fuels from algae biomass. Primary amongst them is the relativelyhigher cost of cultivating and harvesting algae species. Due to this reason,currently, algae based biofuels cost much higher than the biofuels from firstgeneration and second generation biofuel feedstock [2].

Thiswork focuses on modeling and optimization of algae based biodiesel production. Designparameters, for cultivation of algal species, at a geographical location areaffected by that location's climatic conditions such as annual solar irradiance.Hence, geographical location, at which algae is cultivated, plays a vital rolein the biodiesel production. Another significant factor to be considered for producingmaximum amount of biodiesel is the selection of algal species. There are avariety of algal species present throughout the world, however, not all algalspecies are suitable for biodiesel production because of varying amounts of oilcontent and specific growth rates. Hence, selection of algal species is also animportant factor in determining the cost of biodiesel production. There are anumber of technologies for each process involved in the production of biodieselfrom algae biomass but the question of which technology, for which species, andat which location, to attain maximum profits is still an open challenge to theresearchers. Figure 1 presents the tasks that should be completed to producealgae-based biodiesel and options/technologies (which can be used to achievethese tasks) considered in this work. Using Figure 1, a disjunctive programmingmodel is developed to determine the algal species, the location for the cultivationof algae, and the optimum route from growth of algae to production of biodieselby selecting one technology for each process with the objective of minimum costof biodiesel production. The resulting Mixed Integer Nonlinear Programming(MINLP) model is solved using GAMS 23.6.5/SBB solver. With the model parametersused, 8000 tons/year of biodiesel can be produced at a cost of about $153 pergallon for a period of ten years using the selected technologies. Accordingly,Tricornutum species was selected to grow at all the given locations by means ofartificial growing technique-tubular photobioreactor. The cultivated algae wereharvested through flocculation which was then concentrated using centrifugation.The algae biomass was dried by low pressure shelf drying followed by extractionof algal oil by expeller press and finally, the algal oil was sent totransesterification for biodiesel production.

REFERENCES

1. Y. Chisti, Biodiesel from microalgae, *Biotechnol. Adv.* **25** (2007), pp. 294-306.

2. Sheehan, J., Dunahay, T., Benemann, J., Roessler, P. A Look Backat the U.S. Department of Energy's Aquatic Species Program: Biodiesel fromAlgae; Close-Out Report. U.S. Department of Energy (1998).

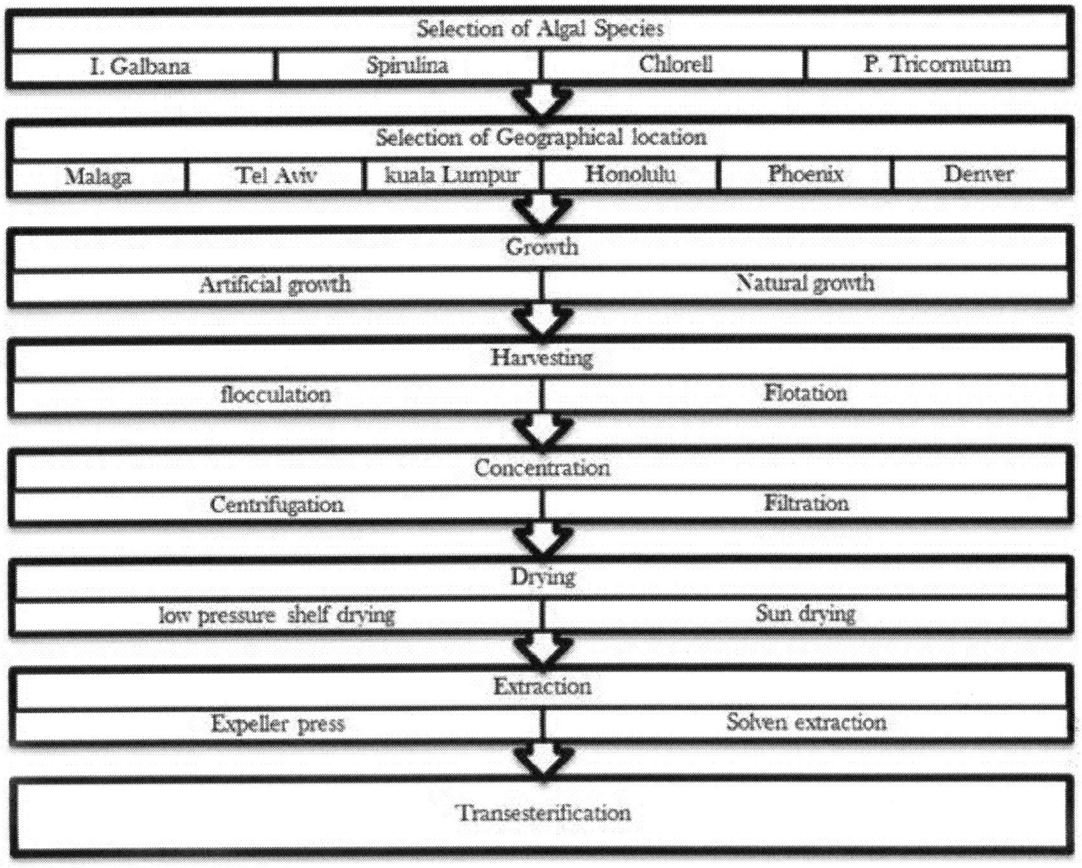

Figure 1. Tasks and corresponding options/technologies for biodiesel production from algae

See more of this Session: Advances In Algal Biorefineries I
See more of this Group/Topical: Sustainable Engineering Forum

(69a) Modeling the Effects of Pt Dispersion During NOx Storage and Reduction on Pt/BaO/ Al_2O_3

Bijesh M. Shakya, Vemuri Balakotaiah and Michael P. Harold, Department of Chemical and Biomolecular Engineering, University of Houston, Houston, TX

A crystallite-scale model is developed and incorporated into a reactor scale model to study the effects of Pt dispersion during the periodic lean-rich operation on a Pt/BaO/Al_2O_3 NOx storage and reduction (NSR) catalyst. The model is an extension of a previously developed regeneration model [D. Bhatia, M. P. Harold and V. Balakotaiah, Catalysis Today, 151 (2010) 314] and accounts for crystallite-scale diffusion limitations in the storage phase. The storage model is based on the concept of NO_x spillover from Pt to BaO and diffusion of stored NO_x in the barium phase. The model predicts the main features of NO_x storage, such as the increase in NOx breakthrough time for increasing Pt dispersion at fixed Pt loading. The predicted NOx storage increase with Pt dispersion is a result of the increase in exposed Pt area which increases the specific NO oxidation activity, and of the increased interfacial perimeter between Pt and BaO, which promotes the rate of spillover. A sensitivity analysis of the stored NO_x diffusivity reveals the importance of this process. The combined storage and regeneration model is used to simulate the entire lean rich cycles to study various cycled-averaged variables such as NO_x conversion and NH_3 selectivity. Simulations show the effectiveness of the catalyst to utilize the storage sites. The results reveal that the low dispersion catalyst (8%) is not able to utilize all of the available barium sites because NOx diffusion is too slow to access barium sites far from the Pt crystallites. The dispersion also affects the extent of axial uniformity of stored NOx during a typical 1-2 minute lean phase. The axial distribution is much more uniform for the low dispersion catalyst whereas a rather sharp storage front is evident for the high dispersion catalyst. Comparisons of the model predictions to experimental data reveal good agreement in the cycle-averaged trends. The model is used to study various storage and regeneration timing protocol which shows that shorter storage time is required for lower dispersion catalysts to achieve higher cycle averaged NO_x conversion. The model is useful in elucidating the complex transient phenomena occurring in the lean NOx trap.

Keywords: NO_x, Hydrogen, Platinum, Barium, NOx storage and reduction, Lean NOx trap, Dispersion, Spillover

See more of this Session: Applied Environmental Catalysis I
See more of this Group/Topical: Catalysis and Reaction Engineering Division

(69b) NOx Storage and Reduction by Multilayer Monolithic Catalysts

Yi Liu, Micheal Harold and Dan Luss, Chemical Engineering, University of Houston, Houston, TX

It is essential to remove NO_x emissions from diesel engine effluents. This can be accomplished by a double-bed combination of a Lean NOx trap (LNT) with a Selective Catalytic Reduction (SCR) system. During the lean operation (i.e., oxygen in excess) the Pt in the LNT catalyzes the oxidation of NO to NO_2 which is captured as nitrates and nitrites by an alkaline-earth metal oxide component, typically comprised of Barium. The stored NOx is then reduced to a mixture of N_2, N_2O, and NH_3 during a short fuel rich period containing a mixture of H_2, CO, and hydrocarbons. The ammonia that is formed and released during the reduction step is adsorbed by the acidic zeolite component in the SCR bed and later reacts with NO_x to form N_2.To improve the operation of this system and to reduce the amount of required expensive Pt it has been suggested to use short in series sequences of LNT and SCR catalysts.

As an alternative, we synthesized and tested catalysts consisting of multilayers of LNT and SCR deposited on a multi-channel monolith to capture and reduce NO_x to N_2 under simulated exhaust lean-rich atmosphere. We investigated which arrangement of the layers yielded the best NO_x removal efficiency. To determine the best balance between the alkalii sites for NO_x storage, the acidic sites (ZSM-5) for NH_3 storage, and the optimal metal loadings on both LNT (Pt) and SCR (Fe or Cu) catalysts, we conducted experiments to determine the NO_x and NH_3 storage on LNT and SCR catalysts separately under different monolith temperatures. The multilayer catalysts led to higher N_2 selectivity and lower NH_3 selectivity during a larger temperature window than those in series LNT or SCR catalysts. The Cu zeolite exhibited a higher redox activity than Fe-zeolite at low temperature. The implications of the findings in terms of the optimal catalyst design will be discussed.

See more of this Session: Applied Environmental Catalysis I
See more of this Group/Topical: Catalysis and Reaction Engineering Division

(69c) The Effect of Accelerated Hydrothermal Aging On NH3-SCR Over Cu-Hbea Catalyst

Norman Wilken[1], Kurnia Wijayanti[1], Krishna Kamasamudram[2], Neal W. Currier[2], Ramya Vedaiyan[2], Aleksey Yezerets[2] and Louise Olsson[1], (1)Chemical Engineering, Chalmers University, Gothenburg, Sweden, (2)Cummins Inc., Columbus, IN

Introduction

Diesel engines have good fuel efficiency, due to thatthey operate with excess of oxygen. However, the excess of oxygen that is fedinto the engine is also present in the exhaust stream. This disables theselectivity of the standard three-way catalyst towards nitrogen oxides (NO_x). It is crucial to reduce the NO_x, due to environmental problems. Oneeffective solution is to use ammonia (NH_3) selective catalyticreduction (SCR). Urea is sprayed into the exhaust, and the urea decomposes andhydrolyses to form ammonia. The ammonia reacts selectively with the NO_x and produces nitrogenand water. Iron and copper based zeolite catalysts have shown very goodperformance, however they can be hydrothermally aged. The objective of thiswork was to examine different reactions involved in the SCR mechanism overCu-Beta and the effect the hydrothermal aging has on these reactions.

Experimental

Thecatalyst was prepared from hydrogen beta zeolite (HBEA) from Zeolyst International. Copper was ion exchanged into theHBEA by an aqueous ion exchange method. A total of 4wt% copper was added to thezeolite, which will be referred to as Cu-BEA. The zeolite powder was coated ona monolith using the incipient wetness technique. One of the monoliths washydrothermally treated at 500°C, 600°C, 700°C, 800°C and 900°C for 3 hourseach. After each aging a set of experiments were conducted to evaluate thespecific change in performance of the SCR mechanism as well as the oxidation ofNH_3 and NO and the adsorption and desorption of NH_3. Theinvestigated temperature range was 150°C-500°C. A total gas flow of 3500 ml/minwas used resulting in a space velocity of 30 300h^{-1}. The gas flowcontained 8%O_2, 5%H_2O and 5%CO_2 as well asdifferent levels of NH_3, NO and NO_2. The gas mixture wasanalyzed with a MKS MultiGas FTIR.

Theadsorption and desorption behavior was further investigated using the powderdirectly in a Setaram SensysDSC. The sample was hydrothermally treated at 500°C, 600°C, 700°C and 780°C andthe coverage dependent heat of adsorption was calculated by performing astepwise adsorption of NH_3 at different temperatures [1].The gas stream was analyzed with a Hiden HPR-20 QUImass spectrometer.

Results and discussion

Theinfluence of hydrothermal aging on the SCR mechanism was studied. The differentSCR reactions were investigated separately. In Figure 1 the results for a fastSCR experiment are shown. A gas flow composed of 5%H_2O, 5%CO_2,8%O_2, 400ppm NH_3, 200ppm NO and 200ppm NO_2 wasused. The temperature was increased in steps from 150°C to 500°C. After eachexperiment the catalyst was hydrothermally treated, which is described in theexperimental section. At 150°C a similar conversion of $NO_x$is observed for aging temperatures up to 800°C. The aging at 900°C caused asignificant drop in activity. When the temperature is increased during theexperiment the conversion increases as well. At 300°C and 400°C the experimentsthat were done when the catalyst was aged up to 700°C reach full conversion.For higher aging temperatures full conversion was not observed. When thetemperature in the experiment is increased to 500°C a drop in conversion isobserved. NH_3 oxidation starts to play a role in that temperaturerange and therefore not all the NH_3 is available for the SCRreaction. The NH_3 storage capacity decreased with higher aging temperaturesas well as the NO and NH_3 oxidation activity.

Thecoverage depended heat of adsorption of NH_3 was investigated in thecalorimeter, using the method described earlier [1]. We could see a trendtowards less strong bound ammonia for higher aging temperatureswhich is in good agreement with TPD experiments that were conducted.

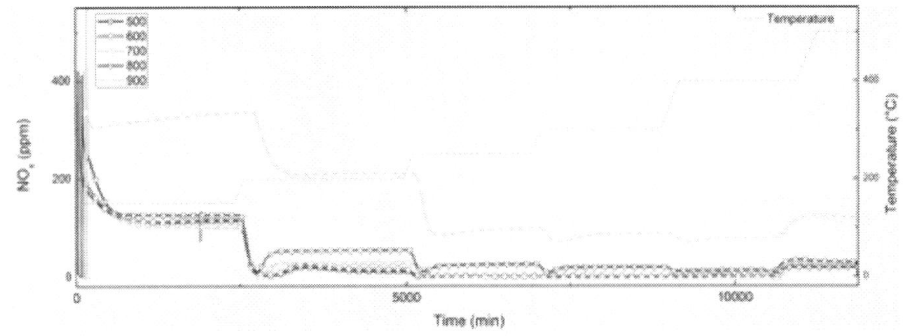

Figure 1. *NH_3 SCRover a Cu-BEA catalyst after different hydrothermal treatment. Conditions: 200ppm NO, 200ppm NO_2, 400 ppm NH_3, 8 % O_2 and 5% H_2, SV = 30 300h^{-1}.*

Conclusions

Ammonia SCR is animportant technique for reducing NO_x fromdiesel and lean burn gasoline engines. Zeolite based catalysts have been provento be a good choice for this reaction, however they can be hydrothermally aged.Since reactions on other components, like diesel particulate filterregeneration, might lead to very high exhaust gas temperatures the objective ofthis study was a detailed study of the catalyst while undergoing hydrothermalaging.

Acknowledgements

This workhas been performed within the Competence Centre for Catalysis and Cummins Inc. Theauthors would like to thank Cummins Inc. for the financial support. One author(Louise Olsson) would also like to acknowledge the Swedish foundation forstrategic research (F06-0006) for additional support. The financial support forthe micro calorimeter from the Swedish Research Council (Contract:621-2003-4149 and 621-2006-3706) and for the FTIR from Knut and AliceWallenberg Foundation, Dnr KAW 2005.0055, is gratefullyacknowledged.

References

1. N. Wilken, K. Kamasamudram, N.W.Currier, J. Li, A. Yezerets, L. Olsson, *Catalysis Today* 151 (2010) 237.

See more of this Session: Applied Environmental Catalysis I
See more of this Group/Topical: Catalysis and Reaction Engineering Division

(69d) Sulfur Poisoning of NOx Storage Reduction Catalysts: Influence the Noble Metal Position of Mono- and Bimetallic Rh and Pt

Robert Bchel, ETH Zurich, Zurich, Switzerland, Sotiris E. Pratsinis, Department of Mechanical and Process Engineering, ETH Zrich, Zrich, Switzerland and Alfons Baiker, Inst. Chemical and Bioengineering, ETH Zurich, Zurich, Switzerland

To limit emission from automobiles NO_x storage-reduction (NSR) catalyst can be used. The exhaust NO_x is trapped first under oxygen rich conditions on an alkali or alkaline-earth metal (like K or Ba) in the form of metal-nitrates and reduced during fuel rich conditions [1]. Catalyst deterioration at high temperatures and sulfur poisoning are currently limiting its application. Previously preferential located Pt on Al_2O_3 or $BaCO_3$ on $Pt/Ba/Al_2O_3$ NSR catalysts was made with a 2 nozzle flame spray pyrolysis (FSP) unit [2]. While Pt located on Ba promoted the Ba-nitrate regeneration [3] the support effect can change the catalytic behavior significantly [4].

Mono- and bimetallic Rh and Pt based NO_x storage-reduction (NSR) catalysts, where the noble metals were deposited on the Al_2O_3 support or $BaCO_3$ storage component, have been prepared using a 2-nozzle flame spray pyrolysis setup. The catalysts were characterized by nitrogen adsorption, CO chemisorption combined with diffuse reflectance infrared Fourier transform spectroscopy, X-ray diffraction, and scanning transmission electron microscopy combined with energy dispersive X-ray spectroscopy. The NSR performance of the catalysts was investigated by fuel lean/rich cycling in the absence and presence of SO_2 (25 ppm) as well as after H_2 desulfation at 750 °C. The performance increased when Rh was located on $BaCO_3$ enabling good catalyst regeneration during the fuel rich phase. Best performance was observed for bimetallic catalysts where the noble metals were separated, with Pt on Al_2O_3 and Rh on $BaCO_3$. The Rh-containing catalysts showed much higher tolerance to SO_2 during fuel rich conditions and lost only little activity during thermal aging at 750 °C.

References

[1] S. Roy, A. Baiker, Chem. Rev. 109 (2009) 4054-4091.

[2] R. Strobel, L. Madler, M. Piacentini, M. Maciejewski, A. Baiker, S.E. Pratsinis, Chem. Mater. 18 (2006) 2532-2537.

[3] M.O. Symalla, A. Drochner, H. Vogel, R. Büchel, S.E. Pratsinis, A. Baiker, Appl. Catal., B 89 (2009) 41-48.

[4] R. Büchel, R. Strobel, F. Krumeich, A. Baiker, S.E. Pratsinis, J. Catal. 261 (2009) 201-207.

See more of this Session: Applied Environmental Catalysis I
See more of this Group/Topical: Catalysis and Reaction Engineering Division

(69e) Mechanistic Transient Kinetic Analysis of the NH3-NO/NO2 SCR Reactions Over a Cu-Zeolite Catalyst for Automotive Applications

Enrico Tronconi, Isabella Nova and Massimo Colombo, Dipartimento di Energia, Politecnico di Milano, Milano, Italy

Tightening NOx emission standards for heavy-duty and passenger car engines require the development of highly efficient exhaust aftertreatment systems. One technology already adopted in Europe for heavy-duty applications to comply with EURO 4/5 regulations is the selective catalytic reduction (SCR) of NOx with NH_3/urea. The new generation of SCR converters is based on zeolite catalysts promoted with transition metals (e.g. Cu, Fe) washcoated onto honeycomb monoliths. Such systems can better exploit the so called "Fast" SCR chemistry resulting from partial conversion of NO to NO_2 over an upstream oxidation pre-catalyst; however little engineering work has been reported so far on the transient kinetics of this complex reacting system over zeolites. Drawing from our previous experience on V_2O_5-WO_3/TiO_2 and Fe-zeolite SCR catalysts, we have investigated by dynamic methods (step-response and temperature-programmed experiments) the intrinsic kinetics of the NH_3-NO/NO_2 reactions over a commercial copper exchanged zeolite powdered catalyst. The results are shown to be qualitatively and quantitatively consistent with a mechanistic model, closely in agreement with the details of the Fast SCR chemistry, which fully explains a variety of observed complex transient features. The accurate quantitative description of such transients is critical for model-based development of SCR converters in view of vehicle applications.

See more of this Session: Applied Environmental Catalysis I
See more of this Group/Topical: Catalysis and Reaction Engineering Division

(69f) NOx Reduction by CO Over Noble Metal-Based FCC CO Emission Control Additives

Behnam Bahrami[1], Vasileios G. Komvokis[1], michael S. Ziebarth[2], Oleg Alexeev[1] and Michael D. Amiridis[3], (1)Department of Chemical Engineering, University of South Carolina, Columbia, SC, (2) GRACE Davison Refining Technologies, Columbia, MD, (3)Chemical Engineering, University of South Carolina, Columbia, SC

Fluid Catalytic Cracking (FCC) is a primary process used to convert crude oil into a variety of higher-value products in modern petroleum refineries. Since crude oil feedstocks incoprorate various compounds containing nitrogen, a portion of nitrogen ends up in the coke formed during the cracking process. While 90 % of the coke-bound nitrogen is converted to N_2 in the FCC regenerator, the remaining part is released in the form of NO_x, HCN, and NH_3. Specific levels of these pollutants depend primarily on the content and the type of the nitrogen compounds present in the feedstock, as well as on the combustion mode used in the FCC regenerator. Since current environmental regulations enforce CO and NO_x emissions from refineries, several new technologies were examined in efforts to control these emissions. Among them, the use of catalytic additives is the most attractive one because it is simple, cost-effective, and applicable to existing FCC units.

In this work, results of *in-situ* FTIR, TPD, and kinetic measurments were combined to clarify the chemistry of NO_x reduction by CO over FCC CO combustion promoters with a general composition M/Ce^{n+}/Na^+/γ-Al_2O_3, where M is Pd, Pt, Rh. Our goal was to understand the structure of these additives under reducing and oxidizing conditions and to examine the role of each component in the NO_x reduction process.

The STEM results indicate the presence of Ce^{n+} cations along with amorphous and crystalline CeO_2 phases on the surface of these additives. XRD and STEM results further suggest that both Ce and Pd can aggregate to a some degree at elevated temperatures in the presence of the reducing environment. This result is consistent with XPS data demonstrating that Pd/Al, Pd/Ce, and Pd/Na ratios decrease under reducing conditions at elevated temperatures. All M/Ce^{n+}/Na^+/γ-Al_2O_3 samples were tested in the NO+CO reaction in the presence and absence of O_2 in the 500–700°C range of temperatures. When O_2 is absent in the feed or present in less than stoichiometric amounts, the complete reduction of NO by CO takes place in this range of temperatures. However, when O_2 is present in large amounts, the reaction of CO oxidation dominates the reaction network. The exposure of all samples to the NO/CO mixture in the 400–600°C temperature range leads to the formation of isocyanate species on the surface. FTIR and kinetic data further suggest that isocyanate species thus formed are reactive towards O_2 and NO, yielding CO_2 and N_2 as the major products. These results suggest that isocyanate species could be intermediates in the CO–NO reaction network that takes place on surfaces of catalytic additives examined.

See more of this Session: Applied Environmental Catalysis I
See more of this Group/Topical: Catalysis and Reaction Engineering Division

(81a) Adsorption and Activation of n-Alkanes On a Pdo(101) Thin Film

Jason F. Weaver[1], Aravind Asthagiri[2], Jose Hinojosa Jr.[3], Can Hakanoglu[1], Abbin Antony[1] and Jeffery M. Hawkins[1], (1)Chemical Engineering, University of Florida, Gainesville, FL, (2)Department of Chemical & Biomolecular Engineering, The Ohio State University, Columbus, OH, (3)Department of Chemical Engineering, University of Florida, Gainesville, FL

The formation of palladium oxide (PdO) is thought to be responsible for the exceptional activity of supported Pd catalysts toward the complete oxidation of alkanes under oxygen-rich conditions. In this talk, I will discuss our recent investigations of the adsorption and activation of *n*-alkanes on a PdO(101) thin film that is prepared by oxidizing Pd(111) in ultrahigh vacuum (UHV) using an oxygen atom beam. We find that alkanes adsorb relatively strongly on the PdO(101) surface by forming σ-complexes along the rows of coordinatively unsaturated (cus) Pd atoms, and that this adsorbed state acts as the precursor for initial C-H bond cleavage. For example, propane and *n*-butane undergo facile C-H bond cleavage on PdO(101) below about 215 K in UHV. I will also discuss density functional theory calculations which show that dative bonding between the alkane and cus-Pd atoms weakens the coordinated C-H bonds, thereby lowering dissociation barriers significantly. Finally, I will discuss recent results on C-H bond selectivity in the initial activation of propane on PdO(101) as well as the activity of binary Pd-M oxides toward alkane activation.

See more of this Session: Computational Catalysis II
See more of this Group/Topical: Catalysis and Reaction Engineering Division

(81b) Multiscale Modeling of the Water-Gas Shift Reaction At the Three Phase Boundary of Pt/TiO2 and Pt/CeO2 Catalysts

Salai C. Ammal[1], Sara Aranifard[2] and **Andreas Heyden**[2], (1)Chemical Engineering, University of South Carolina, Columbia, SC, (2)Department of Chemical Engineering, University of South Carolina, Columbia, SC

For heterogeneously catalyzed reactions with more than one key surface intermediate, it is likely that multiphase catalysts have a significant advantage over conventional monophase catalysts since each phase can potentially be adjusted independently to activate a key reaction step. At the same time, our understanding of bifunctional multiphase systems is relatively poor. It is the objective of this paper to significantly enhance our molecular understanding of heterogeneous catalysis at the three-phase boundary (TPB) of a gas-phase, a reducible oxide surface, and a noble metal cluster. In particular, we intend to illustrate the specific role of the TPB in determining the activity and selectivity of TiO_2 and CeO_2 supported Pt catalysts for the water-gas shift (WGS, $CO + H_2O \rightarrow CO_2 + H_2$) reaction.

We performed an *ab initio* atomistic thermodynamics analysis of the Pt/TiO_2 (110) interface in various gas-phase environments and conclude: (1) In an oxidizing atmosphere, oxygen vacancies in TiO_2 should not play any role and the surface Pt atoms will likely be oxidized. (2) Under WGS reaction conditions, formation of oxygen vacancies at the Pt/TiO_2 interface is thermodynamically favorable and the Pt clusters are surface oxygen free. Furthermore, surface Pt atoms are covered by CO and only interfacial Pt atoms are available for reaction.

Next, we investigated the mechanism of the WGS at the Pt/TiO_2 (110) interface. The redox and carboxyl pathways were studied and the results can be summarized as follows: Pt sites away from the TPB primarily act as a CO reservoir. The oxide support plays an essential role in the O-H bond breaking and H_2 formation process. CO poisoning is reduced on Pt atoms at the TPB. Water activation and dissociation at the TPB becomes exothermic and is facile. Finally, there seems to be no significant advantage of the TPB for COOH dissociation which is another key step in the carboxyl pathway on Pt (111) surfaces.

See more of this Session: Computational Catalysis II
See more of this Group/Topical: Catalysis and Reaction Engineering Division

(81c) Understanding of Environment-Dependent Mechanisms of Alkali Promotion In Heterogeneous Catalysis Using First-Principles Based Monte Carlo Simulation

Hongliang Xin and Suljo Linic, Chemical Engineering, University of Michigan, Ann Arbor, MI

Interactions among adsorbates on solid surfaces play a critical role in heterogeneous catalysis, electro-catalysis, self-assembly, nucleation, growth, and almost every other phenomenon governed by chemical or physical surface processes. For example, alkali promoters (Li, Na, K, Rb, Cs) enhance activity and/or selectivity of various catalytic processes, such as Fisher-Tropsch synthesis, water-gas shift reactions, ammonia synthesis, olefin epoxidation, and automotive three way converters. Although the impact of alkalis in heterogeneous catalysis has been recognized for decades, not much is known about the underlying physical mechanisms that govern the promotion.

We have utilized density-functional theory (DFT) calculations to develop a physically transparent model for the analysis of the interactions between adsorbates on metal surfaces. We have applied this model to investigate underlying mechanisms that govern the interaction between Cs promoters and oxygen adsorbates on the Ag(111) surface. We have also employed an atomistic thermodynamics approach and Monte Carlo simulations to extend the results of DFT calculations to relevant catalytic conditions (T, p). We found that the underlying mechanisms of chemical promotion by alkali promoters change significantly as a function of external conditions.

See more of this Session: Computational Catalysis II
See more of this Group/Topical: Catalysis and Reaction Engineering Division

(81d) Formic Acid Decomposition On Transition Metals: Trends Through Density Functional Theory Studies

Jessica Scaranto[1], Lars C. Grabow[2], Scott D. Tonelli[1], Jeffrey A. Herron[1], Suyash Singh[1], Brandon J. O'Neill[1], James A. Dumesic[1] and Manos Mavrikakis[1], (1)Chemical & Biological Engineering, University of Wisconsin-Madison, Madison, WI, (2)Chemical and Biomolecular Engineering, University of Houston, Houston, TX

Abstract

The investigation of formic acid (HCOOH) decomposition on transition metals is important in order to derive useful insights for vapor phase catalysis involving HCOOH and for direct HCOOH fuel cells. Here we present the main results obtained from periodic, self-consistent Density Functional Theory (DFT-GGA) calculations concerning the gas-phase decomposition of HCOOH on Pt(111), Pd(111) and Cu(111) surfaces. Accordingly, we analyzed the minimum energy paths for HCOOH decomposition into CO_2 and H_{-2} through the carboxyl (COOH) and formate (HCOO) intermediates, along with alternative bond-breaking possible steps in those intermediates. The DFT-derived parameters have been employed in a microkinetic model and then used to compare modeling results with experimental reaction kinetics data. This work will be discussed in the context of thermochemistry trends for the same reaction on a number of other late transition metal closed-packed surfaces.

Acknowledgement

This material is based upon work supported as part of the Institute for Atom-efficient Chemical Transformations (IACT), an Energy Frontier Research Center funded by the U.S. Department of Energy, Office of Science, Office of Basic Energy Sciences.

See more of this Session: Computational Catalysis II
See more of this Group/Topical: Catalysis and Reaction Engineering Division

(81e) Selective Activation and Conversion of Simple Carboxylic Acids and Esters

Ye Xu, Center for Nanophase Materials Sciences, Oak Ridge National Laboratory, Oak Ridge, TN and Lijun Xu, Center for Nanophase Materials Science, Oak Ridge National Laboratory, Oak Ridge, TN

Selective activation and conversion of simple carboxylicacids and esters

Ye Xu,* and Lijun Xu

Center for Nanophase Materials Sciences, Oak RidgeNational Laboratory, Oak Ridge, Tennessee 37831, USA

* xuy2@ornl.gov

The hydrogenation of carboxylic acids and esters to thecorresponding alcohols are industrially important processes, and will likely bean integral pathway in biorefineries for the production of valuable chemicalsand fuels from biomass feedstock because carboxylic acids and esters are commoncomponents of such feedstock. Acids, in particular, are more difficult toreduce than other classes of organic oxygenates. Ester hydrogenation is traditionallycatalyzed by Cu and Cu compounds, which requires high temperature andpressure. Pt group metals can decompose acids and esters at milder conditions butoffer very limited selectivity for alcohols in gas-phase experiments.

Recently Olcay et al. demonstrated that significantselectivity for ethanol can be obtained via the aqueous-phase hydrogenation(APH) of acetic acid on a series of metals, with Ru being the most active andselective at low temperatures.[1] Ru has also been used in the APH of other organic acids with good activity and selectivity.[2] We have performed density functional theory calculations to understand the role of different metal species and surface structure in the activation and conversion of simple carboxylic acids and esters on metal surfaces, using acetic acid and methyl acetate as representative compounds.

Our calculations for the activation and subsequenthydrogenation of acetic acid on several metals suggest that the activity of aceticacid conversion is dictated largely by the intrinsic reactivity of the metals.[1] In agreement with experimental results, the selectivity to ethanol is found to depend on reaction conditions. Dehydrogenation and decomposition are the dominant reaction pathway for acetic acid and methyl acetate at vacuum/low pressure conditions on Ru and Pd, which fundamentally limits the selectivity to ethanol.[3] The activity of C-H and C-O bond dissociations is different on close-packed, open, and under-coordinated Pd sites: Pd step edges and kinks promote both C-H and C-O bond dissociation steps primarily via electronic effects, while (100)-type square surface symmetry is more effective at stabilizing C-O bond dissociation transition states than (111)-type hexagonal symmetry does, thereby differentially promoting C-O relative to C-H bond dissociation.[4]

References

1 H. Olcay, L. Xu, Y. Xu, and G. W. Huber, *ChemCatChem* **2**, 1420 (2010).

2 Y. Q. Chen, D. J. Miller, and J. E. Jackson, *Ind.Eng. Chem. Res.* **46**, 3334 (2007).

3 L. Xu and Y. Xu, *Surf. Sci.* **604**,887 (2010).

4 L. Xu and Y. Xu, *Catal. Today*doi:10.1016/j.cattod.2010.12.021.

See more of this Session: Computational Catalysis II
See more of this Group/Topical: Catalysis and Reaction Engineering Division

(81f) Hydrogenation of Lactic Acid to Propylene Glycol and Propanoic Acid: A Selectivity Challenge

Suyash Singh and Manos Mavrikakis, Department of Chemical and Biological Engineering, University of Wisconsin-Madison, Madison, WI

Periodic, self-consistent Density Functional Theory (DFT) calculations are employed to analyze the hydrogenation of lactic acid[1, 2] to propylene glycol (desired product) and propanoic acid (undesired product). A comprehensive reaction network is proposed and key surface intermediates are identified for both reactions. Detailed adsorption thermochemistry and activation energy barrier calculations are performed to probe the mechanisms on Pt(111) and energetically most favorable pathways for the formation of both products are identified. A thorough investigation of the thermochemistry of all possible elementary steps over several close-packed transition metal surfaces (Rh, Ir, Ni, Pd, Cu, Ag, Au) is then performed to establish reactivity trends among metals. The results from these calculations point to a specific metal which might be an efficient monometallic catalyst.

These DFT calculations, in combination with microkinetic modeling and experiments, will be used to screen for improved (active and selective) alloy catalysts for lactic acid hydrogenation. Our analysis is expected to not only help with identifying improved ways for the manufacture of propylene glycol from biomass-derived lactic acid, but is also expected to capture the qualitative features of the vapor phase hydrogenation of oxygenated biomass-derived feed stocks.

References

1. Cortright RD, Sanchez-Castillo M, & Dumesic JA. Conversion of biomass to 1,2-propanediol by selective catalytic hydrogenation of lactic acid over silica-supported copper. Applied Catalysis B-Environmental **2002**; 39: 353-359.

2. Simonov MN, Simakova IL, & Parmon VN. Hydrogenation of lactic acid to propylene glycol over copper-containing catalysts. Reaction Kinetics and Catalysis Letters **2009**; 97: 157-162.

See more of this Session: Computational Catalysis II
See more of this Group/Topical: Catalysis and Reaction Engineering Division

(81g) Molecular Investigation of Catalytic Isomerization of Glucose to Fructose

Samir H. Mushrif, Nima Nikbin, Stavros Caratzoulas, Vinit Choudhary, Stanley Sandler, Vladimiros Nikolakis and Dionisios G. Vlachos, Chemical Engineering, University of Delaware, Newark, DE

Conversion of carbohydrates, the key components of biomass, to various intermediates, such as 5-hydroxymethyl-furfural (HMF), has gained significant attention in recent years. Even though some success has been achieved in the dehydration of fructose to HMF, there is a pressing need to selectively transform glucose, the most dominant sugar in biomass, to HMF. The pathway to HMF from glucose is believed to involve isomerization of glucose to fructose. In order to convert glucose to HMF, via fructose, in a single reactor system, it is desired to have a catalyst which actively converts glucose to fructose in an aqueous environment. Recent experimental studies at the Catalysis Center for Energy Innovation have demonstrated promising results using various catalysts, such as tin, chromium, titanium. Among them, the tin-beta zeolite catalyst can give considerably good performance. In the present work, we successfully adopt a first principles molecular modeling approach to investigate the fundamentals of the catalytic isomerization reaction, in explicit aqueous medium. Stepwise mechanistic details of the reaction and the dynamics and energetics of individual steps are computed. Results are compared to new experimental data obtained via combining a range of analytical, NMR, and other spectroscopic methods. Molecular details obtained from the present work serve as an input for developing and designing novel catalytic materials for the glucose isomerization reaction, thus providing for the first time a bottom-up approach of catalyst design for biomass conversion reactions.

See more of this Session: Computational Catalysis II
See more of this Group/Topical: Catalysis and Reaction Engineering Division

(104a) Particle Based Multi-Scale Modeling of the Dynamic Response of Hexahydro-1,3,5-Trinitro-s-Triazine (RDX): Constant Energy Dissipative Particle Dynamics with Coarse-Grained Density Dependent Potentials

Joshua D. Moore[1], Sergei Izvekov[1], Martin Lisal[2] and John K. Brennan[1], (1)Weapons and Materials Research Directorate, U.S. Army Research Laboratory, Aberdeen Proving Ground, MD, (2)E. Hala Laboratory of Thermodynamics, Academy of Sciences of the Czech Republic, Prague, Czech Republic

Nanocomposites are inherently heterogeneous, containing several species types of disparate shapes and sizes. For example, energetic materials are often comprised of micro-sized crystallites (or grains) held together by a small amount of polymer binder, and in some instances contain metal nanoparticles. Accordingly, mechanical stimulation (e.g., shock or shear) of these materials can incite responses over a wide range of spatial and temporal scales. Localized regions of elevated thermal energy, or hot spots, often occur near microscale defects in the material such as voids and grain boundaries. However, some energy transfer processes are atomistically governed, for which field-based microscale, mesoscale, and continuum modeling techniques are currently unable to provide the required level of fidelity. Modeling the thermal and mechanical response of these materials atomistically, despite growing computational power and resources, remains a challenge due to the length and time scales required, as often billions of molecules are necessary to be modeled on micron-size scales.

To overcome some of these challenges, we have implemented multiscale techniques to bridge the atomistic and mesoscale levels of description. The multiscale coarse-graining method (MS-CG) [1,2], a *bottom-up* approach based on force-matching, has been utilized to derive particle-based coarse-grained potentials for hexahydro-1,3,5-trinitro-s-triazine (RDX), where one RDX molecule is mapped onto a single interaction site. To improve transferability of the MS-CG model (required to accurately capture thermodynamic responses during mechanical and thermal stimulation), explicit MS-CG mapping from the atomistic to the microscale is made from ambient to high pressures [2]. The resulting density dependent MS-CG model reproduces several atomistic properties within reasonable agreement from ambient to high pressures for the molecular crystal including the lattice structure, melting point, elastic and vibrational properties, and the phase diagram [3]. For the molten state, the 2-body and often multi-body structure are also well described with the model [3].

Despite the successes of the MS-CG model of RDX to describe static properties, it inevitably cannot account for accurate energy and momentum exchange in response to mechanical stimulation using traditional molecular dynamics. This is due to intramolecular degrees of freedom (i.e., intramolecular bonding) being coarse-grained out, and thus, the heat capacity is underestimated by nearly an order of magnitude [3]. To correct this deficiency, we account for momentum and energy transfer under mechanical shock by utilizing the constant-energy Dissipative Particle Dynamics method (DPD-E) [4-6]. Unlike traditional Dissipative Particle Dynamics [7-9], which conserves only momentum, the DPD-E method conserves both energy and momentum by assigning an internal energy to each particle, which for the MS-CG RDX case, helps to minimize the effect of coarse-graining the intramolecular degrees of freedom.

In this talk, we will present results for mechanical and thermal shock loading of our MS-CG model of RDX using DPD-E. Various modeling parameters have been investigated for sensitivity including spatial interaction cutoffs used to determine local densities for the density dependent potentials, the mesoparticle equation of state that accounts for the particle internal energy and the overall dynamics through energy dissipation. Results will be compared directly to atomistic simulation for both mechanical shock and the response due to thermal heating of the material.

This effort is supported by the recently established Institute for Multi-Scale Reactive Modeling of Insensitive Munitions (MSRM), which is a multi-team effort led by the U.S. Army Research Laboratory and the U.S. Army Armament Research, Development, and Engineering Center, involving various other national laboratories and academic groups totaling over 20 scientists. J.D.M. gratefully acknowledges support, in part, by an appointment to the Internship/Research Participation Program for the U.S. Army Research Laboratory administered by the Oak Ridge Institute of Science and Education through an agreement between the U.S. Department of Energy and U.S. ARL (Project ID: 201003211).

[1] S. Izvekov and G.A. Voth, *J. Chem. Phys.*, **123**, 134105 (2005).

[2] S. Izvekov, P.W. Chung, and B.M. Rice, *J. Chem. Phys.*, **133**, 064109 (2010).

[3] S. Izvekov, P.W. Chung, and B.M. Rice, *Submitted* (2011).

[4] J. Bonet Avalos and A.D. Mackie, *Europhys. Lett.*, **40**, 141 (1997).

[5] P. Español, *Europhys. Lett.*, **40**, 631 (1997).

[6] M. Lísal, J.K. Brennan, and J. Bonet Avalos, *Submitted* (2011).

[7] P.J. Hoogerbrugge and J.M.V.A. Koelman, *Europhys. Lett.*, **19**, 155 (1992).

[8] J.M.V.A. Koelman and P.J. Hoogerbrugge, *Europhys. Lett.*, **21**, 363 (1993).

[9] R.D. Groot and P.B. Warren, *J. Chem. Phys.*, **107**, 4423 (1997).

See more of this Session: Multiscale Modeling II

(104b) Temperature-Accelerated Molecular Dynamics Reveals That Insulin Can Undergo Large-Scale Conformational Reorganization On Binding to Its Receptor

Harish Vashisth, Chemistry and Biophysics, University of Michigan, Ann Arbor, MI and Cameron F. Abrams, Drexel University, Philadelphia, PA

Insulin regulates blood glucose levels in higher organisms by specifically binding to and activating a transmembrane glycoprotein known as the insulin receptor (IR). Detailed photo-crosslinking studies have suggested that insulin molecule reorganizes on binding to IR due to the displacement of its flexible B-chain C-terminus in the presence of a tandem hormone binding structural motif called CT-peptide. Although a detailed understanding of how insulin binding leads to receptor activation remains elusive, our recently proposed all-atom structural models of insulin/IR complexes provide information valuable for exploring structure-function repertoire of IR. However, the large-scale conformational change in the insulin was not studied in these structural models due to the absence of then-unresolved CT-peptide, and also the lack of adequate simulation techniques. We have proposed a new conformational sampling algorithm, temperature-accelerated molecular dynamics (TAMD) (2010, *PNAS*, **107**, 4961-4966), that provides an immediate opportunity to use all-atom simulation to directly address these important gaps in our understanding of insulin binding to IR. Using a combination of Metropolis Monte-Carlo (MC) and molecular dynamics (MD) simulations, we have constructed all-atom structural models of insulin/IR complexes in the presence of (now-resolved) CT-peptide. We further apply TAMD to the C-terminus of IR-bound (T and R) insulin molecules at a fictitious thermal energy of 6 kcal/mol, and observe that while insulin remains stably-bound to IR, it can undergo large-scale conformational change in the B-chain C-terminus at ~50-ns time-scale resulting in the exposure of hidden hydrophobic core of insulin. These results are significant because we demonstrate at the atomistic-scale for the first time that insulin molecules can undergo large-scale reorganization on binding to IR in presence of CT-peptide. Moreover, we suggest TAMD as a viable simulation technique to study such large-scale conformational changes in proteins.

(104c) Multiscale Modeling for Phospholipid Bilayer Simulations

Emily Curtis, Chemical & Biomolecular Engineering, North Carolina State University, Cary, NC

Multiscale Modeling for PhospholipidBilayer Simulations

A multiscale modelingapproach was used to develop an implicit-solvent intermediate-resolution modelto simulate the behavior of the lipid, 1,2-dipalmitoyl-*sn*-glycero-3-phosphocholine (DPPC) inwater. By reducing the DPPCrepresentation to 14 coarse-grained sites with unique properties, treatingsolvent implicitly, and employing discontinuous molecular dynamics, a very fastalternative to traditional molecular dynamics, we are able to simulate thespontaneous formation of a bilayer from a random solution of 512 DPPC moleculesin several hours on a fast workstation. Data used to calculate the coarse-grained model parameterswas obtained by running united-atom simulations with GROMACS. Radial distribution functions wereplotted for all pairs of non-bonded coarse-grained sites to estimate thehardsphere and square-well diameters. The interaction energy between each pair of non-bonded coarse-grainedsites was calculated using the Boltzmann inversion scheme. The relative stiffness of each lipid ismaintained by imposing pseudobonds, which limit the bond lengths to fluctuateby the same amount observed in the united atom simulations. In the model each coarse-grained sitehas its own realistic mass. Simulationresults show that the model accurately reproduces structural properties of theDPPC bilayer including the area per lipid, bilayer thickness, bond order andmass density profiles. Currentwork is focused on extending the model to lipid mixtures and simulating thephase separation that occurs as a bilayer composed of DPPC and 1,2-dipalmitoyl-*sn*-glycero-3-phosphate (PA) changes froman initially-homogeneous state (at neutral pH) to a phase-separated state atlow pH.

See more of this Session: Multiscale Modeling II
See more of this Group/Topical: Computational Molecular Science and Engineering Forum

(104d) A Coarse Grained Model of Inhibitors to Scavenger Receptor Uptake to LDL

Michael Tomasini and M. Silvina Tomassone, Chemical and Biochemical Engineering, Rutgers University, Piscataway, NJ

The buildup and oxidation of low-density lipoproteins (LDL) in the walls of arterial blood vessels results in atherosclerotic lesions and arterial damage. The immune response of this arterial damage is to recruit macrophages expressing scavenger receptors specific for the uptake of oxidized LDL. The result is the formation of foam cells and fatty streaks, the first signs of atherosclerotic plaques. One possible method to combat the formation of atherosclerotic plaques is through the use of inhibitors which prevent the uptake of oxidized LDL by the scavenger receptors. Recently, Tian et al. 2004 [1] developed a class of molecules termed amphiphilic scorpion-like macromolecules (AScM) comprised of a mucic acid head group with attached aliphatic chains and a long polyethylene glycol (PEG) tail used specifically to block uptake of oxidized LDL by scavenger receptors. In this work, we use computational simulations to explore the structure and interactions of AscM micelles with the specific scavenger receptor – scavenger receptor A (SR-A) in order to shed light on an optimal inhibitor of oxidized LDL uptake.

In our simulations, we use a coarse-grained model based on the MARTINI force field [2] to describe the AScM molecules as well as the collagenous portion of SR-A where binding is assumed to take place [3]. As the interaction of oxidized LDL is thought to be mediated by electrostatic interactions, we vary the number (0, 1, or 2) of carboxylate groups at different positions on the AScM (near the mucic acid head or at the PEG tail) to determine the effect on both AScM micellation (which it though necessary for SR-A binding) and AScM association with SR-A to determine the best inhibitor.

References:

[1] Tian L, Yam L, Zhou N, Tat H, and Uhrich KE. Amphiphilic Scorpion-like Macromolecules: Design, Synthesis, and Characterization. *Macromolecules*. 2004; **37**: 538-543.

[2] Monticelli L, Kandasamy SK, Periole X, Larson RG, Tieleman DP, and Marrink SJ. The MARTINI Coarse-Grained Force Field: Extension to Proteins. *J. Chem. Theory and Comput.* 2008; **4**(5): 819-834.

[3] Platt N and Gordon SJ. Is the Class A Macrophage Scavenger Receptor (SR-A) Multifunctional? – The Mouse's Tale. *J. Clin. Invest.* 2001; **108**(5): 649-654.

See more of this Session: Multiscale Modeling II
See more of this Group/Topical: Computational Molecular Science and Engineering Forum

(104e) Multiscale Molecular Modeling of Fullerol-Dendrimer Complexes

Seung Ha Kim and Monica H. Lamm, Department of Chemical and Biological Engineering, Iowa State University, Ames, IA

Dendrimers are large, synthetic macromolecules that have regular and highly branched architectures. The multivalency of dendrimers permits them to bind with other species and makes them attractive platforms for a wide range of applications, including medicine, surface science, catalysis, and water treatment. To better understand the physical mechanisms of dendrimer-guest assembly, we have developed a multiscale molecular model for simulating flexible macromolecules in solution, based on the force matching method for coarse graining. We illustrate this approach by simulating the complexation between poly(amidoamine) (PAMAM) dendrimers and $C_{60}(OH)_{20}$ (fullerol) in aqueous solution. Specifically, we systematically examine the effects of concentration, and of dendrimer molecular weight, on the resulting PAMAM-fullerol complexes and compare our model predictions to recent experimental measurements.

See more of this Session: Multiscale Modeling II
See more of this Group/Topical: Computational Molecular Science and Engineering Forum

(104f) Multiscale Modeling of Perfluoropolyether Lubricants with Functional Endgroups

Robert Smith, Pil Seung Chung, Lorenz T. Biegler and Myung S. Jhon, Chemical Engineering, Carnegie Mellon University, Pittsburgh, PA

Physical properties of nanoscale perfluoropolyether (PFPE) films, which are generally used as a perfluoro linear co-polymer lubricant in hard disk drives (HDDs) has been previously studied in disparate molecular/mesoscale simulations. The PFPE molecules must exhibit strong adhesion to the carbon overcoat surface as well as rapid self-healing properties; therefore, a fundamental understanding of nanoscale characteristics of lubricant structure which contribute to desired macroscale performance must be established for successful nanotechnology design.

Due to the hydroxyl functional groups at each chain end of well-known PFPE (e.g., Zdol and Ztetraol), PFPE on an amorphous carbon surface shows peculiar characteristics (i.e., layering structure and entanglement on the surface). Via coarse-grained (CG) molecular dynamics (MD), we qualitatively validated molecular scale phenomena related to nanostructured conformation and dynamic behavior of molecularly thin lubricant film [1, 2]. To validate this coarse-grained model, a procedure in multi-scale modeling [3] which links atomic scale phenomena to CG MD is established thereby providing a form of synergy between the atomistic and molecular scales. Accurate intra / intermolecular force field parameters for PFPE and carbon surfaces are implemented in CG MD to enable improvements in our molecular scale model.

In this study, we use *ab initio* methods to determine intramolecular force field parameters for PFPE molecules and their intermolecular interaction with other PFPEs as well as with various carbon overcoat surfaces. Calculations based on *ab initio* methods are used to determine the intramolecular (stretching, bending, and torsional) force field parameters via solving the *ab initio* Hessian matrix eigenvalue problem. The intermolecular interaction between endgroups in pure and blended PFPEs is quantified using the supermolecular approach to molecular interaction. In addition, we explore novel carbon overcoat technology in HDD by calculating PFPE interaction with graphene. The implementation of atomically thin graphene as a carbon overcoat has the potential for a factor of two improvement in areal recording density. We compare the PFPE interaction with the traditional carbon overcoat versus graphene. We also evaluate PFPE-PFPE and PFPE-carbon surface geometry as a function of the PFPE structures particularly for new-generation lubricants including DDPA. The force field is evaluated by comparing interaction energies and molecular geometries of equilibrium configurations obtained using *ab initio* calculations to those obtained via the classical force field analysis.

[1] Q. Guo, L. Li, Y.-T. Hsia, and M.S. Jhon, "A Spreading Study of Lubricant Film via Optical Surface Analyzer and Molecular Dynamics", *IEEE Trans. Magn.* 42(10), 2528-2530 (2006).

[2] P.S. Chung, H. Chen, and M.S. Jhon, "Molecular dynamics simulation of binary mixture lubricant films", *J. Appl. Phys.*, 103, 07F526 (2008).

[3] D. Kim, P.S. Chung, P. Jain, S.H. Vemuri, and M.S. Jhon, "Multiscale modeling of head disk interface", IEEE Trans. Magn., 46, 2401-2404 (2010).

See more of this Session: Multiscale Modeling II
See more of this Group/Topical: Computational Molecular Science and Engineering Forum

(104g) Nanoparticles At the Water-Decane Interface: Evidence of Emergent Behavior From Equilibrium Multi-Scale Simulations

Heng Fab, School of Chemical, Biological and Materials Engineering, The University of Oklahoma, Norman, OK and Alberto Striolo, School of Chemical Biological and Materials Engineering, The University of Oklahoma, Norman, OK

Pickering emulsions (emulsions stabilized by solid nanoparticles rather than by surfactants) are finding practical applications in several areas, including cosmetics, drug-delivery, and even in-situ upgrade of bio-oil to fungible bio-fuels [Science 327 (2010) 68]. It is known that solid particles adsorb at water-oil interfaces to reduce the contact area between the two immiscible phases. Stable emulsions are obtained when the particles strongly adsorb at the interfaces. We will report here equilibrium molecular dynamics simulation results employed to differentiate the behavior of silica-based nanoparticles used to stabilize either water-in-decane or decane-water emulsions. We will compare nanoparticles with similar surface properties, but different distribution of the surface groups. Particular attention will be given to Janus as opposed to homogeneous nanoparticles [Langmuir 27 (2011) 5264-5274]. The all-atom simulation results have been used to parameterize dissipative particle dynamics simulations. Using the latter tool, we have obtained interesting morphologies depending on nanoparticle concentration, and water-decane molar ratio in the various simulated systems. The simulation results will be discussed in relation to available experimental data.

See more of this Session: Multiscale Modeling II
See more of this Group/Topical: Computational Molecular Science and Engineering Forum

(108a) Dealloyed PtxNi1-x Thin Films for Oxygen Reduction Reaction

Carlos Hangarter[1], Yihua Liu[2], Vladimir Oleshko[2], Leonid Berdersky[2] and Thomas Moffat[2], (1) Metallurgy Division, NIST, Gaithersburg, MD, (2)Metallurgy, NIST, Gaithersburg, MD

Pt cost is a continuing challenge for widespread implementation of PEMFCs, consequently improving durability and increasing Pt mass activity of PEMFCs is essential. Recently, epitaxial Pt_3Ni (111) with a Pt surface and Ni rich subsurface has demonstrated drastic activity enhancements attributed to reduced interaction between Pt and oxidized species resulting from a shift in the d-band structure. This work has motivated low cost efforts to produce high surface area analogs of the alloy which we have explored with electrodeposited Pt_xNi_{1-x} thin films.

In this work nanoporous PtNi thin films over a range of compositions were synthesized by electrodeposition and subsequent dealloying. High mass activity (0.8A/mg) for oxygen reduction reaction was demonstrated by as-deposited $PtNi_3$ with a dealloyed composition of Pt_3Ni. The composition and structure of as deposited and dealloyed Pt_xNi_{1-x} thin films were probed by hydrogen underpotential deposition, electrochemical quartz crystal microbalance, x-ray diffraction, and scanning electron microscopy. The dealloying process was shown to be highly dependent on the initial film, creating trends in composition, porosity, shrinkage, specific surface area, and activity. Select films were further analyzed by small angle x-ray diffraction and transmission electron microscopy to reconcile these differences with structural information and elucidate feature sizes (4-5nm).

See more of this Session: Nanoscale Materials as Catalysts II
See more of this Group/Topical: Catalysis and Reaction Engineering Division

(108b) Autoreduction Kinetics of a Sn-Pt Complex to Pt Metal for the Formation of Pt Nanoparticles

Samuel St. John, Chemical Engineering, University of Cincinnati, Cincinnati, OH and Anastasios Angelopoulos, Chemical and Materials Engineering, University of Cincinnati, Cincinnati, OH

We have recently demonstrated continuous control over Pt nanoparticle size across the cluster-to-crystal transition region (ca. 0.5 – 2.7 nm) using a colloidal synthesis approach that involves autoreduction of a Sn-Pt complex in hydrochloric acid solution. This approach yields particle size distributions less than 10% and permits fine atomic structure contrast in this critical size range. Such uniform structures are needed to characterize the geometric and electronic configurations of nanomaterials used in catalysis and fields like nanoelectronics and optics. Despite this need, the underlying mechanisms leading to nanocrystal formation are poorly understood and often unexplored. Recent attempts at systematic studies to investigate size-controlled NP structures have had limited success especially with particle sizes less than about 1.5 nm leaving size effects occurring at < 1 nm unexplored. This size region is interesting because significant changes from bulk properties can occur here. We have previously demonstrated a 3-fold improvement in the electrocatalytic oxygen reduction activity versus control catalyst for 1.7 nm diameter particles that exhibit densely packed, yet non-crystallographic arrangements of atoms. In the present paper, we describe the kinetics of Sn-Pt autoreduction during colloidal synthesis and the formation of Pt metal using UV-Vis spectroscopy by determining and following spectroscopic features at several wavelengths. These features are used to monitor the generation of Pt-Sn complexes formed when Pt(IV) and Sn(II) are mixed. Broad features extending into the visible spectrum, ca 600 nm, characterize the appearance of the Pt metal. These broad features are induced by the photoexcitation and oscillations of the free electrons in the conduction band that occupy energy states near the Fermi level. Using this information, we propose a mechanism for the autoreduction synthesis scheme that yields precise particle size control. TEM, FFT, and EXAFS are used to characterize the particles observed following reaction completion. Understanding the mechanism of nanoparticle growth and stabilization for Pt will allow us to extend this technique to the other noble metals for both catalytic and non-catalytic applications.

See more of this Session: Nanoscale Materials as Catalysts II
See more of this Group/Topical: Catalysis and Reaction Engineering Division

(108c) Decomposition of Isopropanol As a Probe for Acidic and Basic Sites On Graphite Nanofibers

Randy D. Weinstein, Department of Chemical Engineering, Villanova University, Villanova, PA, Andrew R. Ferens, Department of Chemical Engineering, Villanova University, Villanova, Pa, PA and Robert Giuliano, Chemistry, Villanova University, Villanova, PA

Three types of as-produced and treated graphite nanofibers (GNFs) were used as catalysts in the decomposition of isopropanol to propene and acetone in the presence of oxygen to evaluate the surface chemistry of the fibers. As-produced herringbone fibers were found to produce higher selectivity for propene compared to the as-produced platelet and ribbon fibers at all temperatures explored. Herringbone fibers that had undergone oxidative treatment with nitric acid, phosphoric acid, ruthenium tetroxide or potassium permanganate were also evaluated at a single temperature of 290 °C. BET surface analysis and SEM/EDS were used to assess the effects of oxidation on fiber structure. Selectivity for acetone dehydrogenation product or propene dehydration product could be achieved by the appropriate surface treatment. Nitric acid was the mildest treatment and the treated fibers showed minimal changes. Phosphoric acid treated fibers were found to produce very high conversions and almost pure selectivity for propene. Ruthenium tetroxide did not appear to have a large affect on fiber morphology; however, selectivity for acetone was much higher when GNFs treated with RuO_4 were used as the catalyst. Potassium permanganate was a very harsh treatment that almost completely degraded fiber structure, creating amorphous carbon.

See more of this Session: Nanoscale Materials as Catalysts II
See more of this Group/Topical: Catalysis and Reaction Engineering Division

(108d) Thermal Stability of Nickel-Silica Core-Shell Nanocatalysts

Lu Zhang Whaley, Chemical Engineering Department, University of Pittsburgh, Pittsburgh, PA, Fan Shi, Material Science, National Energy Technology Lab., Pittsburgh, PA and **Gtz Veser**, Chemical and Petroleum Engineering, University of Pittsburgh, Pittsburgh, PA

One of the biggest challenges in the technical application of nanocatalytsts is deactivation due to sintering. While the increasing specific surface area with decreasing particle diameter results in higher catalytic activity, it also increases the thermodynamic driving force for sintering, i.e. the loss of active surface area and hence loss of catalytic activity with time due agglomeration of metal crystallites. To overcome this dilemma, our group has in recent years investigated different approaches towards stabilizing metal nanoparticles via embedding or encapsulation in nanostructured matrices, or via alloying of the active metal phase.

Here, we are presenting the results of a study in which Ni@SiO_2 core-shell materials were systematically investigated to identify structure-stability correlations for this type of metal-oxide core-shell materials. The Ni@SiO_2 materials were synthesized in a straightforward one-pot reverse-microemulsion templated approach. By varying synthesis parameters, we were able to synthesize not only Ni@SiO_2 with carefully controlled metal nanoparticle sizes (~1-7nm), but also to synthesize core-shell nanostructures with a pronounced central cavity (~10 nm). These "nanobubble" materials contain sub-nanometer Ni clusters which decorate the inner wall of the (porous) silica shell. The cavity leaves the surface of the Ni clusters much more accessible than in typical core-shell structures, and is expected to impact the thermal stability of the Ni NPs as it restricts their mobility much less than for the embedded Ni@SiO_2 core-shell materials. The materials were characterized by HRTEM, BET, XRD, TPR, and H_2-chemisorption, and the thermal stability of the different nanocomposite materials were investigated systematically as function of structure and metal NP size over a temperature range from 500°C to 1000°C.

Silica porosity and pore size distribution decreased significantly with increasing calcination temperature above ~800°C, indicating a softening and restructuring of the silica matrix. More significantly, in typical core-shell structures, Ni sintering occurs within the individual silica particles via a transition from multiple Ni cores to a single, larger Ni core, followed by the diffusion of Ni particles out of the silica particles at ~1000°C. This stability is a function of the ratio of Ni NP size to silica pore diameter, where larger Ni NPs induce significant reconstruction of the silica matrix during diffusion. Surprisingly, the very small Ni nanoclusters in the silica "nanobubbles" show excellent stability up to 800°C, presumably due to stronger interaction with the silica support. This enhanced thermal stability of the "nanobubble" catalysts is confirmed in catalytic CO methanation compared to core-shell catalysts as well as conventional Ni/silica catalyst.

See more of this Session: Nanoscale Materials as Catalysts II
See more of this Group/Topical: Catalysis and Reaction Engineering Division

(108e) In-Situ FTIR Spectroscopy to Study the Size Effect of Cobalt Nanoparticles On Carbon Monoxide Oxidation

Bijith Mankidy, Chemical & Biomedical Engineering, University of South Florida, Tampa, FL, Babu Joseph, Chemical and Biomedical Engineering, University of South Florida, Tampa, FL and Vinay Gupta, Chemical Engineering, University of South Florida, Tampa, FL

Catalytic oxidation of carbon monoxide is a topic of significant interest due to its scientific and industrial importance. CO oxidation becomes important in air purification, lowering the toxicity of automobile exhausts, during selective oxidation of the feed stream to a fuel cell, and for several other potential applications. Noble metals are highly active catalysts that are moisture tolerant and stable against deactivation. However, the popularity of these catalysts is affected by their higher cost and limited availability. Recently, CO oxidation using cobalt has been the subject of attention due to the relatively lower cost of cobalt compared to the noble metal counterparts. A systematic understanding of the catalytic activity of cobalt metal nanoparticles, especially the role of the nanoparticle size, has yet to be established. Here, we use a temperature programmed method with an in-situ FTIR spectroscopic characterization to systematically study nanoparticle size effect on catalytic CO oxidation. For a well-defined and model catalyst, stable cobalt nanoparticles of sizes 1 to 14nm were synthesized by thermal decomposition technique in the presence of a surfactant. The nanoparticles were then immobilized on surface tailored SiO_2 colloidal supports. The experimental data was interpreted using kinetic modeling, which allowed a determination of the reactivity as well as the kinetic parameters such as pre-exponential factor and activation energy alongside the impact of nanoparticle size.

See more of this Session: Nanoscale Materials as Catalysts II
See more of this Group/Topical: Catalysis and Reaction Engineering Division

Na-promoted Pt catalysts in core-shell structure for the low-temperature Water-Gas-Shift reaction

***Yuan Wang**, Danny Pierre, Yanping Zhai, Maria Flytzani-Stephanopoulos**

Department of Chemical and Biological Engineering, Tufts University, Medford, MA

Abstract

The water-gas-shift (WGS) reaction is a key step in high-purity hydrogen production. Platinum-based catalysts are active over a wide temperature range and can be stabilized on certain supports, such as ceria and zirconia during WGS operation cycles. Thus, Pt is a good albeit much costlier alternative to Cu-ZnO catalysts for the demanding application to fuel cell systems. The activity of Pt on reducible ceria for low-temperature applications has been attributed to the interaction of Pt with the support and nanoscale ceria with a lot of surface oxygen defects is superior to high-temperature annealed ceria of large crystal size and low defect concentration [1, 2]. Addition of alkali was recently reported to promote the WGS activity of Pt on reducible oxide supports [3], and very recently our group reported that alkali promotion of Pt is possible even on an inert oxide surface such as silica [4]. Moreover, a very small amount of oxidized Pt, atomically dispersed and stabilized by the alkali, Na-O-Pt$(OH)_x$, appeared to contain the active sites. From the practical viewpoint, this is important as it allows the choice of a low-cost support, e.g. silica or alumina for low-temperature shift applications. It also allows one to design a catalyst with potentially trace amounts of the costly Pt metal. From a fundamental viewpoint, understanding the promotion would shed light on the mechanistic aspects and allow rational designs of new WGS catalysts using very small amounts of Pt.

In this work, we investigated the water-gas shift reaction on a differently designed catalyst, comprising a core-shell structure, Pt@SiO_2 and evaluated its activity and stability and compared it to Pt prepared by impregnation on silica surfaces.

The core-shell structured catalyst Pt@SiO_2 was synthesized by the reverse microemulsion (ME) method [5]. Alkali ions were added during the preparation or in a second step by impregnation. In addition to the Pt core, we found that platinum is atomically dispersed throughout the silica shell and sodium ions stabilize the atomically dispersed platinum species in the support. Using XPS, we found that the binding energy of Pt shifts to higher values, and oxidized Pt species were found on the Pt-

Na@SiO_2 sample in addition to metallic Pt. Fig.1 shows steady-state CO conversion data for various Pt@SiO_2 samples. Only the samples containing Na show good low-temperature activity. The surface hydroxyl groups are activated by CO at low temperatures (~150°C) as found by temperature-programmed reaction (TPR) using CO as the reductant (Fig. 2). Cyclic CO-TPR tests show that water activation and hydroxyl regeneration take place on the Na-promoted Pt sites. By contrast, surface hydroxyl groups cannot be recovered on the sodium-free sample after reaction with CO.

In conclusion, the alkali species on the Pt@SiO_2 catalysts stabilize the atomically dispersed Pt and regenerate surface hydroxyl groups. An oxidized Pt-OH_x species associated with alkali ions is the active site for the low-temperature WGS reaction, as was already reported for the open Pt/SiO_2 catalysts [4]. The stability of encapsulated species, however, may be superior in highly reducing fuel gas streams and higher temperatures.

Acknowledgment

The financial support of this work by the DOE/BES (#DE-FG02-05ER15730) and the National Science Foundation (CBET Grant No. 0828666) is gratefully acknowledged.

References

1. Q. Fu, H. Saltsburg, M. Flytzani-Stephanopoulos, "Active nonmetallic Au and Pt species on ceria-based water-gas shift catalysts." *Science* **301** 935 (2003).
2. D. Pierre, W. Deng, M. Flytzani-Stephanopoulos, "The importance of strongly bound Pt–CeOx species for the water-gas shift reaction: Catalyst activity and stability evaluation." *Top. Catal.* **46** 363 (2007).
3. P. Panagiotopoulou, J. Papavasiliou, G. Avgouropoulos, T. Ioannides, D.I. Kondarides, "Water-gas shift activity of doped Pt/CeO_2 catalysts." *Chem. Eng. J.* **134** 16 (2007).
4. Y. Zhai, D. Pierre, R. Si, W. Deng, P. Ferrin, A.U. Nilekar, G. Peng, J.A. Herron, D.C. Bell, H. Saltsburg, M. Mavrikakis, M. Flytzani-Stephanopoulos, "Alkali-Stabilized Pt-OH_x Species Catalyze Low-Temperature Water-Gas Shift Reactions." *Science* **329** 1633 (2010).
5. C.M.Y. Yeung, K.M.K. Yu, Q.J. Fu, D. Thompsett, M.I. Petch, S.C. Tsang, "Engineering Pt in ceria for a maximum metal-support interaction in catalysis." *J. Am. Chem. Soc.* **127** 18010 (2005).

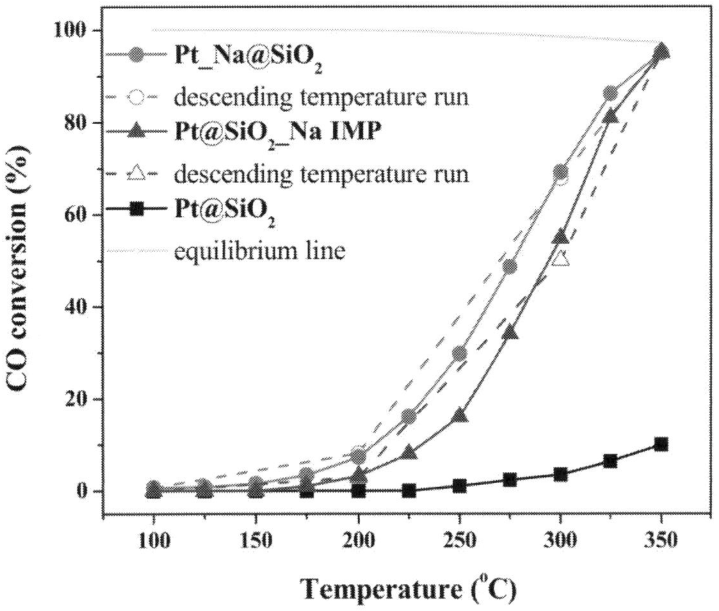

Figure 1. WGS light-off profiles (steady-state data) of Pt@SiO_2 catalysts ($2\%CO-10\%H_2O-He$, 70ml/min, contact time = $0.09 g\cdot s/cm^3$)

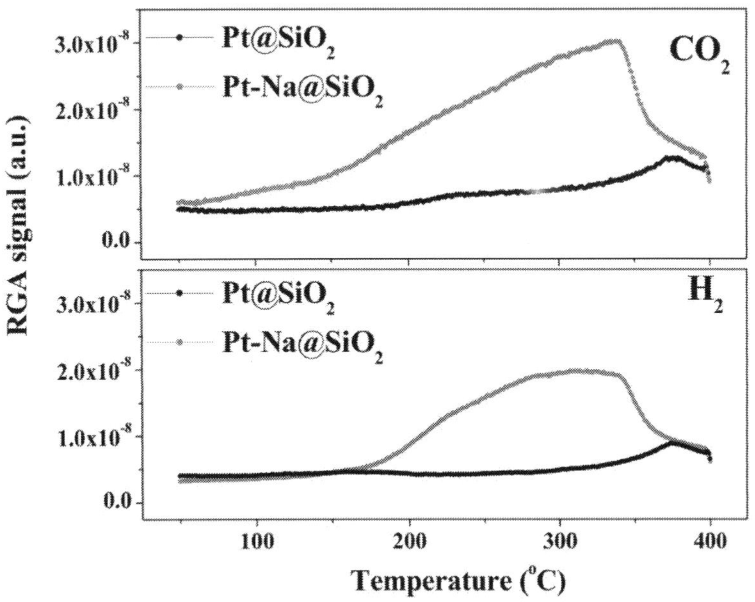

Figure 2. CO-TPR of Pt@SiO_2 catalysts (10%CO-He, 20ml/min, 5°C/min)

Twin Flame Spray Pyrolysis Production of Pt/K/Al_2O_3 NO_x Storage-Reduction Catalysts

Robert Büchel[a,b], Alfons Baiker[b] and Sotiris E. Pratsinis[a]

[a]*Particle Technology Laboratory, Department of Mechanical and Process Engineering, ETH Zurich, Sonneggstrasse 3, CH-8092 Zürich Switzerland.*

[b]*Institute for Chemical and Bioengineering, Department of Chemistry and Applied Biosciences, ETH Zurich, Wolfgang-Pauli-Strasse 10, CH-8093 Zurich, Switzerland.*

Abstract

High surface area Pt/K/Al_2O_3 catalysts were prepared with a 2-nozzle flame spray method resulting in Pt clusters on γ-Al_2O_3 and amorphous K storage material as evidenced by Raman spectroscopy. The powders had a high NO_x storage capacity and were regenerated fast in a model exhaust gas environment. From 300 to 400 °C no excess NO_x was detected in the off gas during transition form fuel lean to fuel rich conditions, resulting in a highly effective NO_x removal performance. Above 500 °C, the NSR activity was lost and not recovered at lower temperatures as K-compounds were partially crystallized on the catalyst.

Introduction

Stricter limits for exhaust emission motivate the development of new catalysts especially for NO_x removal under oxygen rich conditions as encountered in lean burn and direct injection engines. NO_x storage-reduction (NSR) catalysts [1] can trap exhaust NO_x under fuel lean conditions on an alkali or alkaline-earth metal in the form of metal-nitrates. The alkalinity determines the NO_x trap performance in the order: K > Ba >Sr ≥ Na > Ca > Li [2]. Regeneration of the NO_x trap occurs under fuel rich conditions where metal-nitrates are decomposed and the released NO_x is reduced to nitrogen [3]. Potassium- and barium-containing compounds have been studied extensively for their NO_x storage capacity [4]. The best performance of K as storage material, especially at high temperatures, is attributed to its stable nitrates [5]. Furthermore K_2CO_3 is less toxic and cheaper than $BaCO_3$. The main drawback of K is its low resistance to sulfur poisoning whereby regeneration is possible at 700 °C [6]. Catalysts for NSR are typically operated around 400 °C, however, their performance at low temperature is also important. For example, during certification of light duty vehicles, the exhaust stream is most of the time at 200 – 250 °C [7].

Here novel flame-made catalysts are explored with high NO_x removal efficiency as they have inherent advantages compared to wet-prepared ones [8]. More specifically amorphous K/Pt/Al_2O_3 was made by a 2-nozzle FSP setup [9] which allowed simultaneous but separate synthesis of the Al_2O_3 support and K-storage component along with Pt clusters [10] and even preferential Pt deposition on Al- or K-components [11].

Results

With a two-nozzle Flame Spray Pyrolysis unit (FSP) a variety of material combinations have been made. The advantage of having two flames is that each flame can individually produce particles of different composition. The particle-laden streams then mix continuously to form the final product, as shown in **Fig. 1**. Like this, the formation of mixed oxides can be avoided and the noble metal can be selectively deposited e.g when adding noble metal to the Ba precursor (**Fig. 1** left) selective deposition of noble metal on K_2CO_3 is made. When the precursor is mixed with the Al- precursor (**Fig. 1** right) selective deposition on Al_2O_3 is made. The main advantage of using FSP to make such catalysts is that one specific property, e.g. preferential Pt deposition, can easily be changed without (or just marginally) altering other properties of the catalyst.

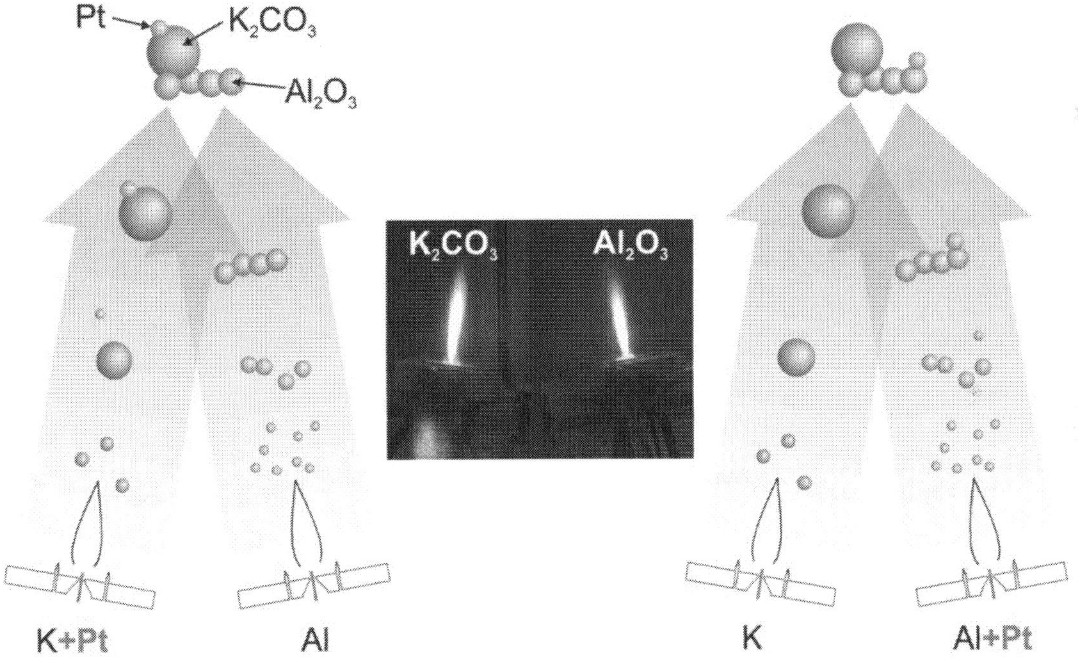

Figure 1: Schematics and photography of the two nozzle FSP setup. Each flame produces individual particles of K_2CO_3 and Al_2O_3, respectively. By adding Pt to the K-containing precursor (left side) preferential Pt deposition on K_2CO_3 was made. Adding Pt to the Al-containing precursor (right side) resulted in preferential Pt deposition on Al_2O_3.

As shown in **Fig. 2**, flame synthesis of $Pt/K/Al_2O_3$ resulted in well mixed powders with crystalline γ-Al_2O_3. As the Pt clusters are smaller than 5nm and less than 3% they are below the detection limit, as well as the amorphous K_2CO_3.

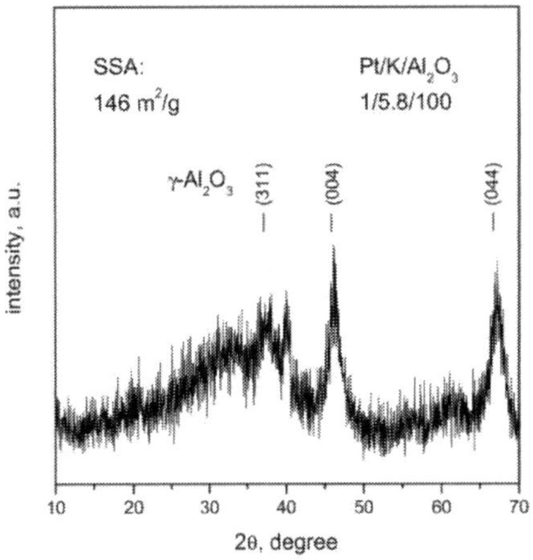

Figure 2: XRD and BET of $Pt/K/Al_2O_3$ catalysts. The amorphous K_2CO_3 cannot be detected with XRD.

The NO_x conversions are shown in **Fig. 3** being >80% in the optimal temperature range of 300 – 400 °C. This high performance could be reached by fast NO_x uptake during the fuel lean and fast regeneration during the fuel rich phase. Additionally the typical overshooting of the NO_x signal during the switch from fuel lean to fuel rich gases was significantly decreased and never exceed the inlet NO_x concentration.

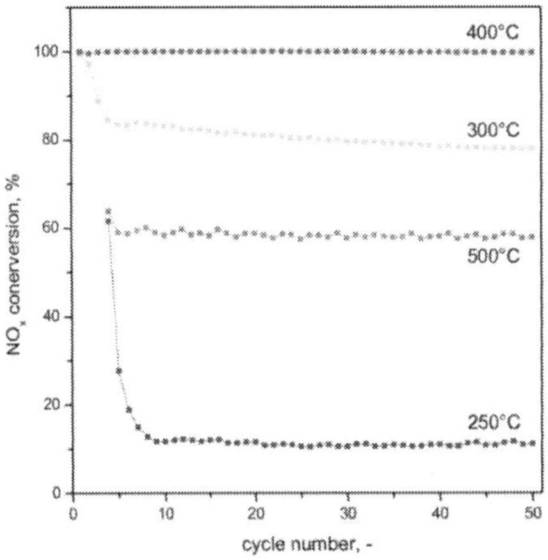

Figure 3: The NO_x conversion at different operation temperatures.
This superior performance was attributed to good K distribution in the catalysts and the amorphous nature of the K species. High temperatures, however, alter the catalyst structure which lowers the catalyst performance, an issue that requires further investigation.

Acknowledgements

We thank for the financial support by ETH Zürich (TH-09 06-2) and the contribution of platinum chemicals by Johnson Matthey PLC is greatly appreciated.

References

[1] N. Takahashi, H. Shinjoh, T. Iijima, T. Suzuki, K. Yamazaki, K. Yokota, H. Suzuki, N. Miyoshi, S. Matsumoto, T. Tanizawa, T. Tanaka, S. Tateishi, K. Kasahara, Catal. Today 27 (1996) 63-69.

[2] T. Kobayashi, T. Yamada, K. Kayano, SAE Tech. Paper 970745 (1997)

[3] A. Fritz, V. Pitchon, Appl. Catal. B 13 (1997) 1-25.

[4] M. Takeuchi, S. Matsumoto, Top. Catal. 28 (2004) 151-156.

[5] L. Olsson, D. Monroe, R.J. Blint, Ind. Eng. Chem. Res. 45 (2006) 8883-8890.

[6] N. Takahashi, A. Suda, I. Hachisuka, M. Sugiura, H. Sobukawa, H. Shinjoh, Appl. Catal., B 72 (2007) 187-195.

[7] H.L. Fang, S.C. Huang, R.C. Yu, C.Z. Wan, K. Howden, SAE Tech. Paper 2002-01-2889 (2002)

[8] R. Strobel, A. Baiker, S.E. Pratsinis, Adv. Powder Technol. 17 (2006) 457-480.

[9] R. Strobel, L. Madler, M. Piacentini, M. Maciejewski, A. Baiker, S.E. Pratsinis, Chem. Mater. 18 (2006) 2532-2537.

[10] R. Büchel, R. Strobel, A. Baiker, S.E. Pratsinis, Top. Catal. 10.1007/s11244-009-9334-x (2009)

[11] R. Büchel, R. Strobel, F. Krumeich, A. Baiker, S.E. Pratsinis, J. Catal. 261 (2009) 201-207.

(111b) A Study On Sulfonated Cross-Linked Chitosan Resin As Solid Acid Catalyst

Yuting Xu and Lichun Dong, Chemical Engineering, Chongqing University, Chongqing, China

Strong acid resins, e.g. Nafion and sulfonated polystyrene are very effective solid acidic catalysts in esterification reaction, ester exchange reaction, etherification reaction, etc., while the high cost limits their extensive application. Sulfonated chitosan has a very similar nature with Nafion and chitosan has the advantages in natural reserves and price, consequently, a lot of efforts have been tried on sulfoanted chitosan as an alternative strong acid resin. However, the thermostability and acidity of sulfonated chitosan prepared by the direct sulfonation of chitosan are not satisfactory. In this study, chitosan was firstly cross-linked by glutaraldehyde before being sulfonated to improve its thermal and acid stability; Moreover, silica particles were added in the molding process to enhance its acidity. The cross-linked chitosan sulfate resin shows good stability and catalytic activity. The DTA-TGA analysis reveals that the effective temperature of the resin is up to 190 °C, at which most organic esterification or etherification reactions can occurs smoothly. The sulfonation ratio of the cross-linked chitosan is about 35% and the retention rate of the sulfonic groups is more than 75% under the endurable temperature for 5h. The addition of silica can increase the sulfonation rate of cross-linked chitosan and improve the stability of the sulfonic groups. The BET specific surface of the resin is around 730m2/g, its sulfonic groups can be fully used since the encapsulation of sulfonic groups can be avoided. Further studies shows that cross-linked chitosan sulfate resin can be used as acidic catalyst for esterification reactions such as synthetizing ethyl nipagin ester and butyl acetate. The esterification ratio can reach up to 96% and the resin can be repeatedly used for more than 5 times.

See more of this Session: Novel Catalytic Materials - II
See more of this Group/Topical: Catalysis and Reaction Engineering Division

(111c) Carbon-Nanotubes/Metal Oxide Hybrid Catalysts As Novel Reactive-Separation System In Water-Oil Emulsions

Jimmy A. Faria, M. Pilar Ruiz and Daniel E. Resasco, School of Chemical, Biological and Materials Engineering, University of Oklahoma, Norman, OK

Catalysis at water/oil interfaces in emulsions stabilized by amphiphilic nanohybrids (carbon nanostructures fused to oxides particles) has been proposed as a novel technology with promising applications, such as bio-oil upgrading and production of specialty chemicals [1,2]. Advantages of using interfacial catalysts in emulsions without surfactants include enhanced mass transfer, easier recovery of the amphiphilic catalyst nanoparticles, enhanced "phase selectivity" (i.e. selective conversion in each phase), and direct partitioning and separation of products, which leads to substantial simplification of the purification stages. Fine control of the hydrophilic/lipophilic balance of these nanohybrids allows us to systematically change type of emulsion, droplet size, and volume fraction of the emulsion [3]. Furthermore, it is possible to control the reactions taking place in each of the phases by selectively depositing active species on the nanohybrids [1,2,4], as well as tailoring metal particle size distributions by grafting functional groups on the carbon structures [2]. The various reactions studied in the biphasic system include aldol-condensation of oxygenates, dehydration of sugars, and hydrogenation of phenolics.

References

1. Crossley, S., Faria, F., Shen, M., and Resasco, D.E. Science 1, 68 (2010).

2. Ruiz, M.P., Faria, J., Shen, M., Drexler, S., Prasomsri, T., and Resasco, D.E. ChemSusChem, In press (2010).

3. Shen, M., and Resasco, D.E. Langmuir 25, 10843 (2009).

4. Faria, J., Ruiz, M.P., and Resasco, D.E. Adv. Synth. Catal. 352, 2359 (2010).

See more of this Session: Novel Catalytic Materials - II
See more of this Group/Topical: Catalysis and Reaction Engineering Division

(111d) Hybrid Sulfonic Acid Catalysts Based On Silica -Supported Poly(Styrene Sulfonic Acid) Brush Materials and Their Application In Ester Hydrolysis

Christopher W. Jones, School of Chemical & Biomolecular Engineering, Georgia Institute of Technology, Atlanta, GA and Wei Long, School of Chemistry and Biochemistry, Georgia Institute of Technology, Atlanta, GA

Catalytic conversions involving water as reactant, product or solvent are of increased importance in biomass conversion into fuels and chemicals. In this context water tolerant solid acids are highly valued. Polymer-oxide hybrid materials based on non-porous silica-supported sulfonic acid-containing polymer brush materials are proposed here as a new class of potentially water-tolerant solid acid catalysis. Atom transfer radical polymerization (ATRP) using both an established and a new ATRP initiator that is designed to improve the hydrolytic stability of the catalyst, is used to create poly(styrene) brushes on the surface of fumed silica. These brushes are sulfonated to produce an acid catalyst akin to an acidic Merrifield resin, but with enhanced accessibility of the active sites. The catalysts are evaluated in the hydrolysis of ethyl lactate, with the polymer brush materials having the same activity as a homogeneous catalyst, p-toluene sulfonic acid, and being substantially more active than an acidic polymer resin (Amberlyst 15). The heterogeneous nature of the catalyst allows for easy catalyst recovery and recycle. The stability of the polymer brush catalysts depends on the nature of the initiator used, with the new alkyl-based initiator introduced here giving enhanced stability relative to the standard, ester-containing initiator that is most commonly used. The activity of the recycled polymer brush catalysts decreased slightly in each cycle due to both desulfonation and the gradual detachment of the polymer chains from the oxide support. Oxide-supported polymer brush materials are suggested to be a promising new architecture for hybrid catalyst materials.

See more of this Session: Novel Catalytic Materials - II
See more of this Group/Topical: Catalysis and Reaction Engineering Division

Oxidative Coupling of Methane Using Novel Catalytic Materials

Ranjita Ghose, Hyun Tae Hwang and Arvind Varma, School of Chemical Engineering, Purdue University, West Lafayette, IN

Since current processes for ethylene production are based on crude oil, there is interest in developing alternative methods. Towards this end, Oxidative Coupling of Methane (OCM) is an attractive opportunity. Methane is the main constituent of natural gas (typically > 95%), for which the reserves are estimated to far exceed those of crude oil. In prior studies, low selectivity and yield have prevented development of a practical catalyst. More recently, mixed metal oxides with complex structures (for example, mixed perovskite with general formula $A_xA'_{1-x}B_yB'_{1-y}O_3$) have generated interest [1]. Solution Combustion Synthesis (SCS) is a novel technique developed in our laboratory, which has previously been used for synthesis of such materials [2-4]. Our research focuses on experimental studies of OCM using nanostructured complex metal oxide catalysts synthesized by the SCS technique, involving systematic variation of the metal components and their ratios, and optimization of reaction conditions, including temperature, flow rates and feed composition, with the goal of maximizing selectivity and yield of ethylene.

References

1. Kondratenko, V. and Baerns, M., Oxidative coupling of methane, Chapter in *Handbook of Heterogeneous Catalysis*, Wiley-VCH Verlag, **2008**.
2. Erri, P., Dinka P. and Varma, A., *Chem. Eng. Sci.* **2006**, 61, 5328.
3. Deshpande, K., Mukasyan, A. and Varma, A., *J. Power Sources*, **2006**, 158, 60.
4. Erri, P. and Varma A., *Ind. Eng. Chem. Res.* **2007**, 46, 8597.

(117a) Aqueous-Phase Reforming of Bio-Oil Model Compounds Over Pt-Re/C

Jie Fu[1], Sikander H. Hakim[2] and Brent H. Shanks[1], (1)Department of Chemical and Biological Engineering, Iowa State University, Ames, IA, (2)Department of Chemical and Biological Engineering, University of Wisconsin-Madison, Madison, WI

While fast pyrolysis bio-oilis an inexpensive liquid fuel derived from lignocellulosic biomass, the qualityof bio-oil is poor partially due to its high content of oxygenated species,which included various acids, alcohols, aldehydes, ketones, sugars, guaiacols, syringolsfurans, and furfurals. Thesespecies can be divided into water soluble and insoluble compounds, with thewater soluble species coming primarily for the carbohydrate portion of thebiomass. Aqueous-phase reforming(APR) is a potential approach to convert these highly oxygenated compounds to $H_2$and mono-functional species that can be subsequently upgraded to hydrocarbons. However, a myriad of reactions willoccur when APR is applied to the soluble portion of bio-oil, so there is a needto perform model compound studies to better understand the reactionsystem. As a starting point, we conducted APR of ethanol over Pt-Re/C at250 °C. H_2, CO_2, CH_4, C_2H_6,acetaldehyde, ethyl ether, ethyl acetate, acetic acid were main products. Additionally, acetaldehyde, acetic acidand formic acid were employed as reactants for APR, to provide information on thereaction pathways over Pt-Re/C. Fluxanalysis on the ethanol reaction system was made to determine the reaction pathwayand relative rates (v_1-v_8) for each step, which is shownin Figure 1.

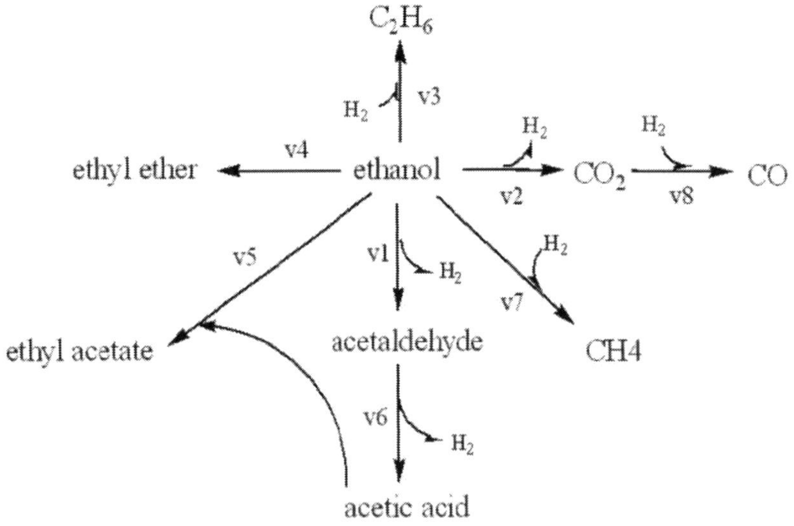

Figure1. The reaction pathwayfor aqueous phase reforming of ethanol over Pt-Re/C.

See more of this Session: Reaction Engineering for Biomass Conversion II
See more of this Group/Topical: Catalysis and Reaction Engineering Division

(117b) A Novel Experimental Technique for Study of Isothermal Pyrolysis of Cellulose

Matthew S. Mettler, Chemical Engineering, Univeristy of Delaware, Newark, DE, Dionisios G. Vlachos, Chemical Engineering, University of Delaware, Newark, DE and Paul Dauenhauer, Chemical Engineering, University of Massachusetts Amherst, Amherst, MA

Fast pyrolysis of biomass converts solid biopolymers into a valuable liquid product (bio-oil) which can be further upgraded to fuels or chemicals. Much effort has been dedicated to developing processes and catalysts which maximize efficiency, throughput and/or bio-oil quality (e.g., [1-2]). While these efforts are important to developing thermochemical biorefineries of the future, the complex chemistry by which the biomass feedstock breaks down is still largely unknown. Global reaction kinetics for the pyrolysis of cellulose have been published (e.g., [3]), but the accuracy of these models is debated [4]. Additionally, recent work has indicated both indirectly (e.g., [5]) and directly [6] that solid biomass pyrolysis proceeds through a high temperature intermediate liquid before volatilizing into the final bio-oil constituents. In order to maximize the accuracy of pyrolysis experiments and prevent device-dependent results, it is important that mass and heat transport rates within the solid biomass and intermediate liquid be faster than the intrinsic reaction kinetics. In this work, we demonstrate an experimental technique whereby biomass pyrolysis can be studied in a kinetically-limited manner. By reducing the size of and controlling shape of biomass, we can eliminate composition and thermal gradients. Additionally, volatiles, gases and char are all characterized in a single pyrolysis run with multi-dimensional analytical techniques. The mechanism of cellulose pyrolysis is investigated by comparing the product distributions of several short-chain cellulose oligomers to cellulose (degree of polymerization equal to 250). We find that oligomers have different product distributions compared to cellulose. Additionally, the production of furans (e.g., HMF, furfural) is maximized for cellobiose. Interestingly, a maximum in char is also observed at cellobiose, indicating that char and furan production are connected.

1. Vispute, T.P., et al., *Renewable Chemical Commodity Feedstocks from Integrated Catalytic Processing of Pyrolysis Oils.* Science, 2010. **330**(6008): p. 1222.

2. Agblevor, F.A., et al., *Fractional Catalytic Pyrolysis of Hybrid Poplar Wood.* Ind. Eng. Chem. Res., 2010. **49**(8): p. 3533.

3. Bradbury, A.G.W., et al., *KINETIC-MODEL FOR PYROLYSIS OF CELLULOSE.* Journal of Applied Polymer Science, 1979. **23**(11): p. 3271.

4. Varhegyi, G., et al., *IS THE BROIDO-SHAFIZADEH MODEL FOR CELLULOSE PYROLYSIS TRUE.* Energy & Fuels, 1994. **8**(6): p. 1345.

5. Boutin, O., et al., *Radiant flash pyrolysis of cellulose - Evidence for the formation of short life*

time intermediate liquid species. Journal of Analytical and Applied Pyrolysis, 1998. **47**(1): p. 13.

6. Dauenhauer, P.J., et al., *Reactive boiling of cellulose for integrated catalysis through an intermediate liquid.* Green Chemistry, 2009. **11**(10): p. 1555.

See more of this Session: Reaction Engineering for Biomass Conversion II
See more of this Group/Topical: Catalysis and Reaction Engineering Division

(117c) Determining Kinetics and Mechanisms of Model Compounds to Understand Chemistry of Algae Liquefaction In High Temperature Water

Shujauddin M. Changi and Dr. Phillip E. Savage, Chemical Engineering, University of Michigan, Ann Arbor, MI

Hydrothermal liquefaction refers to the process of converting biomass to bio-oil by contacting the biomass with water at high temperatures (>200°C) and sufficient pressures to maintain water in the liquid state. Hydrothermal liquefaction of algae producing bio-oil has the advantage of energy efficiency and capability of dealing with wet biomass[1-3]. Although few studies exist, the exact chemistry for bio-oil formation during algae liquefaction is not quite clear. In this project, we try to overcome this gap by studying ethyl oleate (triglyceride), phenylalanine (protein) and phytol (chlorophyll) as model compounds, representing different classes of biomacromolecules in algae. Our ultimate aim is to use the principles of reaction engineering and develop a unique model based on the understanding of these pure model compounds and their mixtures that is capable of predicting conversions given a starting composition of algae.

Recently, we determined the kinetics for ethyl oleate hydrolysis in high-temperature water and for the reverse reaction, oleic acid esterification, in near- and supercritical ethanol[4]. Hydrolysis was clearly autocatalytic. A phenomenological kinetics model is proposed using our experimental data that has thermodynamically and thermochemically consistent Arrhenius parameters. We also developed a unified mechanism for this system. We suggest that hydrolysis of ester is catalyzed by both H^+ and fatty acid, while esterification is only specific acid catalyzed. Both the phenomenological and the mechanistic model provide a good correlation of the data and also exhibit the ability to make quantitatively accurate predictions within and outside the parameter space investigated. This study, a first of its kind for hydrolysis and esterification kinetics in tandem, is especially useful for a multicomponent system that will occur during algae liquefaction. For instance, hydrolysis of proteins in algae can generate organic acids, in turn catalyzing the hydrolysis of triglycerides. Our model can be extended to understand and predict this type of behavior.

References

1) Brown,T.M.; Duan,P.; Savage,P.E. *Energy Fuels*, 24 (6) (2010) 3639-3646.

2) Dote,Y.; Sawayama,S.; Inoue,S.; Minowa,T.; Yokoyama,S. *Fuel*, 73 (1994) 1855 -1857

3) Inoue,S.; Dote,Y.; Sawayama,S.; Minowa,T.; Ogi,T.; Yokoyama,S. *Biomass and Bioenergy*, 6 (4) (1994) 269-274

4) Changi,S.; Pinnarat,T.; Savage,P. *Ind. Eng. Chem.Res,* 50 (2011) 3206-3211

See more of this Session: Reaction Engineering for Biomass Conversion II
See more of this Group/Topical: Catalysis and Reaction Engineering Division

(117d) Catalytic Studies of Reforming of Oxygenates

Sarah A. Tupy[1], Tushar Vispute[2], George W. Huber[2], Jingguang G. Chen[1] and Dionisios G. Vlachos[1], (1)Chemical Engineering, University of Delaware, Newark, DE, (2)Chemical Engineering, University of Massachusetts - Amherst, Amherst, MA

Aqueous phase catalytic reforming is a promising technology for the production of hydrogen from biomass feedstocks. Challenges arise from intrinsic properties of biomass processes such as low thermal stability of reactants and support stability. To rationally design catalysts for biomass conversion to fuels and chemicals, a fundamental understanding of surface processes must be linked to reactor scale performance. Previous studies under ultrahigh vacuum conditions have found that the reforming activity was higher for Ni/Pt than either monometallic Pt or Ni. The objective of this work is to correlate these findings with supported catalysts under aqueous phase reforming conditions. Using ethylene glycol as a model biomass compound, the aqueous phase reforming of ethylene glycol was studied on Pt/γ-Al_2O_3, Ni/γ-Al_2O_3, and NiPt /γ-Al_2O_3 in a packed bed reactor. Under the experimental conditions used, the Ni catalyst deactivates quickly. The bimetallic catalyst NiPt /γ-Al_2O_3 exhibited higher activity than the monometallic Pt catalyst, which is in agreement with literature. The catalysts have been characterized using chemisorption, transmission electron microscopy (TEM), and extended X-ray absorption fine structure spectroscopy (EXAFS).

See more of this Session: Reaction Engineering for Biomass Conversion II
See more of this Group/Topical: Catalysis and Reaction Engineering Division

(117e) Theoretical and Experimental Study of Glycolaldehyde Reaction Pathways On Ni/Pt(111) and Ni/WC Surfaces

Weiting Yu, Mark Barteau and Jingguang Chen, Chemical Engineering, University of Delaware, Newark, DE

Global energydemand growth and green house effect are two of the main drivers for thedevelopment of replacing fossil fuels. Biomass-derived molecules are apromising class of alternative resources because they offer the advantages ofbeing widely available, renewable, and potentially carbon-neutral. Suchmolecules generally contain more oxygen atoms than are found in petroleum-basedfeedstocks. Previously, small alcohols and polyols [1-2] were used asrepresentatives of oxygenates to be studied with gradually increasing thecomplexity of the molecular structure. The result revealed the possibility ofoxygenates reforming to H_2 and CO (syngas) by selectivelycontrolling the C-H, C-C, C-O, and O-H bond scissions. In this work,glycolaldehyde ($HOCH_2CH{=}O$), which contains both OH and ° C=Ofunctionalities similar to many biomass derived molecules, was studied as theprobe molecule for biomass conversion to syngas. The current study ofglycolaldehyde activity utilized density functional theory (DFT) prediction, aswell as experimental verification using temperature programmed desorption (TPD)and high resolution electron energy loss spectroscopy (HREELS).

Bimetallic catalysts are known to often exhibit uniqueproperties different from either of the parent metals, and show potential forbiomass conversion. The Ni/Pt(111) bimetallic system has been extensivelyinvestigated and was therefore studied for glycolaldehyde reactions. As established in previous studies ofalcohols and polyols [1-2], enhanced catalytic conversion of these molecules onthe bimetallic surfaces could be correlated to the binding energies and thesurface d-band center with respect to the Fermi level. The binding energy ofglycolaldehyde was found to increase as the surface d-band center approachedthe Fermi level, with the NiPtPt(111) configuration exhibiting the highestbinding energy and thus predicted to present the highest activity. In order to verify the DFTprediction, glycolaldehyde TPD experiments were performed on Pt(111), NiPtPt(111), PtNiPt(111) surfacesand a thick Ni(111) film. H_2 and CO were observed as the reactionproducts, which confirmed the prediction to produce syngas from glycolaldehyde.The activity of glycolaldehyde on each surface was quantified from the TPD peakareas. Among the four surfaces, the NiPtPt(111) surface showed the highestactivity, consistent with the DFT prediction. HREELS experiments ofglycolaldehyde were employed to show the intermediates in the reactions. At300K, the C-C peak disappeared and a CO peak was observed, demonstrating the C-Cbond cleavage to produce CO, which was consistent with the TPD result.

However, the favorable NiPtPt(111) bimetallic structure is notstable at high temperature [3]; the top monolayer Ni atoms tends to diffuseinto the Pt bulk and stay beneath of the Pt surface. The resulting

PtNiPt(111)subsurface configuration showed significantly lower reforming activity. Since tungstenmonocarbide (WC) [4-6] has been shown to possess similar electronic propertiesto Pt(111), Ni-modified WC surfaces were proposed to replace NiPtPt(111) in theglycolaldehyde study. Parallel DFT glycolaldehyde binding energy on theone-monolayer Ni-modified WC (1ML NiWC) surface was calculated and found to besimilar to that on NiPtPt(111). TPD experiments of glycolaldehyde were performedon WC and Ni-modified WC surfaces. On clean WC surface, ethylene was observedas the product, which was consistent with the HREELS result. On Ni-modified WCsurface, H_2 and CO were produced, exhibiting the same chemistry withNiPtPt(111) configuration. A similar glycolaldehyde reforming activity was alsofound on the 1ML NiWC surface after the quantification of reaction activity. Theseresults suggested that Ni monolayer catalysts supported on WC may be preferableto Ni/Pt bimetallics as active and selective catalysts for biomass reformingwith higher stability and lower cost.

References

1. Skoplyak, O.,Barteau, M. A., and Chen, J. G. *ChemSusChem* 1, 524 (2008).

2. Skoplyak, O.,Barteau, M. A., and Chen, J. G. *J. Phys. Chem. B* 110, 1686 (2006).

3. Kitchin, J. R,Khan, N.A, Barteau,M. A., and Chen, *J. G. Surf. Sci. 544*, 295 (2003).

4. Humbert M.P., MenningC.A., and Chen J.G. *J. Catal.* 271, 132 (2010).

5. Oyama, S.T.Editor. Chemistry of Transition Metal Carbides and Nitrides. *Blackie Academic& Professional*. 1996.

6. Hwu, H.H, Chen,J.G. *Chem. Rev.* 105, 185 (2005).

See more of this Session: Reaction Engineering for Biomass Conversion II
See more of this Group/Topical: Catalysis and Reaction Engineering Division

(117f) The Effects of Water On Sugar Reactions

Xianghong Qian, Department of Chemical Engineering, University of Arkansas, Fayetteville, AR, Mark Nimlos, National Bioenergy Center, National Renewable Energy Laboratory, Golden, CO, David K. Johnson, Bioscience Center, National Renewable Energy Laboratory, Golden, CO and Michael E. Himmel, Chemical and Bioscience Center, National Renewable Energy Laboratory, Golden, CO

Ab initio molecular dynamics and metadynamics simulations were used to determine the free energy surfaces (FES) for the acid catalyzed β-D-glucose condensation reaction. Protonation of C1-OH on the β-D-glucose, breakage of the C1-O1 bond, and the formation of C1 carbocation is the rate-limiting step. The effects of solvent on the reaction were investigated by determining the FES both in the absence and presence of solvent water. It was found that water played a critical role in these reactions. The reaction barrier for the proton-catalyzed glucose condensation reaction is solvent induced because of proton's high affinity with water. During these simulations, β-D-glucose conversion to α-D-glucose process via the C1 carbocation was also observed. The associated free energy change and activation barrier for this reaction were determined.

See more of this Session: Reaction Engineering for Biomass Conversion II
See more of this Group/Topical: Catalysis and Reaction Engineering Division

(117g) Kinetic Study of the Hydrothermal Liquefaction of Microalgae

Peter J. Valdez and Phillip E. Savage, Chemical Engineering, University of Michigan, Ann Arbor, MI

Hydrothermal liquefaction can convert wet algal biomass, at elevated temperatures and pressures, into an energy-dense biocrude without the need for dewatering the algae or extracting the lipids. Presently, there has been limited study to determine the global kinetics of the hydrothermal liquefaction of microalgae. We present results from a kinetic study using microalgae feedstocks with varying carbohydrate, lipid, and protein concentrations. We determined reaction pathways and calculated activation energies and frequency factors from empirical data. We analyzed all of the liquefaction products to obtain data for mass and elemental balances. The elemental analyses were used specifically to examine elemental partitioning amongst the products and how they relate to kinetic parameters. We propose that with the global kinetic analysis we are able to model the outcome of hydrothermal liquefaction based on the biochemical and elemental profile of the feedstock. In addition to kinetic studies, we also examined how processing conditions affect the yield and composition of all of the liquefaction products, not just the biocrude. We completed a systematic study of the processing conditions related to microalgae liquefaction. We examined the effects of biomass preservation, reaction atmosphere, and biomass loading. Our results show that varying these effects does not affect the product yield or composition significantly.

See more of this Session: Reaction Engineering for Biomass Conversion II
See more of this Group/Topical: Catalysis and Reaction Engineering Division

Rules of Thumb for Reactor Design and Process Conditions for Thermochemical Treatment of Chromated Copper Arsenate (CCA) Treated Wood Waste

Lieve Helsen, University of Leuven (K.U.Leuven), Heverlee, Belgium
Frederic Cuypers, University of Leuven (K.U.Leuven), Heverlee, Belgium

Extended abstract

Lab-scale pyrolysis experiments have been performed to gain insight in the mechanisms influencing the metal behavior during carbonization of chromated copper arsenate (CCA) treated wood chips. To this end a fixed bed low temperature pyrolysis reactor and a thermogravimetric (TG) analyser have been used. Besides CCA treated wood chips, mixtures of metal and organic model compounds (As_2O_5, As_2O_3, $CrAsO_4$, $CrAsO_4$ + $Ca(OH)_2$, As_2O_5 + glucose, As_2O_5 + lignin, As_2O_5 + activated carbon, As_2O_3 adsorbed on activated carbon, $CrAsO_4$ + glucose, $CrAsO_4$ + lignin, $CrAsO_4$ + $Ca(OH)_2$ + glucose, $CrAsO_4$ + $Ca(OH)_2$ + lignin) have been pyrolyzed in a controlled way. The aim of these model compound experiments is not to imitate the real CCA wood pyrolysis process, but to isolate some individual mechanisms (without being influenced by a variety of complex interactions) and to study the influence of different process parameters on these mechanisms. The parameters controlled during pyrolysis (and their corresponding range) are: temperature (330°C - 430°C), heating rate (5°C/min – 20°C/min), residence time (10 min – 40 min), reactor pressure (0 bar gage – 5 bar gage), heater gas flow rate (1000 Nl/h – 1800 Nl/h), oxygen concentration (1 vol% - 3 vol%) and steam concentration (5 wt% - 15 wt%).

Based on the experimental observations, reactor design guidelines are derived in order to identify an operating window that minimizes the volatilization of metals during thermochemical conversion of CCA treated wood, while ensuring a good quality of charcoal product.

The main observations related to metal and organic model compound experiments are:

1. The interactions between inorganic compounds (metals and minerals) and organic compounds play an important role during thermal decomposition. On the one hand the presence of organic compounds (wood, carbon, glucose, ...) promotes the reduction of As_2O_5 and subsequent volatilization of As_2O_3. On the other hand formation of complexes or agglomerates and adsorption effects may help to prevent arsenic loss below 400°C and may facilitate metal recuperation.
2. Arsenic adsorption on the solid residue may prevent the volatilization of As_2O_3, and may facilitate the formation of agglomerates of As_2O_3 with wood minerals, giving rise to thermally more stable micro-structures. Furthermore, in a fixed bed or moving bed reactor, small fractions of As_2O_3 that volatilize at the bottom of the reactor can be re-adsorbed by above lying wood or partially pyrolyzed wood layers, giving rise to thermally more stable compounds.
3. Mixtures of metal model compounds and respectively glucose and lignin behave in a different way. Glucose, which is a representative organic model compound for tar formation and secondary char formation, shows foam formation during pyrolysis which hampers the release of arsenic trioxide. Elevated pressure has a pronounced effect on the foam density and the generation of pyrolysis gas, resulting in almost 5 wt% higher arsenic retentions at 5 bar gage compared to 0 bar gage in glucose + As_2O_3 and glucose + As_2O_5 samples. Lignin, which is a representative organic model compound for

char production, shows no foam formation. The pyrolysis residue of lignin helps to stabilize arsenic trioxide.

4. Chromium arsenate forms thermally stable compounds with wood minerals consisting of Cr(III), As(V) and Ca(II), at temperatures below 310°C. Elevated pressure seems to have no significant influence on this process. Chromium arsenate, which had not formed stable compounds, decomposes to Cr_2O_3 (which may form volatile organo-chromium compounds) and As_2O_5, at temperatures below 310°C. Lignin helps to stabilize $CrAsO_4$, and thus counteracts this decomposition.
5. Once As_2O_5 is formed, it is readily reduced to As_2O_3 (following a radicallary mechanism which is favored by elevated pressure) in the presence of reductive pyrolysis vapors. Once As_2O_3 is formed, mass transfer limitations and adsorption on the pyrolysis residue (which are both enhanced by elevated pressure) play an important role in retaining arsenic in the pyrolysis residue.

The CCA wood experiments show that:

1. Percentages of metals volatilized depend strongly on the operating conditions of the process. Therefore, arsenic release can be limited by choosing the operating conditions very carefully.
2. The effect of residence time on arsenic and chromium retentions is limited for the ranges considered.
3. Heating rate mainly affects chromium retentions, primarily through a higher thermal lag of the inner particle temperature at higher heating rates.
4. Reactor temperatures lower than 400°C result in a carbon-like product, having the same matrix as the original wood, only porosity has increased. This allows the minerals and metals to migrate during the thermal process, and to be liberated afterwards since the solid structure does not collapse (in contrast to conventional carbonization processes).
5. Arsenic retentions decrease slightly with increasing temperature till 390°C, with a sharp decrease at higher temperatures, which may be attributed to accelerated desorption of arsenic trioxide at temperatures above 390°C. Chromium retentions show a less pronounced effect of temperature.
6. Arsenic and chromium retentions are affected in a different way by elevated pressure. Arsenic retentions increase about 5% at 5 bar gage compared to 0 bar gage, which is attributed to a higher mass transfer resistance. In the case of chromium, the effect is less clear; the increased mass transfer resistance might be counteracted by enhanced heat transfer at the early stages of pyrolysis.
7. Since pressure increase does not only influence the mode of occurrence of arsenic, but also the char structure and extent of tar reactions (elevated pressure enhances secondary reactions and heat transport), the metal behavior and ease of metal recuperation depends on the reactor pressure.
8. Oxygen concentration mainly affects the reactor temperature (a change of inlet oxygen concentration from 1 vol% to 3 vol% increases the peak temperature from 420°C to 550°C at the bottom of the reactor, and from 390°C to 420°C at the top of the reactor), which results in a decrease in arsenic retention with increasing oxygen concentration.
9. A higher gas flow rate and the addition of steam mainly result in a more homogeneous reactor temperature profile. The net impact on metal retention is limited.
10. Scanning electron microscopy coupled to energy dispersive X-ray analysis (SEM-EDXA) shows that during pyrolysis the metal and mineral compounds form agglomerates, which suggests a relatively easy (mechanical) separation between

metals/minerals and carbon. The mineral agglomerates (already existing in the initial wood before pyrolysis) may assist in the metal agglomeration process.

Although further research is needed to investigate the cross correlation between the various parameters, the experimental observations can be translated in following recommendations for reactor design and process conditions:

1. To ensure a good quality and brittle pyrolysis residue a residence time of at least 20 min and a process temperature of at least 370°C is necessary. However, the process temperature has to be limited to 390°C to limit metal release.
2. A more homogeneous wood input, with less fine particles, will result in a more homogeneous pyrolysis residue and reduced arsenic and chromium volatilization.
3. A longer residence time and lower temperature result in a similar wood conversion while reducing chromium and arsenic volatilization, however a lower plant capacity is obtained.
4. An operating pressure of 5 bar (or even higher) results in 5% higher arsenic retentions and mass retentions. Furthermore, elevated pressures allow to use lower heater gas temperatures to obtain the same wood conversion, resulting in even higher metal retentions.
5. Small gas flow rates are important with respect to:
    a. localizing the release of free and adsorbed water (drying) on the one hand and tar formation on the other hand in distinct phases
    b. minimizing the dilution of combustible vapors
6. Tar problems can be avoided by:
    a. preventing mixing of tar and water compounds (which would result in sticky tar compounds),
    b. facilitating secondary reactions (such as cracking of tar); repeated cracking of organic compounds (stimulated by stratification of pressure and temperature in the reactor column) eliminates the liquid phase.
7. The inlet oxygen concentration has to be strictly controlled below 1 vol% or even 0.5 vol% (at a heater gas flow rate of 1400 Nl/h and without steam addition), which increases both the arsenic retention and mass retention with 5% compared to an oxygen concentration in the range 1-3 vol%.
8. If oxygen control below 1 vol% is not feasible, the heater gas flow rate should be limited to 500 Nl/h to limit arsenic release. However, this could hamper the heating of the reactor bed. Nevertheless, heater gas flow rates should be lower than 1800 Nl/h to reach acceptable charcoal yields.
9. Steam addition results in better reactor temperature control, and as a consequence increased metal and mass retentions. Moreover, thermal lag is reduced which is beneficial with respect to process time. However, significant amounts of steam are needed (40 wt% steam at an initial oxygen concentration of 2 vol% and 70 wt% steam at an initial oxygen concentration of 3 vol% to limit the reactor bed temperature to 390°C).

These recommendations still need to be tested in an industrial scale reactor.

(118b) Biomass Pyrolysis and Gasification: Engineering Rules of Thumb

Michael J. Antal, Hawaii Natural Energy Institute, University of Hawaii, Honolulu, HI

In this presentation I review important findings recorded in the archival literature concerning the thermodynamics, kinetics, mechanisms, and catalysis of biomass pyrolysis and gasification that serve as rules of thumb for reactor design.

See more of this Session: Reactor Engineering for Biomass Feedstocks
See more of this Group/Topical: Sustainable Engineering Forum

(118c) Rule of Thumb for Design of Catalytic Pyrolysis of Biomass In Bubbling Fluid Bed Reactor

Foster Agblevor, Biological Engineering, Utah State University, Logan, UT and Ofei Mante, Biological Systsems Engineering, Virginia Polytechnic Institute and State University, Blacksburg, VA

The production of stable biomass pyrolysis depends strongly on both reactor design and the effective design of a condensation train to collect the pyrolysis products. There are several challenges in the design of the pyrolysis system including solids feeding, char separation from the pyrolysis vapors, condensation of the pyrolysis vapors and compression of the non-condensible gases. In the feeding of solid biomass into the reactor, the feedstock properties such is softening point are critical in the feeding. Bridging properties are also very important and that is dictated by the method of biomass preparation, either knife milling or hammer milling. Hammer milled and disc-milled feedstocks tend to bridge easily whereas knife milled material bridges less. The reactor operating temperature is strongly influenced by the composition of the feedstocks. Biomass feedstocks such as poultry litter which are high in ash tend to form clinker in the reactor and therefore reactor temperature control is very critical for successful operation. The condensation train; especially with recycling of the gases tend to create major problems especially if an electrostatic precipitator is used to condense the aerosols. For systems employing electrostatic precipitator, the cooling temperature in the condensers must be very low because any water vapor escaping the condensers tend to cause electrical short-circuiting in the unit and reduces efficiency of the aerosol capture. Catalytic pyrolysis tends to produce more water and therefore the cooling system must be very efficient. The compressor for the non-condensible gases must be able to handle dirty gases (large clearances) otherwise it is easily clogged with the pyrolysis oils that tend to degrade the performance of the compressor and requires frequent replacements.

See more of this Session: Reactor Engineering for Biomass Feedstocks
See more of this Group/Topical: Sustainable Engineering Forum

(118d) Rule of Thumb for Reactor Design and Process Conditions for Supercritical Water Gasification of Wet Biomass

Yukihiko Matsumura, Division of Energy and Environmental Engineering, Hiroshima University, Higashi-Hiroshima, Japan

The Rule of thumb for reactor design and process conditions for supercritical water gasification of wet biomass is introduced. The important aspects includes heat transfer and heating rate, reaction rate, plugging, gas composition, inorganic element behavior, and heat of reaction. The know-hows were developed by the laboratory scale, and pilot scale reactors.

See more of this Session: Reactor Engineering for Biomass Feedstocks
See more of this Group/Topical: Sustainable Engineering Forum

Rule of Thumb for Reactor Design and Process Condition for Biodiesel Production

Masato Kouzu, Doshisha University, Kizugawa, Kyoto, Japan

Introduction

For the last decade, the planet earth clearly tends toward warming due to the greenhouse gas emissions. One of solutions for the globally environmental problem is to utilize biomass resources, which are generated from CO_2 and H_2O with the aid of photosynthesis, as alternative fuel and chemicals. Plant oil is one of biomass resources, and is used as feedstock of alternative to petroleum diesel oil. The alternative fuel is extensively termed as "Biodiesel". Biodiesel has advantages of not only renewable nature mentioned above but also good fuel properties such as high flush point and excellent lubricity [1-2]. Moreover, it was reported that emissions of both carbon monoxide and particulate were reduced by fueling the engine with biodiesel [3-4]. In 2003, European Community decided to replace at least 5.75 % of the yearly consumed fossil fuels with biofuels, by the year 2010. This decision accelerated the use of biodiesel and its production has been constantly growth. The total of biodiesel yearly produced in the world was 7.75 milion metric tons in 2008 [5].

Biodiesel consists of fatty acids methyl ester (FAME), which was produced by transesterifying plant oil with methanol. For the industrial biodiesel production, plant oil is transesterified with the help of alkali catalyst dissolved in methanol. The alkali catalyst such as sodium hydroxide is so active that the yield of FAME is above 90 % for 0.1 h at 333 K under atmospheric pressure [6]. Glycerol that is by-produced from plant oil is separated from biodiesel by decantation. Then, purification to eliminate impurity such as soap from biodiesel is carried out.

In this presentation, the rule of thumb for reactor design and process condition for biodiesel production is introduced. The important aspect includes type of plant oil used as feedstock, an amount of methanol mixed with plant oil and quality of the produced biodiesel. Recently, much interest has been taken in transesterifying plant oil with the novel manners: enzymatic and heterogeneous catalytic reactions. Also, it is very important to consider the way for recycling glycerol of the by-product.

Type of plant oil

Plant oil consists of triacylglycerides having fatty acid moieties, and contains some minor components. Among the minor components, water and free fatty acids (FFA) inhibit the alkali-catalyzed transesterification that is applied to the industrial biodiesel production. Water brings about the undesirable hydrolysis of triglycerides. It is easy to limit the water content to the trace level by drying, because of the hydrophobic property of plant oil. FFA neutralizes the alkali catalyst. Moreover, soap produced by the neutralization causes formation of stable emulsion with the result that it is difficult to decant biodiesel from glycerol. An industrial way to eliminate FFA is the pre-neutralization with alkali hydroxide. For the pre-neutralization, aqueous solution of alkali hydroxide is stirred into the plant oil. Then, the soap sediment and the FFA-free oil separate out by decantation. Recently, much interest has been taken in pre-esterifying FFA into FAME with the help of acid catalyst such as sulfonated cation-exchange resin [7].

Waste cooking oil is the cost-advantageous feedstock for biodiesel production. In addition to FFA and water, polar compounds are impurities of the waste cooking oil. Heat for food preparation such as frying leads to thermolytic and oxidative reactions of triglycerides having unsaturated fatty acid chains, which produces polar compounds. Polar compounds seem to be of no negative effect on the base-catalyzed transesterification [8].

Furthermore, it should be noted that fatty acids composition of plant oil affects quality of biodiesel. As palm oil is abundant in triacylglycerides consisting of saturated fatty acid chains, the resultant biodiesel is inferior in the low-temperature properties such as cold filter plugging point (CFPP).

Methanol/oil molar ratio

An amount of methanol mixed with plant oil is the most important variable affecting the yield of FAME, because transesterification of plant oil with methanol is a reversible reaction. Due to the chemical equilibrium, excess of methanol is required for achieving the conversion of plant oil into biodiesel. The required methanol/oil molar ratio seems to be 6, while the stoichiometric ratio is 3 [6]. Although an increase in the methanol/oil molar ratio is effective in not only preventing the reverse reaction but also enhancing the conversion rate, a large amount of methanol mixed with the transesterified products makes it difficult to decant biodiesel from glycerol.

The efficient way to achieve the perfect conversion with the minimized amount of methanol is the two step transesterification. In this way, glycerol produced by the primary transesterification is removed in advance of the secondary transesterification. Due to the removal of glycerol, for the secondary transesterification, the chemical equilibrium shifts towards the forward reaction that is useful in maximizing the FAME yield.

Quality of biodiesel

As properties of biodiesel must meet the quality standard such as EN 14214 and ASTM D6751, purification is a very important operation for biodiesel production. Biodiesel collected after the decantation which is carried out after the transesterification contains soap deriving from alkali catalyst. Moreover, it is difficult to achieve the perfect removal of glycerol by only the decantation. Since these impurities are soluble in water, biodiesel is commonly purified by washing with water. As a large amount of soap leads to formation of stable emulsion, an amount of alkali catalyst employed for the transesterification should be optimized. Use of hot water is effective in preventing the emulsion formation. Water mixed with biodiesel for the purification can be eliminated perfectly by decantation followed by drying.

On the other hand, there are some water-free procedures to purify biodiesel. Cation-exchange resin serves adsorbent for eliminating soap and glycerol from biodiesel [9]. Also, it was reported that impurities of biodiesel was eliminated by utilizing amorphous magnesium silicate as the absorbent [10]. Some researchers studied vacuum distillation to collect highly-pure biodiesel [11].

Recycle of glycerol

It is essential for biodiesel production to recycle glycerol of the by-product. Glycerol collected after the transesterification followed by the decantation is contaminated with methanol, water and soap. Soap is decomposed with sulfuric acid, and the resultant alkali-sulfate is removed from stream of the glycerol purification. Fatty acids from soap are decanted from a mixture of glycerol, methanol and water. Finally, methanol and water are topped by heating in order to obtain the crude glycerol used as raw material of chemicals.

Also, it is important to study novel conversion routes of glycerol into chemicals, because intensive production of biodiesel has caused overproduction of glycerol that makes its market price plummeting. Corma et al. showed a variety of the novel catalytic reactions to convert glycerol into chemicals in their research paper

[12]. Hydrogenolysis into propanediol is one of the potential routes for promoting the recycle of glycerol as chemicals.

Catalyst selection

Sodium and potassium hydroxides are the alkali catalyst of commercial availability. Although sodium hydroxide is cheaper than potassium hydroxide, easy handling of glycerol that contains soap deriving from potassium hydroxide is advantageous to operation of biodiesel production process.

For the purpose of improving biodiesel production process, many researchers have focused their attentions on enzymatic and heterogeneous catalytic traansesterifications [13-14]. Both of them bring about not only the water-free purification of biodiesel but also easy recycle of glycerol. For the enzymatic transesterification, plant oil is converted into biodiesel with the help of lipase acting as bio-catalyst. Major characteristic of the enzymatic transesterification is mild reaction conditions: the transesterification plant oil is catalyzed at room temperature. Furthermore, the enzymatic transesterification is insensitive to FFA and water contained in the low-quality feedstock. For the practical use of the enzymatic process, the effect of immobilization on the catalytic activity of lipase has been investigated [15].

Zinc-aluminum mixed oxide was utilized as the heterogeneous catalyst for a biodiesel production process, which was termed as "Esterfif-H™" [16]. The heterogeneous catalyst seemed to be active at the temperature of 483-523 K under the pressure of 3-5 MPa. The catalytic activities of calcium oxide and sulfonated amorphous carbon are close to that of alkali catalyst [8, 17].

Reference

1. Dorado, M. P., E. Ballesteros, J. M. Arnal, J. Gomez and F. J. Lopez (2003), "Exhaust emissions from a diesel engine fueled with transesterified waste olive oil", *Fuel*, 82, pp. 1311-1315.

2. Lin, C. Y., H. A. Lin and L. B. Hung (2006), "Structure and properties of biodiesel produced by peroxidation process" *Fuel* ,86, pp. 1743-1749.

3. Karmee, S. K. and A. Chadha (2005), "Preparation of biodiesel from crude oil of *Pongamia pinaanta", Bioresour. Technol.*, 96, pp. 1425-1429.

4. Szybist, J. P., J. Song, M. Alam and A. L. Boehman (2007), "Biodiesel combustion, emissions and emission control", *Fuel Process. Technol.*, 88, pp. 679-691.

5. Knothe, G (2010), "Biodiesel: Current trends and properties", *Top. Catal.* 53, pp. 714-720.
6. Freedman, B. ,E. H. Pryde and T. L. Mounts (1984), "Variables affecting the yield of fatty esters from transesterified vegetable oils", *J. Am. Oil Chem. Soc.*, 61, 1638-1643.
7. Russbueldt, B. M. E. and W. F. Hoelderich (2009), "New sulfonic acid ion-exchange resins for the preesterification of different oils and fats with high content of free fatty acids", *Appl. Catal. A: Gen.*, 362, pp. 47-57
8. Kouzu, M., T. Kasuno, M. Tajika, Y. Sugimoto, S. Yamanaka and J. Hidaka (2008), "Calcium oxide as a solid base catalyst for transesterification of soybean oil and its application to biodiesel", *Fuel*, 87, pp. 2798-2806.
9. Berrios, M. and R. L. Skelton (2008), "Comparison of purification methods for biodiesel", *Chem. Eng. J.*, 144, pp. 459-465.
10. Kucek, K. T., M. A. F. Cesar-Oliveira, H. M. Wilhelm and L. P. Ramos (2007), "Ethanolysis of refined soybean oil assisted by sodium and potassium hydroxides", *J. Am. Oil Chem. Soc.*, 84, pp. 385-392.
11. Sakai, T., A. Kawashima and T. Koshikawa (2009), "Economic assessment of batch production processes using homogeneous and heterogeneous alkali catalysts", *Bioresour. Technol.*, 100, pp. 3268-3276
12. Corma, A., S. Iborra and A. Velty (2007), "Chemical routes for the transformation of biomass into chemicals", *Chem. Rev.*, 107, pp. 2411-2502.
13. Kulkarni, M. G. and Dalai A. K. (2006), "Waste cooking oil -An economical source for biodiesel: A review", *Ind. Eng. Chem.*, 45, pp. 2901-2913.
14. Yan, S., C. DiMaggio, S. Mohan, M. Kim, S. O. Salley and K. Y. Simon Ng (2010), "Advancements in heterogeneous catalysis for biodiesel synthesis", *Top. Catal.*, 53, pp. 721-736.
15. Oda, M., M. Kaieda, S. Hama, H. Yamaji, A. Kondo, E. Izumoto and H. Fukuda (2005), "Facilitatory effect of immobilized lipase-producing *Rhizopus oryzae* cells on acyl migration in biodiesel-fuel production", *Biochem. Eng. J.*, 23, pp. 45-51.
16. Bournay, L., D. Casanave, B. Delfort, G. Hillion and J. A. Chodorge (2005), "New heterogeneous process for biodiesel production: A way to improve the quality and the value of the crude glycerin produced by biodiesel plants", *Catal. Today*, 106, pp. 190-192.
17. Hara, M. (2010), "Biodiesel production by amorphous carbon bearing SO_3H, COOH and phenolic OH groups, a solid Bronsted acid catalyst", *Top. Catal.*, 53, pp. 805-810.

(118f) Rule of Thumb for Design and Analysis of Biomass Utilization System

Kazuhiro Mochidzuki, IIS, Univ. of Tokyo, Tokyo, Japan and Akiyoshi Sakoda, Institute of Industrial Science, University of Tokyo, Tokyo, Japan

This paper deals with the rule of thumb for design and analysis of overall biomass utilization system, including, the matching of available biomass and conversion technologies, the market and demand of biomass-based energy/products, the evaluation of material and energy balances throughout the system, the environmental impacts, and the sustainability. When laying a plot of biomass utilization, it is important to develop comprehensive strategy based on a reasonable scenario. Nevertheless every plot has different circumstances and viewpoints on it. There are endless varieties of factors regarding to design and analysis of the biomass systems. The problems spread across wide-ranged issues from agriculture/forestry to industry, from political/social sectors to scientific/engineering sectors, from local to worldwide, and so on. To establish a reasonable biomass system, it is necessary to make clear the whole picture of it, organizing the key factors and problems.

We have developed concepts and methods to design biomass utilization systems. Our approach includes the followings: 1) screening the key factors, 2) drawing up a scenario and its modeling, 3) matching the available biomass sources and utilization technologies, 4) analysis/optimization of material and energy balances throughout the whole system, and 5) evaluation of economical and environmental issues. In this paper our recent progress in this field is shown with some results of case studies.

See more of this Session: Reactor Engineering for Biomass Feedstocks
See more of this Group/Topical: Sustainable Engineering Forum

(141a) Ozone Removal At Micro-Second Contact Time for Aircraft Cabin Air Using Microfibrous Entrapped Catalysts

Qiang Gu, Chemical Engineering, Auburn University, Auburn University, AL and Bruce Tatarchuk, Department of Chemical Engineering, Auburn University, Auburn, AL

Microfibrous Entrapped Catalyst (MFEC) manufactured from 8 μ m diameter nickel fibers were engineered into pleated heterogeneous catalytic reactors to improve removal performance from conventional packed bed and monolith reactors, since microfibrous materials can entrap small particles (150-200 μ m) which significantly improves inter phase mass transfer. Conventional reactors usually operate with contact time less than 1s. MFEC reactors were able to reduce the contact time to micro seconds while maintaining similar catalytic performance. This gives MFEC a huge advantage in terms of weight and volume saving, since the conventional meter-long reactors were shortened to millimeter thick material sheets. These unique reactors were targeted at ozone, which was widely recognized as No.1 aircraft cabin air pollutant.

In this research, MFEC reactors were investigated under aircraft bleed air conditions of high temperature (100-200°C) and high face velocity (10-30 m/s) resulting an interlayer contact time of 67-200μ sec. Precious metal (Pd, Ag) and transition metal (Mn) catalysts were impregnated on entrapped particles (e. g. γ-Al_2O_3) using incipient wetness method. Ozone test concentration was set at a high-demanding 1.5 ppmv. Results showed that a high level of ozone decomposition was achieved with a significant reduction of catalyst consumption. Compared with conventional aircraft filters, this can be a huge advantage in terms of material cost and labor. Reaction kinetics analyses were compared for different catalysts. Results showed that precious metal catalyst performs better at a higher temperature while transition metal catalyst maintains similar conversion when changing temperature. XPS, TPR and TGA tests have been used to evaluate performance of catalysts and modify catalysts to improve conversion rate. Also catalyst aging tests were carried according to the frequency of commercial aircraft ozone reactor replacement, which evaluated long-term performance for actual MFEC reactor usage. In addition, pressure drop improvement by pleating MFEC was also evaluated with comparison of traditional low pressure drop monolith reactors. CFD model has also been established for pressure prediction and comparison. It showed that pleated structure enable MFEC to avoid higher pressure drop and eliminate the problem of direct channeling caused by monolith.

See more of this Session: Applied Environmental Catalysis II
See more of this Group/Topical: Catalysis and Reaction Engineering Division

(141b) In Situ and Operando X-Ray Absorption Spectroscopy On Cu-Zeolite Catalysts for Fast SCR by Ammonia

Vincent F. Kispersky[1], Jun Wang[1], Aleksey Yezerets[2], Neal W. Currier[2], Jeffrey T. Miller[3], W. Nicholas Delgass[1] and Fabio H. Ribeiro[1], (1)School of Chemical Engineering, Purdue University, West Lafayette, IN, (2)Cummins Inc., Columbus, IN, (3)Chemical Science and Engineering, Argonne National Laboratory, Argonne, IL

NO_x removal by selective catalytic reduction (SCR) has been studied extensively by the exhaust aftertreatment community. Cu-zeolites, especially Cu on ZSM-5, have been found to be particularly well suited for low temperature SCR [1]. Despite a large body of research on Cu-ZSM-5, the debate about the active Cu species is still ongoing. Our recent measurements of the Cu X-ray Absorption Near Edge Spectra (XANES) in a custom-built, *operando* reactor using glassy carbon tubes have produced intriguing results. *In situ* XANES spectra were collected over a range of Cu-ZSM-5 catalysts, from loadings of 1-7 wt.% Cu, under NO oxidation, dry SCR, wet SCR and pure ammonia gas conditions and a 2.7 wt.% Cu was tested *operando* under wet and dry NO oxidation and wet SCR reaction conditions. During NO oxidation, the Cu was split between isolated Cu^{2+} species and Cu^{2+} in the form of bulk-like CuO, present in small amounts at lower loadings and becoming the dominant species at high loadings where the Cu over-exchange was around 300%. Nevertheless, the Cu was fully oxidized. Upon the introduction of the reducing agent, ammonia, for both dry and wet SCR conditions, a significant portion of Cu reduced to its 1+ oxidation state. Similar to the Cu^{2+}, however, this 1+ species was an isolated Cu^{1+}, entirely different from the Cu^{1+} found in bulk Cu_2O. In fact, there was never any identifiable amount of Cu^{1+} in a bulk-like Cu_2O form during any phase of our experiments. Lower Cu wt.% loadings (50-100% exchanged) had higher proportions of reduced Cu where the lowest Cu loading showed the presence of very little Cu^{2+}. The amount of reduced Cu^{1+} ranged from 5-70% going from high to low Cu loadings. Furthermore, at low weight loadings, some Cu^{2+} was identified as Cu coordinated with NH_3 where the higher wt.% Cu samples appeared to be coordinated with H_2O instead. Implications of the effects of the presence of reduced states of Cu on SCR and NO oxidation kinetics will be discussed.

[1] S. Brandenberger, O. Kröcher, A. Tissler, R. Althoff. Catalysis Reviews, 50:4, 492-531, 2008

See more of this Session: Applied Environmental Catalysis II
See more of this Group/Topical: Catalysis and Reaction Engineering Division

(141c) Catalytic Removal of Volatile Organic Compounds Over the Three-Dimensionally Ordered Mesoporous or Macroporous MOx (M = Co, Fe, Mn, Cr) Catalysts

Hongxing Dai[1], Yunsheng Xia[1], Ruzheng Zhang[1], Lei Zhang[1], Jiguang Deng[1], **Yingshu Liu**[2], **Kai Wang**[3], Hong He[1] and Jian Li[1], (1)Department of Chemistry and Chemical Engineering, Beijing University of Technology, Beijing, China, (2)School of Mechanical Engineering, University of Science and Technology Beijing, Beijing, China, (3)School of Resource and Safety Engineering, China University of Mining & Technology, Beijing, Beijing, China

Catalytic removalof volatile organic compounds over the three-dimensionally ordered mesoporous ormacroporous *MO*$_x$ (*M* = Co, Fe,Mn, Cr) catalysts

HongxingDai [a,*], YunshengXia [a], Ruzhen Zhang [a], Lei Zhang [a], Jiguang Deng[a], Yingshu Liu [b], Kai Wang [c], Hong He [a],Jian Li [a]

[a] College of Environmental and Energy Engineering,Beijing University of Technology, Beijing 100124, China. E-mail address: hxdai@bjut.edu.cn.

[b] School of MechanicalEngineering, University of Science and Technology Beijing, Beijing 100083, China

[c] School ofResource and Safety Engineering, China University of Mining & TechnologyBeijing, Beijing 100083, China

Most of volatile organic compounds (VOCs) are environmentalpollutants. Catalytic combustion is agood way for VOCs removal.The key issue is the availability of an effective catalyst. Recently, we have fabricatedMO_x(M = Co, Fe, Mn, Cr) with a three-dimensionally (3D) or 3Dordered macroporous (3DOM) structureusing nitrate of Co, Fe, Mn or Cr as metal source, and found that these materials (see Table 1) exhibited excellent catalyticactivities for the combustion of formaldehyde, acetone, methanol, and toluene[1].

Theas-fabricated materials were characterized by the XRD, BET, SEM, TEM,XPS, and H_2-TPR techniques. The typical TEM and SEM images of the samples are shown inFig. 1. It is found that the KIT-6- and SBA-16-templating derived samples possessedordered mesoporous architectures with crystalline walls. There was formation of nanovoids within thewalls of the 3DOM-structured Fe_2O_3-2 sample.The surface areas of the mesoporous MO_x (M = Co, Fe, Mn, Cr) samples were above100 m^2/g. XPS analyses revealed the copresence of Co^{2+}/Co^{3+} in Co_3O_4, Fe^{3+}/Fe^{2+} in Fe_2O_3, and $Cr^{3+}/Cr^{5+}/Cr^{6+}$ in CrO_x. Comparedto the corresponding bulk metal oxides, the macro- or mesoporous counterparts exhibited higher surface oxygen species amount and were more reducibleat low temperatures.Theporous materials showed much better catalytic activities (evaluated with the temperature $T_{50\%}$ and $T_{90\%}$ at VOC conversion = 50 and 90%, respectively) than the bulkcounterparts in VOC combustion (Table 1). The unusual catalytic performanceof the porous transition-metal oxides was related to their better low-temperature reducibility, 3D ordered mesoporousor3DOM architecture,highersurface oxygen species concentration, and larger surface area.

Table 1. Fabrication methods, pore structures, surface areas (*S*), and catalyticactivities of the as-prepared catalysts

Catalyst	Fabrication method and calcination conditions	Pore structure	S (m^2/g)	Catalytic activity[a] $T_{50\%}/T_{90\%}$ (°C)			
				HCHO	acetone	methanol	toluene
bulk Co_3O_4	thermal decomposition; 500°C for 3 h	nonporous	10	-	-	142/-	200/-
Co_3O_4	KIT-6-templating; 400°C for 3 h	ordered mesopore	121	-	-	105/139	140/180
bulk Fe_2O_3	thermal decomposition; 500°C for 3 h	nonporous	27	-	235/-	264/-	380/-
Fe_2O_3-1	KIT-6-templating; 400°C for 3 h	ordered mesopore	113	-	151/208	170/204	-
Fe_2O_3-2	P123/PMMA-templating (Fe/P123 molar ratio = 232); 550°C for 3 h	3DOM with mesopore walls	46	-	-	-	240/288
bulk MnO_2	(Beijing Chemical Reagent Co.)	nonporous	10	-	-	-	285/340
MnO_2	ultrasound-aided SBA-16-templating; 450°C for 3 h	ordered mesopore	266	-	-	-	190/240
bulk Cr_2O_3	thermal decomposition; 500°C for 4 h	nonporous	5	152/-	142/-	164/-	190/-

CrO_x-1	ultrasound-assisted KIT-6-templating; 400°C for 4 h	ordered mesopore	124	92/117	75/124	98/130	-
CrO_x-2	solvent-free KIT-6-templating; 240°C for 24 h	ordered mesopore	106	-	-	-	140/234

Reaction conditions: 1000 ppm VOC and spacevelocity = 20,000 mL/(g h).

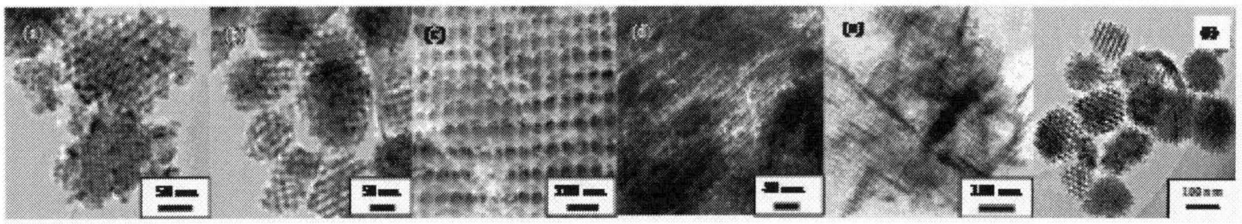

Figure1. TEM andSEM images of (a) Co_3O_4, (b) Fe_2O_3-1, (c)Fe_2O_3-2, (d) MnO_2, (e) CrO_x-1 and (f) CrO_x-2

Reference

1 (a) Y.S. Xia, H.X. Dai, H.Y. Jiang, et al., *Environ. Sci.Technol.* 43(2009) 8355;(b) Y.S. Xia, H.X. Dai, L. Zhang, etal., *Appl. Catal. B* 100 (2010) 229; (c) Y.S. Xia, H.X. Dai, H.Y.Jiang, et al., *Catal. Commun.* 11 (2010) 1171; (d) J.G. Deng, L. Zhang, H.X. Dai, et al., *J. Phys. Chem.* C114 (2010) 2694; (e) Y.S. Xia, H.X. Dai, H.Y.Jiang, et al., *J. Hazard. Mater.* 186 (2011) 84; (f) R.Z. Zhang, H.X. Dai, Y.C. Du, et al., *Inorg. Chem.* 50 (2011) 2534.

See more of this Session: Applied Environmental Catalysis II
See more of this Group/Topical: Catalysis and Reaction Engineering Division

(141d) Role of CeO2 In the Total Oxidation of Toluene Over CuO-CeO2/Al2O3

Unmesh Menon[1], Vladimir V. Galvita[1], Hilde Poelman[1], Vitaliy Bliznuk[2] and Guy B. Marin[1], (1)Laboratory for Chemical Technology, Ghent University, Ghent, Belgium, (2)Department of Material Science and Engineering, Ghent University, Ghent, Belgium

Introduction

Benzene, toluene and xylene(BTX), are considered to be three major carcinogens and their emission in theenvironment has to be controlled. Among the destructive technologies, catalyticoxidation is the most promising approach for the abatement of Volatile OrganicCompounds (VOCs) owing to its advantages including high efficiency, lowoperating temperature and no secondary pollution[1].The performance of the catalyst determines the effectiveness of this technique.

Supported or unsupported noblemetal catalysts are used for complete oxidation of VOCs, but high costs limittheir wide application, thus giving way to transition metal oxides such ascopper oxide. These are less active at lower temperatures but present comparableactivity at higher temperatures and have high catalyst loading capabilities[2-6].

However, a pure copper-basedcatalyst is less active and stable in the presence of water vapor and/or CO_2in the gas mixture. At the same time, copper catalysts promoted by ceria showbetter catalytic performance for the complete oxidation of toluene, propane, benzeneand p-xylene than unpromoted copper catalysts[7,8]as well as improved activity in the presence of oxidation products $CO_2$and water[9]. Ceria as a promoter forsupported CuO has shown several advantages: (a) ceria stabilizes the dispersionof the active component[10,11]; (b) metal/ceriainteractions strongly affect their redox and catalytic properties; (c) ceriaalso acts as an oxygen storing component due to the presence of mixed oxidationstates (3+/4+) of cerium; (d) copper-doped CeO_2 can improve theoxygen storage capacity, diffusivity and redox properties, which are attributedto the formation of structural defects.

However, it is still unclear howcopper and ceria interact with each other and promote catalytic activity. Thequestions addressed in this study are: (1) Assistance of CeO_2 in theoxidation of reduced copper species, (2) the crucial copper species inducingthe coppeR&Deg;Ceria interaction, (3) the active sites of coppeR&Deg;Ceria system fortoluene oxidation, (4) the role of ceria in decreasing the negative effect of thepresence of water and CO_2, (5) the enhancement of catalytic activitydue to coppeR&Deg;Ceria interaction

In this study the binary mixedoxide, $CuO-CeO_2/g-Al_2O_3$, hasbeen investigated in detail in comparison to its corresponding single-oxidecomponents $CuO/g-Al_2O_3$, $CeO_2/g-Al_2O_3$in order to correlate catalytic activity towards total oxidation oftoluene with their physicochemical characterization. A transient responsetechnique with millisecond time scale[12] was used to investigate thecatalytic activity and influence of H_2O and CO_2 on thecatalytic behavior. The structure of the catalysts was investigated by the useof high resolution electron microscopy (HRTEM), selected area electrondiffraction (SAED), X-ray diffraction (XRD) and X-ray absorption Spectroscopy(XAS).

Experimental procedure

The $CuO-CeO_2/\gamma-Al_2O_3$,$CuO/\gamma-Al_2O_3$ and $CeO_2/\gamma-Al_2O_3$catalysts were synthesized via impregnation of $\gamma-Al_2O_3$with $Cu(NO_3)_2 2.5H_2O$ and/or $Ce(NO_3)_4 9H_2O$ precursors, followed by drying at 357K for 8h andcalcination above 973 K for 8h in air. 10mg of 250-500mm particles were used for experiments to determine the catalyticactivity.

The selected areaelectron diffraction (SAED) and high resolution electron microscopy (HREM) ofthe samples were carried out in a transmission electron microscope (TEM) operating at 300 kV (Jeol, JEM-2200FS).

The temporal analysis of products(TAP) experiments are performed in a quartz micro-reactor, with 33mm bed-lengthand 4.75mm inner diameter, evacuated to 10^{-4} 10^{-5} Pa. Avery small amount of reactant molecules (~10^{-9} mol), which is ~5orders less than the active sites in the catalyst, is pulsed into the reactor bymeans of two high-speed pulse valves. The products and reactants at the outletof the reactor are monitored by a quadrupole mass spectrometer. A temperaturerange of 723K -873K is covered. To study the activity of the $CuO-CeO_2/\gamma-Al_2O_3$,$CuO/\gamma-Al_2O_3$ and $CeO_2/\gamma-Al_2O_3$ catalysts towards toluene totaloxidation, experiments were performed over O_2 pre-treated catalystsby pulsing C_7H_8 with and without di-oxygen in the feed. Typically, a stoichiometric ratioof di-oxygen to toluene, 9:1, was used in the mixture $C_7H_8/O_2/Ar$,when experiments were conducted in the presence of di-oxygen.

In situ XAS experiments wereperformed at the SAMBA beam line of the SOLEIL synchrotron. X-ray absorptionnear edge (XANES) spectra were recorded with 1s time resolution during H_2reduction and reoxidation with CO_2 and water at both the Cu K (8979 eV)and Ce L_{III} edge (5723 eV).

Results

XRD analysis of the$CuO-CeO_2/\gamma-Al_2O_3$ catalyst shows thatalumina and ceria have a crystallite diameter of about 5 nm. At the same time CuO presents singlecrystals of about 100nm. From the STEM and SAD images the CuO single crystalunits of ~100nm were confirmed, while CeO_2appeared as 100nm clusters of nanocrystallites. EDX line scans were undertakenthrough specific regions of a STEM frame, displaying the elements present alongthe line (see Fig.1). The line scan proves that in addition to the large CuO crystallites,°CuO is also present inside the CeO_2 clusters as small sizedspecies. In situ XANES measurements at the CeL_{III} and Cu K edgeyield direct evidence that CeO_2 in the $CuO-CeO_2/g-Al_2O_3$ catalystis reduced by H_2 at similar temperature as the CuO phase. Fig. 2shows XANES results which confirms partial reoxidation of ceria by water. Whereas,CuO is not reoxidised by water.

The toluene oxidation experimentaldata indicate that the reaction is carried out according to a redox Mars-vanKrevelen mechanism. Lattice oxygen atoms from the surface of the catalyst areconsumed by toluene and oxygen vacancies are created. These vacancies arefilled by the oxygen atoms that diffuse from the bulk to the surface of thecatalyst or by di-oxygen. It is shown that the degree of reduction of the metaloxide determines to a large extent its performance for VOCs oxidation. Thepresence of water or CO_2 on the catalyst surface decreases the rateof regeneration of the reduced catalyst by di-oxygen. Furthermore, if waterand/or CO_2 are present in the gas phase, the rate of desorption fromthe surface becomes slower due to the high partial pressure of water in the gasphase.

The activity of the$CuO-CeO_2/Al_2O_3$ catalyst increases in the presenceof H_2O or CO_2. On the other hand, CuO/Al_2O_3showed loss of activity in the presence of H_2O. There was nosignificant changes in activity on CuO/Al_2O_3 and $CeO_2/Al_2O_3$when co-feeding toluene with CO_2.

Conclusions

Two CuO phases existin the $CuO-CeO_2/Al_2O_3$ catalyst: large ~100nm crystals, not contacting CeO_2 and small CuO speciesdispersed upon and incorporated into clusters of CeO_2nano-crystallites with size ~5nm.

Theeffect of CeO_2 on the catalytic activity can be explained by twofactors. First, ceria stabilizes the dispersion of the active component, i.e. asmaller crystallite size of the CuO phase on the support becomes one of the keyfactors in determining the chemical reactivity. Second, incorporation of copperinto the CeO_2 lattice can form $Ce_{1-x}Cu_xO_{2-d}$solid solution which also has high activity and reducibility at lowertemperatures compared to CuO or CeO_2 alone. The promoting effect ofceria is found to be more pronounced at low temperatures.

AddingH_2O or CO_2 into the C_7H_8/O_2mixture increases the catalytic activity of $CuO-CeO_2/Al_2O_3$catalyst by producing lattice oxygen species from reoxidation of ceria by CO_2or H_2O. Over the CuO/Al_2O_3catalyst the re-oxidizing role of CO_2 or H_2Oco-fed with toluene is insignificant.

Acknowledgement

This work was supportedby a Concerted Research Action (GOA) of Ghent University and the Long TermStructural Methusalem Funding by the Flemish Government'.

References

[1] J.J.Spivey, J.R. Anderson, M. Boudart, Catalysis: A Specialist Periodical Report,Vol. 8, The Royal Chemical Society, 1989, p. 157.

[2] J.Carpentier, J.F. Lamonier, S. Siffert, E.A. Zhilinskaya, A. Aboukas, AppliedCatalysis A: General 234 (2002) 91-101

[3] Y.Liu, M. Luo, Z. Wei, Q. Xin, P. Ying, C. Li, Applied Catalysis B: Environmental29 (2001) 61-67.

[4] M.Ferrandon, E. Bjornbom, Journal of catalysis 200 (2001) 148-159.

[5] M.C.lvarez-Galvn, V.A. de la Pea O'Shea, J.L.G. Fierro, P.L. Arias, CatalysisCommunications 4 (2003) 223-228.

[6] N.Burgos, M. Paulis, M. Mirari Antxustegi, M. Montes, Applied Catalysis B: Environmental38 (2002) 251-258.

[7] P.M.Heynderickx, J.W. Thybaut, H. Poelman, D. Poelman, G.B. Marin, Journal ofcatalysis 272 (2010) 109-120.

[8] C.-H.Wang, S.-S. Lin, C.-L. Chen, H.-S. Weng, Chemosphere 64 (2006) 503-509.

[9] V.Balcaen, R. Roelant, H. Poelman, D. Poelman, G.B. Marin, Catalysis Today 157(2010) 49-54.

[10] E.C.Su, W.G. Rothschild, Journal of catalysis 99 (1986) 506-510.

[11] B.Harrison, A. F. Diwell, C. Hallett, Platinum Metal Review, 32 (1988) 73.

[12] J.T.Gleaves, J.R. Ebner, P.L. Miils, J.W. Ward, Studies in Surface Science andCatalysis 38 (1988) 633-644.

Fig. 1 EDX line scans through a STEM frame on the CuO-CeO$_2$/Al$_2$O$_3$ catalyst

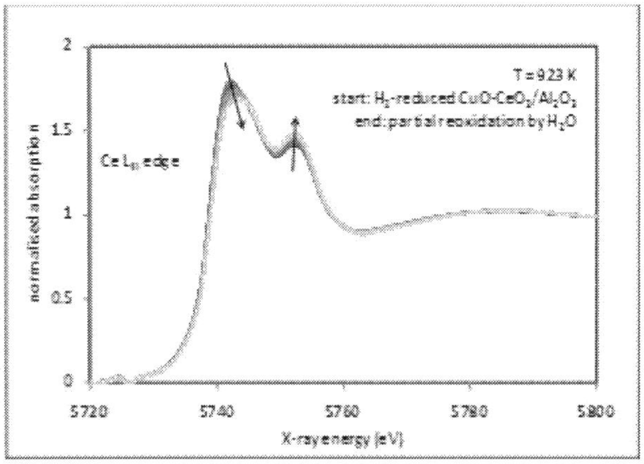

Fig. 2 CO$_2$ yield and ratio of CO$_2$ yields vs reaction temperature for experiments with and without H$_2$O in the feed over CuO-CeO$_2$/Al$_2$O$_3$ catalyst

See more of this Session: Applied Environmental Catalysis II
See more of this Group/Topical: Catalysis and Reaction Engineering Division

(141e) The Effect of Ca2+ and Al3+ Additions On the Stability of Potassium Disilicate (K2Si2O5) As a Soot Oxidation Catalyst

Changsheng Su, Depart of Chemical Engineering, University of Notre Dame, Notre Dame, IN and Paul J. McGinn, Department of Chemical Engineering, University of Notre Dame, Notre Dame, IN

The high activity of potassium-containing catalysts for soot oxidation has attracted much attention over the past decade. Unfortunately, potassium containing catalysts usually tend to degrade after repeated thermal cycles due to the loss of potassium. This is especially the case for K_2CO_3 or KNO_3. This deactivation is thought to occur due to loss of potassium through sublimation. Thus the challenge in using potassium and other alkali metal based catalysts is minimizing, if not eliminating, the loss of the potassium or alkali metal. One approach often used is to incorporate K into a more stable structure (e.g. perovskites)

In many cases attempts to stabilize K were pursued by impregnating potassium species into a support material. This approach has been common over the past decade, with examples such as potassium catalyzed soot oxidation supported by TiO_2, ZrO_2 , Al_2O_3, SiO_2 etc. Among these support materials, Al_2O_3 and SiO_2 do not show any catalytic effect while ZrO_2 and TiO_2 show very low activity in soot oxidation. The potassium-TiO_2 interaction and the improvement of contact conditions between soot and the catalyst, studied by means of thermal programmed reduction (TPR) and X-ray diffraction(XRD), were offered as an explanation for the activity differences for soot oxidation compared to catalysts supported on alumina. K/ZrO_2, with a K/Zr ratio=0.14, synthesized from potassium nitrates was reported as providing the highest activity. The reversible transformation of a bridged NO^{3-} and a monodentate NO^{3-} was suggested as a possible mechanism. Studies on the application of K on silica supports was only recently reported.

Ogura et al. tried to prevent potassium losses by using silica-alumina and zeolites as support materials and impregnating potassium as the active catalytic center. They found that the second test cycle of soot oxidation on K/sodalite showed higher activity than the first run. It appears that the state of potassium was changed under thermal conditions with a minor loss of the crystal structure of sodalite resulting from interaction of Al sites and K in aluminosilicate. However, no extended stability test was performed to reveal the relatively long-term catalytic performance in this study. Relatively extended stability testing was performed by López-Suárez et al. when they studied the effect of copper on potassium stability in $K/SrTiO_3$ catalyzed soot oxidation. According to their 6-cycle repeated TPR results, significant degradation was found between the first and second TPR cycle for all potassium catalysts.

Even though the less active $Cu/SrTiO_3$ catalysts do not degrade, the addition of copper into $K/SrTiO_3$ structure does not affect the potassium stability.

Research results show that catalysts with potassium can lead to high activity for soot combustion. Nevertheless, such catalysts can be readily degraded due to the loss of active potassium. Conversely, an investigation showed that catalysts with immobile K ions that are rigidly bound in the lattice give lower activity and higher soot oxidation temperatures. Neither case is ideally suited for the application on DPF. As an alternative one can consider the use of potassium containing glasses as catalysts. This approach relies on the slow passive release of potassium from a glass to provide for renewal of the catalytic surface activity of the glass. The purpose is to mitigate the effect of loss of the active potassium species by providing new ions over time.

The slow release of ions desired in a catalytic glass is the opposite of what is desired for a durable, weather resistant glass. Several approaches to modeling glass durability have been described, primarily for predicting the durability of glasses for nuclear waste storage. In that application, the goal is to minimize leaching from, or corrosion of, the glass, with researchers hoping to predict glass behavior over thousands of years. The approaches usually involve considering a glass as a mixture of species with known free energies of hydration. Glass durability is then predicted based upon thermodynamic hydration equations, with the contribution of the various species scaling with mole percentage of components present in the glass.

To design durable glasses, glass reactant species are selected based on hydration reactions that are expected to occur between the glass and an aqueous solution (acidic or basic). This is based on expectations as to whether cations in the glass will anionically complex with silica or other oxides. This is determined from their relative anionic force, which reflects their relative field strength. Cation species that might be incorporated into a glass can be classified as to whether they are network formers (i.e. ions with high atomic field strengths (F), calculated as the atomic charge (Z) divided by the square of the ionic radius (r)), network modifiers (i.e. ions with low atomic field strengths), or intermediate cations, that can act as either network formers or network modifiers. Network modifier cations are oxide species which are highly anionically associated with $[SiO_4]^{-4}$ tetrahedra. Potassium and sodium are examples of network modifiers that have low field strengths, so are susceptible to leaching.

The field strengths are considered along with the relative partial molar free energies of the hydration reactions (DG_i) of the cation species that can occur in an aqueous environment. A chemically and electrically balanced hydration reaction can be written for potential components, and the partial molar hydration free energy for each reaction be calculated ($DG_i = DG_{(products)} - DG_{(reactants)}$) for the expected hydration reactions in the environment (e.g. aqueous oxidized basic environment). Jantzen showed that the thermodynamic free energies of hydration (DG_{hyd}) of the silicate and oxide glass components are correlated with their ionic field strengths (F).

The results of these studies provide a starting point for selecting ions for incorporation into a glass with poor weathering characteristics, as is desired for DPF application. Such a glass should incorporate relatively high levels of low field strength, highly negative free energy of hydration ion species (i.e. alkali ions). For the novel application of glasses as soot combustion catalysts it is necessary to balance two characteristics of the glass: melting point and ion delivery. Not only is easy ion exchange desired, but the glass should have a melting point high enough to withstand the diesel exhaust environment without substantial melting or particle coarsening. During soot combustion, it is possible that local temperatures can reach 700-800C for brief periods. Thus a desirable glass possesses a relatively high melting point as well as facile ion exchange.

For designing a catalytic glass, a good starting point is provided by the alkali disilicates (or "phyllosilicates"), of the type $M_2Si_2O_5$, where M is one or more alkali metal ions. The alkali disilicates are of particular interest because while some of them are very susceptible to degradation by moisture, others are unaffected. For example, $K_2Si_2O_5$ is very hygroscopic, while $Li_2Si_2O_5$ is unaffected by moisture. Potassium disilicate ($K_2O.2Si_2O_2$) glass thus can provide a starting point for examining mixed ion effects in soot oxidation catalysts based on glass. In the present study we examine if such mixed ion effects can be observed in terms of soot oxidation activity.

See more of this Session: Applied Environmental Catalysis II
See more of this Group/Topical: Catalysis and Reaction Engineering Division

(141f) On the Activity and Selectivity of Ag/Al2O3 Catalyst During the Lean Reduction of NO by C3H8 Studied At High Space Velocities

M. Eugenia Hernndez-Tern[1], Isidro Mejia-Centeno[2], Sergio A. Gmez[1] and **Gustavo A. Fuentes**[1], (1) Department of Process Engineering, UAM-Iztapalapa, Mexico DF, Mexico, (2)Divisin de Qumica Molecular, Instituto Mexicano del Petrleo, Mexico, DF, Mexico

Ag/Al_2O_3 is considered a promising catalyst for the direct selective lean reduction of NOx by hydrocarbons and oxygenates. It has been reported to have high activity, good selectivity to N_2, stability, durable activity in presence of water and SO_2, moderate operating temperature window and relatively low cost (1-4). Most of the studies have been made at low space velocities, however, and that difficults the analysis of intermediates and the effect of catalyst composition.

In this study we present results obtained at high space velocities (F/W =666-32,000 $cm^3.g^{-1}.min^{-1}$) using a ~2 wt% Ag/Al_2O_3 catalyst prepared by impregnation. The feed gas mixture contains 200 ppm NO, 1750 ppm C_3H_8, 3% O_2 and 6% H_2O diluted in nitrogen as a balance gas. The reactions were carried out in a tubular quartz microreactor equipped with a mass flow controllers and PID temperature control. The gas analyses were done by GC (HP 6890 and Shimadzu GC-12A), on line with an FTIR (Bruker Tensor 27) equipped with 0.75 m path-length infrared gas cell heated at 120ºC to prevent condensation, and a chemiluminescence NO-NOx analyzer (Rosemount 951A).

We observed that the reduction of NO with C_3H_8 occurs in the 375-500°C temperature range in the absence of H_2. By adding a small amount of H2 to the feed the light-off temperature decreases nearly 200ºC. These results are in line with those reported in the literature (2), but operation at high F/W shows that our catalyst is significantly more active in the reduction of NOx. The conversion - W/F behavior, particularly at low temperature, suggests that the mechanism involves consecutive reactions. This issue is being studied and will be discussed further. Characterization of the catalysts using UVVis in situ allows us to understand the role of Ag nanoparticles and AgOx species present under reaction conditions and to dwelve into the reason for the high activity of this class of catalysts.

References

1. Midyadera, T., Appl. Catal. B 2 (1993) 199.
2. Zhu, T., Hao, J., Fu, L., Li, J., Liu, Z., React. Kinet. Catal. Lett. **84**,1(2005) 61.
3. She, X., Flytzani-Stephanopoulos, M., Catalysis Today 127 (2007) 207.

4. DiMaggio, C., Fisher, G., Rahmoeller, K., Sellnau, M., SAE Paper 2009-01-0277

See more of this Session: Applied Environmental Catalysis II
See more of this Group/Topical: Catalysis and Reaction Engineering Division

(141g) Mechanistic Investigation of Ethanol SCR of NOx Over Ag/Al2O3

William Johnson II[1], Josh A. Pihl[1], Galen B. Fisher[2] and **Todd J. Toops**[1], (1)Oak Ridge National Laboratory, Knoxville, TN, (2)Chemical Engineering, University of Michigan, Ann Arbor, MI

Supported Ag on gamma-Al_2O_3 has been shown to be an effective catalyst for selectively reducing NOx to N_2 in a wide temperature range with various hydrocarbons and oxygenates typically found in fuels used in mobile systems. Additionally, this reactivity has been demonstrated at a range of hydrocarbon to NOx (HC_1/NOx) ratios that would not introduce a severe fuel penalty [1]. A particularly effective HC for this selective catalytic reduction (SCR) of NOx is ethanol, which is found at low levels in all gasoline and is becoming more prevalent in concentration and availability in domestic automotive fuels (e.g., E85). Fisher et al. [2,3] have demonstrated that flowing NOx and ethanol vapor (and other HCs) over these same Ag-based catalysts can form NH_3 under lean conditions, which can then be used with a downstream zeolite-based NH_3-SCR catalyst to produce NOx conversions in excess of 90% for realistic exhaust streams in lean burn engine systems. This important finding may offer an alternative to urea dosing currently being implemented with several NH_3-based SCR systems using either zeolite- or vanadia-based catalysts [4]. All of these recent findings point to the importance of studying this chemistry and establishing a predictive model of the reaction in the silver catalyst that may also account for NH_3 formation.

A 2 wt.% Ag/g-Al_2O_3 catalyst was synthesized using the incipient wetness technique with a $AgNO_3$ precursor. Catalysts were calcined at 700 °C for 15 h in flowing air and evaluated at gas hourly space velocities (GHSV) ranging from 30k to 180k h^{-1} in temperatures from 200 to 550°C. Feed gas conditions were 500 ppm NO, 10% O_2, 5% H_2O, and either 750 ppm ethanol at 180k h^{-1} GHSV or 1500 ppm ethanol at 30k h^{-1} GHSV (balance Ar). All gas flows were controlled by mass flow controllers with H_2O and ethanol introduction using impingers submersed in temperature controlled baths. Chemiluminscent NOx analyzers were used in conjunction with a quadrupole mass spectrometer for gas-species analysis.

At 30k h^{-1} and a HC_1/NOx=6 ethanol conversion increases very rapidly from 0% at 200 °C to over 80% at 250 °C; this contrasts with the low NOx conversion at 250°C, less than 20%, despite the high HC_1/NOx ratio of 6. However, this ethanol oxidation does not occur in the absence of NOx; in fact, there is 0% conversion at 250°C when flowing ethanol vapor and O_2 to the reactor. This new result is consistent with NOx initiating the ethanol oxidation pathway, especially considering that Ag/Al_2O_3 is an excellent NO to NO_2 oxidization catalyst, greater than 80% NO_2/NOx at 200°C. This finding suggests that the high ethanol oxidation rate at 250°C may be due to NO_2 initiating the reaction. However, this NO_2 is only reduced to NO, which can be re-oxidized to NO_2, and the cycle can continue. At higher temperatures the NO can be dissociated and its reduction pathway can be followed as evidenced by the higher NOx conversion rates.

To further study the kinetics and mechanisms at work in this catalyst system it is necessary to investigate the reactions in the kinetic regime. Differential conversion of both NOx and ethanol is achieved between 250-290°C at a GHSV of 180k h^{-1}. This allows for the determination of the activation energy for NOx conversion of 63 kJ/mol, which is comparable to the 67 kJ/mol reported by Shimizu et al. for NOx reduction with n-hexane over Ag/Al_2O_3 [5]. Additional efforts are being performed to determine the dependence of the reaction rate on the partial pressure of the reactants. These tests will help with kinetic rate law development that includes the rate of NH_3 formation.

1. Wu, Q., Hong, H., and Yu, Y. Appl. Catal. B: Environmental 61, 67 (2005).
2. Fisher, G. B., DiMaggio, C. L., Trytko, D., Rahmoeller, K. M., and Sellnau, M. SAE Paper 2009-01-2818 (2009).
3. DiMaggio, C. L., Fisher, G. B., Rahmoeller, K. M., and Sellnau, M. SAE Paper 2009-01-0277 (2009).
4. Diewald, R., 2010 Directions in Engine-Efficiency and Emissions Research (DEER) Conference, September, 30, 2010, (2010).
5. Shimizu, K., Shibata, J., Yoshida, H., and Hattori, T. Appl. Catal. B: Environmental 30, 151 (2001).

See more of this Session: Applied Environmental Catalysis II
See more of this Group/Topical: Catalysis and Reaction Engineering Division

(147a) Density Functional Theory Research On Mechanisms and Rate-Determining States of Cyclohexanone Ammoximation Over Titanium Silicalite-1

Feng Xin and **Yanying QI**, School of Chemical Engineering and Technology, Tianjin University, Tianjin, China

Density functional theory research on mechanisms and rate-determiningstates of cyclohexanone ammoximation over Titanium Silicalite-1

Yanying Qi Feng Xin~

School ofChemical Engineering and Technology, Department of Chemical Engineering, TianjinUniversity, Tianjin 300072, China

Abstract

The concept of "rate-determining steps" hasbeen of great significance in deriving reaction kinetics from reactionmechanism for many decades. Nevertheless, more and more articles have got anagreement on identifying the rate-determining states more useful than therate-determining steps for both theoretical and practical researches. Therate-determining states provide the possibility of finding an appropriatecatalyst or improving the existed ones. Foran environmental friendly reaction of cyclohexanone ammoximation on titaniumsilicalite-1 (TS-1), the mechanisms should be confirmed from molecular simulation,and then its kinetics will be established by the rate-determining states andexperiments, which is useful for improving the catalyst performance.

In this paper, the pathways of catalyticammoximation of cyclohexanone in TS-1 and H_2O_2 system wassimulated in a cluster model of TS-1 containing a defect Ti site by using Dmol3 in Material Studios software. Densityfunctional theory (DFT) were utilized with exchange functional of Becke's 1988plus Lee-Yang-Parr's 1988 correlation energy function (BLYP), which was moresuitable to this system compared to other functions in Dmol3. The used atomicorbital basis sets were Double Numerical plus polarization (DNP), whichincluded a polarization p-function on all hydrogen atoms and a polarizationd-function on non-hydrogen atoms. All of the energies were calculated withzero-point energy (ZPE) correlation. For searching transition state, we chose themethod of complete linear synchronous transit and quadratic synchronous transit(complete LST/QST). For acquiring more accurate transition state, we used TSconfirmation tool, which began by approximating the Intrinsic Reaction Path (IRP)with QST and then performed subsequent refinements.

In the beginning,we compared our results of the formation of Ti-OOH with some experimental andcomputational data to correct the method and model. As a result, a novel activecenter was proposed, where TS-1 had Ti(1-OOH) defect active center with a ligand of NH_3.The transition states of all these reaction pathways were explored. A whole catalyticcycle was determined. In the imine mechanism, NH_3 and cyclohexanonereacted firstly to form an intermediate complex C_6H_{10}-OH-NH_2that was transformed to imine with a hydrogen transfer from the nitrogen to theoxygen. The formation of imine was rate-determining step in the above mechanism.While in the hydroxylamine mechanism, we compared the energies of differentreaction routes in the catalytic and non catalytic reactions of hydroxylamineand cyclohexanone. In the end, Campbell'sdegree of rate control was utilized to obtain the rate-determining transitionstates and intermediates.

Keywords:

degree of rate control£££

*Corresponding author. E-mail: xinf@tju.edu.cn

Tel: +86 22 27409533, Fax: +86 22 27892359

See more of this Session: Computational Catalysis III
See more of this Group/Topical: Catalysis and Reaction Engineering Division

(147b) Understanding Hydrocarbon Reactivity In Zeolites with Molecular Dynamics Simulations

Paul M. Zimmerman[1], Martin Head-Gordon[1] and Alexis T. Bell[2], (1)Dept. of Chemistry, University of California at Berkeley, Berkeley, CA, (2)Department of Chemical Engineering, University of California Berkeley, Berkeley, CA

First principles simulations in zeolites reveal the principles of elementary catalytic reactions that are not captured by experiment. One such methodology for molecular dynamics simulations is quantum mechanical/molecular mechanical (QM/MM) simulation, which greatly improves computational efficiency. Although many such techniques exist, our method does so with little sacrifice in accuracy by systematic choice of MM parameters. This methodology is utilized to simulate short time kinetics involved in hydrocarbon conversions in zeolites. These trajectories can provide estimates of carbocation lifetimes, and we will describe dynamics studies of the tert-butyl carbocation in H-FER and H-MFI zeolites. The simulations provide estimates of the lifetime of this important intermediate, and these will be compared to lifetimes obtained from transition state theory.

See more of this Session: Computational Catalysis III
See more of this Group/Topical: Catalysis and Reaction Engineering Division

(147c) A Theoretical Investigation of the Methylation of Olefins by Surface Methyl Groups On Zeolites

Mark N. Mazar[1], Saleh Al Hashimi[2], Aditya Bhan[1] and Matteo Cococcioni[1], (1)Chemical Engineering and Materials Science, University of Minnesota, Minneapolis, MN, (2)Chemical Engineering, The Petroleum Institute, Abu Dhabi, United Arab Emirates

Emerging research in the conversion of methanol to hydrocarbons (MTH) over zeolitic acids suggests that observed catalytic rates and selectivity in the indirect "hydrocarbon pool" mechanism are primarily a result of olefin and arene methylation reaction sequences. In this research, cluster-based and periodic density functional theory studies were used to examine the kinetics and mechanism of olefin methylation by surface methyl groups on CHA, FER, BEA, MOR and MFI zeolite frameworks. The location of the Al atom was chosen such that the bridging O atom had access to the largest open volume of each framework. In each case, the transition state was found to consist of a nearly coplanar CH_3 species that was coordinated with both the original bridging O atom and the olefin suggesting that an sp^2 hybridized carbon species is involved in the transition state. Activation energies of 70, 86, and 98 kJ mol^{-1} for ethylene methylation were found for FER, CHA, and BEA, respectively; in each case, lower than that reported for a mechanism involving the co-adsorption of methanol and ethylene (104 kJ mol^{-1})(Svelle et al. *JACS* **2008**, *131*, 816-825). Observed trends in the activation energies and transition state geometries will be discussed with respect to the zeolite pore size and the local pore configurations.

See more of this Session: Computational Catalysis III
See more of this Group/Topical: Catalysis and Reaction Engineering Division

(147d) Density Functional Theory Study of Propane Ammoxidation Over Mo-V-Te-Nb-O M1 Catalyst

Junjun Yu[1], Muthukumar Kaliappan[1], Vadim V. Guliants[1] and Ye Xu[2], (1)Chemical Engineering and Energy and Materials Engineering, University of Cincinnati, Cincinnati, OH, (2)Center for Nanophase Materials Sciences, Oak Ridge National Laboratory, Oak Ridge, TN

The selective ammoxidation of propane into acrylonitrile catalyzed by the Mo-V-Te-Nb-O mixed metal oxide received considerable attention because of its environmentally friendly nature. This process consists of a series of steps involving propane oxidative dehydrogenation (ODH), ammonia and O_2 activation, and N-insertion into C_3 surface intermediates. However, the limited fundamental understanding of the catalyst has hindered the progress in improving their activity and selectivity required for commercial applications. In this study, density functional theory (DFT) calculations were performed to investigate the mechanism of propane ammoxidation over cluster models of the proposed selective and active sites present in the surface *ab* plane of the M1 phase, in order to understand the roles of the different metal cation species in the reaction. We have calculated the activation energy barriers for the oxidative dehydrogenation (ODH) of propane on vanadyl, molybdyl, and telluryl oxo groups (V=O, Mo=O, and Te=O). V^{5+}=O is more active than V^{4+}=O, in line with the current proposed reaction mechanism. However, Te=O is found to be the strongest adsorption site for H atoms and a significantly more active site for propane ODH than V=O regardless of the oxidation state of V. Radical C_3 species (isopropyl and allyl) energetically prefer to adsorb on V=O and Te=O. We also explored the adsorption of NH_x (x=0-3) species and the N-insertion steps into C_3 intermediates over proposed active sites present in the surface *ab* plane of Mo-V-Te-Nb-O M1 phase. The reactivity trends established for the constituent metal cations and the information about elementary reaction steps for propane ODH found in this study are highly useful for further improvement of the M1 catalyst.

See more of this Session: Computational Catalysis III
See more of this Group/Topical: Catalysis and Reaction Engineering Division

(147e) Active Sites for Olefin Metathesis On WO_3 Catalysts : A Density Functional Theory Study

Zhuo Cheng and Cynthia S. Lo, Department of Energy, Environmental and Chemical Engineering, Washington University, Saint Louis, MO

Industrial demand for polypropylene has spurred research on improved catalysts and mechanisms for propylene production. Recently, WO_3 was reported to exhibit high activity for the metathesis of ethylene and 2-butylene to form propylene. This study investigates active sites for this metathesis reaction on WO_3 surfaces, using first-principles calculations based on density functional theory. For the WO_3 orthorhombic crystal, the (001) surface has been found to be the most stable with sixfold coordinated tungsten sites W_{6c} and fivefold coordinated tungsten sites W_{5c}. The whole process is divided into two stages: 1. Initiation - the formation of active W-carbene species from the W-oxo precursor, and 2. Propagation - the metathesis reaction on these active sites to yield propylene. The oxidation states of W on the surface have a great influence on the adsorption of ethylene and 2-butylene. It is more favorable for W5c sites to adsorb ethylene while it is more favorable for W6c sites to adsorb 2-butylene, which were the compromising results of electrophilic and steric effects. The formations of the W-carbene species from ethylene and 2-butylene were found to follow different pathways. These study results reveal the relationship between WO_3 surface structure and metathesis activities, which is significant in guiding the preparation of more suitable WO3 catalysts for the production of propylene.

See more of this Session: Computational Catalysis III
See more of this Group/Topical: Catalysis and Reaction Engineering Division

(147f) Role of Ceria In Ti0.9Ce0.1O2 Materials for Adsorptive Desulfurization

Michael J. Janik[1], Siddharth Sitamraju[2], Jiahua Guo[2] and Chunshan Song[2], (1)Chemical Engineering, Pennsylvania State University, University Park, PA, (2)Energy and Mineral Engineering, Pennsylvania State University, University Park, PA

Fuel cell systems efficiently convert chemical energy to electrical energy. The reforming and water-gas shift catalysts used to produce hydrogen and the fuel cell electrodes are all sensitive to the sulfur compounds in the fuels. Conventional desulfurization methods, i.e. hydrodesulfurization (HDS), consume large amounts of energy to reach sulfur concentrations low enough for fuel cells. Adsorptive desulfurization (ADS) is a promising method being developed to produce ultra-low sulfur fuels. A $Ti_xCe_{1-x}O_2$ material has been reported by the Song group at Penn State to be a potential adsorbent for ADS due to its good selectivity and regenerability. The ratios of Ti:Ce were optimized to 9:1 for the maximum adsorption capacity. Comparing $Ti_{0.9}Ce_{0.1}O_2$ to TiO_2, the 10% CeO_2 addition significantly increases the capacity, despite only a slight surface area increase. We have applied density functional theory (DFT) calculations to examine the role of Ce addition to TiO_2 in promoting the adsorption of thiophenic molecules. A number of possible contributions of Ce addition were evaluated, including 1) stabilization of TiO_2 surfaces that adsorb thiophene strongly, 2) stronger adsorption to Ce-doped TiO_2 surfaces, 3) stabilization of reduced or over-oxidized sites on Ce-doped TiO_2, and 4) increased oxygen storage capacity to promote formation of adsorbed sulfone-like complexes. We find that Ce-doping of the surface can offer adsorption sites with stronger binding of thiophene, and can stabilize the anatase (100) surface which demonstrates strong adsorption affinity. Ceria may also serve as an oxygen reservoir to promote formation of adsorbed sulfone species to the TiO_2 surface. Thiophene adsorption energies are compared with cyclopentene to probe adsorption selectivity, and substitution of both thiophene and cyclopentene is considered to evaluate variations in selectivity. Calculations using DFT and dispersion corrected DFT, DFT+D, are compared.

See more of this Session: Computational Catalysis III
See more of this Group/Topical: Catalysis and Reaction Engineering Division

(147g) DFT Study of the Hg Oxidation Mechanism Over the V2O5-TiO2(001) Surface

Ana Suarez Negreira, Energy Resources Engineering, Stanford University, Stanford, CA and Jennifer Wilcox, Department of Energy Resources Engineering, Stanford University, Stanford, CA

We present a density functional theory study of Hg oxidation from the interaction between Hg and HCl with the V2O5(001) surface supported on TiO2(001). We investigate the intermediate products of the reaction and calculate the reaction energy barrier for each elementary reaction, taking into account several different adsorption geometries and intermediate products. We propose that Hg oxidation occurs through a Mars-van Krevelen type mechanism consisting of seven steps: adsorption and dissociation of HCl on a surface oxygen vacancy, combination of atomic chlorine with neighboring adsorbed Hg, combination of HgCl with adsorbed Cl, desorption of HgCl2, hydroxyl recombination, water desorption and dissociative oxygen adsorption. The rate-determining step in the reaction is calculated and the results are compared qualitatively to available experimental data.

See more of this Session: Computational Catalysis III
See more of this Group/Topical: Catalysis and Reaction Engineering Division

(158a) Reaction Engineering of New Energy Technologies

Lanny D. Schmidt, Chemical Engineering and Materials Science, U of Minnesota, Minneapolis, MN

While the first routes to producing biofuels involves variations on petroleum refining, the processes will ultimately be very different: probably as different as the whaling industry was from petroleum refining. Biomass is very different from crude oil in its composition and reactivity, and its geographic distribution and energy density require very different processing reactions and reactors. The biorefinery will be much smaller and will produce a very different product spectrum.

Consequently, we will have to rethink strategies for biomass processing. This involves smaller systems, transient operation, lower pressures, reduced recycle, and combined production of heat, power, fuels, and chemicals. Since feedstocks vary widely in composition and energy density, biorefining systems will have to be tightly controlled with all conditions preprogrammed.

We illustrate these differences with examples from catalytic partial oxidation of biomass surrogates in short contact time reactors. Topics considered are novel reactor configurations, new catalysts, solids gasification, and management of inorganic impurities.

See more of this Session: Future Directions In Reaction Engineering
See more of this Group/Topical: Catalysis and Reaction Engineering Division

(158b) A Chemical Perspective On the Conversion of Biomass to Fuels

Aditya Bhan, Chemical Engineering and Materials Science, University of Minnesota, Minneapolis, MN

From a chemical standpoint, the general $[-CH_2O-]_n$ stoichiometry of cellulosic biomass implies the conversion of biomass to fuels - with general stoichiometry $[-CH_2-]_n$ centers on the removal of oxygen. The stoichiometric hydrogen deficiency of biomass is accurately described by the effective hydrogen index (H-2(O))/C proposed by Chen et al., and all chemical processes for the conversion of biomass to fuels must remove the oxygen in biomass and concurrently increase the hydrogen content. Present direct and indirect processes for the conversion of biomass to hydrocarbon fuels use molecular hydrogen to increase the final H:C ratio of the fuel while rejecting oxygen as water. A potentially significant step in biomass-to-fuels conversion lies in the development of a single-step process that does not involve molecular hydrogen.

In this talk, I will briefly discuss our efforts to co-process hydrogen rich alkanes with biomass-derived oxygenates for the production of synthetic fuels. As raw materials for fuels, biomass and light alkanes lie at opposite ends of the chemical spectrum. Light alkanes are inert and their chemical conversion involves the removal of hydrogen and may involve oxygen addition while, biomass-feedstock contains oxygen, the removal of which limits biomass-to-fuels conversion and involves the addition of hydrogen. I will describe our results related to coupling biomass-deoxygenatation with alkane-dehydrogenation pathways over zeolite catalysts so that in essence, alkanes serve as a surrogate for molecular hydrogen for biomass deoxygenation while biomass serves as the oxygen carrier for hydrogen removal from alkanes.

See more of this Session: Future Directions In Reaction Engineering
See more of this Group/Topical: Catalysis and Reaction Engineering Division

(158c) Water-Tolerant Lewis Acids for the Activation of Biomass-Derived Molecules

Yuriy Roman, Chemical Engineering, Massachusetts Institute of Technology, Cambridge, CA

The depolymerization of the constituents of biomass by way of thermochemical or biological routes results in the production of a wide variety of oxygenated intermediates. Although the composition of these intermediates is quite diverse, carbonyl functionalities are prevalent (e.g. aldehydes, ketones, carboxylic acids, etc). In order to transform these functional intermediates into useful chemicals and fuels, robust and active catalysts need to be developed to selectively activate these molecules. Lewis acids are known to form strong complexes with carbonyl groups. A particular instance involves the complexation of the lone pair of electrons from the oxygen to the Lewis acid, effectively polarizing the carbonyl carbon atom, thereby making it prone to nucelophilic attack. Unfortunately, biomass-derived molecules are commonly found in aqueous solutions and, in most cases, Lewis acidity is suppressed in the presence of water, since the hydroxyl group of water also forms strong complexes with Lewis acid sites. The incorporation of metal centers (e.g. tin, titanium, and zirconium) into the framework of pure silica zeolites results in materials that maintain Lewis acidity in purely aqueous environments. In this presentation, several interesting Lewis-acid-mediated reactions of biomass-derived molecules, including isomerizations, condensations, and 4+2 cycloadditions will be discussed.

See more of this Session: Future Directions In Reaction Engineering
See more of this Group/Topical: Catalysis and Reaction Engineering Division

(158d) Future Directions and Research Challenges In Biomass Pyrolysis

Paul Dauenhauer, Chemical Engineering, University of Massachusetts Amherst, Amherst, MA

Recent interest in conversion of lignocellulosic biomass has increased focus on the engineering of pyrolysis of wood fibers and grass for subsequent refining to fuels and chemicals [1]. Pyrolysis of biomass can occur within numerous types of reactors utilizing several different heating, mixing, and quenching methods. Additionally, biomass feedstocks exhibit significant variability in both composition of biopolymers, extractives, and inorganic components (natural catalysts). The ordered combination of these components comprises a microstructure developed from plant cells which also varies with different species and growing conditions. The variability within lignocellulosic materials and reactor types requires a more detailed understanding of the reaction chemistry and transport phenomena within thermally degrading biomass particles. Existing descriptions of reacting wood fibers couple heat transfer and lumped kinetic models capable of describing up to six chemical groups [2]. This approach successfully describes approximate yield of product vapors and particle conversion which is useful for optimizing reactors for bio-oil yield.

The next generation of biomass pyrolysis models must be capable of predicting the molecular-level chemistry occurring during pyrolysis and the impact of transport phenomena on specific chemical reactions. Recent experimental evidence has indicated that the reactive intermediates for carbohydrates, lignin, and lignocellulose exist within a short-lived intermediate liquid which subsequently evaporates to vapors and gases [3]. The entire pyrolysis process occurs through a combination of solid, liquid, and gaseous chemistries which must be independently described. Additionally, mass, heat and momentum transfer between the three reacting phases dictates the residence time distribution for reactions within each phase and ultimately the distribution of products from pyrolysis. These chemistries are also affected by the microstructure of lignocellulosic materials (e.g. wood fibers) which are known to exhibit significant changes in geometry as well as heat and mass transfer throughout the lifetime of the biomass particle. The ability to examine each of these reaction/transport phenomena independently and also the effect of their coupling within particle models will ultimately determine our ability to design pyrolysis reactors which select for optimal fuel precursors.

[1] NSF. 2008. Breaking the chemical and engineering barriers to lignocellulosic Biofuels: Next generation hydrocarbon refineries. Ed. George W. Huber, University of Massachusetts, Amherst. National Science Foundation. Chemical, Bioengineering, Environmental, and Transport Systems Division. Washington D.C. 180 p.

[2] C. Di Blasi, "Modeling chemical and physical processes of wood and biomass pyrolysis," *Progress in Energy and Combustion Science* **34** (2008) 47-90.

[3] P.J. Dauenhauer, J.L. Colby, C.M. Balonek, W.J. Suszynski, L.D. Schmidt, "Reactive boiling of cellulose for integrated catalysis through an intermediate liquid," *Green Chemistry* **11** (2009) 1555-1561.

See more of this Session: Future Directions In Reaction Engineering
See more of this Group/Topical: Catalysis and Reaction Engineering Division

(158e) Modern Catalytic Technologies for Converting Biomass to Renewable Fuels and Chemicals: An Overview of the Catalysis Center for Energy Innovation

Dion G. Vlachos, Chemical Engineering, University of Delaware, Newark, DE

In this talk, an overview of the research thrusts of the Catalysis Center for Energy Innovation (CCEI), one of the Energy Frontier Research Centers (EFRC) will be given. Modern catalytic-based technologies will be presented that may overcome challenges of traditional methods in converting biomass to renewable fuels, chemicals, and electricity. Cross-cutting technologies, including hierarchical multiscale materials and models, will be reviewed. We will then review modern technologies that rely on biomass degradation to simple derivatives, such as sugars, followed by a number of reactions, such as isomerization, acid-based chemistry, and etherification to convert sugars to valuable intermediates, such as furans. Reforming of biomass for hydrogen production will also be discussed.

See more of this Session: Future Directions In Reaction Engineering
See more of this Group/Topical: Catalysis and Reaction Engineering Division

(169a) Online Kinetic Analysis of Reactions In a Microreactor System

Jason S. Moore[1], Christopher D. Smith[1], Norbert Heublein[2] and Klavs F. Jensen[1], (1)Chemical Engineering, Massachusetts Institute of Technology, Cambridge, MA, (2)Chemical Engineering, Technical University of Munich, Munich, Germany

Multicomponent reactions (MCRs) are important to drug discovery by affording complex products in only a single step. By linking two of these MCRs, a Petasis boronic acid-Mannich reaction and an Ugi reaction, six different components can be incorporated in a relatively short time. The kinetics of each reaction and their side reactions, both in isolation and in series, were investigated in an automated silicon microreactor system with online UPLC analysis. The use of silicon microreactors allowed for precise temperature control and rapid changes between setpoints due to the high thermal conductivity of silicon. In addition, the small volume minimized the reagent cost and waste production from these studies. Furthermore, the use of automation significantly increased reaction efficiency and allowed for minimal operator intervention. Online analysis *via* UPLC allowed for quantification of a number of reaction components, including monitoring the formation of side products that were unknown prior to experimentation, which would not be possible with other some online techniques, such as IR.

These kinetic data were then used to perform a model-based optimization of these reactions in series. The optimum from this model was then used as the initial condition for an experimental optimization to test the model prediction and adjust the optimal conditions if necessary. This experimental optimization was performed with a conjugate gradient algorithm that determined the search direction using gradient information calculated from a factorial design of experiments around the initial condition. A line search was then executed with an Armijo-type technique. The efficiency of this method was then compared to an optimization performed with no *a priori* knowledge of the system kinetics using the same experimental optimization procedure.

See more of this Session: Microreaction Engineering
See more of this Group/Topical: Catalysis and Reaction Engineering Division

(169b) Comparison of Catalyst Preparation In Microreactors: Atomic Layer Deposition Vs. Impregnation

Huiliang Shi, Department of Chemical Engneering, Tsinghua University, Beijing, China and **Yi Cheng**, Department of Chemical Engineering, Tsinghua University, Beijing, China

Comparison of CatalystPreparation in Microreactors:

Atomic Layer Depositionvs. Impregnation

Huiliang Shi, XuliZhai,DayongTian, Weixiong Pan and Yi Cheng

Department of ChemicalEngineering, Tsinghua University, Beijing 100084, China

Microreactorshave great potential for fast heterogeneously catalyzed gas phase reactions underhigh temperature due to their superior mass and heat transfer properties.However, two fundamental challenges remain before we could fully exploit thekinetic potential of catalyst in microreactors. Firstly, traditional catalystpreparation methods, such as impregnation, would result in poor dispersion ofactive species on reactor walls. Secondly, severe catalyst fall-off tends to takeplace at high temperature due to the thermal expansion difference between themetallic substrate and the catalyst layer, which would hinder the practical applicationsof microreactors.

In the present study, a new method called "pre-heatedthermal spraying" was adopted to enhance the binding force between the catalystlayer and the metallic substrate of the catalytic plates. As a result, the preparedcatalyst by this means had a fall-off rate that could be neglected during high-temperaturereaction and vigorous ultrasonic treatment. This method has been successfullyapplied in hydrogen production from steam reforming of methane in microreactorsat our lab. Besides, we also employed atomic layer deposition (ALD) to preparecatalyst nanoparticles on the support of catalytic plates. Using ALD we wereable to obtain smaller catalyst particles with a narrower size distribution sothat the dispersion of active species could be improved. The differences in catalyticperformances (i.e. activity and selectivity) using the two preparation methods,ALD and impregnation, were investigated by evaluating the model reactions: ethane/propaneoxidative dehydrogenation (ODH). Besides, the differences in appearances (i.e.size and valence) of active species between fresh and used catalysts were alsoanalyzed. These results revealed that it is possible to use ALD for catalyst preparationin microreactors with high performance in catalytic activity and selectivity.

See more of this Session: Microreaction Engineering

Enzyme hydrolysis of soyabean oil in 3-phase slug flow microreactors

J. Čech, W. Schrott, M. Přibyl, D. Snita

ICT Prague, Department of Chemical Engineering, Technická 5, 166 28 Prague 6, Tel.: +420 2 2044 3239, Fax: +420 2 2044 4320, E-mail: cechj@vscht.cz;

Key words: segmented flow, microfluidic device, microchip, microfabrication, hydrolysis

Abstract

In this study, we deal with a microcapillary/microchannel 3-phase enzyme reaction system that uses a slug/segmented flow to increase the reaction surface. The microdevices are fabricated from polydimethyl siloxane (PDMS) or polymethyl methacrylate PMMA by a combination of microfabrication techniques – UV photolithography, micromachining, hot assembling, casting and UV/O3 treatment. The water-oil emulsion was generated in a plexiglass microchip with an X-cross microchannel arrangement. Both the oil and water were pumped into the microchip by a precise syringe pumps.

In this contribution, we present fabrication of microfluidic systems in which segmented flow will be used for hydrolysis of the vegetable oil, experimental arrangement for continual hydrolysis of vegetable oil. and characterization of segmented flow with comparison batch reaction and reaction in slug flow. Enzyme lipase, which will be used for hydrolysis, is dissolved in water phase.

1. Hydrolysis of soya bean oil

Enzyme lipase breaks triglycerides into free fatty acids and glycerol. Lipase operates at interfacial area. Therefore, it is necessary to create a large interfacial area.

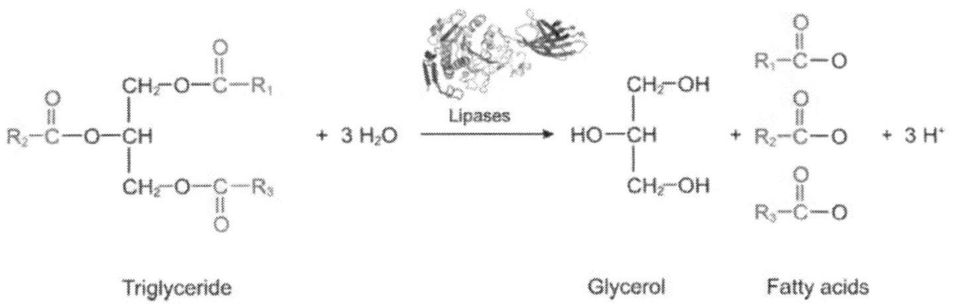

Fig 1: *Scheme of hydrolysis reaction*

2. Conversion evaluation

Product was evaluated by high pressure liquid chromatography. Mixture of acetone and acetonitrile (1:1) was used as a carrier phase [1]. Comparison between samples before and after reaction is depicted in fig. 10.

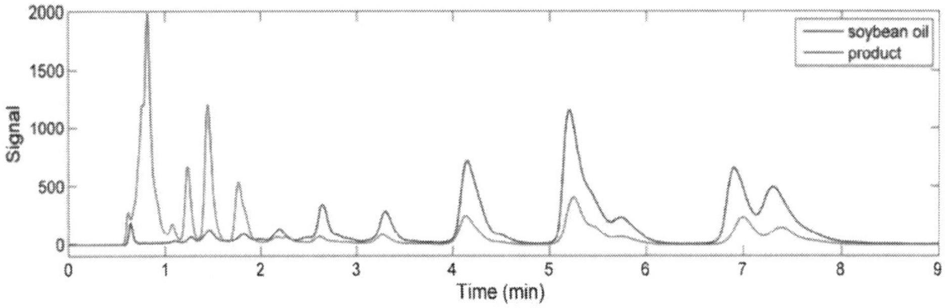

Fig. 2: *Comparison of chromatograms before and after reaction. Chromatogram before reaction is green, chromatogram after reaction is yellow. Reduced peaks represents triglycerides and enlarged peaks are fatty acids.*

3. Batch reaction

Reactor was placed in water bath with temperature 35°C. Mass composition of reaction mixture was oil/buffer/enzyme:20/10/1, phospate buffered saline (pH 7,4; 0,01M) was used as a buffer.

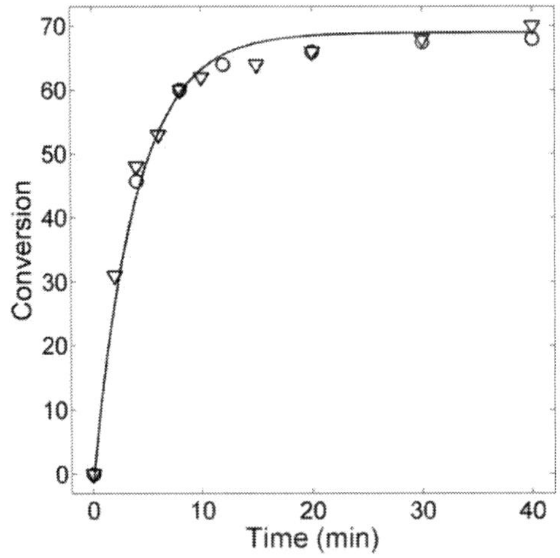

Fig. 3: *Dependence of triglyceride conversion on time in batch reaction.*

4. 3-phase (oil-water) slug flow reactor

Experimental arrangement

The system is designed for continuous hydrolysis of vegetable oil. For 3-phase system is needed to create the smallest beads of water to maximize interfacial area. This can be achieved either by creating a generator with small dimensions of the channels (Fig. 5), or by creating larger drops, which are further divided into drops with a half volume. Then the two phase slug flow is mixed with nitrogen in the L-G slug flow generator and the three-phase slug flow is created. Reactor is a capillary of chosen length and internal diameter 1mm. The L-G separator separates nitrogen from reaction mixture and L-L separator divides product (hydrolysed oil) from water phase.

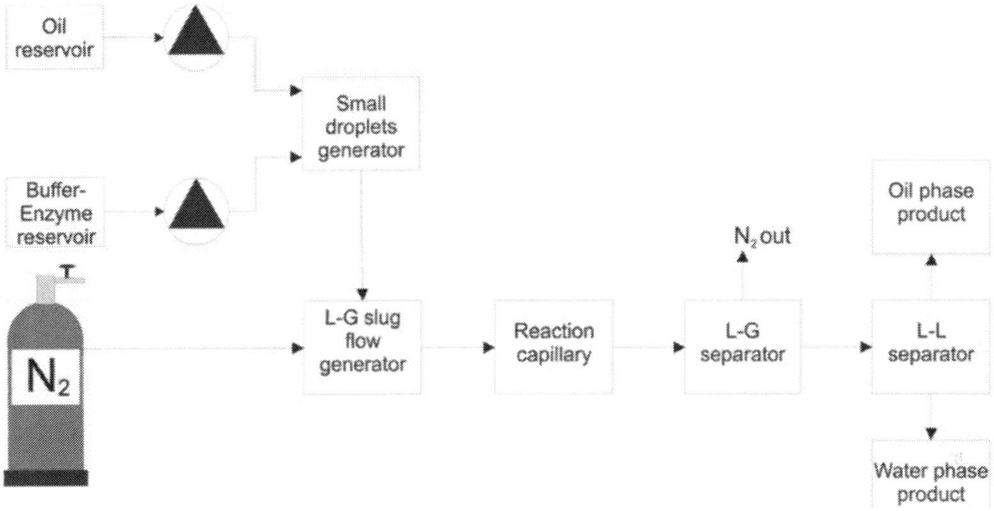

Fig.4 : *Experimental arrangement for 3-phase slug flow reaction*

Fabrication of small droplets generators

A PMMA (polymethylmathacrylate) was used for fabrication of generators. The PMMA was structured by means of micromachining. The chips consisted of two PMMA plates where the first one bore the fluidic structures and the second one openings for liquid delivery. The two PMMA plates were bonded with the use of a solvent (isopropylalcohol) [2]. The whole process of the solvent assisted bonding is as follows:

(i) Isopropyalkohol (IPA) is applied by a pipette on both plates and microfluidic parts.

(ii) Then, the microchip is put together and placed on a brass plate with drilled sucking holes at which position agrees with the position of the inlet and outlet holes in the microfluidic part of the chip.

(iii) The brass plate is then placed on a structured fiberglass mate, which is connected to a vacuum air pump.

(iv) The whole sandwich is placed in a hydraulic press, on an electrically heated plate.

(v) The pressing force is applied and IPA is sucked out of the micro-channels by vacuum.

Temperature of the heated plate is set and controlled by a regulator (Shimaden). Applied force is measured by the force transducer (HBM). As many experiments were carried out the optimal boxing parameters were found (bonding temperature: 65°C, applied force: 2500 N, bonding time: 10 min).

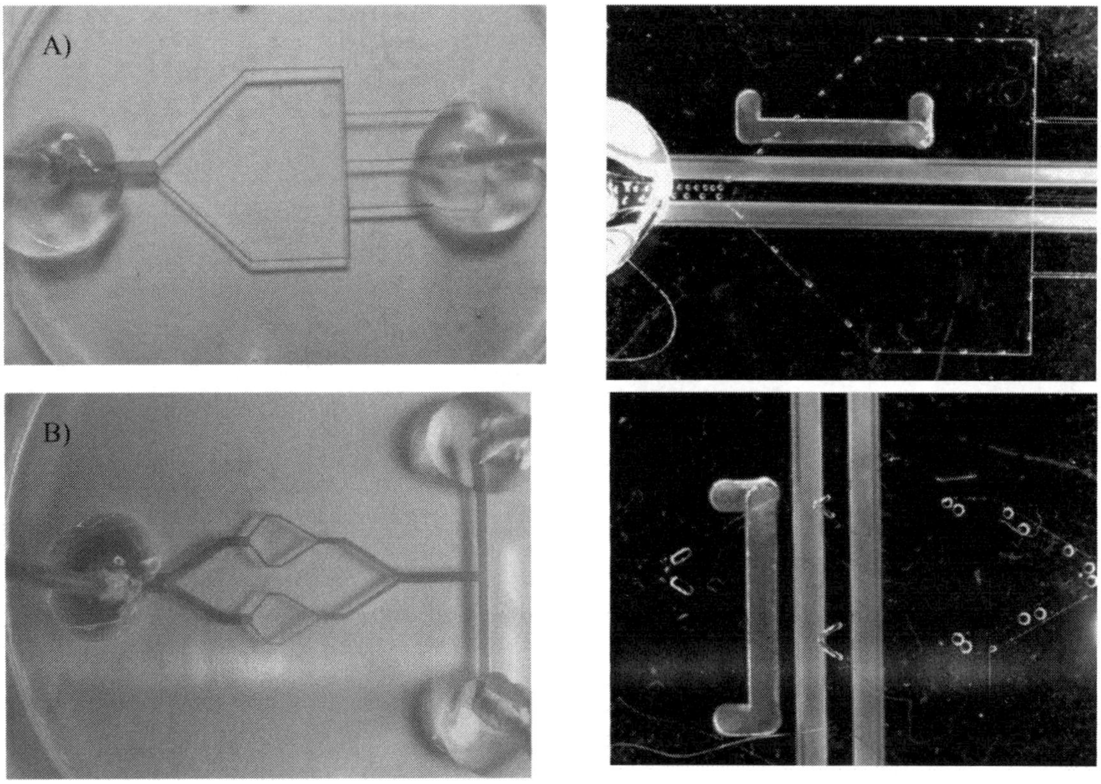

Fig 5: Two types of generators of small droplets: *A) Generator of small water droplets. B) Divider of water droplets.*

Fabrication of L-L separator

PDMS chips are fabricated by means of a casting method where a pre-polymer mixture is poured onto a stamp with the required pattern to be transferred [3]. This pattern is usually made in a thick layer of a SU8 photoresist structured by UV lithography [4, 5]. The PDMS mixture is cured at an elevated temperature and peeled off the stamp. The microchannel network is then enclosed by bonding to a pre-polymerized PDMS slab [6].

The whole process of the SU8 master fabrication is as follows:

(i) Cleaning and etching of a PB substrate in a diluted nitric acid (1 volume part of 68% acid in 2 volume parts of water).

(ii) Washing in DI water and acetone.

(iii) Drying for 30 min in an oven at 100°C.

(iv) Cooling to the laboratory temperature and spin coating a layer of SU8-2025 (MicroChem) on the substrate. The chosen height of the SU8 layer depends on an application.

(v) Pre-baking the SU8 layer at 100°C (the duration depends on the SU8 layer thickness).

(vi) After cooling, exposing through a mask with a desired pattern (the duration depends on the SU8 layer thickness).

(vii) Post-baking at 100°C and development (the duration depends on the SU8 layer thickness).

Fig.6 : *SU8 master on a PB substrate placed in the casting form*

The whole process of casting and sealing can be described as follows:

(i) PDMS (Sylgard 184, Dow Corning) is prepared by thorough mixing the pre-polymer and the curing agent in the weight ratio 10:1.

(ii) After degassing, PDMS is poured onto the SU8 master and degassed again.

(iii) PDMS is cured in a oven for 60 min at 70°C.

(iv) PDMS cast is peeled off the master and the inlet and outlet holes are punched at required locations.

(v) PDMS surface is finally cleaned with isopropyl alcohol and dried with nitrogen.

(vi) Another PDMS slab is prepared on a glass substrate and pre-polymerized (approx. 10 hrs at laboratory temperature or 35 minutes on a hot plate at 60 °C).

(vii) The two PDMS slabs are brought in contact and placed in a oven for 30 min to complete the microchip fabrication.

The microchip consisted of a central channel and an array of small side channels. Due to hydrophobicity of the PDMS substrate, the oil phase preferably entered the side channels. Thus the immiscible phases could be separated.

Fig.7 : *Primary PDMS separator*

Slug flow reaction

Reactants were mixed in the same ratio as in the batch reaction. Inner diameter of capillary was 0,5mm.

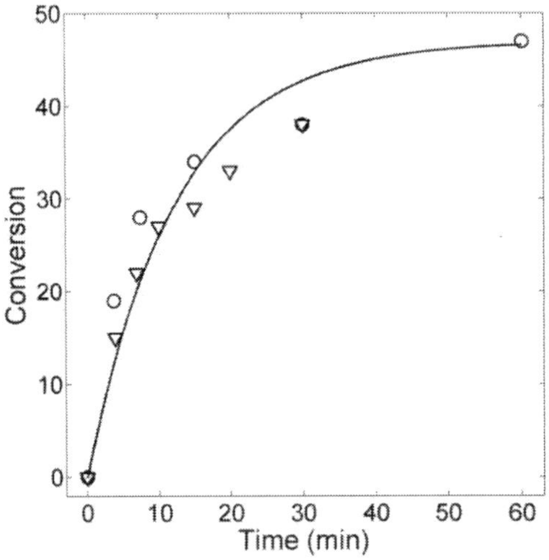

Fig. 9: *Dependence of triglyceride conversion on time.*

Character of flow in capillary

2-phase segmented flow is mixed with nitrogen. This creates a 3-phase segmented flow, where the entire slug flow ensures intensive mixing. Large interfacial surface is guaranteed a small volume of water drops.

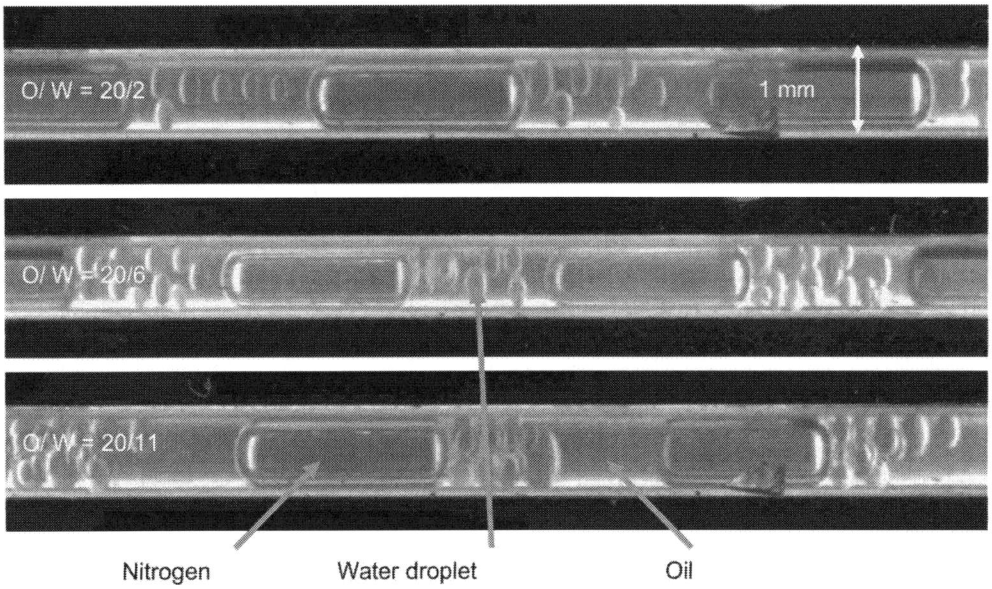

Fig 12: *Character flow in capillary for 3- phase slug flow*

5. The experimental apparatus

Temperature is one of the key parameter of enzyme reaction. The experiments are carrying out in temperature controlled box. The whole apparatus is equipped with computer controlled syringe pumps that provide constant flow rates of liquids. There is also camera and light source in the box.

Conclusions

3-phase slug flow in developed microfluidic was successfully created. The chips were made of PDMS and PMMA. It was found that the flow rate basically influences the size of droplets as well as the frequency at which these droplets are generated. The droplets can be efficiently employed as small reactors. The PDMS phase separator was fabricated by "pre-polymerization" technique. The effect of different parameters on segmented flow separation was examined. Experimental devices were placed into tempered box. Obtained conversion of batch reaction is higher then the conversion of reaction in slug flow, it may be caused by higher surface area in batch arrangement due to intensive mixing. This problem can be probably resolved by using a slug flow liquid-gas system. The obtained results provide a necessary knowledge in further development of microreactors for hydrolysis of vegetable oil by enzyme lipase.

Acknowledgement

The authors thank for the support of the research by the grant of the Czech Science Foundation 104/08/H055 and by the grant of Ministry of Education, Youth and Sport of the Czech Republic (Grants No. ME10036 and MSM 6046137306 and No. 21/2011).

Selected references

1. Ali Türkan, Saban Kalay, *Journal of Chromatography A*, 2006, **1127**, 34-44.

2. Svoboda, M., Slouka, Z., Schrott, W., Červenka, P., Pribyl, M., and Snita D., *Microelectronic Engineering*, 2010, **87**, 1590–1593.

3. Owen, M. J.; Smith, P. J., *Adhes. Sci. Technol.*, 1994, **8**, 1063-1075.

4. Tang Y.; Tan W.K.; Fuh J.Y.H.; Loh H.T.; Wong Y.S.; Thian S.C.H.; Lu L., *Journal of Materials Processing Technology*, 2007, **192-193**, 334-339.

5. Carlberg P.; Montelius L.; Tegenfeldt J.; *Microelectronic Engineering* , 2008, **85**, 210-213.

6. Schrott, W., Svoboda, M., Slouka, Z., Pribyl, M., and Snita D., *Microelectronic Engineering*, 2010, **87**, 1600–1602.

(169d) Oxidative Coupling of Methane (OCM) In a Microchemical Reactor

Sen Liu, Department of Chemical and Petroleum Engineering, University of Pittsburgh, Pittsburgh, PA, Shuang Liang, Department of Chemical Engineering, 1) US DOE-National Energy Technology Laboratory, Pittsburgh, PA; 2) University of Pittsburgh, PA, USA, Pittsburgh, PA and Gtz Veser, Chemical and Petroleum Engineering, University of Pittsburgh, Pittsburgh, PA

Understanding radical behavior is of great significance in homogeneous-heterogeneous reaction systems, such as high-temperature catalytic oxidation of hydrocarbons. Microreactor technology, with its precise control of fluid and temperature fields, improved reactant mixing, and large surface-to-volume ratio, is ideally suited for the study of such reaction systems. Most significantly, the pathways of homogeneous-heterogeneous coupling can be better understood as critical reactor dimensions are reduced into micrometer range, where diffusive flux starts to play an equally important role as convective transport.

Here, we are reporting a novel modular microreactor system based on the use of silicon-based thin-film catalysts, which allows stable operation during thermal cycling up to 800°C. Precise control and adjustment of the critical dimension, i.e. the height of reaction chamber, offers the ability to steer the relative importance of gas- and catalytic-phase chemistries due to transport of reactants between the gas phase and the catalytic walls. Additionally, a moveable thermocouple and quartz-glass capillary (connected to a mass spectrometer) allow the in-situ measurement of temperature and composition profiles in the reaction chamber. By combining these two capabilities, the system thus provides an efficient and sensitive way to investigate the interplay between gas phase and catalytic chemistries, and to evaluate the catalytic performance of thin-film catalysts.

The reactor system was tested using oxidative coupling of methane (OCM) as model reaction. OCM has been studied intensively for many decades and offers a potentially highly efficient path for direct upgrading of methane to higher-value C_2 products. The reaction is also well known to include catalytic steps in the generation of methyl radicals (as well as in undesired methane combustion) and homogeneous reaction steps in which methyl radicals are coupled to form the desired C_2 products (C_2H_6 and C_2H_4). It thus forms an ideal test system for the above described microreactor system.

A La-based thin-film catalyst was deposited onto a silicon chip via dip-coating, characterized, and inserted into the microreactor. The effect of major reactor operating parameters, such as temperature, flow rate, C:O feed ratio, and, most importantly, the surface-to-volume ratio were tested in detailed experimental studies using both reactor outlet and spatially-resolved concentration profiles. The results show a strong decrease of C_2 production rates with decreasing microreactor channel height (680 to 460 and 330), in agreement with the established OCM reaction mechanism. Detailed FEM reactor simulations including gas phase and catalytic reaction steps are currently under way to validate the experiments and gain deeper insights into the coupling pathways.

See more of this Session: Microreaction Engineering
See more of this Group/Topical: Catalysis and Reaction Engineering Division

(169e) Hydrogenation IN Microreactors with Immobilized Noble Nanocatalysts

Rui Lin, Xianfeng Ma and Robert Ofoli, Chemical engineering & Materials Science, Michigan State University, East Lansing, MI

Multi-phase hydrogenation is one of the most important reactions in converting biorenewables to fuels and/or chemicals. The conversion reactions should ideally be run under moderate temperatures and pressures to protect fragile feedstock. However, due to the low solubility of hydrogen in the aqueous phase, they have typically been run at high temperatures and pressures, which make it difficult to determine the intrinsic kinetics. In addition, the large number of catalyst candidates makes rapid and accurate evaluation of catalyst performance very desirable. To address these issues, we have combined nanocatalysis and microfluidics to create a microreactor that can be used to study the kinetics and reaction mechanisms of hydrogenation and other renewables conversions. We synthesized noble nanoparticles (Pd, Pt and Ru) stabilized by n-dodecyl sulfide by thermal decomposition. The nanoparticles ranged from 2 to 5 nm in size with a narrow size distribution, as shown by high resolution transmission electron microscopy (HRTEM) images. The resulting high surface-area-to-volume ratios enhanced hydrogenation at much lower temperatures and pressures than typically used. We fabricated microreactors by soft lithography from polydimethylsiloxane (PDMS) to greatly reduce mass transfer limitations and shorten reaction times. We then developed schemes to immobilize the catalytic nanoparticles *in-situ* in the microreactors to promote interaction of reactants and catalysts, and enable easy catalyst recycling[1,2]. X-ray photoelectron spectroscopy (XPS) and energy dispersive X-ray spectroscopy (EDS) were used to measure the atomic concentrations of nanocatalysts on the surfaces, with the results confirming successful immobilization of the catalysts. The system was assessed by hydrogenation of 6-bromo-1-hexene at room temperature and 1 atm. of hydrogen pressure. The estimated residence time in the microreactor was 4 s, with a selectivity of 100% and complete conversions at lower substrate concentrations[3,4]. Control experiments showed no hydrogenation occurred in blank microreactors. Catalyst leaching was negligible according to ICP-OES analysis, indicating robustness of the immobilization protocol. The catalyst turnover frequency was three orders of magnitude higher than in conventional batch systems[4]. The nanocatalysts were recycled at least 3 times before decreases in conversions were observed. This system has the potential to provide an important tool for efficient and rapid evaluation of catalytic nanoparticles, measurement of intrinsic kinetics, and assessment of reaction mechanisms.

References

1. Nidumolu B. G. and Urbina M.C., “Functionalization of Gold and Glass Surfaces with Magnetic Nanoparticles Using Biomolecular Interactions.” *Bitechnol. Prog.* Vol. 22 (2006), pp. 91-95.

2. Kobayashi J., Mori Y., Okamoto K., Akiyama R., Ueno M., Kitamori T., Kobayashi S., "A

microfluidic device for conducting gas-liquid-solid hydrogenation reactions." *Science*. Vol. 304(2004), pp. (5675), 1305-1308.

3. Lin R., Freemantle R., Fielitz T., Obare S., Ofoli R., "*In-situ* immobilization of palladium nanoparticles stabilized by biotin and investigation of their reactivity in a microfluidic reactor". *Nanotechnology*. Vol. 21(2010), 325605.

4. Lin R., Ma X., Fielitz T. R., Obare S., Ofoli R., "Facile hydrogenation in PDMS microfluidic reactors with immobilized noble nanoparticles." *In preparation*.

See more of this Session: Microreaction Engineering
See more of this Group/Topical: Catalysis and Reaction Engineering Division

(169f) Chemisorption and Physisorption Rates On Catalysts Using Microfluidic ATR-FTIR Spectroscopy

T.J.A. Renckens, Chemical Engineering, Delft University of Technology, Delft, Netherlands, Guido Mul, Faculty of Science and Technology (TNW), University of Twente, Enschede, Netherlands and **Michiel Kreutzer**, Dept. of Chemical Engineering, Delft University of Technology, Delft, Netherlands

We present a method to determine adsorption and desorption rates from liquids onto solid surfaces. In most experiments to determine these rates, diffusion to the surface readily becomes limiting, and the intrinsic rates are hard to determine. This is in stark contrast to high-vacuum surface science, where one can determine such rates using so called Transient Analysis of Products (TAP) reactors. In other words, the present study is a first attempt to create a TAP reactor for liquids.

Diffusion limitations can only be overcome using miniaturization, which we present in this work Even in small channels, Taylor-Aris dispersion limits the resolution of a pulse or step change of concentration over a surface. We overcome this limitation by using a bubble between a concentration step-change, which remains a sharp transition even under flow. We measure the concentration on the surface using attenuated total reflection FTIR spectroscopy. This method has the advantage that both adsorbed and solution concentration can be distinguished and determined simultaneously. We report fast switching for non-adsorbing compounds and show intrinsic adsorption rates for heptylamine and cyclohexanone onto TiO2 catalysts. Especially heptylamine shows interesting non-langmuir behavior.

See more of this Session: Microreaction Engineering
See more of this Group/Topical: Catalysis and Reaction Engineering Division

(207a) Supported Bimetallic Catalysts for the Production of γ-Valerolactone

Jesse Q. Bond, David Martin Alonso, Stephanie G. Wettstein, Drew J. Braden and James A. Dumesic, Chemical and Biological Engineering, University of Wisconsin-Madison, Madison, WI

Transportation fuels may be produced from simple carbohydrates in cascade processes which improve upon the energy density of the feedstock and facilitate C-C bond formation to build the molecular weight of the final product. Such processes generally involve the isolation of intermediate platform chemicals, of which γ-valerolactone (GVL) is a particularly versatile example. Commonly, GVL is produced by hydrogenation of levulinic acid, over supported metals such as Ru/C. Once prepared, it may be used directly as a fuel additive or alternatively processed to yield jet fuel components. To date, the most significant hindrance to large scale demonstration of GVL-centered biorefining is the cost associated with its production, much of which is attributed to recovery and purification of levulinic acid.

Though they have demonstrated appropriate activity, Ru catalysts commonly used in the reduction of levulinic acid suffer a marked sensitivity to residuals from prior stages (sulfuric acid, for example) and have a limited capacity for processing unrefined solutions of levulinic acid. Further, supported Ru catalysts are not entirely selective for carbonyl reduction (they readily saturate carbon-carbon double bonds, for example), which requires that levulinic acid undergo extensive refining to remove residuals prior to GVL production.

In recent research, we have observed that bimetallic catalysts are more robust than traditional monometallic species and may allow relaxed tolerances on levulinic acid purity to facilitate low-cost production of GVL and its many upgrading products. For example, the addition of certain promoters to Ru, such as Re, confers superior stability and activity in the presence of sulfuric acid, offering the possibility of direct hydrogenation of unrefined levulinic acid solutions. Other bimetallic systems enhance the selectivity for reduction of the ketone moiety, allowing intensified strategies by which GVL may be produced selectively in the presence of functionalized extracting solvents without intermediate distillation.

In this session, we shall outline strategies which rely upon the unique capabilities of bimetallic catalysts and may help to establish GVL as a commodity scale platform for fuels production. In addition, we shall present detailed characterization of various bimetallic systems to describe their improved stability, activity, and selectivity in processing lignocellulose derived solutions of levulinic acid.

See more of this Session: Catalytic Processing of Fossil and Biorenewable Feedstocks: Fuels I
See more of this Group/Topical: Catalysis and Reaction Engineering Division

(207b) Catalytic Synthesis of High-Octane Fuels From Biologically-Derived Sources

Samuel Blass, Aditya Bhan and Lanny Schmidt, Chemical Engineering and Materials Science, University of Minnesota, Minneapolis, MN

Autothermal reactors using noble-metal catalysts have been shown to produce α-olefins from higher hydrocarbons such as biodiesel without a need for external heat input or regeneration. We aim to develop a staged autothermal reactor to carry out an additional hydroprocessing reaction and convert α-olefins to saturated branched hydrocarbons. We fabricated a reactor that contained a 0.5 wt % Pt/γ-Al_2O_3 stage with 10 g of zeolite H-BEA situated downstream of a 2 wt % Rh/α-Al_2O_3 stage. Decane was chosen as a surrogate for a higher alkane and was partially oxidized over the Rh stage to form C_{4+} olefins with an 8% carbon yield and oxidation products H_2, CO, CO_2, and H_2O. The H_2 and hydrocarbon products formed in the Rh stage reacted on the downstream bifunctional zeolite stage and the C_{4+} yield increased to 19%. This process also enhanced the percentage of branched isomers in the product distribution. The increase in yield of C_{4+} products and the presence of branched isomers indicates the feasibility of integrating a hydroprocessing catalyst with a partial oxidation catalyst in a single-pipe autothermal reactor.

See more of this Session: Catalytic Processing of Fossil and Biorenewable Feedstocks: Fuels I
See more of this Group/Topical: Catalysis and Reaction Engineering Division

(207c) Improved Pyrolysis Oil Properties Through Catalytic Deoxygenation

John R. Carpenter, Maruthi Pavani, Matthew Von Holle, David Dayton and Raghubir Gupta, Center for Energy Technology, RTI International, Research Triangle Park, NC

Pyrolysis oil prepared by conventional (non-catalytic) fast pyrolysis is not suitable for integrating into the current fuel infrastructure or into a petroleum refinery. Negative properties of conventional pyrolysis oil include: a) thermal instability and high fouling tendency, b) corrosiveness due to high carboxylic acid content (pH 2.2 – 2.4 typically), and c) immiscibility with refinery feedstocks. Many of these properties are hypothesized to result from the high oxygen content (>38 wt%, dry basis) of the pyrolysis oil. RTI is screening a wide variety of catalysts in multiple reactor systems to identify highly active and selective catalysts that reduce the oxygen content of the resulting bio-crude.

A bench-scale entrained flow biomass pyrolysis reactor with a downstream vapor phase upgrading reactor is utilized with a series of over 15 catalysts spanning a range of catalyst functions. Catalyst screening was performed with a nominal biomass feed rate of 100 g/hr and a white oak feedstock at temperatures between 350°C to 500°C, similar to pyrolysis temperatures. Full mass closures are obtained by measuring all liquid, solid, and gas products. Bio-oils are characterized by ultimate analyses, Karl-Fischer titration, total acid number (TAN), pH, and GC/MS. To address thermal instability, a residual testing methodology was developed and a variety of bio-oils and other hydrocarbons have been screened to determine the ability to re-vaporize the condensed bio-oils. The quantity and characteristics of the residual material formed during the vaporization is compared with the bio-oil's characteristics, primarily oxygen content but also acid content.

Selected catalysts are also tested in a bench-scale fluidized bed reactor where the biomass is injected directly into a fluidized bed of catalyst particles to determine product yields for catalytic biomass pyrolysis. Comparisons between vapor phase upgrading and catalytic pyrolysis are made. The most significant difference being lowered gas yields. This screening is directed toward developing a catalytic process to produce a pyrolysis product that is more suitable for upgrading with traditional refining process technology.

See more of this Session: Catalytic Processing of Fossil and Biorenewable Feedstocks: Fuels I
See more of this Group/Topical: Catalysis and Reaction Engineering Division

(207d) Methanol and Ethanol Coupling Reactions to Oxygenates Over Copper Containing Mixed Metal Oxides

Juan J. Bravo-Suarez, Center for Environmentally Beneficial Catalysis, University of Kansas, Lawrence, KS, Bala Subramanian, Center for Environmentally Beneficial Catalysis, Chemical & Petroleum Engineering, University of Kansas, Lawrence, KS and Raghunath V. Chaudhari, Center for Environmentally Beneficial Catalysis, Chemical and Petroleum Engineering, University of Kansas, Lawrence, KS

Currently, there is a tremendous interest in the development of technologies to produce alcohols from lignocellulosic sources for use as fuels and chemical feedstocks for synthesis of olefins, paraffins, and aromatics in a bio-refinery. Use of bio-derived alcohols as fuels is strongly motivated by existing legislation mandating a renewable fuels standard of 36 billion gallons per year by 2022 (EISA 2007, RFS 2). Higher alcohols such as propanol and butanols could be viable candidates to fulfill these mandates as their energy contents are higher than those of methanol and ethanol.

In this work, the synthesis of chemicals and potential fuels (higher alcohols) from methanol and ethanol coupling reactions was studied over copper containing mixed metal oxides. Layered double hydroxides with varying copper contents (prepared by coprecipitation methods and used as parent materials) were treated at high temperatures to yield the corresponding highly dispersed copper mixed metal oxides. Typical products included those from Guerbet reactions (e.g., formaldehyde, acetaldehyde, propanal, propanol, butanals, butanols), esterification of aldehydes with MeOH and EtOH (methyl and ethyl esters), and MeOH decomposition (CO and CO_2).

On CuMgAl mixed metal oxides (Cu/(Cu+Mg+Al)=0-0.45; Mg/Al=1; 250 ^oC, 1 bar, He:MeOH: EtOH=75:20:5, SV=4000 $cm^3\ g_{cat}^{-1}\ h^{-1}$), for example, it was found that increasing Cu content resulted in significant increase in EtOH conversion (typical EtOH conversion > 90%), at the expense of a decrease in C-C coupling alcohols selectivity but with corresponding gain in selectivities towards C-C coupling (e.g., C3+) and non C-C coupling (HAc, EtOAc, MeOAc, $HCOOCH_3$) aldehydes and esters products. The result is higher yields of C-C coupling (~150 g $kg_{cat}^{-1}\ h^{-1}$) and non C-C coupling products (~450 g $kg_{cat}^{-1}\ h^{-1}$) than those observed with Cu free MgAl and related mixed oxides, reported in the literature. The C3+ aldehydes and esters could be easily hydrogenated to the corresponding alcohols for use as fuels, while non C-C coupling esters could be used as bio-derived feedstocks for various chemical conversions. The presence of Cu also resulted in more stable catalysts (EtOH conversion drop < 15% after 24 h) in comparison with the MgAl oxide (EtOH conversion drop > 50% after ~5 h). These results suggest that appropriate manipulation of acid/base, dehydrogenation/hydrogenation properties (e.g., by adjusting Cu content) of mixed metal oxides can be used for the fine tuning of product yields relevant for use as fuels and chemical feedstocks from bio-derived methanol and ethanol mixtures.

See more of this Session: Catalytic Processing of Fossil and Biorenewable Feedstocks: Fuels I
See more of this Group/Topical: Catalysis and Reaction Engineering Division

(207e) Investigation of Benzofuran Deoxygenation In Sub and Supercritical Water

Jacob G. Dickinson and Phillip E. Savage, Chemical Engineering, University of Michigan, Ann Arbor, Ann Arbor, MI

Abstract

Algae have received much attention as a biomass source for second-generation biofuels. The dilute solutions of water that algae grow in provide significant fuel processing challenges. One proposed solution is to process algae in the aqueous phase through hydrothermal liquefaction, to remove the energy intensive drying step. Hydrothermal liquefaction produces a crude bio-oil high in nitrogen and oxygen content. The high oxygen content is likely due to the presence of phenolic and cyclic oxygen-containing compounds. Benzofuran was chosen as a model cyclic oxygen-containing compound.

Our work focuses on the deoxygenation of benzofuran over Pt/C in supercritical water. The effects of hydrogen pressurization, water density, residence time, and reaction temperature were examined. The deoxygenation reaction is highly dependent on amount of hydrogen added to the reactor, but shows little effect when the water concentration is varied. The primary product benzofuran deoxygenation is ethylbenzene. Smaller amounts of benzene and toluene with the accompanying ethane and methane were also observed. Fully hydrogenated products such as straight chain or cyclic alkanes were observed only in minute quantities. The results of the kinetic model will also be discussed.

See more of this Session: Catalytic Processing of Fossil and Biorenewable Feedstocks: Fuels I
See more of this Group/Topical: Catalysis and Reaction Engineering Division

(207f) Increasing the Hydrothermal Stability of Alumina Based Catalysts

Ryan M. Ravenelle[1], Maximilian W. Hahn[2], John R. Copeland[2], Adam Van Pelt[2] and **Carsten Sievers**[1], (1)Chemical and Biomolecular Engineering, Georgia Institute of Technology, Atlanta, GA, (2)School of Chemical & Biomolecular Engineering, Georgia Institute of Technoloy, Atlanta, GA

Aqueous phase processes are expected to play a key role in future biorefineries. This environment leads to new requirements for the stability of heterogeneous catalysts because many of these materials were developed for gas phase reactions. Recently, it was demonstrated that γ-alumina supports are converted into a crystalline boehmite phase [1]. This transformation results in a drastic decrease of the surface and the concentration of accessible Lewis acids sites.

The conversion of γ-alumina to boehmite is significantly retarded when supported Pt or Ni particles are present on the support because the metal particles block surface hydroxyl group. These groups also serve as initiation sites for the hydration of γ-alumina to boehmite. To eliminate the remaining surface hydroxyl groups Pt/γ-Al_2O_3 catalysts were capped. Boehmite formation was not observed for any of the protected samples and a significantly increased hydrogen production was observed when the catalysts were used for aqueous phase reforming of sorbitol. Increased stability was also observed for amorphous silica-alumina supports that were synthesized under well controlled conditions.

Reference:

[1] R.M. Ravenelle, J.R. Copeland, W.-G. Kim, J.C. Crittenden, and C. Sievers, ACS Catalysis 1 (2011) 552-561.

See more of this Session: Catalytic Processing of Fossil and Biorenewable Feedstocks: Fuels I
See more of this Group/Topical: Catalysis and Reaction Engineering Division

(207g) Design of Active and Stable Catalysts for the Hydrodeoxygenation of m-Cresol

Andrew J. Foster, Phuong T. Do, Jingguang G. Chen and Raul F. Lobo, Chemical Engineering, University of Delaware, Newark, DE

Pyrolysis is an emerging technology for the conversion of biomassto liquid fuel products. However, due to the oxygen content of thesebiomass-derived liquids they are acidic, highly viscous, prone topolymerization, and have a low heating value (1). Forthese reasons, oxygenates present in "pyrolysis oils" must be upgraded tohydrocarbons before consideration as potential transportation fuels. Hydrodeoxygenation (HDO) is a potential route for the oxygenateupgrading, but the design of active, stable HDO catalysts that operate withminimal H_2 consumption is still a challenge (2).

Phenolic compounds account for a large fraction of biomasspyrolysis oils. The phenolic C-O bond is particularly strong, and thus HDO mustoccur by hydrogenation of the phenolic aromatic ring followed by dehydration toremove oxygen (3). This path suggests the need for a catalyst active forhydrogenation of aromatics and with acid sites able to catalyze dehydrationreactions. As a model for the phenolics frompyrolysis oil, we have studied the HDO of m-Cresol (3-methylphenol) overPt-based supported catalysts in a lab-scale flow reactor. Modification of thePt catalyst by bimetallic formation with 3d metals has a significant impact onthe overall activity for HDO. Pt-Ni/Al_2O_3 and Pt-Co/$Al_2O_3$are both more active (63% and 60% conversion of m-Cresol, respectively, underconditions tested) for HDO than a monometallic Pt catalyst (38% conversion),but also are more selective towards fully hydrogenated products. Similarly, increasingthe acidity of the Al_2O_3 support by F^- washingincreases HDO activity, and improves selectivity towards the aromatichydrocarbon product. We have further investigated the role of the support by testingalternative supports such as SiO_2 and ZrO_2. We will reportan investigation of the structure-function properties of the metal particles inthe catalysts and their synergy with acid sites on the support to maximize stabilityand selectivity for desired products.

References

1. Zhang Q,Chang J, Wang T, Xu Y. Review of biomass pyrolysisoil properties and upgrading research. Energy Conversion andManagement. 2007;48:87-92.

2. Elliott DC.Historical Developments in HydroprocessingBio-oils. Energy & Fuels. 2007;21(3):1792-815.

3. Massoth FE, Politzer P, Concha MC, Murray JS, Jakowski J, Simons J. Catalytic Hydrodeoxygenationof Methyl-Substituted Phenols: Correlations of Kinetic Parameters withMolecular

Properties. Journal of Physical Chemistry B. 2006;110:14283-91.

See more of this Session: Catalytic Processing of Fossil and Biorenewable Feedstocks: Fuels I
See more of this Group/Topical: Catalysis and Reaction Engineering Division

Ethanol Synthesis From Syn-Gas: How Surface Diffusion of Intermediates Impacts the Product Distributions Predicted for Bimetallic Catalysts

Ming He, James McAliley and ***David A. Bruce****, Chemical and Biomolecular Engineering, Clemson University, Clemson, SC*

Introduction

Ethanol has been receiving great interest for its use as a potential alternative fuel, a potential source of hydrogen, as well as a raw material for the synthesis of many chemicals and polymers.

A major challenge associated with the synthesis of ethanol from syn-gas is an inability to find a low-cost catalyst that promotes the proper combination of CO dissociation and CO insertion steps, so as to yield ethanol as the primary reaction product and inhibit the formation of methane, longer chain alkanes, and other coking reaction products. For this purpose, sub-nanometer size, bimetallic cobalt-palladium (Co-Pd) particles are studied for the desired ethanol reaction. On Co catalysts CO molecules are readily dissociated, followed by C-C coupling reactions to form higher hydrocarbons. On Pd-based catalysts CO molecules adsorb in the molecular state, followed by subsequent hydrogenations to methanol.

DFT simulations and Bronsted-Evans-Polanyi (BEP) relations were used to map out the full reaction mechanism from syn-gas to ethanol. Microkinetic models were built, considering the reaction steps as well as the diffusion of intermediate species between different metal surface sites: Co_3, Pd_3, and mixed CoPd sites.

Result and Discussion

By DFT (B3LYP, LACVP**) calculations, we identified 37 surface reactions in our proposed mechanism for ethanol formation from syn-gas. Based on the most stable morphology of the CoPd cluster, the surface sites can be classified into three categories: Co_3 sites (Co sites), Pd_3 sites (Pd sites) and mixed Co_2Pd or $CoPd_2$ sites (CoPd sites). With three different possible reaction sites and the complex reaction mechanism, there are more than 100 possible surface reaction steps from which to calculate rate constants. To solve this computationally intensive problem, the implementation of scaling methods is critical.

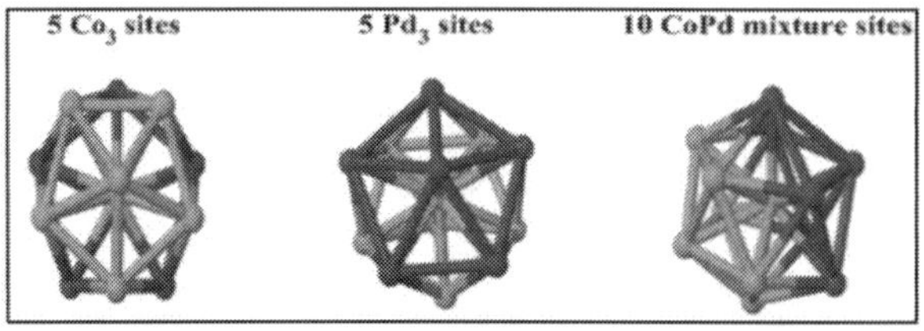

Figure 1. Different sites on the most stable conformation of a Co7Pd6 cluster (Co, blue; Pd, pink)

A widely used scaling method for estimating activation energies is the Bronsted-Evans-Polanyi (BEP) relation. The BEP relation linearly correlates the transition state energy of an elementary step to the reaction enthalpy of that step. It has been successfully applied to many catalytic systems for efficiently locating activation barriers.

To examine the intrinsic nature of the three distinct catalytic sites, the tool of microkinetic modeling was implemented. In this application, the preferred product is ethanol. Other byproducts generated along the process are methane, methanol, and acetaldehyde.

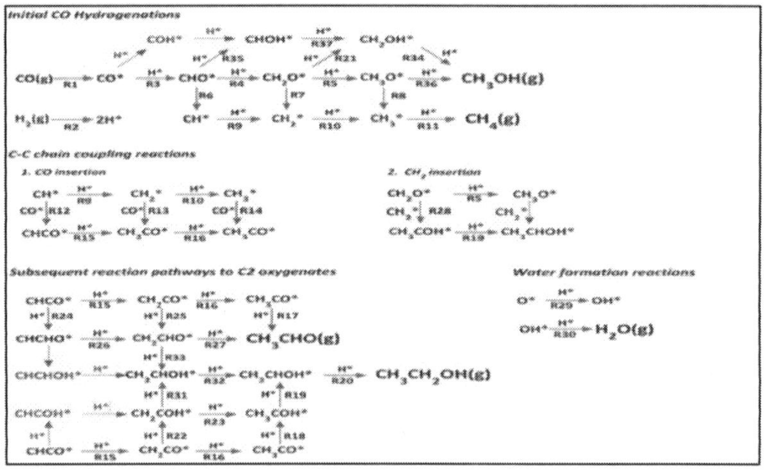

Figure 2. Full reaction mechanism of ethanol formation reacion from syn-gas

The pseudo-steady state hypothesis (PSSH) is applied to calculate surface concentrations of intermediate species. Initially, separate microkinetic models were built, targeting reaction mechanisms on isolated Co sites, CoPd sites, and Pd sites.

On the pure Co and pure Pa catalyst sites, we find good agreement of reaction selectivity between our model and experiments. Methane is the major product on the Co sides, with a selectivity of 98%. The selectivity of methane on a Co catalyst in experiment at similar conditions was 97%. On the Pd side, the microkinetic model predicts that methanol is the major product. Experiments have also shown methanol is the major product on Pd catalysts. It is interesting to note that the C2 oxygenate is formed on the interface sites (CoPd sites). Acetaldehyde is the major product with a selectivity of 58%.

For a cluster of such small scale, surface diffusion of intermediates between different reaction sites should be fairly common during the reaction. Also, because of the existence of heterogeneous reaction profiles, species may take part in distinct reactions on different sites. Thus, a proper description of the intermediates' diffusion behaviors is critical to reveal the overall reaction mechanism on the cluster.

The incorporation of surface diffusion into microkinetic models is very rare due to the complication it brings to the evaluation of kinetic parameters and the extra simulation effort it requires. Surface diffusion can be seen as a reaction between the diffusing species and an empty site. By this meaning, the treatment of diffusion is identical to that of a reaction. Ten species diffusion behaviors are studied here. They are HCO^*, CH_2O^*, CH_3O^*, CH_3CO^*, CH_2CHO^*, CH_3COH^*, CH_3CHOH^*, OH^*, CH_2^* and CH_3^*.

In the 10 species studied for surface diffusion, only the CH_3^* and CH_3CO^* species' diffusions have major effect on overall reactions. The other species have no impact on the reaction mechanism, since the diffusion rates between different sites are much faster than the reaction rates.

Diffusion has two significant effects on the reaction system. First, the overall reaction rate is much faster on Co sites than the other two sites. There is almost no products formation on CoPd sites or Pd sites. Moreover, on Co sites the selectivity of methane is 85%. The ethanol selectivity is 13% after accounting for surface diffusion. These results imply a synergetic effect between Co and Pd favoring the production of ethanol on the bimetallic cluster that would not occur on either of the individual metals.

CO insertion into the metal-carbon bonds of adsorbed methyl groups is the primary pathway to produce C_2 oxygenates. We observed that overall reaction rates are limited by the hydrogenation of CO to methoxy species. In part, this is due to the fact that the CoPd and Pd sites are primarily covered with CO, due to strong CO binding strengths on these two sites. Thus, a significant fraction of product forming reactions occur on the Co sites, which make methane the major product. Some ethanol can also form on Co sites via the diffusion of CH_3 and CH_3CO species between Co sites and CoPd sites, thus facilitating CO insertion reactions (Figure 3).

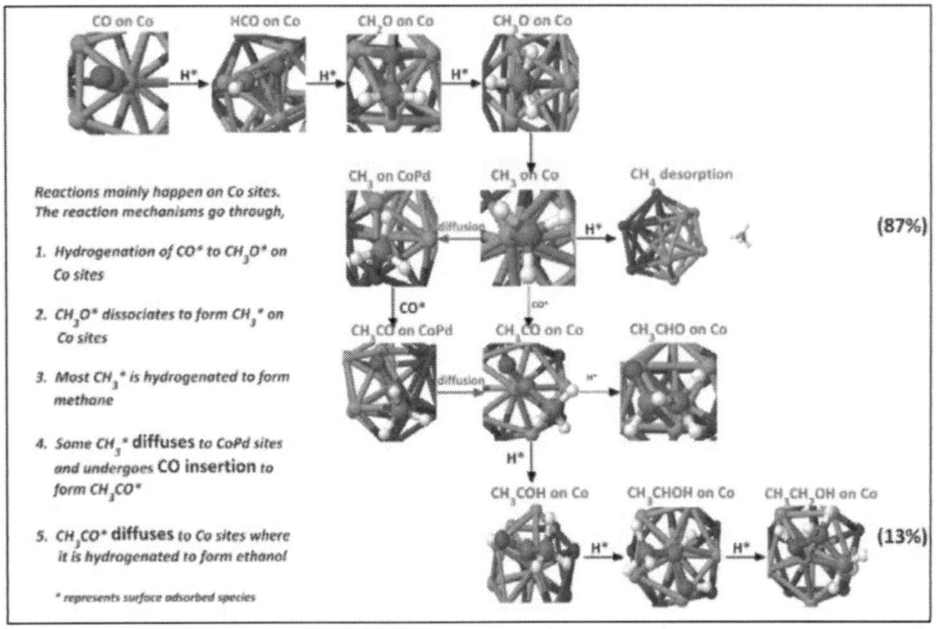

Figure 3. The overall reaction mechanism of ethanol formation from syn-gas on CoPd cluster

Conclusions

By using DFT simulations coupled with BEP methods, we calculate more than 100 intermediate reaction steps. The synergetic effect between Co and Pd is seen on the interface CoPd sites, where C2 oxygenates are the major products.

For this system, CO insertion into the metal-carbon bonds of adsorbed methyl groups is the primary pathway for C_2 oxygenate production. Due to most CoPd and Pd sites being blocked by adsorbed CO, a significant fraction of product forming reactions occur on the Co sites, which leads to methane being the major product (90%). However, the diffusion of CH_3 and CH_3CO between Co and CoPd sites facilitates CO insertion reactions, which ultimately leads to some ethanol (10%) being produced on the Co sites. These results indicate the key for ethanol formation on bimetallic clusters is the presence of intermixture sites, which can effectively reduce the reaction barrier differences of CO insertion reactions and hydrogenation reactions on the CH_3^* species. This feature has to be coupled with fast diffusions of intermediates to accomplish ethanol production.

References

1. Spivey, J. J.; Egbebi, A. *Chemical Society Reviews* 2007, *36*, 1514-1528.
2. Subramani, V.; Gangwal, S. K. *Energy & Fuels* **2008**, *22*, 814-839.
3. Lahtinen, J.; Vaari, J.; Kauraala, K. *Surface Science* **1998**, *418*, 502-510.
4. Cheng, J.; Hu, P.; Ellis, P.; French, S.; Kelly, G.; Lok, C. M. *Journal of Physical Chemistry C* **2008**, *112*, 9464-9473.
5. Neurock, M. *Topics in Catalysis* **1999**, *9*, 135-152.
6. Conrad, H.; Ertl, G.; Latta, E. E. *Surface Science* **1974**, *41*, 435-446.

7. Bligaard, T.; Nørskov, J. K.; Dahl, S.; Matthiesen, J.; Christensen, C. H.; Sehested, J. *Journal of Catalysis* **2004**, *224*, 206-217.
8. Nørskov, J. K.; Bligaard, T.; Logadottir, A.; Bahn, S.; Hansen, L. B.; Bollinger, M.; Bengaard, H.; Hammer, B.; Sljivancanin, Z.; Mavrikakis, M.; Xu, Y.; Dahl, S.; Jacobsen, C. J. H. *Journal of Catalysis* **2002**, *209*, 275-278.
9. Liu, P.; Logadottir, A.; Nørskov, J. K. *Electrochimica Acta* **2003**, *48*, 3731-3742.
10. Wintterlin, J.; Zambelli, T.; Trost, J.; Greeley, J.; Mavrikakis, M. *Angewandte Chemie International Edition* **2003**, *42*, 2850-2853.

(211b) Development and Application of a ReaxFF Description for Formation of Hydrocarbons On Iron and Iron Carbide Fisher-Tropsch Catalysts

Chenyu Zou[1], Adri van Duin[1] and Dan C. Sorescu[2], (1)Mechanical Engineering, Pennsylvania State University, State College, PA, (2)Computational Chemistry, National Energy Technology Laboratory, Pittsburgh, PA

A ReaxFF reactive force field description for iron and iron carbide bulk phases, surfaces and their interactions with water, carbon monoxide and hydrocarbon species has been developed, which puts the foundations together for large scale, fully reactive molecular dynamics simulations of Fisher-Tropsch (FT) catalysis on iron catalysts.

The ReaxFF description in this project will be tested and trained against a suite of ab initio data, and will be validated using dynamical simulations on relatively small systems(50-2500 atoms), enabling us to compare ReaxFF reaction kinetic and diffusion data with experimental observations.

This project aims to close the gap between atomistic-scale simulations and experiment, and the methods developed here will greatly enhance the ability to design and test materials on an atomistic scale. Specifically, the large system sizes (>>1000 atoms) and large simulation time (> 1ns) will enable us to study the ageing, poisoning and regeneration of iron-based FT-catalysts.

See more of this Session: Computational Catalysis IV
See more of this Group/Topical: Catalysis and Reaction Engineering Division

(211c) Investigation of Structure Sensitivity for the CO Oxidation Chemistry On Pt and Au

Michail Stamatakis, Chemical Engineering, University of Delaware, Newark, DE and Dionisios G. Vlachos, University of Delaware, Newark, DE

The CO oxidation is a prototypic catalytic reaction for the study of structure sensitivity on metal surfaces. Yet, experimental results from different groups appear contradictory: studies on single crystal Pt, Ir, Rh and Pd surfaces have shown that CO oxidation is structure insensitive and attribute this effect on the surface being predominantly covered by CO, resulting in the rate being controlled by the CO desorption. Other studies, however, show that steps and terraces exhibit different activity, with the latter being much more active than the former, due to the preferential poisoning of the steps by CO. Thus, it still remains an open question whether the CO oxidation chemistry is structure sensitive and under what conditions.

In this study, we perform multiscale simulations that combine kinetic Monte Carlo with density functional theory based input on two different metals, namely Pt which exhibits strong binding for CO and O, and Au which exhibits weak binding for these species. We consider several surfaces with different ratios of step/terrace site numbers under various conditions, and investigate whether CO oxidation is structure sensitive. We show that the steps in the Au surfaces are much more active than the terraces. However, for Pt surfaces, poisoning effects by CO and/or O are important and may render the steps inactive, thereby resulting in the CO oxidation chemistry appearing structure insensitive under certain conditions.

See more of this Session: Computational Catalysis IV
See more of this Group/Topical: Catalysis and Reaction Engineering Division

(211d) Selective Methane-to-Methanol Oxidation On Bimetallic Transition Metal Surfaces

Kathryn Bjorkman and Linda J. Broadbelt, Department of Chemical and Biological Engineering, Northwestern University, Evanston, IL

Direct methane conversion to useful chemicals is a catalysis problem that has challenged researchers for many years. Methanol is one such desired product because of its potential use in a wide range of industrial applications, including methanol-to-olefin (MTO) processes to form ethylene and propylene, which then may be used as raw material in production of other chemicals. However, part of the challenge in converting methane to more useful products is that the products of those reactions are often more reactive than the methane itself. This results in selectivity problems and higher extents of oxidation. Nanoscale bimetallic catalytic particles have the potential to improve upon existing monometallic micro- and macroscale catalysts by introducing unique electronic and geometric structure effects to selectively convert methane to methanol while avoiding overreaction to complete oxidation products.

Quantum mechanical calculation methods are used to study the binding energies of methane-derived intermediates on monometallic and bimetallic transition metal surfaces in an effort to select the best catalytic sites for selective methane-to-methanol oxidation. Linear scaling relationships between binding energies and calculated activation energy barriers are established to showcase trends for bimetallic surfaces and provide for detailed reaction pathway analysis.

See more of this Session: Computational Catalysis IV
See more of this Group/Topical: Catalysis and Reaction Engineering Division

(211e) Mechanistic Studies of Oxygen Reduction by Hydrogen On PdAg(110)

Carrie A. Farberow[1], Andres Godinez-Garcia[2], Juan Francisco Perez-Robles[2], Omar Solorza-Fera[3] and Manos Mavrikakis[1], (1)Department of Chemical and Biological Engineering, University of Wisconsin - Madison, Madison, WI, (2)Departamento de Materiales, CINVESTAV-IPN, Quertaro, Mexico, (3)Departamento de Qumica, CINVESTAV-IPN, Ciudad de Mxico, Mexico

The reduction ofoxygen by hydrogen is a metal-catalyzed reaction important in the direct synthesisof hydrogen peroxide (HOOH) and at the cathode in proton exchange membrane fuelcells (PEMFCs). Due to its technological applicability and widespread historicalinterest within the catalysis community, many detailed mechanistic studies ofthis reaction have been conducted on monometallic and alloy surfaces. Most recently, bimetallic catalysts containing Pt[1, 2] have garnered much attention forfuel cell applications and PdAu alloys[3] have been the subject ofresearch for hydrogen peroxide synthesis applications. In this study,first principles electronic structure calculations based on periodic,self-consistent, density functional theory (DFT) were utilized to study themechanism for oxygen reduction by hydrogen on an alternative Pd-noble metalalloy surface, namely PdAg(110). Our results demonstrate that the minimumenergy pathway involves the initial formation of a peroxyl (OOH) intermediatefollowed by O-O bond scission, consistent with the minimum energy pathway shownfor oxygen reduction by hydrogen on monometallic Pd and Ag (111) surfaces.[4] The lower activation energybarrier for O-O bond scission in OOH versus hydrogenation of OOH to form HOOHsuggests that PdAg is not an effective catalyst for the direct synthesis ofhydrogen peroxide. The detailed thermochemistry and activation energy barriersof important elementary steps and intermediates in oxygen reduction by hydrogenon PdAg(110) are compared and contrasted with the analogous results recentlyreported for the monometallic surfaces.[4]

1. Stamenkovic, V.R.; Mun, B. S.; Arenz, M.; Mayrhofer, K. J. J.; Lucas, C. A.; Wang, G. F.;Ross, P. N.; Markovic, N. M., Trends in electrocatalysis on extended andnanoscale Pt-bimetallic alloy surfaces. *Nature Materials* **2007,** *6*(3), 241-247.

2. Zhang, J. L.;Vukmirovic, M. B.; Xu, Y.; Mavrikakis, M.; Adzic, R. R., Controlling thecatalytic activity of platinum-monolayer electrocatalysts for oxygen reductionwith different substrates. *Angewandte Chemie-International Edition* **2005,***44* (14), 2132-2135.

3. Edwards, J. K.; Solsona,B.; N, E. N.; Carley, A. F.; Herzing, A. A.; Kiely, C. J.; Hutchings, G. J., Switching Off Hydrogen Peroxide Hydrogenation in the Direct Synthesis Process. *Science***2009,** *323* (5917), 1037-1041.

4. Ford, D. C.;Nilekar, A. U.; Xu, Y.; Mavrikakis, M., Partial and complete reduction of O-2by

hydrogen on transition metal surfaces. *Surface Science* **2010,** *604*(19-20), 1565-1575.

See more of this Session: Computational Catalysis IV
See more of this Group/Topical: Catalysis and Reaction Engineering Division

(211f) A First Principles Analysis of the Selective Hydrogenation of Unsaturated Ketones: Influence of Hydrocarbon Structure

Bing Hao and Matthew Neurock, Department of Chemical Engineering, University of Virginia, Charlottesville, VA

The selective hydrogenation of carbonyl groups in unsaturated ketones (UK) is important in the catalytic synthesis of various pharmaceutical and fine chemical intermediates as well as in biomass conversion strategies [1]. The ability to selectively hydrogenation the C=O group over that of the C=C bond presents a significant challenge as the hydrogenation of the C=C is thermodynamically favored over that of the C=O bond [2,3]. Recent experimental observations reveal that the adsorption and selectivity for the hydrogenation of unsaturated ketones depends strongly on the substituents on the C=C bond [4,5]. One successful example is the selective hydrogenation of the sterically-hindered C=O group in ketoisophorone (KIP) over Pd/Al_2O_3 [6]. Very little, however, is known about the specific role that substitutions play in the elementary hydrogenation mechanisms of unsaturated ketones. To probe this effect, density functional theory (DFT) calculations were carried out to explore the elementary hydrogenation steps for a series of model ketones over different metal surfaces, beginning with the simplest unsaturated ketone, methyl vinyl ketone. Methyl substituents were then added one by one to the C=C bond to establish their influence on the kinetics.

Increasing degree of substitution at the C=C bond was found to weaken the adsorption strength of UK reactants and alter the energetics of hydrogenation steps by increasing the hydrogenation barrier at the substituted carbon, while lowering the barrier to hydrogenate at its nearby carbon. Increasing the degree of substitution notably decreases the activation barrier for the first hydrogenation step of UK thus increasing the likelihood of selectively hydrogenating the C=O bond over the C=C bond.

We have also examined the influence of substituents on the hydrogenation of KIP over Pd. KIP is a much more complex ketone as it is comprised of three reducible groups and as such offers a number of different products that can form. Increasing the degree of substitution on the C=C bond was again found to increase the hydrogenation barrier for C=C while inducing a decrease in barrier to hydrogenate the sterically hindered C=O bond. This is in agreement with the experimental results which show enhanced selectivity to 4-HIP.

Reference

(1) R.A. Sheldon, H. van Bekkum, Fine Chemicals through Heterogeneous Catalysis, Weinheim: WILEY-VCH, (2001).

(2) C. Mohr, P. Claus , Science Progress, 84 (2001) 311-334.

(3) P. Gallezot, D. Richard, Catalysis Reviews-Science and Engineering, 40 (1998) 81-126

(4) C. Milone, R. Ingoglia, A. Pistone, G. Neri, F. Frusteri, S. Galvagno, Journal of Catalysis, 222 (2004) 348-356.

(5) C. Milone, R. Ingoglia, M.L. Tropeano, G. Neri, S. Galvagno, Chemical Communications, (2003) 868-869.

(6) M. von Arx, T. Mallat, A. Baiker, Journal of Molecular Catalysis a-Chemical, 148 (1999) 275-283.

See more of this Session: Computational Catalysis IV
See more of this Group/Topical: Catalysis and Reaction Engineering Division

(211g) Reaction Free Energies In Solutions for Glucose Conversion Into Biomass Intermediates

Vinit Choudhary, Dionisios G. Vlachos and Stanley I. Sandler, Chemical Engineering, University of Delaware, Newark, DE

Sugar dehydration reactions are of great importance in the conversion of biomass into fuels and chemicals. In this study, we have computed reaction free energies in solution for glucose conversion into levulinic acid and levoglucosan (LG). We computed vapor phase energies using the G4 ab initio method in Gaussian and estimated solvation free energies using COSMO-SAC (conductor-like screening model-segment activity coefficient) model. We used water and dimethyl sulfoxide (DMSO) as solvents. Our calculations predict glucose isomerization into fructose and mannose is reversible in both water and DMSO, whereas fructose dehydration into 5-(hydroxymethyl)furfural (HMF) is practically irreversible reaction. We predict that glucose dehydration into LG can be a reversible reaction depending on the reaction conditions as well as the solvent choice. In the vapor phase, at higher temperature, the entropic effect dominates and LG formation becomes favorable. It is found that glucose dehydration into levoglucosan is not thermodynamically feasible below 200°C in aqueous media but can become feasible in organic solvents. These calculations provide guidance into how to selectively control sugar dehydration by changing the reaction conditions. Experiments conducted in our lab and from the literature are contrasted with model predictions and good overall agreement is found.

See more of this Session: Computational Catalysis IV
See more of this Group/Topical: Catalysis and Reaction Engineering Division

(222a) In Situ Spectroscopic Investigation of the Surface Dependence of CO Oxidation Over CeO2 Nanocrystals with Well-Defined Surface Planes

Zili Wu[1], Meijun Li[2] and Steven H. Overbury[1], (1)Chemical Science Division and Center for Nanophase Materials Sciences, Oak Ridge National Laboratory, Oak Ridge, TN, (2)Chemical Science Division, Oak Ridge National Laboratory, Oak Ridge, TN

CO oxidation is a model reaction for probing the redox property of ceria-based catalysts. Despite the intensive experimental and theoretical studies of CO adsorption and oxidation on ceria catalysts, there is still a lack of spectroscopic insights of how CO interacts with different ceria surfaces and what the reaction mechanism is. In this study, CO oxidation was investigated over ceria nanocrystals with defined surface planes (nanoshapes) including rods ({110} + {100}), cubes ({100}), and octahedra ({111}). To understand the strong surface dependence of CO oxidation over these ceria nanoshapes, *in situ* techniques including infrared and Raman spectroscopy coupled with online mass spectrometry, and temperature-programmed reduction (TPR) were employed to reveal how CO interacts with the different ceria surfaces, while the mobility of ceria lattice oxygen was investigated *via* oxygen isotopic exchange experiment. CO adsorption at room temperature leads to strongly bonded carbonate species on the more reactive surfaces of rods and cubes but weakly bonded ones on the rather inert octahedra. Ceria reduction proceeds *via* several channels including CO reduction of lattice oxygen, surface water-gas shift reaction, and CO disproportionation reaction. It reveals that the reducibility of these ceria nanoshapes is in line with their CO oxidation activity, i.e., rods > cubes > octahedra. The mobility of lattice oxygen also shows similar surface dependence. It is suggested that both the reducibility and mobility of ceria lattice oxygen are assisted by the presence of defect sites on ceria, whose nature and amount are intrinsically determined by the surface terminations of ceria nanocrystals. Several reaction pathways for CO oxidation over the ceria nanoshapes are proposed and certain carbonate species, especially those associated with reduced ceria surface, are considered among the reaction intermediates to form CO_2, although the majority carbonate species observed under CO oxidation condition are believed to be spectators.

ACKNOWLEDGMENT: This Research is sponsored by the Division of Chemical Sciences, Geosciences, and Biosciences, Office of Basic Energy Sciences, U.S. Department of Energy. Part of the work was conducted at the Center for Nanophase Materials Sciences, which is sponsored at Oak Ridge National Laboratory, by the Office of Basic Energy Science, U. S. Department of Energy. The research was supported in part by the appointment for M.J. Li to the ORNL Postdoctoral Research Associates Program, administered jointly by ORNL and the Oak Ridge Associated Universities.

See more of this Session: Fundamentals of Oxide Catalysis

(222b) Condensed Phase Ketonization of Bio-Oil Model Compounds: Catalysis by Ceria Nanoparticles

Ryan W. Snell and Brent H. Shanks, Chemical and Biological Engineering, Iowa State University, Ames, IA

Bio-oil, made from the fast pyrolysis of biomass, suffers from low energy density and high acidity due to the large number of oxygen containing functionalized molecules and low molecular weight carboxylic acids. While hydrodeoxygenation could potentially solve these problems, it would be an expensive option. Condensed phase catalytic conversion of these undesirable molecules to increase the carbon-carbon chain length as well as remove oxygen would be a desirable alternative. Ketonization, taking two organic acids and creating a ketone, carbon dioxide, and water, would be an ideal reaction for this application as it is environmentally clean and efficiently removes acidity and oxygen from the oil. Unfortunately high temperatures have been necessary for the ketonization reaction, thus the vapor phase has almost exclusively been used with the coupling of low molecular weight acids. A number of different metal oxides have been found to catalyze this particular reaction, but cerium oxide is particularly active. Despite a significant amount of research, the exact mechanism of ketonization has not been determined and much about the reaction remains unknown.

In order to make condensed phase ketonization of low molecular weight acids more feasible, a significant lowering of the reaction temperature is necessary. Therefore a more active catalyst must be designed. Further understanding of the oxides role in the reaction, particularly in the condensed phase is thus crucial. The work presented here focuses primarily on the condensed phase reactions and potential applications of shape selective ceria nanocatalysts. Pre- and post-reaction characterization has demonstrated that the condensed phase ketonization reaction may proceed through a different mechanism than has been typically proposed in the vapor phase.

See more of this Session: Fundamentals of Oxide Catalysis
See more of this Group/Topical: Catalysis and Reaction Engineering Division

(222c) Microstructural Characterization of Bulk Ab Planes In Mo-V-Te-(Ta, Nb)-O M1 Phase Catalyst by Aberration-Corrected High Angle Annular Dark Field (HAADF) STEM

Jungwon Woo[1], Albina Borisevich[2], Miaofang Chi[3] and Vadim Guliants[1], (1)Department of Chemical and Materials Engineering, University of Cincinnati, Cincinnati, OH, (2)Oak Ridge National Laboratory, Oak Ridge, TN, (3)Materials Science Division, Oak Ridge National Laboratory, Oak Ridge, TN

Microstructural characterization of bulk *ab* planes in Mo-V-Te-(Ta, Nb)-O M1 phase catalyst by aberration-corrected high angle annular dark field (HAADF) STEM

*Jungwon Woo[1] and Albina Borisevich[3], Miaofang Chi[3], and Vadim Guliants[1,2]**

[1]Chemical Engineering, University of Cincinnati, Cincinnati, OH 45221-0012

[2]Energy and Materials Engineering, Cincinnati, OH 45221-0012

[3]Oak Ridge National Laboratory, Materials Science & Technology Division, Oak Ridge, TN 37831-6064

*vadim.guliants@uc.edu

Abstract

There has been growing interest in developing highly active and selective catalysts for direct propane ammoxidation to acrylonitrile (ACN) in order to replace the current industrial technology that relies on conversion of olefins and aromatics. Mo-V-M-O (M= combination of Nb, Te, Sb, and Ta) based mixed metal oxide catalyst consisting of "M1" and "M2" phases is the most promising system being investigated by academic and industrial research groups for the one-step propane ammoxidation. The crystal structures of the Mo-V-Te-(Ta, Nb)-O M1 phases have been reported that were solved by combined application of X-ray and neutron diffraction methods. However, questions still remain regarding the Nb location in the M1 phase lattice and its impact on the distribution of Mo and V cations among several sites in the *ab* planes proposed to contain the active and selective surface sites for propane ammoxidation. These questions arise because Mo and Nb are located next to each other in the

Periodic Table and therefore, are virtually indistinguishable by X-ray, neutron and electron diffraction techniques. The main objective of this study was to employ Ta cations (that are heavier than Nb but chemically similar) as a probe for Nb to investigate its location and impact on the partial occupancy of various Mo and V at sites in *ab* planes by the aberration-corrected high angle annular dark field scanning transmission electron microscopy (HAADF-STEM).

The phase pure M1 catalysts were prepared by two different synthesis methods, standard hydrothermal and microwave-assisted hydrothermal synthesis. Microwave-assisted synthesis resulted in the M1 phase that showed uniform Ta distribution across different M1 crystallites and no significant variation in Ta content from the surface region to the bulk of the M1 phase. Moreover, the M1 samples for STEM characterization were prepared by sectioning epoxy-embedded M1 crystals that produced M1 slices of uniform thickness (ca. 50 nm) which eliminated significant uncertainty in the atomic contrast of various metal lattice sites encountered in previous studies of mechanically crushed M1 crystals. Collected HAADF STEM images were quantified and analyzed statistically in terms of atomic contrast of metal lattice sites, atomic column by atomic column, using the Gatan Digital Micrograph (DM) software running DM scripts. The statistically analyzed intensity of atomic columns suggested that the Mo site 4 in the *ab* planes of Ta-containing M1 has some V occupancy unlike the bulk crystal structure of the M1 phase based on the X-ray and neutron diffraction data. The atomic contrasts of various metal lattice sites confirmed the location of Ta in the pentagonal bipyramidal site 9 and suggested a low level of Ta occupancy at site 11. These results suggested that chemically similar Nb may also partially occupies sites 9 and 11 in the *ab* planes of the Nb-containing M1 phase. The STEM-based partial occupancy data were used to estimate the fraction of the proposed active and selective surface sites that were correlated to experimentally observed performance of this catalyst in propane ammoxidation.

See more of this Session: Fundamentals of Oxide Catalysis
See more of this Group/Topical: Catalysis and Reaction Engineering Division

(222d) Probing A Mixed $Fe_2(MoO_4)_3$-MoO_3 Catalyst to Elucidate the Active Phase and Site for the Oxidative Dehydrogenation of Ethanol

Shane A. Bates[1], Jeffrey T. Miller[2], W. Nicholas Delgass[1] and Chelsey D. Baertsch[1], (1)Chemical Engineering, Purdue University, West Lafayette, IN, (2)Chemical Science and Engineering, Argonne National Laboratory, Argonne, IL

It is known that mixed $Fe_2(MoO_4)_3 - MoO_3$ catalysts have a longer life and are more selective to acetaldehyde in the oxidative dehydrogenation of ethanol when synthesized to contain excess molybdenum. Until now, the exact role of the excess Mo and location of the active phase and site of mixed $Fe_2(MoO_4)_3 - MoO_3$ catalysts have been debated. This study presents work done to determine the active phase and site with a series of experiments probing the bulk and surface of the catalyst during both oxidative and non-oxidative reactions with ethanol, with the latter conditions enabling anaerobic titration of the active redox phase. X-ray absorption near edge structure (XANES) was used to determine the extent of reduction in iron (Fe) and molybdenum (Mo) during anaerobic titration and showed that the active phase was associated synergistically with both species. Extended X-ray absorption fine structure (EXAFS) identified the removable/active oxygen during anaerobic titration as a shared oxygen between Fe and Mo in the lattice of $Fe_2(MoO_4)_3$; thus $Fe_2(MoO_4)_3$ was determined to be the active phase of the catalyst. X-ray photoelectron spectroscopy (XPS) identified surface oxidation states before and after anaerobic titration for Fe and Mo and showed that near surface Mo species stayed mostly oxidized throughout the anaerobic titration, while near surface Fe species were mostly reduced. Based on XANES and EXAFS results, it was expected that Mo species would reduce similarly to Fe, however, there was much more Mo in the form of Mo oxide in the near surface region. This suggests that the only phase reduced during this process is the $Fe_2(MoO_4)_3$ phase and that surface Mo species are kept oxidized by oxygen donation from the $Fe_2(MoO_4)_3$ phase.

See more of this Session: Fundamentals of Oxide Catalysis
See more of this Group/Topical: Catalysis and Reaction Engineering Division

(222e) Surface Speciation and Alkane Oxidation with Fe(III) Sites On Alkali-Modified Oxide Supports

Dario Prieto-Centurion and **Justin M. Notestein**, Chemical and Biological Engineering, Northwestern University, Evanston, IL

Supported iron oxides are active oxidation catalysts whoseefficiency and selectivity can be controlled by the level of dispersion of theFe phase. Here we describe a simple route to highly dispersed, supportedFe (III) catalysts. Well-defined active sites are prepared by incipient wetnessimpregnation of aqueous Fe (EDTA) complexes on neat and alkali-modified SiO_2,Al_2O_3, ZrO_2, and CeO_2. Impregnationis followed by heat treatment under O_2 to remove the EDTA ligand, leaving only the highly dispersed iron oxide centers on the support surface. Thecatalysts are characterized by H_2-TPR, XPS and diffuse reflectanceUV-visible spectroscopy (DRS) and compared to iron catalysts prepared usingmore common, although often more expensive, precursors. DRS results of thesecatalysts are characteristic of isolated Fe^{3+} cations up to 0.5Fe/nm^2 and of highly dispersed Fe^{3+}amorphous layers up to 0.9 Fe/nm^2, greatly exceeding that of typicalprecursors. Additionally, H_2-TPR and XPS show the Fe(III) to Fe(II)transition expected of highly dispersed Fe^{3+} species and no evidenceof bulk iron oxides. Finally, the activity, selectivity, and kinetic behaviorof the catalysts are studied for adamantane oxidation with aqueous H_2O_2as a model reaction for selective C-H bond activation. For this reaction, theinitial TOF of this highly dispersed Fe^{3+} surface species is in theorder of 10 s^{-1}, as compared with TOF <1 s^{-1} for amore traditional catalyst starting with $Fe(NO_3)_3$.

See more of this Session: Fundamentals of Oxide Catalysis
See more of this Group/Topical: Catalysis and Reaction Engineering Division

(222f) Consequences of Composition and Acid Strength for Catalysis On Solid Acids

Robert T. Carr[1], Josef Macht[1], Matthew Neurock[2] and Enrique Iglesia[1], (1)Department of Chemical and Biomolecular Engineering, University of California at Berkeley, Berkeley, CA, (2)Departments of Chemical Engineering and Chemistry, University of Virginia, Charlottesville, VA

Keggin polyoxometalate (POM) clusters and acid form zeolites were used to develop rigorous relations between catalyst structure and function to determine the roles of acid strength and solvation in stabilizing transition states and reactive intermediates in alkanol dehydration and alkene isomerization reactions. Keggin POM clusters with W-addenda atoms are Brønsted acids well suited to build such relations by combined experimental and theoretical studies because their uniform active sites and structures are unchanged over diverse compositions. Keggin clusters have acid strengths, defined as their deprotonation energies (DPE), which vary widely based on their central atom (P, Si, Al, Co) and number of charge-balancing protons. Kinetic and isotopic experiments and *ab initio* calculations led to the formulation of mechanism-based rate expressions needed to measure rate constants of elementary steps from kinetic experiments. In all cases, measured values decreased exponentially with increasing DPE, suggesting that stronger acids lead to more stable ion-pairs at transition states and lower activation barriers. Sensitivities to DPE varied among different reactions, yet activation barriers always changed less than the corresponding change in DPE. Specifically, activation barriers for isomerization reactions were more sensitive to DPE than either 1-butanol or 2-butanol dehydration, whose similar dependences to DPE demonstrated that difficult reactions are not more sensitive to acid strength. In agreement with this conclusion, the effects of DPE on 2-methylpentene (2MP) isomerization rate constants were similar for multiple products with varying rearrangement difficulty; as a result, 2MP selectivities were influenced by acid strength only because facile reactions approach equilibrium more quickly than difficult reactions. Thermochemical cycles of activation barriers that include properties of reactants (proton affinity, adsorption energies), catalysts (DPE), and transition states (electrostatic interactions, charge distribution) demonstrated that the effects of DPE on reactivity are caused by changes in the stabilities of reacting intermediates and transition states with acid strength. Electrostatic interactions at ion-pair transition states, a ubiquitous feature of Brønsted acid catalysis, partially recover the energy required for deprotonation and attenuate activation barriers of all reactions to DPE. Compensation of DPE by these electrostatic interactions is most effective, and the effects of DPE on activation barriers are weakest, when organic cations have localized charges like those at butanol dehydration transition states compared to diffusely charged isomerization transition states. The effects of DPE on rates of methanol dehydration to dimethyl ether (DME) are determined by the prevalent intermediate at active sites under different reaction conditions. Protonated dimer intermediates are ion-pairs whose stabilities change similar to DME formation transition states with DPE, while stabilities of weakly charged monomer intermediates are much less sensitive to DPE. The similar effects of DPE on protonated dimer and DME formation transition state stabilities cause rates to be less sensitive to acid strength when protonated dimer intermediates are prevalent and indicate the role of reacting intermediates and their charges in determining the effects of DPE. Rates of methanol dehydration on zeolites vary less with pore geometry when active sites are occupied by protonated dimers than by monomers and are also closer to DPE correlations developed for Keggin POM. These results arise from greater solvation of transition states and dimers via van der Waals forces than for monomers because both CH_3OH reactant molecules are confined within zeolite pores. Effective solvation of reactants at the transition state in zeolites reduces activation barriers to the greatest extent when monomers occupy active sites and causes differences among POM and zeolites that are unrelated to DPE. This study provides a complete and consistent description of the ways in which reactions sense acid strength and confinement separately and prescribes a framework to rigorously analyze their effects. Specifically, charge distributions in the relevant intermediates and transition states determine the effects of acid strength and the relative stabilization of these species by van der Waals forces determine the effects of confinement.

See more of this Session: Fundamentals of Oxide Catalysis
See more of this Group/Topical: Catalysis and Reaction Engineering Division

(222g) Density Functional Theory Evaluation of Rare-Earth Oxides for Biomass Gasification Effluent Cleanup

Matthew D. Krcha[1], Adam D. Mayernick[2], Michael J. Janik[1] and Kerry.M. Dooley[3], (1)Chemical Engineering, Pennsylvania State University, University Park, PA, (2)Department of Chemical Engineering, Pennsylvania State University, University Park, PA, (3)Chemical Engineering, Louisiana State University, Baton Rouge, LA

Biomass conversion to liquid fuels may be accomplished through gasification to syngas followed by fuel synthesis processes, enabling a renewable energy source of liquid fuels. Prior to fuel synthesis catalysts, the syngas must be cleaned of sulfur and tar species. In a Department of Energy forecast for 2012, approximately 50% of the cost to produce ethanol from biomass is involved in syngas cleanup. Mixed Rare-Earth Oxides (REOs) have shown promise in both desulfurization and hydrocarbon conversion. Our goal is to design a REO catalyst that can reform the large hydrocarbons into CO and H_2 and remove sulfur at high temperatures, thus making biomass gasification-based processes viable for sustainable liquid fuel production. Density functional theory (DFT+U) is used to generate composition-function relationships of mixed REOs for H_2S adsorption and hydrocarbon conversion. Oxygen vacancy sites in the doped oxide serve as the active site for H_2S adsorption and dissociation. Relative rates of the initial H_2S activation step predict trends in experimental H_2S adsorption capacity over a series of dopants in ceria, suggesting surface kinetic rates impact the adsorption capacity. As an indicator of methane reforming and tar cracking activity, the C-H bond activation energy of ceria based catalysts doped with transition metals is correlated with the surface reducibility. The findings from the DFT calculations compare with experimental H_2S adsorption energy and propane reforming activity.

See more of this Session: Fundamentals of Oxide Catalysis
See more of this Group/Topical: Catalysis and Reaction Engineering Division

(227a) Session Introduction

Concetta La Marca, DuPont Engineering Research & Technology, Wilmington, DE

The Catalysis and Reaction Engineering Division is pleased to honor Dr. Robert W. McCabe for his contributions in the areas of catalyst deactivation, catalyst characterization, catalytic reaction kinetics and reaction pathways, catalysts for alternatively-fueled vehicles, and catalysts and aftertreatment systems for lean-burn vehicles. This session will honor his contributions.

See more of this Session: In Honor of the Practice Award Winner Robert McCabe
See more of this Group/Topical: Catalysis and Reaction Engineering Division

(227b) Progress and Challenges in Automotive Exhaust Catalysis

Robert McCabe, Chemical Engineering, Ford Motor Company, Dearborn, MI

The automotive exhaust catalytic converter and associated emission control system has been one of the greatest success stories of heterogeneous catalysis over the past thirty-five years. Tremendous progress has been made, resulting in emissions reductions to less than 1% of pre-regulated levels. This success represents contributions from a broad base of companies supplying various aftertreatment components including the finished catalysts, raw materials, substrates, sensors, and converter assemblies, as well as system-level integration and control strategy provided by the OEMs and supporting companies. In this presentation, I highlight some of the major advances that have allowed catalyst technology to keep pace with a succession of increasingly stringent emission standards, and also consider challenges brought on by phase-in of even lower emission standards. In addition, new and proposed regulations around fuel economy and greenhouse gas emissions present another wave of challenges to the design of exhaust catalysts and emission control systems.

See more of this Session: In Honor of the Practice Award Winner Robert McCabe
See more of this Group/Topical: Catalysis and Reaction Engineering Division

(227c) NOx Abatement for Lean Burn Engines

Patrick Burk, BASF Corporation, Iselin, NJ

xx

See more of this Session: In Honor of the Practice Award Winner Robert McCabe
See more of this Group/Topical: Catalysis and Reaction Engineering Division

(227d) Low-Dimensional Models for Real Time Simulations of Catalytic After-Treatment Systems

Vemuri Balakotaiah, Department of Chemical and Biomolecular Engineering, University of Houston, Houston, TX

It is well known that the performance of the monolith reactors used in catalytic after-treatment systems (such as TWCs, DOCs, LNTs and SCRs) depends on the processes occurring at various length and time scales, e.g. complex chemistry and interactions between the various catalytic components at the crystallite (nanometer) scale, the diffusion of reactants and products thorough the porous washcoat (micrometer scale), the momentum, heat and mass transport processes in the monolith channels (millimeter scale) and the speed and width of various fronts along the monolith bricks (centimeter to meter scale). The design, optimization and control of these systems requires mathematical models that incorporate the processes occurring at all scales but simple enough so that they can simulated in real time or faster than real time. The focus of this talk will be on the development, validation and use of such low-dimensional models.

The low-D models are derived directly by averaging the governing partial differential equations and using the concepts of internal (intra-phase) and external (inter-phase) transfer coefficients. They are expressed in terms of multiple concentration and temperature modes (that are measurable in experiments) and include washcoat or intra-phase diffusional effects without using the concept of the effectiveness factor. The models are validated by simulating the steady state and transient behavior for various test cases and comparing the predictions with detailed solutions. It is shown that these new low-D models are robust, accurate with practically acceptable error, speed up the computations by orders of magnitude, and can be used with confidence for the real time simulation and control of various catalytic after-treatment systems. The application of these models to the transient operation of TWCs and LNTs will be illustrated. The use of the low-D models to estimate kinetic parameters using bench scale data as well as the determination of the controlling regimes as a function of various design and operating conditions will also be discussed.

See more of this Session: In Honor of the Practice Award Winner Robert McCabe
See more of this Group/Topical: Catalysis and Reaction Engineering Division

(227e) Advanced Emission Control Strategies for Gasoline Engines with Special Emphasis on the Development of Pd-Based Technologies

John Nunan, Umicore Autocat USA Inc., Catoosa, OK

In the current paper we will review the most important developments in gasoline emission control. This includes the introduction of advanced support materials and promoters for improved performance, greater durability and ultimately the use of lower Precious Group Metal (PGM) loadings while still meeting stringent emission standards. The development of Pd-based technologies will be discussed in detail and the effective incorporation of Pd into multi-catalyst/converter concepts. The implementation of Pd in Three-Way-Conversion (TWC) catalysts reflects one of the most important developments pioneered by Dr Gandhi and his colleagues at the Ford R&D laboratories. This required a detailed understanding of the detrimental effects of S and Pb on the performance of Pd. The development of specialized support materials that enhance & modify the performance of Pd allowing it to perform the catalytic functions specific to the more expensive PGM - Rh for NOx control will be reviewed. The latest approaches in achieving the more stringent ULEV, SULEV and LEV-III emission standards will be discussed.

See more of this Session: In Honor of the Practice Award Winner Robert McCabe
See more of this Group/Topical: Catalysis and Reaction Engineering Division

(229a) Evaluation of Alternative Fuels to Replace Coal In Cement Manufacturing

Steve R. Duke, Chemical Engineering, Auburn University, Auburn, AL, Anton Schindler, Civil Engineering, Auburn University, Auburn University, AL and Don Stafford, US East Business Unit, Lafarge North America, Alpharetta, GA

A multiyear DOE-supported study was completed to identify, determine, and understand fundamental burn characteristics and properties of alternative fuel sources to replace coal for energy generation, with emphasis on impacts in cement processing. Portland cement manufacturing involves the combustion of solid fuels along with raw materials to produce clinker. Temperatures exceeding 1,400°C are needed to carry out reactions for cement chemistry. Energy needs for a typical manufacturing plant are 12,000 MMBTU/day. Potential alternatives to fossil fuels must provide adequate and economical energy but they must also result in products of combustion that have negligible impact on cement quality and emissions.

An operating Portland cement manufacturing plant (Lafarge North America's Roberta Plant in Calera, Alabama) served as facility for a series of trial burns to replace coal at various substitution rates with bio-based and waste fuels. More than 20 three-day trial burns were conducted from 2007 to 2010 replacing a portion of the coal with scrap tires, waste plastics, woodchips, switch grass, poultry litter, forests trimmings, scrap railroad ties, soy waste products, or waste liquids. Each trial burn contained one or more of the alternative fuels. An elaborate sampling and analysis plan was carried out for each trial burn with close cooperation between the Lafarge plant, the Lafarge testing laboratory, Auburn University, and outside laboratories.

Results will be presented from a comprehensive evaluation of the raw materials, fuels, clinker, cement, emissions, and plant operations from the test burns. The technically viability for each fuel as supplements for coal for direct burn in a manufacturing plant will be discussed. Alternative Solid Fuel (ASF) handling and process operations were established and evaluated for the use of each of the fuels. Replacement ratios (on a BTU basis) were up to 18% for the waste and bio-based fuels and the synergistic burns where two or more of the alternative materials were used together. Impacts on product quality and emissions were negligible and in some cases advantageous.

See more of this Session: Innovations for Energy Intensive Industrial Processes and Building Technologies
See more of this Group/Topical: International Congress on Energy 2011

(229b) Kinetics Study of Calcium Carbonate Decomposition At High CO2 Environment

Yong-Hua Duan[1], **Jian Zhang**[1], Delong Xu[2], Hui Li[3], Le-le Yang[1], Xiao Fan[1], Wen-bin Yang[1], Yongxiang Ren[3] and **Yong Min**[3], (1)Material Institute, Xi'an University of Architecture and Technology,, Xi'an, China, (2)Institute for Powder Eng, Univ. of Archi. & Tech., Xi'an, China, (3)Xi'an University of Architecture &Technology, Institute of Powder Engineering, Xi'an, China

With the concerns of "global warming", CO2 sequestration from cement industry are becoming significant important for the "green processing" for the environments. This is because the cement processing released about 36% of CO2 comparing to the 12% from regular coal-power plant. CaO/CaCO3, a major component and raw material from cement, was investigated in order to design suitable processing to sequestrate the CO2 during the cement production. It was found that the raw material particle size, surface area as well as chemical components were important factors, which will effect on the CO2 absorption efficiency, cycle times, particle activities, and so on. Various analytical equipments such as SEM, XRD, FTIR, TGA, were used to monitor the processing and analysis materials during the processing.

It was found that there are two competitive mechanisms involved during the processing i.e., chemical reaction and physical diffusion. In order to investigate the dominated effect, we selected the different particles as model compounds for the analysis. As a result, we found the small particle (particle size less than 5 micron) is dominated by chemical reaction mechanism, i.e., CO2 can completely diffusion into the core of the particle and result in a complete chemical reaction to form CaCO3; and the larger particle is dominated by the physical diffusion mechanism, i.e., only partial CO2 can diffuse into the CaO particles and formed a "core-shell" particles, where the centre core is CaO and out-shell is CaCO3. In addition, the small particle always accomplished with a high cycle times and more complete chemical reaction, which was recommended to use in the industry.

Based on the cement processing, various chemical looping processes were used to demonstrate the CaCO3 Calcinations / CaO Carbonation Cycle such as, vacuum processing, solids flow processing, fluidization processing, and cocking gas recycling processing. The energy efficiency, raw material activities, and corresponding thermodynamic date will be presented during the presentation.

See more of this Session: Innovations for Energy Intensive Industrial Processes and Building Technologies
See more of this Group/Topical: International Congress on Energy 2011

(229c) Application of Ionic Liquids In the Aluminum Electrorefining

Zheng Yong, Lu Xingmei and Zhang Suojiang, State Key Laboratory of Multiphase Complex Systems, Institute of Process Engineering, Chinese Academy of Sciences, Beijing, China

Application of Ionic Liquids in the Aluminum Electrorefining

Yong Zheng, Xingmei Lu *, Suojiang Zhang *

State Key Laboratory of MultiphaseComplex Systems, Institute of ProcessEngineering, Chinese Academy of Sciences, 100190, Beijing, P. R. China

Corresponding author: xmlu@home.ipe.ac.cn,sjzhang@home.ipe.ac.cn

The Hall-Heroult process for aluminum production has beenwidely employed in the past 100 years. According to this technique, the purityof primary aluminum is usually lower than 99.8% and needs to be improved further.At present, there are three main industrial methods applied in the aluminumrefining: zone melting, three-layer electrolysis and segregation process. However,these methods suffer from some inherent problems, such as high temperature and energyconsumption. Labeled as novel green solvents, ionic liquids have exhibited aseries of attracting properties and excellent performance in low-temperature aluminumelectrodeposition [1,2].The application of ionic liquids in aluminum electrorefiningis worth of investigation.

In our work, ionic liquid 1-butyl-3-methylimidazoliumheptachloroaluminate ([Bmim][Al_2Cl_7]) was prepared and used aselectrolyte in the aluminum electrorefining. The electrorefining processes were examined under different temperatures,potentials and current densities. Experimental results shownthat the main impurity elements, including Si, Fe, Tiand Na could be efficiently separated from primary aluminum in [Bmim][Al_2Cl_7]. Compact and smoothaluminum deposits with high purity (>99.9%) were obtained at 100 ^{0}C, -0.5~-1 V and 10~15 mA/cm^2(see Figure 1 and 2). Furthermore, the current efficiency of aluminum electrorefining increased to 98%. And the energyconsumption was lower than 5 kWh/kg-Al. As for [Bmim][Al_2Cl_7],it could be recycled and reused after purification. In general, ionic liquids area class of promising low-temperature electrolytes which open a new way to thealuminum refining.

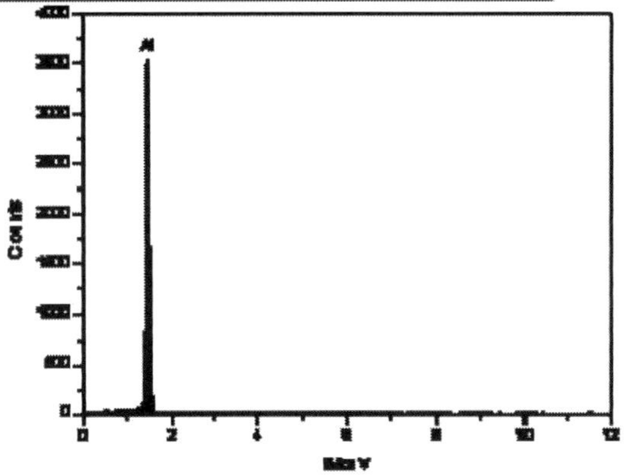

Figure 1. SEM image of the Al deposits   Figure 2. EDAX profile for the Aldeposits

Reference

[1] A. Campion and P. Kambhampati, *Chem. Soc. Rev.*, 1998, **27**, 241.

[2] M. M. Zhang, V. Kamavaram andR. G. Reddy, *JOM*, 2003, **11**, 54.

See more of this Session: Innovations for Energy Intensive Industrial Processes and Building Technologies
See more of this Group/Topical: International Congress on Energy 2011

(229d) Novel Zeolite Membranes for Energy-Efficient Air Dehumidification and Conditioning

Rong Xing[1], Yuxiang Rao[1] and Wei Liu[2], (1)Energy and Environment Directorate, Pacific Northwest National Laboratory, Richland, WA, (2)Pacific Northwest National Laboratory, Richland, WA

Traditional vapor compression air conditioners have very high power consumption for cooling of hot and humid air due to condensation of a large amount of water inside the cooler. We have developed a novel zeolite membrane supported on a thin (~50 micrometer thick) porous metal sheet to remove water vapor from an incoming humid air stream without changing its temperature, i.e., through vapor-phase membrane separation. The membrane dehumidifier reduces air humidity levels well below its due point under the cooling conditions so that only sensible cooling occurs. In this way, the energy consumption for air conditioning in hot and humid climate can be reduced by 50%. The high permeance and high H2O/air selectivity are critical performance parameters. Compared to previous membranes reported in the literature, our membrane exhibits a high H2O/air separation factor (> 100), a permeation flux about 2 or 3 orders of magnitude higher than the literature result, and excellent stability. This presentation will provide systematic discussion on impacts of membrane separation conditions, such as temperature, feed air humidity, and permeate pressure, on the membrane separation performances. The new membrane technology has great potential to be developed as a commercially-viable membrane dehumidifier by obtaining exceptional membrane separation performances and substantially reducing the zeolite membrane fabrication cost at the same time.

See more of this Session: Innovations for Energy Intensive Industrial Processes and Building Technologies
See more of this Group/Topical: International Congress on Energy 2011

An Osmotic Membrane Dehumidifier

Arthur S. Kesten and Jack N. Blechner, Nanocap Technologies LLC, Longboat Key, FL

OSMOTIC MEMBRANE DEHUMIDIFIER

It has long been known that relying on a vapor compression system to cool air down to the dew point in order to dehumidify is expensive to build and uses far too much energy to operate. The ability to dehumidify independently of cooling provides substantial advantage, provided that it can be done using inexpensive materials, uses little energy to operate, and is easily packaged with a separate cooling system.

An osmotic membrane dehumidifier has been developed that uses a flexible, semi-permeable membrane to facilitate capillary condensation of water vapor and the transport of condensed water through the membrane into a salt solution by osmosis. Here a humid gas stream is brought into contact with a semi-permeable membrane, which separates the gas stream from an osmotic (e.g., salt) solution. Some of the pores of the membrane are small enough to permit capillary condensation. Liquid formed within these pores connects with liquid formed in adjacent pores, collectively forming continuous paths of liquid. These 'liquid bridges' extend across the thickness of the semi-permeable membrane and provide paths by which water can travel across the membrane. Because the membrane is so thin, water concentration gradients across the membrane can be large. This can provide a large driving force for water transport between the humid air and the osmotic fluid. The flexibility of the polymeric membrane allows for considerable design flexibility that enhances the potential for retrofit with any cooling system.

BACKGROUND

Conditioning of a gas stream, such as air, in warm weather generally involves removal of moisture and decrease of temperature to make the gas stream suitable for its intended environment. Current dehumidification technology is commonly based on the conventional, refrigerant vapor compression technology (DX technology) or on desiccant substrate capture (DS technology). DX technology requires cooling humid supply air, such as the air within a room and/or outside air, to the water vapor condensation point, with external heat rejection on the compression side. This usually requires the supply air to be cooled below comfortable temperatures, and, thereafter, either reheated or mixed with warmer dry air to raise its temperature to an acceptable level before directing it into the space being dehumidified. Twenty to thirty-five percent of the energy expended in cooling the high humidity air is utilized to remove the latent heat of condensation from the air. Cooling and dehumidification of the air are thus coupled. That makes it impossible to control comfort parameters independently, making the DX cycle less efficient, from an overall system perspective, than a technology that would allow independent control of sensible and latent heat.

In applications where the outside air has both high humidity and temperature and the functional use of the interior space generates high water vapor levels (e.g., populated convention halls, exercise rooms, school buildings, etc.), it may not be possible for the DX technology to maintain the air introduced into the interior space at the correct humidity and temperature for maintaining comfort. The air delivered is cool but 'muggy', since further cooling to remove additional water would result in the air being uncomfortably cool.

In stand-alone dehumidification using a conventional compression cycle, heat reject is in direct contact with room air. As a consequence, the room air becomes more comfortable from a humidity side, but may be too warm. Again, the comfort parameters are coupled.

DS systems are generally applied in central air, ducted systems. Water vapor capture on a solid phase substrate is efficient and rapid. However, removal of the water vapor from the pores requires energy input. It also requires removing the substrate from the high humidity air stream and placing it in an exhaust, water-reject stream, before adding the re-evaporation energy. Alternatively, the substrate remains fixed and the treated air and exhaust streams flow directions interchanged as is done in a parallel bed, desiccant dryer system. DS technology requires, in steady state operation, the addition of this energy at a rate equal to or greater than the latent heat of condensation of water in the desiccant substrate.

The advantage of DS technology is that humidity levels in the outside air and/or re-circulated air can be adjusted independently of the cooling step. The disadvantage is the requirement to move the substrate and treated air stream relative to each other for capture and rejection of the water vapor.

In a liquid desiccant system, direct contact equipment, such as packed-bed towers, allows for simultaneous heat and mass transfer between the air and the liquid. The essential issue here is the requirement for intimate contact of desiccant and air. The conventional technique of using a packed-bed adsorption tower is not energy efficient due to its low contact surface area per unit volume. Moreover, evaporation of some liquid desiccant in a packed-bed tower may cause contamination of the conditioned air stream.

Osmotic dehumidification provides a more compact and efficient option. The process is described in Refs. 1 and 2. In the earlier work, a capillary condenser with pore sizes around 2 to 3 nanometers was used to condense water vapor in air at moderately high relative humidity. The condenser (originally designed and fabricated by Mobil Oil) had been fabricated from a ceramic substrate using a molecular template which was subsequently calcined at high temperature to leave behind a fairly uniform porous structure. A semi-permeable membrane was used to separate the capillary condenser from an osmotic fluid. An osmotic driving force, resulting from a water concentration gradient, transports the condensed water through the condenser and into the osmotic fluid. This type of device is effective in promoting water transport at fluxes in excess of 1 liter/square meter-hour. Under some circumstances, however, the pore structure of the capillary condenser can become unstable and degrade over a relatively short period of time in a humid environment. In addition, the capillary condenser is typically made from a rigid material and is therefore limited to those applications where a rigid body is acceptable.

DEMONSTRATION – PROOF OF CONCEPT

The feasibility of using commercially available membranes for capillary condensation and subsequent transport of condensate into a salt solution was demonstrated first in a laboratory scale experiment. A glass container, filled to the top with a salt solution, was covered with a membrane. A narrow tube attached to the side of the container extends up above the container. Humid air was blown across the top of the membrane and the rate of removal of water from the air was measured by the change in liquid level in the tube. Membranes tested were those used commercially for water purification. Salt solutions that were very effective were lithium chloride and magnesium chloride. The latter salt is inexpensive, benign and readily available. Results suggest that 1 square meter of the membranes that worked most effectively removes about 0.7 liters/hr of water vapor at relative humidity levels between 80 and 90%. The rate of condensation per unit of surface area is considerably smaller than for a uniform pore size capillary condenser. However, the polymer membranes are stable, inexpensive, flexible, and offer simple methods for providing large surface area interfaces between humid air and films of liquid salt solution.

VALIDATION AND SCALABILITY – BENCH TESTS

Subsequent experiments were performed using a larger apparatus in which shallow trays were filled with a salt solution and each covered with a membrane. Trays were arranged in a cafeteria style with narrow spaces between the trays for air flow over the membranes. Air was blown over the membranes and then through an exhaust manifold. An anemometer was used to measure air flow and humidistats were placed upstream and downstream of the trays to measure temperature and humidity. A circulating pump was used to move the salt solution through the trays from inlet to outlet manifolds.

Tests were performed on arrays of trays and on single trays with various spacing and air flow. A single tray, for example, covered with a membrane of 0.13 square meters reduced relative humidity of an air stream flowing at 15 cfm (0.4 cubic meters per minute) by 7%. At 20 cfm, the relative humidity was reduced by 5%. An array of 12 of these trays (occupying a volume of less than 0.02 cubic meters) would remove about 0.5 liters/hr of water from a humid air stream. Reducing volumetric flow rates results in increased relative humidity reduction. However, capillary condensation with current membranes is limited by pore size and relative humidity is generally not reduced below 65-70%.

A mathematical model of the transport of air and removal of water vapor at membrane surfaces has been developed to help to optimize the design of the system. The tray system is but one of many potential configurations.

Ref. 1: Kesten, A.S. and Blechner, J.N., Versatile Dehumidification Process and Apparatus, U.S. Patent 7,758,671 B2, July 20, 2010

Ref. 2: Kesten, et al, Dehumidification Process and Apparatus, U.S. Patent 6,539,731 B2, April 1, 2003

(229f) Ethanol/Water Biomass Feedstock Separation Through Inorganic A-Type Zeolite Membrane on Thin Porous Ni Sheet Support

Yuxiang Rao[1], Rong Xing[1] and Wei Liu[2], (1)Energy and Environment Directorate, Pacific Northwest National Laboratory, Richland, WA, (2)Pacific Northwest National Laboratory, Richland, WA

Bio-Ethanol/water separation by fractional distillation becomes a popular industrial process to remove water from ethanol-water mixture. However, the mixture of 95.6wt% ethanol and 4.4 wt% water is an azeotrope and cannot be further purified by distillation. Alternatively, this can be achieved by membrane separation technology since it is not works on vapor-liquid equilibrium but on pervaporation mode.

In this work, real ethanol feeds (rectifier bottom, 40 wt%, PH=3.99 and rectifier overhead, 95wt%, PH=4.54) from Pacific Ethanol Inc. were chosen to test the performance and stability of A type zeolite membrane (2~3um), which supported on novel thin porous Ni sheet substrate (~50um). The preliminary results shows that NaA zeolite membrane was unstable at operation condition (~100C), mainly due to the strong acidity of real ethanol-water biomass feedstock (PH≈4).

However, after neutralized with 0.1M NaOH solution to PH=8.5 or by pass a guard bed filled with weak base ion exchange resin (Amberlite@IRA-93), this novel membrane shows high performance (Water flux >2 kg/m^2 h, water/ethanol selectivity > 5000) and stability (>100hrs). ICP analysis result also shows that sodium and sulfur ions will remains in the feed side during the separation process, mainly due to their large hydrated diameter (~7.2 Å for Na^+ and ~8.6 Å for SO_4^{2-}) comparing with the pore size of NaA zeolite membrane. (~4Å)

See more of this Session: Innovations for Energy Intensive Industrial Processes and Building Technologies
See more of this Group/Topical: International Congress on Energy 2011

(231a) A Novel SHF Process for Conversion of AFEX-CS to Ethanol Featuring High Ethanol Productivity, Enzyme Recycling, and Yeast Cells Reuse

Ming Jie Jin, Christa Gunawan, Nirmal Uppugundla, Shishir Chundawat, Bruce Dale and **Venkatesh Balan**, Chemical Engineering and Material Science, Michigan State University, East Lansing, MI

There are two major problems in current cellulosic ethanol production process: low ethanol productivity and high enzyme loading. Based on the current Separate Hydrolysis and Fermentation (SHF) process, it takes at least 96h to perform enzymatic hydrolysis to convert AFEX-CS to fermentable sugars at 6% glucan loading. It will then take another 96-168h to ferment those sugars to ethanol. By simply integrating enzymatic hydrolysis and fermentation, which is SSCF (simultaneously saccharification and co-fermentation), it still requires at least 168h to reach comparable results as SHF. Here we will present a novel SHF process. In this process, the easily hydrolyzed part of biomass is hydrolyzed first to fermentable sugars in 24h. The hydrolysate was then separated from residue solids and fermented to ethanol in 24h using high yeast cell inoculation. The residual solids (difficultly hydrolyzed part of biomass (DHPB)), which bound enzymes, are then transferred to the next cycle of hydrolysis. This way, enzymes are largely reused and enzyme loadings in the next cycles can thus be reduced. The yeast cells after 24h fermentation are also recovered and reused for the next cycle of hydrolysate fermentation. The time for each cycle hydrolysis/fermentation is 24h/24h. After five cycles, the difficult hydrolyzed part is collected and enzymatic hydrolysis is performed without adding additional enzymes. Since enzymes are accumulated to a relatively high concentration and part of the decomposition products are removed with liquid during the five cycles, the hydrolysis of DHPB would be easier to achieve high sugar conversion. Profiles of enzyme recycled in each step were characterized.

See more of this Session: Integrated Processes for Biochemical Conversion of Renewable Feedstocks to Fuels and Chemicals I
See more of this Group/Topical: Sustainable Engineering Forum

(231b) Concentration of a Solids-Free Lignocellulose Hydrolysate for Very High Gravity Fermentation: Inhibition Effects

Steven J. Schneiderman, Todd J. Menkhaus and Patrick C. Gilcrease, Chemical and Biological Engineering, South Dakota School of Mines and Technology, Rapid City, SD

A major problem in the efficient conversion of lignocellulose hydrolysates to ethanol is low ethanol concentrations in the fermentation broth due to low initial sugar concentrations. This is undesirable because low ethanol concentrations lead to increased separation costs downstream. Distillation is energy intensive, and constitutes more than 50% of the total energy consumption of a typical ethanol plant. An obvious solution is to start with a high initial sugars concentration, which is referred to as a very high gravity (VHG) fermentation. By increasing the glucose concentration in the hydrolysate from 40 to 100 g/l, the distillation energy consumption would be reduced by 60%. In the first step of our integrated process, polyelectrolyte flocculants facilitate the efficient removal of biomass solids from a pine wood hydrolysate. Once the hydrolysate is solids-free, an evaporation or reverse osmosis step can be readily used to remove water and concentrate the fermentable sugars. VHG fermentations present some challenges of their own, including higher energy costs associated with concentrating the sugars, simultaneous concentration of some or all of the fermentation inhibitors, and osmotic stresses on the fermentation organism at higher substrate concentrations. Because some of the inhibition effects are synergistic, this could become more problematic at VHG conditions.

The goal of this study was to investigate the potential problems associated with fermenting a concentrated lignocellulose hydrolysate, and to determine operating conditions that maximize productivity and minimize costs for the combined sugars concentration-fermentation-distillation process. A batch culture design of experiments was used to quantify the inhibition effects of five common inhibitors on the productivity of our *S. cerevisiae* D_5A strain. The inhibitors studied were glucose, ethanol, furfural, HMF, and acetic acid at concentration ranges of 25 – 300 g/l, 0 – 150 g/l, 0 – 5 g/l, 0 – 5 g/l and 0 – 10 g/l, respectively. Our experimental results were used to regress kinetic model parameters, and define an acceptable operating window for VHG fermentation. The hydrolysate concentration step (evaporation/reverse osmosis) was then tailored to minimize total separation costs (sugars concentration and ethanol distillation) while maintaining a high level of ethanol productivity. Finally, the fermentation of a concentrated pine hydrolysate at optimum conditions was demonstrated at the bench scale.

See more of this Session: Integrated Processes for Biochemical Conversion of Renewable Feedstocks to Fuels and Chemicals I
See more of this Group/Topical: Sustainable Engineering Forum

(231c) Separate Processing of Hemicellulose and Cellulose to Optimize Ethanol Production From Corn Stover

Alexandre Chapeaux[1], Nancy Dowe[1] and Daniel J. Schell[2], (1)National Bioenergy Center, National Renewable Energy Laboratory, Golden, CO, (2)National Renewable Energy Laboratory, Golden, CO

Separate processing ofhemicellulose and cellulose to optimize ethanol production from corn stover

Alexandre Chapeaux, NancyDowe, Daniel J. Schell

Using a series of thermochemical (pretreatment)and biochemical steps, lignocellulosic biomass is converted to sugars that subsequentlycan be made into fuels and chemicals. One such process uses heat and dilute sulfuric acid to hydrolyzes the hemicellulosic fractionof the biomass to monomeric and oligomeric sugars.The removal of hemicellulose and to a lesser extent lignin from the biomass significantlyimproves the performance of the next step in the process, enzymatic hydrolysisof cellulose to glucose. However, enzymatic cellulose hydrolysis can beperformed in the presence or absence of the hemicellulosic sugars producedduring pretreatment. It is well known that removing the hemicellulose sugarsand other compounds produced during pretreatment improves enzymatic celluloseconversion yields. In previous work, we assessed the potential of a processthat removes the hemicellulosic sugars prior to enzymatic cellulose hydrolysis byexploring the interaction between biomass solids loading, yields, and processcosts. However, this process is more expensive because of the additional costfor solid-liquid separation equipment required to remove th-richsugar liquor from the cellulosic solids.

In this study, the goal was to deepen ourunderstanding of the advantages and disadvantages of the separatehemicellulose/cellulose process. The feedstock was dilute-acid-pretreated cornstover produced in a 200 kg dry biomass/d continuous horizontal reactor. Thepretreated slurry was separated into a liquor stream (hemicellulosic sugars)and a washed cellulosic solid stream. Enzymatic cellulose hydrolysis was investigatedat four enzyme loadings (10, 20, 30 and 40 mg protein/g cellulose) and atinsoluble solids concentrations from 9% to 21% (w/w). The resulting glucose wasfermented to ethanol using several yeasts and a glucose-xylose co-fermentingbacterium, *Zymomonas mobilis* 8b. The sugars in the liquorstream were fermented to ethanol with three strains of co-fermenting *Z. mobilis* after dilution of the streamwith water to lower sugar concentrations. Finally, the overall performance ofthe process was assessed after partitioning water between the two phases in away that optimizes overall ethanol yields. We then compared these results to aprocess configuration in which the streams are not separated. In the latterprocess, enzymatic cellulose hydrolysis is performed in the presence of thebackground hemicellulosic sugars produced during pretreatment of the biomass.

See more of this Session: Integrated Processes for Biochemical Conversion of Renewable Feedstocks to

(231d) Ethanol and Furfural Production From Corn Stover Using Hybrid Fractionation Process with Zinc Chloride and Simultaneous Saccharification and Fermentation (SSF)

Chang Geun Yoo[1], **Tae Hyun Kim**[1] and Monlin Kuo[2], (1)Agricultural and Biosystems Engineering, Iowa State University, Ames, IA, (2)Department of Natural Resources Ecology and Management, Iowa State University, Ames, IA

Two-stage hybrid fractionation process was investigated to produce cellulosic ethanol and furfural from corn stover. Zinc chloride ($ZnCl_2$) was used to selectively solubilize hemicellulose in the first stage. During the second stage, the remaining treated solids were converted into ethanol using commercial cellulase and *Saccharomyces cerevisiae* or recombinant *Escherichia coli*, KO11.

Hybrid fractionation process recovered >90% of glucan, ~90% of xylan, and ~75% of lignin at optimal reaction conditions: 1st stage: 5% acidified $ZnCl_2$, 7.5 ml/min, 150 °C (10 min) and 170 °C (10 min); 2nd stage: Simultaneous saccharification and fermentation (SSF). Furfural yield from hemicellulose hydrolysates was 58 %. SSF of $ZnCl_2$ treated solid with *Saccharomyces cerevisiae* resulted in 69-98 % of theoretical maximum ethanol yields based on glucan present in the treated solids. After fermentation, solid residues contained mostly lignin. Based on total lignin in untreated corn stover, the lignin recovery yield was ~75%.

See more of this Session: Integrated Processes for Biochemical Conversion of Renewable Feedstocks to Fuels and Chemicals I
See more of this Group/Topical: Sustainable Engineering Forum

(231e) Fischer Tropsch Synthesis Via Biomass Derived Synthesis Gas

Syed Ali Gardezi, Chemical Engineering, University of South Florida, Tampa, FL, Babu Joseph, Chemical and Biomedical Engineering, Clean Energy Research Center, University of South Florida, Tampa, FL and John T. Wolan, University of South Florida, Tampa, FL

Due to the recent surge in energy costs and uncertainties in fuel prices, there is significant interest in seeking alternative sources of energy particularly from renewable such as biomass. Indirect liquefaction of biomass (via the thermo-chemical conversion to syngas followed by liquefaction via Fischer-Tropsch Synthesis or FTS) offers a commercially viable route for meeting the challenge of producing renewable fungible liquid fuels. Apart from biomass, FTS is also the critical step for converting natural gas and coal to liquid fuels.

A FTS technology has been developed for producing middle distillates (diesel and aviation fuel) by using pine chips as the main feedstock. In this "two step" process, pine chips were subjected to an entrained flow gasification process developed by Pearson et al. [1]. The resulting tar free gas is comprised mainly of carbon monoxide, hydrogen, carbon dioxide, methane and some organic contaminants such as benzene, toluene, and naphthalene along with substantial amount of water. For further liquefaction this gas was cleansed using multiple adsorbents. An inline "Shaw Moisture Meter" was used to continuously monitor the moisture content of the feed to the liquefaction process.

For the production of liquid fuels, eggshell cobalt catalysts were used. Cobalt catalysts have several advantages e.g. they produce linear paraffinic hydrocarbons and minimize undesirable water gas shift reaction. This eggshell design is a well known strategy for producing a narrow hydrocarbon distribution. In this regard previous research work conducted by Iglesia et al. was also consulted [2]. A bench scale fixed bed reactor was used for the conversion of syngas. Modeling of the reactor startup and of adjustment of the reactor packing ensured successful operation resulting in a high conversion and narrow distribution of hydrocarbons in the range of diesel and aviation fuel.

References :

[1] Pearson et al., U.S. Patent 4,857,076, issued Aug 15, 1989.

[2] E. Iglesia, S. L. Soled, J. E. Baumgartner, S. C. Reyes, *J. Catal.* **153**, 108 (1995).

See more of this Session: Integrated Processes for Biochemical Conversion of Renewable Feedstocks to Fuels and Chemicals I
See more of this Group/Topical: Sustainable Engineering Forum

(231f) Integrations of Biosurfactant Production Into Advanced Biorefineries

William Colonna[1], **Mustafa E. Marti**[1], Michelle Pynn[2], Gabriel Reznik[2], Kevin Jarrell[2], Buddhi Lamsal[3] and Charles E. Glatz[1], (1)Chemical and Biological Engineering, Iowa State University, Ames, IA, (2)Modular Genetics, Inc., Woburn, MA, (3)Food Science and Human Nutrition, Iowa State University, Ames, IA

Several classes of biosurfactants are produced by microorganisms. One such surfactant is FA-Glu, which is related to the natural biosurfactant, surfactin. FA-Glu consists of a long-chain β-hydroxy fatty acid coupled to glutamic acid. It is produced by a strain of *Bacillus subtilis* genetically-engineered by Modular, Inc. It has several potential applications including use as a dispersant for oil spills, where biodegradability is desirable.

Economic production and eventual application will be favored if substrate costs can be reduced and low-cost recovery methods developed. Several such opportunities arise in advanced biorefineries. Product value would be further enhanced if such production could achieve favored "green" rating by avoiding solvents other than water and simple alcohols and effect emulsification and low toxicity could be demonstrated. This report focus on production considerations.

Production of FA-Glu by fermentation on glucose is compared to results using lower cost substrates available in advanced biorefineries, including fiber components of soy and glycerol. FA-Glu yields from soybean hulls (SBH) ranged from 29-51% of that from glucose on a substrate weight basis.. When SBH polysaccharides were hydrolyzed (~65%) by pre-treatment with carbohydrases, FA-Glu levels were only ~21-37% of those from glucose. The cells consumed the enzymatically-released sugars; however, very little of this was converted into FA-Glu. Addition of glucose to the enzymatic SBH hydrolysate improved FA-Glu yields, but only slightly. The low FA-Glu yields from SBH could be due to inhibitors released from the hulls by the enzymes or inhibitory substances in the enzyme preparations.

Glycerol, a by-product of biodiesel production, proved to be a suitable carbon source for FA-Glu production. Cells grown on 2% w/v glycerol produced ~25% more FA-Glu than cells grown on 2% glucose. Glycerol is abundant and low in cost. Its use as a feedstock for production of microbial biosurfactants would provide a means for efficient utilization of soy by-products and improving the ag economy.

The FA-Glu was purified by collecting the foam from the aerated broth, clarification by centrifugation or filtration, and adsorption and elution from an anion exchange column to a purity of 97.9%. Final product color can be improved by liquid-liquid extraction, but this may compromixe "green" status. A precipitation based alternative process is being examined as a means of eliminating solvent use.

(247a) Nanoparticle Synthesis Using Reversible Ionic Liquids

Amy Rohan[1], Emily Nixon[2], Rani Jha[2], Manish Talreja[1], Pamela Pollet[2], Charles Liotta[2] and Charles Eckert[1], (1)School of Chemical & Biomolecular Engineering, Georgia Institute of Technology, Atlanta, GA, (2)School of Chemistry and Biochemistry, Georgia Institute of Technology, Atlanta, GA

Applications of metal nanoparticles have attracted significant attention due to their shape and size dependent properties. Current methods for nanoparticle synthesis remain cost-, energy- and time-intensive. This work will present Au nanoparticle synthesis with reversible ionic liquids, formed upon the addition of CO_2 to a silylated amine precursor. The resulting ion pair allows for reverse micelle formation. Heating the solution to a moderate temperature results in the reversal of the ionic liquid and the break-up of the reverse micelles, for subsequent characterization or deposition of the Au nanoparticles. Formation of metal nanoparticles without the requirement of capping agents or surfactants, sometimes in large excess, could result in significant energy and waste reduction. We have tuned a number of process parameters to demonstrate the formation of monodisperse Au nanoparticles.

See more of this Session: Nanostructured Particles for Catalysis
See more of this Group/Topical: Particle Technology Forum

(247b) Synthesis, Characterization and Catalytic Application of Dendrimer-Mediated Nickel and Iron Alloy Nanoparticles Containing An Average of 40 Atoms

Vanessa A. Castillo and John Kuhn, Chemical and Biomedical Engineering, University of South Florida, Tampa, FL

Bimetallic nanoparticles (NPs), composed of two differentmetals, have recently drawn great interest in the field of catalysis because bimetallization makes it possible to obtain catalysts withimproved catalytic activity that may not be achieved with monometallicnanoparticles [1]. In addition, the size of the metal particles plays animportant role in the activity and selectivity of catalysts,and an enhancement in the catalytic properties has been previously shown formonodisperse particles with ~1nm of diameter [2]. Here we report the synthesis and characterizationof different ratios of Fe-Ni bimetallic nanoparticles of < 2nm diametersupported on SiO_2 for the study of the catalytic activity in theethylene hydrogenation reaction.

Thenanoparticles were synthesized within a hydroxyl-terminated fourth generationpoly-aminoamide (PAMAM) Dendrimer (G4-OH), a hyperbranchedpolymer, in aqueous solution and then immobilized by depositing onto aCAB-O-SIL M-5 Silicon dioxide support. UV-Visabsorption spectrophotometry was used to monitor the coordination processbetween $Fe^{(+3)}$ and $Ni^{(2+)}$metals ions and the internal tertiary amine groups of the dendrimers. TransmissionElectron Microscopy and X-ray absorption spectroscopy were used to measure thesize and distribution of the zerovalent metals afterreduction of the coordinated metal ions. Catalytic activity of the supported Fe-Ninanoparticles was studied with ethylene hydrogenation at different temperaturesafter hydrogen reduction pretreatment at 523 K.

Reference

1. J.H. Sinfelt, Acc. Chem. Res., Vol. 20 p. 134, 1987.

2. Huang,W. et al., Nano Letters. Vol. 8, No. 7 p. 2027, 2008.

See more of this Session: Nanostructured Particles for Catalysis
See more of this Group/Topical: Particle Technology Forum

(247c) Scalable Synthesis of Palladium Nanoparticle Catalysts by Atomic Layer Deposition

Xinhua Liang[1], Lauren Blinn[2], Brittany Michael[1] and Alan W. Weimer[1], (1)Chemical and Biological Engineering, University of Colorado at Boulder, Boulder, CO, (2)Chemical Engineering, University of Florida, Gainesville, FL

Atomic layer deposition (ALD) was used to produce Pd/Al_2O_3 catalysts using sequential exposures of Pd (II) hexafluoroacetylacetonate and formalin at 200 °C in a fluidized bed reactor. The ALD prepared Pd/alumina catalysts were characterized by various methods including hydrogen chemisorption, XPS and TEM, and compared with a commercially available 1 wt.% Pd/alumina catalyst, which was also characterized. The content of Pd on alumina support and the size of Pd nanoparticles can be controlled by the number of ALD coating cycles and the dose time of the Pd precursor. One layer of organic component from the Pd precursor remained on the Pd particle surface. The oxidized/reduced ALD 0.9 wt.% Pd/alumina had greater active metal surface area and percent metal dispersion than the commercial 1 wt.% Pd/alumina catalyst. The ALD and commercial catalysts were subjected to catalytic testing to determine their relative activities for glucose oxidation to gluconic acid in aqueous solution. The ALD 0.9 wt.% Pd/alumina catalyst had comparable activity as compared to the commercial 1 wt.% Pd catalyst. No noticeable amount of Pd leaching was observed during the vigorously stirred reaction. ALD can be an efficient method for the preparation of highly dispersed noble metal nanoparticles.

See more of this Session: Nanostructured Particles for Catalysis
See more of this Group/Topical: Particle Technology Forum

(247d) A First-Principle Study of Pt Clusters On Anatase TiO2 (101) Surfaces

Yun Zhou[1], Christopher L. Muhich[2], Charles B. Musgrave[2] and Alan W. Weimer[2], (1)Department of Chemical and Biological Engineering, University of Colorado, Boulder, Boulder, CO, (2)Department of Chemical and Biological Engineering, University of Colorado, Boulder, Bouler, CO

The absorption properties of Pt_n (n is up to 13) clusters on defect free TiO_2 anatase (101) surfaces have been studied using *ab initio* pseudopotential calculations based on density functional theory. The (111) crystal plane was chosen for the Pt atom arrangement. The change of the Pt cohesive energy and the electronic structure were investigated with the change of the Pt cluster size. The Pt cluster properties were also compared with a full monolayer Pt thin film to understand thoroughly the Pt island growth on an anatase TiO_2 (101) surface. The results indicate that Pt prefers to absorb on existed Pt clusters rather than on the TiO_2 surface. For the same size of the Pt cluster, different absorption positions, orientation and cluster configurations were examined. The binding energy difference of the 1 layer configuration and several 2 layers configurations for the same number of the Pt atoms (eg. 10 Pt atoms) were compared. Results showed that the Pt energetically prefers the formation of three-dimensional particles. Different cycles of Pt nanoparticles were also experimentally deposited on TiO_2 nanopowders by use of the atomic layer deposition method. These Pt/TiO_2 nanoparticles were explored via high resolution transmission electron microscopy and the results validated the theoretical study.

See more of this Session: Nanostructured Particles for Catalysis
See more of this Group/Topical: Particle Technology Forum

(247e) Size Dependence of Sulfur Deactivation Using Silica Supported Platinum Nanoparticles

Lyndsey M. Baldyga, University of South Florida, Tampa, FL and J. N. Kuhn, Chemical & Biomedical Engineering, University of South Florida, Tampa, FL

Sulfur poisoning is of large concern in the fossil fuel and the renewable energy industries. A working hypothesis is that an optimal particle size exists because small particles, with their unsaturated coordinated surfaces, will bind sulfur strongly as compared to the surfaces of larger particles and large particles have a low surface to bulk ratio as compared to small particles.

For this project, the sulfur tolerance of four different sizes of silica – supported Pt nanoparticles are tested using ethylene hydrogenation as a probe reaction. These particles were synthesized using colloidal synthesis and immobilized on silica. The catalyts were characterized using transmission electron microscopy, X – ray diffraction, and temperature programmed experiments using a mass spectrometer. The sizes determined as 2.0, 3.4, 4.3, and 6.8 nm. Using mass spectrometry and gas chromatography for analysis, the sulfur tolerance was tested by performing an ethylene hydrogenation reaction with and without sulfur treatment.

See more of this Session: Nanostructured Particles for Catalysis
See more of this Group/Topical: Particle Technology Forum

(247f) Silica-Supported Ruthenium Nanoparticles for Hydrogenation of Pyruvic Acid In the Aqueous Phase

Xianfeng Ma[1], Rui Lin[1], James E. Jackson[2] and Robert Y. Ofoli[1], (1)Chemical Engineering and Materials Science, Michigan State University, East Lansing, MI, (2)Chemistry Department, Michigan State University, East Lansing, MI

Heterogeneous catalysis is an important technology in the chemical industry. Incorporation of nanoscience into catalysis provides one of the most powerful approaches to understanding reaction mechanisms and designing new-generation catalysts more efficienctly[1]. In this work, we present a strategy for design of monodisperse Ruthenium nanoparticles (RuNPs), immobilization of the RuNPs onto ordered mesoporous silica (MSUF), and reactivity assessment of the supported catalyst (RuNPs/MSUF) by aqueous phase hydrogenation of pyruvic acid to lactic acid. The RuNPs were prepared by a polyol reduction method[2] using ruthenium(III) acetylacetonate and poly(vinylpyrrolidone) (PVP) as precursor and stabilizer, respectively. Transmission electron microscopy (TEM) characterization showed that the particles were spherical with a fairly narrow size distribution and an average diameter of 3.5±0.5 nm. RuNPs/MSUF was prepared by immobilizing the RuNPs onto ordered mesoporous silica support MSUF using sonication method. RuNPs/MSUF was prepared by immobilizing the RuNPs onto ordered mesoporous silica support MSUF using sonication. TEM characterization showed well-dispersed particles on the surface of MSUF with no agglomeration and no damage to the framework of the support. Both inductively coupled plasma-atomic emission spectroscopy (ICP-AES) and X-ray photoelectron spectroscopy (XPS) characterization confirmed the attachment of RuNPs to MSUF. The reactivity of RuNPs/MSUF was evaluated by aqueous phase hydrogenation of pyruvic acid. All the reactions were conducted in a multi-batch reactor system (Parr Instrument Co., Model 5000), and the samples were analyzed by HPLC (Waters Co.) with an RI detector. A differential approach was used to compute the initial turnover frequency (TOF) for each reaction[3,4]. Results showed that, even under much milder reaction conditions than reported to-date[5], the initial TOF of the reaction (685.2 h^{-1}) catalyzed by RuNPs/MSUF is more than 25 times higher than values reported in the literature (24.2 h^{-1}) [5]. It is also noteworthy that RuNPs/MSUF shows a significant improvement in TOF compared to commercial Ru catalysts, which is an indication of a higher utility efficiency of Ru in catalysis.

References:

1. Gabor A. Somorjai, Feng Tao, *et al. Top Catal* (2008) 47:1–14
2. Sang Hoon J, Jeong Y. Park, *et al. Nano Lett.*, 2010, 10 (7), pp 2709–2713
3. Zhang, Z.; Jackson, J. E.; Miller, D. J. *Ind. Eng. Chem. Res.*, 2002, 41, 691-696.
4. Mehmet Zahmakýran, *et al. J. Am. Chem. Soc.*, 2010, 132 (18), pp 6541–6549
5. R.. Luque, J.H. Clark. *Catalysis Communications* 11 (2010) 928–931

(247g) Single-Step, Plasma-Based Synthesis of Supported Nanoparticle Catalysts

Travis Koh and Michael Gordon, Chemical Engineering, University of California, Santa Barbara, Santa Barbara, CA

Many established and future technologies rely on the unique chemical and opto-electronic properties of nanometer-scale particles. For example, metal, metal oxide, and semiconductor nanoparticles find application in heterogeneous catalysis, photocatalysis, and photovoltaics. In light of such needs, a robust synthesis route for size and composition-controlled nanoparticles would be highly desirable.

In this work, we present a single-step, plasma-based synthesis route to produce a wide range of surfactant-free nanoparticles (*e.g.* metals, alloys, metal oxides, oxide-supported metals, and semiconductors) in the 1-20 nm size range that can be used in catalytic applications. A hydrodynamically-stabilized, atmospheric pressure, micro-hollow cathode plasma discharge with high density and temperature is used to pyrolyze volatile organometallics and semiconductor precursors to nucleate small crystalline nanoparticles. Particle growth is quenched at small, uniform size because the residence time in the plasma is very short (~microseconds). The end result is an aerosol of surfactant-free nanoparticles that can be spray-deposited on a variety of surfaces.

The talk will focus on the synthesis and characterization of metallic (Pd, Fe, Ni), alloy (Pd-Ag, Fe-Ni), and oxide-supported metallic nanoparticles (Pd/SiO_2) realized via microplasma deposition. *In-situ* FTIR and catalytic testing of Pd and Pd/SiO_2 for selective hydrogenation and CO oxidation were performed to probe activity and metal-support interactions. With respect to hydrogenation, both Pd catalysts exhibited low temperature light-off (< 70 deg C) and high ethylene selectivity at high conversion. Pd catalysts spray-deposited on ITO and graphitic supports were also tested for electrochemical oxidation of methanol and formic acid. Finally, we will highlight how microplasmas can be used to realize a variety of nanostructured materials for various technological applications.

See more of this Session: Nanostructured Particles for Catalysis
See more of this Group/Topical: Particle Technology Forum

(258a) Mechanistic Modeling of Simultaneous Enzymatic Saccharification of Cellulose and Xylan Using Continuous Distribution Kinetics

Ambarish Nag[1], Andrew J. Griggs[2], Jonathan J. Stickel[2], Michael A. Sprague[1] and James J. Lischeske[2], (1)Computational Science Center, National Renewable Energy Laboratory, Golden, CO, (2) National Bioenergy Center, National Renewable Energy Laboratory, Golden, CO

Cost effective production of biofuels and biochemicals often involves enzymatic saccharification, which has been the subject of intense research and development. In this work, we develop a mechanistically based kinetic model for the enzymatic hydrolysis of biomass composed of cellulose, xylan, and lignin. The prototypical biomass particle in our model has a cellulosic core that is partially covered by xylan and lignin that impede the access of glucanase enzymes to the cellulosic core. Our model hence incorporates both glucanase and xylanase enzymatic saccharification of surface xylan and the cellulosic core. For degradation of insoluble cellulose chains we consider the actions of endoglucanase (endo-β-1,4-glucanase,) and exoglucanase (exo-β-1,4-cellobiohydrolase). Likewise for insoluble xylan we consider the actions of endoxylanase (endo-β-1,4-xylanase) and exoxylanase (xylan β-1,4-xylosidase), the latter enzyme hydrolyzing soluble xylan oligomers as well. The enzyme system also includes beta-glucosidase for the hydrolysis of soluble glucan oligosaccharides. The kinetics for the depolymerization of cellulose and xylan chains are described using population-balance equations, which are capable of describing the evolution of the distribution of chain sizes during hydrolysis, without solving equations for all chemical species present in the reacting mixture. This approach affords computationally efficient simulations, enabling the use of the model for process design and optimization. In combination with experiments, the kinetic model is used to study the transformations of the cellulose and xylan chain-size distributions during hydrolysis by an enzyme cocktail of all the enzyme species.

See more of this Session: Reaction Kinetics and Transport Fundamentals for Biomass Conversion
See more of this Group/Topical: Sustainable Engineering Forum

(258b) Kinetic Modeling of Cellulose Hydrolysis with First Order Inactivation of Adsorbed Cellulases

Zhuoliang Ye and R. Eric Berson, Chemical Engineering, University of Louisville, Louisville, KY

The rate of glucose release during enzymatic hydrolysis of cellulose slows down significantly as the reaction proceeds, and the mechanism is not yet clear. It is, therefore, desirable to develop a kinetic model that accurately describes the enzymatic hydrolysis reaction. In this work, a model based on Langmuir binding kinetics that also incorporates first order inactivation of adsorbed cellulase was developed that accurately predicts rates and yields during the hydrolysis reaction for two different cellulose substrates. Substrate concentrations of 20g/L and less were tested in order to remove any potential effects due to product inhibition or mass transfer limitations. A key result of the modeling suggests that Vmax decreases over time due to deactivation of the adsorbed enzyme. Rate constants associated with product formation and enzyme inactivation from our model yield are comparable to reported values elsewhere. The model predicted results within 10% of experimental results for both Solka Floc and Sigmacell substrates. Key model parameters were similar for the two substrates: the apparent hydrolysis rate (k_2) and inactivation rate (k_f) were 33 min^{-1} and 0.38 h^{-1}, respectively, comparable to values reported elsewhere. The rate constants were within 5% for the two different substrates, further strengthening the value of the model and implying that a common but yet undefined rate-limiting step associated with loss of enzyme activity likely exists in the pathway of cellulose hydrolysis. It is proposed that this rate-limiting step may be either dissociation or isomerization of the enzyme.

See more of this Session: Reaction Kinetics and Transport Fundamentals for Biomass Conversion
See more of this Group/Topical: Sustainable Engineering Forum

(258c) Observing and Modeling Cellulosic Substrate Depolymerization by Commercial Enzyme Cocktails Using Confocal Fluorescence Microscopy

Jeremy S. Luterbacher[1], Jose M. Moran-Mirabal[2] and Larry P. Walker[2], (1)Department of Chemical and Biomolecular Engineering, Cornell University, Ithaca, NY, (2)Department of Biological and Environmental Engineering, Cornell University, Ithaca, NY

Understanding and modeling the depolymerization mechanisms of cellulosic and lignocellulosic substrates by cellulase cocktails is a critical step towards further optimizing the production of monosaccharides from biomass. The Spezyme CP cellulase cocktail combined with the Novo 188 b-glucosidase blend was used to depolymerize bacterial microcrystalline cellulose (BMCC), filter paper cellulose and untreated and H_2O-CO_2 pretreated lignocellulosic substrates (hardwood and switchgrass), which were immobilized on a glass surface. The enzyme mixture was supplemented with a small fraction of fluorescently labeled *T. Reseii* CBH I, which served as a reporter to track cellulase diffusion into and binding onto the internal physical structure of the cellulosic substrate. All reported experiments were conducted at 50°C, the optimal temperature for maximum hydrolytic activity of the enzyme cocktail. BMCC was observed throughout degradation by labeling it with a fluorescent dye. Native autofluorescence was used in a similar way to track filter paper and lignocellulosic substrates. This method allowed us to measure the binding of cellulases *in situ* and follow the temporal morphological changes of cellulose and lignocellulosic substrates during their depolymerization by a commercial cellulase mixture.

In the case of BMCC, a kinetic model was developed and fitted to fluorescence intensity data obtained through image processing. This model was successfully used to predict the bulk sugar concentrations that were liberated over time from BMCC in a separate experiment and measured by liquid chromatography. A more complete mass transfer and kinetic model is currently under development for the more complex substrates (filter paper and lignocellulosic biomass).

See more of this Session: Reaction Kinetics and Transport Fundamentals for Biomass Conversion
See more of this Group/Topical: Sustainable Engineering Forum

(258d) Effects of Mixing Quality On the Kinetics of Biomass Liquefaction

David M. Lavenson[1], Emilio J. Tozzi[1], Tina Jeoh[2], Michael J. McCarthy[3] and Robert L. Powell[1], (1) Chemical Engineering and Materials Science, University of Calfiornia Davis, Davis, CA, (2)Biological and Agricultural Engineering, University of Califonrnia, Davis, Davis, CA, (3)Biological and Agricultural Engineering, University of Calfiornia Davis, Davis, CA

Abstract: Biochemical conversion is a promising route to produce fuels and chemicals from cellulosic biomass. A key step in the process is biomass liquefaction, where enzymes are used to break down pretreated cellulosic particles. During this process, drastic changes in the mechanical properties of the cellulosic slurry cause a transition from a solid-like material to a pumpable liquid. A fast liquefaction is desirable to achieve reductions in equipment size and cost. To assess the effect of mixing quality on the speed of liquefaction, cellulases were added to static fiber beds in vials. Different initial mixing conditions were used. One of the conditions was a well-mixed slurry where the enzyme is uniformly distributed in the slurry. The opposite extreme condition was addition of a thin layer of enzymes on one end of the vial, representing a completely unmixed enzyme. Samples were incubated for 96h at 50C. The state of liquefaction of the fiber beds was assessed using a mechanical penetrometer test. Large differences in the force-displacement curves were observed between untreated control slurries and enzymatically treated slurries. More importantly among the treated ones, the liquefaction times were much shorter for well-mixed slurries than for poorly mixed ones. Liquefaction data was also compared with reducing sugars measured throughout the incubation period to correlate with product yields.

See more of this Session: Reaction Kinetics and Transport Fundamentals for Biomass Conversion
See more of this Group/Topical: Sustainable Engineering Forum

(258e) Coupled Fluid-Dynamics and Population-Balance Kinetic Models for Enzymatic Hydrolysis of Biomass

Michael A. Sprague[1], Jonathan Stickel[2] and Andrew Griggs[2], (1)Computational Science Center, National Renewable Energy Laboratory, Golden, CO, (2)National Bioenergy Center, National Renewable Energy Laboratory, Golden, CO

A population-balance kinetic model developed previously for the enzymatic hydrolysis of lignocellulosic biomass is extended to include fluid mixing and transport. The kinetic model is focused on the depolymerization of insoluble cellulose chains by an endoglucanase and an exoglucanase. Population-balance models combined with a continuous-distribution representation provide computational efficiency, which is critical when considering the addition of fluid-mixing effects. The resulting coupled equations are solved with a high-order open-source computational fluid dynamics (CFD) package. The CFD model is validated against experimental and analytical data for a simple vane mixer. Preliminary results examining the effects of fluid mixing on dilute enzymatic hydrolysis are presented and compared to existing experimental data. This work provides a foundation for future efforts to model pilot- and commercial-scale bioconversion-reactors.

See more of this Session: Reaction Kinetics and Transport Fundamentals for Biomass Conversion
See more of this Group/Topical: Sustainable Engineering Forum

(258f) Flow and Transport Properties In High-Solids Biomass Slurries

James J. Lischeske and Jonathan J. Stickel, National Bioenergy Center, National Renewable Energy Laboratory, Golden, CO

Economic biochemical conversion of cellulose to sugars necessitates minimal addition of water to biomass slurries during conversion, resulting in high-solids systems. The resulting slurries are effectively wet-granular materials. These materials presents difficulties for reactor engineering, as their flow and transport properties, along with the mechanical properties that govern them, are poorly understood. This work examines the nature of high-solids flow, along with the heat and mass transfer properties of biomass and related model materials. Methods were developed to measure the volume fraction of solid, liquid and gaseous phases of high-solids slurries. The effect of phase composition on flow was systematically examined in a rotating vessel. Qualitative observations, as well as accessible quantitative information about size distribution of aggregates, were made. Further, a model for measuring bed diffusion in biomass was developed, and experiments were performed to measure the bed diffusion coefficient in model materials and biomass. Finally, heat transfer in a packed bed of biomass was examined, and conductivity coefficients were calculated and related to phase composition.

See more of this Session: Reaction Kinetics and Transport Fundamentals for Biomass Conversion
See more of this Group/Topical: Sustainable Engineering Forum

(265a) New Fischer-Tropsch-Ready Syngas Preparation Strategies From Coal and Natural Gas Using Solid Oxide Fuel Cells

Thomas A. Adams II, Chemical Engineering, McMaster University, Hamilton, ON, Canada and Paul Barton, Chemical Engineering, Massachusetts Institute of Technology, Cambridge, MA

For energy polygeneration processes (the simultaneous generation of multiple energy products such as liquid fuels, chemical products, and electricity in a tightly integrated system), the key defining characteristic is the manner in which syngas is created with an appropriate H_2:CO ratio. Many options are possible, primarily using some combination of coal and/or biomass gasification, natural gas reforming, the water gas shift reaction, hydrogen blending, and the use of absorption or adsorption separation processes. However, the ability to design energy efficient and cost-effective systems depends strongly on the synergies that can be harnessed between differing product mixes, the cleaning needs of each fuel, the market prices of the feedstock and products, and the type of carbon emission regulations.

In this work, we present several promising new process designs to convert coal, biomass, and/or natural gas into a mixture of methanol, naphtha, diesel, and electricity. For example, we consider the combination of natural gas reforming with coal gasification to produce syngas with an H_2:CO ratio of 2:1, ideal for the Fischer-Tropsch and methanol processes. In addition to the traditional approaches, novel heat-integrated systems are proposed in which both steps are performed within the same unit, such that the high-temperature waste heat produced by coal gasification can be use to satisfy the natural gas reforming heat needs directly. This synergistic configuration provides efficiency improvements throughout the plant, as well as interesting environmental benefits. In addition, the use of solid oxide fuel cells (SOFC) as the primary electricity generation source enables significant advantages for cost effective CO_2 capture. With this approach, synergistic efficiency improvements can be made throughout the process, particularly in the areas of sulfur and CO_2 removal. Using detailed flowsheet simulations, the efficiencies, costs, and environmental benefits of these proposed designs strategies are compared with several traditional concepts.

See more of this Session: Syngas Production and Gas-to-Liquids Technology
See more of this Group/Topical: Catalysis and Reaction Engineering Division

(265b) Tri-Reforming of Methane and CO2: A Novel Concept for Catalytic Production of Solid Waste Syngas with Desired H2/CO Ratios for Liquid Biofuels

Devin M. Walker, John T. Wolan, John N. Kuhn and Philip Saraneeyavongse, Chemical and Biomedical Engineering, University of South Florida, Tampa, FL

Tri-reforming of Methane and CO_2: A Novel Concept for Catalytic Production of Solid Waste Syngas with Desired H_2/CO Ratios for Liquid Biofuels

Devin M. Walker, John N. Kuhn, Philip Saraneeyavongse, and John T. Wolan*

Department of Chemical & Biomedical Engineering, University of South Florida

This study focuses on upgrading the Municipal Solid Waste (MSW) syngas for the synthesis of liquid fuels using Fischer-Tropsch synthesis (FTS). The process includes novel gasification of MSW via a tri-reforming process which involves a synergetic combination of CO_2 reforming, steam reforming, and partial oxidation of methane. Typical biomass or MSW derived syngas H_2:CO is 1:1. This innovation allows for cost-effective one-step production of syngas in the required H_2:CO of 2:1 for use in the FTS. To maximize the performance of the tri-reforming catalyst an attempt to control oxygen mobility, thermal stability, dispersion of metal, resistance to coke formation, and strength of metal interaction with support is investigated by varying catalyst synthesis parameters. These synthesis variables include Ce and Zr mixed oxide support ratios and amount Mg and Ni loading. Testing under controlled reaction conditions and the use of various catalyst characterization techniques were employed to better explain the affects of the synthesis parameters. This paper highlights the performance of the tri-reforming catalyst and attempts to explain results through catalyst characterization.

See more of this Session: Syngas Production and Gas-to-Liquids Technology
See more of this Group/Topical: Catalysis and Reaction Engineering Division

(265c) Effects of Biomass Contaminants (N and Cl species) On Fischer Tropsch Synthesis

Fan Shi, Material Science, National Energy Technology Lab., Pittsburgh, PA, Adefemi Egbebi, Material Science Engineering, National Energy Technology Laboratory, Pittsburgh, PA and Ward A. Burgess, National Energy Technology Laboratory, Pittsburgh, PA

Introduction

The production of transportation fuels from biomass, and the use of biomass as a means of carbon dioxide capture and sequestration have been identified as a strategy with potential to address the energy and environmental concerns, including fuel independence, energy availability and reliability, and environmental issues, especially as related to global warming. Most of the contaminants are reduced to an acceptable level by the current coal cleaning processes, i.e. gas scrubbing process. However, the remaining impurities in biomass derived syngas may still negatively affect Fischer Tropsch synthesis catalysts with time on stream, which have not been fully comprehended and present many contradictions. This study has established an understanding of the influence of biomass-derived synthesis gas on the processes involved in syngas conversion to transportation fuels or chemicals.

Materials and Methods

Iron based catalyst containing K and Cu promoters on SiO_2 has been tested for the Fischer Tropsch synthesis using in a 1-liter Continuous Stirred Tank Reactor (CSTR) system. The contaminants in the simulated biomass syngas include nitrogen, and chlorine. The gas products were analyzed by an online micro-GC (Aglient 300A) and liquid/wax samples were analyzed by GC-MS (HP GC 6890, and HP MS 5973).

Results and Discussion

The *ex situ* contaminated Fe catalyst by NH_3 and HCl at 300 °C have been characterized using XRD, TPR, and XPS. Preliminary experiment results from XRD indicated that iron carbides were bulk materials after carburized in CO and then subsequently exposed to NH_3. However, predominant $FeCl_2$ were formed after CO carburization then followed HCl exposure. XPS results also indicated strong binding between Cl species and Fe at catalyst surface. It is also interesting to find that after exposed to NH_4Cl in FT synthesis, the performance of Fe catalyst improved instead of deactivated. Areas of research to ensure the success of biomass conversion in presence of impurities include an understanding of the deactivation mechanism over the catalyst, and the identification of new catalyst materials through experimental and computational exploration.

THERMAL PROCESSING TECHNIQUES TO IMPROVE METAL SULFIDE MIXED ALCOHOL CATALYST PERFORMANCE

Jesse Hensley[a], Martin Menart[a], Kellen Costelow[b], Jason Thibodeaux[a], and Matt Yung[a]

[a]National Renewable Energy Laboratory, Golden CO
[b]Suncor Energy, Denver CO

Abstract

Thermal history has been shown to be an important factor in the performance of bulk-precipitated alkali promoted cobalt molybdenum sulfide (K-$CoMoS_x$) mixed alcohol catalysts. Sulfide catalyst precursors derived from ammonium tetrathiomolybdate were decomposed at various temperatures and dwell times, and with/without potassium promoters. When combined with temperature programmed decomposition studies, results suggested that precursors treated at milder conditions yielded more active and selective catalysts, in general. Precursors decomposed in the presence of alkali promoter yielded highly selective catalysts, with lower activity in general. X-ray diffraction showed important differences in materials decomposed with/without alkali promoter, but failed to provide correlations between catalyst morphology and performance in mixed alcohol synthesis.

Introduction

Catalyst researchers have long understood that there is a significant 'art' in catalyst preparation. There are many examples of industrial catalysts that have been fine-tuned through changes in thermal treatment, including the temperature and atmospheric conditions used for calcination, oxidation, or reduction.[1] Alkali-promoted metal sulfide catalysts are nearing commercialization, but there have been few public studies on the thermal histories of these materials and how they affect performance in mixed alcohol synthesis. The hydrodesulfurization (HDS) community has given consideration to thermal treatments of the metal sulfides, so this is a good place to start. Factors such as sulfidation conditions, thermal decomposition, reduction/regeneration, etc have been studied and the effects on desulfurization activity and catalyst morphology have been reported.[2-7]

Even with the large and growing body of HDS literature, however, one can appreciate that mixed alcohol catalysts are not HDS catalysts: they are asked to perform a much different set of chemical reactions (notably, molecule building instead of destruction to remove S), they operate under different conditions with different feed streams, and they contain alkali metals which significantly changes their functionality. Therefore, we have adopted the hypothesis that preparation strategies optimal to HDS catalysis are not necessarily optimal for mixed alcohol catalysis. The purpose of this presentation is to provide evidence of dependencies between thermal histories and catalyst performances and to convey the need for improved thermal treatment strategies in the synthesis of sulfided mixed alcohol catalysts.

We have chosen for study unsupported cobalt-molybdenum-sulfide (CoMoS) materials, promoted with potassium (K). This choice was made for a number of reasons:

- Bulk sulfides tend to be more selective in mixed alcohol synthesis than supported sulfides such as $K/CoMoS_x/Al_2O_3$
- Supports may add compounding factors to catalyst performance and use of bulk materials allows one to remove these factors for detailed study,
- There has been a resurgence of commercial activity with these materials (mostly startups looking to make ethanol from a low cost feedstock),
- There is a larger body of literature to work from, probably due to the initial disclosure of the catalyst by Dow/UOP in the 1980s[8-11] and subsequent follow-on research from the scientific community, and
- Methods for making this catalyst at commercial scale are into the demonstration stage, and any recommendations we can make on thermal processing should not require a major reworking of those processes.

In particular, we have studied the thermal 'events' in the decomposition of sulfide catalyst precursors, which form the final catalysts. From those events, we have invoked a number of alternate thermal processing strategies and observed changes to catalyst morphology and performance. While these studies are ongoing, the content of this presentation will provide some important insights into the importance of thermal treatments when making high-performance sulfided mixed alcohol catalysts. Manuscripts already submitted and soon to be submitted will provide significant supporting detail.

Experimental

Sulfide catalyst precursors were prepared using a bulk precipitation technique. Pure ammonium tetrathiomolybdate (ATTM) crystal was added to warm distilled water to form a near-saturated solution. The presence of pure ATTM was confirmed by UV-Vis spectrometry using published guidelines.[12] Fine crystals of cobalt acetate tetrahydrate were added to the rapidly stirred ATTM solution with a Co:Mo ratio of 1:3, which caused a black precipitate to form. Additional metal ratios have been considered and the results will be disclosed separately. After complete reaction between Co and ATTM, the pH of the solution was lowered to 3 by adding 2 M sulfuric acid dropwise and observing the solution pH with a pH probe. This caused all ATTM to precipitate out of solution as filtered solution showed no evidence of Mo or ATTM via ICP or UV-Vis analysis, respectively. The resulting precipitate was filtered by vacuum on a Büchner funnel with Whatman 'sharkskin' paper to form a wet 'cake' of precipitates. The surface of the cake was sculpted with a spatula as cracks formed in the cake, to speed the dewatering process. The cake was then washed with three cake volumes of distilled water. In some instances, the final wash was substituted with a potassium salt solution, as a means of adding K promoter to the catalyst before thermal treatment. These samples are hereafter called 'salted' samples.

Thermal treatment of the catalyst precursor was performed in a stainless retort (100 g scale synthesis) or in a quartz reactor tube (2 g synthesis), each located within a controlled furnace. In the larger scale method, washed cake was packed along the sides of a rotary retort and a thermocouple was placed within the cake and in the centerline of the retort. The retort was sealed and a flow of welding grade nitrogen was established over the precipitate. After a period of time sufficient to flush all air from the retort, the furnace was heated to achieve a cake temperature of ~95 °C (boiling point of water at our location), under constant flow of nitrogen. When all water had evaporated from the cake, its temperature rose rapidly to match the temperature of the furnace, at which point the retort was set to rotate at 1-2 rpm and the sample temperature was held at 120 °C for additional drying (when used). The precursor was then heated to one or two temperatures for different periods of time to decompose the precursor to molybdenum disulfide (MoS_2) and cobalt sulfides. The catalyst was then cooled under flowing nitrogen and the retort subsequently plugged, transferred to a nitrogen-purged glove bag, and unloaded in a nearly air-free environment. In most cases the finished material consisted of a coarse powder with chunks the size of button cell batteries.

In the second method, freshly prepared, washed, and filtered catalyst precipitate was frozen under dry ice and crushed to 5×10^{-3} m chunks, then loaded onto a quartz boat which was placed in a horizontal quartz tube. A thermocouple was placed directly above the catalyst to correlate sample temperature with thermal events. A flow of pure nitrogen was established as quickly as possible. The furnace was then set to a temperature of 120 °C and the precursor allowed to melt/dry under flowing nitrogen for 2 h, followed by heating to 700 °C at a rate of 10 $°C\text{-}min^{-1}$ (termed here as 'temperature programmed decomposition, TPD) or by heating to known decomposition temperatures and held until the effluent gas showed no evidence of decomposition reactions (i.e. those that generate a gaseous product). Catalyst precursor decomposition products were monitored with a mass spectrometer (MS). Finished materials were cooled under nitrogen and passivated with industrial grade argon. In all cases the finished material consisted of chunks that were easily pulverized in a mortar and pestle, with a small fraction of curled sulfide flakes that were later identified as cobalt sulfides. In all experiments white and yellow sulfur deposits were found in cooler sections of the effluent tubing, confirming that elemental sulfur is lost on decomposition of the sulfide precursors.

Only the catalysts prepared at large scale were tested in reactors. Samples were prepared by blending finely divided sulfide powder (200-230 mesh) with 325 mesh anhydrous potassium carbonate to form a material with a 1:1 atomic ratio of K:Mo. Blended powders were diluted 3:1 w:w with fine mesh silicon carbide to prevent formation of hot spots and to ensure, to the extent possible, isothermal operation. 'Salted' samples were sized between 200 and 230 mesh, blended with diluent, and used without further manipulation.

Catalyst morphology was examined using X-Ray diffraction. Catalyst performance was measured using an isothermal tubular reactor connected to a process GC which analyzed all hydrocarbons, oxygenates, and permanent gases with FID and TCD detectors. The data presented below were collected at steady state, and all mass and atomic balances closed within 5%.

Results & Discussion

TPD experiments showed different and distinct thermal-chemical events for salted and unsalted precursors as the materials were heated to 700 °C. Figure 1 shows the temperatures at which those events occurred. In previous unpublished experiments, we observed that nearly all water is lost at its normal boiling point, suggesting that water is bulk in nature and simply left over from the batch precipitation. Therefore, an event at 100 °C, precursor drying, is not included in the plot as it is not unique to the system.

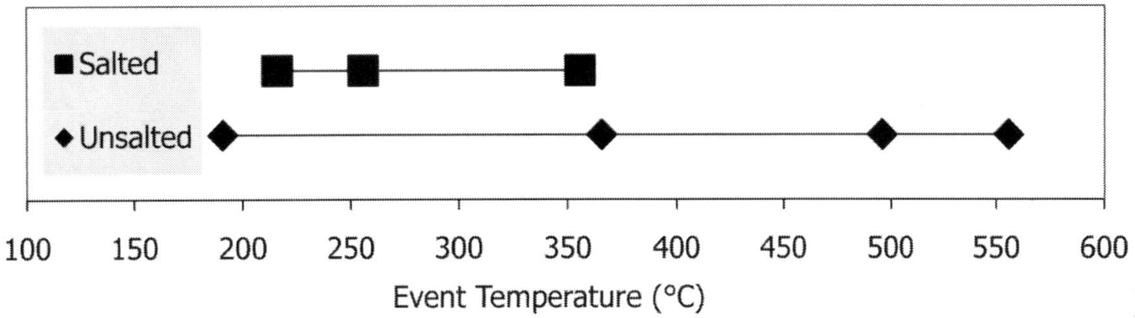

Figure 1: Temperatures at which thermal events were observed on salted and unsalted precursor samples.

Figure 2 shows TPD plots for all ion signals observed at the MS between 1-100 amu for the unsalted precursor. In general we observed the formation of ammonia (m/z = 16, 17) at lower temperatures with cogeneration of some sulfur species (onset 190 °). As the temperature increased, gaseous sulfur evolved from the materials, showing evidence of elemental sulfur at (m/z = 32, 64) and SO (m/z = 48) (onsets at 365, 495, and 555 °C). The presence of SO suggests that at least some O must persist on the precursor up to these higher decomposition temperatures, perhaps as tightly held H_2O or SO_4^{2-} or COO^- leftover from the precipitation and not washed completely from the precipitate cake. Indeed, a small signal for water was observed (m/z = 18) at nearly all thermal events, indicating a small amount of water released with ammonia and sulfur. H_2S was observed in the decomposition product at all events but the signal was orders of magnitude lower than for other species. The TPD results for the unsalted sample are directionally consistent with the thermal decomposition of ATTM as determined by others,[6,7] even though the precursor is expected to consist of sulfides formed through the following reactions, which contain intermediates to ATTM decomposition:

$$MoS_{4\ (aq)}^{2-} + Co^{2+}{}_{(aq)} \rightarrow CoS \cdot MoS_{3(s)} \quad (1)$$

$$MoS_{4\ (aq)}^{2-} + 2H_3O^{+}{}_{(aq)} \rightarrow MoS_{3(s)} + H_2S_{(aq/g)} + 2H_2O \quad (2)$$

The similarities between decomposition of Co-Mo sulfides and ATTM suggests that reactions 1 and 2 may not go to completion (or go at all), and that some ATTM may recrystallize at low pH, a mystery that we're currently investigating.

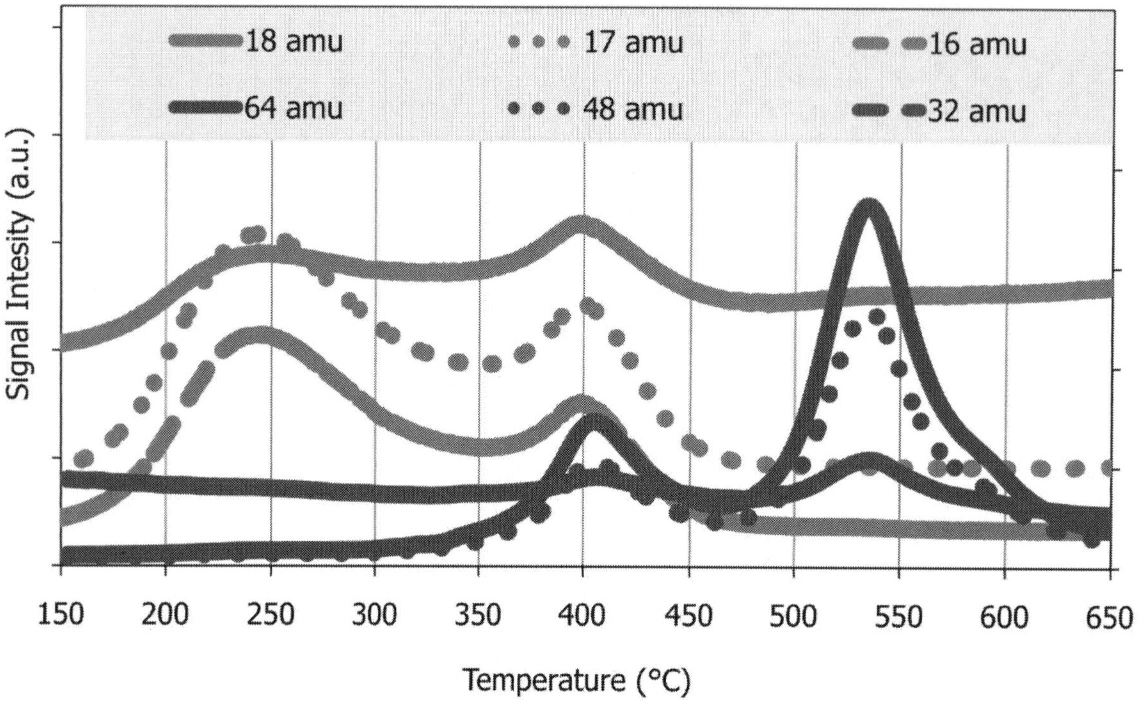

Figure 2: TPD profile of unsalted CoMoS precursor. Peak intensities have been rescaled to view all signals on the same plot.

Figure 3 shows TPD plots for the salted precursor. The species observed on the unsalted sample were observed in similar abundance on the salted precursor, with the addition of CO_2 and C, likely derived from the CO_3^{2-} anion that was added with K^+ in the final precipitate wash. The overall temperature/event results are strikingly different from those observed on the unsalted sample, however. For instance, the salted sample did not exhibit decomposition events above 355 °C, and exhibited fewer events associated with sulfur removal. Further, the salted sample showed no events wherein sulfur and ammonia (or carbon) left the material together. This immediately suggests that a catalyst made via 'salting' a sulfide precipitate needs to be heated much differently than a sample 'salted' after thermal treatment, and this notion is explored further below.

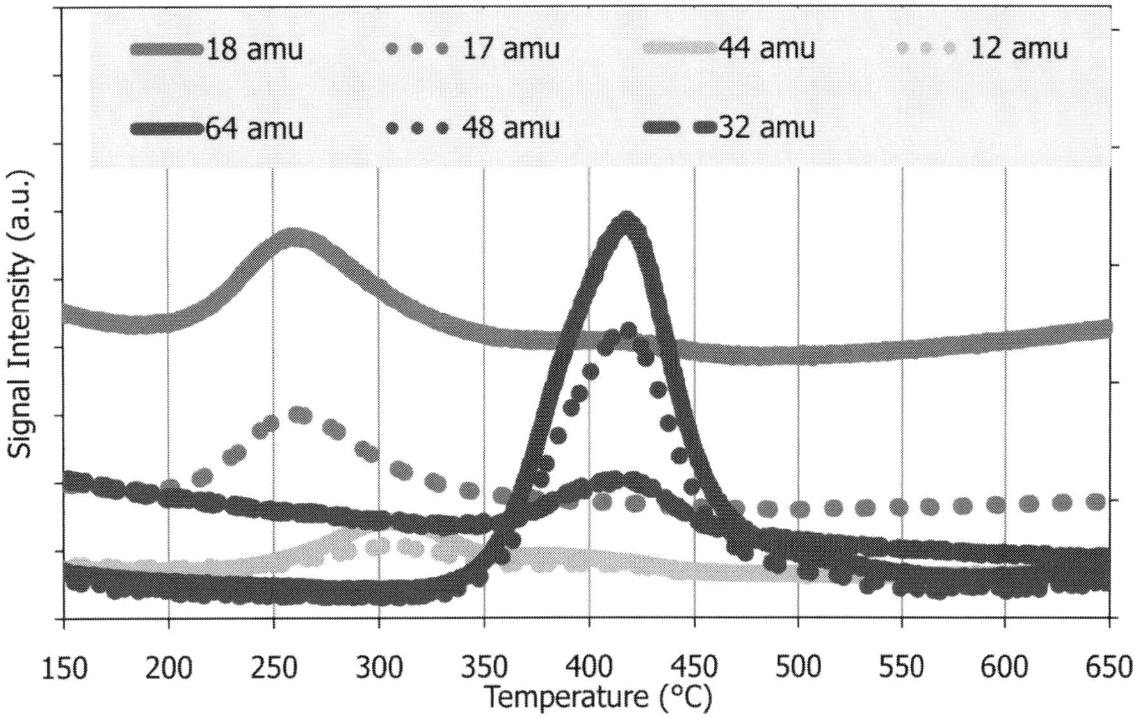

Figure 3: TPD profile of salted CoMoS precursor. Peak intensities have been rescaled to view all signals on the same plot.

Using information from the TPD experiments shown here and previous ones not shown, we prepared a number of catalysts for comparative testing. A first set consisted of those samples created during data collection for figures 2 and 3 and additional samples prepared by holding the salted and unsalted samples at the inflection points of their respective thermal event temperatures (those below 500 °C) until no signals at 12, 16, 17, 18, 32, 44, 48, or 46 amu could be differentiated from the baseline (a so-called 'gentle' thermal treatment). A second set of samples were prepared at large scale with the temperature/soak strategies noted in table 1, and were largely guesses of appropriate treatment temperatures derived from earlier, more ambiguous data.

Table 1: Treatment schedules for mixed alcohol catalysts. All finished materials have a nominal 1:3:3 ratio of Co:Mo:K.

Catalyst Sample	Potassium Addition Method	Drying Conditions	Thermal Treatment #1	Thermal Treatment #2
NREL-1	Dry blend K_2CO_3	Held furnace at 300 °C until solid T > 100 °C	Heat to 500 °C, hold 1 h	none
NREL-3*	Wash precipitate with KOH solution	Evaporate to dryness at 95 °C + 2h at 120 °C	Heat to 250 °C, hold 30 min	Heat to 475 °C, hold 15 min
NREL-4	Dry blend K_2CO_3	Same as NREL-3	Same as NREL-3	Same as NREL-3
NREL-6	Dry blend K_2CO_3	Same as NREL-3	Heat to 250 °C, hold 1 h	Heat to 375 °C, hold 45 min
NREL-7A*	Wash precipitate with K_2CO_3 solution	Same as NREL-3	Same as NREL-3	Same as NREL-3
NREL-8A*	Wash precipitate with $KOCH_3$ dissolved in methanol	Same as NREL-3	Same as NREL-3	Same as NREL-3
NREL-9*	Wash precipitate with K_2CO_3 solution	Same as NREL-3	Heat to 280 °C, hold 2 h	Heat to 380 °C, hold 1 h

*Other potassium salts may be considered but we have limited our testing to those which lose their conjugate anions during activation, a requirement for alcohol synthesis activity as shown by Lee et al.[13]

XRD data are shown for the first set of samples noted above in figures 4 and 5. The data show some interesting features that were not anticipated. Considering the unsalted sample (figure 4) first, we observe a material that is poorly crystalline MoS_2 with some crystalline CoS_2 and an unidentified phase. This shows that even with a ramp to high temperature, the material cannot assume much crystalline order, and that at least some Co and Mo sulfides are segregating, which may or may not be beneficial for mixed alcohol synthesis. The milder temperature, longer treatment resulted in a material that is quite similar, though the CoS_2 assignments are less ambiguous. Considering now the salted sample (figure 5), we observe a material that is significantly more crystalline, with many additional crystalline phases of Co and some crystalline potassium sulfide phases. The sample treated 'gently' with lower final temperatures is less crystalline than the material heated to 700 °C, but still contains many phases. This shows very clearly that potassium has a significant influence on how the catalyst forms its final structure, and suggests a catalyst made by 'salting' the precursor before thermal treatment will behave very differently.

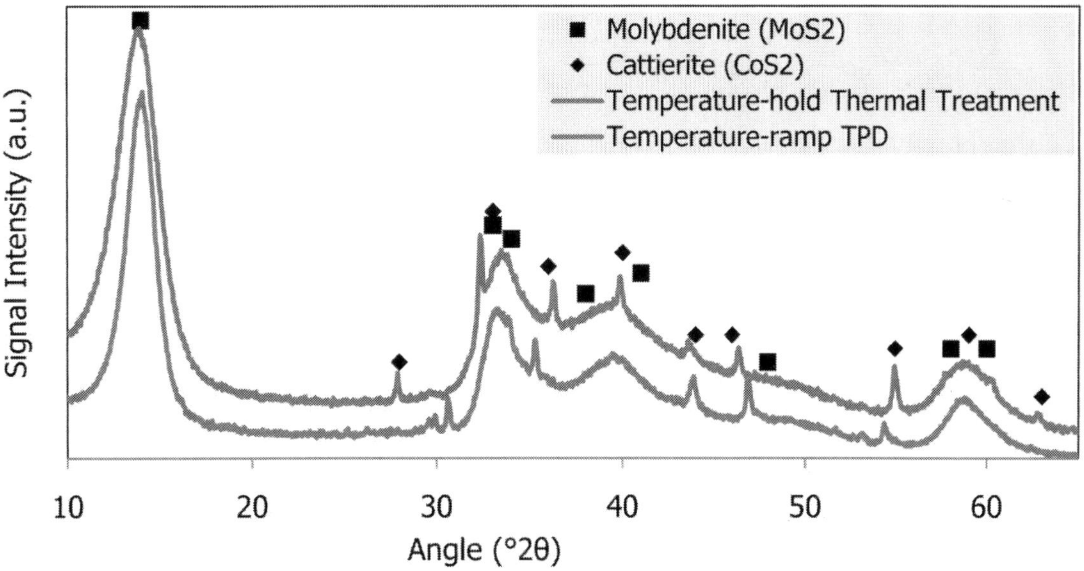

Figure 4: XRD results for unsalted precursor samples subjected to TPD or a temperature-hold 'gentle' treatment. Signals are offset for viewing ease. Symbols above traces indicate observed or expected crystalline phases.

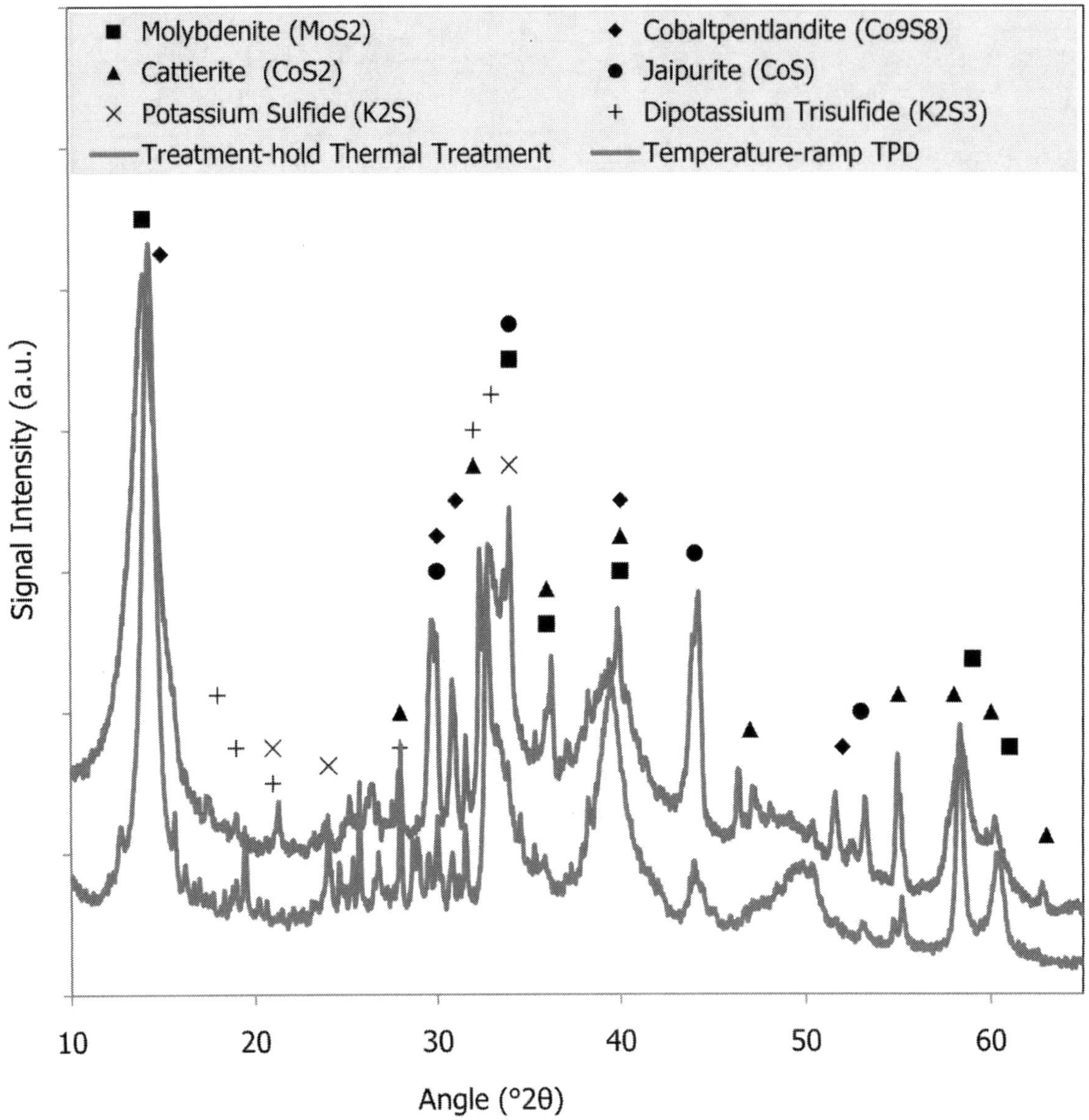

Figure 5: XRD results for salted precursor samples subjected to TPD or a temperature-hold 'gentle' treatment. Signals are offset for viewing ease. Symbols above traces indicate observed or expected crystalline phases.

Figures 6 and 7 show performance comparisons for the catalysts noted in table 1. All catalysts were tested under identical operating conditions; therefore, points at higher CO conversion and/or higher ethanol productivity in figure 6 are more active materials. Points at higher ethanol selectivity *at the same level of CO conversion* in figure 7 are more selective materials. Results for productivities and selectivities of other reaction products are not shown here, therefore, a limited analysis is possible. From the data shown, however, the following is observed:

1. For materials made by salting the catalyst after thermal treatment, milder temperatures result in greater productivity (NREL-6 > NREL-4 > NREL-1),
2. Materials made by salting the catalyst after thermal treatment are slightly more active, in general, than those salted before thermal treatment (NREL-4 >~ NREL-7A),
3. Materials salted with more basic potassium salts are less active (NREL-7A > NREL-8A > NREL-3) but can yield similar selectivity (NREL-3 = NREL-7B > NREL-8A),
4. Materials salted before thermal treatment and subjected to milder temperatures are NOT superior to salted samples treated at higher temperatures (NREL-7A >> NREL-9), and
5. Materials salted before thermal treatment have better ethanol selectivity in general.

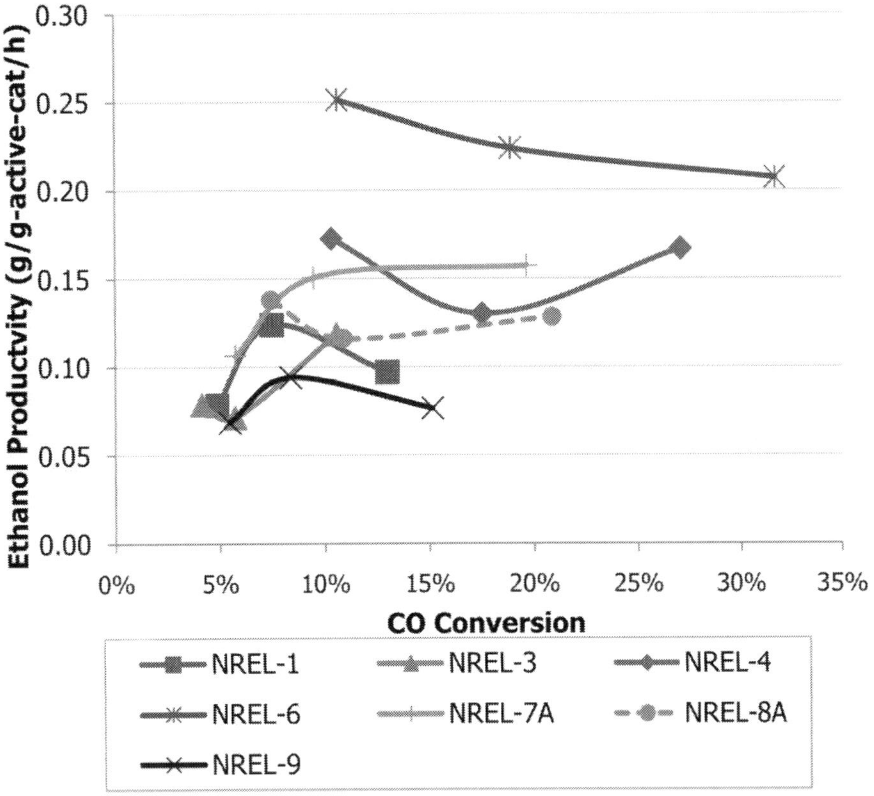

Figure 6: Ethanol productivity at 325 °C as a function of CO conversion for the catalysts in table 1. Data collected at pressures of $p_{H2} = p_{CO}$ = 700 psi at gas hourly space velocities of 6, 10, and 14k NTPL-kgcat^{-1}-h^{-1}.

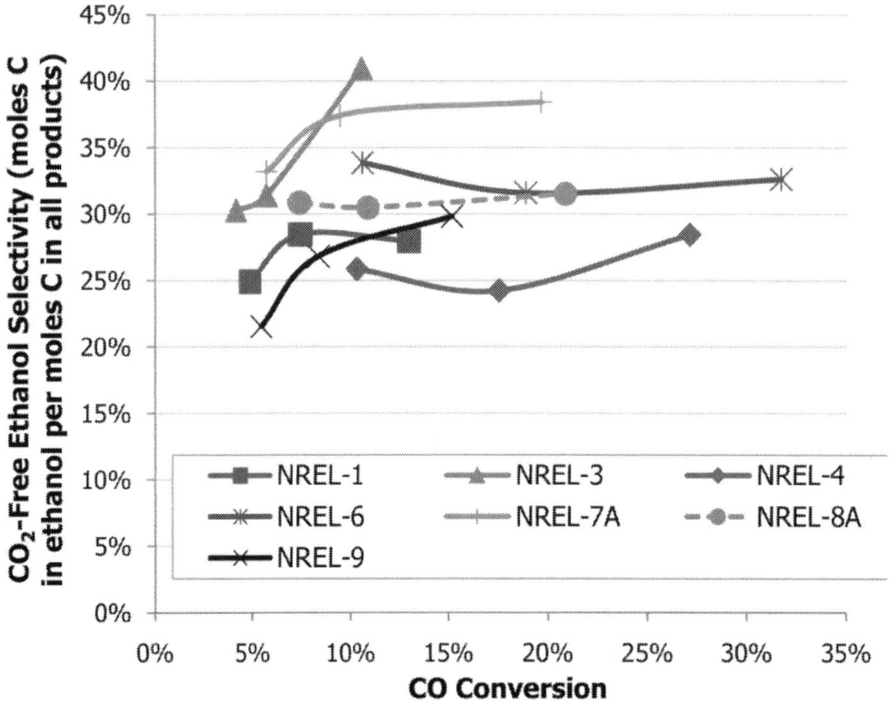

Figure 7: Ethanol selectivity at 325 °C as a function of CO conversion for the catalysts noted in table 1. Data collected at pressures of $p_{H2} = p_{CO}$ = 700 psi at gas hourly space velocities of 6, 10, and 14k NTPL-kgcat^{-1}-h^{-1}.

In general, the data in figures 6 and 7 show that thermal treatments and the time at which salts are incorporated can have very drastic effects on catalyst performance. This suggests that better mixed alcohol catalysts can be made through careful selection and control of thermal processing temperatures/times, and more importantly, that catalysts of reasonable performance can be made by 'salting' the precursor before thermal treatment. The data do not, unfortunately, provide a large population of results for rigorous statistics, as we must rely on good material balances and instrument calibrations to assure ourselves that the catalysts are truly different from a performance standpoint.

A heuristic would be useful at this point. That is, what catalyst properties should one aspire for when finding increasingly optimized thermal treatments for these materials? The data in figures 4 and 5 suggest that XRD might provide some clues. Figure 8 shows XRD signals from the catalysts listed in table 1 and shown in figures 6 and 7. The figure shows that for the samples not salted before treatment, NREL-1 is slightly more crystalline than NREL-4 and NREL-6, but there are no obvious large differences between the materials. Of the salted samples NREL-3, NREL-7A, and NREL-8A, NREL-7A is perhaps the least crystalline and the best in terms of performance in the reactor. NREL-9 is almost completely amorphous however, suggesting that lower crystallinity does not necessarily equate to better catalyst performance. Overall, the data in figure 8 show that XRD is probably not a useful tool in predicting catalyst performance. This is unfortunate, as the only reliable tool we've found to date is reactor testing, which is a costly and time-consuming enterprise.

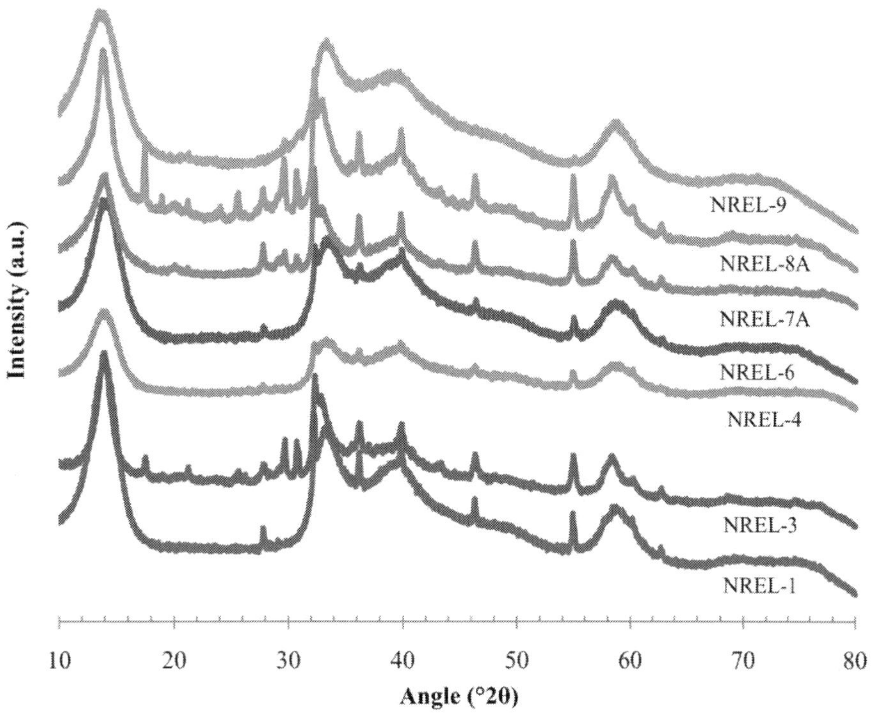

Figure 8: XRD results for the catalysts listed in table 1. Signals are offset for viewing ease.

Other techniques may prove useful, and work is underway in our laboratories to find physical or chemical flags that are good predictors of performance. These include surface characterization (BET, chemisorption, Hg-intrusion) and spectroscopic (FTIR, Raman) techniques, as well as advanced techniques like XANES, EXAFS, and XPS if possible.

From an industrial perspective, exotic synthesis techniques and promoter metals can render a catalyst/reaction system uneconomical. Therefore, when attempting to improve the performance of alkali-promoted metal sulfide catalysts for industrial purposes, it is prudent to target synthesis techniques that do not add cost or complexity to the process. In the case of potassium incorporation, significant savings might be realized if catalysts can be prepared without the need to dry blend sulfide and alkali salt powders, a step that must be performed under inert conditions and one that is complicated even at the bench scale. Thermal processing strategies offer one way to influence catalyst performance, and potentially batch to batch consistency, without changing the process and increasing the cost of making the material.

It is clear from the above results that small changes in thermal treatment can have large effects on catalyst performance. For instance, comparing catalysts NREL-4 and NREL-6, a small drop in decomposition temperature coupled with a slight increase in treatment time resulted in a material with 25-50% higher ethanol productivity and 20-30% better ethanol selectivity, depending on where in the conversion/performance continuum the data is read. Both materials have identical metal ratios, were prepared using identical recipes (and indeed, identical lots of chemicals), were heated in the same furnace with the same temperature control, were promoted identically, and contain both crystalline and amorphous molybdenum and cobalt sulfides. It is also clear that it is possible to make high-performance materials using alternative alkali promotion strategies. Catalysts NREL-4 and NREL-7A were prepared with identical thermal treatments and were promoted with the same type of potassium salt and showed similar activity. NREL-7A appears to be even more selective. As a caution, however, it appears easy to get the 'salted' samples wrong, evidenced by catalyst NREL-9. For both salted and unsalted precursor samples, there is probably room for significant additional improvements in catalyst performance, through thermal treatment adjustments alone.

We are continuing this research in our lab and plan to investigate further the thermal conditions that yield improved catalysts, especially those 'salted' before heating. Considerable efforts are being placed on understanding the effects on catalyst morphology, including changes in surface area, titratable active sites, crystallography, and overall performance in mixed alcohol synthesis. Work is also underway to evaluate these phenomena using different purge gases during thermal events, and using different precursor materials that contain different promoter atoms. Most of this work has been recently submitted for review, and more is soon to come.

Conclusions

The thermal history of K promoted Co-Mo-S catalysts has marked effects on catalyst performance. Specifically, the temperatures and dwell times at which catalyst precursors are decomposed have direct implications on catalyst activity and selectivity to alcohols. The order in which alkali promoter is added to the catalyst appears to have a large impact on catalyst performance and morphology as well. While it appears that similar performance can be derived from materials promoted before or after thermal treatment, the temperatures at which precursors decompose will change, and the correct thermal treatments are not straightforward. Importantly, it is apparent that the synthesis of bulk metal sulfide mixed alcohol catalysts is far from optimized and that there is room for additional innovation.

References

(1) Bartholomew, C. H.; Farrauto, R. J., *Fundamentals of Industrial Catalytic Processes.* 2nd ed.; John Wiley & Sons, Inc.: Hoboken, New Jersey, 2006; pp. 90-106.

(2) Chianelli, R. R.; Berhault, G.; Torres, B., *Catal. Today* **2009,** *147* (3-4), 275.

(3) Daage, M.; Chianelli, R. R., *J. Catal.* **1994,** *149* (2), 414.

(4) Kelty, S. P.; Berhault, G.; Chianelli, R. R., *Applied Catalysis a-General* **2007,** *322,* 9.

(5) Pawelec, B.; Navarro, R.; Fierro, J. L. G.; Vasudevan, P. T., *Applied Catalysis a-General* **1998,** *168* (2), 205.

(6) Prasad, T. P.; Diemann, E.; Muller, A., *Journal of Inorganic & Nuclear Chemistry* **1973,** *35* (6), 1895.

(7) Tsyganenko, A. A.; Can, F.; Travert, A.; Mauge, F., *Applied Catalysis a-General* **2004,** *268* (1-2), 189.

(8) Kinkaid, N. E. US 1984.

(9) Quarderer, G. J.; Stevens, R. R.; Cochran, G. A.; Murchison, C. B. US Pat.#4,825,013, 1989.

(10) Stevens, R. R. US Pat.#4,882,360, 1989.

(11) Stevens, R. R.; Conway, M. M. US Pat.#4,831,060, 1989.

(12) McDonald, J. W.; Friesen, G. D.; Rosenhein, L. D.; Newton, W. E., *Inorg. Chim. A.-Article.* **1983,** *72* (2), 205.

(13) Lee, J. S.; Kim, S.; Lee, K. H.; Nam, I. S.; Chung, J. S.; Kim, Y. G.; Woo, H. C., *Applied Catalysis A-General* **1994,** *110* (1), 11.

(265e) Supercritical Adiabatic Reactor for Fischer Tropsch Synthesis

Ed Durham, Department of Chemical Engineering, Auburn University, Auburn, AL, Mario Richard Eden, Department of Chemical Engineering, Auburn University, Auburn University, AL and Christopher Roberts, Chemical Engineering, Auburn University, Auburn, AL

The Fischer Tropsch reaction is highly exothermic, with an adiabatic temperature rise on the order of 1800°C. Managing the released heat is a major factor in Low Temperature Fischer Tropsch (LTFT) reactor design. The two conventional LTFT reactor designs are fixed bed (with a high gas velocity in a large number of narrow tubes) and slurry bed (with a powder catalyst dispersed in a liquid wax media).

Since 1989, when Kaoru Fjumoto's research group first published the use of a supercritical fluid as the reaction media for fixed bed LTFT, a number of research groups have compared Supercritical LTFT (SC-FTS) with traditional fixed-bed LTFT (GP-FTS), consistently finding that SC-FTS gives lower methane and CO_2 selectivities and enhanced olefin or aldehyde selectivities in the gasoline and heavier fractions. These benefits motivate the possible industrial utilization of SC-FTS. However, the presence of the solvent and the higher reaction pressure lead to cost penalties. In order for SC-FTS to be industrially viable, the costs the "upgrade" must be less than the benefits.

One of the less noted benefits SC-FTS is the decreased adiabatic temperature rise. Simulation based on using hexane as the media at a ratio of 3.5 moles of media per mole of syngas drops the adiabatic temperature rise to 60°C. Consequently, it is feasible to carry out SC-FTS in a series of modules each with an adiabatic reactor and a heat exchanger.

While upgrading Sasol's ARGE reactor for SC-FTS would approximately double its cost, preliminary economic analysis suggests that this system should have a cost comparable to the ARGE reactor, lowering the capital cost of a SC-FTS plant by approximately $1000 per BPD. This reactor system also allows for far easier catalyst replacement and greater reactor control flexibility than the traditional ARGE reactor.

This work does not address the operating cost penalty of SC-FTS or the balance of plant capital cost penalty, though a thorough analysis of these costs is underway.

See more of this Session: Syngas Production and Gas-to-Liquids Technology
See more of this Group/Topical: Catalysis and Reaction Engineering Division

(265f) Methane Conversion towards Liquid Fuel Production Over Promoted Mo/MFI Catalyst: Effect of Zinc, Tungstan, Galuim

Sachchit Majhi[1], Pravakar Mohanty[2] and **K. K. Pant**[1], (1)Chemical Engineering, Indian Institute of Technology, Delhi, New Delhi, India, (2)IIT Delhi India

Methane is the major component, most stable and symmetric organic molecule with four C–H bonds and bond energy = 435 kJ/mol has been a challenge for conversion on the aspect of catalysis science. Among catalytic conversion of methane in indirect processes rely upon the production of synthesis gas, either by reforming reaction or by partial oxidation, followed by FT chemistry. Methane can be converted directly to the desired products, e.g. olefins, aromatics and products with higher added value can eliminate the intermediate costlier step of production of synthesis gas. Direct conversion of methane to higher hydrocarbon can be an effective process to solve the problem of natural gas utilization. In this work, four different catalysts namely 3%Mo/MFI, 3%Mo/1%Zn/MFI, 3%Mo/1%W/MFI and 3%Mo/3%Ga/MFI were synthesized by incipient impregnation method. The effects of metal promoters Zn, W and Ga on methane conversion in the presence of trace amount of methanol over 3%Mo/MFI catalysts has been investigated. The synergistic effect is exhibited in 1%Ga-3%Mo/MFI catalyst on which a high methane conversion of 21.3 % is achieved as compared to other catalyst. Detailed characterization of the catalysts prepared by incipient impregnation method have been analyzed in terms of BET surface area, pore volume, BJH isotherm, TPR, TPD,SEM-EDX, TEM, and XRD. Coke (%) deposition during the course of reaction was determined by TGA/DTA analysis. The activity test of different catalysts was carried out in the temperature range of 650 to 800 ^{O}C.

See more of this Session: Syngas Production and Gas-to-Liquids Technology
See more of this Group/Topical: Catalysis and Reaction Engineering Division

#226402

Thermal Decomposition and Cobalt Species Transformation of Carbon Nanotubes Supported Cobalt Catalyst for Fischer-Tropsch Synthesis

Zhenhua Li, Suli Bai, Chengdu Huang, Jing Lv and Jinyu Han, Key Laboratory for Green Chemical Technology of Ministry of Education, School of Chemical Engineering & Technology, Tianjin University, Tianjin, China

1. Introduction

Currently, the world's fuel production is based predominantly on crude oil. With the decline of crude oil reserve, identifying other alternative energy sources is of great importance. The processes of Gas-to-liquid (GTL), biomass-to-liquid (BTL) and coal-to-liquid (CTL) attracted much attention in past decades for producing transportation fuels. Fischer–Tropsch synthesis (FTS) is known as the key process for producing hydrocarbons from syngas (CO and H_2) derived from coal, natural gas, or biomass. Many efforts have been paid to improve FTS activity and C_{5+} selectivity [1,2] on Co- or Fe-based catalysts [3,4]. However, to our best knowledge, few paper reported the decomposition of Co precursor and Co species transformation on CNTs supported Co catalyst. Carbon nanotubes (CNTs) was reported as a potential support for FTS [5–13] because of its unique properties [14]. In this paper, the thermal decomposition and cobalt species transformation of the Co/CNTs catalyst were investigated in detail. The Co/CNTs catalyst was prepared by the impregnation method. The catalyst was characterized by a range of techniques, and the FTS activity was evaluated in a fixed-bed reactor.

2. Experimental

The 20 wt.% Co/CNTs (20Co/CNTs) catalyst was prepared using the incipient wetness impregnation method. The catalyst was evaluated in terms of its FTS activity (CO conversion) and selectivity of hydrocarbons in a laboratory fixed-bed stainless steel reactor (8 mm i.d.). Prior to the test, the catalyst (3 mL) was activated in H_2 (100 ml/min) at 400 °C for 10 h under atmospheric pressure. The reaction was conducted at 230 °C and 2.0 MPa. Ar was used as an internal standard gas for quantitative analysis. The outlet of the reactor was connected to a hot trap (typically 170 °C) and then a cold trap (-4 °C) both at the system pressure. The liquid product accumulated in hot and cold traps were periodically collected every 5 h and analyzed by gas chromatographs (SP 3420 GC) off-line equipped with an FID. The effluent gas was was monitored with an on-line Agilent 6820 GC with both of TCD and FID. The CO conversion, selectivity of C_{5+} hydrocarbons and chain growth probability (α value) were used to evaluate the FTS performance [15,16]. The characterization procedure is ignored here.

3. Results and Discussion

Figure 1 shows the TG curves of the raw CNTs, dried 20Co/CNTs, and 200 °C

Ar-calcined 20Co/CNTs in Ar or air/N_2 atmosphere. The TG results demonstrate that the CNTs did not lose weight in Ar at up to 650 °C, which means that the CNTs is more stable in an inert gas than in air as CNTs starts losing weight in air from 500 °C [14]. The dried 20Co/CNTs sample has two rapid weight loss stages in Ar, one from 90 °C to 230 °C and the other from 500 °C to 530 °C. The first stage corresponds to the dehydration of $Co(NO_3)_2 \cdot 6H_2O$ and the decomposition of $Co(NO_3)_2$ to Co_3O_4 [17]. The second could attribute to the transformation of Co species, which will be discussed later in detail. The TG curve of 20Co/CNTs after calcined in Ar at 200 °C for 5 h has similar features from 500 °C to 530 °C with that of the dried 20Co/CNTs except that a slight weight loss occurs from 90 °C to 230 °C because of the decomposition of residual $Co(NO_3)_2$. The TG curve of dried 20Co/CNTs in air/N_2 mixture shows two weight loss stages from 90 °C to 230 °C and 560 °C to 570 °C similar with those of the dried 20Co/CNTs sample in Ar. The significant weight loss between these two stages is the oxidation of CNTs in air which began at 230 °C much lower than pure CNTs oxidation temperature 550 °C due to the presence of Co species [14]. The above discussion indicates that calcination of dried 20Co/CNTs experienced different transformations at 200 °C, 500 °C, and higher temperature than 600 °C. To further understand these transformations and the effect of calcination temperature on cobalt species formation and FTS activity, 20Co/CNTs samples calcined in Ar at 200 °C, 500 °C, and 800 °C were characterized by XRD and TPR techniques, and evaluated in the case of FTS activity and selectivity.

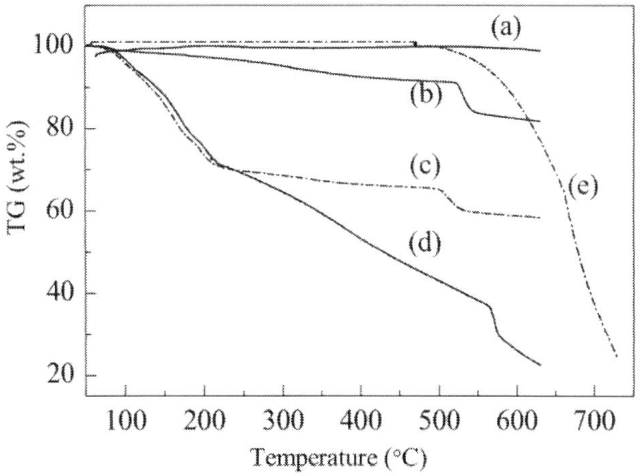

Figure 1. TG curves of the CNTs support and the 20Co/CNTs catalyst calcined in different atmosphere:
(a) CNTs in Ar; (b) 200 °C Ar-calcined Co/CNTs in Ar; (c) dried Co/CNTs in Ar;
(b) (d) dried Co/CNTs in air/N_2; (e) CNTs in Air

The XRD patterns of 20Co/CNTs shown in Figure 2a prove that 200 °C calcination in either air or Ar results in the formation of Co_3O_4 and CoO species. With increasing the calcination temperature to 500 °C, some amount of Co_3O_4 is autoreduced to CoO by CNTs as suggested by the appearance of peaks at 62.5° and 78.0° (JCPDS No. 72–1474). Further increasing the calcination temperature up to 800 °C, significant change occurs from 40° to 58°

which was enlarged and shown in Figure 2b. Besides CoO species, cobalt carbide (Co_2C) species (JCPDS No. 72–1369, 2θ = 37.0°, 41.3°, 42.5°, 45.7°, and 56.6°) also shows up in 500 °C-calcined 20Co/CNTs sample and diminishes greatly with metallic Co formation in 800 °C-calcined 20Co/CNTs sample, which indicates that the autoreduction of cobalt species by CNTs occurs at higher calcination temperature in Ar via two approaches, Co_3O_4 to CoO to Co or Co_3O_4 to Co_2C to Co. The transformation from Co_3O_4 to Co_2C can be supported by the point that Co_2C is formed at 450–500 °C [18]. The transformation from Co_2C to Co is ascribed to the decomposition of Co_2C which occurs at elevated temperatures above 500 °C [19].

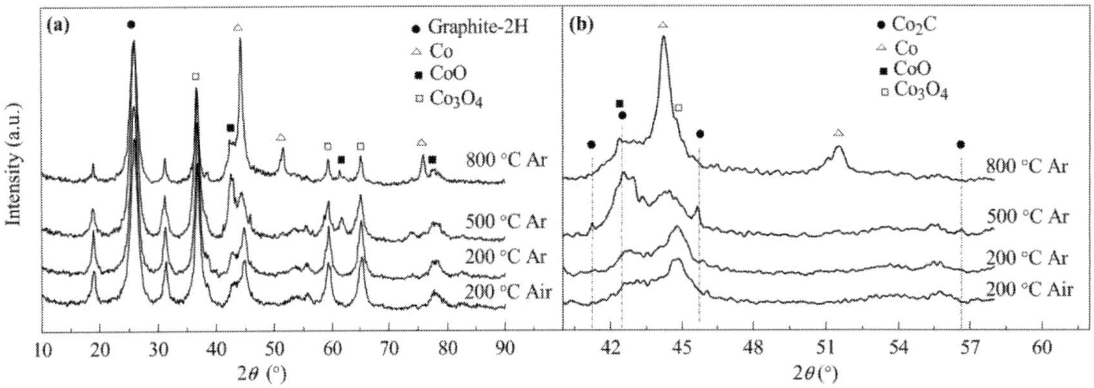

Figure 2. XRD patterns of 20Co/CNTs calcined in Ar at 200 °C, 500 °C and 800 °C: (a) 2θ range of 10°–90°; (b) 2θ range of 40°–58°

Additionally, no metallic Co was found in 500 °C Ar-calcined 20Co/CNTs sample. Further increasing the calcination temperature up to 800 °C results in the formation of metallic Co due to the hard reducibility of Co_2C [20]. This autoreduction behavior is in agreement with Xiong et al. [21] who found that 15 wt.% Co/CNTs can be autoreduced at temperature higher than 480 °C in Ar. The main difference is that no Co_2C was found in his study, which suggests the effect of the autoreduction on the FTS activity is different as discussed later.

Table 1. The FTS performance of 20Co/CNTs catalysts pretreated under different conditions [a]

Catalyst	CO conversion (%)	CO_2 selectivity (%)	Hydrocarbon selectivity (mol%)					α
			CH_4	C_2	C_3	C_4	C_{5+}	
200 °C Air Cal.	91.4	4.3	13.1	0.7	0.9	0.6	84.8	0.89
200 °C Ar Cal.	95.4	7.1	14.1	0.7	0.9	0.6	83.7	0.88
500 °C Ar Cal.	75.4	6.1	17.2	1.6	2.2	1.4	77.6	0.83
800 °C Ar Cal.	65.5	5.8	17.4	1.8	2.3	1.4	77.1	0.83

[a] Reaction conditions: $H_2/CO = 2$, $P = 2.0$ MPa, $T = 230$ °C, GHSV = 1800 h^{-1}

Table 1 shows the catalytic performance of 20Co/CNTs calcined under different conditions. 20Co/CNTs catalysts calcined at 200 °C in Ar or air achieve equivalent FTS activity. Higher calcination temperature resulted in decreased FTS activity. One reason for the decrease is the cobalt particle aggregation resulted from sintering at high temperature and another is that some cobalt oxides were transformed to Co_2C at high temperature which cannot be reduced to metallic Co by H_2 at temperature less than 500 °C [22,23].

4. Conclusions

$Co(NO_3)_2$ supported on CNTs can be decomposed to form Co_3O_4 at 200 °C in either Ar or air atmosphere. Higher calcination temperature resulted in decreased FTS activities because CNTs was consumed by oxidation in air at temperature higher than 230 °C. Cobalt species could be partly transformed from Co_3O_4 to metallic Co in Ar by autoreduction at temperature over 500 °C. The autoreduction route might be from Co_3O_4 to CoO to Co or from Co_3O_4 to Co_2C to Co.

Acknowledgments

The authors gratefully acknowledge the support from the National Natural Science Foundation of China (21076151), New Century Excellent Talents in University Chinese Ministry of Education (NCET-06-0239) and the Programme of Introducing Talents of Discipline to Universities (B06006).

References

[1] Vosloo A C. *Fuel Process Technol*, 2001, **71**: 149
[2] Dry M E. *Appl Catal A*, 2004, **276**: 1
[3] Khodakov A Y, Griboval-Constant A, Bechara R, et. al. *J Catal*, 2002, **206**: 230
[4] Khodakov A Y, Chu W, Fongarland P. *Chem Rev*, 2007, **107**: 1692
[5] Tavasoli A, Abbaslou R M M, Trepanier M, et. al. *Appl Catal A*, 2008, **345**: 134
[6] Tavasoli A, Sadagiani K, Khorashe F, et. al. *Fuel Process Technol*, 2008, **89**: 491
[7] Abbaslou R M M, Tavasoli A, Dalai A K. *Appl Catal A*, 2009, **355**: 33
[8] Tréanier M, Tavasoli A, Dalai A K, Abatzoglou N. *Fuel Process Technol*, 2009, **90**: 367
[9] Tréanier M, Tavasoli A, Dalai A K, Abatzoglou N. *Appl Catal A*, 2009, **355**: 193
[10] Zhang H, Chu W, Zou C J, et. al.. *Catal Lett*, 2011, **141**: 438
[11] Tréanier M, Dalai A K, Abatzoglou N. *Appl Catal A*, 2010, **374**: 79
[12] Abbaslou R M M, Soltan J, Dalai A K. *Appl Catal A*, 2010, **379**: 129
[13] Bahome M C, Jewell L L, Padayachy K, et. al. Coville N. *J Appl Catal A*, 2007, **328**: 243
[14] Serp P, Corrias M, Kalck P. *Appl Catal A*, 2003, **253**: 337
[15] Song D C, Li J L. *J Mol Catal A*, 2006, **247**: 206
[16] Bartholomew C H. *Catal Lett*, 1990, **7**: 303
[17] Ehrhardt C, Gjikaj M, Brockner W. *Thermochim Acta*, 2005, **432**: 36
[18] Nagakura S. *J Phys Soc Jpn*, 1961, **16**: 1213
[19] Fukumiya Y, Haga Y, Nittono O. *Mat Sci Eng A*, 2001, **312**: 248
[20] Tsakoumis N E, Ronning M, Borg O, et. al. *Catal Today*, 2010, **154**: 162
[21] Xiong H F, Motchelaho M A M, Moyo M, et. al. *J Catal*, 2011, **278**: 26
[22] Volkova G G, Yurieva T M, Plyasova L M, et. al. *J Mol Catal A*, 2000, **158**: 389
[23] Cheng J, Hu P, Ellis P, et. al. *J Phys Chem C*, 2010, **114**: 1085

(279a) Control of Isobutane Cracking Selectivity Over H-ZSM-5 by High-Temperature Pretreatments

Jang Ho Yun and Raul F. Lobo, Chemical Engineering, University of Delaware, Newark, DE

Zeolites are crystalline materials that have wide application in industry as solid acid catalyst. Zeolites have Brønsted acid sites that are hydroxyl groups bridged between Al and Si (Al-OH-Si) tetrahedra. The acid sites of zeolites are decomposed at high temperatures, usually above 873 K in a process called dehydroxylation. This high temperature condition is commonly found in the fluidized catalytic cracking where catalyst is recycled forth and back between the riser and the regenerator under an oxidative atmosphere. The dehydroxylation is believed to proceed via a dehydration mechanism of the acid sites, however, hydrogen is also formed during the dehydroxylation process[4]. To explain this observation, the decomposition of Brønsted acid through the formation of electron deficient redox sites, such as $[AlO_4]^0$, in zeolites are proposed. The activation of small alkanes over acid sites has been investigated extensively because of its relevance to technologically important processes such as fluidized catalytic cracking in petroleum refineries, but also because C-H and C-C bond activation is of fundamental scientific interest. Here, we investigate reactivity and selectivity of newly generated sites using isobutane conversion.

We have found that in addition to the catalytic effect of Brønsted acid catalyst, redox sites, which are generated by high temperature for dehydroxylation, also promote alkane cracking process. The specific reaction rates of dehydroxylated samples increase respect to the pristine samples. When conversion is low, the product distribution is limited to the monomolecular cracking of the C-C bond and dehydrogenation of the C-H bond. For isobutane, the cracking-to-dehydrogenation ratio increases by a factor of ~2 after dehydroxylation treatments. We proposed that the presence of redox sites resulted in radical cation chemistry instead of protolytic chemistry in the isobutane cracking process. In addition there are dramatic differences in selectivity: the dehydroxylated catalysts are much more selective toward the cracking of isobutane into propylene and methane than towards dehydrogenation. This is in contrast to propane, where the dehydroxylated samples show preference towards dehydrogenation. The molecular origin of these differences is being investigated and will be discussed.

See more of this Session: Catalytic Processing of Fossil and Biorenewable Feedstocks: Fuels II
See more of this Group/Topical: Catalysis and Reaction Engineering Division

(279b) Exocyclic and Endocyclic C-C Bond Cleavage - Mechanism and Site Requirements

David W. Flaherty[1], Alper Uzun[2] and Enrique Iglesia[2], (1)Department of Chemical Engineering and Biomolecular Engineering, University of California at Berkeley, Berkeley, CA, (2)Department of Chemical and Biomolecular Engineering, University of California at Berkeley, Berkeley, CA

The success of metal-catalyzed hydrocarbon ringopening for fuel upgrading depends on the suppression of dealkylation and multiplehydrogenolysis, both of which form lower value products. Recent reports state thatturnover rates and selectivities for ring opening may be improved by using bimetallicclusters [1] or bifunctional catalysts that promote ring contraction pathways *via*solid acids [2, 3], without rigorous mechanistic interpretations or clearimplications for the design of more active and selective catalyst sites. Here, weshow how the level of saturation of chemisorbed intermediates and the coverageof chemisorbed hydrogen atoms, [H*], determine the probability of breaking eitherexocyclic or endocyclic C-C bonds, and hence, turnover rates and selectivities.We also demonstrate how ring opening turnover rates can be increased onmonofunctional Ir catalysts, with little loss of selectivity, by concurrentincreases in temperature and H_2 pressure to maintain the degree of hydrogenationof the intermediates.

Thehydrogen content of gas-phase cycloalkanes and acyclic alkanes is equilibrated atall conditions demonstrating that the surface is covered with an equilibratedmixture of H* and cyclic and acyclic hydrocarbons whose degree of hydrogenationis determined by chemical equilibrium with the prevalent H_2 pressure.Cyclic surface intermediates undergo initial C-C bond cleavage at endocyclic andexocyclic positions, resulting in the formation of ring opening anddealkylation products, respectively, with individual rate constants for surfaceintermediates at every level of hydrogenation. Multiple hydrogenolysis productsform primarily by subsequent C-C bond cleavage of cyclic, dealkylation products,because acyclic species formed by ring opening are present at very low surface concentrationsdue to low dehydrogenation and adsorption equilibrium constants. Figure 1ashows that turnover rates for 1,3-dimethylcyclohexane (1,3-DMCH) conversion initiallyincrease with H_2 pressure because of a concomitant increase in the hydrogenationof adsorbed species leading to a greater fraction of intermediates withsignificant rate constants. Higher H_2 pressures lead to near-saturatedintermediates which bind to the surface weakly such that H* ultimatelydisplaces all reactive hydrocarbon species and decreases turnover rates for allreaction pathways. As shown in Figure 1a, the selectivity for exocyclic C-Cbond rupture, described here as the ratio ($\beta_{Exo-Endo} = (r_{Exo})/(r_{Endo})$; r_X is the rate of reaction at position *X*),decreases monotonically with increasing H_2 pressure because thearomatic character of highly unsaturated intermediates preferentially stabilizesendocyclic bonds, thus decreasing ring opening turnover rates.

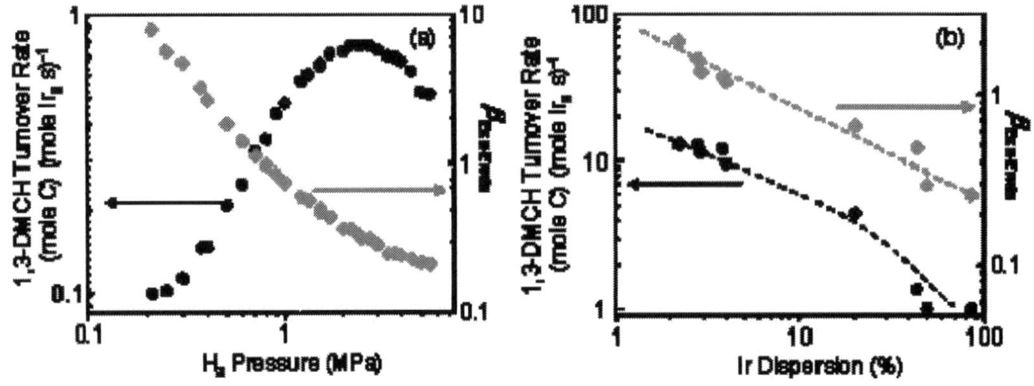

Figure 1. Turnover rates (•) and $\beta_{Exo-Endo}$(♦), the ratio of the rates of exocyclicand endocyclic C-C bond rupture. (a) shows the effect of H_2pressure for small Ir particles, $<d_{TEM}>$ = 0.6 nm, at 20kPa 1,3-DMCH and 593 K. (b) displays the dependence on Ir dispersion at 3.4 MPaH_2, 110 kPa 1,3-DMCH and 593 K.

Turnover ratesand selectivity also depend on the structure of the cycloalkane reactants andthe metal dispersion. The degree of saturation depends on the C-H bondstrength in reactants. With increasing substitution (cyclohexane →1,3,5-trimethylcyclohexane), the weaker C-H bonds at tertiary carbon atoms favordehydrogenation and lead to more unsaturated surface species. Consequently, alkylsubstitution of the ring decreases turnover rates and ring openingselectivities. Figure 1b shows that turnover rates for 1,3-DMCH conversiondecrease with increasing Ir dispersion, however, the selectivity for endocyclicC-C bond cleavage is significantly higher on smaller clusters. Thisobservation correlates with the known tendency of low-index planes, prevalenton large clusters, to cleave terminal C-C bonds which favors exocyclic C-C bondscission. The dependence of selectivity on particle size suggests that low-coordinatesites lead to a disproportionate increase the adsorption constant of saturatedspecies in comparison to highly unsaturated species.

Theauthors acknowledge financial support from the ExxonMobil Research andEngineering Co. and technical discussions with Drs. Stuart L. Soled and GuangCao.

References

1. P. Samoila, M.Boutzeloit, C. Especel, F. Epron, P. Marecot, *Appl. Cat. A* 369, 104(2009)

2. G.B. McVicker *et al., J. Catal.* 210,137 (2002)

3. J.-W. Park, K.Thomas, J. van Gestel, J.-P. Gilson, C. Collet, J.-P. Dath, M. Houalla, *App.Cat. A: General* 388, 37 (2010)

(279c) Catalytic Pyrolysis of Biomass: The Effect of Hydrothermal Treatment of FCC Catalyst and ZSM-5 Additive

Ofei Mante, Biological Systsems Engineering, Virginia Polytechnic Institute and State University, Blacksburg, VA, Foster A. Agblevor, Biological Engineering, Utah State University, Logan, UT and Ron McClung, BASF Catalysts LLC, Iselin, NJ

The process configuration of catalytic pyrolysis of biomass in a continuous circulating fluidized-bed reactor mimics the fluid catalytic cracking (FCC) process used in the petroleum refineries process to upgrade heavy hydrocarbon molecules to more valuable petroleum products. In the FCC process, the spent catalyst is continuously regenerated at high temperatures and pressures, as a result the catalyst undergoes hydrothermal treatment (deactivation). The influence of hydrothermal deactivation of catalyst used in catalytic pyrolysis of biomass is of importance. In this study, fresh FCC catalyst and ZSM-5 additive were deactivated with 100% steam at 1350 °F (732 °C) and 1450 °F (788 °C) for 4 hours in a fluidized state. The catalytic pyrolysis of hybrid poplar with both fresh and steam treated catalysts were conducted in a 50 mm bench scale bubbling fluidized bed reactor at 475°C and a weight hourly space velocity (WHSV) of 2 h^{-1}. The catalysts were evaluated with respect to the product yields and the property of the upgraded bio-oil. The surface area measurements of the catalyst showed a reduction in total surface area of the FCC catalyst with about 40% loss of zeolite surface area after steam treatment at 787.8 °C for 4 hours. However, the steaming of the ZSM-5 additive at both conditions resulted in a slight increase in the total surface area. The significant loss in the surface area within the micropore of the FCC catalyst is indicative of the loss in catalytic activity. The hydrothermal treatment of the FCC catalyst positively affected the catalytic pyrolysis product yields. The organic and gas yields increased and the water and char/coke yields decreased after steam treatment. From the study, the properties of the bio-oils evidently suggested that the FCC catalyst was still active following treatment at 788 °C. The viscosity and the density of the bio-oils from the steamed FCC catalyst were lower than those produced with the fresh FCC catalyst. Steaming at 1350 °F (732 °C) resulted in bio-oil with the highest HHV (28.55 MJ/kg). In the case of the ZSM-5 additive, the extent of deactivation was reflective only in the organic and gas yields. The organic yield increased and the gas yield decreased. Due to the lower tendency of ZSM-5 to form coke, steaming of the catalyst did not show any significant decrease in the char/coke yield as was seen in the FCC catalyst. Additionally, the ZSM-5 additive did not show similar trend in the improvement of the properties of the bio-oils produced with the steam treated FCC catalyst. The bio-oils from the steamed ZSM-5 additives were rather higher in viscosity and density. The GC analysis of the product gases suggested that steam treatment of the catalyst prolonged the catalyst lifetime owing to the fact the concentrations of CO and CO_2 and CH_4 were fairly stable for the steamed catalyst than for the fresh catalyst. The FTIR and ^{13}C-NMR spectra of the bio-oil showed generally that steaming of the catalyst increased the selectivity for the production of aromatic hydrocarbons.

(279d) Supported Palladium Catalysts for Decarboxylation of Fatty Acids

H. Henry Lamb, Jeremy G. Immer and Jeffery P. Ford, Chemical and Biomolecular Engineering, North Carolina State University, Raleigh, NC

Liquid-phase deoxygenation of stearic acid (SA) in dodecane at 300°C and 15 atm was employed to screen supported palladium catalysts for decarboxylation of fatty acids (FAs) to hydrocarbons. Commercial samples of Pd/C (5), Pd/Al_2O_3, and Pd/SiO_2 catalysts and an in-house prepared Pd/SiO_2 catalyst (each containing 5 wt.% metal) were screened under flowing 0, 5, and 10% H_2 (balance He). Under flowing He, most of the catalysts studied failed to achieve 100% SA conversion after 4 h under reaction conditions due to rapid deactivation. The exception was a uniformly impregnated Pd/C catalyst that gave >99% conversion in ~1 h with 99% CO_2 selectivity. All of the catalysts were far more stable under H_2 yielding nearly complete SA conversion after 4 h; however, they differed markedly in their CO_2 selectivities. The Pd/C catalysts were found to be highly active and selective for SA decarboxylation under 5% H_2. In contrast, SA deoxygenation over Pd/SiO_2 catalysts occurred primarily via decarbonylation and at a much slower rates. Pd/Al_2O_3 exhibited high initial SA decarboxylation activity, but deactivated rapidly under 5% H_2. Similar CO_2 selectivity patterns among the catalysts were observed for deoxygenation of lauric and capric acids; however, the initial decarboxylation rates tended to be lower for Pd/C and Pd/Al_2O_3 with these substrates. The most active Pd/C catalyst was used to investigate the influence of alkyl chain length on deoxygenation kinetics for C8-C18 FAs. Generally, as FA carbon number decreases, reaction time and H_2 consumption increase, and CO_2 selectivity and initial decarboxylation rate decrease. We attribute the higher initial decarboxylation rates for longer chain FAs to an increased propensity for FA adsorption on the catalyst support.

See more of this Session: Catalytic Processing of Fossil and Biorenewable Feedstocks: Fuels II
See more of this Group/Topical: Catalysis and Reaction Engineering Division

(279e) Co-Processing Methane and Biomass-Derived Oxygenates Over Mo/ZSM-5 Catalysts for Making Hydrocarbon Fuels

Do-Young Hong and Aditya Bhan, Chemical Engineering and Materials Science, University of Minnesota, Minneapolis, MN

The concurrent deoxygenation of C_2-C_3 oxygenates and dehydroaromatization of CH_4 was accomplished using Mo/ZSM-5 catalysts at 950 K in order to suppress catalyst deactivation by carbon deposition and to minimize depletion of aromatic products due to deep scavenging of CH_x species by oxidants such as CO_2, O_2 and H_2O. A systematic study of C_2-C_3 oxygenates with varying Effective Hydrogen Index (EHI = (H-2(O))/C) and of reactant mixtures with varying oxygenate/CH_4 ratio (0.01-0.1) revealed that higher methane conversion is achieved in presence of oxygenate co-feed and that the formation rate and selectivity to CO increased with increasing oxygenate/CH_4 ratio. The formation rate of C_2 hydrocarbons was not influenced by the identity of the C_2 or C_3 oxygenate at oxygenate/CH_4 ratio of 0.035. In contrast, the rate of formation of benzene varied depending on the carbon number and alcohol, aldehyde/ketone, or acid functionality of the co-reactant. These results show that CH_4 can be directly used for deoxygenation reactions of biomass compounds at high temperature conditions. Structural and chemical characterization of the catalyst by infrared and X-ray absorption spectroscopy as well as outstanding mechanistic questions pertaining to the tendency of the Mo/ZSM-5 system to shed off oxygen as CO, CO_2, or H_2O will be discussed.

See more of this Session: Catalytic Processing of Fossil and Biorenewable Feedstocks: Fuels II
See more of this Group/Topical: Catalysis and Reaction Engineering Division

(279f) Pilot Scale Evaluation of Ni-Based Reforming Catalyst Performance for Syngas Production

Whitney S. Jablonski[1], Steven D. Phillips[1], Kim Magrini[1], Calvin J. Feik[2] and Katherine R. Gaston[2], (1)National Renewable Energy Laboratory, Golden, CO, (2)National Bioenergy Center, National Renewable Energy Laboratory, Golden, CO

Biomass gasification is an economical thermochemical route from plant-based materials to liquid fuels. Product gas from this process is comprised of synthesis gas, tars, olefins, and other hydrocarbon impurities which must be transformed into synthesis gas in order to achieve a suitable feed stream for a fuel synthesis process. Model compounds are often used to evaluate reforming catalysts for achieving a pure synthesis gas stream. However, these model compounds do not capture the severity of the conditions that a catalyst will likely undergo in a real biomass gasification product stream. This study compares results for catalyst deactivation between a 1/2 ton-per-day pilot plant biomass gasification system and a bench scale system using model synthesis gas. A nickel based reforming catalyst (6 wt-% NiO, 2.4 wt-% MgO, 3.9 wt-% K_2O) was developed and compared to several industrial catalysts of varying nickel content for this work. Oak was gasified in an indirectly heated, steam blown gasifier in a half ton per day pilot plant to produce a raw product gas stream. A slip stream of this product gas was fed to a packed bed reactor loaded with catalyst, and the products were analyzed by molecular mass beam spectrometry (MBMS) for tar speciation and gas chromatography for light gas quantification. Post-reaction catalyst samples were analyzed using XRD, SEM, TPR, and coke removal techniques to quantify the amount of carbon deposited on the catalyst. The oak gasification product stream produced in the pilot plant was modeled in a bench scale system using bottled gases. Olefins and alkanes were estimated using ethylene. Product gas was analyzed using gas chromatography. Post bench scale reaction catalysts were characterized in the same way as post pilot scale catalysts. Reaction and characterization results from both pilot scale and bench scale operation will be compared and discussed during this work. It was found that due to the high concentration of hydrogen sulfide in the stream, nickel-based catalysts were deactivated according to the amount of nickel contained within them. As a result, several new catalyst compositions are proposed, and were tested on the bench scale using the modeled oak product gas composition.

See more of this Session: Catalytic Processing of Fossil and Biorenewable Feedstocks: Fuels II
See more of this Group/Topical: Catalysis and Reaction Engineering Division

Accelerated catalytic processing of fossil and biorenewable feedstocks using Avantium's Technology and methodologies

Pieter Imhof and Alex Pérez de Santana

Avantium Chemicals BV,

Alex.Santana@Avantium.com

Zekeringstraat 29, 1014 BV, Amsterdam, The Netherlands)

Introduction

Worldwide specifications for transportation fuels have become more stringent over the years. This has fueled catalyst development in the field of fossil and biorenewable feedstocks. Like current day oil-refineries, it is expected that Biorefineries will be complex facilities which will rely heavily on the integration of various catalytic processes High-throughput catalysis research is beneficial in facilitating an optimal integration of these various processes, and can also be used to address changing feedstock compositions typically encountered with natural product feed streams. Avantium's methodologies and proprietary fixed bed technology, which is used in Avantium's catalyst research services and in the Flowrence^{TM}, allows our customers to accelerate their catalyst and process development. Being able simultaneously study various process steps as well as to test 16-64 catalysts in parallel reduces the time to develop a new catalyst generation and find the most economical process from multiple years to less than a year. Avantium's capabilities allow testing of both extrudates and sieved materials with a discrimination power similar to those normally only observed in pilot test units. Testing both extrudates and sieved materials creates the opportunity to effectively evaluate effect of mass transport limitation and catalyst shape.

The unique capabilities of Avantium's methodologies and technology will be reviewed using real data from hydroprocessing/cracking, Fischer-Tropsch synthesis of waxes from syngas and the selective hydrogenation of 5-ethoxymethylfurfural case studies. The accelerated methodologies used by simultaneously researching various steps towards the development of a new process and catalyst will be presented. Further details on the catalyst can be obtained by comparing the test results of extrudates and sieved material for different catalyst systems, as will be shown in the ULDS and VGO hydrocracking studied discussed. Additionally, the data obtained from Avantium's technology will be compared to those obtained in a pilot plant using same feed and same process conditions.

Experimental

Avantium's gas and trickle flow technology is applied for ULSD, VGO-HDS, hydrocracking, Fischer-Tropsch and a selective hydrogenation reaction. Testing consists of 4 reactor blocks with 4-16 reactors each.

For the hydrotreating/hydrocracking in the ULSD, VGO-HDS study, the each of the reactors was equipped with an individual flow controllers allowing accurate feeding of both the hydrogen and diesel, SRGO and VGO feeds at very low flows with very high accuracy. The unit is operating fully automated 24/7. When required the liquid samples are collected using a 4x16 sample robot makes it possible to sample for multiple days without operator interference. The catalyst volume applied for the test in the unit is between 0.5-1ml. Both sieved materials and complete extrudates have been loaded and tested. In the ULSD, VGO-HDS study extrudates and sieved material of respectively 1.6 mm and 200-400micron have been applied. Two commercial catalysts have been tested in both Flowrence™ unit and pilot plant. In the Flowrence™ unit 4 reactors are loaded with extrudates of Cat A, 4 reactors are loaded with sieved material of Cat A, 4 reactors with extrudates of Cat B and 4 reactors with sieved material of Cat B. In the case of the pilot plant only extrudates have been applied. Catalysts are tested at 80 bar, H2/Oil ratio of 800 and low LHSV at 355, 365, 375 and 385 °C. Both sulfur and nitrogen content of the treated VGO have been analyzed on a daily basis.

All catalysts for the hydrogenation of furanics molecules were synthesized using a wet impregnation procedure [1,2]. The liquid volume was chosen such that all catalysts could be prepared using identical precursor mass concentrations. This gives a rather dilute impregnation slurry, where the volume is determined by the metal precursor with the lowest solubility. In all cases impregnatition time was 24 h. Drying of catalysts was performed at 105 °C, followed by calcination in air at 320 °C. All experiments were performed in trickle flow in a commercial 16-reactor setup between 80 and 120 °C. In a typical experiment, the reactor system is loaded with 15 calcined catalysts at 50 mg each. The remaining reactor is left empty to obtain a reliable measurement of conversion and probe for uncatalysed reactions. The catalysts are first reduced at 300 °C using a mixture of hydrogen and nitrogen. After cooling down to 80 °C, each reactor is supplied with a feed of substrate solution corresponding to a LHSV of 0.032 mol g-1 h-1. The samples are collected in the liquid phase and analyzed offline using GC.

High-Throughput experiments for syngas conversion processes were performed on a 64 fixed-bed tubular reactors operated in parallel and in co-current downstream mode [3]. The so-called 3rd generation Nanoflow consisted of four reactor blocks containing 16 reactors each. The tubular reactors had internal diameters of 1.0, 1.5 or 2.0 mm and were typically used for testing catalyst masses of between 25 and 200 mg. The reactors can operate at a maximum temperature of 550 ºC, a maximum pressure of 100 bar and gas hourly space velocities of up to 100,000 h^{-1}. To enable investigation and optimization of process conditions the pressure, temperature and flow rates may be systematically varied during a run. The reactor holder is of a special design and the reactor effluent can be diluted with liquid or gas thus preventing condensation of products and up to C-15 products can be measured on-line using GC as well as the collection of liquid samples for wax analysis. Using 64 reactors in parallel for over a period of 3 months uninterrupted, the catalysts are compared to one another while accurately being able to combine the online analysis with offline analysis for determining the alpha values(Anderson Schulz Flory distribution).

Results and Discussion

High-Throughput experimentation is a valuable tool to speed-up development, to reduce costs and to improve safety. This requires a combination of high accuracy, increased speed of catalyst testing and least of all high quality data. In three examples these concepts are shown.

Accuracy

Desulphurization of diesel, VGO and other feeds often work at very high conversion levels over 99.9% in order to meet the legislative demands for sulphur emissions by fuels. As a consequence it requires very accurate control of feeds, pressures and temperature and a precise plug flow over the catalyst bed to attain such high conversion levels. When comparing different catalysts one is faced with the challenge of meeting these high standards over duri ng each experiment.

In Figure 1 the performance of catalyst A and B in the Flowrence unit and in the pilot plant are shown for the desulfurization of a VGO feed. As can be seen from figure1 the ranking and behavior of both catalysts are very similar in the Flowrence™ unit and pilot plant. For the HDS performance slightly lower sulfur concentrations are found than in the pilot plant. The small differences in performance between the Flowrence™ and pilot plant are due to small differences in actual reactor temperature (isothermal vs. adiabatic), better plug flow in the Flowrence unit and small difference in hydrodynamics.The accuracy of the Flowrence test data allows discrimination between catalysts that show only 2-3°C difference in temperature required to achieve a certain activity.

Similar data demonstrating the performance of Flowrence technology for HDN and hydrocracking is also presented.

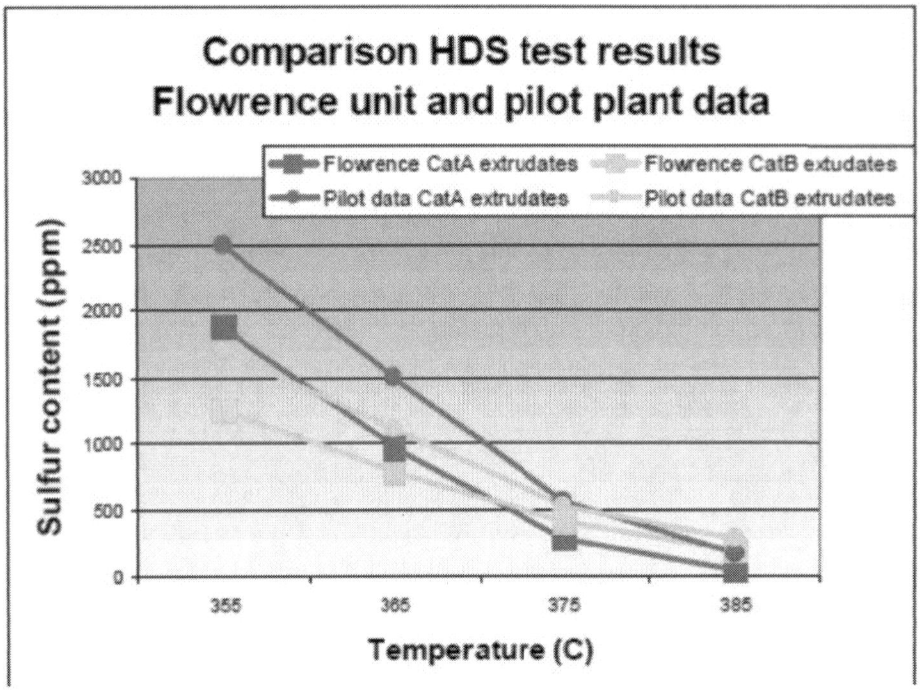

Figure 1 Comparison Sulfur data from Flowrence™ unit with those obtained in Pilot plant.

Speed

Catalytic biomass conversion is a highly competitive area so the speed of development is crucial for any party to remain frontrunner in the development and to guarantee a solid intellectual property position. The following example illustrates the strength of High-Throughput technologies regarding the speed of developments. To develop biofuels from the Furanic intermediates, new catalytic systems for the selective reduction of furanic molecules need to be found. For example, in the hydrogenation of EthoxyMethylFurfural in the route to new bio-fuels hydrogenation of the furan ring, aldehyde moiety and the ethyl ether moiety is possible (see Scheme 1).[1,2] Using a 3-month High-Throughput screening program consisting of 576 experiments, a group of Iridium-based catalysts could quickly be identified as an active and selective metal for the hydrogenation of the aldehyde group. The results for all catalysts tested are shown in Figures 2 and 3. best results of >99% yield to the desired product, 5-ethoxymethylfurfuryl alcohol, are obtained using an iridium/chromium (Ir/Cr) catalyst.

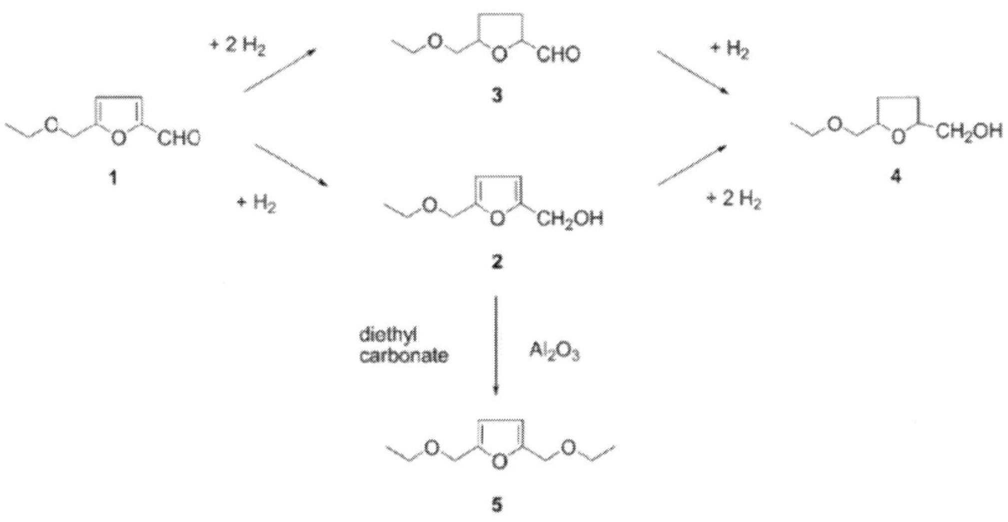

Scheme 1 Hydrogenation pathways for the catalytic hydrogenation of 5-ethoxymethyl furfural

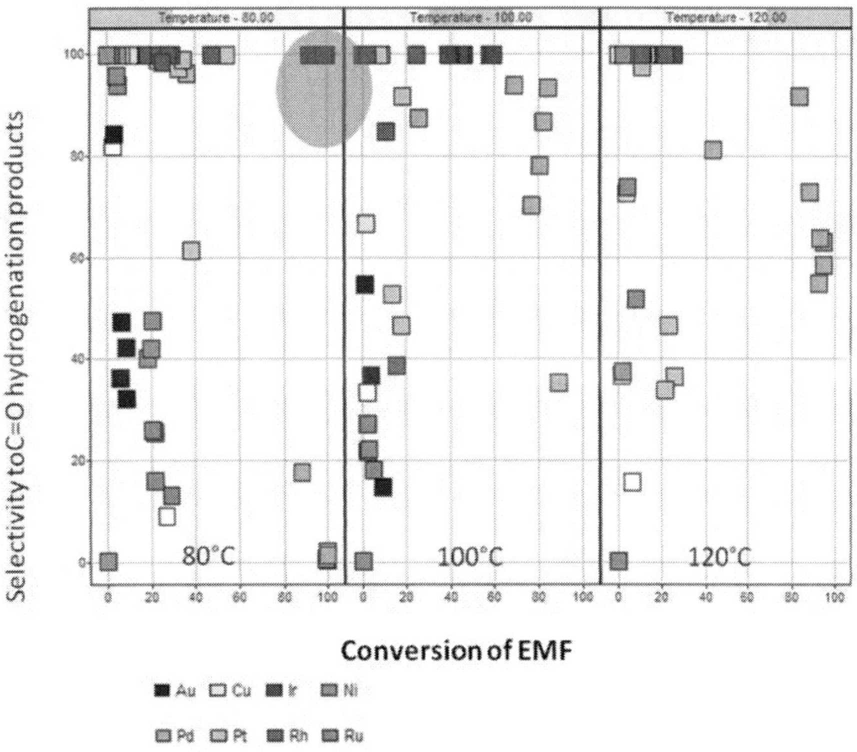

Figure 2: Screening of catalysts for the selective hydrogenation of the aldehyde moiety in EMF. In the circle a group of Ir-based catalysts is highlighted.

In Figure 3 an alternative way of presenting the data is shown, that highlights the influence of process parameters in more detail. The results show that both substrate conversion and product selectivity are sensitive towards temperature changes.

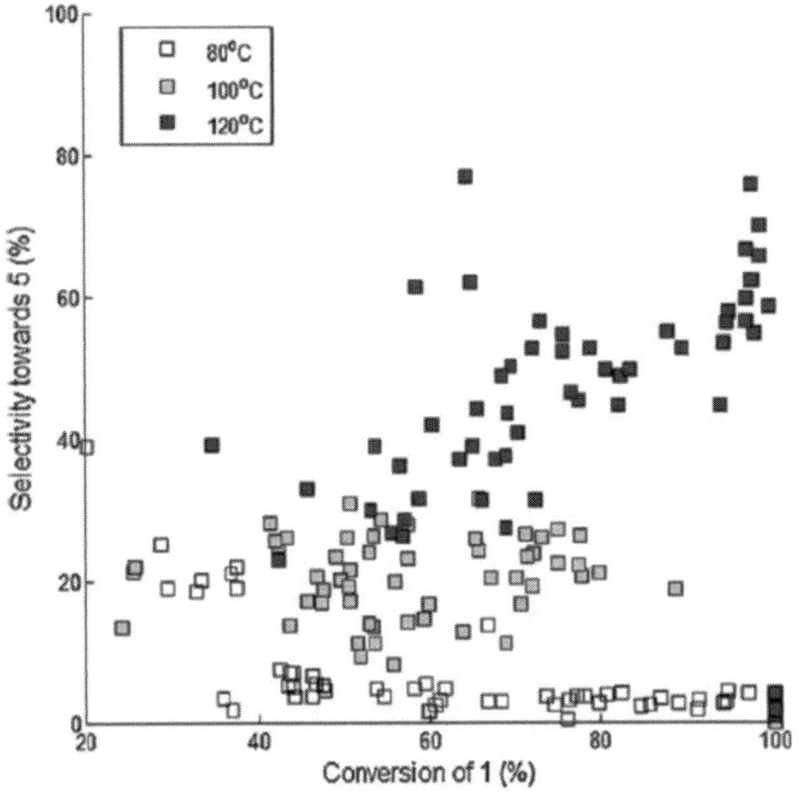

Figure 3 Semilogarithmic plot of the ratio between the selectivities towards the unsaturated alcohol 2 and its the diethyl ether 5 as a function of conversion at three temperatures.

Quality

Obtaining high quality data is often key in a successful catalyst program, especially those that involve length catalysts test, i.e. lifetime studies. Not only is it obvious that a sequential testing of catalysts will be a lengthy process, it is often less clear that small changes in equipment or variation over time of the feed used also play an important role, especially if one wished to discriminate between catalysts with only a very small difference in activity. Using the parallel approach offers significant advantages here as not only will it allow testing more catalysts in a shorter period of time, it also ensures that all the catalysts are tested using the same feed and process conditions.

As an example of this a series of Co-on-alumina catalysts were tested in lifetime studies with a continuous time-on-stream of over 2000 h. Figure 4 shows the CO-conversion over time.[3] The CO feed flow was adapted to maintain a stable conversion. Overall using the Nanoflow parallel reactor technology allowed a 3-month lifetime study for 64 catalysts which would otherwise have taken 16 years on a single reactor thus saving over 98% in experimental time compared to conventional 1-reactor systems.

Moreover, in this study the liquid wax was collected in 64 catch pots, i.e. one for each reactor. The selectivity information from the liquid/wax phase for C20 to C52 showed the same trends and rankings as that from the gas phase (C4 to C12) (See Figure 5). Hence, these observations seem to justify screening catalysts just by analyzing the gas phase, thus simplifying the set-up and limit operational efforts. However, Fischer-Tropsch catalysts may display a carbon-dependent alpha value (as would be visible by a change of slope in the Anderson-Schulz-Flory plot), and for in depth selectivity studies the use of catch pots to collect wax still is advised.

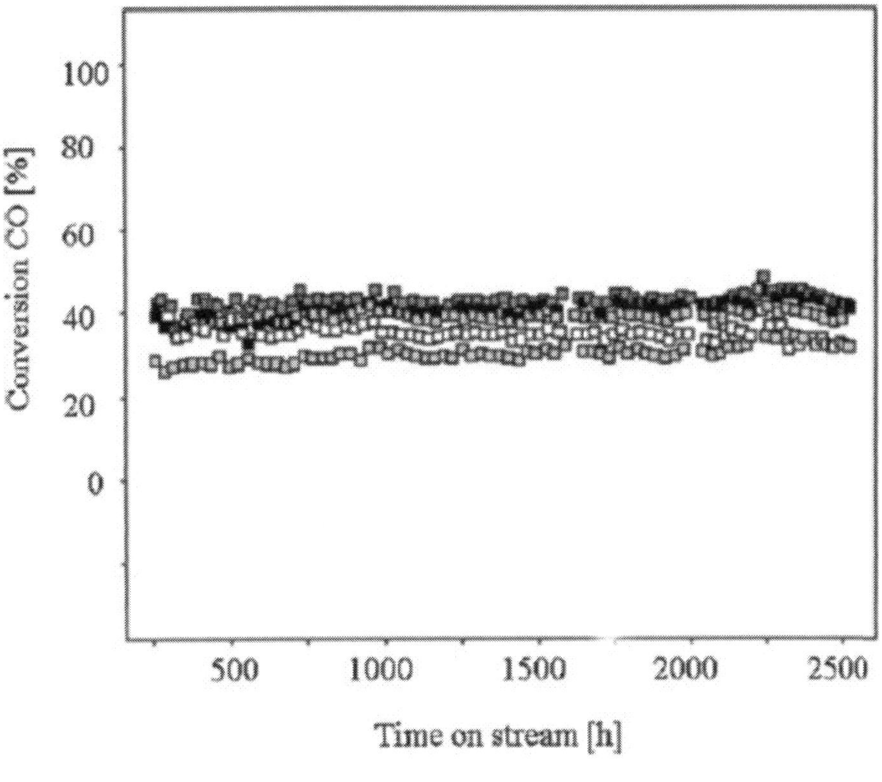

Figure 4 CO conversion for a number of different catalysts in Fischer-Tropsch synthesis during a 3-month test.

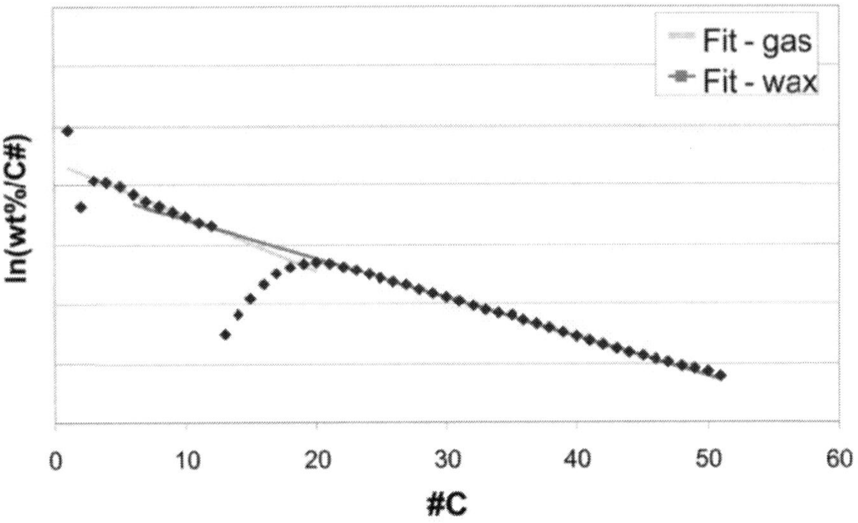

Figure 5 ASF plots of in-situ analyzed gaseous compounds and ex-situ analysed on wax sample collected.

Conclusions

Avantium's methodologies and technologies are very powerful acceleration tools to speed up catalyst and process development efforts in the field of fossil and renewable feed stocks while maintaining high accuracy and high data quality. Examples are shown for trickle flow hydrocracking of a VGO feed and selective hydrogenation of a bio-derived furan and for gas-to-liquid syngas to wax programs, but the advantages of Avantium's methodologies and technologies are applicable for many different types of catalytic chemistries.

The major strength of the quality data as presented by the three examples is that data obtained in Avantium technology relate very well to those obtained in a pilot unit and demonstrate the application flexibility and data accuracy that Avantium can be achieved today.

References

1. E.J. Ras, G. Rothenberg and B. McKay, Understanding catalytic Biomass Conversion through Data Mining Topics in Catalysis, 2010, 53,1202-1208.

2. E.J. Ras, S. Maisuls, P. Haesakkers, G.J. Gruter and G. Rothenberg, Advanced Synthesis and Catalysis. 2009, 351, 3175 – 3185.

3. J.C. van der Waal, G. Klaus, M. Smit, C. M. Lok, Catalysis Today, High-Throughput Experimentation in Syngas-based Research, Catalysis Today, accepted for publication.

Activation studies of an iron-based Fischer-Tropsch catalyst in a slurry reactor

Hao Qinglan [1,2]*, *Bai Liang* [2], *Xiang Hongwei* [2], *Li Yongwang* [2]

1. College of Material Science & Chemical Engineering, Tianjin University of Science & Technology, Tianjin 300457, China

2. State Key Laboratory of Coal Conversion, Institute of Coal Chemistry, Chinese Academy of Science, P.O. Box 165, Taiyuan 030001, Shanxi, China

Abstract: The impact of activation gas on the phase compositions of iron catalysts and on the long-term (>500 h) activity and selectivity for Fischer-Tropsch synthesis (FTS) was explored. The effect of gas composition on catalyst bulk compositions during reduction is more complicated than those of reduction temperature and pressure. Iron phase compositions, and FTS performances of the catalysts were strongly dependent on pretreatment process. Metallic iron is formed when the catalyst is subjected to activation with H_2. However, metallic iron rapidly transforms to other iron phases once exposed to syngas. The amount of carbides in reduced catalyst with syngas is much more than that of activation with CO. The content of carbide in reduced catalysts using syngas decreases with the increase of H_2/CO ratios. Reduction by syngas with high H_2/CO ratios of precipitated iron-based catalysts can achieve the similar activity of the catalyst reduced by CO.

***Keywords*:** Fischer-Tropsch synthesis; activation; spray-dried iron catalyst; slurry reactor; Mössbauer effect spectroscopy

1. Introduction

As the alternative route to synthesize hydrocarbons, Fischer-Tropsch synthesis (FTS) has been the subject of renewed interest for conversion of coal and natural gas to liquid fuels. Before Fischer-Tropsch synthesis (FTS), the catalyst precursor is subjected to an activation treatment [1-5]. Co, Ni and Ru catalyst are almost reduced in flowing H_2 at 200~450 °C to the zero valent metallic state and remain in the metallic state during FTS. However, the pretreatment for the iron catalyst is not so clear. The iron catalyst can be successfully activated with H_2, CO or syngas [1,3-10]. Significant changes in iron catalyst compositions, structure and activity can result from different pretreatment parameters prior to exposure to

FTS conditions.

Proper control of the activation procedure used during catalyst reduction is the key for controlling the catalytic behavior of the spray-dried iron catalyst. This study will provide a better understanding of the relationship between reduction ambient, FTS activity, selectivity and stability of the catalyst used in a slurry phase reactor. The results clearly showed that activation ambient had a significant effect on subsequent FTS activity. And the conclusions from this paper may give us some specific indications on how to use an iron-based catalyst in a slurry phase FTS process.

2. Experimental

2.1 Catalyst preparation

A typical iron-based catalyst was prepared by a combination method of continuous co-precipitation and spray-drying technique. The final obtained fresh catalyst composition is $100Fe/5Cu/6K/16SiO_2$ (mass weight). Detailed description of the extraction has been provided elsewhere [11].

2.2 Reactor system

The FTS tests were carried out in a $1dm^3$ stirred tank slurry reactor (STSR) equipped with a sintered metal filter to allow the removal of wax products from reaction system. Detailed description of the reactor has been provided elsewhere [12].

2.3 Product analysis

H2, CO, CH4 and C1-C8 hydrocarbons (n-paraffins, n-olefins, and branched isomers) in tail gas were analyzed by an Agilent 6890N (HP) gas chromatograph (GC) with a 5A molecular sieve column (Ar carrier) and an Al2O3 column (N2 carrier) equipped with a TCD and a FID. Gas components were determined by the methane concentration correlation method. CO2 in tail gas was measured periodically on-line using an Agilent 4890D GC (HP) equipped with a TCD (H2 carrier) and quantified by the external standard method. The heavy wax and the oil products were analyzed using an Agilent 6890N (HP) GC with UA+-(HT) stainless steel capillary column (FID, N2 carrier) and an Agilent 6890N (HP) GC with DB-1 quartz capillary column (FID, N2 carrier), respectively. Oxygenates in water products were measured by an Agilent 6890N (HP) GC with DB-WAX quartz capillary column (FID, N2 carrier)

2.4 Reduction and reaction conditions

All activations were conducted in situ, using CO, H2, CO followed by H2, CO followed by H2, and syngas (H2/CO=0.67 and 1.0) at 270 oC, 1 NL/g-cat.h and 0.1 MPa. The detailed pretreatment conditions were summarized in Table 1. At the end of the pretreatment, the slurry was cooled down to reaction temperature (250 oC) and the reactor was adjusted to operating pressure (1.5 MPa). After reaching the desired reaction temperature and pressure, the space velocity of syngas (H2/CO=0.67) was increased to 2 NL/g-cat.h for FTS reaction tests for more than 500 h.

Table 1 Pretreatment conditions of the catalyst and test designations

Run No.	Reductant	Duration (h)
A-CO	CO	13
A-H_2	H_2	13
A-H_2+CO*	H_2+CO	6.5+6.5
A-CO+H_2**	CO+H_2	6.5+6.5
A-0.67	H_2/CO=0.67	13
A-2.0	H_2/CO=2.0	13

*The catalyst was firstly reduced by H_2 for 6.5 h, and then treated with CO for 6.5 h.

** The catalyst was firstly reduced by CO for 6.5 h, and then treated with H_2 for 6.5 h.

2.5 Catalyst sample handling

During reduction and reaction tests, catalyst suspended in the liquid paraffin was withdrawn from the reactor through a dip tube connected to a sampling trap. In order to prevent the exposure of the catalyst to oxidizing atmosphere, the sampling system was purged with argon prior to slurry withdrawn. The slurry samples from the sampling trap were treated to separate the catalyst from the heavy waxy products. After the extraction, the catalysts were sealed and stored in a glass holder immerged in ethanol to prevent exposure of the catalysts to air.

2.6 Catalyst characterization

Mössbauer effect spectroscopy(MES) was measured with a CANBERRA series 40MCA constant-acceleration Mössbauer spectrometer (CANBERRA, USA) at room temperature. The radioactive source comprised 25 mCi of 57Co in a Pd matrix. All spectra were analyzed with a set of independent Lorentzian lines and with the help of a non-linear least squares fitting procedure. The number of center channels and the increment of velocity per channel were determined by a standard α-Fe sample at room temperature. The hyperfine parameters, the isomer shift (IS), the quadruple splitting (QS), and the magnetic hyperfine field (Hhf) were used to identify the spectral components. Isomer shifts were reported with respect to α-Fe sample (99.99%, Alfa Aesar). Usually it was assumed that the Mössbauer area ratios are equal to the relative amount of the associated species.

3. Results and discussion

3.1 The catalyst reduction

The CO_2 content in tail gas during reduction process is measured periodically on-line using GC. The changes of CO_2 concentration with time on stream under the different procedures are shown in Fig. 1.

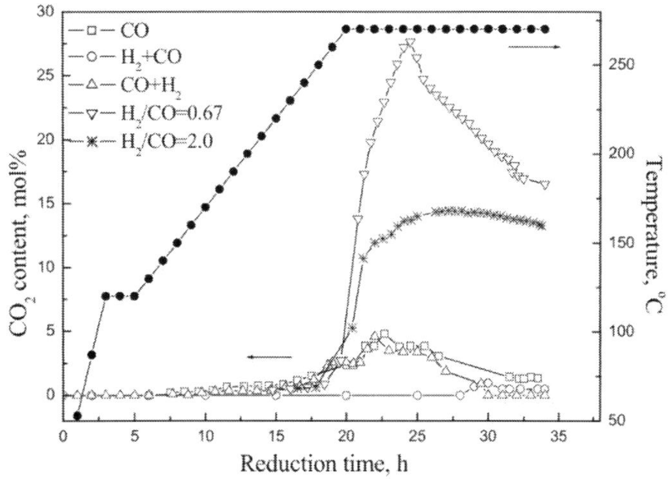

Fig 1 The change of CO_2 content during reduction process

Reduction conditions: 270 °C, 0.1 MPa, 1.0 NL/gcat.h

3.2 Catalyst phase characterization

MES can provide quantitative information about the amount of various iron phases present. Bulk compositions in the catalyst samples after the activation under different reduction ambient are determined by MES. The MES spectra of corresponding contents of the reduced catalysts are summarized in Table 2. The MES results show that the differences in phase compositions of the reduced catalysts are obvious. The effect of activation ambient on the catalyst bulk compositions is more complicated than those of reduction temperature and pressure. α-Fe, Fe_3O_4, iron carbide, and superparamagnetic (spm) phases in the reduced catalyst are coexist. The amount of carbides in reduced catalyst with syngas is much more than that of activation with CO. The content of carbide in reduced catalysts using syngas decreases with the increase of H_2/CO ratios.

Table 2 Iron phases identified by MES in reduced catalysts under different ambient

Run No.	Phase identification	Area (%)
A-CO	χ-Fe_5C_2	15.3
	ε'-$Fe_{2.2}C$	16.2
	Fe_3O_4(A)	3.2
	Fe_3O_4(B)	13.3
	Fe^{3+} (spm)	34.5
	Fe^{2+} (spm)	17.4

$A\text{-}H_2$	α-Fe	9.6
	Fe_3O_4(A)	6.6
	Fe_3O_4(B)	40.9
	Fe^{3+} (spm)	19.6
	Fe^{2+} (spm)	23.3
$A\text{-}CO+H_2$	α-Fe	2.8
	$\chi\text{-}Fe_5C_2$	28.2
	$\varepsilon'\text{-}Fe_{2.2}C$	12.4
	Fe_3O_4(A)	9.6
	Fe_3O_4(B)	6.5
	Fe^{3+} (spm)	24.2
	Fe^{2+} (spm)	16.2
$A\text{-}H_2+CO$	$\chi\text{-}Fe_5C_2$	22.5
	$\varepsilon'\text{-}Fe_{2.2}C$	3.0
	Fe_3O_4(A)	3.8
	Fe_3O_4(B)	32.0
	Fe^{3+} (spm)	22.0
	Fe^{2+} (spm)	16.7
A-0.67	$\chi\text{-}Fe_5C_2$	22.4
	$\varepsilon'\text{-}Fe_{2.2}C$	54.9
	Fe^{3+} (spm)	1.5
	Fe^{2+} (spm)	21.1
A-2.0	$\chi\text{-}Fe_5C_2$	17.6
	$\varepsilon'\text{-}Fe_{2.2}C$	21.7
	Fe_3O_4(A)	1.5
	Fe_3O_4(B)	9.8
	Fe^{3+} (spm)	20.0
	Fe^{2+} (spm)	29.4

3.3 Activity and stability

To systematically investigate the effect of reduction ambient on the FTS performances, a series of the FTS tests for more than 500 h under the baseline conditions (250 °C, 1.5 MPa and 2.0 $NL/g_{-cat}.h$) are carried out. After then, FTS reaction temperature for each test was increased. The FTS activity measured by carbon monoxide conversion is shown in Figure 2. As shown in Figure 2, there have been two areas of the higher catalyst activity ($A\text{-}H_2$ and $A\text{-}H_2+CO$) and the lower catalyst activity A-CO, $A\text{-}CO+H_2$, A-0.67, and A-2.0). Reduction by syngas with high H_2/CO ratios of precipitated iron-based catalysts can achieve the similar activity of the catalyst reduced by CO.

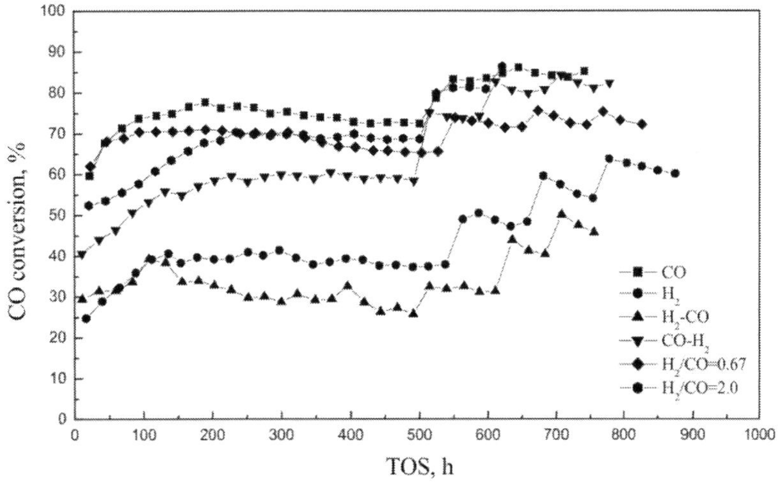

Fig 2 CO conversion and stability of catalysts after reduction at different ambients

Reaction conditions: 250-280 °C, 1.5 MPa, $H_2/CO=0.67$, 2.0 NL/gcat.h

There is a relationship between the catalyst activity and activation ambient. No iron carbide (A-H_2), or iron carbide content less than 30% (A-H_2 + CO) in the reduced catalyst, the FTS activity is very low. However, the iron carbide content in the reduced catalyst achieved greater than 30%, the FTS activity is relative high. Meanwhile, the FTS activity also depended on the phase equilibrium of the iron phases when the iron carbide content in the reduced catalyst increases to certain level.

Some researchers suggested that Fe_3O_4 was the active phase [13] while numerous studies supported that iron carbides were the active phases [14,15] for FTS. Dictor et al. [16] reported that the mixture of χ-, ε'-iron carbides and a small amount of α-iron was the active phase. The results of this study showed that the active phase for FTS was a mixture of carbides and corresponding amounts of superparamagnetic phase. Different preparation method and different pretreatment conditions of the iron-based catalyst led to different understanding results.

4. Conclusions

There is a relationship between the catalyst activity and activation ambient. The carbides formation may be controlled primarily activation ambient. Multi-phase coexistence becomes more apparent. The changes in the bulk compositions resulted in the variation in catalyst activity during FTS. The results of this study showed that the active phase for FTS was a mixture of carbides and corresponding amounts of superparamagnetic phase. The highest FTS activity was obtained when the iron catalyst was reduced using CO. However, the activation with H_2, or with H_2 followed by CO resulted in the low FTS activity.

References

1. Shroff M D, Kalakkad D S, Coulter K E, Kohler S D, Harrington M S, Jackson N B, Sault A G, Datye A K. Activation of precipitated iron Fischer-Tropsch synthesis catalysts. *J Cata.*, 1995, **156**(*2*): 185
2. Bukur D B, Lang X, Ding Y. Pretreatment effect studies with a precipitated iron Fischer-Tropsch catalyst in a slurry reactor.*Appl Catal A*, 1999, **186**(1-2): 255
3. O'Brien R J, Xu L, Spicer R L, Davis B H. Activation study of precipitated iron Fischer-Tropsch catalysts.*Energy & Fuels*, 1996, **10**(4): 921
4. Rao K R P M, Huggins F E, Huffman G P, Gormley R J, O'Brien R J, Davis B H. Mössbauer study of iron Fischer-Tropsch catalysts during activation and synthesis. *Energy & Fuels*, 1996, **10**(3): 546
5. Bukur D B, Okabe K, Rosynek M P, Li C, Wang D, Rao K R P M, Huffman G P. Activation studies with a precipitated iron catalyst for Fischer-Tropsch synthesis. I. Characterization studies. *J Catal*, 1995, **155**(2): 353
6. Bukur D B, Koranne M, Lang X, Rao K R P M, Huffman G P. Pretreatment effect studies with a precipitated iron Fischer-Tropsch catalyst, *Appl. Catal. A,* 1995, **126**: 85
7. Bukur D B, Nowicki L, Manne R K, Lang X. Activation studies with a precipitated iron catalyst for Fischer-Tropsch synthesis. II. Reaction studies. *J. Catal.* **1995**, 155: 366
8. Bukur D B, Lang X, Rossin J A, Zimmerman W H, Rosynek M P, Yeh E B, Li C. Activation studies with a promoted precipitated iron Fischer-Tropsch catalyst, *Ind. Eng. Chem. Res.*, 1989, 28: 1130
9. Bukur D B, Lang X. Highly active and stable iron Fischer-Tropsch catalyst for synthesis gas conversion to liquid fuels, *Ind. Eng. Chem. Res.*, 1999, 38: 3270
10. Luo M, Davis B H. Fischer-Tropsch synthesis: Activation of low-alpha potassium iron catalysts. *Fuel Process. Technol.* 2003, **1672**: 1
11. Xiang H W, Liu F X, Hao Q L, Zhang Ch H, Bai L, Zhang Zh X, Li Y W. CN1583259.2006
12. Bai L, Xiang H W, Li Y W, Han Y Z, Zhong B. *Fuel*, 2002, **81**(11-12): 1577
13 .F. Blanchard, J. P. Reymond, B. Pommier, S. J. Teichner, On the mechanism of the Fischer-Tropsch synthesis involving unreduced iron catalysts, J. Mole. Catal. A Chem. 17 (1982)171

14. N. Sirimanothan, H. H. Hamdeh, Y. Zhang, B. H. Davis, Fischer-Tropsch synthesis: changes in phase and activity during use, Catal. Lett. 82 (2002) 181

15.T. R. Motjope, H. T. Dlamini, G. R. Hearne, N. J. Coville, Application of in situ Mössbauer spectroscopy to investigate the effect of precipitating agents on precipitated iron Fischer-Tropsch catalysts, Catal. Today 71 (2002) 335

16. R. A. Dictor, A. T. Bell, Fischer-Tropsch synthesis over reduced and unreduced iron oxide catalysts, J. Catal. 97 (1986) 121

(281b) Effect of Ammonia In Syngas On the Fischer-Tropsch Behavior of a Precipitated Iron Catalyst

Weinping Ma, Gary Jacobs, Jungshik Kang, Dennis Sparks and Burtron H. Davis, Center for Applied Energy Research, University of Kentucky, Lexington, KY

Effectof ammonia in syngas on the FischeR&Deg;CTropsch behavior of a precipitated iron catalyst

WenpingMa, Gary Jacobs, Jungshik Kang, Dennis Sparks, and Burtron H. Davis*

Center for Applied Energy Research, University of Kentucky, 2540Research Park Drive, Lexington, Kentucky, 40511, USA

Abstract

The effect ofammonia in syngas in the concentration range of 100 ppbw °C 200 ppmw onFischeR&Deg;CTropsch reaction over 100 Fe/5.1 Si/2.0 Cu/3.0 K catalyst was studiedat 260 °C, 175 psig, $H_2/CO = 0.67$ and 10 Nl/g-cat/h in a 1-L slurry phase reactor. Theammonia added in syngas originates from concentrated ammonia and ammonianitrate soution. Three concentrations of ammonia in syngas feed, i.e. 100 ppbw, 20 ppmw and 200 ppmw, wereemployed to examine the sensentivty of the iron catalyst to NH_3. Theexaminion of the impact of each NH_3 concentration on the ironcatalyst FTS performance lasted for 144 or 168 h. The results indicatedthat the iron catalyst was quite resistant to NH_3 poisoningat the FTS conditions used. During 950 h of testing, no changes on the catalystselectivities to CH_4, C_{5+}, and CO_2, theselectivities to C_4 olefin and paraffin, and the selectivities to straight C_4olefin and internal C_4 olefin was observed. The catalystdisplayed a slight increase in CO conversion while adding 100 ppbw solution inthe first 96 h, which might be partly due to a slight increase in the water-gasshift activity for the catalyst. Afterwords, the iron catalyst displayed a stable CO conversion (~56.3%)regardless of the concentrations of NH_3 used.

Graphics

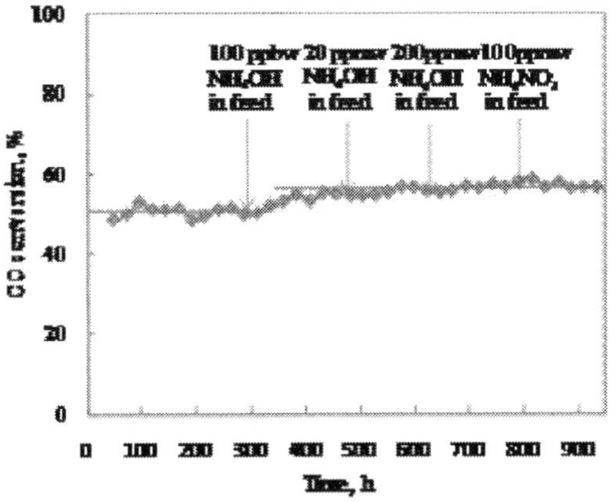

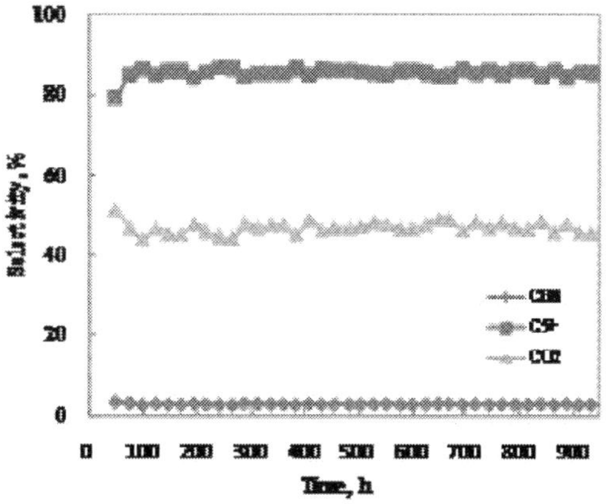

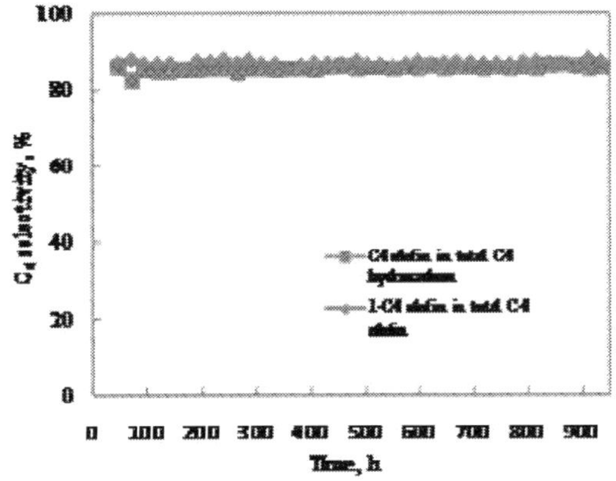

* Corresponding author: BurtronH. Davis, davis@caer.uky.edu,Tel: 859-257-0251.

See more of this Session: CO Hydrogenation I
See more of this Group/Topical: Catalysis and Reaction Engineering Division

(281c) Aldehydes and Ketones From Fischer Tropsch Synthesis

Ed Durham, Sihe Zhang, Rui Xu and Christopher B. Roberts, Department of Chemical Engineering, Auburn University, Auburn, AL

The use of a supercritical fluid as the reaction media for Fischer Tropsch Synthesis (SC-FTS) was first published by Kaoru Fujimoto's research group in 1989. In the years since, a number of groups have compared SC-FTS with traditional fixed bed Fischer Tropsch Synthesis (GP-FTS), finding that SC-FTS suppresses methane and CO_2 formation while enhancing the selectivity of olefins at higher carbon numbers. Recently published work by our group has demonstrated that, under SC-FTS conditions on an iron-based catalyst, 10% to 30% of the diesel-length reaction products are aldehydes, with an additional 5% as methyl-ketones.

We will present our continuing work on the role of process conditions and catalyst composition on aldehyde synthesis.

See more of this Session: CO Hydrogenation I
See more of this Group/Topical: Catalysis and Reaction Engineering Division

(281d) Influence of Support Composition, Base Promoter and Catalyst Passivation On Higher Alcohol Synthesis Over Molybdenum Carbide

Heng Shou and Robert J. Davis, Chemical Engineering, University of Virginia, Charlottesville, VA

Although world-scale catalytic processes have been developed to convert synthesis gas to diesel fuel or methanol, there is a growing need for its direct conversion to higher alcohols for use as fuel additives or chemical precursors. Promoted molybdenum carbide is a promising catalyst for higher alcohol synthesis because of its low cost, slow deactivation, high alcohol selectivity and no requirement for on-stream additives. The current work explores the influences of support composition, base promoter and catalyst passivation on the activity and selectivity of supported-Mo_2C in the synthesis of higher alcohols from syngas at 30 bar and 573 K. Although α-Al_2O_3-supported Mo_2C was more active than MgO-supported Mo_2C, the basic support MgO promoted the formation of alcohols. In addition, higher alcohol synthesis over both α-Al_2O_3-supported and MgO-supported Mo_2C catalysts was substantially improved by adding rubidium as a promoter. The surface of freshly prepared Mo_2C can be highly reactive with oxygen, so the influence of passivation on syngas conversion after carbide preparation was evaluated. Compared to a freshly prepared sample that was never exposed to air, a passivated Mo_2C/α-Al_2O_3 catalyst had lower activity and chain-growth probability. Interestingly, the influence of passivation on Rb-promoted Mo_2C/α-Al_2O_3 was rather minimal. Structural characterization of the catalysts evaluated by X-ray absorption spectroscopy was related to catalytic performance.

See more of this Session: CO Hydrogenation I
See more of this Group/Topical: Catalysis and Reaction Engineering Division

(281e) Mixed Oxide Supports Derived From Layered Metal Hydroxides to Reduce Methanol Selectivity In the Catalytic Synthesis of Higher Alcohols Over Potassium-Promoted Molybdenum Sulfide Catalysts

Michael R. Morrill[1], Nguyen Tein Thao[1], Christopher W. Jones[2], Pradeep K. Agrawal[2], David J. Barton[3], Daniela Ferrari[4], Robert J. Davis[5] and Heng Shou[5], (1)Chemical & Biomolecular Engineering, Georgia Institute of Technology, Atlanta, GA, (2)School of Chemical & Biomolecular Engineering, Georgia Institute of Technology, Atlanta, GA, (3)Dow Chemical Company, Midland, MI, (4)Dow Benelux B.V., Terneuzen, Netherlands, (5)Chemical Engineering, University of Virginia, Charlottesville, VA

Molybdenum sulfide, a well-known hydrodesulphurization catalyst, is being increasingly investigated as a catalyst for conversion of synthesis gas to higher alcohols. Here it is demonstrated that potassium promoted MoS_2 catalysts supported on basic, mixed metal oxides derived from decomposed, synthetic hydrotalcites convert syngas into higher alcohols with reduced methanol and increased n-propanol selectivity relative to bulk molybdenum sulfide or conventionally supported molybdenum sulfide catalysts (activated carbon, basic sepiolite clay, or alumina). This advantage may be attributed to the basic, mesoporous nature of the mixed metal oxide support. X-ray absorption spectroscopy shows that before sulfidation, the molybdenum oxide domains on the supported catalyst are highly dispersed. Treatment of this oxide catalyst at 723 K in 20% hydrogen sulfide in hydrogen results in the formation of molybdenum sulfide moieties that in turn provide a favorable environment for higher alcohol formation.

See more of this Session: CO Hydrogenation I
See more of this Group/Topical: Catalysis and Reaction Engineering Division

(281f) Synthesis of Higher Alcohols From Syngas Over a K Promoted Cu-Co-Zn Catalyst In Supercritical Hexanes

Rui Xu, Sihe Zhang, Ed Durham and Christopher B. Roberts, Department of Chemical Engineering, Auburn University, Auburn, AL

Synthesis of higher alcohols, isobutanol in particular, from synthesis gas (syngas) has been attracting interest due to the potential for producing clean and high octane value fuels. Higher alcohols could be either used directly as fuels, as fuels additives for octane enhancement, or as feed stocks in the chemical industry. The abundance of various carbonaceous sources for producing syngas, including coal, natural gas and biomass, makes this process appealing as an alternative to non-renewable petroleum resources. However, due to the highly exothermic nature of the reactions and the underlying kinetic, this procedure suffers from poor productivity and selectivity towards higher alcohols preventing a large scale and economically feasible commercialized process.

For the last two decades, there has been intensive research centered on shifting the synthesis away from methanol and towards higher alcohols, including; i. the use of modified catalysts, ii. utilizing dual-bed reactors, or iii. slurry phase reactors. In this study, a supercritical medium (i.e. hexanes) was employed to investigate the effect of supercritical solvent on the formation of higher alcohols.

In this work, 0.5 wt% K promoted $CuCo/ZnO/Al_2O_3$ was investigated as a catalyst for the synthesis of higher alcohols from syngas. The catalyst was prepared by coprecipitation methods and tested in a fixed bed reactor using a supercritical hexanes as the reaction medium. The results suggest that the presence of supercritical hexanes improved the heat transfer from the catalyst bed. It has also been observed that the formation of CH_4 was significantly reduced under supercritical hexanes reaction conditions. The results of the catalytic investigations both with and without the supercritical hexane medium demonstrated that supercritical hexane has a significant effect on the selectivity and the productivity towards higher alcohols.

See more of this Session: CO Hydrogenation I
See more of this Group/Topical: Catalysis and Reaction Engineering Division

(281g) Effect of Modifiers On Zirconia-Based Catalysts for Isosynthesis

Nicholas E. McGuire, Narasimharao Kondamudi, Lucia M. Petkovic and Daniel M. Ginosar, Interfacial Chemistry, Idaho National Laboratory, Idaho Falls, ID

This work reports the characterization and performance of modified zirconia catalysts for synthesis of isobutene from syngas (isosynthesis). Several commercial catalysts employing a range of modifier species are considered here. The catalysts examined include tungstated zirconia (W-ZrO_2), sulfated zirconia-alumina (S-ZrO_2-Al_2O_3) and lanthanum zirconia (La-ZrO_2). Addition of such modifiers to zirconia is known to have an effect on the physicochemical properties of the catalyst which in turn affects catalytic performance for isosynthesis. In this work, catalyst properties are characterized via a number of diagnostic techniques including N_2 physisorption, x-ray diffraction, and probe-molecule chemisorption. Isosynthesis activity is evaluated in a packed bed reactor configuration and the results are analyzed in terms of the physicochemical characteristics of the catalysts. Catalyst activity test results indicate that the species of the modifier has a strong effect on carbon monoxide conversion and selectivity to isobutene. The amount of CO conversion decreases in the order W-ZrO_2 > S-ZrO_2-Al_2O_3 > La-ZrO_2. Despite decreased CO conversions, the lanthanum-modified catalyst shows the highest selectivity to isobutene in the hydrocarbon products.

See more of this Session: CO Hydrogenation I
See more of this Group/Topical: Catalysis and Reaction Engineering Division

(284a) Role of Band-Filling, Oxidation State, and Strain On Perovskite Reactivity

John R. Kitchin, Department of Chemical Engineering, Carnegie Mellon University, Pittsburgh, PA

The principles of reactivity for oxide surfaces are not as well understood as metal surfaces. We will discuss the role of band filling, strain and oxidation state on the reactivity of perovskite (ABO3) surfaces, where A is either La or Sr, and B is a 3d transition metal. We will present trends in the adsorption energy of oxygen and the surface vacancy formation energy. The reactivity can be correlated with electronic structure features of the transition metal atom at the adsorption site. The database of reactivities and correlations may help identify oxide materials with desireable properties.

See more of this Session: Computational Catalysis V
See more of this Group/Topical: Catalysis and Reaction Engineering Division

(284b) Energetic Electron Induced Chemical Reactions On Metal Surfaces: First-Principles Based Electron Scattering Model

Hongliang Xin, Phillip Christopher and Suljo Linic, Chemical Engineering, University of Michigan, Ann Arbor, MI

Understanding of the interaction between energetic electrons and molecules adsorbed on metal surfaces is of great significance in fundamental surface science and heterogeneous catalysis. In this study, we have employed an electron scattering model with first-principles calculations to investigate the energetic electron induced activation of adsorbed diatomic molecules over metal surfaces. In this model, the electronic system of the adsorbate and substrate is described by a Newns-Anderson-type Hamiltonian[1] with the linear coupling of unoccupied resonant electronic states and vibrational modes of the adsorbate. When the resonant electronic state of the adsorbate becomes transiently occupied due to electron scattering, the adsorbate confined within the Born-Oppenheimer ground state potential well undergoes a Franck-Condon transition onto the excited state potential energy surface (PES) calculated using linear expansion ΔSCF-DFT method implemented in GPAW.[2] The reaction probability of the energetic electron induced surface reaction can be calculated based on the model using several parameters from density functional theory calculations. The critical feature of the model compared with previous ones[1] [3] is that the effect of substrate temperature has been explicitly included, and the energy transfer between excited adsorbate vibrational state and photon modes of metal substrates has been taken into account using Friction model.

We have applied the model to study energetic electron mediated oxygen activation on plasmonic silver (Ag) nanostructures. We found that the energetic electrons, generated due to surface plasmon resonance (SPR) excitation of Ag nanostructures, can induce a significant enhancement in the rate of O-O bond activation by transient occupation of the O2 $2\pi^*$ anti-bonding orbital. The electron scattering induces nuclear motion along the reaction coordinate of O-O bond, thereby facilitating the dissociation step. The driving force for nuclear motion is characterized by the coupling constant obtained from the excited state PES and the normal mode frequency of O-O bond. We have shown that the model can quantitatively capture the phenomenon of temperature and wavelength dependent rate enhancement of ethylene epoxidation reaction over Ag due to visible light illumination.[4]

[1] J. W. Gadzuk, *Phys. Rev. B* **1991**, *44*, 13466.

[2] J. Enkovaara, C. Rostgaard, J. J. Mortensen, et al., *J. Phys.: Condens. Matter* **2010**, *22*, 253202.

[3] T. Olsen, J. Gavnholt, J. Schiotz, *Phys. Rev. B* **2009**, *79*, 035403.

[4] P. Christopher, H. Xin, S. Linic, *Nat Chem* **2011**, *advance online publication*, DOI 10.1038/nchem.1032.

See more of this Session: Computational Catalysis V
See more of this Group/Topical: Catalysis and Reaction Engineering Division

(284c) Development and Application of a Hybrid QM/MM Method for the Computational Investigation of Reactions At Metal/Water Interfaces

Muhammad M. Faheem, Chemical Engineering, University of South Carolina, Columbia, SC and Andreas Heyden, Department of Chemical Engineering, University of South Carolina, Columbia, SC

Computational investigations of chemical reactions at metal/water interfaces pose a unique challenge of properly accounting for the effect of a liquid-phase environment on free energies of reactants, products and transition states while keeping the computational model simple enough to warrant its practical application. In this paper, we propose and validate a hybrid QM/MM method for computation of the effect of an aqueous-phase environment on the heat of reaction of a metal catalyzed reaction step. By treating only the reaction zone and its immediate environment quantum mechanically (QM) for higher accuracy and the remainder of the metal surface and the bulk of the liquid water at the classical molecular mechanical (MM) level of theory, computational speedup of multiple orders of magnitude can be realized. Using the C-C cleavage in double-dehydrogenated ethylene glycol on the Pt (111) surface as a model reaction, we first show that it is indeed possible to correctly predict the heat of reaction in water through a combination of density functional theory (DFT) calculations on periodic-slab and non-periodic-cluster models. Then, we demonstrate that the same effect can also be described within the framework of periodic embedded electrostatic cluster models (PEECM) coupled with molecular dynamics (MD) simulations of the MM subsystem. This scheme based on a combination of plane-wave (PW), PEECM and MD calculations quickly converges with QM cluster size to within a few kJ/mol of the results of complete ab-initio simulations.

See more of this Session: Computational Catalysis V
See more of this Group/Topical: Catalysis and Reaction Engineering Division

(284d) Connecting Molecular Processes to Macrokinetic Observables Using First-Principles Simulations

William F. Schneider, Chao Wu and David Schmidt, Chemical and Biomolecular Engineering, University of Notre Dame, Notre Dame, IN

Quantitatively connecting observed reaction rates, rate orders, and activation energies with microscopic mechanisms remains one of the grand challenges of heterogeneous catalysis science. Such connections are difficult to make even for simple catalytic reactions at surfaces, where interactions between adsorbates can create a myriad of local reaction environments that collectively determine the observed reactivity. These adsorbate interactions influence coverage, adsorbate arrangements, and even the rates of elementary surface reactions. A proper description of these coverage effects on reaction rates requires a careful, quantitative treatment of adsorbate-adsorbate interactions and their effects on microscopic reaction rates. Here we describe a "basis site" approach to incorporating this coverage dependence that permits ready and accurate calculations of macroscopic kinetic parameters. We use as example NO oxidation by O_2 on the Pt(111) surface. A minimal mechanism for this reaction includes the following steps:

$NO + O^* = NO_2 + *$ (1) equilibrium

$O_2 + 2^* \rightarrow 2\ O^*$ (2) rate controlling

The rate of the latter reaction is a strong function of the available surface sites. We develop a DFT-parameterized cluster expansion to describe coverage-dependent O adsorption. Grand canonical Monte Carlo simulations on Reaction (1) provide equilibrated surface configurations as a function of reaction conditions. The rate of Reaction (2) is described using a DFT-developed Bronsted-Evans-Polyani relationship between activation barrier and final state binding energy. Combining these, we obtain condition-dependent rates suitable for differentiation and comparison with experiment. The approach highlights the subtle links between microscopic mechanism and observed kinetics.

See more of this Session: Computational Catalysis V
See more of this Group/Topical: Catalysis and Reaction Engineering Division

(284e) Optimization-Based Methods for Catalysis: Developing Improved Approaches to Microkinetic Modeling and Their Application to Methanol Synthesis

Patricia B. Rubert-Nason, Chemical and Biological Engineering, University of Wisconsin - Madison, Madison, WI, Lars C. Grabow, Chemical and Biomolecular Engineering, University of Houston, Houston, TX, Lorenz Biegler, Chemical Engineering, Carnegie Mellon University, Pittsburgh, PA, Manos Mavrikakis, Chemical & Biological Engineering, University of Wisconsin-Madison, Madison, WI and Christos Maravelias, University of Wisconsin, Madison, WI

In the face ofrapidly declining fossil fuel reserves and the prospect of global climatechange, it is imperative that we develop renewable fuels and energytechnologies. Methanol is an importantcandidate as it is an industrially important chemical with significant promiseas a transportation fuel. Furthermore, it has the potential to be produced fromthe CO_2, which would make it a carbon neutral fuel. However, current industrial catalysts andprocesses are not optimized to function effectively in the absence of CO andour understanding of the reaction mechanism is incomplete.

To improve ourunderstanding of the reaction mechanism for methanol synthesis and otherreactions we use density functional theory (DFT) and microkinetic modeling. DFTuses the principles of quantum mechanics to predict the properties of materialsat the atomic and molecular levels.Specifically, an approximation of the Schrdinger equation is solved andcan be used to predict the binding sites and binding energies of atoms andmolecules on catalytically active surfaces and the energy barriers to variouspotential elementary reaction steps on the surface. It has provided us with unprecedented insightinto the reactions on the surface of catalytic materials. To use this information most effectively, we needto translate our new understanding of the elementary steps into predictions ofmacroscopic outcomes such as the rate of methanol production. Microkinetic models provide a bridge betweenthe elementary steps occurring at the molecular scale and production rates atthe macroscopic scale. DFT provides anexcellent starting point for parameter values for modeling. However, production rates can be verysensitive to the binding and activation energies (BE and Ea)of the surface species and elementary steps, respectively and the DFT valuesmay contain inaccuracies due to selection of the wrong surface to model, orsurface reconstruction under the reaction conditions. Therefore, parameter estimation is anecessary component of successful microkinetic modeling.

However,microkinetic models are computationally expensive to evaluate and highlynon-linear, rendering optimization and parameter estimation difficult. In the past, we had formulated microkineticmodels as a system of ordinary differential equations (ODEs). This is simple to set up and can be run inreadily available software such as Matlab. However, the ODE version of the model iscomputationally intensive to solve and problematic for parameter estimation asthe gradient is not readily available.Moreover, the optimization does not tend to result in large changes inthe parameter values, so very good initial guesses are required. This requires iterative adjustment of theparameter values by the user and considerable physical insight to obtain goodfits to the experimental data. Overall, parameter estimation with the ODE model is expensive in both computational andhuman time.

We havereformulated the microkinetic model as a system of non-linear equations(NLP). This technique requires carefulformulation and setting of appropriate limits on all variables to producephysically relevant surface coverages and gas phase partialpressures. However, done correctly, itproduces a microkinetic model that is much less computationally intensive tosolve and for which an explicit formula for the gradient is available, whichallows much more effective use of gradient-based optimization techniques. The resulting NLP model requires 3 orders ofmagnitude less computational time for optimization. Furthermore, the problem converges from alarger neighborhood of the local minimum.Together these features allow the search for a good fit to the data tobe substantially automated, removing the need for someone to iteratively adjustthe initial parameter values by hand and dramatically reducing the human timerequired to obtain a comparably good fit to the data.

See more of this Session: Computational Catalysis V
See more of this Group/Topical: Catalysis and Reaction Engineering Division

(284f) Computational Characterization of Pd/Ceria Catalysts Using DFT+U and ReaxFF

Thomas Patrick Senftle[1], Adam Mayernick[1], Adri van Duin[2] and Michael Janik[1], (1)Department of Chemical Engineering, Penn State University, State College, PA, (2)Department of Mechanical and Nuclear Engineering, Penn State University, State College, PA

Catalyst design plays an indispensible role in optimizing processes to meet ever rising demands for efficient and clean energy technologies. Numerous experimental studies have demonstrated the high activity of ceria supported palladium catalysts. Pd/ceria catalysts have the potential for application in catalytic combustion, as automotive three-way catalysts, and as solid-oxide fuel cell (SOFC) anodes. To fully realize this potential, we must determine the nature of active sites on the Pd/ceria interface. In this study, multi-scale computational techniques are used to characterize the structure, stability, and reactivity of the Pd/ceria surface. Density functional theory including on site Coulombic interaction (DFT+U) provides quantum mechanical data describing the energetics of Pd/ceria interactions and hydrocarbon activation over the Pd/ceria surface. The data obtained from DFT+U calculations are used to parameterize a reactive force-field (ReaxFF) capable of describing larger length and time scales via Monte Carlo (MC) and reactive molecular dynamic (RMD) simulations. RMD simulations allow this study to investigate aspects of the system that are computationally intractable for *ab initio* methods, such as the *dyanmic* restructuring of the catalyst surface during reaction. Initial DFT+U results demonstrate that a single Pd atom incorporated into the $CeO_2(111)$ surface significantly lowers the reaction barrier for methane C-H bond activation compared to pure $CeO_2(111)$. Furthermore, *ab initio* thermodynamic approaches indicate that the $Pd_{incorporated}$ structure is stable under high oxygen partial pressures and low temperatures (~5 atm, 250 K).

See more of this Session: Computational Catalysis V
See more of this Group/Topical: Catalysis and Reaction Engineering Division

(284g) Molecular Simulations of the Oxidation of Platinum-Based Alloy Catalysts for Fuel Cells

Rafael Callejas-Tovar and Perla B. Balbuena, Artie McFerrin Department of Chemical Engineering, Texas A&M University, College Station, TX

One of the most important challenges in low-temperature fuel cell technology is improving the catalytic efficiency at the electrode-catalyst where the oxygen reduction reaction takes place. Platinum is the most popular catalyst for this reaction but its high cost and scarcity hinder the commercial implementation of fuel cells in automobiles. Pt-based alloys are promising alternatives to substitute platinum while maintaining the efficiency and life-time of the pure catalyst. However, the acid medium and the oxidation of the surface impact the activity and durability of the alloy catalyst through changes in its local composition. This study applies molecular simulation techniques to characterize the dynamic evolution of the surface composition of platinum-based alloys under reaction conditions. We investigate the complex surface dynamics occurring in the near-surface layers influenced by oxygen and alloy atoms which yields to the dissolution of the catalyst.

Our simulation scheme of the surface oxidation of a supported nanoparticle combines classical molecular dynamics (MD) and density functional theory (DFT) simulations. The electrostatic interactions between the adsorbates and the two topmost layers as a function of the coverage of oxygen are modeled based on the results of the Bader charge analysis from DFT calculations. This approach is able to reproduce the main features of the oxidation phenomena observed experimentally. In addition, we implemented a coarse-grained 3-D Kinetic Monte Carlo method which utilizes thermodynamic and kinetic information obtained from DFT simulations to study the surface segregation of alloy atoms in presence of adsorbed oxygen, such phenomenon causes changes in the local composition of the catalyst affecting its performance under operating conditions.

See more of this Session: Computational Catalysis V
See more of this Group/Topical: Catalysis and Reaction Engineering Division

(297a) Promoting Selectivity In Heterogeneous Catalysis with Self-Assembled Monolayers

Carolyn Schoenbaum[1], Rudy Kahsar[1], Daniel K. Schwartz[1] and **J.Will Medlin**[2], (1)University of Colorado, Boulder, CO, (2)Department of Chemical and Biological Engineering, University of Colorado, Boulder, CO

The utilization of inorganic catalyst modifiers in supported metal catalysts is widespread, but the intentional use of organic modifiers to promote selectivity is rare. Modification of metal surfaces with organic ligands represents a unique opportunity, in that the structure and functionality of the ligands can be controllable varied to optimize performance. Our group has explored several techniques to promote selectivity in hydrogenation reactions, including the modification of supported metal catalysts with organic ligands. Alkanethiols can be deposited on metal surfaces to form organized self-assembled monolayers (SAMs) that alter the selectivity of several catalytic reactions. This attachment strategy was found to enhance the selectivity to the desired product during epoxybutene hydrogenation from 15% on uncoated catalysts to as high as 95% on SAM coated catalysts at equivalent reaction conditions. SAM-coated Pd catalysts also exhibited improved selectivity at comparable conversion for acetylene hydrogenation to ethylene. In this contribution, experiments aimed at identifying the mechanisms behind observed selectivity and activity trends will be described.

See more of this Session: Fundamentals of Supported Catalysis I
See more of this Group/Topical: Catalysis and Reaction Engineering Division

(297b) Novel Oxidation Pathway for Low Temperature Conversion of Methane to Methanol: a Spectroscopic Investigation

Mayfair C. Kung, Sean S.-Y. Lin and Harold H. Kung, Chemical and Biological Engineering, Northwestern University, Evanston, IL

Novel oxidation pathway for low temperature conversion of methane to methanol: a spectroscopic investigation

Cobalt ZSM-5 had been reported to convert methane to methanol at temperatures 200C and below. However, the products had to be extracted as they were purported to adsorb very strongly on the catalyst.[1] Here we report the use of DRIFT spectroscopy to understand the oxidation of CH_4 over Co-ZSM-5. The catalyst was 2 wt.% Co impregnated onto Na-ZSM-5 (Si/Al=25). It was observed that at 200C the cobalt oxo species can react with methane to form surface adsorbed species even in the absence of gas phase oxygen. However, the intensities of the IR bands of the adsorbed products increased with the inclusion of oxygen in the reaction feed. These oxidation products included methoxide, formate and formaldehyde oligomers. It was also observed that after CH_4 oxidation reaction, the prominent Co-formate band at 1578 cm^{-1} continued to increase with time on stream in a He stream. It was proposed that the formaldehyde oligomers underwent a self oxidation-reduction reaction (Cannizzaro reaction) in the presence of water to form adsorbed methoxide and formate. The proposed reaction pathway was verified when a pulse of water after the CH_4 reaction markedly accelerated the formate formation rate. Another observation consistent with this proposed pathway is that IR bands associated with adsorbed methoxide and formate were detected upon heating adsorbed formaldehyde to 200C. The existence of an another viable pathway in addition to direct oxidation of CH_4 is very interesting as it opens up many possibilities both in catalyst formulation and reaction conditions to tune the selectivities of CH_4 oxidation reaction.

1. Benzis et. al., Catal. Lett. 136 (2010) 52

See more of this Session: Fundamentals of Supported Catalysis I
See more of this Group/Topical: Catalysis and Reaction Engineering Division

(297c) Dispersed Ta(V): A Site-Isolated Epoxidation Catalyst and Unexpected Reactivity In Selective Hydrogenation / Hydrodenitrogenation

Justin M. Notestein[1], Natalia Morlanes-Sanchez[1], Mark Bachrach[2] and Tobin J. Marks[2], (1)Chemical and Biological Engineering, Northwestern University, Evanston, IL, (2)Department of Chemistry, Northwestern University, Evanston, IL

Highly dispersed Ta^V species are known, albeitrelatively infrequently studied, epoxidation catalysts with H_2O_2.We recently reported on replacing traditional $TaCl_5$ precursors with calixareneTa(V)complexes on SiO_2, which resulted in epoxidation catalysts of higherH_2O_2 utilization and selectivity for the directepoxidation of cyclohexene (without formation of cyclohexenol) as compared to otherhighly dispersed Ta-SiO_2 catalysts and as compared to the benchmarkTi-SiO_2 catalysts. Calixarene-derived catalysts had initialcyclohexene direct epoxidation turnover rates of 3.9 0.1 10^{-2} s^{-1}unaffected by surface density, demonstrating single-site character.

In parallel, soluble Ta^{III} organometalliccomplexes have been studied as HDN models, showing direct C-N insertion ofpyridine without ring hydrogenation, and highly dispersed, supported Ta areknown to be active in a number of unusual reactions including alkanemetathesis, and direct N_2 activation. As such, we investigated ourhighly dispersed Ta catalysts for HDN of quinoline. Catalysts were used aftercalcination at 550C for 6 h in air to and activation of the calcined catalyst underH_2 at up to 350C, 4 h. HDN reactions were run up to 12h at 40 barand 275-350C, followed by GC/FID with a TR-WaxMS column. Some catalysts wereactivated with small amounts of Pd deposited by incipient wetness.

Table1 compares dispersed Ta vs Pd catalysts. Pd is used here as a typicalhydrogenation catalyst and, as expected, shows much higher HDN conversion (tohydrocarbons), forms essentially only propylcyclohexane (PCH). Although quinolineHDN was slower over Ta catalysts, it proceeded via a less H_2-consumingroute with measurable formation of propylaniline (PA), propylbenzene (PB), andpropylcyclohexene (PCHE). Combining Pd and Ta increases both HDN activity andyields of propylbenzene. The direct cleavage of aniline to benzene was observedover Pd-Ta/Al_2O_3 catalysts by in situ diffuse reflectanceUV-visible spectroscopy-mass spectrometry and gave evidence for a key Ta imidointermediate.

catalyst	Yields

	THQ or DHQ	PA	PCHE	PCH	PB
Pd/Al_2O_3	32	0	0	66	0
Pd/SiO_2	24	0	1	65	0
Ta/Al_2O_3	42	2	4	0	0
Ta/SiO_2	47	1	0	0	1
Ta/SiO_2[a]	56	2	1	0	0
Ta/SiO_2[b]	67	1	0	0	0
0.5%Pd-Ta/SiO_2	30	3	7	47	5
0.5%Pd-Ta/Al_2O_3	24	5	21	29	19
Ta/Al_2O_3+Pd/ Al_2O_3	84	0	5	3	0

See more of this Session: Fundamentals of Supported Catalysis I
See more of this Group/Topical: Catalysis and Reaction Engineering Division

Polymer Anchored Cu(II) and Mn(II) Complex Catalysts for the Oxidation of Styrene

Sweta, Shishir Sinha, Prakash Biswas, Shri Chand*

Department of Chemical Engineering, Indian Institute of Technology Roorkee, India

* Corresponding author. Email: schanfch@iitr.ernet.in; Telephone: +91-1332-285383

Abstract

Liquid phase styrene oxidation was investigated with polymer anchored Cu(II) and Mn(II) complex catalysts prepared by schiff base ligand. The catalysts were characterized by various techniques such as elemental analysis, atomic absorption spectroscopy (AAS), fourier transform infrared spectroscopy (FT-IR), field emission scanning electron microscopy (FE-SEM), energy dispersive X-ray analyses (EDAX), thermo-gravimetric analysis (TGA) and UV-vis spectrophotometer. The effect of various reaction parameters such as temperature, styrene to TBHP mole ratio, catalyst amount and oxygen pressure were studied. The main reaction product was styrene oxide with 78.36% selectivity at a temperature of 70°C with tert-butyl hydroperoxide (TBHP) as oxidant. Benzaldehyde was found as the main reaction product with oxygen. Cu(II) based catalyst showed better catalytic activity than Mn(II) based catalyst for TBHP as well as O_2.

Keywords: Batch reactor, Oxidation, Polymer anchored complex, Styrene

Introduction

Styrene oxidation is of interest from both academic and commercial point of view for the synthesis of styrene oxide which is an important intermediate in the production of large variety of fine chemicals such as phenyl ethanol, phenylaldehyde and perfume etc. Traditionally, oxidation of styrene is carried out by using stoichiometric amounts of organic peracids such as peracetic acid and m-chlorobenzoic acid with permanganate as the oxidizing agent.[1] However, peracids are very expensive, hazardous and create environmental pollution. Therefore, much attention has been paid for the development of new oxidation method which employs safer oxidants and produces little waste. The use of various oxidants such as hydrogen peroxide, tert-butyl hydroperoxide (TBHP) and oxygen are attractive options both on environmental and economic ground.[2-5] Different transition metal complex catalysts have been reported for the catalytic oxidation of styrene. Epoxidation of styrene demonstrated using homogeneous catalysts,[6] titanium silicate (TS-1)[7] and zeolite-encapsulated metal complexes.[8] Recently, the use of polymer-supported catalysts have received considerable attention because they are insoluble, recyclable and generate less solid waste compared to metal supported catalyst, therefore making the overall process more environmental friendly.[9,10]

In the present study, styrene was investigated over polymer anchored Cu(II) and Mn(II) complex catalysts prepared by schiff base tridentate ligand. Tert-butyl hydroperoxide (TBHP) was used as oxidant. The effect of O_2 as oxidant was studied in the pressure range of 1-5 bar.

Experimental

Reaction at atmospheric pressure

The liquid phase styrene oxidation reaction at atmospheric pressure with TBHP as oxidant was carried out in a 50 ml glass flask equipped with a reflux water condenser and placed in a thermostatted oil bath. In a typical experiment, 25 mg of catalyst and 10 mmol of styrene were added to the reaction mixture followed by the addition of TBHP (10 mmol) in 10 ml acetonitrile. The conversion and product selectivity data were taken at the reaction temperature of 70°C after a reaction time of 6h.

Reaction in batch reactor

The catalytic oxidation of styrene using O_2 as the oxidant was carried out in a 300 ml batch stainless steel Parr reactor (4843, USA). The reactor is equipped with temperature and pressure controller, agitator, thermocouple, gas inlet valve and sampling valve. The reactor was charged with 50 mmol of styrene, 50 mg of catalyst in 40 ml of acetonitrile and the reaction was carried out at 60-85 °C at 5 bar pressure. After the reaction temperature and pressure reached at the desired value, the agitator was started and this time was recorded as the starting time of the reaction. The conversion and selectivity data were calculated after a reaction time of 6 h with a stirring speed of 600 rpm.

The products were analyzed with a gas chromatograph (Thermo Scientific, USA) equipped with a flame ionization detector. A capillary column of length 30 m, ID 0.25 mm and film thickness 0.25 μm was used for product separation. The products were also confirmed through gas chromatography-mass spectrometry (GC-MS) (Perkin-Elmer Clarus 500, USA). Iso-octane was used as the internal standard.

Results and discussion

The elemental analysis, AAS and electronic spectra confirmed the formation of Cu(II) and Mn(II) complexes with ligand. FT-IR results showed the coordination bond formation between the polymer ligand and metal ion. The ligand bound and complex bound complexes showed slightly roughening of the top surface as indicated by FE-SEM analysis. TGA revealed the stability of the catalysts upto 250 °C.

The oxidation of styrene with TBHP in the presence of polymer anchored Cu(II) complex was conducted and effect of styrene to TBHP ratio, temperature and catalyst amount was studied. Styrene oxide and benzaldehyde were the main reaction products. At the styrene to TBHP mole ratio of 1:3, the maximum conversion (87.3%) of styrene was obtained at 70 °C with a catalyst amount of 25 mg. At this condition, the product selectivity to styrene oxide and benzaldehyde were 78.36% and 15.71%, respectively. The catalytic activity of Mn(II) complex catalyst was determined at same reaction conditions and the results obtained are shown in Table 1.

Table1

Selectivity to styrene oxide, benzaldehyde and other products obtained over Cu (II) and Mn(II) complex catalysts

Catalysts	Conversion (%)	Selectivity (%)**		
		Styrene oxide	Benzaldehyde	Others
Cu(II) complex	87.3	78.3	15.7	6.0
Mn(II) complex	75.4	68.9	21.30	9.8

**Reaction conditions: styrene: 10 mmol, TBHP: 30mmol, temperature: 70 °C, catalyst: 25 mg, solvent: 10 ml

The styrene oxidation was also conducted in batch reactor at 1-5 bar oxygen pressure in the temperature range of 60 to 85 °C. The maximum conversion (45.6%) of styrene was obtained at 80°C at 5 bar oxygen pressure. The selectivity to styrene oxide and benzaldehyde were 15.5% and 67.7%, respectively. Small amount of benzoic acid was also detected (8.8% selectivity). The activity of Mn(II) complex catalyst was evaluated at same reaction conditions and the results obtained are shown in Table 2.

Table 2

Selectivity to styrene oxide, benzaldehyde, benzoic acis and other products obtained over Cu (II) and Mn(II) complex catalysts

Catalysts	Conversion (%)	Selectivity (%)**			
		Styrene oxide	Benzaldehyde	Benzoic acid	Others
Cu(II) complex	45.7	15.5	67.7	8.8	8.0
Mn(II) complex	36.5	12.4	74.5	6.7	6.4

**Reaction conditions: styrene: 50 mmol, temperature: 80 ^{o}C, pressure: 5 bar, catalyst: 50 mg, solvent: 40 ml

Conclusions

Catalytic oxidation of styrene using TBHP as oxidant showed higher conversion (87.3%) and higher selectivity to styrene oxide as compared to oxygen as oxidant. The effect of oxygen pressure over styrene oxidation reaction showed the maximum styrene conversion of 45.6% at 80 ^{o}C and 5 bar pressure. The major oxidation products were benzaldehyde, styrene oxide and benzoic acid. The Cu(II) based complex showed better catalytic activity and high selectivity (78.36%) to styrene oxide as compared to Mn(II) based complex.

Literature cited

1. Swern D. Organic Peroxides, Wiley- Interscience, New York, 1971.

2. Hulea V, E. Dumitriu E. Styrene oxidation with H_2O_2 over Ti-containing molecular sieves with MFI, BEA and MCM-41 topologies. *Appl.Catal. A: Gen.* 2004; 277:99-106.

3. Rinaldi R, Schuchardt U. Factors responsible for the activity of alumina surfaces in the catalytic epoxidation of *cis*-cyclooctene with aqueous H_2O_2 . *J. Catal.* 2004; 227:109-116.

4. Lu X, Yuan Y. Copper-containing hexagonal mesoporous silicas for styrene epoxidation using tert-butylhydroperoxide. *Appl. Catal. A: Gen.* 2009;365:180-186.

5. Rao SN, Munshi KN, Rao NN.Catalytic oxidation of styrene using *cis*-MoO_2 (L) (solv) [L=salicylidene salicyloyl hydrazine] and its zeolite composite as catalysts in the presence of molecular oxygen. *J. Mol. Catal. A*.2000;156:205-211.

6. Srinivasan K, Michaud P, Kochi JK. Epoxidation of olefins with cationic (salen)Mn^{III} Complexes. The modulation of catalytic activity by substituents. *J. Am. Chem. Soc.*1986;108: 2309–232.

7. Kumar SB, Mirajkar SP, Pais GCG, Kumar P, Kumar R. Epoxidation of styrene over a titanium silicate molecular sieve TS-1 using dilute H_2O_2 as oxidizing agent. *J. Catal.* 1995; 156:163–166

8. Joseph T, Srinivas D, Gopinath CS, Halligudi SB. Spectroscopic and catalytic activity studies of VO (saloph) complexes encapsulated in zeolite-Y and Al-MCM-41 molecular sieves.*Catal. Lett.* 2002;83: 209-214.

9. Chang Y, Lv Y, Lu F, Zha F, Lei Z. Efficient allylic oxidation of cyclohexene with oxygen catalyzed by chloromethylated polystyrene supported tridentate Schiff-base complexes. *J. Mol. Catal. A*.2010;320:56-61.

10. Sherrington DC, Simpson S. Polymer-supported Mo alkene epoxidation catalysts. *Reactive Polymers*.1993;19:13-25.

(297e) A Combined Experimental and Computational Study of the Selective Hydrogenation of Acrolein On Supported Silver Alloy Catalysts

Carolina Gomez[1], Haojuan Wei[1], Neng Guo[2], Tianpin Wu[2], Christopher Marshall[3], Jeffrey T. Miller[4] and Randall J. Meyer[5], (1)Chemical Engineering, University of Illinois at Chicago, Chicago, IL, (2) Argonne National Laboratory, Argonne, IL, (3)Chemical Sciences and Engineering Division, Argonne National Laboratory, Argonne, IL, (4)Chemical Science and Engineering, Argonne National Laboratory, Argonne, IL, (5)Department of Chemical Engineering, University of Illinois - Chicago, Chicago, IL

α, β–unsaturated alcohols are important intermediates in the production of fine chemicals, flavors, perfumes and pharmaceuticals. Selective hydrogenation of α,β–unsaturated aldehydes is a promising way to produce these alcohols. However, selective hydrogenation of the aldehydes to form the alcohols requires tight kinetic control since the hydrogenation of C=C bond is thermodynamically favored. Acrolein (the smallest α,β–unsaturated aldehyde) was chosen as a model compound to study the selectivity between double bond hydrogenation and hydrogenation of the aldehyde functionality.

We have undertaken a detailed kinetic study of acrolein hydrogenation over silica supported Ag and Ag alloy (AgIn, AgAu) catalysts. We have found that particle size is a key controlling factor in determining the selectivity to allyl alcohol. 8 nm Ag particles possessed a 30% selectivity to allyl alcohol at atmospheric pressure whereas 2.5 nm Ag particles showed no alcohol production. Density functional theory studies of acrolein hydrogenation have been carried out over Ag and AgIn alloy surfaces are used to help explain the results.

See more of this Session: Fundamentals of Supported Catalysis I
See more of this Group/Topical: Catalysis and Reaction Engineering Division

(297f) Studies of NH_3 Formation Over Pt/BaO/ Al_2O_3 LNT Monolith In the Presence of Excess Water

Prasanna R. Dasari, Michael Harold and Rachel L. Muncrief, Department of Chemical and Biomolecular Engineering, University of Houston, Houston, TX

Ground level ozone produced by the NO_x emitted from lean burn and diesel engines is the main driving force for the active research in lean NO_x reduction. Several NO_x reduction technologies, such as selective catalytic reduction (SCR) using urea, lean NO_x traps (LNT) etc., are under development. Considerable research is also being done in developing technologies that combine both LNT and SCR after-treatment systems. The LNT/SCR technology utilizes NH_3 produced during the regeneration of the LNT for storage and reaction with NO_x in the downstream SCR. It is therefore, important to understand the NH_3 formation chemistry and kinetics on the LNT catalyst to develop an optimal design and operating strategy for the LNT-SCR system.

The production of NH_3 is known to occur by reaction between NO and H_2 under the rich conditions on precious group metals. However, in the presence of NO, CO and excess H_2O, two additional routes are possible. The first is by the water-gas shift reaction of CO and H_2O to give H_2, which then reacts with NO to give NH_3. A second pathway is through reaction between NO and CO forming support bound isocyanate (-NCO) species, which are readily hydrolyzed to form NH_3. The contribution of both pathways is of significance because each proceeds in the presence of H_2O which is obviously in high concentration in the exhaust (5-10%) but without H_2 in the feed, which is typically low in diesel exhaust. While Cant and coworkers [1-2] have conducted systematic studies in which, they studied the reaction of NO, CO, H_2 and H_2O on different precious metal catalysts and also determined the effect of supports on the production of NH_3. However, the major pathway leading to NH_3 is still unclear and the kinetics of NH_3 formation has not been properly determined.

To this end, comprehensive steady-state experiments of CO-NO, CO-NO-H_2, CO-NO-H_2-H_2O, CO-H_2O, NO-H_2 reaction systems on Pt/BaO/Al_2O_3 monolithic catalyst have been carried out to evaluate the NO_x conversion and product distribution features as a function of the feed composition and temperature. CO reduces NO to N_2, N_2O and NH_3 in the presence of excess water on precious metal catalysts with the product selectivity shifting from N_2O to N_2 and then finally to NH_3 with an increase in the feed concentration of CO, consistent with the stoichiometry. However, an increase in the feed concentration of CO leads to a significant drop in the overall conversion of both NO and CO, which was also the case with the reduction of NO by CO in the absence of water. Differential kinetics measurements showed that the NO+CO reaction is nearly first order with respect to NO and negative first order with respect to CO. This confirms that the strong adsorption of CO on the Pt serves to block sites for adsorption of the other reactants. In addition, support-bound isocyanate (-NCO) species are also expected to form at high CO concentrations in the presence of NO [3]. These isocyanate species readily hydrolyze in the presence of excess water to give NH_3.

The water gas shift (WGS) reaction was observed to be ~ -0.23 order in CO, indicating that CO has a stronger inhibiting effect on the NO+CO reaction than the WGS reaction. This suggests that during the reduction of NO by CO in the presence of a large excess of water, CO is consumed more rapidly by H_2O than NO, thereby producing H_2, so that NO reduction is by reaction with H_2 than with CO. Furthermore, H_2 reduces NO more selectively to NH_3 than N_2O or N_2 irrespective of the stoichiometric requirements, when the catalyst surface is largely covered by CO. Thus, while the isocyanate route cannot be ruled out, the WGS pathway appears to be the main route for the production of NH_3. The results of additional experiments including the use of a spatially-resolved mass spectrometer (SpaciMS) will be presented to establish phenomenological mechanism and to support the kinetics measurements.

References

1. Dumpelmann R, Cant N.W, *Catalysis and Automotive Pollution Control III*, 1995, **96**: p. 123-135.
2. Cant N. W., Chambers D. C., *Journal of Catalysis*, 2001, **204**: p. 11-22.
3. Granger, P, *Journal of Catalysis*, 1998, **177**: p. 147-151.

See more of this Session: Fundamentals of Supported Catalysis I
See more of this Group/Topical: Catalysis and Reaction Engineering Division

(297g) Development of Palladium Oxide Catalysts Supported On Nanoparticle Oxides for C-H Activation and C-C Coupling Reactions

Helena E. Hagelin-Weaver, Luke M. Neal and Justin J. Dodson, Chemical Engineering, University of Florida, Gainesville, FL

A number of palladium oxide catalysts supported on various nanoparticle and conventional oxides were prepared and tested for activity in the C-H activation and C-C coupling of 4-methylpyridine. This is an environmentally friendly, atom economic reaction since waste formation is minimized at the source. However, it is a slow reaction with limited yields. Our research has therefore focused on improving these catalysts to obtain higher product yields.

PdO has been shown to be important for the preparation of an active catalyst in this reaction, possibly due to facilitated H abstraction. In search of catalysts that can outperform the commonly used commercial Pd/C catalysts, a number of different oxides were used as supports for PdO. PdO/n-Al_2O_3 (nanoparticle alumina support) and PdO/p-TiO_2 (porous titania support) were shown to be the most active catalysts in the oxidative coupling of 4-methylpyridine, significantly outperforming the commercial Pd/C catalyst. However, other nanoparticle supports, such as nanoparticle zirconia (n-ZrO_2) and zinc oxide (n-ZnO), can also give reasonable yields despite rather low support surface areas (35-60 m^2/g). Therefore, active catalysts can be prepared by depositing PdO onto very high surface area supports ($\geq$500 m^2/g), or onto supports that will result in strong palladium-support interactions.

Catalyst characterizations reveal that there is no simple correlation between the Pd surface areas (measured after reduction of surface PdO) and the catalytic activities for these reactions. The results indicate that this is a complex reaction system and a high PdO surface area is important but not sufficient to yield a highly active catalyst. The complexity of the reaction system is even more evident when adding a third oxide to the system (PdO/MeO_x/support, where MeO_x is the added metal oxide). The best catalyst reported to date for this reaction is PdO/ZrO_2/n-Al_2O_3 prepared using the coprecipitation method.

See more of this Session: Fundamentals of Supported Catalysis I
See more of this Group/Topical: Catalysis and Reaction Engineering Division

(302a) Bringing Broader Understanding of Step Growth Polymerization

W. Harmon Ray, Chemical and Biological Engineering, University of Wisconsin - Madison, Madison, WI

This paper will discuss how Santosh Gupta's work brought an understanding of Step Growth Polymerization to a wide audience.

See more of this Session: In Honor of Professor Santosh K Gupta I
See more of this Group/Topical: Catalysis and Reaction Engineering Division

(302b) Polyelectrolyte Brushes In Multi-Valent Ionic Media

Matthew Tirrell, Institute for Molecular Engineering, University of Chicago, Chicago, IL

Multi-valent ionic media produce attractions between polyelectrolyte chains, in contrast to mono-valent media in which interactions are always repulsive. Numerous examples from our lab of this interesting and important realm of polymer science will be presented and explained drawing on data from surface force measurement, electrochemical measurement, x-ray scattering, rheology and cryo-transmission electron microscopy. Such electro-statically generated attraction offer new possibilities for self-assembled materials in polymer science.

See more of this Session: In Honor of Professor Santosh K Gupta I
See more of this Group/Topical: Catalysis and Reaction Engineering Division

(302c) Adventures with and Insights Into the Bayesian Design of Experiments In Complex Polymerizations

Alexander Penlidis, Department of Chemical Engineering, Institute for Polymer Research, Waterloo, ON, Canada and Afsaneh Nabifar, Department of Chemical Engineeerign, Institute for Polymer Research, Waterloo, ON, Canada

For many polymerizations, experimental data may be available either from industrial or exploratory laboratory work. In addition, mathematical models usually do exist, albeit often with unreliable and/or highly correlated parameters, and sometimes even unverified mechanistic bases. In complex polymerization systems, these problems are even more prominent. Hence, ideas from the (statistical) design of experiments applied as early as possible can be very beneficial for the clarification of polymerization kinetics. Classical experimental design methods (for example, (fractional) factorial designs) have been used extensively and are useful in optimizing a wide variety of systems. However, these designs usually ignore the prior knowledge available and are often limited when the system digresses from linear behavior. Using more efficient experimental designs, which can accommodate both prior information and nonlinearities, could (hopefully) lead to optimal performance in fewer trials.

Bayesian design is a powerful and largely unstudied (in the polymerization area) experimental design methodology, which can accommodate practical limitations encountered in classical (fractional) factorial designs. In addition, Bayesian design allows the use of a nonlinear (fully mechanistic) model along with experimental information (hence, it is essentially an optimal model-based design of experiments). The approach can shed light on the most uncertain parts of our process understanding, identify the least reliable (less well known) parameters (e.g., uncertain values of kinetic rate constants), and further guide sensitivity analysis studies focusing on key uncertain parameters in one's model.

The Bayesian design approach is an experimental design technique which has many advantages over standard experimental designs. The approach incorporates prior knowledge about the process (or product) into the design to suggest a set of future experiments in an optimal, sequential and iterative fashion. Since for many complex polymerizations prior information is available, either in the form of experimental data or mathematical models, use of a Bayesian design methodology could be highly beneficial. Hence, exploiting this technique could hopefully lead to optimal performance in fewer trials, thus saving time and money. Advantages of the Bayesian design approach will be illustrated via case studies drawn from different complex polymerization systems. However, since this technique is perfectly general, and exploiting the duality between many systems in science and engineering, it can potentially be applied to other chemical engineering processes.

Why Bayesian design? As mentioned above, standard experimental design methods (for example, (fractional) factorial designs) have been employed extensively and are useful in optimizing a wide variety of systems. However, these designs usually suffer from several limitations and cannot handle certain situations, as listed in Table 1. In addition, these approaches do not take direct advantage of the considerable prior knowledge that is available about the reaction system to design experiments. As prior information is already available within existing data, it is logical that it should be used in order to contribute to the optimality of the designed experiments, and hence to improved models and performance of the process in question. Using more efficient experimental designs, like the family of Bayesian designs, can accommodate these restrictions and could lead to optimal performance in fewer trials.

Table 1. Typical Limitations in Standard Designs

Available resources do not match number of trials that can be designed
Handling of impractical treatment combinations
Handling of situations with missing observations
Factor levels change in the middle of experimentation
Factors with several (or combination of) levels
Dropping/adding factors
Not a sequential and iterative approach
Nonlinearities
Incorporation of prior knowledge

The Bayesian experimental design is based on Bayes' theorem which is well established, especially among statisticians. Recent applications more relevant to chemical and process engineering include heat transfer in packed beds [1], emulsion terpolymerization kinetics [2], drug and cell transport kinetics [3], particle size distribution studies in suspension polymerization [4], pharmaceutical kinetics [5], and analyses of catalytic systems [6]. The common characteristic of all these Bayesian applications is that they are concerned with, admittedly very important, (multi)parameter estimation questions (of kinetic rate constants), but not with the issue of the design of experiments (with the exception of [2], [4], and [6]). In other words, looking at the last 15 years or so, whereby many advances have taken place in polymerization, Bayesian design of experiments has not been exploited extensively nor frequently in complex polymerization systems, which could benefit tremendously from its important traits.

What could one gain from the Bayesian design? In our presentation, principles and capabilities/benefits of the Bayesian design approach will be illustrated and its superiority to the currently practised (standard) design of experiments will be presented via case studies drawn from representative complex polymerization processes. These include examples from nitroxide-mediated controlled radical polymerization (both bimolecular and unimolecular), emulsion copolymerization of acrylonitrile/ butadiene (nitrile or NBR) and styrene/butadiene (SBR) rubber, and crosslinking copolymerizations (of styrene and divinyl benzene under controlled radical polymerization conditions). All these case studies address important, yet practical, issues in not only the study of polymerization kinetics but also, in general, in process engineering and improvement. A preview of these important process issues and, accordingly, topics that can be handled efficiently by the Bayesian design approach are listed in Table 2.

Table 2. Overview of Gains in Bayesian Design Approach

Flexible wrt number of trials that can be designed
Changing factor level/range in the middle of experimentation
Accommodating extra trial(s) mid-way through experimentation
Sequential nature (n-trials versus sequences of fewer trials)
Process constraints (and impractical treatment combinations)
Situations with missing observation(s)
Increase of information content (flexibility and cost-effectiveness)
Incorporation of prior knowledge (informative versus non-informative priors)
Flexible wrt source and quality of prior knowledge (screening experiments versus models and/or combinations)
Factors with several (or combination of) levels
Dropping/adding factors
Single versus multi-response scenarios

In addition, the analysis is amenable to a series of statistical diagnostic tests that one can carry out in parallel. These diagnostic tests serve to quantify the relative importance of the parameters (intimately related to the significance of the estimated factor effects) and their interactions, as well as the quality of prior knowledge (in other words, the adequacy of the model or the expert's opinions used to generate the prior information, as the case might be).

The case studies will clearly show that what is novel in the Bayesian approach is the simplicity and the natural way with which it follows the logic of sequential model building paradigm, taking full advantage of the researcher's expertise and information (knowledge about the process or product) prior to the design, and invoking enhanced information content measures (the Fisher Information matrix is maximized, which corresponds to minimizing the variances and reducing the 95% joint confidence regions, hence improving the precision of the parameter estimates).

1. Duran, M. A.; White, B. S. *Chem. Eng. Sci.* **1995**, *50*, 495.
2. Dube, M. A.; Penlidis, A.; Reilly, P. M. *J. Polym. Sci. Part A-Polym. Chem.* **1996**, *34*, 811.
3. Murphy, E. F.; Gilmour, S. G.; Crabbe, M. J. C. *Febs Letters* **2004**, *556*, 193.
4. Vivaldo-Lima, E.; Penlidis, A.; Wood, P. E.; Hamielec, A. E. *J. Appl. Polym. Sci.* **2006**, *102*, 5577.
5. Hermanto, M. W.; Kee, N. C.; Tan, R. B. H.; Chiu, M. S.; Braatz, R. D. *AIChE J.* **2008**, *54*, 3248.
6. Hsu, S. H.; Stamatis, S. D.; Caruthers, J. M.; Delgass, W. N.; Venkatasubramanian, V.; Blau, G. E.; Lasinski, M.; Orcun, S. *Ind. Eng. Chem. Res.* **2009**, *48*, 4768.

(302d) Novel Pathways for Biomass-to-Liquid Fuel Production

Rakesh Agrawal, Dharik S. Mallapragada, Fabio H. Ribeiro and W. Nicholas Delgass, School of Chemical Engineering, Purdue University, West Lafayette, IN

In a limited fossil fuel world, liquid fuels from biomass offer a renewable solution for meeting the transportation sector's fuel needs. In this context, for a given liquid fuel demand, the role of increasing the liquid fuel output per unit of biomass input is critical to reducing the land area requirement for harvesting the feedstock. A grand challenge for chemical engineers is to develop novel processes for converting biomass to high energy density liquid fuels while requiring minimum additional energy input. Moreover, when the additional energy input is derived from solar, increased process energy efficiency would also lead to reduced corresponding land area requirements. Through a systems approach, this presentation will identify some novel biomass-to-liquid fuel pathways based on the efficient use of different forms of supplemental solar energy. In addition, experimental findings pertaining to one particular pathway will also be presented.

See more of this Session: In Honor of Professor Santosh K Gupta I
See more of this Group/Topical: Catalysis and Reaction Engineering Division

(302e) Application of Multi-Objective Optimization In Polymer Reaction Engineering

Ajay K. RAY, Chemical and Biochemical Engineering, University of Western Ontario, London, ON, Canada

Multi-objective optimization involves the simultaneous optimization of more than one objective function. This is quite commonly encountered in Chemical Engineering, particularly in polymer reactors. A considerable amount of research has been reported in this area over the years. The general background of this area will be presented at the beginning, followed by a description of how the results can be described in terms of Pareto sets. We will then present the several methods available for generating these optimal solutions. Applications of optimization in Polymer Reaction Engineering, wherein multiple objectives are encountered, as well as special adaptations of the basic algorithms required to solve these problems, will then be discussed. Most importantly, major contributions of Dr. Santosh K. Gupta in this field will be presented.

See more of this Session: In Honor of Professor Santosh K Gupta I
See more of this Group/Topical: Catalysis and Reaction Engineering Division

(302f) Self-Organized Nano-Lens Arrays by Intensified Dewetting of Polymer Thin-Film

Ankur Verma, Department of Chemical Engineering and DST Unit on Nanosciences, Indian Institute of Technology Kanpur, Kanpur, India and **Ashutosh Sharma**, Chemical Engineering, Indian Institute of Technology Kanpur, Kanpur, India

Dynamics of instabilities and structure formation in thin polymer films havebeen widely studied because of their increasing applications in the areas ofcoatings (thermal and electrical insulation, optical, protective, biologicaletc.), plastic electronics, organic light emitting diodes (OLEDs), membranes,lubricants and adhesives. In many such applications instabilities are undesiredand have to be suppressed for the product to remain functional. However, thestudy of instabilities and their kinetics in ultrathin films (<100 nm) canprovide some important insights into the inter-surface potentials that cannotbe directly measured otherwise. It also can be used to generate useful meso-and nano-scale patterns over large areas (~cm^2). Therefore, thespontaneous dewetting of ultrathin polymer films on a non-wetting substrate hasbeen a subject of great interest owing to its scientific and technologicalimportance.

Dewetting on a flat homogeneous surface starts with the nucleation ofrandomly placed holes at a certain mean separation (λ), which grow andcoalesce with time and result in randomly placed droplets of the polymer.Average diameter as well as the mean separation of dewetted structures are afunction of initial thickness (h) and the interfacial tension of thepolymer film. However, randomness of the dewetted structure limits theusefulness of this method in some applications as a potential soft patterningtechnique. Various strategies have been explored to impose a long range-orderin the dewetted structures. One strategy for the alignment of the dewettedstructure is to combine it with other top-down lithographic approaches such ascontrolled dewetting on topographically or chemically patterned substrates.Dewetting of ultrathin polymers films on physico-chemically patternedsubstrates has also been extensively studied both theoretically andexperimentally. However, despite its promising scientific and technologicaladvantages, the feature-size generated by the self-organized dewetting has twomajor limitations on the pattern resolution and its aspect ratio. The firstlimitation arises from the weak van der Waals destabilizing force and highsurface tension (γ), both of which impose a severe limit on minimumfeature size, which is related to the wavelength, (λ) of the long-waveinstability in spinodal dewetting:
$\lambda = [\ -8\pi^2\gamma/(\partial\varphi/\partial h)]^{1/2}$

Where h is the film thickness and f is the destabilizingintermolecular potential (~ h^{-3} for van der Waalsinteraction). In a model system of polystyrene (PS) thin-film on siliconsubstrate, this limits the spot size to >1m even in case of films as thinas 10 nm.

Other limitation is the very small contact angle (< 10°) and thusthe aspect ratio of dewetted structures in air is rather small.

We have recently developed a novel method which overcomes the limitations onthe feature size and aspect ratio to a great extent. The PS thin (< 100 nm)films were destabilized by immersing into an optimal mixture of solvent(methyl-ethyl ketone and acetone) and water at room temperature. Selectivediffusion of solvent molecules into the polymer matrix brings down its glasstransition temperature (T_g) below room temperature and thuscauses the instability and dewetting of the film. However, water being themajority phase in the bounding media inhibits the dissolution of PS. Roomtemperature dewetting in the liquid media provides the advantage of cleanerenvironment, greater control over the shape of structure and better defectcontrol as compared to thermal annealing or dewetting in air. Further itprevents the sensitive materials from harsh conditions (UV, laser, electronbeam, heating high vacuum etc.) of patterning. Interestingly, dewetting underthe water-solvent mix reduces the droplet diameter by more than one order ofmagnitude compared to dewetting in air. We have also investigated theunderlying mechanisms of intensification of surface instability underwater-solvent mix including the length scale of instability beforefragmentation of the film into droplets and compared the conditions for theformation of ordered patterns on topographically structured substrates in airand under the water-organic mix. Toward these ends, we first systematicallyinvestigate the length scale of instability, droplet diameters andinter-droplet spacing as a function of film thickness in order to clearlydifferentiate these from the well-known aspects of spontaneous dewetting inair. Finally, we explore the conditions for the formation of ordered patternson topographically structured 1-D and 2-D substrates where dewetting occurs inhighly confined but structured spaces. We demonstrate the use this techniquefor fabrication of ordered arrays of polymeric nano-lenses of tunable curvatureby controlled dewetting under the water-organic mix.

We have achieved more than an order of magnitude reduction in the featuresize of the dewetted structures and thus fabricate structures as small as ~60nm. The mean separation of these structures is also reduced close to ~200 nm.Further, the mechanisms of miniaturization of the interfacial instability wasexplored and it was found that the thickness dependence of the characteristiclength scales of dewetted structures is weaker ($\lambda \sim h^{1.51}$)as compared to the instability caused by van der Waals attraction ($\lambda \sim h^2$).Moreover, a far greater control over the shape and aspect ratio of the dewettedstructures is demonstrated as the contact angles can be tuned in the range of40-140°. The ability to tune the curvature of the dewetted structures can beexploited in the fabrication of polymeric nano-lenses. Further, the controlleddewetting under the water-solvent mix on topographically patterned substrateswas also employed for fabrication of sub-micron ordered structures. It was alsoshown that under the same conditions, dewetting in air produces no dewetting,incomplete dewetting or is incapable of producing ordered structures owing to alarge length scale of instability in air. A further miniaturization of lengthscales because of the 2-D confinement is also demonstrated.

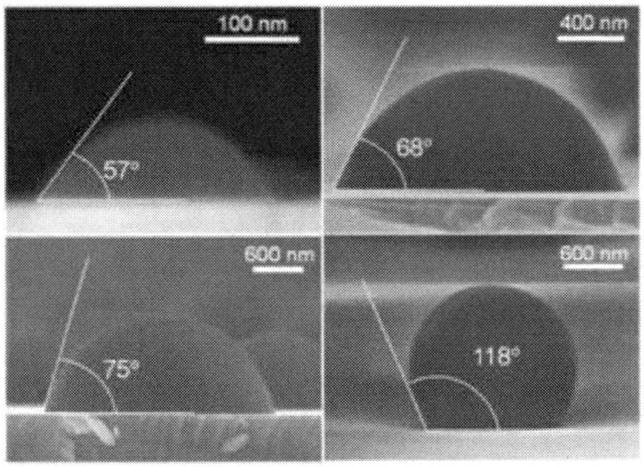

Figure 1. PS droplets of different contact anglesand size visualized in the scanning electron microscope in transverse view.

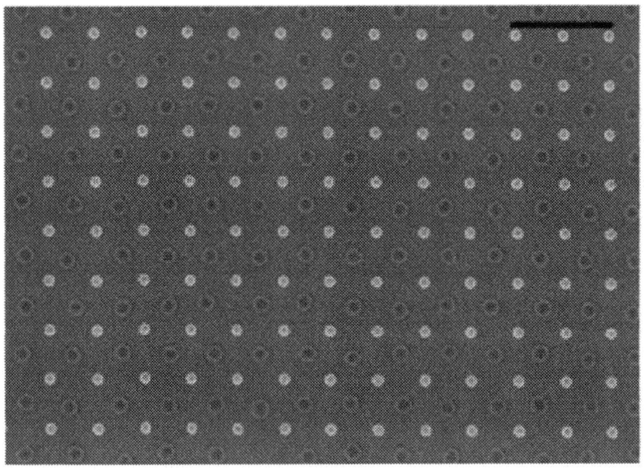

Figure 2. 2-D array of polymeric nano-lenses obtained from the dewettingof PS thin film on the topographically patterned substrate. (Scale bar: 2 mm)

Field-induced self-organized patterning in ultra-thin polymer films producesstructures that are limited to few microns to tens of microns in size becauseof the high energy penalty for the surface deformations on small scales andbecause of weak destabilizing van der Waals forces. We have resolved this longstanding problem of miniaturization of length scales in thin filmself-organization to sub-100 nm scale by reducing the interfacial tension andintensifying the field using a water-solvent mixture for dewetting. This provesto be a simple, powerful, flexible and inexpensive technique for the roomtemperature fabrication of sub-micron polymeric structures such as nano-lensesand their ordered arrays.

In conclusion, dewetting under water-solvent mixture takes the physicalself-assembly of polymer thin-films to its limits by producing sub-micronstructures of various degrees of ordering and tunable shapes. Among otherthings, such patterns have the potential to be used as tunable polymericmicro-lenses and lens-arrays, modifiers for the optical properties, and asdelivery and positioning tools for a variety of functional materials that canbe mixed with the polymer.

References

1. Ankur Verma andAshutosh Sharma, "Enhanced Self-organized Dewetting of Ultrathin Polymer Filmsunder Water-organic Solutions: Fabrication of Submicrometer Spherical LensArrays", *Advanced Materials* **2010**, *22*, 5306-5309.

2. Ankur Vermaand Ashutosh Sharma, "Submicrometer Pattern Fabrication by Intensification ofInstability in Ultrathin Polymer Films under a Water-Solvent Mix", *Macromolecules***2011**, *communicated.*

See more of this Session: In Honor of Professor Santosh K Gupta I
See more of this Group/Topical: Catalysis and Reaction Engineering Division

(312a) Optimization Studies Based On CFD Modeling of a Multiple Tube Reactor for Solar-Thermal Processes

Janna Martinek and Alan W. Weimer, Department of Chemical and Biological Engineering, University of Colorado, Boulder, CO

A numerical model of an indirectly-irradiated solar receiver with five tubes enclosed within a reflective cavity has been developed and experimentally validated for the gasification of carbon black. The radiation model consists of a combination of Monte Carlo and finite volume methods and is coupled with a three-dimensional CFD model treating heat, mass, momentum and species transport for the two-phase fluid/particle mixture. This model is used to investigate the impact of factors describing receiver configuration and operating conditions on overall receiver performance. Thirteen individual factors are considered with four factors describing operating conditions (gas flow rate, carbon feed rate, solar power input, and steam to carbon ratio) and nine two-level factors describing receiver geometry. The geometric factors are defined relative to a base design in which all tubes are arranged in a semicircle around the back cavity wall and allow for variations in cavity size, and number, radius, and arrangement of tubes with both staggered and non-staggered arrangements considered along with factors allowing for offset of all or part of the tube semicircle toward the window. Designs are evaluated in the framework of a 2^{13-8} fractional factorial design on the basis of receiver solar to chemical efficiency and reaction conversion for both absorbing and reflective cavity walls.

See more of this Session: Modeling and Analysis of Chemical Reactors
See more of this Group/Topical: Catalysis and Reaction Engineering Division

(312b) CFD Based Simulation of Heat Transfer In a Fixed Bed Reactor Tube

Mohsen Behnam[1], Anthony G. Dixon[1], Michiel Nijemeisland[2] and E. Hugh Stitt[2], (1)Chemical Engineering Department, Worcester Polytechnic Institute, WORCESTER, MA, (2)Johnson Matthey Catalysts, Billingham, United Kingdom

Low *N* (tube to particle diameter ratio) fixed bed reactors are commonly used in the chemical industry for extremely endothermic and exothermic reactions due to the better heat transfer. In the past, lots of researches have been carried out to simulate the heat transfer in fixed bed reactors by using k_r/k_f and h_w. Most models for k_r/k_f and h_w are based on a direct correlation with *Re* and *N*. Although most models seem to predict k_r/k_f very well, but there is a lot of problem still exists to predict h_w [1-2]. The main objective of this study was to develop a new heat transfer model with emphasis of industrial application for conduction and convection heat transfer in fixed bed tubes without using k_r/k_f and h_w.

Fluid flow in a fixed bed of tube to particle ratio 3.96, 5.96 and 7.99 were obtained for packed bed column by solving the 3D Navier–Stokes and using commercial software FLUENT 6.3.26. The heat transfer was modeled by using pseudo-homogenous energy equation. COMSOL 3.5 was used to simulate heat transfer based on pseudo-homogenous equation in 2D. Effective radial thermal conductivity was calculated from Zehner-Schlünder model as function of bed porosity position, fluid thermal conductivity and solid thermal conductivity [3]. The convection heat transfer was calculated in the 2D flow fluid from the CFD run.

Results were obtained for Reynolds numbers in the range 240–1900. The accuracy of the new model has been validated by analytical solution. The temperature calculated by the pseudo-homogenous energy equation showed reasonable quantitative agreement with values predicted by CFD model.

[1] S. A. Logtenberg, A. G. Dixon, "Computational fluid dynamics studies of fixed bed heat transfer", Chemical Engineering and Processing 37 (1998) 7–21.

[2] A.G. Dixon, "An improved equation for the overall heat transfer coefficient in packed beds", Chem. Eng. Process. 35 (1996) 323.

[3] P. Zehner, E.-U. Schlunder, "Die Effective Warmeleitfahigkeit Durchstromter ugelschuttungen bei Massigen und Hohen Temperaturen", Chem. Ing. Tech. 45 (1973) 272.

See more of this Session: Modeling and Analysis of Chemical Reactors
See more of this Group/Topical: Catalysis and Reaction Engineering Division

(312c) Effect of Spatial Segregation On Commensalistic Cultures - Series Reactors

Satish J. Parulekar, Chemical and Biological Engineering, Illinois Institute of Technology, Chicago, IL

A comprehensive analysis of static and dynamic behavior of a mixed culture in two bioreactors in series is presented considering anaerobic digestion involving acidogens (X) and methanogens (Y) as the example bioprocess. A single continuous culture may operate at up to seven steady states, including up to four coexistence steady states, with only one coexistence steady state being locally stable. The bioreactors in the two-reactor system are identical in terms of composition of and dilution rate with respect to fresh feed. A two-reactor system may admit up to forty nine steady states, which are comprised of up to forty coexistence steady states, at least at very low interaction rates (R). When physically realizable, up to five coexistence steady states are locally stable. The static and dynamic analysis of the two-reactor system is facilitated by appropriate grouping of large number of steady states arising for very low R into nine clusters and grouping of the clusters into three cluster groups. Numerical illustrations reveal the rich steady state structure of the bioprocess for bioreactors in series. The two bioreactors can operate at up to five locally stable coexistence steady states over certain ranges of R. The two-reactor system is operationally more flexible and more robust vis-a-vis single reactor as concerns maintenance of mixed culture. Emergence of additional steady states at intermediate R reveals that series bioreactors are an example of a complex system.

See more of this Session: Modeling and Analysis of Chemical Reactors
See more of this Group/Topical: Catalysis and Reaction Engineering Division

(312d) Modeling the Start-up Phase of Fischer Tropsch Synthesis In a Fixed Bed Reactor: Strategizing the Optimum Operation

Syed Ali Gardezi, Chemical Engineering, University of South Florida, Tampa, FL, Babu Joseph, Chemical and Biomedical Engineering, Clean Energy Research Center, University of South Florida, Tampa, FL and John T. Wolan, University of South Florida, Tampa, FL

To date, very little work has been done to model Fischer Tropsch Synthesis startup. In this study, finite difference approximation is used to model the intra-pellet reaction and heat transport. Finite element analysis was used to model the catalyst inter-pellet heat and mass transport. Vapor-liquid equilibrium was assumed to be attained at each pore, fugacity coefficients were used to develop a quantitative correlation. The goal was to estimate the time required to fill the catalyst pores. This time represents the unsteady transient phase which requires fine temperature control to avoid reaction run away in an actual working packed bed reactor.

In the resulting kinetic model, coupled material and energy balance are used to assess the optimum reactor startup procedures. Another goal is to identify the impact of different types of inert packing used to control heat transfer and conversion in the fixed bed reactor with a view towards avoiding the hot spots and steep temperature gradients. The combined effect of heat transfer and kinetics present a challenging problem for both the bed design and for optimal startup procedure.

See more of this Session: Modeling and Analysis of Chemical Reactors
See more of this Group/Topical: Catalysis and Reaction Engineering Division

(312e) Modeling and Optimization Studies of Combined LNT-SCR Catalyst Systems

Arun S. Kota, Dan Luss and Balakotaiah Vemuri, Department of Chemical and Biomolecular Engineering, University of Houston, Houston, TX

Abstract:

Aone-dimensional two-phase model with global rate expressions for variousreactions is used to study the performance of combined LNT-SCR catalystsystems. The LNT system modeled is that of Pt/BaO/ Al_2O_3catalyst with two different BaO storage sites,designated as fast and slow sites. LNT model includes reactions that describeNO oxidation, storage of NO_2 in the form of nitrates during leanstorage cycle and reduction to N_2 in the subsequent rich cycle withH_2 as a reductant. The SCR systemconsidered is that of Cu-ZSM5 and the model includes NO oxidation, ammoniastorage and oxidation, selective catalytic reduction of NO, NO_2 andformation of N_2O. Model predicts the spatio-temporalconcentration profiles of reactants and products inside the channel, which givesinsight into the performance of each LNT and SCR catalyst during operation. Ithas been found that series arrangement of alternate LNT and SCR catalyst, bydividing them into equal halves, substantially improves NO_xconversion. For fixed total catalyst length as the number of divisionsincrease, NO_x conversion increases andreaches a limit which is theoretically equivalent to that of mixed catalyst. Itis also observed that for fixed length of LNT catalyst better $DeNO_x$ performance is obtained by increasing thelength of SCR catalyst in the alternate arrangement. Spatio-temporalprofiles indicate complete storage of NH_3 on SCR catalyst which canbe efficiently used to reduce NO_x slippingout of LNT. Various catalyst arrangements, the effect of different catalystloading and rich cycle time are investigated in the present work.

See more of this Session: Modeling and Analysis of Chemical Reactors
See more of this Group/Topical: Catalysis and Reaction Engineering Division

(312f) A Parameterized Iron-Zeolite SCR Model Calibrated to Reactor Data

Seth DeLand[1], Gordon Parker[1], **Jason Keith**[2] and John Johnson[1], (1)Department of Mechanical Engineering - Engineering Mechanics, Michigan Technological University, Houghton, MI, (2) Department of Chemical Engineering, Michigan Technological University, Houghton, MI

A numerical model was developed to capture the behavior of the SCR catalyst for varying inlet conditions. The model form was developed based on analysis of reactor data, and simulated over a temperature range reflective of typical exhaust gas temperatures. The experimental protocol was designed to capture the effects of different NH3/NOx ratios and different NOx compositions, as well as ammonia storage and oxidation. Feedback from the modeling effort was used to modify portions of the protocol in order to further investigate key areas. Two different types of storage sites were used to represent the adsorption and desorption of ammonia. An adsorption equation that includes ammonia inhibition was used to capture the slow approach of the catalyst to steady state storage. Two different states were used to represent the gas distribution in the catalyst: a bulk-gas concentration and a surface concentration. Species move between the bulk-gas and surface states via mass transport. Initially, the model used only the Standard (NH3-NO), Fast (NH3-NO-NO2), and Slow (NH3-NO2) reactions to capture the reduction of NOx by stored ammonia. Based on experimental data, reactions were added to extend the model beyond the typical SCR reactions. The single-channel, isothermal model was discretized axially to form a system of ordinary differential equations that could be numerically integrated. An optimization routine was used to determine the reaction rate parameters that provided the best fit to the reactor data, and the resulting model is presented. We discuss changes that were made to the original model form, which allowed the model to capture the dynamics seen in the experimental data.

See more of this Session: Modeling and Analysis of Chemical Reactors
See more of this Group/Topical: Catalysis and Reaction Engineering Division

(318a) Accurate Treatment of Electrostatics During Molecular Simulations In Nanoporous Crystals without Assigning Point Charges to Framework Atoms: Application In Material Selection for Flue Gas Separation

Taku Watanabe and David Sholl, School of Chemical and Biomolecular Engineering, Georgia Institute of Technology, Atlanta, GA

Molecular simulations are now routinely performed to study gas adsorption and diffusion in metal-organic frameworks (MOFs). Traditionally, the electrostatic potential energy surface (EPES) in the pore of a MOF is reproduced by assigning the electrostatic charges to the framework atoms. Despite some recent advance in the charge assignment methods, point charge models are inherently an imperfect representation of the material's true EPES. We have shown how the EPES determined directly from DFT calculations without assigning point charges can be used to accurately describe molecular adsorption in MOFs. Examples will be given including adsorption isotherms and molecular diffusivities of CO_2 and N_2 in selected MOFs including ZIF-8 and ZIF-90. We will also discuss recent advances in using information from ab initio Molecular Dynamics to find useful information regarding the internal flexibility of MOFs without requiring development of classical force fields. We will show how models based on these methods are being combined with experimental efforts by our collaborators to develop robust, scalable MOF-based membranes for carbon capture.

See more of this Session: Molecular and Multiscale Modeling for Renewable Energy Systems
See more of this Group/Topical: Computational Molecular Science and Engineering Forum

(318b) Molecular Dynamics Studies of Diffusion and Solvation of Rubidium Bromide Solutions In Nanoconfinement

Katherine A. Phillips, Jeremy C. Palmer and Keith E. Gubbins, Chemical and Biomolecular Engineering, North Carolina State University, Raleigh, NC

The need for improved energy storage devices has increased due to demands for renewable energy sources. Supercapacitors, or electrochemical double-layer capacitors (EDLCs), have become a viable option as a leading energy storage device due to their high charge efficiency, charge/discharge times, and long cycle life. While EDLCs have many attractive properties, they cannot store large amounts of energy. Carbide derived carbons (CDCs) have shown an improvement of energy storage in EDLC applications due to a dramatic increase in capacitance in pores ranging from 0.7 to 1.1nm[1]. The increase is thought to occur due to the stripping or deformation of the hydration shell as the ions enter the nanopores[1]. In the present work we analyze the hydration of ions in simple nanopore geometries to gain an improved understanding of the solvation structures inside the EDLC.

Using molecular dynamics (MD) simulations, we studied the impact of confinement on the solvation of rubidium and bromide ions in aqueous solutions confined in slit-shaped pores ranging from 0.65 to 1.6 nm in size, which encompasses the range typically found in CDCs and other nanoporous EDLC materials [2]. In the MD simulations, the rubidium and bromide ions were modeled as charged Lennard-Jones particles as prescribed by Koneshan, *et al.*[3]. We investigated 0.1 M, 0.5 M and 1.0 M solutions of these ions using the SPC/E[4] model for water. We found the average number of water molecules hydrating each of the ions and the structure of the solvation shell, which is characterized by the ion-water radial distribution function. The effect on confinement on the dynamics was also investigated by calculating the self-diffusion coefficients of the ions. Finally, the solvation of ions in an amorphous carbon structure was studied in order to determine the propensity of ions to reside in pores of specific sizes. Our results indicate that although materials with smaller pores have significantly increased capacitances, the diffusion, of ions is hindered by the confinement, which is potentially detrimental to enhancing the power density of EDLCs.

[1] Chmiola, J.; Yushin, G.; Gogotsi, Y.; Portet, C.; Simon, P.; Taberna, P. L. *Science*, **2006**, *313*, 1760.

[2] Simon, P.; Gogotsi, Y.; Nature Materials, **2008**, *7*, 845.

[3] Koneshan,S.; Rasaiah, J. C.; Lynden-Bell, R. M.; Lee, S. H. *J. Phys. Chem. B*, **1998**, *102*, 4193.

[4] Berendsen, H. J. C.; Grigera, J. R.; Straatsma, T. P. *J. Phys. Chem.*, **1987**, *91*, 6269.

(318c) An Electronic Structure Based Understanding of Amine-Carbon Dioxide Interactions for CO2 Capture

Anita S. Lee and John R. Kitchin, Department of Chemical Engineering, Carnegie Mellon University, Pittsburgh, PA

Gas scrubbing using amine solvents is one of the most promising first generation CO_2 capture technologies, due to the chemical and process knowledge gained through their long term use in natural gas. However amine scrubbing has high parasitic energy costs due to equipment heating, amine regeneration, and captured CO_2 compression costs. Thus, minimizing this energy requirement is an obvious place to start when developing enhanced amines for CO_2 capture. The energy associated with amine regeneration is a function of the CO_2 capacity of the amine as well as the reaction energy. Early studies of amine-CO_2 interaction explained capacity and bond strength through two acid/base reactions to form carbamates (2:1 amine:CO_2 interaction) and bicarbonates(1:1 amine:CO_2 interaction), with bicarbonates favored for sterically hindered secondary amines and tertiary amines. The picture is not that simple, however, as significant variations in the equilibrium CO_2 capacity among structurally similar amines and amines of similar pH have been reported. Understanding the origin of these variations in the experimental equilibrium capacity is the motivation behind this work. At equilibrium, the capture capacity can be characterized by the Gibbs free energy of reaction. Using Density Functional Theory (DFT) we studied the reactions of CO_2 with structurally similar amines and amines of similar pH to form carbamic acid, carbamate and bicarbonate ions, determined the reaction energies for each amine with CO_2 relate these energies to electronic structure properties of the amine. We show that the reaction energy associated with amine-bicarbonate formation correlates well with the amine electronegativity. Additionally, we identify a means to manipulate amine electronegativity through functionalizing the amine with different functional groups. These results provide us with a fundamental understanding of one of the reactions critical to CO_2 capture and a means to control the reaction energy through modifying the amine itself.

See more of this Session: Molecular and Multiscale Modeling for Renewable Energy Systems
See more of this Group/Topical: Computational Molecular Science and Engineering Forum

(318d) Engineering Protein-Carbohydrate Binding Affinity Via Glycosylation: A General Strategy to Improve Cellulase Performance

Courtney B. Taylor[1], Clare McCabe[1], Lintao Bu[2], Michael E. Himmel[3], Michael F. Crowley[4] and Gregg T. Beckham[5], (1)Chemical and Biomolecular Engineering, Vanderbilt University, Nashville, TN, (2)National Bioenergy Center, National Renewable Energy Lab, Golden, CO, (3)Biosciences Center, National Renewable Energy Lab, Golden, CO, (4)Biosciences Center, National Renewable Energy Laboratory, Golden, CO, (5)National Bioenergy Center, National Renewable Energy Laboratory, Golden, CO

Carbohydrate-binding modules (CBMs) display specificity to mono- and polysaccharides for docking of enzyme complexes to carbohydrates. CBMs are found ubiquitously as components of cellulase and hemicellulase enzymes that degrade cell wall polysaccharides in nature. For this purpose, CBMs have evolved to recognize specific cell wall polysaccharides via hydrophobic interactions and hydrogen bonding. Additionally, CBM binding affinity to carbohydrates has been shown to be proportional to enzyme activity, and thus increasing binding affinity is an important component of engineering improved cellulases. To that end, here we apply molecular simulation and free energy calculations to examine the binding affinity of a well-characterized Family 1 CBM to cellulose. To validate our approach, we quantify the impact of aromatic-carbohydrate interactions on the CBM binding affinity. Our predictions are in excellent quantitative agreement with previous experimental results showing that single-point mutations can alter the CBM binding affinity by approximately 2-fold. Following this validation, we demonstrate that a single, native *O*-linked mannose increases the CBM binding affinity to cellulose by 40-fold, which has not been considered explicitly in previous work. Furthermore, we show that the addition of a single artificial glycan on the CBM can have the same order of magnitude impact on binding affinity. These results have significant implications for protein engineering, in that modifying glycosylation patterns, for example by using heterologous host expression, manipulation of culture conditions, or introduction of artificial glycosylation sites, can drastically alter protein binding affinity to carbohydrate substrates, and may thus be a general strategy to improve cellulase performance.

See more of this Session: Molecular and Multiscale Modeling for Renewable Energy Systems
See more of this Group/Topical: Computational Molecular Science and Engineering Forum

(318e) Multi-Scale Models for Poly (1, 3-cyclohexadiene) (PCHD) Polymer

Qifei Wang[1], David J. Keffer[1], Suxiang Deng[2] and Jimmy W. Mays[2], (1)Chemical and Biomolecular Engineering, University of Tennessee Knoxville, Knoxville, TN, (2)Chemistry, University of Tennessee Knoxville, Knoxville, TN

Abstract

Atomistic and coarse-grained (CG) models of Poly (1, 3-cyclohexadiene) (PCHD) were developed and implemented in Molecular Dynamics (MD) simulations of PCHD chains with different architectures. In the atomistic model, two PCHD chains are cross linked by a sulfur-sulfur bond. Sulfonic acid groups are evenly distributed along the chain. The architecture is specifically aimed for application as a proton exchange membrane used in fuel cells. An atomistic force field for this architecture was tested and applied in the atomistic MD simulation of PCHD for the first time. The atomistic simulations generate density, chain radius of gyration and chain end-to-end distance distribution. To further study the structural properties of longer chain systems, a CG model was proposed. The bonded structural probability distribution functions (PDFs) and nonbonded pair correlation function (PCF) of the CG beads were obtained from the atomistic simulation results. The bonded CG potentials are obtained by simple inversion of the corresponding PDFs. The CG nonbonded potential is parameterized to the PCF using the Iterative Boltzmann inversion (IBI) method. The CGMD simulations of PCHD chains using potentials from above method satisfactorily reproduce the structural properties from atomistic MD simulation of the same system. A comparison of the CG potentials and structural distribution functions from the PCHD melt and a hydrated PCHD membrane is presented.

See more of this Session: Molecular and Multiscale Modeling for Renewable Energy Systems
See more of this Group/Topical: Computational Molecular Science and Engineering Forum

(318f) Band Gap Engineering In Donor-Acceptor Conjugated Copolymers

Ying-Chieh Hung, Department of Chemical Engineering, National Taiwan University, Taipei, Taiwan and Shiang-Tai Lin, Chemical Engineering, National Taiwan University, Taipei, Taiwan

Here we report the principles of the change in the electronic properties (especially, HOMO/LUMO, band gap, and band width) by introducing different amounts of electron donating (D) or electron accepting (A) units to the backbone of a conducting polymer. We performed systematic studies on thiophene (T) based copolymers containing thienopyrazine (TP) or methoxythiophene (OT) at varying compostions using *ab initio* density functional theory (DFT). We found that

(1) The increase of acceptor (TP or OT) content in the mainchain leads to a smaller band gap of conjugated copolymers.

(2) For moderate D-A contrast system (e.g., T-OT), band gap varies linearly with composition at similar D-A ratio region; however, there is little composition effect at high D and A concentration regions.

(3) For strong D-A contrast system (e.g., T-TP), LUMO and band gap drop rapidly when a small amount of acceptor (TP) is added. The band gap become composition insensitive at high acceptor concentration regions.

(4) An extremely narrow bandwidth of conduction band would be observed for strong D-A contrast system especially at low concentration of π-electron acceptor.

Based on these results, we demonstrate an effective molecular design strategy by adjusting D-A composition toward desired electronic properties for donor-acceptor conjugated copolymers.

See more of this Session: Molecular and Multiscale Modeling for Renewable Energy Systems
See more of this Group/Topical: Computational Molecular Science and Engineering Forum

(318g) Structure and Diffusion of Furans and Other Cellulose-Derived Compounds In Solvents Via MD Simulation

Brooks D. Rabideau, Laboratoire de Matriaux et des Structures du Gnie Civil, Champs sur Marne, France and Ahmed E. Ismail, Mechanical Engineering, RWTH Aachen University, Aachen, Germany

There now exists a large push towards the development of sustainable, environmentally friendly energy sources. The conversion of cellulose derived from biomass into biofuels is one such promising route. As part of this process, a variety of intermediary compounds that appear critical to the successful expansion of the process to an industrial scale have been identified. These intermediary compounds consist of either different furans or acids derived from cellulose, as well as the cellulosic strands themselves. We examine the structure and diffusion of these platform chemicals in 1-butyl-3-methylimidazolium chloride ([BMIM]Cl), an ionic liquid, and water via molecular dynamics (MD) simulations. Specifically, we study the solvation properties of these chemicals by examining their pair correlation functions in each solvent. We determine the diffusion constants of these compounds, and explore the dispersibility of these chemicals within these different solvents as well as the hydrogen bonding networks that develop. Finally we explore the effect of the chain length of cellulose on its behavior in these solvents.

See more of this Session: Molecular and Multiscale Modeling for Renewable Energy Systems
See more of this Group/Topical: Computational Molecular Science and Engineering Forum

(356a) Transfer Hydrogenation to Produce Chemicals and Fuels From Lignocellulosic Biomass Using Furfural As a Platform Molecule

Stephanie G. Wettstein, Elif I. Gurbuz, Mei Chia and James A. Dumesic, Chemical and Biological Engineering, University of Wisconsin-Madison, Madison, WI

The conversion of biomass to produce fuels, power, and chemicals involves a balance between production of high-value chemicals with the production of lower-value, but higher-volume transportation fuels, leading to profitable operations to meet energy demand. Furfural derived from ligno-cellulosic biomass has the potential to be utilized as a platform intermediate molecule in the production of renewable liquid fuels and chemicals. We have developed a process to convert the hemi-cellulose fraction of biomass into chemicals and fuel that does not require the use of precious metal catalysts and that allows the remaining cellulose fraction to be processed separately (e.g., for use as pulp or for conversion to biofuels). Thus, this approach offers new opportunities for the effective utilization of renewable ligno-cellulose resources.

In our approach, the hemi-cellulose fraction of biomass is first treated in a dilute aqueous solution of mineral acid to produce xylose, followed by dehydration of the xylose in a biphasic system containing a hydrophobic solvent to produce and concentrate the furfural. The solvent does not extract the mineral acid, which remains in the aqueous phase and can be used for additional hemi-cellulose deconstructions. The furfural can then be converted to other chemicals or fuels. One method we demonstrate is to use transfer hydrogenation with a secondary alcohol over mixed oxide catalysts to convert furfural to furfuryl alcohol, the latter of which can be converted over an acid catalyst to levulinic esters. In addition, we show that levulinic esters can be converted by transfer hydrogenation over mixed oxide catalysts to produce gamma-valerolactone, which can be used as a chemical, fuel additive, or decarboxylated to produce butene that can be upgraded to liquid hydrocarbons fuels.

See more of this Session: Catalytic Processing of Fossil and Biorenewable Feedstocks: Fuels III
See more of this Group/Topical: Catalysis and Reaction Engineering Division

(356b) The Effects of Acetate Anion On Cellulose Dissolution and Reaction In Imidazolium Ionic Liquids

Hongbo Du, Colorado State University, Ft Collins, CO and **Xianghong Qian**, Department of Chemical Engineering, University of Arkansas, Fayetteville, AR

Quantum mechanical calculations were carried out to determine the mechanisms for the superiority of the imidazolium acetate based ionic liquids for biomass processing to the corresponding chloride based ionic liquids. Our results indicate that imidazolium cation can react with the acetate anion to generate carbene, a highly reactive intermediate. Carbene produced then reacts with the reducing ends of cellulose to facilitate its dissolution in the ionic liquid solvents. The mechanisms for imidazolium cation and acetate anion reaction involve the initial ion pairing of the cation and anion via hydrogen bonding and electrostatic interactions. The hydrogen bond formed between the C2-H on the imidazolium cation and COO^- of the anion facilitates the transfer of the H^+ to the anion to form a carbene intermediate.

See more of this Session: Catalytic Processing of Fossil and Biorenewable Feedstocks: Fuels III
See more of this Group/Topical: Catalysis and Reaction Engineering Division

(356c) Upgrading Pyrolysis Bio-Oils by Hydrodeoxygenation Over Non-Sulfided NiMo/Al2O3 Catalysts Using Guaiacol As a Model Compound

Zhong He and **Xianqin Wang**, Chemical, Biological and Pharmaceutical Engineering, New Jersey Institute of Technology, Newark, NJ

Abstract

Pyrolysis oils are considered as reasonable and promising second-generation biofuels. However, the high oxygen content (26-47 wt.%) in bio-oils leads to a lower energy density than traditional fuels and the strong acidity (pH~2-3) makes them enormously unstable. Herein, we evaluated hydrodeoxygenation (HDO) of guaiacol (GUA) as a bio-oil model compound over non-sulfide NiMo/Al_2O_3 under conditions close to those of conventional hydrotreatment process (300 °C and 7.1 MPa), and the results were compared to those obtained with non-sulfided commercial catalyst CoMo/Al_2O_3. NiMo/Al_2O_3 was found to be superior to the commercial catalyst CoMo/Al_2O_3 with respect to conversion of GUA and HDO level in the liquid product mixture (O/C and H/C molar ratios). The main products for HDO of guaiacol were cyclohexane, methane, cyclohexanol, methyl-cyclohexane, and methyl-cyclopentane. The updated liquid products with C/H and C/O molar ratios close to those of gasoline and diesel were produced. Thus, non-sulfided NiMo/Al_2O_3 are promising catalysts for the treatment of bio-oils.

See more of this Session: Catalytic Processing of Fossil and Biorenewable Feedstocks: Fuels III
See more of this Group/Topical: Catalysis and Reaction Engineering Division

(356d) Production of Green Aromatics From Catalytic Fast Pyrolysis of Lignocellulosic Biomass

Jungho Jae, Chemical Engineering, University of Massachusetts, Amherst, Amherst, MA and George W. Huber, Chemical Engineering, University of Massachuseets - Amherst, Amherst, MA

Catalytic Fast Pyrolysis (CFP) is a promising technology for the direct production of gasoline range aromatics from lignocellulosic biomass in a single catalytic reactor. The most challenging part of this process is controlling the complex homogeneous and heterogeneous reaction chemistry to maximize aromatic yield and suppress the production of coke. To gain fundamental understanding of this process, we studied reaction chemistry for CFP of glucose (i.e. cellulose model compound) with isotopic labeling of ^{13}C glucose and ex-situ FT-IR technique in a pyroprobe reactor coupled with GC-MS. It was found that glucose is completely decomposed in one second and converted into dehydrated products, including anhydrosugars and furans. The dehydrated products then enter into the zeolite catalyst pore where they are converted into aromatics, CO, CO_2, and water through a series of dehydration, decarbonylation, decarboxylation, oligomerization, and dehydrogenation reactions. The isotopic labeling studies revealed that aromatic formation reaction proceeds through a common intermediate or "hydrocarbon pool" composed of these dehydrated products. This is a shape selective reaction where the product selectivity is related to the catalyst properties. We have systematically studied CFP with different zeolites including small pore zeolites (ZK-5 and SAPO-34), medium pore zeolites (Ferrierite, ZSM-23, MCM-22, SSZ-20, ZSM-11, ZSM-5, IM-5, and TNU-9), and large pore zeolites (SSZ-55, beta, Y zeolite). The aromatic yield is a function of the pore size and internal pore volume of the zeolite catalyst. Medium pore zeolites with pore sizes in the range of 5.2 to 5.9 Å and moderate pore intersection size such as ZSM-5 and ZSM-11 produced the highest aromatic yield and least amount of coke. The kinetic diameter estimation of aromatic products and the reactants revealed that the majority of these molecules can fit inside the zeolite pores of most of the medium and large pore zeolites, while the polycyclic aromatics form by secondary reactions on the exterior surface of zeolite catalyst. ZSM-5 catalyst, the best catalyst for aromatic production, can be modified further to improve its catalytic performance. These modifications include: (1) adjusting the concentration of acid sites inside the zeolites catalyst; (2) incorporation of mesoporosity into the ZSM-5 framework to enhance its diffusion characteristics, and (3) adding metal oxides promoters to the ZSM-5. Mesoporous ZSM-5 showed high selectivity for monofunctional aromatics, such as benzene, toluene, xylenes, and less for naphthalene. Metal oxide promotion of ZSM-5 increased the aromatic yield by over 40%. A pilot scale reactor was designed and built where CFP was demonstrated that it could produce liter quantities of aromatic products directly from solid woody biomass feeds.

(356e) Bifunctional Materials for the Catalytic Conversion of Cellulose Into Soluble Renewable Biorefinery Feedstocks

Damian Reyes-Luyanda[1], Josseant Florez-Cruz[1], Pedro J. Morales-Prez[1], Luis G. Encarnacin-Gmez[1], Fengyuan Shi[2], Paul M. Voyles[2] and Nelson Cardona-Martnez[1], (1)Chemical Engineering, University of Puerto Rico Mayagez Campus, Mayagez, PR, (2)Department of Materials Science and Engineering, University of Wisconsin Madison, Madison, WI

The development of bifunctional nanostructured materials for the catalytic conversion of cellulose into sugar alcohols offers the potential of a sustainable source of renewable biorefinery feedstock. Supported Ru catalysts were prepared by evaporative deposition on various ordered mesoporous silicas (SBA-15) with different functionalities and characterized using multiple techniques. The catalytic performance of these materials was compared to Ru/C and the corresponding supports. The effects of functional group loading, reaction time and temperature on the activity and selectivity of the bifunctional catalysts were studied by monitoring the cellulose conversion and the production of sugars and sugar alcohols in a high-pressure batch reactor. Sorbitol is the main product obtained by the hydrolysis of cellulose to glucose followed by the corresponding reduction. Secondary products include sugars, ethylene glycol and glycerol. The activity of mesoporous silica catalysts increases with an increase in sample acidity and the addition of Ru allows control of the selectivity towards sugar alcohols. Ruthenium supported on arenesulfonic acid-functionalized mesoporous silica (Ru/SBA-15S) displays the best catalytic performance. Ru/SBA-15S is more hydrothermally stable than SBA-15, but loses a significant fraction of its surface area, crystallinity, and acidity after prolonged exposure to water at 483K. However, even after 72h under these conditions the catalyst retains some of its acidity and the activity is still higher than fresh unfunctionalized Ru/SBA-15 and Ru supported on amorphous silica.

See more of this Session: Catalytic Processing of Fossil and Biorenewable Feedstocks: Fuels III
See more of this Group/Topical: Catalysis and Reaction Engineering Division

(356f) Etherification of Biomass Derived Molecules for Effective Diesel Fuel Production

Eric R. Sacia and Alexis T. Bell, Department of Chemical and Biomolecular Engineering, University of California, Berkeley, Berkeley, CA

Etherification of Biomass DerivedMolecules for Effective Diesel Fuel Production

Eric Sacia and Alex T. Bell*

EnergyBiosciences Institute and Department of Chemical Engineering, University of California, Berkeley, CA 94720

Email: alexbell@uclink.berkeley.edu

The EnvironmentalProtection Agency has estimated that anthropogenic CO_2 emissions dueto the transportation sector in the United States account for nearly 1/3 of netCO_2 emissions (1.7 billion metric tons in the year 2009)[1]. Since the combustion of fossilfuels is considered to be a major contributor to the growth of atmospheric CO_2levels, there is currently significant interest in finding ways to reduce thenet carbon footprint from transportation fuels. One approach is to convertbiomass to both gasoline and diesel range fuels. Toward this aim, the EU hasset a target that 10% of transportation fuels should be biomass sourced by 2020in addition to the DOE's estimate that 20% of U.S. fuels will be sourced frombiomass by 2030.

To produce high qualitydiesel fuels, compounds with high cetane numbers must be formed that canreplace the C11-C22 hydrocarbons derived from petroleum. While biodiesel can beproduced by transesterification of palm, rapeseed, or soybean oil, the supplyof these oils is limited. Further, there are challenges in scalability and in thedevelopment of effective heterogeneous catalysts for transesterifaction[2]. An alternative method is to formdiesel components from sugars generated through the hydrolysis of thecellulosic components of lignocellulosic biomass. The reaction pathways in thiswork have been focused on producing furanic ethers since it has been shownthat such products have high cetane numbers and can be blended intopetroleum-derived diesel up to 17 wt% [3].

Our work has focused on twopaths of utilizing the hydroxymethyl furfural (HMF) that is generated fromdehydration of glucose. First, we have carried out the direct etherification ofHMF with each primary alcohol from methanol through 1-butanol with emphasis onethanol and 1-butanol since these alcohols can already be readily formed frombiomass. The second pathway involves etherification of the hydrogenationproducts of HMF as the aldehyde is reduced to an alcohol, formingbis-hydroxymethyl furan (BHMF), then to a methyl group, forming methylfurfurylalcohol (MFA) [4]. This effort has examined bothliquid and solid acids as catalysts and the effects of solvent on the rates ofetherification. By examining the effect of acid choice, it was determined thatspecific solid acids could be used to increase selectivity beyond thehomogeneous result through catalyst shape selectivity and environment. Underoptimum conditions ether yields in excess of 98% have been achieved.

The kinetics ofetherification have been investigated and found to be first order in thefurfuryl alcohol and the acid catalyst but zero order in ethanol. Further, therate of etherification is highly sensitive to the ring substituents. As thesubstituent on the 2 position of HMF is altered from an aldehyde to a hydroxymethylgroup (BHMF), and finally to a methyl group (MFA), the substituent becomes moreelectron donating. Increasing electron donation to the ring decreases theactivation energy of the etherification and increases the rate significantly. Decreasingsolvent polarity has also been observed to increase reaction rate by lessstabilizing charged reaction intermediates. Rate parameters for each step ofetherification have been measured and the variations in these parameters can beinterpreted in terms of the structure and composition of the reactants and thesolvent. The results of this work will contribute to a better understanding ofthe means for converting biomass-derived sugars into products suitable forblending into diesel.

1. Inventory of U.S. Greenhouse Gas Emissions and Sinks:.Agency, U. S. E. P., Ed. Washington DC, 2011.

2. Huber, G. W.; Iborra, S.; Corma, A., Synthesis oftransportation fuels from biomass: Chemistry, catalysts, and engineering. *ChemRev* **2006,** *106* (9), 4044-4098.

3. Imhof, P.; Dias, A. S.; de Jong, E.; Gruter, G.-J., OA02- Furanics: Versatile Molecules for Biofuels and Bulk Chemicals Applications. *NAMAbstract* **2009**.

4. Chidambaram, M.; Bell, A. T., A two-step approach for thecatalytic conversion of glucose to 2,5-dimethylfuran in ionic liquids. *GreenChemistry* **2010,** *12* (7), 1253-1262.

See more of this Session: Catalytic Processing of Fossil and Biorenewable Feedstocks: Fuels III
See more of this Group/Topical: Catalysis and Reaction Engineering Division

Hydrodeoxygenation of Guaiacol with Rh Based and CoMo and NiMo Catalysts

Yu-Chuan Lin,[*,†] *Chia-Linag Li,*[†] *Hou-Peng Wan,*[‡] *Hom-Ti Lee,*[‡] *Chiung-Fang Liu*[‡]
and Ying-His Chang[‡]

[†]*Department of Chemical Engineering and Materials Science*

Yuan Ze University, Chungli, Taiwan 32003

[‡]*Green Energy and Environment Research Laboratories*

Industrial Technology Research Institute, Hsinchu, Taiwan 30011

[*]Corresponding Author's E-mail: yclin@saturn.yzu.edu.tw

Keywords: guaiacol, hydrodeoxygenation, rhodium, sulfided-nickel-molybdenum,

sulfided-cobalt-molybdenum,

Abstract

Hydrodeoxygenation (HDO) of guaiacol (GUA), a lignin model compound, was examined by two types of catalysts: mono- and bimetallic Rh based catalysts and classical sulfided CoMo and NiMo catalysts. The former possessed greater reactivity than the latter. However, no positive effect could be found when substituting some of the Rh ions by another noble metal (i.e., Pd or Pt). The mechanisms of GUA HDO by these two types of catalysts were also proposed. The first step of Rh based catalysts was hydrogenation of GUA's benzene ring, proceeded by demethoxylation and dehydration. As for the sulfided CoMo and NiMo catalysts, the reaction steps were just the reverse: dehydration and demethoxylation first, followed by further hydrogenation. Future improvements in understanding HDO surface chemistry can possibly lead us to design efficient catalysts for target products.

1. Introduction

Lignin is the most abundant polymer containing organic aromatics in the plant kingdom.[1,2] However, it has been long-term considered as a waste byproduct or used as a low-grade fuel.[2-4] This is mainly because of the challenge of lignin conversion to high grade fuels and chemicals.[2,5] Yet, advanced developments of agriculture and biology in lignocellulosic biomass breeding,[6] together with declining crude oil resources and world-wide environmental concerns have drawn comparatively more attention on lignin conversion than it has ever been.[7,8]

One promising way for lignin conversion is hydrodeoxygenation (HDO). HDO is a process by which hydrogen acts as the reductant for deoxygenation, usually operated in a batch[9,10] or a continuous system[11,12] with catalysts. Under H_2-pressured hydrothermal conditions, oxygen can be removed in the form of water.[13] Therefore, the upgraded products contain less oxygen, rendering then with a higher heating value and chemically more stable than their unrefined state.[13]

Guaiacyl species have been well acknowledged as the major structure in lignin.[2] Significant amount of guaiacyl groups, e.g. guaiacol (GUA, $C_7H_8O_2$) and mono-methoxyl phenols ($C_8H_{12}O_2$), can be produced through various conversion approaches such as pyrolysis[14,15] and liquefaction.[5,16] Therefore guaiacyl species are a popular representative of lignin and its derivatives.[10,17-19] An effective upgrading process for guaiacyl species should facilitate the design of a lignin conversion platform.

The most encountered HDO catalysts are sulfided CoMo- and NiMo on Al_2O_3. In 1970, a Canadian patent reported by Alpert and Shuman[20] revealed the potential of lignin conversion by CoMo catalysts. This inspires follow-up GUA HDO studies, such as the pioneer works by Huff and Klein[21] and the research team of University of Toronto.[22] A recently published review by Zakzesji and collaborators has summarized GUA HDO in Table 4 of their paper.[5] It can be seen that sulfided CoMo and NiMo supported on Al_2O_3 are the major catalysts used in this process at temperatures ranging from 500 to 700 K under pressures of about 3-7 MPa. Six-membered ring compounds such as benzene and phenol are major products. Compared to sulfided CoMo and NiMo on Al_2O_3, however, relatively less work has been conducted using noble metal based catalysts. Despite the high cost, noble metal catalysts possess several advantages over conventional sulfide catalysts: 1) high reactivity at moderate temperature, allowing energy efficient HDO process, 2) free of sulfur stripping, and 3) flexible catalyst design by tailoring the active phase and support.[10,23]

Centeno and coworkers[24] have employed trace platinum (~0.5 wt%) on CoMo sulfide catalyst for HDO of GUA-type molecules. Significant improvement of reactivity was reported. Elliot and Hart[10] have studied catalytic HDO of a bio-oil like compound (a mixture of furfural, GUA, acetic acid and water) with palladium or ruthenium on carbon extrudates. Different hydrogenation abilities, which resulted in diverse hydrodeoxygenation pathways of GUA, have been discovered. Gutierrez and collaborators[23] have explored mono- and bimetallic noble metal catalysts (including Pt, Rh, Pd and their combination) in GUA HDO. By varying the ingredient of the active phase, the reactivity could be changed. In general, Rh containing catalysts showed better performances than other noble metal catalysts.

The objective of this study is to investigate HDO chemistry of lignin derivatives. GUA was used as lignin model compound. Two types of HDO catalysts, Rh based catalysts (Rh, PtRh, and PdRh on ZrO_2) and sulfided CoMo- and NiMo/Al_2O_3 were surveyed in GUA HDO. Catalyst characterization and the catalytic HDO testing were conducted. Based on these outcomes, plausible mechanisms mediated by these catalysts were thereby proposed.

2. Experimental

2.1. Catalyst preparation

All catalysts in this study were prepared by incipient wetness method. For Rh, PtRh, and PdRh catalysts, ZrO_2 (Aldrich, 99.99%) was used as the support. Precursors including $Pt(NH_3)_2(NO_2)_2$ (Aldrich, 3.4 wt% in dilute ammonium hydroxide), $Rh(NO_3)_3$ (Aldrich, 10 wt% Rh in >5 wt% nitric acid), and $Pd(NO_3)_2$ (Aldrich, 99.95%) were used. The Rh content in Rh/ZrO_2 was 1 wt%. For the bimetallic catalysts, an equivalent amount of Pt and Rh or Pd and Rh was used to allow 1 wt% loading. After impregnation, the resulting powder was dried at 80°C overnight followed by calcination at 400°C for 3 hours using a heating rate of 5°C/min.

The synthesis procedure of CoMo supported on Al_2O_3 (Aldrich, 99%) was as follows: alumina was calcined at 600°C overnight prior to the impregnation. Appropriate amount of $(NH_4)_6Mo_7O_{24}$-$4H_2O$ (Alfa Aesar 99%) was first impregnated on alumina, followed by a drying at 80°C overnight and calcined at 400°C for 3 hours using a 5°C/min heating rate. Subsequently, designated amount of $Co(NO_3)_2$-$6H_2O$ (Alfa Aesar 98%) was impregnated on the Mo-containing powder, following the same drying-calcination step mentioned above. By replacing Co precursor to $Ni(NO_3)_2$-$6H_2O$ (Alfa Aesar 98%), $NiMo/Al_2O_3$ could be made with the same synthesis process. The composition of CoMo or NiMo catalyst was about 15 wt% of MoO_3 with 3 wt% of CoO or NiO.

2.2. Catalyst characterization

Surface areas were measured by the Brunauer, Emmet, and Teller (BET) method, using a N_2 automated adsorption system (Micrometrics ASAP 2010). Powder X-ray diffraction (XRD) patterns were collected using a Shimadzu Labx XRD-6000 with a Cu-Kα radiation source. H_2-tempaerature programmed (TPR) and temperature programmed desorption (TPD) of GUA were performed by the same instrumentation described in our earlier publication.[25] In short, the system was assembled by a U-shaped cell and a temperature-controlled furnace. For TPR trials, about 0.1 g of sample was used. H_2 consumption was monitored by a thermal conductivity detector (TCD) in a 10% H_2/Ar stream. As for GUA TPD, approximately 0.1 g of CoMo or $NiMo/Al_2O_3$ was consumed. About 0.2 g of Rh based catalysts were required to exaggerate sufficient signal-to-noise TCD signals. At the outset, the catalyst was reduced at 300°C for an hour with 5°C/min heating rate. After cooling down to ambient temperature, GUA was introduced into the cell overnight by means of a saturator with a Ar stream (30 ml/min) as the carrier gas. Subsequently, the sample was purged with a He stream (30 ml/min) at 220°C for an hour to remove physisorbed GUA. After attaining a stable TCD baseline, the TPD was performed from 220 to 700°C (5°C/min).

2.3. Activity testing

An in-house designed high-throughput system (shown in Figure 1) was employed in catalytic HDO testing. The system consisted of four heating plates, three reactor blocks and a blank block. Heating plates were placed surrounding the reactor and blank blocks with electric heating control. Reactor and blank blocks were aligned in an array. Three 29.8 cm long and 1.1 cm wide quartz reactor tubes (30 ml, shown in Figure 1) were used as the containers for reactants and catalyst. After filling appropriate amount of reactants and catalysts, the reactor tubes were then sealed and placed in the reactor blocks. To allow for good mixing, a rotary motor was equipped and set a 100 rpm rotation speed during the testing.

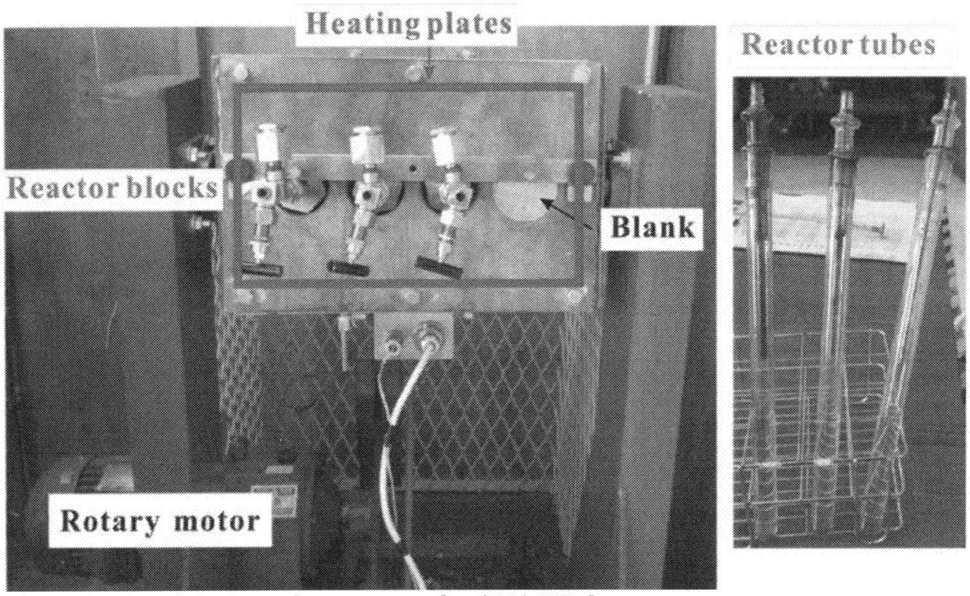

Figure 1. The high-throughput system for GUA HDO.

Approximately 0.3 g of catalyst, with particle sizes ranging from 40-80 mesh, was used per trial. Prior to each test, catalyst pretreatment was conducted for both types of catalysts. The noble metal catalyst was reduced in situ at 300°C under 10 bar of H_2 pressure for 30 min. CoMo and NiMo catalysts were activated by a reduction-sulfidation system. By flowing 5 vol% H_2S/H_2 (110 ml/min), the temperature was raised from room temperature to 360°C at a rate of 0.5°C/min and held for 3 hrs. After cooling back to room temperature, the system was purged with N_2 for 30 mins. Sulfided catalyst was then transferred to the high-throughput system for activity testing. About 0.23 g of GUA and 7.63 g of tetradecane were employed to set the reactant-to-solvent ratio close to 3 wt%. Prior to the experiments, blank runs without catalyst and effects of supports (ZrO_2 and Al_2O_3 used as catalysts) were surveyed. These trials showed negligible influences of homogeneous reaction as well as support's effect.

Speciation of liquid products was conducted by a gas chromatograph-mass spectrometry system (GC/MS, Agilent 6890N and 5973N; with DB1701 capillary column); Quantitative analysis was conducted by a GC/FID (Agilent 6890N) with the same column and programming as the above-mentioned analytical configuration. There were eleven major species identified in this work. For Rh based catalysts, 2-methoxycyclohexanol ($C_7H_{14}O_2$), 2-methoxycyclohexanone ($C_7H_{12}O_2$), 1-methoxycyclohexane ($C_7H_{14}O$), cyclohexanol ($C_6H_{12}O$), and cyclohexanone ($C_6H_{10}O$) were recognized; for sulfided CoMo and NiMo, methoxybenzene (C_7H_8O), methylphenol (C_7H_8O), phenol (C_6H_6O), benzene (C_6H_6), and cyclohexene (C_6H_{10}) were found. Cyclohexane (C_6H_{12}) was observed for both systems. Interpolated calibration was employed by injecting a standard solution of species for product quantification. However, due to the lack of reference reagents, calibrations of 2-methoxycyclohexanol and 2-methoxycyclohexanone were assumed to be the same

FID response as 2-methylcyclohexanone ($C_7H_{12}O$); methoxybenzene and methylphenol were assumed to be equal to guaiacol; cyclohexene was set as cyclohexanone.

The amount of coke was determined by a thermal gravimetric analyzer (TGA, Perkin Elmer Pyris 1). The used catalyst was first soaked in a N_2 stream at 280°C for an hour, followed by changing the sweeping gas to air at 600°C (10°C/min). Due to the high pressure batch-type design, gaseous products were not collected for safety concerns. The conversion of GUA (X), carbon yield of product (Y), and selectivity (S) were defined as follows:

$$X = \frac{(\text{mole of GUA})_{in} - (\text{mole of GUA})_{out}}{(\text{mole of GUA})_{in}} * 100\% \quad (1)$$

$$Y_i = \frac{\text{moles of carbon in product i}}{\text{moles of carbon in GUA}} * 100\% \quad (2)$$

$$S_i = \frac{\text{moles of carbon in product i}}{\text{the sum of carbon moles in products}} * 100\% \quad (3)$$

3. Results and Discussion

3.1. Characterization of catalysts

Table 1 presents surface areas of catalysts in this study. All catalysts showed similar surface areas. Figure 2 shows the XRD patterns of both noble metal and CoMo and NiMo catalysts. For noble metal catalysts, the XRD responses were identical to ZrO_2 support, suggesting well dispersed metal particles on the catalyst surface. For CoMo and NiMo catalysts, a clear peak was detected at $2\theta = \sim 27°$. This indicates the presence of $CoMoO_4$[26] and $NiMoO_4$.[27]

Table 1. Physicochemical properties of catalysts used in this study

	Rh	PtRh	PdRh	CoMo	NiMo
S_{BET} (m^2/g)	6.6	7.0	5.9	1.5	3.0
T_i (°C)	118	100	68	380	382
T_{max} (°C)	175	166	122	668, 855	650, 850

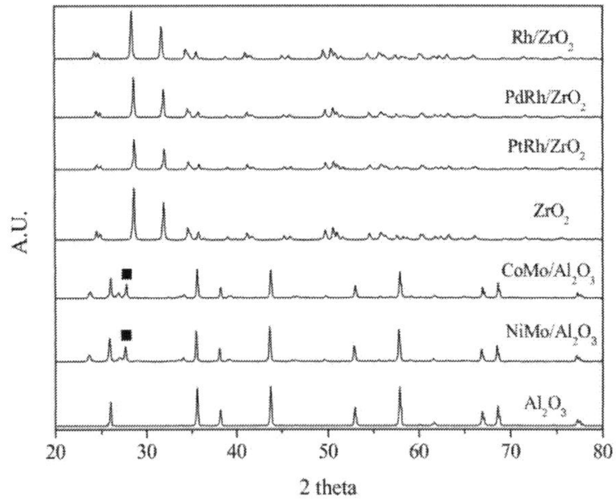

Figure 2. XRD patterns of HDO catalysts.

Figure 3 displays the H_2-TPR profiles. Their corresponding temperatures of onset (T_i) and maximum reduction (T_{max}) were summarized in Table 1. For Rh based catalysts, the order of increasing T_i and T_{max} was: PdRh > PtRh > Rh. In addition, the shapes of the reduction profiles were significantly different: peaks of bimetallic catalysts were broader than that of Rh catalyst. This suggests the chemical interactions between the metals in bimetallic catalyst, and the reducibility of Rh-based catalyst can be improved by introducing another noble metal ion (herein Pt or Pd).[23] For sulfided CoMo and NiMo catalysts, two major peaks were shown in the TPD. The first located at 650-670°C and the second at around 850°C. This is in agreement with earlier studies, where the first peak is assigned as partial reduction of Mo ions (Mo^{+6} to Mo^{+4}) followed by the second peak as deep reduction.[28-30] The influence of Co in CoMo or Ni in NiMo was insignificant in varying the reducibility of catalyst.[28]

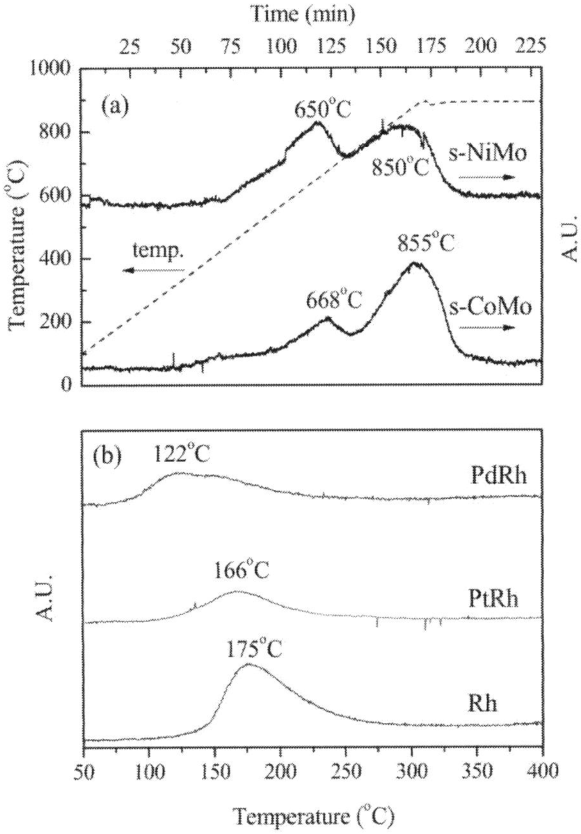

Figure 3. Temperature programmed reduction of (a) CoMo- and NiMo/Al_2O_3 and (b) Rh based catalysts.

Figure 4 displays the GUA TPD patterns. While it is very difficult to interpret TPD profiles with certainty,[31] a comparative study allowed us to approximately evaluate the strength of chemisorbed GUA. Both Al_2O_3 and ZrO_2 can chemisorb GUA molecules. While no significant response was detected for Al_2O_3, at least three peaks could be identified for ZrO_2. In addition, the area of the TPD peaks of ZrO_2 was greater than that of Al_2O_3, suggesting more weakly bonded species could be released from the ZrO_2 surface than that of Al_2O_3. This is in agreement with Popov and coworkers,[32] who showing strong doubly anchored phenates, a potential coke precursor, when GUA interacted with Al_2O_3. By decreasing the support's acidity, the amount of doubly anchored phenates could be suppressed.[32] With incorporating an active phase, CoMo or NiMo/Al_2O_3 displayed newly formed peaks (at ~250, 350, and 450°C). This implies the presence of potential active sites for GUA HDO. As regards to Rh based catalysts, two tiny responses located at around 300-350°C and 400-500°C were observed. With elevating the temperature to 600°C and higher, gradually increasing signals could be seen. Indeed, these two types of catalysts showed different results in GUA TPD, hence, disparate HDO chemistry was expected.

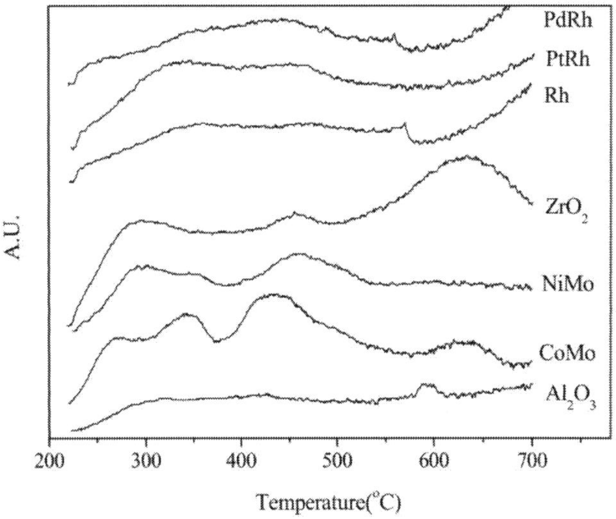

Figure 4. Temperature programmed desorption of GUA.

3.2. Reactivity

Figure 5 (a) displays GUA conversion as a function of reaction time at 400°C. Noble metal catalysts were very active. More than 50% conversion could be achieved at 20 minutes after starting the reaction. The trend then leveled off (100% conversion) at duration time for 40 to 60 minutes. Sulfided CoMo and NiMo catalysts were comparatively inert than the noble metals. Not until a certain time (60 minutes after start-up) could significant amount of GUA be converted (~80%). Figure 5 (b) shows GUA conversion as a function of temperature for a 60-minute reaction time. The conversion is weakly related to the reaction temperature for the noble metals. Rh and PtRh displayed full conversion while PdRh had ~80%. As for sulfided CoMo and NiMo, significant amount of GUA conversion could be observed merely at 400°C. Clearly, Rh based catalysts were more active than sulfided CoMo and NiMo.

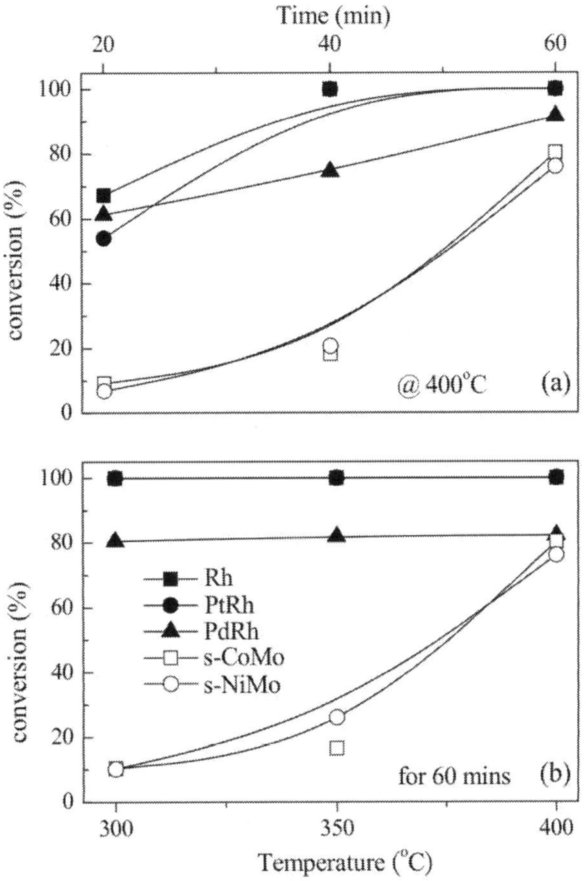

Figure 5. GUA conversion over Rh based catalysts and sulfided CoMo- and NiMo/Al_2O_3 as a function of (a) reaction time and (b) temperature.

Figure 6 shows the carbon yields of Rh based catalysts for GUA HDO, including 1) two-oxygen containing species: 2-methoxycyclohexanol and 2-methoxycyclohexanone, 2) mono-oxygen containing species: 1-methoxycyclohexane, cyclohexanol, and cyclohexanone, and 3) cyclohexane. Note that because of the lack of a mass spectrum database of 2-methoxycyclohexanol, the spectrum of 4-methoxycyclohexanol ($C_7H_{14}O_2$) from NIST Chemistry WebBook was used as a reference.[33] The electron ionization spectrum of 4-methoxycyclohexanol is very similar to our speculated product, i.e. 2-methoxycyclohexanol, showing the strongest ion fragments at m/z equals to 72, 71, and 73. Therefore, it is highly plausible that such a product is 2-methoxycyclohexanol.

With increasing temperature, yields of two oxygen-containing products decreased; while cyclohexane, increased. Among the Rh-based catalysts, monometallic Rh catalyst generated the highest yield of cyclohexane (~42%) at the final stage. Moreover, the sum of yields of mono-oxygen compounds was the highest. The same trend could also be observed by illustrating carbon yields as a function of reaction time. This indicates that the monometallic Rh catalyst possessed the highest HDO reactivity of the Rh catalysts. Although its reducibility was the lowest among the noble metal catalysts as determined by H_2-TPR.

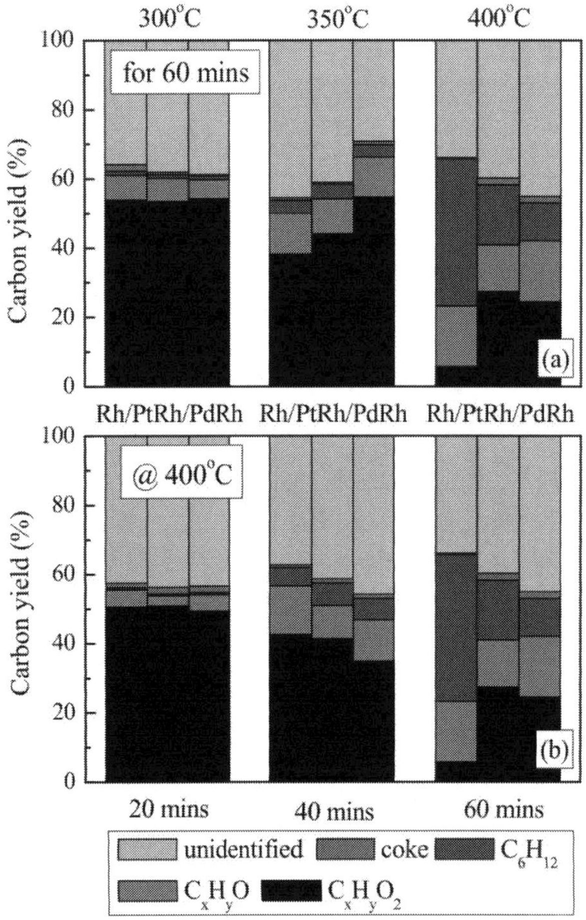

Figure 6. Carbon yields for GUA HDO with Rh based catalysts. *Reaction conditions*: (a) H_2-pressure 50 bar, reaction time 60 minutes; (b) H_2-pressure 50 bar, reaction temperature 400°C. *Key* - unidentified: gray, coke: cyan, cyclohexane: blue, 1-oxygen containing oxygenates ($C_7H_{14}O$, $C_6H_{12}O$, and $C_6H_{10}O$): red, and 2-oxygen containing oxygenates ($C_7H_{14}O_2$ and $C_7H_{12}O_2$): black.

Figure 7 displays the carbon yield of GUA HDO by sulfided CoMo and NiMo as a function of reaction temperature and time. Under 300°C for 60 minutes and 400°C for 20 minutes, no products except coke could be identified. This might be attributed to low conversions of GUA (15% and less), therefore yielding insignificant amount of products. With increasing temperature and time, product yields could be improved associated with a decrease in coke formation. The main products were 1) mono-oxygen containing species including methoxybenzene, methylphenol, and phenol, and 2) hydrocarbons including benzene, cyclohexene, and cyclohexane. The two-oxygen containing compounds generated by noble metal catalysts, were absent. This suggests that the first step of GUA HDO by sulfided CoMo and NiMo catalysts was oxygen removal through dehydration. A detailed examination of the product selectivity is presented in Figure 8. It is interesting to note that selectivity of cyclohexane was significantly high (~80%) for NiMo catalyst under 40 minutes at 400°C and 60 minutes at 350°C, whereas, phenol and methylphenol were the main products for CoMo. This implies that sulfided NiMo had stronger influences in benzene ring hydrogenation than that of sulfided CoMo. It is also worth mentioning that coke formation was more favorable on sulfided CoMo and NiMo catalysts than the noble metals. This might be attributed to the coke precursors, e.g. catechol, formed by the CoMo and NiMo catalysts.[13,34]

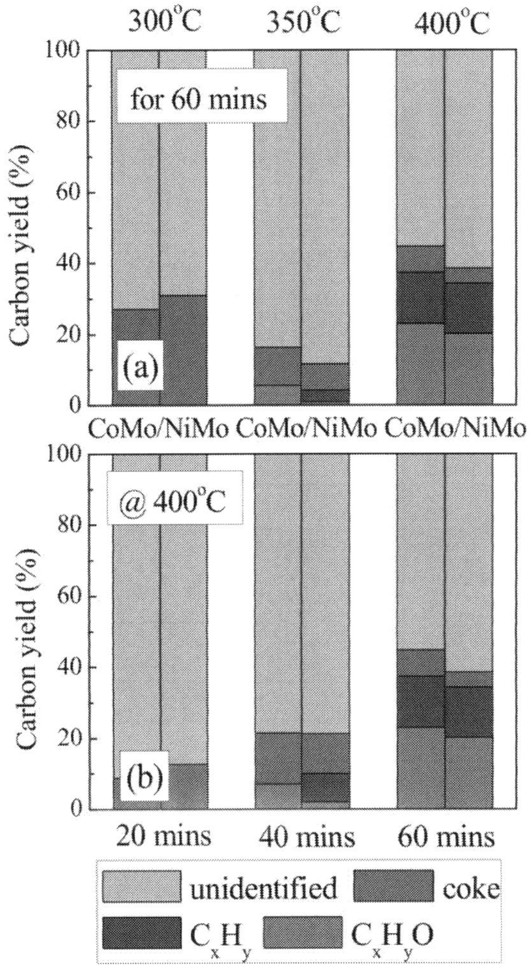

Figure 7. Carbon yields for GUA HDO with sulfided CoMo and NiMo catalysts. *Reaction conditions*: (a) H_2-pressure 50 bar, reaction time 60 minutes; (b) H_2-pressure 50 bar, reaction temperature 400°C. *Key* - unidentified: gray, coke: cyan, hydrocarbons (C_6H_6, C_6H_{10}, and C_6H_{12}): blue, 1-oxygen containing oxygenates (C_6H6O and C_7H_8O): red.

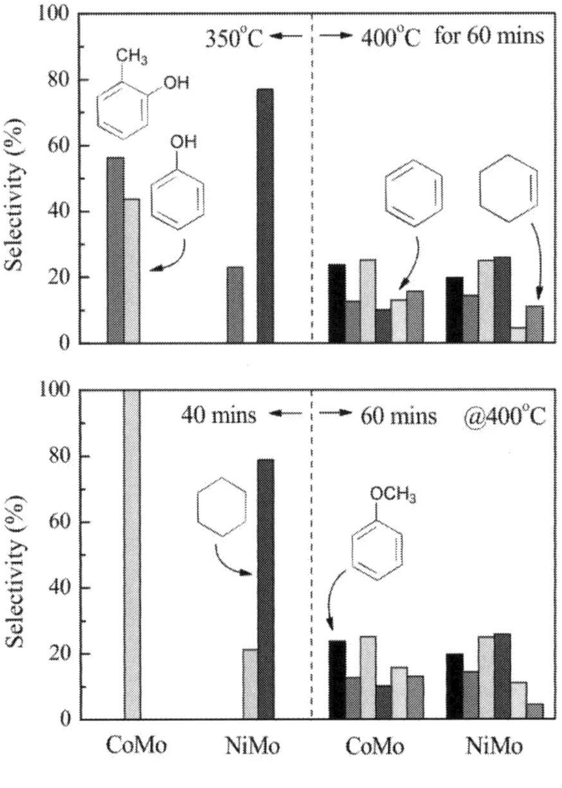

Figure 8. Product selectivity of sulfided CoMo and NiMo catalysts.

3.3. Mechanism

Figure 9 shows the proposed GUA HDO mechanisms for two types of catalysts in this study. The first step of Rh based catalysts is hydrogenation of GUA's benzene ring, yielding 2-methoxycyclohexanol and 2-methoxycyclohexanone as the major products. The second step is demethoxylation and dehydration to form cyclohexane as the final product. In contrast, the first step of sulfided CoMo and NiMo catalysts was dehydration and possibly demethylation of GUA, yielding methoxybenzene, methyl phenol and phenol. Subsequently, these mono-oxygen compounds were further dehydrated to form benzene. Under a hydrogen rich environment, benzene double bonds can be partially and/or fully saturated to form cyclohexene and cyclohexane. It is worth mentioning that catechol ($C_6H_6O_2$), a proposed intermediate for the first step of GUA HDO on CoMo and NiMo catalysts,[21,24,35,36] was not observed. A possible explanation might be that catechol was converted during our sampling period and/or in the handling process. This could be supported by the significant amount of coke[13] observed at initial conditions (300°C, 60 mins and 400°C, 20 mins) and the high selectivities of its derivatives, phenol and methylphenol,[36] at the two moderate conditions (350°C, 60 mins and 400°C, 40 mins). Similar behavior was also reported by Zhao and coworkers when using transition metal phosphides in gas phase GUA HDO in a continuous fixed bed system.[34] With longer contact time, no catechol could be observed.

Rh based catalysts

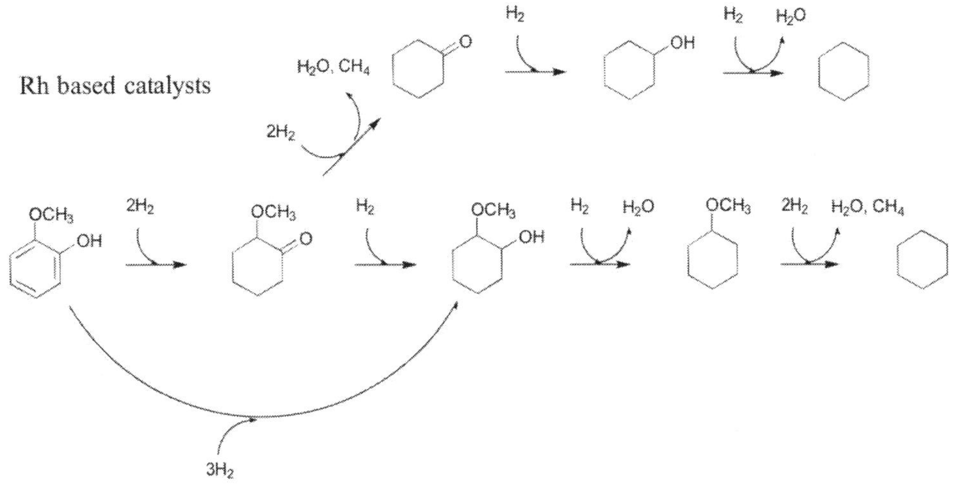

sulfided CoMo and NiMo catalysts

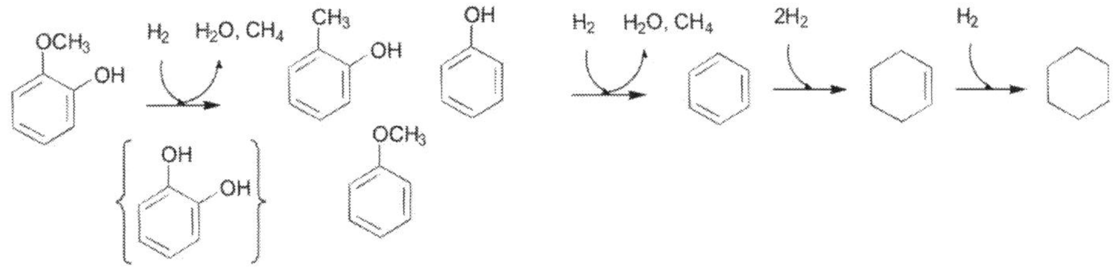

Figure 9. Mechanism of GUA HDO by Rh based and sulfided CoMo and NiMo catalysts.

4. Conclusions

Catalytic behaviors of Rh based and sulfided CoMo and NiMo catalysts in GUA HDO were significantly different. Either monometallic or bimetallic Rh catalysts showed surpassing activity than CoMo and NiMo. No noticeable improvement was found by using bimetallic catalysts. Instead, some adverse effects in activity were observed for the PdRh catalyst. The activity of both sulfided CoMo and NiMo, were almost identical. In regard to product distributions, Rh/ZrO_2 was deemed to be the most effective among the noble metal catalysts, yielding the most cyclohexane at 400^oC for an hour. From the analysis of the products, the mechanism of GUA HDO promoted by Rh based catalysts was proposed to follow two consecutive steps: hydrogenation of the GUA benzene ring, followed by deoxygenation and demethylation of yielded oxygenates. For sulfided CoMo and NiMo, the HDO process was the reverse mechanism, namely: deoxygenation and demethylation first, followed by benzene ring saturation. This likely explains the difference in product distributions of these catalysts.

In summary, the Rh based catalyst had the best HDO activity, but tended to saturate valuable benzene rings; classical sulfided CoMo and NiMo catalysts showed potential to make aromatics, however, they were relatively inert compared to noble metals and produced significant coke. It seems that future advances in understanding HDO surface chemistry both experimentally and theoretically will allow us to design improved catalysts to produce targeted products from lignin and its derivatives.[37,38]

Acknowledgements

Valuable discussion with Prof. Geoff A. Tompsett was deeply appreciated. The authors acknowledge financial support provided by the Bureau of Energy, Ministry of Economics Affairs, Taiwan, (99-D0103).

References

(1) Crawford, R. L. *Lignin Biodegradation and Transformation*; Wiley-Interscience: New York, 1981.

(2) Holladay, J. E.; White, J. F.; Bozell, J. J.; Johnson, D. *Top Value-Added Chemicals from Biomass - Volume II, Results of Screening for Potential Candidates from Biorefinery Lignin*, 2007.

(3) Struszczyk, H. In *Lignin: Properties and Materials*; Glasser, W. G., Sarkanen, S., Eds.; American Chemical Society: Washington, DC, 1989, p 245.

(4) Gosselink, R. J. A.; de Jong, E.; Guran, B.; Abacherli, A. *Ind. Crops Prod.* **2004**, *20*, 121-129.

(5) Zakzeski, J.; Bruijnincx, P. C. A.; Jongerius, A. L.; Weckhuysen, B. M. *Chem. Rev.* **2010**, *110*, 3552-3599.

(6) Huber, G. W. *Breaking the Chemical and Engineering Barriers to Lignocellulosic Biofueld: Next Generation Hydrocarbon Biorefineries*, University of Massachusetts-Amherst, 2008.

(7) Mohan, D.; Pittman, C. U.; Steele, P. H. *Energy Fuels* **2006**, *20*, 848-889.

(8) Huber, G. W.; Iborra, S.; Corma, A. *Chem. Rev.* **2006**, *106*, 4044-4098.

(9) Sealock, L. J.; Elliott, D. C.; Baker, E. G.; Butner, R. S. *Ind. Eng. Chem. Res.* **1993**, *32*, 1535-1541.

(10) Elliott, D. C.; Hart, T. R. *Energy Fuels* **2009**, *23*, 631-637.

(11) Chheda, J. N.; Huber, G. W.; Dumesic, J. A. *Angew. Chem., Int. Ed.* **2007**, *46*, 7164-7183.

(12) Li, N.; Huber, G. W. *J. Catal.* **2010**, *270*, 48-59.

(13) Furimsky, E. *Appl. Catal., A* **2000**, *199*, 147-190.

(14) Dorrestijn, E.; Laarhoven, L. J. J.; Arends, I. W. C. E.; Mulder, P. *J. Anal. Appl. Pyrolysis* **2000**, *54*, 153-192.

(15) Pattiya, A.; Titiloye, J. O.; Bridgwater, A. V. *J. Anal. Appl. Pyrolysis* **2008**, *81*, 72-79.

(16) Kleinert, M.; Barth, T. *Energy Fuels* **2008**, *22*, 1371-1379.

(17) Klein, M. T.; Virk, P. S. *Energy Fuels* **2008**, *22*, 2175-2182.

(18) Hou, Z.; Bennett, C. A.; Klein, M. T.; Virk, P. S. *Energy Fuels* **2010**, *24*, 58-67.

(19) Binder, J. B.; Gray, M. J.; White, J. F.; Zhang, Z. C.; Holladay, J. E. *Biomass Bioenergy* **2009**, *33*, 1122-1130.

(20) Alpert, S. B.; Shuman, S. C. Canadian Patent 851709, 1970.

(21) Hurff, S. J.; Klein, M. T. *Ind. Eng. Chem. Fund.* **1983**, *22*, 426-430.

(22) Kallury, R. K. M. R.; Restivo, W. M.; Tidwell, T. T.; Boocock, D. G. B.; Crimi, A.; Douglas, J. *J. Catal.* **1985**, *96*, 535-543.

(23) Gutierrez, A.; Kaila, R. K.; Honkela, M. L.; Slioor, R.; Krause, A. O. I. *Catal. Today* **2009**, *147*, 239-246.

(24) Centeno, A.; Laurent, E.; Delmon, B. *J. Catal.* **1995**, *154*, 288-298.

(25) Li, C.-L.; Lin, Y.-C. *Catal. Lett.* **2010**, *140*, 69-76.

(26) Hada, K.; Tanabe, J.; Omi, S.; Nagai, M. *J. Catal.* **2002**, *207*, 10-22.

(27) Brito, J. L.; Laine, J. *J. Catal.* **1993**, *139*, 540-550.

(28) Arnoldy, P.; Franken, M. C.; Scheffer, B.; Moulijn, J. A. *J. Catal.* **1985**, *96*, 381-395.

(29) Qu, L.; Zhang, W.; Kooyman, P. J.; Prins, R. *J. Catal.* **2003**, *215*, 7-13.

(30) El Kady, F. Y. A.; Abd El Wahed, M. G.; Shaban, S.; Abo El Naga, A. O. *Fuel*, *89*, 3193-3206.

(31) Gorte, R. J. *Catal. Today* **1996**, *28*, 405-414.

(32) Popov, A.; Kondratieva, E.; Goupil, J. M.; Mariey, L.; Bazin, P.; Gilson, J.-P.; Travert, A.; Mauge, F. *J. Phys. Chem. C* **2010**, *114*, 15661-15670.

(33) NIST Chemistry WebBook, National Institute of Standards and Technology, Gaithersburg, MD, 2005

(34) Zhao, H. Y.; Li, D.; Bui, P.; Oyama, S. T. *Appl. Catal., A* **2010**, *In Press*.

(35) Laurent, E.; Delmon, B. *Appl. Catal., A* **1994**, *109*, 77-96.

(36) Bui, V. N.; Toussaint, G.; Laurenti, D.; Mirodatos, C.; Geantet, C. *Catal. Today* **2009**, *143*, 172-178.

(37) Lin, Y.-C.; Huber, G. W. *Energy Environ. Sci.* **2009**, *2*, 68-80.

(38) Vlachos, D. G.; Caratzoulas, S. *Chem. Eng. Sci.* **2010**, *65*, 18-29.

(359a) Catalysts for Gas-to-Liquids Conversion - Effects of Support Type and Gas Space Velocity

Zhendong Pan, Chemical Engineering, Texas A&M University at Qatar, Doha, Qatar and **Dragomir B. Bukur**, Department of Chemical Engineering, Texas A&M University at Qatar, Doha, Qatar

Dueto high activity of cobalt metal in syngas conversion to higher hydrocarbons,supported cobalt catalysts have been widely used for gas-to-liquids (GTL)conversion. Cobalt is supported on refractory metal oxide, among which Al_2O_3,SiO_2 and TiO_2 have been used most extensively by majorcompanies engaged in development and commercial practice of this technology.

Inthis study we investigated effects support type and gas space velocity on catalyticperformance of 15 wt% Co on three supports (g-Al_2O_3,SiO_2 and TiO_2). Catalysts were prepared by incipientwetness impregnation of supports with aqueous solution of $Co(NO_3)_2$•$6H_2O$.Catalysts were characterized by different methods such as BET, XRD, TPR, and H_2-TPD. Catalytic tests wereperformed in a fixed-bed reactor at the following conditions: 220 °C, 300 psig, H_2/ CO = 2 and gas space velocities of 1 and 3.85 NL/g-cat/h. The catalysts wereactivated in situ withhydrogen at375 ° C for 12 h.

Physicalproperties of calcined catalysts are shown in the following Table.

Catalyst	BET SA (m^2/g)	Pore volume (cm^3/g)	Pore diameter (nm)
15 Co/Al_2O_3	153	0.42	9.8
15 Co/SiO_2	221	1.24	20.1
15 Co/TiO_2	13.9	0.092	11.8

TPRresults show existence of two step reduction of Co_3O_4. Degree of reduction wasabout 93% (H_2-TPD followed by pulse re-oxidation) and dispersion 8.6% (Co/Al_2O_3)catalyst, whereas cobalt was completely reduced on silica and titania supports.Dispersions on these two supports were 2.4% and 3.2%, respectively.

Duringtesting at SV = 3.85 NL/g-cat/h the alumina supported catalyst achieved thehighest CO conversion, 59%, whereas the silica supported catalyst was the leastactive 8.8% CO conversion. During testing at SV = 1 NL/g-cat/h CO conversionincreased and varied from 15.5% (SiO_2 support) to 94% (Al_2O_3support). Turnover frequencies (TOFs) were consistent with CO conversion trendsand ranged from $2.8*10^{-3}$ (s^{-1}) on silica support to $12*10^{-3}$(s^{-1}) on the alumina support. Lower TOFs on the silica and titaniasupports are ascribed to metal support interactions resulting in coverage ofcobalt crystallites with support moieties.

Onalumina supported catalyst we observed decrease of methane and increase of C_5^+selectivity with increase in CO conversion, however this trend was not observedon the silica and titania supported catalysts. Titania supported catalyst hadthe best selectivity (6.5% of CH_4 and 89% of C_5^+hydrocarbons) at SV = 3.85 NL/g-cat/h (CO conversion of 21.2%).

Acknowledgement

Thisresearch was supported by a grant from the Qatar National Research Fund (NPRP28-6-7-32).

See more of this Session: CO Hydrogenation II
See more of this Group/Topical: Catalysis and Reaction Engineering Division

Assessment of Pore Diffusion Limitations for the near Critical and Supercritical Fischer-Tropsch Synthesis

Aswani K. Mogalicherla, and Nimir O. Elbashir*

Chemical Engineering Program
Texas A&M University at Qatar, PO Box 23874, Doha, Qatar

***Corresponding author email address:** nelbashir@tamu.edu

1. Introduction

The catalyst material in the commercial FTS fixed bed reactors is called a pellet, which is 1 to 6 mm in size and contains macro, meso and micro pores. These pores are generally assumed to be filled with liquid hydrocarbon products in conventional gas phase FTS process or with supercritical solvent and hydrocarbon products in supercritical fluid FTS (SCF-FTS). As a result of pore filling, transport of reactants and products to and from the metal surfaces becomes slow and is sometimes the rate controlling step [1]. In such a situation, the heavy molecules diffuse at slower rate compared to lighter molecules. This difference in the transport properties alters the reactant and product concentrations near the active sites and creates spatial variations in the H_2/CO ratio in the catalyst pellet; it also affects both the FTS conversion and product selectivity [2,3,4]. Therefore, it is very important to understand and to develop an appropriate method to estimate the actual pellet behavior for detailed reactor design. In general the supercritical fluid region is very wide whereby the reaction temperature and pressure are the only two variables to manipulate the reaction environment from liquid-like properties (i.e. diffusivities) to gas-like ones. In the process of tuning transport properties, there is a possibility that the reaction moves from a kinetic controlled regime to diffusion controlled regime. Therefore, analyzing only the variations in transport properties (diffusivity, density and viscosity) with respect to operating conditions is not sufficient for optimal reactor design. The relative change in the diffusion rate with respect to the rate of reaction has to be determined to identify the optimal operating regime [3].

This communication represents a continuation of our previous efforts in developing a kinetic model for the non-ideal SCF-FTS environment [5]. The objective of the present work is to assess the diffusional limitations under the gas-phase as well as in the near critical and supercritical FTS conditions in a fixed bed reactor that typically contains catalyst particles of few millimeters (1mm) in diameters. For this a conventional theory of diffusion-reaction in porous pellets is employed. To account for the details of reaction kinetics, an LHHW model developed by Mogalicherla and Elbashir [5] from cobalt catalyst at near critical and supercritical fluid (hexanes) conditions as well as for gas-phase FTS conditions was used. The binary diffusional coefficients of all reactants and products in supercritical solvent were estimated from a correlation proposed by He [6], which accounts for the molecular weights and critical properties of all compounds, reactor operation conditions and reaction mixture density. The phase behavior in the fixed bed reactor in both gas-phase and supercritical phase FTS has been accounted in the effectiveness factor estimation process. The following conditions have been considered in developing this model for SCF-FTS: temperature range from 220 – 250 °C, pressure range from 35-80 bar, and H_2/CO feed ratios of 1-3. The catalyst effectiveness factor and concentration profiles of reactants and products along the pellet radius were estimated in 1 mm size catalyst particles. A single chain growth probability (α-value) model was assumed for both gas-phase and supercritical phase FTS. The other catalyst properties such as porosity, pore volume, surface area, and tortuosity factor were adopted from the properties of commercial Co/Al_2O_3 pellets [7].

2. Theory

Assuming that the catalyst pellet is spherical, the steady state material balance of key component in the catalyst pores can be represented as follows:

$$\left(\frac{1}{r^2}\right)\frac{d}{dr}\left(r^2\frac{df}{dr}\right) = \frac{(83.14*T*\rho_{Cat}*r_{CO})}{D^e} \qquad (1)$$

with boundary conditions

$$\left.\begin{aligned} &\text{at } r = R_p : f = f^{bulk} \\ &\text{at } r=0 : \frac{df}{dr} = 0 \end{aligned}\right\} \qquad (2)$$

where

$$r_{CO} = \frac{K f_{CO}^{1/2} f_{H_2}^{1/2}}{\left[1 + K_1 f_{H_2}^{1/2} + K_2 f_{CO}^{1/2} + K_3 f_{CO}\right]^2} \qquad (3)$$

And f_{CO} , f_{H2} are fugacities of carbon monoxide and hydrogen. r_{CO} is the rate of CO disappearance (mol/$_{gcat}$.min), D^e is the effective diffusivity of key component (cm^2/s), T is the absolute temperature (K) and ρ_{cat} is catalyst density (g/cc). The corresponding kinetic parameters are given Ref. 1.

To solve this problem using finite-differences, the interval [0, R_P] has been divided into twenty five sub-intervals. The derivatives in equation 1 and in boundary conditions have been replaced by central difference formulas to get system of non-linear algebraic equations and they were solved using fixed point iteration method in MATLAB 7.1 for determining the concentration of CO at different locations of the catalyst pellet. From the concentration of CO, the concentration of hydrogen at different locations has been identified as follows:

$$f_{H_2} = f_{H_2}^{bulk} - \frac{D_{CO}^{e} * (3-\alpha)}{D_{H_2}^{e}} \left(f_{CO}^{bulk} - f_{CO} \right) \quad (4)$$

From the estimated concentration profiles, the average CO consumption rate ($\overline{r}_{CO}$) in the catalyst pellet was estimated as follows:

$$\overline{r}_{CO} = \frac{\sum_{R_P}^{0} r_{CO} R_i}{\sum_{R_P}^{0} R_i} \quad (5)$$

where R_i is the position in the catalyst pellet from the center of the catalyst pellet and R_p is the catalyst pellet radius. Finally, the catalyst effectiveness factor determined as follows:

$$Catalysteffectivenessfactor(\eta) = \frac{\overline{r}_{CO}}{r_{CO}^{bulk}} \quad (6)$$

3. Results and Discussion

3.1 Phase Behavior

FTS is a complex multi-phase reaction, in which CO and H_2 enter the fixed bed reactor in the gas phase and some of the reaction products condense in the reactor and form liquidsin the reactor; this liquid phase trickles down the reactor. Caldwell and van Vuuren [8] reported that at the dew point of reaction mixture, the two phase regime (gas-catalyst) of the fixed bed reactor shifts into the three phase regime (gas-liquid-solid). The flash calculations have been conducted in Aspen Plus ® using Soave-Redlich-Kwong equation of state. Typical FTS product distribution has been used in our analysis that's represented by the hydrocarbons products (from C_1 to C_{30}) and water as byproduct. The Aspen simulation results indicated that under the SCF-FTS conditions, the typical reaction mixture (temperatures ranged from 220 to 250 °C, pressures ranged from 35 to 80 bar , the solvent:syngas ratio is 3:1 and the H_2/CO feed ratio is 2, H_2 usage ratio is 2.2 and chain growth probability 0.8) can be represented by a single phase and two-phase regime prevail inside the fixed bed reactors. It is noteworthy to mention that the phase behavior of the reaction mixture in the fixed bed reactor during gas phase FTS is sensitive to CO conversion. At higher CO conversions, the reaction mixtures easily condense because of the increase in heavy product concentration in the reaction mixture. The conversion of CO at which the condensation of the reaction mixture begins in the reactor (critical conversion) varies with the reactor's operating temperature. At 250 °C, the reactor bed is under a two phase regime up to 70% CO conversion, however at 220 °C, the critical CO conversion was only 29%. The model suggested by Caldwell and van Vuuren [8], indicated that a three phase regime prevails in the reactor once the liquid condensing in the catalyst bed voids with complete catalyst internal wetting.

In the two phase regime of a packed bed reactor, the catalyst pellets are surrounded by the reaction mixture in a vapor phase containing heavy hydrocarbons. Inside the catalyst pores, a considerable fraction of the reaction mixture might be liquid due to capillary condensation. The larger pores of the pellet are vapor filled whereas the smaller pores may be filled with the condensed hydrocarbons. The diffusional limitations in FT catalyst pellets are largely affected by the fraction of the liquid in the pellets which is highly depend on the reaction process parameters. It has been assumed in previous works that even in the two-phase regime, the catalyst pores are filled with heavy hydrocarbon products [3]. Dry collected catalyst samples from different positions of a fixed bed reactor and extracted hydrocarbon products from catalyst pellets using a mixture of decalin, xylene and MEK [9]. This analysis indicated that the amount of wax extracted from catalyst pellets at the entrance of the catalyst bed is lower than that at the middle and bottom of the bed: which indicate that the wax formation rate continuously increases along the reactor bed length. A similar assessment has been conducted in this study to identify the size and fraction of catalyst pores filled with heavy hydrocarbons in two-phase regime of the reactor using Kelvin's equation. Ermolaev et al. [10] modified the conventional Kelvin's equation for reacting systems as follows:

$$\ln\left(\frac{p_i}{p_i^s}\right) = \pm \frac{2*\sigma_i V_i}{r_C * R * T} \tag{6}$$

where p and p^s are the partial pressure and saturated vapor pressure, σ is surface tension, V is molar volume of ith compound in the reaction mixture. The physical properties required for Kelvin's equation have been extracted from Aspen Plus ® property data bank. Our finding shows that at 250 °C and critical CO conversion of 70% pores up to pore radius 210 A° are filled with the heavy hydrocarbon products.

3.2 Concentration Profile in catalyst pellet in SCF-FTS:

The typical CO and H_2 concentration profiles in catalyst pellets in SCF-FTS are shown in Figure 1a and 1b. The concentration profiles indicate that nearly 50% of the catalyst volume has been utilized for reaction in SCF-FTS, which is twice of the catalyst utilization reported for gas-phase FTS. The trends of catalyst effectiveness factor with operating parameters indicate that the catalyst particle effectiveness factor is proportional to the fugacity of CO in the bulk phase as well as the effective diffusivity of carbon monoxide. The isothermal catalyst effectiveness factor is sensitive to operating temperature only at low pressures which are nearer to the critical pressure of the solvent.

3.3 Comparison of catalyst effectiveness factor in gas-phase FTS and SCF-FTS

To compare the diffusional limitations in the near critical and supercritical FTS with the gas phase FTS, the isothermal effectiveness factor of the catalyst in gas-phase FTS has been evaluated. Unlike near critical and supercritical FTS, the catalyst effectiveness factor in gas phase FTS depends on the fraction of the active surface covered with hydrocarbon products as well as on the solubility of CO and H_2 in the heavy wax with which catalyst pore are filled. In the three phase regime of the packed bed reactor, catalyst pores always fill with condensed hydrocarbons. As discussed in section 3.1, the catalyst pores in the two-phase regime of the packed bed have partial internal wetting behavior.

The partial internal wetting has been determined as follows [11]:

$$\eta^T_{gas} = (\eta_{wax} * F) + (1 - F) * \eta_{gas} \quad (7)$$

where η_{wax} is the catalyst pellet effectiveness factor when is the catalyst pore completely filled with wax hydrocarbon product, η_{gas} is the catalyst pellet effectiveness factor when it is completely dry, F is the fraction of catalyst surface covered with hydrocarbon products and η^T_{gas} is the catalyst effectiveness factor with partial internal wetting.

The variation of catalyst effectiveness factor with carbon monoxide conversion in both gas-phase and supercritical phase FTS is shown in Figure 2. As shown in the figure the catalyst effectiveness factor is higher in the gas-phase FTS specifically in the initial portion of the catalyst bed. As the conversion increases along the reactor bed, the fraction of catalyst pores filled with heavy hydrocarbon increases and catalyst effectiveness factor decreases drastically as result. Due to single phase operation the active sites in supercritical phase FTS exposed to higher CO concentrations and catalyst effectiveness factor is higher than gas-phase FTS. This is in agreement with the experimental data reported in previous studies [12, 13].

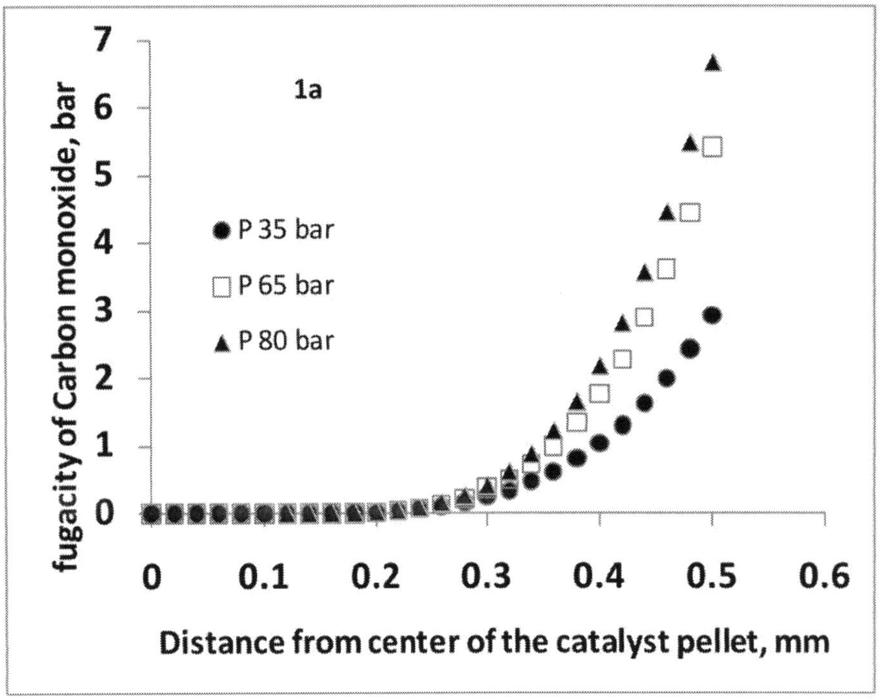

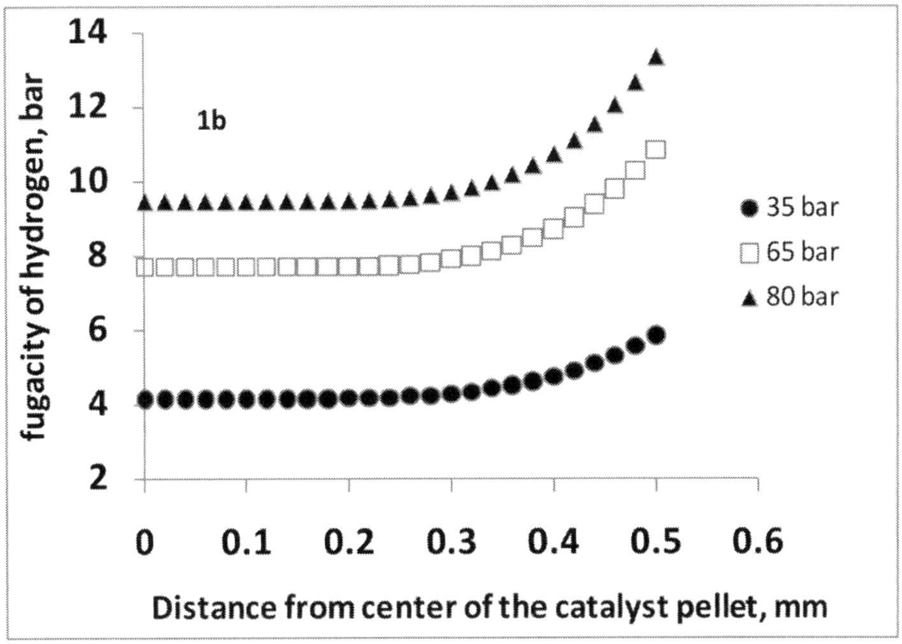

Figure 1: Concentration profile in catalyst pellet at 523 K in 1 mm size catalyst pellet

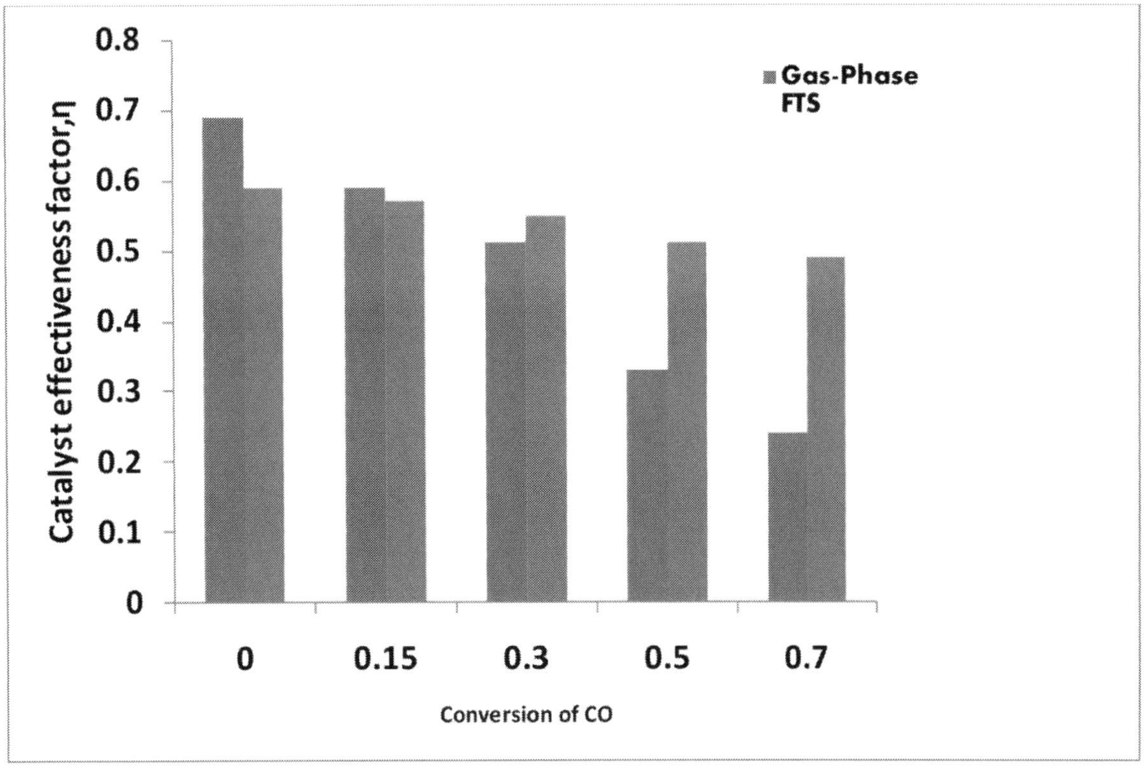

Figure 2: Variation of catalyst effectiveness factor in Gas-Phase FTS and SCF-FTS at 250 °C

4. Conclusions

Our investigations that focused on assessing the diffusional effects in catalyst pores for the SCF-FTS concluded that the fugacity of CO, H_2/CO ratio, the operating pressure and the temperature have significant impact on the catalyst effectiveness factor. For the studied range of operating conditions our findings show that the CO concentration drops rapidly within the initial layers of the catalyst particle. The higher diffusivity of hydrogen creates intense hydrogen rich environment inside catalyst pores, which can lead to undesired reactions, such as the methanation. The trends of catalyst effectiveness factor with operating parameters indicating that the catalyst particle effectiveness factor is proportional to the fugacity of CO in the bulk phase. The isothermal catalyst effectiveness factor is sensitive to operating temperature only at low pressures (equivalent in this case to the critical pressure of the solvent). The catalyst effectiveness factor value in the gas-phase FTS is higher than that of the supercritical phase FTS; however, only at the entrance of the catalyst bed. After certain reactants conversion, the products start to condense and catalyst effectiveness factor drastically decreases in the gas-phase FTS. However, due to the two phase operation in SCF-FTS conditions (i.e. semi single phase for the reaction mixture inside the catalyst bed) coupled with improved diffusivities of reactants in supercritical solvent, the active sites normally exposed to higher CO concentrations relative to gas phase FTS. This phenomenon increases the effectiveness factor of the catalyst particles and suppresses undesired reactions such as methanation and chain termination to light paraffin.

Acknowledgment:

The authors would like to acknowledge the financial support of this work to the Qatar National Research Fund (QNRF), under the National Priorities Research Program grant (NPRP 08-261-2-082).

References

1. D.B. Bukur, X. Lang, A. Akgerman, Z. Feng. Effect of process conditions on Olefin selectivity during conventional and supercritical Fischer-Tropsch synthesis. Ind. Eng. Chem. Res. 36(7) (1997) 2580-2587.
2.Y.-N. Wang, Y.-Y Xu, H.-W. Xiang, Y.-W.Li, B.-J. Zhang. Modeling of catalyst pellets for Fischer-Tropsch synthesis. Ind. Eng. Chem. Res. 40(20) (2001) 4324-4335.
3.X. Zhan, B.H. Davis . Assessment of internal diffusion limitation on Fischer-Tropsch product distribution. Applied Catalysis A: General 236 (2002) 149-161.
4. A. Jung, C. Kern, A. Less. Modeling of internal diffusion limitations in a Fischer-Tropsch catalyst. In Advances in Fischer-Tropsch synthesis, catalysts and catalysis; Davis, B.H.; Occelli., Eds.: Taylor and Francis group, New York, 2010 , 215-227.
5.A.K. Mogalicherla, N.O. Elbashir. Development of a Kinetic Model for Supercritical Fluids Fischer Tropsch Synthesis. Energy and Fuels 25 (2011), 878-889.

6.C. He. Prediction of binary diffusion coefficients for solutes in supercritical solvents. AIChE.J, 43(11) (1997) 2944-2947.
7.C.N. Satterfield, P.N. Calde, Diffusion in commercially manufactured pelleted catalyst. Ind. Eng. Chem. Des. Dev. 7(2), (1968) 256-260.
8. L. Caldwell, D.S. van Vuuren. On the liquid formation and composition of the liquid phase in Fischer-Tropsch reactors. Chemical Engineering Science 41(1) (1986) 89-96.
9.M.E. Dry. Fischer-Tropsch synthesis over Iron catalysts. Catalysis Letters 7 1990,241-252
10. V.S. Ermolaev, V.Z.Mordkovich, I.G.Solomonik. Influence of capillary condensation on heat and mass transfer in grain of a Fischer-Tropsch synthesis catalyst. Theoretical Foundations of Chemical Engineering 44(5) 2010, 660-664.
11.J.Wood; L.F.Gladden.. Modeling diffusion and reaction accompanied by capillary condensation using three-dimensional pore network. Part 1.Fichian diffusion and pseudo first order reaction kinetics. Chemical Engineering Science 57 2002, 3033-3045.
12. N.O. Elbashir, P. Dutta, A. Manivannan, M.S. Seehra, C.B. Roberts. Impact of cobalt based catalysts characteristics on the performance of conventional gas-phase and supercritical-phase Fischer-Tropsch synthesis. Applied Catalysis A: General 285 (2005) 169-180.
13. D.J. Bochniak, B. Subramaniam. Fischer-Tropsch synthesis in near –critical n-hexane: Pressure tuning effects. AIChE J.44 (1998) 1889-1896.

(359c) Development of a Comprehensive and Flexible Model for Fixed-Bed Fisher-Tropsch Reactors

Joseph W. Pratt, Thermal/Fluid Science and Engineering, Sandia National Laboratories, Livermore, CA

Liquid fuels synthesized from renewable CO and H_2 could displace petroleum-based fuels to provide clean, renewable energy for the future. Small scale (<100 bbl/day) reactors for synthetic liquid fuels production (e.g. methanol, Fischer-Tropsch diesel) are an emerging development area that may enable the use of diverse, renewable, carbon-neutral feedstocks, such as biomass. For small scale production of synthetic liquid fuels from syngas, fixed-bed reactors are well-suited and have high potential productivity per reactor volume. To fully realize the potential of these reactors, four key issues must be addressed: 1) minimize temperature gradients due to the highly exothermic reactions, 2) maximize diffusion of reactants and products, 3) prevent heavy hydrocarbons (waxes) from accumulating in the catalyst pellets, reducing conversion and 4) management of lifecycle degradation.

Over the last 90 years, the determination of optimal Fischer-Tropsch synthesis (FTS) reactor conditions has evolved starting with primarily experimentally determined inputs, to increased use of empirical modeling, to incorporation of more theory so that optimal conditions can be predicted and laboratory and pilot plant tests are no longer needed. However, although reactor size and capacity has increased over the years, the same general reactor design today is not much different than that used in the 1940's, especially in the case of fixed-bed reactors. This has led, rightly so, to models that explicitly or implicitly take advantage of this wealth of experience and knowledge with some degree of empiricism. Although they give very accurate results for modeling traditional reactors, the existing models are unable to examine significant design changes and innovations that may be needed to solve the four challenges listed above.

The goal of the model developed in this work is to improve upon the fidelity and flexibility of current models, by reducing the level of empiricism, to enable examination of small-scale reactors or unique designs. Accomplishing this requires validated, coupled, heat and mass transport and chemical kinetics models derived from a component-based, dynamic (time-varying) two-dimensional heterogeneous model that accounts for interfacial and intra-particle gradients. There is no evidence in the open literature of a model of this type being successfully developed.

This paper presents our model development progress in the following areas:

1. Two-phase interparticle (bulk) flow
2. Interparticle (bulk) heat transfer

3. Intraparticle mass transfer (within pore)
4. Intraparticle heat transfer (within solid and pore)
5. FTS chemical kinetics
6. Time-dependent catalyst deactivation by coke deposition, hydrothermal sintering, liquid/wax accumulation, and poisoning.
7. Combination of the various phenomena into a single comprehensive model.

The successful, validated model of this type should enable accurate simulation of innovative and non-traditional reactors.

See more of this Session: CO Hydrogenation II
See more of this Group/Topical: Catalysis and Reaction Engineering Division

(359d) Influence of Gas Flowrate On Behavior of High Thermal Conductivity Microfibrous Entrapped Catalyst for Exothermic FTS

Donald R. Cahela, Min Sheng, and Bruce J. Tatarchuk

Center for Microfibrous Materials Manufacturing, Department of Chemical Engineering, 230 Ross Hall Auburn University, Alabama 36849

Fischer-Tropsch synthesis (FTS) is a highly exothermic reaction with challenging heat transfer issues. Enhanced heat transfer has been realized by using microfibrous entrapped catalysts (MFEC) to perform FTS in fixed bed reactors. Pseudo-homogeneous thermal conductivities greater than 9 W/m-K have been obtained using MFEC made from highly conductive copper fibers. Much larger tubes can be used for FTS using Cu-MFEC than is possible using packed beds of extrudates. 2D (r,z) Simulations and recent experiments indicate that tubes as large as 50 mm ID can safely be used for this highly exothermic reaction without danger of runaway. The pseudo-homogeneous thermal conductivity of MFEC made from copper, nickel or stainless steel without flow correlates with the electrical conduction through the sintered metal matrix in terms of a junction factor. In flowing conditions the pseudo-homogeneous thermal conductivity increases linearly with the Reynolds number. This behavior is investigated by considering the tortuous nature of the sintered fibers. In the wet-lay papermaking process used to prepare preforms [1] the fibers form in stacks randomly oriented in the horizontal plane. Sintering produces contact joints between fibers lying in random directions, so the paths through the metal fibers are tortuous. This effect is accounted for by modeling the tortuous fibers as a separate phase in a segregated temperature, plug flow, fixed bed tubular reactor (ST-PFFBTR) model for FTS. Heat can flow from the main fiber phase that contacts the wall through the gas phase to the extra fiber phase that does not contact the wall. Also, the domain of the influence of this separate fiber phase only extends some multiple of the original length of the fibers used to prepare performs by wet-lay papermaking process. This segregated temperature simulation uses heat transfer correlations to model heat transfer in the reactor without resorting to pseudo-homogeneous thermal conductivity and lumped wall heat transfer coefficients that must be estimated. Flux limiters are applied to make the simulation more robust.

The highly exothermic nature of the Fischer-Tropsch synthesis reaction results in a temperature excursion at the tube centerline in a tubular fixed bed reactor. Using microfibrous entrapped catalyst (MFEC) made from copper fibers has potential to improve selectivity for Fischer-Tropsch synthesis by minimizing temperature excursions, resulting in improved selectivity causing a reduction of the size and weight of a processing system for producing JP-5 type products. Temperature excursions in packed bed reactors can be minimized by recycling some of the products, possibly both gas and liquids. Recycling requires additional process equipment that may be avoided by using MFEC to minimize temperature excursions. MFEC catalysts allow larger diameter reactors to be used without large centerline temperature excursions occurring compared to packed beds. This results in a reduction in the number of reaction tubes required compared to a traditional packed bed of catalyst particles.

Selectivity for the Fischer-Tropsch Synthesis (FTS) reaction is sensitive to temperature, the chain propagation probability dropping by about 0.1 for every 25ºC increase in temperature. For the highly exothermic FTS reaction, a 50ºC temperature increase from wall to centerline has been indicated in two inch diameter fixed bed reactors for FTS employing Iron based catalysts. Cobalt based catalysts are more active than Iron catalysts, such as Ruhrchemie, and experience greater temperature excursions. Minimizing the temperature excursion at the centerline of a tubular reactor for the FTS reaction is important to produce the desired product such as diesel fuel or JP-5. MFEC manufactured using copper fibers demonstrate effective bed thermal conductivities of about 7-10 W/m-K while the thermal conductivity of packed beds is about 0.2 W/m-K, similar to that of the gas phase. This significant increase in the effective thermal conductivity of the MFEC catalyst bed allows a much larger diameter reactor to be used for FTS to produce JP-5 while minimizing the plant footprint and weight.

A 2-D (r,z) model of a tubular reactor packed with catalyst particles or with an MFEC catalyst is solved to estimate the temperature excursion encountered in the FTS reaction. The accentuation of the hot spot by the increased reaction rate due to the hot spot at the centerline of the reactor tube is accounted for in this model. This model uses segregated energy balances on the gas, catalyst and fiber phases. Using this approach avoids estimation of an overall effective thermal conductivity. The heat of reaction is generated inside the catalyst particles and the catalyst temperature will be assumed to be uniform inside the catalyst particles. A material balance on CO is solved for the conversion. A simulation using Euler implicit integration to determine the steady-state is implemented using three-point central finite difference formulas. The complex rate expression is reduced to a first order reaction form by lumping the volumetric reaction rate with the film mass transfer coefficient. A mass balance on CO is written as a 2-D (r,z) cell model. The axial Peclet number is very large, for the relevant cases simulated, indicating that very little axial dispersion of the gaseous components occurs. Comparison of the performance of a packed bed of catalyst particles and MFEC catalysts are made.

Steady-State Heating of a Packed Tube with Flowing Gas

The steady-state temperature profile inside a heated tube with flow through a homogeneous material was measured on a packed bed of alumina powder and a microfibrous copper sample. Segregated 2-D heat balances for the gas, catalyst and fiber phases are written to model the tube heatup. The PDEs describing the heat transfer inside the material are discretized using central finite differences. Flux limiters were applied to the convective flux term and it was also averaged in the radial direction. These equations describing the transient heat transfer were integrated to steady-state using a line implicit Euler method in the axial direction. The simulation was programmed using the Euphoria [1] interpreter. All simulations were performed using 32 and 128 divisions in the radial and axial directions respectively. The derivation of the segregated 2-D model is given as Appendix A. Adding mass balances, with heat generation in the catalyst phase, to the segregated 2-D model would be a simulation of the FTS reaction with uniform velocity at any cross section. The model in Appendix A includes a term describing heat generation in the catalyst phase.

Pfeffer's heat transfer model for muliparticle systems was used for estimation of the gas-catalyst and gas-fiber heat transfer coefficients. The catalyst-wall and fiber-wall heat transfer coefficients were estimated by equation 34 given by Dixon and Cresswell [8]. The solid to wall heat transfer coefficient estimates were adjusted by multiplying by the solid fraction of each component divided by that typical for packed beds (0.6). The gas-wall heat transfer coefficient was estimated by a correlation due to Dixon et. al. [9].

Two materials were tested for this comparison. One is a bed of 80x100 mesh alumina powder. The other is a sintered mesh consisting of equal amounts of 4 and 12 micron copper fibers and the same alumina powder. A two inch section of a 1 ½" O.D. stainless steel tube was filled with the media to be evaluated. The tube was sealed with five thermocouples inside it in the first set of tests. In a second set of heat-up tests eleven thermocouples were used. A diagram of the location of the thermocouples in the packed bed heat-up test is shown as Figure 1. Three thermocouples labeled 3, 7 and 10 were at the middle of the tube located at the centerline, R/2 and at the wall.

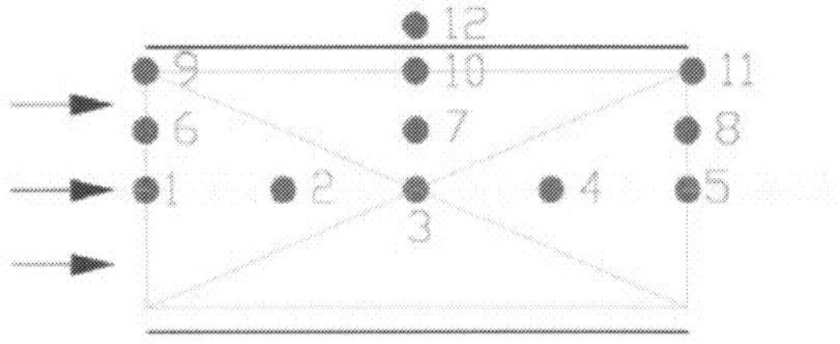

Figure 1. Diagram of thermocouple locations in the packed tube heat-up test.

The data and centerline temperatures predicted by the simulation for the packed bed and for the copper MFEC are shown as Figures 2 and 3. The thermal conductivities of the catalyst phase was taken as the volume average of the alumina in the tube multiplied by a contact efficiency factor. The simulation was converted into a regression routine and a catalyst phase contact efficiency factor of 0.0383 (± 67%) was determined by a Newton-Raphson search using the data on the packed bed heatup The conduction through the catalyst phase depends on particle to particle contacts resulting in a lowering of the thermal conductivity. Similarly, conduction through the copper fiber phase depends on how well the fibers are sintered. A contact efficiency factor of 0.106, determined from electrical resistance measurements, was applied to the volumetric average value of thermal conductivity for copper as an estimate of the radial conductivity of the fiber phase. The MFEC is loaded into the tube as layers, so a reduction of the conductivity of the fiber phase in the axial direction is expected. The axial fiber conductivity reduction of 0.622 (±112%) was estimated by a Newton-Raphson search using the data for the MFEC heatup. The uncertainties indicated are the 95% confidence interval estimation. The heat transfer parameters in this model were predicted by the Nusselt No. correlation of Achenbach [2]. This correlation gave heat transfer coefficients about one-half as large as those predicted by Pfeffer's [3] theoretical model. Figure 2 and 3 show steady state centerline temperature profiles for a 1-1/2" I.D. tube packed with alumina powder and a copper MFEC respectively.

Discussion of Packed Tube Heatup

The simulation of the temperature profile for the packed bed of alumina powder deviates from the data at low flowrates. The simulation of the heatup for the tube packed with copper MFEC follow the data trend. We expect to acquire better data by modifying the setup to prevent heating of the flowing nitrogen before it enters the media. A higher conductivity gas such as Helium could also be used to verify the applicability of this model.

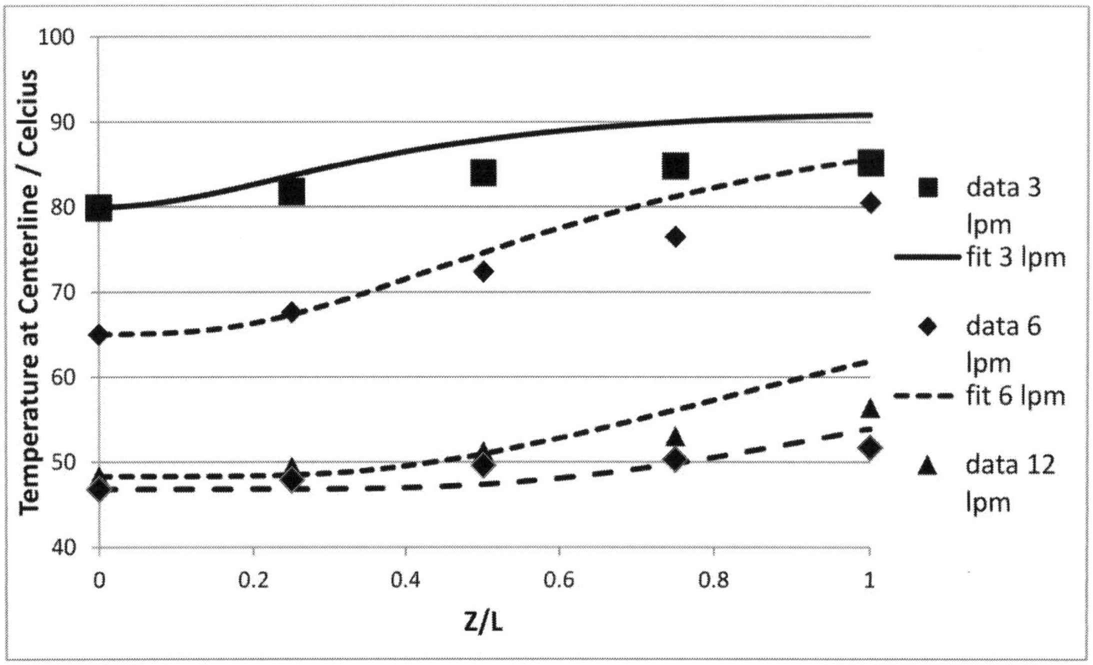

Figure 2. Steady-State data and segregated 2-D simulation result for the heat-up of a tube packed with alumina powder-80x100.

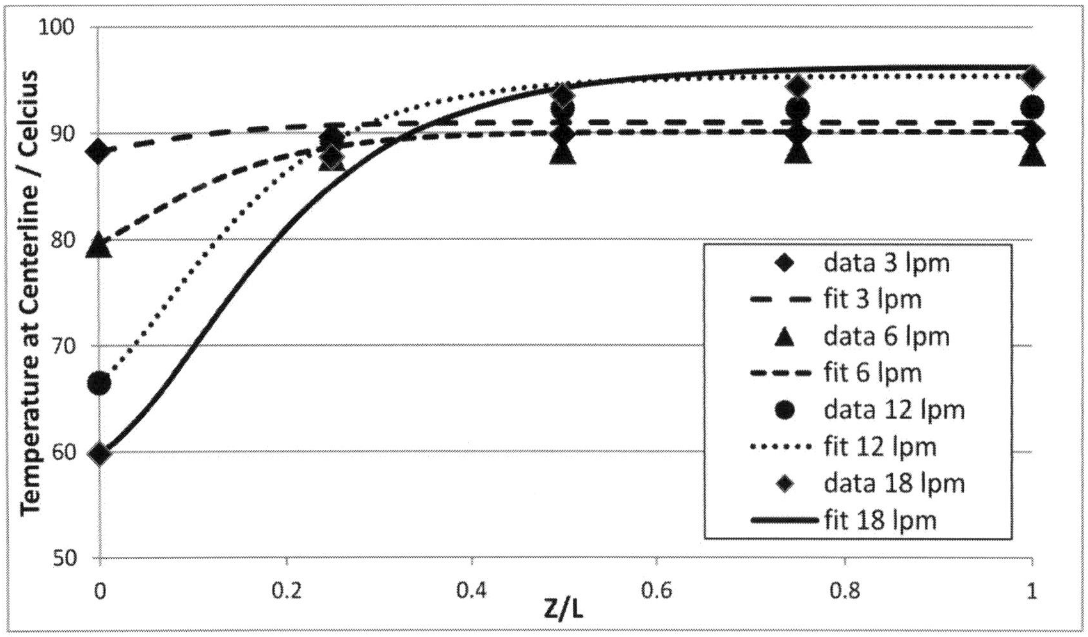

Figure 3. Steady-State data and segregated 2-D simulation result for heat-up of a tube packed with MFEC (Copper fibers and alumina powder-80x100).

ASF Product Distribution for Cobalt /Alumina Catalyst

Chain growth probability for the ASF distribution were determined in our laboratory at temperatures for 225 to 255ºC by analyzing gas, hydrocarbon liquid and wax fractions by gas chromatography. The alpha values as a function of temperature are shown in Figure 4. The slope of the linear fit for the cobalt catalyst is about one-half of that typical for iron catalysts.

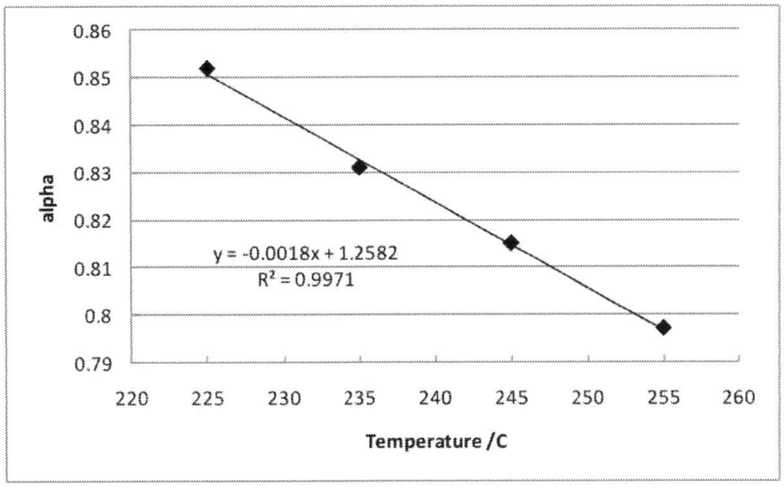

Figure 4. ASF chain growth probability for Fischer-Tropsch synthesis using 15 wt% cobalt/alumina, pressure of 20 bar and H2:CO=2.0.

The temperature dependence indicated by the trend line was used to estimate the product distribution of hydrocarbon fractions, C1-C4, C5-C8, C9-C17, and +C17. The C9-C17 fraction corresponds to the desired synthetic JP-5 product. In the simulation all of the points in the 2-D (r,z) domain were averaged. The gas phase temperature was used to compute the chain growth probability. Then a fractional yield of each of the four groups of paraffins was computed. These fractional yields multiplied by the rate of carbon monoxide consumption were then averaged over the entire domain.

Simulation of Fischer-Tropsch Synthesis using a Segregated 2D Reactor Model

The empirical kinetic model used by Ceramatec[4] was employed in this study to represent the rate of conversion of synthesis gas to hydrocarbon products..

$$r_{C,org} = -r_{CO} = \frac{A_1 e^{-E_{act}/RT} p_{CO}^m p_{H2}^n}{\left(1 + A_2 e^{-E_{ads1}/RT} p_{CO}^x + A_3 e^{-E_{ads2}/RT} p_{H_2}^y + A_4 e^{-E_{ads3}/RT} p_{CO}^{Z_1} p_{H_2}^{Z_2}\right)^2}$$

[1]

The first of the four sets of parameters[4] for cobalt catalysts were used: A1=1.27E+08 1/s-atm^(m+n), A2=A3=0, A4=7.50E-02, Eact=90927 J/mol, Eads1=Eads2=0, Eads3=-6753 J/mol, m=0.5, n=0.75, x=y=0, z1=0.5, z2=0.25. Additionally the loading of cobalt on alumina for the catalyst used in the experimental results shown is fifteen percent. A dispersion of 0.186 for packed bed and 0.053 for MFEC were applied to match the experimental conversions in our 15 mm I.D. laboratory reactor operated at 225ºC and 20 bar and H2:CO ratio of two. The results of this calculation are shown in Table 1. The activation energy in model 1 is the largest of the four sets of parameters, but it is still not high enough to reproduce the conversion versus temperature data we observed in our 15 mm I.D. tubular reactor shown in Table 1.

Table 1. Comparison of experimental CO conversion and centerline temperature excursion with FTS model. 14.8 mm I.D. tube, 50.8 mm catalyst bed, 1gm of catalyst particles, 2.485 gm of 6 m copper fibers, 113 sccm of synthesis gas at 20 bar outlet pressure. Volume fraction of 6 m copper fibers and 60x80 mesh 15 wt. % Co/alumina catalyst are 0.032 and 0.163 respectively.

Wall Temp.	xCO		dTmax	
Celcius	Data	Calc	Data	Calc
225	0.389	0.393	3.2	2.3
235	0.728	0.614	2.7	3.7
240	0.987	0.742	4.1	4.7

Segregated energy balances on gas, catalyst and fiber phases are shown in Appendix A. The partial pressure of carbon monoxide was computed by a transient 2-D (r,z) mass balance.

$$\varepsilon \frac{\partial p_{CO}}{\partial t} = \frac{\partial v_z p_{CO}}{\partial z} + \frac{\varepsilon}{\tau} D_{CO} \left(\frac{1}{r}\frac{\partial}{\partial r}\left(r\frac{\partial p_{CO}}{\partial r}\right) + \frac{\partial^2 p_{CO}}{\partial z^2}\right) + \delta_{CO} r_{C,org} \qquad [2]$$

The rate of the FTS reaction was converted to a pseudo first order rate constant by dividing by the concentration of CO and lumped with the external mass transfer rate to obtain an overall first order rate constant. The first order effectiveness factor for spherical particles was also used employing the pseudo first order rate constant. The lumped first order rate constant is computed as follows:

$$\phi = \frac{R}{3}\sqrt{\frac{\left(\frac{-r_{C,org}}{p_{CO}/_{RT}}\right)}{D_{COe}}} \quad [3]$$

$$\eta = \frac{1}{\phi}\frac{\left(\frac{3\phi}{\tanh 3\phi}-1\right)}{3\phi} \quad [4]$$

$$k_{CO} = \frac{k_f a_c * \eta * \left(\frac{-r_{C,org}}{p_{CO}/_{RT}}\right)}{k_f a_c + \eta * \left(\frac{-r_{C,org}}{p_{CO}/_{RT}}\right)} \quad [5]$$

The mass balance becomes as follows:

$$\varepsilon \frac{\partial p_{CO}}{\partial t} = \frac{\partial v_z p_{CO}}{\partial z} + \frac{\varepsilon}{\tau} D_{CO}\left(\frac{1}{r}\frac{\partial}{\partial r}\left(r\frac{\partial p_{CO}}{\partial r}\right) + \frac{\partial^2 p_{CO}}{\partial z^2}\right) + \delta_{CO} r_{C,org} \quad [6]$$

Flux limiters and radial averaging are applied to the axial convection term. An example of the discretization of this type term is shown in the appendix. The composition of the reacting mixture is computed from the CO conversion and the reaction stoichiometry. A Danckwerts' type B.C. is applied at the reactor entrance. The exit B.C. is obtained by using a backward difference formula for the convective flux term in the PDE and eliminating the axial diffusion term from the PDE. This form of the exit B.C. is referred to as a minimum reduction of PDE B.C. [6]. At the centerline L'Hopital's rule is applied to the radial diffusion term and zero gradient is also applied. At the wall the zero gradient of concentration is applied. The mass balance was solved in a line implicit fashion in the axial direction over the entire radial domain from the centerline to the wall. This was accomplished by incorporating the B.C.s into the discretized forms of the mass balance PDE. All simulations were performed using 32 and 128 divisions in the radial and axial directions respectively.

Figure 4 indicates that the temperature corresponding to the maximum production of C8-C17 with an alpha of 0.86 is 220ºC. Some example computations are done using entrance and wall temperature of 220ºC and outlet pressure of 20bar. The volume loadings of copper fibers and 15 wt% Co/Al_2O_3 catalyst in the MFEC were taken as 5% and 15%. Simulations were integrated until the average change in temperature was less than 10^{-5} ºC. The first set of computations employ a constant gas hourly space velocity of 300 hr^{-1} and the reactor diameter is varied from 20mm to 50 mm. At 60 mm the simulation indicated a runaway condition.

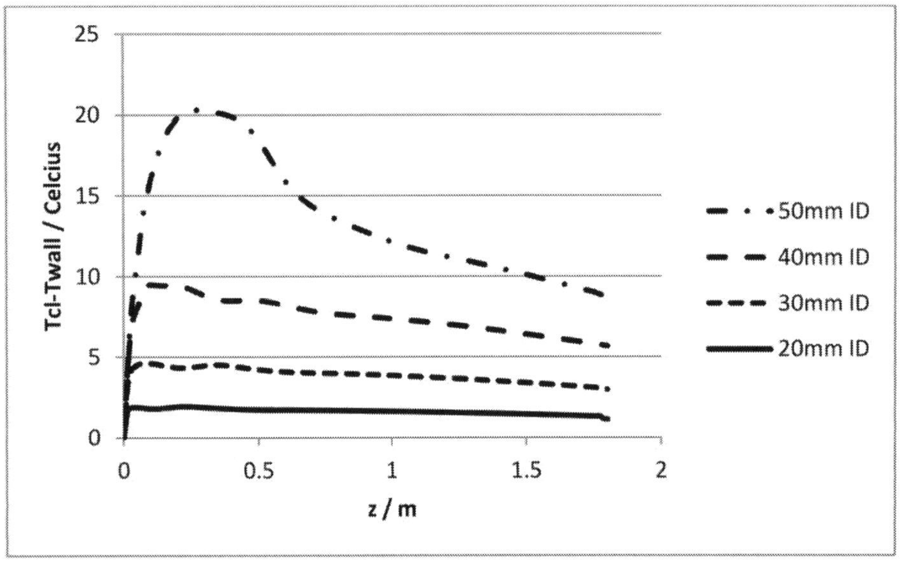

Figure 5. Centerline Temperature Profile for Several Internal Diameters with a GHSV of 300 hr^{-1}. Simulation settings: wall temperature of 220ºC, outlet pressure of 20 bar, length 1.8 m, 5 vol% of 6 micron Cu fibers, 15 vol% of 15 wt% of Co/Al2O3.

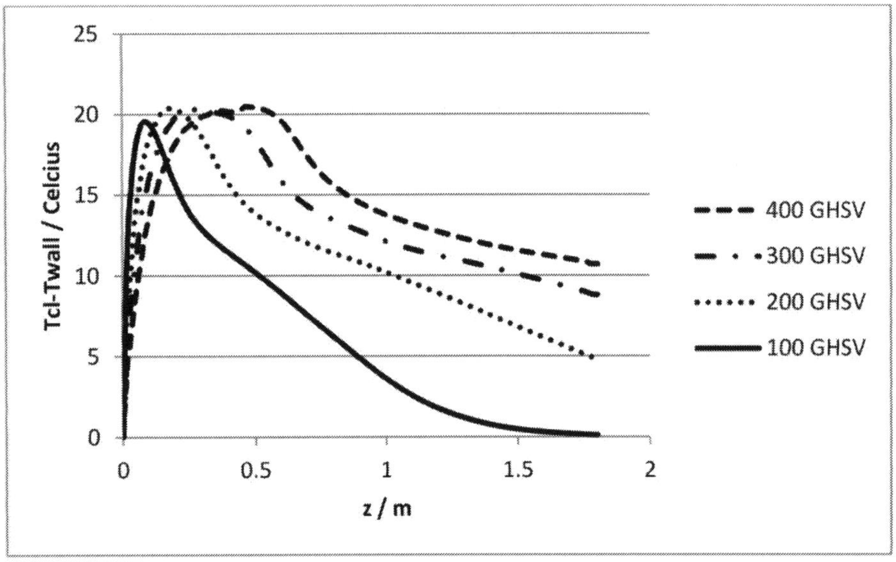

Figure 6. Centerline Temperature Profile for Several GHSV with an iternal diameter of 50 mm. Simulation settings: wall temperature of 220ºC, outlet pressure of 20 bar, length 1.8 m, 5 vol% of 6 micron Cu fibers, 15 vol% of 15 wt% Co/Al2O3.

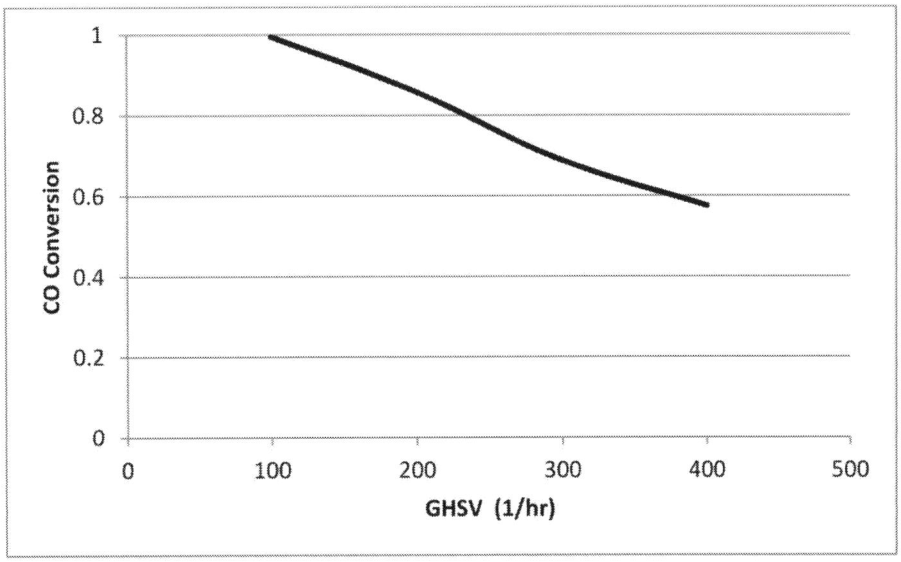

Figure 7. Conversion of carbon monoxide for an Internal Diameter of 50 mm. Simulation settings: wall temperature of 220ºC, outlet pressure of 20 bar, length 1.8 m, 5 vol% of 6 micron Cu fibers, 15 vol% of 15 wt% Co/Al2O3.

Discussion

The simulation results shown in Figures 5-7 indicate that the MFEC used in the steady state heatup, described in Table 1, with nitrogen flowing would allow a reactor tube of 50 mm I.D. to be used without any additional heat transfer enhancements such as inert gas added to the reactor feed, feed gas recycle or recycling hexane and pentane products. These preliminary results indicate that a MFEC with five volume percent copper fibers is sufficient to moderate the exothermic heat effect of the Fischer-Tropsch synthesis in tubes less than two inches upto a GHSV of 400 hr^{-1}. For larger diameters or higher flowrates the volume fraction of copper fibers would need to be higher to increase the thermal conductivity of the MFEC. To increase the reactor tube diameter even further, the volume fraction of copper fibers in the MFEC must be increased to prevent a runaway reaction from occurring. Preparing MFEC with larger volume fractions of highly conductive metal fibers can be accomplished by several techniques. MFEC preforms can be compressed before sintering to increase the volume fraction. Alternately, after sintering, MFEC can be compressed to the thickness required to obtain the desired thickness. The size of the fibers used in MFEC can be chosen so that the pore size in the material is similar to the size of the particles to be loaded into the MFEC material.

The MFEC used in the thermal conductivity test, shown as Figure 3, was prepared at the minimum sintering condition of 650ºC for forty-five minutes. The connectivity of the copper fibers in the MFEC could be improved by sintering the material at a higher temperature and/or for a longer time in the furnace. Also, electroplating and chemical impregnation [7] have been used to increase the thermal conductivity of MFEC.

The advantage from using segregated energy balances in the reactor model is that the thermal parameters are well defined. So, no lumped overall thermal conductivity needs to be estimated as in a pseudo-homogeneous reactor model. Dependable methods for estimating lumped overall thermal conductivity for a mixture of metal fibers and catalyst particles have not been developed although dozens of articles have been published on the subject.

Nomenclature

[to be added]

Acknowlegment

Funding for this research came from the U.S. NAVY.

Project title: Enhanced Fischer Tropsch Catalysis & Synthesis of Naval Logistic Fuels

Award # N00014-09-C-0208

References:

1) http://www.rapideuphoria.com/

2) E. Achenbach, "Review: heat and flow characteristics of packed beds," Experimental Thermal and Fluid Science 1995, 10:17-27.

3) R. Pfeffer, "Heat and mass transport in multiparticle systems," I&EC Fund., 1964, 3:380-383.

4) K.M. Brunner, D.S. Frost, R.P.S. Peguin, C.H. Bartholomew, W.C. Hecker, "A 1-D trickle fixed-bed recycle reactor model for the Fischer-Tropsch Synthesis," ACS Extended Abstract 20090929.

5) D.R. Cahela, http://www.rapideuphoria.com/mat.htm, "Simultaneous Equations-Don Cahela".

6) W.E. Schiesser, "PDE boundary conditions from minimum reduction of the PDE," Applied Numerical Mathematics 20 (1996) 171-179.

7) M. Sheng, D.R. Cahela, H.Y. Yang, W.Y. Yanz, C.F. Gonzalez, D.K. Harris, B.J. Tatarchuk, "Effective thermal conductivity and junction factor for sintered microfibrous materials," manuscript in preparation to be submitted to the International journal of Heat and Mass Transfer.

8) A.G. Dixon and D.L. Cresswell, "Theoretical prediction of effective heat transfer parameters in packed beds," AIChE J., V25, No4, 1979, 663-676.

9) A.G. Dixon, Arias and Wiley, "Wall-to-liquid mass transfer in fixed beds at low flow rates," Chem. Eng. Sci. 58 (2003) 1847-1857.

Appendix: Segregated 2-D energy balances for simulation of heat-up of a packed tube with flowing gas or reaction with heat effects occurring in a packed tube.

The Partial differential equations describing the heat balance between gas, catalyst and fibers inside the tube are as follows:

$$\varepsilon\rho_g C_{pg}\frac{\partial \mathrm{T_g}}{\partial \mathrm{t}} = \varepsilon k_g \frac{1}{\mathrm{r}}\frac{\partial}{\partial r}\left(\mathrm{r}\frac{\partial \mathrm{T_g}}{\partial r}\right) + \varepsilon \mathrm{k_g}\frac{\partial^2 \mathrm{T_g}}{\partial \mathrm{z}^2} - \rho_g C_{pg} v_z \frac{\partial \mathrm{T_g}}{\partial \mathrm{z}}$$

$$-\mathrm{x_c}(1-\varepsilon)\mathrm{a_c h_{gc}}(\mathrm{T_g}-\mathrm{T_c}) - \mathrm{x_f}(1-\varepsilon)\mathrm{a_f h_{gf}}(\mathrm{T_g}-\mathrm{T_f}) \quad \text{[A-1]}$$

$$\mathrm{x_c}(1-\varepsilon)\rho_c C_{pc}\frac{\partial \mathrm{T_c}}{\partial \mathrm{t}} = x_c(1-\varepsilon)k_c \frac{1}{\mathrm{r}}\frac{\partial}{\partial r}\left(\mathrm{r}\frac{\partial \mathrm{T_c}}{\partial r}\right) + x_c(1-\varepsilon)\mathrm{k_c}\frac{\partial^2 \mathrm{T_c}}{\partial \mathrm{z}^2}$$

$$-\mathrm{x_c}(1-\varepsilon)\mathrm{a_c h_{gc}}(\mathrm{T_c}-\mathrm{T_g}) + \mathrm{r_{c,org}}(-\Delta H_{FT}) \quad \text{[A-2]}$$

$$\mathrm{x_f}(1-\varepsilon)\rho_f C_{pf}\frac{\partial \mathrm{T_f}}{\partial \mathrm{t}} = \mathrm{x_f}(1-\varepsilon)\left(k_{fr}\frac{1}{\mathrm{r}}\frac{\partial}{\partial r}\left(\mathrm{r}\frac{\partial \mathrm{T_f}}{\partial r}\right) + \mathrm{k_{fz}}\frac{\partial^2 \mathrm{T_f}}{\partial \mathrm{z}^2}\right) - \mathrm{x_f}(1-\varepsilon)\mathrm{a_f h_{gf}}(\mathrm{T_f}-\mathrm{T_g}) \quad \text{[A-3]}$$

With boundary conditions: $T=T_0$ at z=0. And $T=T_{bath}$ at the outside of the tube. x_c and x_f are the volume fraction of the solid phases occupied by the catalyst and fiber phases ($x_c + x_f=1$). The specific surface areas of the catalyst and fibers are designated by a_c and a_f. The bed voidage is represented by ε. Equation A-3 has been simplified by dividing out the volume fraction of fibers from each term.

A heat transfer coefficient is applied at the interface of the tube wall and the gas, catalyst and fiber phases.

$$-k_g \frac{\partial \mathrm{T_g}}{\partial r} = h_{gw}(T_g - T_w)\ ,\ -k_c \frac{\partial \mathrm{T_c}}{\partial r} = h_{cw}(T_c - T_w) \text{ and } -k_f \frac{\partial \mathrm{T_f}}{\partial r} = h_{fw}(T_f - T_w) \text{ at r=R} \quad \text{[A-4]}$$

The wall temperature was taken as a constant value.

Discritization of the PDE model

Flux limiters were applied to the convective flow terms in the gas phase energy balance and the mass balance. They are calculated as show below:

$$r = \frac{y_j - y_{j-1}}{y_{j+1} - y_j},\quad \emptyset = \frac{1+r^2}{1+3r^4},\quad \frac{\partial y}{\partial z}_j = \emptyset\frac{y_j - y_{j-1}}{\Delta z} + (1-\emptyset)\frac{y_{j+1}-y_j}{\Delta z} = \frac{-\emptyset y_{j-1} + (2\emptyset - 1)y_j + (1-\emptyset)y_{j+1}}{\Delta z} \quad \text{[A-5]}$$

For r=0, ϕ=1 results in the backward difference formula, for r=1, ϕ=1/2 results in the central difference formula and for r=∞, ϕ=0 results in the forward difference formula. The use of flux limiters resulted in a much more robust solution.

Discritization of the tubular geometry is done as follows:

$r_i = (i-1)\Delta r, i = 1,2, \ldots, N+1$ where Δr=R/N and $z_i = (j-1)\Delta z, j = 1,2, \ldots, M+1$ where Δz=L/M.

Radial averaging of the convective terms was done with a commonly used formula that can be derived by volume averaging over a three point stencil in the radial direction.

$$\langle y \rangle_{i-1}^{i+1} = \frac{(2i-3)y_{i-1} + 4(i-1)y_i + (2i-1)y_{i+1}}{8(i-1)}$$

So the convective flow term in the energy balance discretized for solution by axially implicit Euler integration becomes as follows:

$$\frac{\partial T_g}{\partial z}_{i,j} = \frac{\begin{array}{c}(2i-3)\left[-\emptyset_{i,j}^T T_{g\,i-1,j-1}^n + (2\emptyset_{i,j}^T - 1)T_{g\,i-1,j}^n + (1-\emptyset_{i,j}^T)T_{g\,i-1,j+1}^n\right] \\ +4(i-1)\left[-\emptyset_{i,j}^T T_{g\,i,j-1}^{n+1} + (2\emptyset_{i,j}^T - 1)T_{g\,i,j}^{n+1} + (1-\emptyset_{i,j}^T)T_{g\,i,j+1}^{n+1}\right] \\ +(2i-1)\left[-\emptyset_{i,j}^T T_{g\,i+1,j-1}^n + (2\emptyset_{i,j}^T - 1)T_{g\,i+1,j}^n + (1-\emptyset_{i,j}^T)T_{g\,i+1,j+1}^n\right]\end{array}}{8(i-1)\Delta z}$$

The integration of the energy balances over a single time step, two for packed bed or three for MFEC, was solved using a block tridiagonal linear algebra solver that has source code freely available on the web [5]. The centerline B.C. was applied by setting a one-sided three point first derivative formula equal to zero:

$$T_{1,j} = \frac{4T_{2,j} - T_{3,j}}{3}.$$

The wall B.C. was applied by discretizing equation A-4 using backward difference formulas for the first derivatives resulting in a B.C. that is of only first order accuracy. For the FTS simulation a Danquent's B.C. for the gas phase energy balance and transient heat transfer to the solid phases was applied. Minimum reduction of the PDE [6] was used to obtain the exit B.C.

(359e) In Situ FTIR Study for Tri Reforming Process

Yishan Zhang, Juan Cruz and Tracy J. Benson, Chemical Engineering, Lamar University, Beaumont, TX

Tri-reforming process is the conversion of CO_2, H_2O, CH_4, and O_2 to syngas without CO_2 separation from industrial flue gases. Trireforming is a possible process of reducing total carbon dioxide emissions by extending our available carbon resources. This process reduces the chance of catalyst deactivation by carbon formation compared with the dry reforming of methane. However, the interaction of this mixture of gases with catalytically active surface sites of catalyst is still a puzzle that we need to investigate to understand the mechanism of the tri-reforming reaction.

In situ FTIR technique plays an important role in the characterization of heterogeneous catalysts under real reaction conditions with simultaneous study of reaction products. In our lab, the $Ni/Mg_xTi_{1-x}O$ catalyst prepared by incipient wetness and nickel nanoparticle formation via reverse micelles has been studied in a temperature-controlled continuous flow fixed reactor. Using probe molecules as reactants, in situ FTIR spectroscopy was used to investigate the catalyst surface chemistry during a temperature programmed study (room temperature up to 500°C) at 1 atm pressure. Some points will be illustrated: (1) which are the active sites of the catalyst; (2) how the reaction changes due to the different Mg (or Ti) concentrations; (3) how does the adsorbed gases react with the active sites; and (4) what the intermediate species will be generated during the reaction and why does this happen. These results allow one to determine the appropriate catalyst composition and then control of reaction conditions to get the satisfied products through these techniques. In the proposed trireforming process, the syngas, which will be used in Fischer-Tropsch and/or methanol synthesis reactions, must have a $H_2/CO = 2$. In addition, possible mechanism of tri-reforming reaction on $Ni/Mg_xTi_{1-x}O$ will be proposed.

See more of this Session: CO Hydrogenation II
See more of this Group/Topical: Catalysis and Reaction Engineering Division

(359f) The Role of Added Promoters In Reducing the Deactivation of Co Catalyst Used In Fischer Tropsch Synthesis

Nianthrini Balakrishnan[1], Babu Joseph[1], Venkat Bhethanabotla[1] and D. Yogi Goswami[2], (1)Chemical and Biomedical Engineering, Clean Energy Research Center, University of South Florida, Tampa, FL, (2)Clean Energy Research Center, University of South Florida, Tampa, FL

Cobalt catalysts used in Fischer Tropsch synthesis deactivates over time for a variety of reasons. Promoters have been suggested as a way to retard the deactivation of catalysts caused by oxidation or carbidization. These promoters reduce the deactivation of catalyst by making the surface less susceptible to carbon deposition and poisoning caused by trace impurities in the feed. Saeys et al[1], in a combined experimental and DFT study on B promoted Co showed that B and C have similar binding preferences and hence B can block C deposition. The deactivation rate was reduced 6 times without affecting selectivity and activity. Swart et al[2], investigated the possible intermediates in the formation of carbon deposits in cobalt-based FTS and showed that graphene was formed via stable, smaller carbon clusters on the flat Co surface. Ru promoter enhanced Co catalyst stability by preventing filamentous carbon formation[3].

In this study, we investigate if Ru added as a promoter to Co catalysts aids in preventing the deactivation of Co catalyst due to carbidization. A surface alloy model, where the promoter metal is dispersed on the top surface of the catalyst, is studied. The formation energies of dimer, trimer, linear, ring and graphitic carbon structures are calculated on both Ru promoted and unpromoted surfaces. The activation barrier for the diffusion of C on both the surfaces is also calculated to check the mobility of C atoms on the surfaces. If the formation energies of these C structures are reduced on the promoted surface, Ru prevents the deactivation of Co catalyst due to C formation. For the calculations, VASP (Vienna Ab Initio Simulation package) code[4-6] with Perdew–Burke–Ernzerhof (PBE) form of the generalized gradient approximation (GGA)[7] functional is utilized for the exchange and correlation functional. The electron-ion interactions are modeled by the projector-augmented wave (PAW)[8] method. The activation barrier is calculated using the Climbing Image Nudged Elastic Band (CI-NEB) method [9-11].

References:

(1) Saeys, M.; Tan, K. F.; Chang, J.; Borgna, A. *Industrial & Engineering Chemistry Research* **2010**, *49*, 11098.

(2) Swart, J. C. W.; van Steen, E.; Ciobica, I. M.; van Santenc, R. A. *Physical Chemistry Chemical Physics* **2008**, *11*, 803.

(3) Bae, J. W.; Kim, S.-M.; Park, S.-J.; Prasad, P. S. S.; Lee, Y.-J.; Jun, K.-W. *Industrial & Engineering Chemistry Research* **2009**, *48*.

(4) Kresse, G.; Furthmuller, J. *Computational Materials Science* **1996**, *6*, 15.

(5) Kresse, G.; Furthmuller, J. *Physical Review B* **1996**, *54*, 11169.

(6) Kresse, G.; Hafner, J. *Physical Review B* **1993**, *47*, 558.

(7) Perdew, J. P.; Burke, K.; Ernzerhof, M. *Physical Review Letters* **1996**, *77*, 3865.

(8) Blochl, P. E. *Physical Review B* **1994**, *50*, 17953.

(9) Henkelman, G.; Johannesson, G.; Jonsson, H. *Theoretical Methods in Condensed Phase Chemistry* **2000**, *5*, 269.

(10) Henkelman, G.; Jonsson, H. *Journal of Chemical Physics* **2000**, *113*, 9978.

(11) Henkelman, G.; Uberuaga, B. P.; Jonsson, H. *Journal of Chemical Physics* **2000**, *113*, 9901.

See more of this Session: CO Hydrogenation II
See more of this Group/Topical: Catalysis and Reaction Engineering Division

(359g) First Principle Guided Design of Cobalt Fischer-Tropsch Catalysts with Enhanced Stability

Kong Fei Tan[1], Jie Chang[2], Armando Borgna[2] and **Mark Saeys**[1], (1)Chemical and Biomolecular Engineering, National University of Singapore, Singapore, Singapore, (2)Applied Catalysis, Institute of Chemical and Engineering Sciences (ICES), A-STAR, Singapore, Singapore

Supported cobalt catalysts slowly deactivate during Fischer-Tropsch Synthesis (FTS). Various deactivation mechanisms have been reviewed recently [1], and include catalysts sintering, oxidation by water, and deposition of resilient carbon species. To explore the importance of deactivation by carbon deposition, we evaluated the thermodynamic stability of various forms of deposited carbon using Density Functional Theory (DFT) [2]. A surface carbide phase and graphene islands were calculated to be thermodynamically stable under FT conditions. Two forms of deposited carbon were also identified by X-ray Photoelectron Spectroscopy and by temperature programmed hydrogenation after testing supported Co catalysts for 200 hours in a fixed bed reactor [2]. Based on these insights into the deactivation mechanism, boron was proposed as a promoter to enhance the stability of Co catalysts [3]. Calculations indicate that boron and carbon display similar relative binding preferences, and boron is proposed to selectively block the nucleation and growth of resilient carbon species. To evaluate the predictions, supported 20 wt% Co catalysts were promoted with 0.5 wt% boron and tested in a fixed bed reactor. Under enhanced deactivation conditions of 240 °C and 20 bar, boron promotion was found to reduce the deactivation rate six-fold, without affecting selectivity and activity. This finding was subsequently confirmed at more typical FTS conditions.

References

1. Saib, A. M.; Moodley, D. J.; Ciobîc•, I. M.; Hauman, M. M.; Sigwebela, B. H.; Westrate, C. J.; Niemantsverdriet, J. W.; Van de Loosdrecht, J., Catal. Today 154, 271 (2010)

2. Tan K. F.; Xu, J.; Chang, J.; Borgna, A.; Saeys, M., J. Catal. 274, 121 (2010)

3. Tan, K.F.; Chang, J.; Borgna, A.; Saeys, M., J. Catal., 280, 50 (2011)

See more of this Session: CO Hydrogenation II
See more of this Group/Topical: Catalysis and Reaction Engineering Division

(362a) Mass Transfer with Surface Chemical Reaction In Fischer-Tropsch Synthesis: Simulation by a Lattice Boltzmann Method

Mohammad R. Kamali[1], Jurriaan J.J. Gillissen[1], Sankaran Sundaresan[2] and Harry E.A. Van den Akker[1], (1)Dept. of Multi-Scale Physics, Delft University of Technology, Delft, Netherlands, (2) Department of Chemical Engineering, Princeton University, Princeton, NJ

Abstract

In the current study, the interaction of convection, diffusion and a surface reaction is simulated with the help of an advanced two-phase lattice Boltzmann method in the context of a Fischer-Tropsch synthesis. In our case of interest, carbon monoxide and hydrogen are converted into liquid hydrocarbons with the help of a catalyst deposited on the internal structures of a structured reactor. The liquid product and the heat of reaction are carried away from the reacting surface with the help of a fully wetting liquid. A dedicated FTS kinetic model was implemented on the reactive surfaces. Reaction rate and conversion are calculated in terms of the dimensionless Pclet and Damköhler numbers.

See more of this Session: Computational Fluid Dynamics In Chemical Reaction Engineering
See more of this Group/Topical: Catalysis and Reaction Engineering Division

(362b) Micro-Scale Modelling of Flow and Particle Transport In Porous Media Via CFD

Daniele Marchisio[1], Federica Lince[1], Tiziana Tosco[2] and Rajandrea Sethi[2], (1)Dismic, Politecnico di Torino, Torino, Italy, (2)Ditag, Politecnico di Torino, Torino, Italy

Porous media are ubiquitous in chemical engineering. They are involved in many separation processes (e. g., filtration, chromatography) and they play an important role in numerous chemically reacting systems such as in catalysis and in remediation of contaminated aquifers. Very often the fluids involved are characterized by very diverse rheological properties, ranging from classical Newtonian to non-Newtonian behaviors. Although the final applications are very different from each other and are typically characterized by very different features there are some common issues that despite the huge amount of experimental data still need to be addressed. These are generally related to the need of describing the porous medium at the macro-scale as a continuum defined in terms of a set of parameters that quantify its flow properties (e.g., porosity, permeability) and its particle transport characteristics (e. g., collection efficiency).

This work (co-funded by European Union project AQUAREHAB FP7 - Grant Agreement Nr. 226565) aims at tackling these issues from a fundamental perspective by the use of pore-scale (or micro-scale) simulations via computational fluid dynamics.

Porous media with different characteristics have been analyzed and used to extract some representative two-dimensional geometries of a few millimeters in size. Grids were created with Gambit whereas the flow of Newtonian and non-Newtonian (i.e., Carreau and Cross models) fluids through these media was investigated with Fluent 12.0. Data were analyzed in terms of the flow field and pressure drops across the media for different superficial velocities ranging from laminar to turbulent flows. Simulation predictions were then analyzed with different classical approaches (i.e., Darcy, Ergun, Darcy-Forchheimer equations) and the corresponding parameters were eventually extracted. Results show that the considered geometry are large enough to model and represent a fraction of the porous medium and allowed to correct and extend the Darcy-Forchheimer law to non-Newtonian fluids.

In the second part of the work the presence of solid particles suspended in the fluid was considered. The collection efficiency of the different porous media was then quantified with computational fluid dynamics by using an Eulerian approach. The gathered simulation predictions were then compared with classical approaches for the estimation of the collection efficiency of the porous medium (i.e., classical filtration theory). Results show that these theories generally under-predict the overall efficiency and a correction was eventually formulated. Future work includes the use and implementation of these corrections in macro-scale simulations.

ANSYS

Packing Low Tube-to-Particle Diameter Ratio Packed Beds for CFD Simulation

Genong Li,
Aniruddha Mukhopadhyay,
Yi Dai and Chi-Yang Cheng

Ansys Inc.

Sept. 2011

Introduction

ANSYS

- Low D/d ratio packed beds (D/d<5) are widely used in fixed bed reactors.

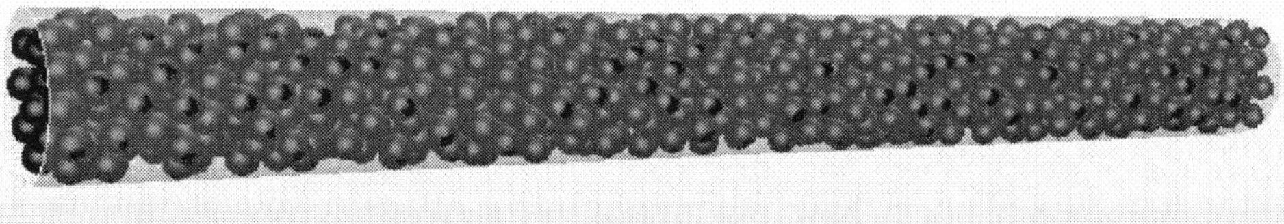

  - Porosity is highly non-uniform, flow channeling
  - Velocity, temperature, species are highly non-uniform
- Limitations of the porous medium model
  - Cannot capture many physics: local hot spots and runaway reactions
  - Characteristics of the packing: Porosity and S/V ratio profile
  - Effective viscosity, effective conductivity ...

3D Fully Resolved Model

ANSYS

- **With the development of high performance computing and advances in CFD, it is possible to conduct fully resolved 3D simulations for packed beds especially at low D/d ratio**
  - Reasonable long reactor segments can be simulated within days time frame
- **Advantages of the fully resolved model**
  - It can provide detailed information of a packed bed reactor. Reactor design can rely more on simulation-based solution.
  - The information extracted from fully resolved simulations can be used to evaluate the performance of other reduced order methods, and provide input parameters for them.

Procedures and Challenges

ANSYS

- **Prepare packed bed geometry**
  - Development of a packing algorithm for random packing
- **Mesh the packed bed geometry**
  - A lot of point-to-point contacts, small gaps
- **Set up and run the model in a CFD solver**
- **Extract useful information**
  - Packing characteristics: porosity profile,...
  - Flow field information

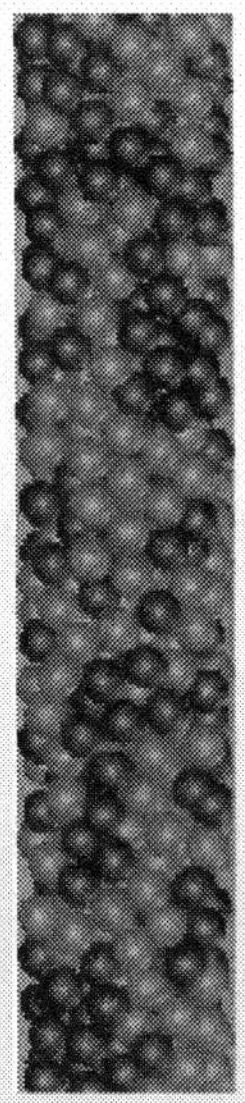

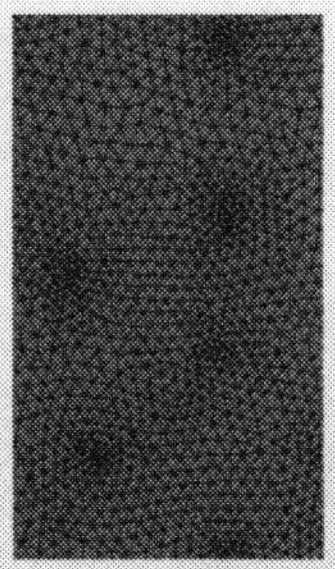

Packed-bed Packing

- **Random packing algorithms fall under two main categories**
  - Dynamic techniques
    - It is usually conducted in a DEM code. The movement of every packing element is tracked till it is settled in the bed according to the Newton's law.
      - Considering more physics, relatively slow
  - Geometric techniques
    - Packing is generated based on purely geometric considerations. The method is very fast.
- **In this work, only cylindrical beds packed with spherical balls are considered. A geometric packing method is developed and used.**

Packing Algorithm

- We will put balls sequentially into the tube.
- The bottom layer of the tube is first filled.
- Given all the balls already in the tube, the next ball's sitting position is limited to a finite number of locations. We will search for all these possible positions.
- We will put the next ball at one of these possible positions (randomly or at the lowest possible position).
  - Note that the packing will be random as a result of this!
- We will continue this process until the whole cylindrical column is filled up.

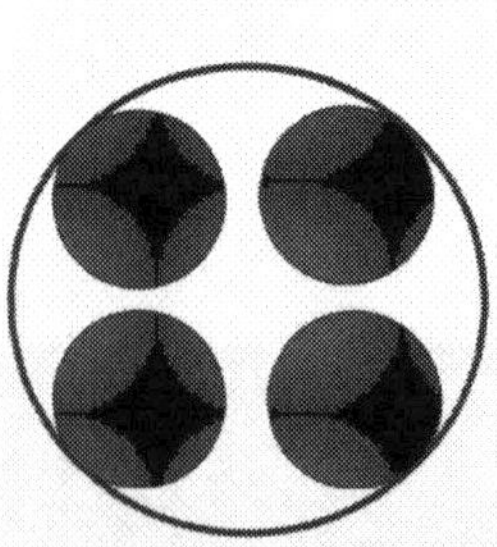

Illustration of the Packing Method

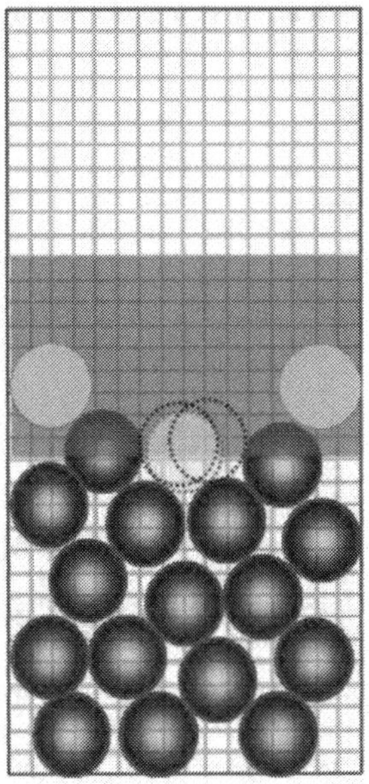

- A structured background mesh is created to facilitate the searching of the next ball's all possible positions.

$$\begin{cases} |\vec{x} - \vec{x_1}| = r + r_1 \\ |\vec{x} - \vec{x_2}| = r + r_2 \\ |\vec{x} - \vec{x_3}| = r + r_3 \end{cases}$$

- Only the stable positions are considered.
- Packing options:
  - The lowest possible position
  - A random position
  - The combination of the two

Several Packing Examples

Mono-dispersed particles

Poly-dispersed particles

Meshing Problem and Solutions

ANSYS

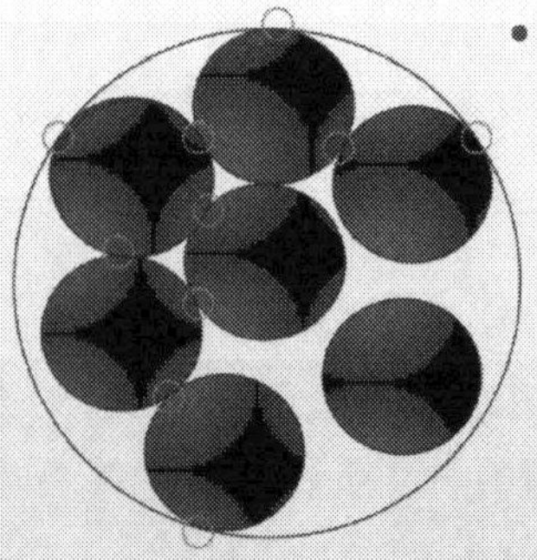

- A packed bed will have a lot of point-to-point contacts, which will cause problem in meshing!
- Solutions:
  - Method I: Packed balls are shrunk by a small percentage so there is always a gap between balls. Method is always robust, but there is no contact among balls. The heat conduction among balls cannot be considered. For the hydraulic simulation, the method should be fine.
  - Method II: Use "welding balls" to connect balls and ball and the wall.
- Size functions are used in meshing

Meshed Packed Bed

ANSYS

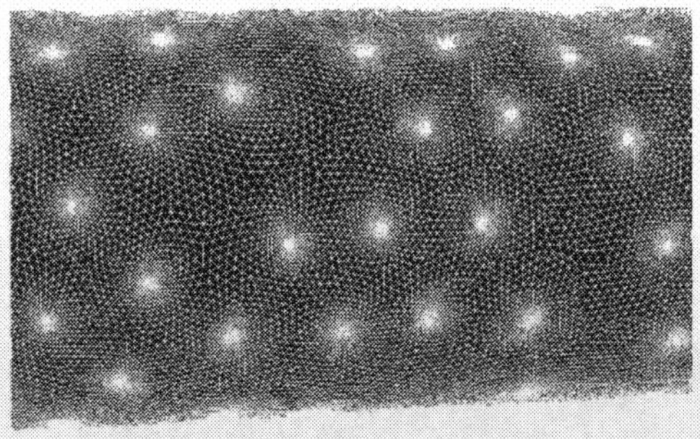

Using size function is critical to get high quality mesh with reasonable mesh count

Characteristics of Packing Beds

- **Once a packing structure is known, packing characteristics can be derived.**
- **Two packing characteristics are considered**
  - Porosity profile
    - Very important for flow distribution and pressure drop
  - S/V ratio
    - Very important for mass transfer and reaction.
- **A Monte Carlo method is developed to compute the above two parameters.**

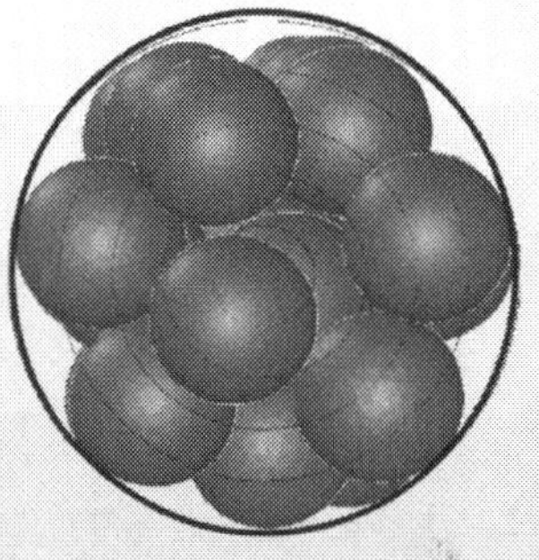

Packing Patterns at Different D/d Ratio

D/d≤2, structured packing; D/d>2, random packing

Porosity and S/V Ratio Profiles

ANSYS

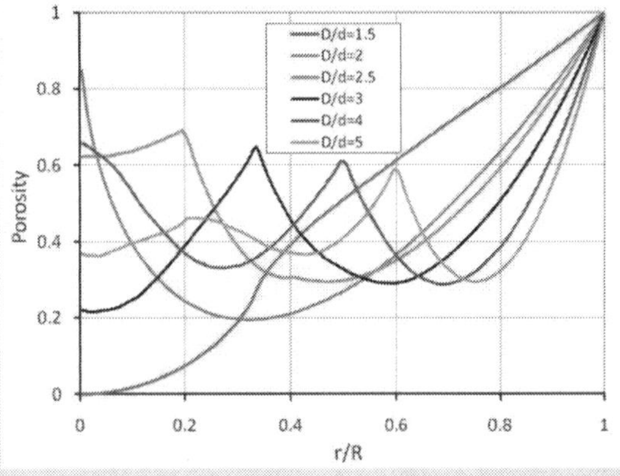

Porosity profiles for packed beds at different D/d ratios

Surface/volume ratio profiles for beds at different D/d ratios

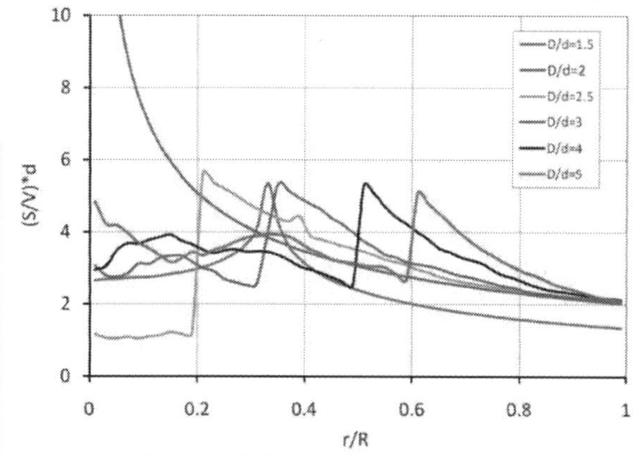

- The profiles highlight the non-uniformity of the packed beds.
- Such information can be used in the reduced order models!

Simulation of a Fully Resolved Bed

ANSYS

- **Dow chemical's plant scale reactor***
  - Tube diameter: D=26.12 mm
  - Tube length: H=2319 mm
  - Ball diameter: d=10 mm
- **Meshing detail**
  - Total number of balls: 916
  - Mesh size: 10M
- **Simulation condition**
  - Material: air
  - Mass flow rate: 3.65, 7.31, 10.96 kg/hr
  - Measured pressure drop: 884, 3491, 7593 Pa

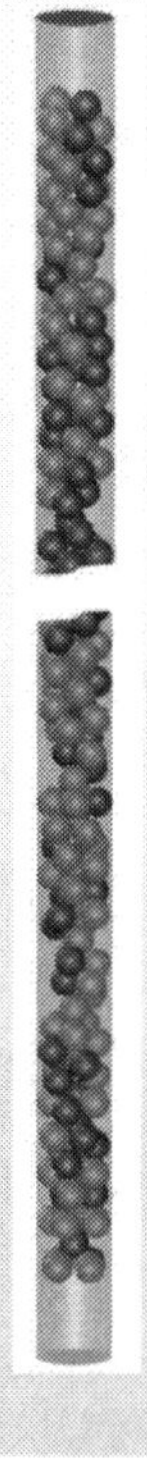

* Bai et al., "A coupled DEM and CFD simulation of flow field and pressure drop in fixed bed reactor with randomly packed catalyst particles." Ind. Eng. Chem. Res. 2009;48:4060-4074.

Some Flow Results

Velocity: 0~20 m/s

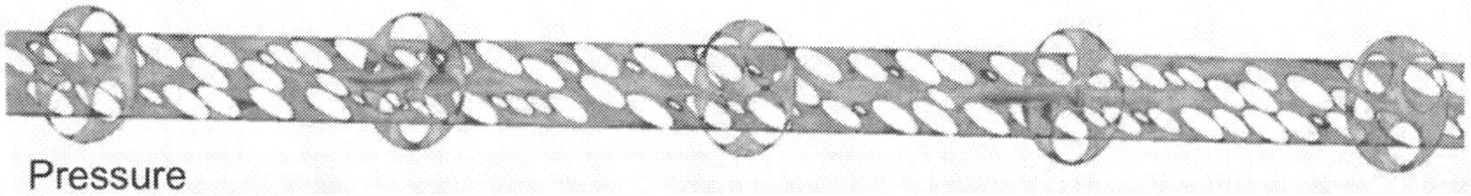

Pressure

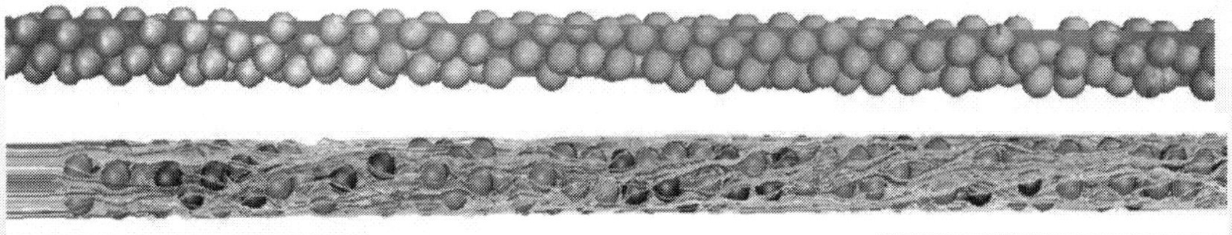

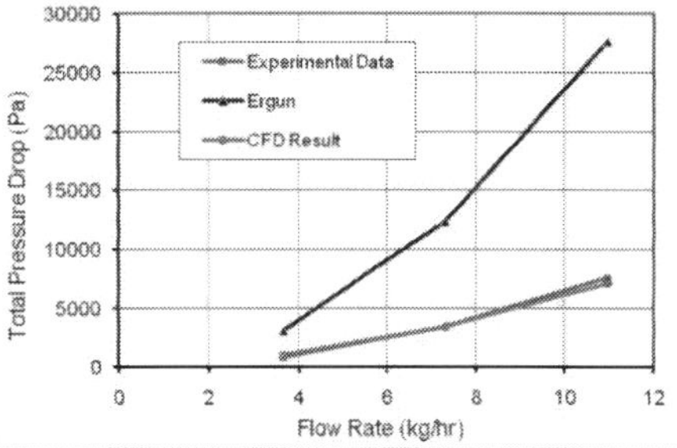

Averaged Velocity Profile

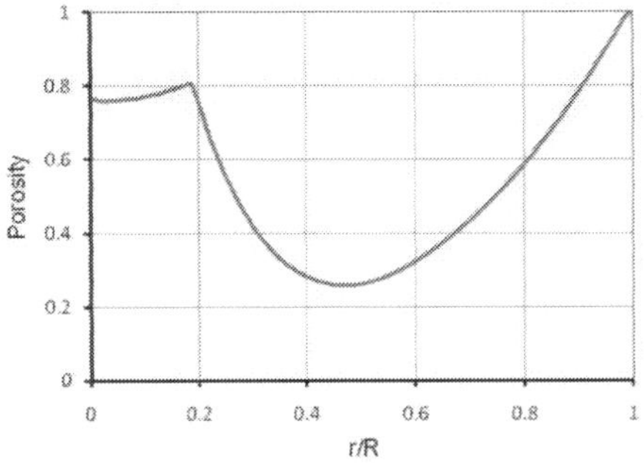

Porosity profile for Dow's plant-scale packed reactor

Averaged axial velocity along the reactor's radius

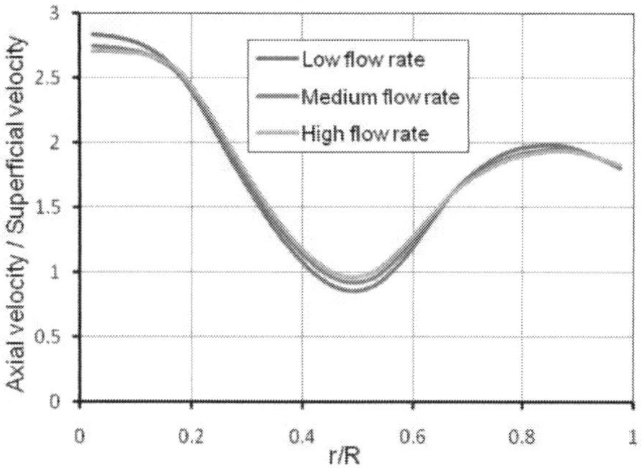

Such averaged profiles can be used to evaluate the accuracy of any reduced order model.

Conclusions

ANSYS

- A packing algorithm was developed to pack spherical balls into a cylindrical column randomly.
- Packing characteristics including porosity and surface/volume ratio were computed for low D/d ratio packed beds.
- Fully resolved simulations were conducted for a plant-scale real reactor. Results are in good agreement with experimental data.
- A fully resolved model provides not only a simulation-based solution in reactor design, but also a very useful tool to validate reduced order models.

(362d) CFD Simulations of An Ethylene Flame and Propylene Flares

Kanwar Devesh Singh[1], Hitesh Vaid[1], **Daniel H. Chen**[1], Helen Lou[1], Kuyen Li[1], Xianchang Li[2] and Christopher Martin[3], (1)Dan F. Smith Department of Chemical Engineering, Lamar University, Beaumont, TX, (2)Mechanical Engineering, Lamar University, Beaumont, TX, (3)Chemistry, Lamar University, Beaumont, TX

Air quality models (CAMx, CMAQ) often significantly under-predict observed peak O3 with the current VOC emission inventories in HGB/BPA. Even though this could be due to the inaccuracy of the photochemical models used, it is very likely that some VOC emissions may be under-reported or unreported due to the assumption of robust destruction efficiencies of industrial flares. From rigorous combustion chemistry, incomplete combustion of hydrocarbons is a likely source of free radicals such as perhydroxyl, radical producing species such as formaldehyde, and highly-reactive VOCs (HRVOCs) such as ethylene and propylene.

CHEMKIN CFD and FLUENT were used for our simulations. The simulation of laboratory McKenna burner is aimed at validating the turbulence model parameters and a 50-species reduced mechanism based on a combination of GRI 3.0 and USC mechanisms for the combustion of C1-C3 light hydrocarbons. The simulation of Tulsa flare tests is aimed at predicting the combustion efficiencies (CE and DRE) and hydrocarbon emissions based on validated CFD models. The simulations can establish the needed data base for finding the relationship between flare efficiencies/emissions and flare operating/ meteorological conditions.

See more of this Session: Computational Fluid Dynamics In Chemical Reaction Engineering
See more of this Group/Topical: Catalysis and Reaction Engineering Division

(362e) Comparative Study of Spouted Bed Hydrodynamics Based On Dimensionless Approach Using Optical Probes and CFD

Shreekanta Aradhya, Chemical and Biological Engineering, Missouri University of Science and Technology, Rolla, MO and Al-Dahhan Muthanna, Chemical & Biological Engineering, Missouri University of Science and Technology, Rolla, MO

@font-face { "Calibri"; }@font-face { "SimSun"; }p.MsoNormal, li.MsoNormal, div.MsoNormal { margin: 0in 0in 10pt; line-height: 115%; font-size: 11pt; ; }p.Paragraph, li.Paragraph, div.Paragraph { margin: 0in 0in 0.0001pt; text-indent: 0.25in; font-size: 10pt; ""; }span.go { }.MsoChpDefault { font-size: 10pt; ; }div.WordSection1 { page: WordSection1; }

Comparative Study of Spouted Bed Hydrodynamics Based on Dimensionless Approach using Optical Probes and CFD

[*]Shreekanta B Aradhya, [#]Xingying Lan and [*+]Muthanna Al-Dahhan

[*]Department of Chemical & Biological Engineering

[+]Department of Nuclear Engineering

Missouri University of Science & Technology, Rolla, MO – 65409, USA

[#]State Key Laboratory of Heavy Oil Processing, China University of Petroleum, P.R China - 102249

Email: sba2p4@mst.edu

Dimensional analysis has been used as a tool to develop a set of scaling parameters for proper design of fluidized beds. He et al. (1997) modified the scaling factors for fluidized beds proposed by Glicksman (1984) to provide a set of scaling parameters for the similarity behavior of spouted beds by adding the internal friction angle and the loose packed voidage. He et al. (1997) proposed the following controlling dimensionless parameters for a spouted bed: , , , , , sphericity of particles (), interfacial angle of particle (), loose peak voidage (), dimensionless particle size distribution, and dimensionless bed geometry. Using different spouted bed conditions that cover ambient and high temperature and pressure conditions and different particle types and sizes, He et al. (1997) demonstrated that close hydrodynamic similarity was maintained in the beds when these scaling parameters were closely matched.

However, the scale-up methodology of dimensional analysis has been evaluated only based on global parameters such as dimensionless spout diameters, dimensionless fountain height, and pressure along the bed height. No detailed hydrodynamics evaluation has been performed or reported. This work will focus on assessing the several match and mismatch conditions (based on dimensionless groups) reported by He et.al (1997) by using advanced optical probes and by CFD simulations. The parameters will be evaluated with the help of CFD simulations and optical probe measurements, which will be compared not only based on global parameters, but also local parameters such as solids holdup, solids velocity and their fluctuations.

See more of this Session: Computational Fluid Dynamics In Chemical Reaction Engineering
See more of this Group/Topical: Catalysis and Reaction Engineering Division

CFD STUDY OF HEAVY OIL GASIFICATION IN AN ENTRAINED-FLOW GASIFIER

*Vikram Sreedharan**, *Dept. of Biotechnology, Chemistry & Environmental Engineering, Aalborg University, Denmark*
Bjørn H. Hjertager, Faculty of Science & Technology, University of Stavanger, Norway
Tron Solberg, Dept. of Biotechnology, Chemistry & Environmental Engineering, Aalborg University, Denmark

**E-mail: vs@bio.aau.dk*

Introduction

Entrained-flow gasifiers are widely used for the gasification of coal and petcoke (fluidized bed gasifiers being another popular choice), especially for the synthesis of chemicals, in power generation and for the production of synthetic fuels. This gasifier type has also been historically associated with the gasification of heavy oil, for instance in the Shell gasification process (SGP) and in Texaco gasification studies. Entrained-flow gasifiers are characterized by high combustion-zone temperatures (in excess of 1400°C) and an operating pressure between 20 and 70bar (Higman & Burgt, 2008). These high temperatures are necessary to ensure high carbon conversion despite the short residence time of the feedstock within the reactor.

With the aim of promoting fuel security, diversification and an increase in thermal efficiency, feeds such as Orimulsion™ (heavy oil) are being considered for use in combined-cycle power plants. Despite a number of studies focusing on the investigation of gasification of solid fuels, only a limited number of studies are found in the literature for the gasification of liquid fuels, most of which deal with the gasification of black liquor. The current study aims to model the physical and chemical processes taking place within the entrained-flow heavy oil gasifier using computational fluid dynamics (CFD). This model needs to account for a number of complex processes such as turbulence, multiphase flow, heat transfer, mass transfer and chemical reactions. In this study, the experimental results of Ashizawa et al (2005) have been compared with the predictions obtained from the CFD model.

The focus of this work is to describe accurately the different chemical processes occurring during gasification, i.e., devolatilization, char combustion and the gas-phase reactions for the heavy oil. Furthermore, turbulent multiphase (dilute) flow with combustion is modeled, with the use of an appropriate model for radiation.

Theoretical Background

When the feed is introduced into the reactor at high temperature, with a "carrier" medium such as air or oxygen, a number of chemical and physical processes occur within the gasifier. In order to obtain the benefit of savings in space and energy, gasification takes place at high pressures. If the feedstock used is heavy oil, the volatile content is consumed during the evaporation/devolatilization phase and the resulting char is burned or gasified.

In an entrained-flow gasifier, the path of the heavy oil droplets would be largely influenced by the gas flow which is represented typically as a dilute flow regime. In this regime, the volume occupied by the droplets and the droplet-droplet interactions are assumed to be negligible. A criterion used in the CFD software FLUENT for dilute flow is that the particle volume fraction should not exceed 10%.

Multiphase Flow

An Euler-Lagrange model known as the discrete phase model (DPM) is appropriate for multiphase flows where the dispersed phases are dilute. The continuous (gas) phase is described in an Eulerian formulation by solving conservation equations for mass, momentum and energy, while the dispersed phase (droplets) is described in a Lagrangian formulation by tracking the exchange of mass, momentum and energy for individual particles.

Continuous Phase

The reacting flow in the continuous gas phase is of a turbulent nature, and can be described by Favre-averaged balance equations for mass, momentum and energy. The governing transport equations can be written in a generalized form using Cartesian tensor notation as

$$\frac{\partial}{\partial t}(\rho\Phi)+\frac{\partial}{\partial x_j}(\rho u_j\Phi)=\frac{\partial}{\partial x_j}\left(\frac{\mu_t}{\sigma_\Phi}\frac{\partial\Phi}{\partial x_j}\right)+S_\Phi+S_{\Phi P} \tag{1}$$

In the above expression, the dependent variable $\Phi = 1, u_i, h, Y_k, k, \varepsilon$ representing variables such as mass, momentum, energy, species, turbulent kinetic energy and turbulent energy dissipation rate, respectively. S_Φ is a standard source term in the governing equation (1) and $S_{\Phi P}$ is a source term taking into account the interaction of the continuous (gas) phase with the dispersed particle (droplet) phase. The turbulent eddy viscosity is given by

$$\mu_t = \rho C_\mu \frac{k^2}{\varepsilon} \tag{2}$$

This eddy viscosity is determined from the Standard $k-\varepsilon$ turbulence model.

Discrete Phase

The conservation of mass equation for the discrete phase (oil droplets) takes the form

$$\frac{dm_p}{dt}=\frac{dm_1}{dt}+\frac{dm_2}{dt}+\frac{dm_3}{dt}+\frac{dm_4}{dt}+\frac{dm_5}{dt} \tag{3}$$

Where, $m_1 - m_5$ are the masses of gas originating from chemical reactions (11) – (15). The trajectory of the discrete phase droplets are determined from Newton's second law of motion. The force balance for a droplet can then be written as

$$\frac{du_{pi}}{dt} = F_D(u_{gi} - u_{pi}) + g_i \frac{(\rho_p - \rho_g)}{\rho_p} + \frac{\rho_g}{2\rho_p}\frac{d}{dt}(u_{gi} - u_{pi}) + \left(\frac{\rho_g}{\rho_p}\right) u_{pj} \frac{\partial u_{gi}}{\partial x_j} \tag{4}$$

Here, the subscripts g and p refer to the gas and particle (droplet) phase, respectively. The droplet inertia is balanced by the drag, the gravity, the virtual mass and the force due to the pressure gradient acting on the droplet. For a rigid spherical particle (droplet), the drag is defined by

$$F_D = \frac{18\mu_g}{\rho_p d_p^2} \frac{\mathrm{Re}\, C_D}{24} \tag{5}$$

The drag coefficient is determined from the Schiller and Naumann model correlation

$$C_D = \begin{cases} 24\left(1 + 0.15\mathrm{Re}^{0.687}\right)/\mathrm{Re} & \mathrm{Re} \le 1000 \\ 0.44 & \mathrm{Re} > 1000 \end{cases} \tag{6}$$

Thermal energy conservation for the discrete phase particle (droplet) can be written as

$$\frac{d(m_p c_p T_p)}{dt} = A_p \frac{\lambda Nu}{d_p}(T_g - T_p) + A_p \varepsilon_a (Q_R - \sigma_B T_p^4) + \sum_{k=1}^{5} \dot{m}_k \Delta h_k \tag{7}$$

In the above expression, the rate of change of thermal energy for a droplet is balanced by the convective and radiative heat transfer as well as the heat due to evaporation, devolatilization and particle surface chemical reactions. The convective heat transfer coefficient is calculated from the Ranz and Marshall's Nusselt correlation

$$Nu = \frac{hd_p}{k_\infty} = 2.0 + 0.6\mathrm{Re}_d^{1/2}\,\mathrm{Pr}^{1/3} \tag{8}$$

Heavy Oil Gasification

The gasification of heavy oil can be broken down into the following simple sub-processes for ease of modeling.

Passive Heating

The heavy oil droplets are heated till the vaporization/devolatilization temperature is reached. There is no mass transfer or chemical reaction during this stage. When the moisture in the heavy oil reaches a certain temperature - say the vaporization temperature, it is released. Meanwhile, energy is taken out from the gas phase to supply the latent heat of vaporization.

Devolatilization

A phenomenological model has to be constructed based on data from experiments used to characterize the heavy oil. This model is then used to predict the yields of some major gas components while preserving a strict elemental balance to determine stoichiometry (Shi et al, 2006). The main species included in the devolatilization model are CH_4, CO_2, CO, O_2, H_2, H_2S, SO_2, N_2 and H_2O. The release of volatiles from the heavy oil can be represented as

$$C_{m01}H_{m02}O_{m03}N_{m04}S_{m05}Ash_{m06} \rightarrow C_{m11}H_{m12}O_{m13}N_{m14}S_{m15} + C_{m21}Ash_{m26} \quad (9)$$

The kinetic devolatilization rate is then determined from the Kobayashi model, which is defined by two competing rates. The weighted rate of devolatilization is given by

$$\frac{dm_v}{dt} = (m_{p,0} - m_a)(\alpha_1 R_1 + \alpha_2 R_2)\exp\left(-\int_0^t (R_1 + R_2)dt\right) \quad (10)$$

$R_1 = A_1 \exp(-E_1/RT_p)$ and $R_2 = A_2 \exp(-E_2/RT_p)$ are the competing rates that control devolatilization over different temperature ranges. The first yield factor α_1 is the fraction of volatiles as determined by proximate analysis of the heavy oil, representing devolatilization at low temperature. The second yield factor α_2 is taken as unity, which is the yield of volatiles at high temperature.

Oxidation of Volatiles, Char Combustion and Gasification

After the complete release of volatiles, its oxidation, char combustion and gasification take place until all the char is consumed or the particles flow out of the reactor. The chemical reactions include volatile oxidation, char combustion, char-steam gasification and char-carbon dioxide gasification, followed by the gas phase reactions. Volatiles oxidation/combustion for oxygen-rich conditions are described by,

$$C_{m11}H_{m12}O_{m13}N_{m14}S_{m15} + \left(m_{11} + \frac{m_{12}}{4} - \frac{m_{13}}{2} + m_{15}\right)O_2 \xrightarrow{k1} m_{11}CO_2 + \frac{m_{12}}{2}H_2O + \frac{m_{14}}{2}N_2 + m_{15}SO_2 \quad (11)$$

Using the expression for the partial oxidation of volatiles given in Chen et al (2000), the chemical reaction for volatiles oxidation/combustion for oxygen-lean conditions, that is, when $\phi < \left(m_{11} + \frac{m_{12}}{4} - \frac{m_{13}}{2} + m_{15}\right)$ is given by

$$C_{m11}H_{m12}O_{m13}N_{m14}S_{m15} + \phi O_2 \xrightarrow{k2} m_{11}(\alpha_1 CO_2 + (1-\alpha_1)CO) + \frac{(1-\alpha_1)}{2}(m_{12} - 2m_{15})H_2 + \frac{\alpha_1(m_{12} - 2m_{15})}{2}H_2O + \frac{m_{14}}{2}N_2 + m_{15}SO_2 \quad (12)$$

Table 1: Stoichiometric coefficient values

Parameter	Value
m11	14.88
m12	27.96

m13	0.072
m14	0.124
m15	0.336

The stoichiometric coefficient values in Table 1 are determined by the proximate and ultimate analysis of heavy oil (Orimulsion) given in Ashizawa et al (2005). If we substitute the parameter values from Table 1 in chemical equations (11) and (12), we get the complete and partial oxidation relations for the released volatiles for heavy oil. Carbon monoxide and hydrogen are produced during the gasification of the char particles contained in heavy oil. The performance of a gasifier is determined primarily by the char gasification reaction. The following heterogeneous (particle-surface) reactions have been included in the model

$$C+\left(1-\frac{\gamma}{2}\right)O_2 \xrightarrow{k3} \gamma CO+(1-\gamma)CO_2 \tag{13}$$

$$C+CO_2 \xrightarrow{k4} 2CO \tag{14}$$

$$C+H_2O \xrightarrow{k5} CO+H_2 \tag{15}$$

The heterogeneous surface reaction rate is determined by a kinetics-diffusion limited scheme given by

$$\frac{dm_p}{dt}=-A_p p_{ox}\frac{D_0 R}{D_0+R} \tag{16}$$

$$D_0=C_1\frac{\left[(T_p+T_\infty)/2\right]^{0.75}}{d_p} \tag{17}$$

and, the kinetic rate for the chemical reactions (13) – (15) is given by

$$R=C_2 e^{-(E/RT_p)} \tag{18}$$

Gas Phase Reactions

The continuous gas phase is modeled using a reduced global reaction scheme to describe gas phase chemistry. Chemical reaction kinetics appropriate for gasification studies have been used for the reaction scheme in this analysis. The following reaction paths have been considered for this model

$$CO+H_2O \underset{k6b}{\overset{k6f}{\rightleftarrows}} CO_2+H_2 \tag{19}$$

$$CH_4+H_2O \underset{k7b}{\overset{k7f}{\rightleftarrows}} CO+3H_2 \tag{20}$$

The gas phase chemistry is modeled by combining laminar finite-rate chemical kinetics and eddy-dissipation turbulence-chemistry interaction. The values of the pre-exponential coefficient and the activation energies for the various chemical reactions considered for the gasification model (to determine the reaction rates) are given in Table 2.

Table 2: Chemical kinetics reaction rates

Eq. #	A_r	E_r (J/kmol)	n_r	$\prod_j [C_{j,r}]^{(\eta'_{j,r}+\eta''_{j,r})}$	Source
k1	2.119 x 10^{11}	2.027 x 10^{8}	0	$[Vol]^{0.2}[O_2]^{1.3}$	FLUENT user guide
k2	4.4 x 10^{11}	1.25 x 10^{8}	0	$[Vol]^{0.2}[O_2]^{1.3}$	Jones & Lindstedt (1988)
k3	1.36 x 10^{6}	1.3 x 10^{8}	0.68	-	Watanabe et al (2002)
k4	2.07 x 10^{7}	2.2 x 10^{8}	0.733	-	Watanabe et al (2002)
k5	2.07 x 10^{7}	2.2 x 10^{8}	0.733	-	Watanabe et al (2002)
k6f	2.5 x 10^{5}	1.38 x 10^{8}	0	$[CO]^1[H_2O]^1$	Maki & Miura (1997)
k6b	9.4 x 10^{6}	4.12 x 10^{8}	0	$[CO_2]^1[H_2]^1$	Maki & Miura (1997)
k7f	9.1 x 10^{7}	1.31 x 10^{8}	0	$[CH_4]^1[H_2O]^1$	Maki & Miura (1997)
k7b	5.5 x 10^{-6}	1.76 x 10^{9}	0	$[CO]^1[H_2]^3$	Maki & Miura (1997)

Model Description

An experimental study of a pilot-scale entrained-flow heavy oil gasifier by Ashizawa et al (2005) provides a framework of comparison for the current work. A simplified schematic of the heavy oil entrained-flow gasifier is depicted in Figure 1. The gasifier is cylindrical with a length of 3m and a diameter of 0.3m. The diameter of the inlet is 0.075m and that of the outlet is 0.15m. The computational mesh consists of 300 x 200 quadrilateral control volumes. The discrete phase method in ANSYS FLUENT 13.0 was used for the CFD simulation.

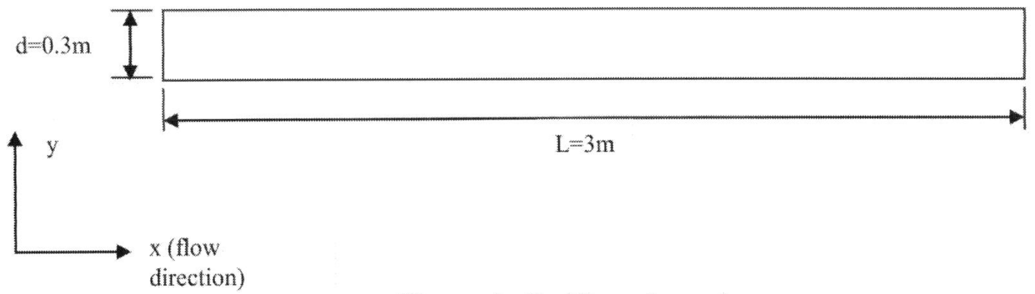

Figure 1: Gasifier schematic

Flow and Heat Transfer Models

The standard k-ε turbulence model was used for the continuous phase, and stochastic tracking was implemented in a steady-state model of the gasifier to simulate particle motion and turbulent dispersion. The SIMPLE algorithm was used to implement the pressure-velocity coupling with the second-order upwind discretization scheme. The oil droplets were assumed to have a Rosin-Rammler size distribution with a minimum size of 4μm and a maximum of 150, with an average of 40μm. The spread parameter was 1.2 and 10 discrete particle sizes were considered. The heavy oil feed rate is 0.0278kg/s and the oxygen feed rate corresponds to an oxygen ratio of 0.39. The operating pressure used in this model is 1.9MPa. Conductive, convective and radiative heat transfer was accounted for between the gases, droplets and reactor walls. The scattering property of the droplets due to radiation is assumed to be isotropic and the standard P1 model has been used for the purpose of modeling radiative heat transfer. This model is based on the expansion of radiation intensity into an orthogonal series of spherical harmonics. The weighted-sum-of-gray-gases-model (wsggm) was used to calculate the absorption coefficient of the gas phase.

Results and Discussion

Using the chemical kinetics scheme with the parameters outlined in Table 2 - comprising heterogeneous reaction rates from Watanabe et al (2002) and the gas-phase reaction rates of Jones & Lindstedt (1988) and Maki & Miura (1997), we can obtain the following contour plot for the temperature and velocity vector plot within the gasifier (Figure 2 and Figure 3).

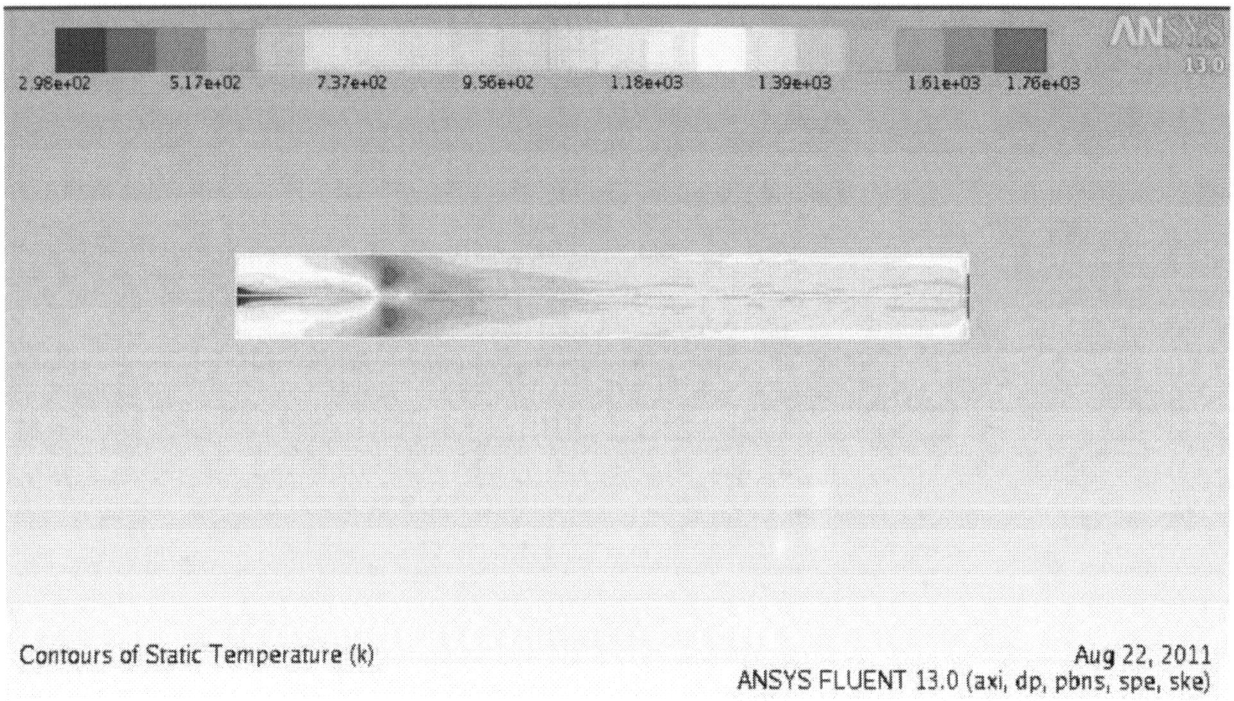

Figure 2: Temperature contour plot

As seen from Figure 2, it can be observed that the maximum temperature within the gasifier is approximately 1760K, and this peak temperature is observed at an axial distance of about 0.6m from the gasifier inlet, which corresponds to the location of the combustion zone within the gasifier. The temperature of the hot combustion gases at the outlet is almost 1330K.

From the velocity vector plot in Figure 3, it is seen that a peak velocity magnitude of approximately 34.4m/s is observed along the flow axis, close to the point of injection of the heavy oil spray.

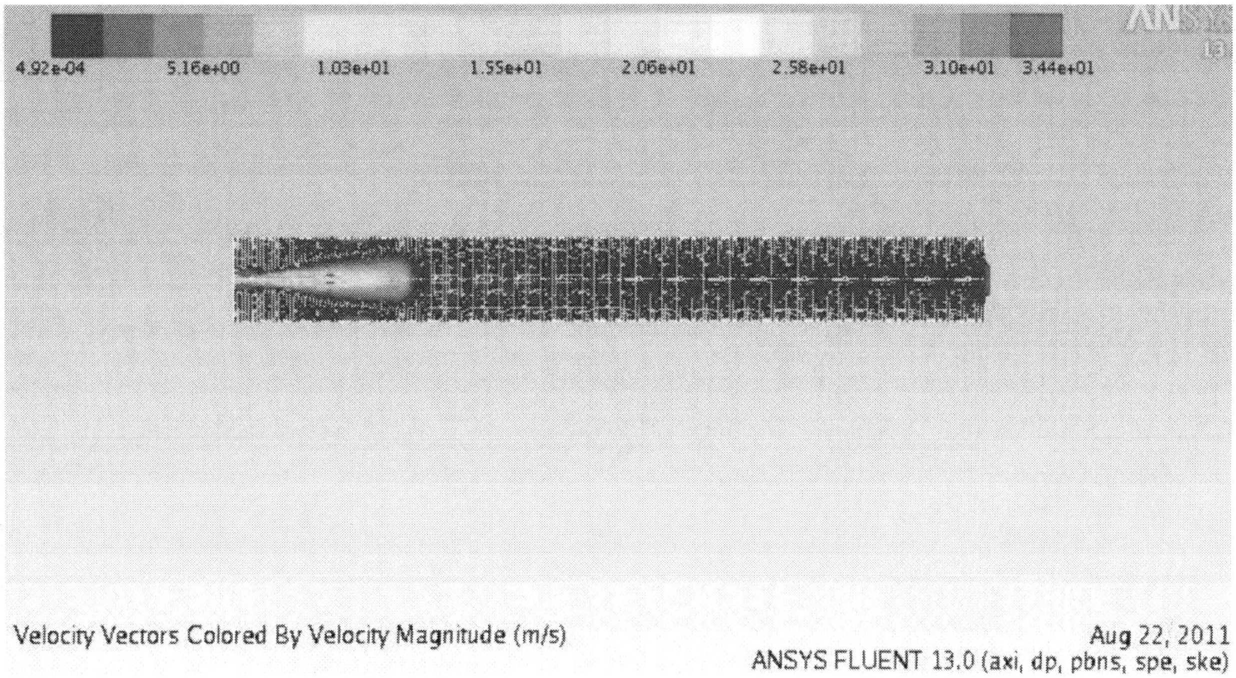

Figure 3: Velocity magnitude vector plot

In Figure 4, the axial temperature profile prediction obtained with the CFD model is compared with experimental data from Ashizawa et al (2005).

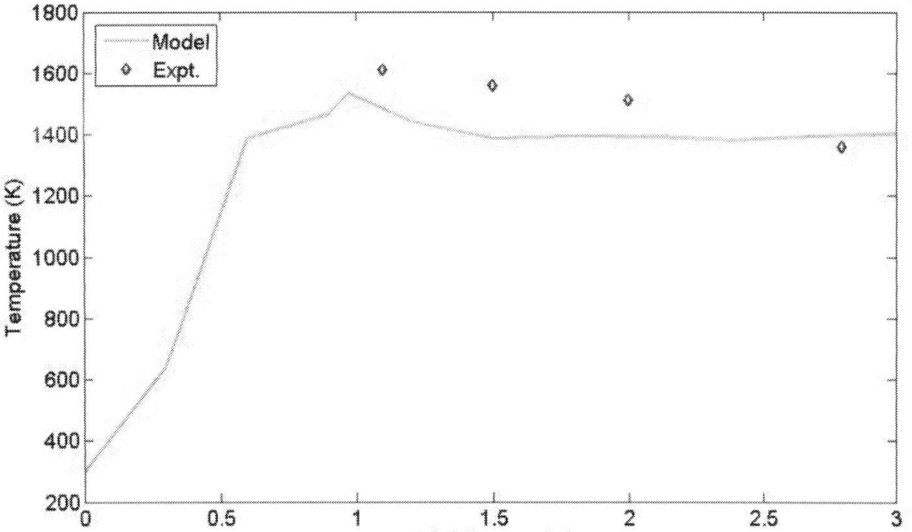

Figure 4: Axial temperature distribution

From Figure 4, it is seen that there is very good agreement between the CFD predictions and experimental results for the axial temperature distribution. The location of the combustion zone as well as the peak temperature is captured well by the model. Furthermore, the temperature trend and magnitude along the axis also shows good agreement with experimental data.

The carbon monoxide and hydrogen species mole fraction model predictions are compared with the experimental results obtained by Ashizawa et al (2005) in Figure 5. Again, it is seen that there is very good agreement with experimental data for both hydrogen and carbon monoxide mole fraction percentages. The correct axial trends are obtained for both the species with the CFD predictions slightly under-predicting the species mole fraction values as compared to the experimental data.

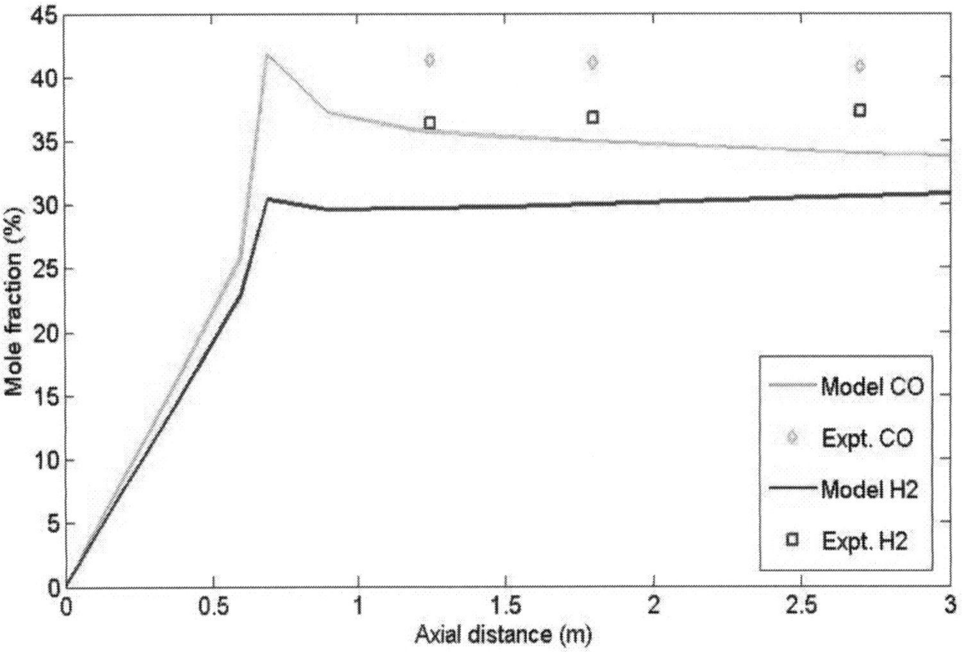

Figure 5: Species mole fraction (%) for carbon monoxide and hydrogen

The carbon dioxide and water vapour species mole fraction model predictions are compared with the experimental results obtained by Ashizawa et al (2005) in Figure 6. It can be observed that the agreement with experimental data for both water vapour and carbon dioxide mole fraction percentages is good. The correct axial trends are obtained for both the species with the CFD predictions slightly over-predicting the species mole fraction values as compared to the experimental data.

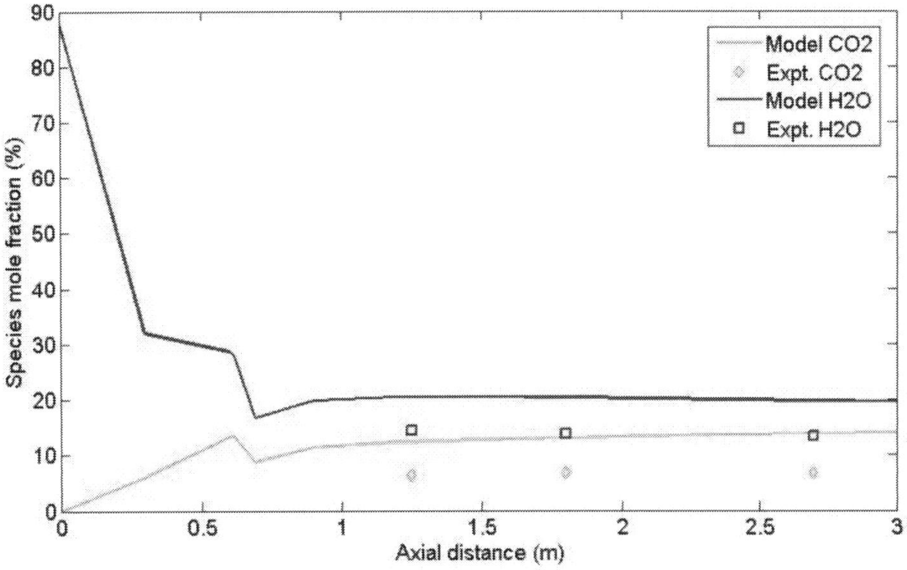

Figure 6: Species mole fraction (%) for carbon dioxide and water vapour

Conclusion

A CFD model for the gasification of heavy oil has been presented in this study which captures the essence of the physical and chemical processes taking place in the gasifier. The accurate chemical composition of the heavy oil (Orimulsion) was obtained from its proximate and ultimate analysis, and the volatiles oxidation chemistry is more detailed than is available in the literature.

From the predictions observed earlier, it is clear that the model captures very well the axial values and trends for temperature and mole fractions of the major species obtained during the gasification of heavy oil. The model slightly under-predicts the absolute values for the axial species mole fractions of carbon monoxide and hydrogen, while slightly over-predicting the values for carbon dioxide and water vapour. Further refinement of the model, including (but not limited to) more detailed combustion modeling for the heavy oil, use of more sophisticated turbulence and radiation models, use of the eddy-dissipation concept, etc. might lead to even more accurate results.

Acknowledgement

The financial support of Statoil Research Centre in Trondheim, Norway is gratefully acknowledged.

Bibliography

ASHIZAWA, M., HARA, S., KIDOGUCHI, K. and INUMARU, J., (2005), "Gasification characteristics of extra-heavy oil in a research-scale gasifier", *Energy*, **30,** 2194-2205.

CHEN, C., HORIO, M. and KOJIMA, T., (2000), "Numerical simulation of entrained flow coal gasifiers. Part I: modeling of coal gasification in an entrained flow gasifier", *Chem. Eng. Sci.*, **55**, 3861-3874.

HIGMAN, C. and VAN DER BURGT, M.,"Gasification", Second Ed. 2008, Elsevier Science.

JONES, W.P. and LINDSTEDT, R.P. (1988), "Global Reaction Schemes for Hydrocarbon Combustion", *Combust. Flame*, **73**, 233-249.

MAKI, T. and MIURA, K. (1997), "A Simulation Model for the Pyrolysis of Orimulsion", *Energy & Fuels*, **11**, 819-824.

SHI, S.-P., ZITNEY, S.E., SHAHNAM, M., SYAMLAL, M. and ROGERS, W.A. (2006), "Modelling coal gasification with CFD and discrete phase method", *J. Energy Inst.*, **79(4)**, 217-221.

WATANABE, H., OTAKA, M., HARA, S., ASHIZAWA, M., KIDOGUCHI, K. and INUMARU, J. (2002), "Modelling and Simulation for Extra Heavy Oil Gasification on Entrained Flow Gasifier", *Proc. IJPGC'02*, 667-674.

SMR3D: AN INDUSTRIAL CFD TOOL FOR LARGE STEAM METHANE REFORMER DESIGN OPTIMIZATION

Dr. Julien Cancès, Air Liquide R&D, Paris, France

Frédéric Camy-Peyret, Air Liquide R&D, Paris, France

Dr. Dieter Ulber, Lurgi GmbH, Frankfurt am Main, Germany

Introduction

The worldwide hydrogen demand has strongly increased in the past years, mostly for refinery needs in order to match environmental requirements such as desulfurization. This trend is expected to maintain for a long period and raise new challenges associated to the design and operation of large scale hydrogen plants. As of today most hydrogen plants are based on steam reforming of natural gas [1]. Steam reforming is also applied to produce CO or synthesis gas for other products.

As a major actor in the field of hydrogen production, and hydrogen plant engineering through its subsidiary LURGI, AIR LIQUIDE has developed advanced in-house simulation tools to understand, predict and optimize the efficiency, compactness and reliability of large scale steam methane reformers.

The present paper aims at presenting the SMR3D modeling tool and its applications to key aspects of a recent LURGI reformer design.

SMR3D presentation

SMR3D consists of a 3D CFD tool, ATHENA, dedicated to large industrial combustion chambers simulations coupled with a 1D SMR tube model.

Combustion chamber model

ATHENA, the AIR LIQUIDE in-house CFD code developed since the 80's, is a robust industrial tool, which includes several established models dedicated to heat transfers in industrial furnaces, including oxy-combustion [2]:

- The averaged three-dimensional flow is solved by use of a RANS approach (Reynolds Average Navier-Stokes). The Navier-Stokes equations are computed averaged on time, while turbulent fluctuations are modeled with using the standard k–ε model [3].
- The solver is based on a cell-centered, structured finite volume discretization of Navier-Stokes equations [4] and on transport equations of passive or active scalars (enthalpy, mass fractions, turbulence quantities of RANS models ...), with a SIMPLE-algorithm segregated solution procedure.
- Radiative heat transfer equations are solved by use of a specific ray tracing method (DTRM: Discrete Tracing Radiative Method [5]), a method which proved to be adapted to complex geometries. The gas radiation properties are taken into account with a Weighted Sum of Grey Gases (WSGG) model with four grey gases tables available for air and oxy-combustion products of different fuel compositions [6], with or without soot. The high radiation contribution to SMR furnaces heat transfer implies the use of an efficient and detailed radiation model, taking into account the geometric complexity of industrial reformers.
- The Magnussen eddy dissipation turbulent combustion model [7] is commonly used for industrial furnaces simulations, because it provides several pertinent tuning parameters thus enabling the use of a coarse mesh suited to such large geometries and avoiding to fully detail the burner geometry.
- The heat losses at the combustion chamber walls are evaluated using a global heat transfer law, as described in the following equation:

  $$\varphi = K_{wall} \cdot (T_{wall} - T_{\infty})$$

  The wall heat transfer coefficients – K_{wall} – were calculated by means of the thermal resistance method using the material properties provided by vendors. The heat conduction is computed inside the tunnel walls using an energy coupling at the interface.

Reformer tubes model

The computation of the flow, heat transfers and chemical endothermic reaction of steam methane reforming inside the catalytic tubes has been carried out using the code developed at Gent University described in [8][9]. The tube is described as a 1D plug-flow reactor in the vertical direction and computes the Hougen-Watson reactions set:

$$CH_4 + H_2O \Leftrightarrow 3 \cdot H_2 + CO$$
$$CO + H_2O \Leftrightarrow H_2 + CO_2$$
$$CH_4 + 2 \cdot H_2O \Leftrightarrow 4 \cdot H_2 + CO_2$$

The reforming model takes into account the intrinsic catalytic kinetics and the limitation through diffusion into the porous catalysts, using reaction effectiveness factors. The effectiveness factors can be computed by the gent code, however these parameters were considered as constant in order to speed-up the simulations and their values were established on the basis of operational data gathered on AIR LIQUIDE operated plants.

The coupling between the combustion chamber and tubular reactor model is performed by iterative computations in the firebox followed by reforming reactions in the tube. The heat flux computed in the combustion part is imposed as a boundary condition of the tube model, which returns the tube temperature by classical energy balance.

Numerical implementation

The SMR3D solver has been developed in order to take advantage of large computers. Two levels of parallelization (MPI and OpenMP) allow running massively parallel computations involving hundreds of million cells with a reasonable restitution time. Therefore, simulation of a full scale large reformer with hundreds of tubes and burners is now at hand. An example of typical SMR3D computation results for a large furnace geometry is illustrated in Figure 1 (Note that due to symmetry, only half of the furnace is computed).

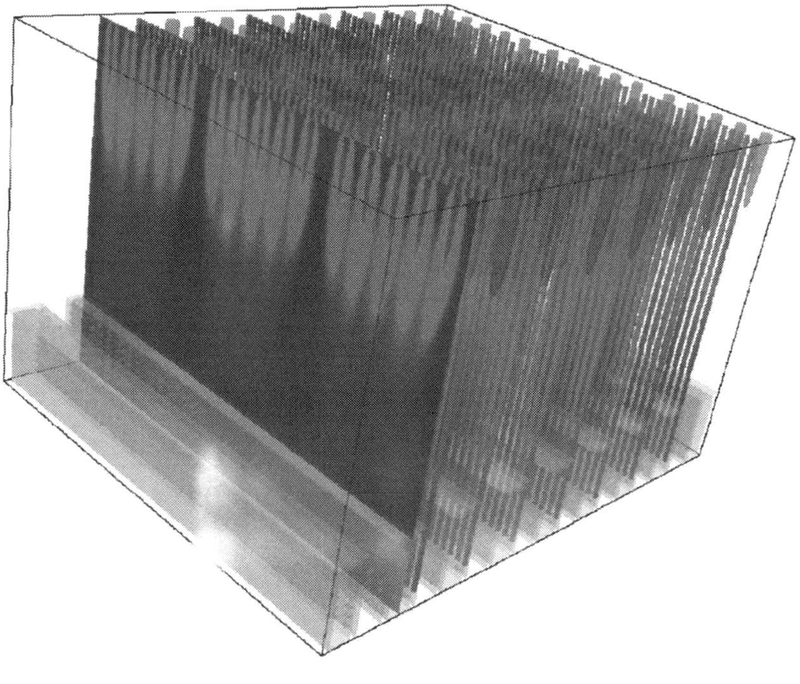

Figure 1: SMR3D computation results illustration for a large SMR furnace. The tubes are colored by temperature. Flames are represented as iso-velocity surfaces at 20 m/s.

SMR3D validation

Before using SMR3D as a design tool, it was necessary to ensure the capabilities of SMR3D to predict accurately the behavior of industrial size steam methane reformers. Therefore, SMR3D has been validated through confrontation with experimental data measured on LURGI design fireboxes worldwide. As an illustration the validation protocol applied to a recent CO plant based on LURGI technology is presented in the present paper.

Mass and Energy balances

The first model validation step is to compare the overall heat and mass balances with the ones gathered on-site using available measurements. However, as each sensor is assorted with a certain uncertainty it is necessary to close balances by applying a post-treatment step to the raw data, called data reconciliation [10], in order to get robust and reliable overall mass and energy balances. The data reconciliation method is a statistical approach combining measurement tolerances with a process simulation, which provides the most probable value - associated with a range confidence plotted as error bars in following diagrams. Thermal losses have been characterized as well during the field test campaign through external temperature measurements which have been analyzed using ASTM 680 standard [11].

Table 1. Comparison of the heat and mass balance for a CO plant experimental (reconciled) data and SMR3D computation results.

	Experimental data		**SMR3D**	
	Mass [kg/s]	Heat [MW]	Mass [kg/s]	Heat [MW]
Injected at burners	75.34	**216. 51** (100.0%)	75.34	**215.01** (100.0%)
Transferred to tubes		**115.12** (53.2%)		**115.14** (53.6%)
Radiation portion				111.92 (97.2%)
Reformed gas temperature		*914°C*		*917°C*
Flue gas	75.34	**99.26** 45.8%	75.34	**98.91** 46.0%
Temperature		*1036°C*		*1020°C*
Heat losses		**2.14 (1.0%)**		
Tube feeders		0.84[1] (0.4%)		
Firebox walls		1.28[1] (0.6%)		**0.95** (0.4%)

[1]: determined using ASTM 680 standard

The overall mass and energy balances are gathered in Table 1 and illustrated in Figure 2, whereas the flue gas and syngas temperature are plotted in Figure 3. A very close agreement between experimental and simulation results is observed. The SMR3D simulation results are inside the confidence intervals. The experimental heat losses evaluation encompasses also the tube feeders' contributions. As the SMR3D value (0.95 MW) corresponds to the furnace walls heat fluxes, it has to be compared to the firebox heat loss estimation (1.28 MW) giving a closer match.

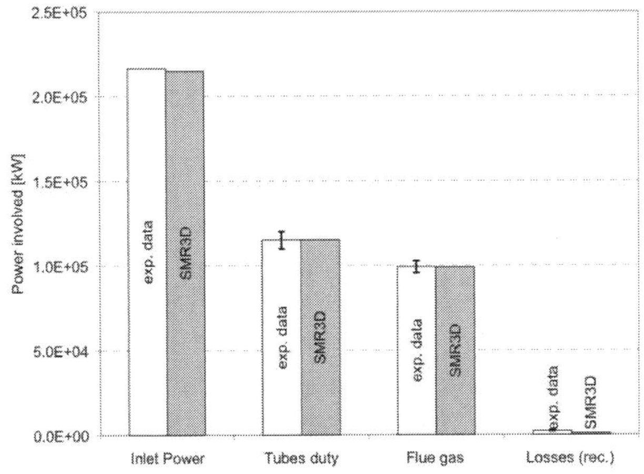

Figure 2. Comparison of the main terms of the energy balance for experimental (reconciled) data and SMR3D computation results.

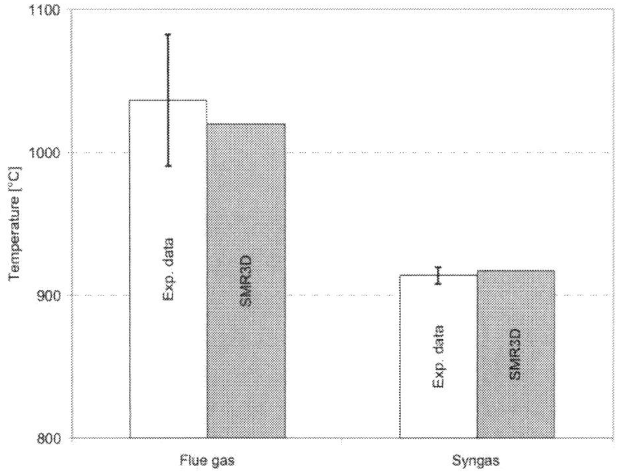

Figure 3. Comparison of the syngas and flue gas outlet temperatures for reconciled data and SMR3D computation results.

A comparison between experimental – i.e. reconciled – and computed syngas composition at the tubes outlet is shown in Figure 4. Both results are very close and comparable one to another, always within the error bars.

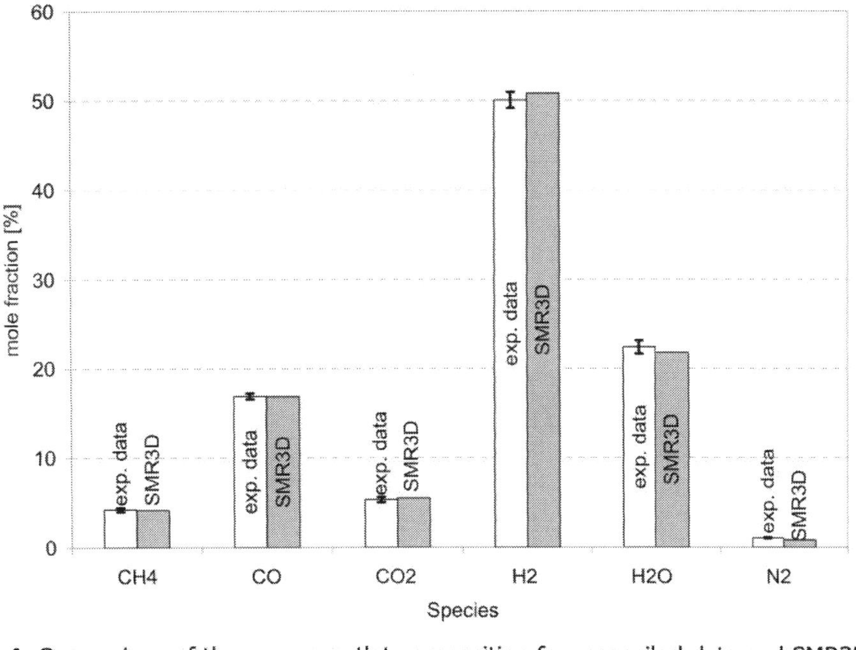

Figure 4. Comparison of the syngas outlet composition for reconciled data and SMR3D computation results.

Tubes temperature profiles

The second step of the validation approach is to compare the energy and flow distribution inside the reformer with experimental data. In the present case of tubular reformer, the tube skin temperatures were measured with a pyrometer applying wavelengths such that the influence of gas radiation is minimized. As this type of device measures and converts the incoming radiative heat flux into a temperature, the measured temperatures require post-treatment of the raw data using a view factors method in order to correct the reflected part from the radiative flux [15].

Computed and corrected experimental tube temperatures are compared in Figure 5. The SMR3D average over all tubes and the tubes with the minimum and maximum temperature profile are shown. The average temperature profiles for both computation and experiments are in close agreement. In addition, the spread of the corrected measurement values is only in a few cases not within the computed min/max envelopes. Note that both SMR3D and experimental data agree well with the maximum temperature taken for the design of the tubes and calculated at design stage with a commercial tool adapted to LURGI's reformer design.

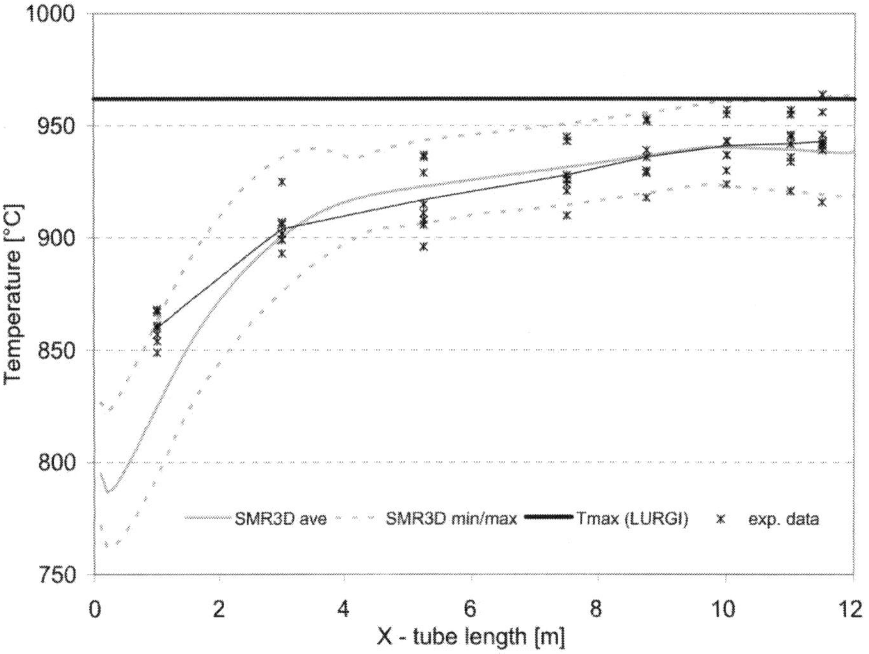

Figure 5. Tube skin temperature profile: corrected experimental data (symbols), SMR3D computation results (average – solid line; min/max envelop – dotted line) and LURGI maximum temperature calculation.

SMR3D applications

Based on the SMR3D predictions and its confrontation with experimental data for overall furnace mass and energy balances and for tube skin temperature profiles as layed down in the preceding paragraph the prediction capability of SMR3D not only allows to describe and understand the governing physical phenomena inside the firebox but also to exploit the detailed coupling effects between the combustion chamber and the reforming tube, with a level of details not reported, to our knowledge, in the literature so far.

A few examples of the application of this tool for design optimization and further improvement of the understanding of the technology are given below.

Heat losses and refractory materials

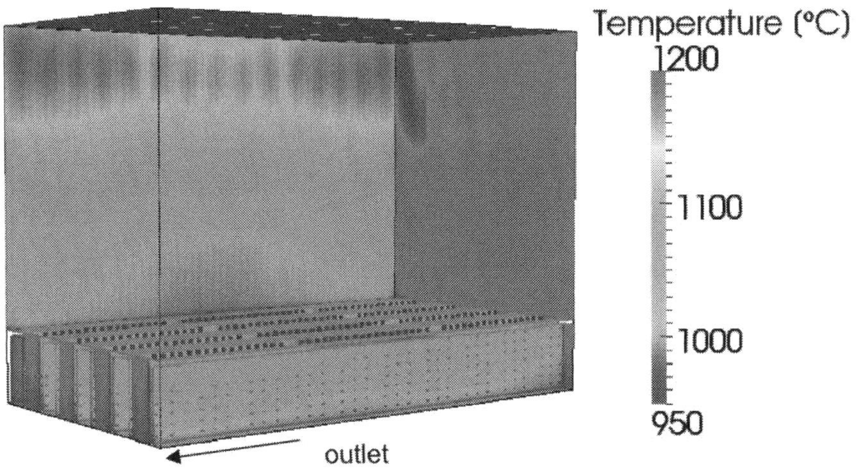

Figure 6. Furnace walls temperature distribution illustration for a large SMR furnace.

A heat flux or temperature mapping at the firebox walls, as illustrated in Figure 6, enables to characterize and map precisely the heat losses of the furnace. The upper part of the reformer walls, subjected to the flames, is the main contributor to heat losses. Additionally, a precise and reliable prediction of the walls temperatures, depending on the operating conditions, helps in the refractory and insulation materials definition during the design phase.

Flow arrangement

The flow arrangement inside the combustion chamber could be characterized and precisely analyzed. In the velocity magnitude field, pictured in **7**, bending of the flame from the external burner rows is observed to a much lower extent than that described in [12].

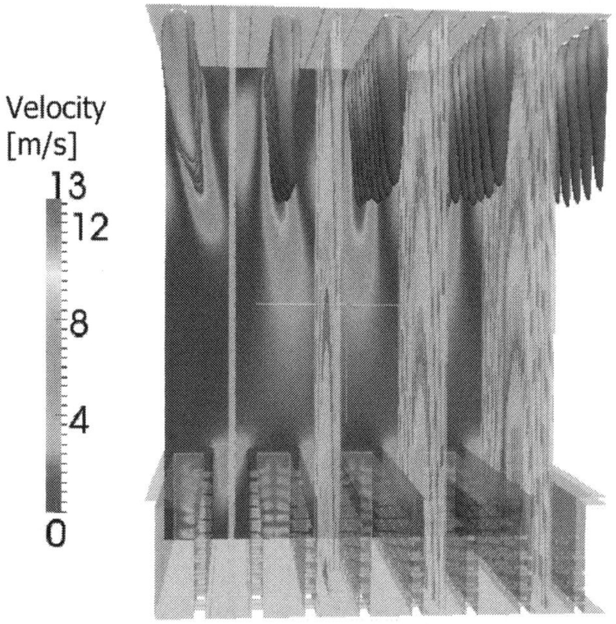

Figure 7. Velocity magnitude field illustration. Flame are represented by iso-temperature surfaces at 1500°C coloured with velocity.

Tube duty polar distribution

The normalized heat flux, $\tilde{\varphi}(\theta,z)$ and temperature variation, $\Delta T(\theta,z)$, circumferential profiles at different altitudes are plotted in Figure 8 for a given 5" tube with 300 mm pitch. Definitions for heat flux normalization and temperature variations used here are given below. The tube is heated on both sides by radiation emitted from hot flue gas and flames, whereas it receives no energy from the neighboring tubes that have locally almost the same temperature. Consequently, the normalized heat flux and temperature variations profiles show two minima in the angular sectors that face other tubes. The ratio between the maximum value for $\tilde{\varphi}(\theta,z)$ of 1.11 computed using SMR3D is close to the API 530 view factor prediction of 1.18, which is a value that doesn't take into account the convection and conduction diffusion influence [13].

The maximum temperature deviation from the average value at the top of the tubes is around 15°C. Indeed, the heat flux received is high in front of the flames, however the syngas and tube wall temperature is low at this location. The temperature difference at the bottom tube section where the maximum tube wall temperature occurs is even lower because of the lower heat flux, an inherent feature of a top-fired reformer enhancing tube lifetime.

$$\tilde{\varphi}(\theta,z)=\frac{\varphi(\theta,z)}{\frac{1}{2\pi}\int_{2\pi}\varphi(\theta,z)d\theta} \qquad \Delta T(\theta,z)=T(\theta,z)-\frac{1}{2\pi}\int_{2\pi}T(\theta,z)d\theta$$

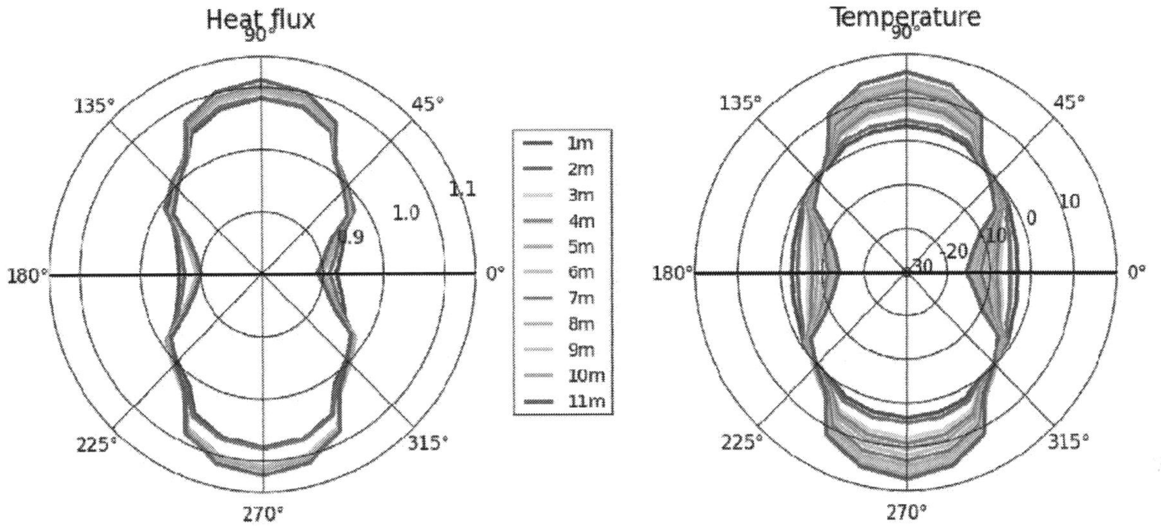

Figure 8. Distributions of the normalized heat flux and temperature difference to the average for a given 5" tube at different altitudes (from the bottom).

Tunnel ports effects on tubes

A potential negative tunnel effect on tube lifetime has been proposed by Farnel on the basis of Monte-Carlo simulations: the radiation intensity could be intensified by multiple reflection inside the tunnel and impact the tubes placed right in front of the opening, inducing a local heat flux increase by 20%, yielding to a local temperature increase of 15°C [14]. The detailed radiation heat transfer model implemented within SMR3D, including finely discretized ray tracing and precise gas radiating properties, has been used in order to characterize this phenomenon. The tunnels area as been highlighted in Figure 9 for temperature fields and radiation contributions (two color codes: tube surface heat flux (W/m^2) and gas radiation source term (W/m^3)). Neither radiation intensification nor temperature peaks have been observed locally on the tubes. On the contrary, the tunnels protect the tubes, as tubes temperature decreases in the tunnel region (see Figure 5). Moreover, it reveals that the tunnels have a small cooling effect on the flue gas entering the tunnels, as illustrated by the temperature gradient between both sides of the tunnel wall in Figure 9. The heat is transferred to the walls and then conducted to the bottom. In terms of radiation, the inside of the tunnel tend to absorb (red) the radiation coming from the outside (green).

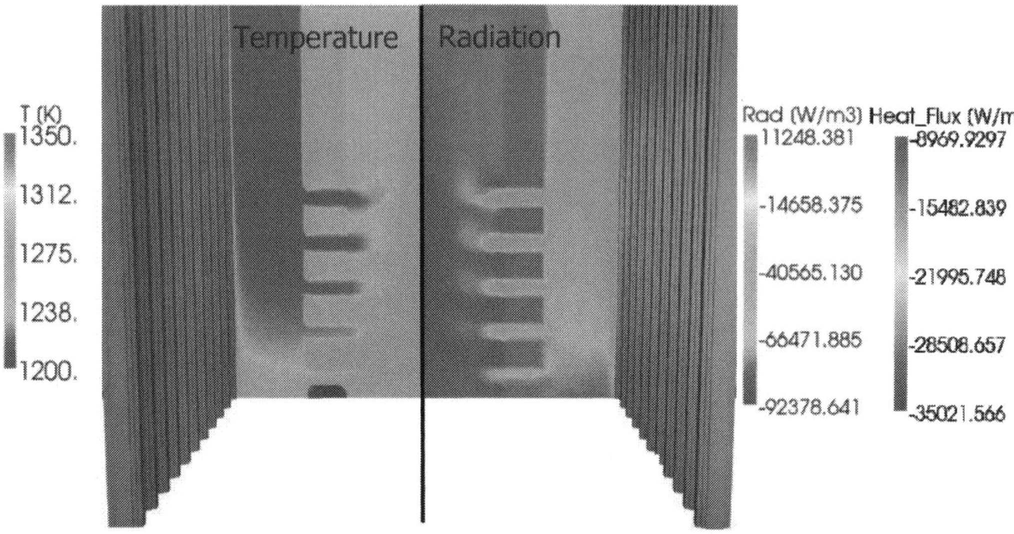

Figure 9. Temperature (left) and radiative contributions fields (right) (two color codes: tube surface heat flux (W/m^2) and gas radiation source term (W/m^3)).

As a result from the detailed coupled computations presented here, the radiation intensification such as in a laser device observed in [14] appears unrealistic in a SMR furnace tunnel. However the cited paper documents several assumptions like the Monte-Carlo approach is not rigorous as detailed in classical references [15][16], the 3D problem is calculated in 2D, spectral aspects from gas, tubes and bricks are neglected, tunnel and tubes walls with specular reflection and uniform gas temperatures assumption is unrealistic and the gas emissivity is twice as big as values given in other literature, e.g. [17].

Design improvement

Finally, the continuous improvement of the commercial LURGI reformer is benefitting from SMR3D simulation capabilities. As an illustration, two of the largest AIR LIQUIDE operated plants, recently designed by LURGI, were simulated and the results are compared (~400 tubes on 8 rows each). The so-called HERACLES plant has been commissioned in 2011 (Figure 10(a)) and was designed 2 years ago, whereas the second project will be started-up in 2013 (Figure 10(b)). Therefore, the latter can benefit from SMR3D inputs to optimize the heat flux and temperature homogeneity and the flue gas flow arrangements. So SMR3D allows significant performance improvements described further below.

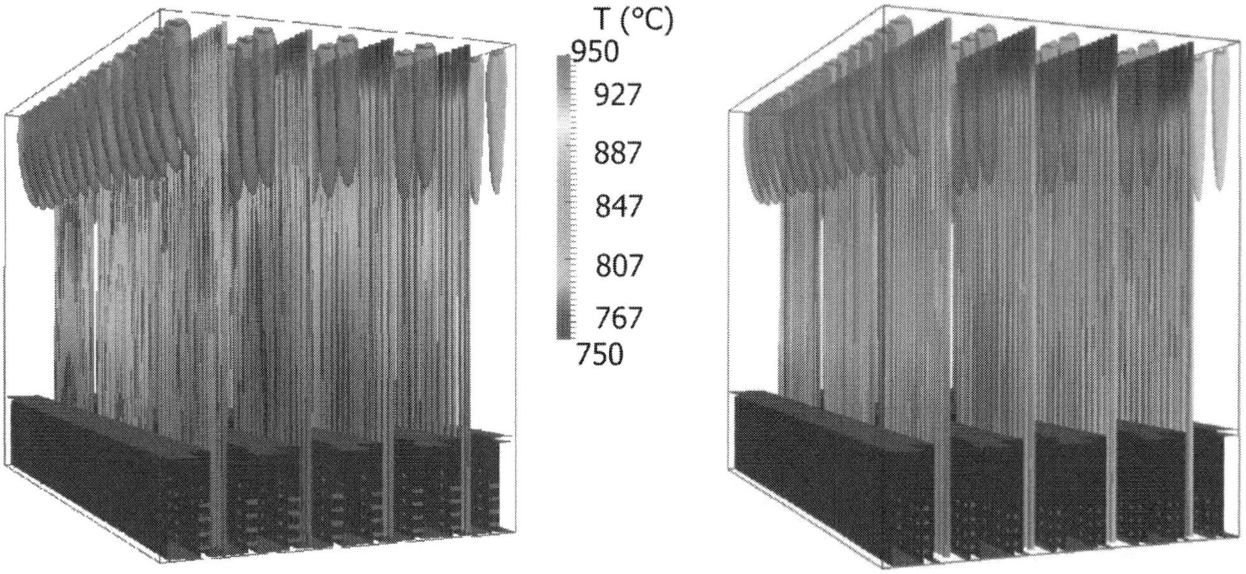

(a) (b)

Figure 10. Illustration of the computed results for two of the largest AIR LIQUIDE operated furnaces: (a) HERACLES, (b) undisclosed project

As the two plants do not have exactly the same design characteristics and operational conditions (e.g. 4″ vs. 5″ tubes, different feedstock, etc.), normalized physical units are compared. The normalized tube duty is plotted in Figure 11 for each tube, row per row. HERACLES data is illustrated using void symbols, whereas the other plant is represented by grey bold symbols. Reforming efficiency rises with higher temperature, while the maximum operating tube temperature is limited by the design for economically optimized reforming conditions and a certain tube lifetime based on creep rupture strength curves. As the hottest tube will limit furnace operation, homogenization of the tube duty at the furnace scale is key in improving furnaces design and operation.

For firebox structural reasons, the tube rows in both furnaces have been split into 3 tubes section. The SMR3D computation results allowed to observe a tube duty maldistribution for HERACLES, which had to be compensated by an adjustment of the firing. Less heat would have been transferred to the central section than that to the extremity ones. This effect was due to different levels of power available per tubes locally, and has been corrected in the latest design.

The average trends have been added to the scatter plot in Figure 11, showing a slight increase of the average duty with an increase of tube number for the HERACLES case. A detailed CFD investigation was performed to identify the reasons for such tendency. It has been found that the flue gas flow arrangement in the firebox was responsible for this inhomogenity. This was corrected in the latest reformer design thanks to an optimized tunnel openings distribution, allowing a homogeneous aspiration of the flue gas in the combustion chamber, while minimizing the pressure induced.

Finally, between these two generations of large reformers, the tube loads have also been significantly homogenized, since the normalized standard deviation has been reduced from 2.3% to 1.2% from HERACLES to the second project.

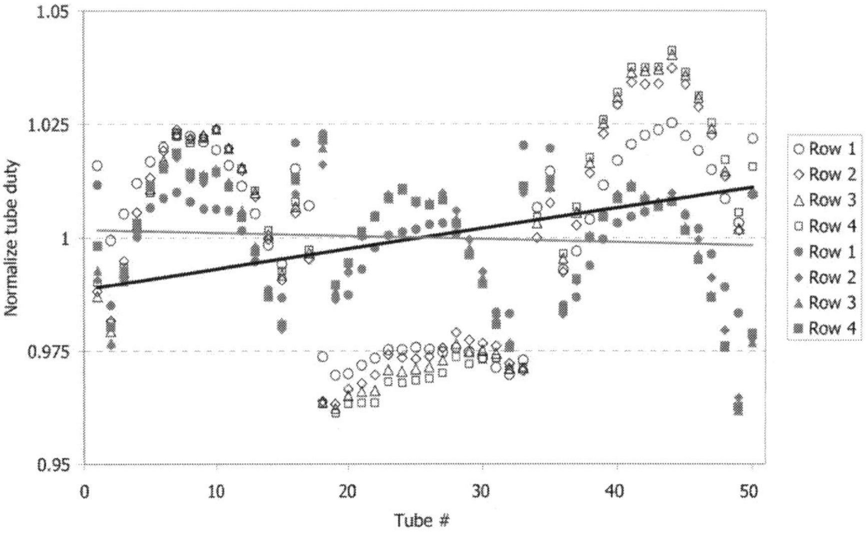

Figure 11. Comparison of normalized tube duty distribution for HERACLES (void symbols) and second project (grey symbols). Row 1 corresponds to the side wall, row 4 is in the furnace middle.

Conclusion

SMR3D is the result of 25 years of continuous efforts for rigorous integration of the most appropriate models, while keeping the solver highly efficient. It is now possible to run simulations of full scale large SMR furnace, including combustion, radiation and the coupling of combustion chamber to individually calculated reforming tubes.

The SMR3D capability to describe different types of reformers (H_2 or CO production plants, 4 or 5 inches tubes) demonstrates its flexibility. The return of experience from SMR3D computations has already provided successful improvement of industrial SMR furnaces.

The understanding of key phenomena involved in reformers operation and design has been greatly improved, resulting in further optimization and benefits in terms of efficiency, reliability and cost optimization.

SMR3D paths the way for larger steam reformer designs, where homogeneity and equal-distribution is essential for the operation of highly reliable and available plants.

References

[1] New York Power Authority; (2005), "Hydrogen fact sheet: Hydrogen production – Steam Methane Reforming (SMR)", www.getenergysmart.org

[2] Champinot, C., Till, M., Haung-Naudin, C., Klug, A., (1996), "Gekoppelte Berechnungen von Glas und Atmosphäre", *Glas Ingenieur*

[3] Jones, W.P., Launder, B.E., (1972), "The prediction of laminarization with a two equation model of turbulence", *Heat and mass transfer*, 15, pp. 310-314

[4] Patankar, S.V. (1980), "Numerical Heat Transfer and Fluid Flow", *Hemisphere*.

[5] Lockwood, F.C. and Shah, N.G., (1980), "A new radiation solution method for incorporation in general combustion prediction procedures", *18th Symposium (International) on Combustion, The combustion Institute, Pittsburg,* pp. 1405-1414

[6] Soufiani, A., Djavdan, E., (1994), "A comparison between weighted sum of grey gases and statistical narrow band radiation models for combustion applications", Combustion and Flame, Vol. 97, pp. 240-250

[7] Magnussen, B.F., Hjertager, B.H., (1977), "On Mathematical Modeling of Turbulent Combustion with Special Emphasis on Soot Formation and Combustion", *16th Symp. (Int.) on Combustion,* pp. 719-729

[8] Xu, J., Froment, G.F., (1989), "Methane Steam Reforming – I Methanation and Water-Gas Shift: Intrinsic kinetics"; *AIChE journal*, 35(1), pp. 88-96

[9] Xu, J., Froment, G.F., (1989), "Methane Steam Reforming – II Diffusional limitations and reactors simulation"; *AIChE journal*, 35(1), pp. 97-103

[10] Kalitventzeff, B., Cransfeld, J.M., Vassilakis, C.E., Dimaratos, J., Delava, P., (1999), "Demonstration of an Energy Saving Process for a Crude Oil Distillation Unit", ESCAPE 9, Budapest, (Hungary), June.

[11] ASTM C680-08, (2008), "Standard Practice for Estimate of the Heat Gain or Loss and the Surface Temperatures of Insulated Flat, Cylindrical, and Spherical Systems by Use of Computer Programs", *ASTM International*

[12] Cotton, W., (2003), "Fluegas flow patterns in top-fired steam reforming furnaces", *Johnson Matthey Catalysts, IMTOF.*

[13] API 530 standard, 2008, (2007), "Calculation of Heater-tube Thickness in Petroleum Refineries", identical to ISO 13704:2007

[14] Farnell, W., (1999), "The Tunnel Port Effect: Validation by Monte Carlo Simulation", *Synetix ICI Group, IMTOF.*

[15] Siegel, R., Howell, J.R., (1981), "Thermal Radiation Heat Transfer – Second Edition", *HPC New York*

[16] Modest, M.F., (2003), "Radiative heat transfer – Second Edition", *Academic Press*

[17] Truelove, J.S., (1976), "A mixed grey gas model for flame radiation", *AERE Harwell*

(364a) Impact of Lignin Extraction and Cellulose III Conversion On the Enzymatic Digestibility of Corn Stover

Leonardo da costa Sousa, Shishir Chundawat, Nirmal Uppugundla, Bruce Dale and **Venkatesh Balan**, Chemical Engineering and Material Science, Great Lakes Bioenergy Center, Michigan State University, Lansing, MI

The benefits of delignification on plant cell wall digestibility by hydrolytic enzymes are well described in the literature. The presence of lignin increases non-productive binding of cellulases, resulting in enzyme inhibition and decrease in the sugar yields. Moreover, lignin acts as an anti-microbial agent, reducing cell viability and ethanol yields.

Recently, additional benefits by modifying the crystal structure of native cellulose (cellulose I) to cellulose III were also reported. After this modification, it was possible to double the rate of enzymatic hydrolysis of cellulose. However, this increase was observed for purified cellulose, native from algae (Cellulose Iα), which has a different crystal configuration than the cellulose native from plants (Cellulose Iβ). Therefore, it is not evident that this modification is possible to be obtained in plant cell walls and that it can represent the same kind of benefits observed for purified cellulose. From the literature, cellulose III is obtained by contacting cellulose I to supercritical ammonia or different types of amines in anhydrous conditions. Additionally, water is known to prevent the conversion of cellulose I to cellulose III, however the water concentration limit that allows cellulose III to be formed is not yet clearly defined. For biorefinery applications, plant biomass will contain a certain degree of moisture that could prevent stable cellulose III formation and therefore, it is important to verify in what extent is possible to convert native cellulose to cellulose III in plant biomass.

In this perspective, our goal was to test the benefits of cellulose III conversion under non-anhydrous conditions, using corn stover, and couple this modification with lignin extraction in a single stage process – Extractive AFEX (E-AFEX). These benefits were compared with the traditional non-extractive ammonia based pretreatment method - AFEX.

See more of this Session: Developments In the Pretreatment of Lignocellulosics for Bioconversion I
See more of this Group/Topical: Sustainable Engineering Forum

(364b) The Effects of Cationic Polyelectrolyte On the Pretreatment and Enzymatic Hydrolysis of Corn Stover

Shaowen Ji and Ilsoon Lee, Chemical Engineering and Materials Science, Michigan State University, East Lansing, MI

Lignocellulose is the most abundant renewable biomass and the potential substrate for ethanol production. The costly pretreatments and large consumption of enzymes are needed to reach high cellulose conversion. Lignin, as one of major cell wall components, is considered as the barrier to enzymatic hydrolysis by blocking cellulase from productively being adsorbed. The removal or effective modification of lignin is required for efficient pretreatment process. In this study, the effects of cationic polyelectrolyte, poly-(dimethyldiallylammonium chloride) (PDAC), on alkaline pretreated corn stover were investigated. Structural study by Scanning Electron Microscopy (SEM) imaging revealed particles morphology on cell walls of pretreated corn stover. These particles were confirmed to contain lignin by Confocal Laser Scanning Microscopy (CLSM) and Transmission Electron Microscopy (TEM) imaging. Chemical compositions of pretreated corn stover at different alkaline and polyelectrolyte concentrations were analyzed. Higher polyelectrolyte concentration in the pretreatment resulted in higher lignin retention. Evaluation of enzymatic hydrolysis showed the enhancement of cellulose and hemicellulose conversion after polyelectrolyte pretreatment. PDAC was proposed to facilitate cellulose disruption and lignin redistribution via its interaction with lignin, which significantly modified biomass morphological properties, such as increased porosity and improved accessibility of cellulose surface.

See more of this Session: Developments In the Pretreatment of Lignocellulosics for Bioconversion I
See more of this Group/Topical: Sustainable Engineering Forum

(364c) Novel Lignocellulosic Biomass Pretreatment Using a Modified Taylor-Couette Mixer for Biofuels and Biomaterials

Wei Wang and Ilsoon Lee, Department of Chemical Engineering and Materials Science, Michigan State University, East Lansing, MI

Bioethanol derived from lignocellulosic biomass can be a good candidate for energy alternative, not only because lignocellulosic biomass is abundant and sustainable, but also bioethanol production forms carbon cycle in a more timely fashion compared to fossil fuel, which leads to reduction of green gas emission. Lignocellulosic biomass pretreatment, which is one of the essential steps in bioethanol production, accounts for the majority portion of the overall bioethanol production cost. Hence an effective and efficient pretreatment method is highly demanded. Although this research area has been deeply investigated several decades ago, it has not attracted tremendous attention until recently. In our research, cost-effective and breakthrough biomass pretreatment was conducted with a novel modified Taylor-Couette mixing system, aiming to overcome the natural recalcitrance of lignocellulosic biomass in a short processing time and facilitate a better accessibility of enzymes to polysaccharides. Compositional analysis proved significant removals in both lignin and hemicellulose and rich glucan content left in the residue only after 2 minutes or less. Parameters like mixing speed, retention time, biomass loading and corresponding enzyme digestibility will be presented.

See more of this Session: Developments In the Pretreatment of Lignocellulosics for Bioconversion I
See more of this Group/Topical: Sustainable Engineering Forum

(364d) Modeling and Scale up of CO2-Water Pretreatment of Guayule Biomass

Ehsan Moharreri and J Richard Elliott, Chemical and Biomolecular Engineering, The University of Akron, Akron, OH

A carbon dioxide – water pretreatment process was developed for guayule biomass. The guayule biomass was mixed with water and then loaded to a reactor where it was exposed to a biphasic environment consisting of CO2 rich vapor phase and water rich liquid phase. Then, the biomass was enzymatically hydrolyzed and the sugar content was measured at the end. The effects of temperature, pressure, moisture content, accessible volume and holding time in the pretreatment step were investigated. One hundred and twenty (120) experimental runs were carried out in the temperature range of 145 C to 210 C, pressure was between 3.4 MPa to 34 MPa, the moisture content ratio was between 0.6 to 6, the accessible volume was between 0.2 to 0.1 L and the holding time was between 20 to 50 minutes. A model was fit to the experimental data based on dissolved CO2 in the liquid phase using CO2-H2O phase envelop and a temperature dependant constant. The model fits to one hundred and twenty (120) data points with an average absolute percent deviation(AAD%) of 13. Then, the same process was implemented to a pilot scale vessel with a 1 liter volume. The process was carried out in batch and semi-continuous methods and twenty (20) data points were produced to confirm the model. The model predicted the pilot plant runs with a 15% average absolute percent deviation (AAD%). In the lab scale of 190 C, 26 MPa, moisture ratio of 1.5, accessible volume of 0.22L, biomass load of 15 grams and 35 minutes in holding time resulted in 78% glucose yield and 66% hemicelluloses yield. In the pilot scale of 200 C, 10MPa, moisture ratio of 2, accessible volume of 0.88, biomass load of 60 grams and 35 minutes in holding time resulted in 78% total sugar yield.

See more of this Session: Developments In the Pretreatment of Lignocellulosics for Bioconversion I
See more of this Group/Topical: Sustainable Engineering Forum

(499g) Effect of Transition Metal Catalyst On Lignin Oxidation During Alkaline Hydrogen Peroxide Pretreatment

Zhenglun Li, Department of Chemical Engineering and Material Science, Michigan State University, East Lansing, MI and David Hodge, Chemical Engineering and Materials Science, Michigan State University, East Lansing, MI

Biomass lignocellulose can be used as feedstock for bioethanol production, and a pretreatment is necessary for achieving high ethanol yield and economical feasibility. As a method to increase enzyme digestibility of biomass by selective delignification, alkaline hydrogen peroxide (AHP) pretreatment has been proposed and studied. The lignin recalcitrance in lignocellulose biomass can be removed by H_2O_2 oxidation and the enzymatic hydrolysis efficiency of the polysaccharides can be greatly improved. However, a big issue with AHP pretreatment is the high cost of hydrogen peroxide during the pretreatment, and the selectivity of the oxidant need to be improved, i.e. to achieve highest pretreatment efficiency with low level of H_2O_2 loading.

In 1890s, Fenton discovered that iron ion can catalyze the oxidation of organic substrate by hydrogen peroxide. Since then, research has been done on the application of transition metal catalysts in catalyzing the phenol contaminant in wastewater and pulp bleaching. It is desired that catalyst can enhance selective electron transfer from lignin substrate to the oxidant and improve the selectivity of delignification reactions to unproductive side reactions.

See more of this Session: Developments In the Pretreatment of Lignocellulosics for Bioconversion I
See more of this Group/Topical: Sustainable Engineering Forum

(364f) Progress Toward Biological Pretreatment - Understanding Plant Cell Wall Degradation by Termite and White-Rot Fungi

Shulin Chen[1], Jing Ke[2], Dhrubojyoti D. Laskar[2] and Jijiao Zeng[2], (1)Department of Biological Systems Engineering, Washington State University, Pullman, WA, (2)Washington State University, Pullman, WA

In an effort developing more efficient and environmentally benign lignocellulosic biomass processing technologies, the authors' laboratory has been studying plant cell wall degradation process in termite and white-rot fungi. This presentation summarizes the results so far obtained from various experimental investigations of the lab. First, termite guts provide both aerobic and anaerobic environments, the former is required for lignin degradation. The termite guts also have close to neutral pH range, which suggest existence of lignin degrading enzymes that are different from that of fungi. The organization of the gut system allows the conversion of biomass in termite to occur as a sequential process in which the pretreatment of the biomass occurs in the foregut and midgut, whereas the hydrolysis of the cellulose occurs mainly in the hindgut. There are synergies between termite and its symbiont in deconstructing plant cell walls. The chewing action of the termite increases the surface area for enzymatic actions, the enzymes produced by the termite tissues modifies the bonds that forms the barriers for cellulase access and symbionts produce the enzymes that are mainly responsible for cutting the cellulose polymer to sugar units. The results shows the actual changes of lignin structure as the biomass material pass through the termite guts. The lignin structure changes are more modification than degradation in nature. Equally import is the deconstruction of the association of lignin with other components of plant cell walls such as hemicelluloses that occurs in the foregut and midgut. It becomes apparent by comparing with other biological systems that the enzyme systems and their actions in biological plant cell wall degradation systems are more complex than what was previously known. A generality has also been observed in that most of the biological systems studied tended to reduce G unit from the original lignin structure. Additionally, termite is a more effective plant cell wall degradation system than white rot-fungus because of the chewing action of the termite and the better-controlled environment in the termite guts. Termite is also more effective than the cow stomach as the latter lacks capability of lignin modification. Developing an artificial system that can surpass the natural biological system in major performance criteria may offer the ultimate solutions to lignocellulosic biomass processing and utilizations.

See more of this Session: Developments In the Pretreatment of Lignocellulosics for Bioconversion I
See more of this Group/Topical: Sustainable Engineering Forum

(375a) Synthesis and Characterization of Bimetallic Overlayer Aqueous Phase Reforming Catalysts

Joseph Holles and Michael Skoglund, Chemical and Petroleum Engineering, University of Wyoming, Laramie, WY

Introduction: A great deal of today's research seeks to create new and sustainable fuels to meet our world's ever growing energy demands. Recently, aqueous phase reforming (APR) has emerged as a process capable of turning renewable feedstocks into hydrogen gas or alkanes. Much of the current hydrogen supply is derived from light hydrocarbons, and a sustainable method of production will prove imperative as demand increases. The majority of past APR studies have used feedstocks that are derived from biomass (methanol, ethylene glycol, glycerol, sorbitol, and glucose) but in order to prove the full potential of the process, further studies need to be done using a non-derived biomass feed. Lactose, a sugar waste product of the cheese and dairy industry, has shown promise as a biomass product capable of undergoing the APR reaction in a single step.

Past APR studies have looked at a variety of different supported metal catalysts for the reaction. These studies have shown that bimetallic Pt/Ni and Pt/Co make ideal catalysts for production of hydrogen through APR. Building upon the success of bimetallic catalysts for APR, we have synthesized pseudomorphic overlayer type bimetallic catalysts which aim to increase the activity of the catalyst while reducing the necessary loading of costly platinum. An ideal pseudomorphic overlayer catalysts consist of a monatomic layer deposited atop a different bulk metal. These catalysts have received widespread attention from computational and single crystal studies because of their notably different adsorption properties when compared to their constituent metals. These changes are likely due to electronic interactions between the two metals causing a shift in the d-band center of the overlayer metal. Computational work has demonstrated a linear relationship between the center of the d-band and heats of adsorption. However, there is still a need to develop effective broad-based synthesis methods for overlayer catalysts on high surface area supports that will empirically demonstrate predicted behavior. Additionally, it would be beneficial to correlate this behavior with reactivity, which cannot be predicted computationally.

In order to capitalize on the behavior of pseudomorphic overlayers, we have previously synthesized Re@Pd (base@overlayer) supported catalysts. A decrease in heat of hydrogen adsorption was seen for Re@Pd catalysts when compared to both the Re and Pd parents; these results were consistent with computational predictions found in literature. Re@Pd overlayer catalysts also demonstrated a decrease in activity for ethylene hydrogenation when compared to the activity of Pd. Thus reactivity was correlated with a computationally predicted parameter. We have built upon these results and have extended the synthesis technique toward creating Ni@Pt and Co@Pt catalysts. Ni/Al2O3, Co/Al2O3, and Pt/Al2O3 monometallic baseline catalysts were prepared using standard techniques. The directed deposition technique was used to selectively deposit the overlayer (platinum) metal on the base metal without it depositing onto the support surface.

Experimental: Hydrogen chemisorption studies for isotherm analysis and to determine hydrogen heats of adsorption were done using a Micromeritics ASAP 2020. Isotherms for each sample were measured from 35 to 300°C and pressures from 1 mtorr to 900 torr. The reactivity of each sample was studied through the ethylene hydrogenation reaction. The ethylene hydrogenation reaction was performed using a plug flow reactor at temperatures from -10 to 300°C at atmospheric pressure with a typical gas flow rate of 750 mL/min. The reactor effluent was monitored using a GCMS.

Results and Discussion: Chemisorption results showed that hydrogen adsorption isotherms were distinct in shape and temperature dependence for each of the Ni/Al2O3 and Co/Al2O3 parent catalysts. Isotherms for Ni@Pt/Al2O3 were extremely similar to those of the Ni/Al2O3; giving no indication of isolated Pt metal formation. Adsorption measurements done at multiple temperatures allowed for the determination of isosteric heat of hydrogen adsorption. Measured hydrogen heats of adsorption for Pt/Al2O3 and Ni/Al2O3 were 57 kJ/mol and 36 kJ/mol respectively at high coverage. Platinum data shows similar results to single crystal and computational predictions. The heat of adsorption for Ni@Pt/Al2O3 was 47 kJ/mol at high coverage. As predicted computationally, the heat of adsorption for hydrogen decreased for the overlayer compared to Pt. The weaker hydrogen adsorption for Ni@Pt when compared to monometallic Pt should aid in increasing APR reaction rates because the less tightly bound hydrogen product will more quickly desorb from the metal surface, thereby freeing up active sites for adsorption of reactants.

Pt/Al2O3 is highly active for ethylene hydrogenation and both of the Ni/Al2O3 and Co/Al2O3 baseline catalysts show lower activities. Addition of a platinum overlayer to make Ni@Pt and Co@Pt overlayer catalysts increased the activity of each catalyst but still showed reduced activity when compared to a pure platinum catalyst. These results agree with predicted d-band shifts that would cause weaker hydrogen adsorption on the metal surface and ultimately reduced activity for ethylene hydrogenation when compared to platinum metal alone. By using the ethylene hydrogenation reaction as a characterization tool we have been able to directly correlate reactivity with hydrogen heat of adsorption. We have previously shown that as hydrogen heat of adsorption decreases, the turnover frequency for ethylene hydrogenation decreases in the Re@Pd system. This activity decrease is consistent with literature results that show ethylene hydrogenation activity is directly related to the amount of hydrogen adsorbed on the metal surface. Thus, the relationship between hydrogen heat of adsorption and ethylene hydrogenation activity has been extended to multiple systems. Additional reactivity studies are underway to further characterize the Ni@Pt and Co@Pt catalysts using apparent activation energies and reaction orders. APR studies will also be initiated for each catalyst and lactose feed in order to determine possible effects that overlayer catalysts can have on the APR reaction.

See more of this Session: Fundamentals of Supported Catalysis II
See more of this Group/Topical: Catalysis and Reaction Engineering Division

(375b) Particle Size Effects On the Selective Dehydrogenation of Cyclohexane Over Pt Particles

Qiang Qian[1], Matthew Neurock[2], Stefan Vajda[3], Lisa D. Pfefferle[4] and Gary L. Haller[4], (1)Department of Chemical Engineering, University of Virginia, Charlottesville, VA, (2)Department of Chemical Engineering and Chemistry, University of Virginia, Charlottesville, VA, (3)Chemical Sciences Division, Argonne National Laboratory, Argonne, IL, (4)Department of Chemical Engineering, Yale University, New Haven, CT

The catalytic dehydrogenation of cyclohexane over transition metals is important in refining reforming [1] as well as in hydrogen production strategies [2]. Nanometer and subnanometer Pt particles demonstrate reasonably high selectivities, lower activation temperatures and higher efficiencies. Recent experimental results [3] show that larger Pt particles (2.15 nm) selectively produce cyclohexene from cyclohexane; while smaller Pt particles (1.32 nm) result in the formation of benzene. We have used theory together with detailed characterization studies to elucidate the particle size effects on the selectivity.

Density functional theory calculations were carried out using the Vienna ab-initio Simulation Package [4] to obtain detailed insights into the mechanism of selective dehydrogenation of cyclohexane on platinum nano-particles. A cubo-octahedral Pt_{55} cluster was employed to model the ~1.3 nm platinum nano-particles. Larger platinum particles with diameter ~2.0 nm were modeled by a truncated cubo-octahedral Pt_{201} cluster.

We examined in detail the reaction pathways for the dehydrogenation of cyclohexane over Pt_{55} and Pt_{201} clusters. The results show that surface species bind more strongly on the smaller cluster (Pt_{55}). The ability to activate the C-H bond increases in the order (111)-facet < (100)-facet < Edge < Corner, which correlates with the degree of coordinative unsaturation at these surface sites.

The product selectivity towards cyclohexene vs. benzene was found to depend on the competitive partitioning between the C-H bond activation process and the desorption process of the cyclohexene intermediates. The computational results were found to be in good agreement with the experimental results.

References

1. Davis, S.M. and G.A. Somorjai, in *The Chemical Physics of Solid Surfaces and Heterogenous*

Catalysis, 1984, Elsevier: Amsterdam.

2. Wang, Y., N. Shah, and G.P. Huffman, *Energy & Fuels,* 2004. **18**(5): p. 1429-1433.

3. Vece, M.D., S. Lee, X. Wang, et al., *Unpublished work,* 2010.

4. Kresse, G. and J. Furthmüller, *Phys. Rev. B,* 1996. **54**: p. 11169.

See more of this Session: Fundamentals of Supported Catalysis II
See more of this Group/Topical: Catalysis and Reaction Engineering Division

(375c) Synthesis, Characterization, and Testing of Shaped PtAg Nanoparticle Catalysts

Louis Jones and Michael Gordon, Chemical Engineering, University of California, Santa Barbara, Santa Barbara, CA

Heterogeneous catalysts play a vital role in many industries such as commodity chemical production, pollution control, fine chemical synthesis, and pharmaceutical manufacturing. In these venues, chemical reactions take place on surfaces that typically consist of supported metal and metal oxide clusters or nanoparticles. Understanding the functionality of these catalysts requires correlating their local electronic, chemical, and morphological structure to catalytic activity and selectivity. It is also well known that dramatic changes in catalytic activity can occur due to finite size effects (*i.e.*, unsaturated atoms at surfaces, edges, kinks, and corners), preferred morphologies, or modification of the nanoparticle's electronic structure by the support. Unfortunately, many traditional catalyst preparation methods (*e.g.* precipitation and impregnation) can result in unknown particle size and shape distributions that make it difficult to establish connections between reactivity and morphology.

In this work, we utilize colloidal synthesis to realize Ag on Pt nanoparticle catalysts with well-defined morphology to critically evaluate the influence of shape (*e.g.* cubes, cuboctahedra, and octahedra) and surface composition (Pt vs. Ag) on C_2 hydrogenation activity and selectivity. Ag on Pt nanoparticles were synthesized using polyvinylpyrrolidone (PVP) and $AgNO_3$ as growth promoters; nanoparticles were subsequently supported on colloidal silica, redox cycled to remove PVP, chemically etched to remove Ag, and evaluated for hydrogenation activity. IR, Raman, and NMR were used throughout the synthesis process to understand how different adsorbates (CO, C_2H_2, C_2H_4, and PVP) associate with specific particle facets as a function of Ag content. This data was used in conjunction with continuous flow reactor experiments to rationalize how shape and surface composition affect catalyst behavior. In particular, it was found that (1) surface-segregated Ag on Pt dramatically decreases hydrogenation activity while increasing selectivity for C_2H_4; (2) Ag removal produces a highly-active, yet non-selective hydrogenation catalyst; and (3) cuboctahedra, as compared to cubic and octahedral particles, tend to favor coke formation. The catalytic behavior of Ag on Pt nanoparticles was also compared with a reverse catalyst of Pt on Ag (nanocubes) synthesized by galvanic exchange. Nanoparticle synthesis, *in-situ* spectroscopy, and catalytic testing results will be highlighted in the talk.

See more of this Session: Fundamentals of Supported Catalysis II
See more of this Group/Topical: Catalysis and Reaction Engineering Division

(375d) Well Defined, Supported MnOx Oxidation Catalysts Via Controlled Decomposition of Coordination Complexes

Andrew Korinda, Nicholas Schoenfeldt and **Justin M. Notestein**, Chemical and Biological Engineering, Northwestern University, Evanston, IL

The low nuclearity metal oxide active sites in several metalloenzymes, including catalase or the oxygen evolving center in photosystem II, have inspired the synthesis of numerous homogeneous mimics. Several of these mimics are active catalysts in their own right for low temperature oxidation. Manganese (IV) triazacyclononane, Mn-tacn, coordination complexes are one such family of examples [1]. Here, we describe our results employing Mn-tacn complexes as a route for creating oxide-supported MnO_x sites of controlled nuclearity and reactivity.

Mono-, di-, and tetranuclear Mn-tacn complexes were supported on silica, titania, and alumina, and calcined to produce the final heterogeneous catalysts. The structures of the supported catalysts were tracked during synthesis by diffuse reflectance UV-visible spectroscopy and XANES/EXAFS, and were consistent with retention of the structure of the original complex. For example, EXAFS showed no higher coordination Mn shells for the supported monomer, and 1 or 3 Mn neighbors for the dimer and tetramer, respectively.

Moderate activity in cyclohexane oxidation with H2O2 at 60°C has been observed for the mononuclear species on silica, while no activity was observed for the higher nuclearities under the same conditions. Further correlations between precursor structure, final supported structure, hydrogen-TPR, and reactivity in alkane oxidation will be described.

[1] Wieghardt, K.; et. al. *J. Am. Chem. Soc.* 110, 22 (1988).

See more of this Session: Fundamentals of Supported Catalysis II
See more of this Group/Topical: Catalysis and Reaction Engineering Division

(375e) Examination of the Effects of the Catalysts Functionality On the Selective Oxidation of Methanol

Erum Qayyum, Chemical & Biomedical Engineering, University of South Florida, Tampa, FL and John Kuhn, Chemical and Biomedical Engineering, University of South Florida, Tampa, FL

Examination of the effects of the catalysts functionality on the selective oxidation of Methanol

E. Qayyum and J.N.Kuhn *

University ofSouth Florida, Tampa, USA

*Department ofChemical & Biomedical Engineering, University of South Florida, Tampa, 813-974-6498 begin_of_the_skype_highlightingend_of_the_skype_highlighting,jnkuhn@usf.edu

Methanol is seen as a viable option as a fuel in futuristic fuel cellvehicle applications and its selective oxidation to numerous high valuechemicals such as dimethyl ether (DME), formaldehyde and hydroformylationproducts demonstrates it as a chemical platform. Both uses of methanol involvedcatalytic chemistry for efficient energy usage.

The focus of the current research is on the selective oxidation of methanol tovarious chemical products using catalytic chemistry. The catalysts were groupedinto two categories, which include bulk catalysts (silica, ceria,silica-alumina, titania, copper, and aluminum phosphate) and supported metalcatalysts (metals immobilized onto the various bulk catalysts). The acidiccatalysts, which include silica-alumina and aluminum phosphate, demonstrated atendency towards DME. Alternatively, ceria, silica, and the supported metalcatalysts showed selectivity towards formaldehyde and combustion products.Currently, catalyst alterations to maximize selectivity for a single route arebeing investigated.

See more of this Session: Fundamentals of Supported Catalysis II
See more of this Group/Topical: Catalysis and Reaction Engineering Division

ADSORPTION AND SUBSEQUENT SURFACE REACTIONS OF BIFUNCTIONAL COMPOUNDS FROM BIO-OIL VAPORS ON OXIDE SUPPORTS

*Chandramouli Vaddepalli and Friederike C. Jentoft**
Chemical, Biological & Materials Engineering, University of Oklahoma
Norman, OK 73019, USA
**fcjentoft@ou.edu*

Introduction

Vapor phase upgrading of short oxygenates in bio-oils (*e.g.*, from fast pyrolysis of biomass) needs development of chemoselective catalysts that convert individual molecules or functional groups while leaving others untouched. The development of chemoselective catalysts requires thorough understanding of the adsorption modes and the reactivity of such bio-oil molecules on surfaces. Likely, some of these catalysts will be supported metals, and the question arises whether oxygenates will solely interact with the metal if the support surface is polar and hydrophilic. Other questions concern the competition of various molecules in a mixture for sites or the preferential mode of adsorption of molecules with more than one functional group.

Hydroxyacetone (HA) is an excellent candidate for a fundamental investigation because it has two functional groups and is contained in bio-oil vapor in significant concentrations [1]. We investigated the adsorption of hydroxyacetone on conventional oxide supports, represented by silica, alumina, titania, ceria and zirconia, using in-situ diffuse reflectance IR spectroscopy (DRIFTS) as the primary method. Pulse thermal analysis (PTA) experiments were carried out to estimate the heats of adsorption. Monofunctional reference compounds, specifically acetone and *n*-propanol, were also adsorbed on these oxides to assess the role of each functionality for reactivity and to generate spectra for comparison.

Materials

The following oxides were obtained from commercial suppliers: silica "Hisil 190" (PGP industries), γ-alumina "Aluminium oxide, gamma phase" (Alfa), and titania "Titanium(IV) oxide, anatase" (Alfa). Zirconia was prepared by calcination of MEL zirconium hydroxide (XZO 1247/01) at 773 K for 1 h. Ceria was prepared by titrating $Ce(NO_3)_3.6H_2O$ aqueous solution into 50 ml ammonium hydroxide solution. The resulting precipitate was washed with de-ionized water, dried overnight in an oven and calcined at 903 K for 1 h. The purities of the organic reagents were as follows: *n*-Propanol, Chromasolv 99.9% (Aldrich), Acetone, Chromasolv 99.9% (Aldrich) and hydroxyacetone 95% (Alfa Aesar). Dried KBr 99+% FT-IR grade (Aldrich) was used for FTIR spectroscopy to record a single channel background.

Methods

The specific surface areas (S_{BET}) were determined by N_2 physisorption using a Micromeritics ASAP 2000 analyzer.

Zirconia and ceria were characterized by X-ray diffraction (XRD) using a Bruker D8 diffractometer with a curved crystal monochromator and system settings of 40 kV and 35 mA. Ceria was found to consist of the cubic phase and zirconia of a mixture of the monoclinic and tetragonal phases.

FTIR spectra of solids were recorded in diffuse reflection geometry. A PerkinElmer Spectrum100 instrument was operated at a resolution of 4 cm^{-1}, and 128 scans were collected per spectrum. A Harrick Scientific DRP-P11 Praying Mantis accessory was placed in the sample compartment of the PerkinElmer spectrometer, and a CHC-CHA-3 reaction chamber was used for measurements under controlled temperature and gas atmosphere. The following protocol was applied, with all flows adjusted by mass flow controllers (Bronkhorst, all values at NTP). Samples were activated by heating in air to a temperature of 673–773 K; a heating rate of 4 K/min was applied and the air flow rate was 20 ml/min. The maximum temperature was held for 20 min; then the samples were cooled to 313 K in air supplied at a flow rate of 20 ml/min followed by purging in N_2 for 30 min supplied at a flow rate of 20 ml/min. The reactants were admixed by bubbling a 20 ml/min stream of N_2 through a saturator filled with the liquid reactant. The saturator temperature was controlled to adjust the vapor pressure; the temperatures and pressures were for *n*-propanol 276 K and 557 Pa, acetone 255 K and 1658 Pa, and hydroxyacetone 298 K and 559 Pa. Spectra were recorded of the saturated state of the surface; however, at the higher reactant partial pressures, gas phase contributions became visible in the spectra. Thus, the reaction chamber was purged at 313 K with 20 ml/min N_2 to remove gas phase and physisorbed species. After 15 min, a heating ramp was started with a rate of 6 K/min and a final temperature of 673 K. Spectra were recorded continuously using Timebase software.

Thermogravimetry (TG) and differential scanning calorimetry (DSC) were performed using a NETZSCH STA 449 F1 Jupiter with evolved gas analysis by a QMS 403 C Aëolos mass spectrometer and a Bruker Tensor 27 Fourier-Transform infrared spectrometer, both connected to the top of the thermobalance by heated capillaries. Solid samples were activated by heating in air to a temperature of 773 K; a heating rate of 4 K/min was applied and the air flow rate was 20 ml/min. The maximum temperature was held for 35 min; then the samples were cooled to 313 K in air supplied at a flow rate of 20 ml/min. Adsorption experiments comprised injecting a pulse of 1-2 μl reactant into the helium stream through a septum. Injections were continued until the weight of the sample did no longer increase.

Results and Discussion

Hydroxyacetone is known to exhibit four different types of conformer in the vapor state [2-3]. Investigations by microwave spectroscopy by Kattija-Ari et al. [4] and Apponi et al. [5] confirmed the most stable form of hydroxyacetone in the vapor phase to be the internally hydrogen-bonded conformation.

Infrared spectra of hydroxyacetone isolated in argon matrix were collected by Coleman et al. [6], who reported the O-H and the carbonyl vibrations at 3540 and 1750 cm^{-1}. Sharma et al. [2], using the same technique, found the same bands located at 3505 cm^{-1} and 1731 cm^{-1}. Jetzki et al. [7] reported the pure gas phase spectrum of hydroxyacetone in the range of 1900-750 cm^{-1}, without specifying band positions. It can be concluded that IR spectra of hydroxyacetone in the literature are inconsistent, and no complete MIR range spectrum has so far been reported.

Gas phase FTIR spectra of hydroxyacetone and acetone (Figure 1) were recorded by vaporizing the organic liquid in the thermobalance and collecting the spectra of the evolving vapor using the connected FTIR spectrometer. Hydroxyacetone has an O-H vibration at 3537 cm^{-1}, whereas propanol, which does not form an internal hydrogen bond, is characterized by an O-H vibration at 3678 cm^{-1}. Internal hydrogen bonding is known to decrease the force constant and the vibrational frequency of the O-H group in (C=O...HO-C) [4]. The red shift in the OH frequency compared to *n*-propanol strengthens this interpretation; hydroxyacetone is predominantly present in an internally hydrogen-bonded conformer, as suggested by various authors in their theoretical and experimental works [2-7].

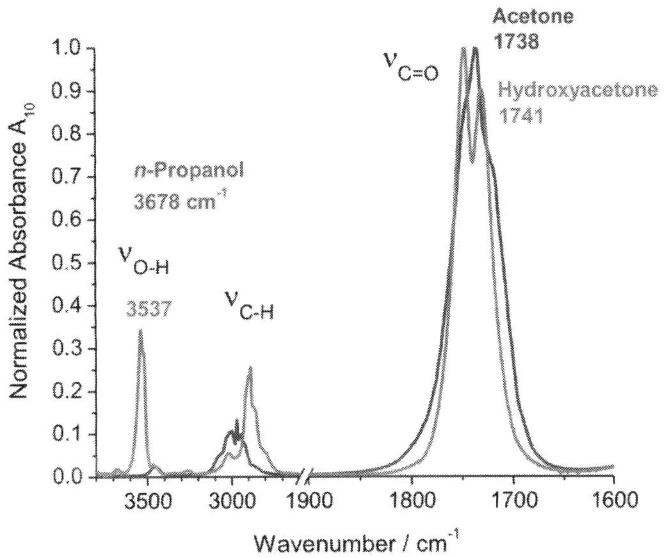

Figure 1. Gas phase IR spectra of hydroxyacetone and acetone.

Hydroxyacetone chemisorbed on all oxides at 313 K. Surface OH groups and metal cations such as Al^{3+} , Zr^{4+}, Ti^{4+} and Ce^{3+}/Ce^{4+}, which are strong Lewis acids, served as adsorption sites as evidenced by the perturbation of OH groups and the red shift of the carbonyl frequency of hydroxyacetone. Alcohols are known to form alkoxides with surface OH groups; because hydroxyacetone possesses both, a hydroxyl group (as propanol does) and a ketonic group (as acetone does), we suspect a chemistry that combines those of the mono-functional molecules. Disappearance of the O-H vibration of hydroxyacetone and formation of a water combination band upon adsorption on several supports suggests that hydroxyacetone may be forming alkoxides on the surface.

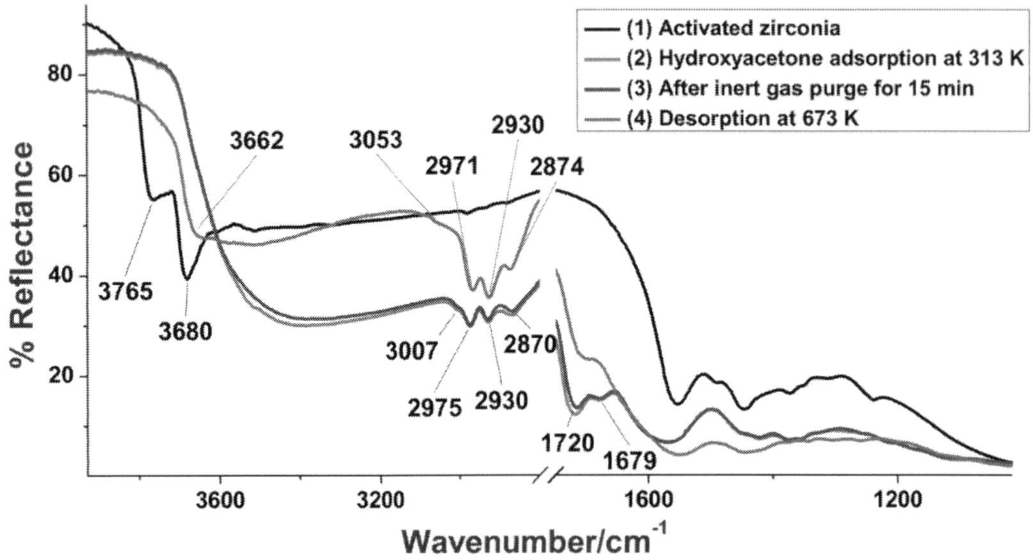

Figure 2. IR spectra recorded during adsorption and desorption of hydroxyacetone on zirconia.

Adsorption-desorption experiments of hydroxyacetone on zirconia are shown in Figure 2. IR spectra of activated zirconia exhibit two surface OH vibrations at 3765 and 3680 cm^{-1}. Upon adsorption of hydroxyacetone both surface OH groups are perturbed forming a broad band centered at 3400 cm^{-1}. After heating to 673 K and partial desorption, the spectrum shows that all basic surface OH groups of zirconia (absorbing at 3765 cm^{-1}) and some acidic surface OH groups (3680 cm^{-1}) have been consumed. We believe these OH groups are involved in complex surface reactions with adsorbed hydroxyacetone, resulting in a product with intense C-H bands that remains on the surface of zirconia.

Desorption of hydroxyacetone was incomplete from all oxides; C-H stretching vibrations were still observable after heating to 673 K. In Table 1, observations regarding adsorption and desorption of hydroxyacetone are summarized. The shift of the carbonyl frequency of hydroxyacetone upon adsorption on the oxides (*i.e.*, relative to the frequency of the free molecule in the gas phase) gives an estimation of the strength of interaction. Because of the internal hydrogen bond in gas phase hydroxyacetone, the carbonyl shifts are calculated relative to an already perturbed carbonyl vibration. The shifts are thus only seemingly smaller than those observed for acetone. In addition, DRIFTS studies show that hydroxyacetone interacts more strongly than acetone and forms surface species stable at 673 K. We suppose that the internal hydrogen bond opens up as the molecule approaches the surface and adsorbs. The correct reference state is then a free conformer, which has a higher carbonyl vibrational frequency than the hydrogen-bonded conformer. On the basis of the work of Jalbout et al. [3], we speculate a conformer obtained through simple opening of the hydrogen bond and without rotation around the C-C bond to be the most likely configuration. Scaling the value of the carbonyl vibration for this conformer with our data, we obtain a free gas phase vibration of the carbonyl group of 1762 cm^{-1}. When the shift of the carbonyl vibration upon adsorption is recalculated with respect to this frequency, much higher values are obtained, consistent with much stronger adsorption of hydroxyacetone on oxides in comparison to acetone.

Table 1

Reactant Catalyst ↓	Propanol	Acetone	Hydroxyacetone	Carbonyl vibration shift (cm^{-1}) relative to gas phase C=O vibration of…		
				Acetone at 1738 cm^{-1}	HA at 1741 cm^{-1}	HA conformer w/o H-bond [3] at 1762 cm^{-1}
Silica	Stable surface species (alkoxides)	Complete desorption	Stable surface species	**-29**	**-11**	**-32**
γ-Alumina	Complete desorption	Stable surface species	Stable surface species	**-36**	**-15**	**-36**
Titania	Almost complete desorptionR eduction of titania	Desorption Reduction of titania	Stable surface species OH- groups of titania are consumed. Reduction of titania	**-29**	**-10**	**-31**
Zirconia	Stable surface species	Stable surface species	Stable surface species. High frequency OH- (basic) groups of zirconia are consumed	**-35**	**-21**	**-42**
Ceria	Almost complete desorption	Stable surface species	Stable surface species	**-41**	**-31**	**-52**

Pulse Thermal Analysis (PTA)

PTA experiments were performed to estimate the heats of adsorption of hydroxyacetone on the selected oxides. The heat is an excellent measure of the strength of interaction of hydroxyacetone with the surface. The TG signal was monitored to determine the chemisorbed amount of hydroxyacetone. Acetone was adsorbed on silica to verify the accuracy of the PTA experiments for estimation of the heat of adsorption. Hair et al. [8] reported a heat of adsorption of 50 kJ/mol for acetone adsorption on Cab-O-Sil. Figure 3 shows acetone adsorption on Hisil. The estimated average heat of adsorption was 51 kJ/mol, which is very close to the value reported in the literature. The TG signal shows that most of the acetone is physisorbed on silica; the carbonyl group of acetone is known to form hydrogen bonding with surface silanols [9]. Hydroxyacetone was adsorbed on alumina at 313 K (Figure 4). The TG curve shows that the first pulse of hydroxyacetone is completely chemisorbed on the surface followed by chemisorption and physisorption in the subsequent pulses. Shoulders on the heat signal and a negative heat signal without associated weight change indicate that adsorption is followed by endo- and exothermic surface reactions.

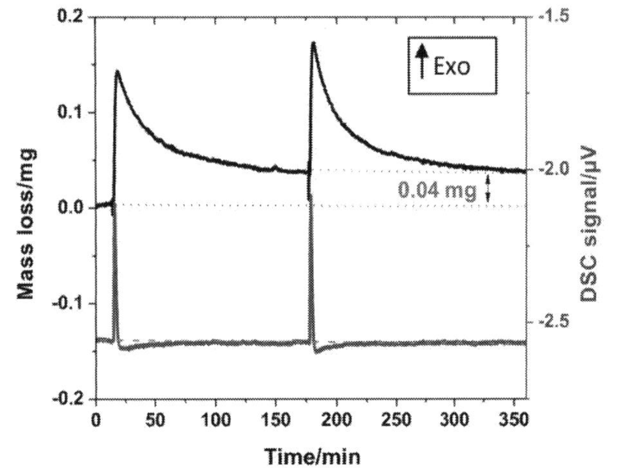

Figure 3. Adsorption of 1 µl injections of acetone on Hisil

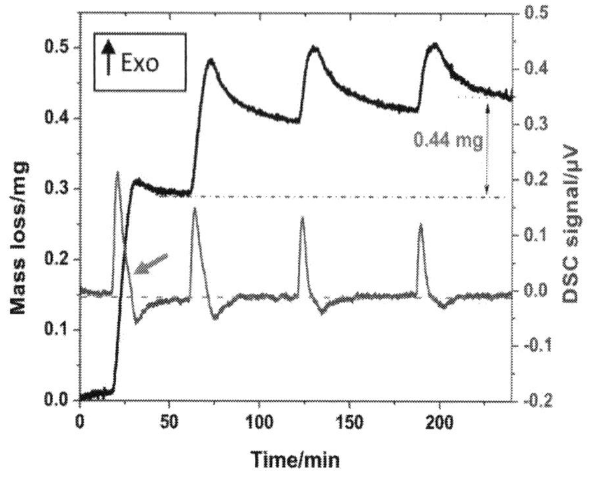

Figure 4. Adsorption of 2 µl injections of hydroxyacetone on alumina

Summary and Conclusions

- Infrared spectra show that hydroxyacetone adsorbs on the polar hydrophilic surfaces of oxides and cannot be thermally desorbed without decomposition.
- Thermal analysis shows that the adsorption of hydroxyacetone on alumina is followed by endothermic and exothermic surface reactions.
- Strong interaction of hydroxyacetone with the polar surfaces of oxides indicates they may be unsuitable as inert supports in catalysts for upgrading of highly reactive bio-oil compounds.
- Short oxygenates present in bio-oil such as hydroxyacetone adsorb and react aggressively, resulting in rapid deactivation of catalysts. Such molecules will have to be removed or converted in the first step of a sequence of catalytic upgrading steps.

References

1. C. A. Mullen and A. A. Boateng, *Energy & Fuels*, **22**, 2104–2109 (2008).

2. A. Sharma, I. Reva, R Fausto*, J. Phys. Chem. A*, **112**, 5935-5946 (2008).

3. A. F. Jalbout, F. F. Contreras-Torres, L. Adamowicz, *Int. J. Quant. Chem.,* **108**, 279-288 (2008).

4. M. Kattija-Ari, M.D. Harmony, *Int. J. Quant. Chem. Quant. Chem. Symp.,* **14**, 443-453 (1980).

5. A. J. Apponi, J. J. Hoy, D. T. Halfen, L. M. Ziurys, M. A. Brewster, *M. A. Astrophys. J.*, **652**, 1787 (2006).

6. W. M. Coleman, B. M. Gordon, *Appl. Spectrosc.*, **43**, 305-310 (1989).

7. M. Jetzki, D. Luckhaus, R. Signorell, *Can. J. Chem.*, **82**, 915 (2004).

8. W. Hertl, M. L. Hair, *J. Phys. Chem.* **72**, 4676-4682 (1968).

9. A. G. Okunev, E. A. Paukshtis, Y. I. Aristov, *React. Kinet. Catal. Lett.*, **65**, 161 (1988).

Population Balance Modeling of Particles with Random Behavior. Application to Gene Regulatory Processes.

Doraiswami Ramkrishna and Che-Chi Shu

Extended Abstract

Population balance models for system of particles with random behavior lead to parabolic partial differential equations with "diffusion" terms acquired from random dynamics of internal coordinates. This talk will demonstrate the origin of such population balance equations, develop their formulation, methods of solution and application to gene regulatory processes in microorganisms. The population balance models are shown to display striking variations from the popular single cell approach that has been in vogue in the literature. Phenomena such as the transfer of drug resistance among bacteria are exciting applications of this approach.

Consider a population of particles of state $\mathbf{x}$ in an environment characterized by vector $\mathbf{y}$ of environmental variables such that the rate of change of $\mathbf{x}$ is determined by the stochastic differential equations

$$d\mathbf{x} = \dot{\mathbf{X}}(\mathbf{x}|\mathbf{y})dt + \mathbf{B}_X(\mathbf{x}|\mathbf{y})d\mathbf{W}_x \tag{1}$$

where $\dot{\mathbf{X}}(\mathbf{x}|\mathbf{y})$ is the average rate of change and $\mathbf{B}_X(\mathbf{x}|\mathbf{y})$ is a matrix and $d\mathbf{W}_x$ is a vector independent differential Wiener (Brownian motion) processes in the state space of internal coordinates describing the stochastic variation of the particle state $\mathbf{x}$ in an environment $\mathbf{y}$. Similarly, the particles may also be assumed to move through physical space in accord with the stochastic differential equations

$$d\mathbf{r} = \dot{\mathbf{R}}(\mathbf{x},\mathbf{r},t)dt + \mathbf{B}_r(\mathbf{x},\mathbf{r})d\mathbf{W}_r \tag{2}$$

where freedom from $\mathbf{y}$ is assumed although not for mathematical reasons. The function $\dot{\mathbf{R}}(\mathbf{x},\mathbf{r},t)$ represents the mean convective velocity and $\mathbf{B}_r(\mathbf{x},\mathbf{r})$ is associated with the diffusion of particles. The derivation of the population balance equation in the expected number density has been given by Ramkrishna (2000). It is given by

$$\frac{\partial n(\mathbf{x},\mathbf{r},t)}{\partial t} + \nabla_x \cdot \dot{\mathbf{X}}(\mathbf{x}|\mathbf{y})n(\mathbf{x},\mathbf{r},t) + \nabla_r \cdot \dot{\mathbf{R}}(\mathbf{x},\mathbf{r}|\mathbf{y})n(\mathbf{x},\mathbf{r},t) = h(\mathbf{x},\mathbf{r},\mathbf{y},t) +$$

$$+\frac{1}{2}\nabla_x\nabla_x : \mathbf{B}_x(\mathbf{x}|\mathbf{y})\mathbf{B}_X^T(\mathbf{x}|\mathbf{y})n(\mathbf{x},\mathbf{r},t) + \frac{1}{2}\nabla_r\nabla_r : \mathbf{B}_r(\mathbf{x},\mathbf{r})\mathbf{B}_r^T(\mathbf{x},\mathbf{r})n(\mathbf{x},\mathbf{r},t) \tag{3}$$

where the first term on the right hand side represents the net rate of producing particles of state $\mathbf{x}$ at location $\mathbf{r}$ by various processes such as breakage, aggregation and so on. Eq. (3) can also be expanded to include spatial coordinates with provisions for the movement of particles through space by convection as well as diffusion. The above equation is also coupled to a balance equation for the environmental variables written as

$$\frac{\partial \mathbf{y}}{\partial t}+\nabla_{\mathrm{r}}\cdot\mathbf{N}_{\mathrm{y}}=\int \mathbf{A}\dot{\mathbf{X}}(\mathbf{x}|\mathbf{y})n(\mathbf{x},\mathbf{r},t)\,d\mathbf{x} \tag{4}$$

where $\mathbf{N}_{\mathrm{y}}$ is the spatial flux of environmental species including convective as well as diffusive contributions.

We address in this presentation an application of the above formulation to the modeling of gene regulatory processes in which a signaling molecule in the cellular environment initiates a set of intracellular reactions culminating in the synthesis of a protein through the expression of a specific gene. Because the number of molecules participating in reaction is small, the behavior of the reaction system is characteristically stochastic in nature. Consequently some cells display high protein levels representing "on" situation of the "gene switch", while other cells with low protein levels represent the "off" state of the switch. The "on" and "off" states often occur with bistability, a scenario that has attracted numerous publications in the literature in which the stochastic analysis of the reaction system provides for bimodal protein level distributions among the cell population, which can also be compared with experiments in a flow cytometer through the use of fluorescent dyes. The modeling is discussed in detail in Shu et al. (2011). The nature of the regulatory process is presented by Chatterjee et al. (2011). In this Extended Abstract, we simply outline the process as follows.

The vector $\mathbf{x}$ represents the concentrations of species connected with gene regulatory process, while $\mathbf{y}$ contains the extracellular concentration of a signaling molecule that induces the expression of a protein tending to produce "on" cells, and an inhibitor produced internally (and exchanged with the environment) that has the opposite effect of preventing protein expression thus producing "off" cells. These details are covered by Chatterjee et al. (2011). The consequence of the stochastic nature leads to a continuous distribution of protein levels among the cell population, some qualifying for on cells and others for off cells.

A common modeling approach in the literature is to focus on a single cell associating with it a portion of the environment without regard to the presence of other fellow cells. The analysis gives rise to a bimodal distribution of protein levels. The population balance approach recounted by Shu et al (2011), on the other hand, accounts for some essential features in this process not included in the single cell model. The cells are assumed to be in a well-mixed environment so that the position coordinate $\mathbf{r}$ is eliminated; growth in biofilms will, however, require the full set of equations (3) and (4). The stochastic nature of individual cell behavior calls for stochastic exchange between cells and the environment, so that the environmental variables are modified by the average exchange from all cells which in turn affects the behavior of single cells also. The net result is that the bimodal behavior from the single cell model is not reproduced by the population balance model. The results are shown below.

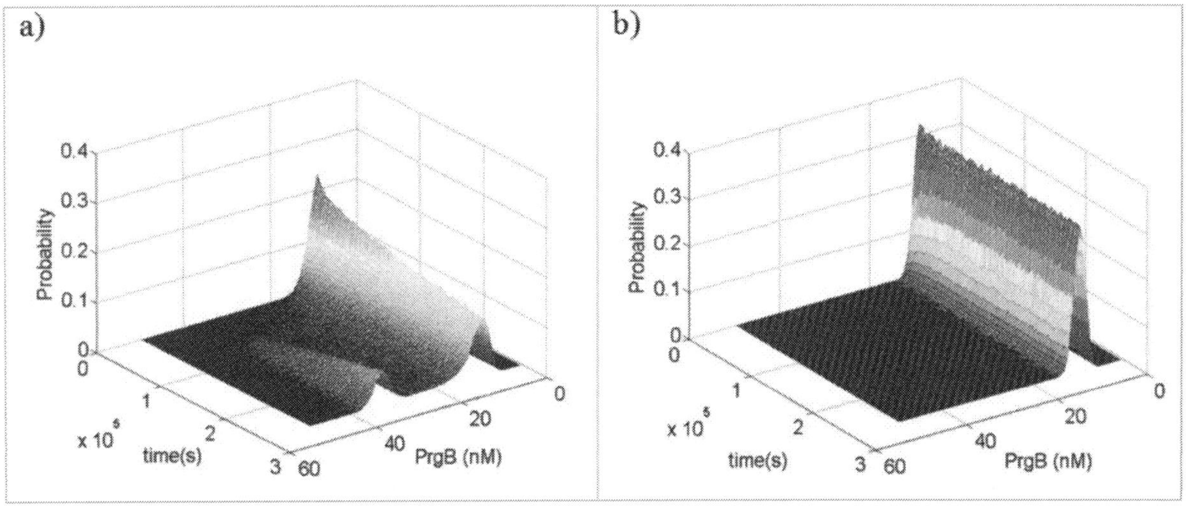

Figure: Single cell model leads to bimodal distribution of protein on the left while the population balance model on the right leads to a unimodal distribution.

In extending such work to transfer of drug resistance, one must also include the conjugation of induced drug resistant cells with those that are not allowing for the transfer of plasmid which bears the drug-resistance trait. Thus the transfer of drug resistance can be studied in various types of growth scenarios such as in planktonic and biofilm environments. The population balance framework proves to be a vital tool in the study of such phenomena.

Ramkrishna, D., 2000. *Population balances: Theory and applications to particulate systems in engineering*. Academic Press San Diego.

Shu, Che-Chi, A. Chatterjee and Wei-Shou Hu, and Doraiswami Ramkrishna, "Modeling of Gene Regulatory Processes by Population Mediated Signaling. New Applications of Population Balances," *Chem. Eng. Sci.,* in press.

Chatterjee, A. Shu, Che-Chi, C. M. Johnson, Y. N. Kaznessis, D. Ramkrishna, G. M. Dunny and Wei-Shou Hu, "Convergent transcription Confers a Bistable Switch in *Enterococcus faecalis* Conjugation," *PNAS,* **108**, 9721-9726, 2011.

Hydrogen Generation From Ammonia Borane for PEM Fuel Cell Applications

Arvind Varma

School of Chemical Engineering, Purdue University, West Lafayette, IN

Ammonia borane (NH_3BH_3, AB) is a promising hydrogen storage material because it contains 19.6 wt% hydrogen. To release hydrogen from ammonia borane, thermolysis, catalytic hydrolysis and heat generated by additional reactive mixtures have been employed. All the current methods, however, have disadvantages which decrease the efficiency of hydrogen storage systems. In this work, we report new approaches to release hydrogen from AB, which do not require any catalyst and provide relatively high hydrogen yield and environmentally benign byproducts. One approach involves nano-aluminium (nAl)/water combustion reaction, which provides heat for AB dehydrogenation and releases additional hydrogen from water [1]. The second approach thermally activates AB hydrolysis in aqueous AB solutions and slurries under modest inert gas pressure [2-3]. The third approach involves AB thermolysis alone, with effective reaction heat management. The combustion-based methods could be used in compact power sources for portable electronic devices, while AB hydrothermolysis or thermolysis are attractive for vehicle transportation applications. For the latter case, we also conducted systematic experiments to compare the various approaches in terms of highest hydrogen yield, lowest ammonia formation, and moderate operating temperature [4], and details will be presented in this work.

References

1. Diwan, M., Hanna, D. and Varma, A., "Method to Release Hydrogen from Ammonia Borane for Portable Fuel Cell Applications," *International Journal of Hydrogen Energy*, **35**, 577-584 (2010).
2. Hwang, H. T., Al-Kukhun, A. and Varma, A., "Hydrogen for Vehicle Applications from Hydrothermolysis of Ammonia Borane: Hydrogen Yield, Thermal Characteristics, and Ammonia Formation," *Ind, Eng. Chem. Res*, **49**, 10994-11000 (2010).
3. Diwan, M., Hwang, H. T., Al-Kukhun, A. and Varma, A., "Hydrogen Generation from Noncatalytic Hydrothermolysis of Ammonia Borane for Vehicle Applications," *AIChE Journal*, **57**, 259-264 (2011).
4. Al-Kukhun, A., Hwang, H. T. and Varma, A., "A Comparison of Ammonia Borane Dehydrogenation Methods for Proton-Exchange-Membrane Fuel Cell Vehicles: Hydrogen Yield, Ammonia Formation and Its Removal," *Ind, Eng. Chem. Res* (*in press*, 2011, DOI 10.1021/ie102157v).

(381c) Understanding Dispersion Polymerizations In Open and Confined Reaction Spaces towards New Applications

Kyu Yong Choi[1], Laleh Emdadi[2], Carla Luciani[2], Sang Yool Lee[2] and In Hak Baick[2], (1)Department of Chemical and Biomolecular Engineering, University of Maryland, College Park, College Park, MD, (2) Chemical and Biomolecular Engineering, University of Maryland, College Park, MD

Dispersion polymerization is a well-known technique to produce monodisperse polymer particles of 1-15 microns. It consists of preparing a solution of a monomer, a solvent, a steric stabilizer and a chemical initiator. Since the polymer is poorly soluble in the solvent-enriched solution, growing polymer chains precipitate in the form of unstable primary particles which agglomerate to produce larger but stable polymer particles.

In general, dispersion polymerizations are carried out in open reaction spaces in the sense that no geometrical confinement is imposed for the polymer to precipitate. As a result, monodisperse polymer particles of several microns in size are generally produced. By varying the reaction conditions, however, other morphologies such as porous polymeric structures can also be developed after inducing the system phase inversion.

The use of confined reaction spaces to carry out dispersion polymerization is interesting because of the possibility of creating micro-particles with complex internal structures. Recently, our research group has developed a micro-dispersive suspension polymerization (MDSP) technique, where the dispersion polymerization takes place in the confined reaction space of a suspended monomer droplet of 10-150 microns in size. Although MDSPs share some characteristics with regular dispersion polymerizations in open spaces, some important difference imposed by the confinement has been identified: due to the presence of water in the surroundings of the confined reaction space, a compositional gradient is induced within the suspended droplets. By electrostatic attractions, the more polar species are forced to preferentially accumulate at the droplet/water interface region, whereas the nonpolar species accumulate at the core region of the droplet. For instance, using a slightly polar monomer such as methyl methacrylate (MMA) and a nonpolar solvent such as *n*-hexane, a pseudo-homogeneous polymerization is promoted at the droplet surface region to produce a dense polymeric structure or shell. In the core region, the solvent concentration is very high thus inducing the precipitation of polymer chains under "starved" conditions. Smaller polymer sub-particles of 300-500 nm in diameter are densely packed within a polymer capsule resulting in the pomegranate-like particle. Similarly to regular dispersion polymerizations, an intra-droplet phase inversion can also occur, thus generating multi-hollow micro-particles.

The applications of the polymer micro-particles from this study are still under investigation. Our research group has developed an in situ micro-encapsulation technique that uses pomegranate-like particles as a sacrifice template to produce ultra-porous silica structures characterized by a huge specific surface area, a wide-open and interconnected internal structure with a unique pore size distribution. More details will be presented at the meeting.

See more of this Session: In Honor of Professor Santosh K Gupta II
See more of this Group/Topical: Catalysis and Reaction Engineering Division

(381d) Early Stages of Crystal Growth In Supersaturated Solutions of Calcium Sulfate Hemihydrate Examined by Time-Resolved Cryogenic Transmission Electron Microscopy

Amitesh Saha[1], **Arijit Bose**[1], Jinkee Lee[2], Anubhav Tripathi[2], Sabrina Pancera[3], Andreas Kempter[3] and Michael Bru[4], (1)Department of Chemical Engineering, University of Rhode Island, Kingston, RI, (2) School of Engineering, Brown University, Providence, RI, (3)BASF SE, Ludwigshafen, Germany, (4) BASF Construction Chemicals GmBH, Trostberg, Germany

ABSTRACT. The early stage of crystal growth from a supersaturated solution of calcium sulfate hemihydrate to form calcium sulfate dihydrate (gypsum) is studied using time-resolved cryogenic transmission electron microscopy (TR-cryo-TEM). We observe, for the first time, amorphous prenucleation clusters that evolve to larger amorphous particles and then to crystalline gypsum domains. The influence of low concentrations of the additives citric acid, potassium sulfate and magnesium sulfate on the evolution from the hemihydrate to gypsum is also investigated. The addition of citric acid significantly delays gypsum crystallization by adsorption and complexation of the carboxylic groups with calcium ions, effectively stabilizing the amorphous phase. The formation of gypsum is expedited when small quantities of potassium sulfate are added. Since the hydration of potassium sulfate is endothermic, addition of this salt accelerates the exothermic transformation of the calcium sulfate hemihydrate to gypsum. In contrast, the hydration of magnesium sulfate is exothermic and addition of small quantities of this salt retards gypsum formation. Results from this work are useful for modulating the rate of gypsum formation in chemical and construction applications.

See more of this Session: In Honor of Professor Santosh K Gupta II
See more of this Group/Topical: Catalysis and Reaction Engineering Division

(381e) Exploiting Chaos: Should Polymerization Reactors Be Chaotic?

Sadhan C. Jana, Polymer Engineering, The University of Akron, Akron, OH and Chang Do Jung, Department of Polymer Engineering, The University of Akron, Akron, OH

It is known that chaotic fluid element trajectories improve distributive mixing of liquids. This study exploited globally chaotic mixing conditions in a chaotic mixer to expedite rate of polymerization in synthesis of thermoplastic polyurethanes by one-step method from of an aliphatic diisocyanate, polyether and polyester polyols, and butanediol. The reactants were hand-mixed with a tin catalyst before pouring into the chaotic mixer. It was found that the tin catalyst not only catalyzed urethane formation but also helped reduce the size of suspended droplets of aliphatic diisocyanate in reaction mixture. The time-scales of mixing (t_{mix}) and polymerization reactions (t_{rxn}) were correlated with the time to reach the maximum torque (t_{max}) and the molecular weight and polydispersity index of the polymer products were analyzed.

See more of this Session: In Honor of Professor Santosh K Gupta II
See more of this Group/Topical: Catalysis and Reaction Engineering Division

(381f) Processing and Characterization of PVDF Films

Devang V. Khakhar, Department of Chemical Engineering, Indian Institute of Technology, Bombay, Mumbai - 400076, India

Poly (vinylidene fluoride) (PVDF) hasexcellent piezoelectric properties and used in applications such assensors, actuators and switches. PVDF has several crystal phases ofwhich the ababb

See more of this Session: In Honor of Professor Santosh K Gupta II
See more of this Group/Topical: Catalysis and Reaction Engineering Division

(411a) Control Emulsions, Suspensions, and Solids Separation with Inline Particle Size Measurements In Dark Crude Oil

Anjan P. Pandey and Des O'Grady, METTLER TOLEDO AutoChem, Columbia, MD

Hydrocarbon pipeline flow is typically a complex multi-phase system comprised of bubbles, droplets, and particles including asphaltenes, waxes, gas hydrates, emulsions, and inorganic particles. To control separations, flow, and process efficiency, analytical measurements are often applied by petroleum engineers and chemists. Yet extreme temperature and pressure conditions make offline sampling and analysis difficult and time consuming. Offline measurements can also alter or destroy multiphase components and cannot be applied to make real-time process optimization and control decisions. With the implementation of inline particle characterization technology, particle or droplet phase behavior can be measured quickly, in real time, without pulling samples.

This paper focuses on the application of real time, in-process particle and droplet measurement tools to avoid processing problems in the following R&D and full scale production applications:

- Controlling asphaltene and wax precipitation in dark crude oil
- Improving oil-water and water-oil separation through in-process droplet measurements
- Visualizing and inhibit gas hydrate formation to ensure consistent flow rates
- Improving oil/sand tailings settling rates and optimizing flocculent and chemical surfactant effectiveness

See more of this Session: Solids-Liquid Separation Processes for the Petroleum Industry
See more of this Group/Topical: Separations Division

(411b) Alternatives to Tailings Ponds for Oil Sand Tailings Disposal

Yuming Xu and Tadeusz Dabros, CanmetENERGY, Natural Resources Canada, Devon, AB, Canada

ABSTRACT

Oil sand resources are playing an increasingly important role in maintaining energy supplies for North America. Bitumen is produced from oil sand, mainly by two processes: in situ recovery by SAGD (steam assisted gravity drainage) for deep reservoirs and open-pit mining operations for shallow deposits. In open-pit mining operations, bitumen is separated from sand grains by a hot-water extraction and flotation process that produces large volumes of tailings. In the existing commercial plants, tailings from the extraction operation are pumped into large settling ponds. The coarse solids settle out quickly as a sand beach while the fine particles settle very slowly and finally form mature fine tailings (MFT), a stable suspension containing about 30% solids. Over the years, large volumes of MFT have accumulated in the ponds and the continuing accumulation is a cause of growing economic and environmental concern.

In order to address these concerns we are actively investigating alternative methods of dry and stackable tailings disposal such as flocculation and thickening, filtration, and centrifugation. Coarse oil sand tailings are flocculated using polymeric flocculants and the sediments are either disposed in the mined-out pit or further filtered. Various factors affecting flocculation and filtration performance were investigated in this work. The results demonstrate that the tailings have much better filterability after appropriate flocculation, and filtration becomes a viable method of producing stackable tailings. Another method for treating fine tailings and MFT is centrifugation. Centrifugal force enhances the separation of fine clays from water and speeds up further consolidation. Pilot-scale test results show that centrifugation of the fine tailings produces consolidated cake that can be disposed of without the need for pond containment. These methods provide potential alternatives for eliminating or reducing the use of tailings ponds and for the recovery of water and the associated thermal energy.

See more of this Session: Solids-Liquid Separation Processes for the Petroleum Industry
See more of this Group/Topical: Separations Division

(411c) Gas Flow Behavior In FCC Disengagers Vessel with Different Outlet Tube Length of Primary Cyclone Separator

Jiangyun Wang, Yu Mao, Juan Wang and Meili Liu, College of Chemical Engineering, China University of Petroleum,Beijing, Beijing, China

To deeply understand the effect of primary cyclone separator's outlet tube length on gas oil flow behavior in fluid catalytic cracking disengager space, the gas oil flow in a full-size disengager is numerically studied on the platform of a commercial computational fluid dynamics (CFD) code Fluent 6.1.22. The Reynolds stress model (RSM) is applied to simulate the gas oil flow field. As no geometry simplification is made in generating the structured mesh for the space both inside and outside the two-stage separator system, the whole flow behavior of the gas oil in the disengager is obtained. Coupled with the RSM, the stochastic tracking mode is used to quantify the gas oil into the secondary cyclone separator. It is shown that the growth of the outlet tube length of primary cyclone separator contributed to oil gas flowing into the inlet of secondary cyclone separator, effectively reducing the gas oil quantity into the disengager and the gas oil residence time, which avoids unwanted overreaction of hydrocarbon.

See more of this Session: Solids-Liquid Separation Processes for the Petroleum Industry
See more of this Group/Topical: Separations Division

(411d) Reclamation of Spent Drilling Fluids by Cross-Flow Microfiltration

Yanling Wu, Environmental Solutions Applied Research, M-I SWACO, A Schlumberger Company, Houston, TX

Oil- and synthetic-based drilling fluids can become contaminated with significant quantities of low-gravity solids, which cannot be removed by conventional solids control equipment and methods. The accumulation of low-gravity solids degrades the performance of costly oil- and synthetic-based drilling fluids in several ways and often generates large quantities of spent drilling fluids not suitable for drilling activities. As a result of rigorous environmental regulations moving toward zero discharge, drilling wastes are the focus of attention in the oil and gas exploration industry.

In this research, pressure driven liquid-phase microfiltration (MF) was investigated for the first time as a potential solution to the removal of low-gravity solids from spent oil- and synthetic-based drilling fluids, where fine solids are retained by the membrane and clean oil passes through the membrane and is recovered as permeate. The results demonstrated that cross-flow MF is capable of breaking the strong emulsions of drilling fluids and recovering clean base oil in the permeate, independent of drilling fluid formulation and composition while minimizing chemical handling and the variability in chemical-based processes. Understanding the impact of each operating parameter on permeate flux will allow us to realize the full potential of cross-flow MF as a spent drilling fluids treatment technology. In this study several operating parameters such as trans-membrane pressure, back-pulse, cross flow velocity and temperature were investigated to optimize permeate flux, minimize membrane fouling and maximize oil recovery using a membrane with a nominal pore size of 0.1 micron. A membrane-cleaning protocol for the fouled membranes was also developed.

See more of this Session: Solids-Liquid Separation Processes for the Petroleum Industry
See more of this Group/Topical: Separations Division

(411e) Technology Development and Demonstration of a Low-Temperature Fluidized-Bed Biomass Dryer

Charles J. Coronella, Mike P Matheus and Victor R. Vasquez, Chemical Engineering, University of Nevada, Reno, Reno, NV

At the University of Nevada, Reno, we initiated a research project to study mechanisms of drying wastewater solids (sludge) in 2007. Preliminary studies led to the development of a new low-temperature fluidized-bed sludge dryer. The sludge dryer has been tested in batch and in continuous studies, and operates at very low temperatures. We report results of a parametric study based on air flow rate and dryer temperature. The dryer produced dried solids at unusually low temperatures, between 125 ° F and 175 ° F, with heat provided at 210 ° F. Collected sludge has been characterized in terms of its moisture content, particle size, and ash content. The moisture of the collected product ranges between 4% and 15%, and can be controlled by proper selection of operating conditions.

The dryer is a novel configuration, and consists of a bed of inert solids operating in the bubbling bed regime. Heat is provided to the bed by internal heat transfer tubes. Wet sludge falls into the warm bed, which is bubbling vigorously. The rapid particle movement causes the wet sludge to break apart, increasing surface area, and thereby enhancing heat transfer. As the sludge dries and fragments, it becomes relatively light dust particles. The small dry particles have specific density 1.2, and are entrained in the fluidizing air. Solid particles are removed from the air in a filter outside the bed, and collected and stored.

The capacity of the dryer is limited by the rate of air flow. Air leaving the bed can remove a finite amount of moisture, up to its saturation of humidity. Increasing too high has a deleterious effect on the performance of the fluidized bed.

We have successfully completed a demonstration of our dryer with a continuous feed of up to 20 lb/hour of wet (85% moisture) sludge onsite at the Truckee Meadows Water Reclamation Facility (TMWRF). We are currently seeking partners for a 200 lb/hour demonstration. Trials and lessons from commercialization with the support of the UNR Tech Transfer Office will be discussed.

See more of this Session: Solids-Liquid Separation Processes for the Petroleum Industry
See more of this Group/Topical: Separations Division

(440a) Sulfur Species During Deactivation and Regeneration of Ni-Based Catalysts Used for Conditioning Biomass-Derived Syngas

Matthew M. Yung[1], Singfoong Cheah[2], Kimberly A. Magrini-Bair[2] and John Kuhn[3], (1)National Bioenergy Center, National Renewable Energy Laboratory, Golden, CO, (2)National Renewable Energy Laboratory, Golden, CO, (3)Chemical and Biomedical Engineering, University of South Florida, Tampa, FL

For syngas conditioning catalysts used for hydrocarbon/tar steam reforming, short lifetimes are an obstacle that leads to high cost of the finished fuel product. While there are many components of syngas that can contribute to catalyst deactivation, hydrogen sulfide (H_2S) is generally considered to be one of the most severe poisons. In order to overcome this deactivation challenge, it would be beneficial to have a catalyst that is i) resistant to deactivation and/or ii) can be readily regenerated. In the present work, we focus on the role of regeneration process on sulfur species found on the catalyst surface. We report sulfur surface chemistry during catalyst regeneration cycles on a Ni tar reforming catalysts used in real biomass-derived syngas with ~ 50 ppm H_2S. These results will be presented in terms of deactivation by sulfur poisoning, as well as other physical methods. In the context of previous results for these samples, we will demonstrate the importance of the regeneration protocols, which we are in the process of developing. The appearance of sulfides that transform into sulfates using a steam/air regeneration will be shown, as well as the persistence of sulfates on the catalyst surface after they are formed. This work will be presented in the context of activity measurements, additional characterization, sulfur reaction mechanisms during regeneration, and schemes to more fully regenerate catalysts will be discussed.

See more of this Session: Catalyst Deactivation
See more of this Group/Topical: Catalysis and Reaction Engineering Division

(440b) Deactivation Behavior of MWCNTs and Activated Carbon-Supported Alkali-Promoted Trimetallic Co-Rh-Mo Sulfide Catalysts for Higher Alcohols Synthesis From Synthesis Gas

Venkateswara Rao Surisetty, University of Saskatchewan, Saskatoon, SK, Canada, Janusz A. Kozinski, Department of Chemical Engineering, College of Engineering, University of Saskatchewan, Saskatoon, SK, Canada and Ajay K. Dalai, Chemical Engineering, University of Saskatchewan, Saskatoon, SK, Canada

Alkali-modified Co-Rh-Mo trimetallic catalysts supported on multi-walled carbon nanotubes (MWCNTs) and activated carbon were used to study the long term deactivation for continuous 720 h of higher alcohols synthesis from synthesis gas. The deactivation behavior of the catalysts were tested for the synthesis of higher alcohols from synthesis gas under similar conditions of 330°C, 9.1 MPa (1320 psig), and 3.8 m^3 (STP)/(kg of cat.)/h using H_2 to CO molar ratio of 1.25. The catalysts were extensively characterized in both oxide and sulfide phases before reaction, together with the spent catalysts. The alkali-promoted trimetallic Co-Rh-Mo catalyst supported on MWCNTs has shown two different deactivation steps: loss of sulfur from the surface and sintering of the catalyst species located on the outer surfaces of the carbon nanotubes. After regeneration, the total activity recovery (about 10%) is close to the total activity loss during the first deactivation step over this catalyst. The % CO conversion decreased from 58% and 43 % to 51% and 30% over the catalysts supported on MWCNTs and activated carbon, respectively. Characterizations of the spent catalyst supported on activated carbon revealed that deactivation occurs due to sintering of metal sulfides, which causes low metal dispersions and high pore blockage of the support.

See more of this Session: Catalyst Deactivation
See more of this Group/Topical: Catalysis and Reaction Engineering Division

(440c) Heterogeneous Adsorbate Based Kinetic Modeling

Max Tirtowidjojo[1], Christina Zarth[2], Liping Zhang[3], Debashis Chakraborty[4] and Naoko Akiya[4], (1) Hydrocarbons, Chemicals & Energy R&D, The Dow Chemical Company, Freeport, TX, (2)Chemicals R&D, The Dow Chemical Company, Zeven, Germany, (3)Chemicals R&D, The Dow Chemical Company, Freeport, TX, (4)Engineering & Process Science R&D, The Dow Chemical Company, Freeport, TX

The heterogeneous catalyzed partial oxidation reaction of hydrocarbons is a commercially important process. Discovery of catalyst formulations and carriers have allowed economical processes with greater than 80% selectivity to the desired products and with a long catalyst lifetime. Many hypotheses have been put forward regarding the kinetics and catalyst active site decay with run time for many partial oxidation processes. In this work, the mechanistic heterogeneous kinetic modeling is considered using elementary reactions involving adsorbates and gas-phase species and intermediates. The approach involves uses of reversible rate of reactions, molar balances of adsorbates, thermodynamic consistency of reaction kinetics involving adsorbates and gas-species, and the use of active site concentration. A model for predicting catalyst lifetime performance will be discussed.

See more of this Session: Catalyst Deactivation
See more of this Group/Topical: Catalysis and Reaction Engineering Division

(440d) Investigate the Impacts of Aging Temperature and Aging Gas On LNTs

Yisun Cheng[1], Robert McCabe[1], Xiaoyin Chen[2] and Johannes W. Schwank[2], (1)Chemical Engineering, Ford Motor Company, Dearborn, MI, (2)Chemical Engineering, University of Michigan, Ann Arbor, MI

LNTshave been widely investigated and developed for NO_xemission control, for both lean-burn gasoline and diesel vehicles. Thermal deactivationhas been a major challenge in LNT development owing to the need to maintainclose proximity of the precious metals to the trapping materials. Aging and deactivation effects are compoundedby sulfur poisoning, which aside from direct poisoning, results in acceleratedthermal deactivation due to the high temperature conditions required to desulfate the trap.

Temperatureand exhaust gas composition are the major factors influencing precious metalsintering and overall NO_x trap performance. Metal-support interactions are alsoimportant, as demonstrated by work reported by Toyota and others. It has been found that high temperature exposurereduces precious metal dispersion; also the LNTs deactivate differently inlean, stoichiometric, and rich gas conditions even atthe same aging temperatures. However, details regarding the combined effects ofaging temperature and gas conditions on LNT deactivation are not clear. Significanttechnical challenges remain for LNT research on developing durable catalystsand understanding their deactivation mechanisms.

Thispaper systematically investigates the impact of aging temperatures and aging gasconditions on LNTs. Rather than using hydrothermal aging as employed widely inmost previous LNT thermal durability studies, the aging conditions representingLNT $deSO_x$ as well as DPF regeneration areapplied in this paper. The testing temperatures, gas composition, and agingconditions were chosen based on the best estimation of real-world conditions. Theaged LNTs were also characterized by TGA, TPR/TPO, H_2/ CO chemisorption, XRD, and XPS to develop furtherunderstandings.

The results showed that theLNTs were aged in very different ways in lean and lean/rich cycling gases. Comparedto aging temperatures, the aging gas conditions had more significant impacts onthe LNTs during the aging. And, lean/rich cycling aging (even at temperaturescharacteristic of normal NO_x storage andregeneration conditions) caused greater deactivation of NO_x,HC, and CO activities than aging in lean or rich environments alone. Theprofiles of NH_3 and N_2O formation were very differentbetween the LNTs aged in lean and in lean/rich cycling gases. Thecharacterization results also showed differences between the LNTs aged in leanand in lean/rich cycling gases, including the state of the NO_xstorage components, OSC, and accessible PGM surface area. The results indicate that there are severaldeactivation mechanisms for LNTs that respond differently to aging conditions.The study provides further insight into LNT deactivation mechanisms, therebyisolating design aspects that must be improved to develop more durable LNTs.

See more of this Session: Catalyst Deactivation
See more of this Group/Topical: Catalysis and Reaction Engineering Division

(440e) Deactivation Model for Commercial Pd-Only TWCs

Sung Bong Kang[1], Seok Jun Han[1], **In-Sik Nam**[1], Byong K. Cho[1], Chang Hwan Kim[2] and Se H. Oh[2], (1) Department of Chemical Engineering, Pohang University of Science and Technology (POSTECH), Pohang, South Korea, (2)Chemical Sciences and Material Systems Laboratory, General Motors Global R&D, Warren, MI

When properly defined and correctly combined with the primary reaction kinetics, the catalyst activity function can be a very useful parameter for predicting the performance of a catalyst by the numerical simulation of a reactor model, especially when the catalyst undergoes a deactivating process.[1] However, few such simulation models that can reliably predict the TWC performance as a function of the noble metal loading and the catalyst mileage have been reported, probably due to the paucity of catalyst samples with a desired range of the catalyst mileages accumulated under realistic driving conditions. Ekström et al. predicted the decline of the catalytic activity of TWC aged under an accelerated engine bench condition by a simple adjustment of the frequency factor of the rate constant in the primary kinetics.[2] Baba et al. developed a TWC deactivation kinetics based upon the change of the frequency factor attributed to both sintering of Pt metal and phosphorus poisoning under the accelerated engine bench condition.[3] They reported that the sintering of noble metal is the primary cause for the deactivation of TWCs due to the decrease of noble metal surface area (MSA). Recently, a simple TWC activity function based upon the alteration of the Pd metallic surface area (MSA) of Pd-only commercial monolith TWCs has been developed to describe the change of the catalytic performance with respect to the Pd metal loading varying from 142 to 283 g/ft^3 and the field-aged catalyst mileage ranging from 4k (stabilized) to 98k miles.[4]

In the present study, the TWC activity function developed previously has been further validated in describing the deactivation behavior of a commercial Pd-only TWC with a low Pd loading of 80 g/ft^3 (Pd80) in the lab-aging program. In a comparative aging study, the TWC activity over the Pd catalyst containing 142 g/ft^3 of Pd (Pd142) aged by the lab-aging program was found to be quite similar to that over an identical catalyst aged by the field-aging program over a wide range of the catalyst mileage from 4k to 98k miles. Given this well-matched TWC activity of the Pd142 catalyst between the lab-aged and the field-aged samples, the Pd80 catalysts were aged for 4k, 20k, 50k, 100k and 150k mile-equivalent by using the lab-aging program. The initial deactivation of the TWC activity over the commercial Pd80 catalyst aged by the lab-aging program was similar to that over the filed-aged catalyst in the low catalyst mileage from 4k to 20k-mile. The change of the Pd MSA of TWC as a function of the catalyst mileage has been well correlated by the 2nd order sintering kinetics for all levels of the Pd loading of the catalyst. The TWC activity function was then incorporated into the primary reaction kinetics developed for the stabilized 4k Pd80 TWCs to predict the deactivation of the Pd-only TWC as a function of the catalyst mileage. The overall reaction kinetic model coupled with the TWC activity function developed in the present study has proven to be capable of describing the alteration of TWC performance as a function of the catalyst mileage, regardless of its metal loading and aging modes – field-aged or lab-aged.

References

1. J. A. Moulijn, A. E. van Diepen, F. Kapteijn, *Appl. Catal. A:Gen.*, 212 (2001) 3.

2. F. Ekström, F. Wallin, A. Fathali, G. Russ, *Top. Catal.*, 52 (2009) 1915.

3. N. Baba, K. Yokota, S. Matsunaga, S. Kojima, K. Ohsawa, T. Ito and H. Domyo, *SAE Technical Paper Series* 2000-01-0214 (2000).

4. S. B. Kang, H. J. Kwon, I.-S. Nam, Y. I. Song, S. H. Oh, *Ind. Eng. Chem. Res.*, (2011) DOI: 10.1021/ie200083f.

See more of this Session: Catalyst Deactivation
See more of this Group/Topical: Catalysis and Reaction Engineering Division

(440f) High Throughput Study of Pt-Pd Catalysts for DOC Applications: XRD and STEM Characterization

Obiefune K., Ezekoye[1], Andrew Drews[2], Robert W. McCabe[2], Robert Kudla[2], George W. Graham[3] and Xiaoqing Pan[3], (1)Materials Science and Engineering, The University of Michigan, Minneapolis, MN, (2)Ford Motor Company, Dearborn, MI, (3)Department of Materials Science and Engineering, Unversity of Michigan, Ann Arbor, MI

Lean-burn gasoline and diesel engines provide significant increase in fuel economy over conventional gasoline engines for transportation applications, but the catalytic after-treatment of exhaust, notably NOx and soot, presents unique challenges. Oxidation of NO plays a prominent role in the catalytic after-treatment of exhaust gas from lean-combustion vehicles. Although alumina-supported Pt is one of the best NO oxidation catalysts, cost and durability must also be considered in selecting a practical catalyst.

At Ford, the search for novel catalysts for DOC applications has been conducted by high-throughput synthesis and testing of a variety of materials, including alumina-supported Pt with partial substitution of Pd. Alloying of Pt with other metals such as Pd has been shown to inhibit particle growth, particularly that of anomalously large particles, under high-temperature lean conditions without excessively degrading NO oxidation activity, but the mechanisms underlying this thermal stabilization remain unknown.

This presentation focuses on high-temperature x-ray diffraction (XRD) of alumina-supported Pt and Pt-Pd bimetallic catalysts. For example, under lean aging conditions, XRD shows that average particle size grows as a power-law in time. The fact that this growth rate follows Arrhenius behavior allows us to combine time and temperature into a single "exposure" variable. These and other findings are considered along with scanning transmission electron microscopy (STEM) results to provide a somewhat more complete picture of the particle coarsening processes at work in these catalysts.

See more of this Session: Catalyst Deactivation
See more of this Group/Topical: Catalysis and Reaction Engineering Division

(440g) Hydrocarbon Deactivation of a Cu/Zeolite SCR Catalyst

William Epling[1], Jin-Yong Luo[2], Cary Henry[3], Krishna Kamasamudram[3], Neal Currier[3] and Aleksey Yezerets[4], (1)Department of Chemical Engineering, University of Waterloo, Waterloo, ON, Canada, (2) Chemical Engineering, University of Waterloo, Waterloo, ON, Canada, (3)Catalyst Technology, Cummins Inc, Columbus, IN, (4)Corporate Research and Technology, Cummins Inc, Columbus, IN

HydrocarbonDeactivation of a Cu/Zeolite SCR Catalyst

A Cu-exchanged zeolite catalystwas evaluated for selective catalytic reduction (SCR) performance before andafter exposure to propylene at different temperatures, ranging from 150 to500°C. A transient response method was used to characterize the standard, fastand NO_2 SCR reactions, as well as NH_3 and NO oxidation,in the presence and absence of propylene.

The results clearly show propyleneinhibition of the standard and fast SCR reactions between 200 and 400°C. Withthe addition of propylene there is an immediate increase in outlet NO_X.According to the DRIFTS data obtained, propylene oxidation intermediates, forexample acrolein-like species, rapidly form on thecatalyst surface upon propylene introduction at these temperatures, explainingthis immediate drop in performance. At 300°C there was a slower decrease inperformance with prolonged exposure to propylene, after the initial rapid loss,which was not observed at 250°C. In the 300 to 400°C range, acroleinspecies were again observed, resulting in the rapid loss in performanceobserved, but coke formation was also observed, and was slow to build-up. Thecoke formation explains the slow decrease in performance noted. At 150°C,inhibition was not observed, nor was any propylene consumption via oxidation,suggesting that some extent of propylene oxidation is required for the onset ofinhibition, further verifying the propylene oxidation intermediates causinginhibition.

In terms of NO oxidation, in alltests, the addition of propylene led to reduced NO_2 formed via NO_2being used as an oxidant for propylene oxidation, and thus reduction of $NO_2$to NO. NH_3 was found to inhibit propylene oxidation, due to therelatively stronger adsorption of NH_3 to the surface. For NO_2SCR, the addition of propylene resulted in improved performance, through HC-SCRas well as reduction of NO_2 to NO, allowing some fast SCR reactionto occur.

See more of this Session: Catalyst Deactivation
See more of this Group/Topical: Catalysis and Reaction Engineering Division

(441a) Hydrogen Generation by Formic Acid Decomposition Over Sub-Nm Gold Species

Nan Yi, Howard M. Saltsburg and Maria Flytzani-Stephanopoulos, Department of Chemical & Biological Engineering, Tufts University, Medford, MA

Hydrogen is a promising source of alternative energy that can be harnessed from simple organic compounds like formic acid and transformed into a practical, industry-usable form. Here we investigated the decomposition of formic acid over gold based oxides (mainly Au-ceria) by a combination of temperature-programmed techniques along with catalytic activity tests. Gold-ceria was prepared by deposition/precipitation method. HCOOH-TPD and HCOOH-TPSR were used to determine reaction onset, reaction intermediates and CO_2 selectivity. These results along catalytic steady-state performance showed that the dehydrogenation pathway was dominant over gold based catalyst. Furthermore, the addition of water or CO did not change the formic acid conversion, demonstrating that not only the absence of the dehydration pathway, but also that the water-gas shift reaction was not involved. Structural analysis and kinetic data will be presented. These findings provide insight into the decomposition of formic acid at ambient temperature.

See more of this Session: Catalytic Hydrogen Generation for Fuel Cell Applications I
See more of this Group/Topical: Catalysis and Reaction Engineering Division

(441b) Robust Nanostructured Noble Metal/ Ceria/ Lanthana Catalysts for Water-Gas-Shift

Shuang Liang, Department of Chemical Engineering, 1) US DOE-National Energy Technology Laboratory, Pittsburgh, PA; 2) University of Pittsburgh, PA, USA, Pittsburgh, PA and Gtz Veser, Chemical and Petroleum Engineering, University of Pittsburgh, Pittsburgh, PA

Noble metal nanoparticles dispersed on oxide supports are widely used as catalysts in fuel processing, chemicals production, and environmental protection. Particularly noble metal/ceria catalyst systems have found much attention due to their excellent activity in partial oxidation and Water-Gas Shift (WGS). However, the noble metal nanoparticles typically suffer from insufficient thermal stability and high sensitivity to poisoning in the presence of S-contamination in the reactor feed. Both sintering due to high temperatures as well as poisoning by S-contaminants result in the loss of active surface area and result in – often irreversible - deactivation.

Here, we are reporting on the synthesis, characterization, and testing of robust nanostructured noble metal catalysts (La-Au-Ce and La-Pt-Ce) in Water-Gas Shift with S-contaminated syngas feeds. The metal nanoparticles are deposited on nanostructured CeO_2 via through deposition-precipitation or impregnation. Subsequently, the supported catalysts were decorated with a La_2O_3 overlayer. This protective overlayer results in enhanced thermal stability of the catalyst, and shields the active metal from S-poisoning while simultaneously enabling the catalyst to capture sulfur, as lanthana is known to be an efficient regenerative S-capturing material in the temperature range of interest in WGS. The nanostructured CeO_2 support then facilitates the fast regeneration of the noble metal nanoparticles in the reducing atmosphere.

The catalyst were characterized (TEM, XRD, BET, and TPO/TPR), and tested in cyclic fixed-bed reactor experiments. WGS activity test in repeated temperature ramps shows greatly improved stability of the active nanoparticle through the surface decoration. The results in repeated sulfidation-regeneration cycles are equally promising, showing fast sulfur uptake kinetics, large S-capturing capacity, and stable regeneration. Overall, the carefully controlled nanostructure hence results in a highly efficient and stable multifunctional catalyst material. Synthesis, characterization, and catalytic activity tests will be discussed in detail in the presentation.

See more of this Session: Catalytic Hydrogen Generation for Fuel Cell Applications I
See more of this Group/Topical: Catalysis and Reaction Engineering Division

(441c) Steam Reforming of Ethanol/Gasoline Mixtures: Deactivation, Regeneration and Stable Performance

Amanda Simson[1], Bob Farrauto[2] and Marco Castaldi[1], (1)Department of Earth and Environmental Engineering (HKSM), Columbia University in the City of New York, New York, NY, (2)R & D, BASF Catalysts LLC, Iselin, NJ

Steamreforming of ethanol/gasoline mixtures: deactivation, regeneration and stableperformance

Introduction

Increasingconcerns about environmental impacts of vehicles with gasoline poweredcombustion engines have increased efforts to find technologies to reduceemissions and fossil fuel consumption. Recently there has been an increasedfocus on using renewable sources to generatetransportation fuels, such as utilizing agricultural waste to generateethanol. Ethanol has a lower vapor pressure relative to gasoline and thusethanol/gasoline blends such as E85 (85% ethanol) are available rather than100% ethanol within the current infrastructure. Thus it is important tounderstand the reforming capabilities of these mixed transportation fuels.There has been extensive research on the reforming of pure ethanol, however,literature on reforming ethanol-gasoline blends is limited. For this study welook at steam reforming of ethanol at less energy intensive conditions: 650Cand steam/carbon ratio of 1.8 with a Rh/Pt catalyst (BASF RM-75ST) washcoatedon a ceramic monolith. At these less energy intensive conditions, however,catalyst deactivation due to either sulfur poisoning or carbon formation ismore favored. The objective of this study is to understand sulfur-inducedcatalyst deactivation and also potential catalyst regeneration methods in orderto develop a process reforming E85 or other sulfur containing liquid fuels atless energy intensive conditions.

Results andDiscussion

Thecatalyst deactivated after exposure to fuel containing 5 ppm sulfur after 22hours on stream whereas the catalyst did not deactivate reforming a sulfur-freefuel for at least 110 hours of continuous operation. Although initially thecatalyst was able to achieve equilibrium product distributions with 100%ethanol and gasoline conversion in both the sulfur-free and the 5 ppm sulfurtest condition the catalyst operating in the presence of sulfur deactivated toan extent unsuitable for industrial hydrogen production, i.e. < 10 mole%hydrogen production and < 21% ethanol conversion after 100 hours ofreforming. After such significant levels of deactivation exposure to air forone hour was able to restore initial catalyst activity. Although initialactivity (100% ethanol and gasoline conversion and equilibrium hydrogenproduction) was restored catalyst deactivation was measured after less time onstream then with the fresh catalyst. Successive air regenerations were alsoperformed after intermediate levels of deactivation as shown in Figure 1.

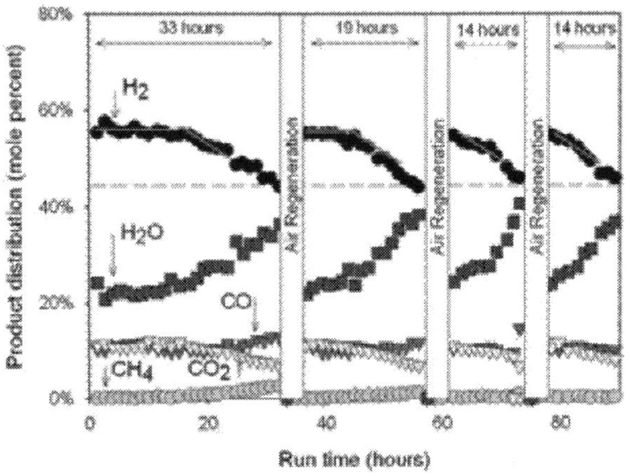

Theseresults show that some deactivation was reversible and thus initial activityafter air regeneration could be restored; however, an aspect of irreversibledeactivation also must have existed which shortened the period of stableactivity following the air treatment. Initial XPS results indicate thatextended periods of time on stream contribute to irreversible changes inprecious metal chemistry which may be attributed to the permanent loss inperformance shown in Figure 1. Coupling reactor studies with characterizationdata it was found that these precious metal chemical changes may be avoidedwith preemptive regeneration. Preemptive regenerations extended the period ofstable performance during reactor studies. These results imply that anindustrial process could be developed utilizing preemptive regenerations toachieve stable activity while reforming a sulfur containing fuel.

Significance

Inindustry, catalyst regenerations are performed more frequently as the catalystages, however, the irreversible aspect of the catalyst deactivation is oftennot understood. The purpose of this study is to understand the difference inthe irreversible and reversible aspects of deactivation in order to optimize anindustrial process for reforming sulfur-containing fuels.

References

1. Swartz,S.L., Matter, P.H., Arkenberg, G.B., Holcomb, F.H., Josefik, N.M., J. PowerSources 188, 515 (2009)

See more of this Session: Catalytic Hydrogen Generation for Fuel Cell Applications I
See more of this Group/Topical: Catalysis and Reaction Engineering Division

(441d) Ambient Pressure PES Study of CeO_2 Supported Catalysts for Water Gas Shift Reaction

M. Cem Akatay[1], Jorge Pazmino[2], Mayank Shekhar[2], W. Damion Williams[2], Anil Mane[3], Eric A. Stach[1], W. Nicholas Delgass[2], Jeffrey W. Elam[3], Fabio H. Ribeiro[2] and Dmitry Zemlyanov[4], (1)School of Materials Engineering and Birck Nanotechnology Center, Purdue University, West Lafayette, IN, (2) School of Chemical Engineering, Purdue University, West Lafayette, IN, (3)Argonne National Laboratory, Argonne, IL, (4)Birck NanoTechnology Center, Purdue University, West Lafayette, IN

Ceria (CeO_2) supported Pd/Fe catalysts for water-gas shift (WGS) reaction were analyzed in situ under 0.5mbar pressure using ambient pressure synchrotron-based Photoemission Spectroscopy (PES). High-resolution core-level spectra were obtained both under reducing atmosphere of H_2 at 300oC and under the WGS reaction mixture consisted of 7% CO, 2% H_2O, 37% H_2, 9% CO_2 at 250oC at total pressure of 0.5mbar. To obtain depth profile, the photon energy of the exciting radiation was varied as to have the kinetic energy of photoelectrons ranging from 120eV to 420eV. The oxidation state of metals under different atmospheres and change of the composition as a function of depth were investigated. Pd/CeO_2 catalyst is promoted by iron. Fe is not much reduced neither under WGS nor reduction environment. Ceria on the surface reduces to Ce_2O_3 both under the reducing and WGS reaction environment. The ratio of the reduced ceria decreases as the sampling depth is increased. Pt concentration on the surface stays around the same value under both environments. The WGS rate of the Fe promoted sample is 6.6 x10-2 and that of un-promoted catalyst is 2.7 x10-2 mole H_2 mole Pt-1 s-1 at 250oC under our standard conditions*, a promotion of a factor of around 2.4, when compared with the un-promoted Pd catalyst. Considering the fact that the ceria supported Pd has the highest rate among different supports (Al_2O_3, ZrO_2), an additional increase of the rate is worth mentioning. Additional systems studied in this work are the ceria supported Pt catalyst promoted with iron and zirconia supported Pd promoted with iron. Based on the findings, models describing the structure of catalysts were developed and compared to images obtained by High-Resolution Transmission Electron Microscopy (HR-TEM).

See more of this Session: Catalytic Hydrogen Generation for Fuel Cell Applications I
See more of this Group/Topical: Catalysis and Reaction Engineering Division

(441e) Evaluation of Rh-Pyrochlore Coated Monolith for the Reforming of Diesel Fuel

Daniel J. Haynes[1], David A. Berry[1], Dushyant Shekhawat[1], Mark Smith[2] and **Matthew M. Seabaugh**[3], (1)National Energy Technology Laboratory, Morgantown, WV, (2)URS Corporation, Washington Division, Morgantown, WV, (3)NexTech Materials, Lewis Center, OH

Evaluation of Rh-Pyrochlore Coated Monolith for theReforming of Diesel Fuel

Daniel J. Haynes[1],David A. Berry[1], Dushyant Shekhawat[1], Mark W.Smith[1,2], Matthew Seabaugh[3]

[1]*National Energy Technology Laboratory/US DOE, Morgantown, WV*

[2] *URS-Washington Division, Morgantown, WV*

[3]*NexTech Materials, Lewis Center, OH*

Through its Solid State Energy Conversion Alliance (SECA)program, the U.S. Department of Energy has been in collaboration withindustrial partners to sponsor the development of solid oxide fuel cells (SOFCs) for clean and efficient energy production from hydrocarbon-based fuels. While the applications for fuel cells arenumerous, auxiliary power units (APUs) for long haul diesel trucks and militaryvehicles have been identified as a possible entry point to introduce thetechnology into the commercial market. On-boardhydrogen production from the reforming of diesel can be used to meet the energyrequirements of the SOFC. However, highlevels of sulfur and aromatic compounds commonly found in diesel fuel limit theformation of synthesis gas (H_2 + CO) because they are poisons totraditional reforming catalysts.

TheNational Energy Technology Laboratory (NETL) has been developing reformingcatalysts based on the pyrochlore crystal structure ($A_2B_2O_7$)to reduce deactivation. Pyrochlore-basedcatalysts are of interest primarily due to their chemical stability in hightemperature reducing and oxidizing environments. They also possess the ability to accommodatethe substitution of a wide variety of active and promoting metals within the oxideframework. This distinguishes them from traditionalsupported metal catalysts, which are more prone to sintering athigh-temperatures and have less control of particle size and next nearest-neighborinteractions. Furthermore,oxygen-conducting supports have been shown to greatly improve hydrocarbonreforming with substituted oxide catalysts such as pyrochlores. In particular, the rate of deactivation isreduced by decreasing the amount of deactivating carbon formed on thecatalytically active phase through oxygen transport from the gas phase, throughthe support to react with strongly adsorbed hydrocarbon species believed to becoke precursors. Researchers at NETLsuccessfully demonstrated a Rh-substituted zirconate pyrochlore catalystsupported onto zirconium-doped ceria (ZDC) to be active and stable for thereforming of commercial diesel fuel for 1000 hrs under oxidative steamreforming conditions. Following thistest, NETL has been collaborating with NexTech Materials in order to develop aneffective process to apply the optimized pyrochlore/ZDC catalyst onto acommercially representative form such as monolith. A square channel alumina monolith structurecoated with an oxygen-conducting support, onto which the active pyrochlorephase was deposited, has been fabricated.NETL has recently completed a successful 93 hour fuel reforming testing ofthe commercially representative monolith using commercial diesel fuel. Synthesis gas yields were stable and nearthose predicted by equilibrium for the 93 hours time on stream. No indicators of catalyst degradation (olefinformation, pressure buildup, etc.) were observed when the run was discontinued.

See more of this Session: Catalytic Hydrogen Generation for Fuel Cell Applications I
See more of this Group/Topical: Catalysis and Reaction Engineering Division

(441f) Sulfur-Resistant Lanthanide Oxysulfide Catalysts for the High-Temperature Water-Gas Shift Reaction

Ioannis Valsamakis, Joseph Lessard and Maria Flytzani-Stephanopoulos, Chemical and Biological Engineering, Tufts University, Medford, MA

We report here on the suitability of the oxysulfides of La and Pr as active and stable catalysts for the high temperature water–gas shift reaction (WGS). The oxygen shuttling mechanism between the oxysulfide and the correspondent oxysulfate occurs at temperatures higher than 600°C, thus making these materials perfect candidates for the high temperature water gas shift reaction. The reaction lights off at around 400°C and at high temperatures CO conversions close to equilibrium are achieved. The redox mechanism is shown to operate under these conditions. The most important feature of these materials is their tolerance to sulfur compounds as confirmed by spiking the feed gas with different amounts of H_2S. Extension of these catalysts into the low-temperature region can be achieved by addition of a precious metal, such as gold. Structural characterization and kinetic analysis will be presented.

See more of this Session: Catalytic Hydrogen Generation for Fuel Cell Applications I
See more of this Group/Topical: Catalysis and Reaction Engineering Division

(442a) Activity and Selectivity of Bimetallic Catalysts In Hydrogenolysis of Sugars and Sugar Alcohols

Xin Jin[1], Bala Subramaniam[1] and Raghunath V. Chaudhari[2], (1)Chemical and Petroleum Engineering, Center for Environmentally Beneficial Catalysis, Lawrence, KS, (2)Center for Environmentally Beneficial Catalysis, Chemical and Petroleum Engineering, University of Kansas, Lawrence, KS

The depleting supply of petroleum has motivated research on developing alternative technologies from renewable resources (biomass) for fuels and chemicals. Hydrogenolysis of biomass-derived sugars and sugar alcohols is an important class of reactions to produce valuable megaton products such as 1,2 or 1,3-propanediol, ethylene glycol, lactic acid, hexanediols and hydrocarbons. Although, hydrogenolysis of glycerol has been extensively studied in recent years[1] , relatively limited work exists on the hydrogenolysis of C_5 and C_6 sugars and corresponding polyols. A major challenge in these processes is to achieve high selectivity to desired products and hence to design appropriate catalyst formulations. For example, hydrogenolysis of C_5 and C_6 sugars/polyols involve a number of consecutive and parallel reaction steps, such as C-C and C-O bond breakage, dehydration, aldol condensation and dehydrogenation. These steps are affected to different degrees by the catalytic components, preparation methods and structural characteristics. Hydrogenolysis of xylitol and sorbitol follows retro-aldol mechanism[2,3] over monometallic catalysts (Ru or Ni) in basic medium[4], which leads to C-C and C-O bond breakage[5]. The complex reaction mechanism underlying the wide product distribution observed in these hydrogenolysis reactions is not yet well understood. In this presentation, the role of mono and bimetallic Ru and Cu based catalysts, acid and base promoters and reaction conditions on hydrogenolysis of sorbitol, xylitol and glucose will be systematically addressed with emphasis on discerning the reaction pathways. Particularly, the activity/selectivity behavior will be discussed.

Our initial results showed that the presence of a solid base enhances the activity of Cu and Ru monometallic catalysts in sorbitol hydrogenolysis. Cu activity is increased by more than two fold by using a solid base promoter. Similarly, the activity (expressed as TOF) of Ru/C catalysts increased from 110 h^{-1} to 214h^{-1} and 145 h^{-1} in the presence of MgO and CeO_2 respectively, at pH=7~9 and 483 K. With bimetallic RuRe and RuCo catalysts, the catalytic activity further increased when CeO_2 was used as a promoter. Secondly, it was observed that the presence of Re also increases the selectivity of C_3 products (glycerol and 1,2-propanediols) by approximately 10% compared to monometallic Ru catalysts while RuCo shows similar product distribution as with Ru/C. This observation provides important information for reaction pathways of sorbitol hydrogenolysis. Further work on reaction pathways, kinetics and mechanism, and catalyst characterization is in progress, and the results of which will be presented.

Reference

[1] A. Torres, D. Roy, et al. *Ind. Eng. Chem. Res.* 49 (2010) 10826-10835.

[2] K. Wang, M. Hawley, et al. *Ind. Eng. Chem. Res.* 34 (1995) 3766-3770.

[3] J. Sun and H. Liu. Green Chemistry (2010) DOI: 10.1039/c0gc00571a

[4] L. Zhao, J. Zhou, et al. *Chemical Engineering Science.* 65 (2010) 30-35.

[5] D. K. Sohounloue, C. Montassier, et al. *React. Kinet. Catal. Lett.* 22 (1983) 391-397.

See more of this Session: Catalytic Processing of Fossil and Biorenewable Feedstocks: Chemicals
See more of this Group/Topical: Catalysis and Reaction Engineering Division

(442b) Hydrodeoxygenation Pathways and Selectivity Descriptors for the Conversion of Propanoic Acid to Alkanes Over Pd(111) Catalysts

Andreas Heyden, Department of Chemical Engineering, University of South Carolina, Columbia, SC and Jianmin Lu, Chemical Engineering, University of South Carolina, Columbia, SC

One of the principle goals of modern catalysis research is to understand reaction mechanisms on solid surfaces to a degree that practical activity and selectivity descriptors can be identified that permit the rational design of new stable catalysts with unprecedented activity and selectivity. High selectivity towards a single reaction product is driven both by economics and the goals of green catalysis, where atom- and energy-efficient processes are required to conserve the world's limited resources.

In this paper we present a computational case study for the determination of activity and selectivity descriptors for the hydrodeoxygenation (HDO) of organic acids to alkanes on transition metal surfaces. In particular, we investigated activity and selectivity issues in the decarboxylation, decarbonylation, and reductive deoxygenation of propanoic acid to propane and ethane on Pd(111) model surfaces.

Decarboxylation: R-COOH -> R-H + CO2

Decarbonylation: R-COOH + H2 -> R-H +CO + H2O

Reductive deoxygenation: R-COOH + 3 H2 -> R-CH3 + 2 H2O

Overall, we identified that a practical activity descriptor for the conversion of propanoic acid (all pathways) is the heat of reaction of the C-OH cleavage step and a selectivity descriptor for the reductive deoxygenation versus decarbonylation/decarboxylation is the heat of reaction of the α-carbon dehydrogenation.

See more of this Session: Catalytic Processing of Fossil and Biorenewable Feedstocks: Chemicals
See more of this Group/Topical: Catalysis and Reaction Engineering Division

(442c) Kinetics and Mechanism of Acetic Acid Esterification with Ethanol On Zeolites

Jeremy Bedard[1], Hsu Chiang[1], Lanny Schmidt[2] and Aditya Bhan[1], (1)Chemical Engineering and Materials Science, University of Minnesota, Minneapolis, MN, (2)University of Minnesota, Minneapolis, MN

Esterification can be used as a mode of biomass deoxygenation and has been studied here using acetic acid and ethanol as biomass surrogates. Steady state, transient, and isotopic-exchange reaction studies of acetic acid and ethanol were used to identify co-adsorbed ethanol–acetic acid dimers as surface intermediates within specific elementary steps involved in the esterification of acetic acid with ethanol on zeolites. A reaction mechanism involving two dominating surface species, an inactive ethanol dimeric species adsorbed on Brønsted sites inhibiting ester formation and a co-adsorbed complex of ethanol and acetic acid on the active site reacting to produce ethyl acetate, was shown to describe the reaction rate as a function of temperature, acetic acid partial pressure, and ethanol partial pressure. Measured differences in rates as a function of zeolite structure and the rigorous interpretation of these differences in terms of esterification rate and equilibrium adsorption constants will be discussed.

See more of this Session: Catalytic Processing of Fossil and Biorenewable Feedstocks: Chemicals
See more of this Group/Topical: Catalysis and Reaction Engineering Division

(442d) A Kinetic and Mechanistic Study of the Selective Oxidation of 5-Hydroxymethylfurfural Over Supported Metal Catalysts

Sara E. Davis[1], Bhushan N. Zope[1], Levi R. Houk[2], Erin C. Tamargo[1], Abhaya Datye[2] and Robert J. Davis[1], (1)Chemical Engineering, University of Virginia, Charlottesville, VA, (2)Center for Micro-Engineered Materials, University of New Mexico, Albuquerque, NM

Interest in the production ofcommodity chemicals from renewable carbon sources instead of from fossilresources continues to grow. Oxidation of alcohols provides one such route fortransformation of biorenewable feedstocksto value-added chemicals. One potential platform alcohol derived from fructoseand glucose is 5-hydroxymethylfurfural (HMF). The HMF oxidation product 2,5-furandicarboxylic acid (FDCA) is a potential replacement forpetroleum-derived terephthalic acid, the monomer usedin production of PET plastic. A high pHis required for HMF oxidation, and oxidation can proceed through multipleintermediate products. Conversion andproduct selectivity is influenced by the choice of catalyst as well as theconcentration of sodium hydroxide in the reactant mixture. The mechanistic roles of sodium hydroxide, dioxygen and water in the oxidation of aqueous HMF are notfully understood and are key to the development ofsustainable methods for biorenewable chemicalproduction.

In the kinetic study, supported Pt, Pd and Au catalysts wereevaluated in the aqueous-phase oxidation of HMF to FDCA at 295 K and high pH ina semibatch reactor.The intermediate reaction product 5-hydroxymethyl-2-furancarboxylic acid(HFCA) was formed in high yield over Au/C and Au/TiO_2 at 690 kPa O_2, 0.15 M HMF and 0.3 M NaOH,whereas the final reaction product FDCA was formed over Pt/Cand Pd/C under identical conditions. However, theinitial turnover frequency of HMF conversion was an order of magnitude greateron Au catalysts compared to either Pt or Pd. Increasing the O_2 pressure and NaOH concentration facilitated the conversion of HFCA toFDCA over the supported Au. The effectof base concentration, O_2 pressure, and metal on turnover frequency,conversion, and product selectivity will be discussed.

In themechanistic study, Pt/C and Au/TiO_2catalysts were utilized in labeling experiments conducted with $^{18}O_2$and $H_2{}^{18}O$. Results from these studies indicated thatwater was the source of oxygen atoms during the oxidation of HMF to HFCA and FDCA, presumably through directparticipation of hydroxide in the catalytic cycle. Molecular oxygen wasessential for the production of FDCA and played an indirect role duringoxidation by removing electrons deposited into the supported metal particles. Adetailed reaction path for HMFoxidation to FDCA will be proposed.

(442e) Production of Gamma-Valerolactone From Corn Stover by Selective Extraction and Hydrogenation of Levulinic Acid

David Martin Alonso, Stephanie G. Wettstein, Jesse Q. Bond and James A. Dumesic, Chemical and Biological Engineering, University of Wisconsin-Madison, Madison, WI

Economic and technically viable conversion of lignocellulosic biomass into chemicals and fuels is an important challenge and several routes have been proposed using intermediate molecules, such as levulinic acid (LA) to maximize the yields. One of the main problems for these routes to be implemented at commercial scale is the high cost of purifying/recovering these intermediates, which are normally produced in dilute solutions of mineral acids. In this presentation we demonstrate the use of a family of high boiling point hydrophobic solvents, to extract LA selectively from water solutions of dilute acids, such as sulfuric acid, and to produce gamma-valerolactone (GVL) over bimetallic catalysts without the necessity to separate/purify the LA. The GVL produced can be used as chemical or fuel additive or it can be decarboxylated to produce biomass derived butene.

The approach described here has 4 main advantages. First, the hydrophobic solvents do not extract sulfuric acid, which remains in the aqueous phase, and can be recycled. Second, LA can be hydrogenated to GVL in presence of the solvent over a bimetallic catalyst, and thus purification of LA is not necessary. Third, the high boiling point of the solvent allows for product recovery at the top of a distillation column without evaporating the solvent. Finally, the organic/aqueous partition coefficient for GVL is higher than for LA, and it is thus possible to increase the final concentration of GVL in the solvent by successive extraction-hydrogenation cycles.

We have carried out this strategy at lab scale for 4 cycles. No sulfuric acid was detected in the organic phase which was recycled to the cellulose deconstruction reactor without further purification with minimal lost in LA yields. The final GVL concentration was increased from 0.4 M in the first cycle to 1.4 M after the 4th. Aspen simulations demonstrate that this increment in GVL concentration reduces the size of the distillation column and the energy required for distillation. Importantly, we demonstrate the aforementioned ideas using a feed produced from real biomass (corn stover) reporting stability for the catalyst during more than 300 h on stream. Solvent impurities in the GVL did not affect negatively the decarboxylation reaction over silica/alumina to produce butene.

See more of this Session: Catalytic Processing of Fossil and Biorenewable Feedstocks: Chemicals
See more of this Group/Topical: Catalysis and Reaction Engineering Division

(442f) Hydrogenolysis of Succinic Acid to GBL, Butanediol : Catalysis & Kinetic Studies

Raghunath V. Chaudhari, Center for Environmentally Beneficial Catalysis, Chemical and Petroleum Engineering, University of Kansas, Lawrence, KS and **Xuchao Li**, Chemical Engineering, University of Kansas, Lawrence, KS

The overall goal of this study is to develop an active heterogeneous catalyst for aqueous phase hydrogenation of succinic acid to value added industrial products such as propionic acid, γ-butyrolactone, 1,4-butanediol, and tetrahydrofuran at relatively milder conditions. It is also proposed to investigate fundamentals of catalysis and kinetics and mechanism of the hydrogenation reactions. This activity is part of a broader goal of utilization of bio-succinic acid for chemicals since succnic acid is a promising bio-based building-block chemical feedstock to replace the current maleic anhydride-C4 platform. Catalytic performance of several ruthenium and palladium based supported bimetallic catalysts, prepared by co-precipitation method, will be reported. A comparison of the bimetallic catalysts with monometallic catalysts with respect to selectivity for gaseous and liquid phase products will also be discussed. Finally, a Langmuir-Hinshwood type kinetic model using one of the best performing catalysts will be established and the correlated interpretation of reaction mechanism will be proposed.

See more of this Session: Catalytic Processing of Fossil and Biorenewable Feedstocks: Chemicals
See more of this Group/Topical: Catalysis and Reaction Engineering Division

(442g) Investigation of Liquid Phase 1,2-Propanediol Dehydration Using Zeolite Catalysts

Vladimiros Nikolakis, Tim Courtney, Stavros Caratzoulas, Raul F. Lobo, Jingguang G. Chen and Dion G. Vlachos, Catalysis Center for Energy Innovation, Chemical Engineering, University of Delaware, Newark, DE

To reduce the dependence of our society in oil, significant effort is devoted to the production of fuels and chemicals from biomass derived sources. Glycerol, which is a major byproduct of biodiesel production, can be selectively reduced to 1,2 propanediol (propylene glycol) and 1,3 propanediol. The dehydration of these low carbon content polyols results in the formation of a number of ketones, aldehydes, or linear or cyclic ethers, which are raw materials for the production of fine chemicals. However, selective dehydration is challenging due to the presence of different of hydroxyl groups with similar reactivity, and due to a complex reaction network (combination of reactions in parallel and in series).

Several zeolite catalysts on the liquid phase dehydration of 1,2 have been studied and compared to of Amberlyst-15. 1,2-diol dehydration using Amberlyst-15 leads to high selectivity to the formation of *cis*- and *trans*- 2-ethyl-4methyl-1,3-dioxolane, which is formed via an acetal reaction of propanal with 1,2-propanediol. This result indicates that Amberlyst-15 catalyzes the dehydration of the secondary OH group of the polyol to form propanal. On the other hand, it has been found that zeolites catalyze the dehydration of either the primary and secondary OH group of 1,2-propanediol as well as etherification reactions. When the primary OH group is eliminated acetone is formed, which can further react with the diol to form 2,2,4-trimethyl-1,3-dioxolane. Etherification reactions lead primarily to the formation di-propylene glycol and tri-propylene glycol. By adjusting the pore size of the zeolite and the strength of its acid sites, it is possible to selectively remove the primary or the secondary OH group of 1,2 propanediol and obtain selective elimination of the primary OH group. The effect of the catalyst properties on the reaction path selectivity and the product distribution will be presented. Furthermore, the experimental results will be compared with *ab-initio* calculations that provide further insight into the reaction mechanisms and help guide the catalyst design.

See more of this Session: Catalytic Processing of Fossil and Biorenewable Feedstocks: Chemicals
See more of this Group/Topical: Catalysis and Reaction Engineering Division

(449a) Density Functional Theory Study of Enhanced One Dimensional Mobility of Oxygen On Strained $LaCoO_3$ (001) Surface

Jeong Woo Han and Bilge Yildiz, Department of Nuclear Science and Engineering, Massachusetts Institute of Technology, Cambridge, MA

Amongst the various types of fuel cells, the solid oxide fuel cell (SOFC) is particularly attractive given its fuel flexibility and high conversion efficiencies. The slow rate of oxygen reduction at the cathode is, however, the main barrier to achieve higher power output in SOFCs at intermediate temperatures (500° C-700°C). Mechanisms by which lattice strain alters the oxygen reduction reaction (ORR) kinetics are important to understand in order to increase the ORR activity of SOFC cathodes. Here we assess the mechanistic and quantitative effects of strain on oxygen diffusion on the $LaCoO_3$(LCO)(001) surface using density functional theory (DFT) calculations. Planar tensile strain reduces the migration barrier of oxygen vacancy anisotropically on the LCO(001) surface, and increases that of the adsorbed atomic oxygen. The reduction in the migration barrier of oxygen vacancy becomes increasingly anisotropic with increasing strain, inducing an enhanced mobility along the [1-10] direction and a suppressed mobility along the [110] direction. The reduction of the energy barrier of migration along the [1-10] is due to the increase in the space around Co that the oxygen traverses with a curved path. The increase in the unstability of oxygen vacancy formation at every other surface lattice oxygen row in the [1-10] direction due to the increasing octahedral distortions with increasing planar tensile strain inhibits the migration of oxygen vacancy along the [110] direction. The increasing energy barrier for the migration of adsorbed oxygen atom is due to the weakening hybridization of the in-plane Co-O bonds, which consequently enhances the atomic oxygen adsorption on the surface. From a qualitative estimate, the significantly lower energy barrier for oxygen vacancy diffusion is expected to dominate the other degrading factors and actually accelerate the ORR kinetics on LCO(001) up to 3% strain. Furthermore, the insights obtained here are useful for designing strategies to control the desired anisotropic and uni-directional oxygen transport along strained hetero-interfaces.

See more of this Session: Electrocatalysis and Photoelectrocatalysis: Fundamentals and Applications I
See more of this Group/Topical: Catalysis and Reaction Engineering Division

(449b) A First-Principles Comparison of Heterogeneous and Electrocatalytic Alcohol Oxidation Over Au and Pt Catalysts

David D. Hibbitts, Bhushan N. Zope, Robert J. Davis and Matthew Neurock, Chemical Engineering, University of Virginia, Charlottesville, VA

AFirst-Principles Comparison of Heterogeneous and Electrocatalytic AlcoholOxidation over Au and Pt Catalysts

DavidD. Hibbitts, Bhushan Zope, Robert J. Davis and Matthew Neurock

Abstract

In the emerging area of biorenewable-basedchemicals, most processes are low-temperature, aqueous phase reactions, instark contrast to the high-temperature, gas phase processes that dominate the currentpetroleum-based chemicals industry. The shift of environments allows formuch-needed synergy in the areas of heterogeneous and electro-catalysis, afield which has already demonstrated the importance of the metal-solvent interface.One area in particular which has demonstrated similarities between electro-catalysisand aqueous-phase heterogeneous catalysis is the oxidation of CO and alcohols.

Recent work[1] on the alcoholoxidation over Au and Pt has demonstrated that the oxidation occurs throughdirect reaction with hydroxyl groups from the alkaline medium, summarized inFigure 1. These hydroxyl groups act as both nucleophiles and Brnsted bases inthe selective oxidation of ethanol to acetic acid and isotopic labeling studieshave confirmed the source of oxygen in the acid products as the solvent, ratherthan the gaseous O_2. The O_2 is still necessary for the reactionto proceed, as it must be reduced to reform the consumed hydroxyl groups andwater. This reduction also removes excess electrons from the metal and closesthe catalytic cycle. Writing the oxidation and reduction separately gives thefollowing half-reactions:

RCH_2O_2+ $4^{18}OH^-$ --> $RCO^{18}OH + 3H_2{}^{18}O$+ $4e^-$ (Eq1)

$(1+x/2)O_2$+ $(2+x)H_2O + 4e^-$ --> xHOOH + $4OH^-$ (Eq2)

Asshown in Equation 2, a byproduct of this oxidation is the production ofhydrogen peroxide, which can catalyze C-C cleavage through a variation of theCannizzaro reaction.

Thismechanism shows that the mechanism during heterogeneous catalysis is nodifferent from a short-circuited electrochemical cell, in which the oxidationand reduction are occurring on the same metal particles. In fact,electro-oxidation of alcohols and CO in alkaline media has been studiedpreviously over Au and Pt electrodes.

Ofcourse, during electro-oxidation, there is an applied potential upon the metalcatalyst, which causes significant changes in the reactivity of the catalyst.In order to further explore the similarities and differences between these twosystems, the calculations performed upon the heterogeneous oxidation wereredone using the double-reference method developed by Filhol, Taylor andNeurock[2]. This allows for independent analysis of the effects of themetal, potential and mechanism.

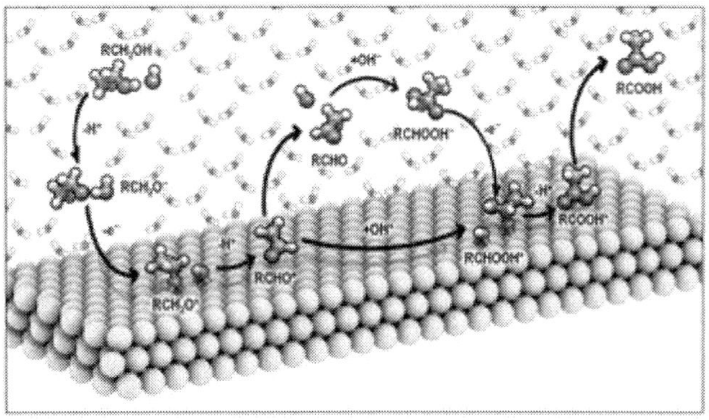

Figure 1. Reaction scheme for the oxidation ofalcohols to acids over a gold surface in water at high pH.

References

1. B.N. Zope, D.D. Hibbitts, M.Neurock, and R.J. Davis. *Science.* **330**. 74-77 (2010)

2. C.D. Taylor, S.A. Wasileski, J-S.Filhol, and M. Neurock. *Physical Review B*. **73**. 165402 (2006)

See more of this Session: Electrocatalysis and Photoelectrocatalysis: Fundamentals and Applications I
See more of this Group/Topical: Catalysis and Reaction Engineering Division

(449c) MoS2 Nanostructures As Efficient, Stable, and Earth-Abundant Catalysts for Hydrogen Evolution In Acid

Zhebo Chen[1], Dustin R. Cummins[2], Benjamin Reinecke[1], Ezra Clark[2], Mahendra K. Sunkara[2] and Thomas F. Jaramillo[1], (1)Chemical Engineering, Stanford University, Stanford, CA, (2)Chemical Engineering, University of Louisville, Louisville, KY

In order to obtain a cost effective alternative to Pt for driving the hydrogen evolution reaction (HER) in acidic media, we need to develop Earth-abundant materials that are highly efficient for the HER and remain stable over long periods of use. Nanostructured MoS_2 has previously shown excellent HER activity with minimal overpotentials of 150-200 mV due to the presence of a large number of highly active edge sites. Driving practically relevant geometric current densities, however, requires a high aspect ratio morphology upon which the edge sites can be vertically integrated. We produce vertically aligned α-MoO_3 nanowires using a hot-wire chemical vapor deposition method followed by sulfidization over a range of temperatures from 150-700°C to create MoS_2 shells of varying thicknesses. At an optimal sulfidization temperature of 200°C, these core-shell nanowires exhibit excellent activity for the HER while remaining stable over thousands of potential cycles in strong acid. The MoS_2 shell further provides a protecting layer for the underlying MoO_3, which is otherwise unstable in strong acid, while the n-type MoO_3 provides a conductive electron pathway to the poorly conducting MoS_2. The understanding of MoS_2 HER activity arising from edge sites provides insight into the possibility of increasing the activity of these nanowires even further. Towards this goal, we also present work on other nanostructures of MoS_2 that also drive efficient HER. Lastly, we explore the optical response properties of these nanowires in solution and their possibility for driving photoelectrochemical water splitting.

See more of this Session: Electrocatalysis and Photoelectrocatalysis: Fundamentals and Applications I
See more of this Group/Topical: Catalysis and Reaction Engineering Division

(449d) Electrocatalysis for CO2 Reduction to Hydrocarbons: Theory-Driven Design Principles Developed From Transition Metals

Andrew A. Peterson, Stanford University, Stanford, CA and Jens K. Nrskov, SUNCAT Center for Interface Science, Stanford University and SLAC National Laboratory, Stanford, CA

The efficient reduction of CO_2 can enable the production of hydrocarbons from renewable and low-carbon sources of energy. We have used density functional theory and a computational hydrogen electrode model in order to understand the reactivity of transition metals in reducing CO_2 to hydrocarbons, carbon monoxide, and hydrogen gas. Using scaling relations to generalize binding energies of the important intermediates of CO_2 reduction, we show how the descriptors of CO_2 reduction can be reduced to key parameters. Our analysis indicates that the key descriptor is the free energy change in the first protonation of adsorbed carbon monoxide. We outline this and other key descriptors to develop catalysts that are efficient and selective in CO_2 reduction to hydrocarbons and alcohols.

See more of this Session: Electrocatalysis and Photoelectrocatalysis: Fundamentals and Applications I
See more of this Group/Topical: Catalysis and Reaction Engineering Division

(449e) Composite Plasmonic Metal/ Semiconductor Photoelectrodes for Overall Water Splitting

David B. Ingram, Department of Chemical Engineering, University of Michigan, Ann Arbor, MI and Suljo Linic, Chemical Engineering, University of Michigan, Ann Arbor, MI

The efficient conversion of solar energy into useful chemical energy, for example through processes such as water splitting and CO_2 reduction, is of critical importance for the development of sustainable, long-term solutions to society's energy problems. These and other photochemical processes require the development of novel photo-catalysts that can efficiently make use of the solar spectrum. Conventional semiconductor photocatalysts suffer from range of issues that limit their ability to efficiently convert solar into chemical energy, including low absorption coefficients, large band gap energies and especially high rates of charge carrier recombination.

In this contribution we show that the photocatalytic activity of a semiconductor can be significantly enhanced by the addition of optically active Ag nanostructures of targeted size and shape to form composite metal/semiconductor photocatalysts. The observed activity enhancement is attributed to the transfer of energy from Ag surface plasmon (SP) states, increasing the steady-state concentration of e/h pairs in the semiconductor, thereby increasing the reaction rate. The ability to predictably tune the energy of Ag SP states by controlling particle shape and size allows for the rational design of optimized metal/semiconductor composite photocatalysts and photo-electro-catalysts wherein the overlap of the metal nanostructure SP states, semiconductor band gap, and energy of the light source is maximized.

See more of this Session: Electrocatalysis and Photoelectrocatalysis: Fundamentals and Applications I
See more of this Group/Topical: Catalysis and Reaction Engineering Division

(449f) Plasmonic Nanostructures As Platforms for Efficient Coupling of Visible Light and Thermal Energy to Drive Chemical Transformations

Phillip Christopher, Department of Chemical Engineering, University of Michigan, Ann Arbor, Ann Arbor, MI, Hongliang Xin, University of Michigan, Ann Arbor, MI and Suljo Linic, Chemical Engineering, University of Michigan, Ann Arbor, MI

Plasmonic nanostructures as platforms for efficient coupling of visible light and thermal energy to drive chemical transformations

Phillip Christopher, Hongliang Xin, Suljo Linic

Department of Chemical Engineering, University of Michigan,

Ann Arbor, Michigan 48109-2136

Heterogeneous catalytic processes play a crucial role in industrial chemical synthesis, energy production and pollution mitigation. Most commercial heterogeneous catalytic reactions are run at relatively high temperatures due to the high activation barriers associated with rate-limiting elementary steps. The high operating temperatures required to achieve reasonable product yields compromise process efficiency and catalyst stability. In this contribution we show that silver plasmonic nanostructures can concurrently use low intensity visible light (on the order of solar intensity) and thermal energy to drive industrially important catalytic oxidation (ethylene epoxidation, CO oxidation and selective NH_3 oxidation) reactions at lower temperatures than their conventional counterparts that use only thermal stimulus (P.Christopher, H.Xin, S. Linic, *Nature Chemistry*, In Press (2011)).

The Ag nanoparticle catalysts exhibit a strong localized surface plasmon resonance due to the nanometer scale spatial confinement and the metals' inherent electronic structure. The excitation of surface plasmons is characterized by the intense concentration of spatially non-homogeneous electric fields at the nanostructure surface. Based on mechanistic experimental studies and density functional calculations, we show that excited plasmons on the Ag surface act to transiently populate O_2 anti-bonding orbitals, thereby depositing vibrational energy in the O—O bond, facilitating the rate-limiting O_2 dissociation reaction and increasing the overall rate of oxidation reaction on Ag surfaces. The results suggest that the nanostructures concentrate the energy of a resonant photon flux in a very small volume at their surface rather than distributing this energy over the entire nanostructure, allowing for a direct and efficient energy channeling in the form of energetic electrons into the reaction coordinate. This unique capacity of plasmonic nanostructure to constructively couple multiple stimuli (the energy of photon flux and the energy delivered by thermally heating the plasmonic materials) opens avenues for the development of a new class of photocatalysts for various reactions.

See more of this Session: Electrocatalysis and Photoelectrocatalysis: Fundamentals and Applications I
See more of this Group/Topical: Catalysis and Reaction Engineering Division

(449g) Mixed-Metal Oxide Electrocatalysts for Oxygen Evolution

Ethan L. Demeter, James R Landon and John R Kitchin, Chemical Engineering, Carnegie Mellon University, Pittsburgh, PA

The electrochemical oxygen evolution reaction could be an important component of future energy systems. If the reaction can be made efficient and cost-effective, it would enable the utilization of renewable energy for hydrogen production, fuel synthesis and oxygen purification from air. Currently there are high overpotentials associated with the O_2 evolution reaction due to large reaction barriers within the complex reaction mechanism. Metal oxides have been identified as potential catalysts for this reaction, but the best performing materials, RuO_2 and IrO_2 are expensive, while other cheaper oxides are not active enough to be practical or are unstable in the conventional acid electrolyte.

Under alkaline conditions many base metals, such as Ni, are stable to corrosion making them potential candidates for the oxygen evolution reaction. We have investigated mixed metal oxides based on Ni for the oxygen evolution reaction. Especially for mixed Ni-Fe oxides, we have found the mixed metal oxide has superior activity compared to the parent metal oxides, even though there is no evidence of bulk mixing by x-ray diffraction and EXAFS. These mixed oxide electrocatalysts show improvements in air separation performance in electrochemical membrane reactors. We will present our synthesis approaches, physical and electrochemical characterizations, and work to integrate the electrocatalysts into membrane electrode assemblies for use in applications.

See more of this Session: Electrocatalysis and Photoelectrocatalysis: Fundamentals and Applications I
See more of this Group/Topical: Catalysis and Reaction Engineering Division

(469a) Landfill Gas Clean-up Using a Flow-Through Catalytic Membrane Reactor

Nitin Narayanan Nair[1], Mirmohammedyousef Motamedhhashemi[1], Fokion Egolfopoulos[2] and Theodore Tsotsis[1], (1)Mork Family Department of Chemical Engineering and Materials Science, University of Southern California, Los Angeles, CA, (2)Aerospace and Mechanical Engineering Department, University of Southern California, Los Angeles, CA

Landfill gas (LFG) is potentially a valuable renewable fuel because of the methane that it contains. But the presence of impurities in landfill gas presents challenges for its effective utilization. Burning the gas for power generation or even when flaring it, if these impurities are not removed, releases gases such as hydrogen chloride, sulfur dioxide, etc., which are toxic and harmful to both humans and the environment. Therefore, there is a strong incentive to develop effective technologies for removing the toxic compounds from landfill gas prior to its utilization as a fuel.

In this project a novel catalytic oxidation technology appropriate for landfill gas clean-up based on the concept of a "Flow-through Membrane Reactor" impregnated with an oxidizing nanocatalyst has been studied. For our experiments we have studied a model LFG stream with a VOC composition which was shown previously by our group to simulate well the behavior of real LFG in field-scale investigations. The effects of feed-stream composition, reactor temperature, and catalyst loading were studied.

Multilayer tubular alumina membranes were used in the research and were rendered catalytic by wet impregnation. Their properties were characterized using gas permeation, surface area measurements (BET), and SEM/EDAX analysis. The pore-structure characteristics of these membranes are important in determining the transport mechanism that prevails through the active layer. For membranes with small enough pores, Knudsen flow prevails, leading to an increase in the number of collisions between the reactants and the catalytic sites inside the pores.

When comparing the behavior of the FTCMR with the more conventional reactors the "yardstick" of success is the ability of the FTCMR to operate under lower temperature for a given level of conversion, and/or attain higher conversion under the same conditions and catalyst loading. For the LFG application, light-off temperature experiments showed promising results when compared to the monolith reactor. Also, no catalyst deactivation was observed during the time-on-stream experiments, proving that the FTCMR is robust towards corrosive by-products (e.g., HCl) produced during the oxidation reactions.

A mathematical model was also developed based on the Dusty Gas formulation of gas transport through the porous membrane being utilized in the FTCMR. The model was used for identifying the advantages of the FTCMR concept compared to the wall-coated catalytic monolith, and also for investigating some of the limitations, which may exist in applying this concept for the aforementioned LFG clean-up application.

See more of this Session: Multifunctional Reactor Design
See more of this Group/Topical: Catalysis and Reaction Engineering Division

(469b) Development of a High-Yield Semi-Continuous Multifunctional Ammonia Synthesis Reactor

Mark Huberty, Chemical Engineering and Materials Science, University of Minnesota, Minneapolis, MN and Ed Cussler, Department of Chemical Engineering and Materials Science, University of Minnesota, Minneapolis, MN

An ammonia synthesis reactor which permits higher conversions than can normally be obtained has been developed and operated at commercial pressures and temperatures. The reactor contains both a commercial magnetite catalyst and an ammonia selective adsorbent. Because the reaction is about the same speed as the absorption, the conversion of nitrogen and hydrogen is much more complete than in a conventional reactor. The mechanism for this unexpectedly rapid adsorption is identified by separate measurements, which show that the absorption isotherm is close to a step function.

This implies that the new reactor can be successfully operated with a pressure swing, and may provide a route to making ammonia for a small group of farms. The nitrogen would be separated from air, and the hydrogen made by wind powered electrolysis of water. Thus this reactor may provide a sustainable, local route to nitrogen based fertilizer.

See more of this Session: Multifunctional Reactor Design
See more of this Group/Topical: Catalysis and Reaction Engineering Division

(469d) La2O3 Modified Ni/Gamma-Al2O3 Catalyst for Steam Gasification of Biomass Model Compounds In a CREC Riser Simulator

Jahirul Mazumder, Department of Chemical & Biochemical Engineering, The University of Western Ontario, London, ON, Canada and Hugo I. deLasa, Department of Chemical & Biochemical Engineering, University of Western Ontario, London, ON, Canada

Biomass is a clean, renewable, CO_2 neutral and abundant energy source. Catalytic steam gasification is a process able to convert low value solid biomass to a uniform gaseous mixture effectively and economically.

To accomplish this and in the context of the present study, Ni catalyst supported on La_2O_3 modified γ-Al_2O_3 was prepared via incipient wetness technique. BET results showed that calcination of γ-Al_2O_3 results in loss of surface area due to thermal sintering via phase transition towards the α-Al_2O_3. Addition of the lanthanum oxides decreased the loss of surface area, as it increases the thermal resistance of γ-Al_2O_3 by reducing the number of sites where nucleation of α-Al_2O_3 is possible. La_2O_3 also improved the reducibility by reducing the nickel-aluminate formation as well as dispersion of active phase as determined by TPR and H_2 chemisorption experiments. The multiple cycles of TPR and TPO experiments proved that the prepared catalysts remains stable under repeated oxidation and reduction cycles, which are the conditions expected in an autothermal fluidized biomass gasifier unit.

Gasification of biomass model compunds glucose (cellulose) and 2-methoxy-4methylphenol (lignin) were conducted using the Ni/La_2O_3-γAl_2O_3 catalyst at different steam/biomass ratios, temperatures and reaction times in a CREC riser simulator. Significant improvements in dry gas yield and carbon conversion compare to the non-catalytic and catalytic gasification using Ni/α-Al_2O_3 catalyst at the same operating conditions confirmed the high activity of the developed catalyst for reforming of tars compounds, coke combustion and WGS reactions. The trends of gasification products (H_2, CO, CO_2, CH_4, etc) with the variation of these parameters are in consistent with the thermodynamics predictions. With the increase in temperature, H_2 and CO concentrations were increased while CO_2 and CH_4 concentrations were decreased. On the other hand, increasing CO_2 and decreasing CO profiles were observed with the steam/biomass ratio and residence time due to WGSR. TOC analysis of spent catalyst showed negligible coke deposition.

(469e) A Hybrid Catalytic Membrane Reactor for Destruction of a Chemical Warfare Simulant

Mirmohammadyousef Motamedhashemi[1], Nitin Nair[1], Majid Monji[1], Fokion Egolfopoulos[2] and Theodore Tsotsis[1], (1)Mork Family Department of Chemical Engineering and Material Science, University of Southern California, Los Angeles, CA, (2)Department of Aerospace and Mechanical Engineering, University of Southern California, Los Angeles, CA

Abstract

The possibility of the use of chemical weapons has been increased in recent years both as a result of potential terrorist attacks and of ongoing international conflicts. The focus of our research is the development of a novel, hybrid catalytic membrane reactor which consists of a flow-through catalytic membrane reactor (FTCMR) integrated with a surface-flow based membrane separator (SFMS) in order to reduce significantly the load on the FTCMR, to achieve complete oxidation of chemical warfare agents (CWAs) at trace levels, applicable in integrated individual protection system (IPS) for civil and military applications. As a part of this research, a catalytic tubular alumina membrane is prepared *via* impregnation with commercial $H_2PtCl_6.H_2O$ solutions, and is utilized in a FTCMR for the catalytic oxidation of dimethyl methylphosphonate (DMMP) in air. DMMP is known as a chemical precursor for the more toxic gas Sarin (GB), and has been widely used to simulate its characteristics.

Experiments are reported for different DMMP feed concentrations (150-1000 ppm) and reactor temperatures (373-573K), which demonstrate the potential advantage of the FTCMR in the complete catalytic oxidation of this important CWA simulant. Complete destruction of low and high concentrations of DMMP was achieved at lower temperatures compared to the values obtained in this study for a wall-coated plug-flow (monolith) reactor containing the same amount of catalytic metal.

A mathematical model has also been developed in order to provide a better understanding of the fundamental transport phenomena underpinning the FTCMR operation. It makes use of the Dusty Gas Model, which incorporates in an appropriate fashion continuum and Knudsen diffusion, and viscous flow as the mechanisms for gas transport through the porous membrane being utilized in the FTCMR. The model is used for identifying the advantages of the FTCMR concept compared to the wall-coated catalytic monolith, and also for investigating some of the limitations, which may exist in applying this concept for the complete oxidation of chemical warfare simulants. The results of the model support the superiority of the FTCMR concept over the more conventional (plug-flow) monolith reactor.

These preliminary experimental and theoretical results indicate that the hybrid membrane reactor concept shows promise for application in the design of individual-protection systems (e.g., gas masks) for soldiers and other military personnel, as well as in the design of collective-protection systems for military armored vehicles and buildings.

See more of this Session: Multifunctional Reactor Design
See more of this Group/Topical: Catalysis and Reaction Engineering Division

(469f) A Novel Structured Catalyst with Enhanced Heat Transfer Characteristics for FTS

Min Sheng[1], Donald Cahela[1], Hongyun Yang[2] and Bruce J. Tatarchuk[3], (1)Department of Chemical Engineering, Auburn University, Auburn, AL, (2)IntraMicron Inc., Auburn, AL, (3)Chemical Engineering, Auburn University, Auburn, AL

Highly exothermic and highly endothermic reactions require catalyst beds with good heat transfer characteristics. A novel structured catalyst, Microfibrous Entrapped Catalyst Structure (MFEC) made of highly thermal conductive metals can significantly improve heat transfer efficiency, compared with traditional packed beds (PB). Firstly, the thermal parameters of metal MFEC were determined experimentally. In a stagnant N_2 gas test, the radial effective thermal conductivity of Cu MFEC is 56-fold of that of alumina PB, while the inside wall heat transfer coefficient is 10 times of that of alumina PB. Compared to PB, even made of pure copper particles, conductive metal MFEC also provides much more effective thermal conductivity and higher inside wall heat transfer coefficient in a flowing gas test. In addition, an application of Cu MFEC in Fischer-Tropsch Synthesis (FTS), demonstrated the improvement of temperature distribution inside the catalyst bed and the increase of product selectivity. Furthermore, unlike monolith catalyst structure, MFEC structure is compatible with pre-manufactured catalyst particles, very flexible and ease to be corrugated. Therefore, the conductive metal MFEC structures have great potential to enhance the intra-bed heat transfer for highly exothermic or highly endothermic reactions, reducing temperature excursions in the reactors.

See more of this Session: Multifunctional Reactor Design
See more of this Group/Topical: Catalysis and Reaction Engineering Division

(478a) Propylene Epoxidation On Supported Silver Catalysts: Insights Into Reaction Mechanism and Effect of Promoters and Butadiene Co-Feed

Apoorva Kulkarni, Marco Bedolla-Panjota, Mark A. Barteau and Raul F. Lobo, Chemical Engineering, University of Delaware, Newark, DE

The gas-phase epoxidation of propylene to propylene oxide (PO) remains one of the most challenging chemical reactions, even though the epoxidation of ethylene by O_2 over Ag/α-Al_2O_3 catalyst has been the preferred industrial synthesis route for decades. The low selectivity for direct propylene epoxidation over silver catalysts is attributed to the vulnerability of the allylic hydrogen to abstraction by oxygen. Epoxidation of propylene and PO decomposition have been studied on unpromoted and Cs-promoted Ag/ α-Al_2O_3 catalyst. Increases in promoter levels of Cs to 1000 ppm resulted in increased selectivity for PO with a simultaneous decrease in conversion. The conversion of propylene and selectivity towards PO decreased with increase in O_2/propylene ratio. Thus the highest conversion (0.95%) and highest selectivity (10.9%) were obtained under propylene-rich conditions (O_2/C_3H_6= 0.4, T= 573 K). On the 1000 ppm Cs promoted catalyst, co-feeding 1,3-butadiene with propylene results in increase in selectivity towards PO from 10.9 % to 36.6% with an increase in conversion from 0.95% to 2.5%. Since relatively little attention has been paid to the network of reactions that may follow oxygen addition to propylene on silver, we have also studied *propylene oxide* reactions on unpromoted and cesium-promoted silver. A more clear picture of the mechanism for propylene epoxidation on silver emerges from these experiments and shows that the reaction network that connects propylene with propylene oxide and other C3 oxygenated products can be explained in terms of reactions of surface oxametallacycle intermediates.

See more of this Session: Reaction Path Analysis I
See more of this Group/Topical: Catalysis and Reaction Engineering Division

MICROKINETIC MODELING OF TOLUENE AND O-XYLENE HYDROGENATION ON PT CATALYST

Tapan Bera, Joris W. Thybaut and Guy B. Marin
Ghent University, Ghent, Belgium.

Abstract

The Single-Event MicroKinetic (SEMK) methodology, which had been successfully applied to benzene hydrogenation, has now been extended towards substituted aromatics, i.e., toluene and o-xylene. In addition to the number of unsaturated nearest neighbor carbon atoms, H-atom addition rate coefficients were assumed to depend on the carbon atom type, i.e., secondary or tertiary. A simultaneous regression of the SEMK model to an experimental data set consisting of 39 toluene and 37 o-xylene hydrogenation experiments resulted in activation energies of H-additions to tertiary carbon atoms that are 11 kJ mol^{-1} higher than to secondary carbon atoms. This can be related to the steric hindrance experienced during H addition to a carbon atom bearing a substituent.

Introduction

Both quality and environmental concerns continue to impose stringent restrictions on the aromatic content of transportation fuels [1]. Hydrocracking of heavier petroleum fractions is one of the most important processes in the refining industry for the production of high quality fuels. Typical hydrocracking feeds contain a significant aromatic fraction. In this work, the hydrogenation of aromatics is studied as part of hydrocracking process.

Kinetic models are very important for process and catalyst design, development and optimization. In this work, the single-event microkinetic (SEMK) concept is used for the determination of aromatic hydrogenation kinetics which ensures feedstock independence of kinetic parameters [2]. The fundamental SEMK model is based on elementary steps without postulating a rate-determining step. In contrast to typical literature reported kinetics models for aromatic hydrogenation, it accounts for the position in which the H atoms are added to the aromatic ring and, hence, for all possible surface species. It provides a unique insight in potential dominant reaction pathways, most abundant surface intermediates as well as the identification of a rate-determining step. While originally applied to thermal [3] and acid catalyzed reactions [2,4], the SEMK methodology has recently successfully applied to benzene hydrogenation [5] and is now being extended towards substituted aromatics, i.e., toluene and o-xylene.

Procedures

Toluene and o-xylene hydrogenation experiments have been performed in a gas phase continuous stirred tank reactor at temperatures in the range of 423–498 K, H_2 inlet partial pressures of 100–600 kPa and aromatic inlet partial pressures of 10–60 kPa on a 0.5%Pt/H-ZSM22 catalyst. The total pressure was in the range 1-3 MPa, while space times from 22 to 90 kg s mol^{-1} have been used. Pt/H-ZSM-22 was selected as catalyst as it is expected to limit secondary acid catalyzed reactions by its shape selective properties. At these conditions intrinsic kinetics are measured as the maximum Weisz moduli were calculated at 10^{-2} [6,7]. The model has been regressed to an experimental toluene and o-xylene hydrogenation data set consisting of 39 and 37 data points respectively.

A Levenberg-Marquardt algorithm [8] as implemented in the ODRPACK routine was used for parameter estimation during simultaneous regression of toluene and o-xylene hydrogenation data [9]. An objective function defined as the sum of squared residuals between the observed and calculated outlet methylcyclohexane and 1,2-dimethylcyclohexane flow rates, is minimized. The minimization occurs through adjustment of the model parameter vector **b** which approximates the real parameter vector **β** at the minimal value of the objective function:

$$SSQ = \sum_{j=1}^{nob} (F_{CA,j} - \hat{F}_{CA,j})^2 \xrightarrow{b} \min$$

Here $F_{CA,j}$ is the experimental outlet molar flow rate for the saturated cycloalkane produced by hydrogenation of corresponding aromatic compound of the jth experiment and $\hat{F}_{CA,j}$ is the corresponding model calculated value.

The cycloalkane outlet flow rate is calculated by solving the non linear equation:

$$\hat{F}_{CA,j} - \hat{R}_{CA,j} \ T, p_t, \hat{F}_{CA,j}, F_A^0 \ W_j = 0$$

i.e., via the mass balance for saturated cycloalkane over the reactor, if the latter is not fed. The saturated cycloalkane net production rate is calculated from the kinetic model, which gives rise to a set of algebraic equations. These algebraic equations are solved simultaneously with the mass balance equation using a hybrid Powell method [10].

Experimental results

The hydrogenation of toluene and o-xylene yielded methylcyclohexane and dimethylcyclohexane as the main product in the investigated range of operating conditions. The effect of the operating conditions on the aromatic hydrogenation behavior is consistent with the literature [5,1114]. In particular a maximum is observed in the hydrogenation conversion as a function of the temperature, see Figure 1. At a certain temperature, the increase of the hydrogenation rate coefficient with the temperature is overcompensated by the decrease of the surface concentrations of the partially hydrogenated species [5,11].

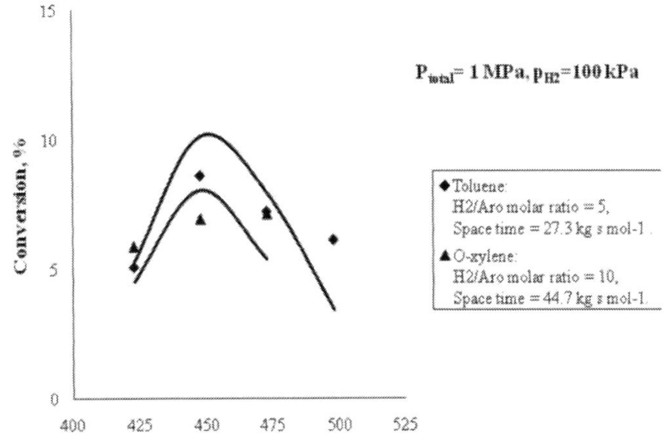

Figure 1. Toluene and o-xylene hydrogenation conversion as a function of the temperature. (symbols: experimental results; lines: obtained from model regression at non-isothermal conditions)

An increase in the hydrogen inlet partial pressure enhances the hydrogenation conversion [5,11]. It results in a corresponding increase of the hydrogen surface concentration and, hence, an enhanced hydrogenation rate. An opposite effect is observed for the aromatic component inlet partial pressure [5,11]. This can be understood if the corresponding increase in surface concentration of aromatic and partially hydrogenated species significantly hinders the hydrogen chemisorption. The expected increase in hydrogenation because of the higher surface concentrations of the partially hydrogenated intermediates is, in that case, overcompensated by a corresponding decrease in hydrogen surface concentration. As a result a decrease in hydrogenation rate is observed with increasing aromatic component inlet partial pressure.

Hydrogenation single-event Microkinetics

Model construction

A characteristic of aromatic hydrogenation kinetic modeling is that many elementary steps occur in the formation of the cycloalkane out of the aromatic component. In steady-state experimentation no direct information is obtained on these reactive intermediates. In order to construct a kinetic model with a reasonable number of adjustable parameters, information with respect to the surface intermediates was obtained from quantum chemical calculations. Quantum chemical calculation of hydrogenation of benzene on a Pt_{22} cluster suggest that hydrogenation takes place by atomic hydrogen addition according to the Horiuti-Polanyi-type mechanism [15,16]. An aromatic ring is transformed into the corresponding cycloalkane by the consecutive addition of H-atoms. The number of unsaturated nearest carbon atoms to the carbon atom to which the H-atom is being added was found to be one of the distinctive features between the rate and equilibrium coefficients involved in the various H-atom additions. A SEMK model for benzene hydrogenation has been constructed defining reaction families on the basis of the saturation state of nearest neighbor carbon atom [5]. In addition to the saturation state of nearest neighbor carbon atom, the rate and equilibrium coefficients for H-addition are also assumed to depend on the secondary or tertiary character of the carbon atom involved.

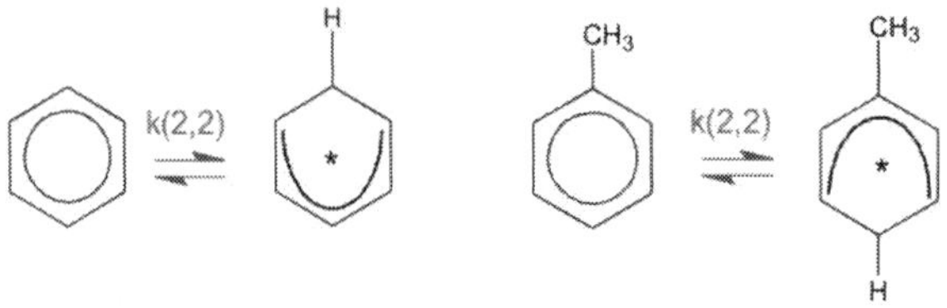

Figure 2. Symmetry effects in hydrogen atom addition.

An additional dependence of the coefficients is related to symmetry effects. This is illustrated in Figure 2, where the first hydrogen atom addition to benzene can occur in 6 equivalent positions whereas the first hydrogen atom addition to toluene has at most 2 equivalent positions. The hydrogenation rate coefficient for an elementary step is written as the product of the number of single-events and the single-event reaction rate coefficient. The latter is assumed to only depend on the saturation state of the nearest neighbors (m= 0,1 or 2) and the degree of branching of the reacting C-atom (n=2 or 3). This results in six different reaction families with specific activation energies and reaction enthalpies.

Unlike in previous Langmuir-Hinshelwood models [1114,16] , no dominant reaction path is considered in this current microkinetic model. This implies that the entire network has to be considered. The reaction network is generated using a computerized algorithm. For benzene hydrogenation by atomic hydrogen, 13 components are interconnected by 20 H-additions and 20 H-abstractions, vide Figure 3. For toluene 40 components are interconnected by 104 H-additions and 104 H-subtractions, while for o-xylene 36 components and 96 H-additions and subtractions are involved. The pseudo-steady state hypothesis is applied for determination of concentration of the partially hydrogenated surface intermediates.

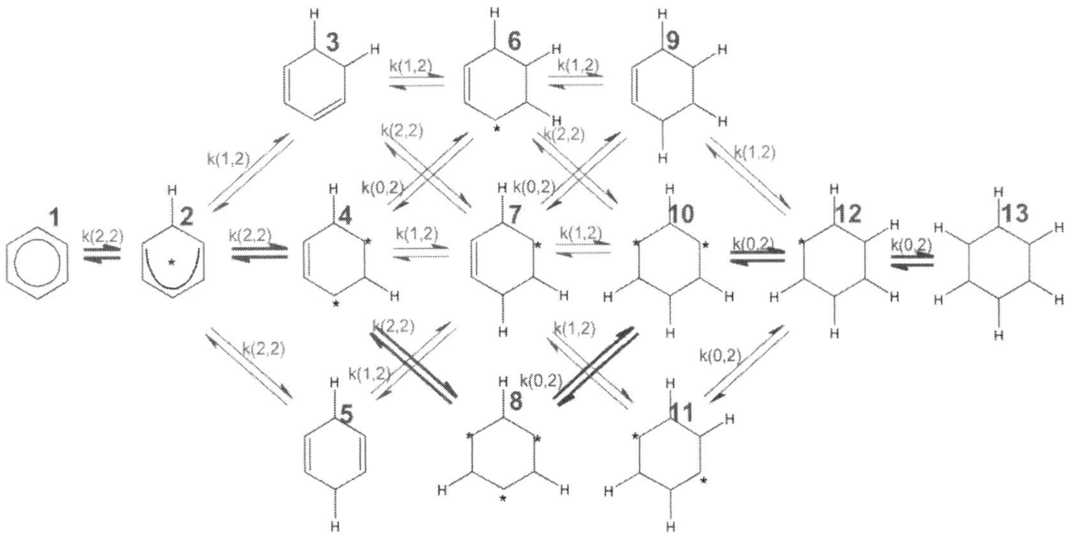

Figure 3. Benzene hydrogenation reaction network indicating the number of the partially hydrogenated intermediates on the catalyst surface as well as the rate coefficients involved in the various hydrogen atom addition reactions [5].

No rate-determining step is assumed. All H-atom additions and abstractions are considered to be non quasi equilibrated. Reactant chemisorption and product desorption was assumed to be quasi equilibrated. The total number of parameters in the model for simultaneous hydrogenation of toluene and o-xylene hydrogenation amounts to 17, i.e., 6 rate coefficients; 6 surface reaction equilibrium coefficients and 5 chemisorption equilibrium coefficients for toluene, methylcyclohane, o-xyne, dimethylcyclohexane and hydrogen. When accounting for their temperature dependence this number doubles to 38, however, order of magnitude calculations have been performed for the preexponential factors of rate, surface reaction equilibrium and chemisorption coefficients. These calculations were based on reasonable assumptions on the mobility of the species involved in the elementary step considered. E.g., aromatic chemisorption was assumed to lead to a significant loss of translational freedom, while the hydrogen mobility on the catalyst surface was assumed to be high. Thermodynamic constraints allow calculating 3 of the surface reaction enthalpies and 2 of the chemisorption enthalpies from the remaining parameter values and the overall reaction enthalpy. Hence 12 adjustable parameters remain to be estimated from regression to experimental data.

The main assumptions in the single-event microkinetic model for aromatic component hydrogenation are summarized below:

1. Competitive dissociative H_2 and molecular aromatic component chemisorption on identical sites.
2. H_2, aromatic component and saturated cycloalkane chemisorption are quasi-equilibrated.
3. No dehydrogenated surface species are considered [17].
4. No rate-determining step or dominant reaction path is assumed.
5. Each reaction step of the network is considered to be reversible.
6. The steady state hypothesis is applied for the partially hydrogenated surface intermediates.
7. The single-event rate coefficients for atomic hydrogen addition only depend on the number of unsaturated nearest neighbor (m) with respect to the carbon atom involved in the hydrogen addition and branching degree (n) of reacting carbon atoms.

The pseudo-steady state hypothesis is applied for the determination of the concentration of the partially hydrogenated surface intermediate. The net production rate (R_j) of any intermediate j is calculated using following expression.

$$R_j = \sum_i r_{i\to j}^{hyd/deh} - r_{j\to i}^{deh/hyd}$$

$$R_j = \sum_i n_{e,i\to j}\left[\tilde{k}_{hyd}(m_i,n_i)C_{H^*} + \tilde{k}_{deh}(m_i,n_i)C_*\right]C_i - n_{e,j\to i}\left[\tilde{k}_{hyd}(m_i,n_i)C_{H^*} + \tilde{k}_{deh}(m_i,n_i)C_*\right]C_j$$

Regression to experimental data

The obtained parameter estimates are in agreement with literature reported values [11,16,12] and lead to a good agreement between experimental observations and model calculations, see Figure 4. The activation energies are in the range of 59 to 73 kJ mol^{-1} and the surface reaction enthalpies correspond to slightly endothermic H addition reactions. This indicates that on Pt the aromaticity is lost upon adsorption of the aromatic component on the surface. Activation energies of the H additions to tertiary carbon atoms were 11 kJ mol^{-1} higher than those to secondary carbon atoms, which can be related to the steric hindrance caused by the methyl group. Chemisorption enthalpies of toluene and o-xylene were estimated very close to each other. As a result, according to the SEMK model, differences in hydrogenation rates between the investigated monoaromatic components are primarily due to differences in H addition rate coefficients rather than in chemisorption strength.

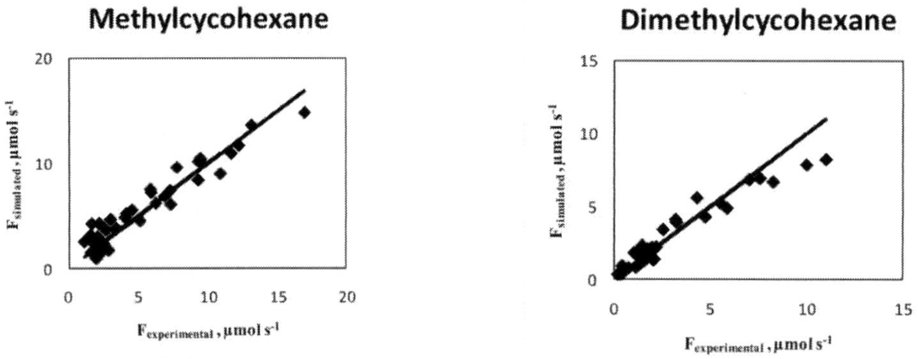

Figure 4. Parity diagram of experimental vs. calculated methylcyclohexane, 1,2-dimethylcyclohexane flow rate. Line: experimental, ♦: obtained from model regression at non-isothermal conditions

As can be observed from Figure 2, the simulated temperature effect is more pronounced than what is actually observed. A refinement in the fundamentals of the pre-exponential factor calculation is expected to further enhance the agreement between model simulations and experimental data.

Conclusions

The single-event methodology has been successfully extended to metal catalysis. The judicious definition of the reaction families has allowed the adequate description of the experimental data using a limited number of adjustable parameters. The single-event hydrogenation rate coefficients are assumed to only depend on the saturation state of the nearest neighbor carbon atoms and the degree of branching of the reacting C-atom. This results in six different reaction families with specific activation energies and reaction enthalpies. Hydrogen additions to tertiary carbon atoms are slower than to secondary carbon atoms due to steric hindrance. Methyl substituent only have a minor effect on the chemisorption characteristics. As a result, differences in aromatic hydrogenation behavior between toluene and o-xylene are mainly governed by differences in hydrogen addition kinetics.

This work demonstrates how single-event microkinetic (SEMK) methodology is a versatile tool in the assessment of hydrocarbon conversion kinetics that can be used in reliable extrapolation towards industrial reactor simulation.

References

[1] Cooper, B. H. and Donnis, B. B. L. (1996), Aromatic saturation of distillates: An overview, *Applied Catalysis A-General*, 137, 203-223.

[2] Vynckier, E., and Froment, G.F., Ed. (1991). Modelling of the kinetics of complex processes based upon elementary steps. *Kinetic and Thermodynamic Lumping of Multicomponent Mixtures,* Amsterdam, Elsevier.

[3] Clymans, P. J. and Froment, G. F.,Computer-Generation of Reaction Paths and Rate-Equations in the Thermal-Cracking of Normal and Branched Paraffins, *Computers & Chemical Engineering*, 8, (1984), 137-142.

[4] Feng, W., Vynckier, E. and Froment, G. F.,Single-Event Kinetics of Catalytic Cracking, *Industrial & Engineering Chemistry Research*, 32, (1993), 2997-3005.

[5] Bera, T., Thybaut, J. W. and Marin, G. B.,Single-Event MicroKinetics of Aromatics Hydrogenation on Pt/H-ZSM22, *Industrial & Engineering Chemistry Research*, In press, doi: 10.1021/ie200541q,

[6] Bos, A. N. R., Lefferts, L., Marin, G. B. and Steijns, M. H. G. M.,Kinetic research on heterogeneously catalysed processes: A questionnaire on the state-of-the-art in industry, *Applied Catalysis A-General*, 160, (1997), 185-190.

[7] Berger, R. J., Stitt, E. H., Marin, G. B., Kapteijn, F. and Moulijn, J. A.,Eurokin - Chemical reaction kinetics in practice, *Cattech*, 5, (2001), 30-60.

[8] Marquardt, D. W.,An Algorithm for Least-Squares Estimation of Nonlinear Parameters, *Journal of the Society for Industrial and Applied Mathematics*, 11, (1963), 431-441.

[9] Boggs, P. T., Byrd, R. H., Rogers, J. E. and Schnabel, R. B.,*NISTIR*, (1992), 92-4834.

[10] Powell, M. J. D., A Hybrid Method for Nonlinear Equations in Numerical Methods for Nonlinear Algebraic Equations, Ed. P Rabinowitz, Gordon and Breach, (1988).

[11] Thybaut, J. W., Saeys, M. and Marin, G. B.,Hydrogenation kinetics of toluene on Pt/ZSM-22, *Chemical Engineering Journal,* 90, (2002), 117-129.

[12] Lin, S. D. and Vannice, M. A.,Hydrogenation of Aromatic-Hydrocarbons over Supported Pt Catalysts .3. Reaction Models for Metal-Surfaces and Acidic Sites on Oxide Supports, *Journal of Catalysis*, 143, (1993), 563-572.

[13] Vanmeerten, R. Z. C., Degraaf, T. F. M. and Coenen, J. W. E.,Gas-Phase Benzene Hydrogenation on a Nickel-Silica Catalyst .3. Low-Field Magnetization Measurements on Hydrogen, Benzene, Cyclohexene and Cyclohexane Adsorption, and Benzene Hydrogenation, *Journal of Catalysis*, 46, (1977), 1-12.

[14] Lindfors, L. P., Salmi, T. and Smeds, S.,Kinetics of Toluene Hydrogenation on Ni/Al2O3 Catalyst, Chemical Engineering Science, 48, (1993), 3813-3828.

[15] Saeys, M., Reyniers, M. F., Neurock, M. and Marin, G. B.,Ab initio reaction path analysis of benzene hydrogenation to cyclohexane on Pt(111), *Journal of Physical Chemistry B*, 109, (2005), 2064-2073.

[16] Saeys, M., Reyniers, M. F., Thybaut, J. W., Neurock, M. and Marin, G. B.,First-principles based kinetic model for the hydrogenation of toluene, *Journal of Catalysis*, 236, (2005), 129-138.

[17] Saeys, M., Reyniers, M. F., Marin, G. B. and Neurock, M.,Density functional study of benzene adsorption on Pt(111), *Journal of Physical Chemistry B*, 106, (2002), 7489-7498.

(478c) Influence of Pt Promoter On Fischer-Tropsch Initiation Pathways Over Cobalt Catalysts

Nianthrini Balakrishnan[1], Babu Joseph[1], Venkat R. Bhethanabotla[1] and D. Yogi Goswami[2], (1)Clean Energy Research Center, Chemical and Biomedical Engineering, University of South Florida, Tampa, FL, (2)Clean Energy Research Center, University of South Florida, Tampa, FL

Determining the chain-growth pathway is very important to understand the Fischer Tropsch Synthesis (FTS) mechanism. FTS consists of the following steps: CO activation, hydrogenation and O removal, chain growth and termination. Three different mechanisms have been proposed depending on how the C-C coupling is achieved. The carbene mechanism involves polymerization of CH_2 intermediates to achieve C-C coupling, the hydroxy carbene mechanism proceeds via dimerization reaction between the adsorbed hydroxyl methylene intermediates and the CO insertion mechanism occurs through insertion of CO into adsorbed alkyl intermediates. However, the carbene mechanism is being supported by many experimental[1-3] and theoretical[4-6] works. There are two types of CO dissociation namely, unassisted CO dissociation and H-assisted CO dissociation. Unassisted CO dissociation involves the dissociation of CO into C and O while H-assisted dissociation does not involve the direct dissociation of CO[7].

Unassisted CO dissociation	H-assisted CO dissociation
CO* --> C* + O*	CO* + H* --> HCO*
C* + H* --> CH*	HCO* + H* --> HCOH*
O* + H* --> OH*	HCOH* --> CH* + OH*
OH* + H* --> H_2O*	OH* + H* --> H_2O*
CO* + O* --> CO_2*	

Both pathways are believed to occur on Fe catalysts, but on Co the H-assisted CO dissociation has been shown to be the preferred route both experimentally and theoretically[7].

Many promoters have been suggested for enhancing the FTS activity of Cobalt. In this paper we investigate how the promoters influence the catalytic pathways. The preferred CO activation pathway in the presence of Pt promoter is studied. We will discuss the energetics of assisted and unassisted CO dissociation pathways. A surface alloy model, where the promoter metal is dispersed on the top surface of the catalyst, is studied. In this work, VASP (Vienna Ab Initio Simulation package) code[8-10] with Perdew–Burke–Ernzerhof (PBE) form of the generalized gradient approximation (GGA)[11] functional is utilized for the exchange and correlation functional. The electron-ion interactions are modeled by the projector-augmented wave (PAW)[12] method. The activation barrier and the transition states are calculated using the Climbing Image Nudged Elastic Band (CI-NEB) method [13-15].

References

(1) Brady, R.; Pettit, R. *Journal of American Chemical Society* **1980**, *102*.

(2) Brady, R.; Pettit, R. *Journal of American Chemical Society* **1981**, *103*.

(3) Van Barneveld, W. A. A.; Ponec, V. *Journal of Catalysis* **1984**, *88*.

(4) Ciobica, I. M.; Frechard, F.; van Santen, R. A.; Kleyn, A. W.; Hafner, J. *Journal of Chemical Physics Letters* **1999**, *311*.

(5) Ciobica, I. M.; Kramer, G. J.; Ge, Q.; Neurock, M.; van Santen, R. A. *Journal of Catalysis* **2002**, *212*.

(6) Liu, Z.-P.; Hu, P. *Journal of American Chemical Society* **2002**, *124*.

(7) Ojeda, M.; Nabar, R.; Nilekar, A. U.; Ishikawa, A.; Mavrikakis, M.; Iglesia, E. *Journal of Catalysis* **2010**, *272*, 287.

(8) Kresse, G.; Furthmuller, J. *Computational Materials Science* **1996**, *6*, 15.

(9) Kresse, G.; Furthmuller, J. *Physical Review B* **1996**, *54*, 11169.

(10) Kresse, G.; Hafner, J. *Physical Review B* **1993**, *47*, 558.

(11) Perdew, J. P.; Burke, K.; Ernzerhof, M. *Physical Review Letters* **1996**, *77*, 3865.

(12) Blochl, P. E. *Physical Review B* **1994**, *50*, 17953.

(13) Henkelman, G.; Johannesson, G.; Jonsson, H. *Theoretical Methods in Condensed Phase Chemistry* **2000**, *5*, 269.

(14) Henkelman, G.; Jonsson, H. *Journal of Chemical Physics* **2000**, *113*, 9978.

(15) Henkelman, G.; Uberuaga, B. P.; Jonsson, H. *Journal of Chemical Physics* **2000**, *113*, 9901.

See more of this Session: Reaction Path Analysis I
See more of this Group/Topical: Catalysis and Reaction Engineering Division

(478d) First Principle Reaction Path Analysis of the Fischer-Tropsch Synthesis Mechanism. Effect of CO Coverage On the Kinetics of the CO Insertion Mechanism

Mingkun Zhuo and **Mark Saeys**, Chemical and Biomolecular Engineering, National University of Singapore, Singapore, Singapore

Fischer-Tropsch Synthesis (FTS), the conversion of CO and H_2 to long chain hydrocarbons, $(CH_2)_n$, and H_2O, involves a sequence of C-O scission, C-C forming, and C and O hydrogenation steps. Different reaction sequences have been proposed. Recent Density Functional Theory (DFT) calculations provide theoretical support for the CO insertion mechanism, where C-C bond formation via a RCH + CO step occurs before C-O scission [1]. Subsequent surface science experiments confirmed that C-O bond breaking is indeed kinetically possible on Co(0001) terraces if a C_xH_y-O moiety can be formed [2], as proposed in the CO insertion mechanism. However, on a clean Co(0001) surface the calculated CO turnover frequency (TOF) for the CO insertion mechanism is several orders of magnitude lower than experimental values.

FTS is performed under a CO partial pressure of about 7 bar. First principle phase diagram calculations indicate that the CO coverage exceeds 1/3 ML under such conditions, in agreement with experimental data [3]. DFT calculations on a CO pre-covered Co(0001) surface further show that the binding energies of the various reaction intermediates decrease by about 10 kJ/mol in the presence of 1/3 ML CO, and the overall reaction energy for the CO insertion mechanism changes from +42 kJ/mol endothermic on a clean surface to -40 kJ/mol exothermic on the CO pre-covered surface. A kinetic analysis using the first principle data furthermore shows that the CO TOF increases from 10^{-8} s^{-1} to 0.1 s^{-1} when the high CO coverage is accounted for.

References

1. Zhuo, M.; Tan, K.F.; Borgna, A.; Saeys, M., J. Phys. Chem. C 113, 1947 (2009); Inderwildi, O. R.; Jenkins, S. J.; King, D. A., PCCP, 11, 11110 (2009)

2. Westrate, C. J.; Gericke, H. J.; Verhoeven, M. W. G. M.; Ciobîc, I. M.; Saib, A. M.; Niemantsverdriet, J. W., J. Phys. Chem. Lett. 1, 1767 (2010)

3. Beitel, G. A.; Laskov, A.; Oosterbeek, H.; Kuipers, E. W. *J. Phys. Chem.* **1996**, *100*, 12494

(478e) Analysis of Reaction Pathways for Ethanol Partial Oxidation On Platinum

Maura A. Koehle, Department of Chemical, Materials, and Biomolecular Engineering, University of Connecticut, Storrs, CT, Angela M. Moreno, Center for Clean Energy Engineering, University of Connecticut, Storrs, CT and Ashish B. Mhadeshwar, Department of Chemical, Materials, and Biomolecular Engineering and Center for Clean Energy Engineering, University of Connecticut, Storrs, CT

Processingof biomass-derived oxygenates, such as ethanol, glycerol, and bio-oil, hastremendous potential to generate H_2 fuel with no net CO_2emissions, providing a clean, renewable, and sustainable fuel source. Whileextensive experimental research has been carried out on oxygenates reforming,there still lacks a comprehensive understanding of the underlying surfacekinetics on an elementary reaction level due to the complexity of reformingchemistry for such large molecular systems. In this work, we will discuss thedevelopment of a comprehensive and predictive microkinetic model for catalytic partialoxidation and reforming of ethanol on Pt, utilizing a combination ofsemi-empirical and first-principles methods as well as experiments to estimate thekinetic parameters for species and elementary reactions. Validation of themicrokinetic model against experimental data and analysis of reaction pathwayswill be presented, along with the approaches for extension of this work tomodel larger oxygenates.

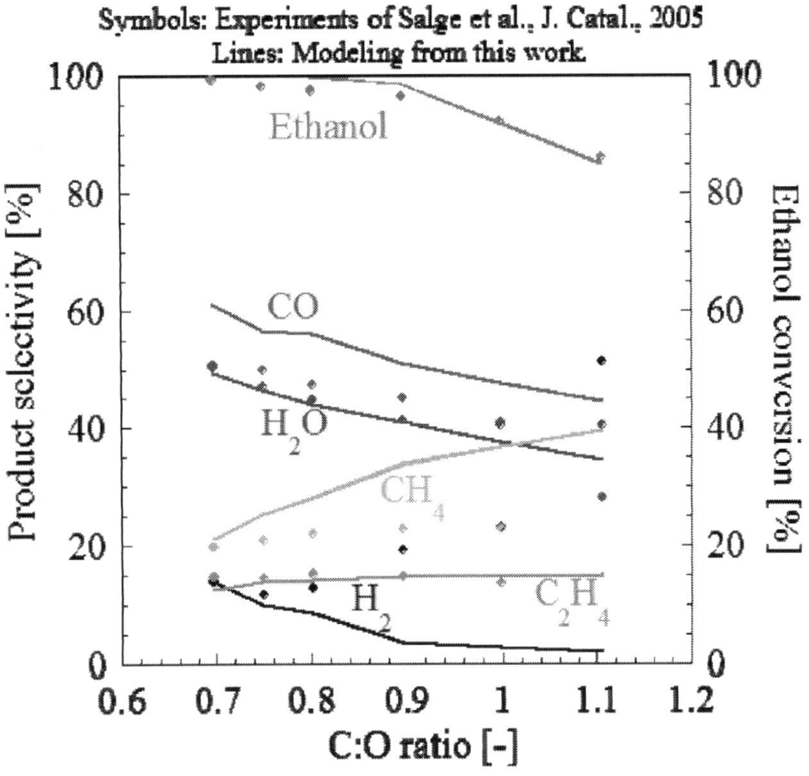

(478f) Magnetic Imaging of AGOs II: Detection of Molecules In the Feed and Relation to Product Yields

Preetinder S. Virk, Department of Chemical Engineering, Massachusetts Institute of Technology, Cambridge, MA

This work reports on a system for the Magnetic Imaging of Pyrolysis Feedstocks, acronym MIPF (pronounced with P silent), that comprizes three facets. First, sample preparation incorporates internal standards into the feedstock oils, to enable precise analysis of the NMR experiments. Second, NMR experiments provide quantitative C13 and H1 spectra, with spectral features elaborated by 1-D and 2-D procedures such as DEPT and HETCOR. Third, data analysis employs (1) an Integral Regions train, which provides coarse but complete information about all types of carbon and hydrogen atoms in a feedstock, and (2) a Canonical Groups train, which provides high-level information about chemical moieties, but detects only ~1/3 of all the atoms in the feedstock. Examples are presented to illustrate how the MIPF parameters of an AGO feedstock might presage its performance in an olefin plant. Overall Gas, RPG and PFO yields are related to the IR train. The CG train is elaborated for two AGO feedstocks to detect a limited number of actual aliphatic and aromatic molecules within them, and to assess their influence on olefin and aromatic product yields from high severity steam cracking.

See more of this Session: Reaction Path Analysis I
See more of this Group/Topical: Catalysis and Reaction Engineering Division

(478g) Gas Phase Methane Oxidative Coupling Studied by Spatial Reactor Profiles and Microkinetic Numerical Simulations

Sardor Mavlyankariev, Oliver Korup, Michael Geske and **Raimund Horn**, Inorganic Chemistry, Fritz Haber Institute of the Max Planck Society, Berlin, Germany

Gas Phase Methane Oxidative Coupling Studied bySpatial Reactor Profiles and Microkinetic Numerical Simulations

Oxidative coupling ofmethane to ethylene (OCM) is a promising one-step reaction pathway to transformmethane into ethylene, which is a valuable intermediate for the chemicalindustry. Unfortunately all research efforts have failed so far to design thishigh-temperature reaction ($T_r \approx 800$ *C*)reaction into a competitive technical OCM process due to insufficient ethyleneyields. Interestingly there seems to be a

virtual upper bound ofabout 25 - 30 % per reactor pass [1] with respect tocombined *C2* yield ($C_2H_4 + C_2H_6$).The kinetic reason for this upper bound is unknown. Since the pioneering workof Keller and Bhasin [2] more than 2700 research articles and reviews and about140 patents have been published on OCM, demonstrating that hundreds ofchemically different materials catalyze OCM, yet there is no convincingexplanation as to why all catalytic data fall roughly on or below aconversion-selectivity trajectory given by $X + S \leq 100$ [3]. The most likely explanation for this indifference of OCM towardsthe nature of the catalytic material is that at a certain temperature methaneoxidation in the gas phase dominates over catalytic oxidation steps, so thatproduct selectivities and yields are not longer determined by the nature of thecatalyst.

In the presentcontribution we use a dedicated high temperature, high pressure spatial profilereactor [4] to measure spatial reactor profiles of species and temperature formethane oxidative coupling in the gas phase. Measurements up to 20 bar pressureand 850 °C were conducted. These profiles provide mechanistic information onhow methane and oxygen are transformed into $C_2H_6$and C_2H_4 and the unwanted by-products $CO$and CO_2 and are ideal datasets for validation ofdetailed microkinetic reaction models. Eight different microkinetic methane gasphase oxidation models were used in a boundary layer simulation and compared tothe experimentally measured profiles. To illustrate the concept Figure 1 showsspecies and temperature profiles measured in the empty quartz tube of the profilereactor described in [4]. This experiment was performed at 8 bar reactorpressure, a typical OCM gas stoichiometry of C/O=4 and a total flow rate of 4000mln min^{-1}(CH_4 = 3200 mln min^{-1}, O_2= 400 mln min^{-1}, Ar = 400 mln min^{-1}). The temperatureprofile of the split furnace surrounding the reactor tube was used as boundarycondition to solve the energy balance in the boundary layer simulation. Thedashed lines in Figure 1 represent predictions by a dedicated OCM gas phasekinetic model, developed by Zanthoff and Baerns for gas phase OCM at elevated pressures[5] comprising 33 species and 192 elementary reactions.

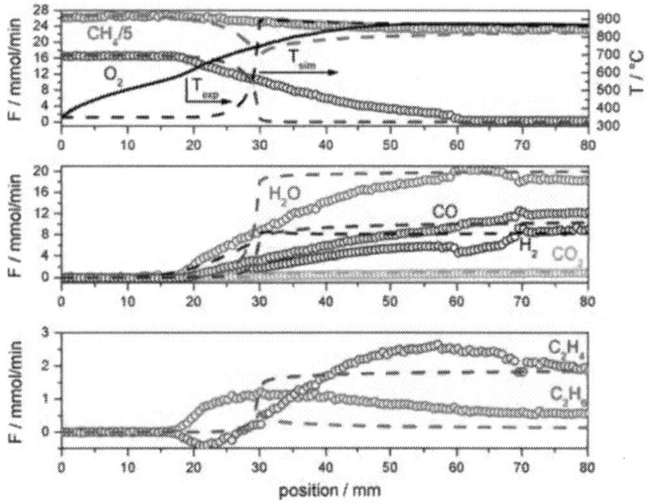

Figure 1: Experimental reactor profiles (scatter) and boundary layer numerical simulations

(dashed lines) of gas phase methane oxidative coupling. Reactor pressure p = 8 bar,

reactant stoichiometry C/O =4.0, CH_4 = 3200 mln min^{-1}, O_2 = 400 mln min^{-1}, Ar = 400 mln min^{-1}, total= 4000 mln min^{-1}.

From the experimentalprofiles it can be seen that CO is the major carbon containing productin gas phase methane oxidative coupling. $C_2H_6$and C_2H_4 are formed in much smalleramounts, and it can be clearly discerned that $C_2H_6$is the primary and C_2H_4 the secondaryproduct. Interesting to note, in particular in view of the predictions by themicrokinetic model, is the crossing of the $C_2H_4$and C_2H_6 profile (here at 35 *mm*)which is reproducibly observed also for other experimental conditions. Thenegative flow rate of the ethylene trace between 16 *mm* and 26 *mm* isan experimental artifact from the mass spectrometric species analysis, which isdue to an isobaric interference on m/z = 30 *amu* by C_2H_6,CH_3OH and CH_2O. As 30 *amu* isused to correct the ethylene peak at 27 *amu* for ethane fragmentation atthis mass, negative peak areas are always obtained when the signal at 30 *amu*is caused by CH_3OH and CH_2O,which are formed in trace amounts at low temperatures. There is basically no CO_2formation by gas phase methane oxidation in contrast to catalytic OCM where $CO_2$is the dominant product (not shown). It can be further seen that there is anignition-delay zone (0-16 *mm*) at the beginning of the free gasphase where no noticeable chemistry occurs. This ignition delay is a combinedeffect of the increasing temperature and the building up of a radical pool.Experimental reactant conversion and product formation begin at about 16 *mm* axialposition and continue until the end of the free gas phase. C_2H_4reaches maximum concentration at around 57 *mm* and is then consumed bysteam reforming. The kinetic model [5] captures the final gas composition atthe reactor outlet sufficiently well, but it does not reproduce the speciesdevelopment inside the reactor. Sufficient reproduction of reactor exitconcentrations has been observed not only for the kinetic model used in Figure 1but also for other models tested, probably because these models have beenoptimized to fit OCM reactor outlet data well. The species development inFigure 10 as predicted by the model is basically confined to a narrow regionbetween 20 and 30 *mm* whereas the experimental profiles develop over amuch longer length between 16 and 81 *mm*. Consequently the numericallypredicted gradients are too steep and not in quantitative agreement with theexperimental data. Also important qualitative features of the experimentalprofiles such as the crossing of the $C_2H_6$and the C_2H_4 profiles are not reproduced bythe model indicating severe deficiencies in terms of the included reactionsteps and/or the kinetic parameters.

References

[1] E.V. Kondratenko, M. Baerns, *Handbook of Heterogeneous Catalysis, 2nd Edition, WILEY-VCH, Ch. 13.17* **2008**, *6*, 3010-3023

[2] G.E. Keller, M. M. Bhasin, *J. Catal.* **1982**, *73*, 9-19

[3] A.M. Maitra, *Appl. Catal. A General* **1993**, *104*, 11-59

[4] R.Horn, O. Korup, M. Geske, U. Zavyalova, I. Oprea, R. Schlgl, *Rev. Sci. Inst.* **2010**, *81*, 064102

[5] H. Zanthoff, M. Baerns, *Ind. Eng. Chem. Res.* **1990**, *29*, 2-10

See more of this Session: Reaction Path Analysis I
See more of this Group/Topical: Catalysis and Reaction Engineering Division

(485a) The Capture of Carbon Dioxide by Ionic Liquid

Suojiang Zhang, 1.State Key Laboratory of Multiphase Complex System, 2.Key Lab of Green Process and Engineering, Institute of Process Engineering, Chinese Academy of Sciences, Beijing, China and Jianmin Zhang, State Key Laboratory of Multiphase Complex System, Institute of Process Engineering, Chinese Academy of Sciences, Beijing, China

Ionic liquid is one kind of organic salts with negligible volatile pressure which might lead to promising substitute candidates for alkanolamines in separation of carbon dioxide. To satisfy this requirement effectively, task specific ionic liquids with alkaline groups, such as imidazolium cation and natural amino acid anions with NH_2-, imidazolium anions with –N(-), or phenate and ethylate with R-O- as anions were designed and synthesized. Intermolecular or intramolecular proton transfer mechanism determines the equilibrium sorption mole ratio. Earlier research focuses on the attempt on displace of functional group within different position in structure of ionic liquid. It is time of significance to look insight into this process from molecular level and illustrate the effect of hydrogen bond net on the sorption mechanism of new designed functional ionic liquids.

See more of this Session: Unconventional Technologies for CO2 Capture, Conversion and Utilization I
See more of this Group/Topical: International Congress on Energy 2011

AUTHOR INDEX

AUTHOR INDEX

AUTHOR INDEX

AUTHOR INDEX

AUTHOR INDEX

AUTHOR INDEX

AUTHOR INDEX

AUTHOR INDEX

AUTHOR INDEX

AUTHOR INDEX

AUTHOR INDEX

AUTHOR INDEX

AUTHOR INDEX

AUTHOR INDEX

AUTHOR INDEX

Catalysis and Reaction Engineering Division

Core Programming Topic at the 2011 AIChE Annual Meeting

Minneapolis, Minnesota, USA
16-21 October 2011

Volume 2 of 2

ISBN: 978-1-61839-739-3

Printed from e-media with permission by:

Curran Associates, Inc.
57 Morehouse Lane
Red Hook, NY 12571

CURRAN ASSOCIATES INC. proceedings.com

Some format issues inherent in the e-media version may also appear in this print version.

Printed by Curran Associates, Inc. (2012)

AIChE
3 Park Avenue
New York, NY 10016-5991

Phone: (203) 702-7660
Fax: (203) 775-5177

www.aiche.org

Additional copies of this publication are available from:

Curran Associates, Inc.
57 Morehouse Lane
Red Hook, NY 12571 USA
Phone: 845-758-0400
Fax: 845-758-2634
Email: curran@proceedings.com
Web: www.proceedings.com

TABLE OF CONTENTS

Volume 1

Volume 2

(485b) Structure and Thermochemistry of Phase-Changing Aminosilicone-Based CO2-Capture Absorbents

Benjamin R. Wood[1], Tiffany E. Westendorf[1], Sarah E. Genovese[1], Robert J. Perry[1], Michael J. O'Brien[1], Matthew L. Meketa[1], Thomas Perry[1], Ravi-Kumar Vipperla[2], Robert M. Enick[3] and Lei Hong[3], (1)Global Research, General Electric, Niskayuna, NY, (2)GE Energy, Greenville, SC, (3) Department of Chemical and Petroleum Engineering, University of Pittsburgh, Pittsburgh, PA

Liquid aminosilicone-based CO_2 absorbents that solidify upon reaction with CO_2 have been developed. These phase-changing absorbents have the potential for superior performance for capturing CO_2 from the flue gas of pulverized-coal power plants. A process has been conceived in which the liquid aminosilicones are reacted with the CO_{2--} in flue gas using a spray column. The solid CO_2-rich powders that form are then collected and sent to a desorber where the powder is heated, releasing high-purity CO_2 and regenerating the liquid aminosilicone, which is recycled back to the spray column. Compared to a conventional monoethanol amine-based CO_2-capture process, phase-changing absorbents have potentially higher CO_2 capacity, greater thermal stability, and simplified separations, due to their low volatility. Studies have been performed to evaluate the particle sizes and morphologies of the solid powders formed in a spray column when these materials are reacted with the CO_2 in simulated flue gas. The crystalline structure of these materials has been studied, and thermochemical methods have been used to study the absorption and desorption of CO_2.

See more of this Session: Unconventional Technologies for CO2 Capture, Conversion and Utilization I
See more of this Group/Topical: International Congress on Energy 2011

(485c) Phase-Changing Absorbents for CO2 Capture

Sarah Genovese[1], Robert J. Perry[1], Michael O'Brien[1], Benjamin R. Wood[1], Tiffany Westendorf[1] and Ravi-Kumar Vipperla[2], (1)Global Research, General Electric, Niskayuna, NY, (2)GE Energy, Greenville, SC

A process to capture CO_2 from flue gas using a liquid material that reacts rapidly with CO_2 to form a solid offers many advantages over conventional liquid-only CO_2 capture methods. The net CO_2 capacity of the system will be increased, the overall system size will be decreased, and the energy required for desorption will be decreased due to the absence of a co-solvent. However, this process also poses significant process design challenges. The material needs to form a solid in the presence of all components of the flue gas, including water. A method for transporting the solids into a pressurized desorber needs to be developed, and the properties of the particles formed in the absorption unit needs to be optimized for maximum absorption and maximum capture of solids from the gas stream as well as for the method of solids transport selected and for the heat transfer into the solids during desorption. Several materials have been developed and have been demonstrated to rapidly form a powdery solid upon reaction with CO_2 in a laboratory-scale spray dryer under various conditions. The physical properties of these materials including the heat of absorption of CO_2, vapor pressure, CO_2 absorption and desorption isotherms, heat capacity, and viscosity have been measured and are used in a process model to predict the impact of this process on the cost of electricity. The design of the process, the materials developed, and the impact on cost of electricity will be discussed.

See more of this Session: Unconventional Technologies for CO2 Capture, Conversion and Utilization I
See more of this Group/Topical: International Congress on Energy 2011

(485d) Extension of Reversible Carbon Dioxide Binding by Frustrated Lewis Pairs to Other Phosphine and Amine Bases

Robert L. Thompson, URS - Site Support Contractor, U.S. Dept. of Energy, National Energy Technology Laboratory, Pittsburgh, PA, Sheila Hedges, U.S. DOE/NETL, Pittsburgh, PA and Damodaran Krishnan, Dept. of Chemistry, University of Pittsburgh, Pittsburgh, PA

Recently, a new approach to metal-free activation of CO_2 using sterically congested Lewis acid/base pairs was discovered which can reversibly bind CO_2 under mild conditions. This reaction was limited to the Lewis base P^tBu_3, which is pyrophoric and difficult to handle in air. A series of alternative Lewis bases, comprised of less-reactive phosphines and amines, was studied in an attempt to extend the CO_2-complexation reaction to other systems. Among the reactions studied, amines tended to form direct acid-base adducts, despite their steric bulk. One system using the amine DBU led to the isolation of crystals which were studied using single crystal X-ray diffraction. Four of the phosphines studied, rather than complexing CO_2, reacted at the para position of one of the borane phenyl rings to form a bridging-C_6F_4 complex, with the lost F atom bonded to B. None of the Lewis bases studied was capable of complexing CO_2 with $B(C_6F_5)_3$, presumably due to the diminished basicities compared to P^tBu_3.

The air and vacuum stability of the $[^tBu_3P\text{-}(m\text{-}CO_2)\text{-}B(C_6F_5)_3]$ complex was studied at ambient conditions. A film of $[^tBu_3P\text{-}(m\text{-}CO_2)\text{-}B(C_6F_5)_3]$ was cast from CH_2Cl_2 solution on CaF_2 and stored both in a vacuum chamber for 6 weeks and in the laboratory air for 12 weeks, both at ambient temperature. Neither film exhibited any change over the span of the experiment, suggesting that the bridging-CO_2 complex is remarkably stable and suggests that $[^tBu_3P\text{-}(m\text{-}CO_2)\text{-}B(C_6F_5)_3]$ could be used as a basis for CO_2-capture and storage. Heating the $[^tBu_3P\text{-}(m\text{-}CO_2)\text{-}B(C_6F_5)_3]$ film under vacuum led to the disappearance of the n(C=O) band at 1694 cm^{-1} in the FTIR spectrum. Heating the $[^tBu_3P\text{-}(m\text{-}CO_2)\text{-}B(C_6F_5)_3]$ film under vacuum in a sealed system led to free CO_2 with a band at 2349 cm^{-1} in the FTIR spectrum. TGA-MS analyses of CO_2-containing products show thermal release of CO_2 between 100 and 160°C.

Finally, variable temperature ^{13}C, ^{31}P and ^{19}F NMR studies of mixtures of both P^tBu_3 and $B(C_6F_5)_3$ with $^{13}CO_2$ in CD_2Cl_2 were conducted. VT NMR analysis indicates that both P^tBu_3 and $B(C_6F_5)_3$

form complexes with $^{13}CO_2$ that can be detected and whose relative amounts vary with temperature. The results from variable temperature scans permit the calculation of the enthalpies of formation for these complexes.

See more of this Session: Unconventional Technologies for CO2 Capture, Conversion and Utilization I
See more of this Group/Topical: International Congress on Energy 2011

2011 AIChE Annual Meeting, Minneapolis, MN

Low-Cost CO_2 Capture Using Enzyme Catalysis

Current CO_2 capture approaches for fossil power plants using solvents give rise to unacceptably high energy losses during regeneration of these solvents. Codexis and CO_2 Solution are jointly developing a technology using the enzyme carbonic anhydrase (CA) that has the potential to allow the economical capture of CO_2 by accelerating the kinetics of solvents with low regeneration. CA has a natural ability to catalyze CO_2 hydration at the rates up to 10^6 molecules per second. In proof of concept lab studies, CA has been shown to accelerate CO_2 hydration in energy efficient but kinetically slow solvents, such as MDEA. CA thus has the potential to provide low-cost capture processes.

Codexis is an innovator in the directed evolution of enzymes for industrial processes and has twice received the Presidential Green Chemistry Challenge Award for development of cleaner, greener pharmaceutical manufacturing processes using optimized enzymes. In 2010, the Department of Energy awarded Codexis $4.7 M under its ARPA-E program to pursue the directed evolution of CA's capable of high activity and stability in carbon capture solvents. CO_2 Solution has been an innovator in the use of CA with CO_2 capture solvents for the past several years and holds issued patents in this field. The presentation will discuss the basis of the technology, initial results and the opportunity to use CA to improve the economics of CO_2 capture.

(485f) Research On Enzyme-Catalyzed Sequential Reduction of Carbon Dioxide

Wenfang Liu, Yanhui Hou, Benxiang Hou and Zhiping Zhao, School of Chemical Engineering and the Environment, Beijing Institute of Technology, Beijing, China

Enzyme-catalyzed reduction of CO2 will not only help to lessen greenhouse effect, but also to obtain valuable chemicals. The sequential reduction of CO2 to methanol under mild reaction conditions has been reported to be feasible by using sol-gel encapsulated or nanoparticle-attached dehydragenases as a biocatalyst, with reduced nicotinamide adenine dinuncleotide (NADH) as a cofactor. However, each enzymatic conversion in multi-step reaction and its influencing factors have not been well studied.

Here we put emphasis on the enzymatic synthesis of formaldehyde with formic acid or CO2 as a substrate, respectively, with a catalyst composing of one or two kinds of dehydrogenases. In the former, reactions were performed in a plastic tube containing formaldehyde dehydrogenase and NADH dissolved in phosphate buffer, and formaldehyde formed was examined by high performance liquid chromatography. The mole ratio of formic acid to NADH plays an important role in the promotion of reaction. The larger the ratio is, the higher the initial velocity and maximum conversion rate are. The yield of formaldehyde was up to 21.3% after reaction for one hour when formic acid was applied as a substrate. The conversion of CO2 to formaldehyde catalyzed by formate dehydrogenase and formaldehyde dehydrogenase with the assistance of NADH also showed a positive result in a preliminary study.

The work will provide a good guidance for further study on the membrane-attached dehydrogenases and their application in the conversion of carbon dioxide to formaldehyde or methanol.

See more of this Session: Unconventional Technologies for CO2 Capture, Conversion and Utilization I
See more of this Group/Topical: International Congress on Energy 2011

(507a) Electrocatalytic Conversion of CO2 to Fuels On Metal Surfaces

Thomas F. Jaramillo, Kendra P. Kuhl and Etosha Cave, Chemical Engineering, Stanford University, Stanford, CA

In order to fully understand the electro-reduction of CO2 on metal surfaces, one must investigate a wide range of metals with differing electronic and geometric structures, and study reaction activity and selectivity under a wide range of applied potentials as mechanistic pathways can change drastically. An experimental setup for running reproducible electrocatalytic CO_2 reduction with on-stream gas analysis and ex-situ liquid analysis is presented; our results lead to a greater understanding of catalytic processes at the metal surfaces. The combination of gas chromatography, proton nuclear magnetic resonance (NMR), and 13C-NMR yields accurate quantitative analysis of H_2, CO, hydrocarbons, formate, and alcohols. Using our optimized CO_2 reduction setup, we proceeded to look at a variety of metal surfaces including Au, Pt, Cu, Ag, Zn, Ni and Fe, measuring all reaction products, major and minor, down to 0.01% current efficiency. As previously reported, these materials can be grouped into three categories: Au and Ag which have selectivity for CO formation, Ni, Fe and Pt which produce primarily H_2 and Cu which yields a high fraction of hydrocarbons. We aim to elucidate the reaction pathways which lead to the observed selectivity, and propose ways to improve selectivity and reduce the required overpotential for production of desirable products, namely hydrocarbons or alcohols.

See more of this Session: Catalysis for CO2 Conversion
See more of this Group/Topical: Catalysis and Reaction Engineering Division

(507b) DFT Studies of the Electrochemical Reduction of CO2 to Alcohols On Cu/ZnO Catalysts

Aravind Asthagiri and Wenjia Luo, Department of Chemical & Biomolecular Engineering, The Ohio State University, Columbus, OH

Within the DOE CALCD EFRC we have an integrated experimental and theoretical effort to develop novel catalysts for the electrochemical reduction of CO_2 to alcohols. There are several challenges to accurately model these systems such as the aqueous electrochemical environment, the changes to the catalyst structure under reaction conditions, the multiple possible reaction pathways and possible catalyst candidates. We have focused our initial modeling work on understanding CO_2 reduction on Cu/ZnO catalysts, which experimental results in our EFRC shows to be selective to methanol with improved activity versus Cu metal and Cu oxides. We have applied density functional theory (DFT) with free energy corrections for the aqueous electrochemical environment to examine CO_2 reduction pathways on various models of Cu on ZnO(10-10). These results will be compared with recent results from the Norskov group on stepped Cu surfaces.While these DFT studies provide insight into possible critical mechanisms on Cu/ZnO, they suffer from the computational cost of electronic-structure calculations. To be able to probe more realistic models of Cu on ZnO in a rapid manner we are developing charge optimized many body (COMB) potentials with Dr. Sinnott's group at University of Florida to describe hydrocarbon chemistry on Cu/ZnO interface. We will present our progress in applying COMB potentials to examine CO_2 chemistry on Cu/ZnO surfaces through direct comparison with DFT studies of the same model surfaces.

See more of this Session: Catalysis for CO2 Conversion
See more of this Group/Topical: Catalysis and Reaction Engineering Division

(507c) Artificial Photosynthesis Utilizing Room Temperature Ionic Liquids

Brian A. Rosen, Department of Chemical and Biomolecular Engineering, University of Illinois at Urbana-Champaign, Urbana, IL and Richard I. Masel, Chemical and Biomolecular Engineering, University of Illinois at Urbana-Champaign, Urbana, IL

Alternative energy and climate change are among the largest concerns facing scientists and engineers at the start of the twenty-first century. The focus of this research is the design of a system that can efficiently capture atmospheric carbon dioxide and reduce it to CO, a precursor for fuel synthesis. Rather than traditional solvents such as water, sulfuric acid, or acetonitrile, room temperature ionic liquids (RTIL's) will be employed as the solvent for CO_2 capture and reduction. RTIL's have many favorable properties such as extended electrochemical windows, high thermal stability, high electrical conductivity, and a 100+ fold increase in carbon dioxide solubility compared to water. Additionally, the absence of bulk water can increase the efficiency of CO production by eliminating the hydrogen and oxygen evolution reactions. Cyclic voltammetry has shown the catalytic reduction of CO_2 at many transition metal cathode surfaces in imidazolium based ionic liquids such as EMIM BF4. Sum Frequency Generation (SFG) spectroscopy was used in identify reduced species and intermediates on the Pt surface. A CO2 conversion flow cell in series with a GC was used to evaluate the steady state performance of CO2 reduction on various metal cathodes.

See more of this Session: Catalysis for CO2 Conversion
See more of this Group/Topical: Catalysis and Reaction Engineering Division

(507d) Photodissociation of CO_2 and Associated Photocatalytic Intermediates In Gas and On Surfaces

Chris Singer and Cynthia Lo, Energy, Environmental & Chemical Engineering, Washington University in St. Louis, St. Louis, MO

Utilizing CO_2 as a feedstock for chemical and fuel production could help in sustainably producing much of what societies demand with reduced environmental impact. Thermal methods such as dry reforming of methane to use CO_2 are too energy intensive to be environmentally beneficial making photocatalytic methods preferable.

We have performed theoretical dynamical photodissociation simulations on CO_2 and similar surface species to determine potential routes towards energetically favorable methods to reactively break down the molecule. This study investigates new methods for predicting CO_2 dissociation rates and products. Semi-classical and fully quantal simulations of wavepacket dynamics were first carried out in the gas phase using monochromatic UV-radiation to confirm experimental results. Photodissociation products and rates are determined as a function of radiation frequency and polarization.

We then performed similar simulations of adsorbed CO_2 and surface intermediate species known to arise from photocatalytic processes on semiconductor surfaces. These molecular results give insight into reactor setups that can improve photocatalytic yield.

See more of this Session: Catalysis for CO2 Conversion
See more of this Group/Topical: Catalysis and Reaction Engineering Division

(507e) Photocatalytic Reduction of CO2 to Fuels by Modified TiO2 Nanoparticles and Nanotubes

Ying Li, Qianyi Zhang and Lianjun Liu, University of Wisconsin-Milwaukee, Milwaukee, WI

Photocatalytic reduction of CO_2 to fuels (CO, methane, methanol, etc) by sunlight is potentially a promising sustainable energy technology that not only reduces greenhouse gas emissions but also produces renewable fuels. However, the low CO_2-to-fuel conversion efficiency has impeded the development of this technology. TiO_2 has been widely used as a photocatalyst due to its low cost, high stability, and environmental benignness. In this work, material innovations have been made to TiO_2 nanostructures to improve the CO_2 conversion efficiency by increasing the surface area, inhibiting photoinduced electron-hole recombination, and enhancing visible light utilization. We have synthesized high-surface-area TiO_2 nanoparticles (TiNPs) and TiO_2 nanotubes (TiNTs) through a hydrothermal process in a Teflon-lined autoclave. The TiNPs and TiNTs are modified by metal and nonmetal dopings to enhance charge separation and narrow the band gap. The catalysts are well characterized by UV-vis, XRD, SEM, TEM, XPS, and BET analysis. The experiments of photocatalytic CO_2 reduction with water vapor using the prepared catalysts are conducted under visible light, UV-visible, and simulated sunlight, respectively. The product yield and selectivity are measured by a gas chromatograph (GC) equipped with thermal conductivity detector (TCD) and flame ionization detector (FID). The effects of the dopings and nanostructures as well as the reaction mechanisms are discussed in this study.

See more of this Session: Catalysis for CO2 Conversion
See more of this Group/Topical: Catalysis and Reaction Engineering Division

(507f) Photocatalytic Conversion of CO_2 to Fuel Using Black TiO_2

Miao Yu, Chemical and Biological Engineering, University of Colorado at Boulder, Boulder, CO

Black TiO_2, prepared by high temperature H_2 treatment, was characterized and used for converting CO_2 into fuel. Visible light absorption by TiO_2 was significantly improved after H_2 treatment, which is expected to generate more charge pairs for surface reactions. In this study, black TiO_2 aggregates were coated with Pt nanoparticles and deposited as thin films for photocatalytic CO_2 conversion. Our preliminary results showed that fuel forming rate increased dramatically by optimizing water vapor pressure in a static photo-reactor and Pt loading. Further studt is conducted to understand fundamental reaction mechanisms.

See more of this Session: Catalysis for CO2 Conversion
See more of this Group/Topical: Catalysis and Reaction Engineering Division

(509a) Water-Gas Shift Catalysis On Supported Au and Pt Catalysts

Mayank Shekhar[1], Jorge Pazmino[1], Wen-Sheng Lee[1], M. Cem Akatay[2], Eric A. Stach[2], Jeffrey T. Miller[3], W. Nicholas Delgass[1] and Fabio H. Ribeiro[1], (1)School of Chemical Engineering, Purdue University, West Lafayette, IN, (2)School of Materials Engineering and Birck Nanotechnology Center, Purdue University, West Lafayette, IN, (3)Chemical Science and Engineering, Argonne National Laboratory, Argonne, IL

The water-gas shift (WGS) reaction ($CO + H_2O \rightarrow CO_2 + H_2$) is an important industrial chemical process for hydrogen production. Our work focuses on determining the active sites, the effect of support and the effect of promoters on supported Au and Pt catalysts. The WGS rate per total mole of Au varies with average Au particle size (d) as $d^{-2.7\pm0.1}$ and $d^{-2.2\pm0.2}$ for Au/TiO_2 and Au/Al_2O_3 catalysts respectively. The variation of reaction rate and apparent reaction orders with particle size was used in conjunction with a physical model of Au clusters and operando IR to show that the active sites are low coordinated metallic corner and perimeter Au atoms. On the other hand, the WGS turnover rates (TOR's), normalized by surface atoms, of Pt/ZrO_2, Pt/Al_2O_3 and Pt/SiO_2 catalysts are independent of average Pt particle size. Unlike for Au catalysts, the apparent reaction orders for Pt/Al_2O_3 catalysts do not vary with particle size. Thus, all surface Pt atoms exhibit the same rate.

The WGS rate per total mole of Au and H_2O order (in parenthesis) varies as Au/Al_2O_3 (~0.6) < Au/CeO_2 (~0.3) < Au/ZrO_2 (~0.0) < Au/TiO_2 (~-0.3) at the same Au particle size at 120°C. Similarly, for Pt catalysts, the WGS TOR and H_2O order (in parenthesis) varies as Pt/Al_2O_3 (0.90) ~ Pt/SiO_2 (0.85) < Pt/TiO_2 ~ Pt/ZrO_2 ~ Pt/CeO_2 (0.65-0.70) at 300°C. We interpret this data to show that the support plays a direct role in activating H_2O.

The effect of promoters like K, Na, Li, Fe and Mo on the catalytic performance on Pt/Al_2O_3 catalyst is determined. A comparison of WGS reaction kinetic parameters and TOR's on supported Au, Pt and promoted Pt catalysts serve as a guideline to determine the descriptors that govern their WGS catalytic performance. These will be discussed.

See more of this Session: Catalytic Hydrogen Generation - General I
See more of this Group/Topical: Catalysis and Reaction Engineering Division

(509b) Nanoarchitectured Ni-ZrO_2 Catalysts for Hydrogen Production From Steam Reforming of Ethanol

Shuirong Li[1], Chengxi Zhang[1], Ping Na[2], Jinyu Han[1], Xinbin Ma[1] and **Jinlong Gong**[1], (1)Key Laboratory for Green Chemical Technology of Ministry of Education, School of Chemical Engineering and Technology, Tianjin University, Tianjin 300072, China, (2)School of Chemical Engineering and Technology, Tianjin University, Tianjin 300072, China

Hydrogen has been considered as an important alternative energy vector and a bridge to a sustainable energy future. Production of hydrogen efficiently and economically from renewable sources, however, remains a challenge. Among diversified hydrogen production technologies, steam reforming of ethanol has been considered as an attractive route for the low toxicity, easy deliverability, and widely availability of bioethanol. This work focuses on the design of nanoarchitectural Ni-ZrO_2 catalysts with high activity in ethanol steam reforming.

The use of nanograined NiO to prepare nanoarchitectural Ni-ZrO_2 catalysts could avoid the use of hazardous reagents during the preparation of nickel metal nanoparticles and simplify the preparation process. The presence of ammonium ion during the precipitation of zirconia precursor led to the production of ZrO_2 species with active sites catalyzing the dehydration of ethanol, thus lowering the catalytic activity and hydrogen yield. However, we have successfully shown that the introduction of Al_2O_3 into the zirconia matrix (Ni-AZ) could greatly improve specific area of the catalyst, the dispersion of nickel metal and catalytic activity and stability. The presence of Al_2O_3 could also effectively block active sites for dehydration of ethanol. The nanoarchitectured Ni-AZ catalyst yielded high activity in both water gas shift reaction as well as the steam reforming of ethanol, at the temperature range of 673-973 K. CO_2, CO and CH_4 were detected as the only byproducts even at low temperature (e.g., 673 K). Yield of hydrogen and selectivities of carbonaceous products were all close to the thermodynamic equilibrium temperatures investigated.

See more of this Session: Catalytic Hydrogen Generation - General I
See more of this Group/Topical: Catalysis and Reaction Engineering Division

(509c) Mechanism for the Water-Gas Shift Reaction On Pt(111) and Catalyst Deactivation

David W. Flaherty[1], Wen-Yueh Yu[2], Zachary Pozun[3], Graeme Henkelman[3] and C. Buddie Mullins[2], (1)Department of Chemical Engineering, University of California at Berkeley, Berkeley, CA, (2) Department of Chemical Engineering, University of Texas at Austin, Austin, TX, (3)Department of Chemistry and Biochemistry, University of Texas at Austin, Austin, TX

The behavior of monofunctional platinum, Pt(111), for the water-gas shift reaction has been investigated using experimental and theoretical methods. Kinetic and isotopic measurements performed from 525 – 675 K are consistent with an associative mechanism for the water-gas shift reaction. The sole kinetically relevant step consists of the unimolecular decomposition of an adsorbed carboxylate intermediate. The turnover frequency of Pt(111) is five-times greater than that observed on Cu(111) under identical conditions (612 K, 26 Torr CO, 10 Torr H_2O); however, Pt(111) loses activity over time due to the formation of carbonaceous deposits, a process not observed in similar studies of Cu(111). Our experimental and theoretical results suggest that CO dissociates *via* two pathways, the Boudouard reaction and through a COH intermediate. Nucleation of carbon at step-edges and subsequent oligomerization deactivates the catalyst. These results provide insight to the synergistic roles of noble metal clusters and supports for the water-gas shift reaction.

See more of this Session: Catalytic Hydrogen Generation - General I
See more of this Group/Topical: Catalysis and Reaction Engineering Division

(509d) Thermodynamic and Kinetic Modeling of the Water Gas Shift Reaction Using Supported Ionic Liquid Phase Catalyst Systems

Johannes Hartmann[1], Alexander Buchele[1], Wolfgang Arlt[1], Sebastian Werner[2] and Peter Wasserscheid[2], (1)Chair of Separation Science and Technology, Erlangen, Germany, (2)Chair of Chemical Reaction Engineering, Erlangen, Germany

Supported Ionic Liquid Phase (SILP) catalyst systems have been successfully applied in continuous gas phase reactions using homogenous catalysts in fixed bed reactors. The homogenously dissolved transition metal complex is immobilized in the thin film of an ionic liquid (IL) dispersed on the high area surface of an inorganic support, thus bridging the gap between homogenous and heterogeneous catalysis.

A technical important reaction is the water gas shift reaction (WGS), what is a key reaction in hydrogen production. Commonly used heterogeneous catalysts only work at high temperatures. However, due to the exothermic reaction, mild conditions are thermodynamically favorable and thus yield higher equilibrium conversions of hydrogen. Werner et al. show the great potential of homogenous catalysis using SILP catalysts for the WGS reaction.

In this work we present a rigorous thermodynamic reaction modeling for this particular reaction. The conversion can only be measured in the bulk gas phase. In order to account for the actual conditions in the reactive IL phase, the phase equilibria of all components are used. The kinetics of the reaction is described by a power law expression, but in comparison to the commonly used concentrations, activities are used for the modeling. Therefore, a consequent thermodynamic description of the kinetics can be achieved. Further, it is possible to predict the equilibrium conversion and use it in the modeling process.

For the validation of the model, we used experimental data sets of two SILP-catalysts systems, [BMMIM][OTf] and [Bu_4P][$MePO_3$] with the homogenous catalyst $RuCl_3$ in the temperature range of 370K – 420K. The CO/H_2O ratio is varied to obtain the partial orders of the reactants. With the derived model, the WGS reaction can be described properly in both IL.

Werner, S.,Szesni, N., Bittermann, A., Schneider, M.J., Haerter, P., Haumann, M., Wassescheid, P.; Appl. Catal. A; 277, 70-75, 2010

See more of this Session: Catalytic Hydrogen Generation - General I
See more of this Group/Topical: Catalysis and Reaction Engineering Division

(509e) Renewable Hydrogen Generation by Steam Reforming of Acetic Acid Over Cu-Zn-Ni Catalyst Supported On Calcium Aluminate

Madhumita Patel, Chemical Engineering, IIT Delhi, Delhi, India and **K. K. Pant**, Chemical Engineering, Indian Institute of Technology, Delhi, New Delhi, India

Catalytic steam reforming of biomass-derived-oil is one of the promising methods for the production of renewable hydrogen. Typically bio oil contains 12-14% by wt. acetic acid therefore it has been chosen as the model compound for the reforming of bio oil. In the present study performance of the Cu-Zn-Ni/ $CaO-Al_2O_3$ catalyst was investigated for hydrogen production with an aim of minimisation of methane and carbon monoxide formation. The calcium aluminate support was prepared by solid-solid reaction between calcium oxide and aluminium oxide (12/7 molar ratio) at 1250^0C temperature for 24 hr. XRD results indicated that $12CaO.7Al_2O_3$ appeared as major phase with $CaO.2Al_2O_3$ and CaO as minor phases in the support. Cu, Zn and Ni were loaded on this support by wet impregnation method. For the preparation of catalyst, initial loadings of Cu, Zn and Ni were kept as 10%, 1% and 7% by wt. respectively. The catalyst was characterized by BET, XRD, SEM-EDX, TEM and TPR to understand the physic-chemical properties of the catalysts. The steam reforming reaction was carried out in a tubular fixed bed reactor (I.D=19mm) at different temperatures ranging from $600\text{-}700^0C$ with 3 gram of catalyst and water to acetic acid ratio of 9:1 (by wt.). The role of Ni metal was to break the C-C bond present in acetic acid and Cu metal promoted the steam reforming reaction on the surface of the catalyst. Incorporation of Zn enhanced the acetic acid conversion and H_2 yield and also reduced deactivation of catalyst. The coke formation on the support surface was insignificant due to the presence of excess oxygen in the $12CaO.7Al_2O_3$ phase. At 700^0C, % hydrogen yield was approximately 92% and acetic acid conversion was approximately 90%.

See more of this Session: Catalytic Hydrogen Generation - General I
See more of this Group/Topical: Catalysis and Reaction Engineering Division

(509f) Colloidal Synthesis and Characterization of Alumina-Supported Cu/Ni Core-Shell Nanoparticles In Water Gas-Shift Reaction

Jiann-Horng Lin, Chemical Engineering, University of Cincinnati, Cincinnati, OH and Vadim Guliants, School of Energy, Environmental, Biological, and Medical Engineering, University of Cincinnati, Cincinnati, OH

Recently, the water-gas shift (WGS) reaction has attracted renewed attention because high purity H_2 is needed for the operation of fuel cells. Metallic Cu and Ni have been predicted as highly promising catalysts for the water-gas-shift reaction based on theoretical (DFT) calculations. Supported bimetallic Cu/Ni nanoparticles with a core-shell structure have not been considered to date as WGS catalysts with enhanced catalytic activity and improved selectivity. However, conventional (support impegnation) approaches show poor control over bimetallic structures. Controlled tailoring of various nanostructures using colloidal methods may lead to new electronic and catalytic properties of nanosized metals. Accordingly, the main goal of the present study is to synthesize a series of stable and well-defined bimetallic Cu/Ni nanoparticle catalysts, namely, (a) Cu core and Ni shell (Cu@Ni), (b) Ni core and Cu shell (Ni@Cu), and (c) Cu-Ni mixed alloy nanoparticles supported on alumina, and to investigate their behavior in the WGS reaction. Another important practical aspect of WGS catalysis is the sensitivity of pure Cu and Ni catalysts to the presence of sulfur-containing impurities in the process streams. These catalytic metals readily chemisorb H_2S and form stable surface metal sulfides that poison WGS activity. Thus, bimetallic catalysts with core shell and alloy structure have been also explored with respect to their sulfur tolerance.

The supported bimetallic Cu@Ni, Ni@Cu, and CuNi alloy nanoparticles were synthesized by a successive chemical reduction, a redox-transmetallation method, and simultaneous reduction, respectively. These catalysts were characterized by scanning electron microscopy (SEM), high resolution transmission electron microscopy (HRTEM), scanning transmission electron microscopy (STEM) equipped with an HAADF detector, X-ray diffraction (XRD), X-ray photoelectron spectroscopy (XPS), thermogravimetric analysis (TGA), CO chemisorption, and UV-vis spectroscopy. The size and shape of supported Cu@Ni and Ni@Cu nanoparticles were determined by HRTEM and STEM. The metal composition in the core and shell regions was characterized by HAADF-STEM. The combination of X-ray diffraction, TGA and UV-vis spectroscopy clearly showed the formation of Cu or Ni core structures distinctly different from CuNi alloy nanoparticles. These catalysts were evaluated for the WGS reaction at 423-673 K under normal atmospheric pressure in a fixed-bed glass reactor connected to a gas chromatograph. Supported Cu@Ni and Ni@Cu nanoparticles showed enhanced WGS activities, as compared to traditional CuNi nano-alloys. Specifically, the Cu core with an approximately 1-2 monolayer thick Ni shell exhibited a greater WGS activity as compared to traditional CuNi nano-alloys.

However, bimetallic CuNi catalysts supported on alumina with core shell or alloy structure showed similar sulfur resistance to that of pure Cu and Ni catalysts. On the other hand, the sulfur resistance of the catalytic system was improved when ceria is employed as a support.

See more of this Session: Catalytic Hydrogen Generation - General I
See more of this Group/Topical: Catalysis and Reaction Engineering Division

(509g) Modified Ferrites As Catalysts for High Temperature Water Gas Shift Reaction

Krishna Reddy Gunugunuri[1], Rajesh Koirala[1], Panagiotis Smirniotis[1] and Punit Boolchand[2], (1) Chemical Engineering Program, School of Energy, Environmental, Biological and Medicinal Engineering, University of Cincinnati, Cincinnati, OH, (2)Department of ECECS, University of Cincinnati, Cincinnati, OH

In recent years, a large number of studies (feasibility, computational and real time analyses) have been performed on water gas shift (WGS) reaction in membrane reactors. The WGS reaction in a membrane reactor (MR) is potentially capable of completing the CO conversion and achieving simultaneous H_2/ CO_2 separation in a single stage operation. A membrane for this reaction typically operates at 450 °C to 550 °C temperatures and pressures ranging from 1-20 bar. Development of catalysts for membrane reactors operating at these high pressures and temperatures have to meet stringent requirements. The present study is aimed at developing various modified ferrite catalysts for high temperature WGS membrane reactor applications. The idea was to stimulate the ferrite formation via doping with certain foreign cations and to promote the Fe^{3+} « Fe^{2+} redox couple. The WGS feed condition were chosen to mimic conditions found in a membrane reactor. For this purpose, various doped modified ferrite catalysts were prepared using ammonia assisted coprecipitation method. The prepared catalysts were characterized by using TPR, Raman, XRD, Mossbauer and X-ray photo electron spectroscopic techniques and evaluated for WGS reaction in the temperature region 400-550 °C and at a stream to CO ratio 3.5 and 1.5 which the conditions used in our membrane reactor. The WGS activity observed was found to be determined by the nature and amount of dopant used in a given case. Our structural and surface characterization results show that dopants exhibit different structural effect in the bulk and on the surface of the ferrite. These results will be presented in this study.

See more of this Session: Catalytic Hydrogen Generation - General I
See more of this Group/Topical: Catalysis and Reaction Engineering Division

(510a) First Principles Analysis of the Selective Hydrogenolysis of Tetrahydrofurfuryl Alcohol Over Rh and ReOx Modified Rh Surfaces

Qiaohua Tan[1], David Hibbitts[1], Mei Chia[2], Yomaira J. Pagn-Torres[2], James A. Dumesic[2], Robert J. Davis[1] and Matthew Neurock[1], (1)Chemical Engineering, University of Virginia, Charlottesville, VA, (2)Chemical and Biological Engineering, University of Wisconsin-Madison, Madison, WI

Abstract

Tetrahydrofurfuryl alcohol (HMTHF), derived from the acid dehydration of xylose, can be converted over supported metal catalysts to either 1,2-pentanediol or 1,5-pentanediol, important chemical intermediates in polymer industry. Recent experimental results show that the hydrogenolysis of HMTHF over supported rhodium particles predominantly produces 1, 2-pentanediol whereas the addition of ReOx to Rh results in 95% selectivity to 1,5-pentanediol. Herein we use ab initio density functional theoretical calculations to explore the differences reported over Rh, ReOx and ReOx modified Rh surfaces.

The results indicate that the hydrogenolysis of HMTHF over Rh as well as other metal surfaces take place via metal catalyzed ring opening paths. The activation predominantly proceeds at the C-O bonds comprised of secondary C atoms thus resulting in the formation 1, 2-pentanediol. The activation of the C-O bond at the more substituted tertiary carbon atom, on the other hand, results in the formation of 1, 5-pentanediol. The activation of the C-O bond appears to proceed via the sequential activation of vicinal C-H bonds followed by the dissociation of the C-O bond to form an oxymetallocycle intermediate. The barrier to break the C-O bond at a secondary carbon atom appears to be linearly correlated to the binding energy of the CH_x intermediate on the metal surface. The barrier to break the C-O bond at a tertiary carbon is about 0.20 eV higher than the barrier to break at the secondary carbon due to the steric effects of the side chain on the tertiary carbon. Higher hydrogen surface coverages that likely form under reaction conditions result in higher activation barriers at both the secondary and tertiary C-O carbon atoms. The selectivity, however, remained about the same. These results which suggest that 1, 2-pentanediol is the predominant product over Rh are in very good agreement with those found experimentally.

The catalytic conversion of HMTHF over a ReOx modified Rh substrate is rather different than that over Rh as the addition of ReOx results in the formation of acid sites and bifunctional hydrogenolysis of HMTHF. The acid sites aid in the activation of the C-O bonds comprised of tertiary rather than secondary character as they result in the formation of the more stable secondary carbenium ions as opposed to the primary carbenium ions. The activation barrier for C-O bond cleavage at a secondary carbon is 0.40 eV higher than over the tertiary carbon, due to the lower stability of the primary carbenium ion that forms C-O bond. The results also show that the hydroxyl group at α-position in HMTHF help to stabilize carbenium ion formation by forming a more stable transition state structure which lowers the activation barrier at the secondary C-O bond. The presence of water further stabilizes the secondary carbenium transition states. The acid-catalyzed mechanism is consistent with the experimental results which show the selective activation of the secondary C-O bond to form 1, 5-pentanediol on the ReO_x modified Rh catalyst.

See more of this Session: Catalytic Processing of Fossil and Biorenewable Feedstocks V
See more of this Group/Topical: Catalysis and Reaction Engineering Division

Oxygen-Free Oxidative Dehydrogenation of Ethane over $VO_x/\gamma\text{-}Al_2O_3$ Catalyst in Riser Reactor Simulator

S. Al-Ghamdi [(1,2)], *H. de Lasa* [(1)] *and M. Volpe* [(3)]

1. *Chemical Reactor Engineering Centre, University of Western Ontario, London, Ontario,(Canada)*
2. *Process Development Group, Research & Development Center, Saudi Aramco Oil Company, Dhahran, (Saudi Arabia)*
3. *Chemical Engineering Department, PLAPIQUI-(UNS-CONICET),Bahia Blanca, Argentina*

INTRODUCTION

Light Olefins, specifically ethylene and propylene, represent the most important building blocks for the petrochemical industry. Steam cracking of hydrocarbon feedstock (ethane, LPG, or naphtha) has been the major source of light olefin accounting for 70% of total olefin production (1). Other methods for olefin production such as fluidized catalytic cracking (FCC) and direct dehydrogenation are also in use. Fluidized catalytic cracking (FCC) is the second most frequently used processes for olefin production accounting for 28% of total olefin production (2). In FCC, ethylene and propylene are produced as co-products during gasoline production with ethylene and propylene in yields of 1-2% and 5%, respectively

Current conventional processes for olefin production suffer from several limitations related to their high energy requirements, coke formation, selectivity control and thermodynamic constrains (1-7). A number of alternative technologies have been investigated for olefin production (8-24), and those are being helped considerably by the increasing demands for ethylene and propylene. Oxidative dehydrogenation (ODH) is an alternate method of olefin production that does not suffer from the drawbacks of traditional methods. selectivity control in ODH reactions is a major challenge. Undesired carbon oxides can be formed either by direct alkane combustion or by deep oxidation of produced olefins. Thus, a catalyst for this reaction should be designed in such as way that it prevents the detrimental reactions leading to carbon oxides (25).

Vanadium-based catalysts have been reported in the technical literature as active and selective catalysts for ethane ODH (4, 9, 11, 12, 26-37). This is due to the vanadium ability to provide the lattice oxygen for abstracting hydrogen from alkanes (25, 38) followed by re-oxidation of the reduced vanadium cations by gas phase molecular O_2 (25, 39). Moreover, the catalytic properties of vanadia can be improved by depositing it on appropriate metal oxide supports such as alumina, silica, titania, zirconia. Variation in catalytic performance of VO_x supported on metal oxides is attributed to the generation of different VO_x species on the surface of the metal oxide support (28, 40-45). However, there is no clear relationship between the structure of surface vanadia species and their reactivity for the ODH of lower alkanes (28). Furthermore, the catalytic activity and selectivity of supported vanadium oxide catalysts is significantly affected by the properties of the support material, the interaction of the surface VO_x species with the oxide support and the vanadium loading (28, 33, 40, 41, 43). Alumina-supported vanadia has been considered in several structural investigations.(12, 28, 31, 40, 46-51) Theses studied showed that at low vanadium loading; a highly dispersed amorphous vanadate phase is formed whose structure changes from isolated tetrahedral vanadium to polyvanadates species at medium loading. However, crystalline V_2O_5 phase was reported at high vanadium loading in addition to amorphous vanadia phases. Variation of the binding strength of surface lattice oxygen with the formation of different VOx surface species structure has been proposed as the main parameter that governs activities and selectivities of alumina-supported vanadia catalysts.

There are no studies in the open literature that have dealt with supported-vanadium catalyst for ethane ODH in dense fluidized beds, risers or downers reactors. Controlled isothermal conditions can be maintained in catalytic fluidized beds are used and hence avoiding hot spots development which can interfere with the reactor performance and damage the catalyst. Moreover, with the need of periodic catalyst re-oxidation, fluidized bed reactors offer the means that enable the transport of reduced catalyst species from the oxidative dehydrogenation zone to the re-oxidation zone; and it is a must in order to implement this technology at the industrial scale. Given the above described facts, the goal for this contribution is to provide an in-depth research on the reactivity and stability of the prepared 10 wt. % VOx /γ-Al_2O_3 catalyst for ethane oxidative dehydrogenation under fluidized bed reaction conditions in CREC Riser Simulator reactor. Ethane was used as a feed for the reaction and air was employed to regenerate the reduced catalyst sample. The prepared oxygen carrier was characterized by different physicochemical techniques, such as temperature programmed reduction (TPR), temperature programmed oxidation (TPO), temperature programmed ammonia desorption (NH_3-TPD), and Powder X-ray diffraction (XRD) analysis.

EXPERIMENTAL

The fluidizable catalyst of this study was prepared by impregnation method with V $(AcAc)_3$ (vanadium III acetyl-acetonate) precursor. Gamma form of alumina support particles were calcined at 500°C for 4 hours under air flow. Then, the support was contacted by V (AcAc)3 solution in toluene at room temperature for a 24 hrs. The impregnated particles were then filtered, dried first at 150 °C and then calcined under air stream at 600 °C for a 6 hrs. Reaction experiments were developed in a novel Chemical Reactor Engineering Center (CREC) mini-fluidized bed Riser Simulator unit. A schematic diagram of the CREC-Riser Simulator reactor body is illustrated in Figure 1. The reaction runs in the CREC riser simulator were carried out with 99.99 % ethane feed composition, and at atmospheric pressure. The reaction temperatures were varied between 550-700 °C. Temperature programmed reduction/oxidation (TPR/TPO) experiments were performed with an Autochem II ASAP 2920 analyzer. To investigate the degree of reduction of the prepared catalyst, the catalyst sample (0.26g) was reduced with a flow of reducing gas mixture (10% H2 in He) at a rate of 50 ml/min. The sample was subjected to a linear temperature heating ramp from ambient to 650°C. The consumption of hydrogen by adsorption/reaction was continuously monitored by a thermal conductivity detector (TCD). The degree of reduction was calculated from the amount of hydrogen consumed.

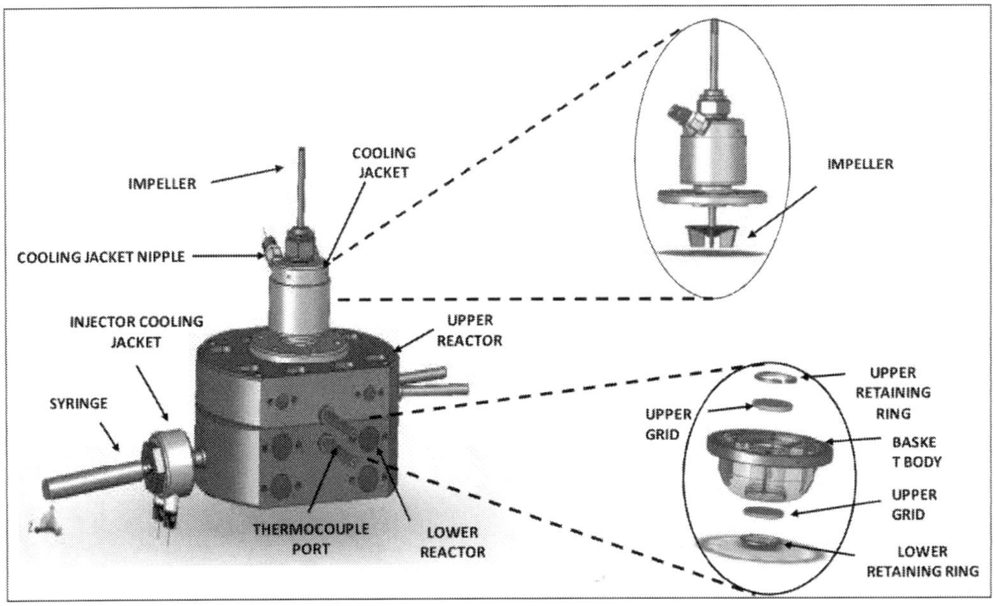

Figure (1): Overview of the CREC Riser Simulator reactor body

RESULTS AND DISCUSSION

TPR and TPO results, Figure (2), show that the prepared γ-alumina supported 10 wt.% VO_x catalyst is stable over repeated reduction and oxidation cycles with reduction temperature around 500 °C. The XRD profile for the prepared catalyst indicates the absence of V_2O_5 bulk surface species and high dispersion of VO_x on the support surface. The VOx phase is primarily present as vanadate or poly vanadate, which is known to be X-ray amorphous.

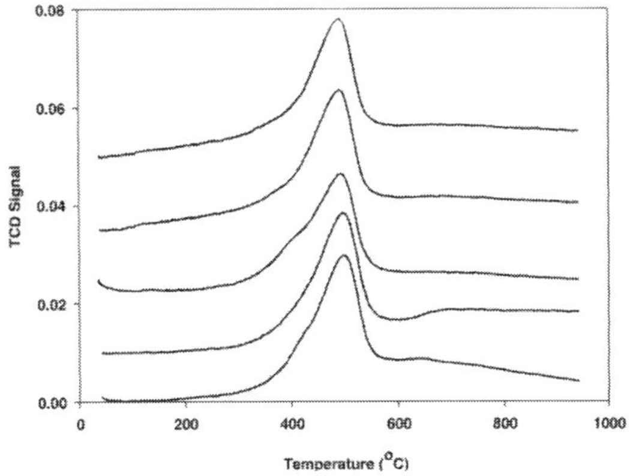

Figure (2): Successive TPR profiles for VOx/γ-Al2O3 catalyst. Pre-treatment: calcination at 773 K.

The effect of the number of TPR/TPO cycles on reducibility of the supported vanadium oxide catalyst was also assessed from the cyclic TPR/TPO experiments. Figure (3) compares the reducibility (% reduction) of the VOx/γ-Al2O3 catalyst sample used in this study under repeated

TPR/TPO cycles. As it can be noticed, approximately 85% reduction of was achieved in each cycle assuming V2O5 as the initial phase of VOx on the catalyst sample. The sequence of TPR/TPO cycles demonstrates consistent percentage reduction of V_2O_5, indicating that the sample is stable over repeated reduction/oxidation. This observation suggests that no phase transformation takes place during the cyclic redox process up to 950 ^{0}C.

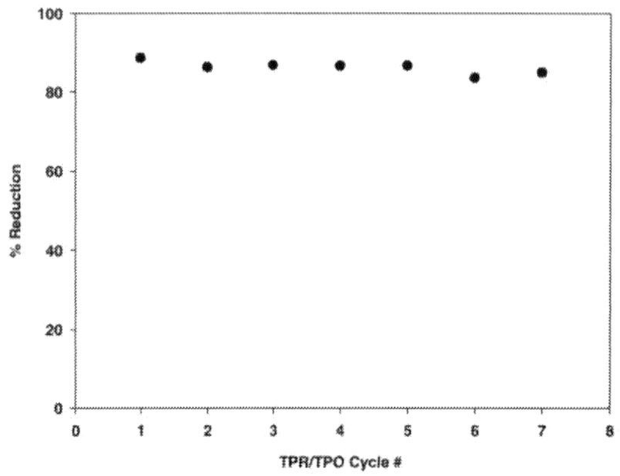

Figure (3): Degree of reduction of Al_2O_3-supported VO_x catalyst sample in multiple TPR/TPO cycles up to 950 °C (heating rate: 10 °C/min; reducing agent: 10 mole% H_2, balance Ar at 50 cm^3.min; oxidizing agent: 5 mole % O_2, balance He at 50 cm^3/min.)

Moreover, XRD resultas, Figure (4), also indicates that no other species is formed due to the interaction between V_2O_5 and Al_2O_3 support. It was shown that the prepared 10 wt. % VO_x/γ-Al_2O_3 catalyst display a low acidity compared to that of the bare alumina support

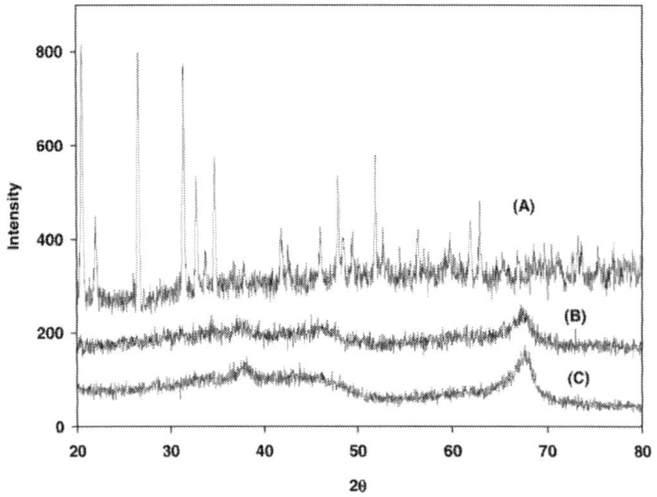

Figure (4): XRD results of (A) Bulk V_2O_5 sample, (B) Fresh 17.8 wt.% V2O5/γ-Al2O3 catalyst sample, (C) Activated Al_2O_3 support sample

Reaction experiments are developed using a CREC Riser Simulator unit over successive reaction-oxidation cycles. Given the results obtained from experiments in oxygen-free environment, it appears that lattice oxygen is involved in the catalytic ODH reaction. It is demonstrated with reactivity tests that the prepared catalyst display promising ethane conversions (18.93% - 25.18%), ethylene selectivity (40.68-55.84%) and catalyst stability during multiple reduction cycles at temperature ranges of 550-600 ^{0}C as shown in Figure (5)

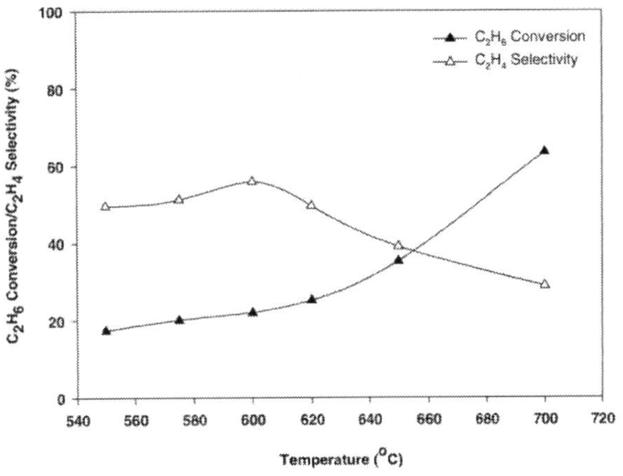

Figure (5): Influence of temperature on ethane conversion and ethylene selectivity (reaction time= 30 sec, C_2H_6 injected= 20 ml, catalyst loaded = 0.76g).

Following the encouraging reactivity results in the CREC Riser Simulator and to be able to assess the catalyst stability, the prepared VOx/γ-A12O3 catalyst was evaluated over multiple ODH cycles/reaction runs. Figure 6-a and 6-b report ethane conversion and different products selectivities as a function of the number of reduction cycles at 600 0C where the catalyst is reduced with ethane feed without catalyst regeneration between runs. It is proven with these results that the VOx/γ-A12O3 catalyst used in this study displays a stable performance in terms of ethane conversion, ethylene selectivity during repeated reduction runs. Following the first injections, the VOx/γ-A12O3 catalyst shows stable ethane conversions (around 21%) with ethylene selectivity (51-58%) at 6000C. Moreover, Figure 6-a and 6-c report ethane conversion and selectivities to various products over repeated reaction-regeneration cycle over the VOx/γ-Al2O3 catalyst at 600 0C. This set of experiments was performed to compare the performance of the catalyst under consecutive reaction-regeneration cycles with those without regeneration steps between reactions run. It is observed that the catalyst shows higher ethane conversion around 27% with cyclic operation, however, selectivities to ethylene are reduced. As it can be seen from Figure 10-c, ethylene selectivity dropped from an initial value of 31% to 2.3% with the repeated reaction-regeneration cycles, while selectivity to CO2 increased from 52.7% to 81 %. Furthermore, selectivity to CO remains rather constant around value of 12% during the reaction-regeneration cycles. This shows that formation of the CO2 can be traced to the oxidation of the ethylene product. Based on these obtained results during oxygen-free ethane ODH reaction/regeneration cycles, it can be concluded that the prepared VOx based catalyst is more active in its reduced form than the fresh oxide /regenerated form. This can be attributed to the excess availability of oxygen when the catalyst is fresh resulting in more oxidation and/or partial oxidation rather than ethane ODH.

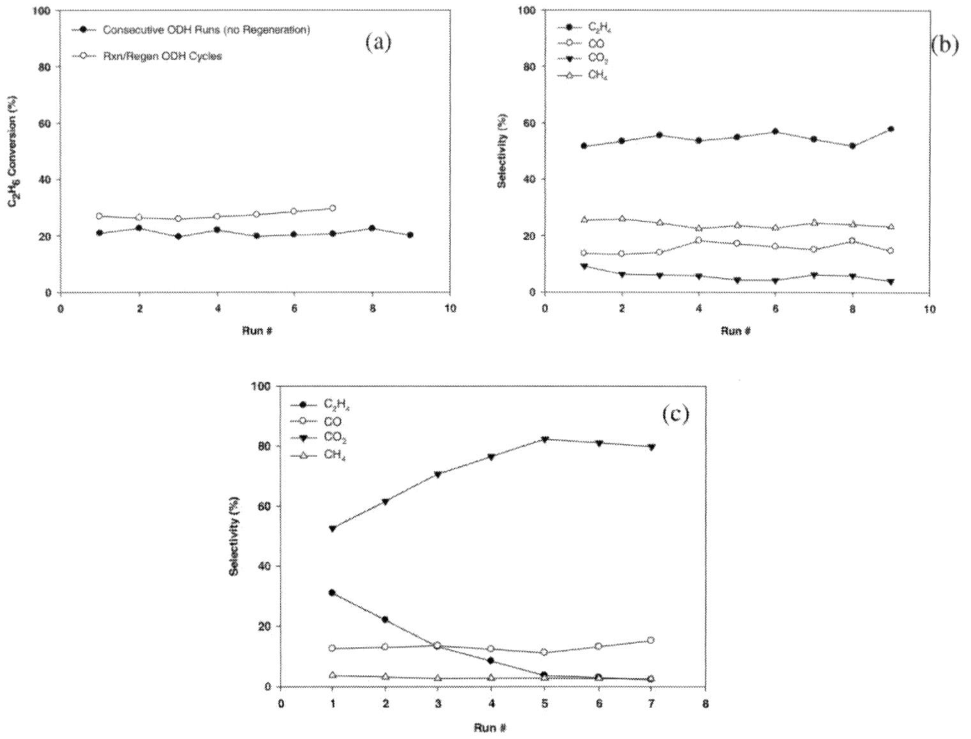

Figure (6): (a) Ethane conversion and (b) ethylene, CO, CO_2 and CH_4 selectivities (consecutive ODH runs), (c) ethylene, CO, CO_2 and CH_4 selectivities (reaction/regeneration ODH Cycles) on $VO_x/\gamma\text{-}Al_2O_3$ catalyst in multiple ethane ODH runs in CREC Riser Simulator. (T= 600 °C, C_2H_6 injected= 20 ml, reaction time=30 s, catalyst loaded = 0.76g).

CONCLUSIONS

The prepared catalyst showed very stable reduction behaviour over repeated reduction and oxidation cycles with reduction temperature around 500°C and 85% reduction of V_2O_5. The XRD profile for the prepared catalyst indicates the absence of V_2O_5 bulk surface. Based on ethane ODH and reaction runs in the CREC Riser Simulator, it was proven that in the absence of gas-phase oxygen, ethane can be converted selectively to ethylene through the participation of lattice oxygen. It was demonstrated with experiments that the prepared catalyst displays promising ethane conversions (18.93% 25%), ethylene selectivity (41-56%) and catalyst stability during multiple reduction cycles at 550-600°C.

REFRENCES

(1) Dharia, D.; Letzsch, W.; Kim, H.; McCue, D.; Chapin, L. *Hydrocarbon Processing*. **2004**, *83*, 61-66.
(2) Ren, T.; Patel, M.; Blok, K. *Energy*. **2006**, *31*, 425-451.
(3) Anon., R. *Oil & Gas Journal*. **2004**, *102*, 50-52.
(4) Bhasin, M. M. *Topics in Catalysis*. **2003**, *23*, 145-149.
(5) Resasco, D. E.; Haller, G. L. *Catalysis: Royal Society of Chemistry*. **1994**, *11*.
(6) Chan, K. Y. G.; Inal, F.; Senkan, S. *Industrial & Engineering Chemistry Research*. **1998**, *37*, 901-907.
(7) Farrauto, R. J.; Bartholomew, C. H. *Fundamentals of Industrial Catalytic Processes*; 1st ed.; Chapman & Hall: London, 1997.
(8) Shen, Z.; Liu, J.; Xu, H.; Yue, Y.; Hua, W.; Shen, W. *Applied Catalysis A: General*. **2009**, *356*, 148-153.
(9) Čapek, L.; Bulánek, R.; Adam, J.; Smoláková, L.; Sheng-Yang, H.; Čičmanec, P. *Catalysis Today*. **2009**, *141*, 282-287.
(10) Shi, X.; Ji, S.; Wang, K. *Catalysis Letters*. **2008**, *125*, 331-339.
(11) Čapek, L.; Adam, J.; Grygar, T.; Bulánek, R.; Vradman, L.; Košová-Kučerová, G.; Čičmanec, P.; Knotek, P. *Applied Catalysis A: General*. **2008**, *342*, 99-106.
(12) Klose, F.; Wolff, T.; Lorenz, H.; Seidelmorgenstern, A.; Suchorski, Y.; Piorkowska, M.; Weiss, H. *Journal of Catalysis*. **2007**, *247*, 176-193.
(13) Danica, B.; Desislava, A.; Mirko, P.; Stefan, H. *Journal of the Serbian Chemical Society*. **2007**, *72*, 183-192.
(14) Haddad, N.; Bordes-Richard, E.; Hilaire, L.; Barama, A. *Catalysis Today*. **2007**, *126*, 256-263.
(15) Karamullaoglu, G.; Dogu, T. *Industrial and Engineering Chemistry Research*. **2007**, *46*, 7079-7086.
(16) Tope, B.; Zhu, Y.; Lercher, J. *Catalysis Today*. **2007**, *123*, 113-121.
(17) Cavani, F.; Ballarini, N.; Cericola, A. *Catalysis Today*. **2007**, *127*, 113-131.
(18) Lemonidoua, A. A.; Heracleous, E. *Catalysis Today*. **2006**, *112*, 23-27.
(19) Solsona, B.; Dejoz, A.; Garcia, T.; Concepcion, P.; Nieto, J.; Vazquez, M.; Navarro, M. *Catalysis Today*. **2006**, *117*, 228-233.
(20) Heracleous, E.; Lemonidou, A. *Journal of Catalysis*. **2006**, *237*, 175-189.
(21) Heracleous, E.; Lemonidou, A. *Journal of Catalysis*. **2006**, *237*, 162-174.
(22) Botella, P.; Dejoz, A.; Lopeznieto, J.; Concepcion, P.; Vazquez, M. *Applied Catalysis A: General*. **2006**, *298*, 16-23.
(23) Nakamura, K.; Miyake, T.; Konishi, T.; Suzuki, T. *Journal of Molecular Catalysis A: Chemical*. **2006**, *260*, 144-151.
(24) López Nieto, J. M.; Botella, P.; García-González, E.; Dejoz, A.; Vazquez, M.; González-Calbet, J. *Journal of Catalysis*. **2004**, *225*, 428-438.
(25) Mamedov, E. A.; Corberfin, V. C. *Applied Catalysis A: General*. **1995**, *127*, 1-40.
(26) Grzybowska, B.; Klisińska, A.; Loridant, S.; Stoch, J.; Gressel, I. *Applied Catalysis A: General*. **2006**, *309*, 17-27.
(27) Grzybowska, B.; Klisińska, A.; Samson, K.; Gressel, I. *Applied Catalysis A: General*. **2006**, *309*, 10-16.
(28) Martinezhuerta, M.; Gao, X.; Tian, H.; Wachs, I.; Fierro, J.; Banares, M. *Catalysis Today*. **2006**, *118*, 279-287.
(29) Grabowski, R.; Sloczynski, J. *Chemical Engineering and Processing*. **2005**, *44*, 1082-1093.
(30) Heracleous, E.; Machli, M.; Lemonidou, A. A.; Vasalos, I. A. *Journal of Molecular Catalysis A: Chemical*. **2005**, *232*, 29-39.
(31) Reddy, E. P.; Varma, R. S. *Journal of Catalysis*. **2004**, *221*, 93-101.
(32) Bozon-Verduraz, F.; Enache, D. I.; Bordes, E.; Ensuque, A. *Applied Catalysis A: General*. **2004**, *278*, 103-110.
(33) Bozon-Verduraz, F.; Enache, D. I.; Bordes-richard, E.; Ensuque, A. *Applied Catalysis A: General*. **2004**, *278*, 93-102.
(34) Navarro, M. T.; Blasco, T.; Panzacchi, B.; Rey, F.; Concepcion, P. *Catalysis Today*. **2004**, *96*, 179-186.
(35) Zhao, Z.; Yamada, Y.; Ueda, A.; Sakurai, H.; Kobayashi, T. *Catalysis Today*. **2004**, *93-95*, 163-171.
(36) Busca, G.; Panizza, M.; Resini, C.; Raccoli, F.; Catani, R.; Rossini, S. *Chemical Engineering Journal*. **2003**, *93*, 181-189.
(37) Bell, A. T.; Iglesia, E.; Argyle, M. D.; Chen, K. *Journal of Physical Chemistry B*. **2002**, *106*, 5421-5427.
(38) Bell, A. T.; Dinse, A.; Schomäcker, R. *Physical chemistry chemical physics : PCCP*. **2009**, *11*, 6119-24.
(39) Cavani, F. *Catalysis Today*. **1995**, *24*, 307-313.

(40) Bell, A. T.; Khodakov, A.; Olthof, B.; Iglesia, E. *Journal of Catalysis*. **1999**, *181*, 205-216.
(41) Banares, M. A. *Catalysis Today*. **1999**, *51*, 319-348.
(42) Banares, M. A.; Martinez-Huerta, M. V.; Gao, X.; Wachs, I. E.; Fierro, J. L. G. *Catalysis Today*. **2000**, *61*, 295-301.
(43) Wachs, I. E.; Weckhuysen, B. M. *Applied Catalysis*. **1997**, *157*, 67-90.
(44) Kung, H. *Advances in Catalysis*. **1994**, *40*, 1-38.
(45) Wan, B.-zu; Kao, C.-yu; Huang, K.-tse *Industrial and Engineering Chemistry Research*. **1994**, *33*, 2066-2072.
(46) Wu, Z.; Kim, H.-sung; Stair, P. C.; Rugmini, S.; Jackson, S. D. *Journal of Physical Chemistry B*. **2005**, 2793-2800.
(47) Sham, E.; Murgia, V.; Gottifredi, J. C.; Torres, E. M. F. *Latin American Applied Research*. **2004**, *34*, 75-82.
(48) Bell, A. T.; Iglesia, E.; Argyle, M.; Chen, K. *Journal of Catalysis*. **2002**, *208*, 139-149.
(49) Arena, F.; Frusteri, F.; Parmaliana, A. *Applied Catalysis A: General*. **1999**, *176*, 189.
(50) Nieto, J. M. L.; Blasco, T. *Applied Catalysis A: General*. **1997**, *157*, 117-142.
(51) Auroux, A.; Bars, J. L.; Forissier, M.; Vedrine, J. C. *Journal of Catalysis*. **1996**, *162*, 250–259.

(510c) Hydrolysis of Poly(1-4-β-glucan) Strands Derived From Cellulose Using Mild Acidity and Temperature

Oz Gazit and Alexander Katz, Chemical and Biomolecular Engineering, University of California Berkeley, Berkeley, CA

Finding a selective synthetic catalyst for the hydrolysis of O-glycosidic bonds in cellulose is crucial for the transformation of biomass to fuels and chemicals. Whereas concentrated mineral acids used under harsh conditions (i.e. 230 °C; 500 psi) can rapidly accomplish this transformation with moderate selectivity, enzymes use a different mechanism to accomplish this transformation under mild conditions of pH and temperature while maintaining higher selectivity, albeit at a much slower rate. In this respect the bioinspired synthetic systems are exceptional because they demonstrate glycosidic bond hydrolysis under mild conditions of pH and temperature in aqueous solution. However, a major practical limitation is that they require the precise positioning of a carboxylic acid functional group adjacent to the glycosidic oxygen: a molecule with carboxylic acid functionality in the *ortho* position has a 13,000-fold faster hydrolysis rate relative to a similar molecule where the carboxylic acid functionality is in the *para* position. To overcome such constraints, we have constructed a system that is capable of circumventing the rigid requirement of intramolecular acid catalyst positioning by essentially providing an array of acid groups in the vicinity of the O-glycosidic bond using a silica surface. Our approach relies on the synthesis and characterization of a new class of materials comprising grafted, isolated poly(1-4-β-glucan) strands within an environment consisting of an ensemble of mildly acidic surface silanols in silica. We demonstrate that by careful control over the grafting density within a chain during synthesis, it is possible to control microenvironment as well as combustion temperature as measured using thermogravimetric analysis. We show that the close proximity of poly(1-4-β-glucan) to the weakly acidic silica surface is able to promote hydrolysis where H_3O^+ activity is considered background.

Finally the extent of hydrolysis of the grafted poly(1-4-β-glucan) strands is demonstrated under mild condition of pH and temperature. This hydrolysis is shown to be highly dependent and controlled by the extent of surface interactions with adjacent silanols.

See more of this Session: Catalytic Processing of Fossil and Biorenewable Feedstocks V
See more of this Group/Topical: Catalysis and Reaction Engineering Division

(510d) Catalytic Hydroprocessing of Fast Pyrolysis Bio-Oil Model Compounds

Haijun Wan, Chemical & Petroleum Engineering, Center for Environmentally Beneficial Catalysis, University of Kansas, Lawrence, KS, R. V. Chaudhari, Center for Environmentally Beneficial Catalysis, Chemical & Petroleum Engineering Department, University of Kansas, Lawrence, KS and Bala Subramaniam, Chemical and Petroleum Engineering, Center for Environmentally Beneficial Catalysis, University of Kansas, Lawrence, KS

Catalytic hydroprocessing of fast pyrolysis bio-oil to produce transportation fuel has attracted much attention in recent years due to the diminishing fossil fuel resources. Given that bio-oil is a complex mixture containing over 300 compounds, hydroprocessing of real bio-oil could involve hundreds of reactions. Model compound studies can reduce the complexity and allow a better understanding of the reaction mechanisms. In this talk, we will present the results of systematic evaluation of various noble metal catalysts with respect to hydroprocessing of model bio-oil substrates such as *p*-cresol, acetic acid and furfural, as well as the binary and ternary combinations of these compounds. As an example, for hydroprocessing of acetic acid as a model carboxylic acid, Ru/C catalyst shows the highest activity among the tested catalysts (including Pt, Pd, Ru on either C or Al_2O_3 support). Further, although hydrogenation of acetic acid to ethanol is favored at lower temperatures and higher H_2 pressures, the selectivity to CH_4 remains high in all reactions with water as a solvent, attributed to rapid hydrogenolysis of ethanol to CH_4. However, with *n*-heptane as a solvent, ethanol hydrogenolysis appears to give way to esterification leading to ≥60% selectivity to ethyl acetate. In addition, the results with binary combination of acetic acid and *p*-cresol on Ru/C catalyst indicated that the hydrogenation of acetic acid is suppressed by the presence of *p*-cresol. In sharp contrast, the presence of acetic acid promotes the HDO of *p*-cresol by dehydration reaction, leading to high selectivity to methylcyclohexane. Such a comprehensive understanding is essential to rationally develop novel multi-functional catalysts and processing strategies for effective deoxygenation of bio-oils.

See more of this Session: Catalytic Processing of Fossil and Biorenewable Feedstocks V
See more of this Group/Topical: Catalysis and Reaction Engineering Division

(510e) Production of Biodiesel by Direct Transesterification of Activated Sludge Using Supercritical Methanol

Adebola T. Coker[1], Rafael Hernandez[1], William T. French[1], Emmanuel Revellame[1], William Holmes[1], Guochang Zhang[1], Alexei V. Iretskii[2] and Mark G. White[1], (1)Dave C. Swalm School of Chemical Engineering, Mississippi State University, Mississippi State, MS, (2)Chemistry, Lake Superior State University, Sault Sainte Marie, MI

Fatty acid methyl esters (biodiesel) can be produced by using activated sludge generated from wastewater treatment facilities. However, drying of sludge prior to oil extraction is a major operating cost of this process. To address this challenge, the authors have conducted a study to determine the yield of biodiesel produced from reacting wet sludge using only supercritical methanol.

Activated sludge was obtained from a wastewater treatment plant in Tuscaloosa, AL and filtered to a composition of 90% water. This was reacted with methanol using varying methanol:solid ratios at 275, 300, and 325 °C for 1 - 9 hours. Two reactor set-ups were used in this study: 1) a 450ml batch Parr ® reactor, and 2) a 46ml batch reactor. Analysis of the product by Gas Chromatography-Mass Spectrometry (GCMS) showed a number of fatty acid methyl esters (biodiesel) like methyl oleate and methyl palmitate. Since sludge contains approximately 5% lipids which varies with time (season) and treatment conditions, a model system of oleagineous yeast - Rhodotorula glutinis was also used to evaluate the production of biodiesel in a system similar to sludge.

The average sludge biodiesel yield obtained on the Parr reactor was 7.1 ± 3.9% based on initial solids while the average sludge biodiesel yield obtained on the 46ml batch reactor thus far was 2.2 ± 0.4%. The average biodiesel yield obtained from Rhodotorula glutinis was 20.7 ± 5.8%. This work shows the results of optimization on temperature, reaction time and methanol volume for the production of biodiesel from sludge and Rhodotorula glutinis with high water content and will also show the proposed mechanism for this reaction.

See more of this Session: Catalytic Processing of Fossil and Biorenewable Feedstocks V
See more of this Group/Topical: Catalysis and Reaction Engineering Division

(510f) Catalytic Production of 1,2-MPG From Glycerin

Sydney D. Sovine, Stephen Dufreche, Rakesh Bajpai and Mark Zappi, Chemical Engineering, University of Louisiana at Lafayette, Lafayette, LA

As biodiesel becomes increasingly important as an alternative fuel its main by-product, crude glycerol, is experiencing a decline in market value. One method of utilizing glycerol is converting this product to propylene glycol using a hydrogenolysis reaction. Glycerol is first dehydrated to form acetol, then with the help of hydrogen gas the acetol is hydrogenised to form 1,2-propylene glycol (MPG). A loosely packed column filled with a copper chromite catalyst with promoters such as barium or manganese is used to help in the conversion of glycerol. Current commercial concerns include selectivity of MPG over other byproducts and catalyst lifetime. Selectivity was modified through addition of organic and inorganic buffers to stabilize conversion. The reaction has been tested with technical grade glycerin to isolate effects of additives. Data presented includes kinetics, selectivity toward 1,2-MPG, and stability of the catalyst.

See more of this Session: Catalytic Processing of Fossil and Biorenewable Feedstocks V
See more of this Group/Topical: Catalysis and Reaction Engineering Division

(510g) Membrane Reactor Technology for Aqueous-Phase Hydrogenation of Glutamic Acid

Mandeep Kular[1], Neha Dhiman[1], Peter Pfromm[2] and Mary Rezac[1], (1)Chemical Engineering, Kansas State University, Manhattan, KS, (2)Department of Chemical Engineering, Kansas State University, Manhattan, KS

Realization of a biorefinery, analogous to a petrochemical refinery depends on the efficient conversion of biomass feedstock into a wide variety of chemical intermediates. One of the biggest hurdle is the high oxygen content of biomass and biomass-derived intermediates from conventional hydrocarbons. Moreover, biomass intermediates are naturally water soluble, which results in high environmental compatibility but it also dramatically impedes the performance of a system with respect to chemical transformations. This difference results in the variation of chemical composition and therefore the variation in important conversion techniques like hydrogenation. Hydrogenation is an important step in transforming chemical intermediates to products. Hydrogenation of biomass derived substrates in aqueous phase is a technical challenge due to the slow rate of delivery of hydrogen through the water phase and to the catalyst surface. Also, biomass-derived substrates have the propensity to undergo undesirable side reactions.

We are investigating hydrogenation of Glutamic acid as a model reaction. In order to improve the availability of hydrogen on the catalyst surface, catalytically active membranes are being looked upon. The Ru metal decorated polymeric membrane allows for high selectivity towards hydrogen and negligible permeability for Glutamic acid.

Results from the initial experiments demonstrate the effect of processing conditions and performance compared to the conventional (high pressure, stirred tank) systems. Using the membrane reactor technology described here, hydrogenation of Glutamic Acid was successfully performed with very high conversion at pressure of less than 40 psig compared to the 2000 psig used in the conventional system. Advantages of this proposed technology include a reduction in the capital cost resulting from mild operating conditions; an increase if ease of use realized by the elimination of high pressure hydrogen, which presents a significant safety hazard; and the potential to influence the product spectrum through process conditions.

See more of this Session: Catalytic Processing of Fossil and Biorenewable Feedstocks V
See more of this Group/Topical: Catalysis and Reaction Engineering Division

Considerations for Research, Development and Commercialization/Scale-up of Biomass Pretreatment Processes

Dale Monceaux[1], *Richard C. Agar*[1] *and Charles E. Wyman*[2], *(1)AdvanceBio, LLC, Milford, OH, (2)College of Engineering-Center for Environmental Research and Technology, University of California, Riverside, Riverside, CA*

Global efforts to cost effectively convert biomass-based polysaccharides to sugars via enzymatic hydrolysis and subsequently to fuels and chemicals have been ongoing since the early discoveries by the US Army Natick Research & Development Command. These, and other early developments related to enzymatic hydrolysis of cellulose were followed by advancements in microbial genetic engineering. Genetically modified microorganisms capable of utilizing xylose-rich hemicellulose fractions of biomass feedstocks in addition to the cellulose derived glucose, promised improved ethanol yields and process economics. The past fifteen years of research has led to other major advancements in cellulase protein chemistry as well as an improved understanding of the microstructure of lignocellulose, moving the world ever closer to commercially viable biomass-based fuels and chemicals.

Early research ultimately revealed the sensitivity of the feedstock, pretreatment, enzyme and organism regime. This, in turn has exposed the need for consistency in methodology as research proceeds from concept to commercial-scale operations.

Much current activities focusing on the commercialization of biomass based fuels and chemicals hinge on the development of cost-effective biomass pretreatment processes. This “first step” in the process of converting biomass to end-products is critical to successful commercialization efforts for a number of reasons. First, it provides reactive cellulose that can be efficiently enzymatically hydrolyzed to oligomeric or monomeric glucose. Second, it efficiently solubilizes hemicellulose fractions of the biomass, producing oligomeric and or monomeric xylose, mannose, galactose, arabinose, etc. Third, it produces minimal reaction by-products that inhibit enzymatic hydrolysis and fermentation processes. Last, it is mechanically and economically scalable.

Many pretreatment technologies employ thermo-chemical reactions that reduce the recalcitrance of ligno-cellulosic matrices, solubilize fractions of the biomass and expose reactive sites to subsequent reactions. Various pretreatment technologies are under development, each with a range of design basis conditions. **Table 1** illustrates general conditions for six common pretreatment chemistries.

Pretreatment	Temp (°C)	Residence Time (min)	Water Loading*	Catalyst	Catalyst Loading**	Catalyst Recyclable
AFEX	150	30	200	Anhydrous NH_3	152	Yes (95-95% of loaded)
Dilute Acid	140	40	891	1% H_2SO_4	9	No
Lime	120	240	1500	$CaOH/O_2$	100/100 psi	Maybe/Ne
Liquid Hot Water	200	10	663	Water	N/A	N/A
SAA	160	60	765	15% Aqueous NH_3	135	Yes, ecess
SO_2	180	10	900	5% SO_2/Steam	5/?	Yes, excess

* Water Loading = L/100 kg of dry biomass input to pretreatment

** Catalyist Loading = kg/100 kg of dry biomass input to pretreatment

Table 1

Typical Thermochemical Pretreatment Chemistries

Reaction kinetics varies significantly with feedstock type, mechanical pretreatment, catalyst type, catalyst loading, water loading, residence time and temperature. Tight control of key parameters is required to maximize conversion efficiency. Elevated catalyst concentration and residence times increase reaction rate but also xylose destruction as illustrated in **Table 2**. Consequently, optimum reaction conditions require mechanical designs that deliver controlled residence times under near-plug flow conditions.

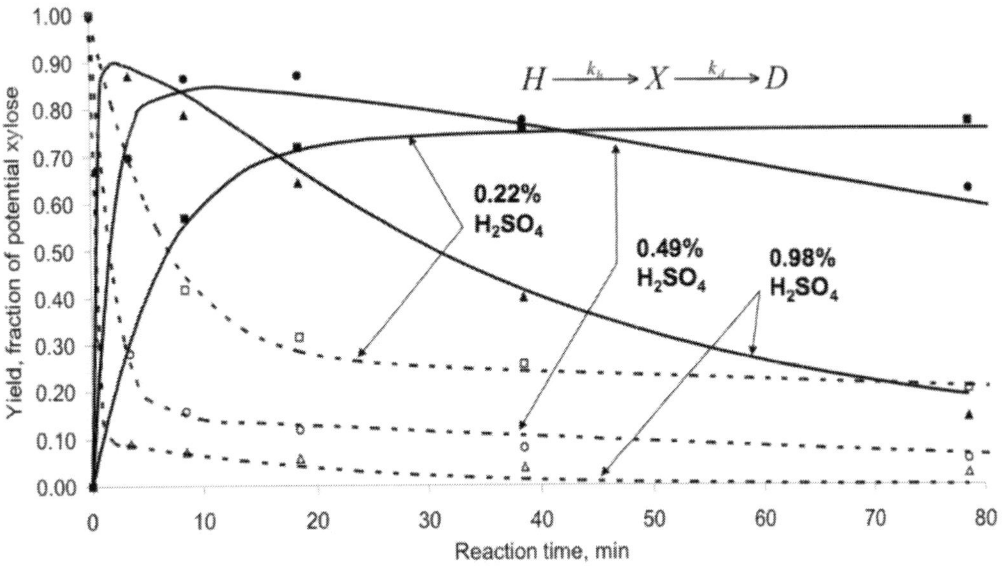

Table 2

Xylose Yield in Pretreatment vs Time at 160°C

Resource limitations, including feedstock supply, workspace, staffing and funding constrain the research and development efforts at each stage. Real-world feedstock properties such as foreign matter content, chemical composition variability, recalcitrance, bulk density, etc. introduce constraints that further complicate research and commercialization efforts. Feedstock selection dramatically impacts capital investment and production cost. Feedstock recalcitrance in conjunction with chemical composition effectively determines yield, and therefore the quantity of feedstock required per unit of fuel or chemical produced. Feedstock chemical composition and bulk density, in conjunction with reactor residence time establishes the size of the major components of the capital and energy intensive pretreatment reactor unit operation, as illustrated in **Table 3**.

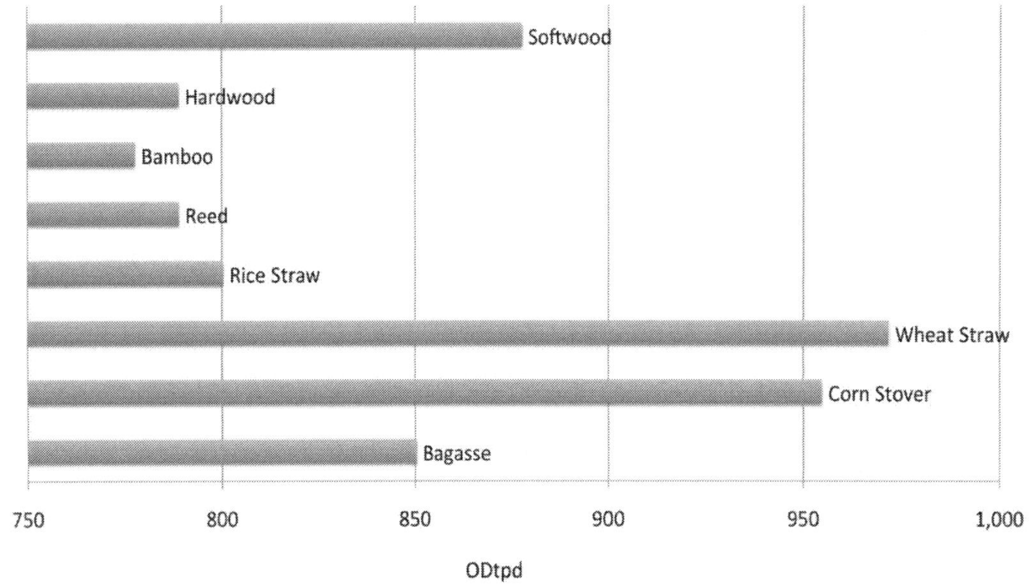

Table 3

Estimated Feedstock Tonnage per 100 Megaliters of Ethanol

High yields are critical to distribute both operating and capital costs. Low operating costs, including chemicals, energy and labor, are essential to providing adequate margins for return on capital. General operating cost categories must be lower for a new process than for established technologies making conventional products since low capital costs are essential to minimize financial exposure. The driving forces to contain capital investment include general technologies and processes that minimize equipment sizes, operate at lower pressures, operate at neutral or higher pHs as well as thermal processes that operate at lower temperatures and biochemical processes that operate at elevated temperatures.

Technical compromises frequently lead to approaches that ultimately expand the research and development program effort, schedule and budget. The impact of decisions made early in the technology development phase of a project may not fully materialize until late in the project after significant design and engineering expenditures have been made. The time period between research decision-making and engineering-based capital cost estimating delays the commercialization or leads to technology dead-ends. Visualization of the expected commercial scale process should occur very early in the technology research and development effort. A process "scale-down" approach to optimize research and technology development, scale-up and commercialization efforts has proven to be and effective strategy to accelerate project development and finance.

(522b) Improving Corn Stover Reactivity In Low Severity Pretreatment Through Deacetylation and Disc-Refining

Xiaowen Chen[1], Joseph Shekiro[1], mary Ann Franden[1], Min Zhang[1], Bonwook Koo[2], Sunkyu Park[2] and Melvin Tucker[1], (1)National Bioenergy Center, National Renewable Energy Laboratory, Golden, CO, (2) Department of Forest Biomaterials, North Carolina state university, Raleigh, NC

The current "state-of-art" dilute acid pretreatment to produce bio-ethanol from corn stover requires severe conditions to overcome biomass recalcitrance, which, in turn, offers 70~80% xylose monomer yield and 5~7% furfural yield. However, high acid loadings and high reaction temperatures harm the economic feasibility of the process. In addition, the amount of acetate and furfural produced during pretreatment is in a range that strongly inhibits cell growth and impedes ethanol fermentation. To decrease the impact of these issues, deacetylation, pretreatment with lower acid concentrations, lower temperatures, and integration of disc-refining was carried out in this work. The results show that deacetylation with dilute NaOH (0.1M) combined with low severitydilute-acid pretreatment increased corn stover reactivity by increasing the xylose monomer yield by 20% in pretreatment while degradation of xylan to furfural remained under 2%. Glucose yield increased by 10% and xylose yield by 20% during dilute solids enzymatic hydrolysis as a result of deacetylation of the feedstock prior to pretreatment. Particle size reduction by mechanical refining after pretreatment is shown to assist achieving> 90% cellulose yield in high solids (20 wt% total solids) enzymatic hydrolysis. Bioscreen C growth assays and mini fermentation at high solids (20wt%) showed less toxicity of deacetylated pretreated corn stover and enhanced xylose utilization by 20%.

See more of this Session: Developments In Biobased Alternative Fuels I
See more of this Group/Topical: Sustainable Engineering Forum

Reducing Process Complexity for Cellulosic Ethanol

Lonnie O. Ingram, Department of Microbiology and Cell Science
University of Florida, Box 110700, Gainesville, FL 32611 Ingram@UFL.edu

Lignocellulosic biomass represents a potential source of carbohydrates (cellulose and hemicellulose) for microbial fermentation into fuels and chemicals. Steam pretreatment with dilute mineral acids serves as an efficient approach to depolymerize hemicellulose and enhance cellulose digestion with enzymes. However, side products (furfural, 5-hydroxymethyl furfural, acetate, phenolics,and others) formed during pretreatment inhibit microbial growth and retard fermentation. Potential engineering solutions to this problem (separation of hemicellulose hydrolysate, counter-current-washing of fibers, and mitigation of toxins) add complexity and increase cost. The level of inhibitors produced during acid pretreatment is affected by the severity of conditions and the choice of acid. Weaker acids such as phosphoric acid produce lower levels of toxic side products and reduce the need for exotic metal alloys. Since mineral acids are not consumed by the pretreatment process, the higher cost of phosphoric acid as compared to sulfuric acid could be offset in part by recovery and reuse as a dilute fertilizer when crops are nearby. Recent progress has been made in the development of more robust *E. coli*-based biocatalyst with increased resistance to dilute acid hydrolysates. Using this biocatalyst and phosphoric acid pretreatment, we have been able to co-ferment sugarcane bagasse hemicellulose hydrolysate and enzyme-hydrolyzed cellulose in a single vessel using a variation of the simultaneous saccharification and co-fermentation process (SScF; Fig. 1). Hexose and pentose sugars from phosphoric acid pretreated sugarcane bagasse were fermented in a single vessel, eliminating process steps for solid–liquid separation and sugar cleanup. After pretreatment, a liquefaction step (L) with cellulase was included to improve mixing and saccharification (L + SScF), analogous to a corn ethanol process. Bagasse slurries containing 10% and 14% dry weight (fiber plus solubles) were tested using pretreatment temperatures of 160–190 C (1% phosphoric acid, 10 min). Enzymatic saccharification and inhibitor production both increased with pretreatment temperature. The highest titer (30 g/L ethanol) was obtained after incubation for 122 h using 14% dry weight slurries of pretreated bagasse (180 C). The highest yield (0.27 g ethanol/g dry bagasse) was 82 gal ethanol per dry ton of bagasse using this L+SScF process and 10% dry weight slurries.

Comparison of conversion processes

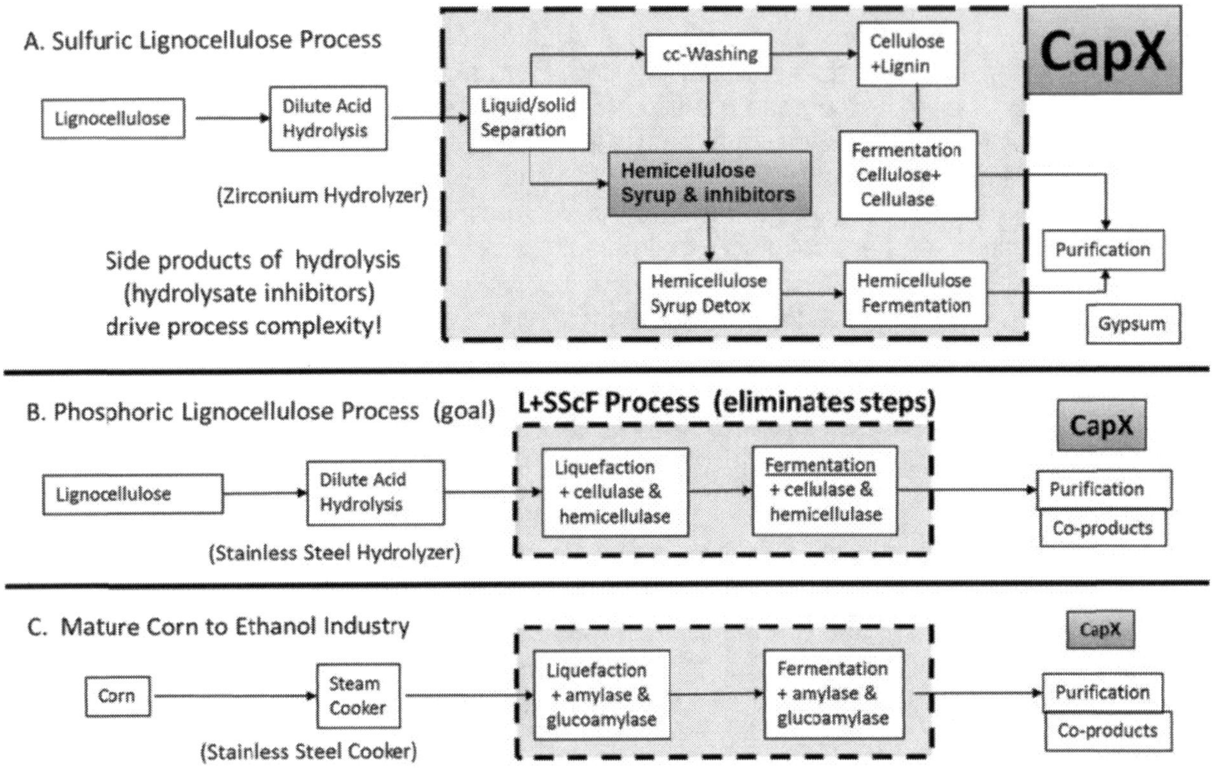

(522d) Co-Utilization of Methane In Steam-Biomass Gasification Using Concentrated Solar Energy

Aaron W. Palumbo, Erica L. Jorgensen and Alan W. Weimer, Chemical & Biological Engineering, University of Colorado at Boulder, Boulder, CO

Presented is a novel gasification application where biomass is steam-gasified and natural gas is steam-reformed simultaneously in a solar reactor. Solar-driven gasification is a proven technology, and can augment the biofuel industry as a domestic fuel source. Gasification performed in an indirectly heated gasifier can achieve temperatures greater than 1150 °C [1], mitigating or eliminating the production of CO_2, tar, char, and light hydrocarbon gases in the product gas. This reduces the need for gas clean-up technology and will not require a downstream water-gas shift reactor to adjust H_2:CO. Thermodynamics favor the formation of only hydrogen and carbon monoxide in the product synthesis gas. Therefore, a predictable H_2:CO can be controlled and tuned for downstream fuel conversion. Since the majority of biomass carbon is converted to CO, more than twice as much biomass is converted to a useable liquid fuel than in a traditional gasification process. Preliminary experiments show that reactor temperatures must exceed 1350 °C in order to reduce the effects of the water-gas shift reaction and reform practically all light hydrocarbon gases, including CH_4. This operating temperature is consistent with realizable concentrated solar energy systems.

1. Piatkowski, N., et al., *Solar-driven gasification of carbonaceous feedstock-a review.* Energy & Environmental Science, 2011. **4**(1): p. 73-82.

See more of this Session: Developments In Biobased Alternative Fuels I
See more of this Group/Topical: Sustainable Engineering Forum

(522e) Poly(oxymethylene) Dimethyl Ethers As Components of Tailored Diesel Fuel - Properties, Synthesis and Purification Concepts

Jakob Burger, Laboratory of Engineering Thermodynamics, Technische Universitt Kaiserslautern, Kaiserslautern, Germany

Poly(oxymethylene) Dimethyl Ethers as components of tailored diesel fuel - properties, synthesis and purification concepts

Jakob Burger [1*], Markus Siegert [2], Eckhard Ströfer [2], Michael Nilles [2], Hans Hasse [1]

[1] Laboratory of Engineering Thermodynamics, University of Kaiserslautern, Germany;

[2] BASF SE Company, Ludwigshafen, Germany
*Corresponding author: Jakob Burger, Laboratory of Engineering Thermodynamics, University of Kaiserslautern, Germany. Email: jakob.burger@mv.uni-kl.de

Keywords: Fuel additive, Soot reduction, Process development

Poly(oxymethylene) Dimethyl Ethers (POMDMEs), oligomers of the general structure CH_3-O-(CH_2-O)$_n$-CH_3, open a new route for tailoring diesel fuels. POMDMEs belong to the group of oxygenates which reduce soot formation in the combustion when added to diesel fuels. They can be produced on a large scale based on Methanol which can be produced from renewable resources. This contribution presents the process chain, possible synthesis routes and some fuel related properties of POMDMEs. Comparisons to other oxygenates, e.g. Dimethyl Ether, are given.

A particularly favourable route for the large scale production of POMDMEs in which they are formed from Methylal and Trioxane is proposed. The chemistry behind this route is given in detail. Measurements of the reaction kinetics and the reaction equilibrium were carried out in a batch reactor using acidic ion exchange resin as heterogeneous catalyst. The parameters of a pseudo-homogeneous model were fitted to the results of these experiments. The formation of side-products was studied experimentally and concepts to minimize them are discussed. Based on the reaction model a process concept was developed. It consists of a reactor and a downstream distillation sequence to yield pure POMDMEs of the most favourable chain length n=3-5.

The use of distillation technology allows an easy scale-up to huge single train plants which can potentially fulfil the need for millions of tonnes of the innovative oxygen carrier POMDME needed for the supply of clean diesel fuel in the future.

See more of this Session: Developments In Biobased Alternative Fuels I
See more of this Group/Topical: Sustainable Engineering Forum

(522f) Synthesis of Green Diesel Fuel From Fatty Acid Feedstocks Via Electrochemical Hofer-Moest Decarboxylation

Cody Wagner, Charles Coronella and Rainer Busch, Chemical Engineering, University of Nevada, Reno, Reno, NV

Production of biodiesel by esterification or transesterification of long-chain fatty acids is a well known process, but the fuel product has characteristics making it not fully compatible with petroleum diesel engines. A better option is to convert fatty acid molecules to hydrocarbons by removing their carboxyl group. Decarboxylation of fatty acid feedstocks was performed by electrolysis using graphite electrodes, which selectively causes a Hofer-Moest (or non-Kolbe)-type reaction. Methanol was used as a solvent for a robust selection of fatty acid feedstocks.

The liquid product created by the electrochemical reaction was composed of mainly alkenes with at least one double bond, ethers, and methyl esters. This fuel product has advantageous cold-flow properties, no sulfur content, heating values close to those of diesel fuels, and is capable of running in current diesel engines as a drop-in replacement for petroleum diesel, unlike biodiesel. The temperature, pH, concentrations of ions, and type of base used were all varied to optimized the current density, which limits the production rate. A separation method was tested for removal of the fuel product from the electrolyte. An efficiency was then calculated for several varieties of fatty acid feedstocks, comparing the electrical energy input to the product generated from the electrolysis. Overall first-law thermodynamics and economics were also evaluated, examining production rates and the large-scale viability of the process.

See more of this Session: Developments In Biobased Alternative Fuels I
See more of this Group/Topical: Sustainable Engineering Forum

Quantitative Comparison of Chemical and Biological Cardboard's Lignin Degradation

E. Isaza
Department of Chemical Engineering, Universidad de Los Andes, Bogotá, Colombia
R. Sierra R.
Advisor, Universidad de Los Andes, Bogotá, Colombia

ABSTRACT: An efficient way to increase biomass digestibility, thereby, make it usable in the production of second generation biofuels, is pretreatment. However, it demands important energy inputs which make it one of the most expensive stages in biofuels processing. Comparisons between chemical (oxidative with lime) and biological (with commercial *Laccase* extract) pretreatments were performed both by experimental measurements and by simulations. All experiments were performed following a full factorial experimental design, with lignin and sugar yields as response variables. The influence of the type of pretreatment (i.e., pretreatment yields) on the process that leads to biogasoline and jet fuel production (MixAlco®) was simulated in AspenPlus®. This simulation showed higher productivity and energy efficiency (15 and 60% respectively) for the chemical pretreatment over the biological pretreatment.

1 INTRODUCTION

First generation biofuels have emerged as an alternative to fossil fuels, whose price and environmental impacts have been subject of intense debate over the last years. From all biofuels, biodiesel and bioethanol are the most popular. They are currently produced using high energy crops such as soy, sugar cane, oil palm and corn starch as feedstocks (U.S. State Department, 2010). These resources, however, are characterized by such high commercial and alimentary value that they become restrictive for biofuels and generate issues of alimentary sustainability (Lee, Biological conversion of lignocellulisic biomasss to ethanol, 1997).

The utilization of more inexpensive feedstock in fermentative process such as lignocellulosic residues is a good baseline for the optimization of this technology (Lee, Biological conversion of lignocellulisic biomasss to ethanol, 1997). Good feedstock resources are agricultural, industrial, and municipal wastes among others. The *MixAlco*® process, created by Dr. Mark Holtzapple in Texas A&M University and patented by Terrabon L.L.C, uses this kind of waste to produce ethanol and higher energy alcohols.

In this process, four main stages may be differentiated: pretreatment, fermentation, chemical dewatering, gasification and cogeneration, thermal conversion involving hydrogen, and final conversion into biofuels. Among these, pretreatment is the most expensive (Gousamme, 2011).

In biological and some chemical pretreatments (such as lime pretreatment), biomass compositional changes are sought. These changes, include but are not limited to delignification. As a result, an increased biomass digestibility is observed. Other effects of pretreatment are: disruption of the crystalline structure of cellulose, increased pore size to enhance enzyme access to carbohydrates, and some degree of separation of the main lignocellulose constituent polymers.

Lignin is a highly heterogeneous tridimensional aromatic polymer, formed by three different units of phenylpropylene: coumaric alcohol, coniferyl alcohol and sinaphyl alcohol; bounded randomly through C-C and C-O-C type bounds (Sylvia, Fuhrman, Hartel, & Zuberer, 1998), (Lee, Biological conversion of lignocellulisic biomasss to ethanol, 1997).

Widely extended delignification processes such as Kraft pulping, (Lachental, Mortha, Sevillano, & Zaroubine, 2004) are controversial due to the use of environmental harmful chemicals. Delignification in alkali with possible oxygen addition is preferred because comparable delignification levels are obtained with less toxic waste products. Pretreatments such as oxidative and non-oxidative lime (i.e., calcium hydroxide ($Ca(OH)_2$), have been reported to promote lignin degradations close to 70% of the compound (Chang, Nagwani M, Kim, & Holtapple, 1999).

On the other hand wild occurring delignification processes are promoted by extracellular enzymes released by fungi, bacteria (Zhang, Kjonas, & Flurkey, 1999) or even plant cells; however, white rot fungi are the degraders by excellence (Lee, Biological conversion of lignocellulisic biomasss to ethanol, 1997). They generate extracellular enzymes such as *Laccase*, Manganese peroxidase and lignin peroxidase. Laccase (EC. 1.10, 32), benzendioloxigenoxidoreductase, is a copper containing enzyme capable of lignin degradation in the presence of oxygen. The only product of this transformation is water (Rodriguez Cuoto & Toca-Herrera, 2006). Because of this reason, this enzyme has been widely studied and used in industrial applications.

Since there are many advantages and drawbacks for both chemical and biological delignifications, it is desirable to formally compare these processes. However, a systematic comparison on the same basis (same feedstock, analytical procedures and general methods) was not found in the literature. This study aimed to make such comparison. To accomplish this goal, a full factorial design was implemented to both: oxidative lime delignification of cardboard (high in lignin content) and delignification of the same material using purified lacasse (provided by Sigma Aldrich)

The experimental data obtained from the best global conversion of carbohydrates was loaded to an existing preliminary simulation of the *MixAlco®* process (Gousamme, 2011) to determine which of the delignification processes is best on the basis of highest productivity and energy efficiency.

2 METHODOLOGY

2.1 *Chemicals and reagents*

2.1.1 Feedstock for pretreatments and substrate for enzymatic hydrolysis

Recycled cardboard is commonly used in Colombia to transport eggs. This biomass is characterized by high lignin content; a desirable feature for a delignification study. Untreated cardboard was the feedstock for pretreatment. The samples were obtained from *Kikes eggs*, Leader Company in the egg market at Colombia. Pretreated cardboard was used as substrate for enzymatic hydrolysis.

2.1.2 *Reagents for enzymatic delignification*

The selected enzyme was *Laccase Suberase* brand *Novartis*. It was activated in a phosphate buffer at pH 3, 4.5, 5, 5.5, 6, 6.5 and 7.

2.1.3 *Chemical pretreatment*

Hydrogen peroxide was obtained from *Panreac* at (%w/w) 0.5, 1, and 2. Calcium hydroxide $Ca(OH)_2$ also was obtained from Panreac.

2.1.4 *Enzymatic saccharification*

It was performed using cellulase 15 FPU obtained from *Sigma Aldrich.*

2.2 *Characterization of the selected biomass*

Quantification of biomass components (i.e., lignin, structural carbohydrates, and ash), was obtained following the laboratory analytical procedures (LAP) produced by the *National Renewable Energy Laboratory (NREL)* for lignocellulosic biomass (Sluiter A. , Hames, Ruiz, Scarlata, Sluiter, & Templeton, 2004). A summary of experimental protocols follows.

2.2.1 *Biomass conditioning*

In accordance to NREL Lap entitled "Preparation of Samples for Compositional Analysis" (Hames, Ruiz, Scarlata, Sluiter, Sluiter, & Templeton, 2008), all biomass was conditioned in a step previous to experimentation, by reducing: particle size (the fraction taken for experimentation was +40/-80 Thyler mesh) and the moisture content was reduced to less than 10%, by drying in a convection oven at 45°C for >24 h.

2.2.2 *Ash content determination*

The ash content of raw biomass was determined by exposing a known dry weight of biomass (7 replicates) to calcination during four hours in a muffle at 575°C. The weights of the resulting ashes were measured and related to the initial dry weight of each sample. (Sluiter A. , Hames, Ruiz, Scarlata, Sluiter, & Templeton, 2008).

2.2.3 *Structural sugars and lignin*

These components were measured in accordance to NREL laboratory analytical procedure entitled "Determination of Structural Carbohydrates and Lignin in Biomass" which involves two acid hydrolysis steps with H_2SO_4. The first stage is concentrated at 72% and the second diluted with 4%. This procedure results in solubilization of structural carbohydrates while lignin is precipitated (Scarlata, Slutier, Termpleton, Crocker,

Hames, & Ruiz, 2008). Sugar concentration is determined by the colorimetric method dinitrosalicilic acid or DNS, and lignin is separated and gravimetrically measured.

2.3 *Chemical pretreatment*

Cardboard (1 g), lime (0,5 g) and 45 mL of hydrogen peroxide were well mixed into a stainless steel reactor (Parr 500). Afterwards, the reactor was tightly closed, and submitted to a pretreatment at the temperature, hydrogen peroxide concentration and time of pretreatment that corresponds to the experiment according to the factorial design 2^3 summarized in Table 1.

Table 1. Factorial design of chemical pretreatment.

Factor	Level 1	Level 2
Hydrogen peroxide concentration%	0.5	2
Operating time (min)	30	120
Operating temperature (°C)	80	150

Pretreated samples were neutralized using 5 N HCl, washed with abundant distilled water, filtered through Whatman #1 filter paper and weighted. Afterwards, the harvested biomass was divided into three parts. The first part was kept frozen to be submitted to enzymatic hydrolysis later. Another part was put to dry in a 45°C convection oven, to be used in lignin and sugar quantification. The last part was used to determine moisture content.

A parameter was calculated to determine total solids yield after pretreatment (Y_T) as follows:

$$Y_T = \frac{\text{recovered biomass (g)}}{\text{biomass before pretreatment (g)}} \quad (1)$$

With this parameter, it is possible to determine structural carbohydrates and lignin yields (Y_s) (Y_L) as defined in equations (2) and (3),

$$Y_S = \frac{\text{recovered carbohydrates (g)}}{\text{carbohydrates before pretreatment (g)}} = Y_T \frac{C_0}{C_t} \quad (2)$$

$$Y_L = \frac{\text{recovered lignin (g)}}{\text{lignin before pretreatment(g)}} = Y_T \frac{L_0}{L_t} \quad (3)$$

Where, (L)=Lignin; (C)= Sugars; (0)=initial; and (t)=at the end of the pretreatment.

For lime pretreatment, kinetics of lignin and sugar degradation were modeled by standard power law (Eq. 4), accounting for two reacting moieties (fast and slow) as proposed by (Sierra, et al., 2011). Model parameters are obtained by minimization of the sum of squared residuals.

$$-r_i = -\frac{dY_i}{dt} = k_{if} {M_{O_2}}^{\beta if} Y_{if} + k_{is} {M_{O_2}}^{\beta is} Y_{is} \quad (4)$$

Where, (i)=Lignin or sugars; (j)=slow or fast reaction; k_{ij}= rate constant for substance; R=universal constant law for ideal gases (8.314×10^{-3} kJ/(mol K); M_{O_2}=oxygen loading (kg initial oxygen/kg dry biomass); and β_{ij}= reaction order for hydrogen peroxide.

2.4 *Enzymatic pretreatment*

2.4.1 *Activity determination*

Activity determination of the enzyme, *Laccase Suberase* brand *Novartis*, was made according to the *Sigma Aldrich* protocol (Aldrich, 1997). This protocol recommends the use of 0.216 mM Syringaldazine solution, water and *Laccase Suberase*, to determine the activity by a continuous spectrophotometric rate determination (Aldrich, 1997). Because experiments were run at several different pH (Table 2), one result of this experimentation was the recommended pH.

2.4.2 *Selectivity*

Enzymatic delignification is supposed to be highly (if not completely) selective. Nevertheless, a selectivity experiment was performed in order to ratify that the enzyme was free of cellulases. For this purpose, the enzyme (20 μL of *Laccase*) was put in contact with 0.1 g of cellulose at pH of 6, 6.5 y 7, and room temperature and 55°C for a day.

2.4.3 *Experimental Design*

A full factorial design 3^3 was made in order to determine enzyme loading, substrate concentration, residence time and temperature that would maximize enzyme performance at optimum pH of 6.5 and optimum temperature of 55°C. (Baldrian, Fungal Laccasses-ocurrance and properties, 2006). It is summarized in Table 2.

Table 2. Factorial design for enzymatic pretreatment

Factor	Level 1	Level 2	Level 3
Enzyme loading (mL/g substrate)	0.0 1	0.1	0.5
Substrate concentration (w/w)	1	5	10
Residence time (min)	60	360	1440

Immediately after pretreatment, enzyme pretreated samples were submitted to thermal shock to denature the enzyme and therefore stop the reactions. Subsequently, they were profusely washed with distilled water.

Based on the obtained results a decrease on the number levels of the factorial design is proposed. Reducing: residence time(0-1 h), enzymatic loading (20,150 μL) and substrate concentration (2 g). For a better compression of the degradation of lignin it is modeled by Michaelis Menten:

$$-r_S = \frac{V_{max} \times (S)}{K_m + (S)} \quad (5)$$

Where (S)=substrate concentration; K_m=affinity constant; and V_{max}=maximum degradation rate of substrate.

2.5 *Enzymatic hydrolysis*

To determine the effect of pretreatment in biomass digestibility, an enzymatic hydrolysis is made quantifying the cellulose fraction that is hydrolyzed by cellulase at optimum conditions.

The procedure is the *NREL Lap* for *Enzymatic Saccharification of Lignocellulosic Biomass* (Brown & Torget, 1996). It consists on exposing pretreated biomass with an excess of cellulase (15 FPU/g biomass) at 50°C and pH 4.8 for 3 days. The protocol is modified adding 1g of biomass instead of 1 g cellulose equivalent in biomass.

The monomers of glucose and xylose are determined later by DNS, and digestibility is calculated with relation (5) and (6):

$$Y_{saccharification} = \frac{\text{digeted celulose (g)}}{\text{added celulose (g)}} \quad (5)$$

$$Y_{saccharification} = \frac{\text{sugars}_e \times V_T}{\text{biomass(g)} \times \text{SST} \times \text{sugars}_p} \quad (6)$$

Where, sugars_e=sugar content of saccharification extract (mg sugars/ml); V_T=total volume of saccharification (10.6 ml), biomass =pretrated biomass= 0.1 g; SST= total solid content in pretreated biomass; and sugars_p =sugar content of pretreated biomass (g sugars/g pretreated biomass).

2.6 *Global conversion*

The combined effect of pretreatment and enzymatic hydrolysis a global recuperation factor is found:

(6)

Mentioned relation allows choosing the conditions of chemical and enzymatic pretreatment to load in the *MixAlco®* process simulation, where higher values of correspond to superior efficiency in pretreatment, therefore, preservation of sugars, breakage of crystalline structure of celluloses and hemicelluloses and higher lignin degradation.

2.7 *MixAlco® process simulation*

The simulation of the *MixAlco®* process, is modified from a existing one made by (Goussame,2011) where experimental data of both pretreatments are loaded. Accomplishing finally, the comparison of the two pretreatments, at a preliminary stage, because the simulation is an ideal modeling of the process.

The energy index was defined by (Gousamme, 2011) as,

(7)

3 RESULTS AND DISCUSSION

3.1 *Biomass characterization*

The obtained results of biomass characterization protocols are, that non pretreated biomass is composed in dry weight: 43% by lignin, 3% by ash, and a 44% of sugars, Illustration 1.

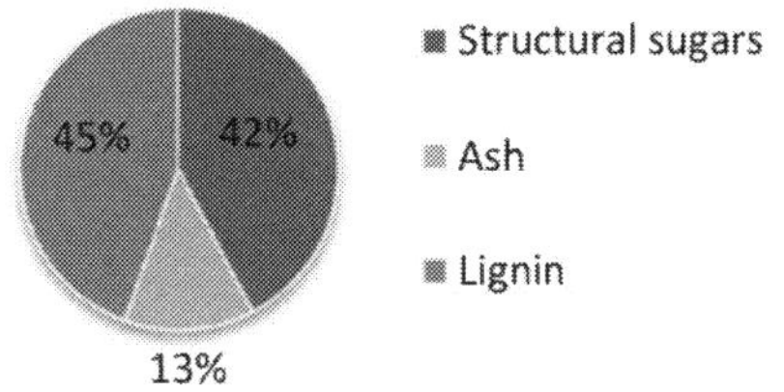

Ilustration1. Composition of cardboard in dry weight.

3.2 *Chemical pretreatment*

3.2.1 *Lignin degradation*

The average of the experimental results obtained for factorial design 2^3, for chemical pretreatment are presented in Illustration 2 with respective standard deviation.

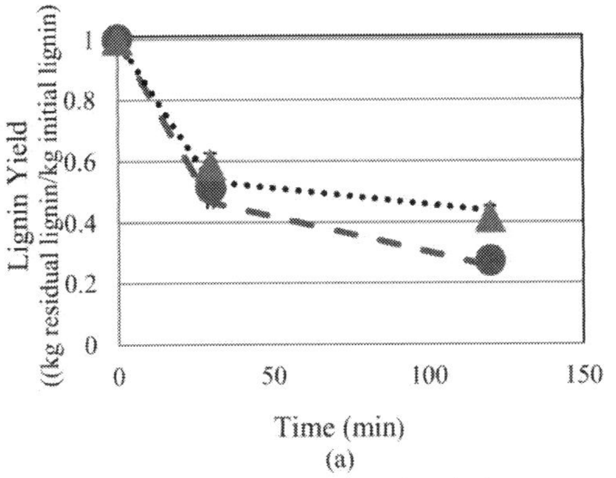

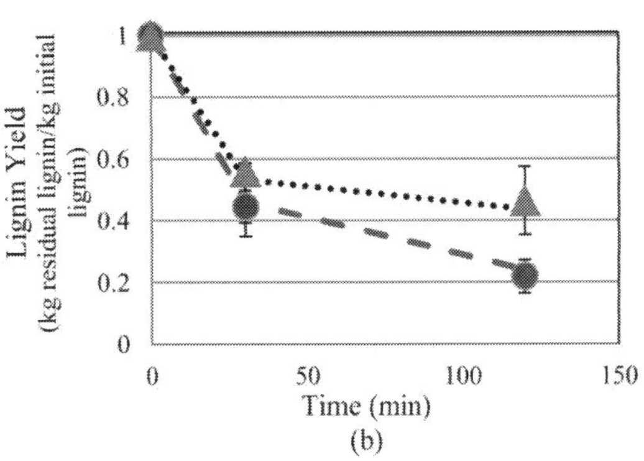

Illustration 2. Lignin yield for chemical pretreatment with lime Ca(OH)2 and respective kinetic model. (a) 80°C(b)150°C. (▲) 2% hydrogen peroxide, (●)0.5% hydrogen peroxide.

According to Illustration 2 at higher peroxide concentration, temperature and time, greater lignin degradation. This can be explained because higher hydrogen peroxide concentration generates more quantity of superoxide ions responsible of breaking C-C and C-O-C bounds.

As shown in Ilustration2, the time that superoxide ions take in breaking the lignin chain ends, where the degradation begins, is half hour. With higher residence time an increase of the lignin degradation is measurable, because the numerical relation between superoxide ions and attack sites become more efficient.

The highest delignification is at two hours, with lime ($Ca(OH)_2$) at 150°C and hydrogen peroxide at 2%.

Calculated lignin yields are presented as cutter lines in the operating rank of each pretreatment. Data is obtained by least squares minimization of the model presented by Sierra ET all, 2009, whose solution generates the parameters show in Table 3.

Table 3. Achieved values for parameters of lignin degradation with lime $Ca(OH)_2$.

Parameter	Value	Units
Y_{if0}	0.545	g residual lignin/g initial lignin
a_{iif}	0.0597	min^{-1}
B_f	0.0257	
E_{iif}	0.0825	kJ/mol
a_{ifs}	4.96E-07	min^{-1}
B_s	2.10	
E_{iis}	0.878	kJ/mol

3.2.2 *Sugar Yields*

Average of experimental data obtained for sugar yield in chemical pretreatment are presented as marks in Illustration 3.

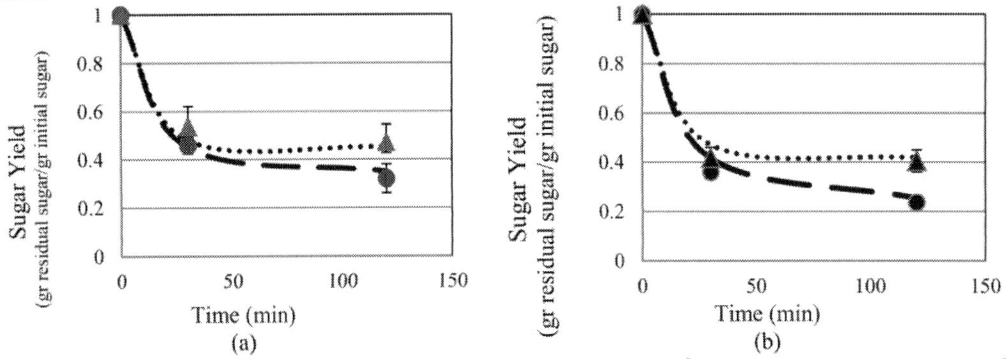

Illustration 3. Sugar yield for chemical pretreatment (a) 80°C, (b) 150°C. (··▲··) hydrogen peroxide at 2% , (–●–)hydrogen peroxide.0.5%

The behavior of sugar yield is similar to lignin yield, because at higher operation conditions lower sugar yield, this according to the presence of superoxide ions and harder thermal conditions.

Comparing part (a) and (b) of the Illustration 3, can be observed that sugars are sensitive to temperature degradation because at every pretreatment with equal time and peroxide concentration and higher temperature lower sugar yield is obtained.

The effect of time in sugar degradation is least at lower hydroxide concentration.

The kinetic modeling is made same as the lignin degradation, the obtained parameters are:

Table 4. Obtained values of kinetic parameters for sugar degradation with lime Ca(OH)2.

Parameter	**Value**	**Units**
Y_{if0}	0,489	g residual sugar/g initial sugar
a_{iif}	0,00156	min^{-1}
B_f	1,02	
a_{ifs}	911×102	min^{-1}
B_s	4,315	
E_{iif}	11,8	kJ/mol
E_{iis}	0,869	kJ/mol

3.3 *Enzymatic pretreatment with Laccase*

3.3.1 *Activity determination and enzyme selectivity*

The higher activity is 864 U ml^{-1} at pH 6.5, Illustration 4.

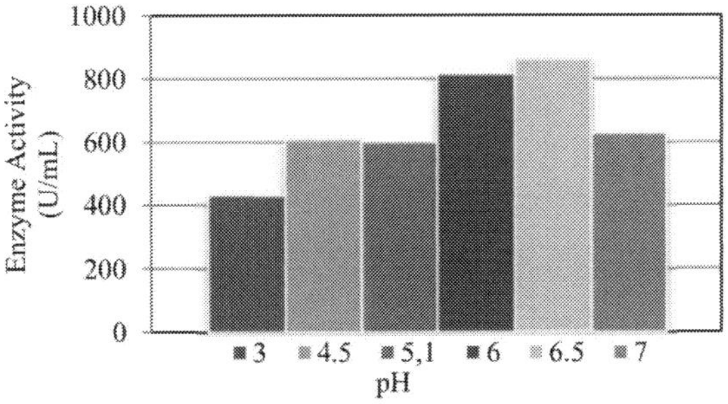

Illustration 4. Activity of the enzyme.

Obtained results for enzyme selectivity showed changes of 10^{-3} g, attributed to moisture content in the samples.

3.3.2 *Enzymatic lignin degradation*

Enzymatic experimental average results of the first substrate level (2 g) of the factorial design 3^3 are presented in Illustration 5. Only this level is presented because the behavior obtained for the other two levels of substrate, 1.5 and 1.0 g are very similar; (where at every enzymatic loading there is an increase of the lignin

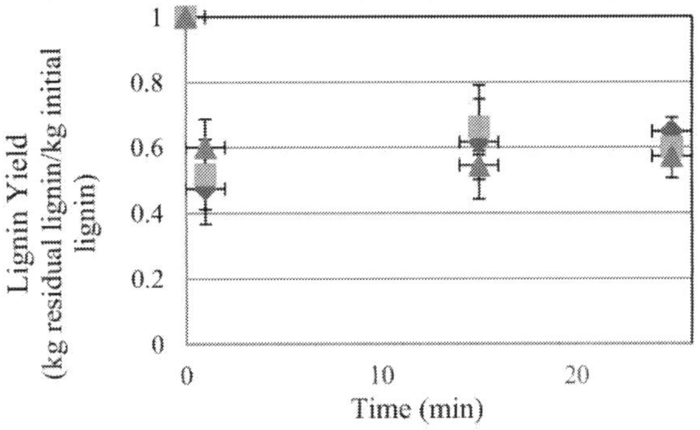

Illustration 5. Enzymatic delignification (◆) 20 µL, (■) 150 µL, (▲) 1000 µL.

For this reason the experimental conditions were cut to: a maximum of an hour, for residence time, and two levels of enzymatic loading, 20 and 150 µL, with a single substrate concentration, 2 g.

In Illustration 6, the results of the last Laccase experimentation are presented where the higher lignin degradation is 0.45(g residual lignin/g initial lignin), accomplish by 150 µL of enzyme at one hour. While the minimum yield is 0.75(g residual lignin/g initial lignin) obtain with 20 µL of Laccase at 30 min.

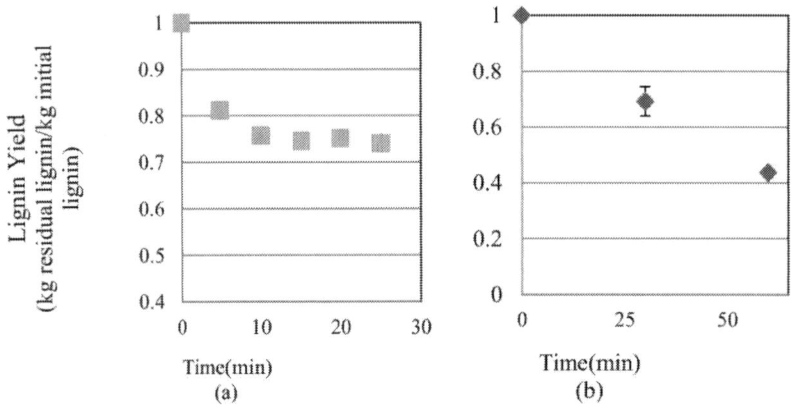

Illustration 6. Enzymatic delignification of 2 g of substrate (a) 20μL enzyme, (b) 150 μL enzyme.

The kinetic modeling for the enzymatic lignin degradation was made by the equation of Michaelis Menten as a first approximation.

The Lineweaver-Burk linearization was used to model the kinetic parameters, for 20 y 150 μl of enzyme data.

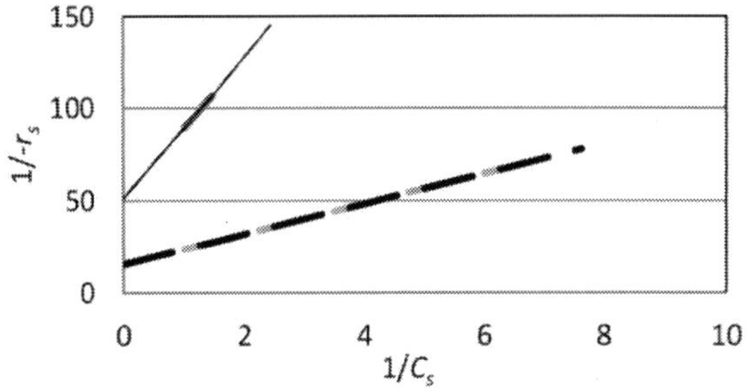

Illustration 7. Linewever-Burk for 2 g of biomass. (- -) 20 μL, ( ) 150 μL.

The values obtained for the kinetic parameters are presented at the Table 5.

Table 5.Obtained values for the enzymatic degradation of lignin.

Parameter	Level	Value	Units
V_{max}	20μl	0,0646	U mg^{-1}
K_m	20μl	0,529	g cm^{-3}
V_{max}	150μl	0,0195	U mg^{-1}
K_m	150μl	0,752	g cm^{-3}

3.3.3 *Sugar Yield*

As shown at Illustration 8, it can be identified that the higher sugar yield is obtained with enzymatic pretreatment at ten minutes and 20 μL of enzyme, while the lowest is also obtained with 20μL at 20 minutes.Thereby time is an important factor at this enzymatic loading level.

At 150 μL of enzyme, sugar yield keeps near 0.6 (g recover sugar/g initial sugar) in the selected time range.

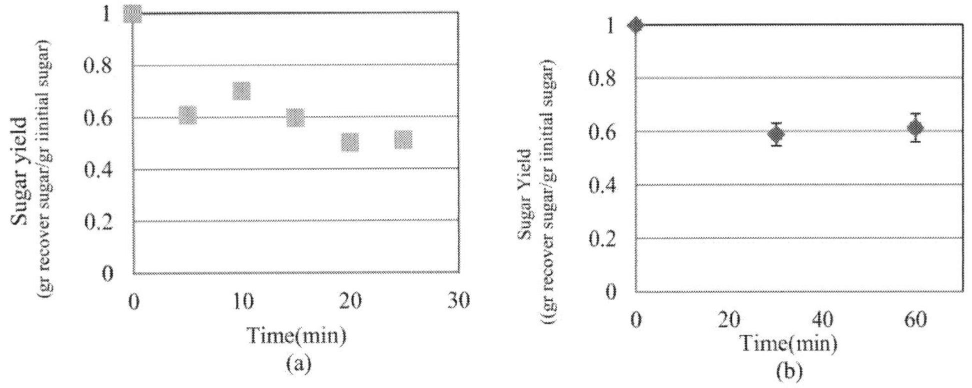

Ilustration8. Sugar yield for 2 g of substrate (a) 20 μL of enzyme (b) 150 μL of enzyme.

3.4 *Enzymatic saccarification*

Higher digestibility values were achieved for chemical pretreatment with the higher level of: time (2 h) and hydrogen peroxide concentration(2%), but since results are similar with different temperatures, Illustration 9, it is necessary to calculate global sugar conversion to select the best chemical pretreatment conditions.

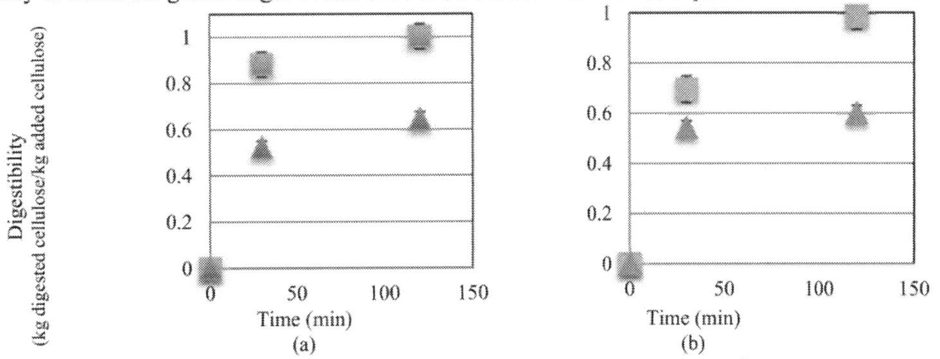

Illustration 9. Digestibility of chemical pretreated biomass.(a) 80°C, (b) 150°C. (▲) 0.5% peroxide,(■)2% peroxide.

Highest digestibility for enzymatic pretreatment is at one hour of pretreatment with *Laccase,* with 150 μL of enzyme loading, Illustration 10.

The biomass, pretreated with 20 μL of *Laccase* obtained the highest digestibility at 10 minutes of pretreatment, which can be justified with the fact that low enzyme concentration triggers quicker enzyme action because there is lower enzyme competition, but also is less efficient compared to 150 μl of enzyme where are more competition but higher lignin degradation.

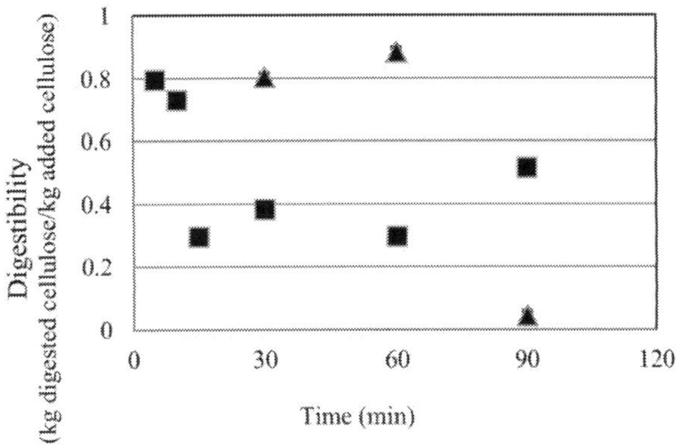

Illustration 10. Digestibility enzymatic pretreated biomass (■) 20μL and (▲)150 μL de enzyme.

The calculation of global conversion is necessary to determine which chemical and enzymatic pretreatment is the proper one to implement at Aspen® simulation of the *MixAlco®* process.

The global conversion results are presented in a contour graph, made using *MiniTab Statistical Software©*.

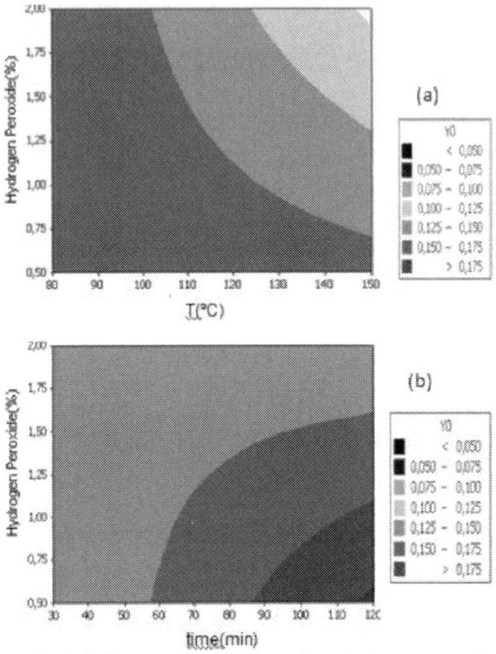

Illustration 11.Contour graph of global conversion for chemical pretreated biomass. (a) %Peroxide vs. temperature(°C),(b) %peroxide vs. time(min).

The global sugar conversion, Y_0, of chemical pretreatment samples are shown in Illustration 11, where at less peroxide concentration, lower temperature y higher time results in the best sugar conversion which can be explained by sensibility of sugar degradation to temperature and lower superoxide ions concentration.

The global sugar conversion for enzymatic pretreated biomass is presented in Illustration 12, where the zone with higher sugar conversion correspond to one hour, 150 µL of enzyme loading which can be explained due to enzyme selectivity to lignin degradation and the inverse relation between enzyme competivity and residence time.

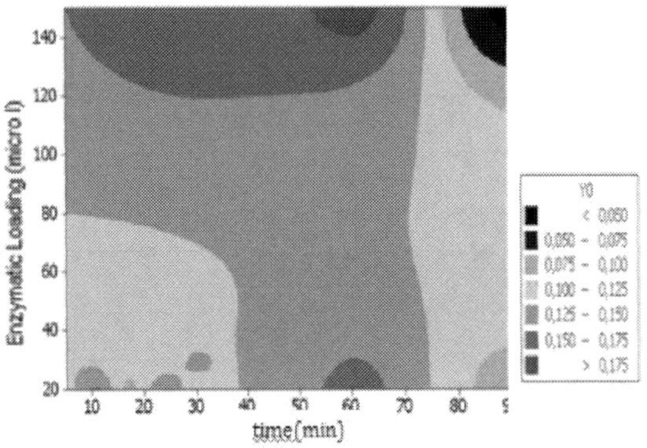

Illustration 12. Contour graph for global sugar conversion for enzyme pretreated biomass.

The best global conversion value was achieved with enzyme pretreated biomass at one hour of pretreatment and 150 µL of enzyme because the recuperation factor Y_t was close to one and the sugar degradation was low compare to chemical degradation recuperation factor close to ,5 and higher sugar degradation.

3.5 *Aspen Plus® Modeling*

The pretreatment conditions chosen to be loaded at the *MixAlco®* process simulation, design by (Gousamme, 2011) are the corresponding to chemical pretreatment 0.5% peroxide, 80ºC y 120 minutes and the corresponding to enzymatic pretreatment with 150 μL and one hour of residence time.

The higher biofuel production was achieved with chemical pretreatment with an energy index of 4,477 vs. 4,189 (kJ out/kJ in) of the enzymatic pretreatment.

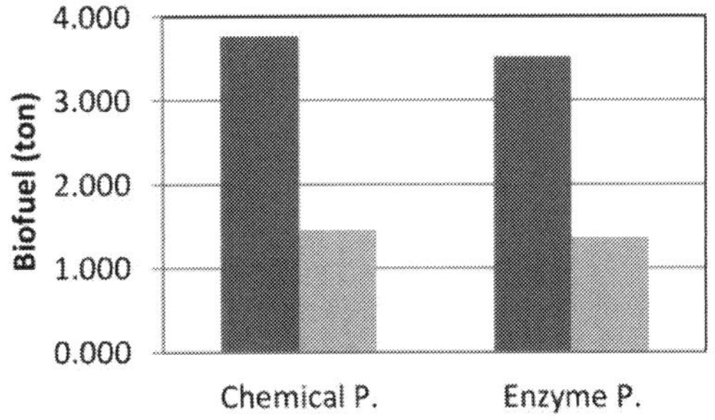

Illustration 13. Bar graph for biofuel production with chemical and enzyme pretreatment. ( ) Jet biofuel, ( ) gasoline.

4 CONCLUSIONS

Comparing chemical with enzymatic pretreatment, the enzymatic is more sensitive to time and pH, which complicates the implementation of this kind of pretreatment in the industry.

Enzymatic pretreatment also achieved the best values for digestibility and global sugar conversion.

The simulation, final step of the comparison between chemical and biological pretreatment, gives better results for the chemical pretreatment in biofuel generation and energy index.

5 BIBLIOGRAPHY

1. Aldrich, S. (1997 11-03). *Enzymatic Assay of LACCASE.* Retrieved 2011 12-03 from http://www.sigmaaldrich.com/etc/medialib/docs/Sigma/General_Information/2/laccase.Par.0001.File.dat/laccase.pdf
2. Baldrian, P. (2006). Fungal Laccasses-ocurrance and properties. FEMS Microbial Review 30 , 215-242.
3. Brown, L., & Torget, R. (1996). *Enzymatic Saccharification of lignocellulosic Biomass.* NREL.
4. Carvajal Montoya, S., & Giraldo Garcia, D. (2010). Simulación del proceso MixAlco en Aspen Plus y análisis de sensibilidad de los parámetros involucrados en las principales etapas del proceso. Bogota D.C: Proyecto de grado, Universidad de los Andes.
5. Chang, V., Nagwani M, M., Kim, C., & Holtapple, M. (1999). *Oxidative Lime Pretreatment of High-Lignin Biomass.* College Station: Texas A&M.
6. David, C., Salvador, S., Dirion, J., & Quintard, M. (2003). Determination of reaction scheme for cardboard thermal degradation using thermal gravimetricanalysis. *Journal of Analitycal an d Applied Pyroysis* , 307-232.
7. Fogler, S. H. (2008). Elementos de ingeniería de las reacciónes químicas. Naucalpan de Juarez, mexico: Pearson Prentice Hall.
8. Gandi, J., Holtzapple, M. T., Ferrer, A., Byers, M. F., Turner, N. D., Nagwani, M., et al. (1997). Lime treatment of agricultural residues to improve rumen digestibility. *Animal Feed and Science Technology* , 195.
9. Gousame, P. (2011). Simulation of the MixAlco Process. *Special Project* .
10. Granda, C. B., & Holtzapple, M. T. (2008). Part 1: Comparison of Three Biomass Conversion Platforms. In *Carboxylate Platform: The MixAlco Process.* Texas: Department of Chemical Engineering Texas A&M University.
11. Hames, B., Ruiz, R., Scarlata, C., Sluiter, A., Sluiter, J., & Templeton, D. (2008). *Preparation of Samples for Compositional Analysis.* Cole Boulevard, Golden, Colorado : National Renewable Energy Laboratory.
12. Hayatsu, R., Winans, R., McBeth, R., Scott, R., Moore, L., & Stuider, M. (1979). Lignin-like polymers in coals. *Nature* , 278.

13. Holtzapple, M., Davison, R. R., Ross, M. K., Aldrett-Lee, S., Nagwani, M., Lee, C.-M., et al. (1999). Biomass Conersion to Mixed Alcohol Fuels Using the MixAlco Process. *Applied Biochementry and Biotechnology* , 609-630.
14. Kaar, W. E., & Holtzapple, M. T. (2000). Using lime pretreatment to facilitate the enzymic hydrolysis. *Biomass and Bioenergy* , 189.
15. Lachental, D., Mortha, G., Sevillano, R.-M., & Zaroubine, M. (2004). Isolation of residual lignin from softwood kraft pulp. Advantages of the acetic acid acidolysis method. *Compites Rendus Biologies* , 911-916.
16. Lee, J. (1997). Biological conversion of lignocellulisic biomasss to ethanol. *Journal of Biotechnoly* , 3-4.
17. Lee, J. (1997). Biological conversion of lignocellulosic biomass to ethanol. *Journal of Biotechnology* , 4.
18. Mandels, M., Hontz, L., & Nystrom, J. (1974). Enzymatic Hydrolysis of Waste Cellulose. *Biotechnology and Bioengineering* , 1471-1493.
19. Morozova, O., Shumakovich, G., Gorbacheva, M., Shleev, S., & Yaropolov, A. (2007). "Blue" Laccases. *Biochemestry (Moscow)* , *73* (10), 1396-1412.
20. Mosier, N., Wyman, C., Dale, B., Elander, R., Lee, Y., Holtappple, M., et al. (2005). Features o promising technologies for pretreatment of lignocellulosic biomass. *Bioresource technology* , 673.
21. Pham, V., Holtazpple, M., & El-Halwagi, M. (2010). Techno-economic analysis of biomass to fuel ocnversion. *Journal of Microbiology and Biotechnology* .
22. Phillips, T. (2006 йил June). Structural Formula of Ethanoic Acid reacting with Propanol to form Propyl-Ethanoate.
23. Robert J. Wooley, V. P. (1996 йил Abril). Development of an ASPEN PLUS Physical Property Database for Biofuels Components. Golden, Colorado, Estados Unidos.
24. Rodriguez Cuoto, S., & Toca-Herrera, J. L. (2006). Lacasses in the textile industry. *Biotechnology and Molecular Biology review* , 115-120.
25. Scarlata, C., Slutier, J., Termpleton, a. D., Crocker, D. A., Hames, B., & Ruiz, R. (2008). *Determination of Structural Carbohydrates and Lignin in Biomass LAP.* Cole Boulevard, Golden, Colorado: National Renewable Energy Laboratory.
26. SE HOON, K. (2004). *LIME PRETREATMENT AND ENZYMATIC HYDROLYSIS.* Texas, USA: Texas A&M University Dissertation.
27. Sierra, R., & Holtzapple, M. T. (2009). Selectivity and Delignification Kinetics for Oxidative and Non-Oxidative Lime Pretreatment of Poplar Wood - Part III: Long-therm. *(Under Review)* .
28. Sierra, R., & Holtzapple, M. T. (2009). Selectivity and Delignification Kinetics for Oxidative Short-therm Lime Pretreatment of Poplar Wood - Part I: Constanr Pressure. *(Under Review)* .
29. Sierra, R., Smith, A., Granda, C., & Holtzapple, M. (2008). Producing fuels and chemicals from lignocellulosic biomass. *Chemical Engineering Progress* , 510-518.
30. Sluiter, A., Hames, B., Ruiz, R., Scarlata, C., Sluiter, J., & Templeton, D. (2008). *Determination of Ash in Biomass (LAP).* Cole Boulevard, Golden, Colorado: National Renewable Energy Laboratory.
31. Sluiter, A., Hames, B., Ruiz, R., Scarlata, C., Sluiter, J., & Templeton, D. (2004). *Determination of Total Solids in Biomass LAP.* Cole Boulevard, Golden, Colorado: National Renewable Energy Laboratory.
32. Smufit Kappa. (2011). *Preguntas frecuentes.* Retrieved 2011 6-Mayo from http://www.smurfitkappa.com.co/DropdownMenu/Contact/FAQ/Recycling/
33. Sylvia, D., Fuhrman, J., Hartel, P., & Zuberer, D. (1998). Decomposition of Naturraly Occuring Organic Materials. In *Principles and applications of soil microbiology.* (pp. 228-234). USA: Prentice Hall.
34. Tchobanoglous, G., Theissen, H., & Eliassen, R. (1982). Producción de desechos sólidos. In *DESECHOS SÓLIDOS PRINCIPIOS DE INGENIERÍA Y ADMINISTRACIÓN* (pp. 67-82). Ambiente y los Recursos Naturales Renovables.
35. Terrabon, L. (2008). *Terrabon Renewable Energy Solutions.* Retrieved 2009 30-Octubre from MixAlco Process: http://www.terrabon.com/mixalco_overview.html
36. Tisma, M., Znidarsic-Plazl, P., Plazl, I., Vasic-Racki, D., & Zelic, B. (2009). *Oxidation of Coniferyl Alcohol Catalized by catalized from Trametes versicolor.* Croacia, Eslovenia.
37. Tuomela, M., Vikman, M., Hatakka, A., & Itavaara, M. (2000). Biodegradation of lignin in a compst enviroment: a review. *Bioresource Technology* , 169-183.
38. U.S. State Department. (2010 йил 15-March). *Raw materials of biofuels.* Retrieved 2011 йил 11-March from http://www.america.gov/st/develop-spanish/2010/March/20100409143620CMretroP3.654116e-02.html
39. UNL. (2004). *Capítulo 2. Aspectos Generales.* From bibliotecavirtual.unl.edu.ar:8180/tesis/.../1/.../07-Capítulo%202.pdf
40. Van Soest, P. J. (1973). Lignin. In *Nutritional ecology of the ruminant* (p. 193). Cornell, USA: Cornell University Press.
41. Yamauchi, K., & Fukushima, K. (2004). The regulation from guaicyl to syringyl lingnin in the differenciating xylem of Robinia pseudoacacia. *Comptes Rendus Biologies* , 791-797.

42. Zhang, J., Kjonas, R., & Flurkey, W. (1999). Does N-hydroxyglycine inhibit plant and fungal laccases? *Phytochemestry* , 775-783.

(523b) Isoconversional Methods Applied to Kinetic Modeling Oxidative Alkaline Pretreatment of Sugarcane Bagasse

Rocio Sierra, Cuellar Gerardo, Simbaqueba Andres, Auza Nicolas and Copete Diego, Chemical Engineering, Universidad de los Andes, Bogota, Colombia

Delignification (decrease of lignin content) has shown to be a valid method to significantly increase lignocellulosic feedstocks digestibility. This operation (widely known as pretreatment) makes these abundantly available and inexpensive feedstocks usable in fermentation processes whose main goal is to produce biofuels and other chemicals. Because pretreatment is one of the most expensive processes in biofuels production plants, it has been extensively studied; however, a throughout literature search shows that development of kinetic models to better understand, optimize and control pretreatment, are very scarce if available at all.

This study uses sugarcane bagasse which is a residue from the sugar industry abundantly produced particularly near and in the Equator zone. This lignocellulosic resource is traditionally used in combustion processes to supply energy to the production plant; however life cycle analysis have shown that if lignocellulose is used as substrate in fermentation processes, the energy yield would be much closer to the theoretically available energy.

In this work, sugarcane bagasse is submitted to an alkaline pretreatment that uses an oxidative agent, systematically applied at widely varied ranges of temperature and time using a full factorial experimental design

Model fitting methods do not achieve a clean separation between the temperature dependent Arrhenius constant and the reaction model. On the contrary, the isoconversional methods evaluate the activation energy values at progressive degree of conversion without modelistic assumptions. This work applies this methodology to calculate activation energy for alkaline pretreatment of sugarcane bagasse and compares with previous work where the traditional curve fitting is employed.

See more of this Session: Developments In the Pretreatment of Lignocellulosics for Bioconversion II
See more of this Group/Topical: Sustainable Engineering Forum

(523c) Alkaline Deacylation of Corn Stover to Reduce Toxicity of Prehydrolyzates and Improve Enzymatic Hydrolysis

Urvi D. Kothari and Y. Y. Lee, Chemical Engineering, Auburn University, Auburn, AL

Hydrolysates from dilute acid pretreatment of biomass contain various biomass degradation products which are inhibitory to enzymes and toxic to microorganisms. One of the main inhibitory compounds is acetic acid which is released from the lignin-hemicellulose matrix in the biomass. Acetic acid at the level of 1 g/L inhibits cellulase activity by 10%. In most pretreatment liquors its concentration is in the range of 3-10 g/L and inhibition of cellulase activity by acetic acid at these concentrations is 20-60%. In this study, sodium carbonate was investigated as a deacylation reagent that can selectively remove part of the acetyl components in corn stover and reduce the acetyl concentration in the pretreatment hydrolysate. Sodium carbonate was chosen since it is a mild alkali that can preserve carbohydrates under the deacylation process conditions. It is inexpensive and can easily be recovered through established Kraft recovery process. Sodium carbonate of 5-10 wt% was used over 30-60ºC for deacylation. A dilute-acid pretreatment was then applied under the conditions of 0.5-1% H_2SO_4 at 60-110ºC in order to assess the effects of deacylation. The two-step process operated under optimized conditions led to reduction of the final acetyl concentration by 50-80%, and minimal loss of hemicellulose at 2-12%. The combined process further improved the glucan and xylan hydrolysis yields of the pretreated corn stover over those of the dilute-acid pretreatment alone.

See more of this Session: Developments In the Pretreatment of Lignocellulosics for Bioconversion II
See more of this Group/Topical: Sustainable Engineering Forum

(523d) Deconstruction of Wheat Straw Lignin by Streptomyces Viridosporus As Insight Into Biological Degradation Mechanism

Jijiao Zeng[1], Department of Biological Systems Engineering, Washington State University, Pullman, WA, Deepak Singh, Edeniq Inc, Visalia, CA, Dhrubojyoti D. Laskar, Washington State University, Pullman, WA and Shulin Chen, Biological Systems Engineering, Washington State University, Pullman, WA

Deconstruction of wheat straw lignin by *Streptomyces viridosporus* asinsight into biological degradation mechanism

JijiaoZeng[1], Deepak Singh[2],Dhrubojyoti Dey Laskar[1]and Shulin Chen[1*]

1. Department of Biological Systems Engineering, Washington State University,Pullman, Washington 99164-6120, USA

2. Edeniq Inc, 2505 N Shirk Road, Visalia, 93291 CA

[*]Corresponding author. Tel: +1 509 335 3743; fax: +1 509 335 2722.

E-mail address: chens@wsu.edu

Abstract:

Lignins are secondary natural organic compositions of plant cell wall besides cellulose and hemicelluse. Structurally lignins are linked with hemicelluloses and cellulose to form recalcitrant complex that resist the environmental stress. Biological systems created by nature over millions of years of evolution can be capitalized for targeted deconstruction of plant cell walls through selective cleavage and structural modification of lignin. Such modification causes changes in cell wall structure for facilitating enzymatic attack to cellulose for sugar release. The study of biological degradation mechanism on lignocellulosic materials could provide the new knowledge to convert plant biomass to biofuel and high value co-products. In this study, we have investigated the effect of wheat straw lignin structural changes by *Streptomyces viridosporus* on downstream enzymatic hydrolysis. ^{13}C CPMAS NMR revealed that the rising of carbonhydrate/lignin ratio accompany with enhanced intensity of methoxyl group in bio-treated wheat straw. The degraded lignin were isolated and analyzed by Fourier Transmission Infrared (FTIR), Pyrolysis Gas Chromatography/Mass Spectrometry (Py-GC/MS) and two dimenssional heteronuclear multiple quantum coherence nuclear magnatic resonance (2D HMQC NMR) to elucidate their structural change. The results indicate that the biological pretreated lignin were associated with critical change and/ or modification on their structure implying selective removal of guaiacyl (G) lignin unit which is more condense than syringyl (S) unit. Moreover, the treated lignin also dramatically reduced the inhibition of cellulose hydrolysis. This information provides new insights for understanding the biological degradation mechanism on lignin for cellulose utilization and exploring a possible route to convert lignin derived co-products.

See more of this Session: Developments In the Pretreatment of Lignocellulosics for Bioconversion II
See more of this Group/Topical: Sustainable Engineering Forum

A New Pretreatment and Processing System for Recalitrant Biomass – Paper 523e

Benjamin Levie, Dwight Anderson, Johnway Gao, Alan Lovas

Catchlight Energy LLC (CLE)

A 50/50 Chevron and Weyerhaeuser joint venture.

Catchlight Energy

Our Mission: To commercialize the large-scale production of liquid transportation fuels from sustainable forest-based resources

A Unique End-to-End Value Chain Solution
Leveraging the strengths of two natural resource leaders

Catchlight Energy (CLE) brings strengths not found elsewhere - feedstocks development and procurement, fuels testing and refining, operations, conversion engineering.

CLE drivers aligned with national interests to develop renewable energy platform, increase energy independence & security and promote rural development.

October 15, 2011

1

Softwood - High Availability/High Recalcitrance

Catchlight Energy

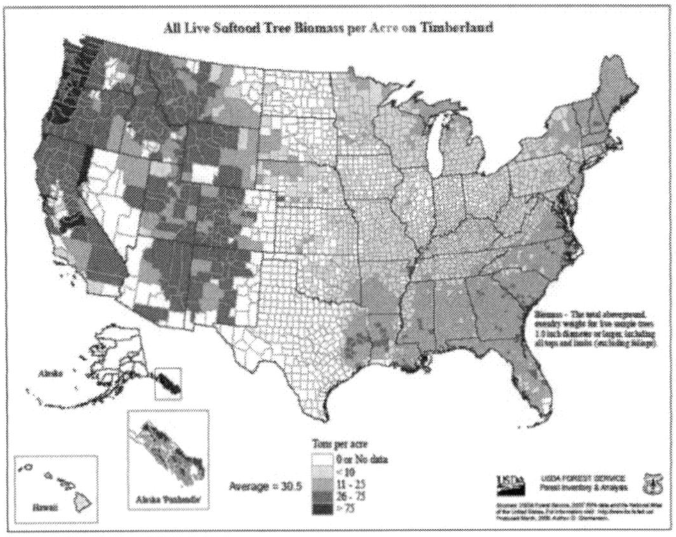

The use of softwood is of high critical value to the biofuels industry, but it remains a difficult feedstock for most pure bio-conversion processes.

October 15, 2011

2

Pretreatment Issues for Softwood and other Recalcitrant Biomass

Catchlight Energy

- **Dilute acid and base systems ineffective at reasonable levels.**
- **Known and proposed "Fractionation" pretreatments are excessively expensive, and not commercial scale.**
- **Pulping processes effective because of removal of lignin, but result in high and unsupportable costs.**
  - ➢ High capex (unless repurposed mill that is obtained at very low cost).
  - ➢ High opex (chemical costs or chemical recovery costs, wastewater treatment).
  - ➢ Fuels value can't compete with pulp value.

October 15, 2011

3

CLE Pretreatment

Catchlight Energy

- **Pulping, especially sulfite pulping, is an excellent but expensive pretreatment**
- **CLE Pretreatment has**
  - ➢ Higher throughput
  - ➢ Lower chemical consumption
  - ➢ Lower inhibitors
  - ➢ Excellent enzymatic hydrolysis results
  - ➢ Higher sugar yield

October 15, 2011

4

CLE Pretreatment Characteristics

Catchlight Energy

- Adaptable to existing infrastructure
- Scalable from pilot operations
- Feedstock flexible

October 15, 2011

5

CLE Pretreatment and Process to Sugar

Catchlight Energy

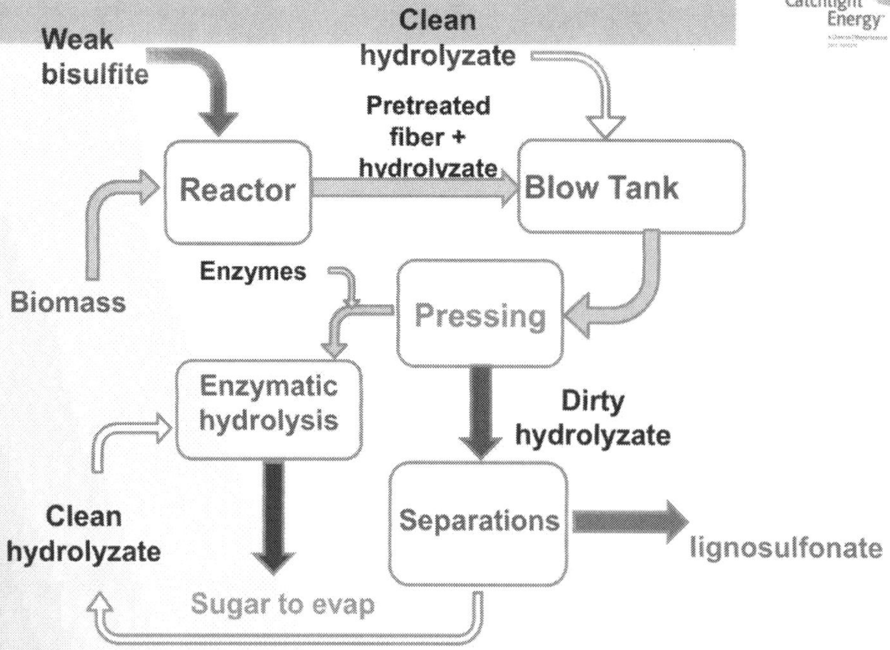

October 15, 2011

6

Softwood Sugar

Pretreated Softwood-Mill Residuals

Catchlight Energy

Softwood Residuals

Calcium Bisulfite Pretreatment

Pretreated Refined Pulp

Pretreated Residual Sugar Composition:

Arabinan	0.21%
Galactan	0.55%
Glucan	49.9%
Xylan	3.42%
Mannan	3.63%
Total	57.7%

Softwood Residual Sugar Composition:

Arabinan	1.01%
Galactan	2.09%
Glucan	39.5%
Xylan	6.05%
Mannan	10.5%
Total	59.2%

Pretreatment Liquor

Liquor Sugar Composition:

Arabinose	0.07%
Galactose	0.17%
Glucose	0.36%
Xylose	0.48%
Mannose	1.10%
Total	2.18%

October 15, 2011

7

Softwood Mill Residual Chips

Pilot Pretreatment Results

Catchlight Energy

Pilot Pretreatment Size: 3 kg OD chips/1 cu ft digester	
Overall Pretreatment Sugar Yield on Wood Sugar	88-95%
Furan Formation on Wood Sugar	<1.5%
Acetic Acid on Wood	~1.0%
Initial Cook Liquor pH	2.0-4.0
Pulp Yield on Wood	65%-80%

CLE pretreatment produces high yield of pulp and ultimately sugar.

October 15, 2011

8

Enzymatic Hydrolysis

Catchlight Energy

- Near 80% conversion of glucan up to 17% solids in the hydrolysis step using low dose of enzymes.
- Overall sugar yields of 87% attained on clean softwood with economic enzyme loading.
- Overall sugar yields of 80% attained on softwood forest residuals.
- Switchgrass and hardwood behave similarly although have higher C5 sugar in the product.

October 15, 2011

9

CLE Pretreated Hydrolysis

Glucose Yield vs. CTec2 Enzyme Dosage

Catchlight Energy

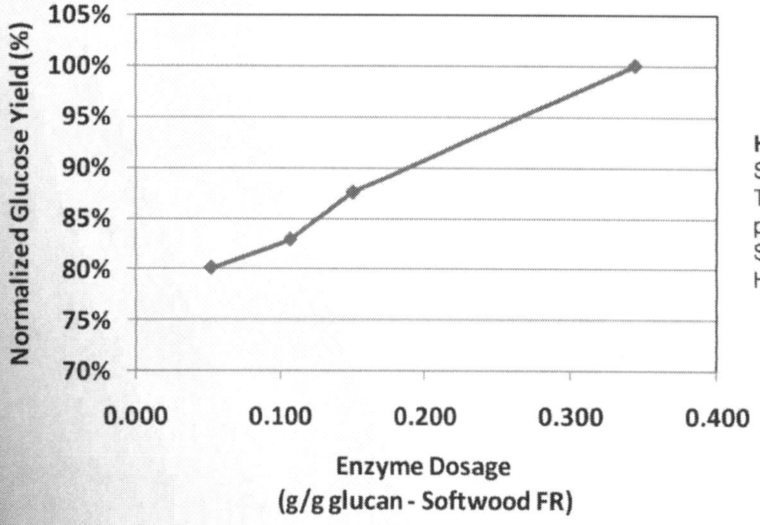

Hydrolysis Conditions:
Solid: 17%, unwashed
Temperature: 50°C
pH: 4.8-5.3
Shaking Speed: 200 rpm
Hydrolysis Time: 96 hrs

October 15, 2011

10

CLE Pretreatment Works on Multiple Feedstocks

Hydrolysis results at 4 days

Catchlight Energy

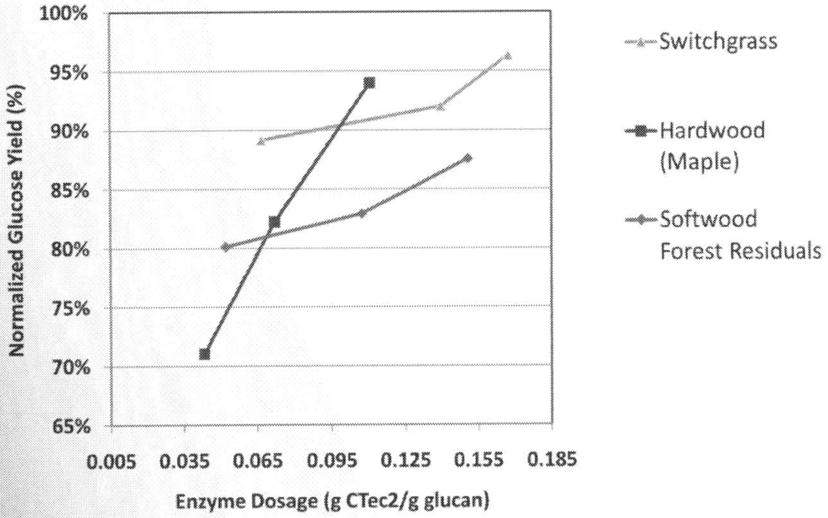

October 15, 2011

11

Commercial Data

CTec2 Dosage on Pretreated Corn Stover (Dilute Acid)

Catchlight Energy

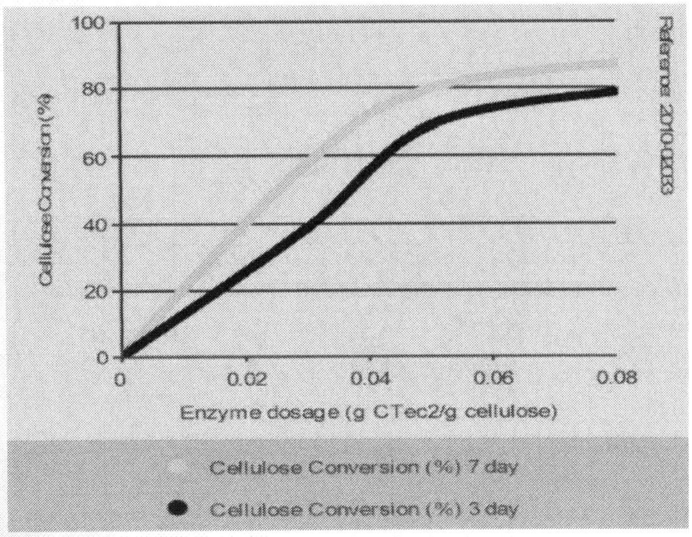

Reference: Corn stover data from CTec2 and HTec2 Application Sheet , www.novozymes.com
Used with permission

October 15, 2011

12

CLE Pretreatment Makes Softwood Hydrolyze Like Acid-Pretreated Corn Stover

Catchlight Energy

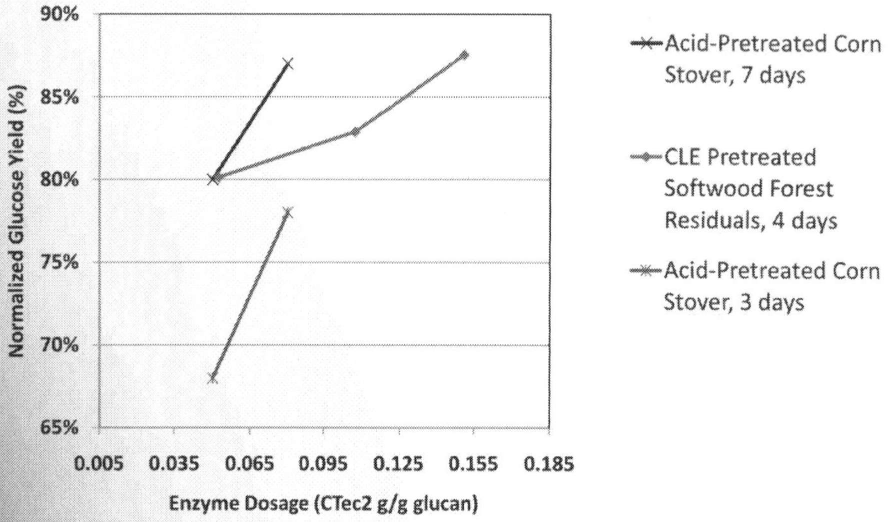

October 15, 2011

13

CLE Sugar: Relative Cost Components

Catchlight Energy

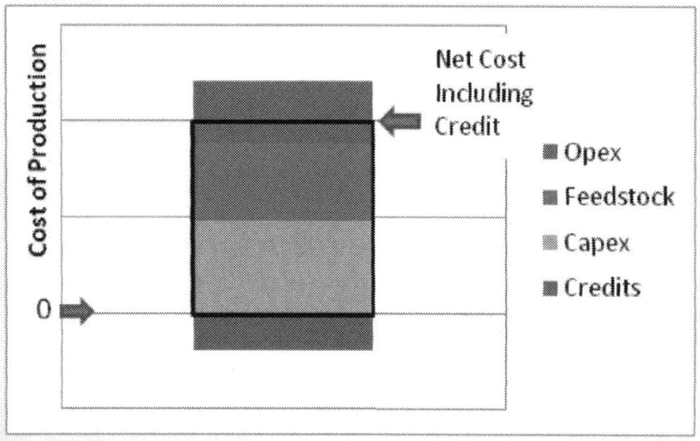

October 15, 2011

14

Conclusions

Catchlight Energy

- CLE's pretreatment technology is advantageous for difficult to hydrolyze feedstocks, such as softwood residuals which are available today at commercial scale.
- The technology is proven, scalable, and cost effective.
  - An engineering design has been developed for a large scale greenfield biofuels with downstream processing and separations for sugar, lignosulfonate, and energy production.
- CLE is actively working with select partners and welcomes interest in the production of sugar, ethanol, and drop-in biofuels.

(523f) Economics of Pretreatment for Biological Processing

Ling Tao, Andy Aden and Richard T. Elander, National Bioenergy Center, National Renewable Energy Laboratory, Golden, CO

Economics of Pretreatment for Biological Processing

Ling Tao[1], Andy Aden[1] and Richard T. Elander[1]

[1] NationalRenewable Energy Laboratory, 1617 Cole Blvd., Golden, CO 80401, USA

Abstract:

A broad understanding of pretreatment economics within anoverall conversion context will be presented. The importance of pretreatmentin the overall economics of the biorefinery is addressed along with the keyparameters that influence the pretreatment economics. A large array ofpretreatment technology options exist and are currently being researched anddeveloped in this study. There appear to be many performances and costtradeoffs for these varying technologies. Only through continued R&D willthese tradeoffs be more fully understood. Process economics, especially thepromising pretreatment technologies studied in the Consortium for AppliedFundamentals and Innovation (CAFI) projects, are analyzed and compared indetails in this work. The six pretreatment technologies vary greatly in termsof their process design and projected total capital investment. Pretreatmentarea capital costs are based on past research, existing databases andengineering judgments, and include not only pretreatment reactor costs, butalso pretreatment chemical recovery and recycle equipment costs for theprocesses that use high pretreatment chemical loadings. More accurate costestimates for each pretreatment process will require specific vendor quotationsbased on pretreatment process conditions and any pretreatment chemical recoveryrequirements. Overall ethanol yield, which is largely based on achieved overallsugar yield achieved in pretreatment and subsequent enzymatic hydrolysis steps,is one of the single-most important factors in determining projected ethanolcosts for each process. Many key variables and parameters can have significantimpact on the pretreatment economics. While many are discussed within this talk, continuing R&D will generate data that further guide the development of thebiomass conversion industry.

See more of this Session: Developments In the Pretreatment of Lignocellulosics for Bioconversion II
See more of this Group/Topical: Sustainable Engineering Forum

(525a) Pt-Free Electrocatalysts for Efficient Oxygen Reduction In Alkaline Fuel Cells: Experimental and Computational Insights

Adam Holewinski and Suljo Linic, Chemical Engineering, University of Michigan, Ann Arbor, MI

The slow rate of the oxygen reduction reaction (ORR) is the major source of overpotential loss in low-temperature hydrogen fuel cells. Expensive Pt-based materials have been found to be the most effective catalysts, and exploration of cheaper alternatives has been hampered by the requirement of operation at low pH, where most metals are susceptible to corrosion. The pH constraint has primarily been due to a lack of electrolyte membranes that are stable under basic conditions, but recent advances in membrane technology have made fuel cell operation at high pH feasible. This is a landscape-changing event which expands the phase space of potential catalytic materials. Most notably, in basic medium Ag is redox stable and shows comparable ORR activity to Pt on a power-per-dollar basis.

In this work, we couple cyclic voltammetry (CV) and rotating disk electrode (RDE) experiments with quantum chemical density functional theory (DFT) calculations to compare ORR mechanisms on Pt and Ag. The calculations account for the potential-dependent interactions of reactive intermediates and co-adsorbed species on the catalyst surface in an aqueous environment. Predicted surface structures agree well with CV adsorption measurements, and predicted overpotentials agree with the onset of catalytic activity measured with the RDE method. Our studies indicate the main tradeoff for an ORR catalyst is between the ability to activate molecular O_2 and the ability to remove adsorbed hydroxyl groups in the form of water. While Ag has difficulty activating O_2, Pt binds hydroxyl groups too strongly. Using these insights, we utilize DFT calculations to identify bimetallic nanostructures of silver and 3d elements (Ni,Co,Fe), which more effectively activate oxygen without binding hydroxyls too strongly. We synthesize these bimetallic structures and demonstrate their enhanced ORR activity. The improvements are discussed in a general framework designating the optimal electronic structure for ORR catalysis.

See more of this Session: Electrocatalysis for PEM Fuel Cells I
See more of this Group/Topical: Catalysis and Reaction Engineering Division

(525b) Electrochemical Characterization of a Platinum Nanotubule Array Electrode Structure Made Via Template Nanofabrication

Eric Broaddus[1], Joel Brubaker[2] and **Scott A. Gold**[2], (1)Institute for Micromanufacturing, Lousiana Tech University, Ruston, LA, (2)Chemical and Materials Engineering, University of Dayton, Dayton, OH

One dimensional nanostructures including nanowires, nanotubes, and nanorods have drawn significant for electrocatalytic applications due to potential advantages that include fewer diffusion impeding interfaces with polymeric binders, more facile pathways for electron transfer, and more effective exposure of active surface sites. Such 1D nanostructures have been fabricated using a variety of techniques including template based methods (wetting and electrosynthesis or electrodepostion), electrospinning, deposition onto nanowire or nanofiber supports, and others. The bulk of the published work on these nanomaterials has focused on demonstrating the viability of the nanofabrication process and describing the morphology, composition, crystal structure and other basic material properties of the nanostructures. The potential of nanomaterials lies in their functional properties which have received less attention, at least in the area of electrocatalysis. For this work a template wetting approach was used to fabricate an array of platinum nanotubules which were examined electrochemically with regard to the electro oxidation of ferrocyanide and formic acid. The former has long been employed as an electrochemical standard in evaluating new electrode structures in comparison to more traditional ones. The latter had long been used as a model compound for other small organics but has also shown great promise in fuel cell applications. For both reactions, arrays of 100 and 200 nm nanotubules were compared to a traditional platinum black catalyst, all of which were found to have similar surface areas. Peak formic acid oxidation current was found to be highest for the 100 nm nanotubule array, followed by the 200 nm nanotubule array and the Pt black, however CO tolerance of the three platinum electrodes was found to be nearly identical, as were the onset potentials of the oxidation and reduction peaks. The higher current response was attributed to enhanced mass transfer in the nanotubule electrodes, likely due to a combination of both the more open nanostructure as well as the lack of a polymeric binder in the catalyst layer. Using the ferrocyanide reaction, electron transfer coefficients were evaluated. Faster electron transfer kinetics were observed with the Pt nanotubules than with the Pt black electrode, likely due to the reduced number of interfaces and the lack of a polymeric binder such as Nafion.

See more of this Session: Electrocatalysis for PEM Fuel Cells I
See more of this Group/Topical: Catalysis and Reaction Engineering Division

(525c) Synthesis of Freestanding Metal-Platinum Core-Shell Structure with Well Defined Facets As Electro-Catalyst for the Oxygen Reduction Reaction

Yanqi Zhang, Qian Xu and Yushan Yan, Chemical Engineering, University of California, Riverside, Riverside, CA

Electro-catalytic oxygen reduction reaction (ORR) has drawn considerable attention because of its slow kinetics and the need for better electro-catalysts with decreased Pt content for proton exchange membrane fuel cells (PEMFCs). One promising approach to solving these problems is to design catalysts with little amount of Pt supported on suitable metal nanomaterials [1].

Here, we report the design and synthesis of freestanding metal-platinum core-shell nanostructures with well defined facets via a simple solution based method. Pd nanocrystal is synthesized by polyol method and used as core for Pt overlayer growth. The surface morphology and composition of the catalyst is characterized by the SEM and TEM techniques. The surface of the nanostructure is bound mainly by [111] and [110] planes. Rotating disk electrode measurements are carried out to analyze the ORR kinetics and activities of the catalysts. Preliminary results suggest that controlling the morphology of Pt nanostructures can improve their catalytic properties and increase their activity both on specific and mass basis compared to the commercially used Pt/C and the first generation catalyst Pt black. The new type of catalysts has a great potential in fuel cell applications. Better performance is expected by optimization of Pt shell thickness. Other substrate can be considered to further reduce the cost of the catalyst.

See more of this Session: Electrocatalysis for PEM Fuel Cells I
See more of this Group/Topical: Catalysis and Reaction Engineering Division

(525d) Double Gyroid Nanostructured Platinum As Highly Durable Oxygen Reduction Reaction Electrocatalyst

Thomas F. Jaramillo, Jakob Kibsgaard and Yelena Gorlin, Chemical Engineering, Stanford University, Stanford, CA

Improving the stability of platinum-based polymer electrolyte fuel cells is a key to facilitate their aspired commercialization [1]. The current industry standard Pt nanoparticles are subject to dissolution, agglomeration and detachment of whole particles which lead to a loss of active catalytic surface area [2, 3]. Here we present on an extended 3D nanoporous double gyroid (DG) Pt network structure that displays a superior durability over commercial Pt nanoparticles. The DG Pt catalyst was synthesized on a glassy carbon rotating disk electrode by means of a silica template created by evaporation induced self-assembly using the commercially available Pluronic P84 surfactant. The high surface area DG Pt shows improved catalytic activity for the oxygen reduction reaction (ORR) as measured by the intrinsic kinetic current density compared to commercial Pt nanoparticles. Accelerated stability test was performed by potential cycling up to 10000 cycles and DG Pt displays a greatly enhanced stability compared to commercial Pt nanoparticles. Whereas Pt nanoparticles loose over half of their electrochemical surface area (ECSA) after 10000 cycles, DG Pt maintains over 80% of the initial ECSA. We attribute the superior durability of the DG structure to a very high mechanical stability. The interlocked continuous DG network structure renders the catalyst sinter-proof and not prone to detach from the carbon support as has recently been demonstrated to be major loss mechanisms [3]. The high stability together with a high catalytic activity ensured by the nanometer scale of the DG network structure and good mass-transport properties render DG Pt an excellent and durable ORR catalyst.

References:

[1] Handbook of Fuel Cells: Fundamentals, Technology, and Applications. W. Vielstich, H. A. Gasteiger, A. Lamm, H. Yokokawa, Eds., John Wiley & Sons Ltd, Chichester, (2009).

[2] Y. Shao-Horn, W. Sheng, S. Chen, P. Ferreira, E. Holby, D. Morgan: Topics in Catalysis, Vol. 46 (2007),

p. 285-305.

[3] K. Hartl, M. Nesselberger, K. J. J. Mayrhofer, S. Kunz, F. F. Schweinberger, G. Kwon, M. Hanzlik, U. Heiz, M. Arenz: Electrochimica Acta, Vol. 56, (2010), p. 810-816.

See more of this Session: Electrocatalysis for PEM Fuel Cells I

(525e) Highly Active and Durable Pt-Bimetallic Electrocatalysts

Chao Wang, Nenad Markovic and Vojislav Stamenkovic, Materials Science Division, Argonne National Laboratory, Argonne, IL

Development of advanced nanomaterials relies on the tuning of material structure at nanoscale. This is in particular manifested in the design of electrocatalysts demanding both high catalytic activity and durability. Here we report the nanoscale architecture of a Pt-bimetallic catalyst with Pt enriched near-surface layers, which shows superior electrocatalytic performance for the oxygen reduction reaction (ORR). The desirable architecture was first established on extended surfaces with customized composition profiles and then realized in nanocatalysts via organic solution synthesis. Electrochemical studies for the ORR demonstrated that the novel Pt-bimetallic catalyst exhibited improvement factors in mass activity of more than six times before, and over one order of magnitude after elongated potential cycling versus conventional Pt catalysts.

See more of this Session: Electrocatalysis for PEM Fuel Cells I
See more of this Group/Topical: Catalysis and Reaction Engineering Division

(525f) Linking Electrode Performance with Structure Using X-Ray Tomography

Molly Jhong[1], Fikile R. Brushett[2] and Paul J.A. Kenis[2], (1)Department of Chemical & Biomolecular Engineering, University of Illinois at Urbana-Champaign, Urbana, IL, (2)Department of Chemical & Biomolecular Engineering, University of Illinois at Urbana Champaign, Urbana, IL

Fuel cells have been extensively investigated as alternative power sources due to their high efficiency, high energy density and low emissions. Unfortunately, broad commercialization of fuel cells has been hampered by prohibitively high costs and insufficient component durability. The complex electrochemical, transport and degradation processes that govern the performance of catalysts/electrodes within an operating fuel cell need to be understood better, so cheaper and more robust electrodes can be manufactured.

The investigation of fuel cell electrode degradation is often done by studying the structural and compositional changes in catalyst layers. However, the structural properties of the gas diffusion layer (GDL) also play an important role. Its structure plays a key role (*i*) in transport of the reactant gas from flow channels to catalyst layer effectively, (*ii*) in draining out liquid water from catalyst layer to flow channels, and (*iii*) in conducting electrons with low resistance. Therefore, a better understanding on the relationship between the electrochemical events in the catalyst layer and the transport phenomena in the GDL is needed to improve the fuel cell performance and durability. To this end, we have employed a microfluidic platform to investigate electrochemical performance characteristics. Also, we have used micro-tomographic imaging (MicroCT) to characterize the 3-D architecture and organization of the whole electrode, information that cannot be accomplished by typical surface imaging method such as scanning electron microscopy (SEM)

In this presentation, we will demonstrate the development of structure-activity relationships for electrodes within operating fuel cells. Specifically, we will demonstrate how MicroCT allows for qualitative and quantitative analysis of the whole fuel cell electrode, for example, as a function of electrode preparation methods.

See more of this Session: Electrocatalysis for PEM Fuel Cells I
See more of this Group/Topical: Catalysis and Reaction Engineering Division

(525g) Anode Catalyst Layers for Direct Liquid Fuel Cells

Cynthia Ann Rice-York[1], Shilpa Beravelli[2] and Akshay Bauskar[2], (1)Department of Chemical Engineering/Center for Manufacturing Research, Tennessee Technological University, Cookeville, TN, (2)Department of Chemical Engineering, Tennessee Technological University, Cookeville, TN

As technology demands for small personal portable electronic devices increases, so does the need for smaller more efficient power sources. Direct liquid fuel cells (DLFCs) have the potential to satisfy this market, due to their small size, portability, ability to 'hot-refuel' permitting 24/7 operation, and their high theoretical power densities. The current technology is limited by poor turnover efficiencies due to surface poison accumulation and fuel crossover inducing parasitic losses. Electrocatalytic efficiency remains a challenge for hydrocarbon-based liquid fuel cells. In this presentation, optimization of the anode catalyst layer for two hydrocarbon-based DLFCs, both methanol and formic acid, will be disused. To increase the electrocatalytic turnover efficiency the catalyst layers have been altered, either by: spontaneous ad-atom adsorption onto commercial catalyst, pore-formers to enhance mass transport, and strategic integration of sulfated zirconia superacid nanoparticles. The catalyst layers were fabricated in-house and a full suite of analytical techniques are employed to unravel DLFC performance enhancement – scanning electron microscopy (SEM), thermal gravimetric analysis (TGA), electrochemical impedance spectroscopy (EIS), cyclic voltammetry (CV), etc.

Acknowledgements:

NSF EPSCoR State-wide Grant and Start-up Funding Provided by the Center for Manufacturing Research

See more of this Session: Electrocatalysis for PEM Fuel Cells I
See more of this Group/Topical: Catalysis and Reaction Engineering Division

(541a) Flow Regime Identification In Different Two-Phase Reactors Based On Extraction of Various Entropies From Gauge Pressure Fluctuations

Stoyan Nedeltchev[1], Shreekanta Aradhya[2], Faraj Zaid[2] and Muthanna Al-Dahhan[2], (1)Department of Chemical and Biological Engineering, Missouri University of Science and Technology, Rolla, MO, (2) Chemical and Biological Engineering, Missouri University of Science and Technology, Rolla, MO

The multiphase reactor hydrodynamics should be studied carefully since the flow structure affects all major parameters. It is essential to have a reliable method for identification of the boundaries of the main hydrodynamic regimes. The multiphase reactors can operate under different flow regimes. The reactor performance can change significantly as a result of flow regime change. Thus, it is very important to know how to identify the boundaries of the different flow regimes since reactor volume productivity, mixing, mass and heat transfer are affected by the prevailing flow regime. In this work, different entropies (Kolmogorov entropy (KE) and maximum information entropy) were derived from the gauge pressure fluctuations (measured by pressure transducers PX309 or PX409, Omega Eng., USA) recorded at different axial positions in three different two-phase reactors (bubble column, fluidized bed and spouted bed). As a criterion for identification of each transition velocity was used the local minimum in the KE profile as recommended by Letzel et al. (1997). The KE algorithm (Schouten et al., 1994) can be used for characterization of flow regimes and transitions between them. The KE is sensitive and robust and it is used frequently for flow regime identification. In general, KE is large in the case of higher degree of turbulence and it is low when more regular behavior is observed.

In a bubble column (0.14 m in ID) equipped with a perforated plate distributor (163 holes, Ø 1×10^{-3} m) and operated with an air-water system three different transition velocities were identified: 0.016, 0.03 and 0.046 m/s, respectively. The first one identified the transition from chain-bubbling regime to bubbling flow regime, the second one identified the onset of the transition flow regime and the third transition velocity marked the onset of the churn-turbulent regime. It is worth noting that the value of the second transition velocity is very close to the prediction of Reilly et al. (1994).

In a fluidized bed (0.14 m in ID) equipped with a porous-plate (pore size: 40 μ m) distributor and operated with an air-glass beads (ρ_s=2500 kg/m^3, d_p=150-210 μ m) system the first well-pronounced drop in the KE profile occurred at 0.121 m/s and it corresponded to the minimum fluidization velocity U_{mf} at which the first bubbles were formed. In the interval $0.151 \le U_g \le 0.377$ m/s the KE profile exhibited a monotonous decrease which was explained with the large bubble formation. Therefore, those were the boundaries of the fast bubble subregime of bubbling fluidization (Trnka et al., 2000). Beyond U_g=0.377

m/s the KE profile practically leveled off and this marked the onset of the slow bubble subregime of bubbling fluidization. Further, at U_g=0.83 m/s a well pronounced minimum was observed which marked the transition to turbulent fluidization regime. The effect of the axial position on the transition velocities was studied, as well.

The dynamic behavior of spouted beds was well described by applying the nonlinear chaos analysis to gauge pressure fluctuations measured by pressure transducer (model PX309, Omega Inc., USA). In the literature hitherto, the flow regimes (packed bed, stable spouting and unstable spouting) in spouted beds have been analyzed based on statistical and frequency analyses of pressure fluctuations, rescaled range analysis and mutual information functions of differential pressure fluctuations. According to Xu et al. (2009) the packed bed and stable spouting are distinguished by the formation of the stable spout/ fountain. Unstable spouting is characterized by swirling and pulsation of the spout with time, distinct regular motion caused by periodic pulsation of the spouting. The signals in the stable spouting regime have more complex nature.

Air was used as a spouting gas, while glass beads (2.1×10^{-3} m in diameter, ρ_s=2450 kg/m^3) were used as a solid material. The measurements were performed in two spouted beds with different diameters (0.076 m in ID and 0.152 m in ID) at two different axial positions (z) above the gas distributor: 0.124 m in the smaller column and 0.24 m in the bigger column. The static bed height in the smaller column was 0.16 m, whereas in the bigger column it was kept at 0.325 m. Both spouted beds were equipped with porous-plate distributors.

In the smaller spouted bed the KE profile exhibited two minima at U_g=0.783 m/s and 0.885 m/s. The first transition velocity delineated the boundary from packed bed regime to stable spouting regime, whereas the second transition velocity identified the boundary from stable to unstable spouting regime. In the bigger spouted bed, these two transition velocities corresponded again to two local minima and were shifted to somewhat lower values (U_g=0.563 m/s and 0.631 m/s). In addition, the KE values in the bigger unit were higher and this implied that the degree of turbulence was higher. The results were validated based on both KE and maximum information entropies (Nedeltchev et al., 1999, 2000 and 2003).

REFERENCES

Letzel, H. M., J. C. Schouten, R. Krishna, C. M. van den Bleek, Characterization of Regimes and Regime Transitions in Bubble Columns by Chaos Analysis of Pressure Signals, *Chem. Eng. Sci.* **52**, 4447 (1997)

Nedeltchev, S., S. Ookawara, K. Ogawa, A Fundamental Approach to Bubble Column Scale-Up Based On Quality of Mixedness. *J. Chem. Eng. Japan* **32**, 431 (1999)

Nedeltchev, S., S. Ookawara, K. Ogawa, Time-Dependent Mixing Behaviors in Both Lower and Upper

Zones of a Bubble Column. *J. Chem. Eng. Japan* **33**, 761 (2000)

Nedeltchev, S., S. B. Kumar, M. P. Dudukovic, Flow Regime Identification in a Bubble Column Based On Both Kolmogorov Entropy and Quality of Mixedness Derived from CARPT Data. *Can. J. Chem. Eng.* **81**, 367 (2003)

Reilly, I. G., D. Scott, T. de Bruijn, D. MacIntyre, The Role of Gas Phase Momentum in Determining Gas Holdup and Hydrodynamic Flow Regimes in Bubble Column Operations, *Can. J. Chem. Eng.***72**, 3 (1994)

Schouten, J. C., F. Takens, C. M. van den Bleek, Maximum-Likelihood Estimation of the Entropy of an Attractor. *Phys. Rev. E* **49**, 126 (1994)

Trnka, O., V. Vesely, M. Hartman, Z. Beran, "Identification of the State of the Fluidized Bed by Pressure Fluctuations", *AIChE J.* **46**, 509 (2000)

Xu, J., X. Bao, W. Wei, H. T. Bi, J. R. Grace, "Chaotic Characteristics of Pressure Fluctuations in a Gas Spouted Bed", *Can. J. Chem. Eng.*.**87**, 252 (2009)

See more of this Session: Multiphase Reaction Engineering
See more of this Group/Topical: Catalysis and Reaction Engineering Division

(541b) Incremental Identification of Reaction and Mass-Transfer Rates In Gas-Liquid Reaction Systems Using Tendency Modeling

Nirav Bhatt[1], Christos Georgakis[2] and **Dominique Bonvin**[1], (1)Laboratoire d'Automatique, EPFL, Lausanne, Switzerland, (2)Systems Research Institute and Chemical and Biological Engineering, Tufts University, Medford, MA

Theidentification of reliable reaction and mass-transfer rates is important forbuilding first-principles models of gas-liquid reaction systems. Theidentification of these rates involves the determination of a model structure(reaction stoichiometry, rate expressions for the reactions and mass transfers)and of the corresponding parameters. The identification of these rate expressionsfrom measured concentrations is a challenging task because of the directcoupling between the reactions and the transfer of reactants and productsbetween the two phases.

The identification task can be performedglobally in one step by choosing the model structure and estimating the modelparameters via the comparison of model predictions and measured data. Theapproach is termed *simultaneousidentification* since all reactions and mass transfers are identifiedsimultaneously. The procedure needs to be repeated for all candidate modelstructures. Hence, the simultaneous identification can be computationallycostly when several candidate rate expressions are available for each reactionand mass transfer. Furthermore, since the global model is fitted so as toreduce the least-squares error, structural mismatch in one rate expression ofthe model will typically result in errors in all the parameters. Finally, it is often difficult to comeup with suitable initial parameter values, which may lead to convergenceproblems.

An *incrementalidentification* approach has recently been proposed, which decomposes theidentification task into the following two steps [1, 2]: (i) computation of theextents of reaction and mass transfer from measured concentrations withoutknowledge of the reaction and mass-transfer rates, and (ii) for each rateindividually, identification of the rate expression and its parameters from thecomputed extents. The fact thateach reaction and mass-transfer rate is treated individually in the incrementalapproach helps reduce considerably the number of model candidates, therebyreducing the computational effort. Although the proposed incremental approachprovides an efficient framework for the identification of gas-liquid reactionsystems, a systematic way of selecting the appropriate rate expressions fromseveral candidate expressions is needed in Step (ii).

Recently, the so-called generalized tendency modeling (GTeMoC) method has been proposed to select appropriaterate expressions from a large number of rate expression candidates [3, 4]. In theGTeMoC methodology, a stepwise linear regression isused as a tool to selectappropriate rate expressions. Moreover, the statistical metrics are developedto discriminate rate expression candidates and avoid collinearity in rate parameters.However, the effect of mass transfer rates is not treated explicitly in the GTeMoC method, and lumped rate expressions containingthe effect of reactions and mass transfers are identified.

This work combines the incremental approach andthe GTeMoC methodology so that the reaction and mass-transfer rates can be identifiedindividually. Hence, the resulting incremental approach proceeds in threesteps: (i) computation of the extents of reaction and mass transfer from measuredconcentrations without knowledge of the reaction and mass-transfer rates, (ii)computation of the reaction and mass-transfer rates through differentiation ofthe corresponding computed extents, and (iii) for each rate individually,identification of the rate expression and its parameters using the GTeMoCmethod. The proposed incremental identification approach combines the strengthsof the incremental approach (can handle each reaction and each mass transferindividually) and the GTeMoC method (can efficiently select the rate expressionfrom several candidate expressions). The approach will be illustrated via thesimulation of the chlorination of butanoic acid.

References

[1] N. Bhatt,M. Amrhein, and D. Bonvin. Extents of Reaction, Mass Transfer andFlow for Gas-Liquid Reaction Systems. *Industrial& Engineering Chemistry Research*, 49(17): 7704-7717, 2010.

[2]N. Bhatt, M. Amrhein, and D. Bonvin. Incremental Identificationof Kinetic and Transport Phenomena Using the Concept of Extents. Submitted to *Industrial & Engineering ChemistryResearch*, 2011.

[3] F. Makrydakiand C. Georgakis. IterativeEstimation of Kinetic Structures of Complex Reaction Systems. To be submitted, 2011.

[4] F. A. Makrydaki.*Optimization of Batch Reactions using Data-Driven and Knowledge-Driven Models -The Case of Asymmetric Catalytic Hydrogenation.* PhD Thesis, TuftsUniversity, Medford, MA, USA, 2010.

See more of this Session: Multiphase Reaction Engineering
See more of this Group/Topical: Catalysis and Reaction Engineering Division

(541c) Local Time-Averaged Heat Transfer Coefficient In a Pebble Bed

Rahman Abdulmohsin and Muthanna Al-Dahhan, Chemical and Biological Engineering, Missouri University of Science and Technology, Rolla, MO

In real pebble bed reactor (PBR) for the 4th generation nuclear energy, thorough understanding the phenomena of the transport of heat generated during nuclear fission from slowly moving hot pebbles to the flowing coolant gas is crucial for safe design and efficient operation. This heat transport can be quantified and characterized in terms of heat transfer coefficient. Therefore, in this work, the local heat transfer coefficient of the gas phase which is used for cooling has been investigated experimentally in a 0.3 m diameter cold-flow pebble bed. Glass bead particles of 5 cm diameter and 2.55 gm/cm^3 have been used in randomly packed bed. The effect of superficial gas velocity on the heat transfer coefficient has been investigated using a wide range from 5 to 60 cm/s that cover both laminar and turbulent flow regimes. A fast-response heat transfer probe (0.02s) was developed to measure the local instantaneous heat flux and surface temperature, simultaneously, at different axial and radial positions in the bed and hence local time-averaged heat transfer coefficient. Experiments results indicate that the increase of superficial gas velocity causes an increase in the heat transfer coefficient. Also, the heat transfer coefficients in the center of the bed are higher than those near the wall of the bed for all the conditions studied.

See more of this Session: Multiphase Reaction Engineering
See more of this Group/Topical: Catalysis and Reaction Engineering Division

Trickle-Bed Reactor Studies for Selective Oxidation of Glycerol to Dihydroxyacetone over Pt-Bi/C Catalyst

Gregory Honda, Wenbin Hu, Wesley Fleming, and Arvind Varma

School of Chemical Engineering, Purdue University
480 Stadium Mall Drive, West Lafayette, IN 47907

Increase in the production of biodiesel has resulted in an oversupply of glycerol, the primary byproduct which forms in a 1:10 ratio with biodiesel. An effective utilization of glycerol (currently ~$0.02/lb) by conversion to the higher value chemical dihydroxyacetone (DHA; currently ~$20/lb) will enhance the economics of biodiesel production. Recently, we have developed a catalyst (3%Pt, 0.6% Bi on activated carbon) capable of selective oxidation of glycerol to DHA at high yield (>40%) in a semi-batch reactor [1]. Development of a continuous-flow process, however, is necessary to increase production. Towards this end, a trickle-bed reactor has been used with cocurrent downflow of oxygen and aqueous glycerol solution over a packed bed. An optimization of the feed solution pH and concentration as well as reactor temperature and pressure will be presented. Additionally, the effects of gas and liquid flowrates will be discussed in light of hydrodynamic characteristics of trickle-bed reactors. The development of a continuous-flow process for conversion of glycerol to DHA increases the economic viability of biodiesel production, while also providing insight into the behavior of trickle-bed reactors.

[1] Hu, W., Knight, D., Lowry, B., Varma, A. Selective Oxidation of Glycerol to Dihydroxyacetone over Pt-Bi/C Catalyst: Optimization of Catalyst and Reaction Conditions. *Ind. Eng. Chem. Res.* 2010, **49** (21), 10876-10882.

(541e) Directed Evolution of Metalloenzymes for Organic Synthesis

Carl A. Denard[1], Lars Martin Jarenmark[2], Ramesh Giri[2], John F. Hartwig[2] and Huimin Zhao[1], (1) Chemical and Biomolecular Engineering, University of Illinois at Urbana Champaign, Urbana, IL, (2) Chemistry, University of Illinois at Urbana Champaign, Urbana, IL

AICHE abstract

Directed Evolution of Metalloenzymes for Organic Synthesis

Carl Denard, Lars Martin Jarenmark, Ramesh Giri, John F. Hartwig and Huimin Zhao, UIUC

Catalytic asymmetric tandem reactions have been reported extensively in synthetic chemistry literature. The use in tandem of catalysts from different disciplines has been reported. However, there are no reports on combining two metal catalysts from completely different fields, i.e., an organometallic catalyst and a metalloenzyme. In the Zhao and Hartwig labs, we have developed a novel catalytic asymmetric tandem reaction to achieve a successive olefin isomerization-enantioselective epoxidation, a reaction that is difficult to envision occurring with a single catalytic system. This process can potentially provide a practical and efficient strategy for the enantioselective preparation of linear terminal epoxides in high *ee* from a mixture of simple substrates. To address the issue of catalyst activity in the biocompatible solvent hexadecane, ruthenium hydride complexes were synthesized that can isomerize several olefins of interest with high activity and interesting selectivities, at room temperature and at enzymatic conditions. Sulfur activation further enhances the isomerization reaction rates. By combining the isomerization catalyst with cell lysates or purified P450 BM3 mutants coupled with a cofactor regeneration system, we report for the first time that the two reactions work in tandem. Starting with a terminal olefin, we were able to detect the corresponding internal epoxide at a ratio of 1.5:1 to 2:1 compared to the terminal epoxide concentration. In addition, we tackled the thermodynamically challenging internal to terminal isomerization in tandem with the epoxidation reaction and were able to detect the terminal epoxide. In the case of 4-methyl-*cis*- and 4-methyl-*trans*-2-pentene, we obtained the most promising result of detecting the terminal epoxide in an equimolar amount to the internal epoxides. This represents the first report that shows that two metal catalysts from very different fields, namely a metalloenzyme and an organometallic catalyst, are capable of working in tandem to tackle an important synthetic challenge. We have further expanded the substrate scope of the reaction system to a number of straight and branched C5-C7 aliphatic olefins. This tandem process should be generally applicable to other high-impact and challenging transformations that require excellent selectivities.

See more of this Session: Multiphase Reaction Engineering
See more of this Group/Topical: Catalysis and Reaction Engineering Division

(541f) Effect of Pebble Sizes On Axial Gas Dispersion In a Pebble Bed

Rahman Abdulmohsin and Muthanna Al-Dahhan, Chemical and Biological Engineering, Missouri University of Science and Technology, Rolla, MO

Despite the recent interest in the pebble bed reactors (PBRs) for the 4th generation nuclear energy that motivates various types of studies; there is still a lack of fundamental understanding of the complex gas flow structure and mixing pattern. In this work, the non-ideal flow behavior of the gas phase which is used for cooling has been investigated experimentally in a 0.3 m diameter cold flow pebble bed. The extent of mixing and dispersion of the gas phase has been qualified in terms of axial dispersion coefficient. The effect of pebble size on the axial dispersion coefficient of gas phase has been investigated with three different sizes of pebbles (1.25 cm, 2.5 cm and 5 cm). The superficial gas velocities were ranging from 5 to 60 cm/s which cover both laminar and turbulent flow regimes. An advanced gaseous tracer technique was developed to measure the gas residence time distributions (RTDs) in the bed. The technique accounts properly for the extra dispersion encountered in different parts of the experimental set-up such as plenum, distributor, sampling lines, etc. Experiments results indicate that the increase of pebble size causes an increase in the axial dispersion coefficient. This can be attributed to an increase in the packing porosity of the bed which yields less pressure drops across the bed. Also, it has been found that the axial dispersion coefficient increases as superficial gas velocity increased.

See more of this Session: Multiphase Reaction Engineering
See more of this Group/Topical: Catalysis and Reaction Engineering Division

(544a) Lignin Depolymerization and Deoxygenation In Ionic Liquids

Blair J. Cox, Thong Q. Ngo and John G. Ekerdt, Chemical Engineering, University of Texas at Austin, Austin, TX

Lignin, an amorphous aromatic biopolymer that comprises up to 30% of plant matter, was dissolved and depolymerized in ionic liquids under acidic conditions. The acidic ionic liquid 1-methylimidazolium chloride (HMIMCl) acts as a combination solvent and catalyst that facilitates degradation of lignin into smaller fragments. This degradation occurs through the catalytic hydrolysis of the β-O-4 ether linkage, the most prominent linkage in the lignin polymer. Studies with guaiacylglycerol-β-guaiacyl ether and veratrylgycerol-β-guaiacyl ether as phenolic and non-phenolic lignin model compounds demonstrated that acid catalyzed degradation of lignin in HMIMCl proceeded efficiently at 150 °C, primarily by eliminating water to produce an enol ether intermediate. This intermediate then undergos hydrolysis of the β-O-4 ether linkage, which is analogous to degradation of the β-O-4 linkage in the lignin polymer. These process conditions were next applied to lignin that was extracted from pine wood biomass through dissolution in the HMIMCl ionic liquid and subsequent precipitation. Analysis with gel permeation chromatography indicates significant decrease in size after treatment in HMIMCl at 150°C for 40 min. After depolymerization, the lignin fragments will be subjected to hydrodeoxygenation (HDO), while in the 1-butyl-3-methyl imidazolium chloride ionic liquid, over supported metal catalysts. Preliminary HDO studies using phenol with ceria- or zirconia-supported copper and nickel catalysts show complete conversion and suggest a viable path to HDO of lignin.

See more of this Session: Novel Catalytic and Separation Process Based on Ionic Liquids
See more of this Group/Topical: International Congress on Energy 2011

(544b) Preparation of Cellulose-Rich Materials From Cornstalks Using Biodegradable Ionic Liquids

Xiaofeng Zhang, Institute of Nuclear and New Energy Technology, Tsinghua University, Beijing, China

In our work, we obtained cellulose-rich materials via utilizing a kind of economic biodegradable IL, choline acetate ([ChO][OAc]), demonstrating the possibility of the partial dissolution and separation of cornstalks after mild grinding. The selected [ChO][OAc] IL presents very low toxicity, due to the benign nature of the cation and its high biodegradation potential which was evaluated under aerobic conditions. Delignification of cornstalks using the [ChO][OAc] IL was successfully achieved that lignin content decreased from 14% of cornstalk raw materials to 2% of undissolved residue under mild conditions. Thus, cellulose-rich materials accounting for 98% could be obtained. Furthermore, the cellulose fibre from [ChO][OAc] solutions can be regenerated to form cellulose membrane on a glass plate. In this way, partial separation of the cornstalk components becomes available with relative efficient, nontoxic and biodegradable ionic liquids.

See more of this Session: Novel Catalytic and Separation Process Based on Ionic Liquids
See more of this Group/Topical: International Congress on Energy 2011

(544c) Preparation of Cellulose-Rich Materials From Cornstalks In Biodegradable Ionic Liquids

Shan Li[1], Minli Zhu[1], Xingmei Lu[2], Zengxi Li[1] and **Suojiang Zhang**[2], (1)College of Chemistry and Chemical Engineering, Graduate University of Chinese Academy of Sciences, Beijing, China, (2) Institute of Process Engineering, Chinese Academy of Sciences, Beijing, China

Agricultural wastes such as bagasse, cornstalks, are relatively cheap sources of lignocellulose consisting of lignin, cellulose and hemicellulose. However, most of them are burned as waste without rational use, even causing severe environment contamination. Lignocellulosic materials have aroused wide public concern due to their potential as starting materials for biofuels and other commodity chemicals.[1]

The unique physicochemical properties of ionic liquids (ILs), such as non-flammability, negligible vapour pressure, high thermal and chemical stability, have attracted increasing interests in their potential applications in various areas. It is showed that lignocellulosic materials can be dissolved in ILs which are almost imidazolium-based ionic liquids (IBILs)[2]. The differences of solubility among the components of lignocellulose make it possible to separate them from each other. However, the scope of possible applications of imidazolium-based ionic liquids is restricted due to their high cost, toxicity and the non-biodegradability of the imidazolium ring[3].

In our work, we obtained cellulose-rich materials via utilizing a kind of economic biodegradable IL, choline acetate ([ChO][OAc]), demonstrating the possibility of the partial dissolution and separation of cornstalks after mild grinding. The selected [ChO][OAc] IL presents very low toxicity, due to the benign nature of the cation and its high biodegradation potential which was evaluated under aerobic conditions. Delignification of cornstalks using the [ChO][OAc] IL was successfully achieved that lignin content decreased from 14% of cornstalk raw materials to 2% of undissolved residue under mild conditions. Thus, cellulose-rich materials accounting for 98% could be obtained. Furthermore, the cellulose fibre from [ChO][OAc] solutions can be regenerated to form cellulose membrane on a glass plate. In this way, partial separation of the cornstalk components becomes available with relative efficient, nontoxic and biodegradable ionic liquids.

Reference

1 Suzie S. Y. Tan, Douglas R. MacFarlane, Jonathan Upfal, Leslie A. Edye, William O. S. Doherty, Antonio F. Patti, Jennifer M. Pringle and Janet L. Scott,Green Chem., 2009, 11, 339-345

2 Ning Sun, Mustafizur Rahman, Ying Qin, Mirela L. Maxim, H°Ctor Rodríguez and Robin D.Rogers, Green Chem., 2009, 11, 646-655

3 Helga Garcia, Rui Ferreira, Marija Petkovic, Jamie L. Ferguson, Maria C. Leitão, H. Q. Nimal Gunaratne, Kenneth R. Seddon, Luís Paulo N. Rebelo and Cristina Silva Pereira, Green Chem., 2010, 12, 367-369

See more of this Session: Novel Catalytic and Separation Process Based on Ionic Liquids
See more of this Group/Topical: International Congress on Energy 2011

(544d) Preparation and Testing of Adsorbents for the Separation of Ionic Liquids From HMF

Yuxiang Rao[1], Alan Cooper[2], Heather Brown[2] and Wei Liu[3], (1)Energy and Environment Directorate, Pacific Northwest National Laboratory, Richland, WA, (2)Pacific Northwest National Lab, Richland, WA, (3)Pacific Northwest National Laboratory, Richland, WA

HMF is an important chemical product since it has the potential to be used as a feedstock to replace petroleum-derived aromatics for production of a range of specialty chemicals and plastics.In this work, we will demonstrate an economically-viable, ionic liquid catalytic process for conversion of biomass-derived raw sugar to 5-hydroxymethylfurfural(HMF). The experimental work includes preparation of adsorbents materials by membrane coating, catalytic reactions in ionic liquid fluids and separation of reacted products by extraction and adsorption, which includes (i) catalytic reaction tests in both a batch and flow reactor system to develop systematic understandings to critical process operation parameters and establish material balances, and (ii) screening of adsorbents and identification of an effective adsorption process for separation of HMF product from reaction mixtures and for complete recovery of the catalyst and ionic liquid.The adsorbent materials tested for this project were characterized by XRD, SEM/EDS and Nitrogen adsorbtion/desorption technology.

See more of this Session: Novel Catalytic and Separation Process Based on Ionic Liquids
See more of this Group/Topical: International Congress on Energy 2011

(544e) Isobutane/1-Butene Alkylation Catalyzed by Ionic Liquid-Superacid Coupled Catalysts

Xueqi Xing, Guoying Zhao, **Suojiang Zhang**, Peng Cui and Hailing Ren, Institute of Process Engineering, Chinese Academy of Sciences, Beijing, China

Abstract-FThe alkylation of isobutane with 1-butene was synergeticly catalyzed by trifluoromethanesulfonic acid and ionic liquids in a batch autoclave. A high-quality alkylates with up to 80% desired trimethylpentane and 95 research octane number were obtained using the coupled catalyst with 33% trifluoromethanesulfonic acid. The coupled catalyst was reused 10 recycles without significant decrease in catalytic performance, compared to the today commercial H2SO4 catalyst. The middle-length cation of ionic liquids, for example C6, was shown to be good for both the quality and yield of alkylates.

See more of this Session: Novel Catalytic and Separation Process Based on Ionic Liquids
See more of this Group/Topical: International Congress on Energy 2011

(544f) Degradation of Poly(ethylene terephthalate) Wastes Catalyzed by Mn-Containing Ionic Liquid

Qian Wang[1], Xingmei Lu[1], Xiangping Zhang[1], **Suojiang Zhang**[1], Chunshan Li[1], Zengxi Li[2], Xueyuan Zhou[1] and Minli Zhu[2], (1)Institute of Process Engineering, Chinese Academy of Sciences, Beijing, China, (2)Graduate University of Chinese Academy of Sciences, Beijing, China

With the increasing of poly(ethylene terephthalate)(PET) consumption in recent years, the effective recycling of PET wastes becomes an urgent affair focused on by many governments and scientific researchers. Ionic liquids (ILs) have been reported as catalysts to degrade PET and most of them have showed some advantages of mild reaction condition and being easily separated from the main product. But the conversion of PET and the selectivity of the monomer bis(2-hydroxyethyl terephthalate)(BHET) are low. Therefore, the exploitation of novel ionic liquids that can efficiently degrade PET is significant.

In this study, Mn-containing ionic liquid was synthesised, characterized by Raman spectroscopy and Electrospray Ionization-Mass Spectrum and used to the catalytic degradation of PET in ethylene glycol. The results shows that the IL exhibits higher catalytic activity, compared with conventional ionic liquids,such as [bmim]Cl, [bmim]H_2PO_4, [bmim]HSO_4. The conversion of PET and the selectivity of the BHET reach up to 100% and 78.9% respectively under the conditions of reaction pressure 1atm, reaction temperature 170°C and reaction time 2.5h. They are both higher than 88.6% and 62.8% catalyzed by 1-butyl-3-methylimidazolium tetrachloroferrate ([bmim]$FeCl_4$) and the reaction time were shortened greatly from 4h to 2.5h(Wang et al, 2010). In addition, the method has advantages that the main product is easily separated from catalyst and has good quality. Finally, the mechanism is proposed and shows that the higher catalytic activity of the IL is owing to the synergistic effect of its cation and anion.

Reference

Wang, H., Yan, R.Y., Li, Z.X., Zhang, X. P., Zhang, S.J.(2010). Fe-containing magnetic ionic liquid as an effective catalyst for the glycolysis of poly(ethylene terephthalate). *Catal. Commun.,* 11, 763-767.

See more of this Session: Novel Catalytic and Separation Process Based on Ionic Liquids
See more of this Group/Topical: International Congress on Energy 2011

Automatic Reaction Mechanism Generation with Group Additive Kinetics

Richard H. West, Joshua W. Allen, and William H. Green
Massachusetts Institute of Technology, Department of Chemical Engineering,
77 Massachusetts Avenue 66-270, Cambridge MA 02139

Abstract

The key challenge in making chemical mechanism development predictive is being able to accurately estimate any possible rate coefficient $k(T)$ even if there are no experimental data. Reaction Mechanism Generator (RMG) is an open-source software project that can build detailed kinetic models for chemical reacting systems (http://rmg.sourceforge.net). It uses a database of rules to propose elementary chemical reactions and to estimate the necessary thermochemical and kinetic parameters. We are modifying the algorithm used to estimate kinetic data to make the estimated reaction rates more reliable and easier to document in cases where they are estimated from sparse data. We present a brief overview of RMG, a discussion of the kinetics estimation options, an explanation of the chosen algorithm, and an assessment of its performance.

Introduction

Kinetic models for gas-phase reacting systems, such as atmospheric chemistry and combustion, often contain thousands of species and reactions. Researchers in these fields have developed a number of tools to help them generate these detailed models [1–5].

Reaction Mechanism Generator (RMG) is an open-source software project that can build detailed kinetic models for reacting systems[4–7]. To estimate reaction rate expressions, RMG uses a group-based approach. The current algorithm works well when the database of rate-estimation rules and associated group values is complete, but performs poorly when kinetic data are sparse. We are modifying the algorithm to make the estimated reaction rates more reliable and easier to document in cases where they are estimated from sparse data.

Automatic kinetic model generation with RMG

Given some starting species (e.g. methane and oxygen) and some reaction conditions (temperature and pressure) it will create a kinetic model of the reaction mechanism consisting of many (up to thousands) elementary reactions between intermediate species. Inside the software molecules are represented as graphs, with atoms as nodes and bonds as edges connecting the nodes. Standard graph-theory methods are used to identify equivalent graphs and ensure uniqueness. RMG uses "reaction families" to generate all the possible reactions that a species can undergo in the presence of the other species in the chemical mechanism. Every reaction family represents a particular type of elementary chemical reaction, such as bond-breaking, or radical addition to a double bond. Each reaction family has a recipe for mutating the graph, and a library of rate expressions for different reacting sites.

Because the model can contain thousands of species and rates, the estimation of thermochemical and kinetic parameters must be very fast. As with most mechanism generating tools, RMG uses a database of known values wherever possible to find thermochemical data for species, but usually the data are unknown and it estimates parameters using a group contribution method. Thermochemistry estimates are based on Benson's group additivity method for standard enthalpies of formation [8, 9]. The functional groups are recognized using a graph-theory matching algorithm. A similar method is used to estimate the rate coefficients for the reactions: functional groups are identified using graph-matching and the rates are estimated from a database of rules.

RMG uses a rate-based termination criterion; the reaction network is expanded until the rates of all reactions going to species not included in the network fall below a certain threshold. This helps to include important pathways without unnecessarily exploring slower pathways, rather than terminating the expansion after a set number of generations [10].

Rate estimation methods

The rate coefficient of a reaction is largely determined by the atoms in the region around its transition state. This region, containing several polyvalent atoms, can be called a "supergroup" [11, 12]. Identifying the supergroup allows one to estimate the reaction rate coefficient.

The supergroup can be decomposed into component groups. For example, in H-abstraction reactions

$$\mathrm{XH} + \mathrm{Y}\cdot \longrightarrow \mathrm{X}\cdot + \mathrm{YH} \tag{1}$$

the component groups would be the abstracting group (Y) and the group from which a hydrogen is abstracted (X).

Currently in RMG, the groups X and Y are used only to locate the transition state supergroup XHY in the database. When a rate expression is not available for XHY, the rates of supergroups close to it in the database are currently averaged using a complicated scheme that can unfortunately lead to poor estimates and obfuscate the source(s) of the final reaction rate expression.

In the new group additive approach, the effect on the kinetic expression from the component groups X and Y are separated and assumed to be independent and additive. For example, the effect of changing Y from a primary to a secondary carbon is independent of the group X [13].

We train our group values with a large database of reaction rates taken from the literature and *ab initio* calculations. We organize the groups in a hierarchical tree structure where child nodes are more specific instances of their parent node. An example is given in Figure 1. The group values for each node are fitted to all the kinetic data that match that node, including those that match its descendants. The goodness of this fit is also stored. When estimating the rate coefficient for a reaction, the most specific instance of each group is identified. If values are missing for that group then its parent node is used, continuing up the tree until a node with data is found. This will allow some indication of the fitting errors at each node and make it clearer how each rate coefficient was estimated in RMG.

This procedure can be made automatic, so that all the group values can be easily refit whenever the user has added new data on individual reaction rates, making it more practical to keep the rate-estimation rules up-to-date with the latest information.

Inspecting the fitted values can suggest modifications to the tree structure. For example, in the bottom left corner of the Y· tree in Figure 1 molecular oxygen $O_2(^3\Sigma_g^-)$ is a sibling of $C_2(X^1\Sigma_g^+)$ although their reactivities are very different.

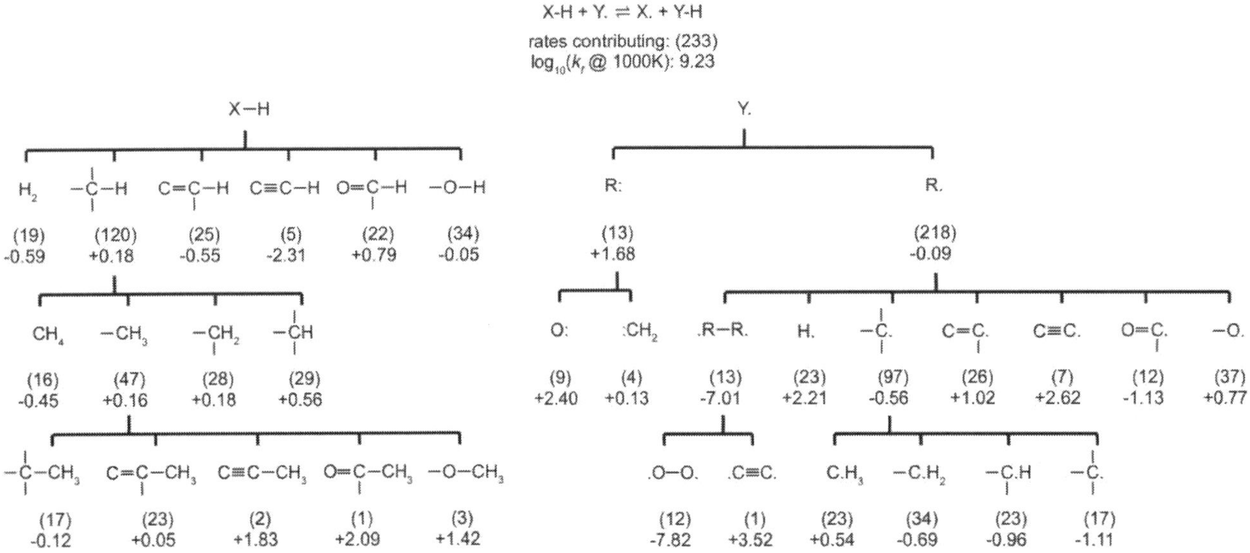

Figure 1: *Part of a pair of trees for hydrogen abstraction reactions, showing the number of reaction rates contributing to the training (in parentheses) and the fitted group value for* $\log_{10}(k_f\ @1000\ K)$

Example

Figure 1 shows part of a pair of trees for the hydrogen abstraction reaction family. In this case the data used are the base-10 logarithms of the forward reaction rate coefficients at 1000 K, per H atom. 223 reaction rate expressions were used in fitting the groups, and the overall average rate was $10^{9.23}$ $cm^3/mol/s$. For the reaction $C_2H_6 + HCO \longrightarrow C_2H_5 + CH_2O$, we can estimate the rate coefficient at 1000 K by identifying the X–H and Y· groups in the tree as follows:

Base		$-C-CH_3$		O=C.		Total
9.23	+	-0.12	+	-1.13	=	7.98

There are 6 equivalent H atoms to abstract so the total rate coefficient is $6 \times 10^{7.98} = 5.6 \times 10^8$ $cm^3/mol/s$, which compares well with a 7.0×10^8 $cm^3/mol/s$ estimate by Tsang *et al.*[14].

Method comparison

To test the methods we extracted 888 rate expressions for hydrogen abstraction reactions of species containing only carbon, hydrogen and oxygen (as covered by our rules) from the PrIMe Kinetics data warehouse[15] and compared them with estimates made using the rules and group values. The test set includes all the available data, not just the curated, checked, and approved values.

For each reaction the reacting functional groups X–H and Y· are identified. Sometimes we have a rule for that exact combination of X and Y, in which case we use it to estimate the rate. The comparison of predicted vs PrIMe k(1000 K) for these cases is shown in Figure 2. The kinetics estimation scheme works quite well. The 95% confidence intervals (shown by the dashed lines) are ± 1.13 in $\log(k_f)$ and most of the outliers are mistakes in the test data from the PrIMe depository. [1]

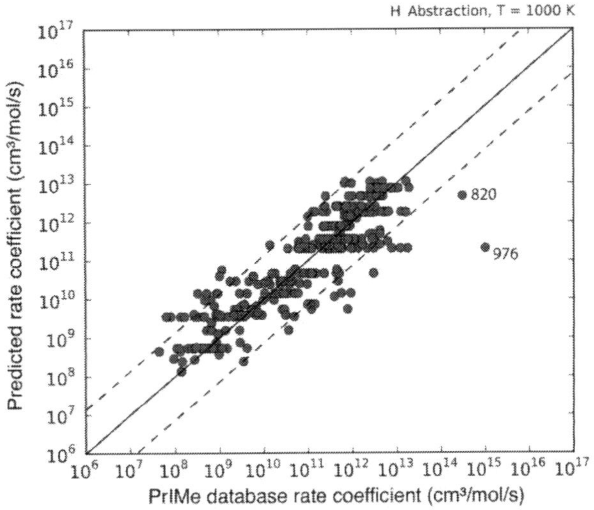

Figure 2: *Parity plot comparing predicted k(1000 K) with data from PrIMe database, for hydrogen abstraction reactions reactions that match a known rule for XHY.*

When there is no rule available for the identified combination of X and Y, the rate must be estimated using the rules that are available. The previous method used in RMG software was to average the rates of rules "nearby" in the trees. When the neighboring pairs of groups are also missing this can lead to complicated expressions which are hard to understand and can give poor estimates. [2] The

[1]Checking the original sources for points 820 and 976 in Figure 2 reveals errors in the activation energies of -6.1 and -9.6 kcal/mol respectively. The outlying point 877 in Figure 3 signifies another mistake in interpreting the PrIMe database: the n in the modified Arrhenius expression represents $(T/298\text{ K})^n$ not $(T/1\text{ K})^n$.

[2]The reaction HC≡C· + H_2O ⟶ HC≡CH + HO· (point 893 in Figure 3) matches the pair of groups (O-pri, Ct-rad), but that rule is unknown. Using the old scheme it is estimated as: (Average of: (Average of: (Average of: (O-pri O2b) & Average of: (O/H/NonDeC O2b) & O-pri H-rad & Average of: (O/H/NonDeC H-rad & O/H/OneDe H-rad) & Average of: (O-pri C-methyl & Average of: (O-pri C-rad/H2/Cs)) & Average of: (O/H/NonDeC C-methyl & Average of: (O/H/NonDeC C-rad/H2/Cs) & Average of: (O/H/NonDeC C-rad/H/NonDeC) & Average of: (Average of: (O/H/NonDeC C-rad/Cs3)) & Average of: (Average of: (H2O2 C4H9O/ c12345 & H2O2 C4H9O/c134(2)5 & H2O2 C4H9O/c134(2)5 & H2O2 C4H9O/c14(2,3)5) & Average of: (H2O2 C3H5/c132)) & Average of: (Average of: (H2O2 C4H9O/c12345 & H2O2 C4H9O/c12345 & H2O2 C4H9O/c134(2)5) & Average of: (Average of: (H2O2 C4H9O/c12345))) & Average

results of using this scheme for the cases when the combination X and Y is not known, are shown in Figure 3a.

Using the new, group additive method to estimate the kinetics for unknown combinations of X and Y is simpler to explain than the averaging scheme. The reaction $HC{\equiv}C\cdot + H_2O \longrightarrow HC{\equiv}CH + HO\cdot$ (point 893 in Figure 3) matches the pair of groups (O-pri, Ct-rad), each of which is trained independently. O-pri was trained from 11 rules (in combination with Y groups other that Ct-rad) and contributes -2.35 to log(k@1000 K). Ct-rad was trained from 7 rules (in combination with X groups other than O-pri) and contributes $+2.53$ to log(k@1000 K). Figure 3b shows the results of using this scheme to estimate the cases when the rules are not available for the matched combination of X and Y. The 95% confidence intervals (dashed lines) are narrower, there is less stratification, and the predicted rates span a larger range than with the averaging method used in Figure 3a.

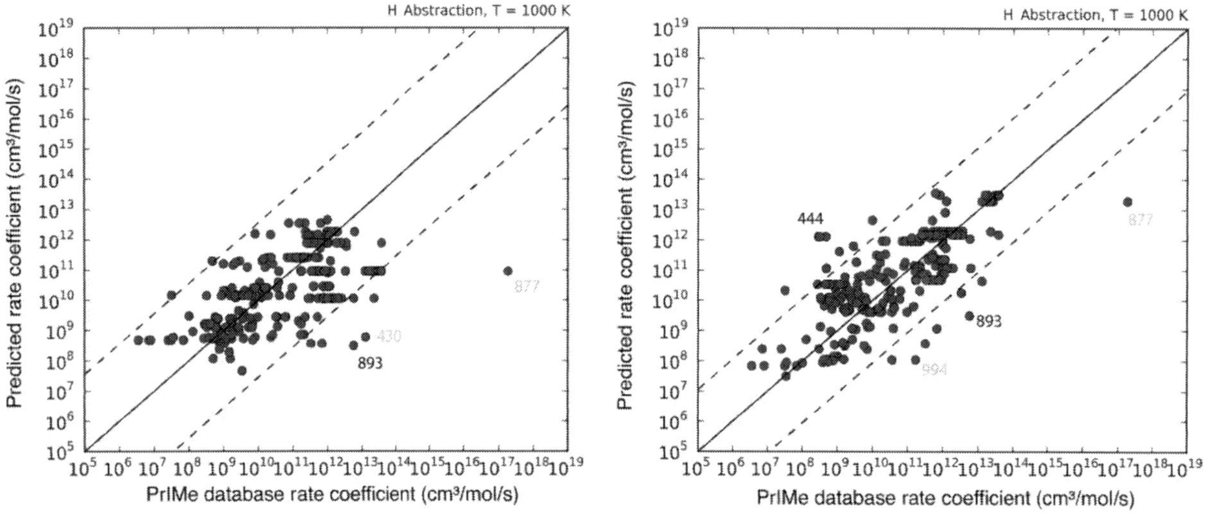

Figure 3: *Parity plots comparing predicted k(1000 K) with data from PrIMe database, for hydrogen abstraction reactions that do not match a known XHY rule. Left (a): old method of averaging "similar" XHY rules. Right (b): new method of estimating from independent XH and Y· contributions.*

Conclusions

The reaction mechanism generation software RMG estimates reaction rate expressions using rules based on the functional groups surrounding the reacting center. A reaction typically involves more than one functional group (e.g. an atom with a hydrogen ligand XH and a radical Y·), which combine to form a "supergoup" XY. When a rule for the supergroup XY is known, it can be used to predict the reaction kinetics with reasonable accuracy. However, when data are sparse and a rule

of: (Average of: (Average of: (H2O2 C4H9O/c134(2)5))) & O/H/OneDe C-methyl) & Average of: (O-pri Cd-pri-rad) & Average of: (O/H/NonDeC Cd-pri-rad & Average of: (H2O2 C4H7/ c1342) & Average of: (H2O2 Cd-rad/NonDeC)) & Average of: (O/H/NonDeC Ct-rad) & Average of: (O-pri CO-pri-rad) & Average of: (O-pri O-pri-rad & Average of: (O-pri O-rad/NonDeC)) & Average of: (O/H/NonDeC O-pri-rad & Average of: (H2O2 O-rad/NonDeO & H2O2 O-rad/OneDe))))) for XY is not known, RMG currently averages 'similar' XY supergroups. For these scenarios we are have tested a group additive method, adding separate contributions from X and Y which are derived from known XY supergroups. The group values can be trained using existing supergroup rules or explicit reactions. The group values can be re-trained when new kinetic data are available or the definitions and hierarchy of the groups are updated. By recording the goodness of fit when the group values are trained, confidence intervals can be calculated on each reaction rate estimated using this method. For the hydrogen abstraction family of reactions, estimates calculated in this manner are better than those estimated using the averaging scheme previously used in RMG software, and their origin is simpler to trace. This approach is now being extended to families of reactions other than hydrogen abstraction.

References

[1] Broadbelt LJ, Stark SM, Klein MT. Computer Generated Pyrolysis Modeling: On-the-Fly Generation of Species, Reactions, and Rates. *Ind Eng Chem Res*. 1994;33(4):790–799. doi:10.1021/ie00028a003.

[2] Grenda J, Androulakis I, Dean A, Green W. Application of computational kinetic mechanism generation to model the autocatalytic pyrolysis of methane. *Industrial and Engineering Chemistry Research*. 2003;42(5):1000–1010. doi:10.1021/ie020581w.

[3] Broadbelt L, Pfaendtner J. Lexicography of Kinetic Modeling of Complex Reaction Networks. *AIChE J*. 2005;51(8):2112–2121. doi:10.1002/aic.10599.

[4] Song J, Raman S, Yu J, Wijaya C, Stephanopoulos G, Green W. RMG: the next generation of automatic chemical reaction mechanism generator. In: *Proceedings of the AIChE Annual Meeting*. San Francisco, CA, USA. 2003; .

[5] Green WH, et al. RMG Reaction Mechanism Generator open-source software package. 2011. URL http://rmg.sourceforge.net

[6] Song J. Building Robust Chemical Reaction Mechanisms: Next Generation of Automatic Model Construction Software. Ph.D. thesis, Massachusetts Institute of Technology, Cambridge MA, USA. 2004.
URL http://hdl.handle.net/1721.1/30058

[7] Van Geem K, Reyniers M, Marin G, Song J, Green W, Matheu D. Automatic reaction network generation using RMG for steam cracking of n-hexane. *AIChE J*. 2006;52(2):718–730. doi:10.1002/aic.10655.

[8] Benson SW. *Thermochemical Kinetics*. New York: Wiley, 2nd ed. 1976.
URL http://www.worldcat.org/oclc/2074434

[9] Cohen N, Benson S. Estimation of heats of formation of organic compounds by additivity methods. *Chem Rev*. 1993;93(7):2419–2438. doi:10.1021/cr00023a005.

[10] Susnow R, Dean A, Green W, Peczak P, Broadbelt L. Rate-based construction of kinetic models for complex systems. *J Phys Chem A*. 1997;101(20):3731–3740. doi:10.1021/jp9637690.

[11] Sumathi R, Carstensen H, Green Jr W. Reaction rate prediction via group additivity part 1: H abstraction from alkanes by H and CH3. *J Phys Chem A*. 2001;105(28):6910–6925. doi:10.1021/jp010697q.

[12] Sumathi R, Green W. A priori rate constants for kinetic modeling. *Theor Chem Acc*. 2002; 108(4):187–213. doi:10.1007/s00214-002-0368-4.

[13] Saeys M, Reyniers M, Marin G, Van Speybroeck V, Waroquier M. Ab initio group contribution method for activation energies for radical additions. *AIChE J*. 2004;50(2):426–444. doi:10.1002/aic.10038.

[14] Tsang W, Hampson R. Chemical kinetic data base for combustion chemistry. Part I. Methane and related compounds. *J Phys Chem Ref Data*. 1986;15:1087.

[15] Frenklach M. Transforming data into knowledge – Process Informatics for combustion chemistry. *Proc Combust Inst*. 2007;31(1):125–140. doi:10.1016/j.proci.2006.08.121. URL http://www.primekinetics.org

(551b) Automation of Composition Modeling for Complex Petroleum Hydrocarbons by Homologous Series

Zhen Hou[1], Steven P. Pyl[2], Craig Bennett[1], Brian Moreno[1] and Michael T. Klein[1], (1)Chemical Engineering, University of Delaware Energy Institute, Newark, DE, (2)Laboratorium voor Chemische Technologie, Ghent University, Gent, Belgium

This article developed a hybrid statistical approach for the modeling of a complex petroleum feedstock by optimizing a molecular representation based on structural attributes to a set of analytical characterizations.

A general petroleum hydrocarbon conversion was qualitatively described as a set of homologous series. Generally, the structures in those series could be determined by user, literature and automated reaction network generation etc.

Consequentially, the quantitative information of a petroleum feedstock (mole fraction or mass fraction) can be obtained via employing an optimization loop. The molecular compositions of a feedstock were described as a set of structural attribute probability density functions (PDF) first. Through constraining the parameters of those PDFs, an optimization loop was then applied based on the juxtapositions of the mapping between molecules and multiple attributes; the objective function was given in terms of available analytical information. A structure property calculation program had been implemented to evaluate this objective function. Finally, the optimal mole fraction of the feedstock was obtained and could serve as the initial concentration values for the kinetic modeling.

A user-friendly program with an Excel-VBA interface called Composition Model Editor (CME) was developed to fulfill this new approach. In order to verify this approach, selected complex feedstocks were modeled by CME and then validation results were shown in this article.

See more of this Session: Reaction Path Analysis II
See more of this Group/Topical: Catalysis and Reaction Engineering Division

(551c) Rule-Based Network Analysis of Complex Reaction Systems In Biomass Conversion

Srinivas Rangarajan, Aditya Bhan and Prodromos Daoutidis, Chemical Engineering and Materials Science, University of Minnesota, Minneapolis, MN

Catalytic upgradation is a method of conversion of lignocellulosic biomass to transportation fuels and chemicals. Such catalytic systems inherently have a large and complex reaction network, elucidation of which is key in reactor design. Moreover, cascades of multifunctional catalytic systems have recently been proposed for upgrading sugar monomers to transportation fuels [1]. Here, we present rule-based systematic computational analyses for: (i) enumerating and analyzing complex catalytic biomass conversion systems, and (ii) formulation of potential synthetic routes to produce valuable biomass-derived molecules. We employ RING, a rule-based network generation and analysis tool with an english-like domain specific user-friendly front-end developed by us earlier [2], in conjunction with ab initio- and DFT-based thermochemistry analysis, as the computational platform for these analyses.

Dehydration to acrolein, in solid acid catalyst systems, is a promising route for value addition of by-product glycerol, obtained from the trans-esterification process to produce biodiesel. Known elementary steps of solid acid catalysis, such as adsoprtion of alcohols, dehydration, desorption of alkoxides (modeled as carbenium ions), hydride shifts, decarbonylation, hydration, deprotonation, protonation of carbonyl group, etc. were input into RING as reaction rules to study the reaction system presented by Corma et al. [3]. The resultant reaction network for this system was more than 70,000 reactions in total. Subsequently, the network was analyzed in terms of extracting reaction pathways leading to products identified in the experimental data at various conditions [3]. The analysis of these pathways further corroborated the observations and discussions made by Corma et al. Specifically, the shortest paths (and paths up to two steps longer) clearly indicated that 3-hydroxypropionaldehyde was an essential intermediate in acrolein production, obtained from glycerol upon dehydration, and hydrogen transfer or enol-keto type rearrangement. Acetone, on the other hand, necessarily required acetol or propane diol as an intermediate, with glycerol undergoing two dehydration steps and one or more rearrangement or hydrogen transfer steps. It was further identified that acetaldehyde production would be accompanied necessarily by the evolution of carbon monoxide. This case study, thus, shows that RING can be employed in the analysis of hypothesized mechanisms of complex reaction systems in a rule-based manner, starting with likely elementary steps and querying for pathways matching experimental observations.

Lignocellulosic biomass upgradation pathways can potentially involve acid catalyzed hydrolysis and dehydration, metal catalyzed hydrogenation and hydrogenolysis, and base catalyzed esterification, ketonization and condensation reactions. Reaction rules describing the overall steps of such a broad spectrum of catalytic chemistries were input into RING with sugar monomers and/ or model oxygenates as reactants. Pathways were traced, in the resultant network, to identify multi-step-catalytic reaction cascades for synthesizing valuable compounds such as long chain alkanes. RING identified proposed synthesis routes in the literature and, in addition, proposed additional potential routes. These pathways can further be analyzed thermochemically to identify feasible/ infeasible and bottleneck steps, and the thermodynamically favorable order of deoxygenation steps.

Application of RING in (a) rule-based systematic and exhaustive analysis of glycerol dehydration pathways in solid acid catalytic systems, and (b) formulating synthesis strategies for biomass upgradation, as well as thermochemical analysis of the identified pathways, will be presented with representative examples.

Alonso, D. M.; Bond, J. Q.; Dumesic, J. A. *Green Chemistry* **2010**, 12, 1493

Rangarajan, S; Bhan, A; Daoutidis, P. *Ind. Eng. Chem. Res.* **2010**, 49, 10459

Corma, A.; Huber, G. W.; Sauvanaud, L.; O'Connor, P. *Journal of Catalysis* **2008**, 257 (1), 163

See more of this Session: Reaction Path Analysis II
See more of this Group/Topical: Catalysis and Reaction Engineering Division

PROGRESS TOWARDS CAPTURING SOLVENT EFFECTS IN AUTOMATIC MECHANISM GENERATION

Amrit Jalan, Richard H. West and William H. Green*
Department of Chemical Engineering, Massachusetts Institute of Technology
77 Massachusetts Avenue, Cambridge, MA 02139

Abstract

Detailed kinetic models provide useful mechanistic insight into a chemical system. Manual construction of such models is laborious and this has led to the development of automated methods for exploring chemical pathways. These methods rely on fast, high throughput estimation of species thermochemistry and kinetic parameters. We discuss methods for extending the idea of automatic mechanism generation to solution phase reactions using linear solvation energy relationships (LSERs) for estimation of solvent effects on thermochemical and rate parameters. Specifically, we employed and validated the Abraham model (using descriptors estimated by group additivity) to estimate the free energy of solvation for several solute and solvent classes. First order temperature dependence of the solvation free energy was modeled by decomposing it into enthalpic and entropic parts. Two approaches to perform this decomposition were evaluated: one derived from a hard sphere model of liquids and another recently proposed empirical model by Mintz *et al.* Ease of implementation, extensibility to different solvents and accuracy of resulting estimates were used as metrics for comparison. A similar approach was adopted for kinetic parameters using existing correlations of solvent effects on elementary reaction rates. Diffusion-limits of various reactions were estimated on the fly using known expressions from literature. These tools were incorporated into the Reaction Mechanism Generator (RMG) software and used to generate kinetic models with automatic solvent effects for tetralin autoxidation in different solvents.

*To whom all correspondence should be addressed

Introduction

Chemical kinetic models are used to simulate and understand many important systems, ranging from the metabolism of bacteria to the performance of new engine designs and fuel candidates. These models typically contain thousands of species and reactions which makes their manual construction tedious and often error-prone. This has led to the development of tools for automatic generation of complex kinetic models in the gas phase. However, many reactive systems of interest to engineers occur in the liquid phase. Tools that could generate detailed kinetic models for liquid phase systems would provide useful mechanistic details about the underlying chemistry of these processes. However, the complexity associated with condensed phase systems makes this a formidable task.

In this contribution, we present and validate an extensible framework for automatic mechanism generation in the solution phase using the Linear Solvation Energy Relationship (LSER) approach for thermodynamics and kinetics, in addition to diffusion limits on reaction rates. For illustrative purposes, we combine this approach with continuum solvation models to understand solvent effects on hydrocarbon autoxidation and kinetic modeling of tetralin autoxidation in different solvents. This framework has been implemented in the latest release of the open source Reaction Mechanism Generator (RMG) tool described below.

Reaction Mechanism Generator (RMG)

Reaction Mechanism Generator (RMG) is an open source (Song, Raman et al. 2003) tool that employs state-of-the-art methods in cheminformatics, thermochemistry and kinetic parameter estimation to build complex gas-phase reaction networks. RMG follows a rate-based algorithm (Susnow, Dean et al. 1997) for the construction of chemical kinetic models for a given set of reactants and initial conditions (temperature, pressure and species concentration).

RMG's algorithm requires estimation of thermochemical and kinetic parameters for thousands of species and reactions, so the methods employed must be both fast and accurate. Because of its high-throughput nature, RMG uses Benson's group additivity for the estimation of gas phase thermochemistry (H, G, S, C_p) along with suitable corrections for open shell species. Rate constants for reactions are generated by estimating the Arrhenius parameters (A and E_a) using *rate rules* for different reaction families that are derived from experimental measurements or *ab initio* calculations.

Extension of RMG to solution phase reactions requires analogous methods for estimating thermochemistry and rate coefficients in the solvents of interest. It is desirable to employ methods that are extensible and allow one to switch between multiple solvents with ease, both for species thermochemistry and reaction rates. In the following sections, we present details of the methods used in RMG towards this objective.

High throughput estimation of solvation thermodynamics

The free energy change associated with the process of transferring a molecule from the gas phase to the solvent phase under standard state conditions is defined as the free energy of solvation ($\Delta G^0_{\mathrm{solv}}$) and many different methods have been developed for computing this quantity from first principles. However, most of these methods are too slow to be used in a high throughput sense. The LSER approach developed by Abraham and co-workers (Vitha and Carr 2006) offers an alternative route to $\Delta G^0_{\mathrm{solv}}$ by first estimating the partition coefficient, K and using the following relation:

$$\Delta G^0_{\text{solv}} = -RT \ln K = -RT \ln \frac{c^{\text{eq}}_{\text{solv}}}{c^{\text{eq}}_{\text{gas}}} \quad (1)$$

The Abraham model uses empirically developed solute and solvent scales to model the partition coefficient of a solute between the gas phase and a solvent of interest as shown below:

$$\log_{10} K = c + aA + bB + eE + sS + lL \quad (2)$$

where the upper case parameters (*A*, *B*, *E*, *L*, *S*) are solute descriptors (unique to a given solute) and the lowercase parameters (*c*, *e*, *s*, *a*, *b*, *l*) are solvent descriptors. Even though the descriptors have been developed to model specific solute-solvent interactions, they can be fairly correlated. The *eE* and *sS* terms mostly account for electrostatic interactions like dipole–dipole and dipole–induced-dipole, the *aA* and *bB* terms account for hydrogen bonds from the solute to the solvent and *vice versa*, and the *lL* term represents the contributions of cavity formation and dispersion (Jalan, Ashcraft et al. 2010).

In order to use the Abraham model in a predictive sense, we need methods for estimation of solute descriptors from their 2D molecular structure. Several methods have been proposed for determining these descriptors using molecular properties, first-principles calculations, or other regression schemes. We use the group additivity method proposed by Platts and co-workers (Platts, Butina et al. 1999) which is similar in concept to Benson's method used for gas phase thermochemistry in RMG. Abraham descriptors for over 50 solvents are available in the literature. Combined with solute descriptors from the Platts group-additivity model, this forms a potential method for high-throughput estimation of ΔG^0_{solv}.

Validation of ΔG^0_{solv} estimates obtained from RMG

Solvation data from the Minnesota Solvation Database (Marenich, Kelly et al. 2009) were used to test the accuracy of the method proposed above. Of the 482 neutral compounds in the dataset, those containing C, H and O were selected giving a test set containing 935 data points across a diverse set of solutes and solvents. The solvation energies in the test dataset were in the 0–14 kcal/mol (negative) range.

The root mean squared error (RMSE) of RMG's estimates over the entire dataset was found to be 0.47 kcal/mol. Most of the major outliers (abs error > 1 kcal/mol) correspond to solute-solvent pairs with significant H-bonding. The RMSE for the subset of H-bonding solvents like alcohols was higher (0.70 kcal/mol) compared to that for alkane solvents (0.30 kcal/mol). The ΔG^0_{solv} estimates obtained from RMG are also in good agreement with those obtained using solute descriptors from a commercial software package (*Absolv* module in ADME ver. 4.95) which uses updated databases to determine the fragments and their contributions.

The Abraham/Platts model only applies to stable species. The approach adopted in RMG for radical intermediates is to use the descriptors of the corresponding saturated compound as a starting point and make corrections based on Platts groups. Using this approximation for the HO. and HOO. radicals in water at 298 K, we obtain ΔG^0_{solv} estimates of -4.7 kcal/mol and -7.5 kcal/mol respectively which are in reasonable agreement with experimentally reported values of (-3.9 kcal/mol for HO. and -6.8 kcal/mol for HOO.)

Temperature Dependence of ΔG^0_{solv}

The Abraham model provides ΔG^0_{solv} estimates only at 298 K since it relies on experimental data which are generally available at room temperature. We model the temperature dependence of ΔG^0_{solv}

by decomposing it into temperature-independent enthalpic (ΔH^0_{solv}) and entropic contributions (ΔS^0_{solv}) and using $\Delta G^0_{\text{solv}} = \Delta H^0_{\text{solv}} - T\Delta S^0_{\text{solv}}$. Two methods for performing this decomposition were explored: (1) Assuming $\Delta S_{\text{solv}} \approx \Delta S_{\text{cav}}$ and using Scaled Particle Theory (SPT) expressions developed by Pierotti (Pierotti 1976) to estimate ΔS_{cav} and (2) Empirical correlations for ΔH^0_{solv} developed by Mintz and co-workers (Mintz, Gibbs et al. 2009).

The solute radii required as inputs for the SPT expressions were approximated from the McGowan volume of each species. The accuracy of this method was tested using the experimental ΔH^0_{solv} values compiled by Mintz and co-workers and using the following approximation:

$$\Delta H^0_{\text{solv}}(\text{SPT}) = \Delta G^0_{\text{solv}}(\text{Platts}) + T\Delta S^0_{\text{solv}}(\text{SPT}) \qquad (3)$$

The estimates obtained from SPT were found to be very sensitive to the molecular radii used in the calculations. Correction factors were employed for both solute and solvent radii to account for the high uncertainties that are typically expected in such estimates. These correction factors were adjusted to minimize errors between theoretical and experimental observations. While this drastically improved the agreement for alkane solvents, significant outliers were still observed for protic solvents as shown in Figure 1. This is indicative of contributions from specific solvation which are not captured in hard-sphere models.

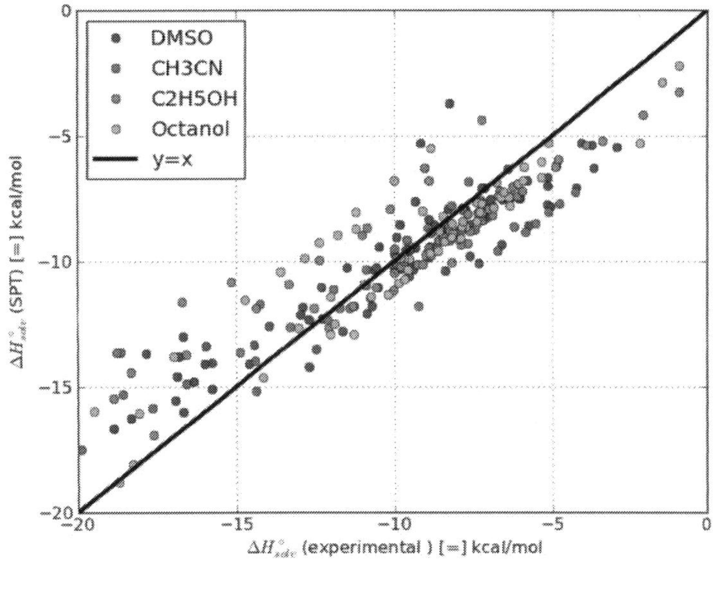

Figure 1. Comparison of $\Delta H^0_{solv}(SPT)$ values with experimental data for protic solvents

The correlations developed by Mintz *et al.* use the same solute descriptors as the Abraham model and also incorporate corrections for H-bonding empirically. The performance of these correlations using RMG-estimated descriptors for the same solvents as Figure 1 is shown in Figure 2. While the Mintz approach gives more accurate estimates, its applicability is limited to about 25 solvents for which correlations are currently available. The present linear approximation for handling the temperature variation of ΔG^0_{solv} may not apply at large deviations from room temperature. The error incurred from second order temperature dependence and approximations to handle these are currently being pursued.

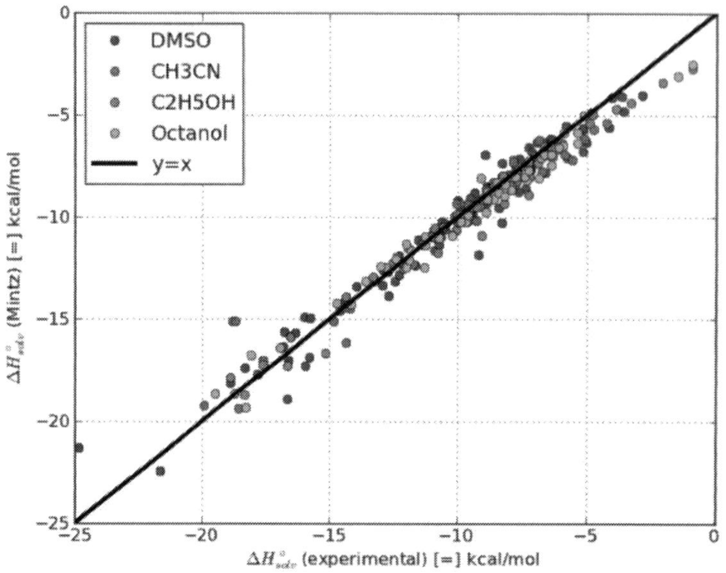

Figure 2. Comparison of $\Delta H^{0}_{solv}(Mintz)$ with experimental data for protic solvents

Solvent effects on kinetics

Solvent effects on reaction rates can be broadly classified into two categories: diffusive limits and kinetic solvent effects. Diffusive limits and expressions for estimating these are well known in the literature. In RMG, we use the expressions due to Smoluchowski to estimate diffusive limits using estimates of effective radii and diffusivity of the reactants and the intrinsic reaction rate.

Kinetic Solvent Effects (KSEs) arise from molecular interactions between the solvent and the reactants, products, or transition state of a given reaction. Even though solvents are well known to affect the rate of ionic reactions, recent studies have pointed out the role that solvents can play in free radical chemistry. KSEs are known for a relatively small number of elementary free radical reactions, popular ones being those for H-abstraction developed by Ingold and co-workers (Snelgrove, Lusztyk et al. 2001) and polar effects observed for beta-scission reactions of tertiary alkoxy radicals.

Case study: Solvent effects on tetralin autoxidation

For illustrative purposes, the solvent effect on AIBN initiated tetralin autoxidation at low temperatures was undertaken as a case study. Experimentally, the rate of hydrocarbon oxidation has been observed to increase with the dielectric constant (Howard and Ingold 1964) of the solvent for a variety of substrates including tetralin. It is well known that peroxy (ROO.) radicals are the major chain carriers in low temperature liquid phase oxidation of alkanes. The oxygen uptake rate in these systems is known to be sensitive to the rate of initiator (AIBN) decomposition, propagation (ROO. + RH), and the terminating self-reactions (ROO. + ROO.) between the peroxy radicals, among other reactions.

AIBN initiation rates in different solvents of interest were obtained from the experimental measurements of Howard and Ingold (Howard and Ingold 1964a) or the empirical correlation developed by Makitra and co-workers using the same data and different solvent scales:

$$\log k_i = -5.574 + 2.09\left(\frac{n^2-1}{n^2+2}\right) + 1.014 \times 10^{-3}B + 6.82 \times 10^{-3}E_T - 0.324 \times 10^{-3}\delta^2$$

where n is the refractive index, B is the Koppel Palm basicity, E_T is the Reichardt electrophilicity index and δ is the Hildebrand solubility parameter. Values of these solvent scales were obtained from the compilation of Abboud and co-workers. Solvent effects on the initiation rate were small and the k_i values were within a factor of two of each other.

Snelgrove et al. (2001) studied H-abstraction reactions in different solvents and found evidence to conclude that the observed solvent effects arise from H-bonding between the substrate and the solvent. The following empirical correlation was proposed to explain the observed solvent effect:

$$\log k_S = \log k_0 - 8.3\alpha_{XH}\beta_S$$

where α_{XH} and β_S are the Abraham parameters for H-bond donation and H-bond acceptance of the substrate (XH) and solvent respectively, and are similar to the A and B parameters employed currently in the Abraham model. The reference rate k_o is the rate of the reaction in a medium with $\beta_S = 0$. Since alkane substrates like tetralin have very little H-bond donating tendencies ($\alpha_{XH} \approx 0$), the propagation rate is expected to vary very little between solvents. This was also confirmed experimentally by Lucarini and co-workers (Lucarini, Pedulli et al. 1998).

The terminating self-reaction between ROO. species can be solvent dependent due to the polar character of the peroxy radicals which is destroyed in the termination step possibly through the tetroxide intermediate. This hypothesis was tested using continuum solvation models (PCM & CPCM) (Tomasi, Mennucci et al. 2005) to estimate the electrostatic solvation energies of methyl and cumyl peroxy radicals and the corresponding tetroxides in different solvents. In both cases the peroxy radicals were found to be more stabilized compared to the tetroxide due to the loss in dipolarity. Based on the calculations for the methyl peroxy radical the termination rate in a solvent was found to vary with its dielectric constant (ε) as given by Equation 4:

$$\mathrm{RT}\log\frac{k_{\mathrm{solv}}}{k_{\mathrm{alkane}}} = 0.38 - 2.02\left(\frac{\epsilon-1}{2\epsilon+1}\right) \quad (4)$$

Predicted variations in termination rate constants for the methyl peroxy radical (with alkane as reference solvent) using Equation 4 also compare well with experimental observations of Niki and co-workers for tetralin autoxidation (Niki, Kamiya et al. 1969) as shown in Figure 4.

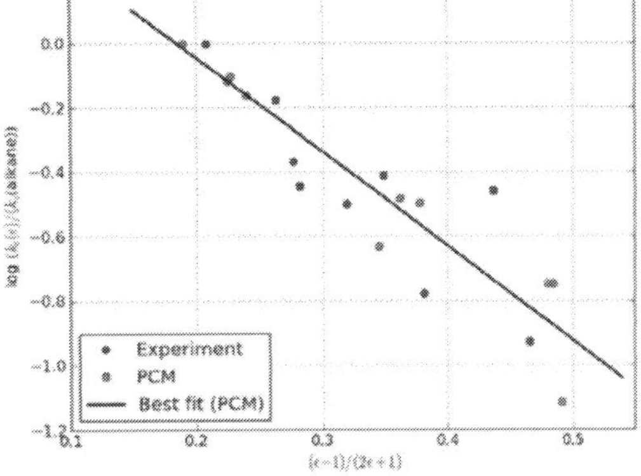

Figure 3. Comparison of predicted solvent effect on termination rates with experimental data at 70°C

The predicted changes in the termination rate were used along with the thermochemistry methods and diffusive limits described earlier to build models of tetralin oxidation in tetrachloromethane (CCl_4), benzene (C_6H_6), acetonitrile (CH_3CN) and dimethylsulfoxide ($(CH_3)_2SO$) solvents. The computer-generated models were found to accurately replicate the solvent effects measured experimentally, and can be used to predict the behavior in new solvents and at different temperatures. In addition to oxidation rates, these models also provide information on the distribution of secondary products and its variation with solvent type.

Conclusions

We have described a systematic approach to incorporating solvent effects in automatic mechanism generation. The Abraham LSER using solute descriptors estimated using the Platts approach was found to give reasonably accurate and fast estimates of ΔG^0_{solv} across a wide range of solutes and solvents. The Mintz correlations were found to provide more accurate and reliable decomposition of ΔG^0_{solv} into ΔH^0_{solv} and ΔS^0_{solv} compared to the analytical SPT model which is sensitive to input solute radii. Kinetic solvent effects on free radical reactions were discussed in the context of solvent effects on tetralin oxidation. The methods and approximations were applied to build and understand the kinetics of tetralin oxidation in different solvents.

Acknowledgments

The authors would like to acknowledge funding from Eni S.p.A.

References

Howard, J. A. and Ingold, K. U. (1964). Solvent Effects on Hydrocarbon Autoxidations. *Canadian Journal of Chemistry-Revue Canadienne De Chimie 42(5,* 1250.

Howard, J. A. and Ingold, K. U. (1964a). Inhibited Autoxidation of Styrene .4. Solvent Effects. *Canadian Journal of Chemistry-Revue Canadienne De Chimie 42(5),* 1044

Jalan, A., Ashcraft, R. W., West, R.H., Green, W.H. (2010). Predicting solvation energies for kinetic modeling. *Annual Reports Section "C" (Physical Chemistry) 106,* 211.

Lucarini, M., Pedulli, G. F. et al. (1998). Do peroxyl radicals obey the principle that kinetic solvent effects on H-Atom abstraction are independent of the nature of the abstracting radical? *Journal of Organic Chemistry 63(13),* 4497.

Marenich, A. V., Kelly, C. P. et al. (2009). Minnesota Solvation Database-version 2009.

Mintz, C., Gibbs, J. et al. (2009). Enthalpy of solvation correlations for organic solutes and gases dissolved in acetonitrile and acetone. *Thermochimica Acta 484(1-2),* 65.

Niki, E., Kamiya, Y. et al. (1969). Solvent Effects in Oxidation of Hydrocarbons .2. Oxidations of Tetralin in Various Solvents. *Bulletin of the Chemical Society of Japan 42(11),* 3224.

Pierotti, R. A. (1976). Scaled Particle Theory of Aqueous and Non-aqueous Solutions. *Chemical Reviews 76(6),* 717.

Platts, J. A., Butina, D. et al. (1999). Estimation of molecular linear free energy relation descriptors using a group contribution approach. *Journal of Chemical Information and Computer Sciences 39(5),* 835.

Snelgrove, D. W., Lusztyk, J. et al. (2001). Kinetic solvent effects on hydrogen-atom abstractions: Reliable, quantitative predictions via a single empirical equation. *Journal of the American Chemical Society 123(3),* 469.

Song, J., Raman, S. et al. (2003). Development of automatic chemical reaction mechanism generation software using object-oriented technology. *Abstracts of Papers of the American Chemical Society 226,* 012-FUEL.

Susnow, R. G., Dean, A. M. et al. (1997). Rate-based construction of kinetic models for complex systems. *Journal of Physical Chemistry A 101(20),* 3731.
Tomasi, J., Mennucci, B. et al. (2005). Quantum mechanical continuum solvation models. *Chemical Reviews 105(8),* 2999.
Vitha, M. and Carr, P. W. (2006). The chemical interpretation and practice of linear solvation energy relationships in chromatography. *Journal of Chromatography A 1126(1-2),* 143.

(551e) Creating Thermodynamically Consistent Microkinetic Mechanisms for Heterogeneous Chemistry Based Upon Gas-Phase Properties

Claude Franklin Goldsmith[1], D. Wayne Blaylock[2] and Raimund Horn[1], (1)Inorganic Chemistry, Fritz Haber Institute of the Max Planck Society, Berlin, Germany, (2)Department of Chemical Engineering, Massachusetts Insititute of Technology, Cambridge, MA

Microkinetic models are useful tools in heterogeneous chemistry. The parameters in these models are based upon physically meaningful quantities – activation energies and frequencies for elementary reactions, plus enthalpy and entropy for adsorbed species – that can be derived from first principles. Unlike global mechanisms, which are postdictive, microkinetic models are predictive. They are not limited to a narrow experimental range, but can be applied over a broad range of conditions, thereby allowing us to narrow the "pressure gap" between UHV surface science and high-pressure industrial application.

Creating microkinetic models, however, is a time-consuming and laborious task, beset with many difficulties. In addition to obtaining all the necessary parameters, one of the challenges is ensuring that they are thermodynamically consistent. Thermodynamic consistency is particularly critical for our intended application: high-temperature and high-pressure adiabatic reactors in which the heterogeneous surface chemistry is coupled with homogeneous gas-phase chemistry.

We present a new method for maintaining thermodynamic consistency in homogeneous/heterogeneous mechanisms. The method is based upon the assumption that the vibrational modes of the gas-phase molecule and the phonons of the active site are conserved upon adsorption. The input parameters are molecular properties of the gas-phase molecule (e.g. vibrational frequencies and moments of inertia) and the heats of adsorptions. These heats of adsorption are obtained either from experiment, Density Functional Theory (DFT), or scaling relations derived from other surfaces. As the gas-phase molecule absorbs, the external translational and rotational degrees of freedom are converted into new surface vibrations. These frequencies can be estimated with suitable accuracy from the input parameters. The partition function of the adsorbed species is then computed using statistical mechanics. Standard thermodynamic parameters, such as enthalpy, entropy, and heat capacity, are obtained.

The estimated vibrational modes for the adsorbed species are tested against detailed DFT calculations obtained from literature. Equilibrium constants obtained from this method are compared against values obtained from other thermodynamically consistent mechanisms in the literature. The results confirm the validity of the assumption of separability of normal modes when computing thermodynamic properties.

See more of this Session: Reaction Path Analysis II

(551f) Efficient Location of Transition State Structures Via the Freezing String Method

Paul M. Zimmerman[1], Andrew Behn[2], Martin Head-Gordon[1] and Alexis T. Bell[2], (1)Dept. of Chemistry, University of California at Berkeley, Berkeley, CA, (2)Department of Chemical Engineering, University of California Berkeley, Berkeley, CA

Transition states along ab initio potential energy surfaces provide important descriptions of molecular reactivity. However, determining these saddle points in highly complex molecular systems remains challenging. Common approaches to finding transition states in chemical systems include string methods, which usually come with significant computational cost. In order to alleviate the cost of such methods, we propose the freezing string method (FSM) to rapidly generate guess transition state structures that can be further refined by standard eigenvector following algorithms. Eigenvector following algorithms are typically reliable only when a suitable approximation to the transition state structure is available. In order to provide this guess, FSM creates a string of structures connecting reactant and product geometries using a combination of linear synchronous transit and gradients from simulation. This method, which will be elucidated in detail, requires many fewer gradient computations than typical string methods. The performance of FSM will be demonstrated for a variety of chemical transformations, including reactions in zeolites.

See more of this Session: Reaction Path Analysis II
See more of this Group/Topical: Catalysis and Reaction Engineering Division

(551g) Reaction Mechanism Analysis Using Periodic Temperature Forcing

Ramanathan Srinivasan and Niket Kaisare, Chemical Engineering, Indian Institute of Technology Madras, Chennai, India

Periodic forcing of the input variables such as pressure has been employed to optimize the selectivity or understand the kinetics of chemical reactions when there is a nonlinear relationship between the reaction rate(s) and the controlled variable. The relationship between the reaction rate constant and the temperatures is also nonlinear. However, temperature control is difficult to implement and hence only a few studies on periodic temperature forcing have been reported. With the advent of microreactors, there is a potential to implement periodic forcing of temperatures during a reaction. In this report, the results of numerical simulations of the effect of periodic forcing of temperatures at various frequencies for a select set of reaction mechanisms are presented. The results are also compared with the analytical solutions in limiting cases. In the study of reaction mechanisms, distinguishability between candidate mechanisms is an important factor. In many cases, the results of isothermal reactions will not be able to differentiate between two mechanisms. We illustrate an example where periodic temperature forcing at various frequencies can help differentiate between candidate mechanisms.

See more of this Session: Reaction Path Analysis II
See more of this Group/Topical: Catalysis and Reaction Engineering Division

(557a) Study On the Capture of Carbon Dioxide with Porous Covalent Triazine-Based Frameworks

Lu Bai, Xiangping Zhang, Guoying Zhao, **Suojiang Zhang**, Yi Nie and Haifeng Dong, Institute of Process Engineering, Chinese Academy of Sciences, Beijing, China

CO_2 emission leads to the serious global warming effect, which has become a seriously global environmental, economic and political issue. CO_2 adsorption technique is a feasible and effective one to capture CO_2 from the industrial gas and eliminate the global warming effect, yet choosing a high-performance adsorbent is the key to the process. Recent years, a novel polymer named covalent triazine-based frameworks (CTFs) could be applied in gas storage because of their high porosities and surface areas (Kuhn et al, 2008). In this paper, CTFs were synthesized by the self-condensation of 1,4-dicyanobenzene (DCB). The samples were characterized by elemental analysis, thermogravimetric analysis and N_2 sorption measurements. Additionally, CO_2 adsorption and desorption were carried out by Intelligent Gravimetric Analyser (IGA-003). The results showed that the surface area of the sample (500 °C, 40h, DCB/$ZnCl_2$=1:5) is 1179m^2/g and the adsorbed amount of CO_2 is 163mg/g at 25°C,15bar. It means that the CTFs show great potential as adsorbents for CO_2 capture and storage. By optimizing the synthesis conditions and raw materials, the CTFs materials with higher surface area and absorbed amounts are obtained.

References

Kuhn, P., Antonietti, M., Thomas, A.(2008). Porous, Covalent Triazine-Based Frameworks Prepared by Ionothermal Synthesis. *Angew. Chem. Int. Ed*, 47, 3450-3453.

See more of this Session: Unconventional Technologies for CO2 Capture, Conversion and Utilization II
See more of this Group/Topical: International Congress on Energy 2011

(557b) Development of Nanoparticle Catalyst for the Trireforming of CO2-Rich Flue Gases

Juan Cruz, Jackie W. Seaman, Yishan Zhang and Tracy J. Benson, Chemical Engineering, Lamar University, Beaumont, TX

Today's chemical industry is very complex with most commodity chemicals being produced from some variation of petroleum feedstocks. From transportation fuels to polymers, petroleum is truly the largest source of basic hydrocarbons for which many of the common household products consumers use. Along with this, stems many industrial and environmental issues which must be addressed for long-term stability of the industry. Upgrading of these crude feedstocks requires hydrogen, generally supplied via steam methane reforming (SMR). Hydrogen is typically produced via methane steam reforming reactions. Carbon dioxide, noted for its greenhouse gas potential, is produced as a byproduct of methane steam reforming. CO_2 sequestration is a growing concern and is at the forefront of many political and environmental policies that are being introduced worldwide. One avenue to combat the growing release of CO_2 into the atmosphere is to convert it to useful chemicals via trireforming.

Trireforming is the conversion of carbon dioxide, methane, and water to produce synthesis gas (syngas). The uniqueness of the trireforming technology is that bulk flue gases can be utilized without the separation of CO_2 and steam. The problems associated with the conversion of CO_2 are overcoming thermodynamic limitations, as carbon dioxide tends to be a very stable molecule. However, through careful catalyst design, a novel catalyst has been synthesized that utilizes reverse micelle technology to create highly active nano-sized nickel particles onto metal-oxide supports. In terms of catalyst reactivity, selectivity, and robustness, a mixed metal oxide consisting of Magnesia and titania oxides is the best performing support. Reverse micelles using sodium dioctyl succinate surfactant allow for metal deposition down to 20 nm sized nickel particles.

Our laboratory has developed some novel trireforming catalysts that are capable of extending our carbon resources with high conversion performance. Characterizations of the catalyst have been performed using Scanning Electron Microscopy, X-ray Diffraction, Temperature Programmed Reaction (and Desorption), and in situ FTIR microscopy. Trireforming reaction temperatures of 800°C have been reported in the literature. According to the results supported by our catalyst characterization efforts, reaction temperatures of 700°C will be sufficient for the H2:CO ratio of 2 required for additional hydrocarbon or methanol synthesis reactions. These findings and a postulated reaction site mechanism will be presented.

See more of this Session: Unconventional Technologies for CO2 Capture, Conversion and Utilization II
See more of this Group/Topical: International Congress on Energy 2011

(557c) Material and Engineering Assessments of Electrochemical Processes for Carbon Capture Technology

Fritz Simeon[1], Michael C. Stern[1], Kristin Vicari[1], Howard Herzog[2] and T. Alan Hatton[1], (1)Chemical Engineering, Massachusetts Institute of Technology, Cambridge, MA, (2)MIT Energy Initiative, Massachusetts Institute of Technology, Cambridge, MA

Integrating carbon capture technology with large source carbon emitters is necessary in the very near future to mitigate the contribution of anthropogenic CO_2 emissions to climate change and global public welfare. However, the implementation of this technology with current separation approaches is a challenging task due to the scale at which CO_2 is produced and the energy consumption to drive the separation process. Thermal-swing separation processes based on chemical sorbents show potential for facilitating large-scale CO_2 capture; however, the energy efficiency in the current thermal-swing processes is unacceptable for their implementation. To overcome this problem, new approaches to gas separation have to be developed.

Electrochemically mediated processes are proposed to facilitate CO_2 capture from a dilute gas mixture by exploiting the significant changes in molecular affinity of certain sorbents for CO_2 molecules when they undergo an electrochemical cycle. The proposed process has the potential for reducing the parasitic energy of the current solvent-based absorption operations since it does not require significant temperature swings to regenerate the sorbents. Molecular optimization of the candidate materials is required to enhance their chemical stability, and implementation requires efficient electron transfer operations for rapid electrochemical switching. This materials optimization is achieved through a combination of electrochemical experiments and quantum chemical simulations. The results from the potentiostatic study of the promising sorbent candidates are used to elucidate the energy required for CO_2 separation from a dilute gas mixture using the proposed technology. A general engineering framework of an electrochemical separation unit is developed to elucidate several key design aspects and parameters that relate the process configuration and properties of the material properties to the energy efficiency of the processes. We conclude with a discussion of the general concepts and the advances needed in both the materials and the engineering aspects for their successful implementation as carbon capture technology of the future.

See more of this Session: Unconventional Technologies for CO2 Capture, Conversion and Utilization II
See more of this Group/Topical: International Congress on Energy 2011

(557d) High Temperature CO2 and Sulfur Removal Using Carbonation-Calcination Reaction (CCR) Process: Results From Computer Simulations and Experiments

William Wang[1], **Nihar Phalak**[1], Shwetha Ramkumar[1], Niranjani Deshpande[1], Yao Wang[1], Robert Statnick[2] and L.-S. Fan[1], (1)The Ohio State University, Columbus, OH, (2)ClearSkies Consulting, Cornelius

Coal-fired power plants generate 50% of the nation's electricity and account for one-third of the anthropogenic CO_2 emissions. Since it is widely anticipated that regulations on CO_2emissions will come into effect in the near future, coal-fired power plants, which are largepoint sources of CO_2, will be needed to adopt CO_2 reduction if the expected CO_2 levels are to be achieved. Several novel technologies are being pursued for this purpose.The Carbonation-Calcination Reaction (CCR) Process, which employs a calcium-based sorbent for CO_2 capture, is being developed at The Ohio State University (OSU). A 90% reduction in CO_2 and 100% SO_2 removal from coal-combustion flue gas, has been demonstrated using the CCR Process at 120 kWth scale. An intermediate reactivation step – hydration – has been integrated in the system to maintain the reactivity of the sorbent over multiple cycles.

Since the CCR Process removes CO_2 and SO_2 at high temperature, there are different process integration options for existing coal-fired power plants. Process analysis from computer simulations shows the CCR Process to be highly effective and efficient in removing both $CO_2$and SO_2 at low energy penalties under realistic conditions. The CCR Process is expected to maintain a constant electrical output with ~25-28% increase in thermal energy.

The feedstock for the CCR Process is naturally occurring limestone. The low-cost and abundance of the feedstock adds to the economic viability of the process. Six different naturally occurring calcium sorbents have been investigated at laboratory-scale over multiple carbonation-calcination-hydration cycles. Results from these tests confirm the suitability of the proposed high temperature steam hydration as a method for sorbent reactivation.

See more of this Session: Unconventional Technologies for CO2 Capture, Conversion and Utilization II
See more of this Group/Topical: International Congress on Energy 2011

(557e) Novel Electrochemical Process for Capture of CO2 From Power Plant Flue Gas

Saurav Datta[1], Yupo J. Lin[1], Michael P. Henry[2], Cynthia S. Millard[3], Jitendra Shah[4], Lisa Wesoloski[4], Rebecca L. Stiles[4], Jianwei Yuan[4], Robert W. Dorner[4], Wayne M. Carlson[4] and Seth W. Snyder[3], (1) Argonne National Laboratory, Argonne, IL, (2)Argonne National Laboratory, Lemont, IL, (3)Energy Systems Division, Argonne National Laboratory, Argonne, IL, (4)Air Protection Technologies, Nalco Company, Naperville, IL

Energy efficient capture of CO_2 released from coal based power plants is a major challenge to reduce greenhouse gas emissions. The existing commercial technologies require substantial energy inputs in the form of heat and vacuum which leads to around 30 % parasitic energy loss. To overcome this problem a novel electrochemical process is developed to efficiently capture CO_2 from flue gas and then release (desorb) pure CO_2 at atmospheric pressure without requiring heat, vacuum, or consumptive chemical usage. The process uses a proprietary resin wafer based Electrodeionization (RW-EDI) technique to shift the pH of the process solution from basic in the capture chamber to acidic in the release chamber. The base/acid pH shift drives the CO_2 capture/release process and catalysts (enzyme or inorganic) are used to enhance kinetics. The goal is to demonstrate 90 % capture of CO_2 from synthetic flue gas (15 % CO_2) and subsequent release of CO_2 with 90 % purity in the released stream. Recent bench-scale experimental results will be presented showing significant progress towards this goal. Water chemistry (CO_2 hydration/dehydration), mass transfer issues (gas-liquid interaction) and kinetic limitations (capture/release rates) associated with the process will be discussed in detail. Funding for this project is awarded by US Department of Energy ARPA-E to Nalco Company and Argonne National Laboratory.

See more of this Session: Unconventional Technologies for CO2 Capture, Conversion and Utilization II
See more of this Group/Topical: International Congress on Energy 2011

(557f) Combating Surface Passivation In the Mineralization of Carbon Dioxide

Edward J. Swanson, Chemical Engineering, Columbia University, New York, NY and Ah-Hyung Alissa Park, Earth and Environmental Engineering and Chemical Engineering, Columbia University, New York, NY

The accelerated weathering of ultramafic rocks is an environmentally benign route to a thermodynamically and kinetically stable form of carbon. The product, solid magnesium carbonate, can be produced with a range of properties depending on the final application. However, before this can be considered, the magnesium must be extracted from mineral particles at a reasonable rate. While mineral dissolution is thermodynamically favorable, the kinetics are prohibitively slow. The silica tends to accumulate on the particle surface creating a passivation layer, through which magnesium and protons must diffuse. In this work, we characterize the effect of silica passivation on the dissolution rate of serpentine minerals. The permeability and dissolution rate of this silica-based passivation layer are shown to ultimately limit the availability of magnesium for further reactions, and a number of silica specific chelating agents are evaluated for their ability to accelerate the removal of the passivation layer and increase the dissolution rate of magnesium.

See more of this Session: Unconventional Technologies for CO2 Capture, Conversion and Utilization II
See more of this Group/Topical: International Congress on Energy 2011

(560a) Evaluation of Modified Feedstocks for Deconstrctionability

Seema Singh, Joint BioEnergy Institute, Emeryville, CA

Today, carbon-rich fossil fuels, primarily oil, coal and natural gas, provide 85 percent of the energy consumed in the United States. Fossil fuel use increases CO2 emissions, increasing the concentration of greenhouse gases and raising the risk of global warming. The high energy content of liquid hydrocarbon fuels makes them the preferred energy source for all modes of transportation. In the US alone, transportation consumes around 13.8 million barrels of oil per day and generates over 0.5 gigatons of carbon per year. This has spurred research into alternative, non-fossil energy sources. Among the options (nuclear, concentrated solar thermal, geothermal, hydroelectric, wind, solar and biomass), only biomass has the potential to provide a high-energy-content transportation fuel. Biomass is renewable resource that is carbon-neutral.

Currently, biofuels such as ethanol are produced largely from grains, but there is a large, untapped resource (estimated at more than a billion tons per year) of plant biomass that could be utilized as a renewable, domestic source of liquid fuels. Well-established processes convert the starch content of the grain into sugars that can be fermented to ethanol. Plant-derived biomass contains cellulose, which is more difficult to convert to sugars. The development of cost-effective and energy-efficient processes to transform cellulose in biomass into fuels is hampered by significant roadblocks, including the lack of specifically developed energy crops, the difficulty in separating biomass components, low activity of enzymes used to deconstruct biomass, and the inhibitory effect of fuels and processing byproducts on organisms responsible for producing fuels from biomass monomers.

The Joint BioEnergy Institute (JBEI) is one of three US Department of Energy Bioenergy Research Centers that will address these roadblocks in biofuels production. JBEI draws on the expertise and capabilities of three national laboratories (Lawrence Berkeley National Laboratory (LBNL), Sandia National Laboratories (SNL), and Lawrence Livermore National Laboratory (LLNL)), two leading US universities (University of California campuses at Berkeley (UCB) and Davis (UCD)), and a foundation (Carnegie Institute of Washington at Stanford University) to develop the scientific and technological base needed to convert the energy stored in cellulose into transportation fuels and commodity chemicals. Established scientists from the participating organizations are leading teams of researchers to solve the key scientific problems and develop the tools and infrastructure that will enable other researchers and companies to rapidly develop new biofuels and scale production to meet US transportation needs, and to develop and rapidly transition new technologies to the commercial sector. This talk will present a summary of the organization and efforts at JBEI, highlight the efforts on the evaluation of modified feedstock for deconstructionability and development of novel biomass pretreatment methods that enable the efficient conversion of biomass into next-generation biofuels.

(560b) Discovery and Characterization of Cellulolytic Enzymes From the Wood Wasp Symbiont Streptomyces Sp. ActE

Taichi E. Takasuka, Adam J Brook, Cameron R. Currie and Brian G. Fox, Great Lakes Bioenergy Center, University of Wisconsin, Madison, WI

One of the major hurdles for producing cost effective biofuels is the enzymatic conversion of lignocellulosic biomass into fermentable sugars. The US Department of Energy has issued the challenge of discovering and designing enzymes with novel biomass-deconstruction capabilities. This study focuses on the discovery of better enzymes and accessory proteins for biomass deconstruction from a novel bacterial species. *Streptomyces* sp. ActE, an isolate from wood-boring wasps of the genus *Sirex*. Its genomic sequence has been determined and initial physiological studies suggested high cellulolytic potential. We analyzed the genome-wide gene expression of ActE grown on insoluble substrates such as cellulose, xylan, chitin and others as the sole carbon source in the culture medium, and compared these expression profiles with glucose-grown control ActE cells. The secretomes obtained from growth on different substrates were analyzed by shotgun LC-MS/MS and enzymatic assays. From the combination of the genomic, transcriptomic, and secretomic analyses, we discovered ~200 genes that have carbon source-specific regulation of gene expression or protein secretion. This work provides the first extensive systems biology characterization of the capabilities of the genus *Streptomyces* for wide variety of polysaccharide utilization.

This work was funded by the DOE Great Lakes Bioenergy Research Center (DOE BER Office of Science DE-FC02-07ER64494).

See more of this Session: Advances In Biofuels: DOE Bioenergy Research Centers II
See more of this Group/Topical: Sustainable Engineering Forum

(560c) Dissecting the Cellulosome Enzyme Complex: Assembly and Function

Yannick J. Bomble[1], Mark Nimlos[2], Michael E. Himmel[1] and Michael F. Crowley[1], (1)Biosciences Center, National Renewable Energy Laboratory, Golden, CO, (2)National Bioenergy Center, National Renewable Energy Laboratory, Golden, CO

Most bacteria and fungi use free enzymes to degrade plant cell walls. However, a small number of bacteria have developed a fundamentally different approach, where enzymes are tethered to a large protein scaffold forming a multi-domain complex known as, the cellulosome. The study of these large protein assemblies is an ongoing research topic that has already yielded numerous breakthroughs. However, the mechanism of assembly of the enzyme subunits onto the natural scaffoldin; as well as the modes of action of the cellulosome and its enzymatic components, are currently not well understood.

In this study, we focused on the cellulosome-integrating protein (CipA) of C. thermocellum; as well as cellulosomal enzymes from families 5, 9 and 48. These three enzymes are representative of the diversity of enzymes secreted by C. thermocellum. This work not only aims at understanding the mechanisms involved in the sequential binding of the cellulosomal enzymes to the CipA scaffold of C. thermocellum, but also the binding of CipA to secondary scaffolds, such as OlpB. We focused on the physical properties of the binding of component enzymes to the scaffoldin. The modularity of the enzymes was found to be one of the main influences on the cellulosome assembly process.

Additionally, we examined several modules from the multi-modular enzyme complex, CbhA. Several of these modules have undefined functions, such as the X1 modules found closely linked to the catalytic domain. We hypothesize that the X modules could serve as tethers between several modules in CbhA to fine-tune the positioning of this enzyme during cellulose deconstruction. We conducted steered molecular dynamics simulations (directly comparable to AFM pulling experiments) to evaluate the energy and the profile for unfolding these modules. We found that the energy required to partially unfold this domain is accessible by the cellulosome and the unfolding pathways are similar in the multiple simulations conducted.

See more of this Session: Advances In Biofuels: DOE Bioenergy Research Centers II
See more of this Group/Topical: Sustainable Engineering Forum

(560d) Construction of Efficient Gram-Negative Biocatalysts for Lignocellulose Conversion

David Keating, Michael Schwalbach, Jeffrey Gardner, Mary Tremaine, Patricia Kiley and Robert Landick, Great Lakes Bioenergy Center, University of Wisconsin, Madison, WI

Research within the Great Lakes Bioenergy Research Center aims to better understand the bottlenecks associated with bioconversion of lignocellulose to liquid transportation fuels, knowledge which will then be used to improve the efficiency of this process. Much of these studies are currently focused on the Gram-negative bacterium *Escherichia coli,* due to its sophisticated genetics, well-understood physiology, and widespread industrial use.

To better understand the limitations associated with microbial biocatalysis of lignocellulosic biofuels, we have engineered *E. coli* strains capable of high-level ethanol production and tolerance. To develop efficient *E. coli* ethanologens, we have constructed mutations that reduce alternative pathways of election flow, and introduced the pyruvate decarboxylase and alcohol dehydrogenase genes from *Zymomonas mobilis*. In addition, we have carried out extensive directed evolution of *E. coli*, which has resulted in strains capable of growth in the presence of 10% ethanol. These strains are currently being subjected to a systems-level microbial analysis during lignocellulosic bioconversion. In collaboration with the Dale laboratory, we have generated enzymatic hydrolysates of AFEX-pretreated corn stover at high solid loadings. Furthermore, we have developed methodology to measure extracellular amino acids, metabolic endproducts, intracellular metabolites, and global transcription. We have used these techniques to characterize the physiological state of our ethanologenic strains during fermentation in lignocellulosic hydrolysates. Our initial analysis has already generated several surprising findings. First, lignotoxins within the hydrolysate appear to strongly limit the production of biomass, but have a more modest effect the conversion of glucose to ethanol. However, these inhibitors appear to strongly limit the conversion of xylose to ethanol. Second, microbial growth in lignocellulosic hydrolysates is a dynamic process, in which global transcription and overall physiology change dramatically during the course of fermentation. Our systems-level analysis of these strains has identified transcriptional and metabolic changes that provide insight into the mechanism of action of the lignotoxins. We are using this information, as well as data from similar experiments with lignotoxin-tolerant mutants, to generate new ethanologens with improved productivities under these conditions. Furthermore, we expect these studies will produce an improved molecular understanding of the current limitations of ethanologenesis in these strains, which we expect will be applicable to diverse ethanologenic microorganisms.

See more of this Session: Advances In Biofuels: DOE Bioenergy Research Centers II
See more of this Group/Topical: Sustainable Engineering Forum

(560e) One-Step Production of Lactate and Isobutanol From Pretreated Biomass by Recombinant Cellulolytic Bacillus Subtilis

Xiaozhou Zhang, Biological Systems Engineering, Virginia Polytechnic Institute and State University, Blacksburg, VA, Chun You, Biological Systems Eng, Virginia Polytechnic Institute and State University, Blacksburg, VA and **Y.-H. Percival Zhang**, Biological Systems Engineering Department, Virginia Polytechnic Institute and State University, Blacksburg, VA

Consolidated bioprocessing (CBP) is low-cost cellulosic biomass bioconversion by integrating cellulase production, cellulose hydrolysis, and sugar fermentation into a single step. Although significant efforts had been made to create recombinant cellulolytic microorganisms, real recombinant cellulolytic microorganism that can produce sufficient secretory active cellulases, hydrolyze cellulose, and utilize soluble sugars for supporting cell growth and cellulase synthesis without any other organic nutrient is not available yet. We demonstrated that over-expression of *Bacillus subtilis* endoglucanase BsCel5 by more than 30-fold enables non-cellulose-utilizing *Bacillus subtilis* to grow on amorphous cellulose and pretreated lignocellulosic biomass as the sole carbon source without addition of costly organic nutrients. After directed evolution and screening on amorphous cellulose, both the expression/secretion level and specific activity of Cel5 and Cel9 had been increased successfully. An advanced recombinant cellulolytic *Bacillus subtilis* strain has been constructed with engineered cellulase and the genes knock out/knock in. Besides these, we also achieved the co-expression and secretion of family 5, 9 and 48 cellulases in a single *B. subtilis* strain. The one-step production of lactate and isobutanol from cellulose and pretreated biomass in the minimal medium without adding any organic nutrient had been demonstrated. We are reconstituting *in vivo* mini-cellulosome on the cell surface of *B. subtilis*. The recombinant cellulolytic B. subtilis would be an ultra-low-cost platform for producing biofuels and other value-added products from non-food biomass.

See more of this Session: Advances In Biofuels: DOE Bioenergy Research Centers II
See more of this Group/Topical: Sustainable Engineering Forum

(560f) Development of Alkene Biofuels

Harry R. Beller, Joint BioEnergy Institute (JBEI), Emeryville, CA and Ee-Been Goh, Fuels Synthesis, Joint BioEnergy Institute (JBEI), Emeryville, CA

Aliphatic hydrocarbons are appealingtargets for advanced cellulosic biofuels, as they arepredominant components of petroleum-based gasoline and diesel fuels and thuswould be compatible with existing engines and fuel distribution systems. Long-chainalkenes are useful as feedstocks that can be crackedto form diesel- or gasoline-range hydrocarbons. We have studied alkenebiosynthesis in the bacterium *Micrococcus luteus*,a close relative of which was reported in the 1960s to biosynthesize *iso-* and *anteiso-*branched,long-chain alkenes. The underlyingbiochemistry and genetics of alkene biosynthesis werenot elucidated in those studies. In fact, until very recently, there were no known sequences forprokaryotic enzymes catalyzing the synthesis of non-isoprenoidaliphatic hydrocarbons ($> C_2$). We have shown that heterologousexpression of a three-gene cluster from *M. luteus*(Mlut_13230-13250) in a fatty-acid overproducing *E. coli* strain resultedin production of long-chain alkenes, predominantly 27:3 and 29:3 (no. carbon atoms: no. C=C bonds). Heterologous expression ofMlut_13230 (*oleA*) alone produced no long-chainalkenes but unsaturated aliphatic monoketones,predominantly 27:2, and *in vitro* studies with the purified Mlut_13230protein and tetradecanoyl-CoA produced the same C_{27}monoketone. Gas chromatography-time of flight (GC-TOF) mass spectrometry confirmedthe elemental composition of all detected long-chain alkenes and monoketones (putative intermediates of alkenebiosynthesis). Negative controlsdemonstrated that the *M. luteus* genes wereresponsible for production of these metabolites. We have proposed a metabolic pathway for alkene biosynthesisstarting with acyl-CoA (or -ACP) thioestersand involving decarboxylative Claisencondensation as a key step, which we believe is catalyzed by OleA. Suchactivity is consistent with our data and with the homology of Mlut_13230 (OleA) to FabH, which catalyzes decarboxylative Claisencondensation during fatty acid biosynthesis. We will discuss how understanding of the alkenebiosynthetic pathway can lead to enhanced production.

See more of this Session: Advances In Biofuels: DOE Bioenergy Research Centers II
See more of this Group/Topical: Sustainable Engineering Forum

(572a) CO2 Methanation with Ru Decorated Carbon Nanotubes-Supported Catalysts

Randy L. Vander Wal[1], Jane Fujiyama-Novak[2], Chung-Hsuan Huang[2], Ganesh Rahul Bhimanapati[2] and Susana Carranza[3], (1)Energy and Mineral Engineering & The EMS Energy Institute, Pennsylvania State University, University Park, PA, (2)The EMS Energy Institute, Pennsylvania State University, University Park, PA, (3)Makel Engineering, Inc., Chico, CA

The sustainable exploration of space requires the minimization of re-supply from Earth by implementation of In-Situ Resources Utilization (ISRU) strategies developed by NASA (National Aeronautic & Space Administration). One of the strategies for the exploration of the Moon is the production of oxygen from the lunar regolith, which is a complex mix of minerals with large oxygen content. In the case of carbothermal based oxygen production, carbon oxides should be converted to methane for reintroduction in the carbothermal system. For that reason, the development of a methanation system that will efficiently convert mixed carbon oxides to methane turn out to be highly necessary.

Many authors have studied the catalytic synthesis of methane and other hydrocarbons from mixture of CO and H_2[1], [2], [3]. It is well established that CO dissociation occurs readily on Ni[4], Co[5] and Ru[5]. Multi-walled carbon nanotubes (MWCT) can be efficiently used as to support metal catalysts for many catalytic heterogeneous reactions[6],[7], [8]. The high specific area of CNTs significantly contributes to improve the final catalytic performance of the system since the reactions are governed by mass and heat transfer phenomena. This reflects the catalytic activity depending on the metal particle distribution as well the particle size on supporting materials.

As a result the key strategy to fabricate a high performance catalytic system is to reach high dispersion levels of metal precursors upon the MWCTs in parallel with high metal loading. Synthesis variables such as the use of surface modified CNTs, impregnation and reduction techniques are critical to both controlling the decoration process and to anchoring the catalyst nanoparticles. In summary, the project objective is the development of a nanostructured catalytic system based on dispersed ruthenium and cobalt particles supported on multi-wall carbon nanotubes. These nanostructed catalysts were deposited within a foam structure to fully expose the decorated MWCT catalysts to the reactant gases for methanation.

Ruthenium and cobalt supported MWCT catalysts were prepared using different metal precursors, ruthenium trichloride ($RuCl_3.xH_2O$), cobalt nitrate hexahydrate [$Co(NO3)_2.6H_2O$], diverse preparation tecnhiques and different MWCT surface chemistry. Results of Thermal Gravimetric Analysis and Transmission Electron Microscopy indicated that the acid treatment by concentrated nitric acid generated additional carboxylic and hydroxyl functionalities on the nanotubes, resulting to a superior amount of metal decoration and better dispersion of the nanoparticles. The effect of chemical reduction process and the addition of cobalt were investigated and compared with a typical single metal catalytic system. The metal/MCNTs were deposited over different density foams for use in the reaction, enhancing heat and mass transfer while maintaining a low-pressure drop. A new microchannel reactor was developed based on the nanofabricated catalysis to evaluate the chemical activity and selectivity towards reduction of mixed carbon oxides.

Ruthenium deposited carbon nanotubes showed superior methane formation comparing to the Co/CNTs. TEM analysis of these catalytic systems illustrated highly dispersed ruthenium nanoparticles on the support, while the cobalt loading on the support has metal particles of 10-25 nm of diameter, 10 times bigger than the ruthenium ones. Different synthesis protocols were tested and the best results will be presented.

References

[1] H. Pichler, A. Hector., *Kirk-Othmer Encyclopedia Chem. Tech.* IV, 446 (1964).

[2] H.-J. Jung, P. L. Walker, Jr., M. A. Vannice, *J. Catal.* 75, 416-422 (1982)

[3] M. A. Vannice, *Cat. Rev.-Sci. Eng.* 14, 153 (1976).

[4] M. Araki, V. Ponec, *J. Catal.* 44, 439 (1976)

[5] J. W. A. Sachtler, J. M. Kool, V. Ponec, *J. Catal.* 56, 284 (1979)

[6] H. Tong, H.-L. Li, X.-G. Zhang, *Carbon* 45:2424-2432 (2007).

[7] J. Garcia, H. T. Gomes, Ph. Serp, Ph. Kalck, J. L. Figueiredo, J. L. Faria, *Carbon* 44:2384-2391 (2006).

[8] W. Li, X. Wang, Z. Chen, M. Waje, Y. Yan, *J. Phys. Chem. B* 110:15353-15358 (2006)

See more of this Session: Catalysis for CO2 Conversion II
See more of this Group/Topical: Catalysis and Reaction Engineering Division

(572c) Unique Processing Considerations for the Trireforming of CO2 to Synthesis Gas

Tracy J. Benson, Juan Cruz, Yishan Zhang and Jackie W. Seaman, Chemical Engineering, Lamar University, Beaumont, TX

Recently discovered natural gas supplies, along with America's increasing desire for fuel independence, has lead to increases in natural gas conversion to hydrogen and to higher fuels and chemicals. However, production of fuels and chemicals from methane feedstocks produces greenhouse gases, such as carbon dioxide, adding to global warming concerns. CO_2 accumulation in the atmosphere is at the forefront of today's responsible chemical industry concerns. This research is aimed at converting CO_2 to syngas using a trireforming process. Trireforming utilizes CO_2 and steam, without separation, from industrial flue stacks. Natural gas and oxygen are added to the mixture to produce H_2 and CO at appropriate ratios for downstream Fischer-Tropsch and methanol synthesis reactions.

Recent literature has alluded to typical processing conditions; however, specific reaction scenarios and conditions have not been reported. The heart of the trireforming process is the heterogeneous nickel catalyst supported on a metal-oxide catalyst. Our group has determined that trireforming ensues via a multi-function reaction site mechanism from the nickel and from oxygen vacancies of the support. Therefore, optimizing reaction conditions (i.e. temperature, pressure, feed ratios, etc.) are important for maximized CO_2 conversions and minimum catalyst coking. Aspen simulation modeling has been used to investigate the trireforming process using reported catalysts and novel nanoparticle catalysts developed in our laboratory. These results and a forecast of future CO_2 conversion technologies will be discussed.

See more of this Session: Catalysis for CO2 Conversion II
See more of this Group/Topical: Catalysis and Reaction Engineering Division

CO_2 METHANATION OVER SUPPORTED BIMETALLIC Ni-Fe CATALYSTS: EFFECT OF SUPPORT AND TOTAL METAL LOADING

Dharmendra Pandey[a], *Goutam Deo*[b]
Department of Chemical Engineering
Indian Institute of Technology Kanpur, Kanpur-208016, India
E-mail: [a] *dpandey@iitk.ac.in,* [b] *goutam@iitk.ac.in*

Abstract

All Group VIII metals were tested for the CO_2 hydrogenation and nickel is preferred due to its high activity, selectivity, long life and relatively low cost. Iron, however, is usually less active and more prone to carbon deposition. In the present work, a series of nickel and iron containing catalysts supported on different supports containing 10 wt. % and 30 wt. % total metal loading were tested for the CO_2 methanation reaction in the temperature range of 448 to 548 K. Bimetallic catalysts with a composition of 7.5 wt. % nickel and 2.5 wt. % iron showed significantly better activity and in some cases also a higher selectivity to methane compared with the monometallic catalysts. It was also observed that amount of enhancement depended on the total metal loading.

Introduction

The hydrogenation reactions of CO and CO_2 to methane and higher molecular weight hydrocarbons are important in purification of coal-derived gases, methanation of ammonia feeds and Fischer-Tropsch synthesis [1]. All Group VIII metals were tested for the hydrogenation of CO_2 and nickel is preferred due to its high activity, selectivity, long life and relatively low cost. Iron, however, is usually much less active and more prone to carbon deposition. Furthermore, it has been shown that besides enhancing surface area for proper dispersion of active metal, the support affects activity and selectivity of the catalysts [2].

It has been suggested [3, 4 and 5] that there are two important properties of the catalyst surface that determine the activity of some catalytic reaction: the barrier for reactant dissociation (E_{diss}) and the stability of the main intermediates on the surface. One of the approaches towards catalyst optimization could then be to construct a surface (active sites) with the desired E_{diss}. This can also be achieved by combining two metals: one with a high and one with low dissociation energy in an alloy catalyst [3]. It is important that both metals are present in a close proximity at the surface so that the active sites of the mixed-metal catalyst have the interpolated properties. This is not always the case since often one of the components can segregate to the surface or there may be alloying. It was furthermore concluded that the specific interactions between metal species in bimetallic catalysts often results in a completely different adsorption properties of the reactants [3]. It was also observed that in many cases, the reduction of bimetallic catalyst, for example Ni–Fe or Co–Fe, proceeds at significantly lower temperatures than for monometallic iron catalysts [6].

In the present work, a number of supported bimetallic Ni–Fe catalysts with different Ni:Fe ratios and total metal loadings were prepared and tested for the CO_2 methanation reaction. The supports considered were Al_2O_3, SiO_2, TiO_2, ZrO_2 and Nb_2O_5. The results were compared with those of the pure Ni and Fe based catalysts.

Experimental

Catalyst Synthesis

Nickel-Iron supported on different supports (Al_2O_3, SiO_2, TiO_2, ZrO_2 and Nb_2O_5) with 10 % total metal loadings was prepared using incipient wetness impregnation method. A nickel-iron supported on Al_2O_3 containing 30% total metal loading was also prepared. Initially, the support was pretreated with the incipient wetness volume of water by making a paste of the support and then it was dried in a dessicator for 6 hrs. The dried paste was heated in a furnace at different temperatures for finite periods and finally calcined at 773 K for 6 hrs. This pretreated support was treated with required amounts of aqueous solution of corresponding nitrates and heat treated in a manner similar to the support.

Catalyst Characterization

The series of catalyst was characterized by X-ray diffraction (XRD). The surface areas of the samples were obtained by the BET method using N_2 adsorption data at 77 K. The instrument used for the measurement was SMART SORB 92/93. The XRD patterns were obtained on a X'Pert Powder Diffractometer using Ni filtered K_α radiation from a Cu target (λ=1.541841 Å). A sweep of 3° min^{-1} and time constant of 3.0 s was maintained. The angles considered were in the range of 20 to 80°. The XRD patterns were obtained for the reduced catalysts. The International Center for Diffraction Data (ICDD) library was used for phase identification.

Experimental Setup

The CO_2 hydrogenation reaction was performed in a down flow tubular quartz reactor operating at atmospheric pressure and the temperatures ranging from 448 to 548 K. For reaction studies, the catalyst was first reduced at 773 K for 4 hrs in a H_2 stream, then cooled to room temperature in a nitrogen stream and then the temperature was increased to reaction condition in a H_2+CO_2 mixture.

Result and Discussion

The surface area values of the support (Al_2O_3 and SiO_2) and the metal supported catalysts containing 10 and 30% total metal loading were determined and tabulated as in Table 1. Also given in the table are the nomenclature, total metal loading and % of total metal loading as Ni and Fe in the catalysts.

X-ray diffraction patterns were obtained for the supports and reduced supported bimetallic Ni–Fe catalysts. The major diffraction features observed over the samples are characteristic of supports. Other possible crystalline phases present in the reduced catalysts could be Ni, NiO, Fe and Fe_3O_4.

Table 1: Specifications and surface area of supports and supported catalysts

Catalysts Nomenclature	Support Used	Total Metal Loading (wt.%)	Ni % in Total Metal Loading	Fe % in Total Metal Loading	Surface Area (m^2/g)
On SiO_2	SiO_2	-	-	-	190
10-100Ni0Fe/SiO_2	SiO_2	10	100	0	153
10-75Ni25Fe/SiO_2	SiO_2	10	75	25	171
10-50Ni50Fe/SiO_2	SiO_2	10	50	50	168
10-25Ni75Fe/SiO_2	SiO_2	10	25	75	159
10-0Ni100Fe/SiO_2	SiO_2	10	0	100	175
On Al_2O_3	Al_2O_3	-	-	-	191
10-100Ni0Fe/Al_2O_3	Al_2O_3	10	100	0	156
10-75Ni25Fe/Al_2O_3	Al_2O_3	10	75	25	163
10-50Ni50Fe/Al_2O_3	Al_2O_3	10	50	50	172
10-25Ni75Fe/Al_2O_3	Al_2O_3	10	25	75	175
10-0Ni100Fe/Al_2O_3	Al_2O_3	10	0	100	157
30-100Ni0Fe/Al_2O_3	Al_2O_3	30	100	0	106
30-75Ni25Fe/Al_2O_3	Al_2O_3	30	75	25	115
30-50Ni50Fe/Al_2O_3	Al_2O_3	30	50	50	119
30-25Ni75Fe/Al_2O_3	Al_2O_3	30	25	75	127
30-0Ni100Fe/Al_2O_3	Al_2O_3	30	0	100	109

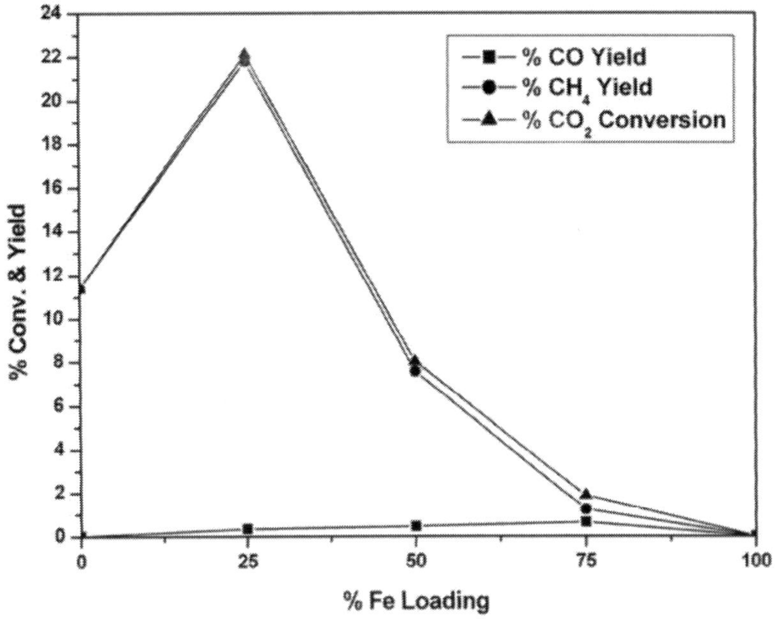

Figure 1: CO_2 conversion and CH_4 yield for the series of 10% Ni-Fe/Al_2O_3 catalysts at 523 K

The pure supports showed negligible CO_2 conversion in the studied temperature range. Reaction results were compared with those of the pure Ni and Fe supported catalysts. Bimetallic catalysts with compositions 7.5 wt.% Ni and 2.5 wt.% Fe supported on Al_2O_3, SiO_2, TiO_2, and ZrO_2 showed better activity and higher selectivity to methane in comparison with the monometallic Ni and Fe supported catalysts. It was also found that the amount of enhancement in activity decreases on increasing the total metal loading from 10 to 30 wt.% for alumina supported Ni-Fe catalysts. For Nb_2O_5 supported catalysts, however, the activity monotonically decreases on increasing the Fe content. In Figure 1, the catalytic activities for CO_2 conversion and CH_4 yield at 523 K are shown as a function of the catalysts composition for 10% total metal loading of a nickel and iron supported on alumina catalysts. Thus, the enhancement in activity depends on the support and total metal loading.

Conclusion

The results of this work suggest that it is possible to increase the activity of supported nickel catalyst by mixing it with Fe. The increase in activity is a function of the support and total metal loading. Additional studies are required to ascertain the reason for this increase in methane yield for the bimetallic system.

References

[1] Weatherbee, G. D. and Bartholomew, C. H. (1981), "Hydrogenation of CO_2 on group VIII metals: I. Specific activity of Ni/SiO_2", Journal of Catalysis, Vol. 68, pp. 67-76.

[2] Chen, S. L., Zhang, H. L., Hul, J., Contescuz, C. and Schwarz, J. A. (1991), "Effect of alumina supports on the properties of supported nickel catalysts", Applied Catalysis, Vol. 73, pp.289-312.

[3] Andersson, M. P., Bligaard, T., Kustov, A. L., Larsen, K. E., Greeley, J., Johannessen, T., Christensen, C. H. and Norskov, J. K. (2006), "Toward computational screening in heterogeneous catalysis: Pareto-optimal methanation catalysts", Journal of Catalysis, Vol. 239, pp. 501-506.

[4] Bligaard, T., Norskov, J. K., Dahl, S., Matthiesen, J., Christensen, C. H. and Sehested, J. (2004), "The Bronsted–Evans–Polanyi relation and the volcano curve in heterogeneous catalysis", Journal of Catalysis, Vol. 224, pp. 206-217.

[5] Norskov, J. K., Bligaard, T., Logadottir, A., Bahn, S., Hansen, L. B., Bollinger, M., Bengaard, H., Hammer, B., Sljivancanin, Z., Mavrikakis, M., Xu, Y., Dahl, S. and Jacobsen, C. J. H. (2002), "Universality in heterogeneous catalysis", Journal of Catalysis, Vol. 209, pp. 275-278.

[6] Tsipouriari, V. A. and Verykios, X. E. (1998), "Catalytic partial oxidation of methane to synthesis gas over Ni-based catalysts", Journal of Catalysis, Vol. 179 pp. 292-299.

(572e) Reaction Mechanism Network of CO2 Hydrogenation to Methanol On Cu(111)

Donghai Mei, Chemical and Materials Sciences Division, Pacific Northwest National Laboratory, Richland, WA

Reaction mechanism network of CO_2 hydrogenation to methanol on Cu(111)

Ya-Fan Zhao[1,2], Yong Yang[2], Charles Mims[3], Charles H. F. Peden[2], Jun Li[1,2,*] and Donghai Mei[2,*]

[1] *Department of Chemistry and Laboratory of Organic Optoelectronics and Molecular Engineering of the Ministry of Education, Tsinghua University, Beijing 100084, China*

[2] *Institute for Interfacial Catalysis, Pacific Northwest National Laboratory, Richland WA 99352, USA*

[3] *Department of Chemical Engineering and Applied Chemistry, University of Toronto, Ontario, M5S 3E5, Canada* Abstract

Methanol synthesis from CO_2 hydrogenation on supported Cu catalysts is of considerable importance in the chemical and energy industries. Although extensive experimental and theoretical efforts have been carried out in the past decades, the most fundamental questions such as the reaction mechanisms and the key reaction intermediates are still in debate. In the present work, a comprehensive reaction network for CO_2 hydrogenation to methanol on Cu(111) is studied using periodic density functional theory calculations. All of the elementary reaction steps in the reaction network are identified in an unbiased way with the dimer method. Our calculation results show that methanol synthesis from direct hydrogenation of formate on Cu(111) is not feasible due to the high activation barriers for some of the elementary steps. Instead, we find that CO_2 hydrogenation to hydrocarboxyl (*trans*-COOH) is kinetically more favorable than formate in the presence of H_2O via a unique proton transfer mechanism. The *trans*-COOH is then converted into hydroxymethylidyne (COH) via dihydroxycarbene (COHOH) intermediates, followed by three consecutive hydrogenation steps to form hydroxymethylene (HCOH), hydroxymethyl (H_2COH), and methanol. This is consistent with recent experimental observations [1], which indicate that direct hydrogenation of formate will not produce methanol under dry hydrogen conditions. Thus, both experiment and computational modeling clearly demonstrate the important role of trace amounts of water in methanol synthesis from CO_2 hydrogenation on Cu catalysts. The proposed methanol synthesis route on Cu(111) not only provides new insights into methanol synthesis chemistry, but also demonstrates again that spectroscopically observed surface species are often not critical reaction intermediates but rather spectator species.

See more of this Session: Catalysis for CO2 Conversion II
See more of this Group/Topical: Catalysis and Reaction Engineering Division

(572f) Mechanism of CO2 Reduction to Hydocarbons Over Ti-SBA-15

Chieh-Chao Yang and **Guido Mul**, Faculty of Science and Technology (TNW), University of Twente, Enschede, Netherlands

Different gas mixtures containing CO, CO_2, H_2O or H_2 were exposed to illumination in the presence of to this date the most effective catalyst for CO_2 reduction, Ti-SBA-15, in order to clarify the route to hydrocarbons in photocatalytic CO_2 reduction over this photocatalyst. A mixture of CO and H_2O led to the highest quantities of CH_4, C_2H_4, and C_2H_6 after 7 hours of reaction, whereas a mixture of CO_2 and H_2 lead to the lowest production rate of these products. H_2O is thus more efficient in activation of CO and CO_2 than H_2. CH_3OH was not detected as significant product, and when fed to the catalyst, did not yield extensive product formation. Formaldehyde was found very reactive over the catalytic system, yielding a product distribution (C_1-C_2) of similar nature as obtained by CO and CO_2 activation. Finally, the backward reaction, hydrocarbon oxidation into CO or CO_2, was found significant over this catalyst system. Based on i) results indicated above, and ii) available literature, a mechanism for photocatalytic CO_2 reduction is proposed involving formation of CO in the initial stages, followed by consecutive formation of formaldehyde, which converts to CH_4, C_2H_4, and C_2H_6, presumably by reaction with photo-activated H_2O (OH radicals). Implications of this study to further improve on the activity of this class of photocatalysts are discussed.

See more of this Session: Catalysis for CO2 Conversion II
See more of this Group/Topical: Catalysis and Reaction Engineering Division

(572g) Mechanistic Study of the Reaction of Ethylene Oxide and CO2 Catalyzed by Salen Complexes

Maria Curet-Arana, Department of Chemical Engineering, University of Puerto Rico, Mayaguez, PR

Few chemical processes have proven successful in using CO2 as a feedstock in chemical reactions. While CO2 is abundant and nontoxic, its stability imposes a challenge in finding routes for its effective use. Some transition metal salen complexes can catalyze the coupling reaction of CO2 with epoxides to produce either cyclic carbonates or polycarbonates via two competing reactions. We have used density functional theory calculations to understand how these reactions proceed and to unravel the role of the catalyst on these systems. We have studied the adsorption of CO2 and ethylene oxide onto Co and Cr salen complexes using the unrestricted OPBE functional along with the effective core potential LANL2DZ for the basis set. Geometrical optimizations were carried out beginning with a variety of different conformations and frequency calculations were used to verify that structures lie in an energy minimum.

See more of this Session: Catalysis for CO2 Conversion II
See more of this Group/Topical: Catalysis and Reaction Engineering Division

(574a) Synthesis, Physical Characterization and Hydrogen Generation by Methanol Dry Reforming Over Silica-Supported Size-Controlled Pd Nanoparticles

Selma Hokenek, Chemical & Biomedical Engineering, University of South Florida, Tampa, FL and John N. Kuhn, Chemical and Biomedical Engineering, University of South Florida, Tampa, FL

In the efforts to create and optimize supported nanoscale catalysts for use in the generation of renewable fuels, the beneficial effects of tuning the sizes and shapes of the catalysts have been actively studied. Due to the impact that the size and shape of the nanoparticle has on the surface energy potentials, and thus the particles' activity, the ability to control the particles' shape and size would be a very beneficial tool in catalyst development.

Here we report the synthesis, physical characterization and catalytic testing of silica-supported size-controlled Pd nanoparticles of 4 to 16 nm average diameter. The nanoparticles were synthesized using a one-pot polyol method and then immobilized by depositing onto a CAB-O-SIL® M-5 Silicon dioxide support. The size of the nanoparticles synthesized was controlled by using two precursors of different oxidation states in varying set molar ratios. Transmission Electron Microscopy (TEM) and X-ray diffraction (XRD) were used to measure the sizes of the nanoparticles. The catalytic activity of the supported particles was studied by running methanol dry reforming under temperature programmed reaction spectroscopy after pretreatment under oxidizing conditions at 623 K. The methanol dry reforming reaction was found to be structure-sensitive on the particles, with the peak TOF going through a maximum over the 6 nm particles.

See more of this Session: Catalytic Hydrogen Generation - General II
See more of this Group/Topical: Catalysis and Reaction Engineering Division

(574b) Nickel-Based Autothermal Reforming Catalysts: Effect of Particle Size On Sulfur Tolerance

Johannes W. Schwank, Joseph M. Mayne, Kevin Dahlberg, Thomas Westrich and Andrew Tadd, Chemical Engineering, University of Michigan, Ann Arbor, MI

Autothermal reforming (ATR) of liquid hydrocarbonfuels over nickel-based catalysts produces an equilibrium-limited mixture of hydrogen-richsyngas that is suitable as feed for solid oxide fuel cells. Autothermalreforming involves exothermic partial oxidation as well as endothermic steamreforming pathways that lead to non-isothermal temperature profiles in thecatalyst bed. The widespread adoptionof Ni-based ATR catalysts is hindered by the vulnerability of the catalysts todeactivate when low concentrations of sulfur compounds are present in the fuel.

In a systematic study ofisooctane ATR in presence of small concentrations of thiophene, we noticed thatcatalyst sites responsible for steam reforming were more vulnerable todeactivation than the sites responsible for partial oxidation. The presence ofthiophene led to an increase in the catalyst bed temperature profile and ahigher yield of intermediate products, such as isobutylene, propylene andethylene. Long-term stable performance of the catalyst could be achieved underreaction conditions that favored nearly complete conversion of thiophene to hydrogensulfide.

However,it was not clear to what extent the nickel particle size affected thesusceptibility of the catalysts to deactivation. Therefore, we conducted a uniqueexperimental study where we attempted to isolate the effect of nickel particlesize from other contributing factors. The total number of active nickel surface sites was removed as avariable in the exploration of particle size effects by selectivelypre-treating catalysts to create a series of samples with different Ni loadingsand particle sizes but identical nickel surface areas. The activities of these catalysts weremeasured for autothermal reforming of sulfur-free and thiophene-containingisooctane. Under sulfur-freeconditions, the isooctane conversion and yield of synthesis gas increased withincreasing Ni particle size. However, in presence of thiophene, catalysts withlarger nickel particles were more susceptible to deactivation. Catalysts with small Ni particle sizes exhibited poorsteam reforming activity, but were less vulnerable to deactivation bythiophene.

There are two possible explanations for the role of Niparticle size. First, it is conceivable that the Ni surface ensembles in thesmallest Ni particles are not large enough for adsorption and carbon-carbonbond breaking of large hydrocarbon molecules. A second possible explanation is thatthe smaller Ni particles are more prone to oxidation under ATR reactionconditions, with the resulting NiO surfaces having diminished hydrocarbon-reformingactivity.

From the standpointof catalyst design, our results clearly show that larger nickel particle sizesare beneficial for sulfur tolerance, with the caveat that sulfur will mainlydeactivate the sites responsible for steam reforming while leaving the partialoxidation sites intact.

See more of this Session: Catalytic Hydrogen Generation - General II
See more of this Group/Topical: Catalysis and Reaction Engineering Division

(574c) Effect of K and P On Rh Catalysts for Autothermal Reforming

Reetam Chakrabarti and Lanny Schmidt, Department of Chemical Engineering and Materials Science, University of Minnesota, Minneapolis, MN

K and P are alkaline and acidic inorganics found in biomass. Rh-based catalysts have been shown to autothermally reform lignocellulosic biomass containing low concentrations of inorganics to syngas with high selectivities at high temperatures (600-1000 °C). The acidity and alkalinity of K and P respectively can have an effect on biomass processing by altering catalyst activity thereby affecting the product distribution. To understand the effect on the chemistry of catalytic gasification over Rh-based catalysts, K and P were added to the catalyst at different loadings and their effect was examined over methane and ethanol. In addition, to understand their effect in presence of each other, the effect of doping the catalyst with potassium phosphate monobasic was also studied. Methane represents a fuel with negligible homogenous chemistry at the temperatures involved whereas ethanol represents a simple surrogate for biomass. For both fuels, the baseline performance was measured with an undoped catalyst and was with the performance of the catalyst after doping with inorganic precursors. Potassium doped catalysts showed volatilization of potassium with temperature whereas phosphorus doped catalyst showed influence of phosphorus chemistry even at higher temperatures. Phosphorus also strongly decreased hydrogen selectivity with both fuels. Catalysts were characterized by SEM and XRD to study changes in microstructure and crystallite formation respectively.

See more of this Session: Catalytic Hydrogen Generation - General II
See more of this Group/Topical: Catalysis and Reaction Engineering Division

(574d) Aqueous-Phase Reforming of n-BuOH Over Ceramic Oxide Based Catalysts

Banasri Roy[1], Hannah Sullivan[1] and Corey A. Leclerc[2], (1)Chemical Engineering, New Mexico Tech, Socorro, NM, (2)Department of Chemical Engineering, New Mexico Tech, Socorro, NM

Hydrogen, the simplest element in the periodic table can be considered the most environment friendly potential fuel for fuel cells application because of the cleanest emission it produces and yet it has the highest energy content per unit weight of all the fuels, three times the energy of same amount of gasoline. Hydrogen, produced from renewable biomass instead of non-renewable fossil fuel sources is an alternative source of environmentally clean renewable energy [**1**]. Recently n-Butanol (n-BuOH) has been considered as a potentially significant source of H_2 , because of it has higher weight % of H content compared to ethanol or methanol, low vapor pressure; so less flammable-easy to handle. Additionally, n-BuOH is a waste byproduct of the classical fermentation process from sugar beet, sugar cane, corn, wheat, lignocellulosic biomass and aqueous fraction of biomass pyrolysis liquids (bio-oil) [**2**]. It can also be produced through the non-fermentative pathways by manipulating metabolic engineering methods [**3**] and from macroalgae or seaweeds [**4**, 5]. There are different methods of producing H_2 from n-BuOH; such as steam reforming [**6, 7**], partial oxidation [**8**], dry reforming [**9, 10**].

Compared to these, aqueous phase reforming (APR) is a single step and low temperature (£ 225 ° C) energy efficient process, which produces hydrogen from water-diluted oxygenated hydrocarbons. The typical operating pressure and low temperature for APR can be helpful for the separation of H_2 and CO_2 from other products that are volatile at atmospheric pressure. Additionally, APR is useful for producing fuel cell grade H_2 with small amounts of CO in a single chemical reactor as a consequence of the water–gas shift (WGS) reaction being thermodynamically favored at lower temperature reaction conditions [**11,12**].

Reaction kinetics studied for aqueous-phase reforming of ethylene glycol over various supported metals indicated that Pt and Pd catalysts are selective for production of H_2 and Pt shows higher catalytic activity, but at a very high cost (~$1700/oz as of 17th Oct, 2010). This coupled with limited availability of Pt make it advantageous to develop catalysts based on less expensive metals, such as Ni (~$10/lb) [**13, 14**]. Studies show that in order to achieve high H_2 selectivity on a catalyst for the APR of an oxygenated hydrocarbon, a high C-C bond breaking rate, a low C-O breaking rate, and a low methanation reaction rate on metal, and low acidic catalyst supports are required [**1**]. For Ni, C-C bond breakage is reported to be high with reasonably good water gas shift activity compared to Co, Pt, Pd, Fe, Ir, and Rh [**15-17**].

High specific surface area-to-volume ratio, homogeneous dispersion of metals, and precise design of mesopore structure (large pore volume and narrow pore size distribution) combined with proper control of acidity/basicity of the support oxides are other important factors for high catalytic performance. There are many different ways of preparing nano metal/Al_2O_3 based catalysts; such as sol-gel, aerosol, co-precipitation, solution combustion synthesis, etc. [**18-23**].

Solution combustion synthesis (SCS) is a fast, simple, and energy efficient technique for the preparation of pure, porous, and small-particle size ceramics generally used as catalysts, phosphors, pigments, etc. [**24-26**]. We have reported APR of EtOH over alumina supported nano-scale nickel catalysts prepared by a SCS method before. Bimbela et al reported the steam reforming of n-BuOH over Ni/Al_2O_3 catalyst prepared by co precipitation method [**7**]. According to the knowledge of the present authors no body so far reported on the formation of H_2 by APR of n-BuOH.

In this paper, we present a study on aqueous-phase reforming of n-BuOH over Ni/CeO_2 and Ni/Al_2O_3 catalysts. Both catalysts were prepared by a SCS route. Over 300hr of run time, the Ni/CeO_2 catalyst lost 25 and 37% of its activity in terms of H_2 and CO_2 yield and 12 and 11.8 % in terms of selectivity to H_2 and CO_2. Ni/Al_2O_3 catalyst lost 41 and 45 % of its activity for H2 and CO_2 yield, and 19.8 and 19.5% in terms of selectivity to H_2 and CO_2. The steady state BuOH conversion and H_2 and CO selectivity increased and CO_2 and alkane selectivity decreased with increasing reactor temperature. The activation energies for H_2 and CO_2 production were 175.3, and 189 kJ/m and 154.4, and 172.7 kJ/m for Ni/CeO_2 and Ni/Al_2O_3 samples, respectively. The catalytic activity of these catalysts was compared at different temperatures, pressures, feed flow rates, and feed concentrations.

References

1. R.R. Davda, J.W. Shabaker, G.W. Huber, R.D. Cortright, J.A. Dumesic, Appl. Catal., B 56 (2005) 171–86.

2. Hipolito CN, Crabbe E, Badillo CM, Zarrabal OC, Morales Mora MA, Flores GP, et al. Bioconversion of industrial wastewater from palm oil processing to butanol by Clostridium saccharoperbutylacetonicum N1-4 (ATCC 13564). Journal of Cleaner Production 2008;16(5):632–8.

3. Atsumi S, Hanai T, Liao JC. Non-fermentative pathways for synthesis of branched-chain higher alcohols as biofuels. Nature 2008;451:86-9.

4. Advanced Research Projects Agency, US Department of Energy. See: http://www.energy.gov/news/8207.htm; Oct 26, 2009.

5. Silvia & Muller, IJHE, 36. 2057-75 (2011)

6. Nahar GA, Madhani SS. Thermodynamics of hydrogen production by the steam reforming of butanol: analysis of inorganic gases and light hydrocarbons. Int J Hydrogen Energy 2010;35:98-109.

7. Bimbela F, Oliva M, Ruiz J, Garcı´a L, Arauzo J. Catalytic steam reforming of model compounds of biomass pyrolysis liquids in fixed bed: acetol and n-butanol. J Anal Appl Pyrol 2009;85:204-13.

8. Wang WJ, Cao YY. Hydrogen-rich gas production for solid oxide fuel cell (SOFC) via partial oxidation of butanol: thermodynamic analysis. Int J Hydrogen Energy 2010;35: 13280-9.

9. Wang WJ. Hydrogen production via dry reforming of butanol: thermodynamic analysis. Fuel, 90 , 1681-88 (2011)

10. Wang & Cao, IJHE, 36, 2887-95 (2011)

11. R.R. Davda, J.W. Shabaker, G.W. Huber, R.D. Cortright, J.A. Dumesic, Appl. Catal., B 43 (2003) 13–26.

12. R. D. Cortright, R. R. Davda J. A. Dumesic, Nature 418 (2002) 964-67.

13. G. W. Huber, J. W. Shabaker, J. A. Dumesic, Science, 300 (2003) 2075-77.

14. New York Spot Price, 30th March, 2011.

15. M.A. Vannice, J. Catal. 50 (1977) 228-36.

16. J.H. Sinfelt, D.J.C. Yates, J. Catal. 8 (1967) 82-90.

17. D.C. Grenoble, M.M. Estadt, D.F. Ollis, J. Catal. 67 (1981) 90-102.

18. J.-H. Kim, D. J. Suh, T.-J. Park, K.-L. Kim, Appl. Catal., A 197 (2000) 191–200.

19. P. Kim, J.B. Joo, H. Kim, W. Kim, Y. Kim, I.K. Song, J. Yi, Catal. Lett. 104 (2005) 181-89.

20. Y. Chen, W. Zhou, Z. Shao, N. Xu, Catal. Commun. 9 (2008) 1418–25.

21. G. Li, L. Hu, J. M. Hill, Appl. Catal., A 301 (2006) 16–24.

22. R. Ran, G. Xiong, W. Yang, J. Mater. Chem. 12 (2002) 1854–59.

23. E. Seker, J.Cavataio, E. Gulari, P. Lorpongpaiboon, S. Osuwan, Appl. Catal., A 183 (1999) 121-34.

24. E.J. Bosze, J. McKittrick, G.A. Hirata, Mater. Sci. Eng. B 97 (2003) 265-74.

25. L.G. Jacobsohn, M.W. Blair, S.C. Tornga, L.O. Brown, B.L. Bennett, R.E. Muenchausen, J. Appl. Phys. 104 (2008) 124303.

26. P. Shuk, H-D. Wiemhofer, U. Guth, W. Gopel, M. Greenblatt, Solid State Ionics 89 (1996) 179-96.

Corresponding author; (575) 835-5412 FAX: (575) 835-5210; Email: leclerc@nmt.edu

See more of this Session: Catalytic Hydrogen Generation - General II
See more of this Group/Topical: Catalysis and Reaction Engineering Division

(574e) Deactivation of Noble-Metal Catalysts During Autothermal Reforming of Jet Fuel

Andrew R. Tadd, Chemical Engineering, University of Michigan, Ann Arbor, MI

Conversion of energy-dense liquid hydrocarbon fuels to syngas is one method for fueling portable fuel cells. Managing carbon deposition and deactivation, especially in high-sulfur feeds such as jet fuel, remains challenging. Catalysts comprising noble metals supported on Ce-Zr mixed oxides have previously been shown as very active and durable for this conversion, with steady state performance demonstrated to 300+ hours. Despite this sustained performance, the catalysts themselves undergo dramatic changes in both physical surface area and chemisorption uptake. During operation, both reversible and irreversible deactivation have been observed. This study summarizes links the observed performance and changing catalyst properties during reforming of clean and sulfur-laden hydrocarbon fuels.

See more of this Session: Catalytic Hydrogen Generation - General II
See more of this Group/Topical: Catalysis and Reaction Engineering Division

(574f) A Review of the Effect of Fuel Type and Sulfur Compound Variation on the Performance of Partial Oxidation Reforming Catalysts

Galen B. Fisher and Johannes W. Schwank, Chemical Engineering, University of Michigan, Ann Arbor, MI

Introduction

Fuel reforming is important for the generation of hydrogen and CO for a range of applications, including solid oxide fuel cells (SOFCs), proton exchange membrane (PEM) fuel cells, and for vehicle emission control systems. The heavy hydrocarbon fuels used in mobile applications contain a range of sulfur compounds. It is important to know the impact of these compounds on reforming performance and how that performance depends on the fuel used. We have measured reformate (H_2+CO) generation by catalytic partial oxidation (CPO) over a Rh-based catalyst for a range of fuels from methane to low S diesel fuel. We have determined the performance impact by sulfur species, sulfur content, and their variation with fuel type. We have shown the reversibility of sulfur addition/removal and have speciated the exiting species.

Materials and Methods

The Rh-based catalysts used in this study were washcoated onto zirconia toughened alumina (ZTA) foams or ceramic monolith supports. Catalysts, typically 1" long, were placed in a 1" OD cylindrical quartz reactor within a furnace. The fuels were methane, propane, isooctane (IO), California Phase II gasoline, dodecane, and Swedish city diesel fuel. For isooctane and several other fuels we have added sulfur compounds up to 2000 wppmS, including SO_2, and several aromatic compounds found in logistic fuels (thiophene [C_4H_5S], benzothiophene [C_8H_6S], and dibenzothiophene (DBT) [$C_{12}H_8S$]). To see the effect of aromaticity alone the structural analog of DBT, fluorene [$C_{13}H_{10}$], was also added. Bench analyzers, a GC, and both an electron-excited and a chemical ionization mass spectrometer measured products. Mass balances were obtained within 1%. The O/C ratio was held at 1.07 for all fuels; typical space velocities were 50,000 h^{-1}, although this was varied in other studies. Catalysts were tested in a de-greened "fresh" state or after baking six hours in air at 1200 °C.

Results and Discussion

For tests with several sulfur compounds, S and hydrocarbon species exiting the reactor were determined. In all cases from SO_2 to dibenzothiophene, the predominant sulfur species was H_2S, in agreement with equilibrium modeling. Our data shows the drop in H_2 concentration from 0 to 500 wppmS of each compound. In our experiments, each data point was collected 10 minutes after a step change in sulfur content. The curve is non-linear with the initial sulfur quantities having the largest effect and saturation occurring after 300 wppm S for all compounds. The impact scales with sulfur content, not with sulfur compound or aromatic concentration. The DBT analog compound, three-ringed fluorene, had no negative effect. When sulfur was removed from the feed, the effect of sulfur was found to be highly reversible. All of the data were consistent with the S concentration being in adsorption-desorption equilibrium at the surface, where it blocks active sites. The exit temperature of the catalyst rose with increasing sulfur content, consistent with its poisoning endothermic steam reforming, one of the main pathways to H_2 formation, after the initial formation of CO_2 and H_2O. These results imply that the S compounds are completely dissociated by the catalyst at these 700-1000 °C operating temperatures and that aromatic compounds, or their cracking products, compete with sulfur for adsorption sites on the catalyst surface. Additionally, aromatic compounds do not inhibit hydrogen-forming reactions, as demonstrated by the results with fluorene addition.

To also examine the effect of fuel variation, we also examined the effect of adding 0 to 100 ppm of SO_2 in the five fuels shown, CH_4, C_3H_8, IO, IO with added aromatics to achieve the C/H ratio of gasoline, and CA Ph II gasoline. Comparisons with analogous reforming research on Rh and other metals will be discussed. What is clear from our data is that increasing both the molecular weight and the aromaticity make the fuel more sulfur tolerant for H_2 generation.

Significance

The effects of fuel sulfur on Rh-based reformer operation are highly reversible, limited at high S levels, and very manageable at current sulfur levels (£ 30 ppm) in transportation fuels.

See more of this Session: Catalytic Hydrogen Generation - General II
See more of this Group/Topical: Catalysis and Reaction Engineering Division

(583a) Production of Concentrated Monosaccharride Solutions From Biomass Using CO2-H2O Mixtures At Varying Temperatures

Jeremy S. Luterbacher[1], Jefferson W. Tester[1] and Larry P. Walker[2], (1)Department of Chemical and Biomolecular Engineering, Cornell University, Ithaca, NY, (2)Department of Biological and Environmental Engineering, Cornell University, Ithaca, NY

Sustainably producing concentrated solutions of monosaccharrides from biomass is a key bottleneck in the conversion of lignocellulosic biomass to biofuels or bioproducts. Most pretreatment and enzymatic hydrolysis processes are run at low-solid concentration (<10 wt%) and use chemical catalysts, while high-solids enzymatic hydrolysis experiments are almost always performed with air-dried pretreatment mixtures.

This study uses a stirred 1 L reactor to study high solids (40 wt%) and high-pressure (200 bar) H_2O-CO_2 pretreatment of mixed hardwood and switchgrass. In addition, two-temperature stage pretreatment was introduced and optimized to take advantage of the temperature dependent biomass depolymerization reaction sequence (a short high-temperature stage at 200-210°C for 0 to 24 min was followed by a long low-temperature stage at 160-170°C for 60 min). Optimal glucan to glucose yields around 80% were obtained after 72 hrs of dilute hydrolysis (1 wt% glucan) for both substrates at similar pretreatment conditions (210°C, 16 min for hardwood and 210°C, 1 min for switchgrass, both followed by 60 min at 160°C). These optimally pretreated substrates were used as feedstock in high-solids enzymatic hydrolysis reactions, which gave glucose yields above 80% for both hardwood and switchgrass after 48 hrs of hydrolysis. Therefore, without additional chemical catalysts or any drying, two-temperature stage H_2O-CO_2 pretreatment coupled with high-solids enzymatic hydrolysis can produce monosaccharide solutions of 185 gr/L (170 gr/L of glucose and cellobiose 15gr/L of hemicellulose sugars) for mixed hardwood and 148 gr/L (115 gr/L of glucose and cellobiose 33gr/L of hemicellulose sugars) for switchgrass. This suggests that H_2O-CO_2 pretreatment is an attractive alternative to chemically catalyzed processes such as dilute acid pretreatment.

See more of this Session: Developments In the Pretreatment of Lignocellulosics for Bioconversion III
See more of this Group/Topical: Sustainable Engineering Forum

(583b) Structural Characterization of AHP-Pretreated Biomass

Muyang Li, Chemical Engineering and Material Science, Michigan State University, East Lansing, MI and David Hodge, Department of Chemical Engineering and Material Science, Michigan State University, East Lansing, MI

Alkaline hydrogen peroxide (AHP) pretreatment is exceptionally well-suited to grasses, yielding high digestibilities at low enzyme loadings, while generating relatively few fermentation innhibitors. For AHP pretreatment, the question of how structural and chemical compositional changes within the plant cell wall correlate to pretreatment effectiveness has not been effectively resolved, while knowledge of how selective modification of lignin during pretreatment may improve digestibility may yield insights into both improving pretreatments and tailoring plant cell walls for deconstruction. This study presents a comprehensive chemical and structural characterization of the changes in plant cell walls associated with switchgrass (*Panicum virgatum* cv. Cave-In-Rock), corn stovers (a commercial hybrid and inbred brown midrib lines bm1 and bm3), sugar maple (*Acer saccharum*) and hybrid aspen (*Populus tremula x tremuloides*) that were AHP pretreated at varied severities. Both the remaining solids and solubilized biomass hydrolyzates were subjected to a number of characterizations including total polysaccharide composition (monosaccharides, acetate, uronic acids), lignin content as both Klason and acetyl bromide lignin, and the ratio of H/G/S monolignols as determined by thioacidolysis GC/MS. Solid-state ^{13}C CP/MAS NMR and HSQC 2D NMR were applied to determine changes associated with intra-lignin, carbohydrate-carbohydrate, carbohydrate-lignin linkages and functional groups. Apparent molecular weight distributions of soluble polymeric lignins and hemicellulose aggregates were determined by HP-SEC. Py-GC/MS was utilized to characterize changes in the volatilized compounds before and after pretreatment implying pretreatment-induced alterations in pyrolysis-labile monolignol linkages.

See more of this Session: Developments In the Pretreatment of Lignocellulosics for Bioconversion III
See more of this Group/Topical: Sustainable Engineering Forum

(583c) Pretreatment of Corn Stover for Bioethanol Production Using Low-Moisture Anhydrous Ammonia (LMAA); Ammoniation and Pretreatment

Chang Geun Yoo, Agricultural and Biosystems Engineering, Iowa State University, Ames, IA, Nhuan-John Nghiem, Crop Conversion Science and Engineering, USDA-ARS-ERRC, Wyndmoor, PA, Kevin B. Hicks, Eastern Regional Research Center, Agricultural Research Service, Wyndmoor, PA and **Tae Hyun Kim**, Department of Agricultural and Biosystems Engineering, Iowa State University, Ames, IA

In the past several decades, various pretreatment methods have been suggested to enhance the enzymatic digestibility and fermentability of lignocellulosic biomass. Although a few of them may be effective, several cost barriers which prohibit scale-up exist including high chemical input and excessive water use. In order to solve these problems, a simpler pretreatment method using anhydrous ammonia was developed. In this method, corn stover with 30~70% moisture was used, which was contacted with anhydrous ammonia in a packed bed reactor. After the ammoniation step, biomass was treated at different reaction temperatures (40-120°C) and reaction times (24-96 h). Pretreated biomass was directly saccharified and fermented "with no additional washing step." Therefore, it is anticipated that this method can reduce energy cost substantially due to the extremely low liquid loading and moderate conditions applied. With the treated solids, we have obtained significantly improved ethanol fermentability (>85%) using *E.coli* KO11 strain.

In this investigation, it is speculated that ammonia swells cellulose structure effectively and ammoniation prevents methoxyl groups on lignin from absorbing cellulase enzymes, thus it enhances the enzyme hydrolysis rate and ethanol fermentation yield of lignocellulosic biomass. Ammoniation pretreatment can also supply assimilable nitrogen (up to 1.25 wt.% of the weight of dry biomass) for microorganism growth in the fermentor.

In this paper, various pretreatment conditions such as moisture levels, residual ammonia concentrations, reaction time, and reaction temperature were explored to reduce the pretreatment cost and its pertinent costs. Enzyme digestibility and ethanol fermentation test results will also be presented.

See more of this Session: Developments In the Pretreatment of Lignocellulosics for Bioconversion III
See more of this Group/Topical: Sustainable Engineering Forum

(583d) Feedstock Mixture Effects On Sugar Monomer Recovery Following Dilute Acid Pretreatment and Enzymatic Hydrolysis

Michael J. Brodeur-Campbell, Jordan Klinger and David Shonnard, Chemical Engineering, Michigan Technological University, Houghton, MI

Feedstock MixtureEffects on Sugar Monomer Recovery Following Dilute Acid Pretreatment andEnzymatic Hydrolysis

MichaelBrodeur-Campbell[1], Jordan Klinger[1], and DavidShonnard[1,2]

[1] Department of Chemical Engineering,Michigan Technological University

[2] Director Sustainable Futures Institute,Michigan Technological University

Increaseduse of lignocellulosic biomass for bioenergyproduction holds promise for improved domestic energy security, reducedgreenhouse gas (GHG) emissions, rural economic development, and developing amore sustainable energy infrastructure [Chum and Overend2001; Perlack, Wright, et al. 2005; U.S. DOE 2006]. Transportation fuels are especially difficultto replace as their replacement has to have a high energy density, be compatiblewith existing fuel distribution infrastructure, and be compatible with existingautomobile technology. In 2009, 97% oftransportation fuels were derived from fossil sources [EIA 2008], andtransportation accounted for one-third of carbon dioxide emissions in 2008 [EIA2009]. Liquid fuels derived from biomassare particularly attractive as a replacement for fossil-derived transportationfuels. Cellulose is the most abundantbiopolymer on Earth, and lignocellulosic feedstocks represent one of the fewnatural resources available in sufficient quantity to replace a significantfraction of liquid transportation fuel.

In acommercial biorefinery feedstocks are likely to come from a variety of sourcesand include many different materials, such as softwoods, hardwoods, agriculturalresidues, and herbaceous or woody energy crops.Much work has been done to understand and optimize the pretreatment andenzymatic hydrolysis conditions for pure feedstocks, but little is knownabout mixture effects on enzymatic hydrolysis.A primary concern in processing woody biomass to produce a range ofbiofuels is to maximize the production of desired intermediates, in this case,determining the quantity of monomer sugars that are recoverable. The biochemical platform as described by theNational Renewable Energy Laboratory (NREL) was the process studied. Briefly, lignocellulosic biomass which iscomposed primarily of hemicellulose, cellulose, and lignin is treated withdilute sulfuric acid to hydrolyze the hemicellulose into soluble sugars. The solid residue remaining after dilute acidpretreatment is then treated with cellulose-degrading enzymes to hydrolyze thecellulose into soluble sugars. Theremaining lignin may be utilized as a high energy co-product stream to provideheat and power for the cellulosic ethanol conversion processes.

This studyseeks to clarify the effects of processing mixtures of feedstocks on overall sugarrecovery. We wish to test the hypothesisthat sugar yields from mixed feedstocks can be predicted by a simple linearinterpolation model based on the behavior of the pure species. Experiments were performed on pure species(aspen, balsam, and switchgrass), and on 50/50 blends of aspen/balsam andaspen/switchgrass. Aspen was chosen as amodel species due to its high suitability for processing in a commercialbiorefinery. Balsam was chosen to mixwith aspen to study the effects of high lignin content on mixed feedstockprocessing, and switchgrass was chosen to mix with aspen to investigate theeffects of a high ash content energy crop.

One-halfgram (0.5 g) of dry sample was weighed and placed into one of a set of 9stainless steel Swagelok cylindrical reactor tubes, and 4.5 ml of dilutesulfuric acid (0.5% w/v) was added to achieve 10% (wt) solids loading in thereactor. Reactors were rapidly heated to160°C in a silicon oil bath (Dow Corning 550 fluid), and three tubes wereremoved from the oil bath at previously determined time intervals. From previous research involving pure samplesof aspen, balsam, and switchgrass (Jensen et al. 2010), the time to achievemaximum yields of monomer and oligomer sugars from combined dilute acidhydrolysis and enzymatic hydrolysis was determined; the "optimum" time. Threereactors were removed at one-half the optimal time, three reactors were removedat the optimal time point, and the last three were removed at two times theoptimal time. Reactor tubes removed fromthe oil bath were immediately submerged in an ice-water bath to quench thereaction. One milliliter of liquid fromeach reactor was removed, filtered, neutralized, and analyzed on an Agilent1200 series HPLC for glucose and xylose concentration. One milliliter of liquid from each reactorwas filtered and then further acidified to 4% (w/v) sulfuric acid, and treatedat 121°C for one hour tohydrolyze any soluble oligomers into monomer form. After neutralization these samples were analyzedby HPLC as above. Results from these twoliquid samples were combined into total soluble (monomer plus oligomer) sugarrecovery from dilute acid pretreatment.Solids were washed with excess distilled water on a glass-fiber filter(VWR) and carried on to enzymatic hydrolysis.

Solidresidues containing cellulose and lignin were subjected to enzymatic hydrolysisusing Accelerase 1500 (Genencore). Each of the three reactors from the samepretreatment time point were loaded with a different enzyme concentration 0.1, 0.25, and 0.5 ml Accelerase 1500 per gram ofcarbohydrate in the feedstock in order to study the interaction betweenpretreatment severity and enzyme loading on sugar yields. Reaction conditions were 50mM citrate buffer,pH 4.8, 50°C,250 RPM orbital shaking. Liquid sampleswere removed at 0, 1, 2, and 3 days, filtered, and analyzed for sugar concentrationsby HPLC as above.

Maximumyields of xylose from pretreatment were seen to vary with both time andfeedstock composition in the expected manner. Maximum glucose yields following enzymatichydrolysis were seen to be strongly affected by both pretreatment severity andenzyme loading. Results are summarizedin Table 1 below.

Table 1: Experimental sugar recoveries for Dilute AcidHydrolysis (DAH), Enzymatic Hydrolysis (EH), and experimental and predictedrecoveries for Total Sugar (TS) yield as percent of theoretical.

	DAH		EH		TS		TS (Predicted)	
	Xylose	**Glucose**	**Xylose**	**Glucose**	**Xylose**	**Glucose**	**Xylose**	**Glucose**
Balsam	72%	7%	0%	2%	72%	9%		
Aspen/Balsam	82%	5%	8%	27%	90%	32%	85%[1]	29%
Aspen (1)	83%	3%	4%	47%	87%	50%		
Aspen/Switchgrass	79%	5%	4%	53%	83%	58%	86%	60%
Switchgrass	69%	12%	5%	59%	74%	71%		
Aspen (2) w/ β-Glucosidase	93%	4%	6%	61%	100%	65%		

[1]Predicted value based on Aspen (2) pretreatment results.

Xyloseyields from pretreatment varied from 69-93% of the theoretical maximum. Pure switchgrass gave the lowest yield andpure aspen the highest. Each purespecies and mixture showed a peak hydrolysis yield near the expected optimaltime. Earlier time points showedincomplete hydrolysis with greater quantities of soluble oligomericsugars in the hydrolyzate. Later timepoints yielded sugar decomposition products such as furfural and hydroxymethylfurfural (HMF) in the hydrolyzate. Longer pretreatment reaction times alwaysresulted in higher glucose yields, but due to the degradation of xylose atlater time points total sugar yields were generally lower for extendedpretreatment times.

Glucoseyields following enzymatic hydrolysis varied from 2-61% with pure balsam givingthe lowest yield and pure aspen with supplemental β-glucosidase giving thehighest. The first aspen experiment wasrepeated due to incomplete pretreatment (indicated by lower than expected sugarrecoveries), and subsequently treated with the highest Acceleraseconcentration plus supplemental β-glucosidase to study the effect ofadditional enzyme loading on yield.

Using asimple linear interpolation model to predict sugar recoveries for mixedfeedstock streams based on the known yields obtainable for pure species gives aresult that is accurate to within about 2% of the experimental data for bothglucose and xylose recovery following dilute acid pretreatment, and to within3% for glucose recovery following enzymatic hydrolysis. Xylose recovery following enzymatichydrolysis for blends containing balsam could not be predicted accurately fromthese data, but since that is only a very small amount of the total sugar yieldthe discrepancy did not affect the model results for total sugar yield. Total sugar yield was predicted to within 5%of experimental results.

References:

Chum, Helena L. and Ralph P Overend (2001). *Biomass and renewable fuels*, FuelProcessing Technology 71, 187-195.

Perlack,R. D., L. L. Wright, et al. (2005). Biomass as Feedstock for a Bioenergy and Bioproducts Industry: The Technical Feasibility of aBillion-Ton Annual Supply. U. S. D. o. A. (USDA) and U. S. D. o. E. (DOE).

U.S. DOE (2006). *Breakingthe Biological Barriers to Cellulosic Ethanol: A Joint Research Agenda*,DOE/SC-0095, U.S. Department of Energy Office of Science and Office of EnergyEfficiency and Renewable Energy (www.doegenomestolife.org/biofuels/).

EIA (2008). *Emissionsof Greenhouse Gases in the United States*. U.S. Energy Information Administration Office of Integrated Analysis andForecasting (http://www.eia.doe.gov/oiaf/1605/ggrpt/carbon.html).

EIA (2009). *Annual Energy Review 2009*. U.S.Energy Information Administration, (http://www.eia.doe.gov/aer/pecss_diagram.html).

Jensen, Jill R., Juan E. Morinelly, Kelsey R. Gossen,Michael J. Brodeur-Campbell, and David R. Shonnard (2010). Effectsof dilute acid pretreatment conditions on enzymatic hydrolysis monomer andoligomer sugar yields for aspen, balsam, and switchgrass. BioresourceTechnology 101, 2317-2325.

See more of this Session: Developments In the Pretreatment of Lignocellulosics for Bioconversion III
See more of this Group/Topical: Sustainable Engineering Forum

(583e) Hydrolysate De-Acidification Using Resin-Wafer Electrodeionization

Yupo J. Lin, Saurav Datta, Michael Henry, Anthony Fracaro and Seth Snyder, Argonne National Laboratory, Argonne, IL

Dilute acid pretreatment of lignocellulosic feedstock offers an economic route for the extraction of fermentable sugars from non-food biomass for biofuel production by bioconversion. However, removal of acids from the hydrolysate is a critical step to ensure optimal conditions for the subsequent bioprocessing/conversion steps, i.e., enzyme hydrolysis of the cellulose to release C6 sugars and for the final fermentation to produce biofuel. The removal (conditioning step) of inorganic and organic acids (released together with C5 sugar molecules from the hemicellulose) after the acid treatment stage poses a great economic challenge to the overall biofuel production cost. To address this challenge, we used a proprietary resin wafer electrodeionization (RW-EDI) technology to remove both the inorganic and organic acids in a single step. Preliminary process economic analysis indicates that the easily scalable and small foot-print unit operation of RW-EDI can offer an economic and viable means of addressing the hydrolysate conditioning challenge. Selective removal of inorganic acids and organic acids was demonstrated when appropriate operation conditions were employed for the RW-EDI operation. This unique feature makes possible the separate recovery of both inorganic acid (e.g., sulfuric acid) and organic acids, which act as fermentation inhibitors, so that the inorganic acid can be recycled to the dilute acid pretreatment stage. The RW-EDI also provides an option to remove other fermentation inhibitor such the furfural and hydroxy methyl furfural (HMF) during the hydrolysate conditioning. In this presentation, process design and performance of hydrolysate de-acidification using RW-EDI will be discussed. Pilot-scale demonstration of technology viability using synthetic hydrolysate will be also presented along with preliminary assessments of economic and process benefits.

See more of this Session: Developments In the Pretreatment of Lignocellulosics for Bioconversion III
See more of this Group/Topical: Sustainable Engineering Forum

(583f) Biomimicry of the Fungal Consolidation of Biomass Pretreatment with Saccharification

Justin T. Kaffenberger and Jonathan S Schilling, Bioproducts and Biosystems Engineering, University of Minnesota, St Paul, MN

Efficient breakdown of ligno-cellulosic material prior to fermentation for bio-fuel production is a key step in optimizing bio-fuel production. Current industrial practice involves chemical pretreatment and acid or enzymatic saccharification to prepare ligno-cellulosic biomass for fermentation. The commercial viability of bio-fuels hinges in large part on production cost. One approach to reducing production cost is process consolidation. Consolidation not only eliminates the capital costs associated with the additional processing steps, but can also trim operational costs by reducing total reaction time and down time between steps.

In nature, the brown-rot fungus *Postia placenta* successfully performs both chemical pretreatment and enzymatic saccharification of biomass in a consolidated manner. As evidenced by its recently sequenced genome, *P. placenta* produces a limited suite of cellulases, lacking genes for exoglucanases. Despite this limitation, it is still capable of rapidly degrading wood. Brown-rot fungi quickly reduce the degree of polymerization of wood with little weight loss through a reduction-oxidation pathway that yields highly reactive hydroxyl radicals through a chelator-mediated fenton reaction. While hydroxyl radicals would readily damage its cellulases, the fungus manages to conduct both reaction types, enzymatic and oxidative, in the same general location. Examination of the spatial and temporal relationship of these two reaction systems as brown-rot fungal degradation of wood progresses will be presented.

See more of this Session: Developments In the Pretreatment of Lignocellulosics for Bioconversion III
See more of this Group/Topical: Sustainable Engineering Forum

(585a) Synthesis, Characterization and Oxygen Reduction Reaction Studies of Monometallic and Bimetallic Palladium-Based Nanoparticles for Proton Exchange Membrane Fuel Cell (PEM FC) Applications

Selasi Blavo, M D Sanchez and J. N. Kuhn, Chemical & Biomedical Engineering, University of South Florida, Tampa, FL

In the search for new PEM FC electrocatalysts materials, Platinum (Pt) nanoparticles supported on carbon feature prominently. However the price of the latter makes commercial feasibility a challenge. Alternative materials and importantly, shapes of these new materials hold the promise of providing useful insight in tailoring the elecrocatalytic properties of synthesized nanoparticles to reproduce or even exceed the oxygen reduction reaction (ORR) results currently obtained with the industry standard.

In recent years, Palladium (Pd) has become a viable alternative to Pt. At a fraction of the price of Pt and, with a similar lattice structure, many studies have looked at monometallic Pd and bimetallic Pt-based nanoparticles to produce electrocatalysts with a reduced Pt load and in some cases no Pt load at all.

Here, we report on the synthesis, characterization and electrocatalytic activity of monometallic and bimetallic Pd-based nanoparticles, varying in shape. Transmission Electron Microscopy and energy dispersive spectroscopy were used to characterize and obtain size distributions of the nanoparticles. Cyclic voltammetry and Linear sweep voltammetry electrochemical experiments were used to determine the electrochemically active surface area and oxygen reduction activity as a function of nanoparticle shapes.

See more of this Session: Electrocatalysis for PEM Fuel Cells II
See more of this Group/Topical: Catalysis and Reaction Engineering Division

(585b) First Principle Insights Into the Electrocatalytic Oxidation of Ethanol Over Pt and PtMo Surfaces

Fei Li, Department of Chemical Engineering, University of Virginia, Charlottesville, VA and Matthew Neurock, Chemical Engineering, University of Virginia, Charlottesville, VA

First Principle Insights into the Electrocatalytic Oxidation of Ethanol over Pt and PtMo Surfaces

Fei Li and Matthew Neurock, Department of Chemical Engineering, University of Virginia, Charlottesville, VA 22904-4741

The electrocatalytic oxidation of ethanol in proton exchange membrane fuel cells is of great interest. The oxidation of ethanol however is met with a number of challenges that result from activating the C-C bond. First-principle density functional theoretical calculations were carried out herein to investigate the oxidation of ethanol over Pt and different Pt-Mo alloy surfaces. The reaction energies and activation barriers for the main pathways involved in the complete oxidation to CO_2 were calculated to establish potential energy profiles and possible rate-limiting steps. The sluggishness of the carbon-carbon (C-C) cleavage reaction and the oxidative removal of poisonous single carbon (C1) intermediates, especially CO, have been probed in detail. It was found that by alloying Pt with Mo, the O-H bond could be selectively activated. The oxidative removal of poisonous C1 intermediates is promoted by adding Mo to the surface as well as to the subsurface. The computational results were calculated for a range of different Pt-Mo alloy surface ensembles and compared with results reported in the literature [2, 3].

[1] Claude Lamy, Alexandre Lima, Journal of Power Sources, 105 (2002) 283-296

[2] D.M. Dos Anjos, K.B. Kokoh, Journal of Applied Electrochemistry, 36 (2006) 1391-1397

[3] A. Oliveira Neto, E.G. Franco, Journal of the European Ceramic Society, 23 (2003) 2987-2992

See more of this Session: Electrocatalysis for PEM Fuel Cells II
See more of this Group/Topical: Catalysis and Reaction Engineering Division

(585c) Electrocatalytic Processing of Polyols Based On Anion Exchange Membrane Fuel Cells

Wenzhen Li, Le Xin and Zhiyong Zhang, Chemical Engineering, Michigan Technological University, Houghton, MI

Biofuels are expected to share a significant portion in our future energy supply landscape. Polyols will be substantially produced with biofuel refinery processing. We will present our recent research on electrocatalytic processing of polyols (including glycerol, sorbitol, etc) based on anion exchange membrane fuel cells. A self-developed solution phase-based nanocapsule method was used to prepare Pt, Pd, Au and their bimetallic catalysts with small particle size of 2-4 nm. The electro-oxidation of polyols on these catalysts was studied in a three-electrode-compartment cell. AEMFCs with commercial Fe-Co based cathode catalysts and self-prepared Pt/Pd/Au anode catalysts have demonstrated exciting performances. For example, the AEMFC with 1.0 mg/cm^2 Fe-Co cathode catalyst and 1.0 mg/cm^2 Pt/C anode catalyst has demonstrated a peak power density of 89 mW/cm^2 at 80°C and 2 atm backpressure of O_2 , which repents 20 times and 3 orders of magnitude higher than the proton exchange membrane fuel cells (with high PtRu and Pt catalyst loadings) and state-of-the-art biofuel cells (with enzymatic catalysts), respectively. The detailed electrocatalytic study of AEMFCs will be presented.

See more of this Session: Electrocatalysis for PEM Fuel Cells II
See more of this Group/Topical: Catalysis and Reaction Engineering Division

(585d) Volcano Behavior of Pd3M@Pd3Pt Core-Shell Oxygen Reduction Electrocatalysts by Gradually Changing the Oxygen Binding Energies

Quang Thang Trinh and **Mark Saeys**, Chemical and Biomolecular Engineering, National University of Singapore, Singapore, Singapore

First principle calculations were used to screen the oxygen reduction activity of a series of $Pd_3M@Pd_3Pt$ core-shell electrocatalysts for direct methanol fuel cells. Density functional theory (DFT) calculations indicate that the subsurface substitution of Pt by a 3d transition metal M improves the activity of the Pd_3Pt catalysts by reducing the oxygen binding energy [1]. Carbon-supported $Pd_3M@Pd_3Pt$ (M=Ni, Co, Fe and Cr) core-shell electrocatalysts with similar particle size, a Pd_3Pt rich surface, and a Pd_3M alloy core were prepared by a galvanic replacement reaction [2] between PdM alloy nanoparticles with a 70:30 Pt:M atomic ratio and an aqueous solution of $PtCl_4^{2-}$. The predicted change in the surface Pt electronic structure was confirmed by comparing the calculated shift in the Pt $4f_{7/2}$ core level binding energies with XPS data, and by comparing the change in the calculated CO binding energies with the shift in the position of the CO stripping peak. Optimal activity close to the maximum of a volcano curve [3] is predicted for $Pd_3Fe@Pd_3Pt$ and $Pd_3Mn@Pd_3Pt$ core-shell catalysts. In agreement with the DFT calculations, optimal activity and high methanol tolerance were observed for $Pd_3Fe@Pd_3Pt$. A lower activity is observed for the $Pd_3Cr@Pd_3Pt$ catalyst with a lower CO binding energy and for the $Pd_3Co@Pd_3Pt$ and $Pd_3Ni@Pd_3Pt$ catalysts with higher CO binding energies. In the presence of 0.1 M methanol, the current density per Pt atom for the optimal $Pd_3Fe@Pd_3Pt/C$ catalyst is 10 times higher than for commercial Pt/C catalysts. DFT calculations further indicate the CO and OH covered $Pd_3Fe@Pd_3Pt$ electrocatalyst is stable towards Fe surface segregation.

References

1. J. Xu, J.H. Yang, J.Y. Lee, M. Saeys, Ind. Eng. Chem. Res., 49, 10251 (2010)

2. J.H. Yang, W.J. Zhou, C.H. Cheng, J.Y. Lee, Z.L. Liu, ACS Appl. Mater. Interfaces 2, 119 (2010)

3. J.K. Norskov, J. Rossmeisl, A. Logadottir, L. Lindqvist, J. Phys. Chem. B, 108, 17886 (2004)

(585e) Computational Modeling of Electrolyte/ Cathode Electrocatalytic Interface In PEM Fuel Cells

Kuan-Yu Yeh, Michael V. Glasspool, Michael J. Janik and Janna K. Maranas, Chemical Engineering, Pennsylvania State University, University Park, PA

Multi-scale modeling, using both density functional theory (DFT) and molecular dynamics (MD), are applied to the electrode/electrolyte interface in a proton exchange membrane fuel cell (PEMFC). The complexity in the electrochemical interface and the difficulties to characterize the interface hinder the fundamental understanding of the oxygen reduction reaction (ORR) under fuel cell operating conditions. Slow ORR kinetics represent a major performance loss in a PEMFC. To capture ORR energetics under different electrochemical conditions, various DFT-based electrochemical models were employed. We determined rate limiting steps in both dissociative and associative ORR paths as a function of electrode potentials and hydration levels. The overall reaction barrier at PEMFC operation potential agrees well with the experimental data. In addition, a solvated model predicts the potential-dependent rate limiting step, qualitatively consistent with the experimental results. However, the influence of the water structure (hydration level) on ORR energetics emphasizes the needs to consider the structure and dynamics of the interfacial water. To extend time and length scales, we performed MD simulations in systems of pure water and 1M H_2SO_4 aqueous solution, which represents a simplified electrolyte solution in PEMFCs, at a potential controlled Pt (111) surface. Results indicate that the electrode potential affects the orientation and the ordering of interfacial water molecules, and the packing of counter ions as well. The structure of the electric double layer is more complicated than that derived from classical-continuum models. Characteristics of the double layer region from MD simulations will provide insights for an advanced DFT-based electrochemical model for a further investigation of the ORR energetics.

See more of this Session: Electrocatalysis for PEM Fuel Cells II
See more of this Group/Topical: Catalysis and Reaction Engineering Division

(585f) Optimization of Electrodes Based On Polymer Properties of Hydrocarbon Polyelectrolyte for PEFCs

Tatsuya Nakajima[1], Takanori Tamaki[2], Hidenori Ohashi[2] and Takeo Yamaguchi[2], (1)Chemical Resources Laboratory, Tokyo Institute of Technology, Yokohama, Japan, (2)Chemical Resource Laboratory, Tokyo Institute of Technology, Yokohama, Japan

Polymer electrolyte fuel cells (PEFCs) are promising future devices for power generation, portable and automotive applications. A membrane electrode assembly (MEA) is one of the key components in PEFC system. The MEA consists of polymer electrolyte membrane and electrodes which made up of a catalyst layer, a diffusion layer, and a carbon paper. Perfluorosulfonic acid (PFSA) such as Nafion has been used as an ionomer in the catalyst layer, and the Nafion-based electrodes have been optimized in a great many studies. Several problems still remain to restrict the application of PFSA to PEFCs, for instance, the high cost, the high environmental load and the difficultly in structure control. On the other hand, hydrocarbon polyelectrolytes have been developed as an alternative to PFSA because of the unique benefits including the low cost and the easiness of structure control. Therefore, the adjustment of the alternative ionomer to the electrode is fundamentally significant. However, none of index could be formed to optimize the structure of catalyst layer up till now. Furthermore, the performance of MEA containing hydrocarbon ionomer in the catalyst layer has been reported to decrease at comparatively-low current density, below 1,000 mA/cm^2, compared with that of Nafion. The followings are considered as main causes of the performance decrement: (a) the inhibition of oxygen diffusion by blocking a secondary pore, which is a pore outside agglomerated catalyst supported carbons (i.e. secondary particles), (b) the resistance of oxygen permeation from the secondary pore to a reaction site through the ionomer layer. In our research, polymer physical properties were investigated to clarify the main cause of performance decrement. The oxygen permeability and the swelling of ionomer were focused on as polymer physical properties since these properties are considered to relate to the above rate limiting steps.

The sulfonated poly(allylene ethel sulfone) (SPES) was synthesized to investigate polymer properties of hydrocarbon polyelectrolyte and to be used to the catalyst layer as the ionomer. Two types of electrodes were fabricated and the introduction weight ratio of Pt supported carbon (Pt/C): ionomer: and polytetrafluoroethylene (PTFE) was set to 60: 25: 15 (SPES-25, Nafion-25) and 80: 5: 15 (SPES-5) at the dry state, respectively. MEAs were fabricated with Nafion NR-212 membrane along with SPES-25 or SPES-5 electrode, respectively (MEA-SPES-25, MEA-SPES-5). The oxygen permeability coefficient [P_{O2}] was measured and calculated from the amount of permeated oxygen through the membrane. The swelling ratio [δ] was estimated from the volume change of the membrane.

P_{O2} of SPES was lower, while δ of SPES was higher than these of Nafion. Firstly, to discuss the effect of ionomer swelling, the porosity of catalyst layer was estimated by considering the volumes of entire catalyst layer, Pt/C, ionomer, and PTFE. The porosity of SPES-25 dropped significantly compared with that of Nafion-based electrode. Besides, the effect of ionomer swelling decreased with decreasing RH. This means that, if the blocking of secondary pore caused by the SPES swelling is the main influence factor, the performance would improve with decreasing RH. Secondly, to discuss a difficulty of the oxygen permeation through the ionomer layer, the resistance of oxygen permeation was calculated from P_{O2}, δ, and the density. The oxygen permeation through the SPES ionomer layer of the SPES-25 is much more difficult than that through the Nafion ionomer layer of the Nafion-25. Furthermore, the effect of resistance increased with decreasing RH. This means that, if the oxygen permeation through the ionomer layer is the main cause of performance decrement, the performance would become lower with decreasing RH.

Therefore, the humidity dependence of performance of MEAs was investigated. The performance of MEA-SPES-25 improved with decreasing RH. These results indicate that the main cause of performance decrement is not the resistance of oxygen permeation but the inhibition of oxygen diffusion by blocking a secondary pore caused by ionomer swelling. Additionally, the performance of MEA-SPES-5 with thin layer of SPES ionomer was high and hardly changed with changing RH. This is because the thinner layer of SPES ionomer is not affected by the inhibition of oxygen diffusion by blocking. The results with MEA-SPES-5 also suggest that the broadened secondary pore enables to maintain the performance of MEA with the hydrocarbon ionomer in the catalyst layer. Therefore, the optimized structure of the catalyst layer will be fabricated by not only using the ionomer with low swelling ratio but also extending the secondary pore of the catalyst layer.

See more of this Session: Electrocatalysis for PEM Fuel Cells II
See more of this Group/Topical: Catalysis and Reaction Engineering Division

(585g) Development of Predictive Models for Screening Multimetallic Electrocatalysts

Hongliang Xin and Suljo Linic, Chemical Engineering, University of Michigan, Ann Arbor, MI

Developing predictive models of chemisorption on metal surfaces is critical for the understanding of surface chemical reactions and rational design of efficient catalysts. It has been shown that the *d*-band model of chemisorption, developed by Hammer and Nørskov, can predict the trends in chemisorption energies of various adsorbates on metal surfaces. [1] The model correlates the central moment of the *d*-band projected on surface atoms (*d*-band center referenced to the Fermi level) with the surface reactivity. It has been used successfully to design novel metal surfaces for various catalytic reactions. In general, for a given adsorption geometry adsorbates bind to the surface of transition or noble metals more strongly if the *d*-band center of the surface atom is higher in energies. The model is very robust and most adsorbates follow the trends predicted by the model.

In this study, we show that there is a family of adsorbate-substrate systems that does not follow the trends in adsorption energies predicted by the *d*-band model. [2] We discuss this exception to the *d*-band model by analyzing hydroxyl (OH) adsorption on a series of Pt and Pd skin alloys. This exception is important since OH adsorption on metals is crucial for the understanding of various catalytic, electro-catalytic and photo-catalytic reactions including oxygen reduction and water splitting reactions. It has been shown previously that Pt and Pd skin alloys are promising alternatives to conventional Pt catalysts in these chemical transformations. A physically transparent model with the incorporation of both the d-band center and the adsorbate-substrate bond length dependent chemisorptions energies has been developed for OH adsorption on Pt and Pd alloy surfaces and used to screen multimetallic electrocatalysts.

See more of this Session: Electrocatalysis for PEM Fuel Cells II
See more of this Group/Topical: Catalysis and Reaction Engineering Division

From Multi-Scale Methods to Virtual Process Engineering --- Efficient and Accurate Simulation of Multi-Phase Complex Systems

Ge Wei, State Key Laboratory of Multi-Phase Complex Systems, Institute of Processing Engineering, Chinese Academy of sciences, Beijing, China

Dynamic simulation of the equipments and reactors in process engineering typically requires very long time and enormous computational cost. The major challenge lies in the multi-scale nature of the structures and behaviors of these systems. The structural consistency between the system simulated, the algorithm employed and the computing hardware used may provide an effective way to speedup the computation and to improve the accuracy, which can be well implemented in a multi-scale discrete computing mode with stability constraints. This presentation will give an overview of the work at our Institute in this direction, focusing on its development and extension in a variety of complex multi-phase systems, and establishment of the corresponding supercomputing software and hardware. The possibility of achieving real-time simulation, that is, virtual process engineering in this computing mode is prospected finally.

(589b) Point of Use Recyclable Bioreactor System for Bio-Agricultural Applications

Matthew A. Robinson and Henry Y. Wang, Department of Chemical Engineering, University of Michigan, Ann Arbor, MI

Farmers have long been respected for their self-sufficiency, acting as primary producers for society. Yet as time progresses, productivity demands drive farmers to require increasing levels of technology in terms of seed strains, fertilizers and pesticides in order to remain competitive. In turn farmers have grown dependent upon manufactured products with rising price tags. Current market prices leave biologically derived fermentation based active ingredients and cultures, like reduced risk certified natural insecticides, uncompetitive against traditional synthetic insecticides placing them out of reach of small farmers, particularly in the developing world. In order to reduce the basic manufacturing cost, we have designed a point of use recyclable bioreactor system capable of fermenting off patent compounds for direct field application eliminating the non-value added concentration step. With recyclable single use inserts, the system can produce an array of bioproducts including nitrogen fixing cultures, phosphate solubilizing cultures, and various natural pesticides without the need for extensive sterilization. The system empowers farmers to do what they do best: cultivate, returning the power of self-sufficiency to the hands of the farmer.

See more of this Session: High-Throughput and Compact Chemical Processing Technologies
See more of this Group/Topical: International Congress on Energy 2011

(589c) The Study of Ultrasonic Irradiation Facilitating Fat Hydrolysis Reaction

Guangming Du, College of Chemical Engineering, Xinjiang Agricultural University, Urumqi, China
See more of this Session: High-Throughput and Compact Chemical Processing Technologies
See more of this Group/Topical: International Congress on Energy 2011

(589d) Autothermal Hydrogen Production From Isobutanol for Fuel Cells

Reetam Chakrabarti, Jacob S. Kruger and Lanny Schmidt, Department of Chemical Engineering and Materials Science, University of Minnesota, Minneapolis, MN

We demonstrate a process which converts isobutanol, a fermentation product of lignocellulosic biomass, to a high selectivity hydrogen stream for fuel cells by autothermal reforming. We use staged reactor configuration, with a Rh-Ce/Al_2O_3 catalyst used to produce a high selectivity syngas stream by partial oxidation and reforming, and an additional Pt-Ce/Al_2O_3 catalyst to consume CO by the water gas shift reaction (WGS). The Rh-Ce/Al_2O_3 catalyst showed high conversion of isobutanol resulting in a product stream consisting mostly of CO, CO_2, H_2 and H_2O. These products react further on the Pt-Ce/Al_2O_3 catalyst. The effect of steam addition to further lower the concentration of CO by promoting WGS has also been investigated. Thermodynamic calculations were performed to optimize the feed compositions and the amount of steam added to obtain a product stream containing minimal amounts of CO for use in a fuel cell.

See more of this Session: High-Throughput and Compact Chemical Processing Technologies
See more of this Group/Topical: International Congress on Energy 2011

(589e) Sulfuric Acid Catalyzed Isobutane-Butene Alkylation Promoted with Ionic Liquids or Trifluoroethanol

Hailing Ren, Guoying Zhao and **Suojiang Zhang**, Institute of Process Engineering, Chinese Academy of Sciences, Beijing, China

Sulfuric acidcatalyzed isobutane-butene alkylation promoted with ionic liquids ortrifluoroethanol

HailingRen[ab], Guoying Zhao[a*], Peng Cui[ab],Jun Huang[b],SuojiangZhang[a*]

[a]Institute of ProcessEngineering, Chinese Academy of Sciences, P. O. Box 353, Beijing, 100190,China

Fax : +86 10 82627080

[b]State Key Laboratory ofMaterials-Oriented Chemical Engineering, College of Chemistry and ChemicalEngineering, Nanjing University of Technology, Nanjing, 210009, China

Fax: +86 2583172263

Abstract

Smallamounts of ionic liquids or trifluoroethanol(TFEA), have been studied as promotersfor sulfuric acid catalyzed alkylation of isobutane/butene. Long-term studiesup to 40+ recycles indicated that the working life of $H_2SO_4$with suitable ionic liquid-based promoter was prolonged at least three timesthat of sulfuric acid. New catalytic active species formed in-situ by the abovepromoter and H_2SO_4 was detected by NMR and ESI-MS analytictechnology. In addition, the promoters are believed to lower the interfacial tensionof hydrocarbon feeds/sulfuric acids dispersion and thus improve the quality ofalkylate.

See more of this Session: High-Throughput and Compact Chemical Processing Technologies
See more of this Group/Topical: International Congress on Energy 2011

(589f) Comparison of Different Process Routes From Coal with Energy and Exergy Analysis

Yuehua Yao, Xiangping Zhang and **Suojiang Zhang**, Institute of Process Engineering, Chinese Academy of Sciences, Beijing, China

Comparison of different process routesfrom coal with energy and exergy analysis

Y. H. Yao[a,b], Y.J. Chen [a,b], X. P. Zhang[a,*], S. J. Zhang [a], B. Z. Ren[b] , J. H. Wang[b]

[a] *Institute of ProcessEngineering, Chinese Academy of Sciences, 100190, Beijing, China*

[b] *School of ChemicalEngineering and Energy, Zhengzhou University, 450001,Zhengzhou, China*

***Corresponding author:** *xpzhang@mail.ipe.ac.cn* **(X. P. Zhang)**

Chinahas plenty coal resource but there are many ways to utilize coal to productfuels and natural gas, et al. Facing the problem of energy shortage in theworld and the target of cutting CO₂ emission, thequestion of sifting the best one from existing process routes of coal utilizationshould be solved at first, it matters whether coal sector of China can satisfythe demand of low carbon economy and that the development form of Chinese coalresource in the future.

Inthis study, considering the existing process routes of coal, models of thedirect coal liquefaction process, the direct coal combustion process, thecoking process and coal gasification process for F-T synthesis oil and naturalgas production were developed. Based on simulation results, combining with theindustrial data, both the conservation of energy and the quality of energy forthe processes were investigated. Exergy loss, CO₂emissions and yield of the four process routes were calculated and compared.For the four systems, the carbon and hydrogen flows were analyzed, and the mainreasons for carbon loss and the hydrogen atom utilization were discussed. Theresults can provide a roadmap for the future coal utilization with low globalwarming effect and high energy utilization.

Acknowledgement

This research was supported by the NationalBasic Research Program of China (973 Program, 2009CB219900) and State Science and Technology SupportProgram (2009BAC64B00)•.

References

Maria Sudiro,Alberto Bertucco. (2009). Production of synthetic gasoline and diesel fuel byalternative processes using natural gas and coal: process simulation and optimization.*Energy*, 34, 2206-2214.

Zhang, X.P.,Christian Solli, Edgar G. Hertwich, Tian, T. (2009). Exergy analysis of theprocess for dimethyl ether production through biomass steam gasification. *Ind.Eng.Chem.Res.* 48,10976-10985.

See more of this Session: High-Throughput and Compact Chemical Processing Technologies
See more of this Group/Topical: International Congress on Energy 2011

(598a) Liquid-Liquid Mass Transfer In a Rotor-Stator Spinning Disc Reactor

Frans Visscher, John van der Schaaf, Mart H.J.M. de Croon and Jaap C. Schouten, Eindhoven University of Technology, Eindhoven, Netherlands

Liquid-liquid mass transfer in arotor-stator spinning disc reactor

FransVisscher, John van der Schaaf, Mart de Croon, andJaap C. Schouten

Laboratory of Chemical ReactorEngineering, Eindhoven University of Technology, TheNetherlands

Introduction

High gas-liquid and liquid-solid mass transfer ratescan be achieved in the rotor-stator spinning disc reactor (SDR) [1, 2]. Here wedemonstrate high liquid-liquid mass transfer rates in this new type of reactorusing the extraction of benzoic acid from n-heptane to water. A benzoic acid inn-heptane flow was continuously contacted in the SDR with a benzoic acid freewater flow. The steady state benzoic acid concentrations in both liquids weremeasured by UV-VIS spectroscopy. The experimental setup is shown in Figure 1.

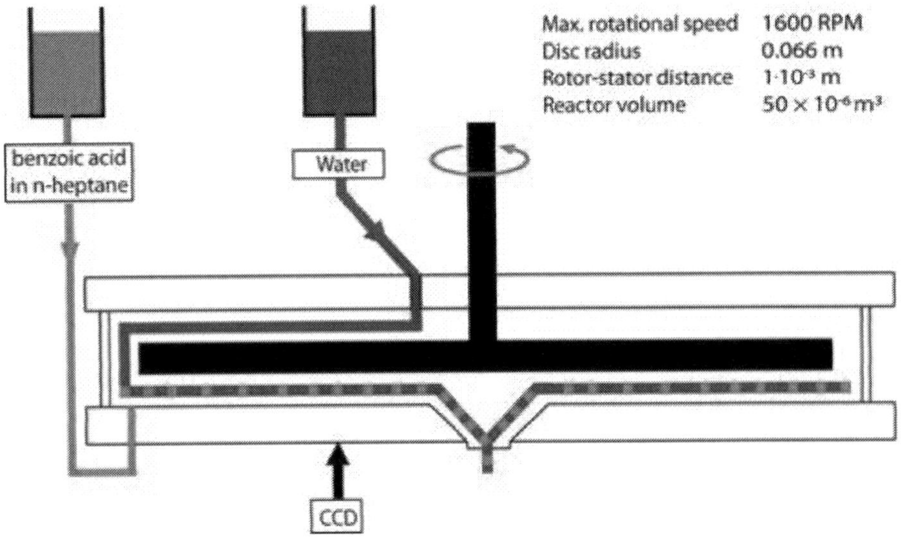

Figure 1.Schematic drawing of the rotor-stator spinning disc set-up.The rotor is located at 1.0 10^{-3} m from the top and bottom stator.Water is fed to the top of the reactor near the axis. n-Heptanewith benzoic acid is fed through an inlet in the bottom stator near the rim ofthe disc.

Results

The overall liquid-liquid mass transfer coefficient, $k_L a_L \varepsilon_{ORG}$,was determined for a total volumetric liquid flow ranging from 4.5 $10^{-6}m^{3-1}$ for an organic to aqueous ratio of 1:1, to 17.2 $10^{-6}m^{3-1}$ at a ratio of 1:7. The multiphase flow pattern was determinedthrough high speed imaging of the bottom stator.

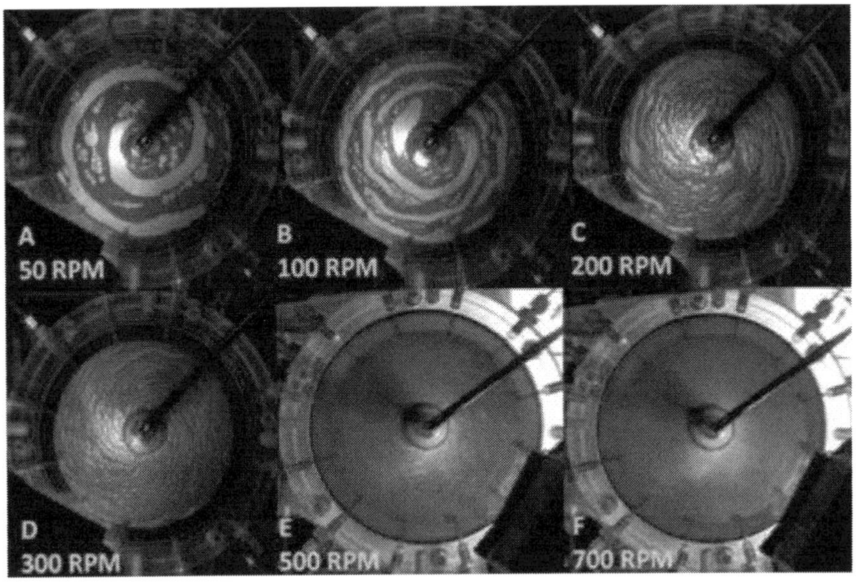

Figure 2. The influence ofrotational disc speed on the holdup inside the reactor. Pictures aremade through the bottom stator, with a total flow rate of 4.710^{-6} m^{3-1}. Water appears blue because of watersoluble ink, n-heptane appears white.

Upon increasing the rotational disc speed the continues n-heptane spiral (in white) breaks up intospiralling droplets due to the enlarged shear force that is present between therotor and the bottom stator. This causes an increase of the interfacial area.

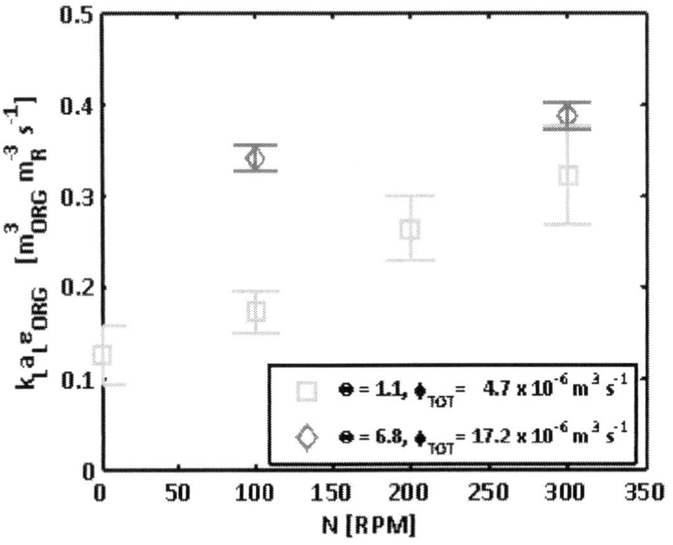

Figure 3 The overall liquid-liquid mass transfer rates in therotor-stator spinning disc reactor are increasing with an increase of the rotorspeed and with the flow ratio. Values are calculated assuming plug flowbehavior for both phases.

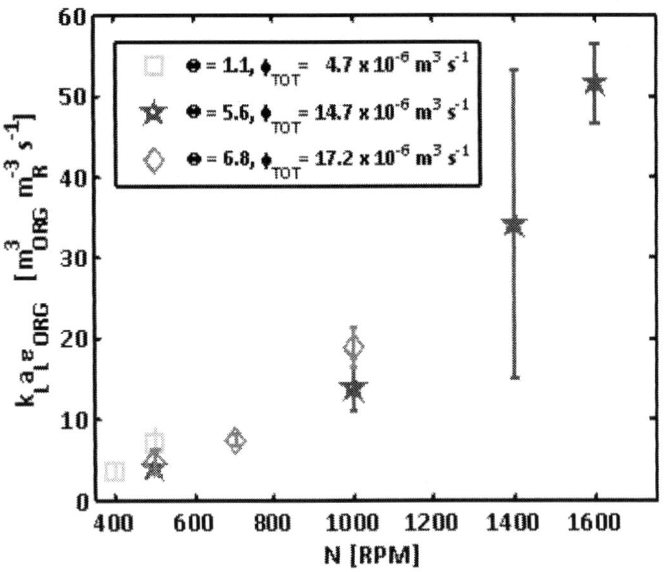

Figure4 At higher rotational disc speed, a stronger shearstress is acting on the liquids between the rotor and the stator. This causesan increase of the overall mass transfer coefficient. The overall liquid-liquidmass transfer rates are calculated assuming ideally mixed behavior.

Conclusion

These liquid-liquid mass transfer rates are at least25 times higher compared to those in packed columns, and at most 15 timeshigher compared to mass transfer rates in state of the art microchannels. Thisimplies that both the process volume and the process time of current equipmentcould be decreased.

References

[1] Meeuwse,et al. (2010) Chemical Engineering Science, 65(1), 466-471.

[2] Meeuwse, et al. (2010)Industrial & Engineering Chemical Research, 49(21),10751-10757.

See more of this Session: Multiphase Reaction Engineering II
See more of this Group/Topical: Catalysis and Reaction Engineering Division

(598b) Study of Effect of Design and Operating Conditions On Spouted Bed Hydrodynamics

Shreekanta Aradhya, Chemical and Biological Engineering, Missouri University of Science and Technology, Rolla, MO and Al-Dahhan Muthanna, Chemical & Biological Engineering, Missouri University of Science and Technology, Rolla, MO

@font-face { "Cambria Math"; }@font-face { "Calibri"; }p.MsoNormal, li.MsoNormal, div.MsoNormal { margin: 0in 0in 10pt; line-height: 115%; font-size: 11pt; ; }span.go { }.MsoChpDefault { font-size: 10pt; ; }div.WordSection1 { page: WordSection1; }

Study of Effect of Design and Operating Conditions on Spouted Bed Hydrodynamics

***Shreekanta B Aradhya, *+Muthanna Al-Dahhan**

***Department of Chemical & Biological Engineering**

+Department of Nuclear Engineering

Missouri University of Science & Technology, Rolla, MO – 65409, USA

Email: sba2p4@mst.edu

Due to their efficiency in contacting gases and coarser particles, spouted fluidized beds have been successfully applied to a wide variety of processes, such as coating, granulation, drying, coal gasification, catalytic reactions, etc. Under proper conditions, the jet penetrates the bed of particles, creating a central spout zone, a fountain above the spout, and an annulus surrounding the spout. Particles entrained in the gas spout form a fountain of particles above the bed surface that disengage from the gases and fall back to the bed surface, thus inducing bed circulation. Different spouted bed configurations have been used and studied, such as conical, cylindrical, cone-based, and slot-rectangular spouted beds (Freitas et al., 2004 a,b; Zanoelo et al., 2004).

One of the spouted bed's features is the possibility of obtaining different flow regimes and flow characteristics with minor variations in geometry or operating conditions. In this work, the effect on hydrodynamics of spouted beds will be studied by changing the design and operating conditions. Local parameters such solids holdup, solids velocity and their fluctuations will be studied with the help of advanced optical probes. The fluctuations produced from optical probe will be analyzed statistically to obtain several different information about the spouted bed system. The time series signals obtained from pressure transducer will also be analyzed using chaotic analysis to deduce useful information. Two different size spouted bed columns have been identified for the present work. Experiments will be performed in these two columns using several different particles at varying conditions to determine the hydrodynamics using advanced optical probes and pressure transducer.

See more of this Session: Multiphase Reaction Engineering II
See more of this Group/Topical: Catalysis and Reaction Engineering Division

(598c) Investigation of the Effect of Fluidized Bed Scale On the Solid Dynamics Using Sophisticated Optical Probe

Faraj M. Zaid, Stoyan Nedeltchev and Muthanna Al-Dahhan, Department of Chemical and Biological Engineering, Missouri University of Science and Technology, Rolla, MO

The gas-solids fluidized-bed reactors have the ability to process large solid volumes and that is why they have found widespread application in chemical industry. It is very important to measure the radial and axial profiles of solids holdups and particle velocities in fluidized beds. In this work, a sophisticated optical probe was used to measure these important parameters in two different fluidized beds (0.14 m and 0.44 m in ID). The radial and axial solids holdup profiles and particle velocities were studied at the same three axial positions in both units. Air was used as a fluidizing gas and glass beads (ρ_s=2500 kg/m^3, d_p=150-210 μ m) as a solids material. The bed aspect ratio was set at two. Both fluidized beds were equipped with a porous-plate (pore size: 40 μ m) distributor. The effect of reactor scale on solids dynamics was studied at several different superficial gas velocities *U*.

By using sophisticated optical probe solid concentrations were first measured and then converted by new method to solid holdups. Their spatial distribution was critically analyzed. The sampling frequency was set at 5 kHz. It was found that at the same dimensionless axial position the solids holdups are higher in the bigger unit. In both fluidized beds the solids holdups increased towards the wall. The same trends were reported earlier by Mabrouk et al. (2005). Additionally, it was found that the radial particle velocity profile was nearly parabolic in both fluidized beds.

In the smaller fluidized bed the solids holdup profile was independent of the axial position z up to 0.09 m. Then, at z=0.14 m it exhibited a pronounced decrease. In the bottom of the column there was a clear trend of solids holdup increase towards the wall, whereas at z=0.14 m the solids holdup profile leveled off towards the wall. At higher gas velocities the solids holdup clearly decreased with the axial height. In the bigger fluidized bed at all *U* values examined always the solids holdup profile decreased with the axial position. In addition, the solids holdup at the wall was always higher than the one in center of the column. Outside the core region, the particle velocities at the different *U* values examined in the smaller fluidized bed were practically the same. In the center of the column, the particle velocity increased with the superficial gas velocity.

Such exhaustive results have not been reported in the literature hitherto. They will be used for preparing a new scale-up strategy.

References

[1] Mabrouk et al., "Scale effects on fluidized bed hydrodynamics" Inter. J. of Chemical Reactor Eng 3, Article A18; 2005.

See more of this Session: Multiphase Reaction Engineering II
See more of this Group/Topical: Catalysis and Reaction Engineering Division

(598d) An Approach for Efficient Conversion of Carbohydrates to Value Added Products Using Metal Coated Polymeric Membranes

Neha Dhiman[1], Peter Pfromm[2] and Mary Rezac[1], (1)Chemical Engineering, Kansas State University, Manhattan, KS, (2)Department of Chemical Engineering, Kansas State University, Manhattan, KS

One of the most important challenges in the development of a biorefinery is to develop a facile conversion technology capable of converting the renewable carbon to high value chemicals. Conventional hydrogenation technology employs slurry-phase systems in which gaseous hydrogen is contacted with a solid catalyst dispersed in a liquid reactant. These three-phase systems suffer from the vanishingly small solubility of hydrogen in liquids. Operation at extreme pressures (frequently in excess of 100 atm) is employed to overcome these thermodynamic limitations. Maintaining high hydrogen coverage at the catalyst surface favors hydrogenation reactions over side reactions.

The research presented here focuses on developing membrane reactor hydrogenation technology capable of operating at modest hydrogen pressures while simultaneously providing rapid hydrogen delivery to the catalyst surface. Specifically, the membrane reactor employed is designed such that one surface of the membrane is decorated with metal catalyst sites. This surface of the catalyst is contacted by the substrate-containing liquid to be converted. In this study, aqueous phase hydrogen of lactic acid, an intermediate of significant importance to the biorefining field was investigated. Hydrogen is supplied on the permeate side of the membrane and is transferred to the catalyst surface by the imposed chemical potential driving force.

The membranes used in the reactor are integrally skinned asymmetric Polyetherimides with thin, defect-free surface layers. Ruthenium was employed as the catalyst for the hydrogenation reaction and applied via spin coating.

This study can thereby help in contributing towards efficient processing conditions for obtaining value added products from renewable carbon.

See more of this Session: Multiphase Reaction Engineering II
See more of this Group/Topical: Catalysis and Reaction Engineering Division

(598e) Characterization of the Degree of Turbulence In Bubble Columns Based On Chaos Analysis of CARPT Data

Stoyan Nedeltchev, Department of Chemical and Biological Engineering, Missouri University of Science and Technology, Rolla, MO and Muthanna Al-Dahhan, Chemical and Biological Engineering, Missouri University of Science and Technology, Rolla, MO

The characterization of degree of turbulence and mixing in bubble columns is very important for the improvement of their performance. The three coordinates (radial, azimuthal and axial) of radioactive tracer particle (Sc-46) were measured by means of noninvasive Computer-Automated Radioactive Particle Tracking (CARPT) technique in four different cylindrical bubble columns (0.05, 0.10, 0.16 and 0.2 m in ID). The facilities were equipped with the same perforated plate distributor (82 holes, 0.4×10^{-3} m in diameter). The bubble columns operated with an air-water system at ambient conditions. Superficial gas velocities covering the range 0.02-0.12 m/s were examined.

The Kolmogorov entropy (KE) algorithm (Schouten et al., 1994) was applied to each coordinate's data in order to extract the degrees of chaos and turbulence in the column. It was found that the degree of turbulence decreased (lower KE values) with the column diameter enlargement. This means that the behavior of the gas-liquid system in the smaller columns is more chaotic and it is less predictable. The same result was obtained from all three coordinates of the radioactive tracer particle moving in the bubble bed.

The main transition velocity (marking the onset of transition regime) was identified by means of the KE algorithm applied to CARPT data and the effect of column diameter on this key parameter was investigated. The results were validated by means of two different entropies (KE and maximum information entropy). Based on the information entropy theory (Nedeltchev et al., 1999, 2000, 2003) the degree of mixing in each column was estimated and a comparison with the degree of turbulence was performed.

REFERENCES:

Schouten, J. C., F. Takens, C. M. van den Bleek, Maximum-Likelihood Estimation of the Entropy of an Attractor. *Phys. Rev. E* 49, 126 (1994)

Nedeltchev, S., S. Ookawara, K. Ogawa, A Fundamental Approach to Bubble Column Scale-Up Based On Quality of Mixedness. *J. Chem. Eng. Japan* 32, 431 (1999)

Nedeltchev, S., S. Ookawara, K. Ogawa, Time-Dependent Mixing Behaviors in Both Lower and Upper

Zones of a Bubble Column. *J. Chem. Eng. Japan* 33, 761 (2000)

Nedeltchev, S., S. B. Kumar, M. P. Dudukovic, Flow Regime Identification in a Bubble Column Based On Both Kolmogorov Entropy and Quality of Mixedness Derived from CARPT Data. *The Canadian J. Chem. Eng.* 81, 367 (2003)

See more of this Session: Multiphase Reaction Engineering II
See more of this Group/Topical: Catalysis and Reaction Engineering Division

(608a) Mechanisms and Energetics for Acid Catalyzed β-D-Glucose Conversion to 5-Hydroxymethylfurfurl

Xianghong Qian, Department of Chemical Engineering, University of Arkansas, Fayetteville, AR

Car-Parrinello based *ab initio* molecular dynamics (CPMD) coupled with metadynamics (MTD) simulations were carried out to investigate the mechanism and energetics for acid catalyzed β-D-glucose conversion to 5–Hydroxymethylfurfurl (HMF) in water, a critical intermediate for biomass conversion to biofuels. It was found that protonation of the C2–OH on glucose, the breakage of the C2–O2 bond and the formation of the C2–O5 bond is the critical rate-limiting step for the direct glucose conversion to HMF without going through fructose, contrary to the wide–spread assumption that fructose is the main intermediate for glucose conversion to HMF. The calculated reaction barrier of 30–35 kcal /mol appears to be solvent induced and is in excellent agreement with experimental observations.

See more of this Session: Reaction Path Analysis III
See more of this Group/Topical: Catalysis and Reaction Engineering Division

(608b) Comparison of Performances for Diesel Fuels and Biodiesel In HCCI Engine Using Detailed Mechanism with On-the-Fly Reduction

Shuliang Zhang[1], Ioannis P. Androulakis[1], Marianthi G. Ierapetritou[1] and Linda J. Broadbelt[2], (1) Department of Chemical and Biochemical Engineering, Rutgers, The State University of New Jersey, Piscataway, NJ, (2)Department of Chemical and Biological Engineering, Northwestern University, Evanston, IL

In recent years, due to the limited resources and possible future shortage of fossil fuels, the production and utilization of alternative energy have been given increasing emphasis. Biodiesel fuels are considered to be very important renewable alternative fuels in the future. To study the combustion characteristics of biodiesel fuels, detailed kinetic mechanisms for several biodiesel surrogates have been recently developed [1, 2]. Fisher et al. [3] developed kinetic model for methyl butanoate which is the first mechanism used for engine simulation of biodiesel. Most recently, Togbe et al. [4] reported a kinetic mechanism for methyl octanoate, and Herbinet et al. reported detailed mechanisms for larger biodiesel surrogate methyl decanoate [5] and mono-unsaturated ester surrogate methyl decenoates [6]. These detailed kinetic mechanisms enable the computational study of combustion simulation using biodiesel.

In order to study the combustion behavior of biodiesel in engine simulations, it is important to use detailed chemistry within the computational fluid dynamics (CFD) model. However, integrating biodiesel surrogate mechanisms in CFD simulation is very computationally expensive due to the large number of species and reactions involved. There are a number of methods that exist in the literature for kinetic mechanism reduction, which are summarized in the review article by Lu et al. [7]. In our previous work, an on-the-fly reduction approach [6] based on element flux analysis was developed and incorporated with CFD simulation [8]. A multi-element flux analysis combining carbon and nitrogen element flux analysis was also employed to describe both the hydrocarbon chemistry and nitrogen chemistry in homogeneous charge compression ignition (HCCI) engine [9], through which the NO_x emission characteristics can be captured.

In the present study, the on-the-fly kinetic reduction approach is extended to incorporate large-scale biodiesel surrogate mechanisms with CFD simulation in HCCI engine. KIVA-3V [10 is used as the CFD framework for HCCI engine simulation. CHEMKIN [11 chemistry solver is incorporated in the KIVA-3V code to handle the chemistry calculation. The multi-element flux analysis is applied to study CO and NO_x emission characteristics. Combustion characteristics and engine performance parameters, including ignition delay, total heat release, work done per cycle and power output, are calculated during the engine simulation. The slightly lower engine power and higher NO_x emission for biodiesel predicted in this study is in agreement with some existing experimental and computational results [12, 13. Comparison between biodiesel and conventional diesel was conducted to evaluate the performance of biodiesel as an alternative diesel fuel.

1. Lai, J.Y.W., K.C. Lin, and A. Violi, *Biodiesel combustion: Advances in chemical kinetic modeling.* Progress in Energy and Combustion Science, 2011. **37**(1): p. 1-14.

2. Pitz, W.J. and C.J. Mueller, *Recent progress in the development of diesel surrogate fuels.* Progress in Energy and Combustion Science, 2011. **37**(3): p. 330-350.

3. Fisher, E.M., et al., *Detailed chemical kinetic mechanisms for combustion of oxygenated fuels.* Proceedings of the Combustion Institute, 2000. **28**(2): p. 1579-1586.

4. Togb, C., et al., *Chemical Kinetic Study of the Oxidation of a Biodiesel–Bioethanol Surrogate Fuel: Methyl Octanoate–Ethanol Mixtures†.* The Journal of Physical Chemistry A, 2009. **114**(11): p. 3896-3908.

5. Herbinet, O., W.J. Pitz, and C.K. Westbrook, *Detailed chemical kinetic oxidation mechanism for a biodiesel surrogate.* Combustion and Flame, 2008. **154**(3): p. 507-528.

6. Herbinet, O., W.J. Pitz, and C.K. Westbrook, *Detailed chemical kinetic mechanism for the oxidation of biodiesel fuels blend surrogate.* Combustion and Flame, 2010. **157**(5): p. 893-908.

7. Lu, T. and C.K. Law, *Toward accommodating realistic fuel chemistry in large-scale computations.* Progress in Energy and Combustion Science, 2009. **35**(2): p. 192-215.

8. He, K., I.P. Androulakis, and M.G. Ierapetritou, *Incorporation of Detailed Chemical Mechanisms in Reactive Flow Simulations Using Element-Flux Analysis.* Industrial & Engineering Chemistry Research, 2010. **49**(21): p. 10471-10478.

9. He, K., I.P. Androulakis, and M.G. Ierapetritou, *Multi-element Flux Analysis for the Incorporation of Detailed Kinetic Mechanisms in Reactive Simulations.* Energy & Fuels, 2009. **24**(1): p. 309-317.

10. Amsden, A.A., *KIVA-3V: A Block-Structured KIVA Program for Engines with Vertical or Canted Valves*, Los Alamos National Laboratory Report LA-13313-MS, 1997.

11. Kee, R.J., et al., *CHEMKIN-III: A FORTRAN chemical kinetics package for the analysis of gas-phase chemical and plasma kinetics*, SAND96-8216, Sandia National Labs: Livermore, CA, 1996.

12. Lin, Y.-f., Y.-p.G. Wu, and C.-T. Chang, *Combustion characteristics of waste-oil produced biodiesel/diesel fuel blends.* Fuel, 2007. **86**(12-13): p. 1772-1780.

13. Um, S. and S.W. Park, *Numerical Study on Combustion and Emission Characteristics of Homogeneous Charge Compression Ignition Engines Fueled with Biodiesel.* Energy & Fuels, 2009. **24** (2): p. 916-927.

See more of this Session: Reaction Path Analysis III
See more of this Group/Topical: Catalysis and Reaction Engineering Division

(608c) Understanding Solvent Effects In the Selective Conversion of Fructose to HMF: A Molecular Dynamics Investigation

Samir H. Mushrif, Stavros Caratzoulas and Dionisios G. Vlachos, Chemical Engineering, University of Delaware, Newark, DE

Selective conversion of fructose to 5-hydroxymethyl-furfural (HMF) involves the participation of high-boiling solvents like dimethyl sulfoxide (DMSO). In order to replace DMSO with low-boiling solvents, it is imperative that we understand the effect of DMSO solvation in protecting HMF from further rehydration reactions and fructose from side reactions other than the dehydration to HMF. In the present work, molecular dynamics simulations of HMF and fructose in water and DMSO mixtures are carried out using the OPLS-AA force field. Radial pair distribution functions, coordination numbers and hydrogen-bond network between the HMF/fructose molecule and the solvent molecules are analyzed. We show preferential coordination of DMSO around certain carbon atoms of the HMF molecule and explain how this has a shielding effect, thus protecting HMF from further rehydration to levulinic acid and formic acid. We also show that, in the case of fructose, the presence of DMSO protects the molecule from side reactions that would lead to condensation or reversion products. However, the local 3-dimensional arrangement of water and DMSO molecules around the fructose molecule also reveals that the presence of DMSO molecules does not hamper the water molecules coming into the contact of the hydroxyl groups of fructose, which is required for the proton transfer step to initiate the dehydration reaction to HMF.

See more of this Session: Reaction Path Analysis III
See more of this Group/Topical: Catalysis and Reaction Engineering Division

(608d) Reaction Control of Transient Systems: Identifying the Dominant Transition and Intermediate States

William D. Michalak, James B. Miller and Andrew J. Gellman, Chemical Engineering, Carnegie Mellon University, Pittsburgh, PA

Elucidating the energetics for all of the states in acomplex reaction pathway, and identifying the kinetically feasible mechanismsfor a chemical process is costly and time-consuming. Many of theintermediate and transition states along the reaction pathway only weaklyinfluence the overall net rate; and therefore, a desirable alternative is todetermine the states that have the most influence and precisely calculate theirproperties. An approach that has been proposed to determine the mostinfluential states is the degree of rate control (C. Stegelmann,*et al.*, *JACS* **131**, 13563 (2009)). While most reportshave only studied stationary (steady-state) processes, there are manyapplications where the reactions operate in a transient mode. In thiswork a variant of the degree of rate control method is introduced for transientsystems; the differences between the transient and stationary approaches arehighlighted. The role that an intermediate species' concentration has onthe rate controlling states is discussed and a method to account for theeffects is described. The transient variation of the degree of ratecontrol is demonstrated for processes with increasing complexity to show howusing the transient degree of rate control provides important insight into eachsystem. The influence of reversibility between intermediate states,energetic interactions between species in the intermediate states, and how theapproach is used to identify the net activation barrier are also discussed.

See more of this Session: Reaction Path Analysis III
See more of this Group/Topical: Catalysis and Reaction Engineering Division

Title : Effects of Point Defects on the Oxygen Precipitation of Si Wafer Grown by Czochralski Method

Paper number : ID#233171:

Author : Do Won Song, Young Hee Mun, Kyo Kim†

LG Siltron Inc, Gyeonggi-do, Korea
† University of Seoul, Seoul, Korea

as part of Catalysis and Reaction Engineering Division
Session: Reaction Path Analysis III
scheduled for Wednesday, October 19, 2011: 3:15 PM-5:45 PM, MCC, 200 B

1. Introduction

1) Defects & Analyzing Method in Silicon Crystal

Various Crystal Defects

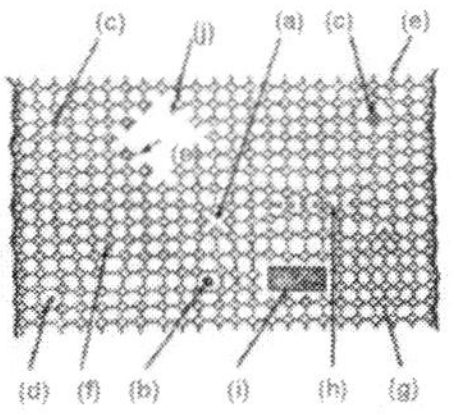

Point defect
- intrinsic
  - vacancy (a), self interstitial atom (b)
- extrinsic
  - substitutional impurity atom (c)
  - interstitial impurity atom (d)
  - dangling bond (e)

Line defect
- dislocation : edge- (f), screw-

Plane defect
- stacking fault
  - interstitial- (g), vacancy-type (h)
- twin, grain boundary

Volume defect : precipitate (i), void (j)

Defects of Si Wafer & Analyzing Method

Species	Vacancy Related Defects			Interstitial Related Defects		
Dimension	Volume Defects	Plane Defects	Point Defects	Point Defects	Plane Defects	Volume Defects
Etching + Visual Light	**COP** FPD					**LDP** EPD
Heat Treat + Etching + Visual Light		**O-band** (OiSF)			**T-zone**	
Heat Treat + Electrical Measurement	V-Cluster	O-Band	**VDP**	**IDP**	T-zone	I-Cluster
Eye	Bubble (Air Pocket)					

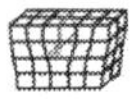

edge-dislocation

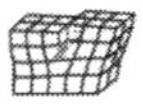

screw-dislocation

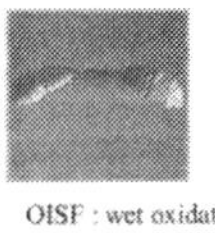

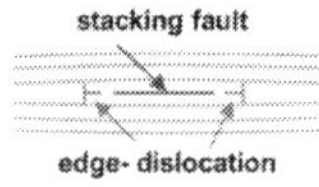

OISF : wet oxidation at 1100℃ for 60min
(Wright 5min, 10,000X, by SEM)

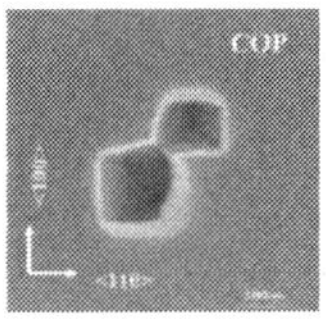

FPD : Flow Pattern Defect, LDP : Large Dislocation Pit, EPD : Etch Pit Defects, OiSF : Oxidation induced Stacking Fault, VDP : Vacancy Dominated Point Defect, IDP : Interstitial Dominated Point Defect

1. Introduction

2) V/G & Point Defects Dynamics

Point Defects Dynamics

Governing Equation :

$$\nabla(-D_V \nabla C_V + \frac{D_V C_V Q_V^*}{k_B T^2}\nabla T) + V\frac{\partial C_V}{\partial z} + \kappa_r (C_I C_V - C_I^{eq} C_V^{eq}) = 0$$

Out Diffusion | **Up-hill Diffusion** | **Pull Speed** | **Recombination**

$\exp(-E_V / k_B T)$ $\quad 1/T = 1/T_0 + G_0 z / T_0^2$

$$C_V^{eq} = C_0 \exp(-E_V G_0 z / k_B T_0^2)$$

$$C_I C_V = C_{Ie} C_{Ve} = C_{i0} C_{V0} \exp(-2z/l) \quad l = 2kT_0^2 / (E_i + E_V) G_0$$

Governing Equation ÷ V, net Flux → 0, then

$$\Rightarrow \frac{V}{G_0} = \xi_c = \frac{E_i + E_V}{2kT_0^2} \cdot \frac{D_i C_{i0} - D_V C_{V0}}{C_{V0} - C_{i0}}$$

Dominant point defect : vacancy $\left(\xi_c - \frac{V}{G}\right) < 0$, interstitial $\left(\xi_c - \frac{V}{G}\right) > 0$

The magnitude of excess point defect (S) $S = f\left(\left|\xi_c - \frac{V}{G}\right|^n\right)$ ξ_c **: critical value**

Crystal Defects Spectrums

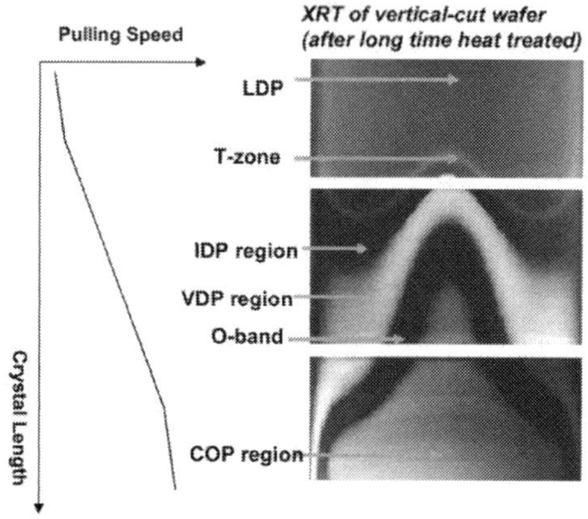

※ V/G(ξ) [=] (m²/s)×(1/T)

V/G(ξ) was proposed by Vladimir V. Voronkov in 1982. The first reasonable theory about the formation behavior of point defects.

2. Intrinsic Gettering

1) Oxygen Precipitates

Oxygen Precipitates

$$\mathrm{Si(s)} + 2\mathrm{Oi} \xrightarrow{k_1} (\mathrm{OP})^*$$

$$(\mathrm{OP})^* + n\,\mathrm{Oi} \xrightarrow{k_2} (\mathrm{OP})^* + \ldots\ldots + (\mathrm{OP})^* \rightarrow \mathrm{BMD}$$

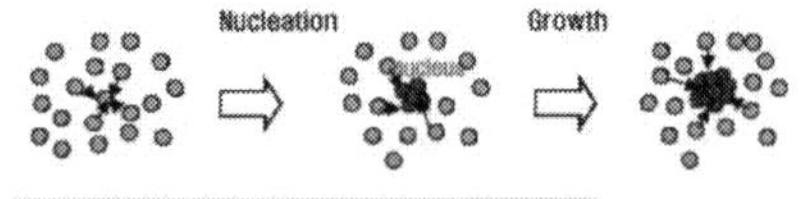

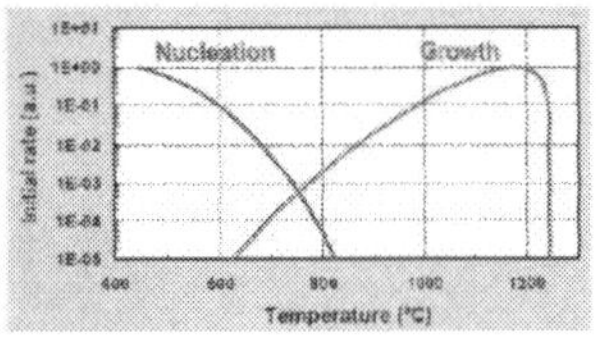

Predominant Temperature Region :
- Nucleation : 450 ~ 700°C
- Growth : 1000 ~ 1200°C

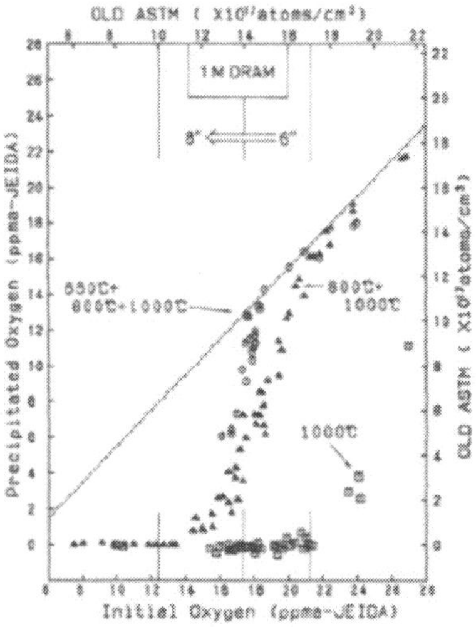

W.A. Tiller
The Science of Crystallization 1991 pp.327-381
"Microscopic interfacial phenomena"

OP : Oxygen Precipitates, BMD : Bulk Micro-Defects

3. Experiments

1) Overall Procedure

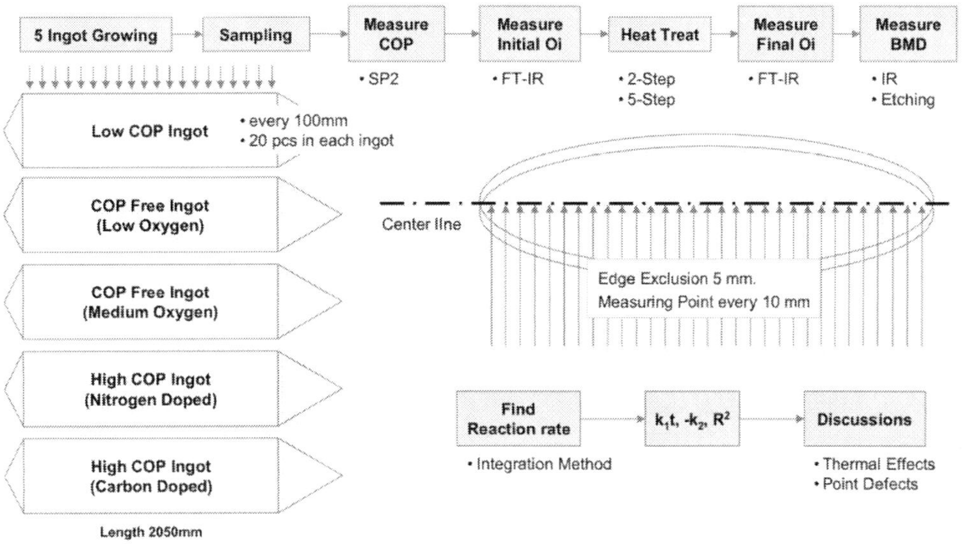

SILTRON

3. Experiments

2) Distribution of Grow-in defects

Usage of ingot	Grow-in defect	Dopant species	Dopant (atoms/cc)	Oxygen (ppma)	Carbon (ppma)	Nitrogen (atoms/cc) (calculated)	Initial vacancy (ea/cc) (calculated)
Polished wafer	Low COP	Boron	9.2E14~14.3E14	9.06~12.24	None	-	2.20E13
	COP free 1	Boron	5.4E14~8.1E14	8.81~10.50	None	-	8.60E12
	COP free 2	Boron	5.4E14~8.1E14	7.82~10.12	None	-	8.60E12
Substrate of EPI	Nitrogen	Boron	9.0E14~14.3E14	10.27~14.36	None	2.8E12~2.5E13	2.95E13
	Carbon	Phosphorous	2.8E14~10.3E14	12.13~17.11	0.26~2.69	-	2.95E13

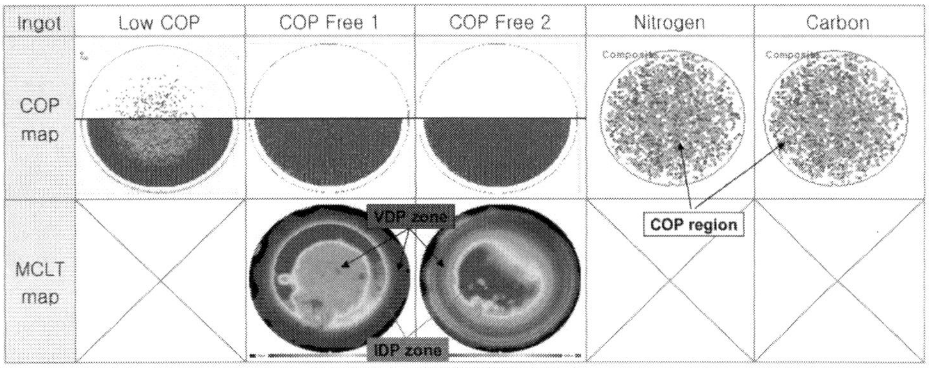

MCLT : Minority Career Life Time

SILTRON

3. Experiments

3) Results : Final Oi & IR-BMD

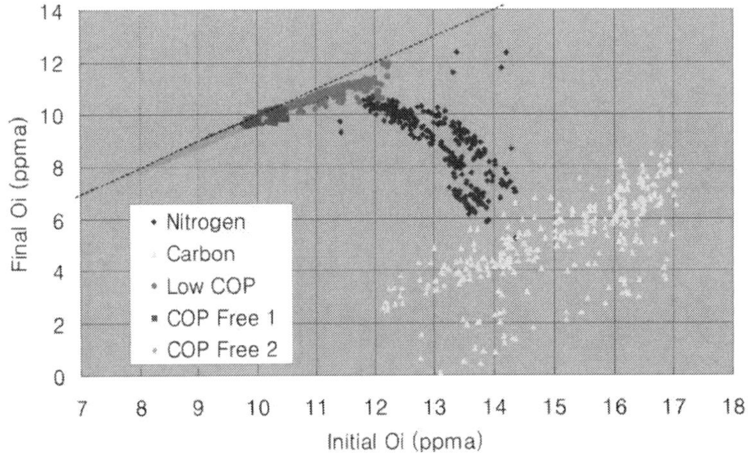

6 / 10

SILTRON

3. Experiments

3) Results : Final Oi & IR-BMD

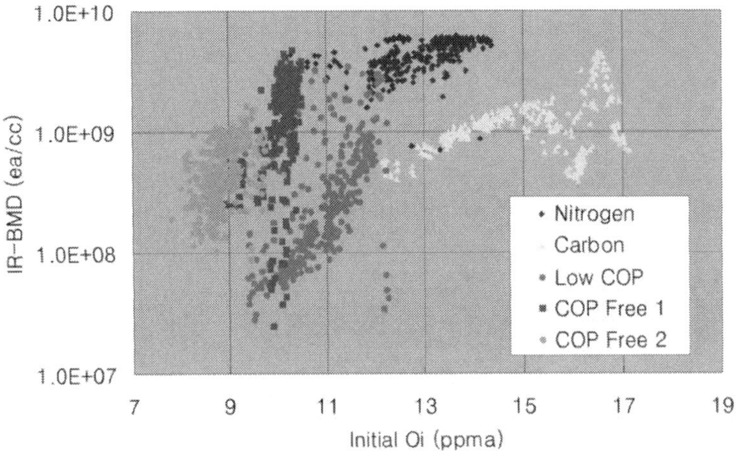

7 / 10

SILTRON

4. Discussion

1) Polished Si Wafer

All Wafers / **Polished Wafers**

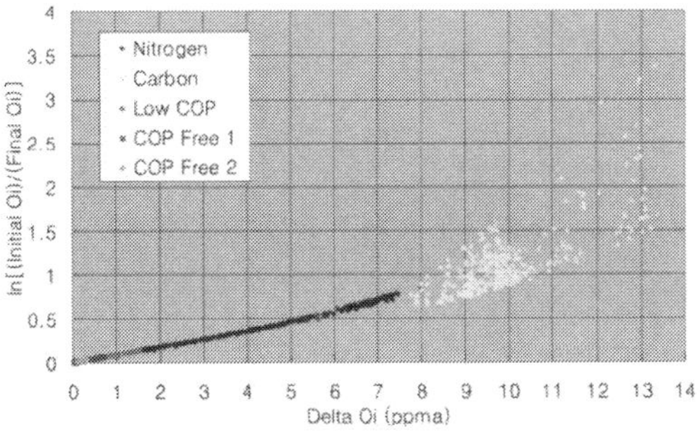

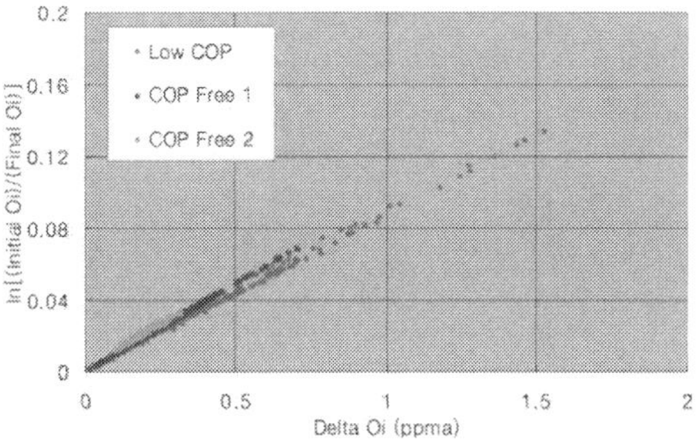

$$r_{\mathrm{BMD}} = -\frac{dC_{\mathrm{Oi}}}{dt} = \frac{k_1 C_{\mathrm{Oi}}}{1+k_2 C_{\mathrm{Oi}}}$$

$$\ln \frac{C^0_{\mathrm{Oi}}}{C_{\mathrm{Oi}}} = -k_2 (C^0_{\mathrm{Oi}} - C_{\mathrm{Oi}}) + k_1 t$$

Delta Oi = (Initial Oi) – (Final Oi)

	k_1t	$-k_2$	R^2	Remarks (H/T)
Low COP	+0.0022	0.08590	99.57%	800℃ 4hr + 1000℃ 16hr
COP Free 1	+0.0006	0.09820	99.72%	
COP Free 2	+0.0028	0.09900	96.98%	

SILTRON

4. Discussion

2) Nitrogen or Carbon Doped Substrates

Nitrogen Doped Wafers

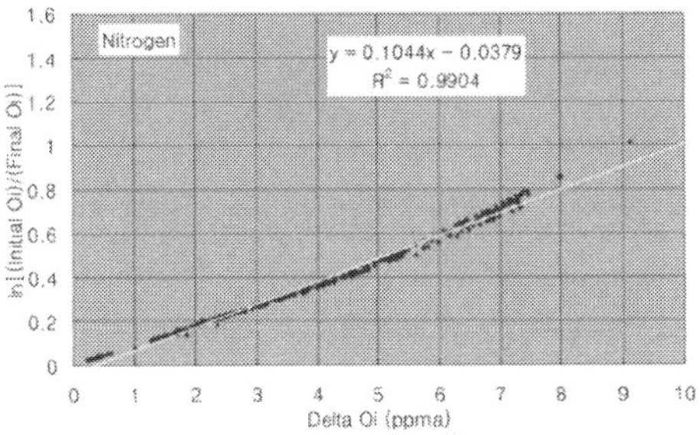

Carbon Doped Wafers

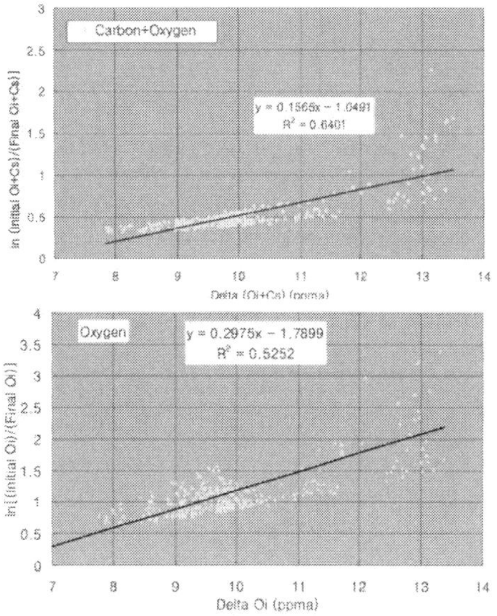

	k_1t	$-k_2$	R^2	Limiting Reactant
Nitrogen	-0.0379	0.01044	99.04%	Oxygen
Carbon	-1.7899	0.29750	52.52%	Oxygen
	+0.0162	0.50150	93.61%	Carbon
	-1.0491	0.15650	64.01%	Carbon + Oxygen

SILTRON

5. Conclusions

1) Order Shifting Autocatalytic Reaction

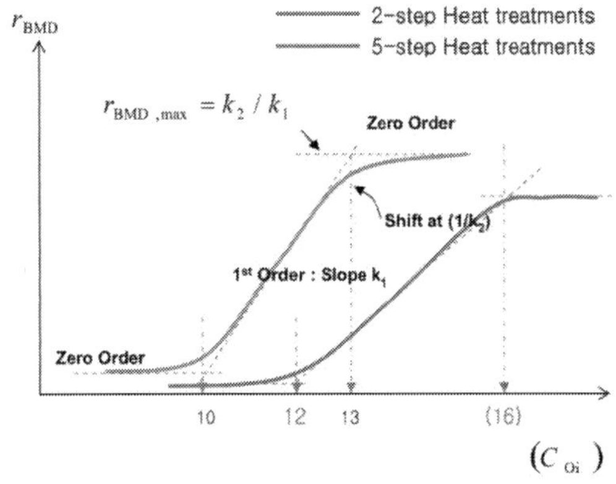

$$r_{BMD} = -\frac{dC_{Oi}}{dt} = \frac{k_1 C_{Oi}}{1 + k_2 C_{Oi}} \qquad \ln \frac{C^0_{Oi}}{C_{Oi}} = -k_2(C^0_{Oi} - C_{Oi}) + k_1 t$$

$$Si(s) + 2Oi \xrightarrow{k_1} (OP)^*$$

$$(OP)^* + nOi \xrightarrow{k_2} (OP)^* + \ldots\ldots + (OP)^* \rightarrow BMD$$

		k_1t	k_2	R^2	Remarks
Substrate Wafer	N-doped	-0.038	0.0104	99.0%	
	C-doped	-1.790	0.2975	52.5%	Oxygen
		+0.016	0.5015	93.6%	Carbon
		-1.049	0.1565	64.0%	Oxygen+Carbon
Polished Wafer	Low COP	+0.002	0.0859	99.6%	
	COP Free 1	+0.001	0.0982	99.7%	
	COP Free 2	+0.003	0.0990	97.0%	

※ Heat Treatments: 800℃ 4hr + 1000℃ 16hr

SILTRON

(608f) Elucidation of Heterogeneous Catalysis Reaction Mechanisms Using a SCT Shock Tube Reactor

Hope H. Connolly and Marco J. Castaldi, Earth and Environmental Engineering, Columbia University, New York, NY

Elucidation of heterogeneous catalysis reactionmechanisms using a SCT shock tube reactor

Hope Connolly,Marco Castaldi

Columbia University, New York, NY 10027 (USA)

hhc2119@columbia.edu

With increasing energy demands and awareness of globalclimate change, it has become necessary to develop methods to reduce the impactof fossil fuels while the hydrogen economy is developed for full-scaleimplementation. Crucial to this new fossil fuel approach are two reactions: 1)catalytic partial oxidation of methane to syngas and 2) reforming of carbondioxide by methane instead of steam. Both of these reactions produce syngas, acombination of hydrogen and carbon monoxide of importance as a hydrogen sourcefor fuel cells. The use of a SCT shock tube reactor at is being used to studythese fast heterogeneous reactions in order to overcome the shortcomings ofprevious studies which have yet to confirm detailed reaction mechanisms. It islocated at ATK/GASL in Ronkonkoma, NY. In the reactor, a fast-acting valve isremotely released, allowing the high-pressure N2 driver gas to propagate a shockwave through the reaction test gas mixture surrounding the SCT catalyzedscreen. A rarefaction wave quickly follows, bringing the temperature andpressure down as quickly as it was increased, and quenching the intermediatesfor detection and identification.

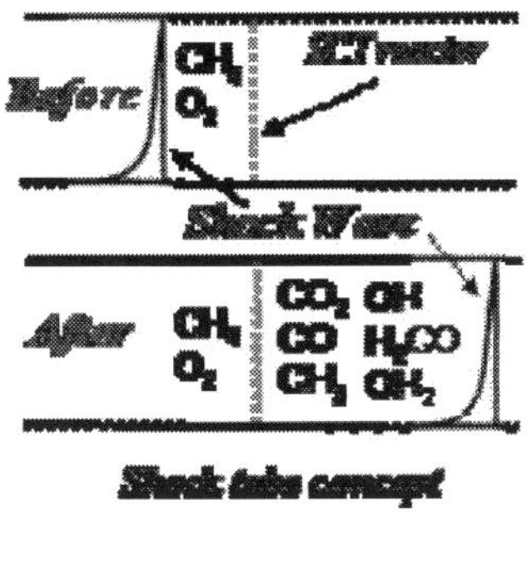

First, this set-up allows analysis ofthe entire catalytic cycle of both surface reactions and gas-phase products andintermediates. In pure surface and continuous flow systems, short livedintermediates released from the surface may react and not be observed in thefinal product gases. Secondly, the reactor looks at the entire surface in aggregate,not limited to a small subsection of the catalyst surface. Furthermore, itoperates under actual reacting temperatures and pressures, enabling insight foreasy scale-up to a flow reactor. Lastly, the SCT shock tube reactor is able tocarry out experiments without transport effects. The microsecond step changeinduced by the shock wave and the $3.64x10^5$ K/s[17] quenching timeeliminates the reaction of unreacted material downstream. Precise control oftemperature, pressure and reaction time allows previously unseen insight intothe progression of the reaction, by quenching the reaction at varying stages ofcompletion. Wiesz-Prater criterion and Damkohler numbers have been analyzed toensure that operations are taking place in a kinetically-controlled regime andwithout external diffusion mass transfer resistance, respectively.

Initial tests have shown 3 promising conclusions.First, a shock wave passed through the uncatalyzed SCT mesh screen elementshowed no signs of attenuation, maintaining its shape and strength throughoutthe interaction. Proof of the robustness of the mesh screen and the absence ofdeterioration of the shockwave were necessary before moving on to furtherexperiments. Next, we passed a test gas mixture (60% nitrogen, 20% oxygen, 20%methane) and an inert gas through both a catalyzed and uncatalyzedPt/γ-alumina SCT element. Only the reactive test gas and the catalyzedscreen combination exhibited a prolonged pressure increase, indicative of thepredicted reaction taking place. Finally, we observed a kinetic time delay of0.005 s, similar to the calculated kinetic time of 0.0051 s. This delay isshorter than a homogeneous ignition delay.

Thispresentation will show results from reforming CH_4-CO_2 mixtures.The mixture range was varied from 20% CH_4 to 70% CH_4(balance CO_2) in reactant concentration at temperatures between 500 and1000K for pressures spanning 1 to 10 atm. Product gas composition as afunction of temperature, mixture concentration and pressure will be discussedwith an emphasis on a reaction sequence understanding.

See more of this Session: Reaction Path Analysis III
See more of this Group/Topical: Catalysis and Reaction Engineering Division

Global Rate Expressions for Reactions in Automotive Catalytic Converters

Nivedita kumar[1], Ravikeerthi Thimmineni[2], Niket Kaisare[2], and Preeti Aghalayam[2*]
[1]Dept. of Chemical Engineering, VIT, Vellore 632014, India
[2]Department of Chemical Engineering, IIT Madras, Chennai 600036, India

In an effort to better understand the underlying chemistry in three-way catalytic converters (TWC), several detailed micro-kinetic models have been developed in the literature [1]. Microkinetic models can capture the observed features at a wide range of operating conditions. They are especially useful under conditions of intrinsic reaction control. However, incorporating detailed microkinetic models, consisting of hundreds of reactions, in computational fluid dynamics simulations for TWC can be prohibitively computationally expensive.

The overall reactions (OR) method [2] provides a convenient means of analysis of microkinetic reaction mechanisms. Its analytical framework lends itself extremely well to the determination of rate limiting steps, and development of corresponding reduced reaction rate expressions. In this work the focus is on analyzing reaction systems of importance in catalytic converters, using the overall reactions method. One of the major gaseous pollutants that is reduced in catalytic converters is nitric oxide (NO). Of the various reactions that occur in the TWC for NO reduction, the NO-CO reactions are important.

In earlier work done in the group, a microkinetic mechanism for the reduction of NO by CO has been proposed and validated [3] against literature experimental data. Furthermore, a sensitivity analysis/partial equilibrium based approach has been recently applied to this reaction mechanism in order to develop simple global reaction rate expressions separately for the formation of the two important products, N_2 and N_2O [4]. In this work, we apply the OR methodology to the same microkinetic mechanism, in order to theoretically derive analytical reduced rate expressions.

The NO-CO reaction is a series-parallel scheme, with the simultaneous formation of N_2 as a main product, and N_2O, as a side product. The rate determining step, and consequently the overall reaction rate expression are expected to vary with the operating conditions (as determined in our earlier work), in this example. In order to successfully apply the OR methodology, we have divided he original NO-CO reaction mechanism to separately reflect the formation of each product (the adsorption of NO & CO, and formation of CO_2, are some of the common steps in both the mechanisms). For both the reaction schemes, a reduced rate expression is to be developed here and validated against the performance of the full microkinetic mechanism, & the reduced expressions based on sensitivity analysis developed earlier [4]. Some of the advantages of the OR-based expressions are that they are derived theoretically without performing full-fledged reactor simulations, and although fairly complex in form and structure, can provide us a wealth of mechanistic understanding.

The validated global rate expressions from this work will also provide appropriate kinetics for reactor-scale modeling of the TWC, and enable further analysis and optimization of the device, for future applications.

REFERENCES

1. R. Burch, J.P. Breen, F.C. Meunier, "A review of the selective reduction of NOx with hydrocarbons under lean-burn conditions with non-zeolitic oxide and platinum group metal catalysts," Applied Catalysis B: Environmental 39 (2002) 283–303
2. S. A. Vilekar, I. Fishtik, R. Datta. "The steady- state kinetics of a catalytic reaction sequence," Chemical Engineering Science 64 (2009) 1968-1979
3. D. B. Mantri , P. Aghalayam ,"Detailed Surface Reaction Mechanism for Reduction of NO by CO," Catalysis Today 119 (2007) 88-93
4. R. Thyagarajan, P. Aghalayam & N. Kaisare, "Microkinetic model for NO-CO reaction: Model reduction," Poster presented at the International Conference of Chemical Kinetics, Cambridge, MA, July 11-14 (2011).

(629a) New Structure of Matter for High Performance Heterogeneous Catalytic VOC Oxidation At Short Contact Time

Sabrina Wahid, Department of Chemical engineering, Auburn University, Auburn, AL and Bruce Tatarchuk, Department of Chemical Engineering, Auburn University, Auburn, AL

A new class of composite materials, microfibrous entrapped catalyst (MFEC); prepared by traditional low cost paper making technique, entrap small (150-230 μ m dia.) catalyst particulates Pd/ γ-Al_2O_3 into sinter-locked networks of metal (e.g. Ni) microfibers (4 and 8 μ m dia.). This MFEC has uniform structure with high and variable voidages. MFEC with pleated configuration (Pleat number 4) enhances significant improvement to the pressure drop reduction and higher catalyst utilization. Also small particle of MFEC improves interphase and intraparticle mass transfer rate. Ensuring comfortable environment in aircraft cabin and similar air handling devices is an issue to be concerned. To provide controlled and improved air quality, VOC removal, by catalytic combustion is of strong interest. Catalytic oxidation of VOC (i.e. n-hexane model compound) was investigated at typical face velocities (ca. 10 to 40 m/s) resulting intrabed/intralayer residence times from 30-200 μsec. This research focuses on unique structures of matter which attain high levels of VOC removal Pleated MFECs (1-2 mm) containing Pd/ γ-Al2O3 are tested at operating conditions similar to those used in air beneficiation. A conversion comparison of n-hexane by noble metal (Pd) and transition metal (Mn, Co) has been made at various flow rates (10 to 30 m/s) and temperature (100 to 200°C). Detailed reaction kinetics has also been studied using the experimental findings.

See more of this Session: Poster Session of CRE Division
See more of this Group/Topical: Catalysis and Reaction Engineering Division

(629aa) Catalytic Evaluation of La1-xSrxCr1-yMnyO3- In Propane Steam Reforming At Intermediate Temperature

Jung Min Sohn and Nak Hyeon Kim, Chonbuk National University, Jeonju, South Korea

In this study, a novel perovskite type anode of $La_{1-x}Sr_xCr_{1-y}Mn_yO_{3-\delta}$ for solid oxide fuel cells (SOFCs) operated under propane has been tested regarding its catalytic activity for propane steam reforming. Samples were synthesized by metal nitrate, citric acid and EDTA using the sol-gel method. Catalytic activity was studied were using fixed-bed quartz reactor and reaction product analyzed by GC. To evaluate the characteristics of physicochemical properties of catalysts, XRD, BET and H_2-TPR were used. The highest performance was obtained using $La_{0.9}Sr_{0.1}Cr_{0.6}Mn_{0.4}O_{3-\delta}$ catalysts of which propane conversion and H_2 yield at 800 °C were 98.7 and 68.3% respectively. The long term stability of $La_{0.9}Sr_{0.1}Cr_{0.6}Mn_{0.4}O_{3-\delta}$ catalyst was confirmed during 24 h of reaction time at 800 °C. Relatively stable activity obtained which propane conversion about 95% and H_2 yield about 55%.

See more of this Session: Poster Session of CRE Division
See more of this Group/Topical: Catalysis and Reaction Engineering Division

(629ab) Behavioral Analysis of Heterogeneous Catalysts Type Zeolite (HZSM-5 and HBETA) on the Transesterification of Soybean Oil Modified with Fatty Acids

Isaac Nava, Ingeniera Qumica, Facultad de Ingeniera Qumica UMSNH, Morelia, Mexico, Horacio Gonzlez, Facultad de Ingeniera Qumica UMSNH, Morelia, Mexico and Alfredo Fuentes Gutirrez, Laboratorio de Ecotecnologas, CIEco, Centro de Investigaciones en Ecosistemas, UNAM campus Morelia, Morelia Michoacn, Mexico

In this work we studied the transesterification of commercial soybean oil enriched with fatty acids (0.25 wt% oleic acid and 0.25 wt% linoleic acid) using HZSM-5 and HBETA zeolite as the catalysts. The purpose of this project is to analyze the behavioral and feasibility of using a solid acid catalyst such as HZSM-5 and HBETA zeolite in the transesterification of modified soybean oil, because the traditional homogeneous catalyst most commonly used (NaOH) is not recommended if the oil has a content greater than or equal to 0.5% by weight of fatty acids, since it produces an emulsion of soap and biodiesel. To carry out the reaction a stirred glass reactor was used. The charge to the reactor consisted of 50 grams of oil, using a molar ratio of methanol/oil = 14 / 1. Commercial HZSM-5 and HBETA zeolite (4 grams) with a $Si0_2/Al_2O_3$ molar ratio of 80 and 75 respectively (Zeolyst) was used in the study. The reactor was operated at 1 atm and temperature of 75 °C. The stirring speed was 1000 rpm and the reaction was carried out within 8 hours. The conversion of fatty acids to methyl esters varied from 2 to 5% at the reaction conditions. However the distribution of product of Fatty Acid Methyl Esters it was diferent those who always obtained with homogeneous catalyst (NaOH). Saturated and Unsaturated chains of Fatty Acid Methyl Ester of more than 18 carbons were the main products obtained with this type catalysts.

See more of this Session: Poster Session of CRE Division
See more of this Group/Topical: Catalysis and Reaction Engineering Division

(629ac) DeNOx Performance of Combined LNT-SCR Catalyst Systems: A Simulation Study

Arun S. Kota, Dan Luss and Balakotaiah Vemuri, Department of Chemical and Biomolecular Engineering, University of Houston, Houston, TX

Abstract:

Transientstudies of combined LNT-SCR catalyst systems are investigated and theirperformance is evaluated against that of individual LNT or SCR systems. Aone-dimensional two-phase model with global rate expressions for various reactionsis used in the simulations. The LNT system modeled is that of Pt/BaO/Al_2O_3catalyst with two different BaO storage sites, designated as fast and slowsites. LNT model includes reactions that describe NO oxidation, storage of NO_2in the form of nitrates during lean storage cycle and reduction to N_2in the subsequent rich cycle with H_2 as a reductant. The SCR systemconsidered is that of Cu-ZSM5 and the model includes NO oxidation, ammoniastorage and oxidation, selective catalytic reduction of NO, NO_2 andformation of N_2O. Model predicts the spatio-temporal concentrationprofiles of reactants and products inside the channel which gives insight intothe $DeNO_x$ performance of combined systems. The effect of variousdesign and operating parameters on $DeNO_x$ efficiency have beenexamined which includes number of LNT and SCR catalysts in series, length ofeach catalyst, catalyst loading, rich cycle time and composition. It hasbeen found that series arrangement of alternate LNT and SCR catalyst, by dividingthem into equal halves, substantially improves NO_x conversion. Forfixed total catalyst length as the number of divisions increase, NO_xconversion increases and reaches a limit which is theoretically equivalent tothat of mixed catalyst. It is also observed that for fixed length of LNTcatalyst better $DeNO_x$ performance is obtained by increasing thelength of SCR catalyst in the alternate arrangement. Spatio-temporal profilesindicate complete storage of NH_3 on SCR catalyst which can beefficiently used to reduce NO_x slipping out of LNT.

See more of this Session: Poster Session of CRE Division
See more of this Group/Topical: Catalysis and Reaction Engineering Division

The Study of the Asymmetric Transfer Hydrogenation of Imines by RuTsDPEN Catalyst – Theoretical and Practical Aspects

Marek Kuzma, Institute of Microbiology Academy of Sciences of the Czech Republic, Prague, Czech Republic

Petr Kacer, Institute of Chemical Technology Prague, Prague, Czech Republic

Jiri Vaclavik, Institute of Chemical Technology Prague, Prague, Czech Republic

Jan Prech, *Institute of Chemical Technology Prague, Prague, Czech Republic*

Libor Cerveny, *Institute of Chemical Technology Prague, Prague, Czech Republic*

Introduction

The asymmetric transfer hydrogenation (ATH) of imines has remained a rather unexplored topic since the first successful approach disclosed by Noyori et al. in 1996, despite its indubitable usefulness, especially in pharmaceutical technology.[1] Noyori showed that the Ru(II)Cl(η^6-*p*-cymene)[*N*-*p*-tosyl-1,2-diphenylethylenediamine] ([RuCl(η^6-*p*-cymene)TsDPEN] – **1**) catalyst in a HCOOH/triethylamine mixture was able to efficiently hydrogenate substituted isoquinolines with great selectivity.[2] Almost simultaneously, the same system was reported to reduce ketones superbly,[3] suggesting that the mechanisms of asymmetric hydrogenation of C=N and C=O bonds should be alike.[4]

In accordance with the latest evidence regarding the ketone and imine reduction mechanistic pathways,[9-11] both posited mechanisms are outlined in Schemes 1 and 2. Scheme 1 focuses on transfer hydrogenation of acetophenone **5** affording chiral 1-phenylethanol **6**, which follows the well-known mechanistic concept published earlier.[5,13] The substrate molecule **5** approaches the Ru hydride **2** and is hydrogenated to **6** by transferring both the proton and hydride from **2**, which leads to formation of a 16e$^-$ Ru complex **3**.

Scheme 2 reflects the process of hydrogenation of 1-methyl-3,4-dihydroisoquinolinium **8** affording chiral 1-methyl-1,2,3,4-tetrahydroisoquinoline **9**. This proceeds by a hydride transfer only, giving the cationic structure of the solvate **4** which is assumed to create an ion pair with a chloride anion present in the solution, while the Ru atom is believed to be solvated.[10]

Scheme 1. Suggested pathways for asymmetric transfer hydrogenation of acetophenone **5** using **1** in the presence of triethylamine (TEA) and formic acid. The chelate phenyl substituents are omitted for the sake of clarity.

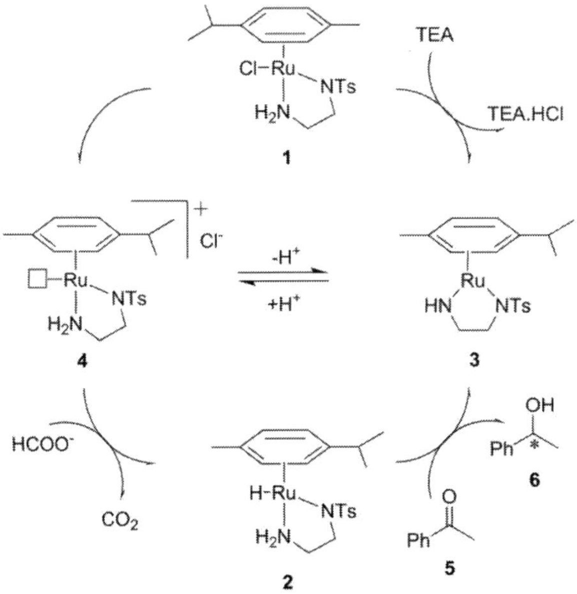

Scheme 2. Suggested pathways for asymmetric transfer hydrogenation of 1-methyl-3,4-dihydroisoquinoline **7** using **1** in the presence of triethylamine (TEA) and formic acid. The chelate phenyl substituents are omitted for the sake of clarity.

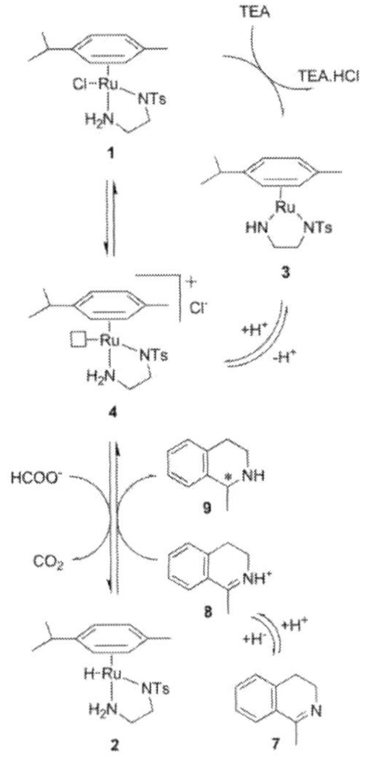

Given the recent evidence for an ionic mechanism of imine hydrogenation and the fact that it has not yet been examined computationally, this paper focuses on a full-scale molecular modelling of the system, mainly exploring the imine pathway and comparing it with a different (ketone) reduction mechanism.

Computational Methods

The geometry optimizations, frequency jobs and single-point energy calculations were performed using the standard Gaussian03 package at the DFT level of theory employing the restricted B3LYP combination of functionals. The standard polarized 6-31G(d,p) basis set was used for all atoms except ruthenium which was described by the LANL2DZ basis set (for 18 valence electrons) together with the LANL2DZ effective core potential (ECP). In addition, single-point energies were computed also at the MP2 level using the previously DFT-optimized structures (an MP2/6-31G(d,p)/LANL2DZ//DFT//6-31G(d,p)/LANL2DZ approach, abbreviated MP2//DFT) to obtain more reliable energy values. All structures were confirmed by frequency analyses to be either true minima (no imaginary vibrations found) or transition states (one imaginary vibration). All intrinsic reaction coordinate (IRC) calculations were carried out using the Gaussian09 package. These were computed separately in the forward and reverse directions, each set up with a maximum of 500 points in order to localize corresponding energy minima. If an IRC calculation failed to reach the potential energy surface minimum, its last available point was used for a manual optimization towards a minimum.

Results and Discussion

The aim of this study was to investigate the currently popular ionic mechanism by means of computational chemistry. The structure optimizations, frequency analyses and energy calculations were conducted using density functional theory (DFT) at the B3LYP level. However, single point energy values were also obtained employing the Møller-Plesset perturbation (MP2) electron correlation method as the energies obtained this way were considered more precise and reliable. Hence, all energies were calculated both at the MP2//DFT and DFT//DFT levels.

Since these in-depth calculations result in demanding hardware requirements, the simplest imine substrate 1-methyl-3,4-dihydroisoquinoline (**7**) was selected for the mechanistic study. The original Noyori mechanism, involving six-membered cyclic TSs and a non-protonated substrate, was first computed. These initial calculations led to a different enantiomer than the one observed experimentally, which supported the opinion that this pathway was not applicable to ATH of cyclic imines. Moreover, the energy differences between the *Si*- and *Re*-TSs were too high (*e.g.*, 43.2 kJ/mol for the Ru-hydride **2** as a catalyst and **7** as a substrate). For these reasons, we decided to abandon the original mechanism and sought to explore the ionic pathway instead.

Scheme 3 shows the resulting MP2//DFT (DFT//DFT) energy profiles of asymmetric transfer hydrogenation of **7** in the **1**/HCOOH-triethylamine system, proceeding *via* an ionic pathway.

Scheme 3. Resulting single-point energy profiles at levels MP2/6-31G(d,p)/LANL2DZ//B3LYP/6-31G(d,p)/LANL2DZ (i.e. MP2//DFT, solid line) and B3LYP/6-31G(d,p)/LANL2DZ//B3LYP/6-31G(d,p)/LANL2DZ (i.e. DFT//DFT, dashed line). The diagrams outline possible reduction pathways proceeding either *via* transition state **fav11a** employing a C(sp^3)H/π interaction (Route A), or *via* **fav11b** stabilized by a C(sp^2)H/π interaction (Route B). DFT//DFT computed Gibbs free energies are also given, showing the differences between Routes A and B. The corresponding structures are listed below, together with their MP2//DFT (DFT//DFT) energies in kJ/mol relative to step I.

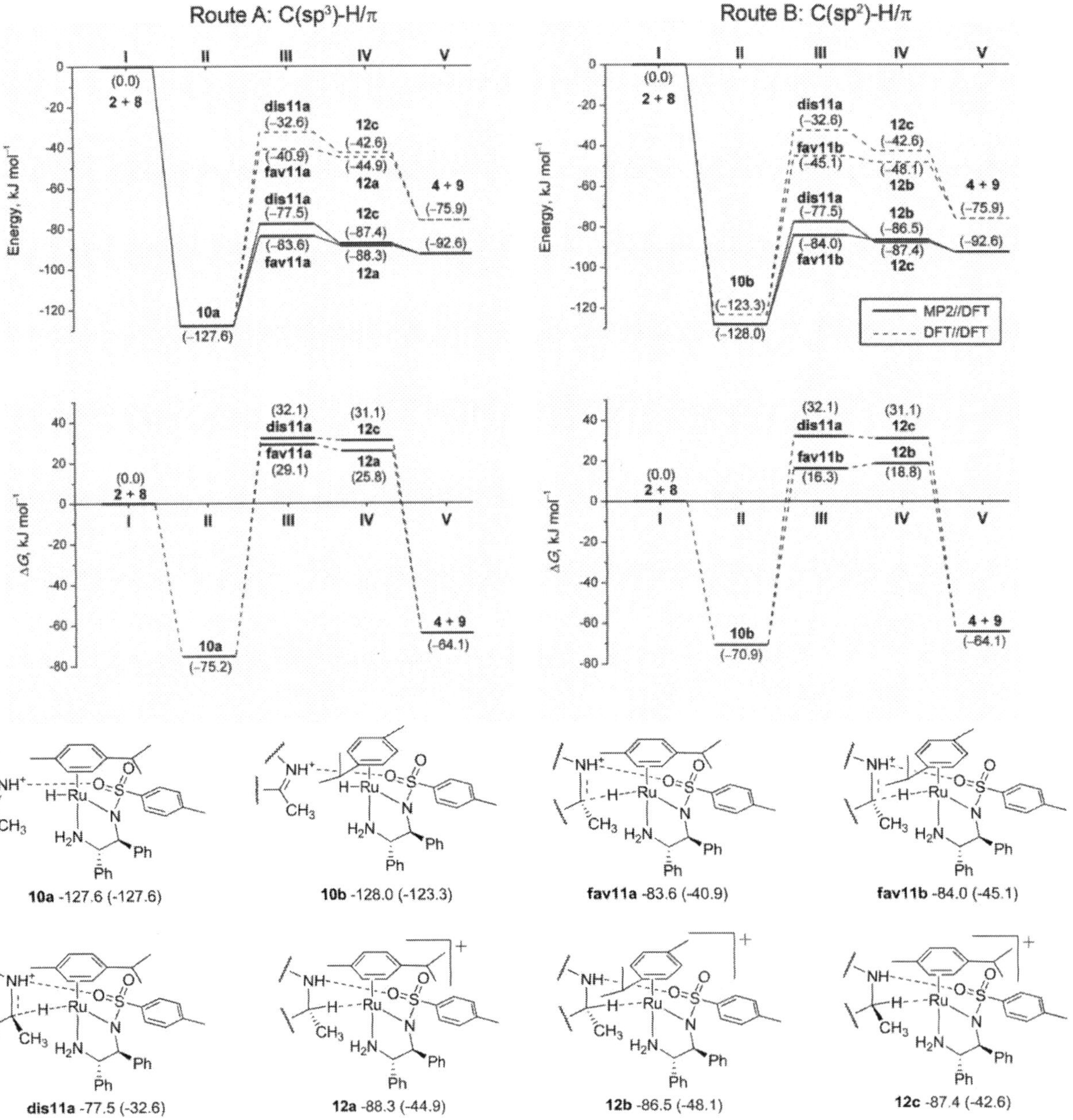

The lower energy (i.e. higher probability of formation) of a favTS is reached by a rather weak, yet critical CH/π interaction between the imine substrate molecule and the η^6-aromatic ligand of **2**.[14] The structure of the η^6-*p*-cymene ligand theoretically allows four different CH/π interactions (Scheme 4) which is inherently connected with a possible rotation of the *p*-cymene moiety. Therefore, two different structures employing $C(sp^3)H/\pi$ interactions were investigated, utilising either the CH_3- group (*A*) or the methine of the 2-propyl group (*D*), and two more $C(sp^2)H/\pi$ interactions were considered involving either of the aromatic hydrogens (*B*, *C*).

Scheme 4. Sites in the η^6-*p*-cymene ligand allowing the formation of a CH/π interaction.

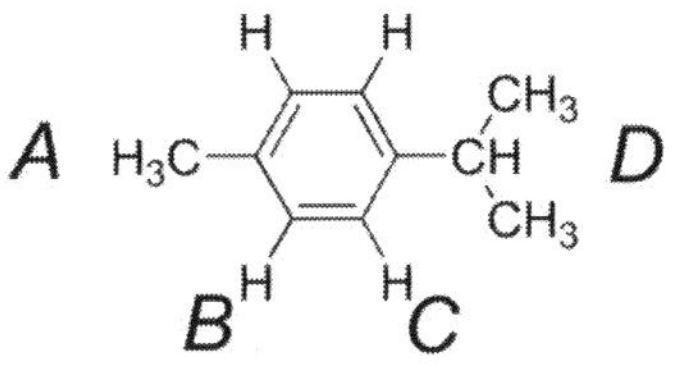

Transition states employing sites *A* and *B* in a CH/π attraction (further denoted as **fav11a** and **fav11b**) were found to be energetically very similar (difference 0.47 kJ/mol, with **fav11b** being the most stable one). The energy of the type-*C* transition state (**fav11c**) was higher than in the case of **fav11b** (4.80 kJ/mol). Unfortunately, the structure of *D*-favTS (**fav11d**) failed to converge in the optimization process and therefore its formation was considered rather improbable. To sum up, the structures **fav11a** and **fav11b** (Figure 3) were chosen as the most probable candidates to be included in the energy profiles shown in Scheme 3 (Route A or B, respectively). In the case of **fav11a**, the total electron density surface was calculated and is involved in Figure 3, graphically emphasizing the CH/π interaction.

Figure 3. Calculated structures of favourable transition states **fav11a**, **fav11b**, **fav11c** and **fav11e**, together with their single-point MP2//DFT (DFT//DFT) energies in a.u. units. Hydrogen atoms, except for the essential ones, are omitted for clarity. Where applicable, CH/π interactions are shown as dotted lines highlighted in green with corresponding interatomic distances in ångström units. The total SCF electronic density surface for **fav11a** is partially displayed with applied MO isovalues of 0.02 and density 0.006.

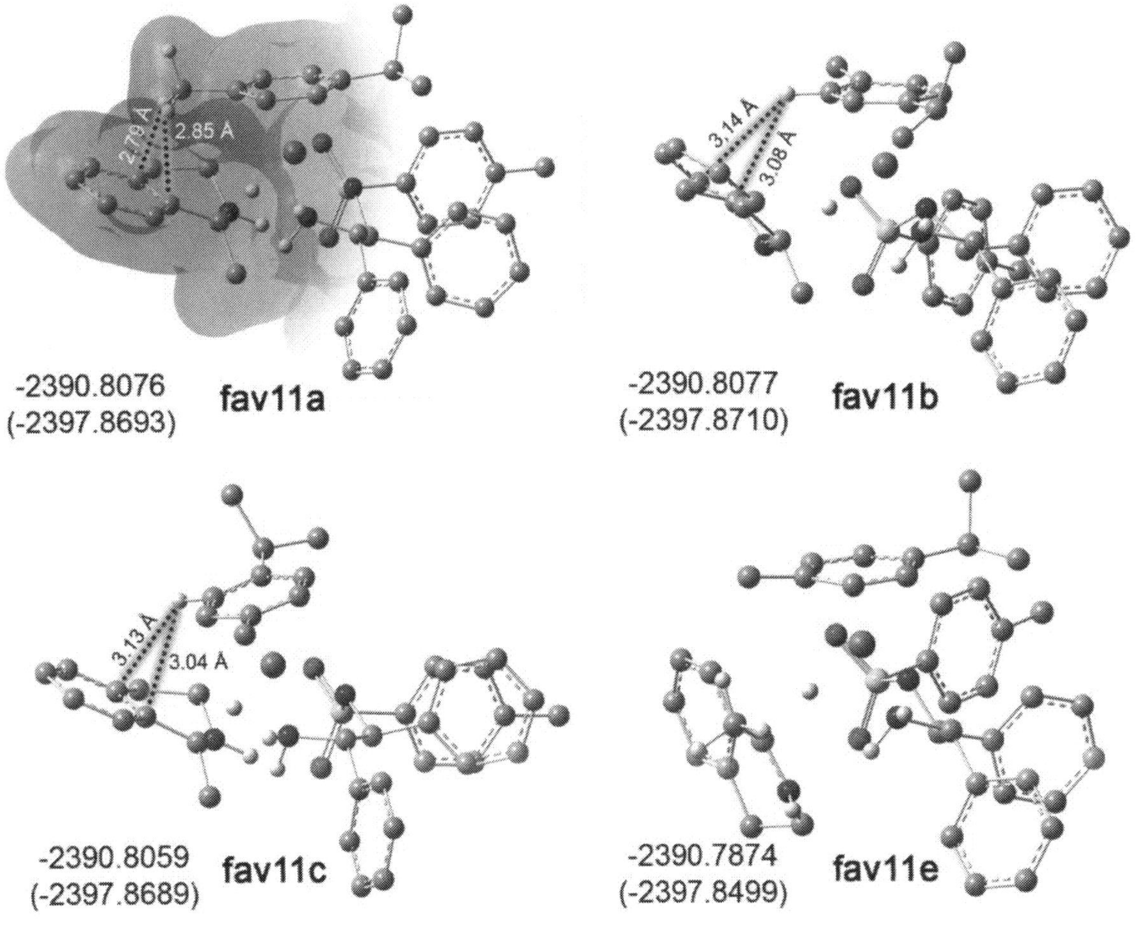

Conclusions

Based on the latest findings concerning the mechanism of asymmetric transfer hydrogenation of imines using chiral Ru(II)Cl(η^6-arene)[(*S*,*S*)-*N*-TsDPEN] complexes, this computational study sought to explore a pathway that was essentially different from the one applicable to ketone reduction. After a detailed analysis of transition state structures and their preceding and following energy minima, two energy diagrams were constructed using the most likely structures. This being the first computational study of such a system using the η^6-*p*-cymene instead of a symmetric η^6-benzene or η^6-1,2,3,4,5,6-hexamethylbenzene ligand, an effort was made to determine whether the C(sp^3)H/π or C(sp^2)H/π interactions took place in the favourable transition states. Interestingly, both possibilities were almost equal in terms of single-point energy, but the pathway involving a C(sp^2)H/π attraction was favoured owing to its lower Gibbs free energy. The interaction between the protonated nitrogen atom of the substrate and the sulfonyl fragment of the TsDPEN ligand was identified as highly important for the stabilization of all transition states and energy minima. This observation represents another key feature of the ionic mechanism which is absent in the original mechanism operating for the ATH of ketones.

Computational studies of this kind provide a valuable insight into the reaction mechanism. Endowed with a deeper understanding of how asymmetric transfer hydrogenations might proceed we are subsequently able to adjust the reaction conditions in order to reach the optimal outcome.

Acknowledgement

This work has been financially supported by the Grant Agency of the Czech Republic (Grant GACR 104/09/1497), the Ministry of Education, Youth, and Sports (MEYS) of the Czech Republic (Grant CEZ: MSM 604 613 7301) and the research intent (AV0Z50200510). The access to the MetaCentrum computing facilities, provided under the programme "Projects of Large Infrastructure for Research, Development, and Innovations" LM2010005 funded by MEYS, is highly acknowledged.

References

(1) Andrews, I.; Cui, J.; DaSilva, J.; Dudin, L.; Dunn, P.; Hayler, J.; Hinkley, B.; Hughes, D.; Kaptein, B.; Kolis, S.; Lorenz, K.; Mathew, S.; Rammeloo, T.; Wang, L.; Wells, A.; White, T.; Xie, C.; Zhang, F. *Org. Process Res. Dev.* **2009**, *13*, 397.

(2) Uematsu, N.; Fujii, A.; Hashiguchi, S.; Ikariya, T.; Noyori, R. *J. Am. Chem. Soc.* **1996**, *118*, 4916.

(3) Fujii, A.; Hashiguchi, S.; Uematsu, N.; Ikariya, T.; Noyori, R. *J. Am. Chem. Soc.* **1996**, *118*, 2521.

(4) Noyori, R.; Hashiguchi, S. *Acc. Chem. Res.* **1997**, *30*, 97.

(5) Noyori, R.; Yamakawa, M.; Hashiguchi, S. *J. Org. Chem.* **2001**, *66*, 7931.

(6) Martins, J. E. D.; Clarkson, G. J.; Wills, M. *Org. Lett.* **2009**, *11*, 847.

(7) Åberg, J. B.; Samec, J. S. M.; Bäckvall, J.-E. *Chem. Commun.* **2006**, 2771.

(8) Blackmond, D. G.; Ropic, M.; Stefinovic, M. *Org. Process Res. Dev.* **2006**, *10*, 457.

(9) Fleury-Brégeot, N.; de la Fuente, V.; Castillón, S.; Claver, C. *ChemCatChem* **2010**, *2*, 1346.

(10) Sandoval, C. A.; Ohkuma, T.; Utsumi, N.; Tsutsumi, K.; Murata, K.; Noyori, R. *Chem. Asian J.* **2006**, *1-2*, 102.

(11) Sandoval, C. A.; Bie, F.; Matsuoka, A.; Yamaguchi, Y.; Naka, H.; Li, Y.; Kato, K.; Utsumi, N.; Tsutsumi, K.; Ohkuma, T.; Murata, K.; Noyori, R. *Chem. Asian J.* **2010**, *5*, 806.

(12) Haack, K.-J.; Hashiguchi, S.; Fujii, A.; Ikariya, T.; Noyori, R. *Angew. Chem. Int. Ed. Engl.* **1997**, *36*, 285.

(13) Yamakawa, M.; Ito, H.; Noyori, R. *J. Am. Chem. Soc.* **2000**, *122*, 1466.

(14) Yamakawa, M.; Yamada, I.; Noyori, R. *Angew. Chem. Int. Ed.* **2001**, *40*, 2818.

ESTERIFICATION OF ISO-AMYL ALCOHOL WITH BENZOIC ACID: A KINETIC STUDY

Yurani Jiménez, Yoshinori Casas, Cristian Otálora, Iván Gil, Julio Vargas and Gerardo Rodríguez[1].

Departamento de Ingeniería Química y Ambiental, Universidad Nacional de Colombia, Bogotá, Colombia.

Abstract

As a result of global ethanol production by fermentation, the by-production of fusel oil used principally as energetic source and coating production, represent low economic benefit, environmental problems and a challenge to the industry. This is composite by higher alcohols, principally iso-amyl alcohol and the implementation of a process to transform it into higher added value products (i.e. esters) presents high potential (Patil, Koolwal et al. 2002).The iso-amyl acetate, propionate, butyrate, benzoate and others as derived ester, than can be used as fragrances, solvents, plasticizers coating, flavoring, cosmetics and others.

This work proposes the esterification of iso-amyl alcohol with benzoic acid due to the add value and its different applications. The iso-amyl benzoate has fruity balsamic odor similar to cacao and cherry, is mainly used in food industry of beverages, candies, sauces and in fragrance industry. In this way, there is a potential market replacing a lot of products used in different applications (Lewis 1989).

In the national market, these products are principally imported or extracted from natural sources, this allows develop different alternative processes to transform by-products in higher added value products and decrease the external dependence. In order to define the desire production process is necessary to known thermodynamics data and kinetic reaction.

This work presents the kinetic study about the esterification reaction between benzoic acid and isoamyl alcohol using p-toluensulfonic acid and a cationic exchange resin. The esterification reaction was carried out at temperatures between 368.15 K and 399.15 K using loading of catalyst at 1.5% in the homogenous case, 1.5 and 2 % in the heterogeneous case and molar ratio alcohol: acid 4 to 6, reaching high values of conversion about 97%.

The experimental test number was discriminate using a superficial design of experiments in both cases. In accordance with the chemical equilibrium limitations of the reactive system, this effect was considered and included in the kinetic model.

The non ideality of the species involved in the mixture reactive was taken into account using the activity coefficients by the UNIFAC group contribution method and due to the thermodynamics data of the system is so limited, calculations were carried-out and compared

---

[1] To whom correspondence should be address. Present address: Universidad Nacional de Colombia Strret 30 No 45-03 Bldg:412-109, Email:ypjimenezr@unal.edu.co, yacasasa@unal.edu.co,

with reported experimental values (i.e. azeotropic data, pure substance properties, etc.). The absolute error obtained was no longer than 3%.

The homogenous kinetic was fitting to a typical non ideal pseudo homogenous kinetic with error value of 9.7%, the estimation of intrinsic parameters of specific constant of reaction rate was fitted using the NelderMead optimization method.

In the heterogeneous case, the kinetic data were recollected below free conditions of the external and internal mass transfer effects. Additionally, the adsorption constants were determinate using the isotherm compound model (Song, Venimadhavan et al. 1998; Orjuela Londoño 2003; Tsai, Lin et al. 2011) , which allow to define the order to the affinity of the compounds in the resin as Water<Alcohol<Ester<Acid. The kinetic data were correlated using the ideal and non ideal pseudo homogeneous model, LHHW (Lagmuir-Hinshelwood-Hougen-Watson) and Eley Rideal models, which predict satisfactorily the experimental data. However, the best model is LHHW with a value error of 1.07%.

Keywords: Esterification, cationic exchange resin, isoamyl benzoate, adsorption constants, heterogeneous kinetic, homogeneous kinetic.

References

- Lewis, R. J. (1989). Food additives handbook New York, Van Nostrand Reinhold.
- Orjuela Londoño, A. (2003). Evaluación de un sistema de destilación reactiva a nivel de planta piloto en reacciones de esterificación. Departamento de Ingeniería Química. Bogotá, Universidad Nacional de Colombia. **Magister en Ingeniería Química**.
- Patil, A. G., S. M. Koolwal, et al. (2002). "Fusel Oil: Compostition, removal and potential utilization." International Sugar Journal **104**(1238): 51-58.
- Song, W., G. Venimadhavan, et al. (1998). "Measurement of Residue Curve Maps and Heterogeneous Kinetics in Methyl Acetate Synthesis." Industrial & Engineering Chemistry Research **37**(5): 1917-1928.
- Tsai, Y.-T., H.-m. Lin, et al. (2011). "Kinetics behavior of esterification of acetic acid with methanol over Amberlyst 36." Chemical Engineering Journal **171**(3): 1367-1372.

(629af) Feedstock to Tailpipe: A Catalysis Approach

Griffin W. Roberts, Department of Chemical & Petroleum Engineering, University of Kansas, Lawrence, KS and Susan Stagg-Williams, Dep't of Chemical & Petroleum Engineering, University of Kansas, Lawrence, KS

The ever growing demand of energy independence has caused significant research in alternative fuels. The work herein is a part of a collaborative effort of developing algae to liquid fuels where this specific work involves developing catalyst for chemical transformation. While the algae production is being scaled up it is necessary to model the feed oils with vegetable oil. Algal oil mostly comprised of triglycerides and free fatty acids similar to seed oils. The catalyst developed is a pillared clay material chosen for textural and chemical properties as far as surface area, thermal stability, pore size, and acidity. Larger pore size is required to allow accessibility of reactant with the active sites. The acidity is of a lower to moderate strength to balance the activity with unwanted side products such as gas and coke. The catalysts has been fully characterized by nitrogen isotherm at 77K, temperature programmed desorption of ammonia, TEM, XRD, and pulsed chemisorption when needed. The catalyst was used in transesterification reaction of 65% of triolein with methanol to compare the results to traditional homogeneous base catalyst (sodium methoxide) and homogeneous acid catalyst (sulfuric acid). The reaction was carried out in a pressurized vessel to ensure that methanol is in the liquid phase. The products were analyzed by means of gas chromatography.

See more of this Session: Poster Session of CRE Division
See more of this Group/Topical: Catalysis and Reaction Engineering Division

(629ag) A Study of the Effect of PVP As Stabilizing Agent In Platinum Nanoparticle Synthesis and the Implications towards Oxygen Reduction Reaction

Selasi Blavo, Chemical & Biomedical Engineering, University of South Florida, Tampa, FL, **Lyndsey M. Baldyga**, University of South Florida, Tampa, FL and John Kuhn, Chemical and Biomedical Engineering, University of South Florida, Tampa, FL

The stability of precious metals under acidic conditions is a potential challenge for several applications including proton exchange membrane fuel cells (PEMFCs). Strategies addressing this challenge have been tested including the addition of organic stabilizing agents such as polypyrrole. Organic stabilizing agents have also been used to synthesize precious metal nanoparticles by assisting in the regulation of the nucleation and growth rates. In this study, well-defined 3 nm Pt nanoparticles were synthesized using a combination of metal precursor, Hexachloroplatinic acid (H_2PtCl_6), surfactant (PVP), alcohol (methanol), and water. The stability of these nanoparticles was evaluated under acidic conditions. Washed (in cycles of ethanol and hexane) and unwashed batches of Pt nanoparticles were characterized using a transmission electron microscope. Further studies to determine the electrocatalytic properties of the particles were performed using cyclic voltammetry. These results were complimented with temperature-programmed experiments after supporting the nanoparticles on silica. A detailed analysis and discussion of these results is made in this presentation.

See more of this Session: Poster Session of CRE Division
See more of this Group/Topical: Catalysis and Reaction Engineering Division

(629ah) Development of a Fluidized-Bed Catalytic Cracking (FCC) System with On-Line Product Analyses for Conversion of Activated Sludge to Green Fuels

William Holmes[1], Emmanuel D. Revellame[1], Rafael Hernandez[1] and William T. French[2], (1)Dave C. Swalm School of Chemical Engineering, Mississippi State University, Mississippi State, MS, (2) Chemical Engineering, Mississippi State University, Mississippi State, MS

The annual production of activated sludges in the United States ranges from 7 – 8 million tons, which are being disposed by land-filling or incineration. These disposal methods are heavily regulated and could create adverse environmental impacts. Activated sludges contain microbial consortiums known to produce lipidic materials and many other organic compounds, which could be used to produce biofuels and specialty chemicals. This study focused on modification of a Parr flow reactor for in situ production of green fuel from activated sludge by fluid-bed catalytic cracking. The modifications include on-line analyses to monitor product formation and reaction completion. The on-line analyses use two gas chromatographs for fixed gases and hydrocarbons.

See more of this Session: Poster Session of CRE Division
See more of this Group/Topical: Catalysis and Reaction Engineering Division

(629ai) Design and Initial Testing of a Spouted-Bed Reactor for Biomass Catalytic Pyrolysis

Shoucheng Du, Julia Valla and George M. Bollas, Department of Chemical, Materials & Biomolecular Engineering, University of Connecticut, Storrs, CT

Spouted-bed reactors are ideal for thecatalytic pyrolysis of biomass due to their ability to handle larger particles,large particle size distributions, and differences in solid densities. Theyalso provide a good mixing and decouple residence times of gas and solids; thus,gas residence times and unwanted secondary biomass reactions are reduced. Inthis study, a bench-scale conical spouted-bed biomass catalytic reactor hasbeen designed, considering different particle properties, geometric factors andimportant hydrodynamic parameters. The design analysis aimed at identifying gasflows capable of generating a stable spouted-bed regime in the conical reactor,while providing flexibility in the catalyst to biomass ratios that the unit canhandle.

The biomass reactor setup is shown inFigure 1. It includes: 1) a conical quartz reactor with a gas distribution gridat the bottom, 2) two biomass feeders providing the capability for batch andcontinuous operation, 3) cooler and condenser for the separation of the liquidproducts, weighing scale for real-time measurement of the liquid product yield,4) filter capable of removing the remaining condensable liquid droplets andfine solid particles, 5) liquid and gas collection systems and 6) a FourierTransform Infrared (FTIR) spectrometer for real time gas composition analysisand the simultaneous use of dynamic simulation and experimentation for theanalysis of reaction kinetics and process optimization. The reactor is capableof converting lignocellulosic and other types of biomass to liquid fuels and/orsynthesis gas, using N_2, H_2O or CO_2 as thefluidization gas. Figure 2 illustrates ranges of stable operation in the relevanthydrodynamic regimes. The regime calculated for the reactor development(red-colored area) demonstrates that conical spouted beds have a wide range ofoperating regimes, which makes them ideal for the variability of biomassdensity and particle diameter.

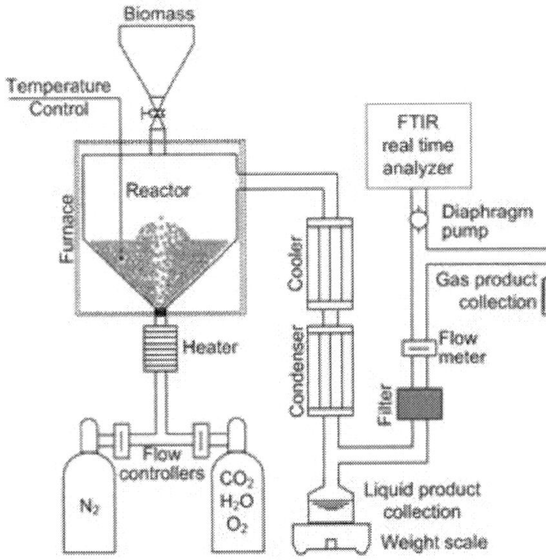

Fig. 1 Flow diagram of the designed biomass reactor

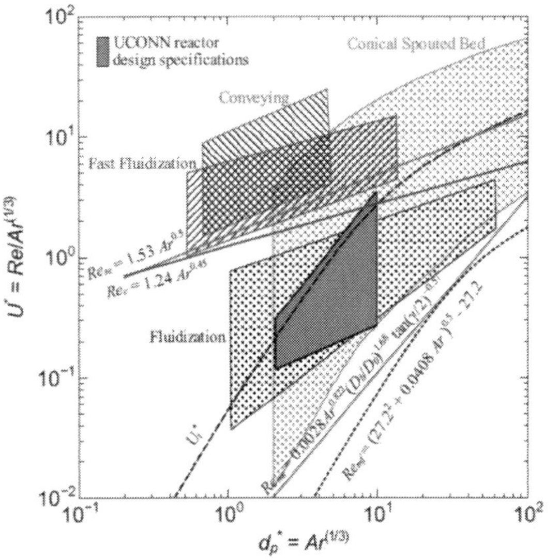

Fig. 2 Hydrodynamic flexibility of conical spouted-bed reactors

In this presentation, the applicabilityof literature correlations on conical spouted bed reactors and the range of hydrodynamicregimes in conical spouted-bed spouted conical fluidization will be shown.Finally results of the biomass catalytic reactor initial testing and acomparison of biomass pyrolysis selectivity under different catalyst to biomassweight ratios, heating rates, and temperatures will be presented.

See more of this Session: Poster Session of CRE Division
See more of this Group/Topical: Catalysis and Reaction Engineering Division

Novel Catalytic Materials for Oxidative Coupling of Methane

Ranjita Ghose, Hyun Tae Hwang and Arvind Varma, School of Chemical Engineering, Purdue University, West Lafayette, IN

Oxidative Coupling of Methane (OCM) is an attractive alternative for the production of ethylene, as compared to current processes based on crude oil. Although low selectivity and yield have prevented development of a practical catalyst, mixed metal oxides with complex structures (for example, mixed perovskite with general formula $A_xA'_{1-x}B_yB'_{1-y}O_3$) have generated interest in recent years [1]. Solution Combustion Synthesis (SCS) is a one-step method for the synthesis of nanostructured complex metal oxides with tailored composition, phase, oxidation state, and surface areas, by variation of tunable synthesis parameters. This technique has been extensively developed and used in our laboratory for synthesis of such materials [2-4]. Our current research involves maximizing selectivity and yield of ethylene in OCM, using catalysts synthesized by the SCS technique, by systematic variation of the metal components and their ratios, and optimization of reaction conditions.

References

1. Kondratenko, V. and Baerns, M., Oxidative coupling of methane, Chapter in *Handbook of Heterogeneous Catalysis*, Wiley-VCH Verlag, **2008**.
2. Erri, P., Dinka P. and Varma, A., *Chem. Eng. Sci.* **2006**, 61, 5328.
3. Deshpande, K., Mukasyan, A. and Varma, A., *J. Power Sources*, **2006**, 158, 60.
4. Erri, P. and Varma A., *Ind. Eng. Chem. Res.* **2007**, 46, 8597.

Selective Oxidation of Glycerol to Dihdroxyacetone in a Trickle-Bed Reactor

Gregory Honda, Wenbin Hu, Wesley Fleming, and Arvind Varma

School of Chemical Engineering, Purdue University
480 Stadium Mall Drive, West Lafayette, IN 47907

The economic viability of biodiesel production can be enhanced by conversion of the primary byproduct glycerol to the higher value chemical dihydroxyacetone (DHA). This can be accomplished by selective oxidation of the secondary hydroxyl group of glycerol by using a 3%Pt, 0.6%Bi on activated carbon catalyst, which has achieved high yield (>40%) in a semi-batch reactor [1]. To increase production, a trickle-bed reactor has been developed with cocurrent downflow of oxygen and aqueous glycerol. Feed solution pH and concentration as well as reactor temperature and pressure have been optimized. The impact of gas and liquid flowrates on yield has also been investigated, with additional consideration given to the influence of the complex hydrodynamics of trickle bed reactors. The development of a continuous-flow process for conversion of glycerol to DHA bolsters the economic capability of biodiesel production, while additionally providing insight into the behavior of trickle-bed reactors.

[1] Hu, W., Knight, D., Lowry, B., Varma, A. Selective Oxidation of Glycerol to Dihydroxyacetone over Pt-Bi/C Catalyst: Optimization of Catalyst and Reaction Conditions. *Ind. Eng. Chem. Res.* 2010, **49** (21), 10876-10882.

(629al) Long CARBON Chain ESTERS FROM Babassu Biodiesel USING Mesoporous Silica AS CATALYST

Monica C G Albuquerque[1], Leandro S Fernandes[1], Maria S. Vale[1], Diana C S Azevedo[1] and **Celio L. Cavalcante Jr.**[2], (1)Universidade Federal do Cear, Fortaleza, CE, Brazil, (2)Chemical Engineering, Universidade Federal do Cear, Fortaleza, CE, Brazil

Long carbon chain esters can be used as lubricants, due to structural similarity with the petroleum derivative. These esters have been obtained using vegetable oil as feedstock and homogeneous catalysts based on pyridine and strong acids. However, such catalysts may not be adequate on industrial scale applications due to safety issues and also because they may require additional washing steps to eliminate reaction by-products and remaining catalyst material. In this study we evaluate the synthesis of long chain esters by transesterification of babassu methyl ester as source material using 2-ethyl-1-hexanol as the transesterifying alcohol and SBA-15 impregnated with CaO (16%wt) as catalyst. The supported catalyst was characterized by N_2 adsorption/desorption isotherms at 77K. The specific surface area was 500 m^2 /g, total pore volume of 0.3946 cc/g and average pore size of 3.14 nm. The reaction parameters were varied according to an experimental factorial design. In all experiments, the alcohol/oil molar ratio was 12:1 and the activation temperature of the catalyst was 783 K. The variable levels were: reaction temperature, 353K and 298K; catalyst concentration, 1.0 and 2.5 wt% relative to liquid phase mass; and magnetic stirring rate, 60 and 1000 rpm. The best performance was obtained at reaction temperature of 353 K, magnetic stirring of 1000 rpm and catalyst concentration of 2.5 wt%, showing conversion values in the order of 80%wt.

See more of this Session: Poster Session of CRE Division
See more of this Group/Topical: Catalysis and Reaction Engineering Division

(629am) Copper Based Plasmonic Catalyst for Efficient and Selective Epoxidation of Propene

Marimuthu Andiappan and Suljo Linic, Department of Chemical Engineering, University of Michigan, Ann Arbor, MI

Propylene oxide (PO) is a valuable commodity chemical to produce important products such as paints, cosmetics and adhesives, etc. It is commercially produced using liquid phase chlorohydrin or hydroperoxide routes. Each of these routes has inherent limitations owing to the undesired byproducts and high operating cost. Because of these limitations in the existing processes, the direct epoxidation of propylene to PO using a heterogeneous catalyst is the most desirable in view of environmental friendliness and operating costs, and it has been one of the most challenging tasks in catalysis. Additionally, the increasing energy demand in near future will force us to design chemical synthesis processes using renewable energy resources. Developing catalysts that can make use of renewable energy resources like solar energy to make desired valuable chemicals is of critical importance to meet this requirement.
In this contribution, we demonstrate that copper based catalyst exhibit high selectivity in epoxidation of propylene to form PO. We also demonstrate that copper nanostructure, characterized by strong plasmon resonance peak in the visible region, can be used to efficiently harvest visible light (with intensity on the order of solar intensity) and utilize it to drive propylene epoxidation at significantly lower temperature than the epoxidation under pure thermal conditions. Our current results show that copper can be used as highly selective and energy efficient catalyst for the direct epoxidation of propylene. These results gives an insight to design energy efficient catalytic processes with a significant fraction of energy input provided in the form of sunlight.

See more of this Session: Poster Session of CRE Division
See more of this Group/Topical: Catalysis and Reaction Engineering Division

HOMOGENEOUS CATALYZED ESTERIFICATION OF *N*-BUTYRIC ACID WITH ISOAMYL ALCOHOL: A KINETIC STUDY

Yohana Páez, Jairo Durán, Iván Darío Gil, Gerardo Rodríguez, Julio César Vargas, Álvaro Orjuela.
Grupo de Procesos Químicos y Bioquímicos, Departamento de Ingeniería Química y Ambiental, Universidad Nacional de Colombia, Bogotá D.C, Colombia.

KEYWORDS: Full factorial design, esterification, isoamyl butyrate, kinetics, homogeneous catalyst.

Isoamyl butyrate, a high-valued ester widely used in fragrances and flavors industries [1-3], has an intern demand of 200000 kg per year [4-5] and is commonly obtained from the esterification of isoamyl alcohol with n-butyric acid. Although that is the main way to obtain this ester, there are works that report other ways such as the condensation between alcohols and aldehydes[6]. In addition to chemical synthesis, some esters are obtained by extraction from natural sources and nowadays some researches are focused on getting esters with enzymatic catalysts [4]. A different way to obtain isoamyl esters is using lipases in organic solvents, which can increase its activity and get higher conversions [1, 4, 7].

Besides the economic potential of this ester, raw materials can be obtained from renewable resources. For instance, isoamyl alcohol may be obtained from separation of fusel oil, a by-product from the bioethanol industry. The quality and quantity of fusel oil generated during alcohol production depends on type and method of preparation of mash used for fermentation, condition and environment under which fermentation proceeds and method of removal of fusel oil [8]. It must be removed during rectification phase of alcohol production and the quantity obtained by a commercial plant could vary between 1 and 11 liter per 1000 liter of alcohol produced (absolute basis) [8].

Determination of the esterification kinetics is the first step necessary to consider isoamyl butyrate production process. In this work, the esterification kinetics experiments using p-toluenesulfonic acid (PTSA) as catalyst is studied.

Butyric acid used were analytical (99.9%) grade purchased from Merck KGaA; isoamyl alcohol was technical grade supplied by F&M Chemicals and p-toluensulfonic acid (PTSA) was supplied by Stephan Co. All the experiences were carried out in a laboratory-scale stirred batch reactor without water removal during the tests. The temperature of the reaction was reached using a water circulating bath where the temperature could be controlled and kept constant within an accuracy of ±0.1K. The stirrer was plate-type and the speed was set at 300 rpm. Furthermore, a total reflux condenser was used to avoid loss of volatile products. At a certain time, liquid samples about 1 ml were taken using a syringe and weighted (accuracy of the balance: ±0.001 g).

During the experimental procedure the catalyst had to be dissolved. The dilution was done in the alcohol, due to PTSA was less soluble in the acid; additional heating was required to increase solubility of the catalyst on alcohol. When the liquid mixture (catalyst and alcohol) in the reactor reached the adjusted temperature, the acid was added (t=0) and the stirrer was started.

In the experiments the acid concentration of the samples was determined using a gas chromatograph (PerkinElmer Clarus 500) equipped with a flame ionization detector (FID) and HP-1 column (polydimethyl siloxane phase, 50m, 0.32 mm ID, WCOT 1.05μm). The acid was the limit reactant, due to this, the concentration of the other compounds was determined from the acid concentration.

A full factorial experimental design (FUFE) based on three levels was used. The influence and interaction of three variables: alcohol/acid molar ratio (1:1, 1.1:1, 1.2:1), catalyst concentration (1, 1.5, 2% based on total mass) and temperature (333.15, 343.15, 353.15K) on the kinetics parameters and isoamyl butyrate yield were evaluated.

In this study the reaction of interest is the direct esterification of isoamyl alcohol and butyric acid to form isoamyl butyrate and water as follows:

$$C_4H_8O_2 + C_5H_{12}O \xrightleftharpoons{\text{Catalyst}} C_9H_{18}O_2 + H_2O$$

As was mentioned above three variables were studied, the catalyst loading, the molar ratio and temperature; mixing speed was not took into account due the homogeneous system, i.e., there are not mass transfer problems [9-10]. The best way to know the effect of catalyst loading on the reaction is with initial rate. **Error! Reference source not found.** shows that there is a direct relation between reaction rate and catalyst loading. As was expected the conversion increases with temperature. The molar ratio of Isoamyl alcohol to butyric acid in this work was close to an equimolar ratio, this effect was assessed among others with a catalyst loading of 2% w/w at 353.15 K. In Figure 2 and Figure 3 the effect of temperature and molar ratio respectively can be observed.

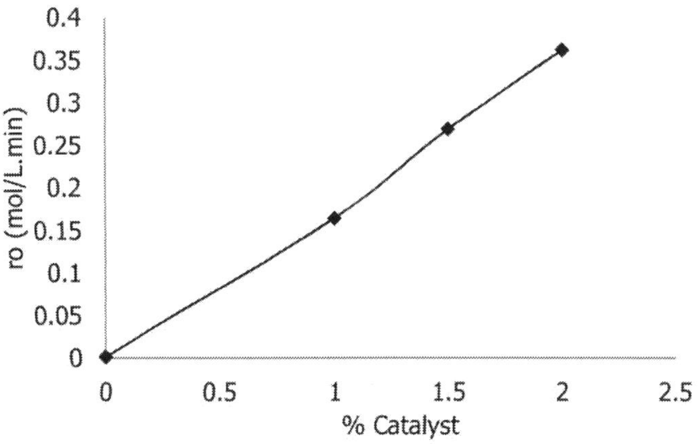

Figure 1. Influence of catalyst loading on initial rate

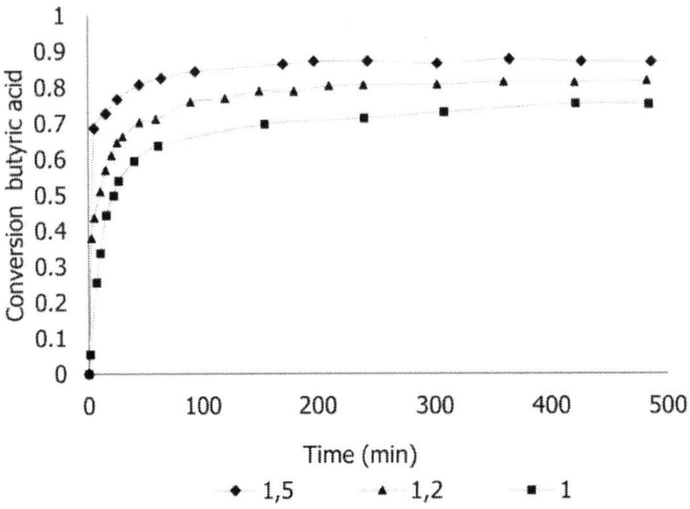

Figure 2. Effect of molar ratio on conversion (353.15K y 2% catalyst).

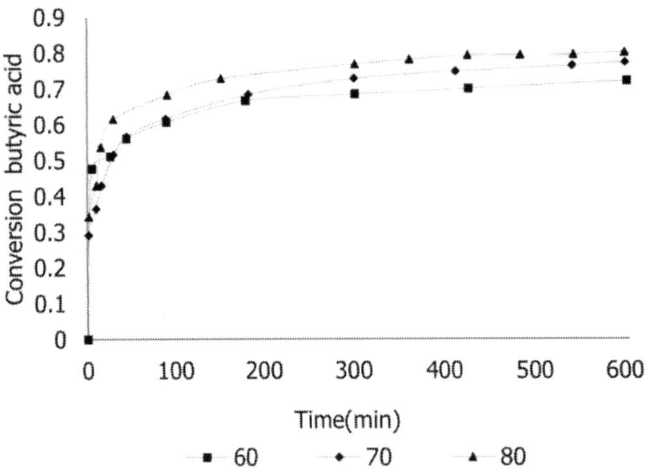

Figure 3. Effect of temperature on conversion of butyric acid (1% catalyst, RM=1.1:1).

The dependency of the reaction rate constant with the temperature was fitted according to Arrhenius equation and the experimental data were used for modeling a pseudo homogeneous kinetic model. The non-ideality of each species in the reacting mixture was considered by using the UNIFAC group contribution method. The activity model coefficients were estimated by using the commercial software ASPEN PROPERTIES®. ANOVA analysis was performed in order to identify the studied variables which have the major influence on fractional conversion. Temperature and catalyst concentration were found to have the major influence in conversion, whereas the molar feed ratio did not show relevant effect. The highest equilibrium conversion (81.6 %) was obtained at 353.15K, molar ratio of 1.2:1 and 2% wt% catalyst concentration. Otherwise, the lowest conversion (74.4 %) was obtained at 333.15K°C, molar ratio of 1:1 and 1% w/w catalyst concentration. Kinetic model parameters, energy of activation (E_a) and pre-exponential factor (A) of the Arrhenius expression were determined and fitted from the experimental data and they are displayed in Table 1.

E_a(J mol^{-1})	68080.25
A (L mol^{-1}min^{-1})	1.2 x 10^8

Table 1. Parameters adjusted for Arrhenius expression.

Conversion as a funtion of time obtained from experimental data and from the parameters adjusted for the model are showed in Figure 4.

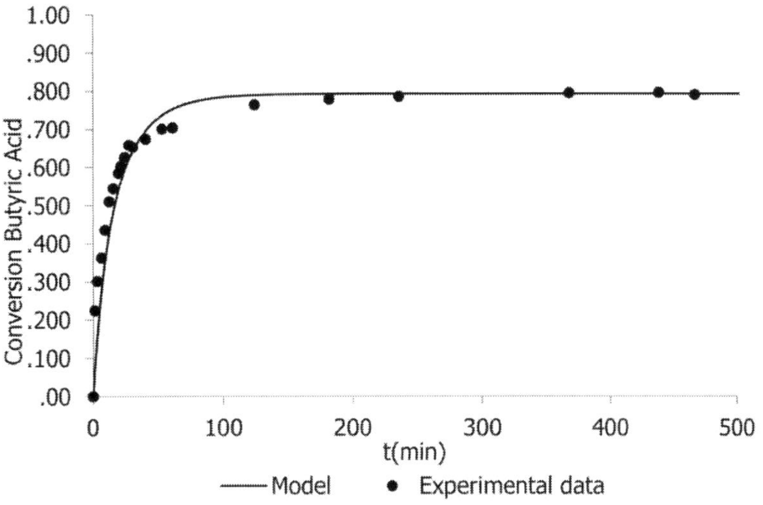

Figure 4. Comparison of Butyric Acid conversion between experimental data an model with estimated parameters

REFERENCES

1. **G.A. Macedo, G.M.P., M.I. Rodriguez, *Optimising the synthesis of isoamyl butyrate using Rhizopus sp. lipase with a central composite rotatable design.* Process Biochemistry, 2004. 39: p. 687-692.**
2. **A. Guvenc, N.K., H. Kapucu, O. Aydogan, U Mehmetogly, *Enzymatic esterification of isoamyl alcohol obtained from fusel oil: Optimization by response surface methodology.* Enzyme and Microbial Technology, 2007. 40: p. 778-785.**
3. **KUCUK Zeki , C.K., *Potential Utilization of Fusel Oil: A Kinetic Approach for Production of Fusel Oil Esters Through Chemical Reaction.* Turk J Chem, 1998. 22: p. 289-300.**
4. **S. Hari Krishna, S.P.a.N.K., *Enzymatic synthesis of isoamyl butyrate using immobilized Rhizomucor miehei lipase in non-aqueous media.* Journal of industrial Microbiology & Biotechnology, 2000. 25: p. 147-154.**

5. Secondini, O., *Handbook of perfumes and flavors.* 1990, New York: Chemical Publishing.
6. Rui Zhao, Y.D., Ahiguang Peng, Xiaolai Wang and Jisuan Suo, *One-Step syntehesis of isoamy butyrate from isoamyl alcohol and n-butyraldehide over TS-1 in air.* Catalys Letters, 2003. 87(81-83).
7. G. Langrand, C.T., J. Baratti, *Lipase catalyzed formation of flavor esters.* Biotechnology Letters, 1988. 10(8): p. 549-554.
8. A.G. Patil, S.M.K., H.D. Butala, *Fusel oil:composition, removal and potential utilization.* Int Sugar Jnl, 2002. 104(1238): p. 51-58.
9. Popken, T., L. Gotze, et al., *Reaction Kinetics and Chemical Equilibrium of Homogeneously and Heterogeneously Catalized Acetic Acid Esterification with Methyl Acetate Hydrolysis.* Industrial & Engineering Chemistry Research, 2000. 39(7): p. 2601-2611.
10. Vieville, C., Z. Moulongui, et al, *Esterification of oleic acid by methanol catalized b p-tolueensulfonic acid and the cation-exchange resins K2411 and K1481 in supercritical carbon dioxide.* Industrial & Engineering Chemistry Research, 1993. 32(9): p. 2065-2068.

(629ao) Deactivation and Regeneration of Commercial SCR Catalyst Used In a Coal-Fired Power Plant

Jung Bin Lee[1], Sung Nam Chun[1], Kwang Beom Hur[1], **Sang Mun Jeong**[2] and Jong Dae Lee[2], (1)Korea Electric Power Research Institute, Daejeon, South Korea, (2)Department of Chemical Engineering, Chungbuk National University, Cheongju Chungbuk, South Korea

The selective catalytic reduction (SCR) of nitric oxides with ammonia as reductant is one of the best-proven and world-wide used method for the removal of NO_X from stationary sources due to its efficiency, selectivity and economics. The De-NO_X process has to be installed to prevent the evolution of NO_X from the coal, oil and orimulsion power plants since the late 1990s in the Republic of Korea. V_2O_5-WO_3 as an active catalyst and TiO_2 as a high surface area support are main components of a commercial catalyst for the SCR unit. When the SCR catalyst is employed in the de-NOx unit, the catalyst would undergo the deactivation during the commercial operation.

Extensive research work has been carried out for better understanding about deactivation of the SCR catalyst during a commercial operation. Alkali metals such as potassium and sodium have been proposed as the major poisoning elements to accelerate the deactivation of SCR catalysts. Among the alkali metals, potassium is the most important element for the deactivation. It was well known that the deposition of the alkali metals can decrease both the number and the strength of Brønsted acid sites, which are largely responsible to the NO_X removal efficiency. Also SCR catalyst plugging can come from the deposition of ammonium sulfates and fly ash. Other reasons of the catalyst deactivation may be a loss of surface area for sintering and rutilization of titania after long-term operation at high temperature and poisoning by arsenic in the case of wet bottom boilers.

The deactivated SCR catalyst has to be regenerated for reuse in terms of process economics and environmental impact of heavy metals contained in the used catalyst. In the present work, the deactivation of a commercial SCR catalyst used in a coal-fired power plant has been investigated and its regeneration has been carried out. The SCR efficiency over the fresh, the deactivated and the regenerated catalysts has been determined at various operating conditions. The SCR catalysts have been characterized by a scanning electron microscope (SEM), a temperature-programmed desoprtion (TPD) and an X-ray diffractometer. An inductively coupled plasma-atomic emission spectrometer (ICP-AES) was used for elemental analyses of the catalyst and a washing solution.

The deactivated catalyst was regenerated by H_2SO_4 solutions with various concentrations (0.005 to 5 wt %). A large amount of deposits was observed on the deactivated catalyst, which resulted in a drastic decrease in the surface area of the deactivated catalyst. The poisoning elements (Ca, K, Na, and S) were detected in the deactivated catalyst. The catalytic activity was recovered by regenerating the catalyst with acidic solutions such as sulfuric acid, hydrochloric acid and so on. A high catalytic activity over the regenerated catalyst came from an increase in the surface sulfate, which could provide a formation of the polymeric vanadate and acid sites on the catalyst surface. Higher concentration of H_2SO_4 brought about dissolution of active material in the catalyst, which caused a decrease of the SCR efficiency and a loss of surface area. Based on the present experimental result, the pilot plant facility for the regeneration will be constructed and demonstrated for reuse of the commercial SCR catalyst.

See more of this Session: Poster Session of CRE Division
See more of this Group/Topical: Catalysis and Reaction Engineering Division

The effect of Zn-Ni-ferrite catalyst for HTS under LNG reformate condition

Myung Suk Lee[a], Joon Yeob Lee[a], Dae-Won Lee[b], Kwan-Young Lee[ac*]*

[a] Korea University, Seoul, Republic of Korea.

[b] Research Institute of Clean Chemical Engineering Systems, Seoul, Republic of Korea.

[C] GREEN SCHOOL, Seoul, Republic of Korea

Introduction

With the exhaustion of fossil fuel, the interesting of fuel cell systems has been increased as one of the energy sources for next generation. Among the several types of fuel cells that have been explored, the PEMFC (Polymer Electrolyte Membrane Fuel Cell) is thought to fit for a small-scale power source such as residential use, vehicle and small plants [1-3]. Hydrogen as source of PEMFC can be produced by taking the steam reforming (SR) of hydrocarbons such as natural gas and LPG. At the steam reforming process, CO is also generated which has to be removed less than 50 ppm because it damage to the Pt electrodes of the PEMFC [4]. The WGSR (Water Gas Shift Reaction) is well-known as the one of the CO reducing processes.

The WGSR has been conducted as two kinds of part: a high-temperature shift (HTS) operated at the temperature range 350-450°C to rapid CO reducing, a low temperature shift (LTS) operated at the temperature range 190-250°C to achieve high equilibrium CO conversion. In the WGSR processing of LNG reformate, Cu-Zn/Al_2O_3 and Fe-Cr-Cu catalysts are generally used as LTS and HTS commercial catalysts, respectively.

Chromium oxide (Cr_2O_3) was introduced to prevent Fe_3O_4 from sintering during the reaction [5]. Even if the substitution of Cr is very powerful to prevent from sintering, the level of about 1 wt% hexavalent chromium (Cr^{6+}) remains into the commercial catalyst and this heavy metal ion very toxic to humans and other organisms [6]. So, many researchers have studied to improve WGSR performance by introducing transition and inner transition metal such as Ni, Mo, Cu, Al, V, Mn, Co, Ce and Zn instead of Cr [7-11].

In this study, the Ni-ferrite catalyst was promoted by adding Zn to restrain methanation and increase WGSR activity. Ni-ferrite catalyst show an effective WGSR performance at 400°C which is similar with the activity of the commercial catalyst, but the significant quantity of CH_4 was produced [7,9,10]. The WGSR activities of the prepared catalysts were investigated under the LNG reformate condition.

Experimental

Catalyst preparation

All of catalysts were prepared by a coprecipitation method which refers to the recipe of Lee et al. [8]. Initially, metal precursors (nitrate hydrate form, Aldrich) were dissolved together in deionized water with appropriate atomic ratio to yield desired metal weight composition. Dissolved sodium hydroxide (Aldrich) in deionized water was added gradually dropwise to the aforementioned mixture solutions until forming a precipitate (pH 8.5) with vigorous stirring. The obtained products were further aged overnight. After filtering, the precipitate was washed several times with sufficient deionized water. The washed cakes were dried at 110°C for 24 h, and calcined at 500°C for 5 h. The rate of heating keeps always at 5 °C/min. The notation and metal composition of the prepared catalysts were described in in Table 1.

Reaction condition and activity test

The feed composition for the HTS reaction refers to previous studies on steam methane reforming (SMR) [9,12,13]. A gas composition of H_2 (56.7 vol%), CO (10 vol%), CO_2 (6.7 vol%) and H_2O (26.7 vol%) was injected into mixed zone. Each gas was controlled by using a mass flow controller (5850E, Brooks), except for steam, which was injected from syringe pump at 160°C. A sieved catalyst (150-250 μm) was packed over a quartz wool bed in the middle of a tubular stainless steel reactor (ID = 12.5 mm) and heated until 400°C and then mixture gases which were preheated at 160°C introduced in a reactor. Unreacted steam was removed from the product gases in a cold-trap (2°C). The outlet gases were analyzed by an on-line gas chromatograph (Acme 6000M, Young Lin). All of isothermal activity tests were performed under a weight hourly space velocity (WHSV) of 30,000 ml/g_{cat}hr.

Results

Catalyst texture properties (XRD, BET)

The XRD patterns of prepared catalysts were shown in Fig. 1. The XRD pattern of fresh F was represented Fe_2O_3, and then after reaction, it was changed to Fe_3O_4 due to the reductive gases of the reaction condition. The XRD pattern of fresh ZF also was seemed to Fe_2O_3, however, actually zinc ion preferred to occupy tetrahedral site in the spinel structure of Fe_3O_4 and when it was introduced enough amount to occupy tetrahedral site, the spinel structure was formed [14,15]. In our case, 5 wt% of Zn was used to substitute and this amount lacked for the formation of spinel structure as much as the detection by using XRD analysis and A. Khan group also reported similar results [11]. At lower Zn/Fe ratios, Zn appears to exist as an amorphous phase of $ZnFe_2O_4$, but it was not detected by XRD analysis [16]. Therefore, invers spinel structure of ZF was produced by reducing with inlet gases like F. In the XRD patterns of fresh NF and ZNF, it was confirm the formation of inverse spinel structure without crystalline phases related to nickel and zinc oxides. On the other hand, the powder XRD results of ZNF and NF catalysts after reaction revealed the appearance of peaks assignable to $FeNi_3$ alloy (JCPDS No. 71-8324). This newly formatted phase ($FeNi_3$) was made from Ni ions which were located in inverse spinel lattices during the reaction. The generated $FeNi_3$ caused a catalytic activity for CO or CO_2 methanation [7,17]

Lattice parameters could be calculated from the main peak of XRD data and the results were shown in Table 2. As substituted from Zn, the expansion of lattice parameters was checked which is attributed to the difference of ionic size of each ion because the larger ionic radius makes the packing of the ions to loose [14,15]. However, the substitution of Ni brought about an opposite phenomenon and it was agreed with the result of other researchers [9,15]. The both case of NF and ZNF showed the expansion of lattice parameters which was occurred from producing of $FeNi_3$ after reaction. Crystallite sizes which were calculated from Debye-Scherrer equation and the BET results were shown in Table 2. Generally, the crystallite sizes were increased after reaction except ZF.

Thermogravimetric analysis of CO

The profiles of CO-TGA for each catalyst were presented in Fig. 2. The results of CO-TGA data reflected the level of reduction of prepared catalysts for CO gas because the weight of samples was changed along with the reduction level. In addition, when its results were represented with the DTG (derivative thermogravimetric analysis) profile, it was definitely distinguished the degree of reduction. Generally, the weight change less than 150°C was thought removal of water which was adsorbed to catalysts or other organic compounds [18,19]. First weight change of F at 316°C appears the reduction of Fe_2O_3 and second change was occurred at 556°C which was the reduction of Fe_3O_4. The third weight reduction at 606°C was represented the reduction of FeO. In the CO-TGA profile of ZF, the three weight diminutions were found at 350, 548 and 600°C. The first reduction temperature means the reduction of Fe_2O_3 and it was shifted to high temperature approximately 50°C. The following weight changes at 548°C and 600°C mean the reduction of Fe_3O_4 and FeO, respectively. Interestingly, the reduction of zinc oxide was not found in the CO-TGA result of ZF and it was identified just slightly from 500°C in Fig. 3(a). This result indicates that zinc oxide can't be reduced by CO gas at least 500°C. In the CO-TGA data of NF, it was confirmed the addition effect of Ni into ferrite. The first weight change appeared the reduction of Fe_2O_3, and in the behind this change the second change was detected slightly around 410°C. This second peak was represented broadly and it was matched with the CO-TGA result of nickel oxide in Fig. 3(b). The following weight changes were the reductions of Fe_3O_4 and FeO and both reduction temperatures were similar with it of F and ZF. From the DTA profile of ZNF, it was found four peaks which were occurred at 378, 423, 560 and 609°C. The order of peak was same with NF's it. The first change means the reduction of Fe_2O_3 and then nickel oxide, Fe_3O_4 and FeO. In the case of NF catalyst, although it was shown the increased reducibility around 400°C, it was not much. However, by adding zinc together, the reducibility of ZNF was dramatically increased compared with NF's reducibility and especially it was remarkable around 400°C.

The catalytic activity for WGSR

The results of WGSR activity for each catalyst were shown in Fig. 4 with the ratio of methanation. The activity of F was measured even lower than the activity of commercial catalyst. Moreover, the activity of ZF showed quite low conversion even less than the activity of F. On the other hand, these catalysts did not produce the methane during the reaction. The commercial catalyst showed the stable and high conversion about 45% compared with F and ZF and also the commercial catalyst did not produce the methane. Although the total CO removal ratio of NF was pretty higher than the commercial's it (total CO removal ratio was not shown), the reaction of NF was simultaneously occurred with WGSR and methanation. If a methanation portion for CO removal of NF was except from total CO removal ratio, the CO conversion by WGSR was obtained and it showed the similar activity with the commercial catalyst. In Fig. 4, the CO conversions by WGSR and by methanation were represented separately. The reaction of ZNF also accompanied with WGSR and methanation at a time and turn up their CO conversions in Fig. 4, respectively. However, the ZNF catalyst manifested the quite improved performance compared with the WGSR activity of NF catalyst. Not only the CO conversion by WGSR of ZNF catalyst was increased fairly, but also the methanation level was lower than the NF catalyst's it. Moreover, the WGSR activity was very stable and almost arrived at the equilibrium conversion.

Conclusions

The Ni-ferrite catalyst was modified by adding Zn to improve their catalytic activity for WGSR and restrict the formation of methane. The NF catalyst shows a good performance for WGSR as much as commercial catalyst, but it accompanied with the formation of methane. The substitution of Zn into the

Ni-ferrite catalyst gave rise to the increased reducibility of catalyst around 400°C for CO gas. It was validated from the CO-TGA results of NF and ZNF catalysts. This increased reducibility for CO gas has something to do with the activity catalyst on WGSR. Moreover, the addition of Zn could reduce the formation of methane because it would interrupt with the formation of Ni cluster which is reason of methanation. Finally, the addition of Zn into Ni-ferrite showed the decreased amount of methane formation and the increased WGSR performance.

References

1. P. Panagiotopoulou and D. I. Kondarides (2004), "Effect of morphological characteristics of TiO_2-supported noble metal catalysts on their activity for the water–gas shift reaction," *J. catal.* ***225*** pp. 327-336.
2. P. Panagiotopoulou and D. I. Kondarides (2007), "A comparative study of the water-gas shift activity of Pt catalysts supported on single (MOx) and composite (MO_x/Al_2O_3, MO_x/TiO_2) metal oxide carriers," *Catal. Today* ***120*** pp. 319-329.
3. H. Iida and A. Igarashi (2006), "Characterization of a Pt/TiO_2 (rutile) catalyst for water gas shift reaction at low-temperature," *Appl. Catal., A* ***298*** pp. 152-160.
4. A.F. Ghenciu (2002), "R eview of fuel processing catalysts for hydrogen production in PEM fuel cell systems," *Curr. Opin. Solid State Mater. Sci.* ***6*** pp. 389-399.
5. C. Rhodes et al. (1995), "Water-gas shift reaction: finding the mechanistic boundary," *Catal. Today* ***23*** pp. 43-58.
6. K. Kochloefl et al. (1997), "Handbook of Heterogeneous Catalysis, vol. 4," *Wiley-VCH, Weinheim,* pp. 1831-1843.
7. K. Watanabe et al. (2009), "High temperature water–gas shift reaction over hollow Ni–Fe–Al oxide nano-composite catalysts prepared by the solution-spray plasma technique," *Catal. Commun.* ***10*** pp. 1952-1955.
8. J. Y. Lee et al. (2009), "Cr-free Fe-based metal oxide catalysts for high temperature water gas shift reaction of fuel processor using LPG," *Catal. Today* ***146*** pp. 260-264.
9. J. Y. Lee et al. (2011), "The CO removal performances of Cr-free Fe/Ni catalysts for high temperature WGSR under LNG reformate condition without additional steam," *Int. J. Hydrogen Energy* ***36*** pp. 8173-8180.
10. A. Khan et al. (2008), "Modified nano-crystalline ferrites for high-temperature WGS membrane reactor applications," *J. Catal.* ***253*** pp. 91-104.
11. A. Khan et al. (2008), "Relationship between temperature-programmed reduction profile and activity of modified ferrite-based catalysts for WGS reaction," *J. Mol. Catal. A: Chem.* ***280*** pp. 43-51.
12. Y.S. Oh et al. (2003), "A highly active catalyst, $Ni/Ce–ZrO_2/\theta-Al_2O_3$, for on-site H_2 generation by steam methane reforming: pretreatment effect," *Int. J. Hydrogen Energy* ***28*** pp. 1378-1392.
13. W.S. Dong et al. (2002), "Methane reforming over $Ni/Ce-ZrO_2$ catalysts: effect of nickel content," *Appl. Catal., A* ***226*** pp. 63-72.
14. L.M. Salah (2006), "Spectroscopic studies of the effect of addition of Y^{3+} on structural characteristics of Ni–Zn ferrites," *Phys. Status Solidi A* ***203*** pp. 271-281.
15. P. Priyadharsini et al. (2009), "Structural, spectroscopic and magnetic study of nanocrystalline Ni–Zn ferrites," *Mater. Chem. Phys.* ***116*** pp. 207-213.
16. E. Iglesia et al. (1998), "Structural and Catalytic Characterization of the Mechanism and Site

Requirements in Fe-Catalyzed Fischer-Tropsch Synthesis," *Technology Development for Iron and Cobalt Fischer-Tropsch Catalysts Quarterly Report* pp. 30-74.
17. W. Wang et al. (2002), "Partial oxidation of ethanol to hydrogen over Ni-Fe catalysts," *Catal. Lett.* ***81*** pp. 63-68.
18. T. Belin et al. (2002), "Influence of Grain Size, Oxygen Stoichiometry, and Synthesis Conditions on the γ-Fe_2O_3 Vacancies Ordering and Lattice Parameters," *J. Solid State Chem.* ***163*** pp. 459-465.
19. X. Bo et al. (2007), "Magnetic diphase nanostructure of $ZnFe_2O_4$/γ-Fe_2O_3," *J. Solid State Chem.* ***180*** pp. 1038-1044.

Tables

Table 1. The notation of the prepared catalysts and the weight ratio of each component

Notation	Description of the prepared catalyst	
	Intended weight ratio	Weight ratio from ICP analysis
F	Fe [100 wt%]	N.M.[a]
N	Ni [100 wt%]	N.M.[a]
Z	Zn [100 wt%]	N.M.[a]
ZF	Zn [5 wt%]/Fe	N.M.[a]
NF	Ni [34 wt%]/Fe	Ni [33.9 wt%]/Fe
ZNF	Zn [5 wt%]/Ni [32 wt%]/Fe	Zn [5.7 wt%]/Ni [31.8 wt%]/Fe

[a] Not measured.

Table 2. The results of BET and XRD data of prepared catalysts

Catalyst		BET analysis		XRD analysis	
		Specific Surface Area [m^2/g]	Pore Volume [cm^3/g]	Lattice parameter [Å]	Crystallite size [nm]
Before reaction (Fresh state)	F	40	0.205	N.M.[a]	27.6
	ZF	11	0.052	N.M.[a]	41.4
	NF	79	0.265	8.339	12.6
	ZNF	55	0.179	8.362	12.6
After	F	N.M.[a]	N.M.[a]	8.384	28.8

reaction	ZF	N.M.[a]	N.M.[a]	8.398	37.9
(3 h, 400°C)	NF	N.M.[a]	N.M.[a]	8.357	19.0
	ZNF	N.M.[a]	N.M.[a]	8.389	22.5

[a] Not measured.

Graphics or Photos

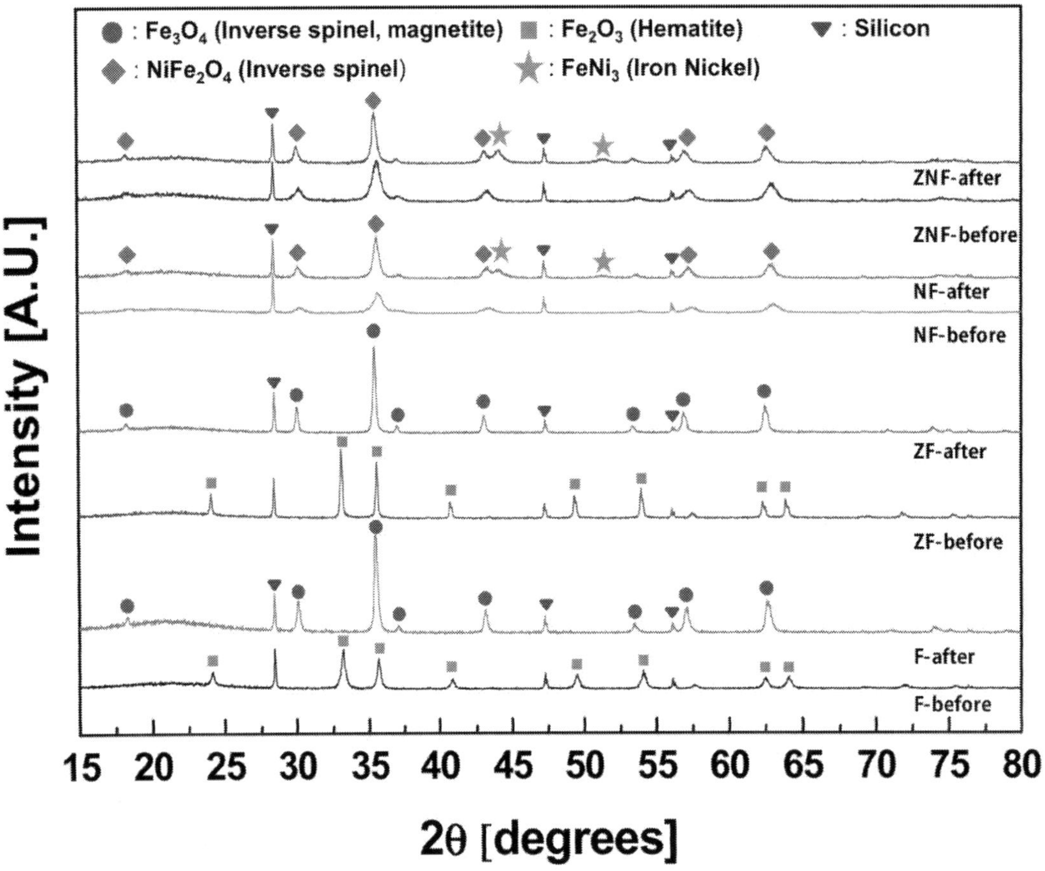

Figure 1. XRD patterns of the prepared catalysts (both after and before the HTS reaction at 400°C for 3 hr)

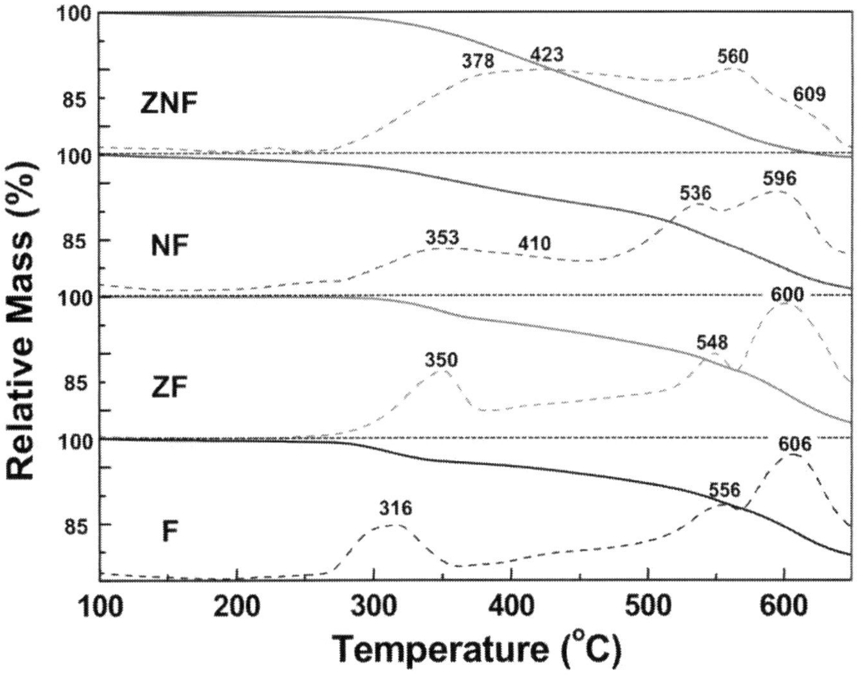

Figure 2. CO-TGA results of the prepared catalysts

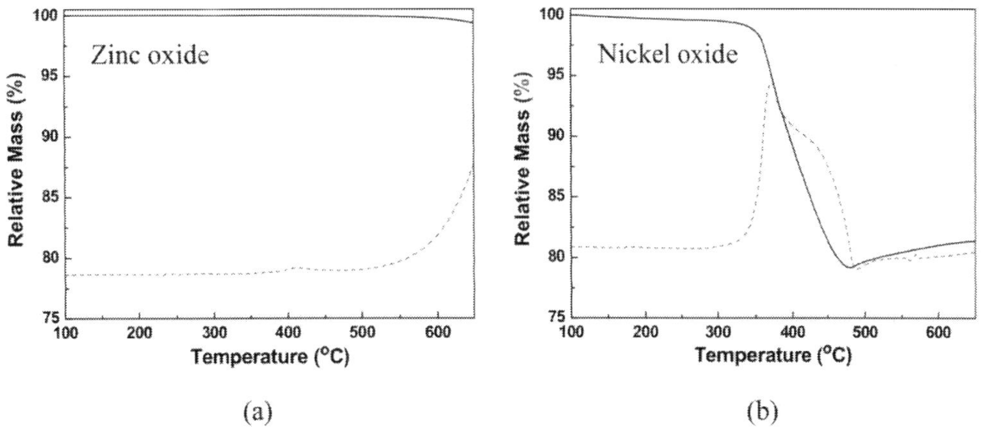

Figure 3. CO-TGA results of zinc oxide (a) and nickel oxide (b)

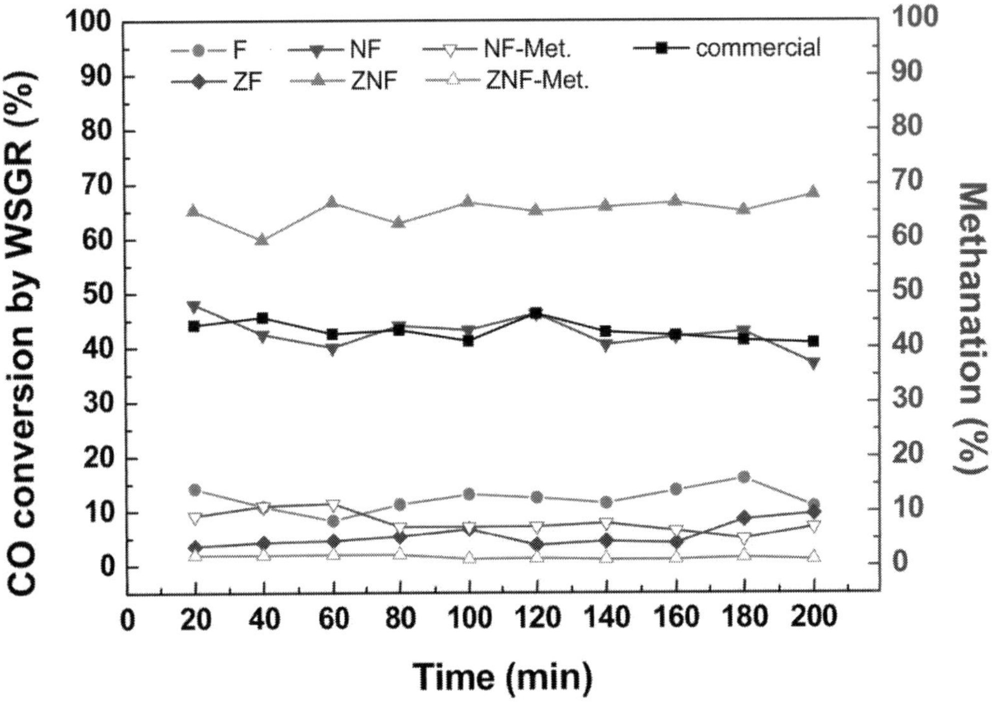

Figure 4. The HTS test results of preapered catalysts under LNG reformate condition (solid symbols for CO conversion and open ones for methanation): R=2, H_2 (56.7 vol%), CO (10 vol%), CO_2 (6.7 vol%) and H_2O (26.7 vol%); Temperature = 400°C; WHSV = 30,000 ml/g_{cat}hr

Two-nozzle flame synthesis of NO_x storage reduction catalysts

Robert Büchel[a,b], Sotiris E. Pratsinis[a], Alfons Baiker[b]

[a]*Particle Technology Laboratory, Department of Mechanical and Process Engineering, ETH Zurich, Sonneggstrasse 3, CH-8092 Zürich Switzerland.*

[b]*Institute for Chemical and Bioengineering, Department of Chemistry and Applied Biosciences, ETH Zurich, Wolfgang-Pauli-Strasse 10, CH-8093 Zurich, Switzerland.*

[1]*Currently with Satisloh Photonics AG, CH-8812 Horgen Switzerland*

Abstract

Catalysts for NO_x storage-reduction (NSR) were made selectively with Pt preferentially on the Al- or the Ba-components without altering significantly the Al_2O_3 or $BaCO_3$ crystal size, Al/Ba weight ratio, specific surface area, porosity, and Pt dispersion using a two-nozzle flame spray pyrolysis (FSP) unit. The NO_x storage performance at 300 °C was best for Pt located near Al_2O_3 as it facilitates the oxidation of NO to NO_2 during the fuel lean period but the reduction rate during the subsequent short fuel rich period was much slower resulting in incomplete regeneration. This contributed to a gradual decrease of the NO_x conversion at increasing cycling. In contrast, Pt on $BaCO_3$ resulted in an initially lower NO_x storage rate but during ten storage-reduction cycles a stable NO_x conversion of about 50% was reached. When using NO_2 instead of NO or higher NO_x oxidation-reduction temperatures (e.g. 350 °C) the Pt location did not affect the NSR performance of the $Pt/Ba/Al_2O_3$ catalysts.

Introduction

Anthropogenic NO_x is formed typically during fuel combustion contributing to smog formation and acid rain [1]. New catalysts are needed to meet stricter emission limits, especially to remove NO_x under oxygen rich conditions of lean fuel and direct injection engines. Among these catalysts, the ones for NO_x storage-reduction (NSR) can trap exhaust NO_x on an alkali- or alkaline earth metal (typically Ba or K) in the form of metal-nitrates [2] without requiring an additional reducing agent, as NH_3 or urea in selective catalytic reduction (SCR). During a subsequent fuel rich period the stored metal-nitrates are reduced to harmless N_2 over a noble metal (typically Pt or Pd) [3]. The storage and the noble metal are supported on a thermally stable carrier material, usually Al_2O_3 or CeO_2 [4]. Pt-free BaO/Al_2O_3 mainly stored NO_2, whereby releasing one NO for three NO_2 stored [5]. In the presence of Pt, NO is oxidized to NO_2 resulting in higher storage capacity and faster NO_x uptake [6], involving NO_2 spillover from the Pt to the Ba sites as proposed from simulations [7] and experimental investigations [8]. A schematic of the process is given in Fig. 1. These catalysts work in alternating cycles; mostly in fuel lean (in Fig. 1 yellow) and for regeneration in fuel rich (in Fig.1 black). In the fuel lean cycle, NO_x is stored on a storage material, typically Ba or K, in the form of a metal nitrate, here $Ba(NO_3)_2$.

In the subsequent fuel rich cycle, the storage material is regenerated and the released NO_x is reduced to N_2. This cycle is repeated continuously during regular operation of the catalyst. For each cycle the amount of NO_x converted can be calculated as: $NO_x\text{conversion} = \frac{NO_{x,in} - NO_{x,out}}{NO_{x,in}} \times 100\%$

Already in 1995 the proximity of Pt to the storage component of NSR catalysts was proposed to strongly influence the performance [9] and was recently investigated by comparing ternary $Pt/Ba/Al_2O_3$ catalysts with binary mixtures of Pt/Al_2O_3 and Ba/Al_2O_3 [10]. For Pt in close contact with Ba a 5-times higher isotopic exchange rate between ^{15}NO was measured, indicative of NO_x forward and reverse spillover from Pt to Ba [10].

In contrast to storage, the reduction (regeneration) is less understood mainly due to parallel occurring phenomena such as the desorption of gaseous NO_x at high temperature [11] and the reaction of the stored nitrate with CO_2 [12, 13]. Nova et al. [14] showed that the reduction of nitrates occurs by Pt-catalyzed surface reactions involving (reverse) spillover processes and concluded that a close proximity of Ba and Pt would be necessary to increase the Ba-nitrate decomposition.

Here, we investigated the importance of Pt being located close to Ba or to Al on NSR catalyst performance. The catalysts were prepared by a two-nozzle flame spray pyrolysis unit as already applied for NSR catalysts [15] with high storage capacity [16]. By this procedure, Pt can be deposited preferentially on Al_2O_3 and/or $BaCO_3$ without altering their structural properties.

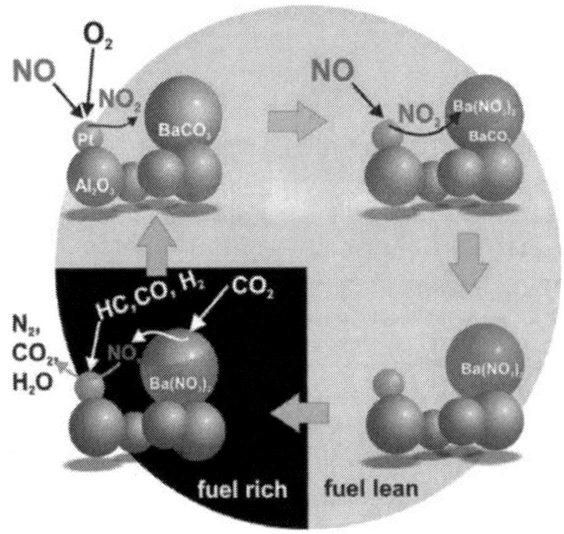

Figure 1:Working principle of the NO_x Storage and Reduction (NSR) catalysts. During fuel lean condition NO_x is stored in the form of metal-nitrates. In the subsequent fuel lean period the stored NO_x is released and reduced to harmless N_2.

Results

With a two nozzle flame spray pyrolysis unit $Pt/Ba/Al_2O_3$ catalysts with controlled deposition of Pt clusters on $BaCO_3$ or Al_2O_3 were prepared for studying the significance of Pt proximity to either Ba or Al in NO_x storage-reduction catalysts. In **Fig. 2** catalysts with Pt preferentially on the Al_2O_3 are shown in A) and with preferential deposition on $BaCO_3$ are shown in B).

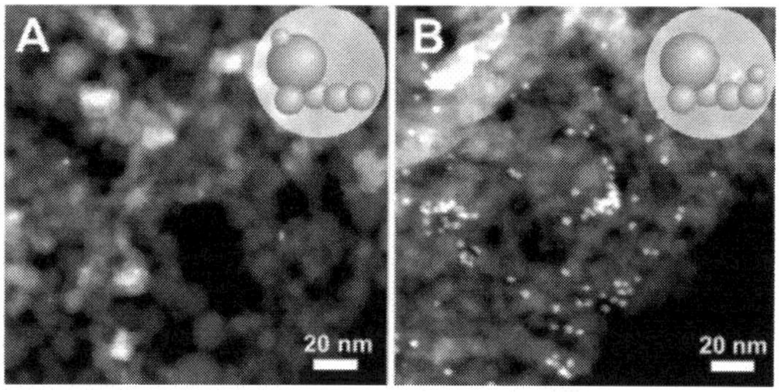

Figure 2: $Pt/Ba/Al_2O_3$ catalysts with Pt preferentially on Al_2O_3 in A) and Pt on $BaCO_3$ in B).

In **Fig.3** the beneficial role of a close interaction of Pt and Ba for NO_x uptake could be confirmed. However, Pt on Al_2O_3 exhibited a better NO oxidation activity, which was the limiting step for the overall NO storage process at low temperatures (300 °C). During reduction, Pt on Ba showed much better activity than Pt on Al_2O_3. This can be attributed to the importance of reverse spill-over and the promotional effect of Ba on the Pt reduction activity. Storage of NO and NO_2 until saturation revealed that the storage capacity is not limited by Ba but by the loss of Pt oxidation activity during catalyst restructuring. We concluded that the spill over is not the limiting step. [17]

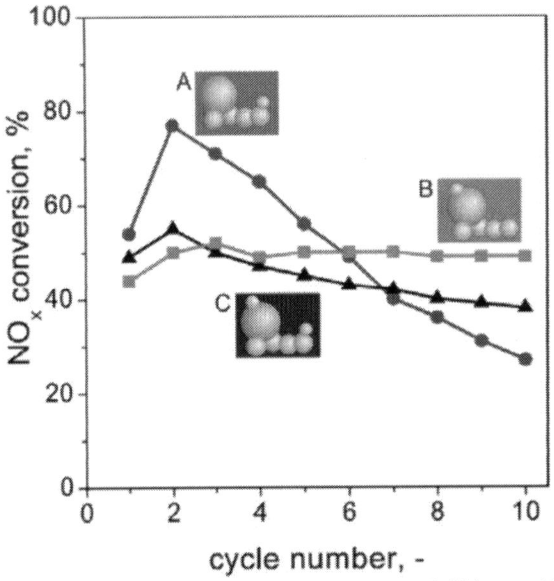

Figure 3: NO_x conversion for Pt preferentially located (A) on Al_2O_3 , (B) on $BaCO_3$ and (C) without preferential deposition. All three catalysts had the same material composition of Pt/Ba/Al_2O_3 the weight ratio as 1/20/100.

Acknowledgements

We kindly acknowledge financial support by ETH Zürich (TH-09 06-2) and thank the Electron Microscopy Center of ETH Zurich (EMEZ) for providing the necessary infrastructure. The contribution of platinum chemicals by Johnson Matthey PLC is greatly appreciated.

References

[1] J. Zeldovich, Acta Physicochimica URSS 21 (1946) 577.
[2] N. Takahashi, H. Shinjoh, T. Iijima, T. Suzuki, K. Yamazaki, K. Yokota, H. Suzuki, N. Miyoshi, S. Matsumoto, T. Tanizawa, T. Tanaka, S. Tateishi, K. Kasahara, Catal. Today 27 (1996) 63-69.
[3] W.S. Epling, L.E. Campbell, A. Yezerets, N.W. Currier, J.E. Parks, Cat. Rev. - Sci. Eng. 46 (2004) 163-245.
[4] A. Casapu, J.D. Grunwaldt, M. Maciejewski, F. Krumeich, A. Baiker, M. Wittrock, S. Eckhoff, Appl. Catal., B 78 (2008) 288-300.
[5] N.W. Cant, M.J. Patterson, Catal. Today 73 (2002) 271-278.
[6] B. Subramaniam, A. Varma, Chem. Eng. Commun. 20 (1983) 81-91.
[7] L. Olsson, H. Persson, E. Fridell, M. Skoglundh, B. Andersson, J. Phys. Chem. B 105 (2001) 6895-6906.
[8] Z.Q. Liu, J.A. Anderson, J. Catal. 224 (2004) 18-27.
[9] N. Miyoshi, S. Matsumoto, K. Katoh, T. Tanaka, J. Harada, N. Takahashi, K. Yokota, M. Sugiura, K. Kasahara, SAE Technical Paper 950809 (1995)
[10] N.W. Cant, I.O.Y. Liu, M.J. Patterson, J. Catal. 243 (2006) 309-317.
[11] K.S. Kabin, R.L. Muncrief, M.P. Harold, Catal. Today 96 (2004) 79-89.

[12] S. Balcon, C. Potvin, L. Salin, J.F. Tempere, G. Djega-Mariadassou, Catal. Lett. 60 (1999) 39-43.
[13] A. Amberntsson, H. Persson, P. Engstrom, B. Kasemo, Appl. Catal. B 31 (2001) 27-38.
[14] I. Nova, L. Lietti, L. Castoldi, E. Tronconi, P. Forzatti, J. Catal. 239 (2006) 244-254.
[15] R. Strobel, L. Madler, M. Piacentini, M. Maciejewski, A. Baiker, S.E. Pratsinis, Chem. Mater. 18 (2006) 2532-2537.
[16] M. Piacentini, R. Strobel, M. Maciejewski, S.E. Pratsinis, A. Baiker, J. Catal. 243 (2006) 43-56.
[17] R. Büchel, R. Strobel, F. Krumeich, A. Baiker, S.E. Pratsinis, J. Catal. 261 (2009) 201-207.

Methanol partial oxidation over palladium-, platinum-, and rhodium-integrated $LaMnO_3$ perovskites

Chia-Liang Li and Yu-Chuan Lin[*]

Department of Chemical Engineering and Materials Science
Yuan Ze Fuel Cell Center
Yuan Ze University, Chungli 32003 Taiwan

Keywords: Perovskite; noble metals; methanol; oxidation

[*] Corresponding author's Tel: (886) 3 463 8800 ext. 3554; Fax: (886) 3 455 9373; email: yclin@saturn.yzu.edu.tw

Abstract

This study investigates methanol partial oxidation over $LaMnO_3$ and $LaMn_{0.95}B_{0.05}O_3$ (B=Pd, Pt, and Rh) perovskites. Their bulk and surface properties were characterized by appropriate physiochemical techniques including BET surface area measurement, XRD, EDS, XPS, H_2-TPR, O_2-TPD, and CH_3OH-TPD. The characterizations showed significant differences of reducibility and oxygen activity between promoted and untainted $LaMnO_3$. Such deviants resulted in diverse catalytic behaviors in methanol partial oxidation. Promoted perovskites showed higher reactivity than pure $LaMnO_3$. Moreover, the redox chemistry was proposed to enhance partial oxidation route to formaldehyde while oxygen activity, particularly surface chemisorbed oxygen, played a central role in combustion. Accordingly, plausible reaction mechanism of methanol partial oxidation over $LaMnO_3$ based perovskites was elucidated.

1. Introduction

Perovskite (ABO_3) has drawn considerable attentions in the last few decades [1]. It is a versatile catalyst which can be applied in numerous fields such as CO oxidation [2], pollution abatement [3], and photocatalysis [4]. This is attributed to perovskite's unique crystalline structure, nonstoichiometry chemistry, and adsorption properties [1].

In some cases, perovskites show surpassing performances and stabilities compared to noble metal supported catalysts, e.g. automotive emission control [5-7]. A successful industrial practice has recently been announced by Daihatsu Motors, who implemented a series of platinum group metal (pgm) promoted perovskites in their exhaust converters [8-12]. The self-regeneration behavior of pgm-doped perovksites in oxidative and reductive environments provides them higher activity and longer durability than conventional noble metal catalysts. Such a breakthrough allows pgm-doped perovskites to be named as "intelligent catalysts [13]."

Another important application of perovskite is partial oxidation of hydrocarbons, especially light alkanes. Partial oxidation of methane is mainly used for syngas preparation [1, 14, 15]. The chemistry of perovskite's A-site and B-site elements is known to play a key role in varying catalytic performance and product distribution. Slagtern and Olsbye investigated B-site effects of $LaMO_3$ (M = Rh, Ni, and Co) and attributed the outstanding reactivity and durability to highly dispersed Rh ions [16]. Lago and coworkers studied A-site effects by a series of $LnCoO_3$ (Ln = La, Pr, Nd, Sm, and Gd). By increasing the lanthanum ionic radius, the activity and syngas selectivity can be improved [17]. Ethane partial oxidation (also called oxidative dehydrogenation) is a potential route to phase out steam cracking in ethylene production [18, 19]. Donsi and collaborators extensively worked on this field by $LaMnO_3$ [20-22] and lately by $Pt/LaMnO_3$ [23]. They proposed that perovskite is intrinsically active because ethyl species, a major intermediate of ethylene, can be formed on its surface [20]. Conner and coworkers investigated propane partial oxidation to acetaldehyde, acrolein, and methanol over complex perovskites, e.g., $Ba_2Bi_{2/3}[\ ]_{1/3}TeO_6$ [24]. Synergistic effects from A-site vacancies and B-sites cations were suggested to be indispensible for its catalytic performance.

Compared to alkanes partial oxidation, however, relatively fewer efforts have been dedicated to methanol partial oxidation (MPO) by perovskites [25, 26]. MPO can be applied to either formaldehyde (Equation 1) or hydrogen (Equation 2) production.

$$CH_3OH + 1/2O_2 \rightarrow HCOH + H_2O \quad \Delta H^o = -159 \text{ kJ/mol} \tag{1}$$

$$CH_3OH + 1/2O_2 \rightarrow H_2 + CO_2 \quad \Delta H^o = -192 \text{ kJ/mol} \tag{2}$$

De and Balasubramanian used $SrVO_3$ in MPO and showed that the lattice oxygen in perovskite's framework can assist methanol conversion [27]. The catalyst is highly selective to formaldehyde (90%) at moderate temperature (320 °C), but the activity is low (~44% conversion). Kuhn and Ozkan studied Sr- and Co-doped $LaFeO_3$ in MPO for hydrogen-rich gas production [25]. A detailed survey of oxygen transfer kinetics and how B-site elements mediate reaction pathways were documented [28-30]. Galenda and coworkers used two manners for $La_{4/5}Sr_{1/5}Ga_{4/5}Fe_{1/5}O_3$ perovskites synthesis and employed them in MPO [31]. Methanol combustion, partial oxidation, and decomposition could proceed simultaneously. In addition, perovskites with different synthesis methods can vary the product distribution. This is ascribed to the difference of catalytic surface caused by preparation [31]. Our group has lately reported that by doping trace palladium into $LaMO_3$ (M = Mn, Fe, and Co), attracting conversion (80% and up) and formaldehyde selectivity (90%) of MPO can be achieved [32]. Such an enhancement is correlated to the improved reducibility and oxygen mobility of Pd-doped perovskites, which promote formaldehyde yield through MPO and methanol dehydration.

The objective of this study is to investigate the effects of B-site-ion replacement by pgms (i.e., Pd, Pt, and Rh) in MPO. $LaMnO_3$ was selected as the host because of its attracting activity and oxygen-excess nature [32, 33]. Both physical and chemical properties of these catalysts were surveyed. The characteristics of perovskites were closely related to their catalytic behaviors. Moreover, an attempt was made to elucidate the reaction scheme under partial oxidation environments. This study is therefore a continuation of our previous work on lanthanum-transition metal perovksites for MPO.

2. Experimental

2.1. Catalyst preparation

A series of $LaMnO_3$ based perovskites, including $LaMnO_3$, $LaMn_{0.95}Pd_{0.05}O_3$, $LaMn_{0.95}Pt_{0.05}O_3$, and $LaMn_{0.95}Rh_{0.05}O_3$ were synthesized by sol-gel method as portrayed elsewhere [32, 34]. For convenience, $LaMnO_3$ and $LaMn_{0.95}B_{0.05}O_3$ (B = Pd, Pt, and Rh) will henceforth referred to as LaMn and LaMnB (e.g., LaMnPd), respectively. Stoichiometric proportions of metal acetylacetonates (acac) (La-, Mn-, Pd-, Pt-, and Rh-$(acac)_3$) were dissolved in a 1-to-4 volumetric ratio mixture of ethylene glycol and methanol (260 mL). Approximately 10 mmol of A-site and B-site metal ions were used. The solution was subjected to vigorous stirring for an hour. Subsequently, excess solvents were removed in a rotary evaporator at 80 °C. The remaining paste was dried in air at 150 °C for 10 hours using a heating rate of 10 °C/min. At last, the dried sample was calcined with the same heating rate from 150 to 700 °C for 2 hours.

2.2. Catalyst characterization

Surface area were estimated by the Brunauer, Emmet, and Teller (BET) method, using an automated N_2 adsorption apparatus (Micromeritics, ASAP 2010).

Powder X-ray diffraction (XRD) patterns were recorded by a Shimadzu Labx XRD-6000 using Cu-Kα radiation (0.15418 nm). Scans were performed using a scanning rate of 4°/min in 20-80° angle range (2θ). The voltage and current were set to 40 kV and 30 mA. Crystallite sizes were measured by employing the Scherrer relation. Instrumental broadening effect was found to be negligible by using the Warren's correction [35].

The surface elemental analysis was carried out by an energy dispersive spectrometer (EDS, Oxford ISIS 310) attached to the scanning electron microscopy (SEM, JEOL JSM-5600). The INCA X-sight software package (Oxford Instrument) was used to quantify surface compositions. Three trials with different locations were averaged for each sample. The acquisition time was 10 min per location. The SEM accelerating voltage was 10 kV. The scanned areas were about 0.12 mm * 0.09 mm.

Elements including O, La, Mn, Pd, Pt, and Rh were determined by EDS from their K-lines.

X-ray photoelectron spectroscopy (XPS) was conducted using a Thermo Scientific K-Alpha equipped with a 180° hemispherical sector analyzer and Al monochromator (Al Kα X-ray sources, 1486.6 eV). The X-ray spot size was 400 μm. The C 1s signal (284.4 eV) associated to adventitious carbon was employed for energy shift correction.

Hydrogen temperature programmed reduction (H_2-TPR), Oxygen temperature programmed desorption (O_2-TPD) and methanol temperature programmed desorption (CH_3OH-TPD) were all conducted in the same system described previously [32, 36]. The system consisted of a U-shaped cell, a temperature-controlled furnace, and a thermal conductivity detector (TCD). Approximately 100 mg of sample was used per trial. Before the experiment, each sample was pretreated in a Ar stream (30 mL/min) at 150 °C (10 °C/min) for 30 min to remove physisorbed water.

In H_2-TPR analysis, a 7.4 % H_2/Ar stream (25 mL/min) was used to reduce the sample by heating 5 °C/min up to 900 °C. O_2-TPD was operated in a He stream (25 mL/min) with a heating rate of 5 °C/min from 100 to 950 °C following by an isotherm at 950 °C for an hour. The amount of desorbed O_2 was quantified by executing the aforementioned program in a thermal gravimetric analyzer (TGA, Perkin-Elmer Pyris 1). The volatile portion was regarded as liberated O_2 during O_2-TPD. CH_3OH-TPD was performed by first saturating the sample with methanol at 60 °C, at which physisorbed methanol and oxidation reactions are suppressed [25]. Pulses of methanol vapor in a He stream was dosed repeatedly through a bubbler and a 6-port valve (VICI) design at room temperature. All tubing was heated to 100 °C to prevent methanol condensation. After cooling and attaining a stable TCD baseline level, CH_3OH-TPD was conducted from ambient temperature to 450 °C (5 °C/min) in a He stream. To inspect oxygen evolution of methanol-exposed perovskite, a series of O_2-TPD was carried out immediately after CH_3OH-TPD with the above-mentioned O_2-TPD program.

2.3. Catalytic activity evaluation

Catalytic performance was measured in a continuous fixed bed reactor (i.d. = 10 mm). The system outlet was connected inline to a gas chromatograph (GC, SRI 8610) equipped with a TCD, a methanizer, and a flame ionization detector (FID). Ultra-high purity He was used as the carrier gas. Two packed columns, 5 Å molecular sieve and Porapak Q, were employed to separate reactants and products. Prior to the experiment, the sample was pretreated in a 10% O_2/N_2 stream at 700 °C for an hour. Identified effluents included N_2, CO, CO_2, H_2, HCOH, and CH_3OH. Nitrogen was served as the internal standard. All products excluding water were calculated relative to GC calibration standards. The amount of water was estimated by the oxygen atom balance. Approximately 10 mg of catalyst with 40 to 80 mesh sizes was used per trial. To maintain an isothermal condition, the catalyst was diluted with 150 mg SiO_2 (40-80 mesh). SiO_2 was inert in each testing. A K-type thermocouple was placed in the middle of catalyst bed to record the reaction temperature.

Methanol was introduced into the system through a syringe pump (KDS-100) with a 1 mL/h injection rate. All tubing of the system was wrapped with heating tape and kept at 150 °C to circumvent condensation. Molar ratio of the feed for MPO was $CH_3OH/O_2/N_2$ = 11/5/84 with a fixed residence time of 0.4 g catalyst * h/mole of CH_3OH. To provide an insight view of MPO mechanism, a set of experiments of methanol decomposition (MD, Equations 3 and 4) and steam reforming (SRM, Equation 5) were conducted under the same reaction platforms as MPO. The feed's molar composition of MD was CH_3OH/N_2 = 11/89; SRM, $CH_3OH/H_2O/N_2$ = 6/10/84.

$$CH_3OH \rightarrow HCOH + H_2 \quad \Delta H^o = +84 \text{ kJ/mol} \tag{3}$$

$$CH_3OH \rightarrow CO + 2H_2 \quad \Delta H^o = +91 \text{ kJ/mol} \tag{4}$$

$$CH_3OH + H_2O \rightarrow CO_2 + 3H_2 \quad \Delta H^o = +50 \text{ kJ/mol} \tag{5}$$

All reported data had carbon and hydrogen atom mass balances within ± 10 % error, mostly less than ± 5%. Two to three trials were recorded after the first steady-state value (about 30 min after start-up of the inline GC system). These multiple trials were applied to estimate 95% confidence intervals. Conversion was calculated as mole of methanol reacted divided by mole of methanol injected; selectivity of product X_i was calculated as 100 * (mole of methanol converted to product X_i)/(mole of methanol converted).

3. Results and discussion

3.1. Physichemical characterizations

Figure 1 shows the XRD patterns of perovskite catalysts. The index reflections of LaMn ([JCPDS: 35-1353]) was clearly defined while diffractions of noble metal oxides were absent. That is, up to 5 mol% of Pd, Pt, or Rh could be implemented in the B-site position with preserved perovskite framework. Another possible explanation may be ascribed to the small cluster size of pgm oxides, which was below the detection limit (~4 nm). A close inspection of XRD patterns within 32.5-33.5^{o} showed that the peaks of precious metal-doped perovskites shifted slightly downward to lower 2θ degree, especially LaMnPt. This indicates the growth of d-spacing by increasing the B-site cation radius.

Table 1 presents surface areas, crystallite sizes, and elemental compositions obtained by EDS analysis. Positive effects were observed for the former two by doping Pd, Pt, and Rh ions into LaMn. Among them, LaMnPt showed the highest surface area and smallest particle size. It is probably because of Pt's ionic radius, which is the largest among noble metals used in this study. According to Goldschmidt's rule [37], LaMnPt should have the lowest tolerance factor, thereby yielding the highly disordered perovskite structure. This is in accord with the shift of XRD peak of LaMnPt shown in Fig. 1. Elemental analysis obtained by EDS showed Mn-to-La ratios were all less than unity, ranging from 0.83 to 0.86. Such a trend can be correlated to the surface enrichment of La. As for contents of noble metal ions, LaMnPd and LaMnPt were closely matched to their stoichiometrics while LaMnRh was about half of its designated value.

Figure 2 displays the XPS profiles of LaMnPd (Pd 3d), LaMnPt (Pt 4f), and LaMnRh (Rh 3d), respectively. For LaMnPd, two major peaks of oxidized Pd^{2+} at 336.7 (Pd $3d_{5/2}$) and 341.9 eV (Pd $3d_{3/2}$) were recognized [38, 39]. Signals of both metallic Pd (335.2 eV) [40] and high-oxidation-state Pd^{3+} and Pd^{4+} (339.0 eV) [38, 39] were insignificant. It is, therefore, greatly plausible that Pd ions exist in LaMnPd as +2 state. The Pt 4f spectrum showed a doublet at 74.5 and 77.7 eV, known as the characteristic binding energies of Pt^{+4} [40, 41]. The contributions of Pt^{+2} (72.9 eV)

[42] and Pt^0 (71.0 eV) [40] were negligible. That is, oxidic Pt species existed in LaMnPt as +4 state. As regards to Rh 3d profile, signals of Rh^{+3} ions appeared at 308.7 and 313.5 eV [43]. Metallic Rh (307.1 eV) [43] were undetectable. This emphasizes the presence of Rh^{+3} ions in LaMnRh. Detailed peak positions and surface elemental contents were summarized in Table 2. Note that surface compositions again displayed high concentrations of La, in agreement with EDS analysis.

Figure 3 illustrates the XPS spectra of O 1s state. By resorting to their deconvolution, two main peaks were identified at 529.0-529.6 eV and 530.5-531.2 eV. The former is associated to lattice oxygen ions in perovskite skeleton; the latter, chemisorbed oxygen [44, 45]. An additional peak identified at 533.2 eV on LaMnPt was assigned to adsorbed water [46, 47]. The presence of this peak may be attributed to the defective surface of LaMnPt. Generally, the shifts caused by noble metal substitution were insignificant, mainly within 1 eV. However, as listed in Table 2, promoted perovskites contained higher lattice oxygen ions than untainted LaMn. This indicates a more reducible surface for noble-metal-substituted perovskites [30].

Figure 4 compares the H_2-TPR results of LaMn based perovskites. Each sample displayed two separated peaks, suggesting a consecutive reduction process. It is generally accepted that a complete reduction of Mn ions to metallic Mn does not happen [48, 49]. Therefore, the first peak can be correlated to Mn^{+4} to Mn^{+3} while the second peak is associated with deep reduction of Mn^{3+} to Mn^{+2} [50]. Some reduction of Mn^{3+} to Mn^{+2} at low temperature region cannot be ruled out [51].

By doping Pd-, Pt, and Rh- ions into LaMn, both of these two peaks shifted to lower temperatures. The first reduction peak declined following the trend of LaMn (425 °C) > LaMnRh (322 °C) > LaMnPt (310 °C) > LaMnPd (224 °C); the second, LaMn (855 °C) > LaMnRh (833 °C) > LaMnPd (821 °C) ~ LaMnPt (820 °C). This is consistent with our XPS analysis of O 1s region, showing doped perovskites are more reducible than LaMn. A probable explanation may be due to the small crystallites and disordered lattice structure [52, 53], or hydrogen spillover [54-56]. It is worth mentioning that the reduction of PdO and Rh_2O_3, both occur at approximately 100 °C [54, 57], were not observed on LaMnPd and LaMnRh. It is also possible that PtO_2 reduction, usually occurs below 0 °C [58], was absence in LaMnPt. This implies well embedded Pd, Pt, or Rh ions in perovskite structure.

Figure 5 depicts the O_2-TPD profiles of freshly prepared samples. The onset temperatures of oxygen desorption increased with the order LaMnPt (495 °C) < LaMnPd (508 °C) ~ LaMn (509 °C) < LaMnRh (590 °C). Moreover, the sum of desorbed oxygen increased in the order of LaMn < LaMnRh < LaMnPd < LaMnPt, as summarized in Table 3. This suggests greater oxygen mobility and capacity of LaMnPt than the rest catalysts.

In general, the evolving profile can be categorized into low (α-O_2) and high (β-O_2) regions, using 700-800 °C as the demarcation [51, 59]. Herein we used 750 °C as the marker to divide these two. α-O_2 is known to be the weakly bonded oxygen species on the surface while β-O_2 is assigned to the lattice oxygen in perovskite skeleton. The former is proposed to be responsible for combustion; the latter, partial oxidation [60]. Table 3 lists the amounts of α- and β-O_2 estimated by O_2-TPD signals. Again, LaMn possessed the least of these two oxygen species while LaMnPt held the most. A close inspection showed quite close β-O_2 concentration for precious metal integrated perovskites. This may account for the B-site replacement, which improve the diffusivity of lattice oxygen from the bulk to the surface, thereby liberating more β-O_2 than LaMn [50].

Figure 6 illustrates the outcomes of CH_3OH-TPD. Two broad desorption peaks were identified at 50-150 °C and 175-325 °C. This corresponds well with earlier studies by Ozkan and collaborators [25, 30], who investigated key desorption species by resorting to their MS fragments. They referred the former to Lewis-bound methanol adducts [25, 30, 61]; the latter, oxidative products, e.g. CO_2 and H_2O [25, 30]. An additional signal was detected at 325-375 °C for LaMn. This indicates stronger chemisorbed species, possibly carbon oxides or formaldehyde [30]. The difference of CH_3OH-TPD patterns between promoted and un-promoted perovskites again revealed the varying chemistry caused by B-site ion substitution.

Figure 7 shows the O_2-TPD after CH_3OH-TPD and Table 3 summarizes the amount of liberated oxygen. This experiment is designed to evaluate the oxygen activity and storage of methanol-reduced perovskites. Compared to Fig. 5, the onset of TPD curve shifted downward to lower temperature. Moreover, irregular oxygen evolving profiles were observed. This suggests the increase of surface inhomogeneity after methanol oxidation (perovskite reduction). Significant growths of α-O_2

accompanying with decreases of β-O_2 were discovered. This indicates that the lattice oxygen (β-O_2) replenished surface chemisorbed oxygen (α-O_2), part of which was spent in methanol oxidation during CH_3OH-TPD [25]. Among the catalysts, LaMnPd liberated the highest amounts of α-, β-, and overall oxygen. Presumably, it should have the greatest oxygen mobility and availability in the reduced form caused by oxidation of chemisorbed methanol.

3.2. Catalytic performances

Figure 8 depicts the temperature-dependence of methanol and oxygen conversions. A strong, positive correlation between these two was acknowledged. The conversions increased steeply within narrow temperature intervals then steadily increased with growing temperature. The order of initiation temperature decreased followed the order: LaMn (260-310 °C) > LaMnPd (160-210 °C) ~ LaMnRh (160-180 °C). As for LaMnPt, the initiation interval was unclear, possibly at lower temperatures than the others. Oxygen was nearly depleted at 285 °C for LaMnRh, 335°C for LaMnPd, and 360 °C for LaMnPt and LaMn, respectively. At these points, methanol conversions declined as: LaMnPt (68%) > LaMnPd (56%) > LaMn (51%) ~ LaMnRh (50%). LaMnPt was the most active in low (< 175 °C) and high temperatures (> 300 °C). LaMn was the least reactive in almost all the trials.

The S-shaped dependence of methanol conversion on temperature implied a stepwise reaction pathway. In light of our feed's stoichiometry, a 5/11 molar ratio of O_2-to-CH_3OH indicates that a total consumption of oxygen could yield ~91% methanol conversion by following partial oxidation route (Equation 1). This is not the case because the maximum methanol conversion for each catalyst was less than 70%. Moreover, a detailed survey on ratios of spent O_2 divided by converted methanol results mainly in the range of 1.2 to 0.6 from 135 to 360 °C. In another word, different amount of O_2 was converted per mole of expensed CH_3OH, receding with increasing temperature.

Figure 9 illustrated product selectivities as a function of methanol conversion. To extrapolate these data, the selectivity pattern was divided into low (< 20%), moderate (20-45%), and high (> 45%) conversion regions. Grossly speaking, HCOH increased with enhancing methanol conversion. The highest HCOH selectivity (~22%) was achieved over LaMnPd at 60% conversion, where HCOH selectivity decreased as: LaMnPd > LaMnRh > LaMnPt > LaMn. In moderate conversion region, LaMnRh became the least selective to HCOH while the rest catalysts still followed the order. As regards to the low conversion region, trace HCOH was formed over each catalyst.

With increasing conversion, CO_2 declined over LaMnPt from 31 to 17% and LaMnPd from 21 to 12%. LaMn showed a reverse trend, increased from 12 to 19% then level off. An inverted U-shape curve was observed by LaMnRh. Its highest CO_2 selectivity was 28% at methanol conversion about 36%. Comparing CO profiles, LaMn produced greater CO than promoted perovskites, declining from 18 to 5%. Less than 2% CO was formed by LaMnPd, LaMnPt, and LaMnRh. Interestingly, LaMnPd and LaMnRh displayed an upward then downward trend. LaMnPt showed a constant CO selectivity (~0.5%) at high conversions.

Precious metal doped perovskites generated two to three folds higher H_2 than LaMn. Not until methanol conversion reached over 40% could LaMn produce detectable H_2. In contrast, LaMn yielded the highest H_2O in moderate and high conversion regions. At low methanol conversion, LaMnRh and LaMnPd showed higher H_2O selectivity than LaMn and LaMnPt. For promoted perovskites, differences of H_2 selectivities were insignificant at methanol conversion below 45% while H_2O selectivities were quite close at 50% and above.

Figure 10 shows methanol conversion and HCOH and CO selectivities as a function of temperature in oxygen-free environments. In the rise of temperature, precious metal incorporated perovskites displayed growing methanol conversions whereas LaMn was inert. All catalysts displayed less than 5% conversions. Among them, LaMnPd and LaMnRh were more active than LaMnPt. Identified products were HCOH, CO, and H_2 (data not shown). In all temperature range, HCOH dominated with trace CO at 300 °C and above. CO receded following the order of LaMnPd > LaMnPt > LaMnRh.

Figure 11 depicts the outcomes of methanol steam reforming. Again, LaMn was inactive while others displayed low activities. Activities of perovskites were very close with respect to their performances in methanol decomposition. The major products were HCOH and H_2; trace CO_2 was generated at high temperatures. According to Eq. 5, converting per mole of methanol through steam reforming should yield the mixture of CO_2 and H_2 to a 1/3 ratio. This suggests that methanol decomposition still dominated the system under steam reforming conditions, yielding HCOH and H_2 as primary products. In another word, methanol steam reforming was trivial under MPO conditions.

3.3. Reaction mechanism

Figure 12 presents possible reactions occur in methanol oxidation [25, 26, 47]. At low methanol conversions, oxygen is still plentiful at this stage. Therefore full or incomplete combustion of methanol should be prominent. This leads CO_2, CO, and H_2O to be the essential products. However, these two reactions evolved diversely over tested catalysts. LaMn exhibited almost two times higher CO than CO_2, suggesting incomplete combustion was favored. In contrast, LaMnPd, LaMnPt, and LaMnRh exhibited great amount of CO_2, the indicator of full combustion.

As methanol conversion increased, relatively less oxygen was available in moderate conversion region. According to the stoichiometrics, the system should gradually transit from combustions to partial oxidation. This is supported by growing HCOH and H_2 together with declining CO_2 and H_2O. Surprisingly, LaMnRh and LaMn exhibited increasing CO_2 in this regime. As shown in Fig. 12, CO_2 generation could be attributed to various reactions including methanol partial oxidation to CO_2 and H_2, HCOH oxidation, CO oxidation, water-gas shift, and steam reforming. Herein the effect of steam reforming could be ruled out. LaMn is known to be highly active in CO oxidation [62]. Moreover, a closely related molar ratio of (increased CO_2)/(lost CO), which nearly equals to unity, suggests that CO oxidation should be more important than the rest reactions. This is not the case for LaMnRh: significant low CO selectivity implied less CO could be converted to CO_2 via CO oxidation or water gas shift. Therefore, methanol and HCOH oxidation may be prominent for LaMnRh at this stage. Worth mentioning that methanol dehydrogenation may proceed under oxidative environments (shown in Fig. 10), thereby contributing certain HCOH and H_2.

In high methanol conversions, oxygen was limited or even depleted. HCOH and H_2 piled up while CO_2 and H_2O declined in this region. That is, contributions of partial oxidation and dehydrogenation of methanol became more and more important. It is interesting to see the deviant of CO selectivity between promoted perovskites: declining CO for LaMnPd and LaMnRh while constant CO for LaMnPt. Remembering the CO distribution pattern in Fig. 10, LaMnPd produced the highest CO under oxygen-free environments. This suggests appreciable CO oxidation reactivity of LaMnPd.

3.4. Influences of surface chemistry and oxygen activity

Based on the aforementioned results, relations between perovskite's physiochemical properties and catalytic behaviors can be corroborated. Substitution of Pt ions into $LaMnO_3$ results in a more defective surface than other perovskites. This leads LaMnPt to possess the highest surface area and lowest O_2 desorption temperature. Furthermore, the electronic state of Pt^{+4} induced a plethora of surface cations. This allows LaMnPt to host excess chemisorbed oxygen species, i.e. α-O_2, to fulfill the electroneutrality. α-O_2 is very active in combustion [50, 60], yielding the highest activity and CO_2 selectivity for LaMnPt in low reaction temperatures. Such a reactivity of α-O_2 can probably be sustained when O_2 is still available, resulting in great methanol conversions of LaMnPt at high temperatures.

The Pd^{2+} ion plays a different role in $LaMnO_3$. It substantially improves the reducibility of perovskite by H_2-TPR observations. The O_2-TPD after CH_3OH-TPD of LaMnPd also showed the lowest desorption temperature and greatest O_2 capacity among all catalysts. This emphasizes the reduction-oxidation (redox) chemistry of LaMnPd. The redox cycle (Equations 6 and 7) is widely accepted to be critical for HCOH production in methanol partial oxidation [63]. The great oxygen storage and mobility of methanol-reduced LaMnPd suggests that lattice oxygen can be efficiently transferred from bulk to the surface, which then participated in the redox cycle. Therefore substantial high selectivity of HCOH can be obtained over LaMnPd.

$$CH_3OH + \text{Catalyst-}O_x \rightarrow HCOH_{(g)} + H_2O_{(g)} + \text{Catalyst-}O_{x-1} \quad (6)$$

$$\text{Catalyst-O}_{x-1} + 1/2\ O_2 \rightarrow \text{Catalyst-O}_x \quad (7)$$

The Rh replaced $LaMnO_3$ seems to be the least reactive and reducible among promoted catalysts. Owing to the relatively close radius of Rh to Mn compared to the pairs of Pd to Mn and Pt to Mn, LaMnRh should have the highest Goldschmidt tolerance factor. In addition, the electronic state of Rh^{+3} equals to most Mn ions. This results in a more stable phase for LaMnRh than LaMnPt and LaMnPd.

4. Conclusions

Substitution trace Pd, Pt, and Rh ions into $LaMnO_3$ can significantly alter the physiochemical properties of perovskite. Improved catalyst reducibility, oxygen mobility, and oxygen storage can be achieved. All these factors allow promoted perovskites to have greater reactivity than untainted catalyst in MPO. Moreover, the reaction pathways can be directed variously.

Over pure $LaMnO_3$, methanol and CO combustions are major pathways with minor contributions of partial oxidation. LaMnPd shows surpassing reactivity of partial oxidation to HCOH compared to other catalysts. Such a trend is correlated to its redox property. Besides, LaMnPd also displays attracting reactivity in CO oxidation. LaMnPt mediates the reaction pathway mainly toward methanol combustion. This is ascribed to the great amount of combustion-prompt α-O_2 on its surface. LaMnRh seems to be the least reactive in promoted catalysts, possibly because of its low reducibility and oxygen activity.

The current work shows that it is possible to manipulate the reaction pathways of MPO with selected promoters in perovskites. Future advances in a molecular level understanding of surface chemistry and reaction mechanism should facilitate the design of $LaMnO_3$ based catalysts.

References

[1] M.A. Peña, J.L.G. Fierro, Chem. Rev. 101 (2001) 1981-2018.

[2] R. Chal, C. Gérardin, M. Bulut, S. van Donk, ChemCatChem 3 (2010) 67-81.

[3] J. Zhu, A. Thomas, Appl. Catal. B 92 (2009) 225-233.

[4] S.G. Ebbinghaus, H.P. Abicht, R. Dronskowski, T. Muller, A. Reller, A. Weidenkaff, Prog. Solid State Chem. 37 (2009) 173-205.

[5] R.J.H. Voorhoeve, J.P. Remeika, P.E. Freeland, B.T. Matthias, Science 177 (1972) 353-354.

[6] J.C. Schlatter, R.L. Klimisch, K.C. Taylor, Science 179 (1973) 798-800.

[7] C.H. Kim, G. Qi, K. Dahlberg, W. Li, Science 327 (2010) 1624-1627.

[8] Y. Nishihata, J. Mizuki, T. Akao, H. Tanaka, M. Uenishi, M. Kimura, T. Okamoto, N. Hamada, Nature 418 (2002) 164-167.

[9] H. Tanaka, Catal. Surv. Asia 9 (2005) 63-74.

[10] H. Tanaka, M. Taniguchi, M. Uenishi, N. Kajita, I. Tan, Y. Nishihata, J. Mizuki, K. Narita, M. Kimura, K. Kaneko, Angew. Chem. Int. Ed. 45 (2006) 5998-6002.

[11] H. Tanaka, M. Uenishi, M. Taniguchi, I. Tan, K. Narita, M. Kimura, K. Kaneko, Y. Nishihata, J. Mizuki, Catal. Today 117 (2006) 321-328.

[12] M. Taniguchi, H. Tanaka, M. Uenishi, I. Tan, Y. Nishihata, J. Mizuki, H. Suzuki, K. Narita, A. Hirai, M. Kimura, Top. Catal. 42-43 (2007) 367-371.

[13] M. Misono, Catal. Today 144 (2009) 285-291.

[14] Y.H. Hu, E. Ruckenstein, Adv. Catal., 2004, pp. 297-345.

[15] B. Christian Enger, R. Lødeng, A. Holmen, Appl Catal. A 346 (2008) 1-27.

[16] Å. Slagtern, U. Olsbye, Appl. Catal. A 110 (1994) 99-108.

[17] R. Lago, G. Bini, M.A. Pena, J.L.G. Fierro, J. Catal. 167 (1997) 198-209.

[18] A.S. Bodke, D.A. Olschki, L.D. Schmidt, E. Ranzi, Science 285 (1999) 712-715.

[19] L.D. Schmidt, J. Siddall, M. Bearden, AIChE J. 46 (2000) 1492-1495.

[20] F. Donsì, R. Pirone, G. Russo, J. Catal. 209 (2002) 51-61.

[21] F. Donsi, R. Pirone, G. Russo, Catal. Today 91-92 (2004) 285-288.

[22] F. Donsi, K.A. Williams, L.D. Schmidt, Ind. Eng. Chem. Res. 44 (2005) 3453-3470.

[23] S. Cimino, F. Donsì, G. Russo, D. Sanfilippo, Catal. Lett. 122 (2008) 228-237.

[24] W.C. Conner, S. Soled, A. Signorelli, Propane Oxidation over Mixed Metal Oxides: Perovskites, Trirutiles and Columbites, International Congress of Catalysis, Tokyo, Japan, 1981, pp. 1224-1238.

[25] J.N. Kuhn, U.S. Ozkan, J. Catal. 253 (2008) 200-211.

[26] K.L. Hohn, Y.C. Lin, ChemSusChem 2 (2009) 927.

[27] K.S. De, M.R. Balasubramanian, J. Catal. 81 (1983) 482-484.

[28] J.N. Kuhn, P.H. Matter, J.-M.M. Millet, R.B. Watson, U.S. Ozkan, J. Phys. Chem. C 112 (2008) 12468-12476.
[29] J. Kuhn, U. Ozkan, Catal. Lett. 121 (2008) 179-188.
[30] N. Lakshminarayanan, H. Choi, J.N. Kuhn, U.S. Ozkan, Appl. Catal. B 103 (2011) 318-325.
[31] A. Galenda, M.M. Natile, L. Nodari, A. Glisenti, Appl. Catal. B 97 (2010) 307-322.
[32] C.-L. Li, C.-L. Wang, Y.-C. Lin, Catalysis Today, in press, DOI: 10.1016/j.cattod.2011.01.038 (2011).
[33] T. Nitadori, S. Kurihara, M. Misono, J. Catal. 98 (1986) 221-228.
[34] S. Youichi, M. Tomohiko, J. Am. Ceram. Soc. 80 (1997) 2702-2704.
[35] B.E. Warren, J. Appl. Phys. 12 (1941) 375-384.
[36] C.-L. Li, Y.-C. Lin, Catal. Lett. 140 (2010) 69-76.
[37] V.M. Goldschmidt, S.N. Videnskaps-Akad, Mater.-Naturvidensk. KI. 8 (1926).
[38] M. Uenishi, M. Taniguchi, H. Tanaka, M. Kimura, Y. Nishihata, J. Mizuki, T. Kobayashi, Appl. Catal. B 57 (2005) 267-273.
[39] E. Tzimpilis, N. Moschoudis, M. Stoukides, P. Bekiaroglou, Appl. Catal. B 84 (2008) 607-615.
[40] D. Briggs, M.P. Seah, Practical Structure Analysis, 2nd ed., John Wiley, Chichester, 1993.
[41] J.P. Dacquin, M. Cabi, C.R. Henry, C. Lancelot, C. Dujardin, S.R. Raouf, P. Granger, J. Catal. 270 (2010) 299-309.
[42] J. Knecht, G. Stork, Fresenius' Journal of Analytical Chemistry 289 (1978) 206-206.
[43] H.J. Gysling, J.R. Monnier, G. Apai, J. Catal. 103 (1987) 407-418.
[44] N. Yamazoe, Y. Teraoka, T. Seiyama, Chem. Lett. 10 (1981) 1767-1770.
[45] M. Machkova, N. Brashkova, P. Ivanov, J.B. Carda, V. Kozhukharov, Appl. Surf. Sci. 119 (1997) 127-136.
[46] M.W. Roberts, Chem. Soc. Rev. 18 (1989) 451-475.
[47] G.I.N. Waterhouse, G.A. Bowmaker, J.B. Metson, Appl. Catal. A 265 (2004) 85-101.
[48] L. Lisi, G. Bagnasco, P. Ciambelli, S. De Rossi, P. Porta, G. Russo, M. Turco, J. Solid State Chem. 146 (1999) 176-183.
[49] S. Royer, H. Alamdari, D. Duprez, S. Kaliaguine, Appl Catal. B 58 (2005) 273-288.
[50] B. Levasseur, S. Kaliaguine, Appl. Catal. A 343 (2008) 29-38.
[51] H. Ziaei-Azad, A. Khodadadi, P. Esmaeilnejad-Ahranjani, Y. Mortazavi, Appl Catal. B 102 (2011) 62-70.

[52] L. Huang, M. Bassir, S. Kaliaguine, Appl. Surf. Sci. 243 (2005) 360-375.
[53] R.M. Navarro, M.C. Alvarez-Galvan, J.A. Villoria, I.D. González-Jiménez, F. Rosa, J.L.G. Fierro, Appl. Catal. B 73 (2007) 247-258.
[54] W.-J. Shen, M. Okumura, Y. Matsumura, M. Haruta, Appl. Catal. A 213 (2001) 225-232.
[55] K. Zhou, H. Chen, Q. Tian, Z. Hao, D. Shen, X. Xu, J. Mol. Catal. A 189 (2002) 225-232.
[56] L. Giebeler, D. Kießling, G. Wendt, Chem. Eng. Technol. 30 (2007) 889-894.
[57] H.F.J. van 't Blik, R. Prins, J. Catal. 97 (1986) 188-199.
[58] A. Jones, B.D. McNicol, Temperature-programmed reduction for solid materials characterization, Marcel Dekker, INC., New York, 1986.
[59] R. Zhang, H. Alamdari, S. Kaliaguine, J. Catal. 242 (2006) 241-253.
[60] T. Seiyama, N. Yamazoe, K. Eguchi, Ind. Eng. Chem. Prod. Res. Dev. 24 (1985) 19-27.
[61] L.J. Burcham, L.E. Briand, I.E. Wachs, Langmuir 17 (2001) 6164-6174.
[62] S. Cimino, S. Colonna, S. De Rossi, M. Faticanti, L. Lisi, I. Pettiti, P. Porta, J. Catal. 205 (2002) 309-317.
[63] C.N. Satterfield, Heterogeneous Catalysis in Industrial Practice, 2nd ed., McGraw-Hill, New York, 1991.

Table 1. Physicochemical properties of catalysts used in this study

Catalyst	S_{BET} (m^2/g)	d_p (nm)	La[a]	Mn[a]	Pd[a]	Pt[a]	Rh[a]	O[a]	Mn/La	Pd/La	Pt/La	Rh/La
LaMn	6.3	16.6	33.6	29.0				37.4	0.86			
LaMnPd	14.1	16.1	35.4	29.4	1.8			33.4	0.83	0.05		
LaMnPt	14.5	15.8	33.8	28.7		1.3		36.2	0.85		0.04	
LaMnRh	11.1	16.2	36.9	30.5			0.6	32.0	0.83			0.02

[a] Obtained by EDS analysis.

Table 2. Surface compositions estimated based on XPS analysis

	LaMn		LaMnPd		LaMnPt		LaMnRh	
	Position (eV)	Surface content (%)	Position (eV)	Surface content (%)	Position (eV)	Surface content (%)	Position (eV)	Surface content (%)
La 3d	833.8	22.8	833.9	25.2	833.3	19.6	833.4	23.9
	837.5	24.5	837.4	23.0	836.9	23.4	837.2	24.6
La total		47.3		48.2		43.0		48.5
Mn 3p	642.1	13.7	642.0	12.2	642.0	13.3	641.2	13.8
	653.3	9.0	653.4	7.5	653.0	9.9	652.7	7.1
Mn total		22.7		19.7		23.2		20.9
Pd 3d			336.7	1.2				
			341.9	1.3				
Pd total				2.5				
Pt 4f					74.5	0.8		
					77.7	-		
Pt total						0.8		
Rh 3d							308.7	1.7
							313.5	1.1
Rh total								2.8
O 1s	529.0	12.0	528.9	14.7	529.6	15.1	529.2	12.5
	530.4	18.0	530.7	14.9	531.2	11.2	530.9	15.3
					533.2	6.7		
O total		30.0		29.6		33.0		27.8

Table 3. Oxygen desorbed from freshly prepared and post CH_3OH-TPD LaMn based perovskites

Catalyst	Amount of oxygen desorbed (μmol/g catalyst)					
	Freshly prepared			Post CH_3OH-TPD		
	α-O_2	β-O_2	Total	α-O_2	β-O_2	Total
LaMn	530	4923	5451	1577	2377	3954
LaMnPd	878	6221	7099	3028	2585	5613
LaMnPt	1112	6375	7487	2434	2417	4851
LaMnRh	665	6335	7000	3024	2093	5117

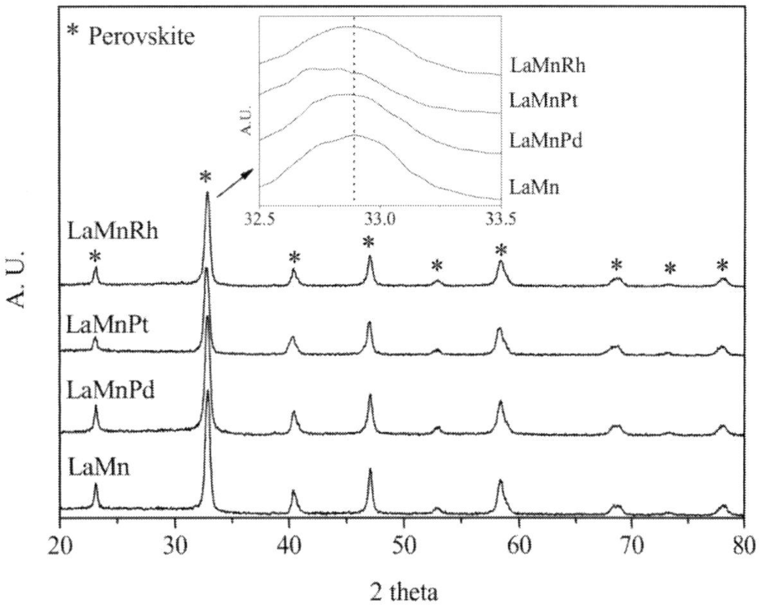

Fig. 1. XRD patterns of LaMn based perovskites.

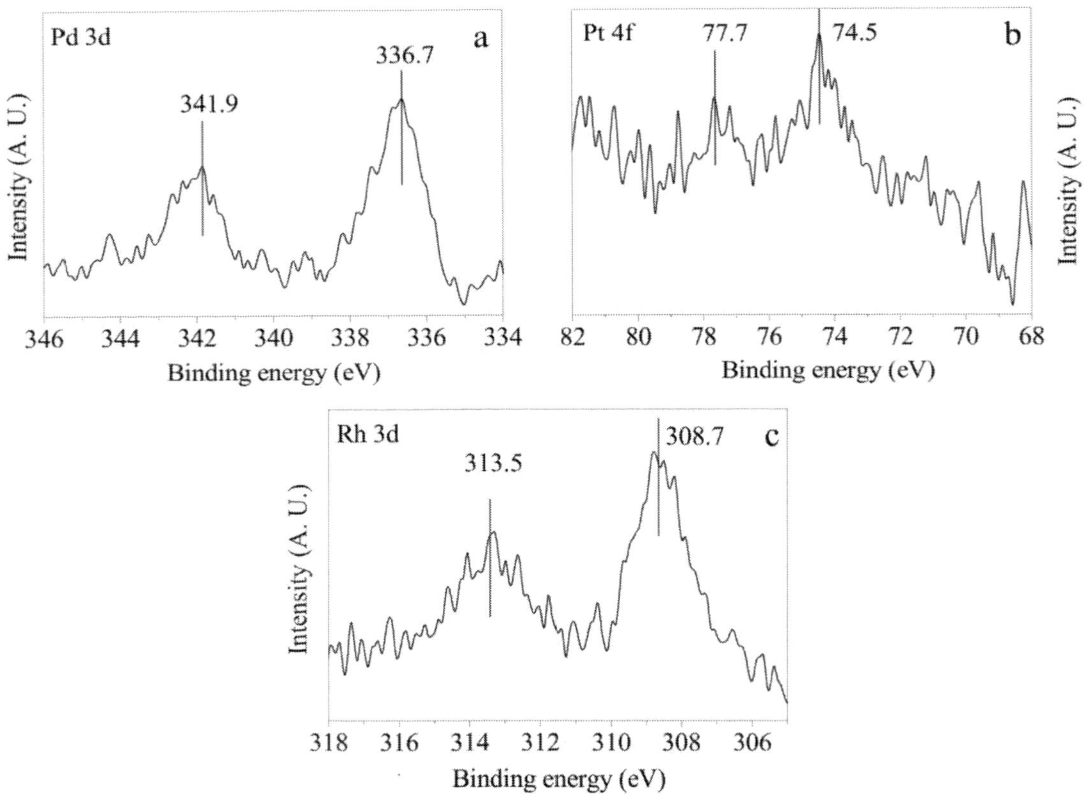

Fig. 2. X-ray photoelectron spectra of (a) Pd 3d, (b) Pt 4f, and (c) Rh 3d of LaMnPd, LaMnPt, and LaMnRh, respectively.

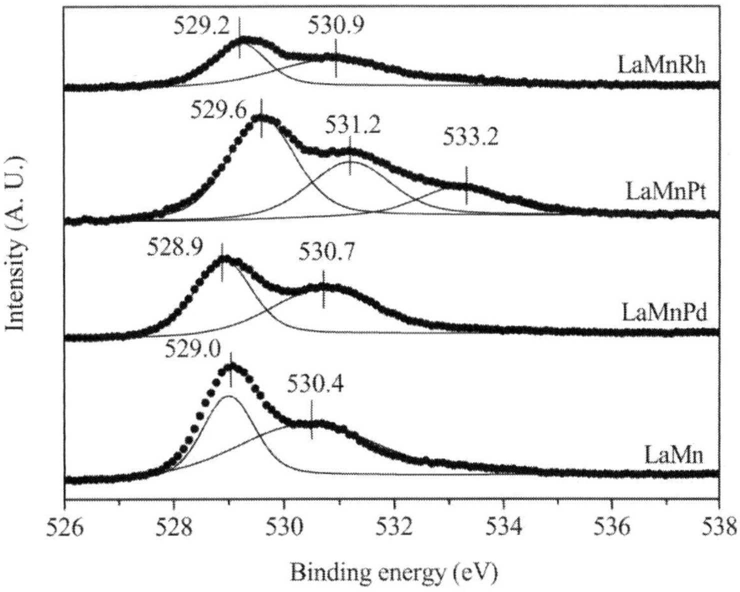

Fig. 3. O 1s region in X-ray photoelectron spectra of LaMn based perovskites.

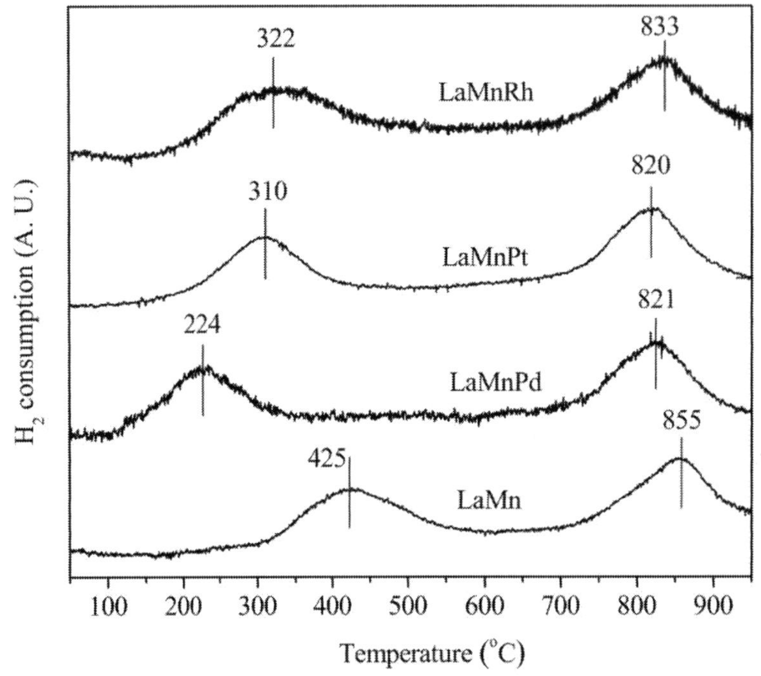

Fig. 4. H_2-TPR of LaMn based perovskites.

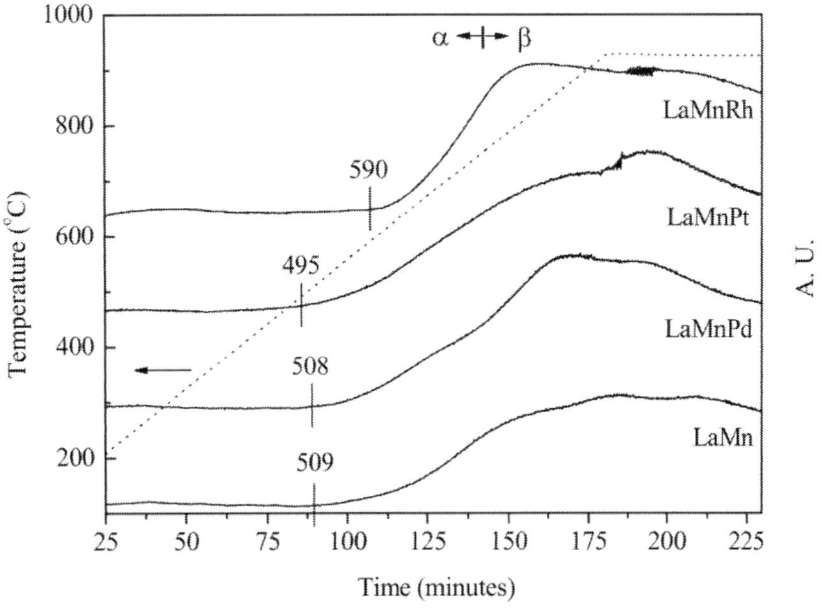

Fig. 5. O_2-TPD of LaMn based perovskites.

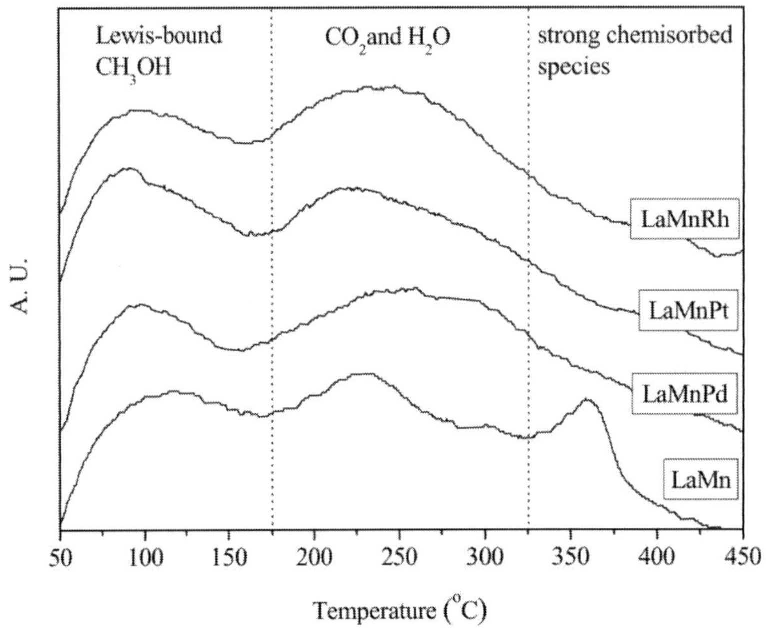

Fig. 6. CH_3OH-TPD of LaMn based perovskites.

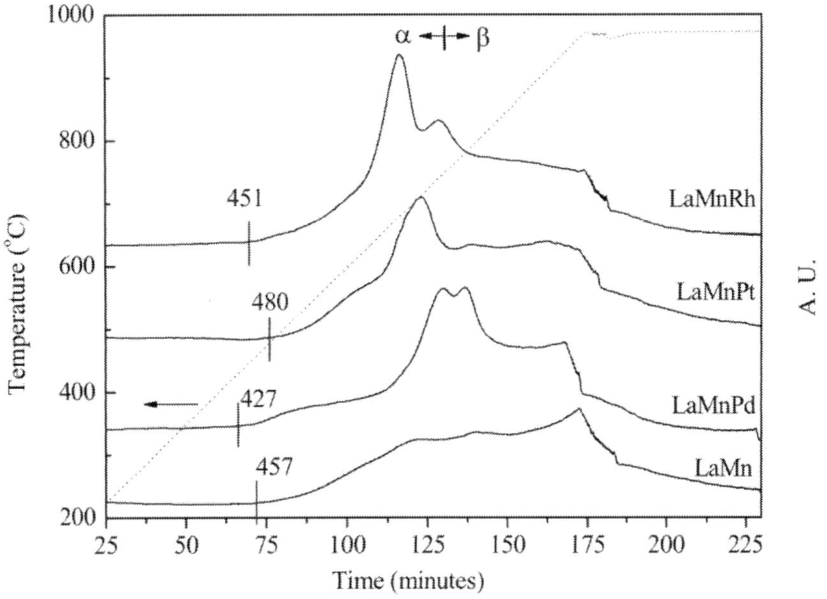

Fig. 7. O_2-TPD of LaMn based perovksites following CH_3OH-TPD.

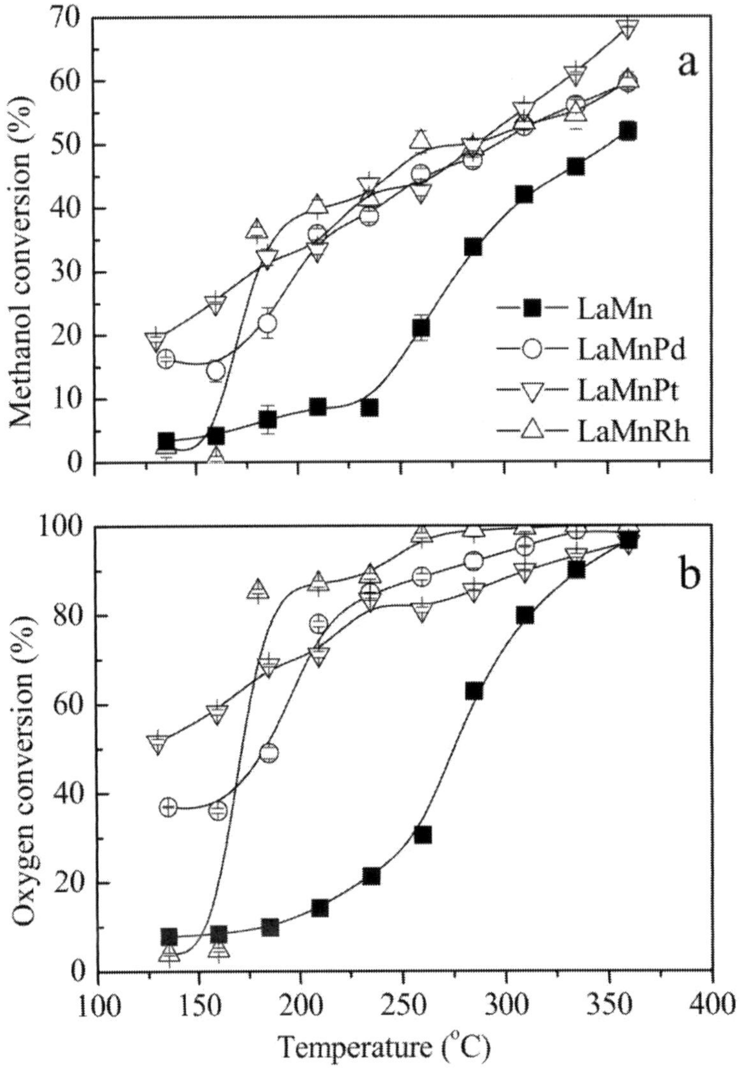

Fig. 8. (a) Methanol and (b) oxygen conversions as a function of temperature.

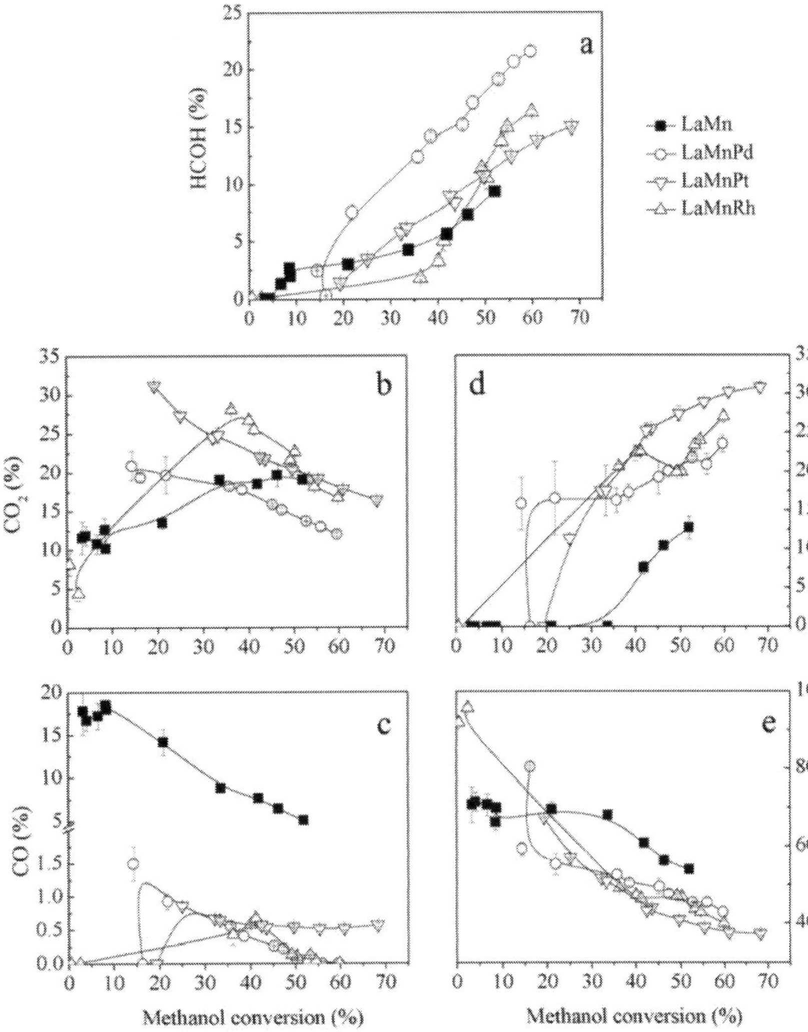

Fig. 9. Product selectivity of (a) HCOH, (b) CO_2, (c) CO, (d) H_2, and (e) H_2O as a function of methanol conversion.

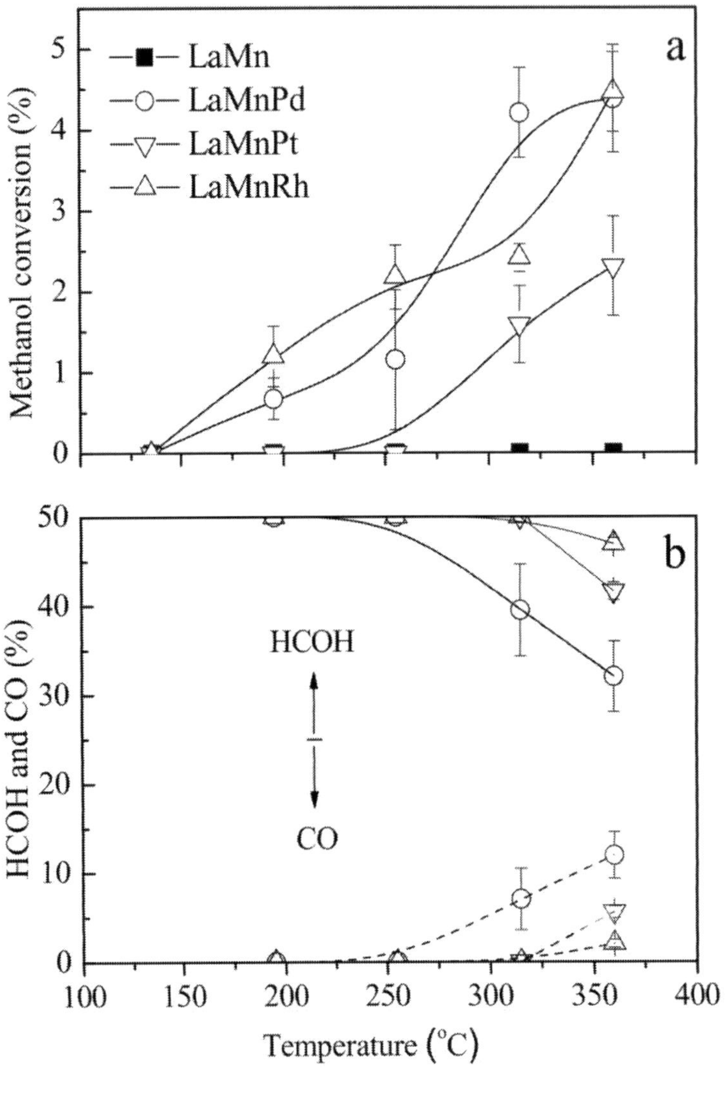

Fig. 10. (a) Methanol conversion and (b) HCOH (solid) and CO (dash) selectivities as a function of temperature in the absence of oxygen. Note that H_2 selectivity was excluded for clarity.

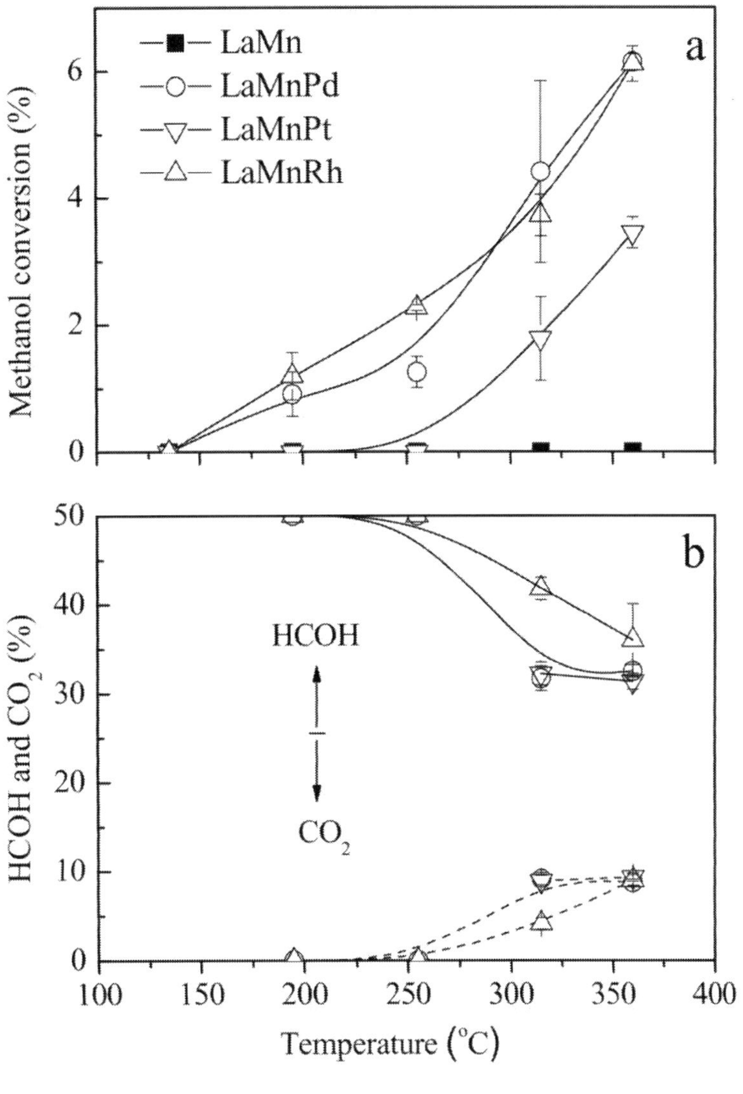

Fig. 11. (a) Methanol conversion and (b) HCOH (solid) and CO_2 (dash) selectivities as a function of temperature in methanol steam reforming. Note that H_2 selectivity was excluded for clarity.

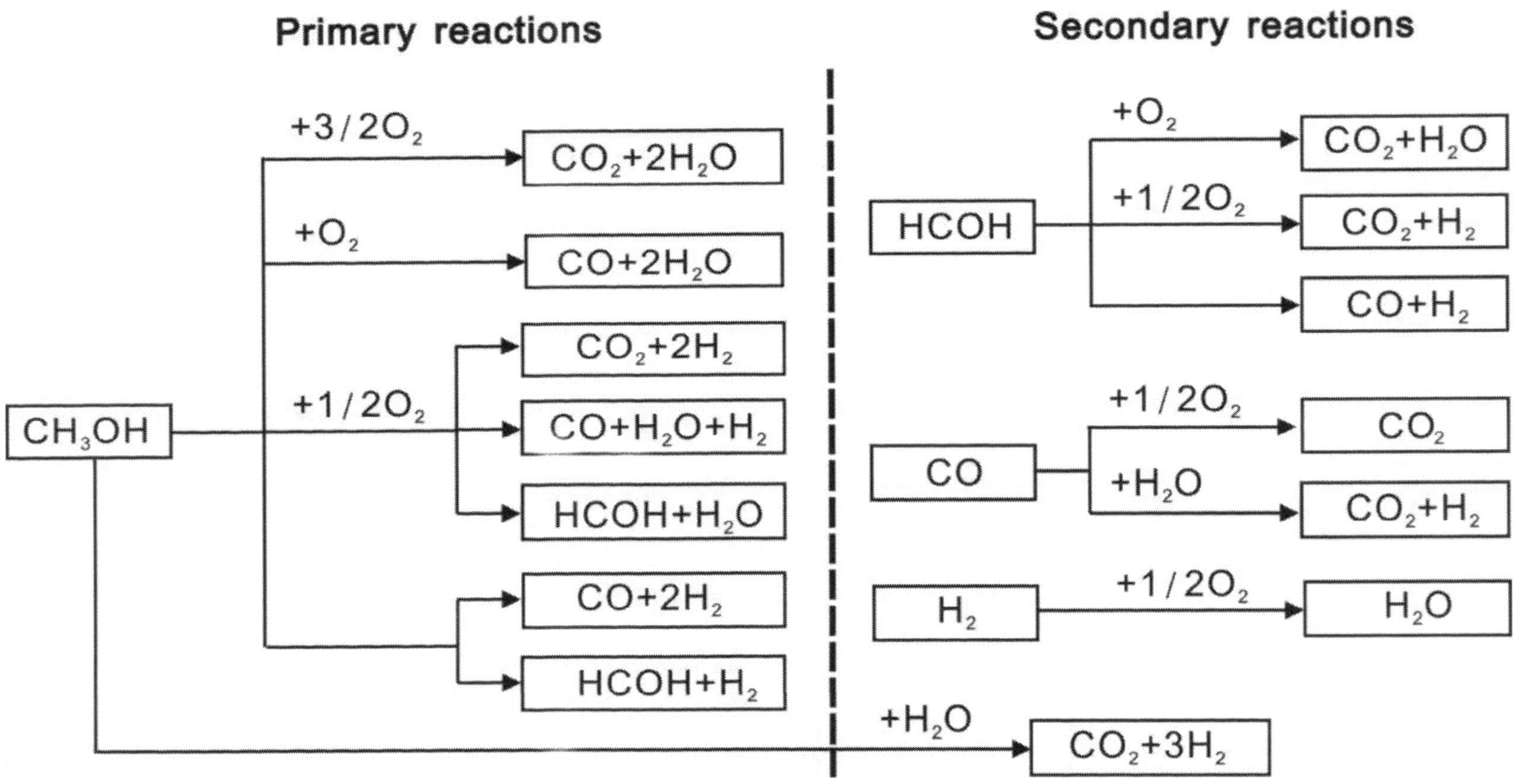

Fig. 12. Reaction pathways of methanol conversion under oxidative environments.

(629b) Synthesis and Catalytic Properties of Nickel Silicide Catalysts for Selective Hydrogenation

Xiao Chen[1], Christopher T. Williams[2] and Changhai Liang[1], (1)School of Chemical Engineering, Dalian University of Technology, Dalian, China, (2)Chemical Engineering, University of South Carolina, Columbia, SC

Hydrotreating research has played an important role in environmental catalysis studies worldwide. This has led to a strong interest in catalytic materials di erent from the classically utilized metal sulfide and noble metal catalysts. Metal carbides and nitrides are initially highly active, but are easily poisoned by a small amount of sulfur. Thermochemical calculations have indicated that metal silicides can tolerate much higher H_2S concentrations than the corresponding carbides and nitrides. Therefore, we explored the possibility of using metal silicides as novel catalysts for heterogeneous catalytic reactions. Nickel-silicon intermetallic catalysts (Ni_2Si, NiSi, and $NiSi_2$) were prepared by direct silification method using SiH_4 as the silicon source. The as prepared nickel silicides were characterized by X-ray diffraction, transmission electron microscopy, magnetic measurements, X-ray photoelectron spectroscopy, temperature-programmed reduction, temperature-programmed desorption measurements. The results show that nickel silicide formation involves the following sequence as a function of increasing temperature: Ni (cubic) → Ni_2Si (orthorhombic) → NiSi (orthorhombic) → $NiSi_2$ (cubic). The insertion of Si atoms into the interstitial sites between Ni atoms resulted in a significant change in the unit cell lattice of nickel. All of nickel silicide materials were ferromagnetic at room temperature, with saturation magnetization values drastically decreased when Si is present. Nickel silicide develops a thin silicon oxide layer during exposure to air, which can be removed by H_2-temperature programmed reduction.

The catalytic hydrogenation of cinnamaldehyde and phenylacetylene over the nickel silicide nanoparticles was investigated. The repulsive force between the electronegative silicon atoms in the nickel silicide intermetallic and oxygen atoms in the C=O bond of cinnamaldehyde inhibit the hydrogenation of C=O bond to some extent. Nickel silicide catalyst presents 77% selectivity to hydrocinnamaldehyde, which is much higher than monometallic nickel catalyst. The nickel silicide catalyst also shows excellent (ca 93%) styrene selectivity in the hydrogenation of phenylacetylene due to the strong modification of electronic structure around the Fermi level caused by the interaction of nickel and silicon. These results indicate that these novel nickel silicides are promising catalysts for the selective hydrogenation of unsaturated hydrocarbons.

References

[1] C. H. Liang, A. Zhao, X. Zhang, Z. Ma, R. Prins, *Chem. Commun.* **2009**, 2047-2049

[2] A. Zhao, X. Zhang, X. Chen, J. Guan, C. Liang, *J. Phys. Chem. C*, **2010**, 114, 3962-3967

[3] A. Zhao, X. Chen, J. Guan, C. T. Williams, C. H. Liang, *Phys. Chem. Chem. Phys.*, **2011**, DOI: 10.1039/c1cp20197b

[4] X. Chen, A. Zhao, Z. Shao, C. Li, C. T. Williams, C. H. Liang, *J. Phys. Chem. C*, **2010**, 114:16525-16533

[5] J. Guan, A. Zhao, X. Chen, M. Zhang, C. Liang, *Stud. Surf. Sci. & Catal.*, **2010**, 175, 259-262

See more of this Session: Poster Session of CRE Division
See more of this Group/Topical: Catalysis and Reaction Engineering Division

(629c) Role of Microwave Heating Strategies In Enhancing the Progress of a First Order Endothermic Reactions

Tanmay Basak, Chemical Engineering, Indian Institute of Technology Madras, Chennai, India and Madhuchhanda Bhattacharya, Independent Researcher, Chennai, India

This work presents a detailed analysis on the role of microwave heating strategies in optimizing the progress of a first order endothermic reaction occurring in a semiinfinite vertical column. Here, microwave heating strategies are analyzed in terms of different intensities of incident radiations from left and right sides of the reactor. Four different configurations of incident radiations are considered such that that they represent the entire spectrum of possible arrangement of microwave sources spanning uniform intensities of radiations from both the sides to highly localized radiations from one side. In all the cases, total intensities of incident radiations from left and right sides are kept constant in order to keep input power constant. The analysis has been performed by considering detail description of microwave propagation within the reactor in presence of varying intensities of incident radiations from left and right sides via Helmoltz equation.

Progress of reactor in presence of heat and mass transfer limitations have been simulated by solving detailed energy and mass balance equations coupled with Helmoltz equation for evaluation of microwave induced heating pattern for various configurations of microwave sources. Efficiency of various strategies of microwave heating are evaluated by compairing the simulated progress of microwave heated reactors with their conventional counterparts, where total heat generated by microwave power absorption in each configuration has been supplied as surface heat flux in its corresponding conventional reactor. The comparison is done for different reactor dimensions corresponding to {\it three} possible regimes of microwave induced heating patterns, namely (1) uniform, (2) oscillatory and (3) exponentially decayed heating from the surface in {\it thin}, {\it intermediate} and {\it thick} regimes, respectively. It has been shown that although all the configurations of microwave heating lead to faster progress of reactor compared to their conventional counterparts, extent of enhancement depend on microwave heating strategy especially in presence of strong diffusion limitations. Progress of reactor in {\it thin} and {\it thick} regimes of microwave heating can be greatly enhanced by configuring equal intensity of radiations on both sides compared to one sided incidences. On the other hand, effect of microwave heating strategies on reactor progress are nontrivial in {\it intermediate} regime of microwave heating, where either one sided or distributed incidences may enhance reactor progress depending on the dimension of the reactor with respect to wavelength and penetration depth of microwave within it. The enhancement of reactor progress in presence of microwave heating has further been quantified in terms of percentage savings of reaction time [$\eta_{\mbox{MW}} = (1 - \tau_{f_{\mbox{\small MW}}}/ \tau_{f_{\mbox{\small Conv}}}) \times 100$]. Variations of $\eta_{\mbox{MW}}$ for all the four heating strategies are determined and shown over the entire range of {\it thin} to {\it thick} regimes. These master curves are extremely essential for selection of the appropriate heating strategy for a given reactor dimension in various regimes of microwave heating especially in {\it intermediate} regime. In this work, the master curves are provided for varying degree of mass transfer limitations showing its effect on the efficiency of various strategies of microwave heating.

Efficiencies of various strategies of microwave heating are also analyzed for different packing materials measured in terms of f_p and f_w. It may be noted that dielectric properties of packing materials as well as packing arrangement can significantly alter f_w and f_p. Effect of f_p and f_w on $ \eta_{\mbox{MW}}$ for each configuration of microwave sources are shown over the entire range of {\it thin} to {\it thick} regimes. These master curves reveal nontrivial variation of optimum heating strategy with f_w and f_p especially in {\it intermediate} regime and show their utmost importance for design of optimum heating strategy to maximize the enhancement of reactor progress for different packing materials.

See more of this Session: Poster Session of CRE Division
See more of this Group/Topical: Catalysis and Reaction Engineering Division

(629d) Significantly Enhanced Reactor-Scale Efficiency and Catalyst Lifetime by Rational Design of the Hierarchical Catalyst Pore Network - Application to Hydrodemetalation

Sanjeev M. Rao and Marc-Olivier Coppens, Isermann Department of Chemical and Biological Engineering, Rensselaer Polytechnic Institute, Troy, NY

Our previous work has shown that the lifetime of nanoporous catalysts may be significantly extended by introducing a broad pore channel network of optimized geometry [1,2]. This is realized by an optimal uniform macroporosity and an optimal constant macropore size; interestingly, broad distributions of pore sizes do not lead to notable improvements. This was demonstrated for the example of hydrodemetalation (HDM) catalysts. However, these results only showed the effects at the catalyst pellet scale. A non-trivial, practically relevant extension that we address in this presentation is the effect of the optimized broad pore network at the reactor scale.

First, we mathematically optimize the uniform macroporosity and uniform broad pore size of a hierarchically structured HDM catalyst, to maximize the total mass of metal sulfides deposited over a given time on stream. Following Mac and Wei, first-order HDM kinetics are assumed [3]. Catalyst deactivation, due to coverage of active sites and pore blockage by metal sulfides, is accounted for through the Random Spheres Model (RSM) [3,4] in both the purely mesoporous regions and the macropores of the hierarchically structured catalyst. The properties of the purely mesoporous regions of the hierarchically structured catalyst are taken from the patent literature [5]. For a time on stream of 3 months, optimization of the broad pore network results in an optimum initial macroporosity of 25% and an optimum broad pore size of 192 nm. Both values are within the corresponding range of macroporosity and broad pore sizes reported in the patent. Optimization for longer times on stream did not significantly increase the total reaction yield, or significantly change the optimal macroporosity and broad pore size. The total reaction yield from the optimal hierarchically structured catalyst is 1.7 times larger than the total reaction yield from the purely mesoporous catalyst. The purely mesoporous catalyst already completely deactivates within 40 days on stream, when the optimized catalyst is still very active.

Next, the optimized HDM catalyst is used in an isothermal, one-dimensional plug flow model of an industrial fixed bed reactor. The purely mesoporous HDM catalyst is chosen as a basis to compare the performance of the optimized HDM catalyst against. The purely mesoporous catalyst bed deactivates completely within 3 months on stream; whereas, the lifetime of the optimized hierarchically structured catalyst bed can be extended to more than 7 months. However, with longer times on stream, the exit conversion from the reactor also decreases, and based on a minimum metals conversion of 60%, a time on stream of 5 months is found to be optimal. When utilizing the reactor for 5 months on stream, the total mass of deposited metal sulfides nearly doubles, compared to the purely mesoporous catalyst. Sensitivity calculations show that varying the initial macroporosity of the catalyst, between 70-200% of its optimized value, in the reactor simulation, does not decrease the total yield from the reactor by more than 15%.

Our work demonstrates the industrial, macroscale opportunities of optimizing the hierarchical catalyst pore network structure in a rational way.

References

1. Rao, S. M.; Coppens, M.-O. Mitigating Deactivation Effects through Rational Design of Hierarchically Structured Catalysts—Application to Hydrodemetalation. *Ind. Eng. Chem. Res.* **2010,** *49,* 11087.

2. Rao, S. M.; Coppens, M.-O. Optimal Broad Pore Networks to Improve Resistance to Catalyst Deactivation—Application to Hydrodemetalation. 102nd AIChE Annual Meeting, Salt Lake City, 2010, paper 48f.

3. Mac, O.; Wei, J. Diffusion in Random Particle Models for Hydrodemetalation Catalysts. *Ind. Eng. Chem. Res.* **1991,** *30*, 909.

4. van Eekelen, H. A. M. The Random-Spheres Model for Porous Materials. *J. Catal.* **1973,** *29*, 75.

5. Abe, S.; Hino, A.; Shimowake, M.; Fujta, K. High-Macropore Hydroprocessing Catalyst and its Use. U.S. Patent 7,186,329, March 6, 2007.

See more of this Session: Poster Session of CRE Division
See more of this Group/Topical: Catalysis and Reaction Engineering Division

(629e) Density Functional Theory Study of External Zeolite Surfaces

Junbo Chen, Chemical Engineering, Worcester Polytechnic Institute, Worcester, MA and **N. A. Deskins**, Department of Chemical Engineering, Worcester Polytechnic Institute, Worcester, MA

Density FunctionalTheory Study of External Zeolite Surfaces

Junbo Chen, N. AaronDeskins; Department of Chemical Engineering;

Worcester PolytechnicInstitute; Worcester, MA 01543

Zeoliteshave been used extensively in the petroleum industry to process crudehydrocarbons into desirable fuel materials. Catalytic cracking,hydro-cracking, and dewaxing(1,2)• are commoncatalytic reactions that take place over zeolites. The reactivity ofzeolite materials is largely attributed to their internal porestructure which leads to a large internal surface area, or highdensity of reactive sites. Large molecules, however, cannot enter thezeolite's pores and therefore do not react within the zeolite, but atexternal surface sites. This phenomena has been well-documented inthe literature. For example one phenomena discussed in the literatureis pore mouth catalysis(3-6)where reactions do not occur in the pore (due to steric hindrance)but outside the pore or at the pore mouth. Another example is thealkylation of biphenyl over several zeolites, which was attributedmainly to external surface reaction sites(7)•.Work on ZSM-5 has also shown the importance of surfaces for largermolecules that cannot enter the pore structure(8,9)•. Several reactionmodels involving large hydrocarbon were shown to agree much betterwith experimental data when surface sites were included in themodels(10, 11)•.

Manydetails however of the surface reactivity of zeolites are largelyunknown, particularly the atomic-level structures and mechanisms ofthese surface reactions. The structural complexity of zeolites hasprecluded many theoretical surface studies in the past due tocomputational limitations. The focus of this work is tosimulate, using density functional theory, the adsorption andreactivity of hydrocarbons over zeolite external surfaces. Various(001) surface terminations of zeolite LTA are considered, at variousSi/Al ratios and with several extra-framework cations. The moststable location of Brnstedacid sites (H^+) was determined, and results show aenergetic preference for the surface, as opposed to bulk internalarea, indicating that the surface may be the most active region ofthe catalyst. The role of surface defects in surface stability isalso presented. Both small and large prototypical molecules (e. g.H_2O, NH_3, benzene, naphthalene) are adsorbed atthe external surfaces in order to determine precursor states toreactivity. Bulk calculations allow a comparison of internal versusexternal catalytic activity. Ultimately, understanding the importanceand role of surfaces of zeolites in hydrocarbon reactivity is the aimof this work.

•References

(1) Vermeiren, W.; Gilson, J.-P. *Topics in Catalysis*. **2009**, *52*, 1131-1161.

(2) Vaughan, D. E. W. *Chemical Engineering Progress*. **1988**, *84*, 25-31.

(3) Souverijns, W.; Martens, J. A.; Froment, G. F.; Jacobs, P. A. *Journal of Catalysis*. **1998**, *174*, 177-184.

(4) Martens, J. A.; Vanbutsele, G.; Jacobs, P. A.; Denayer, J.; Ocakoglu, R.; Baron, G.; Arroyo, J. A. M.; Thybaut, J.; Marin, G. B. *Catalysis Today*. **2001**, *65*, 111-116.

(5) Madeira, F. F.; Gnep, N. S.; Magnoux, P.; Vezin, H.; Maury, S.; Cadran, N. *Chemical Engineering Journal*. **2010**, *161*, 403-408.

(6) Laxmi Narasimhan, C. S.; Thybaut, J. W.; Marin, G. B.; Jacobs, P. A.; Martens, J. a; Denayer, J. F.; Baron, G. V. *Journal of Catalysis*. **2003**, *220*, 399-413.

(7) Aguilar, J.; Corma, A.; Melo, F. V.; Sastre, E. *Catalysis Today*. **2000**, *55*, 225-232.

(8) Röger, H.; Möller, K. P.; O'Connor, C. T. *Microporous Materials*. **1997**, *8*, 151-157.

(9) Röger, H.; Möller, K. P.; O'Connor, C. T. *Journal of Catalysis*. **1998**, *176*, 68-75.

(10) Denayer, J.; Baron, G. V.; Vanbutsele, G.; Jacobs, P. A.; Martens, J. A. *Chemical Engineering Science*. **1999**, *54*, 3553-3561.

(11) Klemm, E.; Scheidat, H.; Emig, G. *Chemical Engineering Science*. **1997**, *52*, 2757-2768.

See more of this Session: Poster Session of CRE Division
See more of this Group/Topical: Catalysis and Reaction Engineering Division

(629f) Kinetics of the Gas-Phase Oxidation of Hexafluoropropylene In a Laminar Flow Reactor

David Lokhat, Deresh Ramjugernath and Maciej Starzak, Chemical Engineering, University of KwaZulu-Natal, Durban, South Africa

The gas-phase oxidation of hexafluoropropylene (HFP) with molecular oxygen has been studied experimentally in an isothermal, laminar flow reactor at a pressure of 4.5 bar and over the temperature range of 453 to 503 K. Feed mixtures comprising 20 to 67% HFP in oxygen were used at total flow-rates of between 150 and 550 cm^3 min^{-1}. Online GC/FID analysis was used for reactant and stable product quantification. The most significant products detected were hexafluoropropylene oxide (HFPO), trifluoroacetyl fluoride (CF_3COF) and carbonyl fluoride (COF_2). Minor products included tetrafluoroethylene (C_2F_4), hexafluorocyclopropane (c-C_3F_6) and carbon dioxide. A kinetic reaction scheme involving 11 species and 7 reactions was developed and used to model the oxidation of HFP. The initial steps in this scheme encompassed the addition of oxygen to the double bond of the fluoro-olefin and decomposition of the resultant dioxetane intermediate to form HFPO and the haloacetyl fluorides. Subsequent steps included the thermal decomposition of HFPO to yield CF_3COF, C_2F_4 and c-C_3F_6, elimination of C_2F_4 through oxidation and decomposition of CF_3COF via C-C bond scission to give COF• and CF_3• radicals, which undergo recombination and oxidation reactions. Rate parameters for the oxidation reactions were determined through a least-squares minimization procedure coupled with two-dimensional numerical integration of the governing equations utilizing the method of lines.

See more of this Session: Poster Session of CRE Division
See more of this Group/Topical: Catalysis and Reaction Engineering Division

(629g) Theoretical Study of Externally Imposed Temperature Profiles In Continuous Reactors and Optimal Profiles for Exothermic Reacions

Dario A. Cruz, Jorge M. Gomez and Watson L. Vargas, Department of Chemical Engineering, Universidad de los Andes, Bogota, Colombia

It is well known that imposing external axial temperature profiles, evidently improves conversion in exothermic reactions that are limited by chemical equilibrium. In the past, this kind of approaches were not possible to implement, due to the difficulties in keeping a reactor at the energetic and reactions conditions required; a situation that demands very precise schemes of control. With technological advances, micro-reactor technologies have increased the possibilities for implementation of externally imposed temperature profiles, in order to reach high conversions in this kind of reactive systems. Three basic archetypal externally imposed temperature profiles have aptly been explored, particularly in micro-structured reactors: convex, gaussian and linear profiles. Results in the literature indicate that depending on reaction conditions, any of the three above-mentioned profiles can lead to higher conversions, with convex been a promising one.

In the first part of this work, three different generic reactions, namely: water gas shift, sulfur dioxide oxidation and amonia synthesis will be tested with each one of this prototypical profiles and their results compared in terms of conversion levels. In the second part of this contribution, an alternative aproximation is proposed. This approach involves a formulation of an optimization problem in which we search for the optimal profile, using optimization tecniques and numerical analysis. The metodology of such aproximation will be illustrated using the water-gas shift reaction as example in order to compare the results with the best temperature profiles described in the literature.

See more of this Session: Poster Session of CRE Division
See more of this Group/Topical: Catalysis and Reaction Engineering Division

(629h) Atomic Layer Deposition for In-Situ Fabrication of Catalyst Supports

Staci A. Van Norman[1], Stefan Strhle[1], Nick Wannenmacher[2], Aldo Steinfeld[3], John Falconer[1] and Alan W. Weimer[4], (1)Chemical Engineering, University of Colorado at Boulder, Boulder, CO, (2) Chemical Engineering, Oregon State University, Corvallis, OR, (3)Department of Mechanical and Process Engineering, ETH Zurich, Zurich, Switzerland, (4)Department of Chemical and Biological Engineering, University of Colorado at Boulder, Boulder, CO

Microchannel catalytic reactors exhibit improved mass and heat transfer properties and are thus often chosen for highly exothermic or endothermic reaction conditions. In conventional microchannel reactors, catalyst is deposited within channels either by wash coating or by packing channels with particles prior to bonding. Both methods exhibit challenges: washcoating can result in poor channel volume usage, while channel packing techniques often result in the crushing and subsequent flaking of particles and clogging or fluid channeling due to differences in the thermal expansion of the materials. A novel *in-situ* fabrication method is presented for the customizable manufacture of microstructured reactor substrates made from ceramics/cermets for use in highly exothermic or endothermic catalytic reaction applications.

Microstructured supports were fabricated by atomic layer deposition on sacrificial polymer templates of polystyrene-divinylbenzene with a surface area of 700 m^2/g, pore size of 30 nm and particles sizes of 50-100 μ m and 80-160 μ m. Atomic layer deposition produces conformal films by two sequential self-limiting surface half reactions of $Al(CH_3)_3$ and H_2O for alumina and similarly for other materials.

Reactions were performed at temperatures below 100°C; once the desired film thickness was achieved the polymer template was volatilized at temperatures above 400°C. The resulting thermal, fluid flow and structural characteristics of these structures are presented. A heat transfer model was developed and is compared to a conventional packed bed of ceramic particles.

See more of this Session: Poster Session of CRE Division
See more of this Group/Topical: Catalysis and Reaction Engineering Division

(629i) Optimal Design In Biodiesel Production From Esterification of Palmitic Acid In Excess Methanol

T.•n Hi•p •ng[1], Yi-Han Chan[1], Bing-Hung Chen[1] and Duu-Jong Lee[2], (1)Department of Chemical Engineering, National Cheng Kung University, Tainan, Taiwan, (2)Department of Chemical Engineering, National Taiwan University of Science and Technology, Taipei, Taiwan

Optimization of esterification of palmitic acid in excess methanol by solid acid catalyst, viz. Amberlite™ IR-120 (H) resin, is studied by combining backward stepwise regression technique with a 2^{5-1} fractional factorial design. Optimum conditions are found as follows: 8:1 in the molar feed ratio of methanol to palmitic acid, a reaction temperature as 61.0 °C, a reaction time of 10.5 h, catalyst loadings at 10 wt% of palmitic acid, and a stirring speed of 600 rpm. At this condition, conversion efficiency of palmitic acid to palmitic acid methyl ester reaction is near 98.9%.

See more of this Session: Poster Session of CRE Division
See more of this Group/Topical: Catalysis and Reaction Engineering Division

(629j) Autothermal Reforming of n-Dodecane Over Promoted Nickel Xerogel Catalysts

M. V. Phanikrishna Sharma, Jale Akyurtlu and Ates Akyurtlu, Chemical Engineering Program, Hampton University, Hampton, VA

Energy demand continues to climb as the world population grows. Meanwhile petroleum reserves are diminishing. Developing new energy sources and more efficient technologies is extremely important. At the same time, the level of emissions and exhaust gases needs to be reduced to minimize their contribution to the greenhouse effect. Environmentally friendly energy solutions are being sought, especially for mobile applications. For example, the auxiliary power units (APUs) can increase fuel economy by displacing engine idling. The combination of an on-board reformer and a SOFC would enable commercial fuels such as jet fuel and diesel to be used as a hydrogen source because of the existing infrastructure and high energy density of these fuels. More popular in transportation, military, and industrial applications, commercial fuels could be a good first step toward a hydrogen-based society.

The conversion of hydrocarbon fuels to hydrogen can be carried out by several different reforming approaches including steam reforming, partial oxidation, and autothermal reforming. Autothermal reforming (ATR) is regarded as the best option since the combustion of some of the fuel supplies the energy required for the reforming reaction and better sulfur tolerance than steam reforming (SR). ATR reactors are smaller, quick-starting and faster-responding than SR reactors, and give higher hydrogen concentration. Faster startup time and better transient response can be attained by tuning the feed stoichiometry. Additionally, the water gas shift reaction, which proceeds simultaneously, reduces the CO content of the hydrogen-rich gas; also, the methane content of the gas is lower than the other methods. The operation variables have to be optimized according to the catalyst employed to minimize carbon formation and methane production.

The present experimental study has been undertaken to investigate the effect of the operation variables on the ATR of *n*-dodecane, which is selected as a jet fuel surrogate to simplify the reaction study and the interpretation of the test results.

Promoted nickel xerogel catalysts were prepared by sol-gel method. Catalysts were characterized by XRD, TPR, TPD of H_2 & CO_2, BET surface area and porosity measurements. The autothermal reforming experiments were performed in a fixed bed micro-reactor system in a temperature range of 600-800°C and space velocity of the order of 100,000 hr^{-1}. The reaction feed containing *n*-dodecane, water, oxygen and carrier gas were sent through a vertical flow quartz reactor containing the catalyst. The product gases were analyzed by two online GCs.

The results from these experiments will be discussed during the presentation in terms of the effect of operation variables on the hydrogen yield and product gas composition.

Acknowledgement: The authors acknowledge financial support from ARO grant (W911NF-10-1-0514).

See more of this Session: Poster Session of CRE Division
See more of this Group/Topical: Catalysis and Reaction Engineering Division

(629k) Comparison of Microwave Drying and Regular Oven Drying of Supported Catalysts

Xue Liu, Rutgers University, Piscataway, NJ, Johannes G. Khinast, Institute for Process and Particle Engineering, Graz University of Technology, Graz, Austria and Benjamin Glasser, Dept. of Chemical and Biochemical Engineering, Rutgers University, Piscataway, NJ

Supported catalysts are used in many industrial processes and applications, ranging from petrochemical and catalytic converters to fuel cells. These catalysts have many advantages, such as a high surface area, a low amount of the often expensive active component (Pd, Pt, etc.) and high mechanical and thermal stability. Clearly, the catalyst design has a pronounced effect on the performance of a catalytic process. With respect to the distribution of the active component in the support materials, four main categories of metal profiles can be distinguished, i.e., uniform, egg-yolk, egg-shell and egg-white profiles. The choice of the desired metal profile is determined by the required activity and selectivity, and tailored for specific reactions and/or processes. Although the development and preparation of supported catalysts have been investigated for many years, many aspects of the various catalyst manufacturing steps are still not fully understood, and in industry the design of catalysts is predominated by trial and error experiments,

Supported catalysts are usually prepared by impregnation, where a porous support is contacted with a liquid solution that contains the desired metal as a dissolved salt. This step is usually followed by the evaporation of the liquid solvent and that is drying. After drying reduction and calcination are carried out. It is generally believed that the metal profile is controlled by the conditions that are applied during impregnation where the metal contacts the solid support for the first time. However, experimental work has shown that drying may also significantly impact the metal distribution within the support. Therefore, to achieve a desired metal profile we need to understand both impregnation and drying steps.

In this work we have carried out impregnation and drying experiments for Ni/Alumina Ni-Cu/Alumina, and Ni-Mo/Alumina systems. Both microwave drying and regular oven drying are used in this work. During microwave drying, the catalyst samples can be heated volumetrically so the hear transport is eliminated and the drying rate increases. Microwave drying can provide rapid and uniform drying within the porous support, leading to uniform evaporation of the liquid solvent, which most likely results in a uniform metal distribution. Usually we hold the impregnation time sufficiently long so a uniform metal distribution can be obtained after impregnation. Given this uniform initial distribution for drying, we find that microwave drying gives us a much more uniform metal distribution than regular oven drying. For a short impregnation time, an egg-shell distribution can be observed after impregnation. The impact of both microwave drying and regular oven drying on this egg-shell initial distribution is also examined in this work. The effect of the energy intensity on the microwave drying process is significant. The drying time required greatly increases with a decrease in the input energy. We have also developed theoretical models to simulate microwave drying and regular oven drying processes and compared the simulations with experiments.

See more of this Session: Poster Session of CRE Division
See more of this Group/Topical: Catalysis and Reaction Engineering Division

(629l) Fundamental Understanding of Adsorption of Aldehydes On Alumina

Zhimin Lu, Cbme, OU, Still Water, OK
See more of this Session: Poster Session of CRE Division
See more of this Group/Topical: Catalysis and Reaction Engineering Division

Catalytic Oxidation of Cyclohexane by Molecular Oxygen Using Polymer Anchored Catalyst

Sweta, Shishir Sinha, Prakash Biswas, Shri Chand*

Department of Chemical Engineering, Indian Institute of Technology Roorkee, India

* Corresponding author. Email: schanfch@iitr.ernet.in; Telephone: +91-1332-285383

ABSTRACT

Catalytic oxidation of cyclohexane was investigated over polymer anchored Cu(II) catalyst prepared by modification of polymer surface by NO_x and subsequent functinalization by amination. The effect of temperature, pressure and catalyst amount on the catalytic activity and product selectivity was investigated. The oxidation of cyclohexane using molecular oxygen has been studied in the temperature range of 100–140 °C. The overall cyclohexane conversion was 15.3 % at 140 °C and cyclohexanone, the major product with 51.7% selectivity.

INTRODUCTION

The cyclohexane oxidation is an important reaction since its oxidation products, cyclohexanol and cyclohexanone, are very important chemical intermediates. They are used as starting materials in the preparation of adipic acid and caprolactam which are further used in the manufacture of nylon-6,6 and nylon-6. Industrially, cyclohexane is oxidized at 423- 453 K and 1-2 MPa air pressure using cobalt naphthenate as catalyst. A low conversion (3-4%) is obtained to avoid the formation of side products and moderate selectivity of 70-85% of cyclohexanol and cyclohexanone is obtained.[1-3]

Although various homogeneous catalysts have been used for this process,[4,5] recent studies have illustrated the use of more efficient heterogeneous catalysts for cyclohexane oxidation.[6-8]

In the present work, we have reported the use of aminated functional group attached to polymer support as precursor for the preparation of heterogeneous catalyst. In this study, Cu(II) metal is loaded on the functional polymer and used for the catalytic activity of cyclohexane oxidation in batch reactor. The product distribution of cyclohexane oxidation is shown in Fig. 1.The various reaction parameters were optimized for maximum conversion and selectivity.

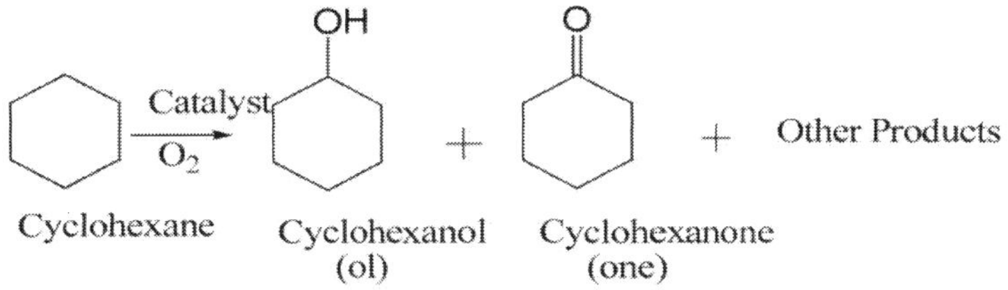

Figure 1. Reaction products of cyclohexane oxidation

Experimental

Preparation of macroporous resin for polymer support

PS-DVB copolymers were prepared by conventional suspension polymerization technique as described in the literature.[9] Direct gas phase nitration of polystyrene–divinylbenzene (PS–DVB) resin was carried out by NO_x method.The resin was subsequent aminated by hydrazine hydrate.[9]

Preparation of catalyst (C-1)

The functionalized resin (2.0g) was added to methanol solution (50 ml) of $CuCl_2.2H_2O$ (1% w/v) and the resulting solution was heated at 60 °C in a thermostated oil bath for 18 h with continuous stirring. The liquid mixture was then cooled and resin was filtered and washed several times with methanol and dried in oven for 12 h.

Oxidation of Cyclohexane

The catalytic oxidation of cyclohexane using O_2 as the oxidant was carried out in a 300 ml batch stainless steel Parr reactor (4843, USA). The reactor is equipped with an inlet for introducing gas and a sampling valve to collect the liquid sample at regular intervals of time. After the reaction temperature and pressure reached the desired values, the agitator was started and this time was recorded as the starting time of the reaction. The reactor was initially charged with 100 ml cyclohexane, 100 mg of catalyst, and 1.0 MPa oxygen pressure and then heated to the required temperature. The extent of the reaction was determined by gas chromatography (Netel, India) equipped with a capillary column (0.25 mm i.d., 30 m long, 0.25 μm thickness) and a flame ionization detector (FID). The oven temperature was programmed at 80 ^{0}C for 1min and then raised to 180 °C at a ramp rate of 10 °C/min.

Results and Discussion

The oxidation of cyclohexane was conducted in the presence of molecular oxygen using polymer anchored complex catalyst. The reaction was carried out in the temperature range of 100-140 °C for 8 h at 1.2 MPa. A large amount of undesired products were formed at temperature >140 °C. The overall conversion increased from 3.0% to 15.0%

when temperature was increased from 100 °C to 140 °C. When the temperature was raised to 140 °C, cyclohexanone and cyclohexanol were the main reaction products as shown in Fig. 5.

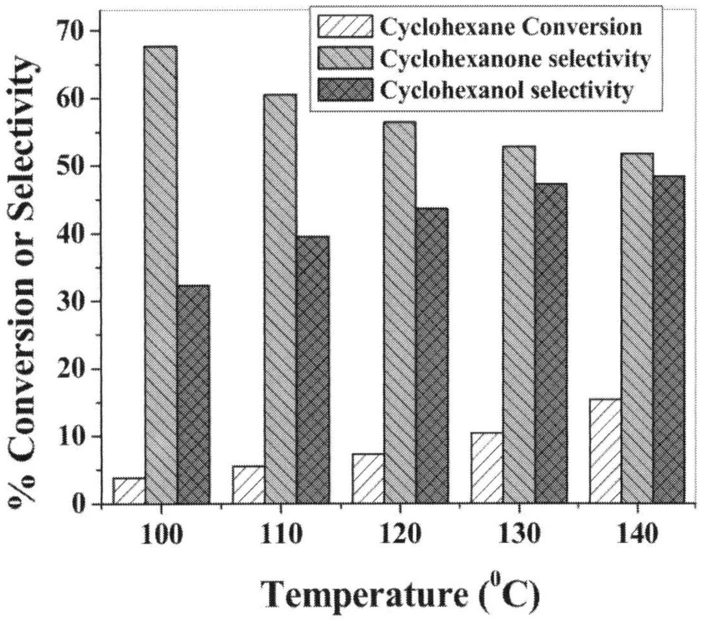

Figure. 2 Effect of temperature on conversion and product selectivity

Conclusion

In the present work, we have reported the formation of polymer anchored Cu(II) complex catalyst by modification of polymer surface by NO_x and further functionalization by amination. The catalyst catalytic activity was investigated for cyclohexane oxidation to cyclohexanone and cyclohexanol using O_2 as oxidant. The oxidation of cyclohexane using molecular oxygen in the absence of promoters and coreactants has been studied in the temperature range 100–140 °C. The maximum conversion of cyclohexane was 15.3%

at 140 °C and 1.2 MPa pressure and cyclohexanone was the major product with 51.7% selectivity.

Literature Cited

1. Sakthivel, A.; P. Selvam, P. Mesoporous (Cr) MCM-41: A mild and efficient heterogeneous catalyst for selective oxidation of cyclohexane. *J. Catal.*2002;211:134-143

2. Anisia KS, Kumar A. Synthesis of heterogeneous copper complex catalyst for oxidation of cyclohexane using molecular oxygen.*Can. Jour. Chem. Eng.* 2008; 86:1054-1061.

3. Moden B, Zhan B, Dakka J, Santiesteban JG, Iglesia E. Kinetics and mechanism of cyclohexane oxidation on MnAPO-5 catalysts. *J Catal.* 2006;239:390-401.

4. Hao J, Wang J, Wang Q, Yu Y, Cai S, Zhao F. Catalytic oxidation of cyclohexane over Ti–Zr–Co catalysts. *Appl. Catal. A: Gen.* 2009;368: 29-34.

5. Pereira R, Rufo M, Schuchardt U. Copper(II) catalyzed oxidation of cyclohexane by tert-butyl hydroperoxide. *J.Braz. Chem. Soc.*1994;5:83-89.

6. Yuan HX, Xia QH, Zhan HJ, Lu XH, Su KX. Catalytic oxidation of cyclohexane to cyclohexanone and cyclohexanol by oxygen in a solvent-free system over metal-containing ZSM-5 catalysts. *Appl. Catal. A: Gen.* 2006; 304: 178-184.

7. Carvalho WA, Wallau M, Schuchardt U. Iron and copper immobilised on mesoporous MCM-41 molecular sieves as catalysts for the oxidation of cyclohexane. *J. Mol. Catal. A.* 1999;144:91-99.

8. Yang D, Gao L, Zhao W. Synthesis, characterization and catalytic activity of dialdehyde starch-schiff base Co(II) complex in the oxidation of cyclohexane. *Catal. Lett.* 2008;126: 84-89.

9. Jayaswal N, Sinha S, Kumar A. Effect of chemical modifications upon exchange capacity of aminated macroporous styrene–divinyl benzene (PS–DVB) copolymer anion exchange resin. *Jour. Applied Polym Sci.* 2001; 79:1735-1748.

(629n) Photocatalytic Conversion of CO_2 to Fuels by Nanostructured CeO_2-TiO_2/SiO_2 Catalysts

Cunyu Zhao and Ying Li, University of Wisconsin-Milwaukee, Milwaukee, WI

Fossil fuel consumption carries with the significant drawback of environment impact – CO_2 emissions that lead to global warming. Conversion of CO_2 to useful chemicals such as methane or syngas under mild reaction conditions has promising application prospects. TiO_2 is well known as an excellent photocatalyst due to its reusable, inexpensive and easily available properties. However, the photocatalytic activity of CO_2 reduction by TiO_2 is low partly due to the fast recombination rate of photo-generated electron-hole pairs. Cerium is a particular lanthanide element which has two oxidation states – Ce(III) and Ce(IV). The oxidation states enable CeO_2 to trap photo-generated electrons and thus reduce the electron-hole recombination rate. On the other hand, surface area and surface hydroxyl groups can also influence photocatalytic activity. SBA-15, a mesoporous silica with 1-D pores has been widely used as an effective support for catalyst. In this work, CeO_2-TiO_2 catalysts with varying compositions were first synthesized via sol-gel preparation method using cerium nitrate and titanium butoxide as precursors. CeO_2-TiO_2/SiO_2 catalysts were finally prepared by adding SBA-15 into the Ce-Ti sol. The catalysts were characterized by XRD, BET, SEM, TEM, UV-vis diffuse reflectance spectroscopy, etc. The photocatalytic activity of CO_2 reduction with water was evaluated under UV-vis irradiation. CH_4 was found to be the major products measured by a GC-TCD/FID. Catalysts with optimal Ce/Ti ratio and catalyst/support ratio were identified in this study.

See more of this Session: Poster Session of CRE Division
See more of this Group/Topical: Catalysis and Reaction Engineering Division

(629o) Ionic Liquid Thin Film Technologies for Advanced Catalytic Processes

Peter Wasserscheid, Chemical Reaction Engineering, Friedrich-Alexander University Erlangen-Nuremberg, Erlangen, Germany

In the last ten years, a large number of new ionic liquid applications have been introduced to catalysis. Many of these build successfully on the unique property profile offered by these low melting salts, such as e.g. extremely low volatility, high solution power for organometallic complexes, tuneable solubility and coordination properties. It has been demonstrated for many catalytic transformations that the use of liquid salts enable processes of much higher efficiency or catalytic materials of much better functionality. Recently, rigorous attempts to minimize the amount of ionic liquid that is required to create a certain, desired "ionic liquid performance" has led to the development of "Ionic Liquid Thin Film Technologies" in catalysis.

The contribution will present preparation methods, characterization techniques and application scenarios of these technologies in catalysis. Special attention will be given to the development and use of "Supported Ionic Liquid Phase (SILP)" materials and "Solid Catalysts with Ionic Liquid Layer (SCILL)" systems. Examples will include selective C-C-linkage reactions and hydrogen production technologies.

See more of this Session: Poster Session of CRE Division
See more of this Group/Topical: Catalysis and Reaction Engineering Division

(629p) Theoretical Investigation of Hydrodeoxygenation of Methyl Propionate On Pd (111)

Sina Behtash, Chemical Engineering, University of South Carolina, Columbia, SC and Andreas Heyden, Department of Chemical Engineering, University of South Carolina, Columbia, SC

The activity and selectivity of various transition metals for the deoxygenation of organic esters derived from biomass has received substantial interest over the recent years. In this study, various gas phase hydrodeoxygenation (HDO) pathways of methyl propionate on Pd (111) have been investigated using periodic density functional theory (DFT). In addition, microkinetic modeling and Campbell's degree of rate control was used to determine rate and selectivity controlling steps and descriptors for the reductive deoxygenation, decarbonylation, and decarboxylation pathways of methyl propionate on Pd (111).

See more of this Session: Poster Session of CRE Division
See more of this Group/Topical: Catalysis and Reaction Engineering Division

(629q) Determination of Controlling Regimes for Various SCR Reactions On Zeolite Based Monolithic Catalysts

Pranit S. Metkar, Michael P. Harold and Vemuri Balakotaiah, Department of Chemical and Biomolecular Engineering, University of Houston, Houston, TX

The selective catalytic reduction (SCR) of NOx with NH_3 is considered to be a highly promising technique for the efficient reduction of highly detrimental NOx (to N_2) emitted from diesel engine vehicles. Fe and Cu-based zeolite catalysts are considered to be highly stable and efficient towards NOx conversion over a broader temperature range. Cu-zeolite has been found to be a very good low temperatures (< 350 °C) NOx conversion catalyst. However, Fe-zeolite has been found to be a better catalyst at higher temperatures (> 350 °C).

An experimental study coupled with the kinetic modeling of key SCR reactions is presented here. This study is focused on the investigation of extent of mass transfer limitations in various SCR reactions. The experiments included NO and NH_3 oxidation, standard SCR ($NO+NH_3$), fast SCR ($NO+NO_2+NH_3$), and NO_2 SCR (NO_2+NH_3).

Monolith catalysts with different washcoat loadings, washcoat thicknesses and lengths were synthesized to study the mass transfer limitations in zeolite catalysts. A steady state experimental study of various SCR reactions showed the presence of washcoat diffusion limitations at intermediate temperature range and hence these diffusion limitations should be considered in any kinetic model and catalyst designs. A detailed analysis of the effect of temperature on the transitions between the controlling regimes (for various SCR reactions) is presented. We investigated the effect of catalyst loading, catalyst aging, monolith dimensions, and space velocity, on the transition between various controlling regimes for all the above reactions.

A kinetic model which includes the effect of washcoat diffusion limitations is developed on the basis of all the above experimental observations. A good agreement is obtained between the experimental observations and the model predictions.

Keywords: Selective Catalytic Reduction, Cu-zeolite, Fe-zeolite, Washcoat diffusion, Mass transfer, Kinetic model.

See more of this Session: Poster Session of CRE Division
See more of this Group/Topical: Catalysis and Reaction Engineering Division

(629r) Computational Study On Dry Reforming Reactions Over Pt/TiO_2 Catalysts

Zhuo Cheng and Cynthia S. Lo, Department of Energy, Environmental and Chemical Engineering, Washington University, Saint Louis, MO

Platinum clusters are potential catalysts for dry reforming reactions between carbon dioxide and methane to form syngas. In this work, titanium anatase (101) and rutile (110) are chosen as supports for platinum nanoclusters. We investigate the adsorption and reaction between CO_2 and CH_4, and its molecular fragments, on TiO_2-supported Pt, using first-principles calculations based on the density functional theory approach. The interaction between the metal nanocluster and the metal oxide substrate are analyzed and used to predict the effect of the photocatalytic substrate on CO_2 dissociation rates.

See more of this Session: Poster Session of CRE Division
See more of this Group/Topical: Catalysis and Reaction Engineering Division

(629s) Synthesis of Mesoporous Aluminosilicates From Commercially Available Ferrierite and Application for Catalytic Reaction

Jong Ki Jeon[1], Nansuk You[2], Hyeonjoo Kim[2], Jin-Heong Yim[3], Seong Jun Lee[4], Young Kwon Park[5] and Do Heui Kim[6], (1)Kongju National University, Cheonan, South Korea, (2)Dept. of Chemical Engineering, Kongju National University, Cheonan, South Korea, (3)Div. of Advanced Materials Engineering, Kongju National University, Cheonan, South Korea, (4)SK Corporation, Daejeon 305-712, South Korea, (5)University of seoul, Seoul, South Korea, (6)Institute for Interfacial Catalysis, Pacific Northwest National Laboratory, Richland, WA

A new synthesis strategy has been reported for ordered mesoporous materials constructed with semi-zeolitic frameworks and high aluminium contents using commercial zeolites such as beta, ZSM-5, USY and mordenite as a source for framework through a simple way combining top-down and bottom-up approaches [1,2]. However, results about the mesoporous substance with ferrierite framework are not known yet.

1-Butene is widely used as a monomer to produce polybutene and a comonomer for copolymerization in production of polyethylene. Recently, its demand has been raised and its price has been increased by price increase of petroleum. Currently, 1-butene is produced through separation processes from C_4 fractions in the naphtha cracking centre of a petrochemical plant and in the refinery plants. When these processes are completed, C_4 raffinate-III containing lots of 2-butene is remained and currently, it is used as a cheap fuel by transforming it into butane through hydrogenation. Therefore, it is necessary to convert 2-butene into more valuable 1-butene by positional isomerization. However, there is yet no commercial process producing 1-butene through positional isomerisation of 2-butene, and there are only a few research results about it. According to the previous reports, the catalysts, which have fewer and weaker Brönsted acid sites as well more Lewis acid sites, are quite useful for the positional isomerisation from 2-butene to 1-butene. Previously, we found that MCM-41, which has a very weak acid site, sufficiently performs a role as a catalyst in positional isomerization of 2-butene [3,4].

In this study, we demonstrate a synthesis method for ordered mesoporous materials constructed with ferrierite framework (MMZ-FER) and the applicability as catalyst for the positional isomerisation from 2-butene to 1-butene. The synthesis method involves combination of top-down and bottom-up approaches using commercially available ferrierite and cationic surfactant as the framework source and the structure directing agent respectively. The mesoporous materials from ferrierite is produced by disassembling ferrierite into unit structure in the presence of alkali solution, adding a surfactant as a templating material and performing hydrothermal synthesis. It also analyzed the effect of diverse variables in preparation of the mesoporous materials from ferrierite on the characteristics of catalyst and performance of isomerization of 2-butene.

When producing MMZ-FER through hydrothermal reaction after dissolving commercial ferrierite in NaOH solution and adding it to CTAB solution, the optimal concentration of NaOH was 0.67 – 1.3 N to contain mesopores and to keep characteristics of FER simultaneously. It was confirmed successful formation of well-ordered meso structure in the manufacturing process of mesoporous zeolite substance by dissolving commercial FER in alkali solution to make precursor solution. This catalyst had decreased amount of Brönsted acid site of FER and increased amount and intensity of weak acid site. In addition, it was identified that MMZ-FER with FER structure and mesopore structure is a superior catalyst for activity, selectivity and lifetime in positional isomerization of 2-butene.

References

[1] H. I. Lee, H. J. Park, Y. K. Park , J. Y. Hur, J. K. Jeon, J. M. Kim, *Catal. Today,* 2008, **132**, 68.

[2] H. J. Park, J. K. Jeon, J. M. Kim, H. I. Lee, J. H. Yim, J. Park, and Y. K. Park, *J. Nanosci. Nanotech.*, 2008, **8**, 5439.

[3] J. K. Jeon, H. Lee, J. H. Yim, Y. S. Kim, S. J. Lee, Y. K. Park, J. K. Shon and J. M. Kim, *Catal. Lett.,* 2007, **119**, 179.

[4] N. You, J. H. Yim, S. J. Lee, J. H. Lee, Y. K. Park and J. K. Jeon, *J. Nanosci. Nanotech.*, 2007, **7**, 3800.

See more of this Session: Poster Session of CRE Division
See more of this Group/Topical: Catalysis and Reaction Engineering Division

Platinum nanoparticles embedded in nanoporous carbon spheres as shape selective catalyst

Maryam Peer [a], *Ali Qajar* [a], *Ramakrishnan Rajagopalan* [b], *Henry C. Foley* [a]
[a] *Department of Chemical Engineering, The Pennsylvania State University, University Park, PA, 16802, USA*
[b] *The Materials Research Institute, The Pennsylvania State University, University Park, PA 16802, USA*

The goal in this study is to synthesize and use nanoporous carbon microspheres within which are embedded even smaller platinum nanoparticles as more efficient shape selective catalyst for liquid phase hydrogenation reactions. Previously, in our group, platinum embedded in carbon was prepared and activity and shape selectivity of the catalyst were shown in liquid phase hydrogenation reactions. The main problem of this as-synthesized catalyst was the low activity due to the long diffusion paths to reach to the active platinum sites. This led to fewer active sites accessible for the reaction. The motivation to make the diffusion length shorter and make more of the platinum accessible led to the synthesis of carbon microspheres with an average particle size of 400 nm. Platinum nano particles were formed by reduction of platinum acetyl acetonate using furfuryl alcohol. With temperature control in reduction step platinum particles as small as 2 nm were formed. The catalyst was synthesized by pyrolysis of preformed platinum/polyfurfuryl alcohol spheres prepared using a micellar solution. Pluronic F127 surfactant was used as the structure directing agent in the solution to form spheres.

Catalyst synthesized using this method showed high stability against sintering even after subsequent high temperature treatments. Activity and shape selectivity of the catalyst were demonstrated for liquid phase hydrogenation of alkenes at different temperatures.

(629u) Effect of the Support of Metal Oxide Supported Ruthenium Catalysts for the Conversion of Cellulose Into Sugar Alcohols

Darlene Z. Galloza-Lorenzo[1], Rafael Mendez-Roman[1], Yomaira J. Pagn-Torres[2], James A. Dumesic[2] and Nelson Cardona-Martinez[1], (1)Chemical Engineering, University of Puerto Rico at Mayaguez, Mayaguez, PR, (2)Chemical and Biological Engineering, University of Wisconsin-Madison, Madison, WI

The increase of the world's energy needs with the combination of several factors including the diminishing of petroleum resources, the increase in the petroleum derivatives prices, as well as the increase of green house gas emissions, have caused great interest of finding new processes to produce renewable fuels, such as the hydrogenation of cellulose by a bifunctional catalyst into sugar alcohols. In this work, we focus on the study of the effect of the support acid-base properties on the catalytic performance during the conversion of cellulose into sugar alcohols, which can be used in a biorefinery platform as a renewable feedstock for the production of fuels and high value chemicals. The series of metal oxide supports studied includes MgO, ZrO_2, TiO_2, two types of Nb_2O_5, Al_2O_3, and SiO_2. Furthermore, binary and ternary metal oxides were also analyzed such as two types of SiO_2-TiO_2, WO_3-TiO_2, Al_2O_3-TiO_3, Fe_2O_3-TiO_2 and SiO_2-TiO_2-WO_3. The materials used display a wide range of acid-base properties. Ru was supported using evaporative deposition on all the supports. The catalytic activity appears to have a correlation with the Sanderson electronegativity for the single metal oxide materials passing through a maximum for Ru/Nb_2O_5 for the supported catalyst and through ZrO_2 for the supports. The catalytic materials were characterized using nitrogen adsorption, X-Ray Diffraction (XRD) and Fourier Transform Infrared Spectroscopy (FTIR). Ru/Nb_2O_5 has a significantly lower surface area than Ru/SBA-15 but displays higher activity. The selectivity of the mesoporous material is equally better than that of Ru/Nb_2O_5. On the other hand, the structural differences between SiO_2 and SBA-15 results in a significant difference in selectivity, and smaller effect over the activity. The binary metal oxide TiO_2-WO_3 had a better activity and selectivity than SBA-15, probably due to its more acidic character. Ru/TiO_2-WO_3 displayed the best cellulose conversion; on the other hand Ru/SiO_2-TiO_2-WO_3 shows better selectivity to sugar alcohols. Also it is seen that the identity of nonporous supports has small effect on cellulose hydrogenation activity.

See more of this Session: Poster Session of CRE Division
See more of this Group/Topical: Catalysis and Reaction Engineering Division

(629w) High Throughput Screening of Catalytic Materials for JP-8 Fuel Reformation

John E. Bedenbaugh, Chemical Engineering, University of Delaware, Newark, DE and Jochen Lauterbach, Chemical Engineering, University of South Carolina, Columbia, SC

Within the past decade, the high-throughput approach hasbecome widespread in the field of catalysis (1-3). In this work, this methodology is applied for thediscovery and optimization of catalytic materials for reformation of militaryjet fuel (JP-8) to lighter hydrocarbons. JP-8 is the single battlefield fuelof NATO and the U.S. Military. These organizations have a critical need forconverting their preferred fuel source into a flexible and readily availablepower supply for portable applications, such as solid oxide fuel cells andliquefied petroleum gas. The objective of this work is to use ahigh-throughput screening approach to discover and optimize novel reformingcatalyst formulations that will enable reforming technology to convert readilyavailable hydrocarbon-based fuels, such as JP-8, directly to a lighterhydrocarbon feed suitable for portable fuel cell applications.

An existing high-throughput reactor system (4) was modified for the JP-8 reformingstudies. The reactor system consists of 16 separate stainless steel reactortubes that are loaded with individual powder catalyst samples, typicallybetween 150 mg and 750 mg, and tested under realistic conditions. Thehigh-throughput reactor system operates at ambient pressure with a temperaturerange between room temperature and 1100 K and space velocities between 3,500(mL/hr * g_{cat}) and 200,000 (mL/hr * g_{cat}). Liquid JP-8is dosed into a heated mixer with high internal evaporative surface area, whereit is vaporized and combined with carrier gas to form the reactant gas feed. Thereactor effluent streams from each of the individual reactors remain inseparate channels throughout the analysis process to prevent the mixing ofreaction products from occurring. The reactor effluent passes through aparallel condenser system to remove any unreacted JP-8 before proceeding foranalysis.

The gaseous product stream then enters a rapid-scan Fouriertransform infrared (FTIR) spectroscopic imaging system. This system enableschemically-sensitive, parallel screening of all 16 gas-phase reactor productdistributions through a gas phase array (GPA), incorporating a 128x128 focalplane array (FPA) detector as the infrared (IR) detecting element. Inaddition, a gas chromatography-mass spectrometry (GC-MS) system is incorporatedvia a computer controlled 16-way switching valve and allows for furtherquantitative analysis of the product composition of the effluent streams. TheGC-MS is capable of identifying and differentiating between specifichydrocarbon molecules and was calibrated to provide compositional informationfor each desired product.

Catalysts are analyzed for their desired properties, such asconversion, selectivity, sulfur tolerance, and coking resistance. Because JP-8may contain up to 3,000 ppmw sulfur, sulfur-tolerant reforming catalysts thatcan withstand coking and demonstrate long-term activity and stability arerequired. Catalysts that can produce a well-defined sulfur-containing productdistribution can be optimized using existing clean coal technology for removalof sulfur compounds. By applying a high throughput methodology to thisprocess, a large number of materials are screened, including several monometallictransition metals, such as Rh, Pd, Pt, Ru, and Ir, as well as bimetalliccombinations involving active metals. In addition,several oxide support materials, including Al_2O_3, SiO_2,CeO_2, ZrO_2, MoO_2, and Nb_2O_5,were synthesized and tested.

The high throughput methodology allows for acomprehensive search of the parameter space affecting catalyst performance,including catalyst composition, support materials, operating conditions, aswell as pre-treatment options.

1. R. J. Hendershot, C. M. Snively, J. Lauterbach,High-throughput catalytic science. *Chemistry-A European Journal* **11**,806 (Jan 21, 2005).

2. W. F. Maier, K. Stowe, S. Sieg, Combinatorial andhigh-throughput materials science. *Angewandte Chemie-International Edition***46**, 6016 (2007).

3. I. Takeuchi, J. Lauterbach, M. Fasolka, in *MaterialsToday*. (2005), vol. 8, pp. 18-26.

4. R. J. Hendershot *et al.*, A novel reactorsystem for high throughput catalyst testing under realistic conditions. *AppliedCatalysis a-General* **254**, 107 (Nov 10, 2003).

See more of this Session: Poster Session of CRE Division
See more of this Group/Topical: Catalysis and Reaction Engineering Division

(629x) Condensed-Phase Cellulose Pyrolysis

Matthew S. Mettler, Chemical Engineering, Univeristy of Delaware, Newark, DE, Dionisios G. Vlachos, Chemical Engineering, University of Delaware, Newark, DE and Paul J. Dauenhauer, Chemical Engineering, University of Massachusetts, Amherst, MA

Fast pyrolysis of biomass converts solid biopolymers into a valuable liquid product (bio-oil) which can be further upgraded to fuels or chemicals. Much effort has been dedicated to developing processes and catalysts which maximize efficiency, throughput and/or bio-oil quality (e.g., [1-2]). While these efforts are important to developing thermochemical biorefineries of the future, the complex chemistry by which the biomass feedstock breaks down is still largely unknown. Global reaction kinetics for the pyrolysis of cellulose have been published (e.g., [3]), but the accuracy of these models is debated [4]. Additionally, recent work has indicated both indirectly (e.g., [5]) and directly [6] that solid biomass pyrolysis proceeds through a high temperature intermediate liquid before volatilizing into the final bio-oil constituents. In order to maximize the accuracy of pyrolysis experiments and prevent device-dependent results, it is important that mass and heat transport rates within the solid biomass and intermediate liquid be faster than the intrinsic reaction kinetics. In this work, we demonstrate an experimental technique whereby biomass pyrolysis can be studied in a kinetically-limited manner. By carefully controlling the size and shape of the biomass sample, we eliminate composition and thermal gradients. Additionally, volatiles, gases and char are characterized using multi-dimensional analytical techniques. Data from our experiments are consistent with recent results [7] that the first several cellulose oligomers (glucose, cellobiose, cellotriose, etc.) have different distributions of volatile organic products. Furthermore, we find that these oligomers differ in the fraction of the biomass which is converted to char and permanent gases. We believe that this novel experimental technique can contribute to the available experimental information on biomass pyrolysis and eventually lead to better chemical understanding and the creation of detailed kinetic models.

1. Vispute, T.P., et al., *Renewable Chemical Commodity Feedstocks from Integrated Catalytic Processing of Pyrolysis Oils.* Science, 2010. **330**(6008): p. 1222.

2. Agblevor, F.A., et al., *Fractional Catalytic Pyrolysis of Hybrid Poplar Wood.* Ind. Eng. Chem. Res., 2010. **49**(8): p. 3533.

3. Bradbury, A.G.W., et al., *KINETIC-MODEL FOR PYROLYSIS OF CELLULOSE.* Journal of Applied Polymer Science, 1979. **23**(11): p. 3271.

4. Varhegyi, G., et al., *IS THE BROIDO-SHAFIZADEH MODEL FOR CELLULOSE PYROLYSIS TRUE.* Energy & Fuels, 1994. **8**(6): p. 1345.

5. Boutin, O., et al., *Radiant flash pyrolysis of cellulose - Evidence for the formation of short life*

time intermediate liquid species. Journal of Analytical and Applied Pyrolysis, 1998. **47**(1): p. 13.

6. Dauenhauer, P.J., et al., *Reactive boiling of cellulose for integrated catalysis through an intermediate liquid.* Green Chemistry, 2009. **11**(10): p. 1555.

7. Patwardhan, P.R., et al., *Product distribution from fast pyrolysis of glucose-based carbohydrates.* Journal of Analytical and Applied Pyrolysis, 2009. **86**(2): p. 323.

See more of this Session: Poster Session of CRE Division
See more of this Group/Topical: Catalysis and Reaction Engineering Division

APPROXIMATE APPROACH FOR KINETIC MODELING OF A COMPLEX CHEMICAL REACTION NETWORK

Jae Ho Lee[1], Sookil Kang[1], Young Kim[2], Sunwon Park[3], Deuksoo Park[1], Seungjun Lee[1] and Sun Choi[1], (1)SK Innovation, Daejeon, South Korea, (2) Korea Institute of Machinery and Materials, Daejeon, South Korea, (3) Korea Advanced Institute of Science and Technology, Daejeon, South Korea

Abstract

The thermal or catalytic cracking of hydrocarbon has very complex reaction network. To build a reasonable model for the chemical reaction network, suitable reaction mechanisms have to be considered. The degree of thermal cracking in the catalytic cracking of light naphtha is investigated. Thermal cracking mechanism becomes more important at temperatures over 650 °C that at temperatures below 650 °C and has to be considered in the kinetic modeling. An approximate approach is newly proposed for the kinetic modeling of complex chemical reaction network systems. This approach is based on the transition state theory and defines the approximate factors which mean the approximate values of entropy change in the thermal and catalytic reactions. Based on the proposed approach, a mechanistic kinetic model with both thermal and catalytic cracking mechanisms is developed and the kinetic parameters are estimated by using the genetic algorithm. The proposed approach is applied to the catalytic cracking of light naphtha and will be effective for the modeling of complex reaction network systems.

Introduction

The catalytic cracking of hydrocarbons involves complex reaction networks with a large number of species, which makes its kinetic modeling difficult. Kinetic modeling of the complex reaction networks has been studied since 1980s in several works. Single event concept[1-5] adopts transition state theory for calculating frequency factors. The structural entropy change of a reaction is calculated by the ratio between the number of configurations, i.e., the number of single events, of the reactant and the activated complex in the transition state. Although its theoretical background is strong, calculating all the configurations of the reactant and activated complex for complex reaction network systems is rather cumbersome. Klein et al.[6-8] suggest a practical approach using linear free energy relationships (LFER) for various cracking reactions with an integrated modeling toolbox. This approach is practical if LFER is known for specific reactions. Structure oriented lumping (SOL)[9,10] is an additive approach for heavy feedstock with huge analytical database focusing the estimation of the properties of products. Some approaches based on basic elementary steps[11-13] are also practical but their theoretical backgrounds are relatively weak and their models consider catalytic cracking reactions only. However more practical approach with less computational load needs to be developed for the high temperature catalytic cracking reactions.

In this work, the mechanisms for the high temperature catalytic cracking reactions for paraffinic naphtha are elucidated by experiments. Subsequently, a kinetic model of the reactions is developed based on a newly proposed approximate approach. This approach defines the approximate factors which mean the approximate values of entropy changes in the thermal and catalytic reactions. Only two types of parameters, the correction factors and activation energy terms, are estimated by a global optimization method. The range of the correction factors are within 0.1~1.5. Activation energies are bounded within literature values. The narrow range of parameters can reduce the stiffness problems in the ODE solving during the parameter estimation.

Experimental

A micro activity test (MAT) was performed without catalyst to measure the degree of thermal cracking. The MAT is widely used to measure the performance of FCC catalysts and it is standardized by ASTM D3907. **Figure 1** shows a schematic diagram of the MAT system used in this work. 1g of naphtha was injected in 75 sec into a 33cc reactor, in the temperature range of 660~700°C. Light straight run naphtha (LSR) supplied from SK Innovation Co. Ltd. was used as a feedstock. The composition of LSR is presented in **Table 1**. The catalytic cracking tests were performed with the same LSR feeds in a 0.94 cm (ID) x 2.4 m (height) circulating fluidized bed riser (CFBR) at 660 to 700°C. The catalyst fed to the riser was a zeolite based catalyst and it was circulated in the reaction system through a stripper and a regenerator. The hydrocarbon product stream was cooled and separated into gas and liquid. Both products were analyzed by GC/FID and hydrogen was analyzed by GC/TCD. The overall mass balances for the experiments were above 95%.

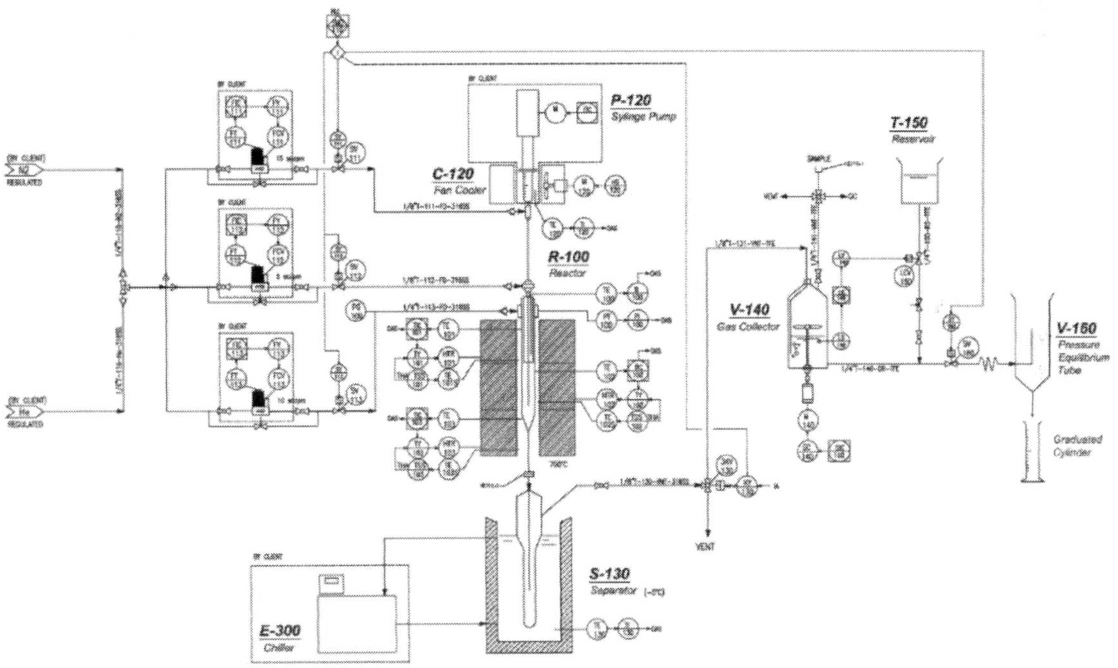

Figure 1. Schematic diagram of MAT equipment

Table 1. Composition of Light Straight Run Naphtha (wt%)

	C_4	C_5	C_6	C_7	C_8	C_9	Sum
Paraffin		16.08	11.53	4.41	0.83	0.04	32.9

Iso-Paraffin		14.60	19.09	11.53	2.40	0.11	47.4
Olefins	1.35	0.27	0.53	0.40			2.6
Naphthenes		1.41	5.78	4.71	1.28		13.2
Aromatics			2.02	1.20	0.42		3.6
Sum	1.3	32.4	38.9	22.2	4.9	0.2	100

Catalytic Cracking Mechanism

Figure 2 shows a schematic diagram for the overall cracking pathways of paraffins.

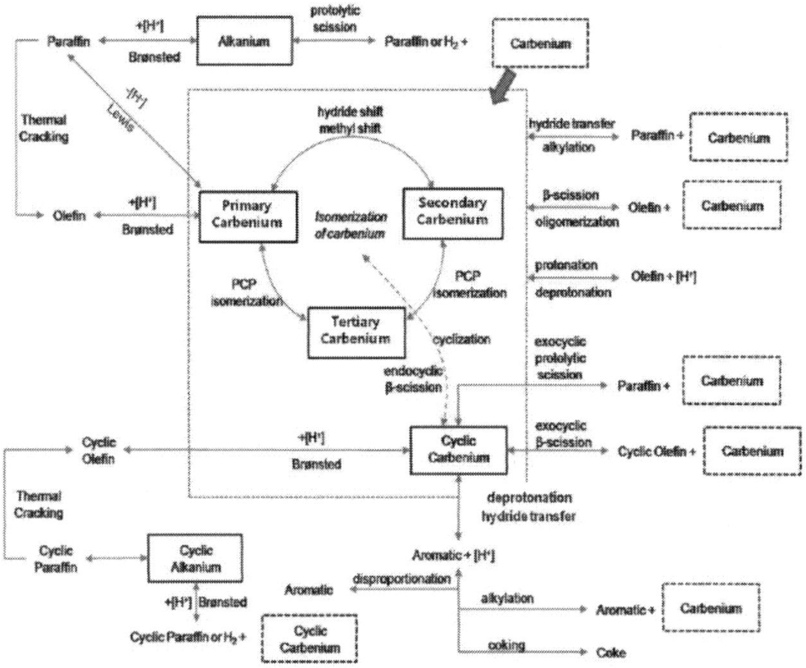

Figure 2. Schematic diagram for cracking pathways of paraffins

Approximate Approach

Kinetic modeling for a complex reaction network involving both thermal and catalytic cracking requires elaborate efforts due to the complexities in mechanisms and huge number of kinetic parameters. To obtain acceptable solutions within a practical calculation time, a systematic approach with reasonable assumptions on the basis of reliable theoretical backgrounds is necessary. In this work, an approximate approach is proposed for kinetic modeling of paraffinic naphtha cracking in a circulating fluidized bed riser system as **Table 2**.

Table 2. Proposed approximate approach

Transition State Theory

$$A_{0,TST} = \left(\frac{k_B T}{h}\right)\left(\frac{\exp(1-\Delta n)}{C^0}\right)\exp\left(\frac{\Delta S^{0*}}{R}\right)$$

energy term — stoichiometry change — entropy term → Quantum Chemical Calculation → Approximate Values

Approximate Factor for Thermal Cracking

$$\exp\left(\frac{\Delta S^{0*}}{R}\right) = \left(\frac{f_r^2}{f_t^3}\right)(\textit{steric factor})$$

Approximate Factor for Catalytic Cracking

$$\exp\left(\frac{\Delta S^{0*}}{R}\right) = \exp\left(\frac{\Delta S^{0,gas}_{trans}\,\theta}{R}\right)$$

Example of Steric Factors = f(degree of freedom)

A B	$[AB]^*$	$Q(A)$ $Q(B)$	$Q^*(AB^*)$	Q'	steric factor[2] F_s
atom atom	linear	f_t^3 f_t^3	$f_t^3 f_r^2 f_v^{3(1+1)-6}$	$(f_r^2 f_t^{-3})$=0.006 @ 300~500K	1
atom nonlinear	nonlinear	f_t^3 $f_t^3 f_r^3 f_v^{3N-6}$	$f_t^3 f_r^3 f_v^{3(N+1)-7}$	$(f_r^2 f_t^{-3})(f_v^2 f_r^{-2})$	$(f_v^2 f_r^{-2}) = 10^{-2}$
linear nonlinear	nonlinear	$f_t^3 f_r^2 f_v^{3N-5}$ $f_t^3 f_r^3 f_v^{3N-6}$	$f_t^3 f_r^3 f_v^{3(N+N)-7}$	$(f_r^2 f_t^{-3})(f_v^4 f_r^{-4})$	$(f_v^4 f_r^{-4}) = 10^{-4}$
nonlinear nonlinear	nonlinear	$f_t^3 f_r^3 f_v^{3N-6}$ $f_t^3 f_r^3 f_v^{3N-6}$	$f_t^3 f_r^3 f_v^{3(N+N)-7}$	$(f_r^2 f_t^{-3})(f_v^5 f_r^{-5})$	$(f_v^5 f_r^{-5}) = 10^{-5}$

Entropy Changes for Catalytic Cracking = f(translational entropy change of gas bulk gas molecules)

$$\frac{\Delta S^{0,gas}_{trans}}{R} = \ln\left\{\frac{RT}{P_t N_A}\frac{(2\pi(MW)k_B T)^{3/2}}{h^3}\right\} + \frac{5}{2}$$

This approach uses the approximate factor, which represents the entropy change between activated complex and reactant(s) in a specific reaction. For thermal cracking reactions, the approximate factors are calculated on the basis of transition state theory. For catalytic cracking reactions, they are calculated on the basis of the translational entropy change from bulk gas to adsorbed species. In the kinetic parameter estimation, only two types of parameters are estimated, which are activation energy and correction factor f_c for pre-exponential factor. The approximate factors p_{therm} and p_{cat} are calculated by the approximate approach. (The kinetic expressions for the approximate approach are shown in **Table 3**.) The activation energies for specific reactions are gathered from the literature and used as the upper and lower bounds in the parameter estimation. All the correction factors f_c are estimated in the narrow range 0.1-1.5. The expected advantages of the proposed approach are as follows:

- The approach is conceptually simple and theoretically reasonable.
- Good performance is obtained in the parameter estimation, as using the narrow range of parameters. Numerical stiffness problems can be reduced in the parameter estimation.
- Both mechanisms of thermal and catalytic cracking can be treated in a model using the same concept.

$$k = A_{0,\mathrm{approx}} f_c \exp\left(-\frac{E_a}{RT}\right)$$

$$A_{0,\mathrm{approx}} = \left(\frac{k_B T}{h}\right) \exp(1 - \Delta n) \exp\left(\frac{\Delta S^{\circ\ddagger}}{R}\right)$$

$$= \begin{cases} \left(\frac{k_B T}{h}\right) \exp(1 - \Delta n) p_{\mathrm{therm}} & \text{for thermal cracking} \\ \left(\frac{k_B T}{h}\right) \exp(1 - \Delta n) p_{\mathrm{cat}} f_{\mathrm{ads}} & \text{for catalytic cracking} \end{cases}$$

$$p_{\mathrm{therm}} = \begin{cases} \left(\frac{f_r^2}{f_t^3}\right) F_{s,\mathrm{tuned}} & \text{for unimolecular reaction of thermal cracking} \\ \left(\frac{f_r^2}{f_t^3}\right) F_s & \text{for bimolecular reaction of thermal cracking} \end{cases}$$

$$p_{\mathrm{cat}} = \begin{cases} \exp\left[\frac{\Delta S_{\mathrm{trans}}^{\circ,\mathrm{gas}}}{R} \theta\right] & \text{for unimolecular reaction of catalytic cracking} \\ \exp\left[\frac{\Delta S_{\mathrm{trans}}^{\circ,\mathrm{A}}}{R} \theta_A\right] \exp\left[\frac{\Delta S_{\mathrm{trans}}^{\circ,\mathrm{B}}}{R} \theta_B\right] & \text{for bimolecular reaction of catalytic cracking} \end{cases}$$

$$f_{\mathrm{ads}} = \begin{cases} C_{\max} \exp\left(\frac{\Delta S_{\mathrm{ads,P}}}{R}\right) & \text{for protonation} \\ 1 & \text{for other catalytic reactions} \end{cases}$$

F_s = steric factor for thermal cracking reaction

f_c = correction factor for pre-exponential factor, $0.1 \le f_c \le 1.5$

θ, θ_A, θ_B = entropy correction factor for catalytic cracking reation

Table 3. Kinetic Expression for the Approximate Approach

Software Development for Integrated Modeling of Thermal and Catalytic Cracking

An integrated modeling software package is developed for complex chemical reaction network systems with a graphic user interface (GUI) for easy application of the developed kinetic model. The package includes various modules, such as an experimental data (ExD) module, an activation energy database (EaDB) gathered from the open literature, a reaction mechanism description (RMD) module, a chemical reaction network generation (CRN) module, as well as an ordinary differential equation (ODE) solver, and a genetic algorithm (GA) module for global optimization. The RMD module is a text-based description with reaction information. The information includes reaction description, reversibility of the reaction, constraints for carbon numbers, initial and boundary values of kinetic parameters, and steric factors depending on reactant types. As it is a text-based descript, the reaction information can be easily added, removed, or updated for other applications.

In the beginning, elementary reactions for thermal and catalytic cracking are selected in the GUI. A chemical reaction network composed of selected reactions and a set of ordinary differential equations are generated in the CRN module. Subsequently, global optimization by genetic algorithm is performed with ODE solver. "ode15s" or "ode23tb" solvers in MATLAB can be selected, as they are known to work well for stiff problems.[14]

Finally, predicted composition data are compared with experimental data using a parity plot in the plot module, and optimized kinetic parameters are obtained. If the fitness of the result is not good, the parameters of the genetic algorithm may be tuned. **Figure 3** shows the schematic diagram of this software, and **Figure 4** shows the GUI.

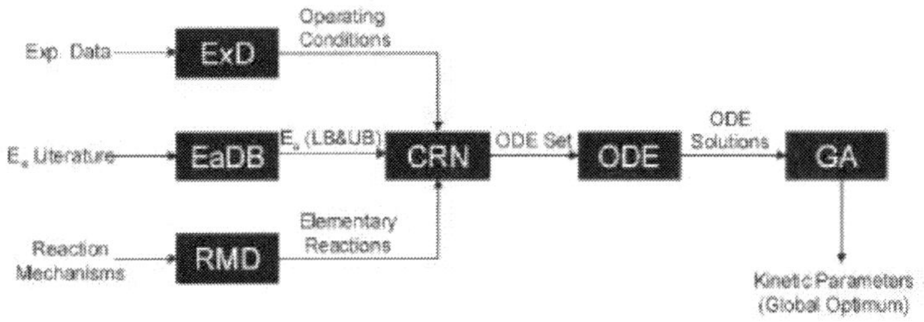

Figure 3. Schematic diagram of the integrated modeling software package for complex chemical reaction network modeling

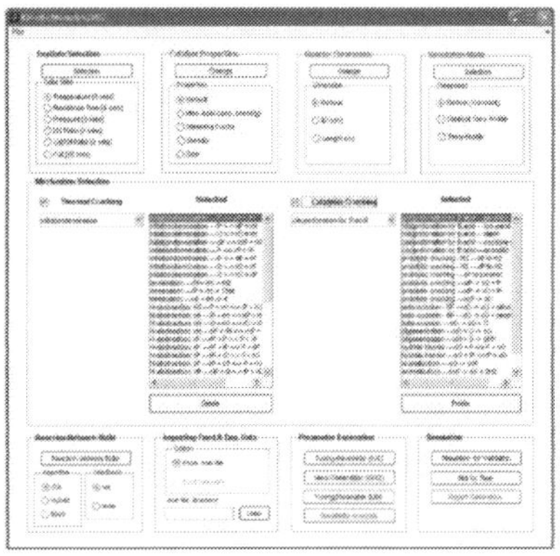

Figure 4. GUI of the developed package

Application

The developed kinetic model based on the approximate approach is applied to a catalytic racking process for paraffin rich naphtha. In the kinetic model, both thermal and catalytic cracking mechanisms are included. The reaction families and species included in this model are described in **Table 4**. In the chemical reaction network generation module, 121 species and 142 reaction families with 1016 elementary reactions are generated.

From the chemical reaction network system, 121 ODEs are generated and solved by ode23tb in MATLAB, which is a reliable algorithm for the problems with stiffness. The object function is to minimize the sum of squares of errors between the experimental and predicted data. A total of 220 kinetic parameters are estimated. 142 correction factors (f_c) for pre-exponential factors bounded in 0.1-1.5; 70 activation energies (E_a) bounded in the upper and lower values found in the literature; 7

coefficients (α for $E_a = E_a + α N_C$) for the protolytic cracking reactions and *β*-scission reactions; and the number of acid sites per gram of catalyst. The amount of acid sites of the catalyst is estimated under the upper limit of 0.85 mmol/g of ZSM-5.50 There are no stiffness problems in the process of parameter estimation.

Table 4. Reaction Families and Species in the Model[a]

(a) Reaction Families	
thermal cracking	catalytic cracking
initiation/termination	protonation/deprotonation by Brønsted acid
hydrogen abstraction	protonation/deprotonation by Lewis acid
radical decomposition/addition	protolytic cracking
radical isomerization	*β*-scission/oligomerization
cyclization	hydride transfer
Diels–Alder reaction for aromatics	cyclization/endocyclic scission
aromatization	alkylation/dealkylation
heavies formation	heavies formation
(b) Species: Total 121 Species	
gas phase (bulk) species	H_2, P, IP, cP, O, cO, D, cD, A, CKP (45 species)
radical species	rH, rP, rIP, rO, rD, rA (34 species)
carbocation (surface) species	sP, sIP, scP, sO, scO, scD, sA (41 species)
acid site	vs (1 species)

[a] Notation: P = paraffins, IP = isoparaffins, cP = cycloparaffins, O = olefins, cO = cycloolefins, D = dienes, cD = cyclodienes, A = aromatics, CKP = C_{10} + heavies, r = radical species, and s = carbocation (surface) species.

The parameter estimation has been performed using two sets of experimental data at 690 and 699 ℃. After obtaining the optimized parameters, this model is applied to another data set obtained at 683 ℃. **Figure 5** shows parity plots of the product wt % yields between experimental data and predicted data using the kinetic model. The coefficient of determination, R^2 ,values are about 0.95-0.97 for the three sets of data.

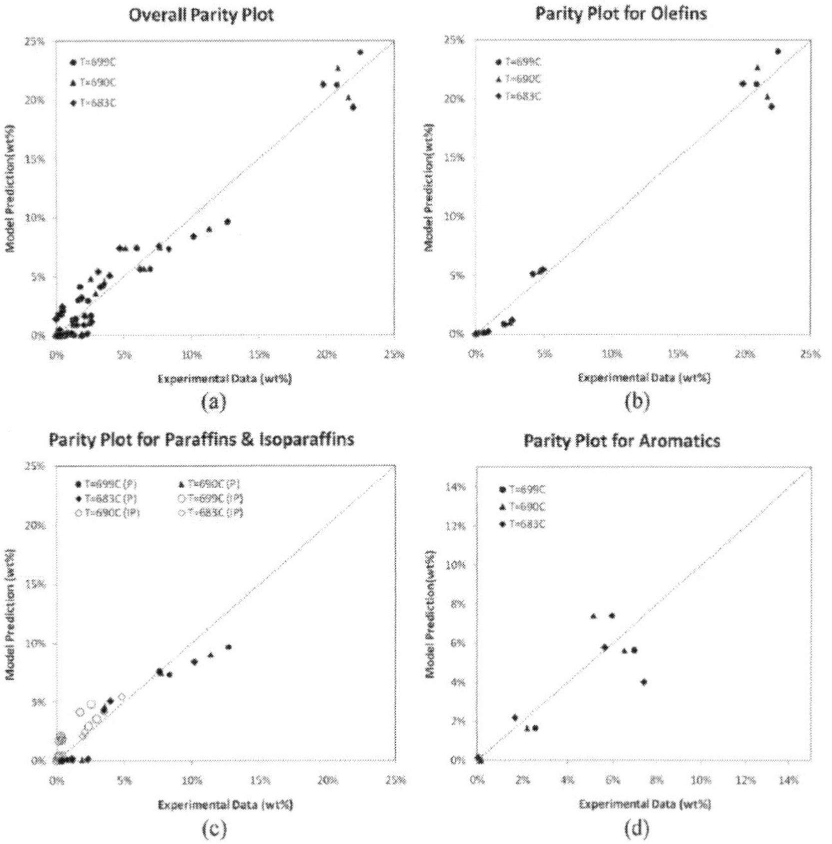

Figure 5. Parity plots for three sets of experimental data: (a) overall parity plot; (b) parity plot for olefins; (c) parity plot for paraffins and isoparaffins; (d) parity plot for aromatics; •, T = 699 ℃; ▲, T = 690 ℃; ◆,T = 683 ℃. Closed symbols are for paraffins, and open symbols are for isoparaffins in part c

Conclusion

The catalytic cracking of paraffin rich naphtha in the circulating fluidized bed is an attractive area in the petrochemical industry. A new approximate approach has been proposed to systematically model the complex chemical reaction network system. According to thermal and catalytic cracking tests using MAT and CFBR units, the degree of thermal cracking (DTC) is more than 50% at 675℃. This implies that both thermal and catalytic cracking mechanisms have to be included in the kinetic modeling to get reasonable results.

The proposed approach has been applied to an industrial problem for the catalytic cracking of paraffin-rich naphtha. An integrated modeling software package with a graphical user interface has been developed for easy application. 142 reaction families and 121 species (45 bulk species, 34 radical pecies,

41 cation species, and acid sites) are automatically generated on the basis of both thermal and catalytic cracking mechanisms through the integrated modeling software package. The kinetic parameters and acid concentration are estimated by a genetic algorithm. It is reasonable to apply the model to yield an estimation of the major olefin products, as the values of R^2 for three sets of experiments are 0.95-0.97.

The developed model can be easily applied to the systems with other acid catalysts if the concentration and their functions of active sites are known. This approach is conceptually simple and theoretically reasonable. A narrow range of the boundary values of kinetic parameters helps to search the reasonable parameters and overcome the stiffness problem in the process of parameter estimation. This approach can be applied to other complex reactions by simple adjustment of reaction stoichiometries or reaction mechanisms.

References

(1) Claymans, P. J.; Froment, G. F. Computer-generation of reaction paths and rate equations in the thermal cracking of normal and branched paraffins. *Comp. & Chem. Eng.* **1984**, 8(2), 137-142.

(2) Feng, W.; Vynckier, E.; Froment, G. F. Single event kinetics of catalytic cracking. *Ind. Eng. Chem. Res.* **1993**, 32(12), 2997-3005.

(3) Quintana-Solorzano, R.; Thybaut, J. W.; Marin, G. B. A single-event microkinetic analysis of the catalytic cracking of (cyclo)alkanes on an equilibrium catalyst in the absence of coke formation. *Chem. Eng. Sci.* **2007**, 62, 5033-5038.

(4) Park, T. Y.; Froment, G. F. Kinetic modeling of the methanol to olefins process. 1. Model formulation. *Ind. Eng. Chem. Res.* **2001**, 40, 4172-4186.

(5) Park, T. Y.; Froment, G. F. Kinetic modeling of the methanol to olefins process. 2. Experimental results, model discrimination, and parameter estimation. *Ind. Eng. Chem. Res.* **2001**, 40, 4187-4196.

(6) Klein, M. T.; Hou, G.; Bertolcini, R. J.; Broadbeldt, L. J.; Kumar, A. *Molecular Modeling in Heavy Hydrocarbon Conversions*, CRC Press, Taylor & Francis, **2006**

(7) Wei, W.; Bennett, C. A.; Tanaka, R.; Hou, G.; Klein Jr., M. T.; Klein, M. T. Computer aided kinetic modeling with KMT and KME, *Fuel Proc. Tech.*, **2008**, 89, 350-363.

(8) Broadbelt, L. J.; Stark, S. M.; Klein, M.T. Computer Generated Pyrolysis Modeling: On-the-Fly generation of species, reactions and rates, *Ind. Eng. Chem. Res.*, **1994**, 33, 790-799.

(9) Quann, R. J.; Jaffe, S. B. Structure-Oriented Lumping: Describing the chemistry of complex hydrocarbon mixtures. *Ind. Eng. Chem. Res.* **1992**, 31, 2483-2497.

(10) Quann, R. J.; Jaffe, S. B. Building useful models of complex reaction systems in petroleum refining. *Chem. Eng. Sci.*, **1996**, 51(10), 1615-1631.

(11) Bhan, A.; Hsu, S. H.; Blau, G.; Caruthers, J. M.; Venkatasubramanian, V.; Delgass, W. N. Microkinetic modeling of propane aromatization over HZSM-5. *J. Catal.* **2005**, 235, 35-51.

(12) Carabineiro, H.; Pinheiro, C. I. C; Lemos, F.; Ribeiro R. Transient microkinetic modelling of n-heptane catalytic cracking over H-USY zeolite. *Chem. Eng. Sci.* **2004**, 59, 1221-1232.

(13) Pinto, R. R.; Borges, P.; Lemos, M. A. N. D. A.; Ribeiro, F. R. Kinetic modelling of the catalytic cracking of n-hexane and n-heptane over a zeolite catalyst. *Appl. Catal. A: Gen.* **2004**, 272, 23-28.

(14) Beers, K. J. *Numerical Methods for Chemical Engineering, Applications in MATLAB*; Cambridge University Press: New York, **2007**.

(629z) 2,4-Di-Ter-Pentyl-Phenol Synthesis On Silica Gel Supported Phosphotungstic Acid Catalyst

Lichun Dong and Nicheng Chen, Chemical Engineering, Chongqing University, Chongqing, China

The synthesis of 2,4-di-tert-amylphenol (2,4-DTAP) by the F–C alkylation reaction of phenol with 2-methyl-2-butene catalyzed by phosphotungstic acid supported on silica gel (PW/SiO_2) was studied. A series of PW/SiO_2 catalysts with various PW loading, prepared by impregnation method, were characterized by XRD, N_2 adsorption, NH_3-TPD, and SEM. The results showed that when PW loading is smaller than 40%, PW can be highly dispersed on SiO_2 surface; 2,4-DTAP has the highest yield when the PW loading of PW/SiO_2 catalyst is 40%. Afterwards, the effects of various operating parameters on 2,4-DTAP was investigated, the optimum reaction conditions were found to be: molar ratio between phenol with 2-methyl-2-butene, 1:2.2; PW/SiO_2 catalyst amount, 0.5g/10mmol phenol; Reaction time, 70°C; Reaction time, 4 h. At the optimum conditions, the yield of 2,4-DTAP can reach up to 76.93%.

See more of this Session: Poster Session of CRE Division
See more of this Group/Topical: Catalysis and Reaction Engineering Division

(640a) Manipulation of the Hydrocarbon Pool for Methanol-to-Hydrocarbons On H-ZSM-5

Samia Ilias and Aditya Bhan, Chemical Engineering and Materials Science, University of Minnesota, Minneapolis, MN

The dehydrative conversion of methanol-to-hydrocarbons over acidic zeolite catalysts has been extensively studied, showing that different classes of hydrocarbons, such as olefins and gasoline-range hydrocarbons, can be formed as a result of the shape selective nature of zeolites, although these reactions often proceed with low selectivity. It is now widely accepted that the conversion of methanol/DME proceeds through a "hydrocarbon pool" mechanism, in which organic molecules present in the zeolite pores act as co-catalysts for olefin formation. In this work, DME was reacted over H-ZSM-5, using propene and toluene as co-feeds to determine the effect of the hydrocarbon pool composition on product selectivity. Co-feeding DME with small amounts of propene (<10 C%) resulted in a product mixture rich in C_4-C_7 aliphatics (47 C%) whereas co-feeding DME with toluene resulted in a product selectivity dominated by C_2-C_3 olefins (36 C%) and aromatics (45 C%). The reaction of DME with a mixed propene/toluene co-feed resulted in an intermediate product selectivity, compared to the single co-feed cases. Varying the olefin:aromatic ratio in the co-feed allowed for the product selectivity to be varied predictably and systematically. The trends did not change for reactions with less than 25 C% conversion. These results show that product selectivity of DME reactions over zeolites can be tuned by varying the composition of the hydrocarbon pool, and thus controlling the relative rates of olefin and aromatic methylation. The effect of temperature, DME pressure, and zeolite crystal size on the hydrocarbon pool evolution and product selectivity will also be discussed.

See more of this Session: Catalysis with Microporous and Mesoporous Materials I
See more of this Group/Topical: Catalysis and Reaction Engineering Division

(640b) Probing the Gold Active Sites In Au/TS-1 for Gas Phase Epoxidation of Propylene In the Presence of Hydrogen

Wen-Sheng Lee[1], M. Cem Akatay[2], Eric A. Stach[2], Fabio H. Ribeiro[1] and W. Nicholas Delgass[1], (1) School of Chemical Engineering, Purdue University, West Lafayette, IN, (2)School of Materials Engineering and Birck Nanotechnology Center, Purdue University, West Lafayette, IN

A single-step, direct catalytic partial oxidation of propylene to propylene oxide (PO) using molecular oxygen has long been desired as a greener and more economical process compared to the current PO production routes. By co-feeding hydrogen with propylene and oxygen and using Au/TiO_2 as the catalyst, one can produce PO with high selectivity (~ 99%) at ambient pressure. However, the low PO rate (~10 gPO/h/kgCat), as well as poor catalyst stability, are short of the requirements for industrial application. The catalytic performance can be significantly improved by using TS-1(titanium silicalite-1) as the support for nano-gold clusters/particles such that the PO rate increases up to ~100 gPO/h/kgCat and the selectivity is maintained at ~ 80% PO. While gold particles with sizes in the range 2-5 nm in Au/TiO_2 are assigned in the literature as the active sites in PO reaction, our previous work suggests that gold clusters smaller than 2 nm are much more active in propylene epoxidation. However, the issue of whether gold clusters inside the TS-1 micropores are active in the PO reaction has not been settled. This work addresses this question and attempts to further understand the gold active sites in Au/TS-1.

A TS-1 support coated with an inert layer of S-1 (silicalite-1) was designed and prepared for probing the activity of the gold clusters inside the TS-1 micropores for the PO reaction. This coated material (S-1/TS-1) was characterized via TEM, XRD, XPS, DRUV-vis and nitrogen adsorption analysis at LN2 temperature to understand the result of the silicalite-1 coating and the corresponding influence on the spatial distribution of

See more of this Session: Catalysis with Microporous and Mesoporous Materials I
See more of this Group/Topical: Catalysis and Reaction Engineering Division

(640c) Bifunctional Mesoporous Silica for Cooperative Catalysis Prepared Through Sequential Grafting of An Organic Base and An Organic Acid

Nicholas Brunelli, School of Chemical & Biomolecular Engineeering, Georgia Institute of Technology, Atlanta, GA, Krishnan Venkatasubbaiah, Chemical & Biomolecular Engineering, Georgia Institute of Technology, Atlanta, GA and Christopher W. Jones, School of Chemical & Biomolecular Engineering, Georgia Institute of Technology, Atlanta, GA

Coupling reactions, such as aldol, nitroaldol, and Knoevenagel reactions that are typically acid or base-catalyzed, can be accelerated through the addition of a second component that acts cooperatively. Depending on the nature of the interaction, the two components can also quench one another, inhibiting the catalytic activity. Mutually reactive and cooperative functional groups can be separated through grafting onto mesoporous silica, permitting cooperativity and enhancing the rate of reaction relative to the single component system. Here we demonstrate the sequential grafting, spectroscopic characterization, and catalytic activity of a cooperative catalyst with an organic acid and an organic base attached to the surface of a mesoporous silica. The acid component was protected using a thermally cleavable protecting group, demonstrating for the first time that two mutually reactive functional groups can be sequentially grafted onto mesoporous silica. The bifunctional catalyst conclusively demonstrates the importance of the strength of the acid component for the aldol condensation. This concept is shown to be broadly applicable to similar C-C bond forming reactions. The observed trend of acid strength has provided direction for future catalyst design.

See more of this Session: Catalysis with Microporous and Mesoporous Materials I
See more of this Group/Topical: Catalysis and Reaction Engineering Division

(640d) Product Selectivity In Catalytic Conversion of Benzyl Alcohol In Mesitylene Using Meso-/ Microporous Zeolites

Dongxia Liu[1], Aditya Bhan[1], Michael Tsapatsis[1] and Saleh Al Hashimi[2], (1)Department of Chemical Engineering and Materials Science, University of Minnesota, Minneapolis, MN, (2)Chemical Engineering Department, Petroleum Institute, Abu Dhabi, United Arab Emirates

Product Selectivity in Catalytic Conversion of Benzyl alcohol in Mesitylene using Meso-/ Microporous Zeolites

Dongxia Liu[1], Aditya Bhan[1], Michael Tsapatsis[1], and Saleh Al Hashimi[2]

1. Department of Chemical Engineering and Materials Science, University of Minnesota, 421 Washington Avenue SE, Minneapolis, MN 55455, USA

1. Chemical Engineering Program, The Petroleum Institute, Abu Dhabi, United Arab Emirates

Meso-/microporous zeolites couple the catalytic features of micropores and the improved access and transport provided by the mesopores. The connectivity between the meso- and micropores could be an important factor determining the performance in catalysed reactions. This talk highlights the catalytic behavior of three acidic meso-/microporous zeolite materials (pillared MWW, pillared MFI, and 3DOm-i MFI) studied using catalytic conversion of benzyl alcohol with mesitylene as probe reactions. As expected, the apparent rate constant for the catalytic conversion of benzyl alcohol in mesitylene on materials processing dual meso- and microporosity was higher than those of the corresponding conventional microporous materials. Interestingly, the selectivity of the catalytic probe reaction in the zeolites possessing meso- and microporosity was comparable to that of the microporous analogue, but remarkably different for each type of zeolite: MWW zeolites favor alkylation of benzyl alcohol with mesitylene, while MFI zeolites favor etherification of benzyl alcohol. Chemical titration using 2,6-di-tert butyl pyridine during the catalytic conversion of benzyl alcohol over the catalysts revealed that alkylation and etherification of benzyl alcohol occurred in mesoporous environments and external surface for both MWW and MFI zeolites; However, due to interconnected micropore and mesopore structures in MFI, the etherification reaction selectively occurred within micropores of MFI zeolite as well. These results indicate that the difference in product selectivity is determined by the different micropore structures of MFI and MWW along with the corresponding differences of meso-/micropore connectivity.

(640e) Effects of Synthesis Conditions and Dispersion On the Propene Oligomerization Activity of Ni-Containing Microporous Catalysts

Anton N. Mlinar and Alexis T. Bell, Department of Chemical and Biomolecular Engineering, University of California - Berkeley, Berkeley, CA

The synthesis of transportation fuels from processes such as fluid catalytic cracking (FCC) and Fischer-Trospch synthesis can produce large quantities of light olefins (C_2-C_4) that cannot be directly blended into the final liquid fuel due to their high vapor pressures. Increasing the molecular weight of the olefins by oligomerization is one method by which these byproducts can be reincorporated into the liquid fuel. Supported heterogeneous nickel materials are known to be highly selective and active ethene oligomerization catalysts producing greater than 94% yield to butene and larger oligomeric products [1], [2]. Work on extending these materials to oligomerize other light olefins, including propene, has been limited due to the relatively poor stability of the catalyst in the presence of olefins larger than C_2. To better understand why these catalysts deactivate in the presence of propene, we have synthesized a series of nickel exchanged zeolite catalysts, namely Na-X, and tested them for activity as a function of time-on-stream. It was observed that the synthesis method can drastically change the type of catalyst produced. When water is used as an ion exchange solvent, with $Ni(NO_3)_2$ as the precursor salt, the zeolite undergoes cation exchange leading to well-dispersed divalent nickel sites that each compensate two Al atoms in the zeolite framework. This catalyst has an initial propene conversion of approximately 5% at 5 bar but decreases to less than 1% conversion after one hour time-on-stream. In contrast, when the catalyst is synthesized using acetone as an exchange solvent, the Ni deposits as NiO on the zeolite. This technique causes the catalyst to possess almost no initial activity, but instead leads to an induction period where maximum activity of 5% is observed after 1.5 hours time-on-stream. After the 1.5 hour activation period, the catalyst deactivates in a similar fashion to the water exchanged catalyst. This suggests a common deactivation mechanism that we attribute to adsorbed hydrocarbons on the nickel sites instead of irreversible reduction of the nickel sites as was seen in another study [3]. Impregnation of silica with nickel nitrate and nickel citrate leads to a catalyst that possesses lower activtiy but a qualitatively similar activity profile to the Na-X sample exchanged in acetone. Temperature-programmed reduction experiments with all catalysts suggests that partial exchange with either the exchange sites on Na-X or the silanol groups on silica could correlate with the activity of the catalyst. These results imply that Brønsted acid sites or Ni-O-Al linkages may not be required for catalyst activity, as was suggested by other studies [3],[4]. Furthermore, these experiments imply that the dispersion of nickel may be the critical factor in determining the activity of a heterogeneous nickel-based oligomerization catalyst and work is ongoing to validate this conclusion.

[1] Heveling, J.; Nicolaides, C.P.; Scurrell, M.S.; *Appl. Catal. A.***1998**, *173*, 1-9.

[2] Corma, A., Iborra, S. In *Catalysts for Fine Chemical Synthesis, Microporous and Mesoporous Solid Catalysts*; Derouane, E.G., Ed.; Wiley-Interscience, **2006**; Vol. 4; p. 132-135.

[3] Brückner, A.; Bentrup, U.; Zanthoff, H.; Maschmeyer, D.; *J. Catal.* **2009**, *266*, 120-128.

[4] Spinicci, R.; Tofanari, A.; *Mater. Chem. Phys.* **1990**, *25*, 375-383.

See more of this Session: Catalysis with Microporous and Mesoporous Materials I
See more of this Group/Topical: Catalysis and Reaction Engineering Division

(640f) Co-Feeding Lignin-Derived Compounds and Hydrocarbons On Different Acid Zeolites

Anh T. To, Teerawit A. Prasomsri, Lei Nie and Daniel E. Resasco, School of Chemical, Biological and Materials Engineering, University of Oklahoma, Norman, OK

Abstract for the AIChE 2011 Annual Meeting

The conversion of pure anisole and guaiacol, a model compound of lignocellulosic biomass-derived oils, and its mixtures with tetralin, propylene, *n*-decane, or benzene, has been investigated over zeolites (HZSM-5 and HY) in continuous flow and pulse reactors. The dominant reaction of pure anisole and guaiacol is transalkylation, which produces catechols, phenol, cresols, xylenols, and methyl anisoles as main products. From the pulse experiment, it is suggested that coking and adsorption of the products of anisole conversion and guaiacol are more severe on HZSM-5 catalyst, and leads to a stronger catalyst deactivation as compared to HY. While with pure anisole in the feed a fast deactivation is observed, the addition of tetralin, which is an effective hydrogen donor, to the feed results in lower amounts of carbon deposits and much higher catalyst stability. Hydrogen transfer from tetralin helps desorbing and dehydrating phenolic products from anisole conversion. However the role of hydrogen transfer is more effective in HY than HZSM-5. The other hydrocarbons with weaker hydrogen transfer capability have lower effect (e.g. n-decane), negligible effect (e.g. benzene), or even a detrimental effect (e.g. propylene) on activity. Transalkylation reaction was also observed when anisole was co-fed with tetralin. Experiments with subsequent pulses of anisole and tetralin were also carried out to elucidate the transalkylation reaction pathway between anisole and tetralin. Dissociative pathway occurs dominantly over HZSM-5, while both dissociative and non-dissociative bimolecular pathways are possible over HY. The alkylation reaction of m-cresol with propanols and propylene over HBeta were also tested. Propanols were dehydrated fast to propylene and then reacted with m-cresol. Propylene formed oligomers which deactivated the catalyst. Water can reduce the oligomers and might generate new weak Bronsted acid sites thus enhance the activities.

See more of this Session: Catalysis with Microporous and Mesoporous Materials I
See more of this Group/Topical: Catalysis and Reaction Engineering Division

ETHYLENE OLIGOMERIZATION CATALYST OPTIMIZATION USING FUNDAMENTAL KINETIC MODELING

K. Toch[1], J.W. Thybaut[1], M.A. Arribas[2], A. Martínez[2], and G.B. Marin[1], (1) Laboratory for Chemical Technology, Ghent University, Ghent, Belgium, (2) Instituto de Tecnología Química, UPV-CSIC, Valencia, Spain

Introduction

Homogeneously catalyzed ethylene oligomerization is an established industrial process, mainly resulting in even carbon atom numbered alpha alkenes. Heterogeneous catalysis is preferred, however, e.g., because of environmental reasons and for avoiding a catalyst separation step from the product. Moreover, heterogeneous catalysis offers more opportunities to tune the product distribution and corresponding yields to the market demands, i.e. even-numbered, linear α-alkenes versus a high octane fuel blend hydrocarbon mixture [1].

The present work aims at developing an active, stable and selective catalyst for ethylene oligomerization. This challenge is addressed via fundamental modeling using Single-Event MicroKinetics (SEMK) [2]. Model parameters are determined which can be classified as kinetic and catalyst descriptors. The first type of parameters is specific to the reaction family considered and independent from the catalyst, while the latter specifically accounts for the effect of catalyst properties, such as Si/Al ratio, metal ion site concentration… on the kinetics, e.g. via chemisorption enthalpies, sterical constraint factors,… Through simulation, the catalyst descriptors can be varied in order to identify optimal catalyst behavior defined in terms of product selectivity or yield.

When brought into contact with an acid catalyst, ethylene is relatively stable at mild conditions because it can only form primary carbenium ions [3]. Therefore, typically a metal ion function is introduced to activate ethylene and dimerize it to butylene. Consecutive ethylene insertions are possible on metal ion sites but also on the acid sites, because of the involvement of more stable secondary carbenium ions. In the present work, for kinetic modeling purposes, the kinetics occurring on the metal ion sites are aimed at, i.e. in the absence of acid catalyzed oligomerization.

Experimental Investigation of Ethylene Oligomerization

Experimental ethylene oligomerization data have been gathered on three different catalysts, i.e. 1.8wt% Ni-SiO_2-Al_2O_3, 2.7wt% Ni-Beta and 1.7wt% Ni-USY. At 393 K, a total pressure of 3.5 MPa and a space time equal to 48 kg_{cat} s mol_{eth}^{-1}, see Figure 1, stable behavior up to 9h time on stream was exhibited by Ni-SiO_2-Al_2O_3 and Ni-Beta, whereas Ni-USY suffered a significant activity loss. Ni-SiO_2-Al_2O_3 was more active than Ni-Beta and the product distributions obtained indicated that on the Ni-Beta acid catalyzed oligomerization was responsible for the formation of mainly C_8 and C_{10} components, while this did not occur on the Ni-SiO_2-Al_2O_3. Hence, the 1.8wt% Ni-SiO_2-Al_2O_3 was selected as the benchmark catalyst for modeling purposes in this work. As observed in Figure 2, the obtained reaction products mainly comprised even-numbered alkenes in the C_4-C_{12} range. Only traces of odd-numbered alkenes were found at the reaction conditions studied. A different product distribution was observed over the silica-alumina-based and microporous Beta-based catalysts. While a typical Schulz Flory distribution was obtained with the Ni-SiO_2-Al_2O_3 catalyst, indicative of a single growth and termination mechanism, the product distribution on the Ni-Beta catalyst exhibited a minimum at C_6. The C_{10}+ fraction (diesel products) was significantly higher on the Ni-Beta compared to the Ni-SiO_2-Al_2O_3 catalyst, i.e. 38 wt% versus 17 wt %. The acid catalyzed ethylene insertion in heavier alkenes and dimerization of butene on the Ni-Beta catalyst may explain these observations.

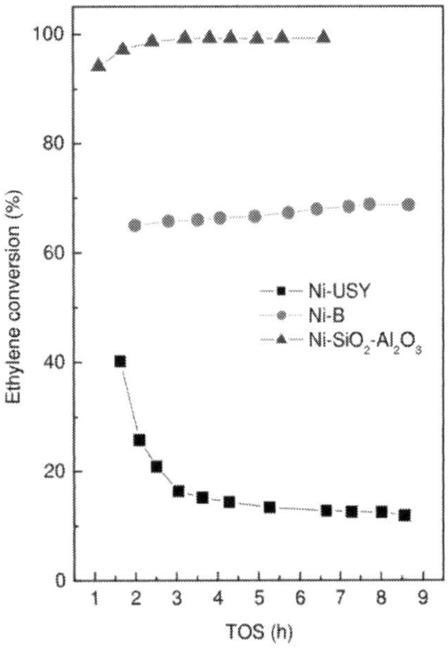

Figure 1: Ethylene conversion on 1.8wt% Ni-SiO_2-Al_2O_3, 2.7% Ni-Beta and 1.7 wt%Ni-USY as function of time on stream at 393 K, 3.5MPa and 48.0 kg_{cat} s mol_{eth}^{-1}

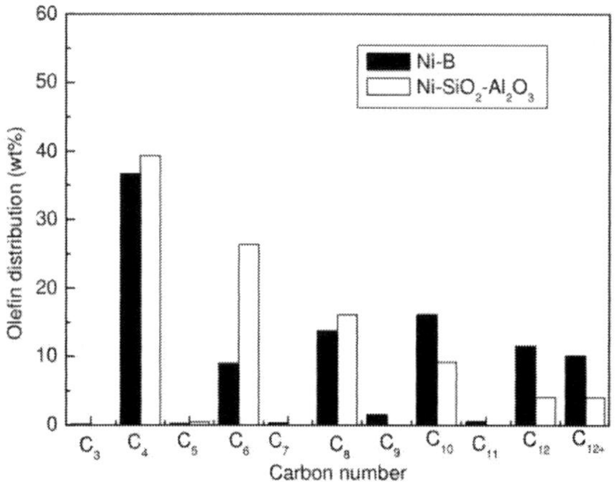

Figure 2: Olefin distribution obtained on 1.8wt% Ni-SiO_2-Al_2O_3 and 2.7% Ni-Beta at 393 K, 3.5MPa and 48.0 kg_{cat} s mol_{eth}^{-1}

A more elaborate experimental study has been performed on the 1.8wt% Ni-SiO_2-Al_2O_3 within a temperature range from 323 K to 503 K, pressure range from 1.5 to 3.5 MPa, with a molar ethylene content in the feed of 10% to 60% and a space time ranging from 4.8 to 45.0 kg_{cat} s mol_{eth}^{-1}. As previously reported in the literature [1,4], two distinct activity regions were observed as a function of the temperature, see Figure 3. At low temperatures mainly linear α-alkenes are obtained through a coordinated ethylene insertion mechanism on an active Ni cation site. Between 373 and 403 K, in the presence of ethylene, an irreversible transformation of Ni occurs which leads to a significantly lower activity of these Ni sites [1]. At higher temperatures, i.e., above 423 K, the Ni cation sites becomes more active again, resulting in an increase of the ethylene conversion. The absence of odd carbon numbered alkenes in the product spectrum shows that no cracking occurred, see Figure 2.

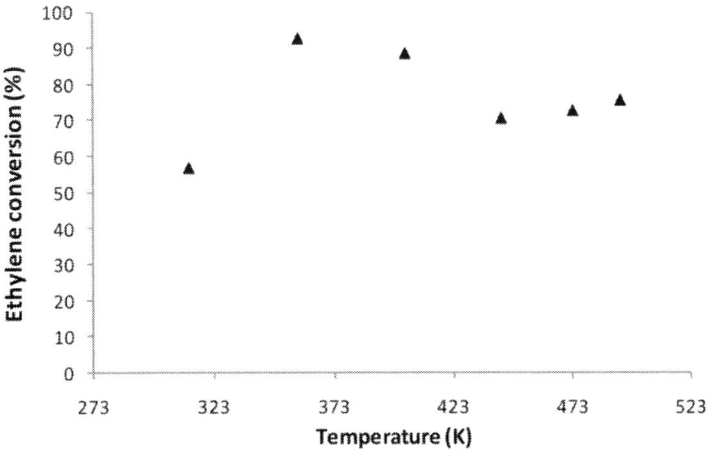

Figure 3: Ethylene conversion on 1.8wt% Ni-SiO_2-Al_2O_3, as function of temperature at 3.5MPa, molar ethylene content in the feed of 60% and 45.0 kg_{cat} s mol_{eth}^{-1}

In the experimental dataset, it was aimed at obtaining a kinetic dataset for which intrinsic kinetics were measured, i.e. in the absence of any transport limitations. Also, a large number of experiments were done at the higher end of the temperature range, i.e. from 443 to 503 K, since higher operating temperatures offer more possibilities for tuning the product distribution. In total, 51 experimental points were included in this range of which 10 are repeat experiments. The kinetic behavior was as expected, i.e. increasing temperature, (partial) pressure and space time all lead to an increase of ethylene conversion, see Figure 4 and Figure 5.

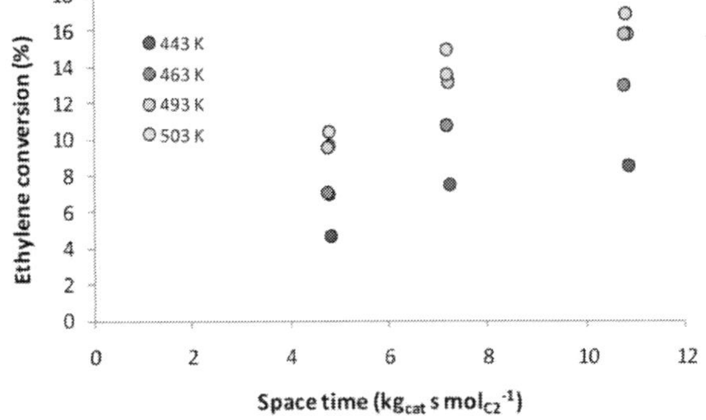

Figure 4: Ethylene conversion on 1.8wt% Ni-SiO_2-Al_2O_3, as function of space time at 3.5MPa, molar ethylene content in the feed of 10% and different temperatures

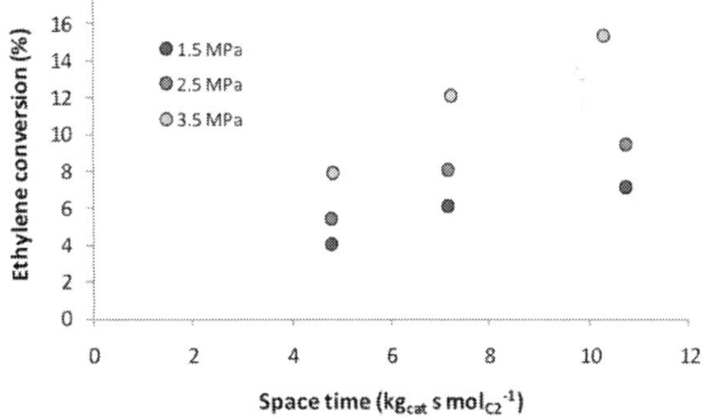

Figure 5: Ethylene conversion on 1.8wt% Ni-SiO_2-Al_2O_3, as function of space time at 473K, molar ethylene content in the feed of 10% and different pressures

The product yields are varying linearly with the ethylene conversion, see Figure 6, indicating that product selectivities are independent from reaction conditions. Generally, the product stream consisted out of (mol%) ±85% butene, ±12% hexene and ±3% octene. No odd carbon numbered products were detected, in correspondence with the first set of experiments. However, no components heavier than octene were present in the reactor outlet, mainly due to the low molar fraction of ethylene in the reactor feed. Also, the temperature at which high molar mass components were detected was within the low-temperature range, i.e. up to 393K, which exhibits higher activity than the high temperature range, see Figure 3. Additional experiments at more severe conditions, i.e. 543K and 43.9 kg_{cat} s mol_{eth}^{-1} confirmed these observations. NH_3-TPD measurements revealed only weak acid sites on the catalyst with NH_3 desorption occurring between 473 – 523 K. All these observations indicated that acid catalyzed oligomerization does not occur on the 1.8wt% Ni-SiO_2-Al_2O_3.

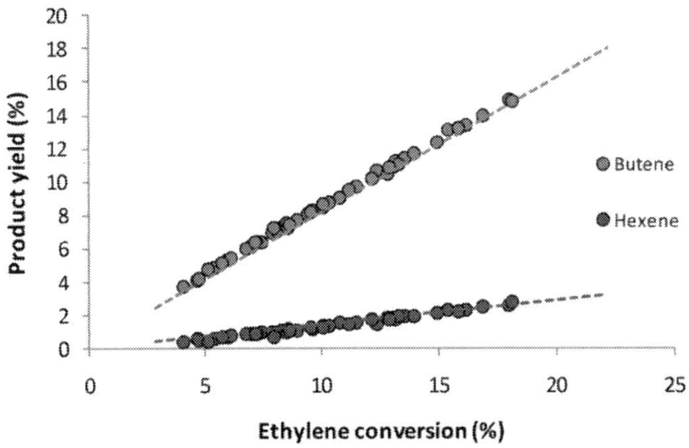

Figure 6: Product yield of butenes and hexenes (octenes not shown) on 1.8wt% Ni-SiO_2-Al_2O_3 as function of ethylene conversion with reaction conditions ranging from 443 to 503 K, 1.5 to 3.5 MPa, a molar ethylene content in the feed from 10 to 30% and a space time between 4.8 and 14.4 kg_{cat} s mol_{eth}^{-1}.

Kinetic Modeling

The reaction network considered in the SEMK model was limited to molecules with a maximum carbon number equal to 8 since no heavier components were observed experimentally. The modeling tools are developed as such that the use of larger networks involving heavier components is straightforward. The considered network consists of 82 alkenes and 60 carbenium ions. Although not evident from the data that are currently assessed, acid catalyzed reactions, e.g. protonation, acid catalyzed oligomerization and isomerization, are already included in the network for future purposes. In total, 403 elementary steps are considered interconverting the considered species. Physical adsorption of the components on the catalyst surface, prior to any chemical reaction was also accounted for using the Langmuir hypothesis. With increasing maximum carbon number of the molecules considered, the reaction network grows exponentially. For a maximum carbon number of 12 over 3000 species and over 10000 elementary steps are involved.

The kinetics for metal ion oligomerization are expressed in terms of 4 elementary steps, see Figure 7, based upon similarity with homogeneous catalysis [5]: initiation of the metal ion by an ethylene molecule, chemisorption of ethylene on the active site, insertion of ethylene into the growing chain and termination, i.e. desorption of the product after ethylene insertion. The initiation is assumed to be in quasi-equilibrium and insertion and termination are assumed irreversible. (de-)Protonation ensures double bound isomerization and is assumed to be in quasi-equilibrium too.

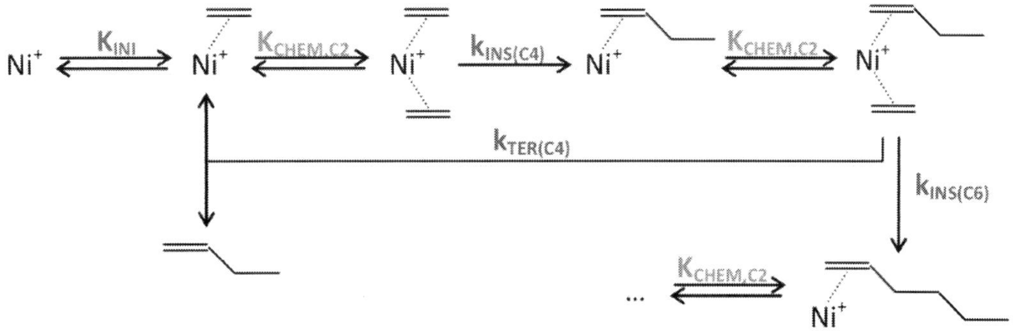

Figure 7: Proposed reaction mechanism for metal ion oligomerisation.

Due to the absence of acid catalyzed oligomerization, carbon skeleton rearrangement and cracking on the catalyst studied, the model parameters to be determined are two equilibrium coefficients, K_{INI}, K_{CHEM}, two rate coefficients k_{INS}, k_{TER} and the components' physisorption coefficients. The latter are calculated using literature data [6]. Applying the SEMK concept, only 4 pre-exponential factors and reaction enthalpies/activation energies are to be determined. The pre-exponential factors are calculated based on statistical thermodynamics, i.e. assuming a certain degree of entropy loss/gain during transition state formation. An equal loss of degree of freedom during transition state formation for initiation and chemisorption is assumed due to the similar nature of the elementary steps and was equal to a net loss of chemisorption entropy of ethylene. For insertion, no change in entropy is expected because prior to ethylene insertion, the two chemisorbed components already have lost their translation entropy via physisorption and chemisorption, and only by rotational movement, the transition state is formed [5]. For termination, one might expect an increase of entropy equal to the chemisorption entropy of the component released. However, based upon quantum mechanical calculations, the formation of the transition state itself does not contain any difference in entropy [5], and, thus this entropy change is set equal to zero too. The remaining four reaction enthalpies/activation energies for, respectively, initiation (ΔH_{INI}), chemisorption ($\Delta H_{CHEM,C2}$), insertion ($E_{a,INS}$) and termination ($E_{a,TER}$) are estimated by regression to the experimental data. The results are given in Table 1. The global regression was globally significant while the kinetic model is adequate. All parameters have a clear physical meaning. The activation energies of ethylene insertion and termination being close to each other is in line with the quasi-independency of the product distribution on the reaction conditions and the temperature in particular. The calculated molar outlet flow rates of ethylene, butenes and hexenes are in very good correspondence with the experimentally observed molar outlet flow rates, see Figures 8 to 10. The residuals for these responses are not dependent from reaction conditions (not shown).

Table 1: Estimated parameters and statistical test for the SEMK model for ethylene oligomerization

Estimated parameters	Values (kJ mol^{-1})
ΔH_{INI}	-103.1
$\Delta H_{CHEM,C2}$	-26.4
$E_{a,INS}$	89.8 ± 0.2
$E_{a,TER}$	87.0 ± 0.7
F-value (regression significancy): 8585 (F_{tab}: 3.59)	
F-value (model adequacy): 1.93 (F_{tab}: 2.55)	

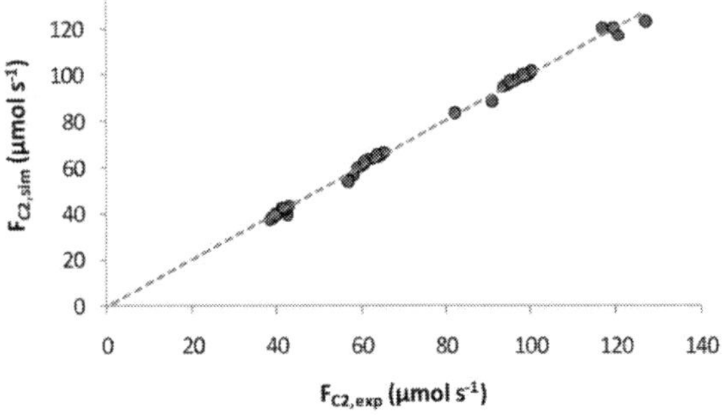

Figure 8: Parity diagram of the molar outlet flow rate of ethylene.

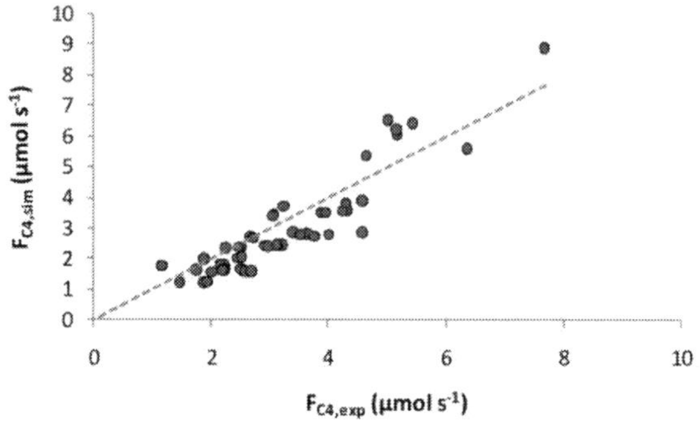

Figure 9: Parity diagram of the molar outlet flow rate of butenes.

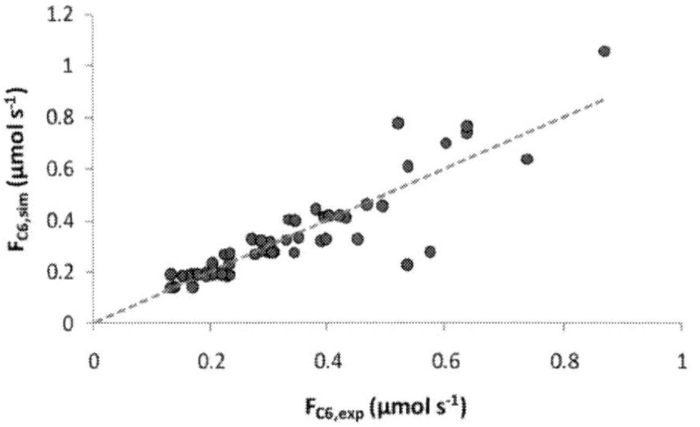

Figure 10: Parity diagram of the molar outlet flow rate of hexenes.

While acid catalyzed oligomerization parameters are not determined with this experimental dataset, the kinetic model is capable of describing these effects using literature reported values for a catalyst containing stronger acid sites [7]. The following simulations are done with a reaction network with molecules having a maximum carbon number of 12. Metal ion catalyzed oligomerization leads to an Anderson Schulz Flory-like product distribution, while the acid catalyzed oligomerization reactions disturb this ASF-like distribution, see Figure 11, left. This is more pronounced with increasing temperatures, see Figure 11 right. Even when acid catalyzed oligomerization occurs, the C_4 fraction obtained is entirely built up out of linear alkenes. C_6 and heavier fractions also contain branched isomers. The decrease with the temperature of the linear alkene content in the C_8 fraction from 43.2 % at 443 K to 25.2 % at 503 K, also illustrates the increase of the relative importance of acid catalyzed reactions at higher temperatures.

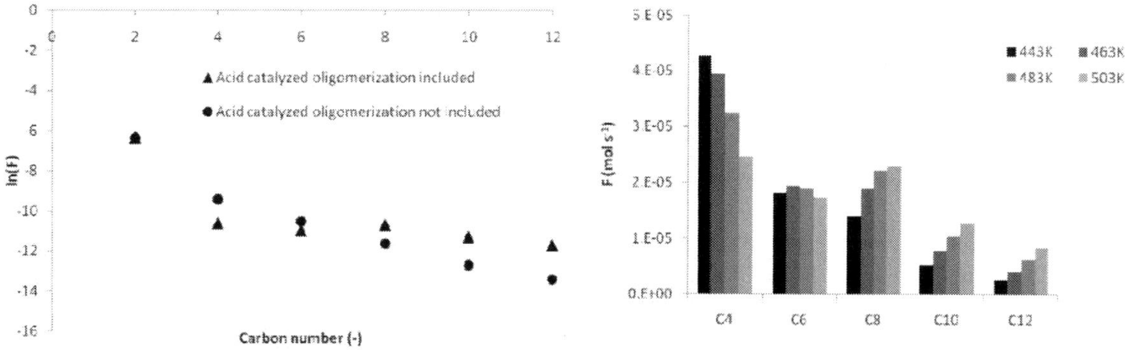

Figure 11: Left: natural logarithm of the molar outlet flow rates as function of the carbon number; Right: product distribution as a function of temperature, acid catalyzed oligomerization included at 3.5 MPa and 6.7 kg_{cat} s mol_{eth}^{-1}.

Conclusions

In the present work, three catalyst candidates were screened for activity and stability in ethylene conversion to oligomers. 1.8wt% Ni-SiO_2-Al_2O_3 and 2.7wt% Ni-Beta showed stable behavior, while 1.7% Ni-USY exhibited a fast initial deactivation. 1.8wt% Ni-SiO_2-Al_2O_3 was the most active catalyst and did not exhibit any acid catalyzed oligomerization. A more extended data set was acquired on this catalyst in order to accurately determine the oligomerization kinetics on Ni ion sites.

Using the SEMK concept, a microkinetic model was constructed and the corresponding kinetic parameters for metal ion oligomerization were estimated. The kinetic model was able to adequately describe the experimental observations. The parameters were estimated significantly with a clear physical meaning. The effect of acid catalyzed oligomerization was investigated by model simulations. It was found to disturb the Anderson Schulz Flory distribution and lead to alkene isomerization while the extent of cracking was still negligible. These simulations illustrate how the SEMK model can be used in catalyst design. In future work the acid catalyzed oligomerization will also be assessed experimentally and corresponding SEMK model parameters will be determined.

Acknowledgements

The research leading to these results has received funding from the European Community's Seventh Framework Program FP7/2007-2013 under grant agreement n° 228953.

References

1. J. Heveling, C.P. Nicolaides, and M.S. Scurrell, Applied Catalysis a-General 173 (1998) 1-9.
2. J.W. Thybaut, I.R. Choudhury, J.F. Denayer, G.V. Baron, P.A. Jacobs, J.A. Martens, and G.B. Marin, Topics in Catalysis 52 (2009) 1251-1260.
3. J. Heveling, C.P Nicolaides, and M.S. Scurrell, Applied Catalysis a-General 248 (2003) 239-248
4. J.R. Sohn, Catalysis Surveys from Asia 8 (2004) 249-263.

5. L. Fan, A. Krzywicki, A. Somogyvari, and T. Ziegler, Inorganic Chemistry 35 (1996) 4003-4006.
6. J.F.M. Denayer, and G.V. Baron, Adsorption Journal of the International Adsorption Society 3 (1997) 251-265
7. R. Van Borm, A. Aerts, M.F. Reynier, J.A. Martens, and G.B. Marin, Industrial & Engineering Chemistry Research 49 (2010) 6815-6823

(641a) Direct Conversion of Lignocellulosic Biomass to Drop-In Hydrocarbon Fuel

Xuejun Pan and Li Shuai, Biological Systems Engineering, University of Wisconisn, Madison, Madison, WI

This research reports an innovative process (HDA) for producing fuel-grade hydrocarbon from lignocellulosic biomass. The process directly converts polysaccharides in lignocellulosic biomass to furan-based precursors for producing fuel-grade hydrocarbons in one step under mild conditions without prior pretreatment and saccharification of the feedstock biomass. Specifically, in the presence of acetone, halide salt, and small amount of mineral acid, the lignocellulosic biomass undergoes hydrolysis of polysaccharides (both cellulose and hemicellulose) to monosaccharides, dehydration of the monosaccharides to furfural (from pentoses) and hydroxymethylfurfural (HMF, from hexoses), and aldol-condensation of HMF and furfural with acetone, and ends up with the formation of furan-derived oxygenates (furfural- and HMF-acetone adducts) with chain length of C5–C15. The resultant furan-based precursors can be subsequently converted to fuel-grade hydrocarbons via hydrogenation. The process integrates the three reactions (Hydrolysis-Dehydration-Aldol condensation) in one step, and is abbreviated as HDA. At the conditions of 120-140°C and 1-2hour, cellulose and hemicellulose in lignocellulosic material (e.g., ground or chipped wood) were completely hydrolyzed into sugars, and more than 80% of the sugars were selectively converted to the furan-acetone adducts. Meanwhile, lignin in the biomass is extensively depolymerized and extracted. Preliminary evaluation indicated the HDA lignin has high purity, low molecular weight with narrow distribution, and good reactivity. The lignin is expected to have great potential in co-products development.

See more of this Session: Chemical and Catalytic Conversions and Processes for Renewable Feedstocks
See more of this Group/Topical: Sustainable Engineering Forum

(641b) A Method for Rapid Screening of Catalysts for Hydrodeoxygenation of Biomass Intermediates

David K. Johnson, Biosciences Center, National Renewable Energy Laboratory, Golden, CO

The focus of biofuels development is shifting from alcohol fuels to advanced biofuels that would be drop-in replacements for petroleum-derived hydrocarbons. There is a wide range of intermediates that can be produced from biomass by chemical or biological routes, but virtually all require further deoxygenation before they can be used as fuel components. Potential routes for decreasing the oxygen content of biomass components include dehydration, hydrodeoxygenation and decarboxylation (possibly after controlled oxidation). Deoxygenation by HDO is preferred over dehydration and decarboxylation reactions, as it is likely to result in higher yields through conservation of carbon, retention of hydrogen, and the production of more energy dense products. Hydrodeoxygenation (HDO) is a hydrogenolysis process that removes oxygen by converting it to water, however, HDO is often performed at high temperatures and pressures that significantly increase the cost of the transformation. We need to develop HDO catalysts that more efficiently convert biomass-derived components into fuel products that are compatible with the existing fuel distribution infrastructure. The feedstocks for the chemical transformations can include sugars, lignins, bio-oils, and chemicals obtained by biological transformations of sugars and biomass. It is clear from previous research that in order to design commercially viable catalysts for HDO upgrading of biomass-derived intermediates, the catalyst system must (1) exhibit high activity under mild conditions (<200°C), (2) contain low-cost materials, (3) demonstrate high stability in an aqueous environment, and (4) have high selectivity for the desired products. To accomplish this, a microwave heated (to 250°C), pressurizable (to 250 psi), automated reaction system has been assembled to rapidly screen catalyst systems using model compounds. To allow automated operation, hydrogen is provided in-situ from a hydrogen donor material, such as formic acid, *i*-propanol or decalin. The rapid screening not only allows the rapid testing of catalysts, but also of reaction conditions, including temperature, reaction time, solvent, and hydrogen donor. The results from tests with model compounds such as acetovanillone and hydroxymethyl furfural, representing typical lignin and carbohydrate derived intermediates, with heterogeneous catalysts, will be described.

See more of this Session: Chemical and Catalytic Conversions and Processes for Renewable Feedstocks
See more of this Group/Topical: Sustainable Engineering Forum

A Novel Reactive Distillation Approach for Producing 1,3-Dihydroxyacetone from Glycerol

Xi Hong, Carl T. Lira, and Dennis J. Miller
Department of Chemical Engineering and Materials Science, Michigan State University
East Lansing, Michigan 48824

Introduction

Glycerol is a viable renewable feedstock that has emerged as a consequence of triglyceride conversion to biodiesel. In addition to glycerol conversion to commodity chemicals such as propylene glycol and epichlorohydrin, formation of higher-value specialty chemical products such as 1,3-dihydroxyacetone (DHA) will provide income for the biorefinery.

DHA is the active ingredient in sunless tanning products. Currently, DHA is industrially produced by fermentation of glycerol, but this microbial approach is challenging because of low production rates and high recovery costs [1]. As a result, the high selling price of DHA limits access of sunless tanning products in the general population.

Recently, scientists have attempted conventional chemical oxidation processes to make DHA from glycerol using metal catalysts with air or molecular oxygen. The goal is to achieve direct oxidization of the secondary hydroxyl group in glycerol to a carbonyl bond. Unfortunately, this reaction is always accompanied with primary alcohol oxidation and a variety of products (i.e. glyceroaldehyde, glyceric acid, etc.) are produced. The highest yield obtained from this method is in the range of 40-60% [2-4].

We present here a novel chemical process to convert glycerol to DHA that greatly improves the efficiency and reduces cost. The essential feature of this process is the selective formation of a 6-membered acetal of glycerol via reactive distillation, followed by selective oxidation of the secondary alcohol to its corresponding ketone product using ruthenium metal and a chemical oxidant. The ketone product is then hydrolyzed to obtain DHA.

A continuous process of making DHA from glycerol has been designed and will be presented. Reactions involved in the process are rapid and can be carried out under mild conditions. Yields of individual process steps are all in excess of 90%, suggesting that this process has good potential to reduce DHA production costs.

Experiments

Materials

Anhydrous glycerol (99.5%), 1-propanol (99.9%) and 1-butanol (99.9%) were obtained from J.T. Baker. Ethanol (99.5%), 1,1-diethoxyethane (99%) and sodium hypochlorite (15% aqueous solution) were purchased from Pharmco and Aldrich, respectively. Sodium carbonate (> 99%), sodium bromate (> 99%), potassium bromate (> 99.8%), calcium hypochlorite (> 98%), sodium periodate (> 99%), sodium iodide (> 99%) were purchased from Sigma-Aldrich. Amberlyst-15 cationic exchange resin (16-50 mesh), purchased from Sigma-Aldrich, was investigated as the catalyst. Before being used in reactions, it was pretreated at 90°C in a vacuum oven for 2 hours. Ruthenium dioxide hydrate (99.9% trace metals basis, 28% water content) and ruthenium trichloride hydrate (reagent plus, 46% Ru content) were obtained from Aldrich. Hydrogen (99.999%) and helium (99.999%) were used for GC analysis; both were purchased from Linde Gas, Inc.

Analysis

Gas Chromatography Analyses

1-Butanol and 1-propanol were used as internal standard and solvent, respectively, for this analysis. To prepare samples for analysis, 0.16 ml 1-butanol and 0.2 ml of reaction sample were added to a volumetric flask and the mixture was diluted with 1-propanol to 10 ml of solution. A Perkin-Elmer Sigma 2000 gas chromatograph with FID and a DB-Wax capillary column (15m × 0.53mm × 1.0 μm) was used for analysis. Ultra high purity helium was used as the carrier gas; the column temperature was ramped from 40°C (3min) at a rate of 30°C/min to 220°C and held for 1 min. The total program time was 10 min. On column injection was applied to analyze the samples. Injector and detector chamber temperatures were set at 60°C and 300°C, respectively.

GC-MS analysis

The structure of some products was identified using GC-MS analysis (GC-Agilent 6890, MS-Agilent 5973). A 1μL sample was separated in a DB-wax capillary column (30m × 0.25mm × 0.25 μm). The initial column temperature was 40 °C (for 5min), the temperature was then increased at 20 °C/min to 250 °C and held for 10 min. Total program time was 24.5 min. Injector temperature was set at 240 °C. Electron impact (EI) mass spectra were monitored from 20 to 300 m/z. The MS source was set at 230 °C.

Results and discussion

Synthesis of glycerol acetal

GA isomers were synthesized at 45 °C by mixing glycerol (3.1 kg, 33.9 mol), DEE (4.0 kg, 33.9 mol), and Amberlyst-15 cationic exchange resin (120 g) and stirring for 4 h. Following reaction and removal of volatile species (DEE and ethanol) by rotary evaporation, GC-MS analysis of the product mixture showed the presence of four different structures corresponding [5, 6] to the four GA isomers displayed in Table 1.

Table 1. Structure and IUPAC name of GA isomers

	Structure	IUPAC Name	Designation
a)	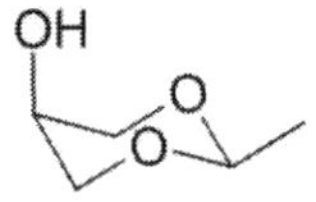	cis-5-hydroxy-2-methyl-1,3-dioxane	Cis-6
b)	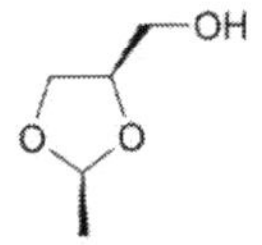	cis-4-hydroxymethyl-2-methyl-1,3-dioxolane	Cis-5
c)	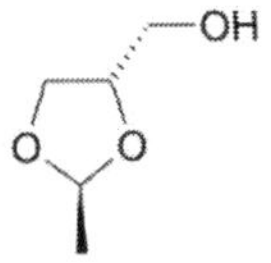	trans-4-hydroxymethyl-2-methyl-1,3-dioxolane	Trans-5
d)		trans-5-hydroxy-2-methyl-1,3-dioxane	Trans-6

GA isomerization equilibrium

In the presence of acid catalyst [7, 8], the four GA isomers interconvert and reach an equilibrated composition. To demonstrate this equilibration, three samples obtained from the distillation experiments below with widely differing isomer compositions (cis-6 from 0% to 84%, cis-5 from 2% to 54%, trans-5 from 2% to 41%, and trans-6 from 0% to 63%) were stirred for 5 hr at 25 °C in the presence of 5 wt% (relative to GA) Amberlyst-15 cationic exchange resin. Identical isomer compositions were observed in the products from the three samples: cis-6, cis-5, trans-5 and trans-6 were 37±1%, 17±1%, 13±1%, and 33±1%, respectively. The molar percentages of 6-membered isomers (cis-6 + trans-6) and 5-membered isomers (cis-6 + trans-5) are approximately 70% and 30%, respectively, at equilibrium.

Rapid isomerization requires vigorous agitation (to ensure mass transfer between solid catalyst and solution) and elevated temperature. For example, Sample 1 took 12 h to reach equilibrium at 0 °C, but only one hour at 80 °C. Increasing temperature also shifts the equilibrium composition. For Sample 1, increasing temperature from 0 °C to 80 °C decreased the sum of cis-6 + trans-6 isomers from 78% to 54%. However, the total quantity of the cis- form isomers (cis-6 + cis-5) and trans- form isomers (trans-6 + trans-5) isomers remained essentially constant over all temperatures at 55±2% and 45±2%, respectively.

Production of the cis-6 GA isomer

A glass vacuum distillation column was constructed to separate the four GA isomers. The column (height = 3.6 m, i.d. = 50 mm) was packed with 0.6 cm 316 Stainless Steel PRO-PAK distillation packing (Scientific Development Co.). The column reboiler, consisting of a two-liter round-bottom flask, was heated with a 750 W electric heating mantle. Vapor temperature was measured at the top of the column.

In the initial distillation, 896 g (800 ml) of a GA mixture containing 35% cis-6, 18% cis-5, 15% trans-5, and 32% trans-6 isomers was placed in the reboiler. To avoid isomer interconversion, the solution was maintained at a weakly basic state by adding 50 g of sodium carbonate to the mixture. This experiment was carried out at a pressure of 20.5 ± 1.0 mm Hg (except for Sample 1, where the pressure was temporarily reduced to 9.0 mmHg).

Twelve 60-mL distillate samples were collected over the course of the distillation and analyzed by GC. Ninety percent of the initial charge to the still pot was recovered as distillate in these samples (Table 2). The material remaining in the still pot consisted primarily of the trans-6 isomer and glycerol; inclusion of this material gave essentially complete (100±2%) closure of the material balance for the experiment.

Table 2. Summary of GA isomers recovered in the initial distillation

	Initial GA charge		Product recovery in distillate		Percent recovered in distillate (%)
	Mass (g)	Composition (wt%)	Mass (g)	Percent of total	
Cis-6	311	35	288	36	93
Cis-5	166	18	152	19	92
Trans-5	132	15	120	15	91
Trans-6	287	32	246	30	86
Total	896	100	806	100	90

Figure 1 gives the reboiler liquid temperature, vapor temperature at the top of the column, and composition (via GC) of the distillate samples taken during the distillation. The results clearly show the order of decreasing volatility of the GA isomers as cis-6, cis-5, trans-5, and finally trans-6. As the more volatile isomers are removed from the still pot, liquid and vapor temperature gradually increase.

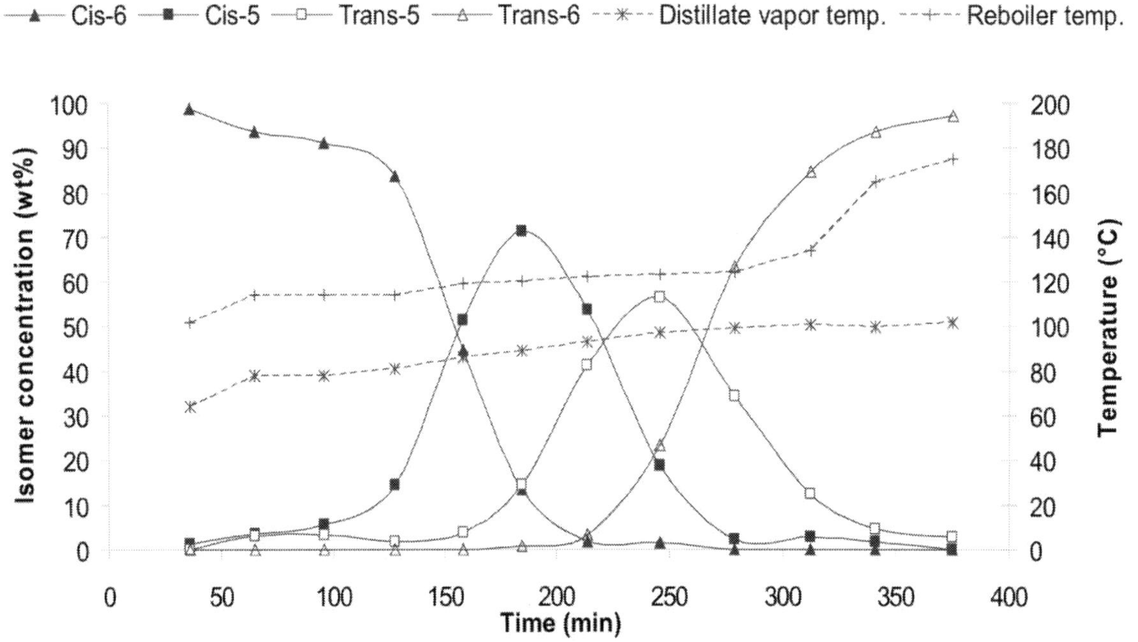

Figure 1. Initial distillation of GA isomer. Column pressure was 20.5±1.0 mmHg (except for the first sample, where the pressure was 9.0 mmHg).

Based on the result of the initial distillation and the fact that GA isomers can interconvert to each other and reach an equilibrated isomer composition in the presence of acid catalyst, a continuous reactive distillation experiment was then carried out to produce the cis-6 isomer as the sole product. In this run, 504 g (450 mL) of a mixture of 36 wt% cis-6, 16 wt% cis-5, 14 wt% trans-5, and 34 wt% trans-6 GA isomers was charged to the reboiler pot of the glass column along with 50 g of Amberlyst-15 cationic exchange resin. For this experiment, the column pressure was maintained at 2.9±0.2 mmHg. Once the distillate product started to condense at the top of the column, a continuous feed of the same GA isomer mixture was introduced to the column reboiler to maintain its liquid level constant. The feed rate was adjusted to match the distillate rate from the column, which was approximately 1.8 mL/min. During the distillation, 16 distillate samples were collected with an average volume of 59 mL.

As shown in Figure 2, this experiment produced samples with high concentration of cis-6 isomer (from 90% to 96%) continuously during operation of the column. The reboiler temperature and distillate vapor temperature were essentially constant during operation, indicating that the reboiler composition was constant and that the column was operating at steady state. The total quantity of distillate obtained was 1028 g, more than twice the amount of the initial charge to the reboiler, demonstrating that continuous production of the cis-6 isomer from the GA acetal mixture is feasible. Based on these results, it is fully expected that an appropriately designed commercial-scale column with sufficient separation capacity would provide essentially pure cis-6 isomer.

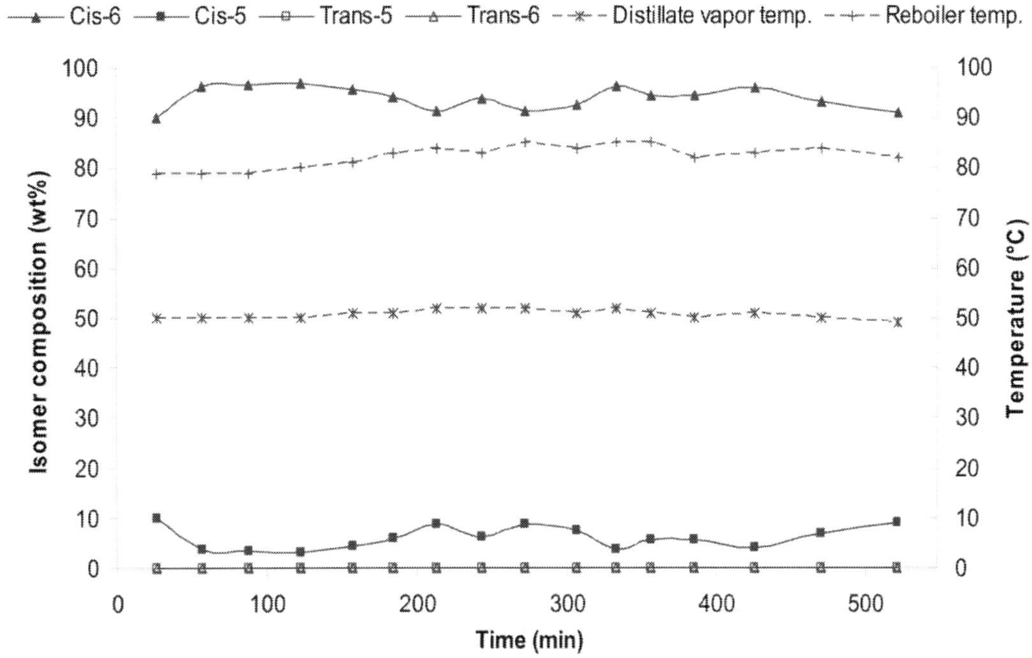

Figure 2. Continuous reactive distillation of GA isomers. Column pressure = 2.9±0.2 mm Hg.

Oxidation of cis-6 to the precursor of DHA

With the availability of cis-6 isomer, a new synthesis pathway can be developed to obtain valuable chemicals. For instance, by oxidizing the remaining free hydroxyl group in cis-6 and hydrolyzing the corresponding ketone product (2-methyl-1,3-dioxan-5-one, G-tone), 1,3-dihydroxyacetone (DHA) can be produced [7].

For this purpose, a variety of oxidation reagents including pyridinium chlorochromate (PCC), ruthenium tetroxide, hydrogen peroxide, molecular oxygen, and ozone were tested to convert cis-6 into its ketone product.

The most effective method is the ruthenium tetroxide-catalyzed oxidation. The reagents used in this reaction involves a ruthenium species (i.e. $RuCl_3$ and RuO_2) and another oxidizing salt (i.e. NaClO, $NaBrO_3$, $NaIO_4$, $Ca(ClO)_2$, $KBrO_3$, etc.) During reaction, the ruthenium species is first oxidized to RuO_4, and then RuO_4 oxidizes cis-6 to the corresponding ketone product. The co-product is a black precipitate of RuO_2, which can be re-oxidized to RuO_4 by the oxidizing salt in the aqueous solution. The regenerated RuO_4 then enters the organic phase again for another cycle of oxidation. In this way, the alcohol substrate can be continuously oxidized to the desired product with catalytic amounts of RuO_4 while the secondary oxidant is consumed. When the reaction is finished, the residue RuO_4 species in the organic phase can be easily recovered by adding isopropanol to make it RuO_2 precipitate [9].

Table 3 presents the results of cis-6 oxidation experiments catalyzed by ruthenium-based substances. It is clear that this type of reaction very efficient. In some reactions over 90% yield of G-tone was achieved within 30 min.

Table 3. Ruthenium tetroxide catalyzed oxidation of cis-6 at 25°C, 1.0 atm.

Reaction	Cis-6 (g)	Water (g)	Catalyst ($RuO_2 \cdot xH_2O$) loading (g)	Secondary oxidant solution/mass (g)/concentration	Time (min)	G-tone yield (%)
1	0.41	10	0	NaClO/4/10%	30	0
2	0.42	10	0	$KBrO_3$/10/saturated	1440	0
3	4.0	50	0.40	NaClO/100/10%	30	80
4	0.39	11	0.31	NaClO/4/10%	30	76
				NaClO/5/10%	60	95
5	0.45	10	0.05	$Ca(ClO)_2$/10/ saturated	20	87
6	0.40	10	0.30[a]	$NaIO_4$/10/saturated	75	88
7	0.42	10	0.26	$KBrO_3$/10/ saturated	45	> 99
8	0.43	10	0.32[a]	$KBrO_3$/10/saturated	75	95
9	4.1	10	0.40	$KBrO_3$/100/saturated	20	97
10	0.30	10	0.41	$NaBrO_3$/10/10%	20	96
11	10.0	10	0.51	$NaBrO_3$/100/20%	40	90
12	3.0	10	0.10	$NaBrO_3$/12.5/20%	20	95

a. $RuCl_3 \cdot xH_2O$ was used as catalyst.

Hydrolysis of G-tone to DHA

Hydrolysis of G-tone to DHA is very fast and can be carried out at mild conditions (Scheme 1). In one experiment, 5 wt% (weight to substrate) of Amberlyst-15 cationic exchange resin was added to the final mixture of Reaction 12 in Table 3. When the mixture was heated to 30°C, complete hydrolysis of G-tone was achieved in 10 min.

Scheme 1. Processes of catalytic conversion of glycerol into useful chemicals.

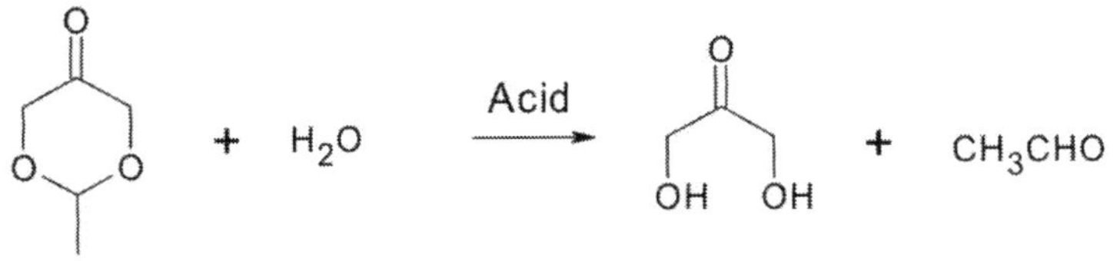

A novel process of making DHA from glycerol

Using cis-6 as an intermediate provides a relatively quick, environmentally safe and economical route to 1,3-dihydroxyacetone (DHA) from glycerol. Based on the experimental results obtained, a novel process of making DHA from glycerol is developed and described here.

Figure 3 presents the overall scheme of a novel DHA production process based on acetaldehyde or 1,1-diethoxyethane (DEE) as the glycerol acetal forming reagent. The process may also work with other aldehydes or ketones. Selection of the acetal forming reagent is dependent on price, availability and process feasibility.

In Figure 3, glycerol and acetaldehyde are introduced to Reactor 1 to form cyclic glycerol acetals. The reaction mixture, which contains four isomers of glycerol acetal, is then evaporated to get rid of water and unreacted acetaldehyde for recycling. After that, the mixture of glycerol acetal isomers is introduced to a vacuum distillation column, in which the cis-6 GA isomer is formed, distilled, and thus obtained from the top of the column. At the same time, other isomers, which have higher boiling points, react in the reboiler to re-establish isomerization equilibrium. The continuous recycling of the less-volatile glycerol acetal isomers through the equilibration-distillation procedure can transform all of the products into the desired 6-membered structure.

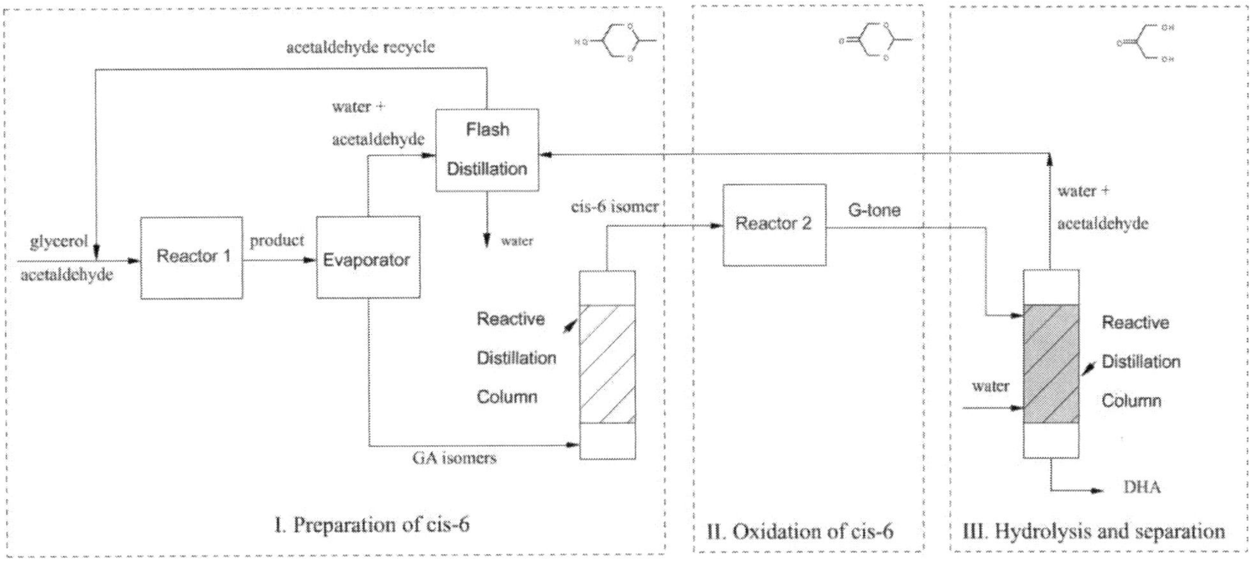

Figure 3. A novel process to produce DHA from glycerol.

In the second stage of this design, cis-6 GA isomer is introduced to the oxidation Reactor 2 to form 2-methyl-1,3-dioxan-5-one (G-tone). A preferred operation for this stage is to use ruthenium tetroxide-related oxidation.

In the last stage, G-tone and an excess amount of water are reacted, perhaps in a reactive-distillation column, to hydrolytically decompose G-tone to the desired product DHA and acetaldehyde, which can be recycled or reformed into DEE. The possible application of reactive distillation along with a large amount of water will overcome any chemical equilibrium limitations of this reaction, such that a high yield of DHA can be achieved from this process.

Summary

A novel process to make 1,3-dihydroxyacetone (DHA) from glycerol is presented in this paper. The process involves three major steps: 1) formation of glycerol acetal (GA) and reactive distillation to get the 6-membered GA isomer 5-hydroxyl-2-methyl-1,3-dioxane (cis-6); 2) oxidation of cis-6 GA isomer to the corresponding ketone product 2-methyl-1,3-dioxan-5-one (G-tone); 3) conversion of G-tone to DHA via hydrolysis. Experimental results show that in each step, over 90% yield of the desired product can be achieved. Thus, the new method holds a good potential for commercial success.

References

1. Mishra, R, Jain, S. and Kumar, A. (2008). Microbial production of dihydroxyacetone. Biotechnology Advances. 26, 293-303.
2. Garcia, R, Besson, M. and Gallezot, P. (1995). Chemoselective catalytic oxidation of glycerol with air on platinum metals. Applied Catalysis A: General. 127, 165-176.
3. Hu, W, Knight, D, Lowry, B. and Varma, A. (2010). Selective Oxidation of Glycerol to Dihydroxyacetone over Pt-Bi/C Catalyst: Optimization of Catalyst and Reaction Conditions. Industrial & Engineering Chemistry Research. 2010, 49, 10876–10882.
4. Kimura, H. (1993). Selective oxidation of glycerol on a platinum-bismuth catalyst by using a fixed bed reactor. Applied Catalysis A: General. 105, 147-158.
5. Da Silva Ferreira, A, Barbe, J. and Bertrand, A. (2002). Heterocyclic acetals f from glycerol and acetaldehyde in port wines: Evolution with aging. Journal of Agricultural and Food Chemistry. 50, 2560-2564.
6. Câmara, J, Marques, J, Alves, A. and Da Silva Ferreira, A. (2003). Heterocyclic acetals in Madeira wines. Analytical and Bioanalytical Chemistry. 375, 1221-1224.
7. Deutsch, J, Martin, A. and Lieske, H. (2007). Investigations on heterogeneously catalysed condensations of glycerol to cyclic acetals. Journal of Catalysis. 245, 428-435.
8. Aksnes, G, Albriktsen, P. and Juvvik, P. (1965). Studies of cyclic acetal and ketal isomers of glycerol. Acta Chemica Scandinavica. 19, 920-930.
9. Tojo, G. and Fernandez, M. (2006). Oxidation of alcohols to aldehydes and ketones. NY: Springer.

(641d) Catalytic Oxidative Dehydration of Butanol Isomers: 1-Butanol, 2-Butanol and Isobutanol

Ivan C. Lee, US Army Research Laboratory, Adelphi, MD

The catalytic production of economically important four-carbon olefins was achieved through catalytic oxidative dehydration of 1-butanol, 2-butanol, and isobutanol using a millisecond contact time reactor. Both alumina foam and rhodium-alumina foam catalysts convert these four-carbon alcohols into four-carbon olefins, with contact time equal to 25 ms over a wide range of equivalence ratios (f) from 0.176 to 6.338 (or C/O_{air} ratio from 0.0659 to 2.11) . The cracking of the carbon backbone was found to be minimal. A mixture of butene isomers was obtained due to the isomerization of the carbocation intermediates. The degree of this isomerization was determined by analyzing the ratios of butenes/(total olefins), 1-butene/2-butene, cis-2-butene/trans-2-butene, and isobutene/(linear butenes). In addition, a reaction mechanism of the catalytic oxidative dehydration of butanol in short contact times was proposed. In this mechanism, the C-O bond of the majority of butanol was broken to form olefinic species on the alumina surface. These surface olefinic species would undergo successive oxidation to form CO and/or CO_2, depending on the availability of oxygen.

See more of this Session: Chemical and Catalytic Conversions and Processes for Renewable Feedstocks
See more of this Group/Topical: Sustainable Engineering Forum

(641e) Effects of Reaction Conditions On the Acid-Catalyzed Hydrolysis of Miscanthus Dissolved In An Ionic Liquid

Sean J. Dee and Alexis T. Bell, Chemical and Biomolecular Engineering, University of California - Berkeley, Berkeley, CA

Miscanthus, a perennial grass, is an attractive biofuel feedstock compared with corn and poplar due to its high energy yield per hectare and low agricultural inputs such as water and soil nutrients. Recent studies have shown the three primary components of miscanthus, cellulose, hemicellulose, and lignin, can be dissolved in ionic liquids (ILs) at levels approaching 20 wt%. An interesting question is then whether the cellulosic and hemicellulosic components of the dissolved miscanthus can be hydrolyzed to sugars in high yield. While acid-catalyzed hydrolysis of cellulose and hemicellulose has been studied extensively in ILs, the conversion of biomass is not as well understood. Recent literature has claimed that effective chemical conversions of biomass in ILs requires either thermochemical delignification which can destroy both hemicellulose and cellulose, or multistage processing that is inefficient and makes IL recycling energy intensive. This study investigated the acid-catalyzed hydrolysis of miscanthus dissolved in the ionic liquid 1-ethyl-3-methylimidazolium chloride ([Emim][Cl]). The effects of varying temperature, acid concentration, miscanthus concentration, and the gradual addition of water were undertaken. The similarities and differences between the rate of hydrolysis of microcrystalline cellulose (Avicel) and amorphous hemicellulose (Xylan from birchwood) compared to the cellulosic and hemicellulosic components of miscanthus were determined. The influence of miscanthus delignification on the hydrolysis rate was also explored. The present work has demonstrated that 90% biopolymer conversion from miscanthus could be achieved, with 93% selectivity to saccharine products in a single stage hydrolysis without pretreatment.

See more of this Session: Chemical and Catalytic Conversions and Processes for Renewable Feedstocks
See more of this Group/Topical: Sustainable Engineering Forum

(641f) Secondary Thermochemical Xylo-Oligomer Hydrolysis of High-Solids Dilute-Acid Pretreatment Slurries

Erik M. Kuhn[1], Joseph Shekiro[1], Nick J. Nagle[2] and Richard T. Elander[1], (1)National Renewable Energy Laboratory, Golden, CO, (2)National Bioenergy Center, National Renewable Energy Lab, Golden, CO

Dilute-acid pretreatment is a critical step in the biochemical conversion of biomass to fuels and chemicals, releasing hemicellulose as oligomeric and monomeric xylose and rendering the cellulose fraction readily digestible by enzymes. It has been difficult to achieve very high yields of monomeric xylose during pretreatment, since acid hydrolysis also converts xylan to various oligomers and non-fermentable sugar degradation products. An alternative strategy for the production of high yields of monomeric xylose is to perform secondary thermochemical hydrolysis on the dilute-acid pretreated material. This secondary hydrolysis, typically performed at lower temperatures than the primary dilute-acid pretreatment step, liberates additional monomeric xylose while minimizing the yield of sugar degradation products. This work presents the results of laboratory and pilot scale primary and secondary thermochemical hydrolysis experiments over a wide range of conditions. Our results suggest that reduced severity conditions for primary dilute-acid pretreatment provide the highest overall yields of monomeric xylose after secondary thermochemical hydrolysis for the conditions tested. The results of our LC-MS investigations suggest that higher-molecular weight xylo-oligomers are converted to xylose during secondary thermochemical hydrolysis, but that a small fraction of the xylan in corn stover is converted to stable but non-fermentable compounds, such as xylo-oligomers, during both primary and secondary hydrolysis.

See more of this Session: Chemical and Catalytic Conversions and Processes for Renewable Feedstocks
See more of this Group/Topical: Sustainable Engineering Forum

(654a) Single Molecule Spectroscopy for Characterization of Acid Acid Catalysts

Xiaojiao Sun[1], Alec Kirkeminde[2], Daniel A. Higgins[2] and Keith L. Hohn[3], (1)Chemical Engineering, Kansas State University, Manhattan, KS, (2)Chemistry Department, Kansas State University, Manhattan, KS, (3)Chemical Engineering Department, Kansas State University, Manhattan, KS

Mesoporous silicas are promising solid acid catalysts since their pore sizes can be tailored form two to tens of nanometers and they can be modified by incorporation of metals and organic functional groups. Strong acidity can be achieved by functionalizing silica with sulfonic acid groups. Researchers have shown extensive interest to mesoporous materials for their potential to be used as heterogeneous catalysts. However, our understanding of where the acid sites are incorporated into the matrix and how their properties vary from site to site is incomplete. Conventional characterization methods provide only limited information of the distribution of the functional sites and acid strengths. Our research will use single molecule spectroscopy to study the locations and properties of the acid functionalities in silicas-based materials. In this technique, probe molecules at nanomolar concentration are introduced into a material during its preparation. These molecules interact with acid sites, and are converted from unprotonated to protonated forms. These forms fluoresce at different wavelengths, providing a way to probe individual sites.

Two pH sensitive probes (C.SNARF-1 and ACISA) have been employed in our laboratory to provide information of the distribution of acid site strengths, the location of acid sites relative to the materials surface, and ultimately, the catalytic rates of certain model reactions. The acidity of alumina-silica catalysts with varying compositions was studied using C.SNARF-1. In addition, aligned porous silica structures were studied, where diffusion of molecules into the pores can be followed using single molecule spectroscopy. From these studies, we can obtain valuable information of acid sites location and acid strength variability.

See more of this Session: In Situ and Operando Spectroscopy of Catalysts I
See more of this Group/Topical: Catalysis and Reaction Engineering Division

(654b) Spectroscopic Investigation of CO Adsorption On Pt(100) At near-Atmospheric Pressures Using PM-IRAS

John E. Bedenbaugh, Chemical Engineering, University of Delaware, Newark, DE and Jochen Lauterbach, Chemical Engineering, University of South Carolina, Columbia, SC

In molecular-level catalyticinvestigations, discrepancies that exist between surface science observationsunder ultra-high vacuum (UHV) conditions and industrial catalytic performanceat higher pressures are referred to as the "pressure gap." For example,changes in the population of adsorption sites and variation in reactionmechanisms on catalyst surfaces have been observed as pressure increases aboveUHV conditions (*1-2*). This work addresses this pressure issue throughthe investigation of surface adsorption behavior of model catalytic systemsover the range from UHV to atmospheric pressures using a polarizationmodulation infrared reflection absorption spectroscopy (PM-IRAS) systemconstructed in our laboratory. In order to study catalytic systems above UHVconditions, investigative techniques capable of functioning in higher pressureenvironments without undue interference from gas phase molecules are required.This PM-IRAS system is capable of removing contributions from gas phasemolecules to yield surface vibrational spectra.

Adsorption of CO on Pt(100) wasinvestigated at near-atmospheric pressures using PM-IRAS measurements. At asample temperature of 325 K, a linear C-O stretch (~2090 cm^{-1}) wasobserved. Peak sharpening and a frequency shift were observed for this COadsorption band at higher pressures. At 325 K, the frequency shift increasedwith exposures between 1 and 200 Torr CO by up to ~6 cm^{-1}.A corresponding decrease in the linewidth of the IR band was observed over thesame range. These results suggest that dipole-dipole coupling effects play animportant role in understanding the surface adsorption behavior in this systemat higher pressures.

A dipole-coupling model (*3*)was applied to these experimental results. The model predicted CO surfacecoverages on the Pt(100) surface increasing from ~0.7 at 1 Torr CO to >0.9at 200 Torr CO. These results indicate that at higher pressures the CO surfacecoverage on Pt(100) is much greater than similar measurements obtained underUHV conditions. The predicted increases in surface coverage at higherpressures were verified through analysis of the integrated peak areas of themeasured absorption bands. The calculated areas were observed to increase byup to 50% in magnitude over the pressure range from 1 Torr CO to 200 Torr CO. Measurementsobtained during reduction from a high-pressure environment indicate thathigh-pressure adsorption behavior is a mix of reversible and irreversibleprocesses.

Measured PM-IRAS spectra exhibitsignificant broadening and decreasing frequency with increasing sampletemperature. These effects are consistent with phonon dephasing models foradsorbed CO (*4*).

Disappearance of the IR band at higher sampletemperatures is attributed to CO dissociation, resulting in carbon deposition onthe Pt(100) surface. Subsequent spectra obtained after exposure of the systemto oxidative conditions reveal the return of the absorption band correspondingto adsorbed CO. Adsorbed CO measured in a CO oxidation reaction environmentexhibit reversible adsorption/desorption processes below CO desorptiontemperatures, in contrast to the experiments conducted in a pure COenvironment.

1. G. Rupprechter, C. Weilach, *Nano Today* **2**, Issue 4, 20-29(August 2007)

2. D. Stacchiola, A.W. Thompson, M. Kaltchev, W.T. Tysoe, *J. Vac. Sci.Technol. A* **20** (6), 2101-2105 (2002).

3. J. Lauterbach, R.W. Boyle, M. Schick, W.J. Mitchell, B. Meng,W.H. Weinberg, *Surface Science* **350**, 32-44 (1996).

4. B.N.J. Persson, F.M. Hoffmann, R. Ryberg, *Phys. Rev. B* **34**,2266 (1986).

See more of this Session: In Situ and Operando Spectroscopy of Catalysts I
See more of this Group/Topical: Catalysis and Reaction Engineering Division

(654c) A Novel Multilayer Infrared Absorption Spectroscopy (MEIRAS) Technique for Infrared Polarization Dependent Operando Study CO Adsorption and Oxidation On Thin Film and Nanowire Catalysts

Prashant Deshlahra, Chemical and Biochemical Engineering Department, University of Notre Dame, Notre dame, IN and Eduardo E. Wolf, Chemical and Biomolecular Enigineering Department, University of Notre Dame, Notre dame, IN

We have developed a new multilayer enhanced infrared reflection absorption spectroscopy (MEIRAS) technique for study of adsorbates, on catalyst such as thin films and low area supported nanoparticles or nanowires [1]. The MEIRAS technique utilizes the optical interference effects in metal-dielectric-metal multilayer substrate to obtain a large signal enhancement by quarter-wave matching of dielectric layer thickness to desired infrared wavelength. These effects have been studied using optical simulation and a thorough theoretical analysis. The sensitivity enhancement through multilayer reflection, combined with the experimental setup developed here, provides unique capability for high pressure FTIR study of adsorption and reactions on low area model catalysts. Unlike commonly used techniques for these studies (such as PM-IRAS and SFG), this technique does not require additional optical instrumentation apart from a standard FTIR setup. Moreover, it is sensitive to both surface normal and tangential vibrational modes [2]. This is in contrast to IRAS techniques on reflecting metal substrates which only detect normal modes.

In the present work, we perform CO adsorption and oxidation study using MEIRAS on TiO_2 supported Pt thin film and nanofabricated parallel monodisperse nanowires under reaction conditions. The effect of the width of nanowires on CO infrared absorption band position and surface ignition behavior of CO oxidation reactions is studied. With incident infrared polarized either along the Pt nanowires or normal to nanowires, signal can be selectively collected from CO molecules adsorbed either on top of the wires or on the edges of the wires. On analyzing these spectra we find that infrared absorption peak on CO adsorbed on the edges of wire, closer to the Pt/TiO_2 interfaces is slightly red shifted compared to the signal from top of the wire, indicating the role support effects. These experiments highlight the role of metal-support interfaces in catalysis and provide a way to selectively probe adsorbates at or near these interfaces.

References:

[1] P Deshlahra et al. *Surf. Sci.* **604**, 2 (2010); P Deshlahra et al. *Appl. Catal. A* **391**, 22 (2011).

[2] P Deshlahra and EE Wolf *J. Phys. Chem. C* **114**, 16505 (2010).

See more of this Session: In Situ and Operando Spectroscopy of Catalysts I
See more of this Group/Topical: Catalysis and Reaction Engineering Division

(654d) Spatially Resolved Raman Spectroscopy In Catalytic Packed Bed Reactors

Michael Geske, Oliver Korup and **Raimund Horn**, Inorganic Chemistry, Fritz Haber Institute of the Max Planck Society, Berlin, Germany

Spatially Resolved Raman Spectroscopy in Catalytic Packed Bed Reactors

Measuring spatiallyresolved species and temperature profiles in catalytic reactors has recentlygained attention in the literature and several reactor designs with or withoutsampling probes have been developed [1,2,3,4]. As reviewed by Urakawa and Baiker [5], also the measurement of spatially resolvedspectroscopic data in catalytic reactors becomes more and more important ascatalysts are dynamic systems that continuously adapt to the local chemicalpotential in the reactor. However, in situ spectroscopy cells are usuallyadapted to the spectroscopic technique and in most cases do not reflect theflow and temperature field in a technical reactor. Measuring spectroscopicprofiles along the centerline of a catalytic fixed bed reactor, e.g. by fiberoptics, would be an approach closer to industrial reality because fixed bedreactors are frequently used for chemical processes. Furthermore no thermalgradients are induced by radiation losses which can be the case in commercialor self-designed spectroscopic cells equipped with windows for optical access.

In the presentpaper we demonstrate how spatially resolved Raman spectroscopy using fiberoptic probes can be applied to monitor the state of the catalyst in a fixed bedalong the flow direction in a catalytic partial oxidation reaction. Theoxidative dehydrogenation of ethane to ethylene on MoO_xsupported on g-Al_2O_3spheres is used as demonstration example. After determining spatialresolution and offset of the used fiber optics by means of mapping a sharptransition between an excellent Raman scatterer(sulfur) and a poor Raman scatterer (graphite)arranged as cylinders in the reactor tube, the reactor is filled with thecatalyst and a simultaneous spatially resolved measurement of gas species andRaman spectra is conducted. In situ measurements are complemented by ex situXRD and Raman measurements on the used catalyst extracted layer by layer fromthe reactor tube. The latter Raman measurements were conducted under a confocal microscope. Two reaction zones were observed inthe catalyst bed. At the entrance, ethane reacts with gas phase oxygen toethylene, carbon monoxide, carbon dioxide and water. Depending on the loading, MoO_x is present in this oxidation zone either ascrystalline MoO_3 or as polymolybdate. Thetemperature maximum is observed at the end of the oxidation zone where about500° C are reached. Upon complete O_2 consumption the color of thecatalyst changes from white or gray respectively to violet, whichwas identified by ex situ XRD and Raman spectroscopy to be MoO_2.The lattice oxygen removed by this reduction preferentially oxidizes $C_2H_4$to CO_2. Interestingly C_2H_6 is much less proneto deep oxidation than C_2H_4 and no CO but only $CO_2$is formed by reaction of C_2H_4

with lattice oxygen. Towardsthe end of the catalyst bed a mixture of polymolybdateand MoO_2 is observed indicating that the catalyst bed had notreached steady state yet and was still in the process of being reduced to MoO_2.

References

[1] R. Horn,O. Korup, M. Geske, U. Zavyalova, I. Oprea, R. Schlgl, *Rev. Sci. Inst.* **2010**, *81*, 064102

[2] J. Sxa, D. L. A. Fernandez, F. Aiouacheet al., *Analyst* **2010**, *135*, 2260

[3] M. Bosco, F. Vogel, *Catal.Today* **2006**,*116*, 348

[4] M. Reinke, J. Mantzaras,R. Schaeren et al., *Comb. Flame* **2004**, *136*, 217-240

[5] A.Urakawa, A. Baiker, *Top. Catal.* **2009**, *52*, 1312-1322

See more of this Session: In Situ and Operando Spectroscopy of Catalysts I
See more of this Group/Topical: Catalysis and Reaction Engineering Division

(654e) Steady State Isotopic Transient Kinetic Analysis of Supported Pt and Au Nanoparticles for the Water Gas Shift (WGS) Reaction Using a Novel Operando Transmission Fourier Transform Infrared (FTIR) Reactor

Jun Wang[1], Vincent F. Kispersky[1], Jorge Pazmino[1], Mayank Shekhar[1], Wen-Sheng Lee[1], W. Damion Williams[1], M. Cem Akatay[2], Jeffrey T. Miller[3], W. Nicholas Delgass[1] and Fabio Ribeiro[1], (1)School of Chemical Engineering, Purdue University, West Lafayette, IN, (2)School of Materials Engineering and Birck Nanotechnology Center, Purdue University, West Lafayette, IN, (3)Chemical Science and Engineering, Argonne National Laboratory, Argonne, IL

An *operando* FTIR reactor was developed for simultaneous collection of kinetic data and monitoring of adsorbed species under reaction conditions. The reactor is made of stainless steel and can be easily loaded and unloaded. It can be operated up to 450 °C and pressure range from 1 to 5 atm. Due to a low dead volume, a 95% signal decay time of less than 2 s for inert gas switching was achieved at a flow rate of 50 ml/min. The apparatus enables steady state isotopic transient kinetic analysis to be performed by monitoring the effluent gaseous species by mass spectroscopy while the adsorbed species are monitored simultaneously by time-resolved transmission FTIR. The reaction rates and kinetics of the low temperature water gas shift (WGS) reaction on Pt and Au catalysts measured by this reactor were comparable to the ones measured in a regular plug flow reactor (PFR).

Details of Pt and Au nanoparticle catalysis of WGS reactions were investigated by *Operando* FTIR, isotopic transient experiments. For a Pt/CeO_2 catalyst, the total amount of surface intermediates, determined by isotopic transient experiments, was found to be approximately 85% of the surface Pt, which is comparable to the CO surface coverage on Pt during WGS reaction [1]. Simultaneous time resolved IR data showed that the normalized IR peak intensities of ^{13}CO adsorbed on Pt best correlate with the normalized MS response of $^{13}CO_2$, while the normalized IR peak intensities of ^{13}CH, from formates, do not. These results suggest that metallic surface Pt atoms are the dominant active sites for WGS reaction on Pt/CeO_2 catalysts.

Bromine poisoning effect, *Operando* FTIR, and isotopic transient experiments were also used to investigate the active sites of Au/TiO_2 catalysts for the low temperature WGS reaction. With addition of only 4 mol% Br to Au, the reaction rate decreased about 50%; while adding 8mol% Br to Au led to an undetectable reaction rate at 200 °C. TEM and kinetics results indicate that the poisoning effect of Br on

Au/ TiO_2 is primarily due to Br blocking the active sites. *Operando* IR data show that normalized peak area of CO adsorbed on Au^0 at 2100 cm^{-1} best correlates with the WGS reaction rate on these catalysts. The fraction of the total moles of Au found to be the active sites on Au/TiO_2, Au(4mol%Br)/TiO_2 catalysts, is 0.60±0.04% and 0.30±0.02% respectively. The turnover rate (TOR) calculated based on the number of active sites from the isotopic transient experiment for Au(4mol%Br)//TiO_2 is 1.7±0.1 s^{-1}, which is similar to that for Au/TiO_2, 1.5±0.1 s^{-1}. These results suggest that metallic corner Au atoms are the dominant active sites for the WGS reaction on Au/TiO_2 catalysts.

References

1. Guo, N., Fingland, B.R., Williams, W.D., Kispersky, V.F., Jelic J., Delgass, W.N., Ribeiro, F.H., Meyer, R.J., and Miller J.T., *Phys. Chem. Chem. Phys*, 12, 5678, (2010).

See more of this Session: In Situ and Operando Spectroscopy of Catalysts I
See more of this Group/Topical: Catalysis and Reaction Engineering Division

(654f) Ambient Pressure PES Study of Al2O3 Supported Catalysts for Water Gas Shift Reaction

M. Cem Akatay[1], Jorge Pazmino[2], Mayank Shekhar[2], W. Damion Williams[2], Anil Mane[3], Eric A. Stach[1], W. Nicholas Delgass[2], Jeffrey W. Elam[3], Fabio Ribeiro[2] and Dmitry Zemlyanov[4], (1)School of Materials Engineering and Birck Nanotechnology Center, Purdue University, West Lafayette, IN, (2) School of Chemical Engineering, Purdue University, West Lafayette, IN, (3)Argonne National Laboratory, Argonne, IL, (4)Birck NanoTechnology Center, Purdue University, West Lafayette, IN

Alumina (Al_2O_3) supported Pt/TiO_2 catalysts for water gas shift (WGS) reaction were analyzed in situ under 0.5 mbar pressure using ambient pressure synchrotron-based Photoemission Spectroscopy (PES). High-resolution core-level spectra were obtained both under reducing atmosphere of H_2 at 300°C and under the WGS reaction mixture consisted of 7% CO, 2% H_2O, 37% H_2, 9% CO_2 at 250°C at total pressure of 0.5mbar. To obtain depth profile, the photon energy of the exciting radiation was varied as to have the kinetic energy of photoelectrons ranging from 120eV to 420eV. Oxidation state of metals under different atmospheres and change of the composition as a function of depth were investigated.

A reduced state of titania was identified under both environments on 3.6 wt% Pt/Al_2O_3 sample promoted with 10 wt% TiO_2. Reduced titania was supposed to distribute around and under the Pt nanoparticles under H2 whereas under WGS reaction mixture the reduced titania state was most likely on the boundary around Pt nanoparticles. The depth profiling allowed us to suppose agglomeration of TiO_2. No diffusion of Pt into the support was observed and Pt was not oxidized under these conditions. Oxidized states of Fe are found on 0.5 wt% Pt/Al_2O_3 sample promoted with 11 wt% Fe. The surface of the support is covered uniformly with the promoter and agglomeration is not as widespread as compared to the titania promoted sample. The WGS rate of the titania promoted sample is $1.3x10^{-2}$ and that of Fe promoted sample is $4.5x10^{-2}$ mole H_2 mole Pt^{-1} s^{-1} at 250°C under our standard conditions, a promotion of a factor of 17 and 60, respectively when compared with the un-promoted Pt catalyst. Based on the findings, models describing the structure of catalysts were developed and compared to images obtained by High-Resolution Transmission Electron Microscopy (HR-TEM).

See more of this Session: In Situ and Operando Spectroscopy of Catalysts I
See more of this Group/Topical: Catalysis and Reaction Engineering Division

(654g) Multiple Inline Spectroscopy for Multiphase Reaction Monitoring At High Pressure

Marcel A. Liauw and Sonja Hardy, Itmc, RWTH Aachen University, Aachen, Germany

The characterisation of reactive systems under reaction conditions requires non-invasive analytics with a high time resolution that can be implemented into a high pressure setup. In situ spectroscopy with fibre-optical ATR probes (both mid-IR and UV-vis) in autoclaves obviously are a very potent tool for kinetic and thermodynamic investigations. Starting from general considerations, the design, construction, testing and usage of an autoclave is presented that permits both spectroscopies in both the upper and lower phase (liquid-liquid or gas-liquid). The combination with transmission spectroscopy permits the monitoring of nanoparticle catalyzed reactive systems. Various examples for in situ monitoring will be presented.

See more of this Session: In Situ and Operando Spectroscopy of Catalysts I
See more of this Group/Topical: Catalysis and Reaction Engineering Division

CFD SIMULATION OF IN-SITU DROP TUBE FURNACE

Esam Jassim[+], Wayne S. Seames[++], Steven A. Benson[++]

[+]Department of Mechanical Engineering, Prince Mohammad Bin Fahd University (PMU)

Mobile: +966 501470909, Fax.: +966 3 8964566, ejassim@pmu.edu.sa

[++]Department of Chemical Engineering, Harrington Hall Room 323, 241 Centennial Drive Stop 7101, Grand Forks, ND 58202-7101

Abstract

A computational fluid dynamics code for calculating three dimensional reacting flows is used to predict the combustion behaviour of pulverized coal inside a Drop Tube Furnace (DTF). The CFD model has linked the eddy dissipation concept with the finite rate chemistry model in calculating the near flame zone field and the far field regions of the flame.

Furnace model is based on the numerical solution of three-dimensional differential equations for conservation of mass, momentum and energy. The simulation is employing with a comprehensive representation of the features of gas phase fluid dynamics, combustion of coal particles, evolution of trace elements, heat transfer including radiation and chemical reactions to investigate the velocity profile, flame temperature, and product emissions at various locations in the combustion zones.

Coal particles are treated as non-interacting spheres with full coupling of mass, momentum and energy with the gaseous phase. Moreover, the eddy dissipation model is coupled with Arrhenius type expressions for devolatilization, char combustion and COx production.

The combustion sub-models treat particle devolatilization, char oxidation and additional gas phase reactions. The model uses the Lagrangian approach for tracking the particulate phase, the Eulerian approach for formulating the gas phase time-averaged conservation equations, the k - ε turbulence model, the eddy dissipation model for the gas mixture phase, particle to turbulence interaction, particle dispersion by turbulence and the discrete transfer model for particle radiation. The char remaining after devolatilization is considered to be pure carbon and its reaction is governed by external diffusion of oxygen to the particle surface.

The particle tracking includes laws for inert heating, drying (wet combustion model), devolatilization, and char burnout. Chemical reactions were modeled using the Eddy-Break-Up model. The reaction mechanism was based on a 2-step mechanism with carbon monoxide as intermediate combustion species.

Introduction

Combustion of fossil fuels is still the backbone of the power generation, manufacturing and transportation industries. However, problems caused by evolution of inorganic elements during burning fossil fuels continue to manifest, even though substantial progress has been made in understanding the physical and chemical process of these elements during combustion [1].

Considerable attention has therefore been directed toward describing the behavior of certain trace metals during the course of combustion. Most inorganic elements in the periodic table can be found in coals although only a small number occur in significant concentrations to cause operational or environmental problems **[2]**. Among lots of elements, trace metals such **as** mercury. Arsenic, sodium, calcium, and selenium are of concern.

All coal blends contains in their structure numerous inorganic minerals and toxic elements. In pulverized coal, mineral grains can be either included in coal particles or excluded from coal particles [3]. Some alkali and alkaline elements may also be founded in organically associated form in the coal [4]. Many of these elements vaporize upon combustion and experience reaction and condensation at varying degrees depending upon the characteristics of the element and other constituents in the gas stream. Moreover, the ability to model the transformations of trace elements during the combustion process and emissions control equipment is very important due to their emissions and impactions on system performance.

The manuscript highlights the research effort on the releasing of trace elements during pulverized coal combustion and the incorporation of the model into CFD simulation of 3D Drop Tube Furnace.

Research Background on the Trace element in fossil fuel

The ability to model the transformations of trace elements during the combustion process and emissions control equipment is very important due to their emissions and impactions on system performance. The elements considered in the investigation include arsenic, selenium and antimony [5]. These elements vaporize upon combustion and will react and condense to varying degrees depending upon the characteristics of the element and other constituents in the gas stream.

The vaporized trace element can react and condense during gas cooling and be concentrated in the submicromet ash size fractions or may remain in the vapor phase.

A mathematical model was developed to study the vaporization of trace elements during the combustion of individual coal and pyrite particle [6]. The initial modeling has focused on the release of As, Se and Sb associated with pyrite since it is the host site for these elements. The model considers the release of the trace elements from pyrite included in coal particles and excluded from coal particles. The model predicts the amount of elements vaporized as a function of coal particle size, mineral grain size, mineral grain association, flame temperature, and residence time. The model considers the combustion of the coal char and the decomposition, fragmentation, melting, and coalescence processes of the pyrite particles.

CFD simulation of in-situ Drop Tube Furnace

A computational fluid dynamics code for calculating three dimensional reacting flows was used to predict the combustion behaviour of pulverized coal inside a Drop Tube Furnace (DTF).

The furnace model is based on the numerical solution of three-dimensional differential equations of conservation of mass, momentum and energy. The simulation is using a comprehensive representation of the features of gas phase fluid dynamics, combustion of coal particles, evolution of trace elements, heat transfer including radiation and chemical reactions to investigate the velocity profile, flame temperature, and product emissions at various locations in the combustion zones [7].

Coal combustion is calculated by combining a particle transport model of the coal particles with an eddy dissipation model for the combustion of volatiles gases in the gas phase. Coal particles are treated as non-interacting spheres with full coupling of mass, momentum and energy with the gaseous phase. Moreover, the eddy dissipation model is coupled with Arrhenius type expressions for devolatilization, char combustion and COx production.

Coal devolatilization is described by a first order 'single-reaction' model [8], while the char remaining after devolatilization is considered to be pure carbon and its reaction is governed by external diffusion of oxygen to the particle surface.

The particle tracking includes laws for inert heating, drying (wet combustion model), devolatilization, and char burnout. Chemical reactions were modeled using the Eddy-Break-Up model. The reaction mechanism was based on a 2-step mechanism with carbon monoxide as intermediate combustion species.

Furnace Layout

The overall dimensions of the furnace shown in Figure (1) are: 6" ID, 8" OD, and 52" long) for the combustion zone and about 188" length of the post-combustion zone. The primary air-coal mixture enters from the top of the furnace through a 1/2" water cooled pipe and the preheated secondary air flows perpendicular 2" from the top. The secondary air undergoes a 3" thick flow straightener to absorb vortices before mixing with the coal/primary air. However, mixing of the coal/primary air with the secondary air occurs by ~7" down the top of the furnace. The pulverized primary coal air mixture enters the furnace at a temperature of 60 C and the secondary air enters at a temperature of 450 C.

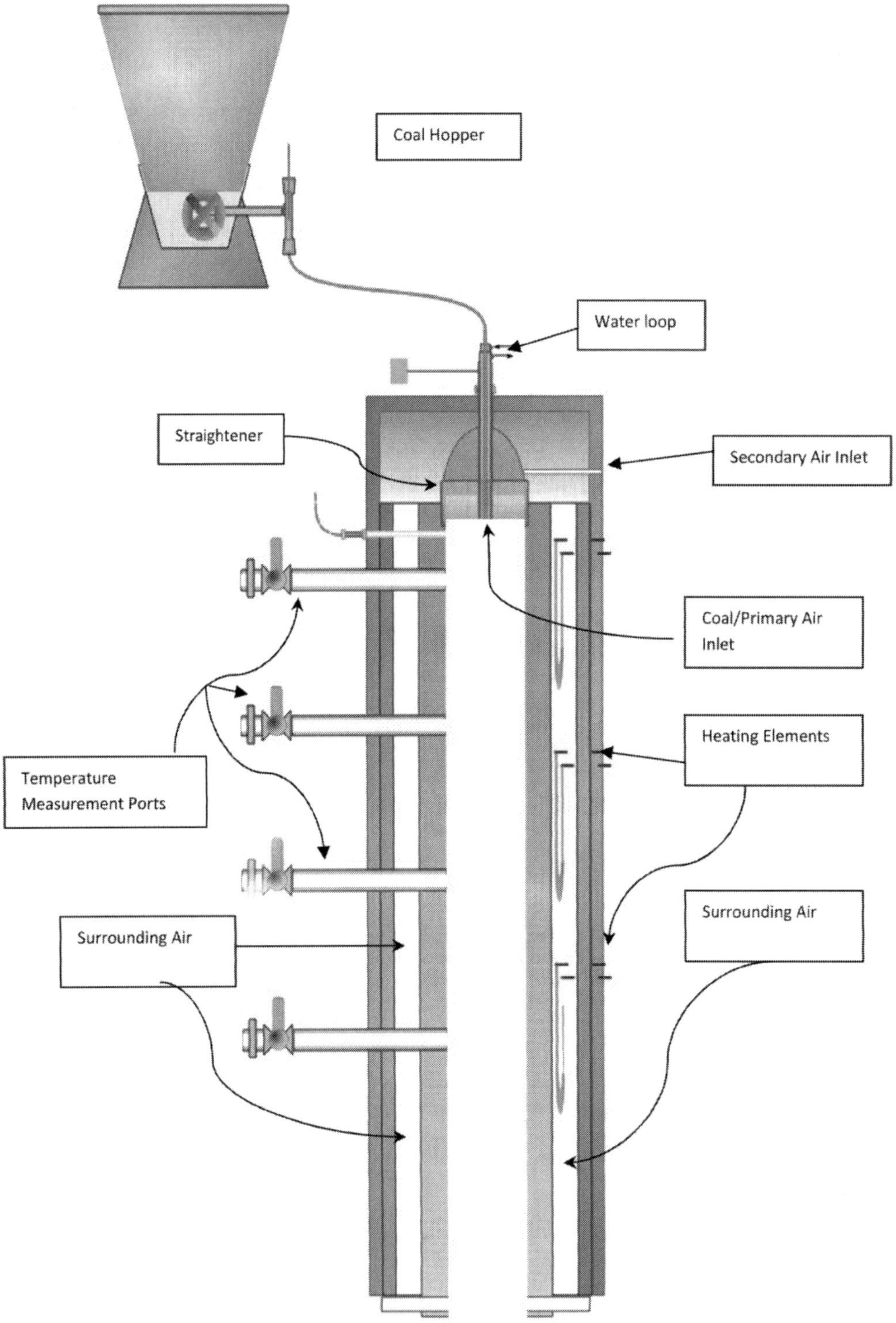

Figure (1) Layout of the Drop tube furnace built in Chemical engineering Department UND

Modeling overview

The DTF combustor was designed using three-dimensional CFD modeling. FLUENT CFD software from Fluent, Inc., was used for the study [7]. Different configurations are planned to be compared based on the required operating parameters, which include temperature level and temperature uniformity inside the combustion zone, uniformity of mixing patterns of the combusted solids, uniformity of trajectory patterns of the injected particles inside the combustor, and sufficient residence time for the particles' moisture release and devolatilization. One of the main requirements for the design was to minimize the interaction of the solids inside the combustion chamber and the chamber walls by optimizing the flow pattern and trajectories of the particles.

Geometry and Mesh

Figures 2 and 3 show the outline of the combustion zones as well as the inlet ports for the computed cases. Meshes with approximately 42780 tetrahedral cells have been generated (Figure 4) for the modeled cases. The computational domain is filled with an unstructured tetrahedral and hexahedral mesh. A refined mesh has been employed around the inlet ports to allow the cell size to grow from the areas with the high strain rates to the rest of the domain. With this approach, it was possible to resolve recirculation pattern of air flow and possible recirculation patterns of the solid particles. The mesh was further refined during the calculations in the areas with high rates of devolatilization of solids.

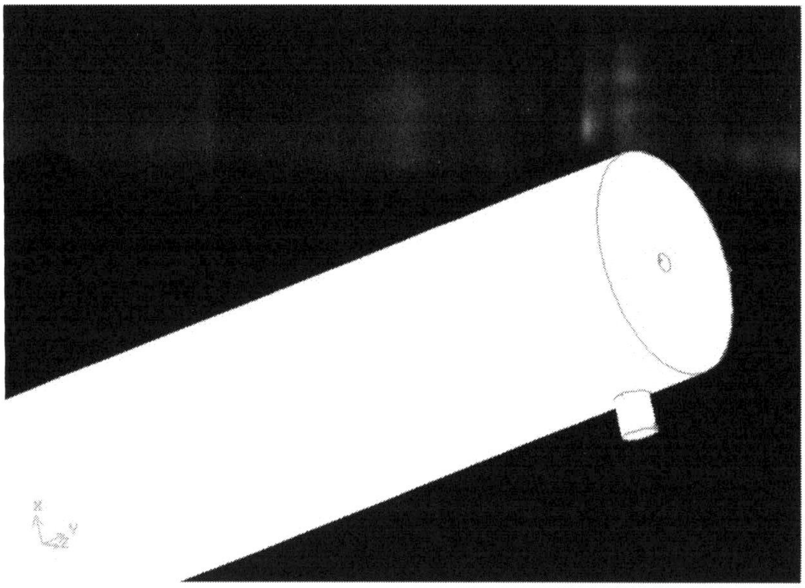

Figure (2) Furnace configuration

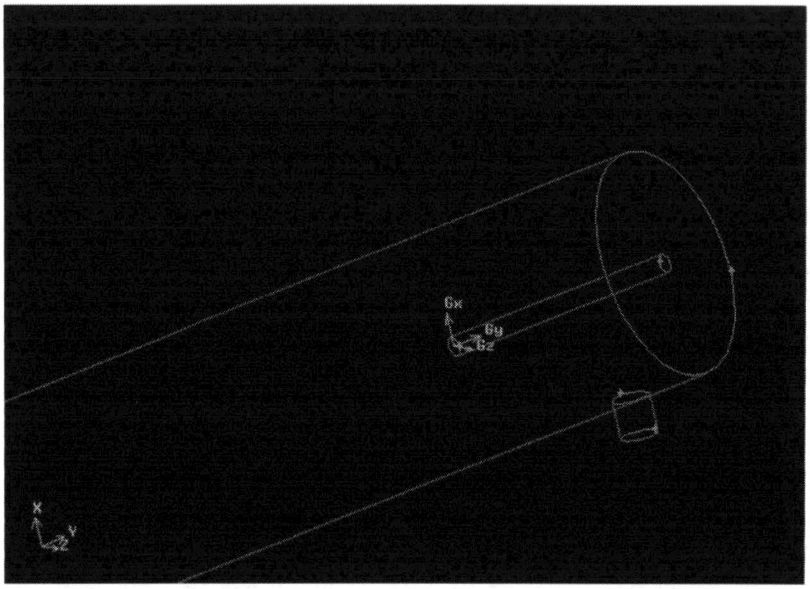

Figure (3) 3-D CFD Geometry of DTF

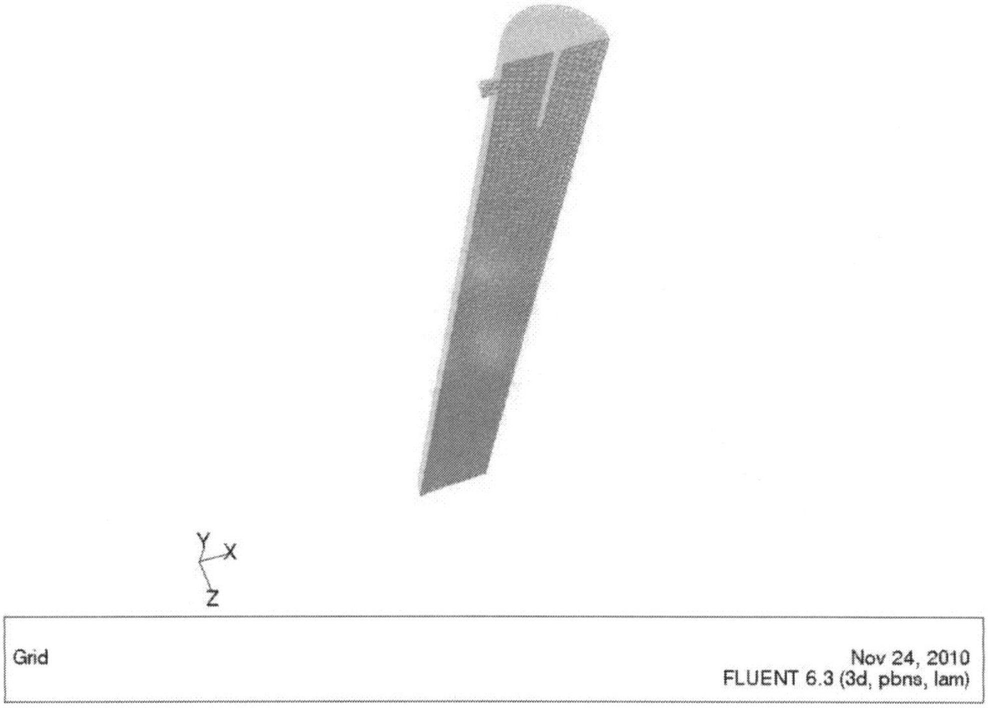

Figure (4) Axisymmetric meshing scheme of DTF

Preliminary CFD results

Figure 5 presents an insight on the flame and flow inside the combustion chamber for optimized design. Flow path lines (see Figure 6) show no strong recirculation pattern and that is advantage as recirculation could remain the coal particles and fly ash particulates in the combustion zone or bring them back to the coal/air inlets.

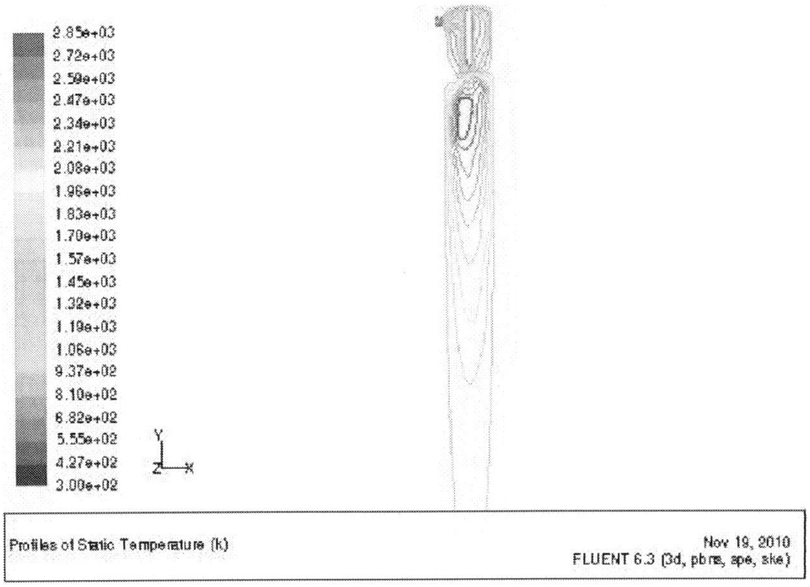

Figure (5) Flame shape at the centre axis of the furnace

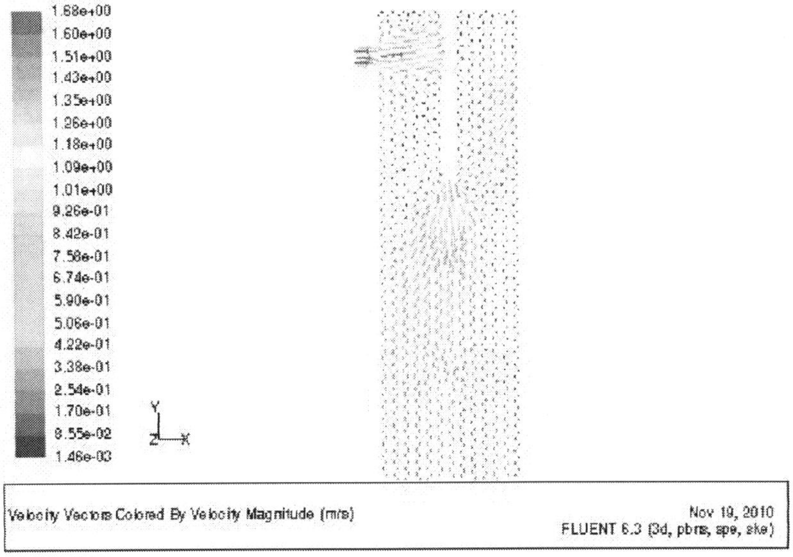

Figure (6) Flow stream pattern near inlets

Temperature, velocity and species contours

Regions of very high temperatures close to the centre prove that no coal has been shuttered toward the wall. Temperature contours predicted by CFD simulation for typical DTF are depicted in Figures (5). The mean temperature of the flame is ~ 2800 K. The temperature contours depict the flame to be approximately ***0.75-1m*** long. The phenomena of flame liftoff may also be seen from the 3D contours of temperature in Figure 7. Flame root position is an important parameter in the design of low NO*x* combustors. If the flame root is closer to the burner then it produces less NO*x* compared to the flame root away from the burner. This may be due to the fact that if the flame is detached from the burner then the concentration of oxygen near the flame-envelope increases the temperature locally-thereby increasing thermal NO*x*. Also, the excess (or unshielded from the flame) O_2 near the flame-envelope results in production of fuel NO*x*.

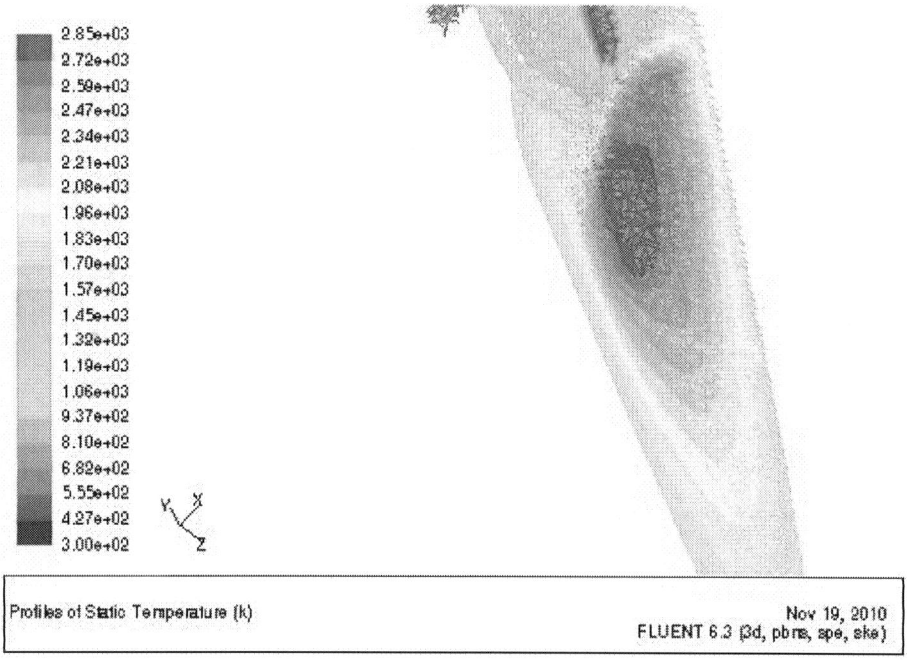

Figure (7) 3-D Temperature contours near the flame

The velocity vector plots in Fig. 6 clearly demonstrate a smooth flow that preventing pushing the plumes towards the wall. The flow vectors show that uniform combustor plume penetrates throughout the system despite the vortex in the top of the furnace. There is a bit shifting of the flow stream to the left at the mixing spot near the coal/primary air inlet as a result of secondary air vortex. Although this small shifting causes no alteration in the flame stability and laminar, the experiment apparatus has straightener just below the secondary air inlet chamber to keep uniformity of the stream at the mixing position.

Contours of O_2 mass fraction provides a clear indication to the fact that complete combustion is predicted during the course of combustion. Figure (8) shows that considerable amount of oxygen at combustion zones exit confirming excess air/fuel ratio has been fed to the system. The zero oxygen concentration in and around the flame suggests complete consumption of the oxygen in such region as well.

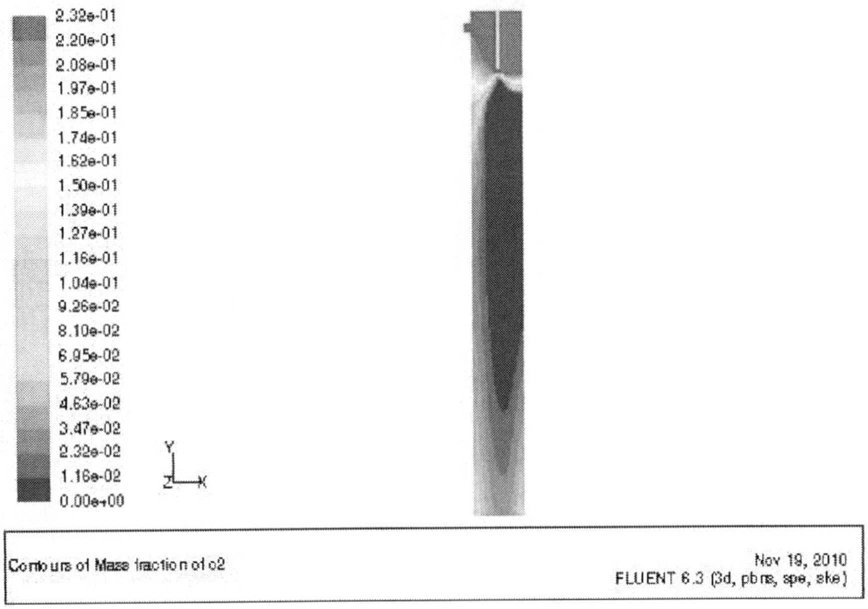

Figure (8) Concentration profile of Oxygen in the combustion zone

Considering CO mass fraction contours illustrated in Figs. 9, the case of coal combustion is characterized by a CO peak in zone1 region corresponding to a temperature peak, because of the strong volatile releasing.

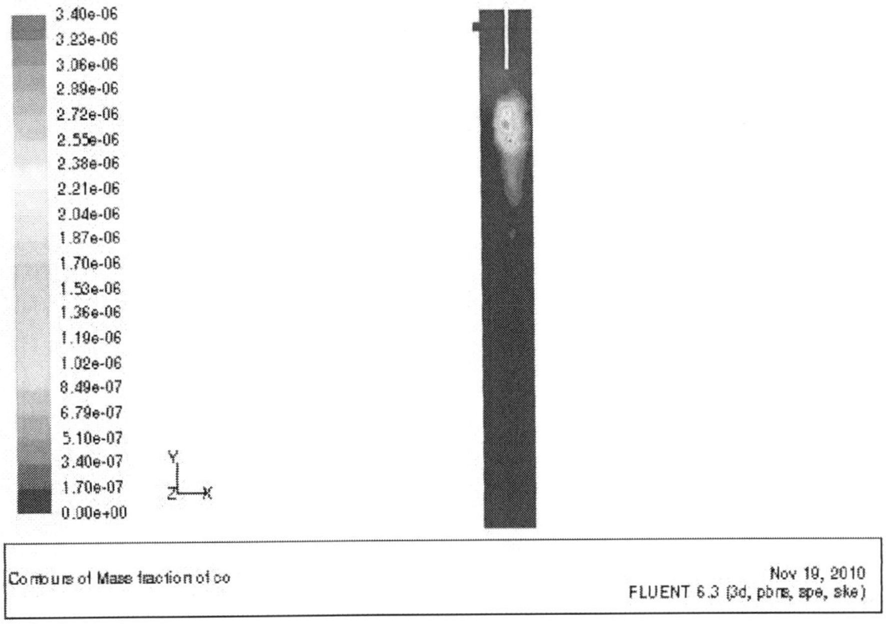

Figure (9) Mass fraction of CO

Another measuring tool to the complete combustion can be concluded using the CO_2 contours. Figure (10) demonstrates that high concentration of carbon dioxide at outlet, confirming that carbon is converted to CO_2 during the process of combustion despite the assumption of two steps reaction mechanism with carbon monoxide as intermediate combustion species.

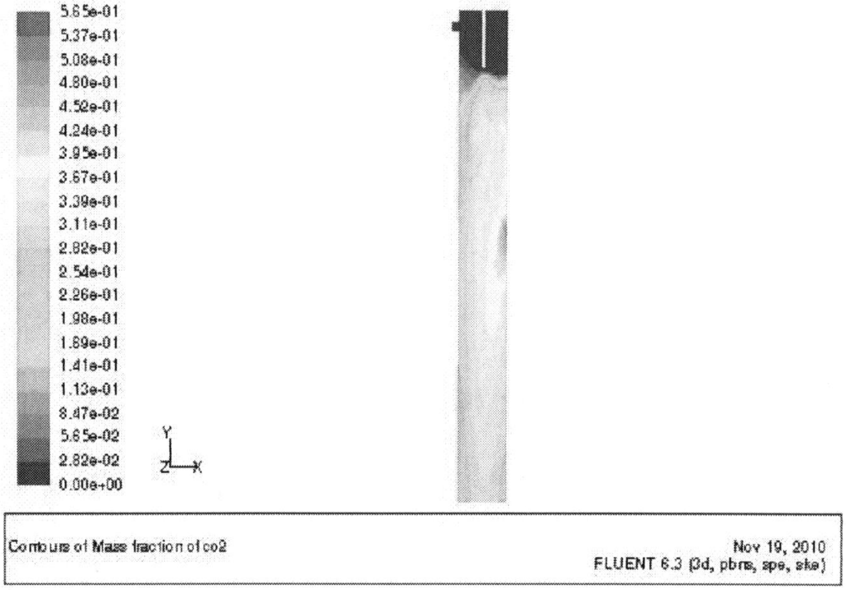

Figure (10) Mass fraction of CO_2

Since the objective of the project is to show how the trace element behaves during combustion, we plugged in the mathematical model of arsenic evolution that developed and discussed in the previous task into the CFD simulation. The assumption we made is the chemical reaction of arsenic with oxygen to form trioxide base on the following expression:

$$As + \frac{3}{4}O_2 \rightarrow \frac{1}{2}As_2O_3 \qquad \text{(R4)}$$

Contours of arsenic shown in figure (11) gives an insight of the spot of arsenic vapor releasing, which is mostly done in the high temperature zone. In turn, trioxide arsenic spread out through the combustion zones as seen in figure (12).

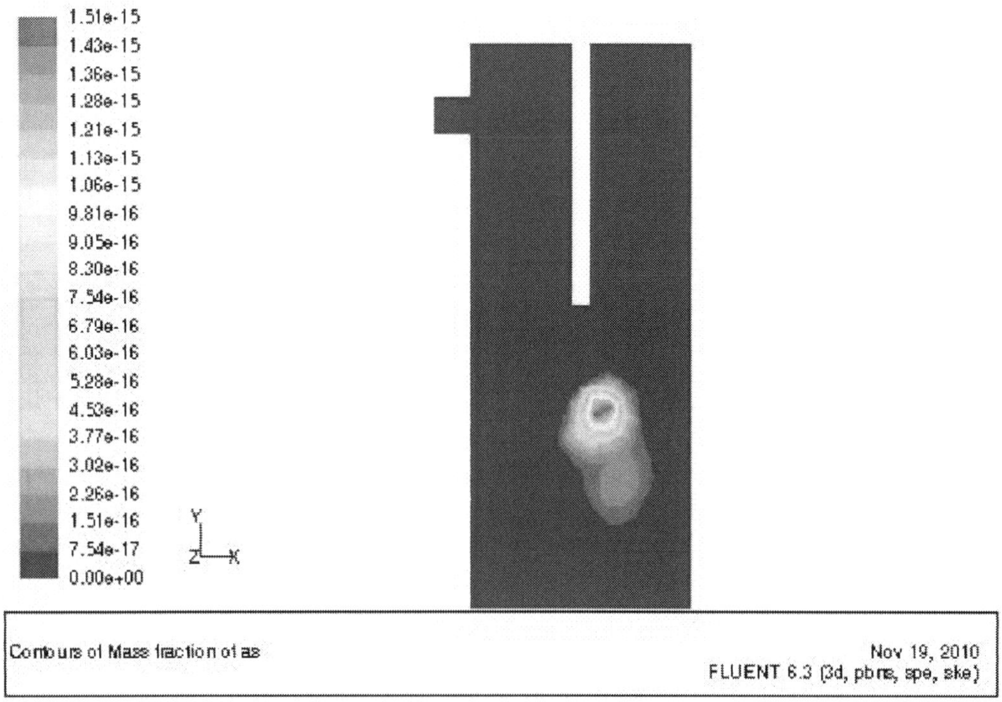

Figure (11) Mass fraction of As

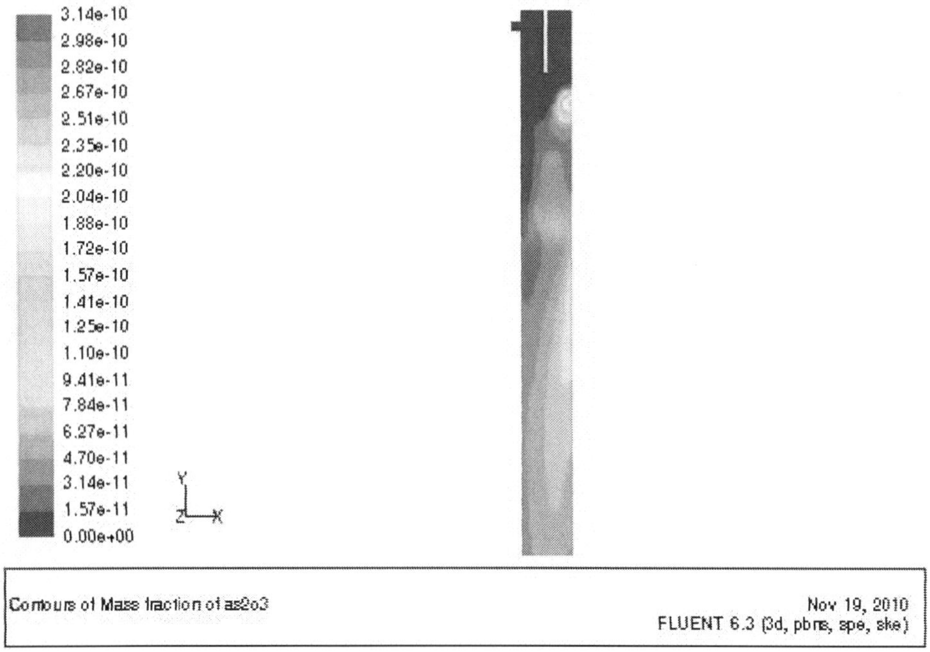

Figure (12) Mass fraction of As_2O_3

Conclusion

The evolution of three trace elements released during coal combustion has been modeled using the criteria of three consecutive transportation processes.

Formulated expressions obtained from the modeling have been incorporated into CFD simulation using Fluent software. In-situ Drop Tube Furnace has been numerically simulated and real experiment data has been used to describe the transformation processes that occur during course of combustion. Profiles of temperature, velocity, product gases, and trace element vaporization and reaction have been presented.

The simulation predicts good design in term of flame progress since the flame base is shown to be close to the fuel source. Flow recirculation is very lean resulting in no coal and/or ash particulates near the furnace walls, keeping the gases stream smoothly flow toward the exit. The simulation also shows that complete combustion is occurred in the combustion zone and less NO_x gasses are emitted. The conclusion gained from the simulation is planned to be compared with the experiment measuring data for agreement and further simulations to include evolution of other trace elements and vaporization of alkali and alkaline are currently under attention.

References

[1] J.R. Qiu, F. Li, C.G. Zheng, Mineral Transformation during Combustion of Coal Blends, Int. J. Energy Res. 23 (1999) 453-463.

[2] Inorganic in fuel: A Century of Scientific and Engineering Progress. Constance **L.**, Physical Sciences, Inc., 20 New England Business Center, Andover, MA 01886 USA

[3] Seams W.S. An initial study of the fine fragmentation fly ash particle mode generated during pulverized coal combustion. *Fuel process Technology*, 2003, 81, 109-125.

[4] Z. Ma, F. Iman, P. Lu, R. Sears, L. Kong, A.S. Rokanuzzaman, D.P. McCollor, S.A. Benson, A comprehensive slagging and fouling prediction tool for coal-fired boilers and its validation/application, Fuel Processing Technology 88 (2007) 1035–1043.

[5] W. Seames, J. Wendt, Regimes of Association of Arsenic and Selenium during Pulverized Coal Combustion, Proc. Combust. Inst. 31 (2007) 2839-2846.

[6] Wayne S. Seames, Esam I. Jassim, Steven A. Benson, "Modeling Trace Element Release from Pyrite During Pulverized Coal Combustion", proceedings in The 27th Annual International Pittsburgh Coal Conference, Istanbul, TURKEY, October 11 - 14, 2010.

[7] FLUENT 6.3., Getting Started Guide, 2006.

[8] Anthony D.B., Howard J.B., Coal Devolatilisation and Hydro-gasification, A.I.Ch.E

Journal, Vol. 22, No. 4, 1976.

(670b) The Thermal Decompositions of Benzyl and o-Xylyl Radicals

Raghu Sivaramakrishnan[1], Meng-Chih Su[1] and **Joe V. Michael**[2], (1)Chemical Sciences & Engineering Division, Argonne National Laboratory, Argonne, IL, (2)Argonne National Laboratory, Argonne, IL

The thermal decompositions of benzyl- ($C_6H_5CH_2$), benzyl α,•-d_2- ($C_6H_5CD_2$), and o-xylyl-radicals (o-$CH_3C_6H_4CH_2$) have been studied at high temperatures with the reflected shock tube technique using H- and D-atom ARAS. The experiments were performed at high-T (1437-1801 K) at nominal pressures ≈ 0.3-1.3 atm.

Using the ultra-sensitive H- and D-atom ARAS technique, the present experiments reveal that there are three channels that contribute to benzyl decomposition: H-atom removal from the ring, H/D-atom removal from the side-chain, and a non-atom producing process (postulated to be cyclopentadienyl + acetylene). The experimental data was used to obtain rate constants for the three channels and also for total benzyl decomposition.

We then initiated experimental studies on o-xylylbromide (o-$CH_3C_6H_4CH_2Br$) decomposition. Br-ARAS experiments used with the shock tube confirm that at high-T (>1200 K) o-xylyl radicals (o-$CH_3C_6H_4CH_2$) are generated instantaneously. The high sensitivity H-atom ARAS detection technique has been subsequently used to obtain quantitative measurements of the H-atom yields, and rate coefficients, for o-xylyl decomposition. These H-ARAS experiments span a T-range, 1267-1597 K, and pressure range, 0.3-1.0 atm. The results from these studies and comparisons to earlier studies are discussed.

This work was supported by the U. S. Department of Energy, Office of Basic Energy Sciences, Division of Chemical Sciences, Geosciences, and Biosciences, under Contract No. DE-AC02-06CH11357.

See more of this Session: Reaction Engineering for Combustion and Pyrolysis I
See more of this Group/Topical: Catalysis and Reaction Engineering Division

ADVANCES IN UNDERSTANDING ACETYLENE COMBUSTION FROM MBMS EXPERIMENT AND MODELING

Wenjun Li[1], Phillip R. Westmoreland[1], Tina Kasper[2], Nils Hansen[2], Bin Yang[3], Terrill A. Cool[3], Katharina Kohse-Höinghaus[4]*
(1) Chemical and Biomolecular Engineering, North Carolina State University, Raleigh, NC,
(2) Combustion Research Facility, Sandia National Laboratories, Livermore, CA,
(3) Department of Applied Physics, Cornell University, Ithaca, NY,
(4) Department of Chemistry, Bielefeld University, Bielefeld, Germany.

Introduction

Flame chemistry from acetylene combustion has been a recurring research target of flat-flame molecular-beam mass spectrometry (MBMS), especially near the sooting limit. In the long history since this research theme began, progress has advanced from empirical observation of species to modeling and precise species identification.

Results and Discussion

In this work, we present experimental detailed flame structures for two premixed laminar flat flames having feed mixtures with "fuel-equivalence ratios" of 0.95 and 2.41 (fuel-oxygen ratios relative to ratio that forms CO_2+H_2O stoichiometrically, so fuel-lean and fuel-rich mixtures here). These data were obtained with Vacuum-Ultraviolet Photo-Ionization Molecular-Beam Mass Spectrometry (VUV-PI-MBMS) in our apparatus at Lawrence Berkeley National Laboratory []. Temperature profiles are measured with type-R thermocouples.

New insights are also made possible by modeling these VUV-PI MBMS from our apparatus and recent measurements in a sister apparatus at the National Synchrotron Radiation Laboratory in Hefei, China. Current modeling reveals good fundamental understanding of C_2H_2 chemistry as well as areas where further advances are needed.

An updated reaction mechanism is used for flame simulation, giving generally good agreement with the experiment. Oxidation and molecular-weight-growth steps in acetylene combustion are then investigated under fuel-lean and fuel-rich conditions using reaction pathway and flux analyses.

Bibliography

1. A. Lucassen, N. J. Labbe, P.R. Westmoreland, K. Kohse-Höinghaus, "Combustion chemistry and fuel-nitrogen conversion in a laminar premixed flame of morpholine as a model biofuel," *Combustion and Flame* **158** (2011) 1647-1666; DOI: 10.1016/j.combustflame.2011.02.010
2. W. Li, M. E. Law, P. R. Westmoreland, T. Kasper, N. Hansen, K. Kohse-Höinghaus "Multiple Benzene-Formation Paths in a Fuel-Rich Cyclohexane Flame," Combustion and Flame (accepted 5/2011), http://dx.doi.org/10.1016/j.combustflame.2011.03.014.
3. A. Frassoldati, T. Faravelli, E. Ranzi, K. Kohse-Höinghaus, P. R. Westmoreland, "Kinetic modeling study of ethanol and dimethyl ether addition to premixed low-pressure propene-oxygen-argon flames," *Combustion and Flame* **158** (2011), 1264-1276.

4. J. Wang, M. Chaos, B. Yang, T. A. Cool, F. L. Dryer, T. Kasper, N. Hansen, K. Kohse-Höinghaus, P. Oßwald, P. R. Westmoreland, "Composition of reaction intermediates for stoichiometric and fuel-rich dimethyl ether flames: Flame-sampling mass spectrometry and modeling studies," *Phys. Chem. Chem. Phys.* **11**(9) 1328-1329 (2009).
5. N. Hansen, T. A. Cool, K. Kohse-Höinghaus, P. R. Westmoreland, "Recent Contributions of Flame-Sampling Molecular-Beam Mass Spectrometry to a Fundamental Understanding of Combustion Chemistry," *Progress in Energy and Combustion Science* **35**(2) (2009) 168-191.

Contact information

Prof. Phillip R Westmoreland (*speaker)
Executive Director, NCSU Institute for Computational Science and Engineering
Professor, Department of Chemical and Biomolecular Engineering
North Carolina State University
2036 Engineering Building I, Box 7905
911 Partners Way; Raleigh, NC 27695-7905
Phone: (919) 515-7121; Fax: (919) 515-3465
Email: phil.westmoreland@ncsu.edu

(670d) On the Role of O2 + QOOH In Low-Temperature Ignition of Alkanes

Claude Franklin Goldsmith III, Inorganic Chemistry, Fritz Haber Institute of the Max Planck Society, Berlin, Germany, William H. Green, Department of Chemical Engineering, Massachusetts Institute of Technology, Cambridge, MA and Stephen J. Klippenstein, Chemical Sciences and Engineering Division, Argonne National Laboratory, Argonne, IL

The kinetics of the reaction of molecular oxygen with hydroperoxyalkylperoxy radicals have been studied theoretically. These reactions – often referred to as second O_2 addition, or O_2 + QOOH reactions – are believed to be responsible for low-temperature chain branching in hydrocarbon oxidation. The O_2 + propyl system was chosen as a model system. High-level *ab initio* calculations of the $C_3H_7O_2$ and $C_3H_7O_4$ potential energy surfaces are coupled with RRKM master equation methods to compute the temperature- and pressure-dependence of the rate coefficients. Variable reaction coordinate transition state theory is used to characterize the barrierless transition states for the $O\text{-}_2$ + QOOH addition reactions as well as subsequent $C_3H_6O_3$ dissociation reactions. A simple kinetic mechanism is developed to quantify the extent to which the second O_2 addition increases the number of radicals. Based upon the simulations, chain branching increases with pressure. For the initial conditions considered here, chain branching is maximized in the temperature range of 700 to 900 K. Low-temperature chain branching is significant only when the isomerization from a peroxy radical to a hydroperoxyalkyl radical proceeds via a transition state with six or more members. The formally direct (or well skipping) reaction O_2 + QOOH → OH + OH + oxy-radical and the corresponding sequential reactions O_2 + QOOH → OOQOOH → OH + ketohydroperoxide → OH + OH + oxy-radical increase the total number of radicals. An analysis of the influence of 1D, separable models for hindered internal rotation on the phenomenological rate coefficients suggests that the uncertainty in the rate constants due to the hindered rotor model is less than a factor of two. The results confirm that n-propyl is the smallest alkyl radical to exhibit the low-temperature combustion properties of larger alkyl radicals, but n-butyl is perhaps a truer combustion archetype.

See more of this Session: Reaction Engineering for Combustion and Pyrolysis I
See more of this Group/Topical: Catalysis and Reaction Engineering Division

Oxy-Coal Combustion Flamelet Dynamics From High-Speed Video Analysis

***Terry A. Ring,** Husam el Gendy and Eric G. Eddings*
University of Utah
Institute for Clean and Secure Energy
Chemical Engineering Department
50 S. Central Campus Drive, MEB 3290
University of Utah
Salt Lake City, UT 84112
Phone: 801-581-6915, Fax: 801-585-9291
ring@eng.utah.edu

Abstract:
At 3,000 frames per second and shutter speeds of 1/250,000 of a second, the action in a turbulent flame in an oxy-coal combustion system is stopped. Analysis of these images shows the instantaneous temperature field and how it changes due to the mixing and reaction of eddies that contain volatiles that have evolved from coal in the radiation field of the furnace and oxygen. The soot generated by this reaction is observed with the video camera. Using image analysis, the macroscopic velocity field and the local flamelet velocity, flamelet heating and flamelet cooling rates are measured. The results show that flamelet temperatures have a probability density distribution that depends on the local composition. Maximum flamelet temperatures approach the adiabatic flame temperature for the coal. The flamelet velocity ranges from 1 to 10 m/s corresponding to sub sonic combustion or micro deflagrations. An analysis of the flamelet heating and cooling allows the size of the soot particles generated in the flamelet to be determined.

Introduction:

The objective of this experimental program is to gather data that will allow comparison to high-fidelity coal combustion simulations. These computational models provide high temporal and spatial information for temperature, pressure and gas and coal and soot particle velocities, as well as chemical species of various types, in a highly turbulent environment. These models are expensive to run taking multiple days to weeks on 1,000 cpu computer clusters. The challenge to experimentalists is to provide experimental data of a similar high-fidelity nature for comparison with these simulations. In this work, we provide high-fidelity experimental data – high temporal and spatial resolution soot temperature data. A challenge to both experimentalists and simulators is how to condense the massive amount of high-fidelity data into a form that can be used for direct comparisons between experiment and theory. Here we suggest a method: population distribution functions (PDF).

Experimental Setup

A 610 cm inside diameter, 1.219 m tall down-flow, axial-flame oxy coal combustion system utilizing various ratios of oxygen to CO_2 for combustion has been used, see Figure 1. The experimental apparatus is described in detail in a companion paper[1]. The walls in the main burner zone, 76 mm thick, are insulated with Insboard 2600 ceramic board material, and the downstream sections are insulated with a castable refractory material. In the main burner zone, twenty four 840 W electrical heaters are placed within the Insboard material to maintain the wall temperature at 1301±10 K using K-type themocouples (Omega) and individual controllers. The heaters enable the 100 kW system to approximate the near-burner conditions of larger boilers in the field. Illinois bituminous coal with the analysis shown in Table 1 has been used for combustion after it was pulverized and classified to 80% minus 200 mesh (74 μm). The coal was burned at a nominal flow rate of 5.26 kg/hr with a nominal 5.5% excess oxygen, and the incoming oxidant had an oxygen mole fraction of 40%, the balance being CO_2 delivered from a liquid storage tank. The burner consisted of co-annular flow of coal and transport gas in the primary stream through a 15.8 mm ID tube and oxidizer gas only in the secondary flow through the annular region between the OD (21.34 mm) of the primary and the ID (35.05 mm) of the secondary tube. The primary flow temperature was 305 K and that of the secondary was 544±2 K. For these experiments, the mole fraction of oxygen in the primary flow was varied from 0 to 0.209 keeping the total oxygen mole fraction for both the primary and secondary the same at 0.401. The coal was metered with a screw feeder into the primary

---

[1] Rezaei, D.M., Eddings, E.G., Kelly, K.E., Zhang, J-W, Shou, Y-G and Wendt, J.O.L., "The Effect of Coal Composition on Ignition and Flame Stability in Co-axial Oxy-fuel Turbulent Diffusion Flames," Proceedings Pitts burgh Coal Conference, Istanbul, Turkey, October 11-14, 2010.

gas stream. To keep the primary flow at a constant so that the feed rate of the coal could be kept constant, additional CO_2 flow was added to maintain the velocity of the primary constant at 6.338 m/s. The secondary flow compensates for variations in the O_2 content of the primary, but since the flow of the secondary is much larger (typically 3:1), it only varies by a small amount keeping the ratio of gas velocity between the secondary and the primary gas flows at a constant of 2.4. The specific experimental conditions used in this work are given in Table 2. These experiments are a subset of those performed in another paper[1]. The results of the other paper suggest that the conditions under study here show detached (A3), partially attached (A4) and fully attached (A6) flame ignition behavior.

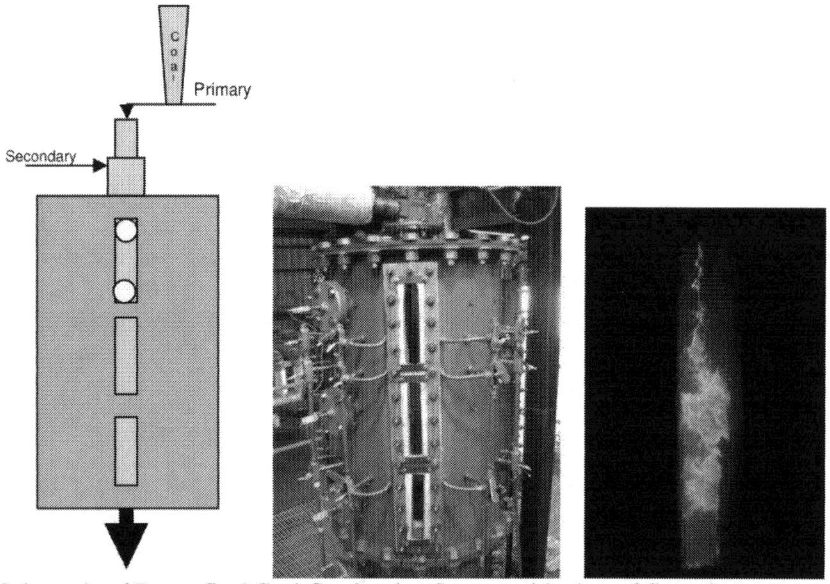

Figure 1 Schematic of Down-fired Coal Combustion System with view of flame in top-most windows indicated by vertical blue rectangles.

Table 1 Analysis of Illinois Bituminous Coal

LOD@ 105°C	Ash@ 750 °C	C	H	N	S	O	Volatile Matter	Fixed Carbon	HHV (BTU/lb)
9.65%	7.99%	64.67%	5.59%	1.12%	3.98%	16.65%	36.78%	45.58	11598

Ash Composition (wt %)

Al_2O_3	CaO	Fe_2O_3	MgO	MnO	P_2O_5	K_2O	SiO_2	Na_2O	SO_3	TiO_2
17.66%	1.87%	14.57%	0.98%	0.02%	0.11%	2.26%	49.28%	1.51%	2.22%	0.85%

Table 2 Experimental Conditions for Oxy Combustion Experiments. Secondary Temperature is 544±5 K and wall Temperature is 1301±10 K and the coal (80% -200 mesh Illinois Bituminous) had a flow rate of 5.26±0.3 kg/h and the total oxygen mole fraction was always 0.401.

A	Primary P_{O2} Mole fraction.	Primary Velocity m/s	Secondary Velocity m/s	Primary O_2 kg/s	Primary CO_2 kg/s	Secondary O_2 kg/s	Secondary CO_2 kg/s
1	0.000	6.338	14.909	0.00E+00	1.90E-03	3.07E-03	4.42E-03
2	0.055	6.338	14.908	7.56E-05	1.80E-03	3.00E-03	4.53E-03
3	0.100	6.338	14.908	1.39E-04	1.71E-03	2.94E-03	4.61E-03
4	0.146	6.338	14.909	2.02E+04	1.63E-03	2.87E-03	4.70E-03
5	0.178	6.330	14.913	2.46E-04	1.56E-03	2.83E-03	4.76E-03
6	0.209	6.338	14.908	2.90E-04	1.50E-03	2.78E-03	4.82E-03

High Speed Video Analysis

Video analysis was performed with a Photron high-speed, 1024x1024 pixels, black and white camera [Itronx Imaging Technology, Westlake Village, CA] at 3,000 f/s and a shutter speed of 1/2,500,000 of a second. This very fast shutter speed essentially stops the turbulent action of the flame and prevents saturation of any of the pixels in the field of view allowing full resolution of the image. A 70 - 300mm macro zoom lens [Tamron USA, Inc., 10 Austin Blvd., Commack, NY 11725] set at 300 mm (macro), an f-stop of 4 and a focal length of 1.3 m was used. The depth of field for these settings was ~3 cm. At these settings a 5 cm x 5cm field of view at the center line of the combustion chamber was in focus. At these camera settings each pixel observes the light from a 50 μm x 50 μm section of the camera's view. The camera was positioned to observe at the center line of the burner two different axial locations below the tip of the burner – one 7.6 to 12.6 cm and the other 30.6 to 35.6 cm from the burner tip. The intensity of light captured is converted to an electrical current by the amplifier attached to the photo detector. Each pixel in this camera can detect a range in the number of electrons generated by the photo detector. These electrons pass through an on-pixel amplifier and into the pixel's resistor producing a voltage that is measured –the pixel intensity scales linearly with the measured voltage and the number of electrons generated in the photo detector. The spectral resolution for the camera is between 400 nm and 1,000 nm with various quantum efficiencies at each wavelength. Using the spectral resolution obtained from the camera manufacturer under confidentiality agreement, the pixel intensity can be scaled albeit non-linearly with temperature using Planck's black body radiation equation assuming an emissivity of 1.0. Work done by Lou, et. al.[2] using two-color pyrometry of similar flames has shown that the emissivity is typically 0.99 to 1.0. Spectroscopic measurements for visible wavelengths fit to Planck's grey body radiation model gave a flame emissivity of 0.989. As a result, the assumption of black body radiation (emissivity of 1.0) is not a bad assumption. The resulting pixel calibration curve is given in Figure 2. In this figure the equilibrium flame temperature [2468 K] for Illinois coal in this atmosphere was also plotted. Typically 1024 frames of video are taken covering a time frame of 0.35 s and creating 1.07 Gb of data for each experimental run. All runs were also performed in triplicate.

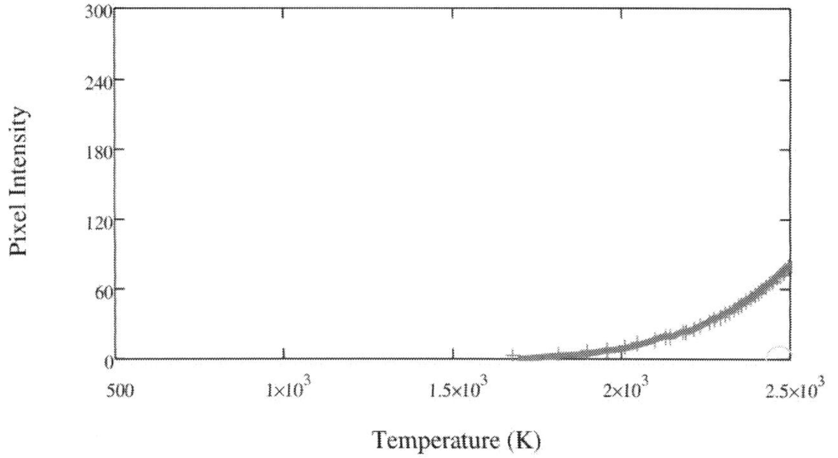

Figure 2 Temperature Calibration based upon Pixel intensity for 1/500,000 s shutter speed and specific camera setup. Green half circle is the equilibrium flame temperature (2468 K) for Illinois coal at stoichiometric conditions. Blue + symbols correspond to a fit function given by T= 2840 K $[I/255]^{0.111}$.

Measurements were made at an axial distance from the burner of 7.6 cm and 30.1 cm corresponding to the top and bottom of the first window indicated by the yellow circles in Figure 1. The windows were cleaned

[2] Lou, C. Zhou, H-C, Yu, P-F, Jiang, Z-W., "Measurements of the flame emissivity and radiative properties of particulate medium in pulverized coal fired boiler furnaces by image processing of visible radiation," Proceedings of the Combustion Institute, 31(2007), 2771-78.

with compressed air just prior to taking high-speed video images, and the images were recorded after the disturbance due to cleaning cleared the combustion system.

Results

Viewing with a traditional video camera at 30 f/s, we find that the image of the flame, while not static, is rather constant in its color and intensity, see single color frame in Figure 1. An example of three frames from the 3,000 f/s high speed video for a 5 cm x 5 cm viewing area at 30.1 to 35.1 cm downstream from the burner tip, where the flame is well established for experimental conditions A6 is shown in Figure 3. The high speed image is far from constant in intensity but consists of discrete wispy bright areas of various shapes and roughly circular spots of light. The circular spots of light are from less than 100 μm to 300 μm in size, with the smaller ones emitting less light and are likely to be coal particles or clusters of coal particles (larger ones) heated to radiating conditions. The whispy bright areas cover a broad range of sizes from 100's of microns to centimeters in length. Given the bright edges of these curved shapes, we can surmise that reaction proceeds at the interface between a turbulent eddy containing volatiles generated from the heated coal, with another turbulent eddy rich in oxygen. The reaction between coal volatiles and oxygen forms soot which is most likely responsible for the light emissions observed. Depending upon the local mixing conditions, the fuel to oxygen ratio will vary significantly in these flamelets giving rise to various flamelet temperatures. Observing the individual images in a sequence, the bright zones tend to get hot over a period of 2 to 10 images or about 2 ms and cool over a period of 15 to 25 images or about 7 ms. Flame velocity can also be assessed using several individual frames of this high speed video flame velocities from 1 to 10 m/s.

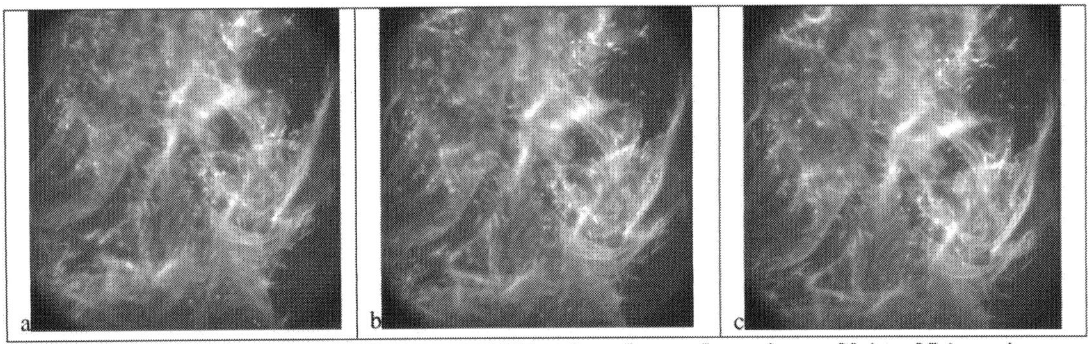

Figure 3 Three Frames of 3,000 f/s High-Speed Video for a 5 cm x 5 cm view at 30.1 to 35.1 cm down stream from the burner tip for experimental conditions A6.

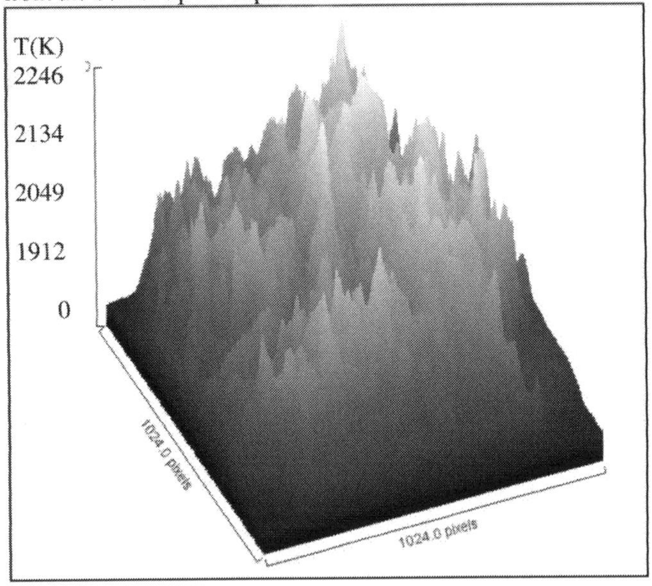

Figure 4 Temperature distributions for the image shown in Figure 3a.

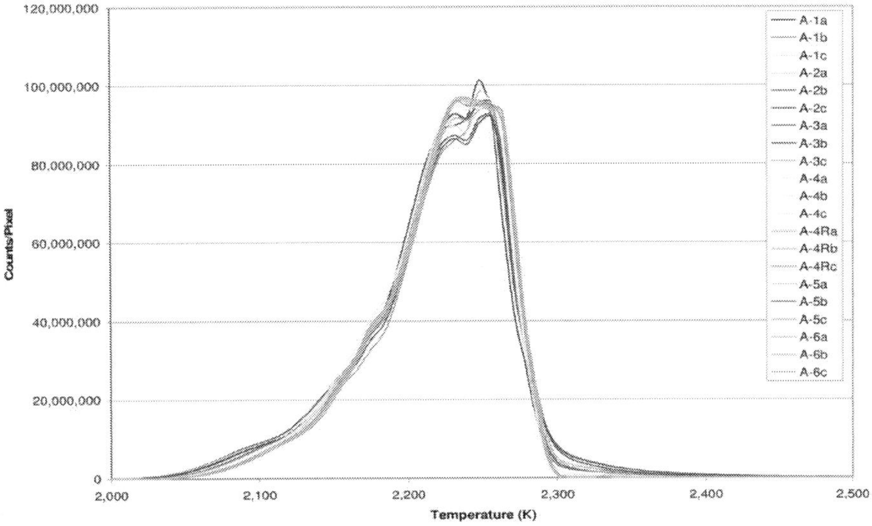

5A) Data from All Experimental Conditions at 7.6 to 12.6 cm from burner tip

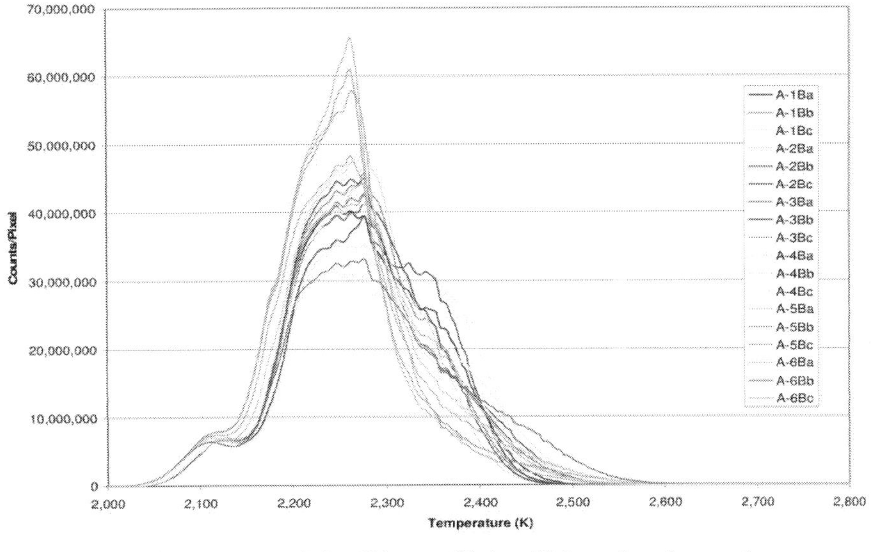

5B) Data from All Experimental Conditions at 30.1 to 35.1 cm from burner tip.

Figure 5 Temperature PDFs for all experimental conditions given in Table 2. Figure 5A corresponds to experimental conditions at 7.6 to 12.6 cm from burner tip, Figure 5B corresponds to all experimental conditions at 30.1 to 35.1 cm from burner tip.

Image Analysis

Using the temperature calibration in Figure 2, converting pixel intensity to temperature, we can determine the temperature distribution for any image. An example of a temperature distribution plot is shown in Figure 4 where the vertical axis is temperature. As we can see the temperature profile ranges from 1900 K to 2200 K which is lower than the equilibrium flame temperature (2468 K) for the coal at stoichiometric conditions. During heat up the coal undergoes drying then devolatilization. The mass released is initially water, followed by coal volatiles, which consist of varying sizes of hydrocarbon molecules that are released as the temperature of the coal particle increases. When a turbulent eddy containing coal volatiles collides and mixes with eddies having a range of oxygen concentrations, then the adiabatic flamelet temperature for those various collisions will vary. If the local mole fraction of oxygen due to eddy mixing is 0.2 and not that of the overall mole oxygen fraction of 0.4 then the equilibrium flame temperature decreases to ~1900 K. In addition, after the flamelet has reacted the soot produced will cool back down to the reaction-increased gas temperature as time proceeds. It is for this reason that we have such a broad distribution of flamelet temperatures in Figure 4.

Taking the temperatures as in Figure 4 for all locations in all 1024 frames of video taken, a temperature probability distribution function (PDF) can be developed for each experimental condition. Temperature PDFs will allow direct comparison between high-fidelity experiments and high-fidelity simulations. The temperature PDFs for all experimental conditions for 7.6 to 12.6 cm (Top) and 30.1 to 35.1 cm (Bottom) from the burner tip are shown in Figure 5. These results show that the Top PDFs are very similar to one another for all the experimental conditions. Using all the experimental data given in Figure 5A, the mean temperature at the Top is 2,226.07K. The high temperature tail of the Top PDF shows higher temperatures for samples with lower Primary P_{O2} values suggesting that higher combustion temperatures are obtained when secondary O_2 mixes with coal volatiles. The results for the Bottom PDFs shown in Figure 5B show a broader temperature distribution than the Top PDFs. The mode of the Bottom PDFs centered at 2250K is similar to the mode in the Top PDFs but some of the intensity is lost to lower temperatures, a second mode to the distribution at 2,100K indicating that some flamelets have cooled 150K in 22.5 cm between the Top and Bottom measurements and to higher temperatures approaching the equilibrium flame temperature at 2,468 K. In some cases, a high temperature mode at 2,300K is observed which in general result for lower Primary P_{O2} values. Taking all the Bottom PDF data together the average temperature is 2,291.16K which is 65 K higher than the average temperature up stream at the Top.

Conclusions

High speed B&W video has been used to measure the soot temperature in an oxy-coal combustion system operating at 40 mole % oxygen. The video images have been converted to temperature using Plank's black body radiation equation assuming an emissivity of 1.0. The temperature profiles show that individual coal particles and clusters of coal particles are present as well as clouds of soot produced by the reaction of oxygen with volatiles generated from the heating of coal. Experimental conditions consist of differing amounts of oxygen in the primary burner gas flow containing the coal. The range of experimental conditions provides for detached, partially attached and fully attached flames. From the high-speed video data temperature PDFs were determined. With detached flames the temperature PDF is initially of low, near constant temperature and increases in breadth to higher temperatures moving away from the burner. With the attached flame the temperature PDF is high and broader in temperature, relative to the detached flame.

Acknowledgement

This material is based upon work supported by the Department of Energy under Award Number DE-NT0005015. Help in running the experiments was provided by Dadmehr Rezaei.

MODELING THE PYROLYSIS OF METHYLDECENOATE AND METHYLDECANOATE

Kevin M. Van Geem[1], Steven Pyl[1], Nick Vandewiele[1], Zhiming Zhou[2], Marie-Françoise Reyniers[1] and Guy B. Marin[1]

[1] Laboratory of Chemical Technology, Ghent University, Belgium
[2] East China University of Science and Technology, Shanghai, China

Introduction

In comparison to hydrocarbons the chemistry of combustion and pyrolysis of oxygenates is quite poorly understood[1]. The available kinetic models developed for combustion of mixtures of hydrocarbons and oxygenated compounds do a reasonable job in predicting auto-ignition, flame propagation parameters, and the formation of the main regulated pollutants. However they do quite poorly for predicting the formation of minor products (alkenes, dienes, aromatics, aldehydes) and soot, which have a deleterious impact on both the environment and on human health. For example most of the kinetic models simply do not include formation of aromatics, let alone soot. Moreover, the validation for toxic aldehydes such as formaldehyde is often impossible because these components are quite difficult to measure quantitatively.

The pyrolysis of methyl esters remained largely unstudied since the early work of Hurd and Blunk[2]. During the last decade the research was mainly focused on shorter methyl esters such as methyl butanoate.[3] More recently, detailed kinetic models describing the oxidation and pyrolysis of methyl decanoate have been developed. The kinetic model developed by Herbinet et al.[4] was built in a systematic way using the reaction rules proposed by Curran et al.[5, 6]. The kinetic parameters and thermodynamic properties in this kinetic model have been taken from literature data. A kinetic model developed by Glaude et al.[7] was generated automatically using EXGAS. This program, previously used to generate reaction mechanisms for the oxidation of alkanes[8-10], was extended to account for the specificities in the reactions and rate constants particular for saturated methyl esters. Thermodynamic properties and kinetic data were calculated automatically using the programs THERGAS[11] and KINGAS[9] respectively. In 2011, EXGAS was used by Herbinet et al.[12] to generate a kinetic model specifically for methyl decanoate pyrolysis. In comparison to the model developed by Glaude et al.[7], some reactions were added to account for the decomposition of unsaturated methyl esters.

Although the first two kinetic models were developed to model the oxidation of methyl decanoate, pyrolysis reactions still play an essential role. The processes involved in pyrolysis are of key importance, especially for the treatment of the formation of poly aromatic hydrocarbons (PAH), since these components are important soot precursors. From an experimental point of view it is difficult to study oxidation at high temperatures without a thorough understanding of the competing pyrolysis reactions[13]. Hence, both pyrolysis and oxidation data are required to perform validation of detailed kinetic models. Therefore in this contribution we focuse on the pyrolysis of methyl decanoate and methyl decenoate, two ideal surrogates for large methyl esters found in biodiesel. New experimental data at high and moderate dilutions are obtained from a bench scale pyrolysis set-up[14, 15]. The currently available kinetic models, mentioned above, for methyl decanoate oxidation and pyrolysis are validated using this dataset.

Experimental

The bench scale pyrolysis set-up has been discussed at length previously[14-18]. During the currently discussed set of experiments the reactor is operated almost isothermally. To compensate for heat losses near the outlet of the reactor, an extra heater is placed at the bottom of the furnace. The pressure drop over the reactor was found to be negligible.

The analysis section of the pyrolysis set-up enables on-line qualification and quantification of the entire product stream, i.e. a wide boiling mixture containing permanent gasses (H_2, CO, CO_2, etc.), alcohols, aldehydes and ketones, esters, and hydrocarbons ranging from methane to polyaromatic hydrocarbons (PAH). As discussed previously[14], three different gas chromatographs are required: a refinery gas analyzer (RGA), a light oxygenates analyzer (LOA) and a GC×GC-FID/(TOF-MS). Response factors (RF_i) of all permanent gasses and light hydrocarbons (C_2-C_4) were determined with the aid of a gaseous calibration mixture provided by Air Liquide. The response factors of all C_{5+} hydrocarbons and methyl esters were determined using the effective carbon number concept, relative to methane.

The effect of temperature on the product distribution was determined for two dilution regimes for methyl decanoate, i.e. 10 mol_{N2}/mol_{MD} and 0.6 mol_{N2}/mol_{MD}, and for a single dilution for methyl decenoate, i.e. 10 mol_{N2}/mol_{MD}. The flow rates were chosen to obtain a residence time of approximately 0.5s in both dilution regimes. The outlet pressure was 1.7 bar in all experiments. For each studied condition, at least 3 repeat experiments were performed. Only experiments that resulted in a mass balance between 95 and 105 wt% were retained. Subsequently, the component weight fractions were normalized to 100 wt%, in order to enable straightforward interpretation of the observed trends. After normalization elemental balances (C, H, O) were checked, all of which deviated less than 1 %.

Results and discussion

Product distribution

Pyrolysis of MD leads to a very complex reactor effluent with strongly varying composition depending on the conversion of MD. Because of the long alkyl chain in MD, the observed pyrolysis products have some similarity with those obtained during the thermal decomposition of long chain alkanes. Obviously the presence of the methyl ester group results in a larger variety of products than those observed during pyrolysis of n-decane for example. Significant amounts of large linear olefins and unsaturated esters were observed at lower temperatures. The amounts of these components decrease from a given temperature in favor of permanent gasses (CO, CO_2, CH_4, H_2), light olefins and, at the highest temperatures, aromatics. As shown in Figure 1a, the linear α-olefins and unsaturated esters, i.e. the primary decomposition products, are clearly visible in the GC×GC-FID chromatogram of the on-line sampled product at a reactor temperature of 720°C.

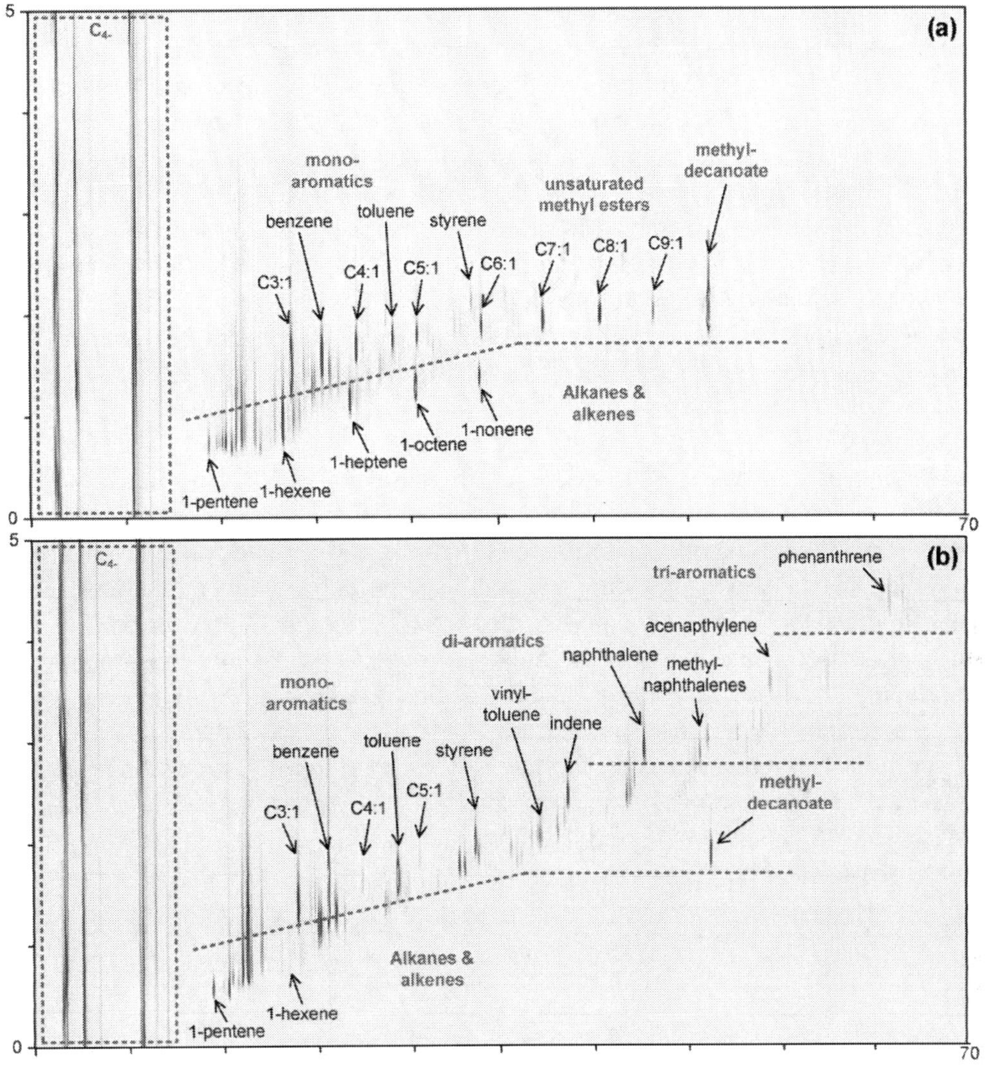

Figure 1: GC×GC-FID chromatograms of the on-line sampled MD pyrolysis effluent; (a) T=720°C; (b) T=780°C (e.g. C7:1 represents the methyl ester with 7 carbon atoms and one double bond at the end of the fatty acid chain, i.e. methyl-6-heptenoate)

The two dimensional separation is crucial for properly identifying and quantifying the product yields. This is especially so for the olefins and unsaturated esters, since the chromatogram in Figure 1a shows that these components have similar 1st dimension retention, and would therefore remain unresolved using conventional 1D gas chromatography.

Figure 1b shows the GC×GC-FID chromatograms of the on-line sampled effluent obtained at a reactor temperature of 780°C. Again, a distinction can be made between several bands of components. These include olefinic and saturated components, followed by mono-aromatics with increasing alkyl substitution, starting with benzene, toluene, styrene, etc. Naphtheno-aromatics such as indene and alkyl substituted indenes have slightly higher second dimension retention. Di-aromatic components, i.e. naphthalene followed by methyl-naphthalenes, show even higher second dimension retention. The next bands of components are made up of, respectively, naphtheno-di-aromatics such as acenaphthylene and even traces of tri-aromatics such as phenanthrene. The GC×GC chromatograms in Figure 1 also clearly illustrate the effect of temperature on the conversion of the FAME mixture. At 720°C a lot of methyl esters are still present, while at 780°C the amount of methyl esters has reduced drastically.

On-line GC×GC-(TOF-MS) confirmation allowed to identify approximately 150 components in these effluents. Besides the main components mentioned above, also small amounts of alkanes and saturated methyl esters, smaller than methyl decanoate, were detected. As well as some dienes and species with the double bond at a different location than the α-position. Moreover, using the Light Oxygenates Analyzer (LOA) it was possible determine the amounts of formaldehyde and water in the effluent. Using the different analytical instruments a detailed composition of the entire product stream was determined at each of the investigated process conditions, which can be found in Supporting Information.

Methyl decanoate conversion

The conversion as a function of average reactor temperature is shown in Figure 2. At low dilution, conversion ranges from 11 wt% at 560°C to full conversion at 800°C. At high dilution, conversion ranges from 5.6 wt% at 600°C to full conversion at 800°C. The conversion at high dilution is consistently lower for a given temperature over almost the entire conversion range although the residence time remained the same. This observation can be explained based on the free-radical mechanism. In analogy with alkane pyrolysis, methyl decanoate is mainly consumed by hydrogen abstractions, producing alkyl and alkyl-ester radicals. Hydrogen abstractions by hydrogen and methyl radicals are the most predominant, but also ethyl, vinyl and resonance stabilized radicals such as allyl are important hydrogen abstracting radicals. The rate of these bimolecular abstractions is considerably reduced at higher dilution, explaining the reduced conversion. However, at the highest temperatures, unimolecular scissions of C-C and C-O bonds become increasingly important, explaining the reduced effect of dilution at these temperatures.

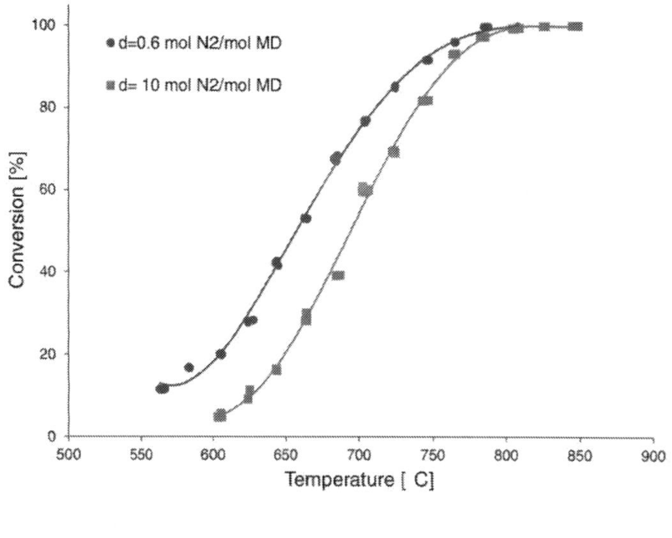

Figure 2: Methyl decanoate conversion as a function of the average reactor temperature (● - low N_2 dilution, ■ - high N_2 dilution)

Simulated conversions, obtained using the detailed kinetic models published by Herbinet et al. (2008), Glaude et al. (2010) and Herbinet et al. (2011), were compared with the experimental data. These simulations were performed using the PFR model of CHEMKIN-PRO (release 15101) and by imposing the measured process gas temperature profiles, which are available in Supporting Information. As shown in Figure 3, all models are quite able to predict the observed trend, although the model of Herbinet et al. (2008) slightly overpredicts the measured conversions, while underpredicted by the model of Glaude et al. (2010). The model of herbinet et al. (2011) still underpredicts the experimental data, but outperforms the model of Glaude et al. (2010). These simulation results clearly show the limitations/uncertainties in the presently available kinetic models and indicates that accurate experimental data is crucial to improve the kinetic models developed for methyl ester pyrolysis and combustion.

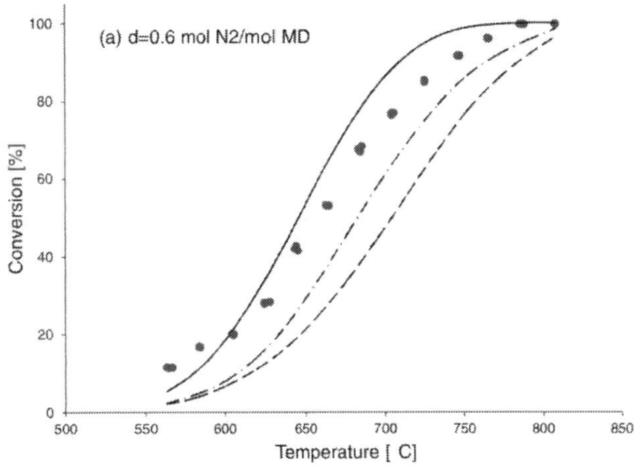

Figure 3: Comparison of the experimental and simulated methyl decanoate conversion as a function of temperature for low N_2 dilution (symbols – experimental, full lines – Herbinet et al. (2008) model, dashed lines – Glaude et al. (2010) model, dash-dot lines – Herbinet et al. (2011) model)

Formation and disappearance of long olefins

Figure 4 shows the experimental yields of the linear α-olefins and unsaturated esters at low dilution. The initial production of these α-olefins and esters is counteracted by their own disappearance via for example retro-ene reactions, resulting in a maximum in the observed yields as a function of MD conversion. Typically, the maximum shifts towards higher conversion/temperature with decreasing carbon number, resulting in a monotonically increasing yield of the smallest olefin, i.e. ethylene. Overall, high light olefin yields were measured, as well as relatively high methyl-2-propenoate yields. Similar trends were observed at high N_2 dilution. In Figure 5 the measured yields at high dilution for some main components are compared with the simulated yields using the three kinetic models mentioned before. The yields of 1-octene and ethene are predicted fairly well by all models. The agreement between the experimental yields of methyl-6-heptenoate and the yields predicted by the model of Herbinet et al. (2008) is also satisfactory. However, the simulation results using the model of Glaude et al. (2010) show that the secondary decomposition of this primary product is to slow. Consequently, the methyl-2-propenoate is underpredicted by the model of Glaude et al. (2010). In the model of Herbinet et al. (2011) unimolecular bond scissions and retro-ene decomposition reactions were added to the mechanism for unsaturated methyl esters. As a result, the predicted methyl-6-heptenoate yield decreases more rapidly at higher MD conversions and the methyl-2-propenoate yield is higher. On the other hand, the model of Herbinet et al. (2008) significantly overpredicts the methyl-2-propenoate yield.

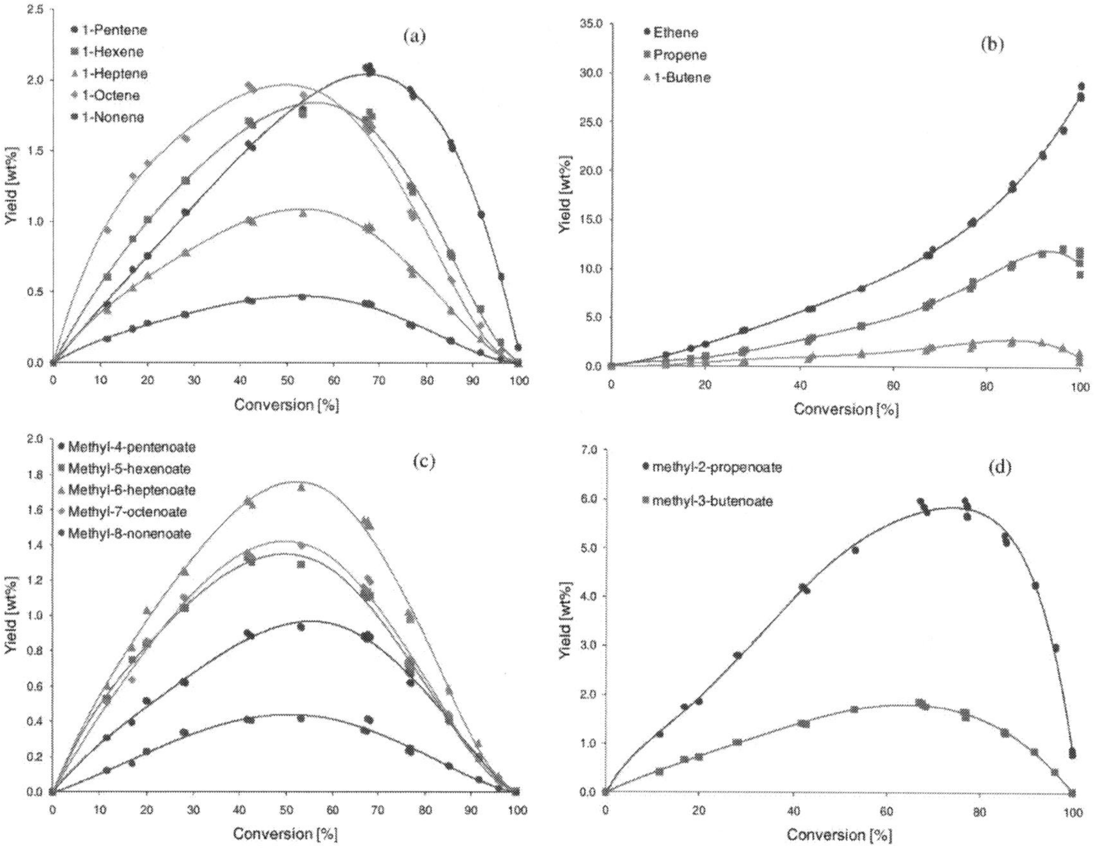

Figure 4: Yields of (a), (b) α-olefins, and (c), (d) unsaturated esters as a function of outlet conversion at low N_2 dilution

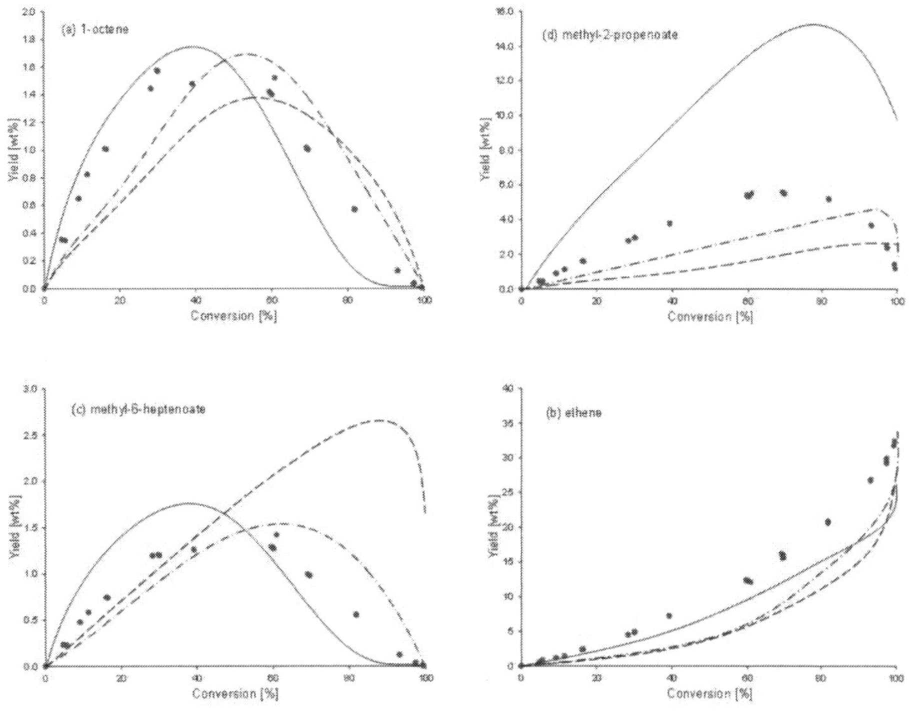

Figure 5: Comparison of the experimental and simulated yields at high dilution; (a) 1-octene, (b) ethene, (c) methyl-6-heptenoate, (d) methl-2-propenoate (symbols – experimental, full lines – Herbinet et al. (2008) model, dashed lines – Glaude et al. (2010) model, dash-dot lines – Herbinet et al. (2011) model)

Formation of aromatics

Although the experimental data shows that significant amounts of aromatic components are produced at high conversion, neither of the studied kinetic models include reactions for their formation. The yield profiles shown in Figure 6 clearly indicate that aromatics such as benzene and toluene are secondary products, produced only at high conversion and in higher amounts at low dilution.

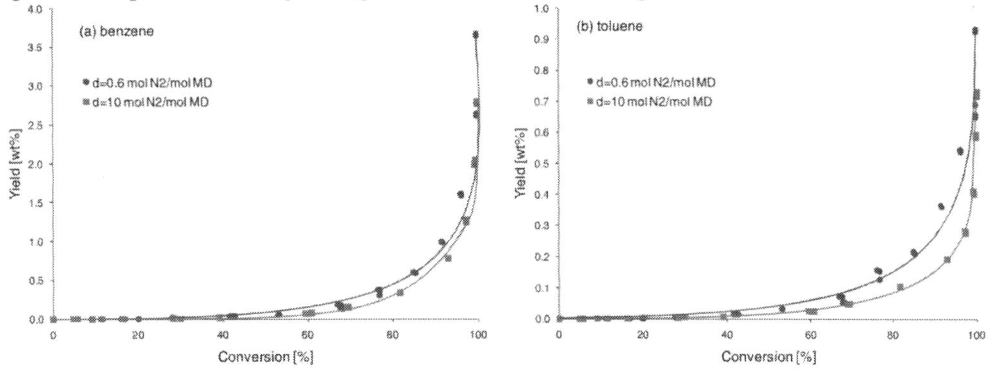

Figure 6: Comparison of the yields at high (■) and low dilution (●) for (a) benzene, and (b) toluene

High conversions also correspond to high yields of light olefins, which are important precursors for the formation of aromatics. In the temperature range discussed here, and when high amounts of ethene and 1.3-butadiene are present, aromatics are mainly produced by addition of olefinic radicals such as vinyl and allyl to butadiene[19]. As shown in Figure 7, this addition produces a resonance stabilized radical that undergoes subsequent cyclization, producing a cyclic olefinic radical that rapidly decomposes into cyclo-hexadiene by splitting of an allylic hydrogen. Subsequent dehydrogenation, again by abstraction and scission of allylic hydrogens, leads to benzene. A similar pathway starting with the addition of the propenyl radical to butadiene leads to toluene, although the latter can also be produced by ipso-addition of methyl radicals on benzene. In several kinetic models the propargyl radical recombination is considered to be dominant path towards benzene[20, 21]. However, as discussed by Sabbe et al. for ethane pyrolysis[19], using the recombination rate coefficients reported by Miller and Klippenstein[22], the recombination of propargyl radicals does not contribute significantly according to a rate of production sensitivity analysis[19]. It only starts contributing to the benzene rate of formation at a temperature of 1130 K.

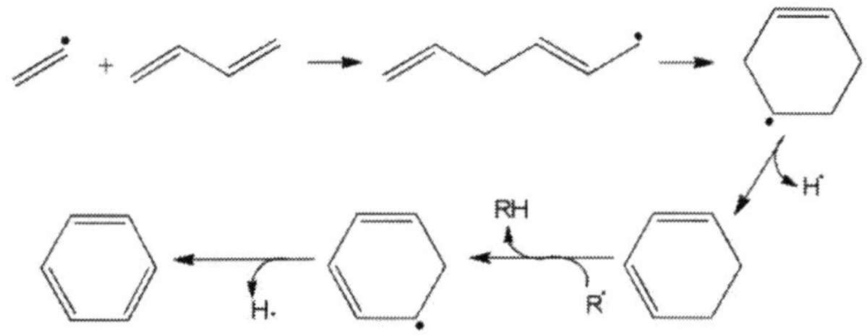

Figure 7: Main reaction path for the formation of benzene[19]

Conclusions

The pyrolysis of methyl decanoate and methyldecenaote have been studied in a bench scale set-up equipped with a dedicated on-line analysis section. The results showed that at lower temperatures significant amounts of large linear olefins and unsaturated esters are formed, the amounts of which decreased from a given temperature in favor of permanent gasses (CO, CO_2, CH_4) and light olefins. At the highest temperatures considerable amounts of mono- (benzene, toluene, styrene) and di-aromatics (naphthalene, methylnaphthalene, biphenyl) were measured. These results show that there are significant similarities between the thermal decomposition of methyl decanoate and methyldecenoate and alkane pyrolysis, implying that previously developed tools [23-25] to model the primary decomposition of long chain molecules can be straightforwardly applied to methyl esters. Nevertheless, the results also show that the presence of the ester group has a significant effect on the kinetics, and consequently, product yields.

This new dataset has been used to validate combustion mechanisms developed for methyl decanoate by Herbinet et al. and Glaude et al. and revealed that overall a reasonable agreement is obtained between experimental and simulated conversion. Without any modifications of the reaction rate coefficients in the kinetic model the agreement obtained between simulated and experimental yields for several of the major products is quite satisfactory. Major differences are observed for methyl propenoate, water, and CO_2. Rate of production analysis together with ab-initio calculations revealed that these differences can be attributed to either the need for more accurate kinetics or simply some missing reactions or reaction families.

References

1. Metcalfe, W.K., J.M. Simmie, and H.J. Curran, *Ab Initio Chemical Kinetics of Methyl Formate Decomposition: The Simplest Model Biodiesel.* Journal of Physical Chemistry A, 2010. **114**(17): p. 5478-5484.
2. Hurd, C.D. and F.H. Blunck, *The pyrolysis of esters.* Journal of Americain Chemical Society, 1938. **60**(10): p. 2419-2425.
3. Farooq, A., et al., *An experimental and computational study of methyl ester decomposition pathways using shock tubes.* Proceedings of the Combustion Institute, 2009. **32**: p. 247-253.
4. Herbinet, O., W.J. Pitz, and C.K. Westbrook, *Detailed chemical kinetic oxidation mechanism for a biodiesel surrogate.* Combustion and Flame, 2008. **154**(3): p. 507-528.
5. Curran, H.J., et al., *A comprehensive modeling study of n-heptane oxidation.* Combustion and Flame, 1998. **114**(1-2): p. 149-177.
6. Curran, H.J., et al., *A comprehensive modeling study of iso-octane oxidation.* Combustion and Flame, 2002. **129**(3): p. 253-280.
7. Glaude, P.A., et al., *Modeling of the oxidation of methyl esters-Validation for methyl hexanoate, methyl heptanoate, and methyl decanoate in a jet-stirred reactor.* Combustion and Flame, 2010. **157**(11): p. 2035-2050.
8. Glaude, P.A., et al., *Modeling of the oxidation of n-octane and n-decane using an automatic generation of mechanisms.* International Journal of Chemical Kinetics, 1998. **30**(12): p. 949-959.
9. Warth, V., et al., *Computer based generation of reaction mechanisms for gas-phase oxidation.* Computers & Chemistry, 2000. **24**(5): p. 541-560.
10. Biet, J., et al., *Experimental and modeling study of the low-temperature oxidation of large alkanes.* Energy & Fuels, 2008. **22**(4): p. 2258-2269.
11. Muller, C., et al., *THERGAS - A computer program for the evaluation of thermochemical data of molecules and free-radicals in the gas-phase.* Journal de Chimie Physique et de Physico-Chimie Biologique, 1995. **92**(5): p. 1154-1178.
12. Herbinet, O., et al., *Experimental and modeling study of the thermal decomposition of methyl decanoate.* Combustion and Flame, 2011. **158**(7): p. 1288-1300.
13. Tsang, W. *Thermal Stability of Hydrcarbon Fuels* in *AIChE Annual Meeting*. 2010. Salt Lake City, UT, USA.
14. Pyl, S.P., et al., *Rapeseed oil methyl ester pyrolysis: On-line product analysis using comprehensive two-dimensional gas chromatography.* Journal of Chromatography A, 2011. **1218**(21): p. 3217-3223.
15. Harper, M.R., et al., *Comprehensive reaction mechanism for n-butanol pyrolysis and combustion.* Combustion and Flame, 2011. **158**(1): p. 16-41.
16. Van Geem, K.M., et al., *Accurate High-Temperature Reaction Networks for Alternative Fuels: Butanol Isomers.* Industrial & Engineering Chemistry Research, 2010. **49**(21): p. 10399-10420.
17. Chen, Q. and G.F. Froment, *Thermal-Cracking of Substituted Aromatic-Hydrocarbons .1. Kinetic-Study of the Thermal-Cracking of I-Propylbenzene.* Journal of Analytical and Applied Pyrolysis, 1991. **21**(1-2): p. 27-50.
18. Chen, Q. and G.F. Froment, *Thermal-Cracking of Substituted Aromatic-Hydrocarbons .2. Kinetic-Study of the Thermal-Cracking of Normal-Propylbenzene and Ethylbenzene.* Journal of Analytical and Applied Pyrolysis, 1991. **21**(1-2): p. 51-77.
19. Sabbe, M.K., et al., *First Principles-Based Simulation of Ethane Steam Cracking.* AIChE Journal, 2010.
20. Matheu, D.M. and J.M. Grenda, *A systematically generated, pressure-dependent mechanism for high-conversion ethane pyrolysis. 1. Pathways to the minor products.* Journal of Physical Chemistry A, 2005. **109**(24): p. 5332-5342.
21. Herbinet, O., et al., *Thermal decomposition of n-dodecane: Experiments and kinetic modeling.* Journal of Analytical and Applied Pyrolysis, 2007. **78**(2): p. 419-429.
22. Miller, J.A. and S.J. Klippenstein, *The recombination of propargyl radicals and other reactions on a C6H6 potential.* Journal of Physical Chemistry A, 2003. **107**(39): p. 7783-7799.
23. Clymans, P.J. and G.F. Froment, *Computer-Generation of Reaction Paths and Rate-Equations in the Thermal-Cracking of Normal and Branched Paraffins.* Computers & Chemical Engineering, 1984. **8**(2): p. 137-142.
24. Hillewaert, L.P., J.L. Dierickx, and G.F. Froment, *Computer Generation of Reaction Schemes and Rate Equations for Thermal Cracking.* Aiche Journal, 1988. **34**(1): p. 17-24.
25. Van Geem, K.M., M.F. Reyniers, and G.B. Marin, *Challenges of modeling steam cracking of heavy feedstocks.* Oil & Gas Science and Technology-Revue De L Institut Francais Du Petrole, 2008. **63**(1).

(670g) Combustion Chemistry of Fischer-Tropsch and Hydrotreated Renewable Fuels

S. Mani Sarathy, Charles K. Westbrook, William J. Pitz and Marco Mehl, Lawrence Livermore National Laboratory, Livermore, CA

Fischer-Tropsch (FT) and Hydrotreated Renewable (HR) fuels derived from biomass are potential renewable hydrocarbon fuels. The bulk physical chemical properties of these are similar to those of conventional petroleum derived diesel and jet fuels, so they can readily be used in compression ignition engine and gas turbine applications. However, the chemical composition of FT and HR fuels is different than petroleum fuels, thereby affecting important combustion performance parameters (e.g., ignition and emissions formation). FT and HR fuels typically contain higher concentration of n-alkanes and lightly branched alkanes with one are two methyl branches. This research project presents a detailed chemical kinetic mechanisms for 2-methylkalkanes, 3-methyalkanes, and symmetrical dimethylalkanes. The model is validated against experimental data from a variety of fundamental combustion devices. This new model is used to show how the presence of a methyl branches and their location affects ignition, extinction, flame speed, and species formation.

See more of this Session: Reaction Engineering for Combustion and Pyrolysis I
See more of this Group/Topical: Catalysis and Reaction Engineering Division

(671a) Supported Manganese Triazacyclononane Cis-Dihydroxylation Catalysts

Nicholas Schoenfeldt, Andrew Korinda and **Justin M. Notestein**, Chemical and Biological Engineering, Northwestern University, Evanston, IL

Manganese complexes of1,4,7-trimethyl-1,4,7-triazacyclononane are highly active and selective for thelow temperature oxidation of alkenes and other substrates with environmentallybenign oxidants such as H_2O_2. Previous investigations onthe homogeneous catalyst revealed that carboxylate co-catalysts could beemployed to raise catalyst productivity by reducing the inherent catalaseactivity of the materials, as well as tune the reaction selectivity towardeither epoxidation or *cis*-dihydroxylation. *Cis*-dihydroxylation isa relatively unusual but powerful reaction, primarily catalyzed by expensive Rhcomplexes or extremely toxic OsO_4. There is thus an incentive forthe development of novel catalysts, especially solid catalysts. These Mncomplexes use mild reaction conditions (room temperature and below), benignoxidants (H_2O_2) and less toxic metal centers (Mn),demonstrating potential utility for the organic transformations necessary in pharmaceuticalsynthesis.

We have found that functionalization of common catalystsupports with carboxylate moieties allows the *in-situ* assembly andactivation of the Mn complex onto the support under oxidation reactionconditions. The immobilized catalyst requires orders of magnitude lesscarboxylate co-catalyst than the soluble analog and requires up to 8x lowerresidence times for a given conversion. The immobilization technique also removescumbersome synthetic requirements typically required to create amine ligandsthat can be covalently attached to a support. For example, we have found thatthe cis-diol selectivity can be increased from <50% to ~75% by changing onlythe nature of the supported carboxylate. Supporting characterization by X-rayabsorption fine structure and diffuse reflectance UV-visible spectroscopy helpsus further understand the nature of this supported catalyst.

See more of this Session: Reaction Engineering In Pharmaceuticals and Fine Chemicals
See more of this Group/Topical: Catalysis and Reaction Engineering Division

(671b) Reaction Development for Bench Scale, Continuous Synthesis of a Drug Substance

Patrick L. Heider[1], Soubir Basak[1], Louis Buchbinder[2], Rachael C. Hogan[1], Ketan Pimparkar[1], Aaron D. Wolfe[2] and Klavs F. Jensen[1], (1)Chemical Engineering, Massachusetts Institute of Technology, Cambridge, MA, (2)Department of Chemical Engineering, Northeastern University, Boston, MA

Recently the pharmaceutical industry has shown an increased interest in implementing continuous manufacturing processes. While this movement has been supported by the Food and Drug Administration, there have been limited reports of continuous production systems. We have worked within the Novartis-MIT Center for Continuous Manufacturing to develop a fully integrated, bench scale process for taking chemical intermediates to a coated tablet. This talk will highlight the reaction development and operation for the two chemical transformations performed. Translating unit operations from batch to continuous required the development of different reaction conditions. Reaction times were significantly decreased and solvent usage was reduced or even eliminated. The increased heat transfer available to continuous processes was utilized to perform a rapid, exothermic quench not possible in batch operation.

Operating a continuous synthesis process at the bench scale required addressing several issues not present in smaller, isolated systems. A thorough understanding of mixing was important for operation at the bench scale where neither the short diffusion times of microreactors nor the turbulent flow of pilot plants were present. This was addressed in multiple ways including modeling the residence time distribution and making use of *in situ* gas formation to promote mixing. Monitoring of chemical composition was required to maintain reaction stoichiometry and was achieved using a variety of online measurements including UV and IR spectroscopy and density. Lastly, the bench scale system provided a challenging scale up for membrane based liquid-liquid separators originally developed for microscale continuous multistep synthesis (1).

References:

1. H. R. Sahoo, J. G. Kralj, K. F. Jensen, Angew. Chem. Int. Ed. 2007, 46, 5704.

See more of this Session: Reaction Engineering In Pharmaceuticals and Fine Chemicals
See more of this Group/Topical: Catalysis and Reaction Engineering Division

(671c) Practical and Selective Oxidation and Amination of Alcohols In Flow

Klaus Hellgardt[1], K.K. Hii[2], N. Zotova[3] and F. Roberts[3], (1)Chemical Engineering, Imperial College London, London, United Kingdom, (2)Chemistry, Imperial College London, London, United Kingdom, (3)Imperial College London, London, United Kingdom

The development of clean, selective and atom-efficient reactions has become the new paradigm in organic synthesis, driven by a desire to reduce economic, as well as environmental costs of chemical production. To this effect, the oxidation of alcohols to aldehydes and ketones may be considered one of the most challenging reactions, ubiquitous in organic chemistry, yet least efficient. The lack of chemoselective and atom-economical oxidants often make these reactions unattractive, particularly for the large scale synthesis of complex molecules. In this work, we demonstrate a number of new ways of oxidising alcohols using either H_2O_2 or O_2 as an oxidant by performing catalysis in a continuous flow high temperature/pressure reactor, using supported Au and Ru cataysts systems, thus fulfilling several criteria for practical and 'green' synthesis. We further demonstrate the utilisation of the intermediate aldehyde in the alkylation of amines without preactivation of alcohols.

See more of this Session: Reaction Engineering In Pharmaceuticals and Fine Chemicals
See more of this Group/Topical: Catalysis and Reaction Engineering Division

(671d) Rotating Foam Reactors for Fine Chemical Synthesis

T. Alexander Nijhuis, M.A. Leon Mattheus, R. Tschentscher, J. van der Schaaf and J.C. Schouten, Chemical Reactor Engineering, Eindhoven University of Technology, Eindhoven, Netherlands

The production of fine chemicals and pharmaceuticals is commonly performed in batch reactors, using slurry catalysts, due to their high flexibility. However, the catalyst particles need to be filtered from the slurry when the reaction is finished, which is a costly and time consuming operation and makes re-using the catalyst difficult. In this presentation we will be showing the development of new three-phase reactor configurations based on the application of solid foam packings. These materials have a good mechanical stability, a high porosity, and a high specific surface area, which offers great potential for their use as catalyst packings. The good mechanical strength, allows us to use the material as both catalyst support and stirrer at the same time. The foam surface area can be further increased by deposition of a catalyst wash coat or using sol-gel methods.

We will be showing that the solid foam stirrer reactor has an efficient mixing behavior and high mass transfer rates. Additionally, the reaction mixture can be drained off easily, without the need to filter the catalyst particles. Using different cell-density foams (different geometrical surface areas) and different washcoat thicknesses it is possible to optimize the performance of the reactor in cases where the catalyst activity and mass transfer are effecting the selectivity of a reaction, for example in the selective hydrogenation of an alkyne to an alkene. We will be showing the performance of this novel reactor type for different gas-liquid reactions catalyzed by a heterogeneous catalyst coated on the stirrer. We will be comparing its performance to conventional slurry catalysts in a reactor equipped with a conventional Rushton stirrer.

See more of this Session: Reaction Engineering In Pharmaceuticals and Fine Chemicals
See more of this Group/Topical: Catalysis and Reaction Engineering Division

(671e) Economic and Process Engineering Evaluations of Polymer-Grade Terephthalic Acid Production In A Spray Reactor

Meng Li, Department of Chemical and Petroleum Engineering, University of Kansas, Lawrence, KS, Fenghui Niu, Center for Environmentally Beneficial Catalysis, University of Kansas, Lawrence, KS, Xiaobin Zuo, Department of Chemistry, University of Kansas, Lawrence, KS, Daryle H. Busch, CEBC/ Chemistry, University of Kansas, Lawrence, KS and Bala Subramaniam, Chemical and Petroleum Engineering, Center for Environmentally Beneficial Catalysis, University of Kansas, Lawrence, KS

There is emerging interest in developing renewable plastics derived from feedstocks (such as renewable *p*-xylene) from plant-based biomass. In such cases, it is imperative that green conversion technologies are also employed. CEBC researchers have been investigating a spray process concept as a greener alternative to the conventional Mid-Century (MC) process for terephthalic acid (TPA) production from *p*-xylene. Compared to the MC process in which the air is dispersed into the stirred liquid phase containing *p*-xylene and the Co/Mn/Br based catalyst dissolved in acetic acid, the spray process uses a nozzle to disperse the reaction mixture as fine droplets into a gas phase dominated by saturated acetic acid vapor and also containing O_2 (in slight excess over the stoichiometric amount needed for *p*-xylene oxidation to TPA) and CO_2 (as carrier for O_2). The increased gas/liquid interfacial area was aimed at eliminating O_2 starvation in the liquid phase, thereby avoiding incomplete oxidation to achieve improved selectivity and purity of the TPA product. In the MC process, the crude TPA from the stirred reactor contains significant amount of impurities such as approximately 1400 ppmw 4-carboxybenzaldehyde (4-CBA) and 600 ppmw *p*-toluic acid due to the incomplete oxidation. To obtain polymer-grade TPA, a purification process by hydrogenation associated with a significant number of additional processing steps is required. Employing a semi-continuous spray process (*p*-xylene sprayed continuously but no continuous product withdrawal), it was confirmed by reliable analytical methods that the solid TPA product meets the polymer-grade purity specification of 25 ppmw 4-CBA or less, implying the hydrogenation step is not required. Comparative economic analyses show that the total capital investment and utility costs associated with the spray process are roughly half of that in the MC process, thus significantly reducing the cost of the polymer grade TPA product. Preliminary experimental results from continuous operation of the spray reactor show that *p*-xylene was being converted mostly to TPA and that the purest solid TPA product obtained in the reactor contains 3 ppmw 4-CBA. Based on reactor modeling, factors affecting the control of reactor temperature and open-loop stability of the reactor are also elucidated.

See more of this Session: Reaction Engineering In Pharmaceuticals and Fine Chemicals
See more of this Group/Topical: Catalysis and Reaction Engineering Division

(672a) Model Studies of Supported Catalyst Preparation - Pd Deposition On Iron Oxide Films From the Liquid Phase

Martin Sterrer, Huifeng Wang, Rhys Dowler, William Kaden and Hans-Joachim Freund, Department of Chemical Physics, Fritz-Haber-Institute of the Max-Planck-Society, Berlin, Germany

In this contribution we present results of surface science studies related to the preparation of supported metal catalysts on well-defined oxide surfaces applying procedures used in real catalysis. While in most UHV model studies of supported metal catalysts the metal is deposited onto a clean surface by physical vapor deposition, the standard methods in real catalyst preparation, e.g. impregnation or deposition-precipitation, are wet chemical routes with the important processes occurring at the solid-liquid interface, followed by calcination and reduction to transform the catalyst precursor into the active compound. The aim of the present work is to apply those preparation procedures to well-defined oxide surfaces using thin oxide films as substrates that allow standard surface science characterization techniques to be applied. We present Scanning Tunneling Microscopy (STM), X-ray Photoemission Spectroscopy (XPS), Temperature Programmed Desorption (TPD) and Infrared Reflection Absorption Spectroscopy (IRAS) results for Pd deposition from $PdCl_2$ containing solutions of different pH onto $Fe_3O_4(111)$ films.

STM showed that uniformly distributed Pd nanoparticles can be prepared by using the deposition-precipitation method. XPS was used to follow the evolution of Pd species at each preparation stage and to study the influence of Pd precursor solution pH on the speciation of Pd complexes on the surface. The Pd particles were further characterized by CO chemisorption with TPD and IRAS. Interestingly, different chemisorption properties of Pd/Fe_3O_4 were found after CO or H_2 reduction at 500 K. CO-reduced Pd samples showed normal CO adsorption behavior, while a striking suppression of CO adsorption was brought about by H_2 reduction. XPS results indicated a change in the electronic structure of Pd after H_2 reduction which can be recovered by reoxidation and CO reduction. The results of these experiments will be compared with the respective results for Pd clusters prepared by physical vapor deposition on $Fe_3O_4(111)$.

See more of this Session: Science and Engineering of Catalyst Preparation I
See more of this Group/Topical: Catalysis and Reaction Engineering Division

(672b) Quantitative Understanding of the Aqueous Solid-Liquid Interface During Catalyst Synthesis

Jason Binz, Pennsylvania State University, University Park, PA and **Robert M. Rioux**, Chemical Engineering, Pennsylvania State University, University Park, PA

The synthesis of heterogeneous catalysts has primarily been considered more of an art than a science. Systematic experimental studies have been performed to characterize the influence of drying, calcination and reduction on the nanoparticle size and distribution; however, very little quantitative work focuses on the initial adsorption of the transition metal complex (TMC) on the metal oxide surfaces [1]. There has been very little work connecting the influence of the initial adsorption conditions on the final size and distribution of particles [2]. Seminal work by Brunelle introduced ion-exchange as a highly controlled adsorption method because the bulk pH solution conditions can be used to manipulate the surface charge on an oxide surface through protonation/deprotonation of surface hydroxyl groups [3]. More recently, Regalbuto and co-workers have introduced the concept of strong electrostatic adsorption, which can be described by a rather simple interfacial potential model [4,5]. Within this model, electrostatics is the dominant driving force for association between TMC and a charged support surface. At this stage, the chemical identity of the TMC ligands has no influence and only the size of the ligand influences the maximum adsorption uptake. In this work, we introduce isothermal titration calorimetry as a quantitative technique to understand catalyst synthesis at solvated neutral and charged solid-liquid interfaces. Initial experiments demonstrate that the electrostatic adsorption of chloroplatinic acid (CPA) on γ-Al_2O_3 at pH 2.11 is ~70 kJ mol^{-1} exothermic with an equilibrium constant of 10^5 M^{-1}. Additional data will be presented for the adsorption of other cationic and anionic Pt TMCs on alumina and silica in order to demonstrate trends in adsorption enthalpy and maximum adsorption capacity. This talk will focus on a study of the influence of adsorption conditions on the final particle size distribution for Pt-based catalysts synthesized from the adsorption of CPA from aqueous solutions onto g-alumina. Control of the surface charge of γ-Al_2O_3 relative to its PZC is the primary driving force for adsorption, which influences not only the final particle size and distribution, but also the kinetics of TMC dissocation and the onset of particle formation. Synchrotron-based high-energy wide-angle scattering has been employed to follow the *in-situ* growth of Pt nanoparticles under reducing conditions. Temperature and temporal dependent pair distribution function analysis of catalysts synthesized under different solvated interface conditions demonstrates the structure of treated catalysts is dependent upon the initial solvated interface conditions during the initial adsorption step.

1. Liu, X., J.G. Khinast, and B.J. Glasser, *Ind. Eng. Chem. Res.*, 49, 2649 (2010).
2. Lambert, J.F. and M. Che, *J. Mol. Catal. A: Chem.*, 162, 5 (2000).
3. Brunelle, J.P., *Pure Appl. Chem.*, 50, 1211 (1978).

4. Hao, X., W.A. Spieker, and J.R. Regalbuto, *J. Colloid Interface Sci.*, 267, 259 (2003)
5. Park, J. and J.R. Regalbuto, *J. Colloid Interface Sci.*, 175,

See more of this Session: Science and Engineering of Catalyst Preparation I
See more of this Group/Topical: Catalysis and Reaction Engineering Division

(672c) Preparation of Ni/Al2O3 Catalysts for Aqueous Phase Reforming of Ethanol by Selective Electrostatic Adsorption Method

Kavi Geetharani Loganathan, Materials Engineering, New Mexico Institute of Mining and Technology, Socorro, NM and Corey Leclerc, Chemical Engineering, New Mexico Institute of Mining and Technology, Socorro, NM

Preparation of Ni/Al_2O_3 catalysts for aqueous phase reforming of ethanol by selective electrostatic adsorption method

Kavi G.L* and Corey Leclerc

New Mexico Institute of Mining and Technology

Socorro, NM

Commercially available alumina, solution combustion synthesized (SCS) alumina and plasma modified SCS alumina were used as supports for Ni catalysts. Nitrogen plasma modification of the SCS alumina as achieved using RF plasma reactor prior to metal deposition. Ni/Alumina catalysts are prepared by electrostatic adsorption of nickel onto the alumina using pH solutions ranging from 9-13. The effect of substituting hydroxyl groups on alumina versus nitrogen containing groups from the plasma has been tested using a cationic nickel complex. The point of zero charge (PZC) of alumina and SCS alumina are found to be 8.1 and 13 respectively. Surface loading of 20,000 m^2/L and 1000 m^2/L alumina were used to determine PZC and pH of strongest interaction in all the experiments. The metallic dispersion of all the catalysts have been measured using hydrogen pulse-chemisorption, while the activity of the catalysts has been compared in the aqueous phase reforming of ethanol.

See more of this Session: Science and Engineering of Catalyst Preparation I
See more of this Group/Topical: Catalysis and Reaction Engineering Division

(672d) Atomic Pair Distribution Study of the Growth of Metal Structures In Zeolites

Liliana Gamez, Chemical Engineering, University of Puerto Rico, Mayaguez, PR and Maria Martinez-Inesta, Chemical Engineering Department, University of Puerto Rico, Mayaguez, PR

Supported metal catalysts in zeolites are used commonly in many petroleum refinement processes. It is known that during synthesis the metal precursor goes through different stages that affect the final dispersion of the metal clusters but what is known about this process is limited. Experimental total scattering of high energy X rays analyzed as atomic pair distributions has been used here to obtain structural models of both the zeolite and the zeolite metal complex during synthesis. These models help understand the evolution of the metal species in the zeolite.

See more of this Session: Science and Engineering of Catalyst Preparation I
See more of this Group/Topical: Catalysis and Reaction Engineering Division

(672e) Modification of Carbon Supports for Novel Platinum-Based Water-Gas Shift Catalysts

Branko Zugic[1], Yanping Zhai[1], Howard M. Saltsburg[2] and Maria Flytzani-Stephanopoulos[1], (1) Chemical and Biological Engineering, Tufts University, Medford, MA, (2)Department of Chemical & Biological Engineering, Tufts University, Medford, MA

Recent studies have shown that platinum-based water-gas shift catalysts supported on metal oxides can be promoted by the addition of alkali metals. Controlling the Group VIII metal-alkali metal interaction during the preparation of these materials, however, is difficult by standard catalyst preparation methods. As a consequence, isolating and understanding these complexes has been a challenge. Here we present a novel approach to controlling the deposition of the alkali metal and platinum using modified carbons. The stability of carbon relative to metal oxides allows for the introduction of oxygen groups on to the surface, which in turn provides a template for the addition and dispersion of alkali metals and platinum. The ability to control the functionality allows for a better understanding of the surface chemistry at work in these materials. The preparation and characterization of these catalysts by x-ray absorption and electron microscopy techniques will be presented.

See more of this Session: Science and Engineering of Catalyst Preparation I
See more of this Group/Topical: Catalysis and Reaction Engineering Division

ID No 225808

Catalytic study of Sulphated Zirconia from Conventional and Modified Conventional Methods for Fatty Acid Methyl Ester

Elizabeth J. Eterigho[a]*, *Jon G.M. Lee*[a], *Karen Wilson*[b] and *Adam P. Harvey*[a]
[a] *School of Chemical Engineering and Advanced Materials, Newcastle University, NE1 7RU, UK*
[b]*School of Chemistry, Cardiff University, CF10 3AT, UK*
Corresponding author. e-mail: elizabeth.eterigho@ncl.ac.uk

Abstract

Sulphated zirconia was synthesized using different volume of sulphating agent in order to optimise its catalytic property: (1) in the first case the "conventional" 15 ml H_2SO_4 to 1g of ZrOH for impregnation was used in wet-precipitation method and the catalyst designated 'wp' and (2) same procedure but with modification in volume of acid; 2 ml of H_2SO_4 was used for 1g of ZrOH, 'mwp'. The properties of uncalcined and calcined catalysts were examined by various techniques: EDX, XPS, XRDP, Py-DRIFTS and BET nitrogen adsorption techniques. The sulphur content in the uncalcined 'mwp' and 'wp' catalysts was 0.07 mol each whereas the calcined catalysts were 0.07 mol and 0.05 mol for 'mwp' and 'wp' respectively. BET surface area was determined to be 65.0 m^2/g for wp and 101m^2/g for 'mwp'. The effect of the variation of sulphating agent on the catalyst activity was investigated in catalytic cracking of triglycerides for fatty acid methyl esters. Sulphated zirconia from the modified conventional method was found to perform better than the conventional wet-precipitated sulphated zirconia. Furthermore, 'mwp' retained approximately 74 % of its sulphur content after 3 h reaction whereas 'wp' catalyst lost 100 % during the reaction as indicated by SEM-EDX. Interesting the results show variation in characterization and in their selectivity to different types of fatty acid methyl esters under the same reaction conditions. The 'mwp' catalyst showed higher catalytic activity for methyl ester yield of 40 % as well as higher selectivity for saturated methyl esters

Keywords: Conventional method; Sulphated zirconia; Characterization; Saturated FAMEs; Catalytic cracking;

1. Introduction

Presently transesterification of triglycerides using monohydric alcohols in the presence of homogeneous alkali catalyst is the most common process for biodiesel production. However the process is associated with a number of limitations, due to the use of homogeneous catalysts and alcohol. The most popular type of feedstock is virgin vegetable oils such as soybean oil, rapeseed oil, palm oil and linseed (Srivastava and Prasad, 2000). However, non-edible oil, waste vegetable oil and waste animal fat could be used; except that this feedstock will need to undergo a pre-treatment process called esterification before they can be used successfully in transesterification. This is because of their high level of free fatty acid that could result into the formation of soap instead of the desired biodiesel. With The increased awareness of the economical and environmental implications of homogeneous catalysts in the production of biodiesel production, other approaches using solid acid catalysts are being investigated. Already solid acid catalysts has been reported as an alternative to the homogeneous counterpart in transesterification (Melero et al., 2009a); however their usage in catalytic cracking is reported to be an ideal alternative to transesterification,(Taufiqurrahmi and Bhatia, 2011). This is because it would remove the need for alcohols altogether, reducing downstream processes, and simplified flow sheet, thereby removing a substantial capital cost. In our previous work, (Eterigho et al 2011) and other authors; it has been shown that catalytic cracking is a good process for biodiesel production. Presently, the research frontier for catalytic cracking of triglycerides is finding solid acid catalyst that is active, selective, and stable under the process conditions that could facilitate highly selective conversion of economically viable feedstocks to the desired products in the existing infrastructure. Theses dominate the key concern obstructing the real potential of the process, (Chew and Bhatia, 2008). Among the many heterogeneous solid acid catalysts, zirconia doped with sulphate has recently attracted great attention due to their strong acid characteristics and their resulting potential as solid acid catalysts for selective hydrocarbon isomerisation and several other acid-promoted reactions (Ardizzone et al., 1997; Arata et al., 1990). Thus far, sulphated zirconias has been heavily reported as super-acid catalyst with acidity 10^4 times stronger than 100% sulphuric, (Katada et al., 2000; Hino et al., 1979). However it has been found to possess relatively low surface area and known to suffer leaching in the presence of polar solvents such as methanol (Hino and Arata, 1994). Most authors reported sulphated zirconia as having more Lewis acid sites than Bronsted acid site (Morterra et al., 1994). Several authors have reported a variety of investigation on the use of sulphated zirconia in transesterification using feedstock with variety of free fatty acid, (Melero et al., 2009b). However, few research studies are reported on the use of sulphated zirconia in catalytic cracking of triglycerides. Nevertheless, its commercial implementation still need more research in terms of increased stability of acid sites to avoid leaching, thermal stability, enhancement of mass transfer to avoiding diffusional limitations.

The catalytic properties of sulphated have been shown by several authors as string function of its preparation method, (Yadav and Nair, 1999; Davis et al., 1994; Yamaguchi, 1990). An acid catalyst should be designed to exhibit combined activity and selectivity to have a positive impact on the biodiesel synthesis or any chemical reaction. Therefore, its preparation has drawn much attention among researchers; sulphated zirconias have been synthesized through different routes and conditions in order to improve its catalytic performance. Schuchardt et al., 2008 prepared sulphated zirconia by calcining the ammonium sulphate and the zirconium source. They reported a dramatic effect of the method on the catalytic activity in transesterification, however recycling test revealed activity decay. It is well known that an acid solid catalyst should have moderate to high concentration of strong acid sites available for reactants to proceed to products. The conventional way of preparing sulphated zirconia is by wet-precipitation method; this method has been reported to produce sulphated zirconia with small surface area and low activity (Yadav and Nair, 1999). The objective of this work is to investigate how the conventional method can be improved upon to synthesize an active sulphated zirconia with improved properties. Hence in the present work the conventional wet-precipitation method was modified in terms of the volume of acid used for sulphation. For comparison purpose, a conventional sulphated zirconia was also prepared and their catalytic properties were investigated. Both were characterized for sulphur content, surface area, chemical state, phase composition and crystallinity. The overall aim is to study the effect of the volume of sulphating agent during sulphation on the catalyst activity.

2. Experimental

2.1. Catalysts synthesis

In the first instance sulphated zirconia was prepared by conventional wet-precipitation method as described by Yadav and Nair, 1999; using zirconium oxychloride octahydrate (99.5%, Sigma-Aldrich) as a precursor. A solution was prepared by mixing zirconium oxychloride (20 g) in 200 ml of distilled water. Then ammonium hydroxide (25%) was added drop wise to the mixture and maintained at a pH 9, under constant stirring for 4 h. When the reaction was completed, the solution was filtered and thoroughly washed to remove chloride ion from the zirconium oxide. The presence of chloride ion was determined with $AgNO_3$ test. The oxide was dried by evaporation in an oven for 24 h at 100°C after which it was impregnated with 1 M H_2SO_4 (97.5%, Sigma-Aldrich) under constant stirring for 2 h. The mixture was filtered and the resulted gel was further dried in the oven at 100°C for 24 h and finally calcined in air at 650°C for 3h to produce sulphated zirconia, denoted as 'wp' (wet-precipitated sulphated zirconia). In the case of the modified form the sample was synthesized using same procedure but different volume of acid for impregnation; 2 ml instead of 15 ml and it is designated 'mwp' (modified wet-precipitated sulphated zirconia). For the sake of comparison both calcined and uncalcined catalysts were subjected to characterization using surface techniques, and their catalytic activities were tested using triglycerides oil in a batch reactor at 270°C.

2.2 Sample characterization

The X-ray diffraction pattern was performed using a Panalytical X'Pert Pro Multipurpose Diffractometer (MPD) fitted with an X'Celerator and a secondary monochromator to determine the nature of the sample before calcination as well as the nature of the phases after calcinations. The diffractograms were recorded from using Cu Kα radiation with a wavelength of $\lambda = 1.54$ Å generated at 40 kV and 40 mA from 2°C to 100°C. Their crystallites size was determined using X'pert data viewer software on X-ray diffractogram to generate the necessary data that were applied in Scherrer equation. The surface area measurements were obtained from N_2 adsorption/desorption isotherms determined at 77K using the Coulter™ (SA 3100™ series) by applying the BET equation. The samples were outgassed for 2 h at 200°C prior to the analysis. Thermal analysis was performed on both calcined and non-calcined samples in order to obtain information on the loss of sulphur involved in the calcination process. The TG analysis was conducted on a Perkin-Elmer (Pyris STA 6000 Model) instrument. The sample was placed on a platinum crucible and heated at a heating rate of 5°C/min in flowing helium of 30 ml/min at STP. Spectra were recorded from 30 to 900°C and the amount of sulphate before and after calcination was determined. IR spectra of samples were performed in a Varian 800 (Scimitar series) spectrometer. The spectra were produced between 4000cm^{-1} and 400cm^{-1} using a Pike Technologies diamond crystal plate ATR. The XPS spectra were performed using CasaXPS 2.3.15 software on Kratos analytical system, equipped with a monochromator AlK$\alpha_{1,2}$ X-ray sources of 1486.6 eV and 0.85 eV widths. The reference used was by setting the CHx peak maximum in the resolved C 1s spectra to binding energies of 284.6 eV. Elemental analysis was investigated using energy dispersive X-ray (EDX). The acids site on the catalysts was determined using Diffuse Reflectance Infrared Fourier Transform Spectroscopy (Pyridine- DRIFTS) after pyridine adsorption and desorption. The measurements were carried out using self-supported wafer; a dilution of 5% KBr was used to avoid absorption saturation at low wave numbers.

2.3 Catalytic tests

Both catalysts were tested in a batch reactor (Model: 4560) equipped with a magnetic stirrer and an external heating mantle. The feed, rapeseed oil was heated from ambient room temperature to

270°C before injecting the catalyst. The operating conditions were controlled by WinISO software and conversion monitored against time. The amount of catalyst used was 2 % weight of the feed. Sampling was performed at every 15 minutes for a reaction time of 3 h. The FAME(s) determination and quantification was according to BS EN 14103:2003. This was performed on Hewlett Packard 5890 Series II gas chromatograph with an FID detector. The column temperature was maintained at 210°C while the FID and injector were operated at 250°C each. Detailed composition of the liquid product was carried out using Perkin Elmer GCMS (Clarus) 600/560D) equipped with a capillary column (15m×0.25mm, i.d. 0.25 μm film thickness), and helium as the carrier gas in order to determined the glycerides present according to BS EN 14105: 2003. No alcohol was used as feed in the catalytic cracking reaction.

3. Results and discussion

The XRD measurement revealed the phase composition of the two sulphated zirconias ('mwp' and 'wp') as crystalline; the wet-precipitated sulphated zirconia ('wp') was tetragonal in nature, interestingly the modified wet-precipitated sulphated zirconia ('mwp') exhibited not only the characteristic peaks of tetragonal phase but in addition, monoclinic ZrO_2 phase was observed as shown in Figure 1. The XRDP diffractograms of 'mwp' and 'wp' catalysts were amorphous before calcination. This shows the effect of calcinations temperature on the catalysts as earlier observed by. Interestingly during the preparation of the modified catalyst, 2 ml of sulphuric acid was used for 1 g of ZrOH compared to the wet-precipitated catalyst where 15 ml to 1g of ZrOH was used. The appearance of the monoclinic phase in 'mwp' which is absent in 'wp' is obviously an important observation and is due to the difference in the volume of sulphate used during the preparation, as reported by Farcas,iu et al., (1997). Their crystallite size was 17.51 nm and 10.42 nm for 'mwp' and 'wp' respectively.

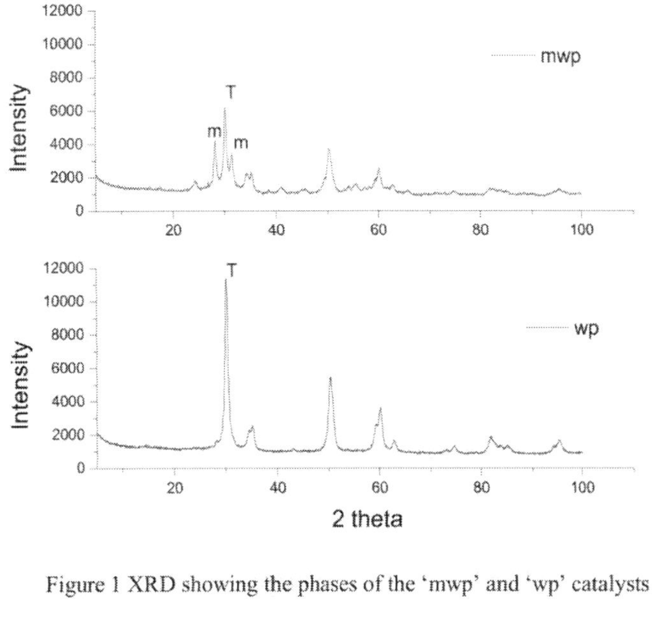

Figure 1 XRD showing the phases of the 'mwp' and 'wp' catalysts

These distinct structural differences are indication that excess sulphate affects the textural properties and may not be necessary for the preparation of an active catalyst. The IR spectra of the catalysts exhibited strong absorptions of sulphate ions coordinated to the zirconium cation between 1297 cm^{-1} and 896 cm^{-1} region. The frequencies from both catalysts were similar, suggesting that the sulphur species in both samples are similar; however the intensity of absorption for 'mwp' catalyst is higher 3% more than the 'wp' catalyst. This is an indication that more sulphate ion was retained on the surface of the 'mwp' and homogeneity was achieved. The elemental analysis (EDX) and X-ray photoelectron spectroscopy reveal similar results of higher retention of sulphur on the 'mwp' catalyst (13 wt %); almost double the weight percentage found on the 'wp' catalyst. The XPS spectra of O 1s of 'mwp' revealed a 1:1 ratio of sulphate and oxide at 532.4 eV and 530.3 eV respectively. This indicates 50 % each of oxide oxygen and sulphate oxygen whereas 'wp' had 42 % sulphate oxygen and 58 % of oxide oxygen, Figure 2. The EDX also indicate sulphur content in the uncalcined 'mwp' and 'wp' catalysts was 0.07 mol each whereas the calcined catalysts were 0.07 mol and 0.05 mol for 'mwp' and 'wp' respectively.

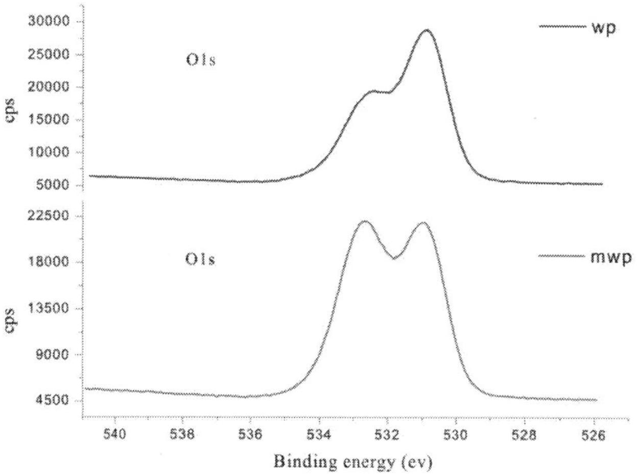

Figure 2 XPS spectra of O 1s of 'mwp' and 'wp' catalysts

It is seen that the peak at 532.4 eV showing the sulphate oxygen in 'wp' catalyst is reduced compared to 'mwp'. This is an indication of approximately 11% more $Zr(SO_4)_2$ on the surface of 'mwp' catalyst. This feature exhibited in 'mwp' is an important observation that has not been reported in any literature to the best of our knowledge. However the S 2p spectra are similar with each other and showed that the sulphate species are S^{6+} of SO_4^{2-} . This is in agreement with Hino et al., (2006). The infrared spectra of the adsorbed pyridine on both catalysts, revealed bands characteristics of Bronsted and Lewis acid sites at 1540 cm^{-1} and 1450 cm^{-1} respectively. However the 'wp' has more Lewis acid site as revealed on the catalyst; (75%) and less of Bronsted acid sites (25%). The reverse is the case with the 'mwp' catalyst, approximately 54% of Bronsted acidity was observed, Figure 3; despite lesser amount and same source of acid was used for impregnation of 'mwp' catalyst coupled with the fact that the same method of preparation was adopted in contrast to (Hong et al., 1999) report. This is an interesting result; the 'mwp' not only has Bronsted acid sites but about 8% more than its counterpart, Lewis acid site. We also observed that despite the large difference in the amount of sulphuric acid used; approximately 1:8 for 'wp' and 'mwp' respectively during preparation, the same mole of sulphur was retained on the surface of the gel before calcination; interestingly the 'mwp' retained its sulphur content after calcination, but 'wp' lost 29%

Figure 4. Apparently, therefore, there was no reduction of sulphate during the post-treatment (i. e. calcination) of the 'mwp'. This goes further to confirm the uniqueness of the modified method of preparation.

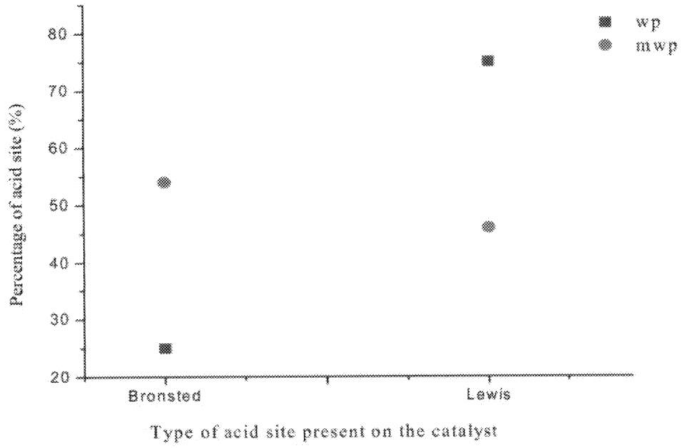

Figure 3 Percentages of Bronsted and Lewis acid site in 'wp' and 'mwp'

The conventional wet-precipitated sulphated zirconia has enhanced catalytic activity which obviously is attributed to the controlled amount of sulphate used during preparation rather than the method in this case as compared to other authors, (Hino et al., 2006; Farcas¸iu et al., 1997; Corma et al., 1994)

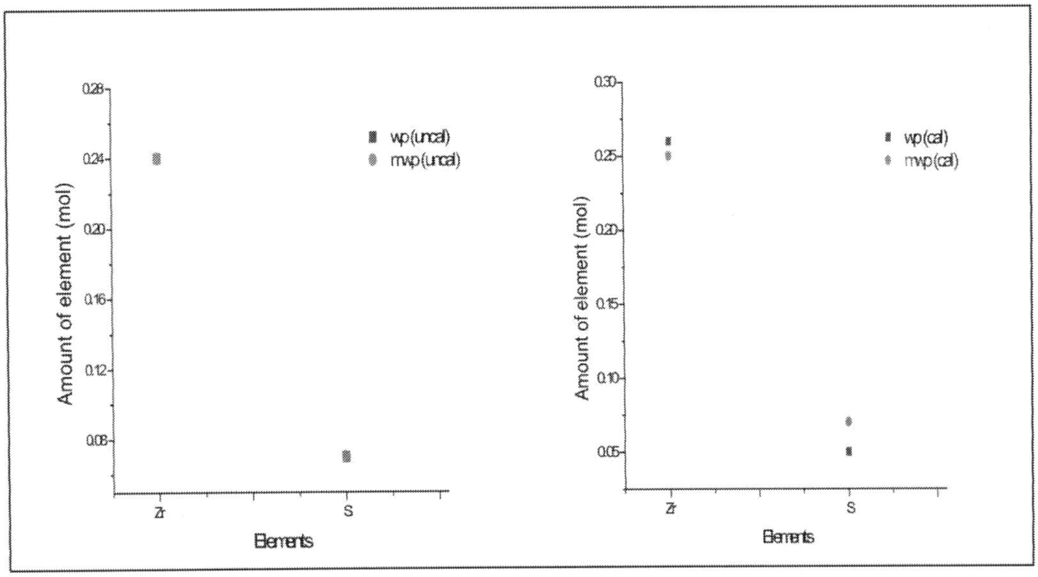

'wp(uncal)' and 'mwp(uncal)' are uncalcined wp and mwp;
'wp(cal)' and 'mwp(cal)' are calcined wp and mwp

Figure 4 EDX results of sulphur in mole before and after calcinations

The BET surface areas were $65m^2/g$, and $101m^2/g$, for 'wp' and 'mwp' respectively. The surface area and acid sites play a great role in the catalytic reactions as it is the area and amount of site available for reactions. The sulphate on the surface of the uncalcined catalysts was monitored by thermogravimetric analysis (TGA). Notably three main mass loss regions were observed: at 100°C due to physisorbed water, the second between 150 – 390°C indicating dehydration and dehydroxylation and finally at 580 – 800°C associated with sulphate decomposition. The 'mwp' experienced a greater sulphate loss (0.17mol) than the 'wp' (0.11mol). This no doubt confirmed the presence of more sulphate ion to zirconium cation been formed on the surface of 'mwp' catalyst during preparation. Both catalysts cracked triglycerides with a conversion of 58 % and 49 % after 3 h for 'mwp' and 'wp' respectively. Not only was 'mwp' conversions greater, it exhibited greater selectivity for total fatty acid methyl esters (FAMEs) approximately 40 % after $2^1/_4$ h and 27 % for 'wp' in less than 2 h as seen in Figure 5. We found interestingly the fatty acid methyl esters profile was more selective for saturated methyl esters compared to unsaturated, Figure 6. However there was higher yield of the saturated methyl esters with 'mwp'

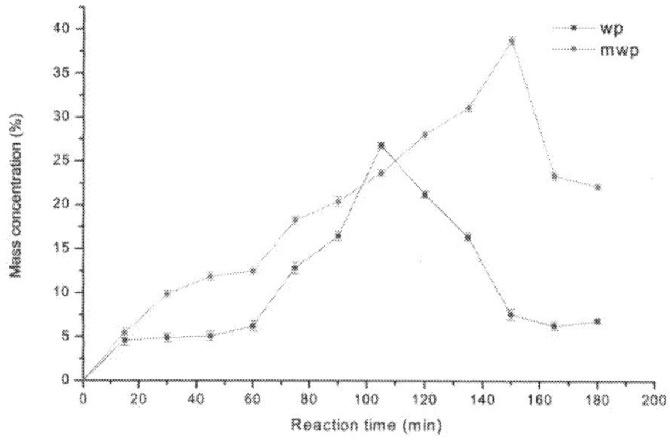

Figure 5 Total yields of methyl esters over 'wp' and 'mwp' catalysts

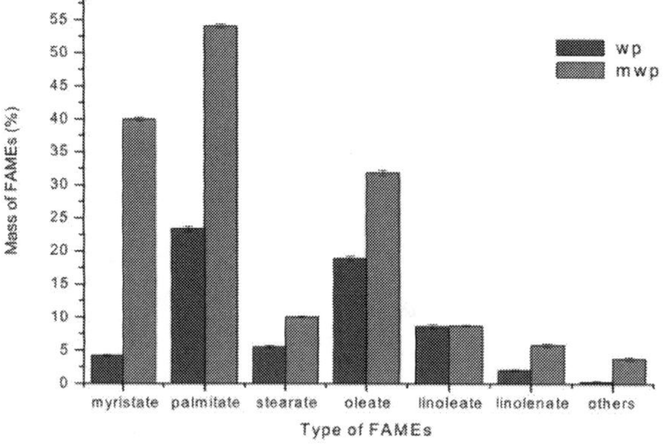

Figure 6 FAMEs profile showing saturated, mono and poly unsaturated methyl ester

The presence of Bronsted acid sites in the conventional wet-precipitated sulphated zirconia enhanced its catalytic activity which obviously is attributed to the reduced amount of sulphate used during preparation rather than the method in this case as compared to other authors, (Hino et al., 2006; Farcasiu et al., 1997; Corma et al., 1994) who reported sulphated zirconia catalyst as only Lewis acid sites. Furthermore, the 'wp' was observed to suffer severe leaching; it lost 100 % of its sulphur content after 3 h of reaction whereas 'mwp' lost only 26 % (1.17 wt %). The weight percent of sulphur on the regenerated 'wp' catalyst was 0 wt %, while that of 'mwp' catalyst was 3.29 wt % as indicated by EDX as shown in Figure 7. This aspect of sulphate leaching needs further improvement.

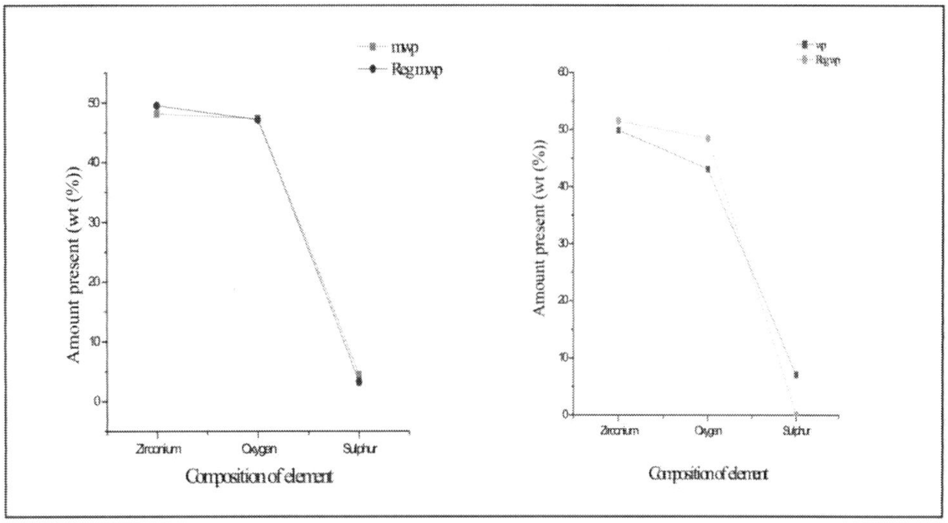

Figure 7 Comparing the sulphur content of both catalyst after regeneration with the calcined catalysts

4. Conclusions

The variation in the volume of the sulphating agent used during preparation led to sulphated zirconias that exhibited different properties in terms of specific surface areas, acid sites, thermal stability and surface sulphate. Both catalysts were catalytically active for triglycerides cracking for fatty acid methyl esters in a catalytic reaction. The 'mwp' catalyst was more active with higher conversion of triglycerides and higher FAMEs yield, approximately 40% after $2^1/_4$ h, as compared to the conventionally wet-precipitation method (27%). The selectivity towards saturated esters could only be explained by the higher activity of 'mwp' which is as a result of the presence of higher sulphate on the surface of the catalyst as revealed by the XPS and EDX studies as well as the increased Bronsted acid site. The activity and selectivity of 'mwp' for saturated esters is an advantage and could be of interest. We can safely conclude that the conventional wet-precipitation method if modified by reducing the "conventional" volume (15 ml) of sulphating acid; could be used to synthesized active sulphated zirconia catalyst with improved surface area and increased Bronsted acid site. Importantly we observed in this work and consequently agree with katada et al., (2000) that excess sulphating agent tends to suppress acidity. Therefore, the conventional wet-precipitation method is still viable for preparing sulphated zirconia with improved activity, structural and textural properties.

Acknowledgement

The authors would like to acknowledge the Nigerian Government for their financial support through the Petroleum Trust Development Fund, Dr. Karen Wilson of School of Chemistry, Cardiff University. Maggie White, and Pauline for characterization of the catalysts, Newcastle University, UK. The service of ACMA is acknowledged www.ncl.ac.uk/acma.

Reference

Arata, K., Hino, M. and Yamagata, N. (1990) 'Acidity and catalytic activity of zirconium and titanium sulfates heat-treated at high temperature. solid superacid catalysts', *Bulletin of the Chemical Society of Japan,* 63, (1), pp. 244-246.

Ardizzone, S., Bianchi, C. L., Cattagni, W. and Ragaini, V. (1997) 'Effects of the precursor features and treatments on the catalytic performance of SO4/ZrO2', *Catalysis Letters,* 49, (3), pp. 193-198.

Chew, T. L. and Bhatia, S. (2008) 'Catalytic processes towards the production of biofuels in a palm oil and oil palm biomass-based biorefinery', *Bioresource Technology,* 99, (17), pp. 7911-7922.

Corma, A., Fornés, V., Juan-Rajadell, M. I. and Nieto, J. M. L. (1994) 'Influence of preparation conditions on the structure and catalytic properties of SO42-/ZrO2 superacid catalysts', *Applied Catalysis A: General,* 116, (1-2), pp. 151-163.

Davis, B. H., Keogh, R. A. and Srinivasan, R. (1994) 'Sulfated zirconia as a hydrocarbon conversion catalyst', *Catalysis Today,* 20, (2), pp. 219-256.

Eterigho, E. J., Lee, J. G. M. and Harvey, A. P. (2011) 'Triglyceride Cracking for Biofuel Production using a Directly Synthesised Sulphated Zirconia Catalyst ', *Bioresource Technology,* in Press.

Farcas,iu, D., Li, J. Q. and Cameron, S. (1997) 'Preparation of sulfated zirconia catalysts with improved control of sulfur content II. Effect of sulfur content on physical properties and catalytic activity', *Applied Catalysis A: General,* 154, (1-2), pp. 173-184.

Farcasiu, D., Li, J. Q. and Kogelbauer, A. (1997) 'The mechanism of conversion of hydrocarbons on sulfated metal oxides .4. Kinetics of the reaction of methylcyclopentane on sulfated zirconia', *Journal of Molecular Catalysis a-Chemical,* 124, (1), pp. 67-78.

Hino, M. and Arata, K. (1994) 'Synthesis of highly active superacids of SO4/ZrO2 with Ir, Pt, Rh, Ru, Os, and Pd substances for reaction of butane', *Catalysis Letters,* 30, (1), pp. 25-30.

Hino, M., Kobayashi, S. and Arata, K. (1979) 'Reactions of butane and isobutane catalyzed by zirconium oxide treated with sulfate ion. Solid superacid catalyst [7]', *Journal of the American Chemical Society,* 101, (21), pp. 6439-6441.

Hino, M., Kurashige, M., Matsuhashi, H. and Arata, K. (2006) 'The surface structure of sulfated zirconia: Studies of XPS and thermal analysis', *Thermochimica Acta,* 441, (1), pp. 35-41.

Hong, Z., Fogash, K. B. and Dumesic, J. A. (1999) 'Reaction kinetic behavior of sulfated-zirconia catalysts for butane isomerization', *Catalysis Today,* 51, (2), pp. 269-288.

Katada, N., Endo, J. I., Notsu, K. I., Yasunobu, N., Naito, N. and Niwa, M. (2000) 'Superacidity and catalytic activity of sulfated zirconia', *Journal of Physical Chemistry B,* 104, (44), pp. 10321-10328.

Melero, J. A., Clavero, M. M., Calleja, G., Garcìa, A., Miravalles, R. n. and Galindo, T. (2009a) 'Production of Biofuels via the Catalytic Cracking of Mixtures of Crude Vegetable Oils and Nonedible Animal Fats with Vacuum Gas Oil', *Energy & Fuels,* 24, (1), pp. 707-717.

Melero, J. A., Iglesias, J. and Morales, G. (2009b) 'Heterogeneous acid catalysts for biodiesel production: Current status and future challenges', *Green Chemistry,* 11, (9), pp. 1285-1308.

Morterra, C., Cerrato, G., Pinna, F., Signoretto, M. and Strukul, G. (1994) 'On the Acid-Catalyzed Isomerization of Light Paraffins over a ZrO2/SO4 System: The Effect of Hydration', *Journal of Catalysis,* 149, (1), pp. 181-188.

Srivastava, A. and Prasad, R. (2000) 'Triglycerides-based diesel fuels', *Renewable and Sustainable Energy Reviews,* 4, (2), pp. 111-133.

Taufiqurrahmi, N. and Bhatia, S. (2011) 'Catalytic cracking of edible and non-edible oils for the production of biofuels', *Energy & Environmental Science,* 4, (4), pp. 1087-1112.

Yadav, G. D. and Nair, J. J. (1999) 'Sulfated zirconia and its modified versions as promising catalysts for industrial processes', *Microporous and Mesoporous Materials,* 33, (1-3), pp. 1-48.

Yamaguchi, T. (1990) 'Recent progress in solid superacid', *Applied Catalysis,* 61, (1), pp. 1-25.

(682a) Methanol Dehydration to Dimethyl Ether In An Autothermal Millisecond Residence Time Reactor

Hui SUN, Chemical Engineering and Material Science, University of Minnesota, Minneapolis, MN and Lanny D. Schmidt, Chemical Engineering and Materials Science, U of Minnesota, Minneapolis, MN

Methanol dehydration to dimethyl ether (DME) has been carried out in a two-stage autothermal millisecond residence time reactor in which top stage is noble metal coated monolith and bottom stage is zeolite catalyst. DME has attracted a lot of attention since it is a clean diesel substitute; it is environmental friendly with zero ozone depletion potential and it is a key intermediate for other important chemicals. The advantage of this reactor is: (1) external heat is not required since the reactor is autothermal; (2) The residence time of the reaction is on the order of milliseconds resulting in a smaller reactor and catalyst volume. The best result was obtained when methanol was fed from the side of the reactor between two stages in terms of maximizing the methanol conversion and DME selectivity. The experiment has been carried out using 1wt% Pt/α-Al_2O_3 catalyst on the top stage where heat is released by exothermic reactions and H-ZSM-5 (Si/Al=25) on the second stage where the heat is used to drive endothermic reactions. 1.2 g of H-ZSM-5 is used and the typical methanol dehydration temperature is between 523K and 603K. Different noble metal (Pt and Rh) coatings with different monolith lengths were tested to optimize the top stage of autothermal reactor. Two reactor configurations have been tested: (1) Methanol fed from the top of the reactor with hydrogen as sacrificial fuel and (2) methanol fed from the side of the reactor with methane as sacrificial fuel which is partially oxidized in the noble metal catalyst layer. For the first configuration, reaction temperature and product distribution of the second stage was adjusted by varying different fuel to oxygen ratio or varying distance between two stages. Second configuration results in higher methanol conversion and DME selectivity than those of the first one. The highest yield to DME is 76.11% at C/O ratio of 1.01.

See more of this Session: Alternative Fuels I
See more of this Group/Topical: Catalysis and Reaction Engineering Division

(682b) Pd-Ni Electrocatalysts for Efficient Ethanol Oxidation Reaction In Alkaline Electrolyte

Le Xin[1], Zhiyong Zhang[1], Kai Sun[2] and Wenzhen Li[1], (1)Chemical Engineering Department, Michigan Technological University, Houghton, MI, (2)Department of Materials Science and Engineering, University of Michigan, Ann Arbor, MI

Abstract: Lignocellulosicbiomass-derived ethanol is among the most promising fuel candidates to supplyfuture energy needs. The development of anion exchange membrane renders thepossibility of ethanol oxidation by using non-platinum catalysts with highefficiency, which significantly reduces the cost of direct ethanol fuel cellsand makes it attractive for large-scale fuel cell applications. It has beenfound that Pd has a competitive ethanol oxidation reaction (EOR) as Pt in highpH media. The theoretical calculation predicts that Pd-Ni could have high EORactivity due to the synergetic effect of Ni element. Here, we report asolution-phase reduction approach to prepare carbon supported Pd-Ni nanoparticles with different Pd-Ni compositions and characterizethem by XRD, TEM, HR-TEM, TGA, ICP-AES. Thecharacterizations reveal that these Pd-Ni catalysts have small diameters ofaround 3 nm and narrow size distributions. The catalytic activities towards EORare carried out in 1.0 M NaOH + 1.0 M $C_2H_5OH$by cyclic voltammetry and chonoammeperometry,and to our knowledge, the highest EOR mass activity and stability on PdxNiy/C catalysts have been demonstrated. We also find theEOR activity is strongly correlated with Pd-Ni composition and structure,especially the Pd-Ni interaction/contact on catalyst surface. Our latestprogress in this area will be presented as well.

See more of this Session: Alternative Fuels I
See more of this Group/Topical: Catalysis and Reaction Engineering Division

Effects of ammonia substitution on stability limits and emissions of premixed hydrogen/air flames

J.M. Joo[a], D.H. Um[a], O.C. Kwon[a,*]
[a]School of Mechanical Engineering, Sungkyunkwan University
Suwon, Gyeonggi-do, Korea

Abstract

The potential of partial ammonia (NH_3) substitution to improve the safety of hydrogen (H_2) use was evaluated, using premixed NH_3/H_2/air flames at normal temperature and pressure generated on a tube type burner. The effects of NH_3 substitution on the stability limits and nitrogen oxide (NO_x) emissions of premixed H_2/air flames and the effects of inert (nitrogen, N_2) coflows on the NH_3-substituted H_2/air flames were investigated. The lower stability limits due to heat losses are observed for all the present test conditions, while the upper stability limits due to insufficient residence times of injected mixture jet are observed for limited test conditions. Results also show reduction of the stability limits with NH_3 substitution in H_2/air flames, supporting the potential of NH_3 as a carbon-free, clean additive for improving the safety of H_2 use in premixed H_2/air flames. Opposite tendencies of the upper stability limits with and without the coflow are observed since nonpremixed flames near the burner exit cannot be generated with the coflow. Ammonia substitution enhances the NO_x formation in general; however, the NO_x emission index that is defined as the fraction of the mass of the produced NO_x per the mass of the provided NH_3 and H_2 is almost constant with the enhanced ammonia substitution. At fuel-rich conditions, the NO_x emission index is reduced with increasing burner exit velocities of the injected mixtures.

Keywords: Hydrogen, Ammonia, Premixed flames, Stability limits, Carbon-free

* Corresponding author. E-mail: okwon@skku.edu

(682d) Low-Pressure Hydrogenolysis of Vapor-Phase Glycerol Over Heterogeneous Catalysts

D. Evan Piephoff[1], Jeremy G. Immer[2], M. Jason Kelly[2] and H. Henry Lamb[2], (1)Materials Processing Center/Center for Materials Science and Engineering, Massachusetts Institute of Technology, Cambridge, MA, (2)Department of Chemical and Biomolecular Engineering, North Carolina State University, Raleigh, NC

The recent increase in biodiesel production has resulted in a global glut of glycerol, the major byproduct of biodiesel transesterification, which has decreased the value of crude glycerol and rendered the product detrimental to biodiesel economics. This study presents a feasible process for the production of value-added products from glycerol via hydrogenolysis of vapor-phase glycerol over various heterogeneous catalysts at atmospheric pressure. Dynamic system behavior was observed via on-line gas chromatography. In order to determine the effect of H_2 concentration on catalyst activity and product selectivity, reactions were typically conducted under H_2 followed by 10% H_2 (balance He) flow. Under all conditions tested, catalyst activity was greater under H_2 than 10% H_2 flow. Cu catalysts were found to be selective for hydrogenation products, such as 1,2-propanediol (1,2-PDO), under H_2 and dehydration products, such as acetol and acrolein, under 10% H_2 flow. Acrolein, a known thermal reaction product of glycerol, was produced most selectively when acetol was injected over blank quartz wool and glycerol was dehydrated over blank SiO_2. A network of reaction pathways for glycerol hydrogenolysis over Cu catalysts was proposed based on the observed temporal product selectivity behavior. Supported Ni catalysts were found to be highly active for C-C bond hydrogenolysis, generally producing C1 products, such as methanol, more selectively under 10% H_2 and C2 products, such as ethylene glycol, more selectively under H_2 flow. When reactions were performed over Ni/Al_2O_3 under 5% H_2 flow, methanol was produced with highest selectivity, which is significant because methanol can be reused as a reactant in biodiesel transesterification.

See more of this Session: Alternative Fuels I
See more of this Group/Topical: Catalysis and Reaction Engineering Division

(682e) Understanding Pt-Re Catalysts Used for Aqueous Phase Reforming of Biomass

Prashant Reuben Daggolu, Yong Wang and Zhehao Wei, The Gene and Linda Voiland School of Chemical Engineering, Washington State University, Pullman, WA

Aqueous phase reforming (APR) is a process by which a substrate molecule is reacted in aqueous phase to produce hydrogen. When applied to biomass, this technique can be used to produce hydrogen from oxygenated hydrocarbon such as glucose, glycerol and acetic acid. Extensive research in the literature has shown the efficacy of Pt-Re bimetallic catalysts for this reaction for various substrate molecules. However, not much characterization is done in understanding this catalyst with respect to APR. Hence we studied the influence of supports, acidity, Pt:Re ratio on the APR process. We studied this using CO-TPD, TPRS, pyridine –DRIFTS and some batch reactions. We have shown that CO-Metal bonding is stronger on silica than on zirconia supports. Also water addition strengthens this bonding. Py-DRIFTS showed the emergence of Brønsted acidity with water addition. This acidity seems to arise from Pt more than from Re and the support. Lewis acidity was seen consistently on the surfaces. More work is underway including a correlation between the observed properties and APR product distribution.

See more of this Session: Alternative Fuels I
See more of this Group/Topical: Catalysis and Reaction Engineering Division

(682f) Pulse and Temperature Programmed Reaction Study of Acetic Acid Conversion to Hydrocarbon On TiO2/ZSM-5

Prashant Reuben Daggolu, The Gene and Linda Voiland School of Chemical Engineering, Washington State University, Pullman, WA and Mark G. White, Dave C. Swalm School of Chemical Engineering, Mississippi State University, Mississippi State, MS

Acetic acid is a major component of pyrolysis oil produced from Biomass. It is also a byproduct when syngas is converted to ethanol on Rh and Mo catalysts. Conversion of acetic acid to hydrocarbon would greatly enhance the viability of biomass conversion to fuel processes. However, acetic acid is shown to coke when reacted on H^+/ZSM-5 without much hydrocarbon formation. This is because acetic acid is not amenable for cracking reactions on acid catalysts due to its poor H/C ratio. Hence we studied using titania (TiO_2) in conjunction with the zeolite, H^+/ZSM-5 to convert acetic acid to liquid hydrocarbon.

Our strategy utilizes the ability of titania to catalyze condensation reactions where acetic acid can be converted to acetone and acetone can be converted to mesitylene via mesityloxide formation. The condensation products then can react on the zeolite to form wider range hydrocarbon.

Initially, acetic acid was reacted in a pulse reactor connected to a GC-MS on Titania P25 catalyst. The reactions were performed between 300 ^oC and 450 ^oC. The results confirmed our hypothesis when acetone and mesitylene were seen in the product stream. Then 10% TiO_2 impregnated on H^+/ZSM-5 was tested which showed the formation of several other hydrocarbon like substituted benzenes and naphthalenes. Hence the combination of titania with the acid zeolite seem to be the way to move forward in converting acetic acid to hydrocarbon. This is further tested in a Temperature programmed reaction system to understand various intermediate formation. Also the catalyst is characterized using pyridine DRIFTS and XRD. Some of the later work is in progress.

See more of this Session: Alternative Fuels I
See more of this Group/Topical: Catalysis and Reaction Engineering Division

The Effects of Carbon-to-Oxygen ratio upon Supercritical Water Reformation for Hydrogen Production

Jared Bouquet[1], Ryan Tschannen[2], Aaron Gonzales[2], Jason Picou[1], and Sunggyu Lee[2]
1: Missouri University of Science and Technology, Rolla, MO
2: Ohio University, Athens, OH

Abstract

With the continually growing energy demands of today's society, finding innovative ways to produce renewable energy is increasingly important. One such process is the supercritical water reformation of hydrocarbons for hydrogen production. Supercritical water reformation is a versatile non-catalytic process that can utilize a wide variety of commonly available raw materials. Such raw materials include methanol, ethanol, glycerin, sucrose, diesel, and jet fuel. With such a diverse feedstock, determining which factors have a substantial impact on the supercritical water reformation is crucial. One such factor is the carbon-to-oxygen ratio of the feed material. To determine the effect of carbon-to-oxygen ratio, an experimental study was performed using isopropanol, propylene glycol, and glycerin solutions. The experimental study was conducted using a 0.1 L Haynes Alloy 282 reactor at a constant pressure of 24.13 MPa and a reactor space time of 45 seconds. The feed solution for each experiment contained an 8:3 water-to-carbon ratio, and the effect of carbon-to-oxygen ratio was observed at 550, 600, and 650°C.

Introduction

Supercritical water reformation is a reaction that utilizes water above it critical point (374°C and 22.06 MPa) to break down organic molecules into hydrogen and carbon monoxide. The resulting hydrogen can be used for alternative energy applications. When water is in the supercritical region, non-polar molecules become much more soluble in it. This provides a highly energetic, non-catalytic reaction medium for reformation.

One of the biggest strengths of supercritical water reformation is the number of different fuel feedstocks it can be used with. Possible reformation materials range from small molecules like methanol, to larger molecules like diesel and jet fuel, to ringed molecules such as sucrose. The process can also take mixtures of fuels, such as is the case of crude glycerin. Crude glycerin is a mixture of methanol and glycerin and a substantial byproduct of biodiesel production.

With such a wide range of potential fuels that can be used in supercritical water reformation, determining the effect of certain functional groups on reformation is vital. One such functional group that should be explored is the amount of hydroxyl groups in the molecular structure of the fuel. Based on the previous work of Picou et. al, there are many differences in the results reforming a fuel with many hydroxyl groups and none.[1,2]

To give insight into how the amount of hydroxyl groups effect reformation, a study was done using a number of three-carbon organic molecules with differing numbers of hydroxyl groups. For this study, supercritical water reformation was done on isopropanol, propylene glycol, and glycerin. In order to determine the effect of increasing carbon-to-oxygen ratio in the fuel, a series of experiments were conducted at 24.2 MPa, a reactor space time of 45 seconds, and water-to-carbon molar ratio of 8:3. The effect of increasing carbon-to-oxygen ratios were evaluated at temperatures of 550, 600, and 650°C.

Chemical Reactions

There are many reactions that can occur at the conditions needed for supercritical water reformation. Besides supercritical reformation there is also pyrolysis and the water gas shift reaction. Also due to the oxygen present in these molecules, it is possible that they undergo a reaction similar to methanol decomposition rather than direct reformation. The general equation for supercritical water reformation is shown below.

Equation 1 Supercritical Water Reformation: $C_mH_n + mH_2O \rightarrow (m + \frac{N}{2})H_2 + mCO$

Supercritical water reformation is an endothermic reaction and involves breaking a hydrocarbon down into hydrogen gas and carbon monoxide. Reforming a compound saturated with hydroxyl groups, like glycerin, would produce carbon dioxide instead of carbon monoxide since the hydrocarbon is already oxygenated. The production of carbon monoxide is more likely the result of decomposition, detailed below, because in that scenario water is not a reactant and only used as a reaction medium.

The reformation of hydrocarbons is commonly performed with the aid of a noble metal catalyst. The experiments presented in this paper were performed non-catalytically, with no catalyst loaded into the reactor, except potential monolithic catalytic effects from the reactor wall made of high-nickel alloy of Haynes® Alloy 282.

Equation 2 Methanol Decomposition: $CH_3OH \leftrightarrow 2H_2 + CO$

While no methanol was used in these experiments, it is possible glycerin and the other alcohols undergo similar decomposition reactions, or some mixture of reformation and decomposition. The best way to tell whether reformation or decomposition is happening is to look at the amount of carbon monoxide being produced verse the amount of carbon dioxide being produced

Equation 3 Water-Gas Shift: $CO + H_2O \leftrightarrow CO_2 + H_2$

The water-gas shift (WGS) reaction is exothermic with a heat of reaction of -41 kJ per mol.[3] The reaction is reversible in nature, and involves the combination of water and carbon monoxide to make hydrogen and carbon dioxide. The forward reaction is thermodynamically favored at temperatures of 815°C or below, which is true for all of the experiments discussed in this paper.[4]

Pyrolysis describes a group of reactions that involve the dismantling of glycerol into smaller hydrocarbons, such as methane, ethylene, and ethane. The ethane produced can still be reformed to make hydrogen or pyrolyzed even further to produce methane. While this reaction does not consume hydrogen, it does competitively consume the reactant for reformation, limiting hydrogen production.

Another important reaction is isopropanol dehydrogenation, where isopropanol is oxidized into acetone and hydrogen is produced. This reaction commonly competes with isopropanol dehydration to make propylene, though the presence of water enhances selectivity to produce acetone.[5] This reaction can produce hydrogen with no carbon coming out in the gaseous effluent, making it a reasonable assumption for situations where there is a lot of hydrogen production with little gaseous carbon conversion.

Apparatus and Chemicals

The chemicals used for the experiments were deionized water, 99.7% pure glycerol from thechemistrystore.com, 99% extra pure propylene glycol from Acros Organics, and 70% Isopropanol from Fisher Scientific. Glycerin above 80 wt-% in purity will absorb ambient water vapor from the air; therefore the bulk glycerin was diluted to 75.0 wt-%. The water, glycerin, propylene glycol and isopropanol for the experiments were then mixed into solutions that had water-to-carbon molar ratios of 8:3. The physical properties of glycerol, propylene glycol, isopropanol and water may be found in Table 1. Note that with the exception of the experiment with glycerin at 550°C the reaction conditions were above the critical points of the fuel.

Table 1. Physical Properties of Glycerin, Propylene Glycol, Isopropanol, and Water [6-9]

Property	Glycerin	Prop. Glycol	Isopropanol	Water
Molecular Weight (g/mol)	92.09	76.09	60.09	18.02
Density at 25 °C and 1 bar (g/cm^3)	1.26	1.032	0.786	0.99
Melting Point (°C)	17	-60	-88	0
Boiling Point (°C)	287	189	82	100
Critical Temperature (°C)	577	353	236	375
Critical Pressure (atm)	74.02	60.2	48.35	217.7

The supercritical water reformation system used for these experiments consists of a water and a fuel feed system, reactant preheater, Haynes® Alloy 282 reactor with a heater, liquid collection system, gas sampling system, and a data acquisition and control system. Figure 1 is a representative process flow diagram of the supercritical water reaction system. The Haynes® Alloy 282 reactor has an internal volume of 101 mL, and is an alloy of mostly nickel, chromium, iron, and molybdenum, along with other elements.[8] The samples collected of the gaseous effluent at ambient conditions were analyzed using a HP 5890 Series A gas chromatograph with a thermal conductivity detector. The gas chromatograph is calibrated to detect hydrogen, nitrogen, carbon monoxide, methane, carbon dioxide, acetylene, ethylene and ethane.

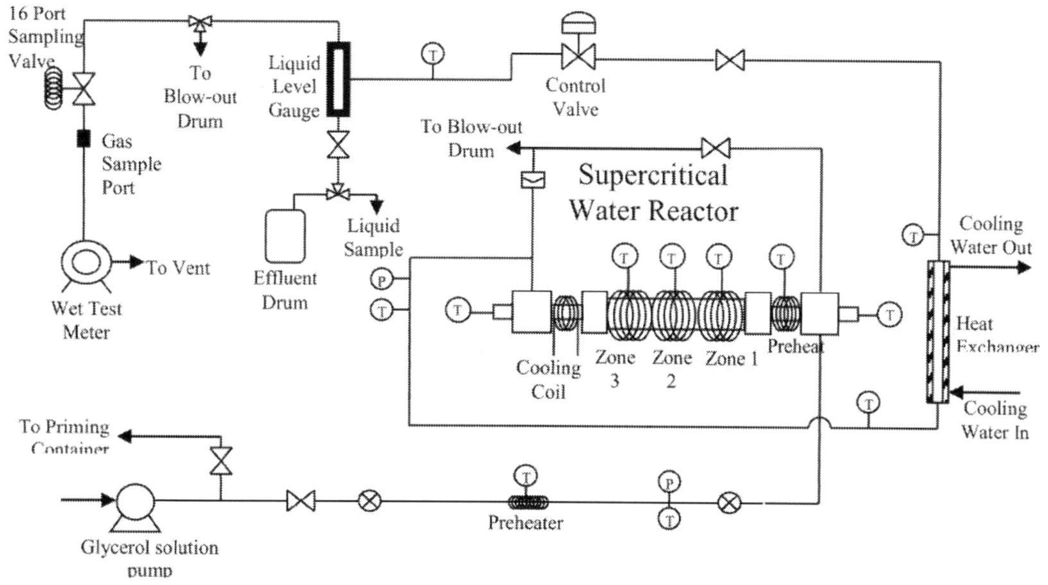

Figure 1The Multi-fuel reformation system at the SEAM Laboratory at Ohio University

Experimental

Experiments using solutions of isopropanol, propylene glycol, and glycerin were run in run at the same conditions. The experiments were performed non-catalytically at a constant pressure of 24.13 ± 0.33 MPa, solutions containing a molar ratio of 8:3 water-to-carbon, and space times of 45 seconds. The temperatures used in these experiments were 550 ± 1°C, 600 ± 2°C, and 650 ± 2°C. The space time was determined using the Peng-Robinson equation of state with Wang-Sandler mixing rules and calculated in Aspen simulation software. Table 2 lists the temperature, flow rate, pressure, water/carbon (W/C) ratio, fuel, fuel weight percent, and methanol weight percent of each experimental run.

Table 2: Experimental Run Summary

Run No.	Temp (oC)	Pressure (psig)	Flowrate (g/min)	Res Time sec	Fuel	OH/C ratio	Fuel wt%	W/C ratio
WI-132M	549	3501	12.3	45.0	Isopropanol	0.33	29.42	2.67
WPG-129M	549	3510	13.5	45.0	Prop. Gly.	0.67	34.55	2.67
WG-126M	549	3498	17.1	45.0	Glycerin	1.00	38.97	2.67
WI-133M	598	3485	11.1	45.0	Isopropanol	0.33	29.42	2.67
WPG-130M	599	3480	12.2	45.0	Prop. Gly.	0.67	34.55	2.67
WG-127M	599	3485	14.4	45.0	Glycerin	1.00	38.97	2.67
WI-134M	649	3499	10.6	45.0	Isopropanol	0.33	29.42	2.67
WPG-131M	648	3456	11.2	45.0	Prop. Gly.	0.67	34.55	2.67
WG-128M	648	3452	13.0	45.0	Glycerin	1.00	38.97	2.67

Results and Discussion

The results of the experiments shown in Table 2 are shown below in Table 3. The results show the amount of carbon gasification as well as the amount of product gas being produced. The amount of each species coming out in the gaseous effluent was measured as moles of the gaseous species produced per mole of carbon fed. Carbon not accounted for in the gaseous effluent either comes out of the reaction system in the liquid effluent, or becomes a solid coke that stays in the reactor.

Table 3 Experimental Gas Results

Run No.	Gas Yield (mol gas/mol carbon fed)							Carbon
	H2	N2/O2	CO	CH4	CO2	C2H4	C2H6	conversion
WI-132M	0.18	0.00	0.00	0.03	0.02	0.00	0.00	3.6%
WPG-129M	0.08	0.00	0.02	0.01	0.03	0.00	0.01	5.9%
WG-126M	0.08	0.00	0.16	0.03	0.02	0.00	0.00	22.7%
WI-133M	0.42	0.00	0.04	0.22	0.07	0.01	0.02	25.3%
WPG-130M	0.31	0.00	0.24	0.17	0.14	0.01	0.11	65.8%
WG-127M	0.22	0.00	0.35	0.12	0.05	0.00	0.03	59.4%
WI-134M	0.25	0.00	0.12	0.74	0.20	0.07	0.13	97.2%
WPG-131M	0.28	0.00	0.30	0.36	0.16	0.00	0.16	96.2%
WG-128M	0.30	0.00	0.49	0.22	0.10	0.00	0.08	97.9%

The graphs showing the effect of water-to-carbon ratio at 550, 600, and 650°C are below. Each graph shows the effect of increasing oxygen content, represented as OH/C ratio, at a certain temperature. This means that data points at 0.33 have isopropanol as their fuel, data points at 0.66 use propylene glycol, and data points at 1.0 correspond to glycerin.

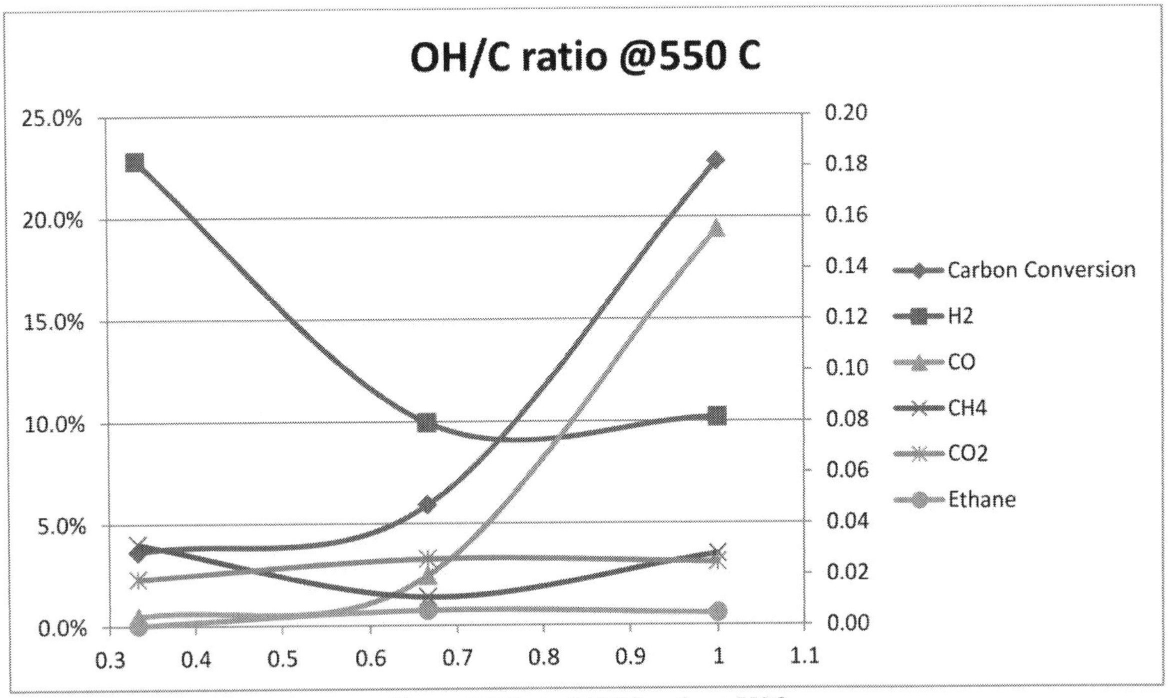

Figure 2 Effect of OH/C ratio at 550C

At 550 °C very little of the isopropanol and propylene glycol are reacting, with carbon conversions of only 3.6% and 5.9% respectively. The only hydrocarbon to have significant carbon conversion is glycerin, with 22.7% conversion. Figure 2 also shows that with increasing oxygen in the fuel, there is an increase in the amount of carbon monoxide produced. In fact the trend in carbon conversion and carbon monoxide seem almost identical. There is relatively little change in the amount of carbon dioxide, methane, or ethane with changing carbon to oxygen ratio.

An interesting result shown in Figure 2 is the relatively large amount of hydrogen gas being produced from isopropanol with so little carbon conversion. This is most likely do to the dehydrogenation reaction taking place and making acetone and hydrogen gas. The resulting hydrocarbons then leave the reaction system in the liquid effluent.

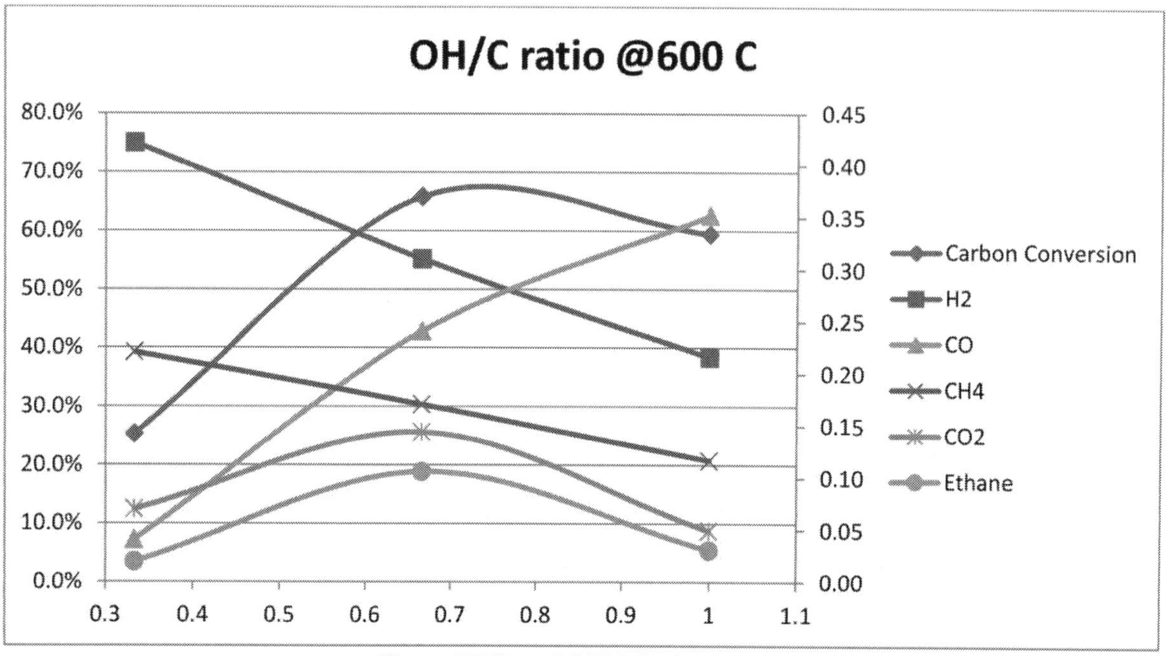

Figure 3 Effect of OH/C Ratio at 600C

Figure 3 has increasing carbon conversion going from isopropanol to propylene glycol, but the conversion going from propylene glycol to glycerin stays about the same. This means that going from two to three hydroxyl groups has little effect on the amount of carbon gasification at 600°C.

With an increasing number hydroxyl groups in the fuel there is a decline in hydrogen production, as well as methane production. This indicates that with increasing oxygen content a decline in pyrolysis reactions as well as any decomposition reactions that produces hydrogen but no gaseous carbon. Carbon monoxide production increases with increased hydroxyl groups similar to the way it did at 550 °C.

Carbon dioxide and ethane both increase going from isopropanol to propylene glycol, but decreases from propylene glycol to glycerin. The initial rise is likely a result increased carbon conversion and general increase in the amount of all gases produced. The decline is the result of decreased pyrolysis, since the conversions for propylene glycol and glycerin were almost the same.

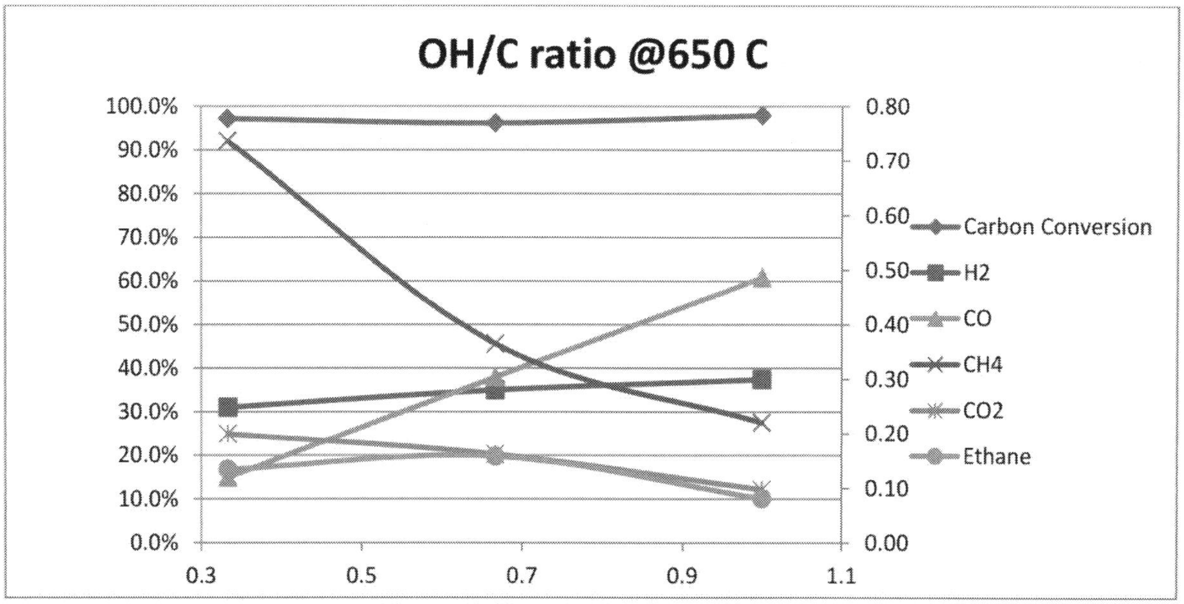

Figure 4 Effect of OH/C Ratio at 650C

Figure 4 displays the results of the experiments of all three fuels done at 650 °C. At these conditions there is almost complete carbon gasification for each fuel, meaning the only thing to change with hydroxyl content is gas composition. Also because almost all of the carbon is accounted for in the gas, there are no reactions producing a liquid carbon product such as acetone.

As with the other graphs, increasing oxygen content also shows an increase in carbon monoxide production. There is also a decrease in methane with increasing hydroxyl content in the fuel. It is also interesting to note the very high amount of methane produced by isopropanol in Figure 4. This suggests that when decomposing due to pyrolysis, the two end carbons of isopropanol break down into methane. Favoring methane production over ethane in pyrolysis could also explain why ethane slightly increases going from isopropanol to propylene glycol. The decrease in ethane going from propylene glycol to glycerin is due to the decreasing pyrolysis.

Figure 4 is also the only graph to have increasing hydrogen production with increasing the amount of hydroxyl groups. This could be due to an increase in the forward water-gas shift reaction, which also explains the decreasing carbon dioxide. Also the reactions producing liquid carbon and hydrogen in Figure 2 and Figure 3 are not present at 650°C, which could also explain why hydrogen trends differently.

Conclusion

The graphs above show that the carbon-to-oxygen ratio of the fuel undergoing supercritical water reformation has a large influence on the final product of that reformation. The most obvious influence is the increase of carbon monoxide with increasing hydroxyl groups in the fuel. The large amount of carbon monoxide means that glycerin decomposes in a way similar to methanol decomposition rather than reacting with the water molecules for supercritical water reformation, since that would result in more carbon dioxide. Isopropanol and propylene glycol may also undergo decomposition rather than reformation, but it is difficult to tell whether the lack of carbon monoxide is a result of decomposition, the fuel favoring pyrolysis over reformation, or a mixture of both.

The increasing the number of hydroxyl groups reduces the amount of pyrolysis going on, and also causes fewer reactions that produce hydrogen and liquid hydrocarbon products. More hydroxyl groups also appear to increase carbon conversion at lower temperatures. Increasing the amount of oxygen also may slightly increase the amount of water gas shift, though it is hard to tell at the lower temperatures with the changing carbon conversion.

In future work, liquid samples will be analyzed to detect the presence of acetone. Also repeats of experiments will be done to ensure repeatability. The addition of experiments using n-propanol will be conducted and compared to the isopropanol to see if the position of hydroxyl groups has an impact.

Acknowledgements

The work presented in this paper was partially funded by Korea Institute of Science and Technology (KIST)

References

(1) Picou, J.W.; Wenzel, J.; Stever, M.; Bouquet, J.; Lee, S. Synthesis gas production by noncatalytic reforming of glycerin in supercritical water: Effect of temperature. Presented at the AIChE Fall National Meeting, Nashville, TN, November 2009.
(2) Picou, J.W.; Putta, S.; Stever, M.; Lee, S. First-order Kinetic Study of the Reformation of Jet Fuel in Supercritical Water. Presented at the AIChE Fall National Meeting, Philadelphia, PA, November 2008.
(3) Boukis, N.; Diem V.; Habicht, W.; Dinjus, E. Methanol Reforming in Supercritical Water. *Ind. Eng. Chem. Res.* **2003,** *42,* 728-735
(4) Lee, S. *Alternative Fuels*, Taylor & Francis, Philadelphia, **1996**; pp.140-151.
(5) Zhang, J.H.; Zhou, X.L.; Wang, J.A. Water Promotion or Inhibition Effect on Isopropanol Decomposition Catalyzed with a sol-gel MgO-Al_2O_3 Catalyst. *Journal of Molecular Catalysts.* **2006**, 247, 222-226
(6) Glycerin Phase Change Properties. National Institute of Standards and Technology. http://webbook.nist.gov/cgi/cbook.cgi?ID=C56815&Units=SI&Mask=4#Thermo-Phase
(7) Water Phase Change Properties. National Institute of Standards and Technology. http://webbook.nist.gov/cgi/cbook.cgi?ID=C7732185&Units=SI&Mask=4#Thermo-Phase
(8) Isopropanol Phase Change Properties. National Institute of Standards and Technology. http://webbook.nist.gov/cgi/cbook.cgi?ID=C67630&Units=SI&Mask=4#Thermo-Phase
(9) Propylene Glycol Physical Properties. Chem Group. http://www.chem-group.com/services/pg.tpl
(10) *Inconel® Alloy 625*, Special Metals. http://www.specialmetals.com/documents/Inconel%20alloy%20625.pdf accessed 9/12/10

(684a) Tailoring the Morphology and Structure of Zeolite Catalysts Through the Use of Molecular Modifiers

Alexandra I. Lupulescu and Jeffrey D. Rimer, Chemical and Biomolecular Engineering, University of Houston, Houston, TX

Propertiessuch as exceptional thermal stability, unique shape-selectivity, and highacidity place zeolites among the most frequently used industrial catalysts. Theinability to *a priori* control singlecrystal growth, however, often yields materials with limited catalyticperformance due to long, tortuous internal diffusion pathlengths, as well as restrictedaccess to surface active sites. Rational design approaches capable ofselectively tailoring zeolite morphology and structure can dramatically improvecatalyst activity and lifetime [1,2]; and given the potentialapplication of zeolite catalysts for alternative fuels production (e.g.biofuels) and mobile emissions technologies, there exists a critical need to expandfundamental understanding of zeolite growth as well as robust design schemesfor optimizing zeolite catalysts. We will present a novel synthesis techniquethat draws inspiration from nature wherein molecular modifiers are employed to mediateanisotropic growth rates of zeolite crystals [3]. By selectivelytuning modifier-zeolite molecular recognition, we can achieve remarkablecontrol of crystal morphology, size, and surface architecture. This versatile,facile approach was used to synthesize several zeolite framework types. Resultsof these studies will be discussed, highlighting the influence of molecularmodifiers on zeolite aspect ratio and surface area. In addition, we willdiscuss interfacial studies that employ scanning probe microscopy to monitorreal time in situ growth of zeolites in actual reaction conditions. These studiesreveal interesting trends in step density, surface defects and growthmechanisms, and growth kinetics in the presence of molecular modifiers.

[1] Choi, M., Na,K., Kim, J., Sakamoto, Y., Terasaki O., Ryoo, R.,*Nature* 461 (2009)246-249

[2] Corma, A., *J. Catalysis* 216 (2003) 298-312

[3] Rimer, J.D.,An, Z., Zhu, Z., Lee, M.H., Goldfarb, D.S., Wesson, J.A., Ward, M.D., *Science*

330 (2010)337-341

See more of this Session: Catalysis with Microporous and Mesoporous Materials II
See more of this Group/Topical: Catalysis and Reaction Engineering Division

(684b) Synthesis and Characterization of Tin(IV) MFI: Sodium Inhibits the Synthesis of Phase Pure Materials

Nataly Garcia-Vargas[1], Scott Stevenson[2] and Daniel F. Shantz[1], (1)Artie McFerrin Department of Chemical Engineering, Texas A&M University, College Station, TX, (2)Sabic Technology Center, Sabic Americas, Inc., Sugar Land, TX

The synthesis and characterization of tin(IV) containing MFI zeolites is reported. Using a variety of characterization methods, it is shown that via conventional hydroxide-mediated preparations that include sodium hydroxide it is difficult to make phase-pure Sn-MFI materials with more than approximately 3 wt % tin in the framework. While at low tin contents and high pH values it is possible to make materials that appear more uniform, it is generally observed that the presence of sodium hydroxide in the synthesis has a deleterious effect on the quality of the material obtained. This is manifested in several ways including the presence of a residual surface phase observed via scanning electron microscopy, a discrepancy between the bulk and surface Si/Sn ratios as determined by X-ray fluorescence (XRF) and X-ray photoelectron spectroscopy (XPS) respectively, and a systematic decrease in the micropore volume as determined by nitrogen physisorption. Moreover, it is observed that the tin/sodium rich overlayer formed is recalcitrant towards methods to dissolve it, including ion-exchange processes in acidic or basic media. In contrast, Sn-MFI samples made in the absence of sodium hydroxide do not display these trends. Rather, they appear to be phase pure and more homogeneous based on the methods outlined above. The current work lays out a different approach to the synthesis of tin zeolites that appear to be highly uniform and should be of interest for the catalysis community. Preliminary catalytic studies of these materials will also be described.

See more of this Session: Catalysis with Microporous and Mesoporous Materials II
See more of this Group/Topical: Catalysis and Reaction Engineering Division

(684c) Delamination of Layered Zeolite Precursors Under Mild Conditions

Isao Ogino[1], Michael Nigra[1], Son-Jong Hwang[2], Einar Eilertsen[1], Jeong-Myeong Ha[1], Thomas Rea[3], Sheila Yeh[3], Stacey I. Zones[3] and Alexander Katz[1], (1)Department of Chemical and Biomolecular Engineering, University of California, Berkeley, Berkeley, CA, (2)Division of Chemistry and Chemical Engineering, California Institute of Technology, Pasadena, CA, (3)Chevron Energy Technology Company, Richmond, CA

Delaminated zeolite precursor materials expand the range of reactions that zeolites catalyze, by providing access for larger reactant molecules, as demonstrated by Corma and his coworkers with the discovery of materials such as ITQ-2, ITQ-6, and ITQ-18. Syntheses of these materials requires a high pH medium during precursor material swelling, typically in the pH range of 13.5–13.8, which is known to lead to partial amorphization of the zeolite layers during delamination. Here, we report a new class of delaminated zeolite catalysts consisting of materials that are synthesized by delaminating layered zeolite precursors at near-neutral pH conditions, using fluoride/chloride anions. We demonstrate that the new method efficiently delaminates layered zeolite precursors such as MCM-22 (P) with retention of high structural order of the zeolite framework in the delaminated layers. We provide evidence of delamination by powder X-ray diffraction (PXRD), transmission electron microscopy (TEM), and nitrogen gas adsorption characterization of the synthesized material, and compare with conventional delaminated materials. The new materials demonstrate high structural integrity of the exfoliated layers using ^{29}Si MAS NMR and FT-IR characterization. Furthermore, results from pyridine chemisorption on the new delaminated materials show features that suggest these materials to be active for acid catalysis applications.

See more of this Session: Catalysis with Microporous and Mesoporous Materials II
See more of this Group/Topical: Catalysis and Reaction Engineering Division

(684d) Synthesis and Characterization of Tungsten Containing Ultra Large Pore Mesoporous Silicate KIT-6

Anand Ramanathan, Center for Environmentally Beneficial Catalysis, The University of Kansas, Lawrence, KS, Bala Subramaniam, Department of Chemical and Petroleum Engineering, Center for Environmentally Beneficial Catalysis, University of Kansas, Lawrence, KS, Rajamanickam Maheswari, Department of Chemistry, Anna University, Chennai, India and Ulf Hanefeld, Department of Biotechnology, Faculty of Applied Sciences,, Delft University of Technology, Delft, Netherlands

Synthesis and characterization of tungsten containing ultra large pore mesoporous silicate KIT-6

Anand Ramanthan[a], Bala Subramaniam[a], Rajamanickam Maheswari[b] and Ulf Hanefeld[c]

[a]Center for Environmentally Beneficial Catalysis, The University of Kansas, Lawrence, KS 66047, USA

[b]Department of Chemistry, Anna University, Chennai – 600025, India

[c]Gebouw voor Scheikunde, Technische Universiteit Delft, Julianalaan 136, 2628 BL Delft, The Netherlands

Tungsten was incorporated into ultra large pore mesoporous silicate, KIT-6[1], via hydrothermal synthesis method using a Pluronic P123 triblock copolymer as the structure directing agent and n-butanol as additive. Four W-KIT-6 samples with Si/W ratio of 200, 100, 40 and 20 were prepared. Calcined W-KIT-6 samples were characterized by XRD, elemental analysis, N_2 sorption, HR-TEM, FTIR, DR-UV-Vis and NH_3-TPD. Cubic *Ia3d* structure was confirmed by presence of 211 reflections around 0.9° (2θ) in Low angle XRD. The incorporation of tungsten is inferred by a slight decrease in unit cell parameter with tungsten content [2,3]. The presence of crystalline WO_3 was observed in high angle XRD for samples with Si/W ratio £ 40. Elemental analysis measured by instrumental neutron activation analysis (INAA) revealed that most of the tungsten in the synthesis gel was retained in to the KIT-6 silica matrix. W-KIT-6 samples showed a very high specific surface area of 780 – 930 m^2/g with pore volume of 1.19 – 1.45 cm^3/g and possessed a narrow pore size distribution of 6.4 – 6.7 nm. The BET surface area and pore volume decreased with increase in tungsten content. The presence of well-ordered cubic 3D mesopores of W-KIT-6 samples was seen by High resolution TEM studies. The average thickness of the wall is found to be 4–5 nm and the pore diameter is around 6-7 nm. This is in agreement with the results of the N_2 adsorption study. The incorporation of tungsten was further evident by FTIR spectra which showed a band at 960 cm^{-1} [4] representative for Si-O-M linkages (M =

heteroatom). An intense band at 216 nm observed in Diffuse reflectance UV-Vis spectra confirm this findings, which is due to ligand to-metal charge transfer in isolated [WO_4] tetrahedral species [5] and is a direct proof for the framework incorporation of tungsten species in silica framework. Existence of partially polymerized W species in octahedral coordination with low nuclearity is also observed in these spectra. NH_3-TPD study revealed presence of low-medium acid strength sites in these samples.

References

[1] T.W. Kim, F. Kleitz, B. Paul and R. Ryoo, *J. Am. Chem. Soc.* 127 (2005) 7601.

[2] O. Klepel, W. Bohlman, E.B. Ivanov, V. Riede and H. Papp, *Micropor. Mesopor. Mater.* 76 (2004) 105.

[3] P. Trens, V. Stathopoulos, M. J. Hudson and P. Pomonis, *Appl. Catal. A* 263 (2004) 103.

[4] D. Zhao, A. Rodriguez, N.M. Dimitrijevic, T. Rajh and R.T. Koodali, *J. Phys. Chem. C* 114 (2010) 15728.

[5] M. S. Morey, G. D. Stucky, S. Schwarz and M. Froba, *J. Phys. Chem. B* 103 (1999) 2037 –2041.

See more of this Session: Catalysis with Microporous and Mesoporous Materials II
See more of this Group/Topical: Catalysis and Reaction Engineering Division

(684e) Synthesis of Oxide Nanocavities for Size-Selective Catalysis

Christian P. Canlas[1], Natalie Ray[2], Junling Lu[3], Sungsik Lee[4], Randall Winans[4], Jeffrey Elam[3], Peter C. Stair[2] and **Justin M. Notestein**[1], (1)Chemical and Biological Engineering, Northwestern University, Evanston, IL, (2)Chemistry, Northwestern University, Evanston, IL, (3)Energy Systems Division, Argonne National Laboratory, Argonne, IL, (4)X-ray Science Division, Advanced Photonic Source, Argonne National Laboratory, Argonne, IL

Enzymes have been the inspiration for catalyst design and areknown for their reactant and product selectivity. For reactions not able to becarried out in zeolites or other microporous solids, translating this size- orshape-selectivity onto heterogeneous catalysts is a challenge. In this study,we present first results on a route to incorporating size selectivity in anarbitrary oxide catalyst. This approach involves templated atomic layerdeposition (ALD) on the surface of an existing catalyst, where bulky moleculessuch as p-tertbutylcalix[4]arene or adamantanecarboxylic acid are immobilized onthe oxide catalyst surface and are used as templating agents. ALD's highlyconformal coverage and angstrom-level control over film thickness is ideal for thisapplication. After typically depositing less than 1.5 nm of an inert barrieroxide, the subsequent removal of the molecular template generates nanoscale cavitiesin the ALD layers, which we call nanocavities. Unlike previously reportedtemplated ALD where generated features are hundreds of nanometers to microns indimension, this templating process generates features that are 1-2 nm or less,the range where size- and shape- selective catalysis can be carried out. Thecatalysts are characterized throughout the synthesis process using DRUV-vis,TGA, and N_2 physisorption. The existence of the nanocavities is demonstratedby QCM studies during synthesis, SAXS and TEM. Size selectivity of thecatalysts is demonstrated through selective photooxidation of benzyl alcohol,1-hexanol and 2-adamantanol over what is normally a non-selective titania photocatalyst.Results showed that 2-adamantanol accesses relatively little of the titaniasurface when nanocavities deeper than ~0.5 nm are in place. In contrast, benzylalcohol and 1-hexanol are still relatively reactive with these nanocavities, allowingfor selective reaction of mixtures. This selective photocatalytic oxidation ofalcohols demonstrates how nanocavities can discriminate size and opens uppossibilities in designing multifunctional, all-oxide catalysts with propertiesthat begin to mimic enzymes, or bulk oxide catalysts with zeolite-likeselectivity.

See more of this Session: Catalysis with Microporous and Mesoporous Materials II
See more of this Group/Topical: Catalysis and Reaction Engineering Division

Synthesis and characterization of microporous carbon spheres based shape selective catalyst for liquid phase hydrogenation reactions

Maryam Peer [a*], *Ali Qajar*[a], *Ramakrishnan Rajagopalan* [b], *Henry C. Foley* [a]

[a] *Department of Chemical Engineering, The Pennsylvania State University, University Park, PA, 16802, USA*

[b] *The Materials Research Institute, The Pennsylvania State University, University Park, PA 16802, USA*

*mop5104@psu.edu

The goal in this study was to synthesize and use platinum embedded in microporous carbon spheres as catalyst in liquid phase hydrogenation reactions. These microporous carbon spheres benefit from high surface area and small particle size, which affects the diffusion length and mass transport of reactants to the active platinum sites. Using furfuryl alcohol as the monomer and a micellar polymerization process, carbon spheres with particle size of 400 nm were synthesized. Size of the spheres was controlled by changing the surfactant concentration in the solution. In addition, the synthesized carbon spheres were also used as catalyst supports and impregnated with platinum precursors to produce carbon supported platinum catalysts. The mass transport effects were studied in detail by performing liquid phase hydrogenation reactions using various alkenes on both types of catalysts. It was found that changing the porosity of the support has a significant effect on both activity and selectivity of the catalyst.

The synthesized catalysts were fully characterized using different characterization methods. Pore size distribution of the porous carbon spheres was measured using nitrogen adsorption in a Micromeritic ASAP 2020 instrument. Adsorption isotherm indicated that the carbon is completely microporous. X-ray Diffraction was used to determine the platinum particle size and Transmission Electron Microscopy applied to show how good is the dispersion of the platinum particles and also to confirm the platinum particle size.

All hydrogenation reactions for different alkenes were performed in a Parr low pressure reactor equipped with temperature and pressure control to set the temperature and pressure at the desired value. Reaction products were analyzed using a Varian GC and a capillary column. The reaction results for impregnated catalyst showed high activity comparable to commercial catalyst. The product analysis results in the case of embedded catalyst verified shape selectivity effect of the catalyst.

Catalytic hot-gas cleaning of biomass derived syngas by CaMgO nanocrystals

Ali Rownaghi, Ajay Kumar, Raymond Huhnke

Biosystems and Agricultural Engineering Department
Oklahoma State University, Stillwater, 74078, OK

ABSTRACT

Gasification of biomass is a promising source of renewable fuels and chemicals [1]. However, the removal of contaminants such as tars and other impurities from the syngas is required [2]. Tars can present a number of process challenges, including coking of catalysts and condensation on downstream piping, filters, and other equipment [3], and the cost of tar removal can be as much as the overall process cost to produce ethanol [4]. Moreover, too much tar in product gases will reduce the biomass utilization efficiency. Therefore, the reduction or decomposition of tar in biomass-derived synthesis gas is one of the biggest obstacles in its utilization for fuels, chemicals and power generation. The gas clean-up technology that offers many advantages is catalytic hot gas conditioning downstream of the gasifier [5, 6]. Dolomite with general chemical formula $CaMg(CO_3)_2$, is generally used as a highly efficient catalyst for removing tar from the product gases of gasifier. Because dolomite possesses a relatively low surface area, catalytic activity is greatly limited [6, 7]. From this point of view, the performance of dolomite can be enhanced by using nanostructure catalyst particles. CaMgO nanocrystals have been prepared with varius CaO/MgO ratios by the coprecipitation, hydrothermal and solvothermal methods [8, 9]. In order to investigate the effect of physico-chemical properties on the catalytic activity, catalysts were characterized using various techniques. Compared to dolomite, the CaMgO nanocrystals exhibited significantly higher surface area. The cracking reactions of toluene, a model tar compound, were carried out in fixed bed reactor under atmospheric pressure to evaluate the catalytic resistance to coke poisoning. In addition, the effects of catalysts particle size, reaction temperature, catalyst amount on the reaction rate have been investigated to evaluate the catalytic activity of all catalysts.

References:

1. D.L. Carpenter, R.L. Bain, R.E. Davis, A. Dutta, C.J. Feik, K.R. Gaston, W.Jablonski, S.D. Phillips, M.R. Nimlos, Ind. Eng. Chem. Res., 2010, 49, 1856-1871.
2. Z. A. El-Rub, E. A. Bramer, G. Brem, Ind. Eng. Chem. Res. 2004, 43, 6911-6919.
3. T. A. Milne, N. Abatzoglou, R. J. Evans, National Renewable Energy Laboratory (NREL) Technical Report, Golden, CO, 1998; Report NREL/TP 570-25357.
4. D. Dayton, National Renewable Energy Laboratory (NREL) Technical Report, Golden, CO, 2002; Report NREL/TP-510-32815.
5. M. M. Yung, W. S. Jablonski, K. A. Magrini-Bair, Energy & Fuels, 2009, 23, 1874-1887.
6. M. A. Gerber, PNNL-16950, Prepared for the U.S. Department of Energy DE-AC05-76RL01830.
7. R.N. Andre, F. Pinto, C. Franco and M. Dias, Fuel 84, 2005, 1635-1644.
8. A. Rownaghi, Y.H. Taufiq-Yap, F. Rezaei, Chemical Eng. Journal, 155, 2009, 514-522.
9. A. Rownaghi, Y.H. Taufiq-Yap, Ind. & Eng. Chem. Res., 49, 2010, 2135-2143.

(711b) Tuning Oxygen Basicity and Metal Reducibility In Bismuth Vanadate-Molybdate Catalysts

Andrew (Bean) Getsoian, Zheng Zhai and Alexis T. Bell, Department of Chemical and Biomolecular Engineering, University of California - Berkeley, Berkeley, CA

The industrial use of bismuth molybdate-based catalysts dates back to 1955, when Standard Oil of Ohio obtained a patent for conversion of propylene to acrolein and acrylonitrile over bismuth molybdates doped with phosphate – a technology that is still in use today. Yet even with the large body of research that has been done on bismuth molybdates in the subsequent decades, there is still not a full understanding of the structure of the active site(s), nor of the details of the chemical steps that link reactants to products on these catalysts [1]. Such understanding is even more lacking in the case of bismuth molybdate doped with vanadate, which shows promise for direct conversion of propane to acrolein [2] and which is also more selective for conversion of propylene to acrolein than undoped bismuth molybdate [3]. In order to better understand experimentally observed trends in selectivity vs. composition in these catalysts, our group has undertaken detailed physical and chemical characterization of a series of materials with composition $Bi_{1-x/3}V_{1-x}Mo_xO_4$, where the ratio x = Mo/(Mo+V) ranges from 0 to 1. Powder x-ray diffraction has been used to establish that a single-phase solid solution exists at all values of the ratio x. Catalytic studies have shown that optimal selectivity of propylene to acrolein occurs near x=0.2: the apparent activation energy for acrolein formation is lower over bismuth vanadate (14.9 kcal/mol) than over bismuth molybdate (19.5 kcal/mol), but greater vanadium content also facilitates further oxidation to CO_2. UV-Vis and x-ray absorption near edge (XANES) spectroscopies have shown that while only molybdenum centers are appreciably reduced under reaction conditions, molybdenum becomes more reducible with increasing vanadium content. Raman spectroscopy, x-ray photoelectron spectroscopy (XPS), and probe molecule adsorption studies have shown that the catalyst also becomes more Lewis basic with increasing vanadium content. Over bismuth molybdate, an initial hydrogen abstraction step is known to be rate limiting [4]. Since this step requires transfer of a proton to an oxygen and reduction of a molybdenum center, greater Lewis basicity and greater metal reducibility both facilitate the rate-limiting step, in agreement with the observed reduction in apparent activation energy in more basic, more reducible vanadium-doped materials. Developing a fuller understanding of the interplay between oxygen Lewis basicity, metal reducibility, and catalyst composition in mixed metal oxide systems like the one studied here will enable rational formulation of the more selective partial oxidation catalysts that greener, more efficient industry requires.

[1] Jung J.C.; Lee H.; Song I.K. *Catalysis Surveys from Asia* **2009**, *13*, 78 – 93.

[2] Yang H.P.; Fan Y.N.; Feng L.Y.; Qiu J.H.; Lin M.; Xu B.L.; Chen Y. *Acta Chimica Sinica* **2002**, *60*

(6), 1006 – 1010.

[3] Ueda W.; Asakawa K.; Chen C.L.; Moro-Oka Y.; Ikawa T. *Journa of Catalysis* **1986**, *101*, 360 – 368.

[4] Grasselli R.K.; Burrington J.D. *Industrial & Engineering and Chemistry Product Research and Development* **1984**, *23*, 393 – 404.

See more of this Session: Rational Catalyst Design I
See more of this Group/Topical: Catalysis and Reaction Engineering Division

(711c) Gold Catalysis Using Calixarene-Bound Nanoparticles

Michael M. Nigra, Jeong-Myeong Ha and Alexander Katz, Department of Chemical and Biomolecular Engineering, University of California Berkeley, Berkeley, CA

Gold catalysis represents a promisingand emerging area that has been used to enhance the selectivity and activity ofchemical reactions, which are often difficult or impossible to accomplish usingother means. Much of gold catalysis relies on active metal nanoparticles andhinges on the use of either inorganic and/ or organic modifiers. This study aims to use calixarene-basedmodifiers as ligands for gold and uses these ligands as tools to elucidate mechanisms of catalysis. Becauseof its size and steric bulk, the calixarene does notcompletely block reactants' access to the surface atoms. Instead we have shown it previously to synthesizepockets of accessibility on the metal surface, which allows reactants access. We have also developed methods to measure theamount of accessible surface area on calixarene-bound gold nanoparticles byfluorescence previously using 2-napthalenethiol (2-NT) binding.[1]

The catalytic performance of homogeneouscalixarene-modified Au nanoparticles was demonstrated using 4-nitrophenol(4-NP) hydrogenation with sodium borohydride as reductant. Synthesisis used as a tool to enable comparative studies employing different surface ligands on the gold nanoparticles. Au nanoparticles that have accessible surfacearea with bound calixarene molecules are bound with 2-NT to remove surface areathat would otherwise be accessible to reactants such as 4-NP. Even without this accessible surface area, inthe combined calixarene-2-NT bound particles, the hydrogenation rates aresimilar to the particles with only the calixarene bound. However,the organic ligand plays a distinct and unique role in controlling theinduction time before onset of catalytic activity. These results provide a unique perspectiveinto the identity of the active sites of these catalysts.

References

1. (a)Ha, J.-M., Solovyov, A., and Katz, A. *Langmuir* 25, 10548(2009). (b) de Sliva, N., Ha, J.-M., Solovyov, A., Nigra, M. M., Ogino, I., Yeh, S. W., Durkin K.A., and Katz, A. *Nat. Chem.* 2, 1062(2010).

See more of this Session: Rational Catalyst Design I
See more of this Group/Topical: Catalysis and Reaction Engineering Division

(711d) A Novel Approach to Designing Supported Metal Oxide Catalysts

Navaneetha Krishnan Nandakumar and Edmund G. Seebauer, Department of Chemical and Biomolecular Engineering, University of Illinois at Urbana-Champaign, Urbana, IL

Supported metal oxides are used as catalysts in a variety of applications including the partial oxidation of alcohols, selective catalytic reduction (SCR) of NO_x, oxidative dehyrogenation of alkanes and hydrodesulfurization. The role of the support in influencing the reactivity of the overlayer, beyond just the provision of mechanical stability and high surface area, has been recognized. The specific nature of the support and its properties, such as the electronegativity of the support cation, has been proposed to affect the catalytic activity of the overlayer. It has also been shown that modification of the support of the overlayer, by doping, alters the catalytic performance. Considering that several transition metal oxides that are typically used as catalysts, including V_2O_5 and TiO_2, are (wide band-gap) semiconductors, we present a novel approach to understanding the performance of and designing supported catalysts - from the perspective of semiconductor heterojunctions. This approach also allows us to assess the effect of doping the catalyst layers with altervalent ions on the reactivity of the supported catalyst. The reactivity of the surface is known to depend on the surface Fermi level position. The supported catalyst is considered as a heterostructure with a thin overlayer. Heterojunction physics is used to estimate the Fermi level position at the surface of a supported catalyst for various doping levels of the substrate and for various thicknesses of the overlayer. The Fermi level position is then linked to the reactivity through empirical correlations involving surface acidity values. V_2O_5/TiO_2 and partial oxidation of methanol to formaldehyde were chosen as the test catalytic system and test reaction respectively. Our calculations indicate that for the test system, higher n-type doping of the TiO_2 substrate and lower thickness of the V_2O_5 layer lead to higher methanol oxidation rates, as can be expected. Fitting of experimental data to the model gives the value of a parameter that is indicative of the ionic / covalent nature of the reaction.

See more of this Session: Rational Catalyst Design I
See more of this Group/Topical: Catalysis and Reaction Engineering Division

A Computational Approach for the Rational Design of Bimetallic Syn-Gas to Ethanol Catalysts

*Ming He, James McAliley and **David A. Bruce**, Chemical and Biomolecular Engineering, Clemson University, Clemson, SC*

Introduction

A major challenge associated with the synthesis of ethanol from syn-gas is an inability to find a low-cost catalyst that promotes the proper combination of CO dissociation and CO insertion steps, so as to yield ethanol as the primary reaction product and inhibit the formation of methane, methanol, longer chain alkanes, and other coking reaction products. Given the complexity of this reaction system, a simple trial-and-error approach to catalyst design is fraught with difficulties, which could severely limit efforts to identify an ideal catalyst material. Therefore, we used quantum mechanical simulations for the rational design of bimetallic catalysts that are optimally suited for the production of ethanol from syn-gas. Specifically, dispersion corrected Density Functional Theory (DFT) simulations and Bronsted-Evans-Polanyi (BEP) relations were used to map out the full reaction mechanism from syn-gas to ethanol for bimetallic clusters that range in size from 13 to 38 metal atoms. These efforts were accelerated by the identification of key reaction descriptors that enabled the elimination of bimetallic systems that were less likely to be effective catalysts for the desired reactions. For the more promising bimetallic catalysts, microkinetic models were built, considering all necessary adsorption and reaction steps as well as the diffusion of intermediate species between varying metal surface sites. These simulations indicate the nature and stability of the various bimetallic nanocatalysts and more importantly identify specific metal combinations that are ideally suited for ethanol production.

Result and Discussion

We identified 37 surface reactions in our proposed mechanism for ethanol formation from syn-gas, as shown in the figure below. Additional consideration was given to the different active sites available on the heterogeneous surface profiles of bimetallic clusters. With different possible reaction sites and the complex reaction mechanism from syn-gas to ethanol, there are usually more than 100 possible surface reaction steps to evaluate on bimetallic clusters. To solve this computationally intensive problem, the implementation of scaling methods is critical.

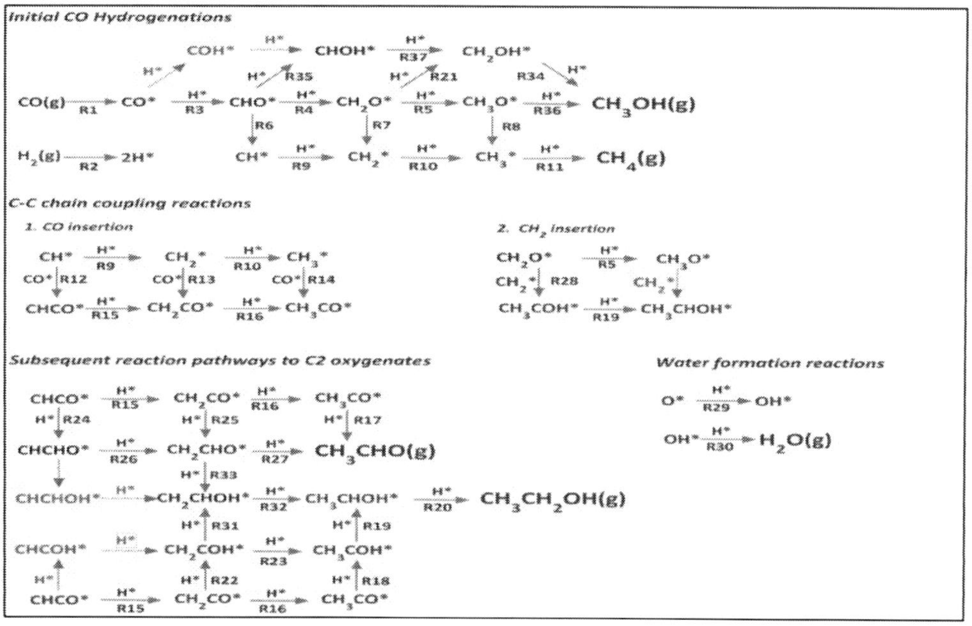

Figure 1. Full reaction mechanism of ethanol formation reacion from syn-gas

A widely used scaling method for estimating activation energies is the Bronsted-Evans-Polanyi (BEP) relation. The BEP relation linearly correlates the transition state energy of an elementary step to the reaction enthalpy of that step. It has been successfully applied to many catalytic systems for efficiently determining activation barriers.

Rational Design of catalysts

Bimetallic catalysts, in which one metal promotes hydrocarbon production and the other oxygenate production, may exhibit a synergistic effect that can facilitate the formation of ethanol. From literature, the catalytic behaviors of various transition metals are identified, as shown in Figure 2. Fe, Co and Ru are Fischer-Tropsch catalysts leading to the formation of long-chain hydrocarbons, while Ni catalysts are used in methane production. CO readily dissociates to form active carbon species on these catalysts. On the other hand, metals marked in blue (Cu, Pd, Ir and Pt) promote the formation of oxygenates, such as methanol. With these catalysts, CO participates in the reaction in its molecular-state.

The 16 different combinations of metals from these two categories are possible candidates to promote the proper combination of CO dissociation and CO insertion steps, so as to yield ethanol as the primary reaction product. Fundamental theoretical investigation is essential to shed light onto the complex reaction mechanism from syn-gas to ethanol for providing guidelines for experimental synthesis as well as computationally screening potential active combinations.

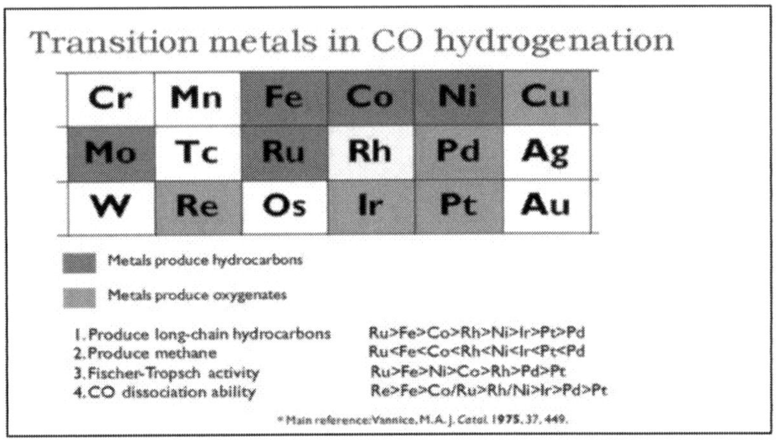

Figure 2. Transition metals in CO Hydrogenation

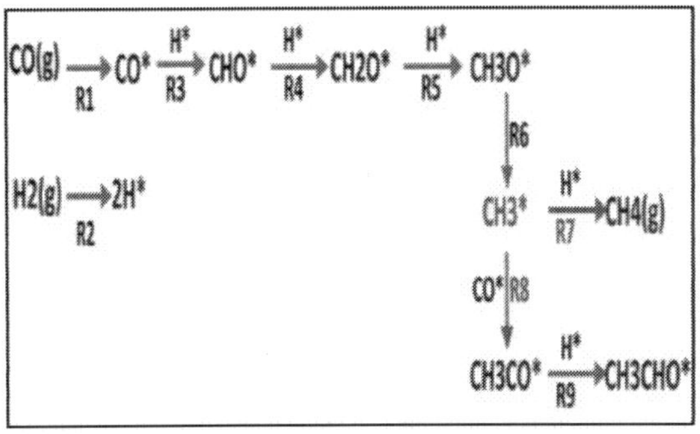

Figure 3. Reaction mechanism leading to C2 oxygenates

From our study of a Co_7Pd_6 icosahedra cluster, the reaction mechanism leading to C_2 oxygenates are deduced, as shown in Figure 3. In this pathway, the dissociation of hydrogenated CO species only happens for CH_3O^*, producing CH_3^* as the only CH_x^* species. A competition exists between the formation of methane and C_2 oxygenates, since CH_3^* serve as a common reactant. Thus, the selectivity toward C_2 oxygenates is affected greatly by the barrier to CH_4 formation via hydrogenation of CH_3^*. A similar reaction mechanism toward C_2 oxygenates has been proposed for a Rh(111) catalyst surface. It is speculated that similar mechanistic steps may be followed on other bimetallic clusters.

To this end, a possible energy descriptor for ethanol formation is the barrier differences of CH_3+H reaction and CH_3+CO reaction, $\Delta(E_{a,CH3\text{-}CO} - E_{a,CH3\text{-}H})$. A narrow gap is preferred for ethanol synthesis. This barrier difference can be effectively lowered due to synergetic effects in bimetallic catalysts. Combined with fast, open surface diffusion of intermediates from/to the binary mixed sites, this effect helps to produce ethanol.

We start the computational screen process by exam all possible mixing patterns for each 50/50 mixed bimetallic cluster to identify the most energetically favorable configuration (Figure 4). On the intermixed sites of these clusters, which are the most likely to promote the synergetic effects between two metals, reaction barriers of CO insertion and hydrogenation of CH3* are located with rigorous quantum mechanical methods to obtain $\Delta(E_{a,CH3\text{-}CO} - E_{a,CH3\text{-}H})$.

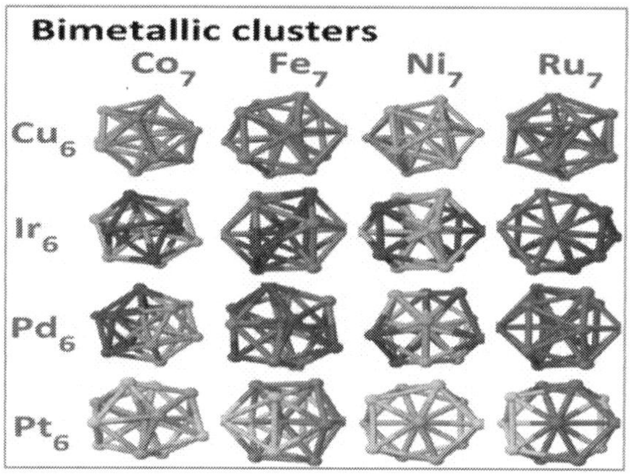

Figure 4. 16 13-atom bimetallic clusters

This reaction descriptor is predicated on two assumptions. First it assumes the mechanisms of ethanol formation reactions on all 13-atom bimetallic clusters are similar with the exact same key reaction steps (R7 and R8 in Figure 3). Also, since the barrier of methane production is used as a reference, it is assumed methane is a dominant product. For the promising candidates identified by this descriptor, the validity of these two assumptions needs to be tested. The following work is a rigorous study of the overall complex reaction mechanism (Figure 1) considering multiple possible reaction sites and surface diffusion of intermediates with the tools of microkinetic modeling. These simulations verify and improve the energy descriptor, which allows fast and efficient identification of specific metal combinations that are ideally suited for ethanol production.

References

1. Spivey, J. J.; Egbebi, A. *Chemical Society Reviews* 2007, *36*, 1514-1528.
2. Subramani, V.; Gangwal, S. K. *Energy & Fuels* **2008**, *22*, 814-839.
3. Cheng, J.; Hu, P.; Ellis, P.; French, S.; Kelly, G.; Lok, C., *Topics in Catalysis* **2010,** *53* (5), 326-337.
4. Fisher, I. A.; Woo, H. C.; Bell, A. T., *Catalysis Letters* **1997,** *44* (1), 11-17.

5. Fukuoka, A.; Kimura, T.; Kosugi, N.; Kuroda, H.; Minai, Y.; Sakai, Y.; Tominaga, T.; Ichikawa, M., *Journal of Catalysis* **1990,** *126* (2), 434-450.
6. Matsumura, Y.; Shen, W.-J.; Ichihashi, Y.; Okumura, M., *Journal of Catalysis* **2001,** *197* (2), 267-272.
7. Suh, Y.-W.; Moon, S.-H.; Rhee, H.-K., *Catalysis Today* **2000,** *63* (2-4), 447-452.
8. Bligaard, T.; Nørskov, J. K.; Dahl, S.; Matthiesen, J.; Christensen, C. H.; Sehested, J. *Journal of Catalysis* **2004**, *224*, 206-217.
9. Nørskov, J. K.; Bligaard, T.; Logadottir, A.; Bahn, S.; Hansen, L. B.; Bollinger, M.; Bengaard, H.; Hammer, B.; Sljivancanin, Z.; Mavrikakis, M.; Xu, Y.; Dahl, S.; Jacobsen, C. J. H. *Journal of Catalysis* **2002**, *209*, 275-278.
10. Liu, P.; Logadottir, A.; Nørskov, J. K. *Electrochimica Acta* **2003**, *48*, 3731-3742.
11. Choi, Y.; Liu, P., *Journal of the American Chemical Society* **2009,** *131* (36), 13054-13061.

(711f) Catalytic Conversion of Methane and Carbon Dioxide to Higher Value Products

Vesna Havran, Milorad P. Dudukovic and Cynthia S. Lo, Energy, Environmental and Chemical Engineering, Washington University in St. Louis, Saint Louis, MO

The possibilities for chemical utilization of methane and carbon dioxide to higher value products at mild conditions are investigated. Although the abundance of these two greenhouse gases makes them attractive raw materials for fuels and chemical synthesis, most of their reactions require significant energy inputs as well as properly designed catalytic systems that lower kinetic barriers in their direct conversion. Based on molecular modeling analysis, bifunctional system, comprised of well-shaped Pt nanoclusters deposited on ceria support is chosen as potential candidate for this study. The presence of low-coordinated sites, such as edge and corner sites on transition metal active centers has been shown to facilitate the methane conversion by reducing the energy barriers and stabilizing reaction intermediates. Tetrahedral colloidal platinum nanoparticles that contain numerous surface steps and kinks, are obtained by wet chemical preparation by carefully adjusting the concentration ratio between the metal precursor and the capping agent. Ceria nanorods which expose more reactive (110) and (100) planes are prepared by hydrothermal method. Influence of different cluster and support morphologies on direct conversion is investigated by reacting methane and carbon-dioxide at mild temperatures (< 350°C) in small lab scale packed-bed reactor. These studies should advance the fundamental understanding of the methane adsorption mechanisms on different cluster morphologies as catalytic active centers. It will also be studied whether improved oxygen storage capacity of ceria nanorods plays a role in CO_2 adsorption.

This integrated approach to the catalyst development, based on theoretical considerations and including catalyst synthesis and characterization, and finally, its testing in a lab scale reactor, should help in determining the key features needed for catalyst design.

See more of this Session: Rational Catalyst Design I
See more of this Group/Topical: Catalysis and Reaction Engineering Division

(712a) Thermochemistry, Reaction Paths and Kinetics On the Isooctane Radical Reactions with O2: Kinetic Study At High Pressures

Itsaso Auzmendi-Murua[1], Suarwee Snitsiriwa[2] and Joseph W. Bozzelli[2], (1)Chemical, Biological, and Pharmaceutical Engineering, New Jersey Institute of Technology, Newark, NJ, (2)Chemistry and Environmental Science, New Jersey Institute of Technology, Newark, NJ

The increasing regulations to control the emissions of pollutants from gasoline-powered vehicles has lead to the need of implementing improvements in the vehicles with the aim of optimizing the combustion, improving fuel economy and reducing emission of several pollutants. One of the innovations is the use of gasoline direct injection (DI) engines, where the fuel is compressed to higher pressure for coupled improvements in efficiency pollutant reduction. Moreover, if homogeneous charge DI engines are implemented, where high-pressure fuel injector sprays fuel directly into the combustion chamber early enough in the cycle to promote homogeneous fuel-air mixing, apart from higher fuel efficiencies, a reduction in pollutant emissions is achieved. Iso-octane, 2,2,4-trimethyl pentane holds an octane rating of 100 and is a major spark ignition engine fuel component; it is widely used as a model compound for detailed reaction mechanisms tha are applied to modeling ignition and combustion in spark and HCCI ignition internal engines. Isooctane is a highly branched molecule, where its thermochemistry and reaction kinetics are affected by its structure. In this study the thermochemical properties of isooctane and its radicals corresponding to loss of hydrogen atoms, along with the peroxy and hydroperoxide-alkyl radicals formed by $^3O_2$ reactions with isooctane radicals have been evaluated.

Calculations on transition state structures for the peroxy radical reactions have been performed using B3LYP and CBS-QB3 methods to identify differences from smaller chemical species model systems which are used for estimation of the higher molecular weight hydrocarbon fuels. The kinetics for reaction paths of the elementary reactions of isooctane radicals with O_2 have been constructed with use of chemical activation analysis for this association using quantum RRK analysis for k(E) and Master Equation analysis for fall-off. The isooctane radical systems have been evaluated over a pressure range up to 125 atm. in order to determine effects of pressure.

See more of this Session: Reaction Engineering for Combustion and Pyrolysis II
See more of this Group/Topical: Catalysis and Reaction Engineering Division

REACTION KINETICS FOR TMEDA AS AN ALTERNATIVE HYPERGOLIC ROCKET FUEL

Nicole J. Labbe, Chemical Engineering, University of Massachusetts Amherst, Amherst, MA
Phillip R. Westmoreland, Chemical and Biomolecular Engineering, North Carolina State University, Raleigh, NC

Introduction

Kinetics of the hypergolic rocket fuel TMEDA (tetramethylethanediamine) has been studied because it is less toxic and non-mutagenic by contrast to hydrazine and monomethyl hydrazine [1]. Hypergolic fuels (fuels that ignite spontaneously upon contact with an oxidizer) have been used in many propulsion and rocket applications. Historically, hydrazine rocket fuels such as monomethyl hydrazine and hydrazine have been used effectively as hypergolic rocket fuels. However, their negative feature is the carcinogenic, highly toxic nature of hydrazines.

Results and Discussion

To explore the combustion chemistry of TMEDA, a reaction set has been developed based on using an oxidizer of red fuming nitric acid (RFNA), a mixture of approximately 84% HNO_3, 13% N_2O_4, and 3% H_2O. As with the skeletal combustion mechanisms for traditional hydrocarbon fuels, it is proposed that TMEDA combustion is first dominated by hydrogen abstraction chemistry. The resulting TMEDA radicals then undergo a series of β-scissions and chemically activated reactions until the parent TMEDA molecule is broken down into various small C/H/N/O species. Due to the nature of the oxidizer, OH and NO_x chemistry had been expected to play a significant role.

Thermochemistry parameters were calculated for all the species in the proposed TMEDA skeletal mechanism using Gaussian 09 [2] at a CBS-QB3 [3] level of theory. Rate constants were estimated by direct analogy to appropriate hydrocarbon reaction rates or by transition-state optimizations combined with master equation theory. The assembled reaction set and thermochemistry database were used in adiabatic flame simulations using a modified CHEMKIN Premix [4-7] code. Flame conditions include pressures greater than or equal to atmospheric pressure and feed conditions ranging from slightly lean to slightly rich.

Acknowledgements

We gratefully acknowledge support of this research through the US Department of Defense under MURI contract W911NF-08-1-0171 (Ralph A. Anthenien Jr., project manager) and a National Defense Science and Engineering Graduate Fellowship (N. Labbe).

References

[1] G. Reddy, J. Song, M.S. Mecchi, M.S. Johnson, *Mutation Research* **700** (2010) 26-31.

[2] Gaussian 09, Revision A.1, M. J. Frisch, G. W. Trucks, H. B. Schlegel, G. E. Scuseria, M. A. Robb, J. R. Cheeseman, G. Scalmani, V. Barone, B. Mennucci, G. A. Petersson, H. Nakatsuji, M. Caricato, X. Li, H. P. Hratchian, A. F. Izmaylov, J. Bloino, G. Zheng, J. L. Sonnenberg, M. Hada, M. Ehara, K. Toyota, R. Fukuda, J. Hasegawa, M. Ishida, T. Nakajima, Y. Honda, O. Kitao, H. Nakai, T. Vreven, J. A. Montgomery, Jr., J. E. Peralta, F. Ogliaro, M. Bearpark, J. J. Heyd, E. Brothers, K. N.

Kudin, V. N. Staroverov, R. Kobayashi, J. Normand, K. Raghavachari, A. Rendell, J. C. Burant, S. S. Iyengar, J. Tomasi, M. Cossi, N. Rega, J. M. Millam, M. Klene, J. E. Knox, J. B. Cross, V. Bakken, C. Adamo, J. Jaramillo, R. Gomperts, R. E. Stratmann, O. Yazyev, A. J. Austin, R. Cammi, C. Pomelli, J. W. Ochterski, R. L. Martin, K. Morokuma, V. G. Zakrzewski, G. A. Voth, P. Salvador, J. J. Dannenberg, S. Dapprich, A. D. Daniels, Ö. Farkas, J. B. Foresman, J. V. Ortiz, J. Cioslowski, and D. J. Fox, Gaussian, Inc., Wallingford CT, 2009.

[3] J. A. Montgomery, Jr., M. J. Frisch, J. W. Ochterski, G. A. Petersson, *J. Chem. Phys.* **110** (1999) 2822-2827.

[4] R. J. Kee, J. A. Miller, T. H. Jefferson, CHEMKIN II version 2.5: *A General-Purpose, Problem-Independent, Transportable, Fortran Chemical Kinetics Code Package*, SAND80-8003, Sandia National Laboratories, Livermore, California, 1980. Modification of CHEMKIN version 3.9, 1992.

[5] R. J. Kee, G. Dixon-Lewis, J. Warnatz, M. E. Coltrin, J. A. Miller, A Fortran Computer Package for the Evaluation of Gas-Phase, Multicomponent Transport Properties, SAND87-8246, Sandia National Laboratories, Livermore, California, 1986. Modification of TRANLIB version 1.6, 1990.

[6] R. J. Kee, J. F. Grcar, M. D. Smooke, J. A. Miller, *A Fortran Program for Modeling Steady Laminar One-Dimensional Premixed Flames*, SAND85-8240, Sandia National Laboratories, Livermore, California, 1985. Modification of PREMIX version 2.5, 1991.

[7] R.J. Kee, F.M. Rupley, J.A. Miller, *The Chemkin Thermodynamic Database*, SAND87-8215, Sandia National Laboratories, Livermore, California, 1987.

Contact information:

Nicole Labbe (speaker)
University of Massachusetts Amherst
Department of Chemical Engineering
686 N. Pleasant St., 159 Goessmann Laboratory
Amherst, MA 01003-3110
Phone: 413-545-6078
Fax: 413-545-1647
Email: labbe@ecs.umass.edu

Prof. Phillip R. Westmoreland (PI)
Executive Director, NCSU Institute for Computational Science and Engineering
Professor, Department of Chemical and Biomolecular Engineering
North Carolina State University
2036 Engineering Building I, Box 7905
911 Partners Way; Raleigh, NC 27695
Phone: (919) 515-7121; Fax: (919) 515-3465
Email: phil.westmoreland@ncsu.com

(712c) Theoretical Studies of HO2 Radical Reactions of Relevance to Ethanol Combustion

Raghu Sivaramakrishnan, Chemical Sciences & Engineering Division, Argonne National Laboratory, Argonne, IL

A detailed chemical kinetic mechanism has been assembled for the combustion of ethanol. While there are no surprises in the mechanistic initiation and propagation routes, the uncertainties in the rate coefficients and branching ratios for many of these primary routes have inhibited an accurate description of the high temperature combustion of ethanol. One specific class of propagation reactions that is particularly important at high pressures are reactions of HO_2 with the fuel and intermediates, in the present case ethanol and acetaldehyde. These are reactions that cannot be experimentally isolated, particularly at combustion temperatures (>700 K), and consequently we have initiated theoretical studies on these reactions.

Ab-initio electronic structure theory has been used to characterize the energetics for the various abstraction channels in Ethanol + HO_2 and Acetaldehyde + HO_2. The energetics for the saddle points were characterized at the CCSD(T)/cc-pV∞Z level of theory using B3LYP/6-311++G(d,p) geometries. Rate coefficients for the various abstraction channels were estimated using Transition State Theory. These results and their implications for ethanol combustion modeling will be discussed.

This work was supported by the U. S. Department of Energy, Office of Basic Energy Sciences, Division of Chemical Sciences, Geosciences, and Biosciences under contract no. DE-AC02-06CH11357 as part of the Argonne-Sandia Consortium on High-Pressure Combustion Chemistry, FWP# 2009 ANL 59044.

See more of this Session: Reaction Engineering for Combustion and Pyrolysis II
See more of this Group/Topical: Catalysis and Reaction Engineering Division

(712d) Themogravimetric Analysis of Press Mud for Energy Generation

I.M. Mishra, Department of Chemical Engineering, Indian institute of Technology Roorkee, Roorkee, India

Press mud is a solid waste obtained from the cane sugar mills in India. It has the potential for energy generation. The paper presents physico-chemical properties of press mud and its thermal degradation in nitrogen atmosphere. The thermal degradation was carried out in a thermogravimetric analyzer from room temperature to 900 °C at heating rates of 20 and 40 °C/min. The thermogravimetric, derivative thermogravimetric and differential thermal analyses were carried out to determine the rate of volatiles evolution, the effect of heating rates on the thermal degradation characteristics and to determine the kinetics of thermal degradation. The thermal degradation was found to occur in several distinct phases: each phase giving volatile evolution in an independent parallel lump. Each decomposition phase was modeled by a single irreversible reaction with respect to the solid mass. The kinetic analysis has also been done for the entire decomposition process, as if occurring in one single step.

See more of this Session: Reaction Engineering for Combustion and Pyrolysis II
See more of this Group/Topical: Catalysis and Reaction Engineering Division

ADVANCES IN THE SIMULATION OF DETAILED CHEMISTRY IN REACTING FLOWS

Geoffrey M. Oxberry
Yu Shi
Raymond L. Speth
Paul I. Barton
William H. Green
Massachusetts Institute of Technology, Cambridge, MA, USA

Introduction

Along with theory and experiment, simulation of reacting flows has played an important role in advancing combustion science. It has proven especially helpful in explaining experimentally observed behavior in homogeneous charge compression ignitions (HCCI) engines, developing better explanations of unburned hydrocarbons in spark ignition (SI) engines [1], and determining the main cause of stabilization in a jet-lifted flame, among other applications [2]. However, existing algorithms and computational resources for reacting flow simulations are limited in the number of species that can be tracked. As a result, it is difficult to simulate reacting flows with detailed chemistry in all but the simplest systems. Furthermore, it is known from experience that simplified chemistry can lead to inaccurate physical results in some situations, such as the failure to predict negative temperature coefficient (NTC) behavior in the ignition delay of hydrocarbons when low-temperature chemistry is omitted or oversimplified. Such simplified chemistry, sometimes simplified without error control, also can fail to yield quantitative predictions of practical systems, which would be of use to experimentalists and industrial designers alike [3, 4]. In order to continue to advance combustion science through simulation, methods must be developed to enable simulations with higher fidelity chemical models that have quantifiable errors. Here, three approaches to tackling these issues are presented: restructuring algorithms to take advantage of graphics processing units (GPUs) through task parallelism, error-controlled model reduction methods that relate the extent of reduction to metrics similar to local truncation errors in ODE solvers, and error quantification of reduced models using a combination of *a priori* bounds and *a posteriori* estimates.

Parallelization and graphics processing units

One promising avenue for speeding up combustion simulations is the use of parallelization in heterogeneous (CPU + GPU) computing environments [2, 5, 6]. Due to thermal and power limitations, CPU chip manufacturers can no longer increase the clock speed of CPUs, and are instead placing more CPU cores on chips [7, 8]. Researchers *must* parallelize their codes in order to take advantage of these extra cores for performance gains [9]. Already, many reacting flow simulations use CPU-only parallel computing to carry out large-scale computations [1]. The best illustrative algorithmic example of this parallelization is an operator splitting solver [10--12]. By decomposing a single simulation time step into separate transport and chemistry time steps, the chemistry at each individual grid point can be evaluated and integrated independently. The idea of distributing many independent tasks among multiple processors is called *task parallelism* [13]; evaluating the chemistry at each grid point on different processors is an example of this concept. By identifying additional task parallel portions of reacting flow simulations, it is possible to extract further speed gains from multicore CPUs that are already out on the market. GPUs present an additional opportunity for researchers because they have many more cores (often a hundred or more) than a state-of-the-art CPU (around 4-8 cores), and have a much lower cost per core than CPUs (see, for instance, the cost of Intel processors [14] and NVIDIA GPUs [15]). Provided that researchers can take advantage of the specialized hardware of GPUs, it is possible to reap significant performance improvements at modest costs [16].

Two opportunities for task parallelism in combustion codes are the evaluation of reaction rates and the integration of the chemistry source term. These tasks are particularly important to parallelize because they account for the bulk of computational time in reacting flow simulations [17]. Reaction rate calculations were parallelized by rewriting the CHEMKIN-II [18] Fortran 77 [19] subroutine CKWYP in C [20] and interfacing it with NVIDIA's CUDA libraries [16], also in C, so that the rate of each reaction can be calculated on independent cores of an NVIDIA GPU. Integration of the chemistry source term was parallelized by writing CUDA interfaces to the portions of both DVODE [21] and DASAC [22] that use LU decomposition [23, 24] (since LU decomposition, and more generally, linear equation solves, are task parallel and compute-intensive [25, 26]). LU decomposition is performed using the CULA library [27] (LAPACK [28] ported to CUDA); MAGMA [29] is a recent free alternative that can also be used. All calculations were carried out in double precision. For reaction mechanisms the size of n-pentane [30] (385 species, 1896 reactions) or larger, use of the GPU for reaction rate evaluation and LU decomposition decreases the wall clock time of an isochoric simulation in SENKIN [31] (relative to a CPU-only calculation). For mechanisms with 2000 species or more, GPU parallelization decreases the wall clock time by a factor of 6-10 and decreases the cost of LU factorization by a factor of 20. However, since there is some overhead when using a GPU because data must be transferred between the CPU and GPU [16], the additional costs of data transfers exceed the savings from GPU parallelization for smaller mechanisms such as GRI-Mech 3.0 [32] (53 species, 325 reactions) [33].

In multi-dimensional reacting flow simulations, these smaller mechanisms (less than 100 species, less than 500 reactions) are of greater interest due to the prohibitive computational expense of simulating larger mechanisms. To better take advantage of the GPU for smaller mechanisms, a hybrid BDF/exponential ODE solver was developed for using in simulations of HCCI engines using the internal combustion engine CFD simulator KIVA-CHEMKIN [34--37] (for references on BDF methods, see [38, 39], for references on exponential methods, see [40]). The main idea behind the solver is to partition the grid points in an HCCI simulation at a given time step into two sets: grid points where the conditions the chemistry source term is stiff, and conditions where the chemistry source term is non-stiff. Stiffness of the chemistry source term is assessed using the initial step size estimate of the exponential integrator. If the estimated initial step size is below a threshold value, the chemistry source term at that grid point is considered stiff; otherwise, it is considered non-stiff. Once the set of grid points is partitioned into stiff and non-stiff sets, the BDF ODE solver DVODE is used on a CPU to integrate the chemistry of the stiff grid points, and the exponential integrator (called a QSS method by the authors) CHEMEQ2 [41] is used on a GPU to integrate the chemistry of the non-stiff grid points in parallel. The threshold value for the stiff/non-stiff cutoff is determined by a load-balancing scheme [42] that allocates grid points among the CPU and GPU by assuming that the rates at which grid points are solved on each processing unit at the current time step are the same as the rates at which grid points were solved on each processing unit during the previous time step. These rates are used to estimate the number of grid points that should be allocated to each processing unit so that they spend roughly equal amounts of time on integrating the current time step. Based on these numbers of grid points, the stiff/non-stiff time step estimate cutoff is adjusted accordingly. By applying this load-balancing scheme, the amount of time either processing unit spends idle is reduced compared to a case without load-balancing. In the ideal case, the idle time would be minimized. Applying the hybrid BDF/exponential solver to HCCI case studies in KIVA-CHEMKIN with two smaller mechanisms (39 species, 131 reactions [43] and 117 species, 499 reactions [44], respectively) yields reductions in simulation wall clock time by a factor of 10-41, demonstrating the efficacy of the approach for HCCI simulations.

Error-controlled model reduction methods

Another avenue for speeding up combustion simulations is the use of error-controlled model reduction methods. Here, *error-controlled* refers to methods that set bounds on the local truncation error due to model reduction (for ODE or DAE integration [39, 45, 46]) or bounds on the residual error (for nonlinear equation solvers [47--49]) due to model reduction. By controlling model reduction error in this manner, it is possible to make meaningful comparisons in error tolerances among model reduction methods and numerical solution methods because the two sets of error tolerances measure quantities with similar interpretations. Previous work in error-controlled model reduction includes computational singular perturbation (CSP) [50, 51], reaction elimination [52, 53], and simultaneous reaction and species elimination [54], among others. Of particular note is work on reaction elimination over ranges of reaction conditions [53], which has enabled users to bound explicitly the approximation error incurred by using a model reduced by this method in a steady-state reacting flow simulation.

However, in order for these bounds to be useful, the reduced model calculated by reaction elimination over ranges must have a steady-state solution within the range of conditions used to reduce the model, and an initial guess within the region of convergence containing the correct root must be supplied to the steady state nonlinear solver, both of which are nontrivial conditions. To compensate for these limitations, the exact steady-state adaptive chemistry (ESAC) method has been developed. In this method, the steady-state problem is solved repeatedly using an adaptive chemistry algorithm [55] coupled to an error-controlled model reduction method. The first reduced models used in a steady-state solve iteration are calculated using large model reduction error tolerances, and error tolerances for the model reduction method are successively decreased, using the solution of the previous iteration as the initial guess for the current iteration. Once a steady-state solve iteration has been reached where the model reduction error tolerances are below a threshold value, the original, full chemistry model is used in the steady-state solve to obtain the desired steady-state solution. When ESAC is applied to KIVA-CHEMKIN simulations of methane (GRI-Mech 3.0) and ethylene [56--58] (75 species, 529 reactions) 2-D laminar steady-state flames, the simulation wall clock time decreased by a factor of 3.5 to 4 with no loss in solution accuracy.

In addition to modifying the chemistry source term in reacting flow simulations, species transport terms may also be modified in an error-controlled fashion. Similar to previous work on reaction elimination and simultaneous reaction and species elimination, an optimization problem was formulated to calculate reduced models by eliminating as many transport terms as possible, subject to tolerances on the model reduction error in the resulting reduced transport term, evaluated at user-selected reaction conditions. Since each grid point and each species transport term is independent (to first order), the optimization problem reduces to evaluating simple inequalities and executes quickly. When coupled to an adaptive transport algorithm [55] and applied to a 1-D strained flame ignition calculation using the UCSD methane reaction mechanism [59, 60] (40 species, 180 reactions), the simulation showed very slight (roughly 0.5%) performance improvements, since the bulk of the simulation time (roughly 90%) is spent on evaluating and integrating the chemistry source term. However, for problems in which transport contributes a larger fraction of the total simulation time, the method is expected to yield greater performance gains.

GPU parallelization and model reduction are two independent approaches that each speed up combustion simulations. The two approaches can be combined to achieve increased performance gains. A multi-approach algorithm combining on-the-fly model reduction using directed relation graphs with error propagation (DRGEP) [61], GPU parallelization methods (mentioned above) [33], and an adaptive multi-grid chemistry (AMC) method [42] was used to enable the efficient simulation of multi-dimensional HCCI and premixed charge compression ignition (PCCI) engines with highly detailed chemistry. The resulting algorithm simulates a single-zone HCCI engine using a detailed methyl decanoate reaction mechanism [62, 63] (2877 species, 8555 reactions) in KIVA-CHEMKIN with the DVODE solver 220-300 times faster than the original KIVA-CHEMKIN simulation without any significant loss in accuracy. The dramatic performance increase achieved by this multi-approach algorithm enables a researcher to carry out a 2-D HCCI simulation using the same mechanism in KIVA-CHEMKIN/DVODE in only 13 hours on a PC with an NVIDIA Tesla C2050 GPU, which is less time than it would take to solve the corresponding 0-D model using the existing full chemistry CPU-only approach.

Bounding and estimation of reduced model approximation errors

Although model reduction is a popular approach to decreasing the computational cost of reacting flow simulations, it also raises concerns about the accuracy of results obtained using this approach. For instance, it is known that many model reduction methods enforce some means of error control using parameter that, when decreased to zero, recover the original, full model and thus decrease the model reduction approximation error to zero. Until recently, there was no way to bound the error due to model reduction *a priori* to obtain some idea of the error in simulation results unless an orthogonal projection-based model reduction method (such as POD) was used in a time-varying simulation [64], or reaction elimination over ranges was used in a steady-state simulation [53]. (For a definition and examples of projection-based model reduction methods in reacting flow simulations, see [65].) These *a priori* bounding results were extended to apply to oblique projection-based model reduction methods (such as CSP and the linearized quasi-steady state approximation, LQSSA [66]) as well as non-projection-based methods (such as reaction elimination), using results similar to those used to prove a priori bounds on errors in the numerical solution of ODEs [67--69]. Much like those *a priori* bounds on numerical errors, the *a priori* error bounding results for reduced model ODEs often drastically overestimate the error due to model reduction because such bounds must account for the error at any physical or non-physical set of reaction conditions.

Despite yielding overly pessimistic error bounds, the *a priori* error bounding results both give insight into factors affecting the error due to model reduction and inspire *a posteriori* error estimation methods similar to previous work on *a posteriori* error estimation for orthogonal projection-based model reduction [70], *a posteriori* estimation of error in reduced models under perturbations in parameters [71], *a posteriori* estimation of error in operator decomposition methods for both ODEs [72, 73] and PDEs [74, 75], *a posteriori* error estimation for numerical solutions of reaction-diffusion equations [76], sensitivity analysis [22, 77--81], and global error control of numerical solutions of ODEs [82--85]. This substantial corpus of work can be synthesized to calculate *a posteriori* error estimates for reduced model ODEs obtained via projection-based or non-projection-based model reduction methods using a number of algorithms that bear strong resemblance to familiar algorithms for sensitivity analysis used in chemical kinetics. These *a posteriori* error estimation algorithms require the model reduction error to be sufficiently small; put another way, the reduced model must be sufficiently accurate for the error estimates to be meaningful. (One way to ensure sufficiently accurate reduced models is to use an error-controlled model reduction technique such as the ones described above.) Provided that the reduced model is sufficiently accurate, error estimates obtained using these algorithms on a number of test problems are accurate to within an order of magnitude, and in some cases, can even achieve quantitative accuracy.

Conclusions

The methods above demonstrate the potential for further improvements in the state-of-the-art of simulations of reacting flows. Leveraging GPU hardware, error-controlled model reduction algorithms, and quantification of reduced order model errors enables the use of higher fidelity chemical models on today's computing resources with known approximation errors. By carrying out reacting flow simulations with more detailed chemistry and known errors, it brings the state of combustion science closer to the goal of quantitative prediction of outcomes in practical systems, helping experimentalists, theoreticians, and designers better understand combustion phenomena to build better engines and combustors to address the environmental and energy problems facing society in the future.

References

[1] C. K. Westbrook, Y. Mizobuchi, T. J. Poinsot, *et al.* Computational combustion. *Proceedings of the Combustion Institute*, 30(1):125--157, 2005. URL http://linkinghub.elsevier.com/retrieve/pii/S0082078404003340.

[2] J. H. Chen. Petascale direct numerical simulation of turbulent combustion---fundamental insights towards predictive models. *Proceedings of the Combustion Institute*, 33(1):99--123, 2011. URL http://linkinghub.elsevier.com/retrieve/pii/S1540748910003962.

[3] C. K. Law. Combustion at a crossroads: Status and prospects. *Proceedings of the Combustion Institute*, 31(1):1--29, 2007. URL http://www.sciencedirect.com/science/article/B7GWS-4M21T9Y-1/2/b30939e7071095b42cc1cbdc289bb741http://www.sciencedirect.com/science?_ob=ArticleURL&_udi=B7GWS-4M21T9Y-1&_user=501045&_rdoc=1&_fmt=&_orig=search&_sort=d&_docanchor=&view=c&_acct=C000022659&_version=1&_urlVersion=0&_userid=501045&md5=b7e481433bfad34948a41d368a26e83dhttp://www.sciencedirect.com/science?_ob=MImg&_imagekey=B7GWS-4M21T9Y-1-2W&_cdi=20466&_user=501045&_orig=search&_coverDate=01%2F31%2F2007&_sk=999689998&view=c&wchp=dGLbVl.

[4] T. Lu and C. K. Law. Toward accommodating realistic fuel chemistry in large-scale computations. *Progress in Energy and Combustion Science*, 35(2):192--215, 2009. URL http://linkinghub.elsevier.com/retrieve/pii/S036012850800066X.

[5] The advantages of first-generation heterogeneous computing on the Cray XT5h. *SciDAC Review*, Summer 2008:42--49, 2008.

[6] K. Spafford, J. Meredith, J. Vetter, *et al.* Accelerating S3D: A GPGPU case study. In *HeteroPar 2009*. 2009. URL http://books.google.com/books?hl=en&lr=&id=-U8ORGDF0OgC&oi=fnd&pg=PA122&dq=Accelerating+S3D:+A+GPGPU+Case+Study&ots=M2hfwxlWVL&sig=kvGeGut7OxDPVL1kPIMYtUW0k10.

[7] W. Knight. Two heads are better than one. *IEE Review*, September, 2005. URL http://books.google.com/books?hl=en&lr=&id=KzXC_qfemngC&oi=fnd&pg=PP5&dq=Two+heads+are+better+than+one&ots=1DbOyDgWPH&sig=DaDlEBq-KXCtuPnLsp6CGZCo2qY.

[8] J. Shalf. The new landscape of parallel computer architecture. *Journal of Physics: Conference Series*, 78:012066, 2007. URL http://stacks.iop.org/1742-6596/78/i=1/a=012066?key=crossref.b8e0329a5a0ac1eef3667fab1253ce6c.

[9] K. Asanovic, J. Wawrzynek, D. Wessel, *et al.* A view of the parallel computing landscape. *Communications of the ACM*, 52(10):56--67, 2009. URL http://portal.acm.org/citation.cfm?doid=1562764.1562783.

[10] S. K. Godunov. A difference method for numerical calculation of discontinuous solutions of the equations of hydrodynamics. *Matematicheskii Sbornik*, 47(89):271--306, 1959. URL http://www.mathnet.ru/php/archive.phtml?wshow=paper&jrnid=sm&paperid=4873&option_lang=eng.

[11] G. Strang. On the construction and comparison of difference schemes. *SIAM Journal on Numerical Analysis*, 5(3):506--517, 1968. URL http://www.jstor.org/stable/2949700.

[12] D. A. Schwer, P. Lu, W. H. Green, *et al.* A consistent-splitting approach to computing stiff steady-state reacting flows with adaptive chemistry. *Combustion Theory and Modelling*, 7:383--399, 2003. URL stacks.iop.org/CTM/7/383.

[13] M. J. Quinn. *Parallel programming in C with MPI and OpenMP*. McGraw-Hill Education, 2008.

[14] Intel. Intel server and workstation chipsets. URL http://www.intel.com/content/www/us/en/chipsets/server-chipsets/server-workstation-chipsets.html.

[15] NVIDIA. NVIDIA products. URL http://www.nvidia.com/page/products.html.

[16] NVIDIA CUDA C programming guide version 4.0. Technical report, 2011. URL http://developer.download.nvidia.com/compute/DevZone/docs/html/C/doc/CUDA_C_Programming_Guide.pdf.

[17] S. R. Tonse, N. W. Moriarty, N. J. Brown, *et al.* PRISM: Piecewise reusable implementation of solution mapping. An economical strategy for chemical kinetics. *Israel Journal of Chemistry*, 39:97--106, 1999.

[18] R. J. Kee, F. M. Rupley, and J. A. Miller. Chemkin-II: A Fortran chemical kinetics package for the analysis of gas-phase chemical kinetics. Technical Report SAND89-8009, Sandia National Laboratory, 1989.

[19] Fortran 77 Standard. Technical Report X3J3/90.4, American National Standards Institute, 1978.

[20] B. Kernighan and D. Ritchie. *The C programming language*. Prentice Hall, 2nd edition, 1988.

[21] P. N. Brown, G. D. Byrne, and A. C. Hindmarsh. VODE: A variable-coefficient ODE solver. *SIAM Journal on Scientific and Statistical Computing*, 10(5):1038--1051, 1989. URL http://link.aip.org/link/SJOCE3/v10/i5/p1038/s1&Agg=doi.

[22] M. Caracotsios and W. E. Stewart. Sensitivity analysis of initial value problems with mixed ODEs and algebraic equations. *Computers & Chemical Engineering*, 9(4):359--365, 1985. URL http://www.sciencedirect.com/science/article/pii/0098135485850146.

[23] L. N. Trefethen and D. Bau. *Numerical linear algebra.* Society for Industrial and Applied Mathematics, 1997.

[24] G. Strang. *Introduction to linear algebra.* Wellesley Cambridge Press, 4th edition, 2009.

[25] J. Dongarra, I. Duff, D. C. Sorensen, *et al. Numerical linear algebra on high-performance environments and tools.* Society for Industrial Mathematics, 2nd edition, 1998.

[26] J. W. Demmel, M. T. Heath, and H. A. van der Vorst. Parallel numerical linear algebra. *Acta Numerica,* 2:111, 2008. URL http://www.journals.cambridge.org/abstract_S096249290000235X.

[27] CULA. URL http://www.culatools.com/.

[28] E. Anderson, Z. Bai, C. Bischof, *et al. LAPACK users' guide.* Society for Industrial and Applied Mathematics, 3rd edition, 1999.

[29] M. Baboulin, J. Demmel, T. Dong, *et al.* MAGMA. 2011. URL http://icl.cs.utk.edu/magma/software/index.html.

[30] H. J. Curran, W. J. Pitz, C. K. Westbrook, *et al.* Oxidation of automotive primary reference fuels at elevated pressures. *Symposium (International) on Combustion,* 27(1):379--387, 1998. URL http://linkinghub.elsevier.com/retrieve/pii/S0082078498804268.

[31] A. E. Lutz, R. J. Kee, and J. A. Miller. SENKIN: A Fortran program for predicting homogeneous gas phase chemical kinetics with sensitivity analysis. Technical Report SAND-87-8248, Sandia National Laboratory, 1988.

[32] G. P. Smith, D. M. Golden, M. Frenklach, *et al.* GRIMech-3.0. 1999. URL http://www.me.berkeley.edu/gri_mech/.

[33] Y. Shi, W. H. Green, H.-W. Wong, *et al.* Redesigning combustion modeling algorithms for the graphics processing unit (GPU): Chemical kinetic rate evaluation and ordinary differential equation integration. *Combustion and Flame,* 158(5):836--847, 2011. URL http://linkinghub.elsevier.com/retrieve/pii/S0010218011000393.

[34] A. A. Amsden, P. J. O'Rourke, and T. D. Butler. KIVA-II: A computer program for chemically reactive flows with sprays. Technical Report LA-11560-MS, Los Alamos National Laboratory, 1989.

[35] A. A. Amsden. KIVA-3: A KIVA program with block-structured mesh for complex geometries. Technical Report LA-12503-MS, Los Alamos National Laboratory, 1993.

[36] A. A. Amsden. KIVA-3V: A block-structured KIVA program for engines with vertical or canted valves. Technical Report LA-13313-MS, Los Alamos National Laboratory, 1997.

[37] S. Kong, C. D. Marriott, R. Reitz, *et al.* Modeling and experiments of HCCI engine combustion using detailed chemical kinetics with multidimensional CFD. *SAE Technical Paper,* (724):01--1026, 2001. URL http://subscriptions.sae.org/content/2001-01-1026/.

[38] C. F. Curtiss and J. O. Hirschfelder. Integration of stiff equations. *Proceedings of the National Academy of Sciences of the United States of America*, 38(3):235--243, 1952. URL http://www.ncbi.nlm.nih.gov/pmc/articles/PMC1063538/.

[39] U. Ascher and L. Petzold. *Computer methods for ordinary differential equations and differential-algebraic equations.* SIAM: Society for Industrial and Applied Mathematics, 1998.

[40] M. Hochbruck and A. Ostermann. Exponential integrators. *Acta Numerica*, 19:209--286, 2010. URL http://www.journals.cambridge.org/abstract_S0962492910000048.

[41] D. R. Mott, E. S. Oran, and B. van Leer. A quasi-steady-state solver for the stiff ordinary differential equations of reaction kinetics. *Journal of Computational Physics*, 164(2):407--428, 2000. URL http://linkinghub.elsevier.com/retrieve/pii/S0021999100966051.

[42] Y. Shi, S. Kokjohn, H. Ge, *et al.* Efficient multidimensional simulation of HCCI and DI engine combustion with detailed chemistry. *SAE Technical Paper*, (01--0701), 2009. URL http://web.mit.edu/~yushi/www/download/T11.pdf.

[43] A. Patel, S. Kong, and R. Reitz. Development and validation of a reduced reaction mechanism for HCCI engine simulations. *SAE Technical Paper*, (01--0558), 2004. URL http://subscriptions.sae.org/content/2004-01-0558/.

[44] Y. Ra and R. D. Reitz. A combustion model for IC engine combustion simulations with multi-component fuels. *Combustion and Flame*, 158(1):69--90, 2011. URL http://linkinghub.elsevier.com/retrieve/pii/S0010218010002075.

[45] E. Hairer, S. P. Nørsett, and G. Wanner. *Solving ordinary differential equations I: Nonstiff problems.* Springer, Berlin, Germany, 2nd edition, 2000.

[46] E. Hairer and G. Wanner. *Solving ordinary differential equations II: Stiff and differential-algebraic problems.* 2nd edition, 1996.

[47] W. Rheinboldt and J. Ortega. *Iterative solution of nonlinear equations in several variables.* Society for Industrial Mathematics, 1987.

[48] C. T. Kelley. *Iterative methods for linear and nonlinear equations.* Society for Industrial Mathematics, 1987.

[49] D. P. Bertsekas. *Nonlinear programming.* 2nd edition, 1999.

[50] S. H. Lam. Using CSP to understand complex chemical kinetics. *Combustion Science and Technology*, 89(5):375--404, 1993.

[51] S. Lam and D. Goussis. The CSP method for simplifying kinetics. *International Journal of Chemical Kinetics*, 26(4):461--486, 1994. URL http://citeseerx.ist.psu.edu/viewdoc/download?doi=10.1.1.18.1725&rep=rep1&type=pdf.

[52] B. Bhattacharjee, D. A. Schwer, P. I. Barton, *et al.* Optimally-reduced kinetic models: Reaction elimination in large-scale kinetic mechanisms. *Combustion and Flame*, 135(3):191--208, 2003. URL http://linkinghub.elsevier.com/retrieve/pii/S0010218003001597.

[53] O. O. Oluwole, P. I. Barton, and W. H. Green. Obtaining accurate solutions using reduced chemical kinetic models: a new model reduction method for models rigorously validated over ranges. *Combustion Theory and Modelling*, 11(1):127--146, 2007.

[54] A. Mitsos, G. M. Oxberry, P. I. Barton, *et al.* Optimal automatic reaction and species elimination in kinetic mechanisms. *Combustion and Flame*, 155(1--2):118--132, 2008.

[55] D. A. Schwer, P. Lu, and W. H. Green. An adaptive chemistry approach to modeling complex kinetics in reacting flows. *Combustion and Flame*, 133(4):451--465, 2003. URL http://linkinghub.elsevier.com/retrieve/pii/S0010218003000452.

[56] H. Wang, A. Laskin, Z. M. Djurisic, *et al.* A comprehensive mechanism of C2Hx and C3Hx fuel combustion. In *Fall Technical Meeting of the Eastern States Section of the Combustion Institute*. 1999.

[57] H. Wang and A. Laskin. Optimization of a comprehensive detailed chemical kinetic model for simulating high-speed propulsion. Technical report, University of Delaware, 1999.

[58] H. Wang and A. Laskin. A Comprehensive Reaction Model of Ethylene and Acetylene Combustion. 1999.

[59] J. Hewson. Reduced mechanisms for NOx emissions from hydrocarbon diffusion flames. *Symposium (International) on Combustion*, 26(2):2171--2179, 1996. URL http://linkinghub.elsevier.com/retrieve/pii/S0082078496800439.

[60] San Diego Mechanism. 2005. URL http://web.eng.ucsd.edu/mae/groups/combustion/cermech/sandiego20051201_CK.txt.

[61] P. Pepiot-Desjardins and H. Pitsch. An efficient error-propagation-based reduction method for large chemical kinetic mechanisms. *Combustion and Flame*, 154(1-2):67--81, 2008. URL http://linkinghub.elsevier.com/retrieve/pii/S0010218007003264.

[62] O. Herbinet, W. Pitz, and C. Westbrook. Detailed chemical kinetic oxidation mechanism for a biodiesel surrogate. *Combustion and Flame*, 154(3):507--528, 2008. URL http://linkinghub.elsevier.com/retrieve/pii/S0010218008000771.

[63] O. Herbinet, W. J. Pitz, and C. K. Westbrook. Detailed chemical kinetic mechanism for the oxidation of biodiesel fuels blend surrogate. *Combustion and Flame*, 157(5):893--908, 2010. URL http://linkinghub.elsevier.com/retrieve/pii/S0010218009003058.

[64] M. Rathinam and L. R. Petzold. A new look at proper orthogonal decomposition. *SIAM Journal on Numerical Analysis*, 41(5):1893--1925, 2004. URL http://citeseerx.ist.psu.edu/viewdoc/download?doi=10.1.1.80.389&rep=rep1&type=pdf.

[65] G. M. Oxberry, W. H. Green, and P. I. Barton. Three representations of projection-based model reduction in combustion. *In Preparation*.

[66] T. Lu and C. Law. Systematic approach to obtain analytic solutions of quasi steady state species in reduced mechanisms. *Journal of Physical Chemistry A*, 110(49):13202--13208, 2006. URL http://www.ncbi.nlm.nih.gov/pubmed/17149834.

[67] T. H. Gronwall. Note on the derivatives with respect to a parameter of the solutions of a system of differential equations. *The Annals of Mathematics*, 20(4):292--296, 1919. URL http://www.jstor.org/stable/1967124.

[68] G. Dahlquist. *Stability and error bounds in the numerical solution of ordinary differential equations*. PhD Thesis, Stockholm University, 1958.

[69] G. Söderlind. The logarithmic norm. History and modern theory. *BIT Numerical Mathematics*, 46(3):631--652, 2006. URL http://www.springerlink.com/index/10.1007/s10543-006-0069-9.

[70] C. Homescu, L. R. Petzold, and R. Serban. Error estimation for reduced-order models of dynamical systems. *SIAM Review*, 49(2):277--299, 2006. URL http://citeseerx.ist.psu.edu/viewdoc/download?doi=10.1.1.74.3252&rep=rep1&type=pdf.

[71] R. Serban, C. Homescu, and L. R. Petzold. The effect of problem perturbations on nonlinear dynamical systems and their reduced-order models. *SIAM J. Sci. Comput.*, 29(6):2621--2643, 2007. URL http://www.siam.org/journals/sisc/29-6/62527.html.

[72] D. Estep, V. Ginting, D. Ropp, *et al.* An a posteriori-a priori Analysis of multiscale operator splitting. *SIAM Journal on Numerical Analysis*, 46(3):1116--1146, 2008. URL http://citeseerx.ist.psu.edu/viewdoc/download?doi=10.1.1.119.6050&rep=rep1&type=pdf.

[73] D. Estep, V. Ginting, and S. Tavener. A posteriori analysis of a multirate numerical methods for ordinary differential equations. Technical report, 2010. URL http://citeseerx.ist.psu.edu/viewdoc/download?doi=10.1.1.145.797&rep=rep1&type=pdf.

[74] D. Estep, S. Tavener, and T. Wildey. A posteriori analysis and improved accuracy for an operator decomposition solution of a conjugate heat transfer problem. *SIAM Journal on Numerical Analysis*, 46(4):2068--2089, 2008. URL http://citeseerx.ist.psu.edu/viewdoc/download?doi=10.1.1.114.5111&rep=rep1&type=pdf.

[75] D. J. Estep. Error estimates for multiscale operator decomposition for multiphysics models. In *Multiscale methods: bridging the scales in science and engineering*, 305--388. Oxford University Press, USA, 2009. URL http://books.google.com/books?hl=en&lr=&id=DX6_gXHflhkC&oi=fnd&pg=PA305&dq=Error+Estimates+for+Multiscale+Operator+Decomposition+For+Multiphysics+Models&ots=bRQ5gYmgAn&sig=PycmQpeSS8uzZDzd5hLxrr_-LDQ.

[76] D. J. Estep, M. G. Larson, and R. D. Williams. *Estimating the error of numerical solutions of systems of nonlinear reaction-diffusion equations.* 696. American Mathematical Society, 2000. URL http://books.google.com/books?hl=en&lr=&id=OnLU1bM-3PcC&oi=fnd&pg=PA1&dq=Estimating+the+error+of+numerical+solutions+of+systems+of+nonlinear+reaction-diffusion+equations&ots=tl4DH2Sg2M&sig=Ot3krBHCE-k_qOVT_BeorgNnCN8.

[77] R. P. Dickinson and R. J. Gelinas. Sensitivity analysis of ordinary differential equation systems -- A direct method. *Journal of Computational Physics*, 21(2):123--143, 1976. URL http://www.sciencedirect.com/science/article/pii/0021999176900073.

[78] E. P. Dougherty and H. Rabitz. A computational algorithm for the Green's function method of sensitivity analysis in chemical kinetics. *International Journal of Chemical Kinetics*, 11(12):1237--1248, 1979. URL http://doi.wiley.com/10.1002/kin.550111203.

[79] T. Maly and L. R. Petzold. Numerical methods and software for sensitivity analysis of differential-algebraic systems. *Applied Numerical Mathematics*, 20(1-2):57--79, 1996. URL http://linkinghub.elsevier.com/retrieve/pii/0168927495001174.

[80] W. F. Feehery, J. E. Tolsma, and P. I. Barton. Efficient sensitivity analysis of large-scale differential-algebraic systems. *Applied Numerical Mathematics*, 25(1):41--54, 1997. URL http://linkinghub.elsevier.com/retrieve/pii/S0168927497000500.

[81] Y. Cao, S. Li, L. Petzold, *et al.* Adjoint sensitivity analysis for differential-algebraic equations: The adjoint DAE system and its numerical solution. *SIAM Journal on Scientific Computing*, 24(3):1076--1089, 2003. URL http://scitation.aip.org.libproxy.mit.edu/getpdf/servlet/GetPDFServlet?filetype=pdf&id=SJOCE3000024000003001076000001&idtype=cvips&prog=normal.

[82] P. E. Zadunaisky. On the estimation of errors propagated in the numerical integration of ordinary differential equations. *Numerische Mathematik*, 71(1):20--39, 1976.

[83] R. Skeel. Thirteen ways to estimate global error. *Numerische Mathematik*, 48(1):1--20, 1986. URL http://www.springerlink.com/index/Q1J2706402477717.pdf.

[84] D. Estep. A posteriori error bounds and global error control for approximation of ordinary differential equations. *SIAM Journal on Numerical Analysis*, 32(1):1--48, 1995. URL http://link.aip.org/link/SJNAAM/v32/i1/p1/s1&Agg=doi.

[85] Y. Cao and L. Petzold. A posteriori error estimation and global error control for ordinary differential equations by the adjoint method. *SIAM Journal on Scientific Computing*, 26(2):359--374, 2004. URL http://citeseerx.ist.psu.edu/viewdoc/download?doi=10.1.1.104.6136&rep=rep1&type=pdf.

Pyrolysis kinetics of glucose using model compounds

Vikram Seshadri, Jordan Keith, and Phillip Westmoreland
North Carolina State University, Raleigh, NC 27695-7905

Introduction

Glucose is the monomer unit of cellulose, and its pyrolysis should provide useful chemical-pathway analogies to cellulose pyrolysis. A mechanism for pyrolysis of glucose is proposed in this work using quantum-chemistry calculations. A possible extension to pyrolysis mechanism of cellobiose is also proposed.

Results and Discussion

Glucose can isomerize between boat, chair, and linear conformation. These isomerization reactions can take place by unimolecular and OH-assisted bimolecular pathways. Rate constants for the gas-phase reactions were calculated using CBS-QB3 quantum-chemistry calculations [1] and classical transition-state theory. Simplifications were made to the glucose structure by replacing –OH at carbon 2, 3 and 4 positions with –H to achieve higher computation speed. This replacement did not have any significant effect on the reactions that were analyzed in this work. The unimolecular boat-to-chair isomerization has activation energy of about 6 kcal/mol, boat-to-linear about 33 kcal/mol, and chair-to-linear about 47 kcal/mol. A water-assisted bimolecular reaction pathway reduces the activation energy of boat-to-linear and chair-to-linear isomerizations significantly to 23 kcal/mol for both reactions. The bimolecular pathway has a pericyclic transition-state structure. Glucose can also isomerize between α- and β- forms through its linear aldose isomer.

Levoglucosan is a dehydration product of β-D-glucose and is one of the important components of biomass pyrolysis oil. The leaving OH group can be either at carbon 1 (type 1) or at carbon 6 (type 2) in glucose. It was found that OH at carbon 1 is the most feasible pathway because of the proton transfer through promoter mode vibration, where the acceptor lone pairs are delocalized into the antibonding orbital of a donor σ bond, leading to reduction in activation energy [2]. Natural-bond-orbital analysis was used to confirm this finding. The unimolecular type 1 and type 2 reactions for levoglucosan formation have activation energies of 52 and 74 kcal/mol, respectively. A water-assisted type 1 bimolecular reaction has activation energy of 46 kcal/mol.

Similar analysis was performed by substituting the –OH group at carbon 1 with $-OCH_3$, $-OC_2H_5$, $-O(iC_3H_7)$, $-OC_2H_4OH$ and $-OC_6H_{12}O_3$ (cellobiose). This analysis can help in the understanding of levoglucosan formation from polymers of glucose. The bimolecular pathways assisted by methanol, ethanol and isopropanol were also analyzed because of the presence of –OH in cellulose.

The reactions discussed above were also performed in solution with water as solvent. A conductor-like polarizable continuum model was used to model solvent-phase reaction. In general, the activation energy was found to be lower in solution compared to gas-phase reactions. The structure of the transition state did not change significantly in solution.

These pathways support some of the hypotheses of Sanders et al. [3] on the mechanism of formation of levoglucosan and isomerization to linear form. The α- to β- isomerization is found to go through the linear aldose form, while Sanders et al. proposed a direct pathway for this reaction. It is also hypothesized that the five-membered furanose structures are formed from the linear form, and they undergo further dehydration to produce the furans that are commonly found in bio-oil.

References

1. Gaussian 09, Revision A.1, M. J. Frisch, G. W. Trucks, H. B. Schlegel, G. E. Scuseria, M. A. Robb, J. R. Cheeseman, G. Scalmani, V. Barone, B. Mennucci, G. A. Petersson, H. Nakatsuji, M. Caricato, X. Li, H. P. Hratchian, A. F. Izmaylov, J. Bloino, G. Zheng, J. L. Sonnenberg, M. Hada, M. Ehara, K. Toyota, R. Fukuda, J. Hasegawa, M. Ishida, T. Nakajima, Y. Honda, O. Kitao, H. Nakai, T. Vreven, J. A. Montgomery, Jr., J. E. Peralta, F. Ogliaro, M. Bearpark, J. J. Heyd, E. Brothers, K. N. Kudin, V. N. Staroverov, R. Kobayashi, J. Normand, K. Raghavachari, A. Rendell, J. C. Burant, S. S. Iyengar, J. Tomasi, M. Cossi, N. Rega, J. M. Millam, M. Klene, J. E. Knox, J. B. Cross, V. Bakken, C. Adamo, J. Jaramillo, R. Gomperts, R. E. Stratmann, O. Yazyev, A. J. Austin, R. Cammi, C. Pomelli, J. W. Ochterski, R. L. Martin, K. Morokuma, V. G. Zakrzewski, G. A. Voth, P. Salvador, J. J. Dannenberg, S. Dapprich, A. D. Daniels, Ö. Farkas, J. B. Foresman, J. V. Ortiz, J. Cioslowski, and D. J. Fox, Gaussian, Inc., Wallingford CT, 2009.

2. B. Saritha and M. Durga Prasad, "An electronic structure perspective of the promoter modes in proton transfer reactions." *Journal of Physical Chemistry A* **115** (2011) 2802-2810.

3. E. B. Sanders, A. I. Goldsmith, and J. I. Seeman. "A model that distinguishes the pyrolysis of D-glucose, D-fructose, and sucrose from that of cellulose. Application to the understanding of cigarette smoke formation." *Journal of Analytical and Applied Pyrolysis* **66** (2003) 29-50.

Contact information

Prof. Phillip R Westmoreland (speaker)
Executive Director, NCSU Institute for Computational Science and Engineering
Professor, Department of Chemical and Biomolecular Engineering
North Carolina State University
2036 Engineering Building I, Box 7905
911 Partners Way; Raleigh, NC 27695
Phone: (919) 515-7121; Fax: (919) 515-3465
Email: phil.westmoreland@ncsu.edu

(712g) Evaluation of Soots for Diesel Particulate Filter Development Studies

Changsheng Su, Depart of Chemical Engineering, University of Notre Dame, Notre Dame, IN and Paul J. McGinn, Department of Chemical Engineering, University of Notre Dame, Notre Dame, IN

Diesel soot consists of agglomerates with diameters of up to several hundred nanometers and a fine structure of spherical primary particles. Unlike the well understood formation of NO_x in internal combustion engines, formation of soot is still more ambiguous because the process is more complicated and difficult to examine. For example, the characterization of diesel particulates from a particular engine at different operation conditions may vary considerably due to variations in combustion parameters and diesel fuels. Laboratory studies often utilize diffusion flames in burners, because the soot formation processes in burner diffusion flames are thought to be fundamentally similar to those in diesel engines with sequential nucleation, growth and agglomeration steps taking place.

Until recently the impact of soot nanostructure on reactivity has not drawn much attention. Vander Wal et al. found by means of transmission electron microscopy (TEM) that the structural differences of soot/ carbon at a nanoscale level (i.e. dimension and orientation of the graphitic layers) affect its oxidation reactivity. Other recent studies showed that the intrinsic oxidation rate of pyrolysis soot might vary from different fuels (such as benzene, ethanol, and acetylene) resulting in differences in soot nanostructure. Moreover, it has been speculated that peculiar nanostructures, such as amorphous and shell/core formations, are present in the soot formation process. Amorphous nanostructures are related to soot particles and have been identified mainly in less developed soot particles and condensed species. The study of soot nanostructure is important for both correlating reactivity and simulating non-catalytic/ catalytic soot oxidation in exhaust systems.

Comparisons of the kinetic parameters of non-catalytic soot oxidation with those of catalytic soot oxidation have been made by many researchers to evaluate the effectiveness of soot oxidation catalysts, and the obtained kinetics information can also reveal the mechanism of soot oxidation reactions. However, the reported activation energy for non-catalytic soot oxidation ranges from 92 to 211kJ/mol since the employed experimental techniques and conditions vary from one group to another. Yezerets et al. compared the oxidative reactivity of diesel soot and carbon black, and the activation energies were found to be 92 and 117 kJ/mol, respectively. In another study of soot samples from different engine conditions, the activation energies for the soot with ash contents of 14 and 6.5% were found to be 126 and 146 kJ/mol, respectively. A recent study by Kim et al. showed that the activation energies for diesel soot and flame soot are 110 and 162 kJ/mol, respectively. Hence, it seems the kinetics parameters for soot oxidation cannot be simply compared from one study to another due to the differences in the nature of the soot and the experimental conditions of the reactors (i.e. the mass and heat transfer limitations).

In soot oxidation studies, both well defined model soots and a reliable means to simulate realistic contact conditions with catalysts are crucial. In catalytic soot oxidation studies, the "contact condition" refers to the nature of the contact between the soot and catalyst, with the contact range varying from "tight" to "loose". The realistic contact condition obtained by in-situ collection on a DPF in an exhaust system is considered as "loose" contact. Replicating this condition in laboratory studies can be challenging. In lab-scale studies, thermogravimetric analysis (TGA) has usually been used to characterize soot oxidation behavior, permitting relatively straightforward examination of kinetics during heating. When the catalysts under study are in a powder form, loose contact can be simulated by gently mixing the soot and catalyst powders with a spatula or gently shaking the sample container with catalyst/soot mixture. However, these methods are not appropriate for study of catalysts incorporated onto complex support material shapes (e.g. porous cordierite/SiC filters).

In the present study, it was anticipated that producing flame soot with a Tiki torch oil lamp and capturing the soot directly onto a catalyst/support structure would yield a contact condition simulating realistic conditions. The properties of this flame soot were examined by means of XRD and TEM for structure analysis, BET for surface area analysis, and TGA for reactivity and kinetics analysis. The flame soot is thought to be more uniform and controllable than similar samples of collected diesel particulates. For validation purposes, catalytic oxidation of Tiki soot using the simulated contact condition was compared with the diesel particulates collected from a real diesel engine exhaust system.

See more of this Session: Reaction Engineering for Combustion and Pyrolysis II
See more of this Group/Topical: Catalysis and Reaction Engineering Division

(713a) Effect of Compressed Gas Addition On H_2/CO Tunability and Hydroformylation Catalysis In Gas-Expanded Liquids (GXLs)

Zhuanzhuan Xie, Chemical & Petroleum Engineering, University of Kansas, Lawrence, KS, Kirk Snavely, Cebc, University of Kansas, Lawrence, KS, Swarup Maiti, Center for Environmentally Beneficial Catalysis, University of Kansas, Lawrence, KS, Jon Tunge, Chemistry, Univ. Kansas, Lawrence, KS and Bala Subramaniam, Chemical and Petroleum Engineering, Center for Environmentally Beneficial Catalysis, University of Kansas, Lawrence, KS

Previously,we had reported on the beneficial effects of carbon dioxide expanded liquidsfor hydroformylation[1, 2]. In this talk, we present systematicinvestigations of the comparative effects of compressed gases ($CO_2$and light alkanes) on the relative miscibilities ofCO and H_2 in GXLs and also the corresponding effects on Rh-basedhydroformylation catalysis with monodendate (triphenylphosphine) and bidendateligands. It is shown that the H_2/CO ratio in the gas-expanded liquidphase (containing substrate and dissolved catalyst complex), a key determinantof the turnover frequency and regioselectivity, canbe easily tuned with GXLs to optimize these performance measures at relativelymild pressures[3]. Continuous hydroformylation with bulky phosphite ligands is demonstrated in a GXL-based stirredreactor equipped with nanofiltration membranes. It is shown that the Rhleaching from these reactors is on the order of a few tens of ppb, which easilyexceeds the economic viability criterion[4, 5].

1. H.Jin, A. Ghosh, J. A. Tungeand B. Subramaniam, *AIChEJournal*, 52(7), 2575-2591 (2006)

2. H.Jin and B. Subramaniam, *Chem. Eng. Sci.*, 59,4887, (2004)

3. Z. Xie, W. Snavely, A. Scurto, and B. Subramaniam, *J.Chem. Eng. Data*, 54 (5), 1633, (2009)

4. J.Fang, H. Jin, T. Ruddy, K. Pennybaker, D. Fahey andB. Subramaniam, *Industrial and EngineeringChemistry Research*, 46, 8687-8692 (2007)

5. J.Fang, R. Jana, J. Tunge, B. Subramaniam,*Applied Catalysis, A: General*, 393(1-2), 294-301 (2011)

See more of this Session: Reactions In near-Critical and Supercritical Fluids
See more of this Group/Topical: Catalysis and Reaction Engineering Division

(713b) Ionic Liquid Production In CO_2-Expanded DMSO

Sylvia O. Nwosu and Aaron M. Scurto, Chemical and Petroleum Engineering, University of Kansas, Lawrence, KS

Novel applications for ionic liquids (ILs) are continually being reported in the fields of reactions, extractions, engineering fluids, CO_2-capture, material synthesis, etc. As these new technologies mature, large volumes of ILs will be needed at low-cost and with a low environmental impact. Previously, our group has shown that DMSO (a relatively environmentally-benign organic solvent) yields the highest kinetic rate for the synthesis of the IL, 1-hexyl-3-methylimidazolium bromide ([HMIm][Br]). However, due to the high boiling point (189°C) and heat of vaporization of pure DMSO, the energy of separations and energy-associated pollution beset the ultimate life-cycle analysis despite high kinetic rates and direct low human and environmental impact. However, CO_2-expanded DMSO may help overcome these issues. Gas-expanded or CO_2-expanded liquids (GXLs/CXLs) are a promising solvent media that reduce the use of organic solvents and operate at pressures often lower than conventional supercritical fluid technologies. The temperature and pressured may be adjusted to provide rapid reactions and a phase split between the IL and DMSO to provide facile separations. Here, we will demonstrate the phase equilibrium thermodynamics, kinetics, and transport properties of the production of [HMIm][Br] in CO_2-expanded DMSO.

See more of this Session: Reactions In near-Critical and Supercritical Fluids
See more of this Group/Topical: Catalysis and Reaction Engineering Division

(713c) Selective Removal of Protecting Groups Using Water At Elevated Temperatures

Wilmarie Medina-Ramos[1], Mike Mojica[2], Pamela Pollet[2], Rani Jha[2], Elizabeth D. Cope[2], Charles L. Liotta[2] and Charles A. Eckert[1], (1)Chemical and Biomolecular Engineering, Georgia Institute of Technology, Atlanta, GA, (2)Chemistry and Biochemistry, Georgia Institute of Technology, Atlanta, GA

Protecting groups are frequently used in multistep syntheses of chemicals and generally require the utilization of strong acids or bases for the removal of these protecting groups. Hot water is an ideal solvent for the development of sustainable, environmentally green processes for the removal of protecting groups. At temperatures and pressures higher than ambient (150-275°C), the physical and chemical properties of water change drastically compared to common polar organic solvents. This change in the properties is due to the decrease of hydrogen bonding, which ultimately results in enhanced solubility of nonpolar organic species. In addition the auto-ionization constant (K_W) of water increases by three orders of magnitude compare to ambient conditions, making hot water a source of hydronium and hydroxide ions. These ions are available to act as catalysts for an acid- and/or base-catalyzed reaction; these currently require additional, mineral acid or base, and the subsequent catalyst neutralization creates a problem of salt disposal. By employing hot water as the reaction solvent and catalyst simultaneously, the need for environmentally hazardous solvents and acid catalysts is eliminated. The efficient water-mediated selective removal of removal of protecting groups will be discussed for a variety of organic substrates.

See more of this Session: Reactions In near-Critical and Supercritical Fluids
See more of this Group/Topical: Catalysis and Reaction Engineering Division

THE EFFECTS OF SPACE TIME AND GLYCERIN TO METHANOL RATIO UPON THE SUPERCRITICAL WATER REFORMATION OF CRUDE GLYCERIN SOLUTION FOR HYDROGEN PRODUCTION

Ryan Tschannen[1], Jared Bouquet[2], Aaron Gonzales[1], and Sunggyu Lee[1]

1: Ohio University, Athens, OH
2: Missouri University of Science and Technology, Rolla, MO

Abstract

Biodiesel is a renewable energy with great potential due to the fact it can be used in existing diesel engines. The transesterification of triglycerides used to produce biodiesel also produces a large amount of glycerin byproduct. Most biodiesel plants need to sell their excess glycerin to remain profitable. This poses a challenge because the crude glycerin solution must be purified of excess methanol reactant before it can be sold and the supply of glycerin greatly exceeds the demand. An alternative to this is the supercritical water reformation of the crude glycerin solution into hydrogen for energy applications. This non-catalytic process utilizes the crude glycerol solution and removes the need for an energy intensive purification step. Reformation of crude glycerin also increases the amount of usable transportation fuel that may be produced from triglycerides in biodiesel production. The effect of space time and glycerin to methanol ratio on the supercritical water reformation of a crude glycerol solution was evaluated in an experimental study. The experimental study was conducted using a 0.1 L Haynes® 282® Alloy reactor[1] at a constant pressure of 3500 psig with a water-to-carbon molar ratio of 4.22:1. The space time was varied between 30 and 60 seconds at temperatures of 550°C, 600°C, and 650°C using a crude glycerin solution of 70 wt% glycerin and 30 wt% methanol. The crude glycerin solution was varied between 60 wt% glycerin and 40 wt% methanol to 80 wt% glycerin and 20 wt% methanol at a temperature of 600°C and a space time of 45 seconds.

Background

As biodiesel gains more popularity, crude glycerin byproduct will need more applications. Reacting crude glycerin in a supercritical water medium yields the potential for electricity generation and transportation fuel applications. At appropriate conditions, crude glycerin can be reacted to produce a hydrogen rich effluent gas or varying compositions of synthesis gas. One application for this effluent gas is to separate and purify the hydrogen for electricity generation in fuel cells. Another application is to use the synthesis gas as a basis for upgrading to long chain hydrocarbons for transportation fuel.[2]

Supercritical water makes for an excellent medium for these experiments. The dielectric constant of water drops significantly once it reaches critical conditions. At the critical point, the hydrogen bonding of water is half of what it is at ambient conditions.[3] This causes water to act as a better organic solvent than as an inorganic solvent. Water is also very energetic at critical conditions. This allows hydrocarbons to reform efficiently without the use of a catalyst. The physical properties of each chemical species used in these experiments are presented in Table 1.

Table 1. Physical Properties of Water, Methanol, and Glycerin[4-6]

Species	Units	Water	Methanol	Glycerin
Molecular Weight	(g/mol)	18.02	32.04	92.09
Density @ 25°C, 1 bar	(g/cm3)	0.997	0.786	1.26
Normal Boiling Point	(°C)	100	64.5	290
Critical Temperature	(°C)	374	240	577
Critical Pressure	(psig)	3200	1175	1088

Apparatus and Procedure

The experiments were performed using a custom built system capable of reforming a wide range of chemicals. It consists of a liquid feed system, integrated heat exchanger, preheater, 0.1 L Haynes® 282® Alloy reactor[1], pneumatic control valve, sample collection ports, and a data acquisition system. A process flow diagram of the experimental apparatus is shown in Figure 1. Analysis of the effluent gas was performed using a wet test meter for volumetric flow rate and a HP 5890 Series A gas chromatograph with a thermal conductivity detector for gas composition. The gas chromatograph is calibrated to detect hydrogen, nitrogen, carbon monoxide, methane, carbon dioxide, ethene, and ethane.

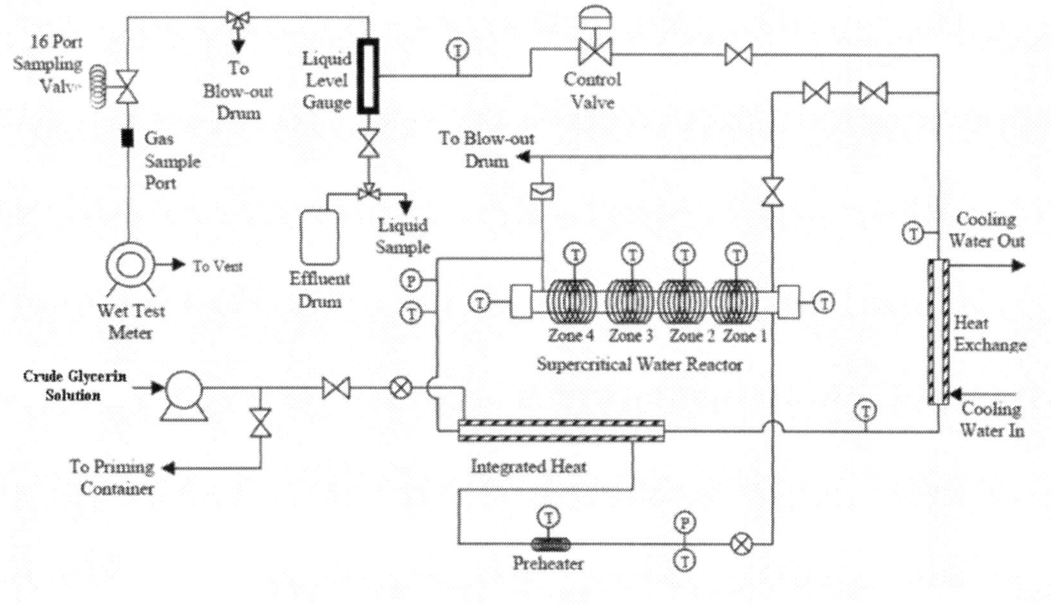

Figure 1. Process Flow Diagram

Feed solutions for the experiments were prepared using deionized water, histological grade methanol from Fisher-Scientific, and 99.7% pure glycerin from The Chemistry Store. After the glycerin containers were opened the glycerin was diluted to a 75wt% glycerin, 25wt% water solution because glycerin solutions of 80wt% and higher readily absorb moisture from the atmosphere.[6] Methanol and glycerin were mixed with water in separate containers to achieve a diluted methanol solution and a diluted glycerin solution with the desired water to carbon molar ratios. Those solutions were then mixed together to attain the appropriate glycerin to methanol weight percentages for the feed solution. Space times were calculated using the Peng Robinson equation of state and based on the reactor entrance conditions

The supercritical water reformer was brought up to temperature and pressure using deionized water. Once the reactor achieved the desired temperature and pressure the deionized water was replaced with the crude glycerin feed solution. Gas samples were taken in approximately 40 minute intervals. The runs were concluded when the gas composition of three consecutive gas samples matched up to indicate steady state.

Chemical Reactions

The purpose of this study is to determine how glycerin and methanol react in a supercritical water medium. Hydrocarbons can react with water to form carbon monoxide and hydrogen. This reaction is known as reformation. The traditional form of this chemical reaction is shown in Equation 1.

$$C_mH_n + mH_2O \rightarrow mCO + (m + \tfrac{n}{2})H_2 \qquad (1)$$

However, since glycerin and methanol are such oxygenated compounds, they can undergo a decomposition reaction without consuming water as a reactant. Equation 2 and equation 3 show the glycerin and methanol decomposition reactions respectively.[6, 7]

$$C_3H_8O_3 \rightarrow 3CO + 4H_2 \qquad (2)$$

$$CH_3OH \rightarrow CO + 2H_2 \qquad (3)$$

Carbon monoxide in the presence of water can undergo a reaction known as water gas shift. This reaction produces carbon dioxide and hydrogen. This is a reversible reaction and is thermodynamically favored in the forward direction at temperatures below 815°C. The water gas shift reaction is shown in Equation 4.[7]

$$CO + H_2O \leftrightarrow CO_2 + H_2 \qquad (4)$$

The combination of Equation 2 and Equation 3 with Equation 4 yields an overall reaction that looks similar to the reformation reaction. These equations are used to determine the maximum theoretical hydrogen yield. They are shown in Equation 5 and Equation 6.

$$C_3H_8O_3 + 3H_2O \rightarrow 3CO_2 + 7H_2 \qquad (5)$$

$$CH_3OH + H_2O \rightarrow CO_2 + 3H_2 \qquad (6)$$

Since glycerin is a three chain carbon molecule it can also undergo pyrolytic decomposition where smaller compounds are formed such as ethane, ethane, methane, methanol, etc. These smaller compounds can then proceed to decompose further or reform into carbon monoxide and hydrogen. A generic form of pyrolytic decomposition of glycerin is shown in Equation 7.

$$C_3H_8O_3 \rightarrow C_xH_yO_z + C_aH_bO_c \qquad (7)$$

Another possible reaction that can occur is the methanation reaction. Carbon monoxide reacts with hydrogen to form methane and water. This reaction is reversible and can be detrimental to the hydrogen yield as hydrogen is consumed in the reaction. The methanation reaction is shown in Equation 8.[6]

$$CO + H_2 \leftrightarrow CH_4 + H_2O \qquad (8)$$

Reactions such as the Boudouard reaction, dehydrogenation, dehydration, and others are possible reactions occurring in this system, but they are not the main focus of this study.

Experimental

The space time was varied at different temperatures to determine its effects of on crude glycerin gasification. Experiments were conducted at space times of 30s, 45s, and 60s. These space times were varied at temperatures of 550 ± 2°C, 600 ± 2°C, and 650 ± 2°C. The pressure, crude glycerin composition, and water to carbon ratio were kept constant during these experiments at 3500 ± 85 psig, 70wt% glycerin, and 4.22:1 respectively. The weight percent of glycerin and methanol in the mixture was also varied to determine the effects of methanol and glycerin concentration on crude glycerin gasification. Experiments were conducted using 60wt%, 70wt%, and 80wt% glycerin in the crude glycerin solution (water free basis). These experiments were performed at a constant temperature, pressure, space time, and water to carbon ratio of 600 ± 2°C, 3500 ± 27 psig, 45s, and 4.22:1 respectively. Table 2 shows the conditions for each experiment used in this study.

Table 2. Experimental Matrix for Space Time and Glycerin to Methanol Ratio Effects on Crude Glycerin in Supercritical Water

Run No.	Temp	Pressure	Feed Flow Rate	Space Time	Glycerin wt%	MeOH wt%	W/C ratio
-	(°C)	(psig)	(g/min)	(s)	-	-	(mol/mol)
WCG-108M	548	3497	19.6	30	20.3	8.7	4.22
WCG-105M	549	3513	12.8	45	20.3	8.7	4.22
WCG-101M	549	3472	9.8	60	20.3	8.7	4.22
WCG-106M	598	3486	17.8	30	20.3	8.7	4.22
WCG-102M	599	3468	11.5	45	20.3	8.7	4.22
WCG-112M	598	3503	11.8	45	20.3	8.7	4.22
WCG-97M	598	3469	8.6	60	20.3	8.7	4.22
WCG-98M	598	3486	8.8	60	20.3	8.7	4.22
WCG-115M	648	3501	16.2	30	20.3	8.7	4.22
WCG-103M	648	3464	10.3	45	20.3	8.7	4.22
WCG-99M	648	3415	8.0	60	20.3	8.7	4.22
WCG-113M	648	3511	7.8	60	20.3	8.7	4.22
WCG-139M	598	3463	11.6	45	17.5	11.6	4.22
WCG-102M	599	3468	11.5	45	20.3	8.7	4.22
WCG-112M	598	3503	11.8	45	20.3	8.7	4.22
WCG-138M	599	3483	11.8	45	23.1	5.8	4.22

Results

The results of the experiments summarized in Table 2 are shown in Table 3; the percent carbon gasification (mol C in effluent/mol C fed) and the yield of each gaseous species in the effluent stream (mol gas/mol C fed) are given.

Table 3. Results for Space Time and Glycerin to Methanol Ratio Effects on Crude Glycerin in Supercritical Water

	Product Gas Yield (mol gas/mol carbon fed)						Carbon
Run No.	H2	CO	CH4	CO2	C2H4	C2H6	Gasification
WCG-108M	0.25	0.12	0.01	0.04	0.00	0.00	18.2%
WCG-105M	0.31	0.18	0.02	0.05	0.00	0.00	25.5%
WCG-101M	0.36	0.24	0.03	0.05	0.00	0.00	34.3%
WCG-106M	0.52	0.31	0.05	0.09	0.00	0.01	48.2%
WCG-102M	0.79	0.46	0.11	0.15	0.00	0.03	77.4%
WCG-112M	0.85	0.52	0.11	0.14	0.00	0.02	81.1%
WCG-97M	0.97	0.48	0.13	0.23	0.00	0.03	91.6%
WCG-98M	1.17	0.46	0.13	0.28	0.00	0.03	92.3%
WCG-115M	0.81	0.51	0.11	0.12	0.00	0.03	80.7%
WCG-103M	1.18	0.50	0.16	0.28	0.00	0.04	101.2%
WCG-99M	1.44	0.22	0.19	0.57	0.00	0.03	104.8%
WCG-113M	1.27	0.36	0.15	0.39	0.00	0.04	97.5%
WCG-139M	0.84	0.37	0.10	0.21	0.00	0.03	73.6%
WCG-102M	0.79	0.46	0.11	0.15	0.00	0.03	77.4%
WCG-112M	0.85	0.52	0.11	0.14	0.00	0.02	81.1%
WCG-138M	0.57	0.38	0.12	0.17	0.00	0.04	74.5%

Figure 2 through Figure 5 graphically depict the results given in Table 3. Figure 2 through Figure 4 show the effects of space time on crude glycerin in supercritical water at 550°C, 600°C, and 650°C. Figure 5 shows the effect of glycerin weight percentage on crude glycerin in supercritical water at 600°C. Each graph has the percentage of carbon conversion on the left side of the graph and the product gas yield on the right side. Gas yields were divided by the moles of carbon fed to normalize the results based on different feed flow rates. The lines used in the graphs are for visual aid only and help the reader see the trends in the data. Gas yield of ethene was not included in the graphs because it was present in such small amounts compared to the other gases or not present at all.

Figure 2 shows that longer space times yield higher carbon conversion and higher gas yields. Larger increases in hydrogen and carbon monoxide than other gases indicate an increase in glycerin and methanol decomposition. The slight increase in methane and ethane formation shows that longer times in the reactor promote slightly more pyrolysis. Little change in the carbon dioxide yield indicates that water gas shift has either reached equilibrium or the compounds don't remain in the reactor long enough to start the reaction.

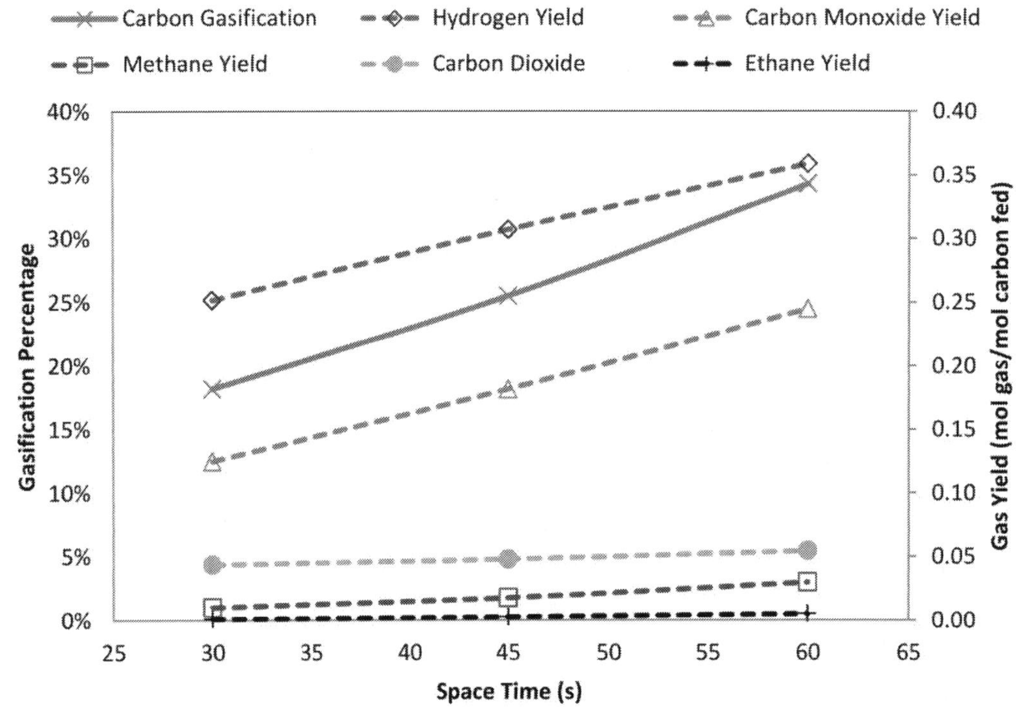

Figure 2. Effect of Space Time on Crude Glycerin in Supercritical Water at 550°C

Figure 3 shows that carbon gasification is strongly dependent on space time at 600°C. Approximately 40% more carbon is converted with an additional 30 seconds in the reactor. Hydrogen yield increases with increasing space time indicating increased glycerin and methanol decomposition. Carbon monoxide yield increases and then begins to level off at higher space times. Carbon dioxide yield increases gradually and then more rapidly at higher space times. The trends in carbon monoxide and carbon dioxide indicate that at higher space times the reactants are able to undergo more water gas shift at 600°C. Slight increases in methane and ethane yield again indicate that longer space times lead to small increases in pyrolytic activity.

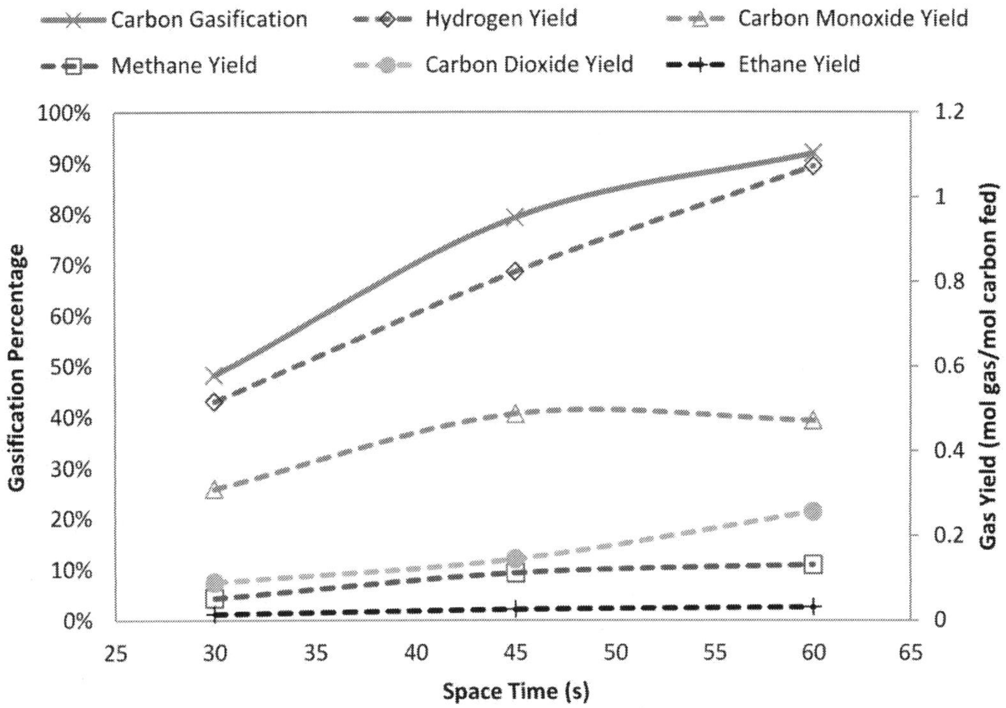

Figure 3. Effect of Space Time on Crude Glycerin in Supercritical Water at 600°C

Figure 4 shows that increases in space time at 650°C leads to an increase carbon conversion, but past a certain point increasing space time does not increase carbon conversion. Hydrogen yield continues to increase even when the carbon conversion levels off. This shows that after the glycerin and methanol have undergone decomposition, water gas shift can continue to produce hydrogen gas. The increase in hydrogen, decrease in carbon monoxide, and increase in carbon dioxide all show that space time has a large effect on the water gas shift reaction at 650°C.

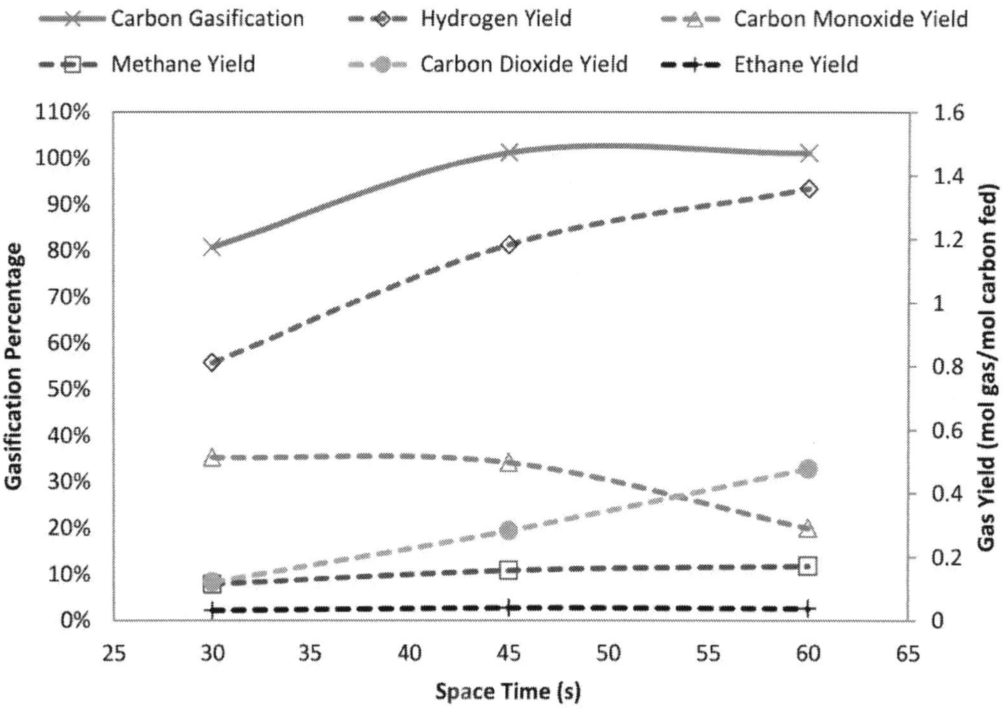

Figure 4. Effect of Space Time on Crude Glycerin in Supercritical Water at 650°C

The composition of crude glycerin from biodiesel production can vary. It can range from approximately 60.34 wt% to 80.71 wt% glycerin depending on the triglyceride feed stock.[8] Figure 5 shows the effect of varying the wt% of glycerin in the crude glycerin feed. Carbon conversion seems to remain constant at approximately 76% conversion. More data needs to be collected verify that there is no trend. Hydrogen yield decreases with increasing glycerin wt%. Carbon monoxide yield increases and then decreases with increasing glycerin wt% whereas carbon dioxide decreases and then increases slightly. The trend from 60 wt% to 70 wt% could be a result of reverse water gas shift consuming some hydrogen and carbon dioxide to produce carbon monoxide and water. Both methane and ethane yields increase with increasing glycerin wt%. The trend from 70 wt% to 80 wt% could occur from either increased methanation activity or an increase in pyrolysis activity. Carbon monoxide and hydrogen being consumed to make methane and water could explain the drop in both hydrogen and carbon monoxide yield. Higher glycerin wt% means more moles of glycerin are being fed into the reactor. Since there is a larger quantity of three carbon molecules, this could encourage more pyrolysis. The hydrogen consuming free radical carbons generated from pyrolysis might account for the drop in hydrogen and carbon monoxide as well as the increase in methane and ethane. To verify the trends, more experiments should be performed and an extensive liquid composition analysis should be conducted to determine if other chemical reactions should be considered relevant.

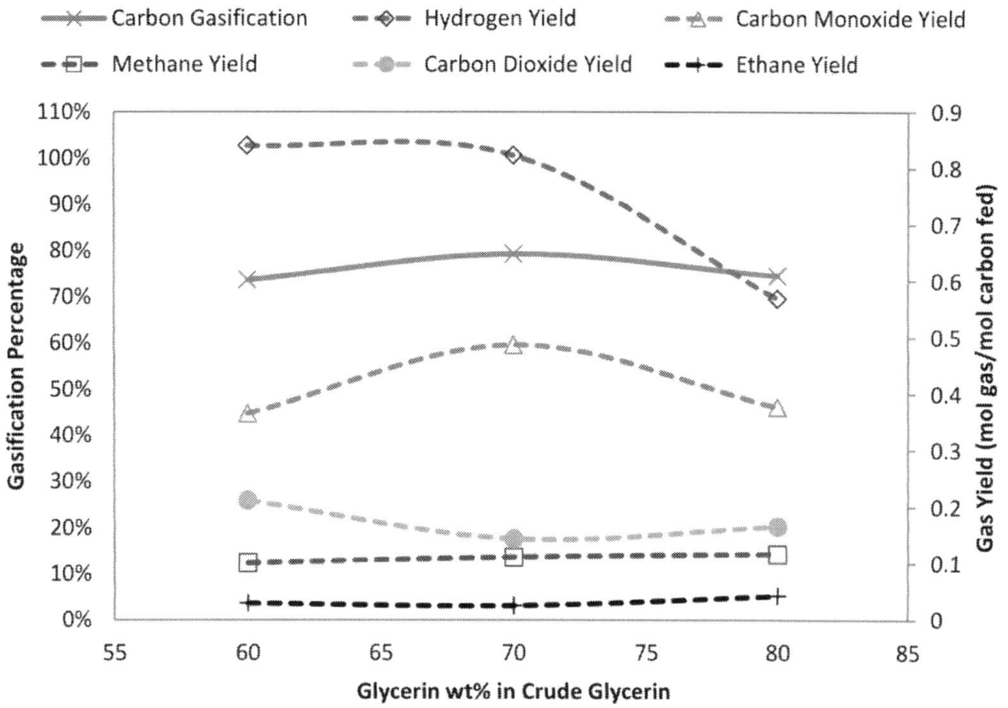

Figure 5. Effect of Glycerin Weight Percentage in Crude Glycerin at 600°C

Conclusion

Space time has a significant effect on the gasification of crude glycerin in supercritical water at both 550°C and 600°C. The carbon conversion increases as the space time increases at both temperatures. At 650°C space time provides an increase in carbon conversion until it reaches 100% conversion. At this point, space time has an effect on the reactions occurring inside the reactor. At higher space times the reactants are able to undergo more water gas shift and effectively increase the hydrogen gas yield and decrease the carbon monoxide yield. These results are applicable to maximizing hydrogen production for fuel cells and generating certain synthesis gas compositions for transportation fuel upgrading. Varying the glycerin weight percentage in the crude glycerin feed stock does not seem to have much of an impact on carbon conversion. However, at higher glycerin concentrations, there are more C_3 hydrocarbons available for pyrolysis. This effectively reduces the hydrogen gas yield. Another possible contributor to decreases in hydrogen production is an increase in methanation activity.

Further analysis needs to be performed on the liquid samples obtained from the experiments. Total Organic Carbon (TOC) analysis will be used to determine the concentration of carbon atoms in the effluent liquid. This will provide a more accurate carbon gasification measure and allow us to close our carbon balance. The composition of the liquid effluent also needs to be analyzed. Glycerin can react with water at the experimental conditions to form compounds such as acrolein and acetaldehyde.[9] Liquid composition analysis would allow us to better understand the reactions that occur in our system and to determine which compounds in the crude glycerin undergo the most gasification.

Acknowledgements

This research work is partially supported by a grant from the Korea Institute of Science and Technology (KIST).

References

(1) Haynes International, Inc. HAYNES® 282® alloy. http://www.haynesintl.com/Haynes282alloy/282HaynesAlloyPF.htm (accessed 8/22/2011, 2011).

(2) Lee, S.; Speight, J. G.; Sudarshan K. Loyalka In *Handbook of alternative fuel technologies;* CRC Press: Boca Raton, 2007; , pp 552.

(3) Adam Joseph Byrd Hydrogen Production in Supercritical Water, Auburn University, 2011.

(4) National Institute of Standards and Technology. Water Phase Change Data. . http://webbook.nist.gov/cgi/cbook.cgi?Name=Water&Units=SI&cTP=on#Thermo-Phase.

(5) National Institute of Standards and Technology. Methanol Phase Change Data. . http://webbook.nist.gov/cgi/cbook.cgi?Name=methanol&Units=SI&cTP=on.

(6) Jason Picou; Jonathan Wenzel; Michael Stever; Jared Bouquet; and Sunggyu Lee In *In Synthesis Gas Production by Noncatalytic Reforming of Glycerin in Supercritical Water: Effect of Reaction Temperature.* AIChE National Conference; 2009; .

(7) Jared Bouquet; Jason Picou; Michael Stever; Jonathan Wenzel; Ryan Tschannen; and Sunggyu Lee In *In The Effects of Temperature Upon the Supercritical Water Reformation of Alcohols for Hydrogen Production;* AIChE National Conference; 2010; .

(8) J. C. Thompson; B. B. He CHARACTERIZATION OF CRUDE GLYCEROL FROM BIODIESEL PRODUCTION FROM MULTIPLE FEEDSTOCKS. *Applied Engineering in Agriculture* **2006**, *22(2)*, 261-265.

(9) Watanabe M*.; Iida T., A. Y.; Aida T.-M.; Inomata H. ACROLEIN FORMATION FROM GLYCEROL IN SUPERCRITICAL WATER. *Research Center of Supercritical Fluid Technology, Tohoku University* .

Elongation of Alanine induced by Discharged Plasma Under Hydrothermal Condition

Atsushi Nagira[1], Kohichi Nagafuchi[1], Hiroshi Watanabe[1], Mtsuru Sasaki[1], Motonobu Goto[2], and Kunio Kawamura[3]
[1]Guraduate School of Science and Technology
[2]Bioelectrics Research Center
Kumamoto University, Kumamoto, JAPAN
[3]Hiroshima Shudo University, Hiroshima, JAPAN

Introduction

The term "plasma" is generally defined as a partially ionized gas composed of ions, electrons and neutral species. Recently, plasma-water interaction has attracted growing interest as it may provide experimental chemists with a quite unique reaction medium.[1-3] Plasma are normally generated by supplying energy to neutral gas, causing the formation of charge carriers. These neutral molecule collisions transfer energy to molecules and form a variety of active species which may include photons, metastables, individual atoms, free radicals, molecular fragments, monomers, electrons and ions. Electrons and ions are produced in the gas phase when electrons or photons with sufficient energy collide with neutral atoms and molecules in the feed gas (electron-impact ionization or photoionization). Charge carriers accelerated in the electric field couple their energy into the plasma via collisions with other particles. Electrons retain most of there energy primarily in inelastic collisions. Figure 1 illustrates a simple plasma generation method consisting of a voltage source that drives current through a low-pressure gas between two parallel conducting plates or electrodes.[4)]

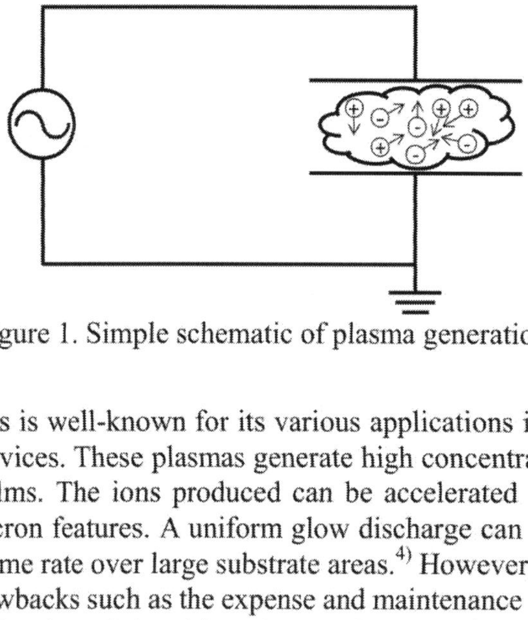

Figure 1. Simple schematic of plasma generation.

Low-pressure plasmas is well-known for its various applications in material processing and its key role in manufacturing devices. These plasmas generate high concentrations of reactive species that can etch and deposit thin films. The ions produced can be accelerated towards a substrate to cause directional etching of submicron features. A uniform glow discharge can be generated so that material processing proceeds at the same rate over large substrate areas.[4)] However, operating plasma at reduced pressure presents several drawbacks such as the expense and maintenance requirements of low-pressure systems and limitations on the size of the object that can be treated due to the size of the reactor or chamber. In this work, water under high argon atmospheric pressure was used as a media with pulsed discharge plasma for chemical synthesis of amino acids. Pulse discharge plasma under high inert atmospheric pressure, such as argon, might serve as an intense source of ultraviolet radiation.[5] Argon is one of the most studied and utilized chemical elements, and dense argon is a promising reaction media alternative. Like carbon dioxide, argon processes easily obtainable critical points and it is also non-polluting (T_c = 150.87 K, P_c = 4.898 MPa).[6-8] Argon is also inexpensive and its optical transparency is superior to that of carbon dioxide since it is monatomic. While pulse discharge plasma at atmospheric and high pressure conditions eliminates the disadvantage of the low-pressure operation the requirement of high voltages for gas breakdown and difficulty of sustaining pulse plasma under these conditions present new challenges. Water is commonly considered a begin solvent due to its non-toxicity and abundant natural occurrence. The first common advantage of using water as solvent for clean technology is the opportunity to directly utilize water-soluble compounds and renewable materials.[9-13] They reported that high temperature water provided the possibility of directly modifying water-soluble and renewable materials in water without the necessary protection-deportation sequences that are commonly associated with conventional chemistry, thus reducing the overall synthetic steps. Therefore, in order to accomplish the chemical reactions that can tolerate water and reactive functional groups, oligomerization of amino acids was conducted in water.

It is well known that some essential organic compounds such as amino acids are synthesized and composed in deep sea due to natural occurrence of supercritical water conditions in volcanic and hydrothermal vents. Elongation of amino acids under hydrothermal conditions had also been reported, however, the yields of the direct formation of oligopeptides from amino acid monomers in these systems were very low at 0.1-1%.[14] Other than the hydrothermal vent system, there could be other factors responsible for the more efficient generation of oligopeptides. Discharge plasmas which can be naturally generated during lightning could be one likely reason for this. In this study, the surface plasma at various conditions; change of temperatures, discharge times frequency, concentrations, and starting materials.

Experimental

Materials

D,L-Alanine (98%, Wako Pure Chemical Industries, Ltd., Japan) (Ala) and L-alanyl-L-alanine (98%, Wako Pure Chemical Industries, Ltd., Japan) ($(Ala)_2$) was used as a starting material and distilled water as a solvent. This compound had high solubility in water (15.8 g/100 mL) at room temperature. Monosodium phosphate, 1-Hexanesulfonicacid, and 2, 5-dihydroxybenzoic acid (DHB) and analytical reagents such as methanol with purities exceeding 98.0 % were purchased from Wako Pure Chemical Industries, Ltd., Japan. All chemicals were used without further purification.

Experimental Procedures

Figure 2 shows a schematic diagram for discharge plasma over a water surface under high pressure argon conditions. The reactor contained two inspection windows to monitor the performance of plasma production and was made of stainless steel (SUS-316, Taiatsu Techno, Japan) with a volume of 15 mL and a maximum temperature of 573 K and pressure of 30 MPa. The discharge plasma over the water surface was initiated by a conical anode composed of copper wire 0.1 cm in diameter which was set at a distance of 0.2 cm from the water surface. A power lead was introduced through the center of a long bushing made of polytetrafluoroethylene (PTFE), and the annular space was sealed with an O-ring placed around its outer surface to prevent gas leakage. The distance of inside wall reactor as a cathode was 3 cm from the water surface. To begin the experiment, feed containing 0.1 mol/L of Alanine (5 mL) was loaded to the reactor manually. The reactor was repeatedly (three times) purged of air using argon gas with a pressure of 0.5 MPa and then directly pressurized with argon gas to the desired pressure, controlled using a pressure regulation; it was then heated up to 333 K. This reactor internal pressure was defined as the initial pressure of the experiment. A thermocouple and needle valve were used to control the temperature and the pressure of the reactor, respectively. After the temperature and pressure achieved the desire operating values, pulse power was applied to produce electric discharge. A voltage was applied to the positive electrode using an MPC (Magnetic Pulse Compressor, MPC2000S, Suematsu Electronics Co. Ltd., Japan) charged by a high-voltage stabilization DC power supply. The breakdown voltage and the current were respectively measured using a high-voltage probe (EP100K, Pulse Electronic Engineering Co. Ltd., Japan) and a current transfer (4997, Person Electronics, Inc., USA) that were connected to a digital oscilloscope. After the reaction time had elapsed, the electric current was cut off and the reactor was opened. The products were directly identified by liquid chromatograph mass spectrometer (LC-MS), and quantified by high-performance liquid chromatography (HPLC).

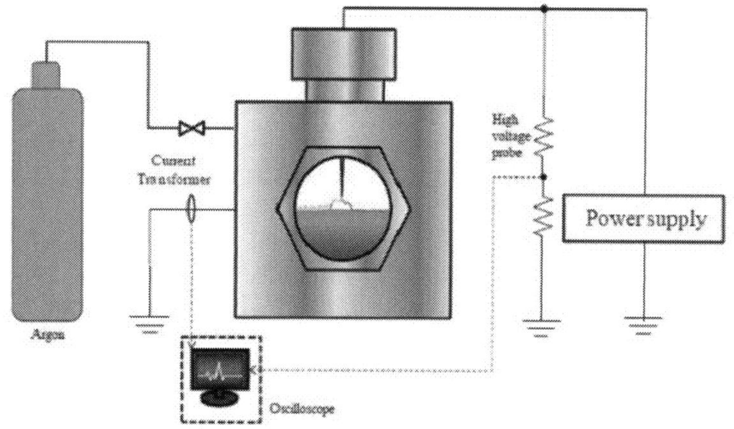

Figure 2. The schematic diagram for discharge plasma over a water surface under high pressure

Analytical Methods

The product was identified by LC-MS. Reversed-phase HPLC (RP-HPLC) was carried out at a wavelength of 220 nm with an ODS column (Inertsil®ODS-3 (u 4.6 x 250 mm)); GL Sciences Inc. Japan); it was carried out 35 C° with a gradient of 5 mM NaH_2PO_4 and 3.6 mM $CH_3(CH_2)_5SO_3Na$ at pH 2.65 with 10 mM NaH_2PO_4 and 7.2 mM $CH_3(CH_2)_5SO_3Na$ in 50% methanol at pH 2.70 using a time program of 30% (t = 0 min) - 60% (15 min) - 60% (25 min) at a flow rate of 0.7 mL/ min. Naturally, amino acids and peptides are stable during the RP-HPLC analyses.

Result and Discussion

Figure 3 shows that the reaction behavior of (Ala) under hydrothermal conditions was investigated by using batch reactor system. These studies in high temperature have shown that large amounts of diketopiperazine (DKP) formed from amino acid monomers.[15-19] This pathway is regarded as a stumbling block from the viewpoint concerning either the prebiotic formation of oligopeptides in hydrothermal systems or the synthetic organic chemistry of oligopeptides in aqueous solution. $(Ala)_2$ is immediately converted to DKPs.[14] To solve this problem, the reaction under hydrothermal discharge plasma have been investigated using starting materials Ala and $(Ala)_2$.

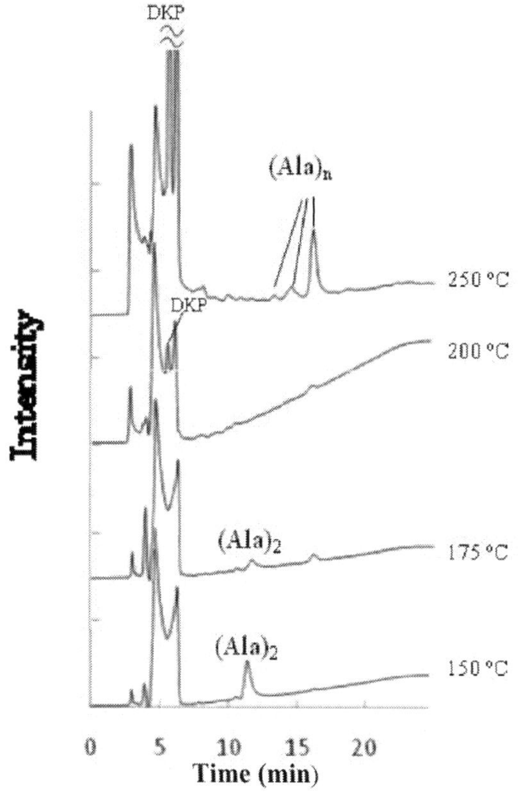

Figure 3. HPLC chromatogram for the products from Ala under hydrothermal condition

To evaluate the reaction of Ala after treatment by pulse discharge plasma, the liquid products were analyzed by HPLC with diode array detector (DAD); this is a useful tool for studying electropolymerization of alanine and its derivatives, since many intermediate products as well as of the resulting polymer products are known to be strongly colored. Parts of the spectra are given inFigure 4, the alanine as a starting material was detected clearly and other products, such as $(Ala)_2$ were also detected in small amount. It could be explained that the reaction of alanine into dimer or trimers increased with the existence of the electric field which produced during discharge plasma process. In the case of $(Ala)_2$, dimer or trimers increased by using discharge plasma. Depending on the reaction region, pulse discharge plasma can be conveniently grouped into localized and extended effects. When pulse discharge plasma occurs, the following individual effects occurred simultaneously: overpressure shock wave, strong electrical field, production of various free radicals, intense ultraviolet radiation, and ozone production with oxygen bubbling. Localized effects occur in the immediate vicinity of the plasma channel, whereas extended effects primarily result from UV radiation and the intense shock wave, which radiates out into the bulk of the solution. These effects play various important roles in different application regions in the liquid. Under these conditions, various chemical reactions occur to cause fluctuation in the reaction of alanine and its derived compounds. However, the temperature between 30 and 60 °C did not give high influence in the reaction process.

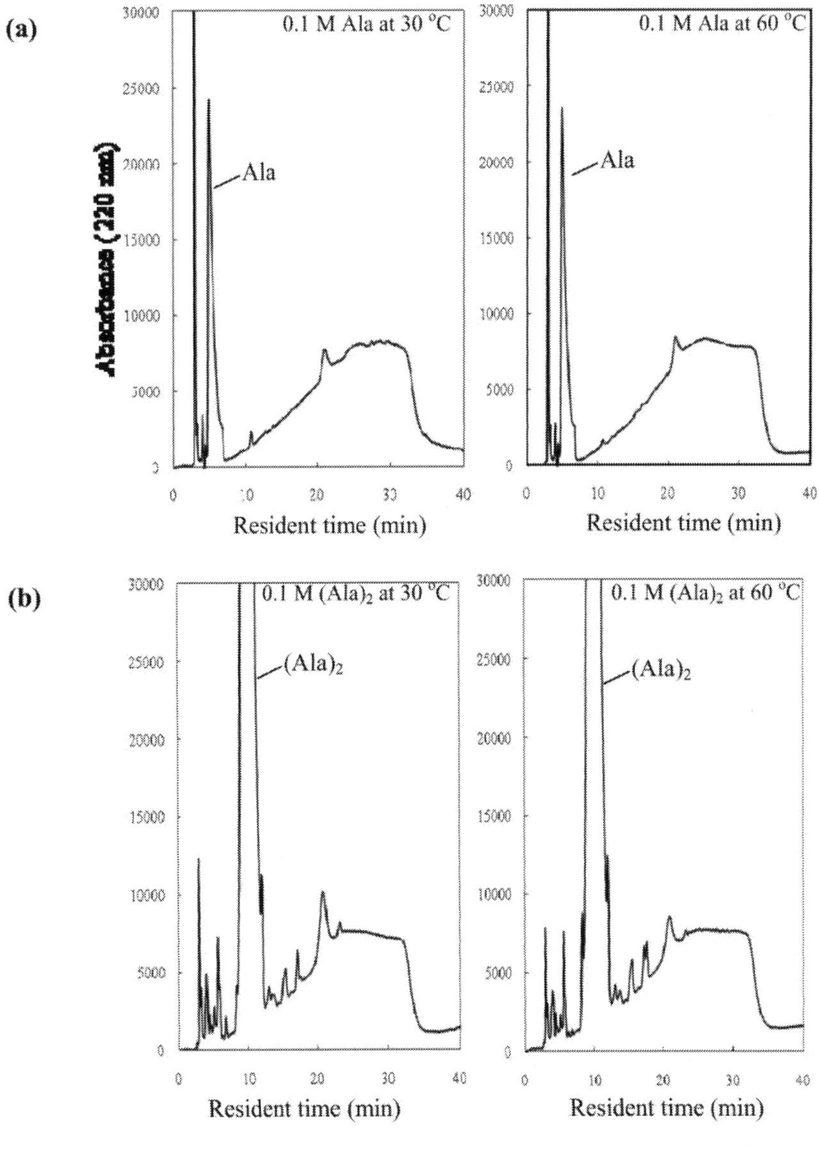

Figure 4. HPLC chromatogram for the reaction of (a) Ala and (b) $(Ala)_2$ under discharge plasma

Figure 5 showed the chromatograms of LC-MS for the elongation of $(Ala)_2$ under discharge plasma at temperature of 60°C. In the products from $(Ala)_2$, LC-MS analyses showed that products possess molecular weights of 90.06 $[M^+H]^+$, 143.16 $[M^+H]^+$, and 321.18 $[M^+H]^+$, which consistent with those of Ala, DKP, and $(Ala)_4H_2O$. It shows that main products from $(Ala)_2$ were Ala, $(Ala)_4$ H_2O and DKP. The materials do not affect the reactions at these temperatures. Additionally, it was surprising that

$(Ala)_3$ was smaller than $(Ala)_4$ in the products from $(Ala)_2$. $(Ala)_4$ was directly synthesized from $(Ala)_2$ without forming $(Ala)_3$. These results suggested that the elongation of $(Ala)_2$ increased slightly with the existence of discharge plasma at the same temperature of reaction.

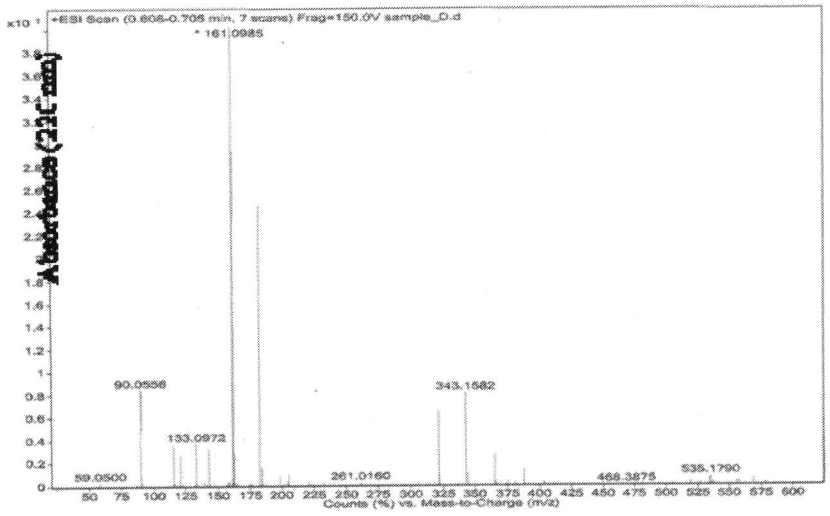

Figure 5. LC-MS chromatogram for the elongation of $(Ala)_2$ under discharge plasma at 60 °C

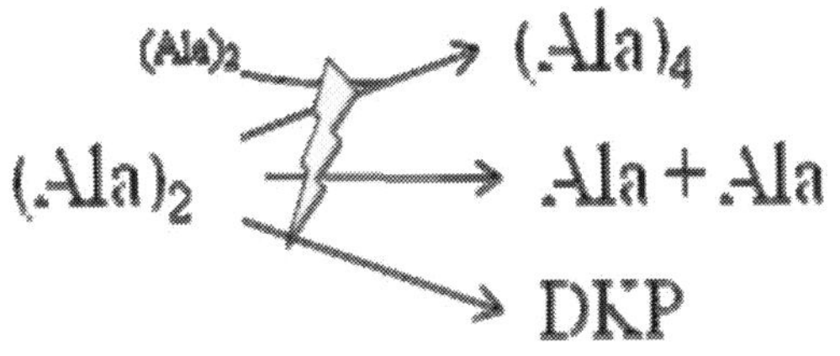

Figure 6. Possible reaction pathway for $(Ala)_2$ induced by pulse discharge plasma under hydrothermal conditions

Based to the results of the above experiments, it is possible to propose the reaction pathway for alanine elongation under pulse discharge plasma conditions. Figure 6 shows the elongation process of alanine in under pulse discharge plasma. Initially, the charged species sustained locally around the anode in contact with the solution surface bombarded water molecules to dissociate them therein. The active radicals, especially hydroxy radical attacks alanine. Otherwise, they recombine with alanine derived compound in a competitive manner.

Conclusion

The preliminary study of alanine reaction behavior under hydrothermal conditions and under pulsed discharge plasma over water surface in a batch-type reactor was performed at 150-250 °C and 30-60 °C, respectively. Under hydrothermal conditions, alanine was converted into its derived compounds. However, the HPLC result could not show clearly. When the reaction was conducted under pulsed discharge plasma, $(Ala)_2$ could be converted into the its derived compounds with higher molecular weight. Furthermore, this reaction method would be applicable for non-condensing agent synthesis of oligopeptides within aqueous solutions combined using the pulsed discharge plasma over water surface reactor system.

Acknowledgment

This work was supported by Kumamoto University Global COE Program "Global Initiative Center for Pulsed Power Engineering".

References

1. N. Sano, T. Kawashima, J. Fujikawa, T. Fujimoto, T. Kitai, and T. Kanki. (2002), *Ind. Eng Chem. Res*. 41, 5906-5911.
2. Y. Wen, and X. Jiang. (2000), *Plasma Chemistry and Plasma Processing* 20, 34.
3. P. Lukes, and B. R.Locke(2005), *Ind. Eng. Chem. Res*. 44, 2921.
4. M. A. Lieberman and A.J. Lichtenberg. (2005), *Principles of Plasma Discharges and Materials Processing, 2nd ed.*,p.6.
5. I. V. Kosinskaya, and L. P. Polozova. (1969), *Zhurnal Principles of Plasma Discharges and Materials*, 1151.
6. Ch. Tegeler, R. Span, and W. Wagner. (1999) , *J. Phys. Chem. Ref. Data*, 28, 3, pp.779-850.
7. S. Liang and D. C. Tilotta. (1998), *Anal. Chem*., 70(3), 616-622.
8. H.M. Roder, R.A. Perkins and C.A. Nieto de Castro. (1989), *Intern. J. Thennophys*. 10, 1141.
9. R. W. Shaw, T. B. Brill, A. A. Clifford, C. A. Eckert and E. U. Franck. (1991), *Chem. & Eng. News,* 69, 26.
10. M. Sasaki, B. Kabyemela, R. Malaluan, S. Hirose, N. Takeda, T. Adschiri and K. Arai, (1998), *J. Supercrit. Fluids,* 13, pp. 261-268.
11. M. Sasaki, Z. Fang, Y. Fukushima, T. Adschiri, and K. Arai.(2000) *Ind. Eng. Chem. Res*., 39(8), 2883-2890.
12. Wahyudiono, S. Fujinaga, M. Sasaki, and M. Goto,(2006), *Chem. Eng. Technol*., Vol. 29, p. 882.
13. Wahyudiono, M. Sasaki, and M. Goto. (2006), *J. Mater. Cycles Waste Manag*. 9, 882.
14. K. Kawamura, T. Nishi, and T. Sakiyama. (2005) "Consecutive Elongation of Alanine Oligopeptides at the Second Time Range under Hydrothermal Conditions Using a Microflow Reactor System", *J. Am. Chem. Soc*., 127 (2), 522-523.
15. E. Imai, H. Honda, K. Hatori, A. Brack, K. Matsuno, (1999), *Science*, 283, 831-833. (b) (1997) *ViVa Origino*, 25, 191-204.
16. E. Imai, H. Honda, K. Hatori, K. Matsuno, (1999), *Origins Life EVol. Biosphere*, 29, 249-259. (b) Y. Ikushima, (2002), *Koukai Tokkyo Kouhou*, JP2002- 37799, JP2002-53594.
17. K. Kawamura, M. Yukioka, (2001) *Thermochim. Acta*, 375, 9-16.
18. S. M. Steinberg, J.L. Bada, (1981) *Science*, 213, 544-545.
19. Qian Y. Engel, M. H, Macko. S. A, Carpenter, S.and Deming J. W. (1993), *Geochim. Cosmochim. Acta*, 57, 3281-3293.

(713f) Continuous-Flow Reforming of Hydrocarbons Underground: Kinetic Analysis

Yousef Alshammari, Imperial College London, London, United Kingdom and **Klaus Hellgardt**, Chemical Engineering, Imperial College London, London, United Kingdom

Water-producing oil wells are potential candidates for clean and economic production of hydrogen, syngas and low sulphur fuels. This may be achieved by transforming the passive production well into a high-pressure, high-temperature, continuous-flow reactor, eliminating the need for above surface reforming steps, and reducing the carbon footprint. As a part of this study, thermodynamic analysis of hydrothermal gasification of hexadecane, a heavy-saturate oil model was carried out and presented in the AIChE meeting 2010. The present work reports the experimental analysis of the hydrothermal gasification of hexadecane at various reaction residence times , temperatures, and H_2O_2 concentrations.

The aim is to develop an uncatalysed continuous-flow hydrothermal reactor model for hydrogen generation from hydrocarbon resources. While some researchers have studies hydrothermal reactions of hydrocarbons in batch-type reactors, none has determined the kinetic data of heavy-saturate hydrocarbons in continuous flow systems (Arai *et al,* 2000, Watanabe *et al,* 2001, Watanabe *et al,* 2000, and Tsuzuki *et al,* 1999). Continuous flow systems offer high potential for commercial scale-up compared with batch systems. Online gas analysis was carried out using a quadruple mass spectrometer while produced oil-water residue was analysed using GC-MS. Moderate yields of syngas, and cracked n-alkanes/1-alkenes (C_9-C_{15}) were produced at 300 – 565 °C and 250 bar. The conversion of hexadecane has been found to be inversely proportional to the residence time, and directly proportional to the water density. In addition, increasing the concentration of H_2O_2 enhanced the gasification of hexadecane as observed by the significant rise in the gas flowrate. On the other hand, coke formation due to hexadecane thermolysis was experienced at 565 °C and 250 bar. It was also found that increasing the pressure at a constant temperature and H_2O_2 concentration reduces the formation of n-alkanes/1-alkenes products while it negatively causes the undesirable oligomerisation of hexadecane. Results produced were compare with previously reported catalytic batch systems (Arai *et al,* 2000, Watanabe *et al,* 2001). Additional improvements have been introduced into the experimental system including mixing the reactants at the desirable reactor temperature, and decomposing the H_2O_2 in a separate reactor prior to oxidising the hexadecane feed. This experimental analysis is also complimented with heat, mass and momentum balance using COMSOL enabling direct characterisation of the hydrothermal reactor.

See more of this Session: Reactions In near-Critical and Supercritical Fluids
See more of this Group/Topical: Catalysis and Reaction Engineering Division

(714a) The Role of Cellulose Accessibility On Enzymatic Saccharification of Lignocelluloses

Junyong Zhu, USDA Forest Products Laboratory, Madison, WI, Zhiguang Zhu, Biological Systems Engineering, Virginia Polytechnic Institute and State University (Virginia Tech), Blacksburg, VA, Y.-H. Percival Zhang, Biological Systems Engineering Department, Virginia Polytechnic Institute and State University, Blacksburg, VA, Yonghao Ni, 3Limerick Pulp and paper Center, University of New Brunswick, New Brunswick, Canada, ZiBin He, Limerick Pulp and Paper Centre, Virginia Polytechnic Institute and State University (Virginia Tech), Blacksburg, VA and Xiaolin Luo, South China University of Technology, Guangzhou, China

So much research has been dedicated to understand the effects of various factors on enzymatic hydrolysis of lignocelluloses. It has been understood that removal of hemicelluloses is very important to improve lignocellulosic substrate digestibility. It has also been identified that improving enzyme accessibility to cellulose is more important than delignification to enzymatic saccharification. This study is attempted to develop a comprehensive understanding of various factors on enzymatic hydrolysis of lignocelluloses by taking a progressive approach. In the first phase, a set of lignocellulosic substrates were produced by drying to different degrees from the same never dried sample. Therefore, these dried substrates have exactly the same chemical composition and structure. However, drying induced fiber hornification produces irreversible change of cell wall structure when rewetting, i.e., fibrils are tightly packed together through hydrogen bonding to produce permanent fiber pore shrinkage and loss of pores in cell wall. As a result, the enzyme accessibilities of these samples were significantly reduced comparing with the never- dried sample and varied with the degree of drying. This set of samples is an excellent candidate to verify the effect of enzyme accessibility on enzymatic hydrolysis. In the second phase, two sets of samples, one produced from fractionation of the same sample, and another from disk milling of the same chemically pretreated biomass to different degrees, were used to determine the effect of substrate size/specific surface or degree of size reduction on enzymatic hydrolysis. Finally, different pretreatment methods were used to explain the effect of chemical composition on enzymatic hydrolysis. Both enzymatic hydrolyses and measurements of cellulase accessiblity to cellulose (CAC) were carried out to quantify the contributions of various factors to enzymatic hydrolysis of lignocelluloses.

See more of this Session: Recalcitrance of Woody Biomass
See more of this Group/Topical: International Congress on Energy 2011

(714b) Structural Studies On Recrystallized Ionic-Liquid (IL) Treated Lignocellulosic Biomass

Indira Samayam[1], B. Leif Hanson[2], Paul Langan[3], Christopher J. Barr[1] and **Constance Schall**[1], (1) Chemical and Environmental Engineering, University of Toledo, Toledo, OH, (2)Instrumentation Center, The University of Toledo, Toledo, OH, (3)Biosciences Division - Los Alamos National Laboratory, Los Alamos, NM

Terrestrial lignocellulosic biomass is composed of three major components; cellulose, a linear polymer of glucose (a hexose sugar); hemicellulose, a complex branched polymer of xylose (a pentose sugar) and other sugar derivatives; and lignin, a polyphenyl propanoid macromolecular assembly that is covalently cross-linked to hemicellulose. These components can serve as a source of carbon based feedstock for fuel and chemical production in much the same way that crude oil serves as the carbon feedstock in petrochemical refineries.

The deconstruction of lignocellulosic biomass into simple sugars constitutes a core barrier for producing value added products from the sugar platform. The cellulosic component has a crystalline and oriented fibrillar structure in biomass, whereas the hemicellulosic and lignin components have less ordered structures that encrust the cellulose fibers. The crystalline structure of native cellulose fibers, called cellulose I, can be a major impediment to its hydrolysis to monomeric sugars. Pretreatment can disorder the crystalline cellulosic component and result in structural changes in the cellulose fibers. These structural changes in cellulose fiber can be achieved by modifying the pretreatment conditions. Depending upon the type of pretreatment, the conversion of cellulose fibers to other crystal forms, such as cellulose II, cellulose III_{II} and cellulose III_{I}, or amorphous forms, can greatly improve their susceptibility to hydrolysis. To investigate the effects of varying the parameters of IL pretreatment on the structure of their cellulosic components, X-ray powder diffraction (XRD) and X-ray fiber diffraction from biomass samples was used with unoriented and partially (fiber) orientated cellulosic components . A comparative study on enzymatic hydrolysis and sugar analysis related to the observed changes in cellulose structure to biomass digestibility was also performed. The improvement in hydrolysis is hypothesized to be primarily due to structural changes in the cellulose constituent.

See more of this Session: Recalcitrance of Woody Biomass
See more of this Group/Topical: International Congress on Energy 2011

(714c) Kinetics and Mechanism of Cellulase Hydrolysis of Nanocrystalline Cellulose

Xiao Zhang, Chemical Engineering and Bioengineering, Washington State University, Richland, WA and Xiaohui Ju, Washington State University, Richland, WA

Kinetics and Mechanism of Cellulase Hydrolysis of Nanocrystalline Cellulose

Xiaohui Ju, Xiao Zhang

School of Chemical Engineering and Bioengineering, Centre for Bioproducts and Bioenergy, Washington State University,

Abstract

Decrystallization of cellulose is a major bottleneck in enzymatic hydrolysis of lignocellulosic biomass. The kinetic and mechanism of enzyme degradation of nanocrystalline cellulose prepared from bleached kraft pulp were investigated. The efficiency of enzymatic degradation of nanocrystalline cellulose by three commercial enzymes was determined. All these cellulase preparations demonstrated the ability to degrade nanocrystalline cellulose. The mechanism of enzyme decrystallization of cellulose was investigated by XRD and NMR. This presentation will discuss some of the new findings in cellulose crystal structural changes during enzymatic hydrolysis.

See more of this Session: Recalcitrance of Woody Biomass
See more of this Group/Topical: International Congress on Energy 2011

(714d) Incremental Disassembling of Woody Biomass and Production of Biochemicals

Shijie Liu, Paper and Bioprocess Engineering, SUNY College of Environmental Science and Forestry, Syracuse, NY

Woody biomass is a reliable renewable source of raw materials for energy and chemicals. Woody biomass is structurally composed of cellulose, lignin, hemicelluloses and other minor components. The compositions vary from species to species. The complexity / heterogeneity of woody biomass presents challenges to conversion technologies, which is perceived as recalcitrance. Pretreatment and detoxification have been two unfavorable but widely discussed processes todate. At SUNY ESF, a strategy of incremental disassembling of woody biomass is adopted in which hemicelluloses and extractives are separated first through hot-water extraction, cellulose and lignin are next. Hot-water extraction uses water as solvent without any other added chemicals. Acetyl groups in the woody biomass dissolve into the liquor and serves as catalyst. In each step of the separation, high value applications are examined. For example, hemicelluloses are hydrolyzed and fractionated via membrane for xylo-oligomers, xylose, acetic acid, methanol, and aromatics, etc. The residual solid mass containing mostly cellulose and lignin has wide applications: fiber board, wood fuel pellet, pulp and paper, and sugars / aromatics. Incremental disassembling is emphasized for value preservation of various components. A clever integration of the processes makes the conventional pretreatment and detoxification processes obsolete. The sugar stream produced at SUNY ESF has been successfully fermented to ethanol, butanol, lactic acid, and PHA. There is no deliberate detoxification that generates waste byproducts.

See more of this Session: Recalcitrance of Woody Biomass
See more of this Group/Topical: International Congress on Energy 2011

(714e) Effect of Steam Explosion On Degradability and Accessiblity of Loblolly Pine In Bioethanol Applications

Yuzhi Kang, Sang Beom Kim, Prabuddha Bansal, Dr Matthew Realff and Andreas S. Bommarius, School of Chemical and Biomolecular Engineering, Georgia Institute of Technology, Atlanta, GA

Abstract

Among the various biomass pretreatment methods, steam explosion is one of the most widely used due to its cost-effective disruption of the plant structure and hence enhanced enzymatic accessibility. The operational parameters including temperature, residence time and acid concentration have pronounced effect on crystallinity (CrI) of the feedstock loblolly pine. CrI was determined from X-ray diffraction pattern with the method of least squares and it was found to decrease as pretreatment severity increase.

Three different substrates, Avicel, phosphoric acid swollen cellulose (PASC), and SO_2 steam exploded Loblolly pine (SELP), were subjected to enzymatic hydrolysis and subsequent adsorption study. Both SELP samples (2.9-419-5 and 5-417-5) were found more readily degraded by the CelluclastÒ enzyme system than Avicel. Despite the complexity of both substrates and enzyme mixtures, the SELP adsorption isotherm can be fitted to a Langmuir isotherm with $R^2 > 0.99$. The saturation point was reached at ~6.24 mg/ml total enzyme loading. Adsorption data was correlated with initial rates and a linear relationship between initial rate and adsorption amount was confirmed.

See more of this Session: Recalcitrance of Woody Biomass
See more of this Group/Topical: International Congress on Energy 2011

CHANGE OF STRUCTURE OF CORNCOB WITH DILUTE ACID AND LIME PRETREATMENT AND A POTENTIAL PROCESS FOR XYLOSE CO-PRODUCTION OF CELLULOSIC ETHANOL

Ji-hong Li[1], Shi-Zhong Li[1], Zhi-pei Yan[1] , Qing Li[2], Liang-cai Peng[3]

[1]Institute of New Energy Technology, Tsinghua University, Beijing, China

[2]Colledge of Science, Huazhong Agricultural University , Wuhan , China

[3]Colledge of Plant Science & Technology, Huazhong Agricultural University , Wuhan , China

Abstract

Degree of polymerization (DP) of cellulose is considered an important factor to effect on the conversion of Lignocellulose. The higher DP of cellulose, the more difficult conversion of lignocellulose is. In this study, corncobs were pretreated with dilute acid process for production of xylose, and then the residues were pretreated with lime. The enzymatic hydrolysis and the structures involved the composition, DP and crystallinity index of these two types of feedstock were determined. Comparison of structures resulted from these two pretreatment revealed that DP was not the crucial factor to determined efficiency of enzymatic hydrolysis. Lime pretreatment could improve the conversion of cellulose although the residues with lime pretreatment had higher DP and percentage of crystallized cellulose. Moreover, the results also confirmed lignin was not the key factors for enzymatic hydrolysis of lignocellulose because the residues with dilute acid pretreatment had the similar percentages of lignin to that with lime pretreatment. The crucial effect of lime pretreatment was to remove the inhibitor for cellulase and increase the percentages of cellulose. With obtained xylose, the residues with further lime pretreatment had expected efficient enzymatic hydrolysis of cellulose. So this process could be a potential process for xylose co-production of cellulosic ethanol to decrease the cost of cellulosic ethanol.

(714g) Green and High Efficient Enzyme Hydrolysis Process for Sweet Sorghum Bagasse

Xinshu Zhuang, Wen Wang, Qiang Yu, Wei Qi, Qiong Wang, Xuesong Tan, Yu Zhang and Zhenhong Yuan, Guangzhou Institute of Energy Conversion,CAS, Guangzhou, China

In order to improve enzymatic hydrolysis efficiency, several approaches including liquid hot water (LHW) pretreatment, adding surfactant, magnetic stirring and fed-batch hydrolysis were investigated, in which Sweet sorghum bagasse (SSB) was chosen as the feedstock. The experiment results showed that with the LHW pretreatment, the SSB conversion ratio reached 35.91% , nearly 75% xylan and 33% lignin were removed. Using the LHW-pretreated SSB, glycan conversion ratio raised from 16.6% to 37.9% by adding surfactant Tween80 with the optimum dosage of 0.175ml/g DM. The virulence test showed that Tween80 nearly had no influence to the growing of ethanol fermentation yeast- *Saccharomyces cerevisiae*. Comparing stirring and shaking mode for the enzyme hydrolysis, the glycan conversion rate was 57.3% in stirring mode, higher than 44.4% obtained from shaking mode. The image of Scanning Electron Microscope (SEM) showed that LHW pretreatment could destroy the tight structure of SSB and the comminution of pretreated material could smaller the particles to enhance the hydrolysis process. Finally the LHW-pretreated SSB, Tween80 and stirring mode were applied to fed-batch process, the concentration of glucose, cellobiose and xylose reached 88.8g/l, 30.7g/l and 18.9g/l respectively, and the glycan conversion ratio was 89.3% under the optimum condition. These research works lay the foundation for the construction of a green and high efficient sugar platform to facilitate the cellulosic ethanol development.

These research work are financially supported by the National Key Technology R&D Program (2011BAD22B01), National High Technology Research and Development Program of China (2010AA101606) and the Knowledge Innovation Project of the Chinese Academy of Sciences (KSCX1-YW-11-A3, KSCX2-YW-G-063)

[1]Corresponding author. yuanzh@ms.giec.ac.cn (Z. Yuan)

See more of this Session: Recalcitrance of Woody Biomass
See more of this Group/Topical: International Congress on Energy 2011

(716a) Metal Distribution During Impregnation and Drying of Ni/Alumina and Ni-Mo/Alumina Catalysts

Xue Liu, Rutgers University, Piscataway, NJ, Johannes G. Khinast, Institute for Process and Particle Engineering, Graz University of Technology, Graz, Austria and Benjamin Glasser, Dept. of Chemical and Biochemical Engineering, Rutgers University, Piscataway, NJ

Supported catalysts are essential components in a variety of industrial processes, ranging from catalytic converters to production of new drugs. They are generally required because of their high surface area and high mechanical and thermal stabilities. The performance of a catalytic process is intimately related to the catalyst design - uniform, egg-yolk, egg-shell and egg-white metal profiles. Although catalyst preparation and catalytic processing have been investigated for many years, many aspects of catalyst manufacturing are still not fully understood and in industry the design of catalysts is usually based on trial and error, which is expensive and time-consuming, and does not offer assurances on the final results.

It is generally believed that the metal profile is controlled by the conditions that are applied during impregnation where the metal contacts the solid support for the first time. However, experiments have shown that drying may also significantly impact the metal distribution within the support. Therefore, to achieve a desired metal profile we need to understand both impregnation and drying. Controlling the impregnation and drying conditions can enhance catalyst performance, and minimize the production of useless batches that have to be disposed, or recycled.

In this work we have developed theoretical models to simulate the impregnation and drying processes of Ni/Alumina and Ni-Mo/Alumina systems. The initial input of the drying model comes from the results from the impregnation simulations. Therefore, we can combine impregnation and drying together to analyze their effects on the preparation of supported catalysts. Of particular interest is to compare simulation results and experimental measurements to determine the key parameters that control the impregnation and drying processing. Experimental work has also been carried out for impregnation and drying of Ni/Alumina and Ni-Mo/Alumina systems and the simulations are generally able to capture the behavior observed experimentally. The effects of impregnation time, initial metal concentration, drying temperature and solution surface tension on the metal distribution have been examined. By adjusting these parameters, uniform, egg-shell, and egg-white profiles can be obtained. To investigate the impact of support shape and support size on the metal distribution, three high surface area γ-alumina carriers have been examined, including rings with 7.9mm diameter, pellets with 3.2mm diameter and trilobes with 1.4mm diameter.

Our simulations and experiments have allowed us to better understand the fundamental mechanisms that occur during impregnation and drying, and to develop a strategy that can generate desired metal profiles.

The models used in the present work can capture the essential physics of impregnation and drying while still maintaining a level of generality. Although the results presented are based on a particular metal/support system, they serve to provide physical insight into the fundamentals of the impregnation and drying processes.

See more of this Session: Science and Engineering of Catalyst Preparation II
See more of this Group/Topical: Catalysis and Reaction Engineering Division

(716b) Simulations and Experiments of Dry Catalyst Impregnation for Improved Content Uniformity

Frank Romanski[1], Atul Dubey[2], Yangyang Shen[1], Arthur W. Chester[1] and M. Silvina Tomassone[1], (1) Chemical and Biochemical Engineering, Rutgers University, Piscataway, NJ, (2)Chemical and Biochemical Engineering, Rutgers University, Piscataway, NJ, Rutgers University, Piscataway, NJ

The dry impregnation of catalysts is widely used in the preparation of heterogeneous catalysts, however, there is very little work done at the computational level. In this work, Discrete Element Method (DEM) simulations coupled with a novel algorithm allowing the transfer of fluid to and between particles were used in combination with geometrically equivalent experiments to model dry catalyst impregnation. Specifically, the effects of flow rate, volumetric fill level, metal solution type, particle dimensions, and morphology were explored within a double cone geometry blender, where the process parameters were optimized for the best overall mixing and content uniformity within the granular bed. While the developed model and corresponding algorithm have shown good agreement with experiments, the double cone blender itself does not establish an optimal content uniformity alone, and as a consequence, various improvements were introduced and subsequently analyzed within the system including baffles and various nozzle configurations. It was found that the addition of baffles to the system greatly improves the mixing and fluid content uniformity during dry impregnation, and additionally, baffles allow for greater volumes of catalyst support to be processed than traditionally acceptable using the double cone geometry. The culmination of this work has been the development of simulations capable of analyzing dry catalyst impregnation in limitless parameters, geometries, and conditions for improving overall mixing, content uniformity, and product quality.

See more of this Session: Science and Engineering of Catalyst Preparation II
See more of this Group/Topical: Catalysis and Reaction Engineering Division

(716c) Preparation, Characterization and Evaluation of Group IB-Pd Bimetallic Catalysts Prepared by Electroless Deposition

Yunya Zhang, Jayakiran Rebelli, Christopher T. Williams and John R. Monnier, Chemical Engineering, University of South Carolina, Columbia, SC

Preparation, Characterization and Evaluation of Group IB-Pd Bimetallic Catalysts Prepared by Electroless Deposition

Yunya Zhang, Jayakiran Rebelli, Christopher Williams, and John R. Monnier

Department of Chemical Engineering, University of South Carolina, Columbia SC 29208, USA

Research has shown that bimetallic catalysts often enhance activity, selectivity and stability relative to their corresponding monometallic catalysts because of interactions between these two metals. Co-impregnation and successive impregnation are two widely used methods in industry to synthesize bimetallic catalysts. However, these two methods commonly produce both monometallic and bimetallic catalyst particles. This complex mixture of monometallic and bimetallic particles (of variable composition) results in poor control of the final catalyst composition and makes it very difficult to characterize these catalysts and correlate catalyst performance with bimetallic catalyst composition.

Electroless deposition (ED) is an alternative, industrially-feasible method to produce a wide range of truly bimetallic catalysts. This methodology uses a controlled chemical reaction to deposit a second metal onto a pre-existing (primary) metallic surface from a solution containing a reducible metal salt and a reducing agent. Because the reducing agent is catalytically activated by the pre-existing metallic surface, the second metal is deposited only on the primary metal. Since the ED process is kinetically-controlled, the final composition of a particular bimetallic catalyst can be controlled to give rather precise combinations of the two metallic components.

We have used ED to prepare an extensive series of Au-, Ag-, and Cu-Pd/SiO_2 bimetallic catalysts having controlled, incremental surface coverages of each group IB metal. Electroless deposition baths were developed using the bis-cyano metal (1+) salts of each Group IB metal as precursors and hydrazine (for Au and Ag) or dimethylamine borane (for Cu) as the reducing agents.

The Group IB bimetallic catalysts have been characterized by atomic absorption spectroscopy, selective H_2 chemisorption, and Fourier transform infrared spectroscopy (FTIR) of CO. Because Cu, Ag, and Au do not chemisorb H_2 or CO at room temperature, chemisorption and FTIR are specific for the exposed Pd component of the bimetallic catalysts. The analytical results demonstrate that electroless deposition is an effective method to prepare bimetallic catalysts with controlled coverages of the IB metal on the Pd surface. These bimetallic catalysts have been evaluated for selective hydrogenation of acetylene in the presence of ethylene, an important industrial process. The kinetic results indicate superior performances for high coverages of the IB metal on the Group IB-Pd bimetallic catalysts. Explanations will be given for the enhanced performances of the compositions.

See more of this Session: Science and Engineering of Catalyst Preparation II
See more of this Group/Topical: Catalysis and Reaction Engineering Division

(716d) Investigation of Gold-Metal Oxide Interface: Active Site for Selective Oxidation

Neema A. Mashayekhi, Mayfair C. Kung and Harold H. Kung, Chemical and Biological Engineering, Northwestern University, Evanston, IL

Gold nanoparticles supported onto metal oxides have potential for selective oxidation of alkanes, but catalytic performance has to be improved for practical applications. A logical point of attack is the active site, but both the specific active site of gold and the role of the support are unresolved. However, it is commonly agreed that the location of the active site is at the interface of gold (Au) and the support (metal oxide) for oxidation reactions. Two important parameters will be investigated at the interface: nuclearity and nature of the metal-oxide at the gold-metal oxide interface. In order to study this interface a synthesis scheme to control local environment of the Au particles must be developed.

We have developed a scheme where polymethylhydrosiloxane (PMHS) is modified with an amine precursor, N-methyl-aza-2,2,4-trimethylsilacyclopentane, which was used to bind and stabilize the Au particles. Using the modified PMHS, Au nanoparticles were prepared by addition of Au precursor, chloro(tetrahydrothiophene)gold (I), which were reduced using hydrosiloxane. At a ligand:Au ratio of 10, particles less than 4nm in size were obtained with high loading (>7%). Titanium-oxo precursors were also incorporated onto amine functionalized PMHS at different loadings to study the effect of the presence of the metal oxide for selective oxidation of propane. For future work, different types of metal-oxo species and different nuclearities will be incorporated onto PMHS and their catalytic effects would be studied.

See more of this Session: Science and Engineering of Catalyst Preparation II
See more of this Group/Topical: Catalysis and Reaction Engineering Division

(716e) Design and Preparation of Nanoengineered Multi-Component Catalysts for Fuel Cells and LithiumAir Batteries

Chuan-Jian Zhong, Department of Chemistry, State University of New York at Binghamton, Binghamton, NY

The ability to engineer the size, shape, composition, phase and surface properties of nanoscale catalysts is important for many electrochemical energy devices. Electrochemical energy devices such as fuel cells and lithium-air batteries represent a clean energy vector to address the energy and environmental problems today. A major area of challenge to the commercialization of these electrochemical energy devices is the poor performance of catalysts for the cathode reactions, calling for design of multifunctional and stable electrocatalysts. To address this challenge, we have been developing advanced approaches to nanoengineering multi-component catalysts with controlled size, shape, composition, and alloying/composite structures. In addition to synthesis and processing of multi-component nanoparticles, alloys, oxides, and core-shell nanostructures, a key emphasis is placed on establishing the correlation of the electrocatalytic activity and stability with nanostructural parameters of the catalysts. Recent results in an effort to achieve the better design and fabrication of functional, stable and low-cost cathode catalysts for fuel cells and lithium-air batteries will also be discussed.

See more of this Session: Science and Engineering of Catalyst Preparation II
See more of this Group/Topical: Catalysis and Reaction Engineering Division

(716f) The Effect of Pretreatment On Electrostatically Adsorbed Noble Metal Precursors

Manuel Nieto, David Childers and **John R. Regalbuto**, U. Illinois at Chicago, Chicago, IL

One of the most important concerns in catalyst synthesis is the metal particle size. After impregnation of the support with the metal precursor, the pretreatment conditions of the catalyst play a crucial role in determining the final metal particle size. This correlation has been indirectly investigated by various studies, but hasn't been systematically examined. This research project deals with the influence of temperature and heating rate during the reduction process in H_2 flow on Pt and Pd particle size on various supports, both low PZC (oxidized carbon, TiO_2, SiO_2) and high PZC (carbon, Al_2O_3). The influence of humidity during reduction is also dealt with for the Pt catalysts. The catalysts are synthesized by dry impregnation (DI) and strong electrostatic adsorption (SEA). Particle size analyses are done by XRD and STEM imaging.

See more of this Session: Science and Engineering of Catalyst Preparation II
See more of this Group/Topical: Catalysis and Reaction Engineering Division

(726a) Biodiesel Production From Waste Cooking Palm Oil In a Continuous Reactive Distillation Column Catalyzed by Superacid Heteropoly Acid: Optimization Using Response Surface Methodology (RSM)

Iman Noshadi[1], Richard Parnas[1], Nor Aishah Saidina Amin[2], Alireza Zarei[2], Hadi Hezaveh[2], Sanaz Hesamedini[3] and Samad Doostdar Somarin[4], (1)University of Connecticut, Storrs, CT, (2)University Technology Malaysia, Johour, Malaysia, (3)Shiraz University, Shiraz, Iran, (4)Institute of Graduate Studies, University of Malaya, Institute of Graduate Studies, University of Malaya, Kuala Lumpur, Malaysia

This study aims to develop an optimal continuous process to produce fatty acid methyl esters (biodiesel) from waste cooking palm oil in a reactive distillation column catalyzed by a heteropolyacid, $H_3PW_{12}O_{40}.6H_2O$. The conventional production of biodiesel in the batch reactor has some disadvantage such as excessive alcohol demand, short catalyst life and high production cost. Reactive distillation combines reaction and separation to simplify the process operation. The reaction catalyzed by $H_3PW_{12}O_{40}.6H_2O$ overcomes the neutralization problem that occurs in conventional transesterification of waste cooking oil with high free fatty acid (FFAs) and water content. Response surface methodology (RSM) based on Box-Behnken was used to design the experiment and analyzed four operating parameters: total feed flow, feed temperature, reboiler duty and catalyst concentration. The optimum conditions were determined to be 115.52 (mol/h) total feed flow, 29.11°C feed temperature, 1.22 kW reboiler duty, and 10 catalyst concentration. The optimum and actual WCO conversion was 95.1% and 94.8%, respectively, which shows that the RSM is an accurate method for the current procedure.

See more of this Session: Alternative Fuels II
See more of this Group/Topical: Catalysis and Reaction Engineering Division

(726b) Production of Green Diesel From Brown Grease Via Decarboxylation

Elvan Sari[1], Craig DiMaggio[2], Manhoe Kim[2], Steven O. Salley[2] and K. Y. Simon Ng[2], (1)Chemical Engineering and Materials Science, Wayne State University, Detroit, MI, (2)Department of Chemical Engineering and Material Science, Wayne State University, Detroit, MI

Brown grease (BG) is a mixture of waste fats and oils collected from wastewater traps. The free fatty acid (FFA) content of BG can range between 50-100%. While BG is a potential low cost and non-food competing feedstock source for biofuels, it is a challenging feedstock due to its high FFA content. However, FFAs are potentially good candidates for decarboxylation reactions where the oxygen is removed as carbon dioxide, producing hydrocarbons referred to as "green diesel." This process is a low cost alternative to hydrogenation, where excess hydrogen is required along with high pressure conditions in order to remove oxygen as water to form straight chain hydrocarbons in the diesel fuel boiling range. Approximately 75% of BG consists of unsaturated FFAs with oleic acid representing the largest portion (around 50%) [1]. Studies on the decarboxylation of oleic acid have shown that the hydrocarbon selectivity is low because the unsaturated carbon-carbon double bond (C=C) in the oleic acid is saturated first under the reaction conditions in the presence of H_2, and therefore decarboxylation of oleic acid yields stearic acid (saturated FFA) as an intermediate and n-heptadecane as final product[2]. In the absence of H_2, the key challenge is catalyst deactivation. We explored the conversion of brown grease to green diesel via decarboxylation with minimal hydrogen consumption. The purpose of this study is to develop a mechanism for the conversion of FFAs to n-paraffins, in the diesel fuel distillate range, to provide a better understanding of the processing conditions required and optimization of the reaction conditions in order to further develop the current green diesel process. Experiments were conducted using a 100-mL stirred Hanwoul high pressure batch reactor and a 5wt.% Pd catalyst on an activated carbon support (Aldrich). Results demonstrated that all the BG compounds can be fully converted in 9 hours to straight chain hydrocarbons using mild reaction conditions (15 bar and 300 °C) and an inert gas with 10 vol% H_2. However, the decarboxylation reaction rate was observed to be very slow due to the presence of unsaturated FFAs in the BG. To further investigate the effects of unsaturated FFAs in the feedstock on the reaction rate, a pre-hydrogenation step was also performed before decarboxylation as part of study.

References

1. Kim, M.; DiMaggio, C.; Yan, S.; Wang, H.; Salley, S.O.; Ng, K.Y.S., Performance of heterogeneous ZrO2 supported metal oxide catalysts for brown grease esterification and sulfur removal. *Bioresource Technology* **2010**, doi:.1016/j.biortech.2010.10.105

2. Immer, J. G.; Kelly, M. J; Lamb, H. H., Catalytic reaction pathways in liquid-phase deoxygenation of C18 free fatty acids. *Applied Catalysis A:General* **2009**, *375*, 134-139

See more of this Session: Alternative Fuels II
See more of this Group/Topical: Catalysis and Reaction Engineering Division

Title: *Reversed-Phase Ternary Phase Diagram for Commercial Biodiesel - Glycerol - Methanol*

Authors: Stephen N. Csernica and James T. Hsu

In the present work, the liquid-liquid ternary phase diagram for a commercial biodiesel/glycerol/methanol system was experimentally determined at 20 °C. Proper knowledge of the phase equilibrium behavior of these systems is required for the design of efficient separation processes. As is common in biodiesel/glycerol/methanol systems, a given mixture generally separates into two liquid phases. The top phase is generally rich in biodiesel while the bottom phase consists of glycerol. Due to these components respective solubility parameters, most of the methanol is found in the glycerol phase. At certain compositions, however, the orientation of the two liquid phases reverse themselves, resulting in the biodiesel phase on the bottom and the glycerol phase on the top. The validity of the experimental data was checked through an Othmer-Tobias plot. The Othmer-Tobias plot was also used to predict the plait point, which was determined to be 60.2% biodiesel, 39.1% methanol, and 0.7% glycerol (by mass).

(726d) Kinetics of Biodiesel Production From Non-Edible Vegetable Oils

A. Chadha[1], **R. Ravi**[2] and G Ravikumar[2], (1)Biotechnology, Indian Institute of Technology Madras, Chennai, India, (2)Chemical Engineering, Indian Institute of Technology Madras, Chennai, India

A comparative study of the kinetics of production of biodiesel from the transesterification of three non-edible vegetable oils is undertaken. A three step mechanism was used to model the kinetics in the case of Pongamia oil which was found to be a mixture tri-, di- and monoglycerides. On the other hand, an one-step model was sufficient for Jatropha and Mahua oils both of which contained largely triglycerides. The rate of transesterification at comparable temperatures and catalyst concentrations was found to be the highest for Jatropha oil followed by Mahua and then Pongamia oil. The effect of cosolvent on the rate of transesterification was noticeably higher for Mahua oil than for the other two oils but still much lower than that reported in the literature for both soybean and canola oil. Kinetics of transesterification of oils to yield biodiesel thus depends on the composition of the oil under consideration.
Therefore studies are under way to examine the fine structure of the triglycerides contained in the various oils and establish a correlation with the kinetics of transesterification.

See more of this Session: Alternative Fuels II
See more of this Group/Topical: Catalysis and Reaction Engineering Division

Production of Green Jet Fuel from Edible and Non-edible Oils

Harvind ,Prafulla Patil, Tapaswy Muppaneni , Shuguang Deng, Tanner Schuab, Peter Cooke

A surrogate of jet fuel range hydrocarbons was produced from renewable feedstock such as Camelina oil, palm oil. The fatty acid methyl esters (FAME's) derived from these oils were used as initial feedstock to produce jet fuel range hydrocarbons. The process includes two steps involving hydrocracking and reforming in presence of the catalyst. This work involves the different reaction conditions by taking temperature, pressure, time of reaction and amount of catalyst as variable parameters. Different lab prepared catalysts were used along with an industrial catalyst. The feedstock, intermediate product and final product were analyzed by using ATR-FTIR, TGA and GC-MS.

Keywords: Green jet fuel; Renewable feedstock; Hydro cracking; Bi-functional catalyst

BIODIESEL PRODUCTION BY TRANSESTRIFICATION USING HETEROGENEOUS BASE CATALYST AND EFFECT OF CO-SOLVENT

Arun Kumar Gupta[a], *Goutam Deo*[b]
Department of Chemical Engineering
Indian Institute of Technology Kanpur, Kanpur-208016, India
E-mail: [a]*guptaak@iitk.ac.in*, [b]*goutam@iitk.ac.in*

Abstract

The transesterification reaction for biodiesel production with a several alkali and alkaline metal alkoxides, such as NaOH, KOH, $Ca(OH)_2$, CH_3ONa, CH_3OK, $(CH_3O)_2Ca$, $(CH_3O)_2Mg$ were studied using refined soybean oil. The metal hydroxide and alkoxides of Na and K were found to be the most active. The ethanolysis reaction was also studied for homogeneous catalyst and favourable transesterification of soybean is observed. The influence of Na loading on a γ-Al_2O_3 support for the base heterogeneous catalysts was considered. An optimum NaOH concentration of the catalysts was observed. The yield for the methanolysis reactions reached an optimum for the 10 wt % Na/ γ-Al_2O_3 and 3.5 g catalysts per 100 g of soybean oil. It was observed that the surface area monotonically decreases with an increase in NaOH loading. In the presence of co-solvent the reaction rate is as fast and similar to homogeneous catalysed transesterification and reaches the same final conversion as without co-solvent. The yield of biodiesel was analysed by 1H NMR spectroscopy method. The catalysts were characterised for their surface area and by determining their XRD patterns.

Introduction

Biodiesel is being given serious consideration as a potential source of energy and as an attractive alternative to replace or be blended with petroleum fuel [2]. It is normally produced from the reaction of oils and fats with alcohols. Chemically, the oil and fats consist of triglycerides molecules of three long chain fatty acids that are bonded to a single glycerol molecule. The stiochiometric reaction for biodiesel production requires the transesterification of 1 mol of a triglyceride and 3 mol of the alcohol, however, an excess of the alcohol is used to increase the biodiesel (alkyl esters) yields and to allow its phase separation from the glycerol formed. Several aspects, including the type of catalyst, alcohol, edible/non-edible oil, purity of the reactants (mainly water content) and free fatty acid content have an influence on the course of the transesterification reaction [1].
In this work, the transesterification of soybean oil is studied using several homogeneous and heterogeneous base catalysts. Quantitative analysis of bio-diesel formation was achieved by ^1H NMR spectroscopy. The transesterification activity of soybean was studied by using homogeneous and heterogeneous catalysts. The homogeneous catalysts contained a methanol solution of the hydroxides of sodium, potassium or calcium, or methoxides of sodium, potassium, magnesium or calcium. The performance of heterogeneous sodium-based basic

catalysts was compared with that of conventional homogeneous NaOH. The deficiencies of the heterogeneous base catalysed reaction, namely, a slower initial reaction rate, was addressed by the use of a co-solvent. The heterogeneous base catalysts were characterised for their surface area and by the obtaining their X-ray diffraction patterns.

Experimental Procedure

The experiment for biodiesel production was performed in two steps. In the first step, the pre-treatment of oil with acid catalyst (H_2SO_4, 1% w/w) was done to reduce the free fatty acid (FFA) concentration below 1%. Reduction of FFA was necessary to avoid the saponification reaction, which could occur during the transesterification of vegetable oil. In the acid catalysed step, 100 gm of oil was placed in the reactor and heated to the desired temperature (60°C). Methanol in 13:1 ratio with free fatty acid content (oleic acid) of the oil was mixed separately with 1 % (w/w of oil) H_2SO_4 [2]. The mixture was then added to the reactor. The esterification reaction was carried out at 60°C for 2 hours. The reaction mixture was kept for 12 hr for separation of reacted FFA content and treated oil. The oil phase of the reaction mixture was separated from the mixture and washed with hot double distilled water at 70°C to remove the acid and methanol. The washed oil phase was then subjected to a second step alkali transesterification.

In the second step, the oil layer from the acid esterification step was placed in the reactor and then heated to the desired temperature. Measured amounts of methanol (6:1 molar ratio) and NaOH (1.5 % w/w) of oil were mixed separately at room temperature to make a homogenous solution. This solution was then added to at the bottom of an agitated reactor (750 rpm) to avoid the loss of methanol. The sample was then allowed to settle for 12 hours followed by the washing of the upper biodiesel layer with 70°C heated distilled water and mild shaking to remove the NaOH catalyst and excess methanol. The waste was taken out from the bottom of a separating funnel [3]. The upper ester layers of the samples were analysed by H^1 NMR spectroscopy for the determination of the conversion/yield using published procedures [5].The same steps were followed for different catalysts and alcohol with the corresponding molar ratios.

Heterogeneous Catalyst Preparation

A series of heterogeneous base catalysts (Na/Al_2O_3) were prepared by the incipient wetness (IWM) method. Prior to catalyst preparation, the support was pretreated by mixing with incipient wetness volumes of water. The paste of the support was heated through a pre-defined temperature cycle. The pretreated support was loaded with required amounts of NaOH solution in incipient wetness volumes and followed by the same pre-defined temperature cycle [4]. The reaction with the prepared catalyst was done in the same reactor using the reagents in the same ratios as the homogeneous catalyst.

Catalyst Characterization

The prepared catalysts were characterized for their surface area and their X-ray diffraction (XRD) patterns were determined,

The surface area of the catalysts samples were measured using a SMART SORBS 92/93, which was based on the BET single point equation using nitrogen absorption.
The XRD patterns of the selected samples were obtained with a 180 Debye Flex-2002 X-ray diffractometer. Nickel filtered K_{α} radiation of a copper target (λ=1.541841 A°) was employed covering 2θ angles up to 90°, data were recorded in the step mode at 2θ sweep of 3° min^{-1} and a time constant of 3.0 s was maintained. The data were collected in the range of 10 to 75°.

Results and Discussion

The catalytic activity of the methanol solution of the hydroxides of sodium, potassium or calcium, and methoxides of sodium, potassium, magnesium or calcium, for the transesterification of soybean oil was investigated. Comparing the different homogeneous catalysts it was observed that sodium hydroxide, sodium methoxide, potassium hydroxide and potassium methoxide showed high and comparable conversion (90-100%). In contrast, calcium hydroxide and calcium methoxide showed less conversion (15-20%) and magnesium methoxide showed no conversion in soybean oil methanolysis under the present conditions. As a test the kharanja oil with high FFA content showed 65% conversion using sodium methoxide catalyst. Different NaOH percentage loadings on Al_2O_3 were tried as heterogeneous catalysts and the results are shown in Figure 1(a). It was observed that the conversion increase and then decrease as the NaOH % increased. The best catalyst under the present operating conditions was 10% Na/γ-Al_2O_3. Furthermore, the optimum amount of 10% Na/γ-Al_2O_3 to be used with 100 g oil was determined and the results are shown in Figure 1(b). From Figure 1(b) it was observed that 3.5 g of 10 Na/Al_2O_3 is sufficient to provide high yields.
To understand the reason for this optimum loading the surface area of the catalyst were determined and the results are presented in Table 1. From Table 1 it is observed that the surface area monotonically decreases with an increase in sodium (Na) loading. It appears that the increase in the amount of Na in the heterogeneous catalyst is not necessarily beneficial since the surface area available for the reaction continuously decreases. Combination of these two factors may give rise to the optimum NaOH loading.

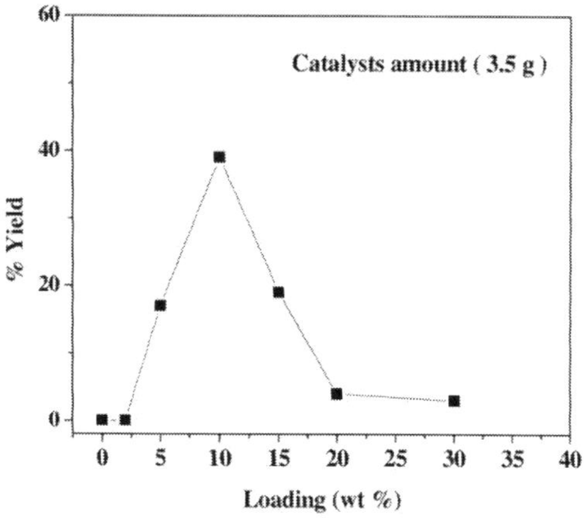

Figure 1 (a) Optimum weight % loading of Na on Al_2O_3 using 3.5 g catalyst

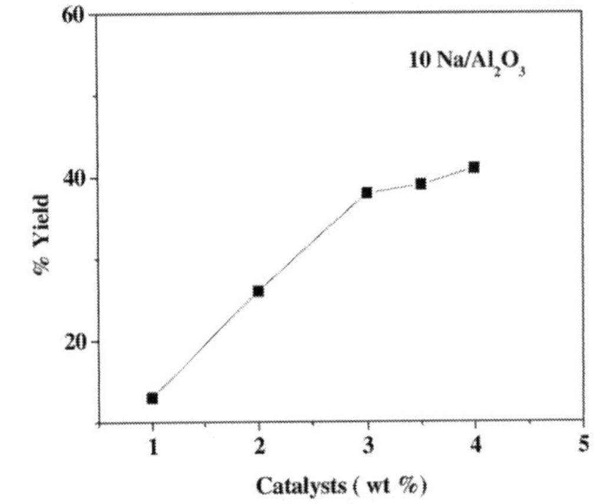

Figure 1(b) The optimum amount of 10 Na/Al_2O_3 catalyst for transesterification reaction

Reaction Conditions: Time = 1 hr, T = 60 °C, rpm = 750, Oil = 100 g, Methanol: oil = 6:1

Table 1 Characterization of catalysis with BET area method

Sample Code	Na/Al_2O_3 (% loading)	BET Area (m^2/g)	Yield (%) (3.5 g Cat./100 g oil)
0 Na/Al_2O_3	0	145	0
2 Na/Al_2O_3	2	137	0
5 Na/Al_2O_3	5	110	17
10 Na/Al_2O_3	10	86	39
15 Na/Al_2O_3	15	32	19
20 Na/Al_2O_3	20	12	4
30 Na/Al_2O_3	30	3	3

Co-solvent effect

The transesterification of soybean oil in the presence of heterogeneous catalyst is a three phase reaction. The three phase reactions may be affected by mass transfer resistance. To overcome the mass transfer resistances between the oil and methanol phases, different co-solvent, such as hexane, heptane and tetra-hydrofuran (THF), were tested. Among these co-solvents THF was found to be most beneficial. Even in the presence of THF as a co-solvent the final yield in the presence of a heterogeneous catalyst is lower than the homogeneous catalyst (1.5 wt%). These results suggest that the presence of THF only enhances the solubility of phases and not final yield value. The results for the homogeneous and heterogenous catalysts systems, with and without THF, are shown in Figure 2.

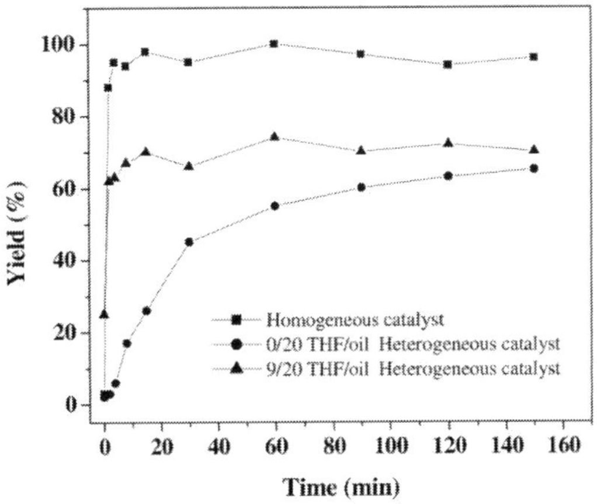

Figure 2 Co-solvent effect on heterogenous base transesterification reaction at Conditions: Methanol: oil = 6:1, THF (ml)/oil(g) = 9 ml/20 g , Catalyst wt. (10 Na/Al_2O_3) = 3.5 g/100 g oil, T = 40 °C, rpm = 250

Conclusions

In the present study several methanol solutions of alkali and alkaline hydroxide and alkali and alkaline methoxides are studied for the transesterification reaction. The sodium and potassium hydroxide solutions or methoxide were found to be the most active. A series of supported sodium on Al_2O_3 heterogeneous catalysts were also tested for the transesterification reaction. An optimum sodium loading was obtained. THF as a co-solvent increased the solubility and the initial reaction rate. Additional studies are required for understanding the effect of sodium loading and the presence of leaching, if any.

References

[1] Schuchardt, U., Serchelt, R. and Vargas, R. M., "Transesterification of vegetable oils", J. Braz. Chem. Soc., 9 (1998) pp. 199-210.

[2] Meher, L. C., Vidya Sagar, D. and Naik, S. N., (2006) "Technical aspects of biodiesel production by transesterification", *Renewable and Sustainable Energy Reviews* pp. 248–268.

[3] Biswanath, H., Garg, S., M.Tech. Thesis, Deptt. of chemical engineering, Indian Institute of Technology Kanpur, June, 2007.

[4] Arzamendi, G., Campo, I., Arguinarena, E., Sanchez, M., Montes, M. and Gandıa, L. M., "Synthesis of biodiesel with heterogeneous NaOH/alumina catalysts comparison with homogeneous NaOH", Chem. Engg. Journal 134 (2007), pp. 123–130.

[5] Dalcin, F. M, Maertini, D. D., Reiznautt, Q. B., Nicolau, A., Pedrotti, F. and Samios, D. (2009) "A transesterification double step process- TDSP for biodiesel preparation from fatty acids triglycerides" *Fuel Processing Technology* 90, pp. 599-605.

(726g) Coal-to-Biomethane Potential of Microbial Consortia Native to Coal Seam Formation Waters

Samuel Papendick[1], Gene Tyson[2], Joan Esterle[3], Suzanne D. Golding[3], Victor Rudolph[4] and Patrick C. Gilcrease[1], (1)Chemical and Biological Engineering, South Dakota School of Mines & Technology, Rapid City, SD, (2)Australian Centre for Ecogenomics & Advanced Water Management Centre, University of Queensland, St. Lucia, QLD, Australia, (3)School of Earth Sciences, University of Queensland, St. Lucia, QLD, Australia, (4)School of Chemical Engineering, University of Queensland, St. Lucia, QLD, Australia

Laboratory culture experiments have demonstrated that microbial consortia native to coal seam formation waters are capable of producing methane given native coal as the primary substrate. The field application of such a process has the potential to generate new natural gas reserves *in situ* in real time. Given the diversity in coal rank, type, and depositional setting, a fundamental understanding of this natural microbial process is needed to develop site-specific stimulation treatments. Multiple samples taken from the Surat Basin of Queensland, Australia were collected and screened for methanogenic activity using native Walloon coal as the sole carbon and energy source. *In vitro* methanogenic rates and yields were 0.24 m^3/ton/day and 6.1 m^3/ton, respectively; if our average laboratory rate could be extrapolated to the field, the Surat Basin mean gas content of 3.5 m^3/ton could be replenished in two weeks. Various coal weights in a constant growth medium volume were subjected to culture tests; results indicate that methanogenesis yields in our system were limited by a coal substrate, as opposed to a supplemental nutrient. While the interior coal matrix holds additional biomethane substrate, only the coal surface is bioavailable in the short term. With this in mind, experiments designed to enhance microbial surface interactions and coal bioavailability showed that both physical and chemical treatment of coal enhances methanogenesis rates and yields. Microscopic imaging has revealed microbial coal surface interactions, which may play a critical role in coal-to-methane biodegradation. The CO_2 reduction pathway was active for these consortia, which may have implications for CO_2 sequestion in these coal seams. Our Walloon coal substrate was separated into vitrain and dull lithotypes, and the coal-to-methane activity was quantified as a function of maceral type; the results were used to hypothesize potential coal substrates and possible scale-up limitations. Finally, a genomic study using pyrotag sequencing from both raw and enriched methanogenic communities was used to identify the most active and critical consortium members.

See more of this Session: Alternative Fuels II
See more of this Group/Topical: Catalysis and Reaction Engineering Division

(731a) Synthesis and Characterization of Nanocrystalline Zeolites for Catalysis

Sarah C. Larsen, Chemistry, University of Iowa, Iowa City, IA

Nanocrystalline zeolites, beta and ZSM-5, with hierarchical porosity were synthesized in one step from single template systems. By systematically tuning the reaction conditions, such as the concentration of the reacting species, temperature, synthesis duration and pH, the formation of isolated nanocrystals or mesoporous aggregates of zeolites beta and ZSM-5 was achieved. Depending on the synthesis conditions, zeolite nanocrystals of approximately 20 to 150 nm in diameter or mesoporous aggregates of approximately 100 to 200 nm in diameter were prepared. The zeolites were characterized by electron microscopy, solid state NMR, nitrogen adsorption and powder x-ray diffraction. The crystallinity of the samples determined from pXRD patterns is similar to that of commercially available micron-sized zeolites. The ability to systematically control the zeolite morphology has important implications for catalysis.

See more of this Session: Catalysis with Microporous and Mesoporous Materials III
See more of this Group/Topical: Catalysis and Reaction Engineering Division

(731b) Kinetics and Mechanism of Olefin Methylation Reactions On Zeolites

Ian M. Hill[1], Saleh Al Hashimi[2] and Aditya Bhan[1], (1)Chemical Engineering and Materials Science, University of Minnesota, Minneapolis, MN, (2)Chemical Engineering Department, Petroleum Institute, Abu Dhabi, United Arab Emirates

The effect of zeolite topology on steady-state dimethyl ether (DME) methylation of ethylene and propylene has been studied for zeolites H-FER, H-MOR, H-MFI, and H-BEA. Reactant pressure dependence studies over these materials at low conversions (<0.5%) and high DME:olefin ratios (30:1) show a first-order rate dependence on olefin pressure and a zero order rate dependence on DME pressure in accordance with a zeolite surface predominantly consisting of activated DME-derived species reacting with olefins. Temperature dependence studies in low temperature regimes (<430K) yield elementary-step activation energies and rate constants. Systematically lower activation barriers and higher rate constants were observed for propylene methylation in comparison with ethylene methlyation on all frameworks studied. Kinetic and isotopic studies probing the involvement of surface methyl groups in olefin methylation cycles and the relative rates of propagation of olefin methylation cycles in methanol to hydrocarbons catalysis on different zeolites will be presented.

See more of this Session: Catalysis with Microporous and Mesoporous Materials III
See more of this Group/Topical: Catalysis and Reaction Engineering Division

(731c) Optimization of Reaction Networks In Zeolites

Eric L. First, Chrysanthos E. Gounaris, James Wei and Christodoulos A. Floudas, Department of Chemical and Biological Engineering, Princeton University, Princeton, NJ

Zeolites are useful materials for shape-selective catalysis due to their regular microporous structure. Difficulty with comprehensive experimental study has prompted interest in computational approaches to systematically study the pore structure of zeolites. With nearly 200 known zeolite framework types [1] and millions of hypothetical zeolite-like structures [2,3], a quick screening method is desirable to discover those structures with the potential to promote desired chemical pathways.

We recently developed an automated method for the three-dimensional characterization of zeolite porous networks [4,5,6,7]. The method requires only crystallographic data as input and identifies channels and cages that connect to form the network of pores. The results of our characterization approach are used to calculate quantities of interest, including pore size distribution, accessible volume, surface area, largest cavity and pore limiting diameters, and probability of pore occupancy. To facilitate use of the automated method by the scientific community, we have also developed a web tool and online database of structure characterizations, ZEOMICS, which is freely available at: http://helios.princeton.edu/zeomics/

In this presentation, we introduce a novel computational approach for the modeling of reaction networks in zeolites. Large reaction networks are particularly important for the conversion of biomass and can be readily generated as input [8]. Typical biomass reactions of interest include deoxygenation of biofuels, conversion of furfural to its derivatives, and dehydration of carbohydrates. Starting from a prespecified network of reactants, products, and intermediates, transition state structures are identified for each reaction step through quantum calculations [9]. We then consider the relative mobility of all species within a zeolite to determine plausible pathways within the reaction network. The contribution of each pathway to the overall reaction yield is weighted by energetic and occupancy factors. We use our approach to screen databases of structures and identify those that exhibit the highest favorability toward generation of desired products.

References:

[1] Baerlocher, C., McCusker, L.B., 2010. Database of zeolite structures: (http://www.iza-structure.org/databases/).

[2] Deem, M.W., Pophale, R., Cheeseman, P.A., Earl, D.J., 2009. Computational discovery of new zeolite-like materials. Journal of Physical Chemistry C 113, 21353-21360.

[3] Pophale, R., Cheeseman, P. A., and Deem, M. W., 2011. A database of new zeolite-like materials. Physical Chemistry Chemical Physics, in press.

[4] Gounaris, C.E., Floudas, C.A., Wei, J., 2006. Rational design of shape selective separation and catalysis--I: Concepts and analysis. Chemical Engineering Science 61, 7933-7948.

[5] Gounaris, C.E., Wei, J., Floudas, C.A., 2006. Rational design of shape selective separation and catalysis--II: Mathematical model and computational studies. Chemical Engineering Science 61, 7949-7962.

[6] Gounaris, C.E., Wei, J., Floudas, C.A., Ranjan, R., Tsapatsis, M., 2009. Rational design of shape selective separations and catalysis: Lattice relaxation and effective aperture size. AIChE Journal 56, 611-632.

[7] First, E. L., Gounaris, C. E., Wei, J., and Floudas, C. A., 2011. Computational characterization of zeolite porous networks: An automated approach. Physical Chemistry Chemical Physics, submitted for publication.

[8] Rangarajan, S., Bhan, A., and Daoutidis, P., 2010. Rule-Based Generation of Thermochemical Routes to Biomass Conversion. Industrial & Engineering Chemistry Research 49, 10459-10470.

[9] Frisch, M.J. et al., Gaussian 09, Revision A.02, Gaussian, Inc., Wallingford CT, 2009.

See more of this Session: Catalysis with Microporous and Mesoporous Materials III
See more of this Group/Topical: Catalysis and Reaction Engineering Division

(731d) Direct Conversion of Ethylene to Propylene On Ni/MCM-41 - Insights Into Catalyst Structure

Tino Lehmann[1], Tanya Wolff[2], Christof Hamel[1] and Andreas Seidel-Morgenstern[2], (1)Institute of Process Engineering, Otto-von-Guericke-University, Magdeburg, Germany, (2)Max-Planck-Institute for Dynamics of Complex Technical Systems, Magdeburg, Germany

Propylene demand – mainly driven by polypropylene production – has been steadily growing in the past. The supply of light olefins like propylene is traditionally provided by either catalytic or steam cracking processes. These will, however, fail in the long run to cope with the changed demand situation since they have been historically designed to maximize ethylene output [1] and cannot be adjusted at will towards higher C_3 yields. Therefore, significant research activities have been aimed at the development of on-purpose propylene production processes.

One of the possible approaches in this direction is the recently found conversion of ethylene (or ethanol) to propylene on Ni/MCM-41 [2]. It has been proposed that the net transformation 3 $C_2H_4 \rightarrow 2\ C_3H_6$ proceeds via a reaction cascade comprised of ethylene dimerization, butene isomerization and finally cross-metathesis between ethylene and 2-butenes. A particularly fascinating aspect is the hypothesis of nickel-catalyzed olefin metathesis since nickel has not been found active for this kind of reaction in the past. One of the open issues for this catalytic system is the chemical identity of the nickel compound on the catalyst surface.

Catalysts with varying nickel contents (0.46 – 5.72 wt%) have been prepared following the template-ion exchange approach reported by Iwamoto and Kosugi [2]. The catalyst sample loaded with 3.18 wt% nickel gave a maximal ethylene conversion of 82 % combined with a propylene selectivity of 51 % (process conditions: packed-bed reactor, feed: 10 % ethylene in nitrogen, W/F = 0.7 kg h/m³, 400 °C, atmospheric pressure). This being said, we will focus in the present contribution on investigations regarding the nature of the supported nickel phase as well as the catalyst structure. Different techniques were used to elucidate the state of nickel on the catalyst. Increasing nickel loadings resulted in the formation of a lamellar phase at the expense of the mesoporous support as evidenced by nitrogen physisorption and XRD. TEM micrographs showed irregularly folded and crumpled sheets covering the MCM-41 particles. These foil-like structures could either present some kind of nickel phyllosilicate or turbostratic nickel hydroxide. However, XPS investigations proved the silicatic nature of the nickel surface compound. These findings are in line with the preliminary study of Ikeda et al. [3], who speculated about the existence of a nickel silicate based on their EXAFS results. Our own TPR studies revealed a single silicatic species in all samples, which turned out to be of the same kind regardless of nickel loading. Finally, FTIR analyses were employed to establish the specific type of nickel silicate: in all cases a badly crystallized 2:1 phyllosilicate was formed.

On the grounds of nickel silicate formation chemistry as well as some auxiliary arguments, it is proposed that nickel silicate is situated exclusively at the external surface of the MCM-41 particles. This is in contrast to Iwamoto [4], who assumed the nickel compound to be present within the pores of the mesoporous support. Furthermore, we suggest that the template-ion exchange approach of Iwamoto and coworkers is more adequately described as a template-protected deposition-precipitation.

References

[1] J. S. Plotkin, *Catal. Today* (2005), **106**, 10-14.

[2] M. Iwamoto, Y. Kosugi, *J. Phys. Chem. C* (2007), **111**, 13-15.

[3] K. Ikeda, Y. Kawamura, T. Yamamoto, M. Iwamoto, *Catal. Commun.* (2008), **9**, 106-110.

[4] M. Iwamoto, *Catal. Surv. Asia* (2008), **12**, 28-37.

See more of this Session: Catalysis with Microporous and Mesoporous Materials III
See more of this Group/Topical: Catalysis and Reaction Engineering Division

(731e) Structure-Property Relationships Between Fractal Description of Pore Structure and Catalyst Effectiveness In V-SBA-15

Alexander Zoelle and Michael A. Smith, Department of Chemical Engineering, Villanova University, Villanova, PA

Structure-property relationships between fractaldescription of pore structure and catalyst effectiveness in V-SBA-15

Alexander Zoelle,Thomas Baldassare, Michael A. Smith*,

Department of Chemical Engineering, VillanovaUniversity Villanova, PA, 19085 *michael.a. smith@villanova.edu

SBA-15is a template-synthesized mesoporous silicate that has found extensive use as amodel support for studies in supported catalysis. Thorough structural analyses clearlydescribe the dual micropore-mesopore structure with a broad distribution ofmicropore sizes; alternatively, it has been shown that the microporosity may alternativelybe understood as a manifestation of the fractal nature of the mesopore surface. Moreover, one can use calcination temperaturefrom 300-900 °C to adjust the degree of microporosity or fractal character inthe SBA materials, where calcination to higher temperatures effectively annealsout the micropores without destroying the mesopore topology of the SBA-15. Webelieve the effect of surface roughness or fractal character has anunderappreciated effect on catalyst performance. In this work, we hypothesizethat catalytic effectiveness of vanadia supported on SBA-15 should depend uponthe fractal character of the original SBA-15.

SBA-15was prepared following a protocol adapted from Zhao et al. We use calcinationtemperature of 500°, 700° and 850 °C to adjust the degree of fractal characterin the SBA materials. Nitrogen sorption isotherms were collected to assessfractal character. VO_x -SBA-15 was prepared using a modifiedgrafting technique to yield loadings corresponding to $^1/_8$,$^1/_4$, $^1/_2$ and full monolayercoverage. Samples are evaluated using partial oxidation of methanol in amicroreactor under differential conditions between 300° and 400° C. Additionalcharacterization includes SAXS and XRD.

Preliminary results for two catalysts, each with thesame nominal vanadia loading but with much different parent SBA-15 fractalcharacter indicate that as fractal dimension or surface roughness of the parentdecreases, catalytic activity increases.We will show results of ongoing efforts to support these findings.

(731f) Consequences of Solvation and Acid Strength for Catalysis by Faujasite Zeolites

Rajamani Gounder, Andrew J. Jones, Robert T. Carr and Enrique Iglesia, Department of Chemical and Biomolecular Engineering, University of California at Berkeley, Berkeley, CA

The effects of extraframework Al (Al_{ex}) and of Na^+ titrants in FAU zeolites on turnover rates and acid strength have been contradictorily interpreted in previous studies. Here, we examine partially-Na^+-exchanged samples of steam-dealuminated FAU (USY) and USY dealuminated chemically (CD-USY) to remove Al_{ex} species. Protons were measured by titration with 2,6-di-tert-butylpyridine during methanol dehydration catalysis, with Na^+ and with dimethyl ether. On both samples, the number of H^+ sites was smaller than the number of framework Al atoms (Al_f) given by ^{27}Al MAS nuclear magnetic resonance spectroscopy and X-ray diffraction. Monomolecular isobutane cracking and dehydrogenation rate constants (per H^+; 763 K) and their ratios were independent of Na^+ content on USY and CD-USY. These kinetic data, together with infrared spectra, indicate that Na^+ stoichiometrically replaces H^+ sites of uniform reactivity. Rate constants inappropriately normalized by Al_f atoms, however, decreased monotonically with Na^+ titration, giving the appearance that Na^+ selectively titrates more reactive protons. Methanol dehydration rate constants confirmed that acid sites on USY and CD-USY are equivalent in strength but that solvation effects are slightly stronger within USY supercages, because of the larger Al_{ex} content and smaller void spaces than in CD-USY. These findings demonstrate that thermal and chemical treatments do not change acid strength but can subtly influence the solvation effects of confinement within voids of molecular dimensions.

See more of this Session: Catalysis with Microporous and Mesoporous Materials III
See more of this Group/Topical: Catalysis and Reaction Engineering Division

(733a) Renewable Fuel From Activated Sludge Using Fluidized-Bed Catalytic Cracking (FCC) Process

Emmanuel D. Revellame, William Holmes, Rafael Hernandez and William T. French, Dave C. Swalm School of Chemical Engineering, Mississippi State University, Mississippi State, MS

Annual production of activated sludges in the United States ranges from 7 – 8 million tons. Currently, disposal technologies of these sludges include incineration and landfilling, both of which are heavily regulated and could create adverse environmental impacts. A fraction of the activated sludges can be used as fertilizer once it complies with regulations for Class A biosolids. Activated sludges contain microbial consortia known to produce lipidic materials and many other organic compounds, which could be used to produce biofuels and specialty chemicals. Several studies showed that reasonable yields (2 – 6% weight) of biodiesel could be produced from activated sludge. At these yields however, activated sludge biodiesel is less economical than petroleum diesel. This study describes the production of renewable fuel from activated sludge using Fluidized Catalytic Cracking (FCC). This process could convert most of the compounds present in sludge into fuel, and thus, increase fuel yields and reduce cost.

Several catalysts, sludge to catalyst ratios, sludge moisture content, fluid flow rates and temperatures were tested. Initial results using ZSM-5 at 500°C with air showed considerable production of aromatic compounds, from benzene to methylnaphthalene.

See more of this Session: Conversion of Solid Wastes to Energy and/or Product
See more of this Group/Topical: International Congress on Energy 2011

(733b) Simultaneous Extracellular Polymeric Substance (EPS) and Lipid Production by Activated Sludge Via Fermentation of Glucose

Patrisha J. Pham, Rafael Hernandez, Hien Nguyen, Andro H. Mondala and W. Todd French, Dave C. Swalm School of Chemical Engineering, Mississippi State University, Mississippi State, MS

The use of activated sludge microflora to convert sugars, derived from the hydrolysis of lignocellulose biomass, into lipids for biofuel production is of current research interest in our group. Lipid accumulation in this system has been enhanced by manipulating the carbon-nitrogen (C:N) ratio and initial sugar loading on activated sludge bioreactors with glucose as the main carbon source. Activated sludge is a flocculated suspension composed of microorganisms that can also produce extracellular polymeric substances (EPS) *in situ* during fermentation. Microbial EPS which are secreted largely by bacteria in harsh environments are a matrix rich in polymers including polysaccharides, glycoproteins, proteins, nucleic acids and phospholipids. Thus, studying EPS production along with lipid formation is essential for understanding the technical and economic value of the whole process. In this study, samples were obtained from aerobic batch fermentation inoculated with 20 % (v/v) activated sludge. The total incubation time for the batch fermentation experiment was seven days. EPS from the collected samples were extracted using a Dowex cation exchange resin in phosphate buffered saline solution and precipitated with 95% v/v cold ethanol and dialyzed. Fourier transform-infrared spectroscopy of freeze-dried EPS revealed characteristic IR peaks for polysaccharides and proteins. Total polysaccharide and protein concentrations were estimated in each EPS sample using phenol-sulfuric and modified Lowry methods, respectively. An increase in lipid % (g/g cell dry weight) and a decrease in specific EPS % (g/g cell dry weight) was observed from day 0 to 7 in all treatments. Increasing the carbon loading while maintaining the same C:N ratio increased lipid % and decreased specific EPS %. However for a low C:N ratio, lipid % remained constant while specific EPS % decreased. These results will be useful for manipulation of EPS and lipid production and consequently enable development of valuable strategies for improving the process.

See more of this Session: Conversion of Solid Wastes to Energy and/or Product
See more of this Group/Topical: International Congress on Energy 2011

(733c) The Hydrolysis of Nylon6 In Ionic Liquid to Recycle Nylon Waste

Xiao-Li Xia and Hai-Jun Wang, Jiangnan University, Wuxi, China

To date, several processes for Nylon6 depolymerization have been put forward. Herein, we first examined several ionic liquids for the hydrolysis of Nylon6. These ionic liquids are Brønsted acidic and they have been used both as a catalyst and environmental solvent for the hydrolytic reaction of Nylon6. The final result showed that using 1-butyl-3-methylimidazolium chloride ([Bmim]Cl) as solvent and catalyst for hydrolysis of Nylon6 ,the high conversion of Nylon6 and yield of Aminocaproic acid were obtained. In order to confirm the product, FT-IR and ^{1}H NMR were performed. After the experimental, theoretically, we also used high level ab initio methods with the 6-31G(d) basis sets to calculate the interaction between [Bmim]Cl and Nylon6. Based on our calculated interaction, we conclude that hydrogen bonds exist between [Bmim]Cl and Nylon6 which explain that Nylon6 dissolve in [Bmim]Cl.

See more of this Session: Conversion of Solid Wastes to Energy and/or Product
See more of this Group/Topical: International Congress on Energy 2011

(733d) Process Development for Efficient Utilization of Waste Plastic to Liquid Product for Rural Development In Nigeria: Gasification Option

Bamikole Amigun, Environmental Biotechnology and Conservation Department, National Biotechnology Development Agency (NABDA), Abuja, Nigeria, Christie Onyia, Environmental Biotechnology and Conservation Department, National Biotechnology Development Agency, Abuja, Nigeria and Bamidele Ogbe Solomon, Director General's Office and consultancy, National Biotechnology Development Agency (NABDA), Abuja, Nigeria

Waste production is a very important problem in all countries especially in developed countries and Nigeria is no exception. Waste plastics create a very serious environmental challenge because of their huge quantities and their disposal problems, due to their non-biodegradable attributes (or the biodegradation process is very slow). Plastic waste poses a serious environmental problem that is not addressed by recourse to landfill. Open dumping has been the predominant solid waste disposal option in Nigeria. Their destruction by incineration poses serious air pollution problems due to the release of airborne particles and carbon dioxide into the atmosphere. Only a limited amount of plastics can be recycled, while most plastics could be used, for example in energy production. The management of plastic waste using gasification in fluidized bed technology is a technically feasible option. Apart from eliminating the waste, a chemical recycling is produced during this process, obtaining a gas (syngas) which can be used in different applications such as ethanol through bacteria conversion of the syngas intermediate. Ethanol can be further converted to ethanol gel fuel to substitute the use of paraffin and traditional firewood in rural communities. This is doubly environmentally friendly since it will reduce the volume of plastic waste being disposed of in landfill whilst producing green fuel without generating green house gases. This application could give some prospect of self-reliant energy supplies at local levels, with potential economic, ecological, social, and security benefits; thereby improving the socio-economic and political wellbeing of the rural people.

See more of this Session: Conversion of Solid Wastes to Energy and/or Product
See more of this Group/Topical: International Congress on Energy 2011

(733e) An Anaerobic Bacterial Treatment of Corn Stover to Prepare a Biorefining Feedstock

James MacLellan, Biosystems and Agricultural Engineering, Michigan State University, East Lansing, MI

Anaerobic Bacterial Treatment of Corn Stover to Prepare a Biorefining Feedstock

James MacLellan, Zhengbo Yue, Robert Kraemer, Wei Liao*

Department of Biosystems and Agricultural Engineering, Michigan State University

East Lansing, Michigan 48823

Abstract

The objective of this study is to apply an anaerobic bacterial consortium to uptake the hemicellulose in corn stover to prepare a relatively cellulose-enriched feedstock for biofuel production. Previous research demonstrated that submerged anaerobic bacterial culture is capable of converting hemicellulose in raw lignocelluloses to methane gas. The in-depth study is needed to optimize the culture, enhance the hemicellulose utilization efficiency, and improve the quality of treated cellulose as a biorefining feedstock. An anaerobic bacterial consortium from animal manure will be screened and enriched as the seed for this study. Two feedstocks of raw corn stover and steam-treated corn stover will be used to compare the effects of structure on the anaerobic culture of hemicellulose uptake. A completely Randomized Design (CRD) on seed-to-feedstock ratio, culture temperature, and culture time will be carried out. Fiber composition, methane production, and microbial consortia will also be analyzed. An optimal condition can then be established based on the quality and quantity of cellulose in the treated corn stover. The ethanol production on the treated corn stover will be fulfilled in order to elucidate the effects of anaerobic bacterial treatment on biofuels production.

*: Corresponding author

Dr. Wei Liao

Department of Biosystems & Agricultural Engineering

Michigan State University

202 A.W. Farrall Hall

East Lansing, MI 48824-1323

Tel: 517-432-7205

Fax: 517-432-2892

E-mail: liaow@msu.edu

See more of this Session: Conversion of Solid Wastes to Energy and/or Product
See more of this Group/Topical: International Congress on Energy 2011

(733f) Towards Developing Anaerobic Digestion Based Biorefinery - Research On Food Wastes As Feedstock

Shulin Chen[1], Liang Yu[2], Jingwei Ma[2] and Craig Frear[2], (1)Department of Biological Systems Engineering, Washington State University, Pullman, WA, (2)Washington State University, Pullman, WA

Anaerobic digestion (AD) is a proven technology for waste treatment. However, as the organic material becomes more valuable and being treated as resources, traditional AD process is no longer adequate. The Bioprocessing and Bioproduct Engineering Laboratory at Washington State University has been devoting its effort in developing the second generation of AD technology with the concept of biorefinery, featured by producing various products from improved AD processes. Key to an AD biorefinary is innovative design of the digestors. This presentation reports the recent results from a study aimed at evaluating a two stage-digester designed for treating food wastes operated within a solid concentration range of 15-20%. The digester had a unique two-stage configuration that resulted in high biogas production rate than the existing technologies based on the laboratory data. The experimental observations also led to a simple mixing design that has low capital and operational costs. The information provides some new insights in AD research and development related to municipal solid waste applications.

See more of this Session: Conversion of Solid Wastes to Energy and/or Product
See more of this Group/Topical: International Congress on Energy 2011

(739a) Metal-Modified Tungsten Carbide (WC) for Catalytic and Electrocatalytic Conversion of Alcohols

Thomas G. Kelly and Jingguang G. Chen, Department of Chemical Engineering, University of Delaware, Newark, DE

Direct alcohol fuel cells (DAFCs) utilizing methanolor ethanol have been proposed as an alternative to hydrogen fuel cells. Alcoholsare easier to transport and more readily obtained from biorenewable sourcesthan other fuels. The anode material most widely studied for DAFCs iscurrently a PtRu alloy; however, Pt is scarce and expensive. Additionally, Ptis not active towards scission of the C-C bond in ethanol. Recently,researchers reported using a Pt/Rh/SnO_2 alloy that was more activetowards ethanol electrooxidation than Pt, and Rh was determined to be crucialfor effective C-C bond cleavage [1]. Tungsten monocarbide (WC) has been shownto have Pt-like properties and is active towards methanol electrooxidation [2,3].Also, metal-modified WC can be used in place of bimetallic catalysts that areless stable and more expensive, as was shown recently in using Ni/WC to replaceNi/Pt for ethanol decomposition [4]. In the current study, we have extendedour previous surface science studies for methanol on Pt-modified WC [5] to C1-C3alcohols on WC modified with Ni, Rh, and Au. Furthermore, an understanding ofthe bond scission sequence of alcohols on metal-modified WC should also provideinsights into the utilization of carbides for catalytic reforming reactions ofalcohols.

Density Functional Theory (DFT) was used to calculatethe binding energies of alcohols and relevant intermediates on WC and metal-modifiedWC, which showed the possibility to correlate reactivity with binding energy.Temperature-programmed desorption (TPD) was used with a WC polycrystalline foilmodified with different metals to quantitatively determine the activity of thesurfaces and their selectivity towards C1-C3 alcohol decomposition. Highresolution electron energy loss spectroscopy (HREELS) identified differentreaction intermediates on WC and metal-modified WC surfaces. The first step inalcohol decomposition on these surfaces was O-H bond cleavage to form anadsorbed alkoxy species. In the case of methanol, the predominant reactionproduct on clean WC was methane; adding small amounts of Ni or Rh promoted theC-H bond scission and shifted the selectivity towards CO and H_2. Ausimply acted as a site blocker with decreasing activity with increasing Aucoverage [6]. Similar experiments have been conducted for ethanol and propanoldecomposition on Rh/WC to investigate the problem of C-C bond scission. Theselectivity on clean WC was for C-O bond scission to produce ethylene. AddingRh led to C-H and C-C bond cleavage; HREELS showed that the C-C bond in ethanolis broken by 200 K.

Of further interest is the bond scission sequence ofethanol and propanol on these surfaces. On most transition metals, afterinitial O-H bond scission, these alcohols decompose through aldehydeintermediates by cleavage of the α C-H bond. However, on Rh(111), theβ C-H bond breaks to form an oxametallacycle intermediate that rapidlydecomposes [7-9]. HREELS experiments on WC and Rh/WC were conducted forethanol, acetaldehyde, propanol, and propanal to investigate the reactionpathways. On WC, ethanol followed a decomposition pathway similar to acetaldehyde;the spectra suggested that both molecules formed a di-σ species thatfacilitates C-O bond cleavage. This behavior was mirrored by propanol andpropanal on WC. On Rh/WC, ethanol and acetaldehyde again followed similarpathways, except that the pathway in this case is through C-C bond scissioninstead of C-O scission. For both molecules, the C-C bond was broken by 200K. A slightly different result was observed for propanol and propanal onRh/WC. By 200 K, the C-C bond was broken in propanal, but the propanol spectrastill showed peaks due to propoxy. The difference in spectra between propanoland propanal suggested that the molecules may have followed differentdecomposition pathways.

The results of alcohol decomposition for C1-C3alcohols on WC and metal-modified WC hold promise for applications in heterogeneousreforming and electrochemical applications. Futhermore, butanol may be anattractive prospect since it can be produced through biomass fermentation. Futurework will focus on electrochemical experiments on metal-modified WC for alcoholelectrooxidation in an acidic medium.

References

1. Kowal, A., *et. al. Nat. Mater.* **8**(2009) 325.

2. Levy, R.B. and Boudart, M., *Science***181** (1973) 547.

3. Weigert, E.C., *et. al. J.Phys. Chem. C.* **111** (2007) 14617.

4. Ren, H., *et. al. ACSCatalysis* **1** (2011) 390.

5. Stottlemyer, A.L., *et. al. J.Chem. Phys.* **133** (2010) 104702.

6. Kelly, T.G., *et. al. J. Phys.Chem. C.* **115** (2011) 6644.

7. Houtman, C.J. and Barteau, M.A., *J.Catal.* **130** (1991) 528.

8. Brown, N.F. and Barteau, M.A., *Langmuir***8** (1992) 862.

9. Brown, N.F. and Barteau, M.A., *Surf.Sci.* **298** (1993) 6.

See more of this Session: Fundamentals of Surface Reactivity I
See more of this Group/Topical: Catalysis and Reaction Engineering Division

(739b) Autocatalytic Decompostion of Chiral Aspartic Acid On Cu(110) Surface

Bharat S. Mhatre, Department of Chemical Engineering, Carnegie Mellon University, Pittsburgh, PA and A.J. Gellman, Chemical Engineering, Carnegie Mellon University, Pittsburgh, PA

Surface explosion is a phenomenon in surface science wherein an autocatalytic increase in the reaction rate during heating results in its occurrence over a very narrow temperature range. We have previously demonstrated that this phenomenon can be successfully exploited to attain extremely high enantioselectivity in the case of tartaric acid decomposition over naturally chiral Cu(643)R&S surfaces. Prior work by others has focused majorly on developing models that describe this phenomenon, however, little efforts have been directed towards understanding the factors in the mechanism of surface explosion that contribute towards observed enantioselectivity. In this regard, tartaric acid as a probe molecule has several limitations, the most important being the fact that, isotopically labeled enantiomers of tartaric acid are not commercially available.

In our most recent study, we have successfully identified aspartic acid as a probe molecule for studying autocatalytic decomposition on Cu(110) surface. Also, we have identified the differences in the kinetics of formation of these products as manifested in the temperature programmed reaction spectroscopy experiment. Using isotopically labeled species, we are now in a position to map the appearance of the atoms in the products from their location on the decomposing parent molecule.

In due course, various isotopically labeled species will be used for further investigation of this reaction to develop our understanding of the surface explosion phenomenon. Preliminary experiments on chiral Cu surfaces indicate that the aspartic acid should also prove to be an ideal probe for investigating the origin of extremely high enantioselectivity as observed in the case of tartaric acid decomposition on Cu (643)R&S surfaces. The ultimate objective of the work is to apply the experimental and parameter estimation techniques developed during the course of study on Cu(110) surface, towards the study of chiral aspartic acid enantiomer decomposition on chiral Cu surfaces. The idea is to be able to identify the mechanism, develop models that best describe the mechanism and finally, to identify the steps in the mechanism that lead to enantioselectivity i.e. to figure out the rate constants that depend on the chirality of the surface.

See more of this Session: Fundamentals of Surface Reactivity I
See more of this Group/Topical: Catalysis and Reaction Engineering Division

(739c) High Throughput Study of Catalytic Surface Chemistry Over $Cu_xAu_yPd_{1-x-y}$ Alloys

A.J. Gellman[1], James B. Miller[1], Petro Kondratyuk[2] and Deepika Priyadarshini[1], (1)Chemical Engineering, Carnegie Mellon University, Pittsburgh, PA, (2)Dept. of Chemical Engineering, Carnegie Mellon University, Pittsburgh, PA

A suite of tools has been developed and is now being applied for the high throughput study of catalytic surface chemistry on composition spread alloy films (CSAFs) and on surface structure spread single crystals (S^4Cs).

A fairly simple offset filament tools is being used for deposition of $Pd_xCu_yAu_{1-x-y}$ CSAFs. These alloys have been chosen for study because of their use as dense metal membranes for hydrogen separation and purification from coal gas. The $Pd_xCu_yAu_{1-x-y}$ CSAFs are being used as libraries for the study of surface segregation in both binary and ternary alloys. Segregation impacts the ability of these $Pd_xCu_yAu_{1-x-y}$ alloys to dissociatively adsorb hydrogen, the first step in hydrogen transport through their bulk. X-ray photoemission spectroscopy (XPS) and low energy ion scattering (LEIS) are being used to determine the near surface and the topmost surface compositions as functions of the bulk film composition. Surface energetics favors the segregation of Cu and Au to the alloy surfaces. In parallel with these efforts, a multichannel microreactor has been developed for high throughput measurements of catalytic surface chemistry. It is being used to study to composition dependence of the kinetics of H_2-D_2 exchange on the $Pd_xCu_yAu_{1-x-y}$ alloys.

See more of this Session: Fundamentals of Surface Reactivity I
See more of this Group/Topical: Catalysis and Reaction Engineering Division

(739d) Enantioselective Hydrogenation of α, β-Unsaturated Carboxylic Acid Over Cinchona-Modified Pd/Al2O3: a Spectroscopic and Kinetic Study

Shuai Tan and Christopher T. Williams, Chemical Engineering, University of South Carolina, Columbia, SC

Enantioselective synthesis plays a key role in the industrial production of agrochemicals, fragrances, and pharmaceuticals. However, studies of C=C hydrogenation with modified catalyst are lagging behind. In this work, the adsorption of methyl pentenoic acid (MPeA), a prochiral α, β-unsaturated carboxylic acid on Pd catalyst, was studied. In addition, the hydrogenation of this molecule with/without the chiral modifier cinchonidine is explored, with the aim of understanding the reaction mechanism. The adsorption of MPeA has been studied by *in-situ* ATR-IR spectra on Pd/Al_2O_3 and Al_2O_3 surface. Several peaks appear upon exposure to acid solution, and these peaks remain after purging with pure solvent, indicating an irreversible adsorption on the Pd/Al_2O_3 surface. Detailed analysis of these vibrational spectra suggests that the acid adsorbs on surface both as monomeric and dimeric molecular species, and as a bridging bidentate carboxylate species. In addition, vibrational studies of cinchonidine-acid interactions in solution and on the surface were carried out. Corresponding kinetic studies of the hydrogenation of MPeA in several solvents in a batch autoclave reactor were carried out. Methanol and dioxane exhibit good performance in terms of activity and selectivity, while dichloromethane (CH_2Cl_2) provides nearly no reactivity, indicating the reaction is strongly solvent-dependent. The data suggest that the reaction is first order with respect of H_2 under low pressure (<7.5 atm), while becoming H_2-independent for pressures larger than 20 atm, consistent with Langmuir-Hinshelwood type adsorption. The reaction order of the substrate MPeA up to around 40-50% conversion is close to 1, but increases to 2 as the reaction progresses further. The reaction rate was also found to decrease with increasing initial product/reactant ratio when product is added initially, suggesting that the product is strongly adsorbed on the surface. In the case of enantioselective hydrogenation, the highest e.e. value was found to be ca 30%. Furthermore, the presence of modifier had no effect on reaction mechanism (i.e., reaction orders of reactant and H_2 remained the same). However, the reaction rate decreased compared with racemic reaction, presumably due to blockage of active sites by the cinchonidine modifier.

See more of this Session: Fundamentals of Surface Reactivity I
See more of this Group/Topical: Catalysis and Reaction Engineering Division

(739e) Au/Ni near-Surface Alloys As Potential Direct H2O2 Synthesis Catalysts: A DFT Study

Fuat E. Celik and Manos Mavrikakis, Chemical and Biological Engineering, University of Wisconsin-Madison, Madison, WI

In high-volume industrial catalytic processes, increasing catalyst selectivity can have a potentially higher impact on process margins than increasing activity. The search for metal catalysts possessing high selectivity and activity has led to the evaluation of many types of metal alloys, hoping to combine the high activity of one metal with the higher selectivity of another, less-reactive metal catalyst. Near-surface alloys (NSAs) are a particular subset of alloys where the alloying is restricted to the top one or two layers of a metal surface. Au atoms substituted into the surface of Ni (111) have already been shown to possess higher steam reforming selectivity by impeding the formation of carbon [1].

In the present study, periodic, self-consistent Density Functional Theory (DFT) calculations were used to evaluate Au/Ni NSAs as potential oxygen reduction catalysts for the direct synthesis of hydrogen peroxide from O_2 and H_2, thereby avoiding the current anthraquinone process. While Au may have higher O-H bond formation activity, it is a poor O_2-dissociation catalyst, and likewise Ni is very effective at O_2-dissociation but not oxygen-hydrogenation [2]. The Au-substituted Ni (111) shows properties intermediate between pure Au and Ni, such as H atom binding energies.

[1] F. Besenbacher, I. Chorkendorff, B.S. Clausen, B. Hammer, A. M. Molenbroek, J. K. Nørskov, I. Stensgaard, Science 279 (1998) 1913.

[2] D.C Ford, A.U. Nilekar, Y. Xu, M. Mavrikakis, Surf. Sci. 604 (2010) 1565.

See more of this Session: Fundamentals of Surface Reactivity I
See more of this Group/Topical: Catalysis and Reaction Engineering Division

(739f) Synergistic Effects In Reactions of Functionalized Alcohols On Pd(111)

Simon H. Pang[1], Michael Griffin[2], Rhea Williams[2] and **J.Will Medlin**[3], (1)Department of Chemical and Biological Engineering, College of Engineering and Applied Science, University of Colorado-Boulder, Colorado Center for Biorefining and Biofuels, Boulder, CO, (2)Chemical and Biological Engineering, University of Colorado, Boulder, CO, (3)Department of Chemical and Biological Engineering, University of Colorado, Boulder, CO

The surface chemistry of multifunctional oxygenates, such as unsaturated alcohols, is of interest due to the prevalence of these compounds as intermediates in the transformation of biomass to fine chemicals and fuels. This contribution focuses on recent experimental and computational studies of highly functional oxygenates such as furfuryl alcohol and furfural – furanic compounds that can readily be derived from biomass – on Pd(111). It is observed that the surface chemistry of these compounds cannot be understood by the reactivity of the individual functional groups alone. Rather, there is a synergistic effect that allows these molecules to undergo deoxygenation in addition to decarbonylation. Theoretical studies were conducted to investigate the complete decomposition mechanism for furfural and furfuryl alcohol, and to relate this chemistry to that of similar probe molecules such as 1-propanol and allyl alcohol. Related investigations of the chemistry of polyols and halogenated alcohols point to a complex synergy between alcohols and additional functional groups in reactions on surfaces.

See more of this Session: Fundamentals of Surface Reactivity I
See more of this Group/Topical: Catalysis and Reaction Engineering Division

A Combined Theoretical and Experimental Study of the Adsorption of $Rh(CO)_2(acac)$ on $TiO_2(110)$

Homa Khosravian[1], Zhu Liang[2], Alexander Uhl[2], Michael Trenary[2], and Randall Meyer[1]

[1] Department of Chemical Engineering, University of Illinois at Chicago, 810 South Clinton Street, Chicago, IL 60607, United States

[2] Department of Chemistry, University of Illinois at Chicago, 845 West Taylor Street, Chicago, IL 60607, United States

Introduction

Organometallic precursors have been recently used to create supported metallic cluster catalysts with unique reactivity. Gates and co-workers have shown that precursors such as $Rh(CO)_2(acac)$ can be decarbonylated by heating in helium (or H_2), and intact metal cluster catalysts can be created on a variety of supports[1]. Although there is significant evidence that hydroxyl groups play an important role in the synthesis and activation of these clusters, the precise interaction of these precursors with supports is currently still largely unknown. In this combined DFT and STM study, we examine the adsorption of $Rh(CO)_2(acac)$ on a $TiO_2(110)$ single crystal substrate in an effort to understand the synthesis of small Rh clusters.

Materials and Methods

All the deposition experiments were done at the University of Illinois at Chicago in ultra-high vacuum with a variable temperature scanning tunneling microscope (VT-SPM, Omicron) system consisting of a STM chamber and a separate preparation chamber. The r-$TiO_2(110)$ single crystal (PKEM) used in the sequential $Rh(CO)_2(acac)$ deposition experiments was prepared by repeated cycles of Ar^+ ion sputtering and subsequent annealing at 630°C. Exposure of the sample surface to $Rh(CO)_2(acac)$, rhodium(I) dicarbonylacetylacetonate (Strem, 99%), was performed via a leak valve, through a 1 cm diameter stainless steel directed doser. During the dosing, the distance between the opening of the stainless steel tube and the sample surface was about 1 mm in order to increase the dosing efficiency with minimum background exposure. All deposition experiments were performed with the $TiO_2(110)$ crystal at room temperature. STM images were collected with electrochemically etched homemade W tips with a sample bias of 1 V and a constant tunneling current of 0.1 nA. All images were generated at room temperature. Density Functional Theory (DFT) calculations were performed using the VASP package with a 2 × 2 unit cell of the $TiO_2(110)$ surface (11.8 × 13.0 Å).

Results and Discussion

After exposing $Rh(CO)_2(acac)$ to the surface at room temperature the formation of small particles is observed, with diameters of 1 nm or less, and a height of 2.5 Å (i.e., monolayer height) as shown in Figure 1a. After depositing the precursor at room temperature, the sample was exposed to 3.5 × 10-7 torr of CO for 10 min while holding the sample at 373 K. As shown in Fig 1b, the size of the particles grew dramatically after exposure to CO. Previously we had established that annealing the sample to 373 K resulted in minimal sintering. The features seen in Fig 1b have an average height of 0.8 nm and an average diameter of 2.5 nm, which we ascribe to small metal carbonyl clusters (but perhaps larger than $Rh_6(CO)_{16}$) This sample was then exposed to H_2 at 3×10^{-7} for 10 min at 373 K, which results in a decrease in particle height (to 0.5 nm) but an increase in particle diameter (to 3 nm) (shown in Fig. 1c). This dynamic restructuring of the particles from hemispheres to flatter disc-like structures is not without precedent (see Hansen *et. al.* [2]). However, in this case, the gas exposures were done under UHV conditions at 373 K, indicating a remarkable sensitivity to the environment. Furthermore, these structural changes were revealed to be at least somewhat reversible as demonstrated by further experiments involving cycles of CO and H_2 exposures. Supporting DFT calculations indicate that the species imaged following exposure must be some partially deligated species ($Rh(CO)_2$) as the intact molecule will not bind with the surface. DFT calculations also suggest that the presence of defects is necessary for dissociative adsorption. If the acac ligand is cleaved, its subsequent hydrogenation to hydroxypetenone is exothermic (ΔE = −0.29 eV) and the product may easily desorb from the surface (Eads =−0.41 eV). $Rh(CO)_2$ then may be a precursor to the small clusters observed when CO is exposed to the surface. Calculations of the kinetics of the deligation and hydrogenation steps are also done.

Significance

The use of organometallic precursors to create clusters in the size range of ~10 atoms gives us an avenue to a potentially new class of nanocatalysts. However, the routes to synthesize these clusters are not well understood on a molecular level. Our work shows that these transformations can take place under surprisingly mild conditions. In addition, DFT calculations indicate that defect sites are critical to the adsorption and deligation and assembly processes. As such support engineering may be required to create and stabilize cluster catalysts prepared through these routes.

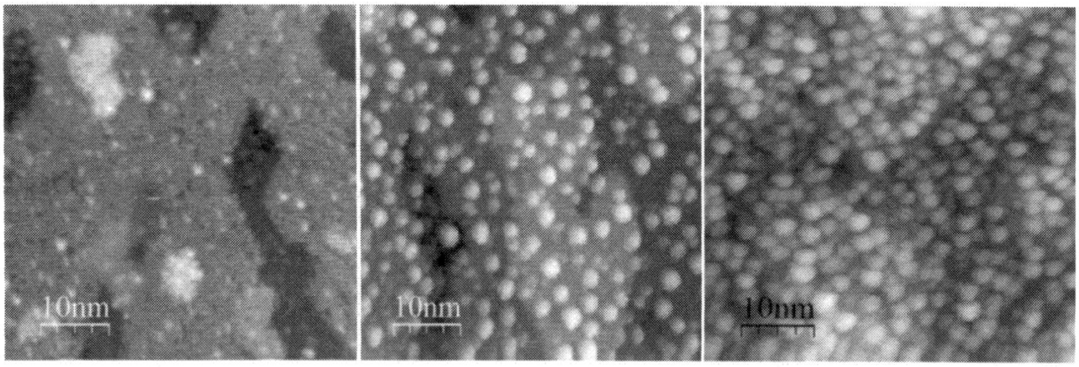

Figure 1. STM images of (a) $Rh(CO)_2(acac)$ deposited on $TiO_2(110)$ (b) $Rh(CO)_2(acac)/TiO_2(110)$ after exposure to CO at 373 K and (c) $Rh(CO)_2(acac)/TiO_2(110)$ after exposure to CO at 373 K, followed by additional exposure to H_2 at 373 K. All images 50 × 50 nm^2

References

[1] J.F. Goellner, B.C. Gates, J. Phys. Chem. B 105 (2001) 3269-3281.
[2] P.L. Hansen, J.B. Wagner, S. Helveg, J.R. Rostrup-Nielsen, B.S. Clausen, H. Topsoe, Science 295 (2002) 2053-2055.

(749a) Polylactic Oligomers Degradation Kinetics

Fabio Codari[1], Stefano Lazzari[1], Miroslav Soos[1], Davide Moscatelli[2] and Massimo Morbidelli[1], (1) Institute of Chemical and Bioengineering, ETH Zurich, Zurich, Switzerland, (2)Chimica, Materiali ed Ingegneria Chimica G. Natta, Politecnico di Milano, Milano, Italy

Polylactic acid (PLA) is a biodegradable and biocompatible, thermoplastic polymer produced from lactic acid (LA) as derived from renewable sources. Degradation studies have been reported in literature for both large scale applications, such as composting processes, and niche applications, such as in drug delivery systems. In the latter case, the degradation kinetics has a direct impact on the rate of drug release and system morphology and, therefore, on the final performance. The aim of the present work is to investigate experimentally the degradation kinetics of low molecular weight PLA as a function of selected key parameters, such as polymer chain length, pH, temperature and chain chiral composition.

PLA oligomers with different chiral composition were produced by polycondensation of DL-LA and L-LA and collected by semipreparative HPLC method. Hydrolysis experiments at different conditions, ranging from pH 1 to pH 7.4 and from 30 up to 120°C, were carried out in batch and characterized by HPLC analysis.[1] Following the time evolution of the concentration of both initial oligomer species and all degradation products it was possible to identify different reaction mechanisms among the different explored conditions. Finally, a comprehensive kinetic scheme has been developed and the corresponding rate coefficients were estimated by minimizing the error between model predictions and experimental data. In particular, it was found that hydrolysis occurs preferentially through chain end-scission and backbiting mechanism at acid and basic pH, respectively, and that the overall degradation constant presents a minimum at the pKa of the carboxylic acid. Moreover, LL-oligomers were found to have larger reactivity compared to DL ones.

See more of this Session: Polymer Reaction Engineering Kinetics & Catalysis
See more of this Group/Topical: Materials Engineering and Sciences Division

(749b) Pseudo-Solid State Polymerization In Amorphous Polymer Micro-Layers: A Novel Route to Produce Ultra-High Molecular Weight Polycarbonate

In Hak Baick[1], Carla Luciani[1], Woojic Yang[1] and Kyu Yong Choi[2], (1)Chemical and Biomolecular Engineering, University of Maryland, College Park, MD, (2)Department of Chemical and Biomolecular Engineering, University of Maryland, College Park, College Park, MD

A novel polymerization technique is presented to produce ultra-high molecular weight bisphenol A polycarbonate via step-growth polymerization in micro-layers using low molecular weight polymer precursors produced by melt polycondensation of bisphenol A and diphenyl carbonate. The polymerization is carried out at low pressures and at temperatures close but below the polymer melting point to ensure the reaction proceeds in amorphous state. It was found that the polymerization rate is much faster than in a conventional solid-state polymerization using pre-crystallized prepolymer particles. Transparent polycarbonates of molecular weights as high as 700,000 Da were obtained. Since the reacting polymer micro-layers are neither in the solid state nor in the melt phase, the proposed polymerization is called *pseudo-solid state polymerization (p-SSP)*. The relatively high mobility of polymer chains in the amorphous state, the efficient removal of the condensation by product (*i.e.*, phenol) from the confined geometry, the free radical-induced reactions via thermal decomposition of the residual casting solvent or Fries rearrangement are hypothesized to be the main responsible for the departure of the proposed polymerization technique from the conventional solid state polymerization. Experimental and theoretical investigations to elucidate the key mechanisms governing the p-SSP in amorphous polymer micro-layers will be presented at the meeting.

See more of this Session: Polymer Reaction Engineering Kinetics & Catalysis
See more of this Group/Topical: Materials Engineering and Sciences Division

(749c) Computational Study of Chain Transfer to Monomer Reactions In Thermal Polymerization of Methyl Acrylate

Nazanin Moghadam, Chemical and Biological Engineering, Drexel University, Philadelphia, PA, Masoud Soroush, Department of Chemical and Biological Engineering, Drexel University, Philadelphia, PA, Sriraj Srinivasan, Arkema, Philadelphia, PA, Andrew M. Rappe, Makineni Theoretical Laboratories, Department of Chemistry, University of Pennsylvania, Philadelphia, PA and Michael C. Grady, DuPont Marshall Lab, Philadelphia, PA

North American regulations [1, 2] have required volatile organic content (VOC) of coatings to be less than 300 ppm by 2010. These and similar regulations [3] have been responsible for the changes in the paint and coatings industries in the last sixty years. Low temperature (50-80°C) polymerization processes producing high molecular weight, low solids polymer resins have been replaced with high temperature (>100°C) polymerization processes producing low molecular weight, high solids polymer resins [4]. It has also been reported that secondary reactions, such as backbiting, β-scission and chain transfer reactions, have stronger impacts on the dynamics of the processes at higher temperatures.

Previous studies [5-9] have shown the successful use of computational quantum chemistry to better understand self-initiation and propagation reactions in polymerization systems. To the best of our knowledge, no computational study of chain transfer to monomer reactions in polymerization of alkyl acrylates has been reported as of yet.

This paper presents a computational study of chain transfer to monomer reactions in self-initiated high-temperature polymerization of methyl acrylate. Density functional theory calculations are used to predict energy barriers and rate constants of the chain transfer to monomer reactions. The effect of polymer chain-length and radical centers (e.g., tertiary vs. secondary radical) on the energy barriers and rate constants of the reactions are studied. Intrinsic reaction coordinate calculations are carried out to determine the pathways for the mechanisms of interest.

References:

[1] Superintendent of Documents, Title 1, US Government Printing Office, Washington, DC, 1990, p.1.

[2] Superintendent of Documents, Clean Air Act Amendments of 1990, Title 111, US Government Printing Office, Washington, DC, 1990, p.236.

[3] VOC's Directive, *EU Committee of the American Chamber of Commerce in Belguim*, ASBL/VZw, Brussels, July 8, 1996.

[4] Adamsons, K.; Blackman, G.; Gregorovich, B.; Lin, L., Matheson, R.; Oligomers in the Evolution of Automotive Clearcoats : Mechanical Performance Testing as a Function of Exposure, *Progress in Organic Coatings* 1998, 34, 64-74.

[5] Heuts, J.P.A.; Gilbert, R.G.; Radom, L.; A Prior Prediction of Propagation Rate Coefficients in Free Radical Polymerizations: Propagation of Ethylene, Macromolecules, 1995, 28, p. 8771-81.

[6] Khuong, K.S.; H. Jones, W.H.; Pryor, W.A.; Houk, K.N.; The Mechanism of Self-Initiated Thermal Solution of a Classic Problem, J. Am. Chem. Soc. 2005. 127, 1265-77.

[7] Yu, X.; Pfaendtner, J.; Broadbelt, L.J.; Ab Initio Study of Acrylate Polymerization Reactions: Methyl Methacrylate and Methyl Acrylate Propagation, J. Phys. Chem. A. 2008, 112, 6772-82.

[8] Srinivasan, S.; Lee, M.W.; Grady, M.C.; Soroush, M.; Rappe, A.M.; Computational Study of the Self-Initiation Mechanism in Thermal Polymerization of Methyl Acrylate, J. Phys. Chem. A 2009, 113, 10787-94.

[9] Srinivasan, S.; Lee, M.W.; Grady, M.C.; Soroush, M.; Rappe, A.M.; Self-Initiation Mechanism in Spontaneous Thermal Polymerization of Ethyl and n-Butyl Acrylate: A Theoretical Study, J. Phys. Chem. A 2010, 114, 7975-83.

See more of this Session: Polymer Reaction Engineering Kinetics & Catalysis
See more of this Group/Topical: Materials Engineering and Sciences Division

(749d) Synthesis and Characterization of Hyperbranched Polyacrylamide Using Semi-Batch RAFT Copolymerization of Acrylamide and N,N'-Methylenebisacrylamide

Dunming Wang[1], Wen-Jun Wang[1], Bo-Geng Li[1] and Shiping Zhu[2], (1)Department of Chemical & Biological Engineering, Zhejiang University, Hangzhou, China, (2)Department of Chemical Engineering, McMaster University, Hamilton, ON, Canada

Abstract Highly branched acrylamido polymers have been widely used in drug delivery systems, catalyst supports, viscosity modifiers, biomedicines, etc. In this work, hyperbranched polyacrylamides (b-PAMs) were prepared via a semi-batch reversible addition-fragmentation chain transfer (RAFT) polymerization of acrylamide (AM) using N,N'-methylenebisacrylamide (BisAM) as branching agent (BA) and 3-benzyltrithiocarbonyl propionic acid (BCPA) as chain transfer agent (CTA). Water soluble b-PAMs were successfully produced at low [BCPA]/[BisAM] ratios of 1/20 ~ 1/50 in comparison with gel formation in RAFT batch copolymerization of AM and BisAM at the same condition, indicating the crosslinking reaction of BisAM was effectively suppressed at low BCPA concentration using the semi-batch technique.

A systematic kinetic study on the semi-batch RAFT copolymerization of AM with BisAM was conducted. Effect of [BCPA]/[BisAM] ratios and feeding rates on the evolution of molecular weight (MW) and distribution (PDI), branching density (BD), and cyclization density (CD) in the polymerization were investigated. It was observed that the b-PAM samples collected at different time during the RAFT copolymerization had similar BD, suggesting that the b-PAMs prepared via the semi-batch RAFT copolymerization technique having homogeneous branching structure. High BisAM feeding rate or BisAM usage benefitted to branching formation. The cyclization reaction was suppressed at low instantaneous BisAM concentration in the copolymerization system.

See more of this Session: Polymer Reaction Engineering Kinetics & Catalysis
See more of this Group/Topical: Materials Engineering and Sciences Division

(749e) Photoinitiators As Copper(II) Reductants: A New Approach to the Copper-Catalyzed AzideAlkyne Cycloaddition

Brian Adzima, Dept. of Chemical and Biological Engineering, University of Colorado, Boulder, CO, Christopher J. Kloxin, Chemical Engineering and Materials Science and Engineering, University of Delaware, Newark, DE and Chistopher N. Bowman, Department of Chemical and Biological Engineering, University of Colorado, Boulder, CO

The copper catalyzed azide–alkyne cycloaddition (CuAAC reaction) click reaction is utilized in a wide range of synthetic strategies, from dendrimer and block copolymer synthesis to bioconjugation. Traditional methods for catalyzing the CuAAC reaction include the direct addition of Cu(I) salts, the reduction of copper sulfate via sodium ascorbate, and the comproportionation reaction of copper metal and Cu(II). Recently, we have demonstrated that the CuAAC reaction can be catalyzed using conventional photoinitiators to reduce Cu(II). This approach allows for spatial control of the CuAAC reaction, as demonstrated by patterning of hydrogel formation and labelling of hydrogels with features as small as 25 microns. This photochemical approach not only simplifies patterning surfaces, compared to traditional approaches such as dip pen lithography, but also allows the CuAAC reaction to be used for spatially selective coupling reactions within a material. Herein, we discuss recent kinetic experiments suggesting that undesirable catalyst consuming reactions, such as disproportionation, further reduction, and oxidation are suppressed, and that the ligands present play a key role in the reaction. Accordingly, like the conventional approach of reducing Cu(II) using sodium ascorbate this process is robust and does not require purging to eliminate oxygen. However, unlike conventional approaches this methodology readily allows for spatial control of the reaction using conventional photolithographic techniques.

See more of this Session: Polymer Reaction Engineering Kinetics & Catalysis
See more of this Group/Topical: Materials Engineering and Sciences Division

(749f) Conversion Control In Acrylate-Epoxide Hybrid Photopolymerizations with Hydroxyl-Containing Acrylates

Gbenga I. Ajiboye and Julie L.P. Jessop, Chemical & Biochemical Engineering, University of Iowa, Iowa City, IA

Photopolymerization uses light, instead of heat, to initiate reactions that convert monomer to polymer. The technique is increasingly used in industry due to its high levels of spatial and temporal control, low reaction temperatures, fast curing speeds, and virtually no emission of volatile organic compounds. Free-radical and cationic photopolymerizations have been implemented in many applications, such as coatings, electronics, biomedicine, and adhesives. Although free-radical polymerizations of acrylates are most common, inhibition by oxygen remains a problem. Oxygen reacts with the excited-state photoinitiators and the free-radical active centers to prevent chain propagation. Cationic ring-opening reactions of epoxides, while not inhibited by oxygen, generally exhibit low reaction rates and are sensitive to moisture and alcohol. Epoxide-acrylate hybrid systems promise to reduce the sensitivity of the reaction to oxygen and moisture by combining the advantages inherent to each reaction. However, the much faster acrylate reaction interferes with epoxide conversion because the epoxide chain growth takes place via the relatively slow active chain end (ACE) mechanism. In order to increase epoxide conversion, the faster activated monomer (AM) mechanism, a chain transfer reaction, is promoted in this research by pairing hydroxyl-containing acrylates with a diepoxide monomer.

Hybrid formulations with varying concentrations of 2-hydroxyethyl acrylate (HEA, acrylate) and 3,4-epoxycyclohexane carboxylate (EEC, epoxide) were photopolymerized using the cationic photoinitiator diaryliodonium hexafluoroantimonate (DAI). During photolysis, DAI also produces free radicals capable of initiating the acrylate polymerization, thereby removing the need for a separate free-radical photoinitiator. Conversions of acrylate and epoxide functional groups were obtained using real-time Raman spectroscopy. As the acrylate monomer concentration increased from 50 to 80%, the epoxide conversion increased from 55 to 85%, indicating that the hydroxyl group did indeed facilitate the AM mechanism. In a control experiment to verify that the increased epoxide conversion was not due to a dilution effect, HEA was replaced with a similar non-hydroxyl-containing acrylate, ethylene glycol methyl ether acrylate (EGMEA). The maximum EGMEA and EEC conversion obtained for an 80:20 mixture after 5-minutes illuminations was 86% and 35%, respectively. These results further confirm that the hydroxyl group in HEA promotes the AM mechanism. Physical and mechanical properties (e.g., glass transition temperature, modulus, and phase separation) for these systems are being investigated using dynamic mechanical analysis to determine how the AM mechanism affects the final hybrid polymer properties.

See more of this Session: Polymer Reaction Engineering Kinetics & Catalysis

IN LINE MONITORING OF RESIDUAL MONOMER BY NIR SPECTROSCOPY DURING STYRENE-DIVINYLBENZENE SOLUTION POLYMERIZATION REACTIONS

Dennis Chicoma Lara, María Verónica Carranza Oropeza,
Leandro Gonçalves de Aguiar and Reinaldo Giudici,

University of São Paulo, São Paulo, SP, Brazil.

Abstract

In this work, the sensitivity of NIR spectroscopy was studied for in-line, in-situ monitoring of residual monomer concentrations in solution copolymerization of styrene and divinylbenzene (S-DVB). Reactions were carried out in a glass reactor (stirred tank) to gather data used to develop a NIR calibration model. Experimental data were collected via off-line measurements of monomer conversion by gravimetry and they were previously validated with a kinetic model of the process. These data were correlated with the NIR spectra collected by an immersion probe (transflectance mode) during the reaction at the same sampling times. Spectral changes in the ranges 4460-4750 cm^{-1} and 6212-6187 cm^{-1} were related to the variations of monomer concentration. The results show good agreement between the experimental data and in-line monitoring predictions from the NIR spectra and the calibration model.

Keywords

Chemical Reactions, In Line Monitoring, NIR Spectroscopy, Polymers.

Introduction

Solution polymerization processes have been widely used for the production of polymer resins and other commercial products. In these processes, monomer, initiator, solvent and polymer are all soluble in the reaction medium, leading to the formation of a homogeneous reacting mixture. This technique show, as advantages, the easy-control of temperature and the solution can be directly used (according to the desired end-product). However, the solvent reduces the polymer molecular weight and reaction rate, promoting some control over these variables. It is well known that the main limitation for the efficient control of polymerization processes is the lack of robust sensors, which provide quick and reliable information on the properties of the polymer being produced [1,2]. Actually, the NIR spectroscopy is enabling a great development in both pharmaceuticals and food industries, as well as, in agriculture area and cosmetic industry. In the polymer field, this technique is used successfully in monitoring in bulk [3,4], suspension [5-7] and in emulsion [8,9] polymerization process. For the solution polymerization, there are also some published works, such as, Long et al. [10], Fontoura et al. [11], Chefri and Févotte [12], Chefri et al.[13] and Beyers et al.[14]. The success of NIR spectroscopy is possible through the acquisition of spectra assisted with chemometric techniques that relate the spectral information with the reference variable (residual concentration of comonomers, for example). The application of these techniques in polymerization process has been studied, with special attention to the analytical and quantitative results due to spectral variations between the peaks of the monomers and solvent during the reaction. As such, the final spectrum displays broad, overlapping peaks and bands with presence of shoulders, indicating that, although "rich" in information, the spectrum does not allow a direct measurement of the variables of interest, thus requiring appropriate calibration. In the present work, in-line and in-situ monitoring using a NIR probe coupled with fiber optics was used to monitor styrene and divinylbenzene (S-DVB) residual monomer concentrations in solution copolymerization reactions. The aim of this work was to produce copolymers with pre-specified properties that are able to attend the increasingly exigent demands of the polymer market.

EXPERIMENTAL

The following reactants were used in the solution copolymerization reactions: industrial grade styrene (BASF-Brazil) and divinylbenzene (Sigma Aldrich) monomers, toluene as solvent and AIBN as initiator. Methanol was used as precipitation agent at the end of the reaction. Polymerization reactions were carried out in a jacketed glass stirred-tank reactor using different ratios between styrene and toluene (see Table 1).The temperature of the reactor content was controlled within 85±1 °C and the agitation speed was kept at 200 rpm. All NIR spectra were collected with an IFS 28/N spectrometer (Bruker Optics), equipped with a quartz bean splitter, using a probe (Hellma 661.622 NIR, with a transflection system with overall light path equal to 1mm), immersed into the reaction medium. Off-line measurements of monomer conversion were obtained by gravimetry and they were previously validated with a kinetic model of the process. These measurements were used as reference for the development and validation of the multivariate NIR calibration model. This model was developed based on Partial Least Squares (PLS) using the software of the equipment (OPUS, program QUANT2).

Table 1. Polymerization recipes

Reactant	Run 1 (g)	Run 2 (g)	Run 3 (g)	Run 4 (g)
Toluene	200.36	280.58	481.83	440
Styrene	193.5	115.73	309	346.6
Divinylbenzene	7.453	4.467	11.911	13.4
AIBN	3.4	3.472	6.765	6.744

In order to obtain the global conversion by gravimetry, an empty aluminum capsule was weighted, then 3 mL of methanol (responsible for the precipitation of the polymer) was added and weighted again. This content was weighted after receiving 1.5 mL of reaction sample. This procedure was done periodically throughout the reaction and the capsules were immediately conducted to an oven at 70 °C, proceeding a drying process during 24 hours. The experimental measurements were validated with the results predicted by the mathematical model of the process based on the work of Ortiz et al. [15] and Gonçalves et al. [16]. In the model, the following assumptions are considered: pseudo-steady state for the radicals, each polymer molecule has only one radical center; pseudo-homopolymerization approach; similarity between the average size of the chain for dead and dormant polymer during the process. Figure 1 shows the comparison between the experimental and the predicted results of global monomer conversion evolution. As can be seen, there is a good agreement on all reactions. Based on these results, the model is used to estimate the concentration of residual monomer during the reaction and used as reference set for the NIR calibration model. For the elaboration this NIR model, the following known steps were performed [17]: identification and selection of the spectral region representative of the conversion of monomer into polymer, spectral pre-processing (necessary to remove unwanted variations such as noise and fluctuations in the baseline), calibration set (evaluation parameters) and validation (internal and external).

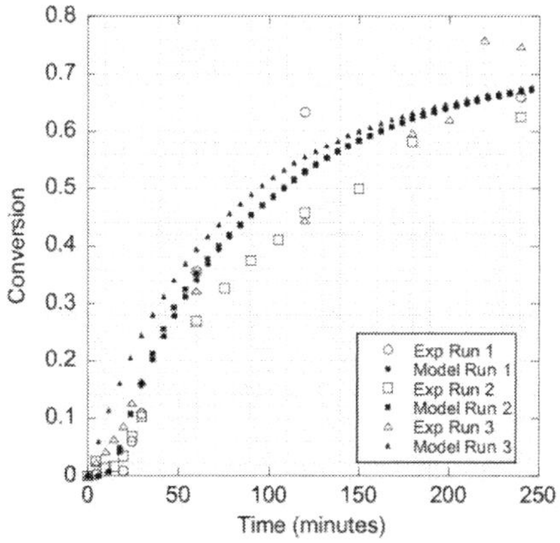

Figure 1. Comparison of the evolution of global conversion between the experimental results (Exp Run) and the ones predicted by the kinetic model (Model Run).

RESULTS

Figure 2 shows the NIR spectral evolution during the polymerization process. It can be observed that the decreasing of the peaks at 6134, 4720 and 4487 cm^{-1} are assigned to stretching vibrational motions of the vinyl group and region of combination bands C–H + C=C, respectively.

As such, the spectral regions chosen for residual monomer were the combination (4460-4750cm^{-1}) and the first overtone region (6212-6187cm^{-1}), wherein the bands are attributed to the monomer concentration. Spectral pre-processing was performed using second derivative with 25 points smoothing and the internal validation was made by using conventional "leave-one-out" cross-validation. In this procedure, parts of the calibration set are left out of the calculation used to fit the model (with a certain number of latent variables –LVs) to the data. Then the model is used to estimate the values of the left-out samples. This procedure is repeated several times for different sets and the root mean square error (RMSECV) between the estimated and real values is calculated. The number of LVs is chosen as the minimum of the plot of RMSECV versus the number of LVs. The resulting calibration model uses three principal components (latent variables) and presents a high correlation coefficient (0.986).

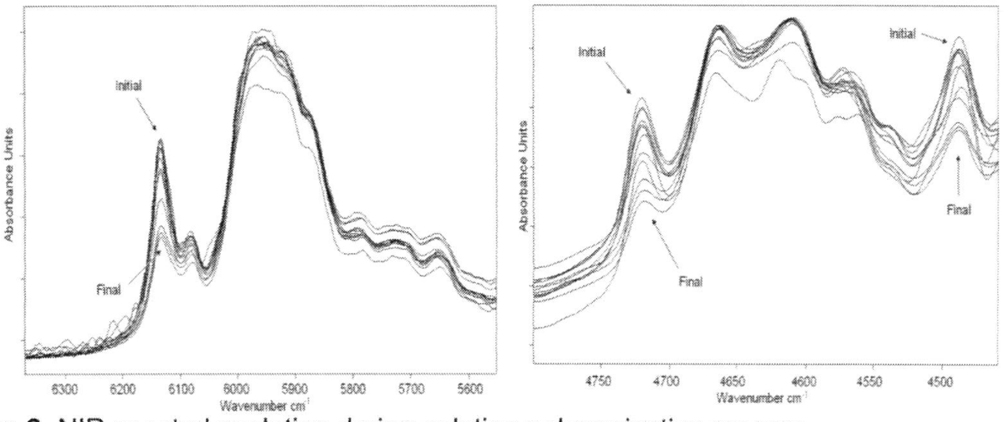

Figure 2. NIR spectral evolution during solution polymerization process.

Figures 3a, 3b and 3c show the comparison for monomer content between the off-line measurements by reference methods and the predictions from the on-line measured NIR spectra with the calibration model (internal validation). These results exhibited good agreements. Then, the NIR calibration model was tested with a new reaction data that were not used in calibration step (external validation), the results of prediction showed good agreement between mathematical model data and the NIR prediction data.

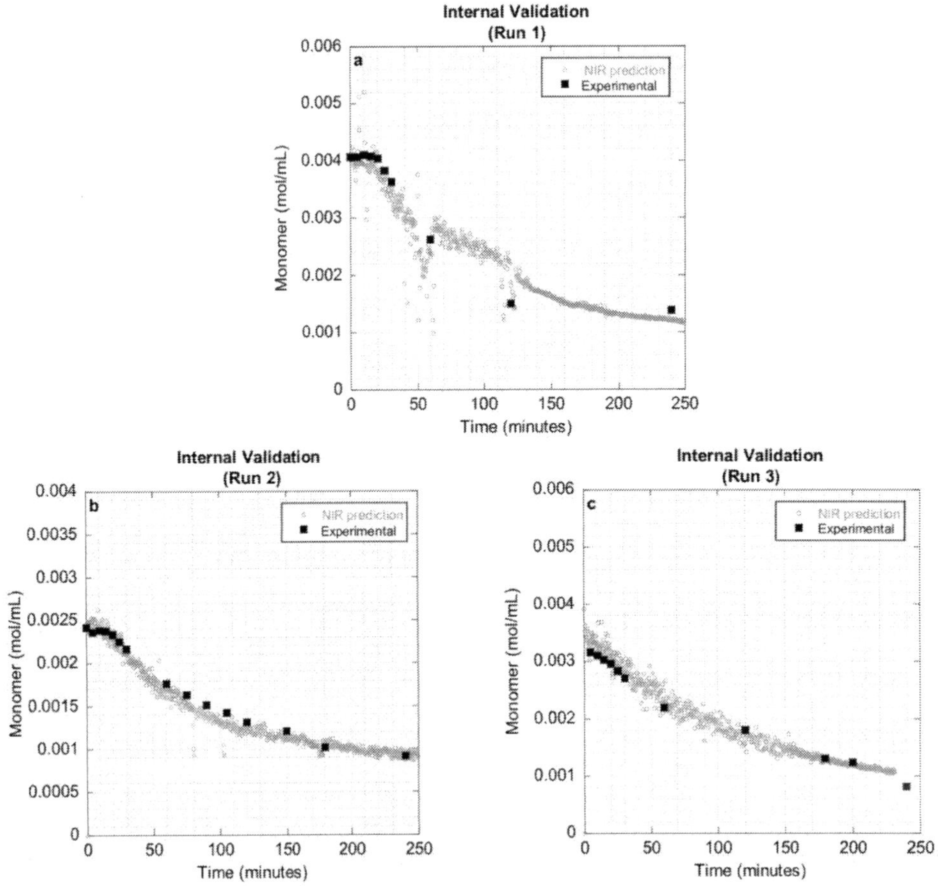

Figure 3. Comparison between the estimates from the NIR spectra and the experimental monomer concentration by internal validation. (a) Run 1, (b) Run 2 and (c) Run 3.

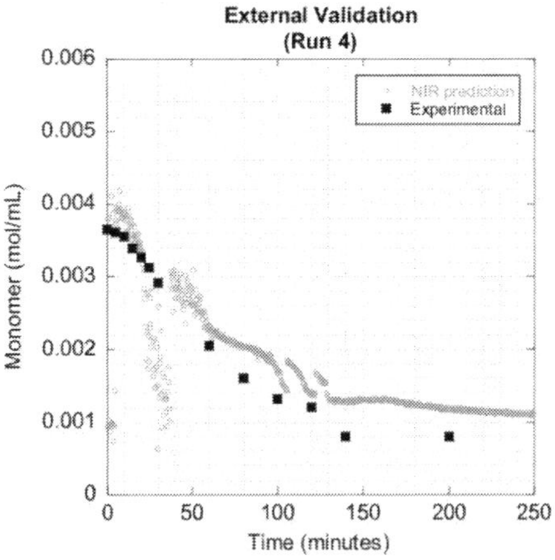

Figure 4. Comparison between the estimates from the NIR spectra and the experimental monomer concentration by external validation.

CONCLUSION

The present approach for the in-line and in-situ monitoring of monomer concentration during solution copolymerization of styrene and DVB was based on a NIR calibration model. This model was developed and validated with NIR spectra (from external validation), off-line gravimetric measurements and a kinetic model of the process. The results showed good agreement between the experimental data and the in-line monitoring predictions from the NIR spectra. The calibration model can enable, in a near future, the implementation of closed-loop control strategies and real-time decision-taking during polymerization processes.

ACKNOWLEDGEMENTS

The financial support from FAPESP (Fundação de Amparo à Pesquisa do Estado de São Paulo) and CNPq (Conselho Nacional de Desenvolvimento Científico e Tecnológico) is gratefully appreciated. The authors also thank BASF Brazil for providing monomer.

REFERENCES

1. MacGregor, J. F.; Penlidis, A.; Hamielec, A. E. (1984), "Control of Polymerization Reactors. A Review", *Polym. Proc. Eng.*, v. 2 (2-3), pp. 179-206.

2. Giudici, R. (2000), "Polymerization Reaction Engineering. A Personal Overview of the State of Art", *Latin American Applied Research*, v. 30(4), pp. 351-356.

3. Aldridge, P. K.; Kelly, J. J.; Callis, J. B.; Burns, D. H. (1993), "Non Invasive Monitoring of Bulk Polymerization Using Short-Wavelenght Near Infrared Spectroscopy", *Anal. Chem.*, v. 65, pp. 3581-3585.

4. Lousberg, H. H. A.; Boelens, H. F. M.; Le Comte, E. P.; Hoefsloot, H. C. J.; Smilde A. K. (2002), "On Line Determination of the Conversion in a Styrene Bulk Polymerization Batch Reactor using Near-Infrared Spectroscopy", *Journal of Applied Polymer Science*, v. 84 (1), pp. 90-98.

5. Santos, A. F.; Lima, E. L.; Pinto, J.C. (1998), "In-line Evaluation of Average Particle Size in Styrene Suspension Polymerization Using Near-Infrared Spectroscopy", *J. Appl. Polym. Sci.*, v. 70, pp. 1737-1745.

6. Santos, A. F.; Lima, E. L.; Pinto, J.C. (2000), "Control and Design of Average Particle Size in Styrene Suspension Polymerization Using NIRS", *J. Appl. Polym. Sci.*, v. 77, pp. 453-462.

7. Machado, F.; Lima, E. L.; Pinto, J. C. (2007), "Uma Revisão Sobre os Processos de Polimerização em Suspensão", *Polímeros*, v. 17, 2, pp. 166-179.

8. Reis, M. M.; Araujo, P. H. H.; Sayer, C.; Giudici, R. (2004[b]), "In situ near infrared spectroscopy for simultaneous monitoring of multiple process variables in emulsion copolymerization", *Ind. Eng. Chem. Res.*, v. 43, pp. 7243-7250.

9. Chicoma, D.; Sayer, C.; Giudici, R. (2011), "In-Line Monitoring of Particle Size during Emulsion Polymerization under Different Operational Conditions using NIR Spectroscopy", *Macromolecular Reaction Engineering*, v. 5, Issue 3-4, pp. 150–162, April 13.

10. Long, T. E.; Liu, H. Y.; Schell, B. A.; Teegarden, D. M.; Uerz, D. S. (1993), "Determination of Solution Polymerization Kinetics by Near-Infrared Spectroscopy 1: Living Anionic Polymerization Processes", *Macromolecules*, v. 26 (23), pp. 6237–6242.

11. Fontoura, J. M. R.; Santos, A. F.; Silva, F. M.; Lenzi, M. K.; Lima, E. L.; Pinto, J. C. (2003) "Monitoring and Control of Styrene Solution Polymerization Using NIR Spectroscopy", *Journal of Applied Polymer Science,* v. 90, pp. 1273-1289,.

12. Cherfi, A.; Févotte, G. (2002), "On-Line Conversion Monitoring of the Solution Polymerization of Methyl Methacrylate Using Near-Infrared Spectroscopy", *Macromol. Chem. Phys.*, v. 203, 9, pp. 1188-1193.

13. Cherfi, A.; Févotte, G.; Novat, C. (2002), "Robust On-Line Measurement of Conversion and Molecular Weight Using NIR Spectroscopy During Solution Polymerization", *Journal of Applied Polymer Science*, v. 85, pp. 2510-2520.

14. Beyers, C. P.; Boelens, H. F. M.; Klumperman, L.; Westerhuis, J. A. (2004), "In-line reaction monitoring of the methyl methacrylate and N,N-dimethylacrylamide copolymerization reaction employing near infrared spectroscopy", *Applied Spectroscopy*, v. 58, 7, pp. 182A-208A.

15. Ortiz, J.C.H.; Lima, E.V.; Lona, L.M.F.; Mccmanus, N.T.; Penlidis, A. (2009), "Modeling of the Nitroxide-Mediated Radical Copolymerization of Styrene and Divinylbenzene", *Macromol. React. Eng.*, v. 3, pp. 288–311.

16. Gonçalves, M.A.D.; Trigo, I.M.R.; Dias, R.C.S; Costa, M.R.P.F.N. (2010), "Kinetic Modeling of the Molecular Architecture of Cross-Linked Copolymers Synthesized by Controlled Radical Polymerization Techniques", *Macromol. Symp.*, v. 291, pp. 239–250.

17. Silva, W.; Chicoma, D.; Giudici, R. (2011), "In-situ Real-time Monitoring of Particle Size, Polymer, and Monomer Contents in Emulsion Polymerization of Methyl Methacrylate by Near Infrared Spectroscopy". *Polymer Engineering Science*. doi: 10.1002/pen.22100.

(753a) Fundamentals of Aerosol Reactor Design for Catalysts

Sotiris E. Pratsinis, Particle Technology Laboratory, Institute of Process Engineering, Department of Mechanical and Process Engineering, ETH Zurich, Zurich, Switzerland

Aerosol manufacturing routes are advantageous over wet chemistry ones as the former offer fewer process steps, easier particle collection and no liquid by-products requiring costly cleaning[1]. In addition, aerosol-made particles and films have unique morphology and high purity (e.g. optical fibers) and can even form metastable phases (e.g. low temperature $BaCO_3$ for NOx storage-reduction catalysts[2]). So they easily form mixed oxides, salts and even pure metals and highly porous (98%) films resulting in novel catalysts, micropatterned sensors, phosphors, battery electrodes, dental prosthetics and even nutritional supplements[1].

The emphasis then shifts is on the fundamentals of flame synthesis of particles. In principle, clusters are formed by oxidation of appropriate precursors and grow by coagulation and sintering to ramified fractal-like aggregates and eventually agglomerates of nanosize primary particles. Primary particles in an aggregate may have a narrow size distribution as it is predicted by theory[3] and verified by experiments at Cabot[4].

Recent understanding of the dynamics of these non-spherical particles shows that at high particle concentrations, classic coagulation theory may not be sufficient to describe the ensuing aerosol dynamics, especially if fractal-like particles are formed[5]. In fact, a transition takes place from dilute to concentrated aerosol dynamics as there is less "free gas" volume per agglomerate. They deviate from the kinetic theory of gases especially during instantaneous coalescence: completely inelastic particle-particle collisions, the so-called *cooling of granular gases*[6]. In the transition regime, the coagulation rate of highly concentrated aerosols is progressively higher than that for dilute aerosols as growing particles enter the continuum regime where coagulation rates are 2 - 30 times higher than that of Smoluchowski theory[5]. Then fractal-like particles are likely to coagulate till gelation as has been observed in batch formation of soot. In aerosol flow reactors, however, shear forces such agglomerates to restructure or break-up given the employed rather high (10^5-10^6) Reynolds numbers[7].

1. B. Schimmoeller, S.E. Pratsinis, A. Baiker, "Flame Aerosol Synthesis of Metal Oxide Catalysts with Unprecedented Structural and Catalytic Properties", *ChemCatChem* in press (2011).

2. R. Buchel, S.E. Pratsinis, A. Baiker "Mono- and bimetallic Rh and Pt NSR-catalysts prepared by controlled deposition of noble metals on support or storage component", *J. Catal.*, in review (2011).

3. M.C. Heine, S.E. Pratsinis, "Polydispersity of Primary Particles in Agglomerates made by Coagulation and Sintering" *J. Aerosol Sci.*, **38**, 17-38 (2007).

4. D. Boldridge, Morphological Characterization of Fumed Silica Aggregates. *Aerosol Sci. Technol.* **44,** 182-186 (2010).

5. M.C. Heine, S.E. Pratsinis, "High Concentration Agglomerate Dynamics at High Temperatures", *Langmuir*, **22**, 10238-10245 (2006).

6. B. Buesser, M.C. Heine, S.E. Pratsinis, "Coagulation of highly concentrated aerosols" *J. Aerosol Sci.*, **40**, 89-100 (2009).

7. Y. Xiong, S.E. Pratsinis, "Gas Phase Production of Particles in Reactive Turbulent Flows", *J. Aerosol Sci.*, **22**, 637-655 (1991).

See more of this Session: Rational Catalyst Design II
See more of this Group/Topical: Catalysis and Reaction Engineering Division

(753b) Structure Sensitivity of Dimethyl Ether Electro-Oxidation

Jeffrey A. Herron, Peter Ferrin and Manos Mavrikakis, Chemical and Biological Engineering, University of Wisconsin - Madison, Madison, WI

Dimethyl Ether (DME) is an attractive chemical for both on-board hydrogen generation as well as a fuel in direct electro-oxidation in fuel cells.[1, 2] Experimental studies of the electro-oxidation have demonstrated significant structure sensitivity on Pt – specifically when comparing the (111) and (100) facets.[3] Using periodic, self-consistent density functional theory calculations, we evaluate the thermodynamics of dimethyl ether (DME) electro-oxidation on a number of model (100) and (111) facets of monometallic fcc catalysts, including Au, Ag, Cu, Pt, Pd, Ni, Ir, Rh. Using the results, we predict the most-probable reaction mechanism on each of these catalysts. Based on these proposed mechanisms, we predict the energy efficiency and catalytic activity of each of these metals. Furthermore, by comparing results on the (111) and (100) facets, we attempt to elucidate the origins of the experimentally observed structure sensitivity of DME electro-oxidation.

[1] Semelsberger, T. A.; Borup, R. L.; Greene, H. L., *Journal of Power Sources* 2006, *156*, 497-511.

[2] Kerangueven, G.; Coutanceau, C.; Sibert, E.; Leger, J. M.; Lamy, C., *Journal of Power Sources* 2006, *157*, 318.

[3] Lu, L.; Yin, G.; Tong, Y.; Zhang, Y.; Gao, Y.; Osawa, M.; Ye, S., *Journal of Electroanalytical Chemistry* 2010, *642*, 82.

See more of this Session: Rational Catalyst Design II
See more of this Group/Topical: Catalysis and Reaction Engineering Division

(753c) Structure-Activity Correlation for Relative Chain Initiation to Propagation Rates In Single-Site Olefin Polymerization Catalysis

Thomas A. Manz[1], Shalini Sharma[2], Khamphee Phomphrai[2], Grigori Medvedev[3], Kendall T. Thomson[3], Mahdi Abu-Omar[2], W. Nicholas Delgass[3] and James M. Caruthers[3], (1)School of Chemical Engineering, work performed as a Ph.D. student at Purdue University (present address: Georgia Institute of Technology), Atlanta, GA, (2)Chemistry Department, Purdue University, West Lafayette, IN, (3) School of Chemical Engineering, Purdue University, West Lafayette, IN

Using experiments and density functional theory (DFT) calculations, we systematically studied factors controlling relative chain initiation to propagation rates in homogeneous olefin polymerization catalysis. These catalysts typically contain Ti or Zr bound to several organic ligands. The active part of the catalyst is a cation coordinated to a weakly bound anion (aka counterion). This ion pair can exist in various forms: (a) in the inner sphere ion pair, the counterion is in direct contact with the counterion, (b) in the outer sphere ion pair, the counterion is in the metal's outer coordination sphere but there are no solvent molecules between the cation and counterion, and (c) in the solvent separated ion pair, the cation and counterion are separated by two or more solvent layers. Since partial sepation of the counterion from the cation is required for the monomer to enter the docking site, catalysts with lower ion pair separation energies have higher reactivity. Twenty-seven such catalysts containing mixed cyclopentadienyl/aryloxide ligation were synthesized and used to polymerize 1-hexene in toluene, bromobenzene, or 1,2-dichlorobenzene solvents. For some of the catalysts, the rate of first monomer insertion (i.e. chain initiation) was comparable to the rate of subsequent monomer insertions (i.e. chain propagation). For other catalysts, the chain initiation rate was three orders of magnitude slower than the chain propagation rate. A comprehensive set of DFT calculations was performed to discover the underlying cause of slow versus facile chain initiation. First, DFT computations were used to identify the lowest energy ion pair form for each catalyst in various solvents. Then, appropriate electronic and steric descriptors were identified and correlated to chain initiation and propagation rate constants. Comparison of DFT calculations to experimental data revealed the underlying cause of slow versus facile initiation is the difference in steric congestion between the initiation kinetically dominant ion pair and the propagation kinetically dominant ion pair. These descriptors were used to construct a quantitative structure-activity relationship (QSAR) that accurately predicts whether chain initiation is facile or slow for each catalyst. Transition state analysis performed for representative catalysts confirmed the underlying chemical mechanism controlling facile versus slow initiation. This underlying chemical mechanism was consistent with the developed QSAR.

See more of this Session: Rational Catalyst Design II
See more of this Group/Topical: Catalysis and Reaction Engineering Division

(753d) Rational Catalyst Design for Oxygenate Reforming

Michael Salciccioli and Dion G. Vlachos, Chemical Engineering, University of Delaware, Newark, DE

Understanding and controlling bond breaking sequences of oxygenates on transition metal catalysts can greatly impact the utilization of biomass feedstocks for fuels and chemicals. Specifically, the reforming of biomass and biomass derivatives to synthesis gas can act as a first step towards this transformation to value added products. In this study, the decomposition of ethylene glycol, as the simplest representative of biomass-derived polyols, was studied via density functional theory (DFT) calculations to identify the differences in reaction pathways between Pt and the more active Ni/Pt bimetallic catalyst. Temperature programmed desorption (TPD) experiments were conducted with deuterated ethylene glycols for comparison with DFT results. Insights from these fundamental studies were used to parameterize a semi-empirical based microkinetic model to probe for active and selective catalysts for ethylene glycol decomposition to synthesis gas. Finally, this model was extended to other C1 and C2 oxygenates and optimal catalysts were tested experimentally.

See more of this Session: Rational Catalyst Design II
See more of this Group/Topical: Catalysis and Reaction Engineering Division

(753e) Shape- and Size-Specific Chemistry of Ag Nanostructures In Catalytic Ethylene Epoxidation

Phillip Christopher and Suljo Linic, Chemical Engineering, University of Michigan, Ann Arbor, MI

Shape- and Size-Specific Chemistry of Ag Nanostructures in Catalytic Ethylene Epoxidation

Phillip Christopher and Suljo Linic

Department of Chemical Engineering, University of Michigan,

Ann Arbor, Michigan 48109-2136

Catalytically active metallic nanoparticles typically exhibit surface structure dominated by the most thermodynamically stable surface facet of the metal. This is a result of catalyst synthesis procedures that allow thermodynamics to control the resulting size and shape of the metal particles. Consequently, it has remained challenging to tune the catalytic function of metallic nanoparticles by engineering the exposed surface facet. Recent progress in solution-based synthesis of metallic nanoparticles (exploiting kinetics to control nanoparticle growth) coupled with advances in molecular scale modeling of catalytic processes is allowing, for the first time, the rational design of catalytic metallic nanostructures with targeted geometric (size and shape) features.

In this work, shape and size controlled synthesis of silver (Ag) nanoparticles was utilized in concert with Density Functional Theory (DFT) calculations to design Ag catalysts that are more selective in the oxidation of ethylene to form ethylene oxide (EO) than Ag catalysts prepared using conventional methods. DFT calculations showed that the Ag(100) surface facets should be inherently more selective than the Ag(111) facet towards EO. The Ag(111) facet is the most abundant surface facet on spherical Ag particles, that are the standard form of Ag catalysts used in ethylene epoxidation. Ag nanoparticles dominated by the (100) facet (Ag nanocubes and nanowires) and the (111) facet (nanospheres) were synthesized using a modified polyol process. Steady-state reactor studies showed that the shape and size of catalytic Ag nanoparticles play important roles in governing the selectivity to EO, as predicted by the DFT calculations. Furthermore, these studies have allowed us to develop a physically transparent model, which describes the effect of shape, size and external conditions on activity and selectivity in the industrially important ethylene epoxidation reaction. (Christopher, P., Linic, S., *ChemCatChem*, **2010**, 2, 78. Christopher, P., Linic, S., *JACS*, **2008**, 130, 11264.)

See more of this Session: Rational Catalyst Design II
See more of this Group/Topical: Catalysis and Reaction Engineering Division

(753f) Structure-Activity Relationships In Group IV Bis (Phenoxide) Amine and Diamine Olefin Polymerization Catalysts: A DFT and Kinetic Analysis

Kendall T. Thomson[1], Xiong Silei[2], W. Nicholas Delgass[2], James Caruthers[2], Mahdi Abu-omar[3] and Jeff Switzer[2], (1)School of Chemical Engineering, Purdue University, West Lafayette, IN, (2)Chemical Engineering, Purdue University, West Lafayette, IN, (3)Chemistry, Purdue University, West Lafayette, IN

While olefin polymerization is a well established industrial process, very little kinetic data is available to characterize these catalytic processes. Even less is know about the relationships between catalyst structure and polymerization kinetics. Here we present a detailed electronic DFT-based analysis of a variety of Group IV bis(phenoxide) amine and diamine catalysts used for the polymerization of 1-butene. By comparing both electronic and structural descriptors, we reveal several important insights into the effect of salen ligand character on (1) initiation, (2) propagation, and (3) chain-transfer kinetics. These DFT results, combined with a unique quantitative kinetic modeling approach involving multi-response data (i.e. the time evolution of the following species: monomer consumption, molecular weight distribution, etc.) provide a consistent characterization of this class of polymerization catalysts, and opens the way to full model-based prediction of polymerization activity and permormance.

See more of this Session: Rational Catalyst Design II
See more of this Group/Topical: Catalysis and Reaction Engineering Division

(766a) Enantioselective Hydrogenation of C=C Bond Over Cinchona-Modified Pd Catalyst

Shuai Tan, Chemical Engineering, University of South Carolina, Columbia, SC
See more of this Session: Fundamentals of Surface Reactivity II
See more of this Group/Topical: Catalysis and Reaction Engineering Division

(766b) Adsorbate Coverage Effects On Catalytic Reactivity At a Low-Symmetry, Kinked Surface

Jason M. Bray and William F. Schneider, Chemical and Biomolecular Engineering, University of Notre Dame, Notre Dame, IN

Adsorbate coverage effects on catalytic reactivity at alow-symmetry, kinked surface

Jason M. Bray, William F. Schneider

Department of Chemical and Biomolecular Engineering,University of Notre Dame, Notre Dame, IN, USA

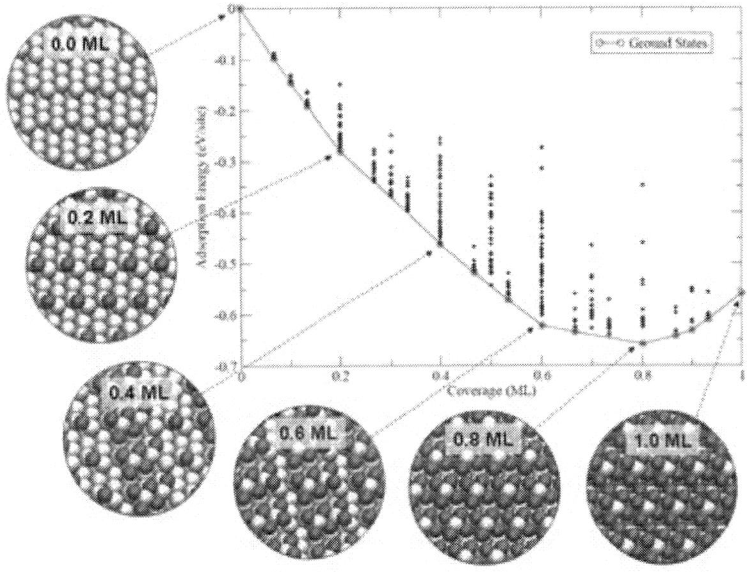

Adsorbate coverage can have a large impact on the number and types ofsites that are available for reactions at a surface. Such coverage effects arecritical, for instance, in understanding the catalytic oxidation of NO to NO_2over Pt. While computational approaches to modeling these coverage effects onhigh symmetry surfaces, such as (111) facets, are fairly well developed, lowersymmetry surfaces remain a challenge. In this work, we use DFT simulations andcluster expansions to understand oxygen coverage effects during catalytic NOoxidation at the (321) surface of Pt, a system that has been shown to have surprisinglylittle structure sensitivity. The chiral (321) facet exposes kinks and stepedges that offer a variety of potential adsorption and reaction sites. In thelow coverage limit we characterize all potential O and O_2 adsorptionsites as well as a number of pathways for O_2 dissociation and Odiffusion, and we use these results to explain the surprisingly faciledissociation of O_2 at low coverage observed experimentally as wellas the related vibrational spectrum of adsorbed O and O_2. While ason the (111) surface, O adsorbates interact repulsively at low coverage, athigher coverages they exhibit complex interactions that cause O atoms tocongregate around single Pt kink sites in distinctive 4-fold configurations.To quantify these O interactions, we use DFT simulations to parameterize anIsing-type cluster expansion of O on Pt(321) for coverages up to 1 ML. Themodel allows us to predict the ground state configurations of adsorbed O,determine coverage-dependent O binding energies, and, with GCMC simulations,predict equilibrium adsorbate coverages as a function of reaction environment, inagreement with experimental observation. Further, GCMC provides informationabout the number and types of reaction sites that are available at *operando*coverages, allowing us to rationalize the similarities in (111) and (321)activity.

See more of this Session: Fundamentals of Surface Reactivity II
See more of this Group/Topical: Catalysis and Reaction Engineering Division

(766c) Hydrogen-Release Kinetics From Pd(100): The Influence of H In the Surface, Subsurface, and Bulk States

William D. Michalak[1], James B. Miller[1], Dominic Alfonso[2] and Andrew J. Gellman[1], (1)Chemical Engineering, Carnegie Mellon University, Pittsburgh, PA, (2)National Energy Technology Laboratory, U.S. Department of Energy, Pittsburgh, PA

Understanding the interactions of hydrogen atoms on thesurface and within the subsurface regions of Pd is critical to the developmentof advanced energy technologies for hydrogen storage andseparations, as well as catalytic processes. While many of the physical,chemical, and electronic properties of the H_2-Pd system are known,the kinetics and thermodynamics during absorption into the bulk, transport backto the Pd surface, and desorption at low temperature remain unclear. In this work,the H_2 release kinetics from Pd were measured and modeled over arange of exposure pressures and temperatures using temperature programmeddesorption. To simulate the observed kinetic behaviors, a continuum-basedmodel was extended from other previously published work (M. Mavrikakis, *etal., J. Chem. Phys* **105**, 8398, 1996) to include activation barriersfor desorption and transport that are dependent on H concentration. The use ofconcentration dependent barriers improves the model's ability to predict theexperimental trends across temperatures ranging from 100 - 600 K. With themodel, the transition and intermediate states that control the net release rateare also identified using a transient version of the degree of rate control (C. Stegelmann, *et al., JACS* **131**, 13563 (2009)). It is shownthat many states are involved in the release process depending on thetemperature and hydrogen distribution in the Pd system, so that the netactivation barrier is defined by either a single reaction step or by statesthat are at distant locations in the reaction coordinate.

See more of this Session: Fundamentals of Surface Reactivity II
See more of this Group/Topical: Catalysis and Reaction Engineering Division

(766d) Surface Structural Evolution During the Thermal Decomposition of a Pdo(101) Thin Film

Jason F. Weaver, Chemical Engineering, University of Florida, Gainesville, FL and Jose Hinojosa Jr., Department of Chemical Engineering, University of Florida, Gainesville, FL

The oxidation of late transition metal surfaces plays a central role in several applications of oxidation catalysis, including the catalytic combustion of natural gas, exhaust gas remediation in automobiles and fuel cell catalysis. In this talk, I will discuss results of recent experiments in which we investigated the isothermal decomposition kinetics of a PdO(101) thin film, and characterized the structural changes using scanning tunneling microscopy (STM).

We find that the isothermal decomposition rate of a PdO(101) film exhibits an autocatalytic regime in which the rate of oxygen desorption increases as the oxide decomposes. Our STM results show that reduced sites created during oxide decomposition promote further PdO decomposition, and reveal strong kinetic anisotropies in the decomposition process. For example, the kinetic anisotropies cause one-dimensional reaction fronts to propagate preferentially along the atomic rows of the PdO(101) surface, generating long Pd-atom chains which subsequently promote oxygen recombination in neighboring rows of the oxide. We also find that reduced domains at the surface are re-oxidized by underlying PdO(101) layers to produce domains of well-defined PdO(101) islands. The re-oxidation process acts to sustain the autocatalytic decomposition kinetics, and appears to be the dominant mechanism by which oxygen atoms in the subsurface of the PdO film migrate to the surface and ultimately desorb.

See more of this Session: Fundamentals of Surface Reactivity II
See more of this Group/Topical: Catalysis and Reaction Engineering Division

(766e) First Direct Demonstration of the Schwab Effect Via a Novel External Voltage Induced Control of Electron Transfer to Surface-CO Bond On Pt/TiO2/Au Multilayer Catalytic Nanodiode

Prashant Deshlahra and **Eduardo E. Wolf**, Chemical and Biomolecular Engineering, University of Notre Dame, Notre Dame, IN

Electronic effects are known to play an important role in catalysis since first proposed by Schwab [1]. Charge transfer is often postulated to play a role in metal-titania [2] and other reducible oxide supported catalysts (sometimes referred as SMSI effect). However these electronic effects are often coupled with geometric factors such as decoration effects or higher concentration of low coordination sites on small metal particles, which are difficult to decouple. Furthermore, in dispersed systems, there is no electrical connectivity so no direct demonstration of electronic effects in catalytic junctions has been possible. An early attempt in this direction was made in our group about 15 years ago by microfabricating Pt particles on a flat silica support for butadiene hydrogenation reactions [3]. While external voltage effects were not detected, the concept of controlling the surface chemical activity by a bias voltage was reported in a related patent [4]. Recently, Somorjai and coworkers prepared a catalytic nanodiode with similar characteristics as described in this work and reported a chemicurrent during CO oxidation [5].

In this work we present results on a nanofabricated Pt/TiO_2 catalytic diode modified with a multilayer Pt/TiO_2/Au structure. The latter was introduced to obtain multilayer enhanced infrared reflection spectroscopy (MEIRAS) results of CO adsorbed on Pt, a novel technique [6] recently developed in our laboratory. Using this technique, we directly demonstrate for the first time that Pt-CO bond can be modified by controlling electron transfer at the metal-support junction through an external bias voltage. The results are consistent with DFT simulations [7] of the effect of electric fields on adsorbed CO which predicted variations in the IR frequency due to polarization of the adsorbate. Pt/TiO_2/Au diodes consisting of Pt thin films as well as nanofabricated Pt nanowires have been studied. By studying the effect on infrared polarization, we have demonstrated that using MEIRAS technique and we can selectively probe adsorbates on edges of nanowire which are closer to the Pt/TiO_2 junction. This effect of proximity to the junctions on observed voltage dependent frequency shifts will be analyzed using incident infrared polarized along the nanowires and normal to the nanowires.

References: [1] GM Schwab *Adv. Catal.* **27**, 1 (1978). [2] SJ Tauster et al. *JACS* **100**, 170 (1978); DW Goodman *Catal Let.* **99**, 1 (2005). [3] AC Krauth et al. *Catal. Lett.* **27**, 43 (1994); [4] EE Wolf, US Patent 5,688,474 (1997). [4] XZ Ji et al. *Nano Lett.* **5**, 763 (2005). [5] P Deshlahra et al. *Surf. Sci.* **604**, 2 (2010); P Deshlahra et al. *Appl. Catal. A* **391**, 22 (2011); P Deshlahra and EE Wolf *J. Phys. Chem. C*

114, 16505 (2010). [6] P Deshlahra et al. *J. Phys. Chem. A* **113**, 4125 (2009).

See more of this Session: Fundamentals of Surface Reactivity II
See more of this Group/Topical: Catalysis and Reaction Engineering Division

(766f) Surface Characterization Studies of Silver-Titania Adsorbents

Zenda D. Davis and Bruce J. Tatarchuk, Chemical Engineering, Auburn University, Auburn, AL

Adsorptive desulfurization using silver-titania adsorbents is effective at selectively removing sulfur containing heterocyclic compounds from JP5 fuel. Molecular information on the surface reactivity of the active sites involved in adsorptive separation is necessary to understand the mechanism of adsorption. A thorough understanding of the surface structure, adsorbates, and the bonding interactions will assist in the development of an adsorbent with a greater capacity for sulfur removal. It is hypothesized that the surface adsorption site involves a silver titanate complex. This theory is supported by XPS studies of the atomic ratio of Ag to Ti which shows Ti is a function of Ag in monolayer equivalent of Ag. TPD/TPRS was also performed to elucidate the desorption pathway of the adsorbates. The information gained will be discussed and is useful to optimize multi-cycle regenerability.

See more of this Session: Fundamentals of Surface Reactivity II
See more of this Group/Topical: Catalysis and Reaction Engineering Division

(766g) Characterization and Surface Chemical Reactions of Model Cu-ZnO/Silica Sorbents Upon H2S Adsorption and Regeneration

Divya Repala, Chemical engineering, Auburn University, Auburn, AL and Bruce J. Tatarchuk, Chemical Engineering, Auburn University, Auburn, AL

Zinc oxide sorbents have been considered to be favorable for H2S removal. The dopant Cu plays significant role in improving the performance of of ZnO sorbents.Well prepared model sorbents are necessary to study the changes in oxidation states,diffusion,thermal induced sintering, surface chemical reactions upon Sulfidation and regeneration.

Model sorbents have been prepared by RF-DC magnetron sputtering .Varying compositions of Cu-Zinc oxides are co-sputtered on to RCA cleaned Silicon wafers using the Argon and oxygen as the inlet gases. Sulfided samples are prepared exsitu by passing H2S.Stuctural changes of the calcined vs sulfided sorbents are investigated by SEM-EDX and XRD,AFM. Oxidation states and atomic ratios are confirmed by XPS. Regeneration of "sulfided" sorbents was done in-situ by dosing oxygen and changes in the oxidation states determined. With the help of the temperature-dependent XPS changes in oxidation states of sulfur and the metal ions present in the sorbents, as well dynamics of ionic diffusion, depending on the temperature and oxygen partial pressure are explained. Also changes in the crystal structure due to annealing are studied with XRD

See more of this Session: Fundamentals of Surface Reactivity II
See more of this Group/Topical: Catalysis and Reaction Engineering Division

Dye-Sensitized Photocatalyst – a Breakthrough In Green Energy and Environmental Applications

Pankaj Chowdhury, Hassan Gomaa and Ajay K. Ray
Chemical and BioChemical Engineering, University Of Western Ontario, London, ON, Canada.

Introduction

Solar energy is abundant; the sun creates its energy through a thermonuclear process that converts hydrogen to helium every second. The earth receives approximately $1.5\text{x}10^{18}$ kWh per year, from the solar energy, which is far beyond the energy consumed by human beings (Galvez et al., 2003).

In photocatalytic process photons are utilized to excite the semiconductor material producing electron/hole pair which ultimately involves in the degradation of pollutants and/or producing hydrogen from aqueous solution. These processes can be considered as green or semi green process as they uses solar energy (photons) to mineralize pollutants or to generate green fuel (hydrogen) without any CO_2 formation.

Till date people have mainly used high energy photons (UV-A or UV-B) with different photocatalysts such as TiO_2, $TiSrO_3$, ZnO, etc. for environmental applications. At the same time photocatalytic water splitting showed remarkable progress in last decade under UV light which was started with the pioneering work of Fujishim and Honda (1972). Due to the high cost and environmental impacts of utilization of artificial UV light in photocatalytic process, there has been a great deal of interest in the application of solar energy.

The main drawback with these catalysts lies in their visible light inactivity ($\lambda \geq$ 400-700 nm). A lot of efforts have been made to improve the photocatalytic activity in the visible range. Semiconductor photocatalyst were modified to expand the photo-response to visible region in several ways such as doping with cation/anion (Rehman et al, 2009; Ling et al., 2008), sensitizing with dye, coupling with another small band gap semiconductor and implantation of metal ion (Ni et al., 2007).

In this work, we are using Eosin Y dye as the sensitizer of semiconductor catalyst (TiO_2). Dye is adsorbed chemically onto the semiconductor surface by conventional adsorption process and the chemisorbed dye molecules play the role of spectral sensitizer, which upon excitation with visible light inject electron into the conduction band of the semiconductor. The main focus of this paper is to discuss the photo-degradation of phenol and photocatalytic hydrogen generation under visible solar light and to elaborate the effect of different parameters such as electron donor concentration, amount of co-catalyst and solution pH, in dye-sensitized photocatalysis.

Experimental

Photocatalytic reactions were carried out in a Pyrex glass reactor with a flat window at top for illumination. Figure 1 shows a schematic diagram of the photocatalytic reactor for both hydrogen generation and phenol degradation. The light source was equipped with AM 1.5 G as well as 420 nm UV cut off filter. Liquid samples were analyzed for phenol with HPLC analyzer whereas gas samples were analyzed with GC through TCD detector.

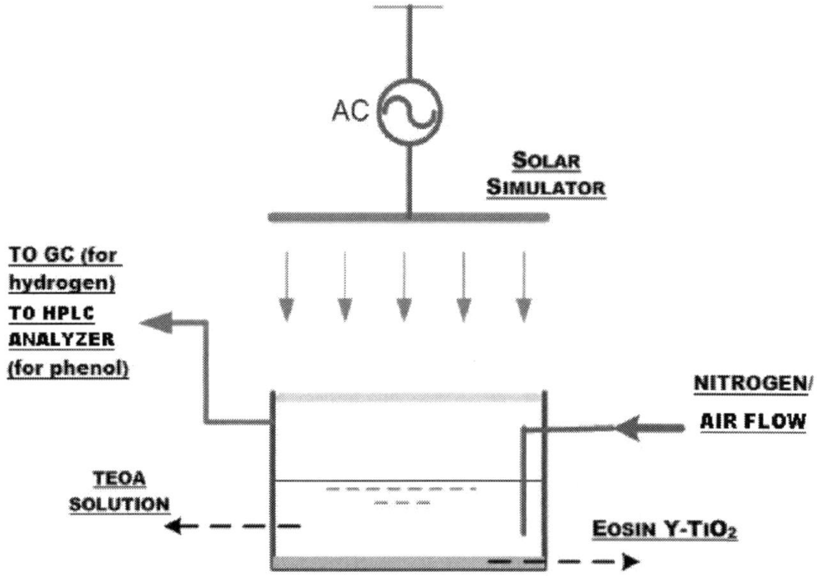

Figure 1. Photoreactor for phenol degradation /hydrogen generation

Results and Discussion

The degradation mechanism of phenol and its derivatives were well explained by Dingwang and Ray (1999) under UV radiation which was mainly through band gap excitation. In dye-sensitized phenol degradation under visible solar light (λ>420 nm) the process starts with excitation of dye molecule from its ground state to the excited state and finally transfer an electron from dye excited state to semiconductor conduction band. Dye-sensitized hydrogen generation also obeyed similar excitation mechanism. In phenol degradation the active species were superoxide ion and hydroxyl radical where as in hydrogen generation, the conduction band electron played the main role. In dye-sensitized process considerable effects were observed with varying the platinum content (in TiO_2), solution pH and electron donor concentration for both hydrogen generation as well as phenol degradation. Neutral pH was found to be optimum for both the processes.

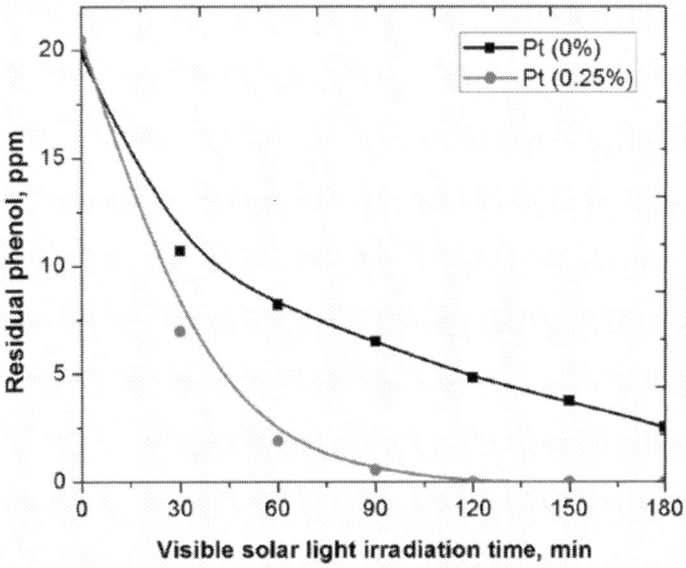

Figure 2. Phenol degradation under visible solar light.

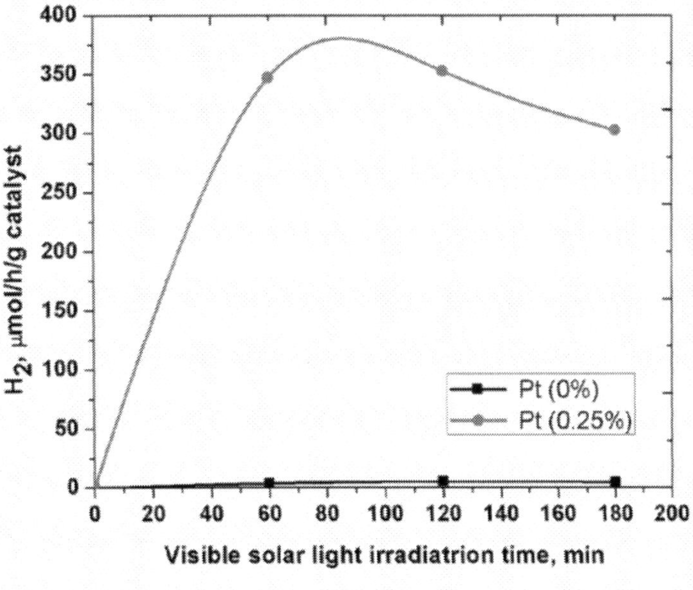

Figure 3. Hydrogen generation under visible solar light.

References

1. Galvez, J. B. and Rodriguez, S. M. (2003), "Solar Detoxification," *United Nations Educational Scientific and Cultural Organization*, pp. 1-246.
2. Fujishima, A and Honda, K (1972), "Electrochemical Photolysis of water at a Semiconductor Electrode," *Nature*, 238, pp.37-38.
3. Rehman, S., Ullah, R. and Gohar, N. D. (2009), "Starategies of making TiO_2 and ZnO visible light active," *J Hazard Mater*, 170, pp. 560-569
4. Ling, Q., Sun, J. and Zhou, Q. (2008). "Preparation and Characterization of visible light driven titania photocatlyst co-doped with boron and nitrogen," *Applied Surface Science*, 254, pp.3236-3241.
5. Ni, M., Leung, M. K. H., Leung, D. Y. C. and Samathy, L. K. (2007), "A review and recent developments in photocatalytic water splitting using TiO_2 for hydrogen production," *Renewable and Sustainable Energy Reviews*, 11, pp.401-425.
6. Chen, D. and Ray, A. K. (1999), "Photocatlytic kinetics of Phenol and its derivatives over UV irradiated TiO_2," *Applied Catalysis B: Environmental*, 23, pp. 143-157.

(769b) The Role of Surface Deposited Pt on the Photoactivity of TiO2

Yun Zhou[1], Christopher Lawrence Muhich[2], Charles B. Musgrave[2] and Alan W. Weimer[2], (1) Department of Chemical and Biological Engineering, University of Colorado, Boulder, Boulder, CO, (2) Department of Chemical and Biological Engineering, University of Colorado, Boulder, Bouler, CO

The role of Pt clusters on defect free anatase TiO_2 (101) surfaces has been studied using *ab initio* pseudopotential calculations based on density functional theory. The calculated electronic structure of Pt clusters supported on (101) anatase shows that excited electrons and holes are trapped by Pt nearly equally (upon addition of an extra electron 0.57 e- localizes on the Pt cluster, while the addition of a hole causes the loss of 0.59 e- from the Pt cluster). Furthermore, Pt states completely bridge the band gap and act as electron-hole recombination centers, which is deleterious to the photoactivity of TiO_2. In addition, calculations show that O_2 will not adsorb on the TiO_2 surface unless an extra electron or subsurface defect exists. On the other hand, the Pt cluster was found to significantly enhance the adsorption of O_2, stemming from the high O_2 adsorption energy of -1.69 eV and the many adsorption sites a Pt cluster presents. This results in enhanced photocatalytic performance because O_2 can scavenge electrons more efficiently from Pt than from TiO_2, which increases the interfacial electron transfer rate. The initial increase and subsequent decrease in TiO_2's photoactivity with increasing Pt loading can be explained by the competition between increased O_2 adsorption and electron-hole recombination. At Pt loadings above the optimum level the increased electron-hole recombination outweighs the increased O_2 adsorption, which is limited by its concentration and diffusion in solution.

See more of this Session: Photo, Microwave and Ultrasound Catalysis
See more of this Group/Topical: Catalysis and Reaction Engineering Division

(769c) Efficiency Factors In Photocatalytic Reactors for Air Treatment

Juan M. Garcia Hernandez, Faculty of Engineering, University of Western Ontario, London, ON, Canada, Benito Serrano, Chemical Engineering, Universidad Autonoma de Zacatecas., Zacatecas Zacatecas., Mexico and Hugo I. de Lasa, Chemical & Biochemical Engineering, University of Western Ontario, London, ON, Canada

Efficiency Factors are key parameters for assessing photocatalytic degradation efficiency and photocatalytic reaction pathways in air treatment units. While the *PTEF* (Photochemical Thermodynamic Efficiency Factor) and QY (Quantum Yield) has been considered for photocatalytic conversion of organic pollutants in water (1), definition of efficiency factors in air treatment reactors is a new initiative of CREC researchers. Efficiency calculations are performed in the present study, using data for acetone and acetaldehyde photocatalytic degradation in air with low concentration of water. Efficiency factors are established in two Photo-CREC-Air reactor scales of 14.7 liters and 60 liters capacity (2,3,4). On the basis of the number of $OH^{\bullet}$ consumed, the enthalpy of $OH^{\bullet}$ formation and the radiation absorbed in the mesh supported photocatalyst, it is shown that quantum yields both for acetone and acetaldehyde supersede in cases the value of 1 (equivalent to 100%) while *PTEFs* remaining below the level of 1 as required by thermodynamic constrains. On this basis, a chain reaction mechanism can be supported for photocatalytic conversion of organic pollutants in air.

1. Serrano B., Ortiz A., Moreira J., de Lasa H., "Photocatalytic Thermodynamic Efficiency Factors (PTEF). Practical Limits in Photocatalytic Reactors" Vol. 9, N5, 6824-6832 Ind.Eng.Chem.Res. (2010).
2. Ibrahim H., de Lasa H., "Kinetic Modeling of the Photocatalytic Degradation of Air Borne Pollutants", AIChE J., Vol.50, 1017-1027 (2004).
3. Romero-Vargas S., Ibrahim H., de Lasa H." Flow Field Investigation in a Photocatalytic Reactor for Air Treatment (Photo-CREC-air). Chem,.Eng.Sci. Vol. 61, 3343-3361 (2006)
4. Garcia Hernandez J, Serrano B., de Lasa H.,"The Photocatalytic Energy Efficiency Factors (PTEF) in Photocatalytic Reactors for Air Treatment" Chem.Eng.J. Vol.165, 891-901 (2010).

See more of this Session: Photo, Microwave and Ultrasound Catalysis
See more of this Group/Topical: Catalysis and Reaction Engineering Division

(769d) Visible Light Enhanced Selective Propylene Epoxidation Over Copper Based Catalyst

Marimuthu Andiappan, University of Michigan, Ann Arbor, MI and Suljo Linic, Chemical Engineering, University of Michigan, Ann Arbor, MI

Developing energy efficient, environmentally friendly chemical transformation processes is of paramount importance for combating dwindling natural resources and growing environmental concerns. Propylene oxide (PO) is an extremely valuable commodity chemical that is still commercially produced using expensive, environmentally harmful routes, due to the difficulties associated with developing heterogeneous catalytic materials that can directly produce PO using oxygen or air as an oxidant. The development of an environemtnally friendly and selective direct heterogeneous process for producing PO represents a major potential advancement.
In this contribution, we demonstrate that copper based catalysts exhibit high selectivity in the direct epoxidation (using O2 as an oxidant) of propylene to form PO. We also demonstrate that the nanostructured copper catalyst can efficiently couple low intensity visible light with thermal energy to drive propylene epoxidation at significantly lower temperature than epoxidation under pure thermal conditions. The critical feature of the copper catalysts that allows it to effectively channel photon energy into chemical energy is its strong surface plasmon resonance. The effect of illumination on PO selectivity and process kinetics will be discussed in the context of the molecular mechanisms that govern the photocatalytic process.

See more of this Session: Photo, Microwave and Ultrasound Catalysis
See more of this Group/Topical: Catalysis and Reaction Engineering Division

(769e) High-Temperature Photocatalytic Ethylene Oxidation Over TiO2

Thomas Westrich[1], Kevin A. Dahlberg[1], Massoud Kaviany[2] and Johannes W. Schwank[1], (1)Chemical Engineering, University of Michigan, Ann Arbor, MI, (2)Mechanical Engineering, University of Michigan, Ann Arbor, MI

Photocatalytic oxidation of ethylene was conduced at higher temperatures than previously reported (60 °C > T > 520 °C) in a novel two-dimensional packed bed reactor. Heat treated Degussa P25 TiO2 was used as a photocatalyst; each catalyst contained a different phase composition, surface area, and surface hydroxyl concentration. Maximum photocatalytic oxidation rates were observed between 100 °C to 200 °C. Further, oxidation rates greater than 75% of the maximum oxidation rate were observed over a large temperature range: 60 °C > T > 300 °C. Decreased oxidation rates at elevated temperatures (T > 300 °C) were attributed to the non-radiative, multi-phonon recombination of photo-generated charge carriers. Regression analysis of a nonlinear, multi-phonon recombination model and simplified chemical rate law were used to determine apparent reaction rate law parameters for the photocatalytic oxidation of ethylene and recombination parameters for the TiO2 photocatalyst. Insights for the development and operation of novel high-temperature photocatalysts are also discussed.

See more of this Session: Photo, Microwave and Ultrasound Catalysis
See more of this Group/Topical: Catalysis and Reaction Engineering Division

(769f) Transesterification of Soybean Oil with Microwave Heating Using Homogeneous and Heterogeneous Catalysts

Christopher J. Gilbert, George W. Huber and W. Curtis Conner, Chemical Engineering, University of Massachusetts-Amherst, Amherst, MA

The finite supply of petroleum-based fossil fuels has prompted research into the production of fuels from renewable feedstocks (such as biomass). One of the fuels that are obtained from a biomass-based feedstock is biodiesel, which is produced by the transesterification of triglycerides or the esterification fatty acids. This fuel can be blended with diesel fuels and used in a conventional diesel engine, or it can be used alone as fuel in a properly modified engine.

The transesterification reaction starts with a triglyceride molecule (a three-chain fat, commonly found in oils), and removes the fat chains and converts them into biodiesel esters (by reaction with methanol) in successive steps. It is typically base catalyzed using homogeneous catalysts, the two most common industrial catalysts being sodium methoxide and sodium hydroxide. Unfortunately, the use of homogenous catalysts for this reaction requires significant neutralization and washing of the product stream. As an alternative to these catalysts, the heterogeneous catalyst strontium oxide is also examined. In this work, we consider how these catalysts behave under microwave irradiation and compare this heating method to conventional heating.

As well as examining the effects of heating on both homogeneous and heterogeneous catalyst, we will perform a kinetic fitting of the system. We will develop a model that incorporates the three individual steps of the reaction, and fit experimental data to determine the rate constants and activation energies of each of the steps. The ratios between these rate constants for different catalyst will also be considered. Lastly our model will also incorporate the inherent starting biphasic nature of the system (methanol and triglycerides are effective immiscible).

See more of this Session: Photo, Microwave and Ultrasound Catalysis
See more of this Group/Topical: Catalysis and Reaction Engineering Division

(769g) Influence of Microwave Band Irradiation On Catalytic Reforming Systems Operating Under Deleterious Conditions

Steven Edmund and Johannes W. Schwank, Chemical Engineering, University of Michigan, Ann Arbor, MI

In catalyst systems microwave radiation has been shown toincrease reaction rates, lower the bulk temperature at which reactions occur,and assist in the desorption of sulfur. Past research into zeolite and catalytic systems has examined theeffects of microwaves on adsorbate equilibriums withthe goal of altering equilibrium states. Despite beneficial effects, the large-scale application of microwaves inadsorption and catalytic systems has remained limited. Significant gaps exist in theunderstanding of microwave interactions in complex catalytic systems.

The present study on Ni-based reforming catalysts examinesthe influence of irradiation in the microwave band (2.45GHz, λ=12.1cm) oncarbon deposition. Catalytic reforming of hydrocarbon fuels is frequentlyplagued by catalyst deactivation via carbon deposition, sulfur poisoning, andoxidation of metal surface sites. To probe interactions with electromagneticradiation a model catalyst system was chosen consisting of cordierite monolithto minimize internal field attenuation, washcoatedwith a Ni/ $Ce_{0.75}Zr_{0.25}O_{-2}$ catalyst. Autothermal reforming of a mixture ofpropane and ethylene was used in this study with O/C and H_2O/C ratioschosen to create reaction conditions where significant deactivation due tocarbon deposition would be expected. Experiments were performed in a single modemicrowave cavity varying the source power from 0-900Watts with continuousmonitoring of the effluent using an electron ionization mass spectrometer. The results of microwave band catalystirradiation on: (i) Carbon deposition and morphology,and (ii) changes in the product distributions and fuel conversion arediscussed.

See more of this Session: Photo, Microwave and Ultrasound Catalysis
See more of this Group/Topical: Catalysis and Reaction Engineering Division

(773a) Autothermal Staged Reactors for Biomass Upgrading

Michael Skinner, **Samuel Blass**, Hui Sun, Reetam Chakrabarti, Aditya Bhan and Lanny Schmidt, Chemical Engineering and Materials Science, University of Minnesota, Minneapolis, MN

We demonstrate an autothermal single-pipe staged reactor system that combines multiple catalytic functions for biomass upgrading reactions. The reactants are fed to an upstream noble metal catalyst consisting of Rh or Pt on 1.3 mm ID α-Al_2O_3 spheres, forming catalytic partial oxidation (CPO) products. These products react further over a downstream HZSM-5 zeolite catalyst composed of crushed pellets that are 0.7 mm ID. The energy released from upstream CPO drives the downstream endothermic zeolite reactions. The staged reactor setup has been used to produce high yields of ethylene and dimethyl ether from common biomass surrogates such as ethanol and methanol. We also demonstrate a side-feed configuration with methane as a sacrificial fuel and the alcohol introduced as a feed between the noble metal stage and zeolite stage. In this configuration, the alcohol is preserved for dehydration reactions on the zeolite stage rather than be consumed by CPO in the noble metal stage. The staged reactor configuration has also been used to form gasoline-range compounds from decane, a non-oxygenated surrogate for biodiesel, by partial oxidation and partial depolymerization in the noble metal stage. The hydrocarbon products react further over a bifunctional Pt/HBEA zeolite with hydrogen produced by upstream catalytic partial oxidation. The bifunctional zeolite catalyzes hydrocracking and hydroisomerization reactions to form branched gasoline-range hydrocarbons. A staged reactor configuration with multiple catalysts can greatly expand the chemical functionality of autothermal reactors for biomass upgrading.

See more of this Session: Structured Catalytic Reactors: Monoliths and M mbranes
See more of this Group/Topical: Catalysis and Reaction Engineering Division

(773b) Hydrogen Production by Catalytic Partial Oxidation of Methane On Reticulated Rhodium and Platinum Foam Catalysts

Oliver Korup, Claude Franklin Goldsmith, Michael Geske and **Raimund Horn**, Inorganic Chemistry, Fritz Haber Institute of the Max Planck Society, Berlin, Germany

Hydrogen Production by Catalytic Partial Oxidation of Methane on ReticulatedRhodium and Platinum Foam Catalysts

Catalytic partial oxidation (CPO) ofmethane is an alternative to steam reforming for industrial production ofsynthesis gas or hydrogen. Since Hickman and Schmidt showed in their pioneeringwork [1] that equilibrium yields of synthesis gas can be obtained inmillisecond contact times by methane CPO on autothermallyoperated noble metal coated foam catalysts this reaction has received a lot of attentionin academia and industry.

Bymeasuring species and temperature profiles through Rhand Pt coated a-Al_2O_3foam catalysts during methane CPO [2] it was shown that H_2 and COare being produced partly in presence of gas phase O_2 by partialoxidation (Eq. 1) and partly by steam reforming after gas phase $O_2$is fully consumed (Eq. 2).

$$CH_4 + 1/2O_2\ CO + 2H_2\ D_rH = -36\ kJmol^{-1} (1)$$

$$CH_4 + H_2O\ CO + 3H_2\ D_rH = +206\ kJmol^{-1} (2)$$

Whereasthe steam reforming pathway to synthesis gas can be rationalized because $CH_4$and H_2O are present at reaction temperatures around 1000°C, theco-existence of CO and in particular H_2 with O_2 over anoble metal surface at these high temperatures comes as a surprise raising thequestion why they are not burned to CO_2 and H_2O. Oneinterpretation that is pushed forward in the literature is that $H_2$and O_2 do only co-exist because methane CPO is fully film transportlimited leading to an effective O_2 concentration close to zero atthe catalyst surface [3]. This argumentation is based on numerical andexperimental studies which show without doubt that methane CPO on reticulated Rhodiumfoam catalysts is indeed fully film transport limited [4].

Inthe present paper it will be shown that in contrast to Rhfoam catalysts methane CPO on Pt foam catalysts is much slower and largelykinetically controlled probably due to formation of oxidation resistant carbondeposits as revealed by in situ Raman spectroscopy. To verify kinetic controlon Pt foam catalysts spatial reactor profiles were measured and analyzed interms of the achieved O_2 conversion rate which is clearly below ofwhat can be expected in the film transport limit. Furthermore flow rate andpressure variations were conducted whose results are also in line with kineticcontrol.

Ifcatalytic data on Rh and Pt foam catalysts are nowcompared it can be shown that up to 50% H_2 selectivity is obtainedin the oxidation zone on a Rh foam catalyst underfull film transport control but that already 30% selectivity is achieved on a Ptfoam catalyst under kinetic control. Therefore film transport has indeed astrong beneficial effect on the achievable H_2 selectivity but it isnot the sole explanation of the co-existence of H_2 and O_2as argued in the literature. In fact Pt and most certainly also Rh produce a non-vanishing selectivity to H_2 evenunder high temperature and high pressure conditions making them interestingcatalysts for industrial application.

References

[1] D.A. Hickman, L. D. Schmidt, *Science* **1993**, *259*, 343-346

[2] R. Horn, K. A. Williams, N. J. Degenstein etal., *J. Catal.***2007**, *249*, 380-393

[3] A. Donazzi, M. Maestri, B. C. Michel et al., *J. Catal.* **2010**, *275*, 270-279

[4] D. Dalle Nogare, N. J. Degenstein, R. Horn etal., *J. Catal.***2008**, *258*, 131-142

See more of this Session: Structured Catalytic Reactors: Monoliths and Membranes
See more of this Group/Topical: Catalysis and Reaction Engineering Division

COUPLING DETAILED HETEROGENEOUS AND HOMOGENEOUS KINETICS WITH MASS AND HEAT TRANSFER IN CATALYTIC REFORMING OF LOGISTIC FUELS

*Lubow Maier[1], Marco Hartmann[2], Steffen Tischer[1], Olaf Deutschmann[1,2]**
[1]*Institute for Nuclear and Energy Technologies, Karlsruhe Institute of Technology (KIT), Campus North, Hermann-von-Helmholtz-Platz 1, Eggenstein-Leopoldshafen 76344 (Germany)*
[2]*Institute for Chemical Technology and Polymer Chemistry, Karlsruhe Institute of Technology (KIT), Campus South, Kaiserstraße 12, Karlsruhe 76128 (Germany)*
**lubow.maier@kit.edu*

Introduction

A considerable long term interest in hydrogen as a fuel in relation to the scarcity of fossil fuels and the associated pollution problems excites the development of systems using catalytic partial oxidation and steam reforming for the production of hydrogen-rich synthesis gas from conventional fuels. Two examples of such systems being currently of great technological interest are the Solid-Oxide Fuel Cell (SOFC) [1] when operated with non-pure hydrogen fuels, e.g. partially reformed logistic fuels, and short-contact time reactors for reforming gasoline and diesel fuels [2], e.g., as first stage of an on-board auxiliary power unit (APU). The non-linear coupling of complex homogeneous and heterogeneous chemical reaction kinetics with heat and mass transfer in such systems matters for reactor behavior, often even superimposed by transient modifications of the active catalytic phase, e.g. by oxidation and coking

In this work, we will present an experimental, modeling, and simulation study on a catalytic reformer for the production of hydrogen-rich synthesis gas from the gasoline surrogate iso-octane. Reforming of this model fuel exhibits all features mentioned above: complex homogeneous and heterogeneous reaction schemes, mass and heat transfer effects, catalyst deactivation. Present work is related to the coupling of models of these phenomena, and their computational implementation to explain the impact of residence time on fuel conversion and hydrogen production and to optimize the reactor performance.

Experimental and Modeling Methods

The model couples elementary-step based reaction mechanisms with a two-dimensional parabolic description of the flow field in a representative number of monolith channels and heat transport in the entire solid structure of the reactor including catalyst, heat shields, insulation, and reactor wall (Fig. 1). This approach is realized in the computer code $DETCHEM^{MONOLITH}$, which uses the code $DETCHEM^{CHANNEL}$ for the simulation of flow field in the individual channels [3].

The concept is applied to analyze conversion, selectivity, and temperature profiles in partial oxidation of iso-octane, a gasoline surrogate, over a rhodium/alumina monolithic catalyst. Although more complex fuels have already been studied experimentally [4], data from our experimental study of a single-component reference fuel iso-octane (2,2,4-trimethylpentane) [5] are chosen as reference because detailed reaction mechanisms of CPOX of iso-octane over Rh has recently been developed and coupled with homogeneous reaction schemes [4, 6]. Both homogeneous und heterogeneous mechanisms are applied in the current study without further modifications.

Results and Discussion

The numerical simulation predicts the two-dimensional temperature profile of the three monoliths of the reactor as function of axial and radial position (as shown in Fig. 1) as well as species profiles and product distribution in the single channel of monoliths for the different flow rates (2 – 6 slpm) studied at 0.8 and 1.0 C/O ratios.

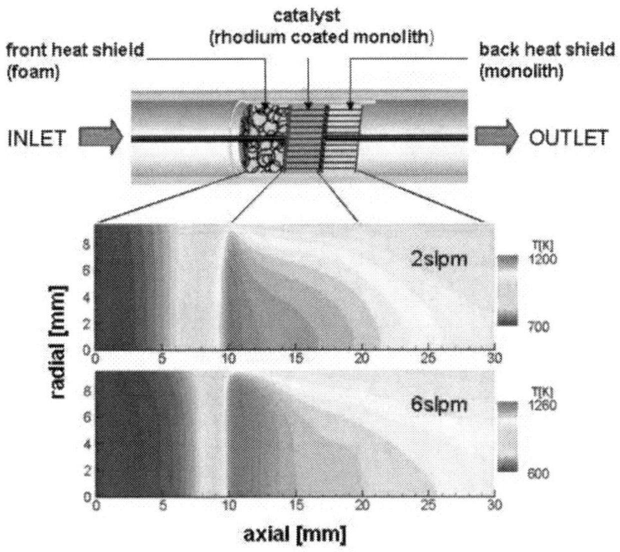

Figure 1. Sketch of the catalyst section of experimental setup with two heat shields simulated (top) and numerically predicted steady-state monolith temperature at C/O = 1.0 and at flow rates of 2 slpm and 6 slpm (bottom). The symmetry axis of the monolith is at radial dimension of zero.

The impact of the mass flow rate on conversion, selectivity, and operating temperature in an autothermal short-contact-time reactor was studied at atmospheric pressure. The simulation shows, that upstream heat transfer leads to a significant temperature increase of the downstream section of the front heat shield, which increases with decreasing flow rate. The residence time of the reactants within this uncoated monolith however is too small to initiate any fuel conversion in the gas-phase as clearly shown in Figure 2. The solid temperature at the entrance of the front heat shield is already notably higher than the temperature of the incoming gas being 463 K. Depending on the flow rate the solid temperature at the entrance decreases from 660 K to 570 K for increasing flow rate from 2 to 6 slpm due to the increasing convective cooling effect of the incoming gas at C/O = 1.0 (Fig. 3b). Convective cooling has a much stronger impact on the front temperature of the front heat shield than upstream heat conduction, because the maximum temperature of the catalytic monolith is highest at 6 slpm but the front temperature exhibits the lowest value at this condition (Fig. 3b). The radial temperature distribution in the front heat shield is rather flat. Interestingly, the temperature slightly increases towards the reactor wall showing the impact of heat conduction in the dense insulation and in the quartz tube.

Within the catalytic monolith radial and axial temperature variations can clearly be seen. The monolith is cooled by heat loss at the reactor insulation and quartz tube; the effect is stronger at low flow rates, because the total amount of heat released by chemical conversion is small and the maximum temperature does not change significantly with flow rate at C/O = 1.0. Even though the temperature exhibits a stronger dependence on flow rate at C/O = 0.8, the same effect was observed showing that the flow rate strongly determines the temperature profile and also the chemical conversion via the ratio of absolute chemical heat release to absolute physical heat loss.

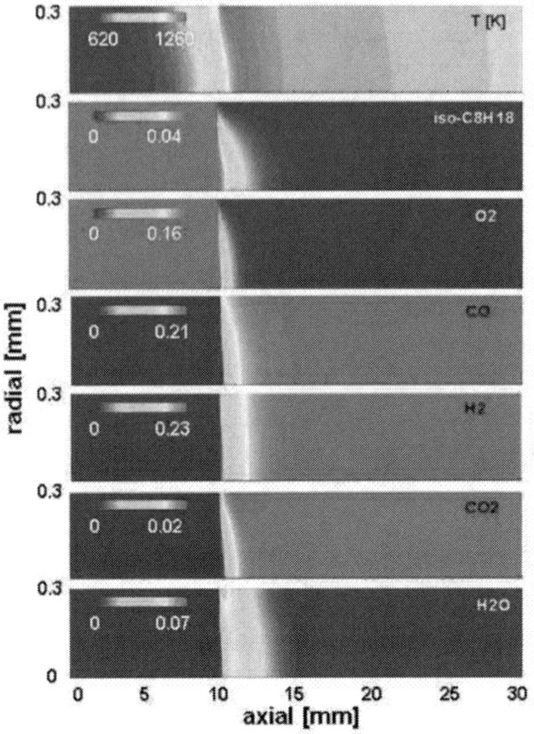

Figure 2. Numerically predicted gas-phase temperature and mole fractions of isooctane, O_2, CO, H_2, CO_2, and H_2O in the catalytic channel at C/O = 1.0 and 5 slpm. The symmetry axis of the channel and the gas-wall interface are at radial position of 0 and 0.3 mm, respectively.

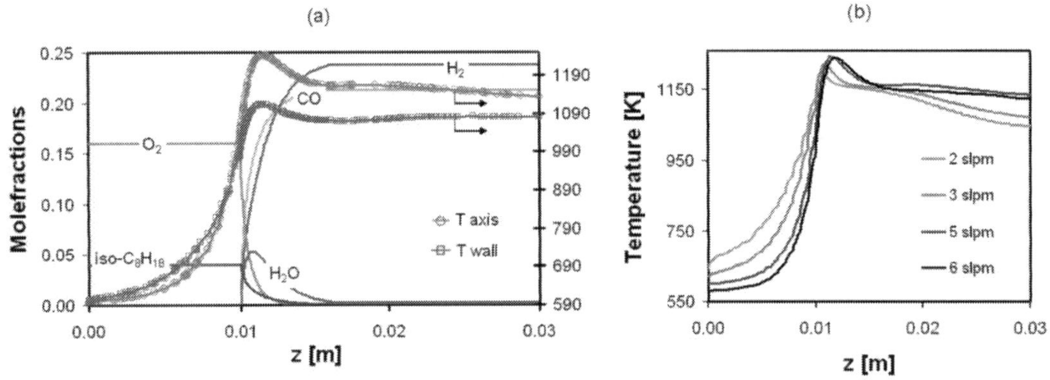

Figure 3. Numerically predicted axial species profiles (in the center of the single channel) in the centerline of the monolith and temperature profiles in the centerline (axis) and at the outer wall of the monoliths, C/O = 1.0, 5 slpm (a); effect of flow rate on temperature profile taking in the centerline of monoliths, C/O = 1.0 (b).

In axial direction, the temperature decreases rather quickly within the catalytic section after reaching its maximum at the entrance region. This behavior is well known in CPOX reactors for all hydrocarbons studied, from natural gas to diesel, and has extensively been described in literature. To make it short (Fig. 2.): in the entrance region of the catalytic monolith total oxidation of the fuel towards CO_2 and H_2O occurs until oxygen is almost completely consumed, at least on the catalyst, leading to the temperature maximum. Further downstream, steam reforming of the remaining fuel leads to the high hydrogen yields. At stoichiometric conditions (C/O = 1.0) not only the oxygen but also the fuel is completely (>98%) consumed within the catalytic monoliths both at low (2 slpm) and high (6 slpm) flow rates. CO is already partially produced in the total oxidation zone, dry reforming with CO_2 is not significant. Within few milliseconds in time or millimeters in space catalytic reforming of the fuel iso-octane is basically completed.

Since steam reforming is endothermic and the total heat consumption by this process depends on the total amount of fuel reformed, the temperature decreases faster for higher flow rates in axial direction in the centerline of the catalytic monolith; compare the 2 slpm vs. 6 slpm case in Figures 1 and 3b. However, closer to the wall of the reactor, the larger heat loss effect at slower flow rates exceeds the smaller (2 slpm vs. 6 slpm) chemical heat consumption effect and leads to more rapid cooling in axial direction at low flow rates. Increasing flow rates lead to larger temperature gradients in axial direction and smaller ones in radial direction.

In the downstream back heat shield, the cooling effect of the surroundings also explains the more rapid cooling for lower flow rates. As shown in Fig. 3b, the axial temperature profiles of varying flow rates in the centerline of the reactor cross each other just before entering the catalyst and again around axial center of the catalytic monolith. The highest temperature maximum is obtained at the highest flow rate studied, which is even much more pronounced at C/O = 0.8 (not shown). On the scale used in Figs. 2 and 3a, no variation of reactants' concentration (vanished) and any major products' concentration can be recognized.

Figure 4 compares the numerically predicted temperature at the outlet of the catalytic section of the reactors with the experimentally measured temperature as function of flow rate. In addition, numerically predicted outlet temperatures of a single channel simulation with adiabatic boundary conditions are shown. The exit temperature increase with increasing flow rate is well-computed with the accurate heat balance model, while the adiabatic model fails. The increase in temperature with increasing flow rate observed in this study can be understood by the effect of heat losses. The total amount of heat released by the reaction almost linearly increases with flow rate, because fuel is fully converted in the first zone of the catalyst. The adiabatic behavior is approached at the highest flow rate as a result of predominance of chemical heat release to thermal heat loss.

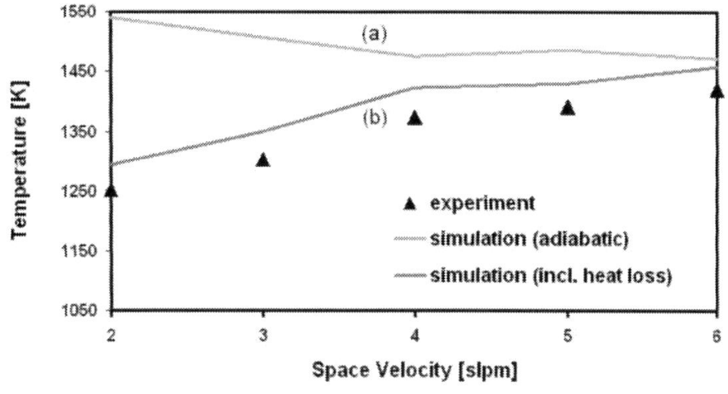

Figure 4. Catalyst outlet temperature as function of space velocity at a given C/O = 0.8; experimental data versus simulation, a) adiabatic single channel simulation without heat loss, b) monolith simulation including heat loss.

The experimentally determined independence of the hydrogen selectivity on the flow rate at C/O = 0.8 is well-reproduced by the numerical simulation using the complex model as shown in Fig. 5(a), the deviations are within the experimental error [5]. At C/O = 1.0 shown in Fig. 5(b), the hydrogen selectivity significantly increases with flow rate up to 4 slpm and is rather constant at higher flow rates, though the temperature increases only slightly. A variety of effects (heat balances, thermodynamics, kinetics, flow rate/residence time) overlap to produce this levelling-out of the hydrogen selectivity between 3 and 4 slpm.

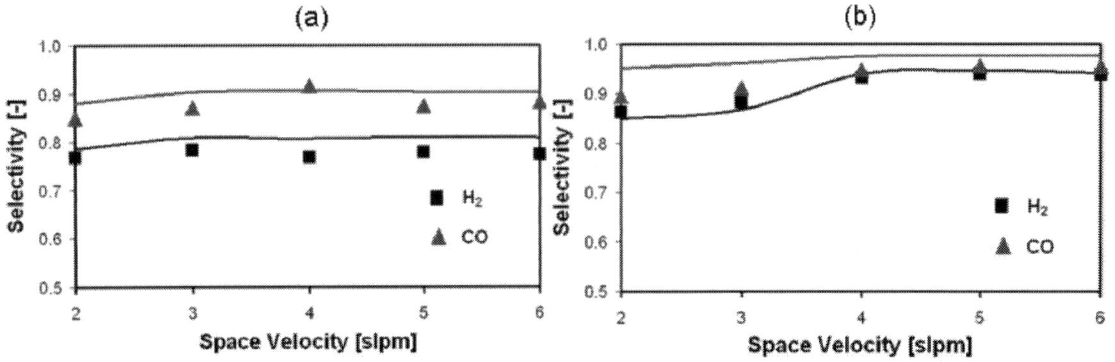

Figure 5. H_2 and CO selectivity as function of space velocity at C/O = 0.8 (a) and C/O = 1.0 (b); experimental data (symbols) versus simulation (lines).

It was found that the major objective of the reactor, i.e. production of high hydrogen yields at minimal formation of coke-precursors, can be achieved at C/O ratios close to 1.0 and sufficiently but not extremely high flow rates. The counter-intuitive flow rate effect on hydrogen yield is explained by the ratio of chemical heat release to physical heat loss. Coking propensity is related to the C/O ratio, the flow rate, and the occurrence of homogeneous fuel conversion.

The implementation of detailed chemical kinetics in a two-dimensional parabolic flow model for individual monolith channels and the coupling with heat balances of the catalytic monolith as well as heat shields, insulation, and reactor wall provides a simulation tool that is able to analyze the behavior of structured CPOX reactors in great detail. The simulation can provide guidance to reactor design and optimization of the operating conditions such as flow rate and fuel/oxygen ratio.

References

1. Hecht, E.S., Gupta, G.K., Zhu, H.Y., Dean, A.M., Kee, R.J., Maier, L., Deutschmann, O. *Appl. Catal. A. General 295*, 40 (2005).
2. Thormann, J., Maier, L., Pfeifer, P., Kunz, U., Deutschmann, O., Schubert, K. *Int. J. Hydrogen Energy* 34, 5108 (2009).
3. Deutschmann, O., Tischer, S. et al. DETCHEM^{TM} vers. 2.3 (2010)www.DETCHEM.com
4. Hartmann, M., Maier, L., Deutschmann, O. *Combustion and Flame* 157, 1771 (2010).
5. Hartmann, M., Maier, L., Deutschmann, O. *Appl. Catalysis A: General* 391 144 (2011) .
6. Kaltschmitt, T., Maier, L., Deutschmann, O. *Proceedings of the Combustion Institute* 33 3177 (2011).

(773d) In-Situ Fabrication of a Novel Microstructured Reactor by Atomic Layer Deposition

Staci A. Van Norman[1], Stefan Strhle[1], Nick Wannenmacher[2], Aldo Steinfeld[3], John Falconer[1] and Alan W. Weimer[4], (1)Chemical Engineering, University of Colorado at Boulder, Boulder, CO, (2) Chemical Engineering, Oregon State University, Corvallis, OR, (3)Department of Mechanical and Process Engineering, ETH Zurich, Zurich, Switzerland, (4)Department of Chemical and Biological Engineering, University of Colorado at Boulder, Boulder, CO

Microchannel catalytic reactors exhibit improved mass and heat transfer properties and are thus often chosen for highly exothermic or endothermic reaction conditions. In conventional microchannel reactors, catalyst is deposited within channels either by wash coating or by packing channels with particles prior to bonding. Both methods exhibit challenges: washcoating can result in poor channel volume usage, while channel packing techniques often result in the crushing and subsequent flaking of particles and clogging or fluid channeling due to differences in the thermal expansion of the materials. A novel *in-situ* fabrication method is presented for the customizable manufacture of microstructured reactor substrates made from ceramics/cermets for use in highly exothermic or endothermic catalytic reaction applications.

Microstructured supports were fabricated by atomic layer deposition on sacrificial polymer templates of polystyrene-divinylbenzene with a surface area of 700 m^2/g, pore size of 30 nm and particles sizes of 50-100 μ m and 80-160 μ m. Atomic layer deposition produces conformal films by two sequential self-limiting surface half reactions of $Al(CH_3)_3$ and H_2O for alumina and similarly for other materials.

Reactions were performed at temperatures below 100°C; once the desired film thickness was achieved the polymer template was volatilized at temperatures above 400°C. The resulting thermal, fluid flow and structural characteristics of these structures are presented. A heat transfer model was developed and is compared to a conventional packed bed of ceramic particles.

See more of this Session: Structured Catalytic Reactors: Monoliths and Membranes
See more of this Group/Topical: Catalysis and Reaction Engineering Division

(773e) Monolith Reactor Platform for Refinery Fuel Gas Utilization In High Value Applications

Vasilis Papavassiliou, Raymond Drnevich, Ramchandra Watwe, John Scalise and Perry Pacouloute, Praxair Technology Center, Tonawanda, NY

Monolith Reactor Platform for RefineryFuel Gas Utilization in High Value Applications

Raymond F.Drnevich, Vasilis Papavassiliou, & Perry Pacouloute, Praxair, Tonawanda,NY; John Scalise Burr Ridge IL; Ramchandra Watwe, Praxair, Houston TX

Anew reactor platform has been developed that can condition refinery fuel gas soit can be used not only as fuel for refinery heaters and boilers but in highervalue applications such as:

feed for hydrogen production via steam methanereforming

feed forgas turbines with dry low NOx (DLN) combustors

substitute natural gas for export to natural gaspipelines

Refinery fuel gas management hasbeen recognized as a key element for optimum refinery operation. As higherquality fuels are mandated, refineries face pressure to increase theirprocessing intensity in order to remain profitable. In addition, refineries areseeking to increase their ability to process heavy and sour crude in order toachieve needed operational flexibility. These two main trends cause increased refinerygas (i.e. H_2, C_1-C_5) production, which exceedsin many cases the capability of the refinery to use it as fuel, forcing therefiner to change refinery operations to meet fuel gas constraints, produce andvent unneeded steam, flare fuel gas, or sell fuel at a discount. Another indirect benefit of better fuel gasmanagement is that it is an enabler for other energy reduction projects. Theseprojects generally lower the demand for fuel gas in various refineryoperations, but the lack of an outlet for the "saved" fuel gas may become animpediment in the implementation of the energy conservation projects.Conversely, such projects achieve higher benefits if the "saved" fuel gas isutilized in a productive manner.

Usingrefinery gas as a feedstock for hydrogen production has high potential butavailable pre-treatment technologies are generally based on natural gasprocessing and are unable to cope with the quality and characteristics inherentto refinery fuel gas. Refinery gas is typically highly variable with higholefin, C_2+ and sulfur content which makes it difficult to processin SMRs. Similarly, refinery fuel gas isunsuitable as a feed to gas turbines with Dry Low NOx (DLN) combustors. Thetypical DLN burner fuel specifications for hydrogen and C_{2+} are 10%and 15% maximum, respectively. Refineryfuel gas likely has hydrogen and C_2+ content that exceeds theselimits. Furthermore, olefins contained in the fuel gas tend to form soot and aminimal compositional variation is required to maintain NO_x performanceof the DLN combustors. A common approach to get around these limitations is touse fuel gas as a supplemental fuel to the heat recovery steam generators(HRSG) in gas turbine cogeneration plant or as a fuel to the SMR furnace. However, this approach limits the amount offuel gas that can be used and/or the value obtained from use of the fuelgas.

Toaddress some of these challenges, a new reactor technology has been developed (RefineryGas Processor or RGP) based on short contact time catalytic monolith technologythat has been so far been targeted for catalytic partial oxidation (CPO)applications. CPO was pioneered by LannySchmidt's group[1] at the university of Minnesota and usesceramic or metallic monoliths washcoated with precious metal catalysts such aPt or Rh. RGP is based on similarcatalytic monoliths but it expands the use of such monoliths to also coverhydrogenation reactions, thus, RGP can operate in two modes to address thedifficulties in treating refinery gas.Firstly in hydrogenation mode (no oxygen) the reactor converts olefins inRFG to paraffins with the contained or supplemental hydrogen but with a muchwider operating temperature window compared with conventional hydrotreater technology. Surprisingly in hydrogenation mode thecatalyst exhibits similar reaction times as those that have been demonstratedin catalytic partial oxidation operation.Given the inherent temperature stability of CPO catalyst RGP permitsutilization of refinery gas streams with high olefin content and high olefinvariability. Secondly with the additionof small amounts of oxygen (up to 10% of RGP feed) and steam (up to 1:1 steamto carbon ratio) the reactor can operate in a prereforming mode that reducesthe amount of hydrocarbons with two or more carbon atoms in addition toreducing olefin levels. Both olefinhydrogenation and hydrocarbon reforming occur simultaneously in this mode. By tuning the oxygen consumption, therefinery fuel gas composition variations can be reduced thus improving the reliabilityof downstream operations. The samereactor can be used in both operational modes and no shut-down is required totransition between modes. The dualoperation can expand the type of refinery gas composition that can used andallow the same technology to be used for different refinery gas applications. In hydrogen production RGP can replace twounit operations, a hydrotreater and a prereformer.

Extensivelaboratory testing was undertaken with simulated refinery gas to performparametric analysis and develop efficient operating conditions, selectappropriate catalyst and test the ability to operate in hydrogenation mode orprereforming mode. Since it is notpossible to simulate all of refinery gas characteristics in the laboratory a demonstrationunit was also designed and built at a refinery location. The details of this new reactor design,laboratory results and demonstration unit results will be presented.

1. Hickman D. A., Schmidt L.D., Synthesis gas formation by direct oxidation of methane over Pt monoliths, *Jr. of Catalysis*, 1992, 138, 267.

See more of this Session: Structured Catalytic Reactors: Monoliths and Membranes
See more of this Group/Topical: Catalysis and Reaction Engineering Division

(773f) Hydrogenation of Dimethyl Oxalate to Ethylene Glycol On a Cu/SiO_2/Cordierite Monolithic Catalyst: Enhanced Internal Mass Transfer and Stability

Hairong Yue[1], Li Zhao[1], Yujun Zhao[1], Shengping Wang[1], Yongli Sun[2], Jun Wang[2], Jinlong Gong[1] and **Xinbin Ma**[1], (1)Key Laboratory for Green Chemical Technology of Ministry of Education, School of Chemical Engineering and Technology, Tianjin University, Tianjin 300072, China, (2)School of Chemical Engineering and Technology, Tianjin University, Tianjin 300072, China

Ethylene glycol (EG) is an important chemical used in a variety of consumer and industrial applications. At present, ethylene oxidation is a universal industrial approach to produce EG. However, as crude oil resource shrinks, synthesis of EG from syngas attracts increased interest. This indirect synthesis process includes two steps: the coupling of CO with nitrite esters to dimethyl oxalate (DMO), and the hydrogenation of DMO to EG. Copper-based catalysts have been widely used for the hydrogenation of DMO. Notably, internal diffusion has been considered as the rate-limiting step. In order to eliminate the internal diffusion limit and radically solve the contradiction between the grain size and bed resistance, A catalytic structured reactor for hydrogenation of DMO has been introduced in this work.

The structured reactor consisting of a cordierite monolith is based on a non-precious metal, thus cost-effective for a widespread application in industry. The washcoat of the monolithic catalyst was dip-coated with highly dispersed Cu/SiO_2 slurry prepared by ammonia evaporation method. This structure could guarantee high copper species diffusion within the mesopores of silica matrix in the form of copper phyllosilicate. These features render a catalyst with high activity in the reaction of hydrogenation of DMO, achieving a 100% conversion of DMO and more than 95% selectivity of EG. Notably, the conversion of DMO over monolith is significantly higher than compared to a packed bed Cu/SiO_2 catalyst in the form of pellet and extruded granular grains. It is primarily due to the short diffusion course of the thin wash-coat layer and regular channel structure of the monolithic catalyst. Calculations indicated that the internal mass transfer is dominated on the extruded granular catalyst. Moreover, the monolithic catalyst possessed excellent stability compared to packed bed catalysts upon heat treatment, which is attributed to the decreased probability of formation of hotter spots and less dead zones.

See more of this Session: Structured Catalytic Reactors: Monoliths and Membranes
See more of this Group/Topical: Catalysis and Reaction Engineering Division

AUTHOR INDEX

AUTHOR INDEX

AUTHOR INDEX

AUTHOR INDEX

AUTHOR INDEX

AUTHOR INDEX

AUTHOR INDEX

AUTHOR INDEX

AUTHOR INDEX

AUTHOR INDEX

AUTHOR INDEX

AUTHOR INDEX

AUTHOR INDEX

AUTHOR INDEX

AUTHOR INDEX